Comprehensive

CHEMISTRY

(For Senior School Classes and Undergraduates)

By

Dr I.V. Oppong
Dr N.K. Verma
Dr Walter Affo
Prof. S.K. Khanna
Dr B. Kapila

UNIVERSITY SCIENCE PRESS

(An Imprint of Laxmi Publications Pvt. Ltd.)

NEW DELHI, INDIA • BOSTON, USA

ISBN: 978-93-80386-94-2
First published 2011; Second Revised Edition 2014

PRINTED AND BOUND IN INDIA
Published by
UNIVERSITY SCIENCE PRESS
(An Imprint of Laxmi Publications Pvt. Ltd.)
113, Golden House, Darya Ganj, New Delhi-110 002, INDIA

Phone : 91-11-43 53 25 00
Fax : 91-11-43 53 25 28

www.laxmipublications.com
info@laxmipublications.com

UCH-9506-750-COMP. CHEMISTRY-VER C—
Typeset at: Goswami Associates, New Delhi, India
printed at : Repro Knowledgecast Limited, Thane

CONTENTS

SHS 2

SECTION 1: INTRODUCTION TO CHEMISTRY

SECTION 2: ATOMIC STRUCTURE

SECTION 3: CHEMICAL BONDS

Comprehensive

CHEMISTRY

SECTION 2: INORGANIC CHEMISTRY

SECTION 3: CHEMICAL KINETICS AND EQUILIBRIUM

SECTION 4: ACIDS AND BASES

SECTION 5: REDOX REACTIONS AND ELECTROCHEMISTRY

SHS 4

SECTION 1: CHEMISTRY INDUSTRY AND ENVIRONMENT

PREFACE

We feel immense pleasure in presenting '*Comprehensive Chemistry*' for students of **Senior High Schools**. The text is strictly based upon the guidelines and syllabus issued by the Ministry of Education. While developing the text, a sincere attempt has been made to keep the language and presentation simple and free from confusion.

The salient features of the book are:

- The book has been designed in such a way that the students as well as the teachers will feel confident in achieving the general and specific objectives recommended in the syllabus designed by the Curriculum Research and Development Division (CRDD) of the Ministry of Education, Science and Sports.
- The book closely follows the order in which the curriculum for the Senior High School Chemistry has been organised. This will ensure that all aspects of the syllabus are covered.
- Each unit begins with a list of specific objectives so that the student knows what goals are expected to be achieved by a careful study of the unit.
- The language level and depth in the text reflect the year group for which the text is written. This facilitates easy reading and understanding.
- The text is well and attractively illustrated with large, simple, clear and fully labelled diagrams to aid quick understanding and to arouse interest.
- A large number of different types of numerical problems and conceptual questions have been solved as Examples wherever required.
- Each unit ends with a set of multiple choice and discussion questions entitled *Evaluate yourself*. This is to enable the student check his or her understanding of the text. These questions cover the full range of the aims and objectives of the syllabus.
- Answers to the objective questions, fill in the blanks and numerical problems are provided at the end of each self-evaluation exercise.

- Important terms and statements have been given in **bold** or *italic* letters.
- A wide range of **Activities** have been suggested to motivate the students to develop practical skills and to aid better teaching and learning.
- Important facts, things to be kept in mind while going through the text, and some relevant and interesting material for reference have been highlighted with easily identifiable icons.
- A **summary** of important terms and concepts has been given in each unit for quick revision.
- WAEC adapted IUPAC names of inorganic compounds and some oxoanions have been given in Appendix-1
- IUPAC system of naming organic compounds has been given in detail in Appendix-2.

We believe that the book in the present form will be appreciated by teachers as well as students. We shall be glad to receive suggestions for further improvement of the book.

– Authors

SHS 2

SECTION

1

INTRODUCTION TO CHEMISTRY

General Objectives

After studying this section, the student should be able to:

1. recognize the centrality of chemistry among science subjects and the role it plays to solve the numerous problems facing society.
2. recognize that science depends heavily on measurement and that one's ability to measure accurately and precisely is a prerequisite for learning chemistry.
3. understand that different forms of matter behave differently and know how to handle them.
4. identify and describe careers related to chemistry and contribution of scientists including Ghanaians in chemistry related fields.

In this section, we shall study

Unit 1: Chemistry as a Discipline

Unit 2: Measurement of Physical Quantities

Unit 3: Basic Safety Laboratory Practices

SECTION-1

UNIT-1

CHEMISTRY AS A DISCIPLINE

Specific Objectives

After studying this unit, the student should be able to:

- describe chemistry as a subject and as a central science discipline
- describe the various branches of chemistry
- outline some careers in chemistry and their importance.

The human mind has always been very curious to make investigations and know about various activities/phenomena occurring around him. This curiosity has led him to collect information through experiments and observations. The curious mind has also been responsible for the research activities of various people all over the world. The knowledge and data acquired through this is then systematized in a way that mankind can make maximum benefit from. This knowledge base is known as science. **Science** may, thus, be broadly *defined as systematized knowledge gained by mankind through observations and experimentation.* Science has been further classified into *different branches due to* its enormous expansion and diversified fields. Some examples are: Chemistry, Physics, Biology, Geology, etc. *Chemistry* is one of the most important discipline of science to which this present book is devoted.

1.1 CHEMISTRY AS A SUBJECT AND AS A CENTRAL SCIENCE

Chemistry may be defined as the branch of science *which deals with the study of matter, its composition, its properties and the changes which it undergoes in composition as well as in energy during various processes.*

The word chemistry has been derived from the word ***alchemy***, which means '*study of metals*'. Alchemy itself might have come from ***al chem*** meaning '*country of dark soil*' and is the ancient name for Egypt. The chemical industries which flourished in Egypt about 6000 years ago are thought to mark the beginnings of chemistry. The Egyptians were successful in their industrial tasks because they developed their dyeing, pottery, glass-making and other areas through experiments.

The journey from *alchemy* to the modern subject of *chemistry* took centuries and involved untiring work by scientists. The recent advancements in the field of chemistry has created a marked effect on our present day life. Chemistry has helped us to meet all our requirement for a better living. The continuous research in the field of chemistry has resulted in the production of useful materials such as, clothes, drugs, artificial foods, plastics, rubbers, fertilizers, insecticides, weed killers, life supporting products, etc., which have revolutionised our life.

Our life would have been very dull and dreary without the knowledge of chemistry. In fact, *we can say, that chemistry is everywhere.* It is *in what we eat, in what we breathe, in how we live* and *even in what we are.*

CHEMISTRY—A CENTRAL SCIENCE

Modern chemistry is an abstract subject whose study presents great intellectual challenges and rewards. It is a practical field at the hub of man's future.

Modern chemistry is a *CENTRAL DISCIPLINE,* which correlates almost all branches of science. It is used to study biological, physical, medical as well as environmental phenomena. For example, a ***chemist*** works with:

- ***biologist*** to understand life processes and metabolic activities,
- ***physicist*** to understand the properties of matter and to develope new sources of energy,
- ***geologist*** to probe outer and inner space,
- ***physician*** to design new drugs and medicines,
- ***ecologist*** to make improvement in environment,
- ***engineers*** and ***technical managers*** to provide material and energy for better life.

Chemistry, thus responds to all social needs. It plays a critical role in any attempt to: *discover new processes*; *tap new energy sources*; *develop new materials*; *feed the people properly; improve health and conquer disease*, *monitor* and *protect our environment.*

1.2 BRANCHES OF CHEMISTRY

Chemistry can be broadly devided into **Pure Chemistry** and **Applied Chemistry.**

A. PURE CHEMISTRY

Pure chemistry deals with the attempt to get a better understanding of nature.

Pure chemistry is further divided into three main branches. *Organic chemistry*, *Inorganic chemistry* and *Physical chemistry.* These main branches have been further divided into a large number of sub-sections. The main branches are described briefly as follows:

1. **Organic Chemistry.** This branch of chemistry deals with the study of the structure, chemical composition and characteristics of compounds of carbon and hydrogen (Hydrocarbons) and their derivatives.
2. **Inorganic Chemistry.** This branch deals with the study of the structure, composition and behaviour of *inorganic compounds, i.e.*, the compounds other than hydrocarbons or their derivatives. Such compounds are found in the crust of the earth and constitute non-living matter.
3. **Physical Chemistry.** This branch deals with the study of the fundamental principles governing various chemical transformations and chemical systems. It is primarily concerned with the laws and theories of the different branches of chemistry.

B. APPLIED CHEMISTRY

Applied chemistry deals with the application of the knowledge of chemistry for the benefit of mankind. The different branches of applied chemistry are:

1. **Analytical Chemistry.** This branch involves a collection of techniques which allows for the exact determination of the composition of a material. It has been further divided into two categories:
   (*a*) *Qualitative analysis.* It deals with the identification of the constituent particles (atoms, ions, molecules) present in a material.
   (*b*) *Quantitative analysis.* It deals with the estimation of the various constituents in a material.
2. **Industrial Chemistry.** This branch deals with the chemistry involved in different industrial processes such as the manufacture of various chemical substances.
3. **Biochemistry.** This branch deals with the study of metabolic pathways and enzymology pertaining to living organisms. It deals with molecular, cellular and chemical activities of living organisms.
4. **Geochemistry.** This branch deals with the chemical processes occurring on earth such as the metamorphism of rocks, the formation of petroleum, etc. It also deals with the composition of soils and rocks.
5. **Petrochemistry.** It is the branch of chemistry which deals with the transformation of crude oil (petroleum) and natural gas into useful products and raw materials.
6. **Radiochemistry.** It is a branch of chemistry which deals with the study of radioactive materials, both natural as well as man-made. It also involves the use of radioactive materials to study the pathways/mechanism of ordinary chemical reactions.

7. **Biotechnology.** It refers to the technological applications which uses biological systems, living organisms or their derivatives to make or modify products or processes for specific use. Biotechnology, infact, combines various disciplines like genetics, molecular biology, biochemistry, embryology and cell biology for developing techniques for beneficial purposes.
8. **Medicinal or Pharmaceutical Chemistry.** It is a scientific discipline at the intersection of chemistry and pharmacology, which is involved with designing, synthesizing and developing pharmaceutical drugs. Medicinal chemistry involves the identification, synthesis and development of new chemical entities suitable for therapeutic use. Medicinal chemistry is a highly interdisciplinary science that combines *organic chemistry* with *biochemistry, pharmacology, *pharmacognosy, molecular biology, statistics* and *physical chemistry.*
9. **Environmental Chemistry.** It is a branch of chemistry which deals with the scientific study of chemical and biochemical phenomena that occur in natural places. Environmental chemistry is also an interdisciplinary science that includes *atmospheric, aquatic,* and *soil chemistry* along with *analytical chemistry, environmental studies* and *other areas of science*.

1.3 CAREERS IN CHEMISTRY

CAREER OPPORTUNITIES

Chemistry is a central science because of its multidisciplinary nature. Therefore chemistry students can pursue careers in the field of *industry, education, research, government agencies* and other *non-traditional fields*. Some of the career opportunities that a student with a degree in chemistry can have are as follows:

1. **Industries.** *Chemical industries* employ about 66% of all chemists. The majority of them find opportunity in *research and product development (R and D), sales*, or *marketing*. Many of them work in *quality control analysis* and *testing* products. Others find work in areas like *industrial hygiene* and *safety* or *regulatory work* for *environmental compliance.*
2. **Academic Institutions.** Educational institutions employ about 26% of chemists. Ph. D. degrees are required for most academic positions at colleges and universities. Some chemists degrees in chemistry education take up teaching assignments in high schools.
3. **Government Agencies.** Government employs about 7% of all chemists. Government agencies hire chemists for a variety of jobs including *basic research, testing work* required to enforce government regulations, *technical program managers, authors/editors* of technical documents and government regulations.
4. **Non-traditional Fields.** A small percentage of chemists (about 1%) find work in *non-traditional fields*. They get opportunities to become *patent lawyers, science writers, information specialists, technical librarians, technical consultants* or *business owners*.

*pharmacognosy is a study of medicines derived from natural sources.

Group Discussion

Identify two applied chemistry professions and explain the chemistry they practice.

Hints: For reference, the two important professions associated with applied chemistry are discussed below:

1. **Analyst:** Analysts find jobs in the chemical industry, food industry and pathological laboratories. In chemical industries, they control the quality of the final product through chemical analysis of the product. In food laboratories, they analyse the food items to detect adultration. In pathological labs they carry out chemical tests on blood or urine to help a doctor diagnose a disease.
2. **Research scientist.** Research scientists find jobs in pharmaceutical companies where they can use their knowledge of chemistry in developing more convenient and economical methods for the synthesis of drugs. They can also help in designing new drugs.

Different career options in chemistry are summarised in Fig. 1.1.

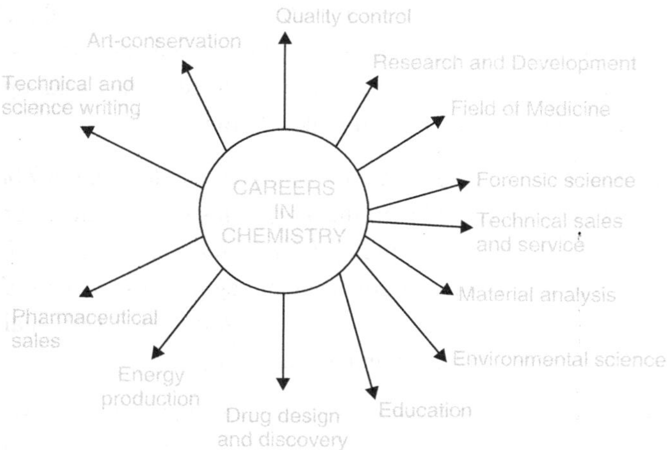

Fig. 1.1. Various careers associated with chemistry.

SUMMARY

- **Chemistry** is the study of matter, its composition, its properties and changes which it undergoes in composition as well as energy during various transformations. Chemistry is a ***central science discipline*** which correlates various important branches of science. Chemistry can be divided into pure and applied chemistry.
- **Pure chemistry** has three main branches viz organic, inorganic and physical chemistry while **applied chemistry** has branches *namely* biochemistry, analytical chemistry, radiochemistry, geochemistry, petrochemistry, environmental chemistry and biotechnology.

EVALUATE YOURSELF

I. Objective Type Questions

Select the most appropriate choice from the options given as (a), (b), (c) and (d) after each question:

1. Which branch of pure chemistry deals with the study of compounds associated with non-living sources?
   (*a*) Physical chemistry (*b*) Organic chemistry
   (*c*) Biochemistry (*d*) Inorganic chemistry.
2. Which of the following is not applied chemistry?
   (*a*) Geochemistry (*b*) Biochemistry
   (*c*) Radiochemistry (*d*) Inorganic chemistry.
3. In order to design a new drug, a chemist has to seek the help of
   (*a*) Engineer (*b*) Geologist
   (*c*) Zoologist (*d*) Physician.
4. The branch of chemistry which deals with the study of hydrocarbons is called
   (*a*) Organic chemistry (*b*) Inorganic chemistry
   (*c*) Radiochemistry (*d*) Nuclear chemistry.

II. Fill in the Blanks

5. Complete the following sentences by supplying appropriate words:
   (*i*) Ecologist and chemist work together to ..... .
   (*ii*) The branch of pure chemistry which deals with the study of fundamental laws and principles is called ..... .
   (*iii*) Qualitative analysis deals with ..... .
   (*iv*) Radiochemistry deals with the study of ..... substances.
   (*v*) The phenomenon of metamorphosis of rocks is studied by ..... branch of chemistry.

III. Discussion Questions

6. Define chemistry and its various disciplines.
7. Comment on the statement that chemistry is a central science discipline.
8. Write the names of the various disciplines of applied chemistry. Define any two of them.
9. Name and define the various branches of pure chemistry.
10. Give a brief account of the various career options of a degree holder in chemistry.

Answers

I. Objective Type Questions

1. (*d*) **2.** (*d*) **3.** (*d*)
4. (*a*).

II. Fill in the Blanks

5. (*i*) make improvement in the environment
(*ii*) physical chemistry
(*iii*) identification of constituents
(*iv*) radioactive substances
(*v*) geochemistry.

SECTION-1

UNIT-2

MEASUREMENT OF PHYSICAL QUANTITIES

2.1 SCIENTIFIC MEASUREMENTS AND THEIR IMPORTANCE IN CHEMISTRY

One important part of scientific work is the *measuring of physical quantities* such as *mass, volume, length, energy, temperature*, etc. It is interesting to note that all these quantities can be expressed in terms of a few basic units. For example, in everyday life, we come across a number of measurements like **kilograms** (*for mass*), **litres** (*for volume*), **metres** (*for length*) etc. In addition to these common measurements we need to measure a number of other quantities such as *concentration, temperature, pressure, density, amount of electrical charge* etc.

"*All such quantities which we come across during scientific studies are called physical quantities*".

A **physical quantity** thus refers to the result of measurement operation. It involves the comparison of the quantity to be measured with some fixed standard. For example, if we are interested in knowing the length of a pencil, the operations would be:

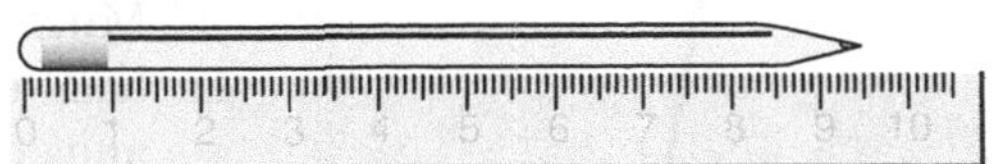

(*i*) to lay the pencil parallel to a centimetre scale and

(*ii*) to count the number of markings on the scale.

Suppose the number of markings on the scale is 9.5. It makes no sense to say that length of the pencil is 9.5. However, if we attach *centimetre*, a unit of length, to the numerical figure, *i.e.*, 9.5 cm, it will become more appropriate. Thus, a measured physical quantity is expressed in two parts, a **numerical coefficient** and a **unit.** Either of them is meaningless without the other.

Specific Objectives

After studying this unit, the student should be able to:

- describe the importance of scientific measurements to the study of chemistry
- measure some physical quantities using appropriate instruments
- demonstrate the correct handling and use of laboratory equipment
- differentiate between base and derived units of measurement
- outline the scientific method.

In the measurement of the length of the pencil, 9.5 is a *numerical figure* whereas centimetre is a *unit*.

The study of experimental science thus depends upon the quantitative measurement of properties. Every measurement gives a numerical result that has three important aspects:

(*i*) *Numerical magnitude,*

(*ii*) *Accuracy or precision with which the number is expressed* and

(*iii*) *Indicator of scale i.e., unit employed to express.*

The measurement of physical quantities becomes scientifically more correct and relevant if all the three aspects as given above are reported. Thus,

"*A unit may be defined as the standard of reference chosen to measure or express any physical quantity*".

2.2 MEASUREMENT OF PHYSICAL QUANTITIES USING APPROPRIATE INSTRUMENTS

MEASURING MASS

Mass is a measure of how heavy something is *i.e.*, it is the amount of matter present in it. It can be determined in the laboratory by using analytical or electrical balances as shown in Figs. 2.1, 2.2 and 2.3.

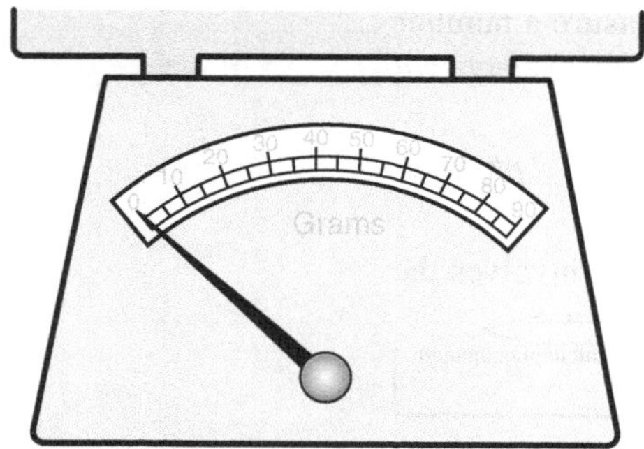

Fig. 2.1

- Mass is measured in *grams* (g), *kilograms* (kg) and *tonnes*. These are known as metric units of mass.

  1 kg = 1000 g

  1 tonne = 1000 kg

- Ounces and pounds are old units of mass. These are known as *imperial* units.

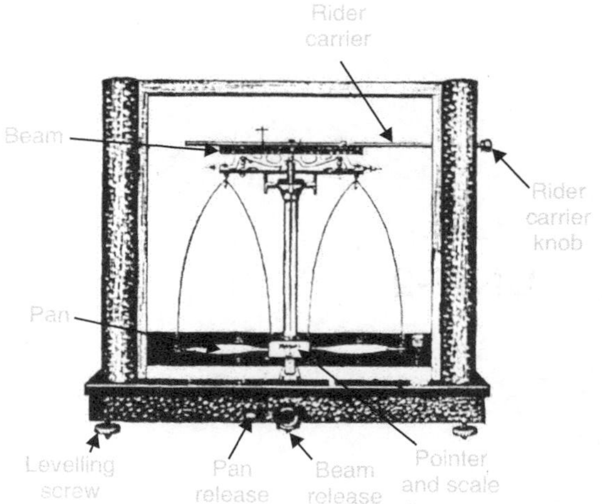

Fig. 2.2. Analytical balance.

- There are 16 ounces in a pound.
- An ounce is roughly equal to 25 grams.
- A pound (454 g) is equal to just under half a kilogram (500 g).

Fig. 2.3. Electrical balance.

MEASURING LENGTH

Length is a measure of how long or wide something is. Rulers and tape measures can be used to measure length.

- Length is measured in *millimetres* (mm), *centimetres* (cm), *metres* (m) or *kilometres* (km). These are known as metric units of length.

  1 cm = 10 mm

  1 m = 100 cm

  1 km = 1000 m

- Miles, feet and inches are old units of length. These are known as imperial units of length.

  There are 12 inches in a foot.

  An inch is roughly equal to 2.5 centimetres.

  A foot is roughly equal to 30 centimetres.

  A mile is roughly equal to 1.5 kilometres.

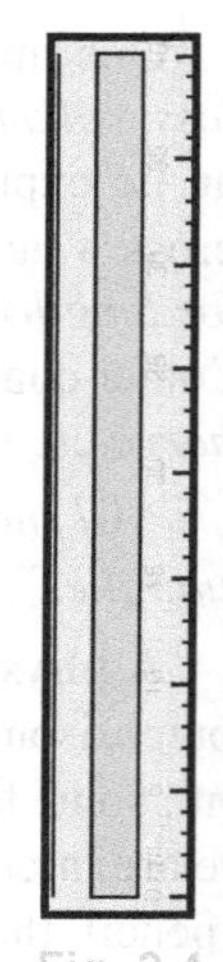

Fig. 2.4

MEASURING CAPACITY OR VOLUME

Capacity or volume is a measure of how much space something takes up. Measuring spoons or measuring jugs can be used to measure capacity.

- Capacity is measured in *millilitres* (mL) and *litres* (L).

  1 L = 1000 mL

- Pints and gallons are old units of capacity (imperial units).

  There are 8 pints in a gallon.

  A pint is equal to just over half a litre.

  A gallon is roughly equal to 4.5 litres.

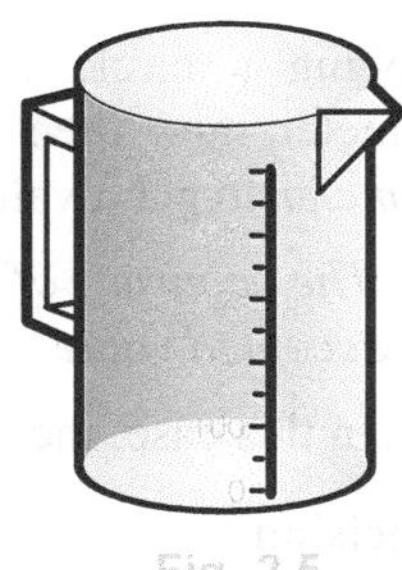

Fig. 2.5

In the laboratory, volume of liquids or solutions can be measured with a graduated cylinder, burette, pipette, volumetric or measuring flask etc. A volumetric flask is used to prepare a known volume of a solution. These measuring devices are shown in Fig. 2.6.

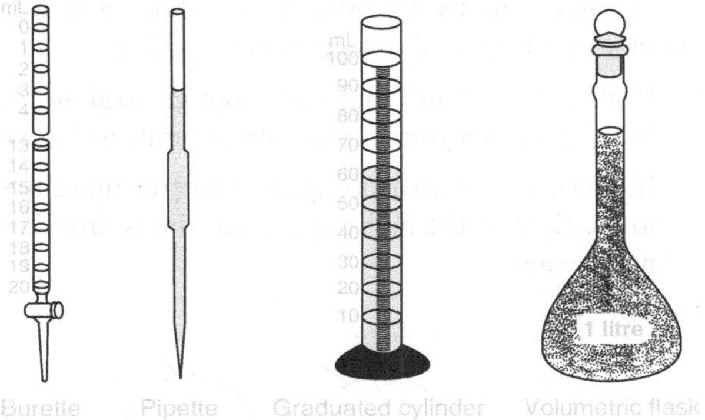

Fig. 2.6. Some volume measuring devices

Reading Scales

To read a scale, first work out how much each mark or division on the scale represents.

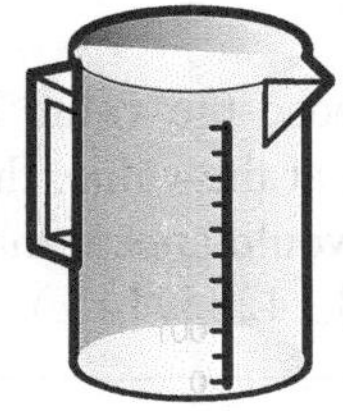

There is one mark between each 100 mL.
So each 100 mL is divided into 2 parts.
100 ÷ 2 = 50.
So each mark must represent 50 mL.

Fig. 2.7

MEASURING TIME

Time is measured with a simple clock, digital clock, etc. (Figs. 2.8 (*a*), 2.8 (*b*))

- AM (am) is morning time (all times between 12 midnight and 12 midday).
- PM (pm) is afternoon and evening time (all times between 12 midday and 12 midnight).
- An analogue or a 12 hour clock is shown in Fig. 2.8 (*a*). An analogue clock is one with a face and hands. It is showing the time twenty past five (Fig. 2.8 (*a*)).

  If it were twenty past five in the morning, it would be written as 5:20 am (Fig. 2.8 (*b*)).

  If it were twenty past five in the afternoon, it would be written as 5:20 pm (Fig. 2.8 (*b*)).

Fig. 2.8 (*a*)

means it is twenty past five in the morning.

means it is twenty past five in the evening.

Fig. 2.8 (*b*)

- Nowadays we have digital or 24 hour clocks. They are also showing the time twenty past five.
- The 24 hour time is the same as the analogue time in the morning (except for the 0 at the beginning for numbers under 10). So 8:45 am becomes 08:45.
- But in the afternoon, you need to add 12 to convert an analogue time to a 24 hour time. So 8:45 pm becomes 20:45.
- Midday on a 24 hour clock is shown as 12:00.
- Midnight on a 24 hour clock is shown as 00:00.

Units of Time

1 minute = 60 seconds

1 hour = 60 minutes

1 day = 24 hours

1 week = 7 days

1 fortnight = 14 days

1 year = 12 months = 52 weeks = 365 days

1 leap year = 366 days

Remember "30 days has September, April, June and November. All the rest have 31. Except for February alone, which has 28 days clear but 29 each leap year".

MEASURING TEMPERATURE

The thermometer is used to measure temperature.

There are three common scales to measure temperature—°C (degree celsius), °F (degree fahrenheit) and K (kelvin).

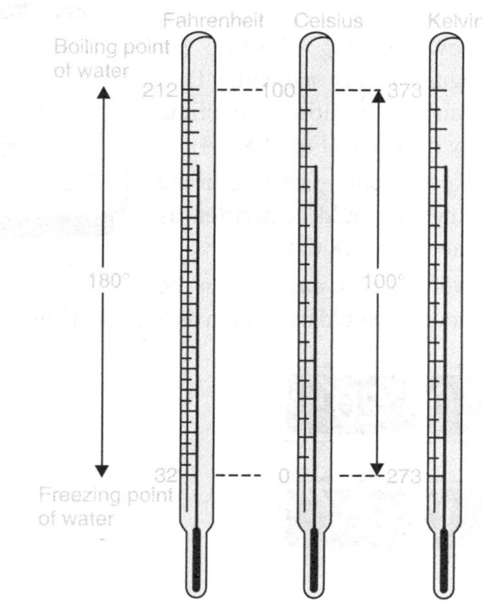

Fig. 2.9

Here, K is the SI unit. The thermometers based on these scales are shown in Fig. 2.9. Generally, thermometers with celsius scale are calibrated from 0° to 100° where the two temperatures are the freezing point and the boiling point of water respectively. The fahrenheit scale is represented between 32° to 212°.

The temperatures on the two scales are related to each other by the following relationship:

$$°F = \frac{9}{5}(°C) + 32°$$

The kelvin scale is related to the celsius scale as follows:

$$K = °C + 273.15$$

It is worthwhile to note that the temperature below 0°C (*i.e.,* negative values) are possible in celsius scale but in kelvin scale, negative temperature is not possible.

ACCURACY AND PRECISION

Accuracy

Accuracy is how close a measured value is to the **actual (true) value.** Accuracy describes how close an approximation is to a correct answer. Thus, accuracy "*is the measure of the difference between the experimental value or mean value of a set of measurements and the true value.*"

Accuracy = *Mean value – True value*

The smaller the difference between the mean value and *the true value the higher the accuracy.*

For *example,* suppose your math textbook tells you that the value of pi (π) is 3.14. You do a careful measurement by drawing a circle and measuring the circumference and diameter, and then you *divide the circumference by the diameter* to get a value for pi (π) of 3.16.

The accuracy of your answer is how much it differs from the accepted value.

In this case, the accuracy is 3.16 – 3.14 = **0.02.**

Precision

Precision is how close the measured values are **to each other.** Precision describes how many digits we use to approximate a particular value. In simple words, it is the difference between the measured value and the arithmetic mean value for a series of measurements, *i.e.,*

Precision = *Individual value – Arithmetic mean value*

The idea is illustrated by the following examples:

(*i*) **Target shot by arrows.** If you shoot a quiver of arrows at a target, several outcomes are possible.

- If all of the arrows that you shoot go straight to the bull's eye, then your aim is both accurate and precise.
- If, however, the arrows cluster in an area immediately to the right of the bull's eye, your aim is precise, but not accurate.

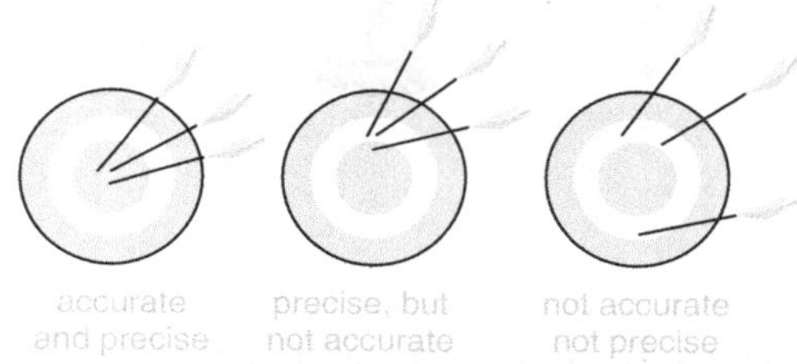

- If your aim is bad and the arrows hit positions scattered all over the target, then your aim is neither accurate nor precise.

(*ii*) **The value of π (pi) is 22/7.** Suppose the value is written as 3.1417 and 3.1392838. Looking at the value, the second number has higher precision, but it would appear that the first is more accurate. (Actual value is 3.142857143.)

Bias (don't let precision fool you !)

So, if you measure something several times and all values are close, they **may** still all be wrong if there is a "**Bias**".

Bias is a systematic (built in) error which makes all measurements wrong by a certain amount.

Examples of Bias

- The scales read "1 kg" when there is no weight on them.
- You always measure your height wearing shoes with thick soles.
- A stopwatch that takes half a second to stop when clicked.

(*iii*) **Measurement of a pencil on two different scales.** How long is the pencil? The best you can say looking at the scale I, is 'about 9 centimetres'. You might guess and say 'about 9.5 centimetres', but the decimal place is just a guess. Because the smallest unit on the ruler you are using is one centimetre, the precision of your measurement is to the nearest centimetre.

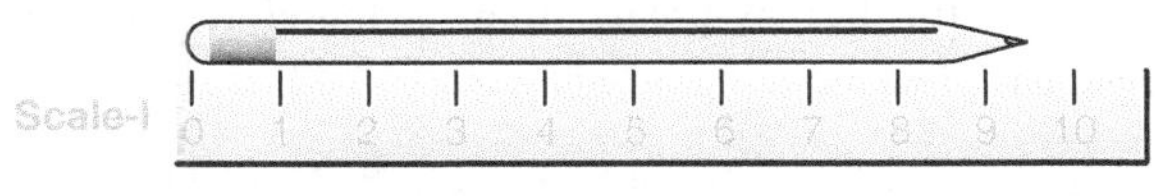

Fig. 2.10

Now look at the scale-II. Here we are using a different ruler to measure the pencil.

How long is the pencil? The best you can say is 'about 9.5 centimetres'. Again, you might guess and say 'about 9.51 centimetres, but the second decimal place is just a guess. Because the smallest unit on the ruler you are now using is one millimetre (one tenth of a centimetre), the precision of your measurement is to the nearest millimetre, or tenth of a centimetre.

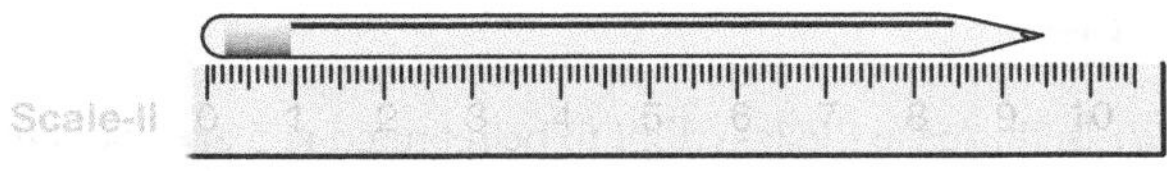

Fig. 2.11

This second measurement is more precise, because you used a smaller unit to measure with.

- The *precision* of a measurement is the size of the unit you use to make a measurement. The smaller the unit, the more precise the measurement.
- The *accuracy* of a measurement is the difference between your measurement and the accepted correct answer. The bigger the difference, the less accurate your measurement.

UNCERTAINTY IN MEASUREMENT

The study of chemistry involves both theoretical as well as practical aspects which further deals with qualitative and quantitative measurements. While dealing with calculations, the meaningful way is to handle the numbers conveniently and present the data realistically with certainty to the extent possible.

SIGNIFICANT FIGURES

Experimental measurements have some uncertainty associated with them. However, one would always like the results to be precise and **accurate.** These aspects depend on the accuracy of the **measuring device** and the **skill of the operator.**

A convenient method of expressing the uncertainty in measurement is to express it in terms of *significant figures.* In this method, it is assumed that all the digits are known with certainty except the last digit which is uncertain to the extent of ± 1 in that decimal place. Thus, a measured quantity is expressed in terms of a number which includes all digits which are certain and a last digit which is uncertain. The total number of digits in the number is called the number of **significant figures.**

The number of significant figures in a measurement is the number of figures that are known with certainty plus one that is uncertain, beginning with the first non-zero digit.

In order to determine the significant figures in a measured quantity the following rules should be applied.

1. ***All non-zero digits are significant.***

   For example, 165 cm has **three** significant figures; 0.165 also has **three** significant figures. Similarly, 2006 has four significant figures, 9.05 has three significant figures, etc.

2. ***Zeros to the left of the first non-zero digit in a number are not significant.***

   For example, 0.005 g has only **one** significant figure, 0.026 g has **two** significant figures.

3. ***Zeros between non-zero digits are significant.***

   For example, 2.05 g has **three** significant figures.

4. ***Zeros to the right of the decimal point are significant.***

   For example, 5.00 g, 0.050 g, 0.5000 g have **three, two** and **four** significant figures respectively.

5. ***If a number ends in zeros that are not to the right of a decimal, the zeros may or may not be significant.***

   For example, 1500 g may have two, three or four significant figures.

   The ambiguity in the last point can be removed by expressing the number in **scientific notation.**

In scientific notation the number is written in the standard exponential form as $N \times 10^n$.

N = a number with a single non-zero digit to the left of the decimal point.

n = an integer.

For example, a mass of 1500 g can be expressed in scientific notation in the following forms depending upon whether it has two, three or four significant figures.

1.5×10^3 g (**Two** significant figures)
1.50×10^3 g (**Three** significant figures)
1.500×10^3 g (**Four** significant figures)

In these expressions all the zeros to the right of the decimal point are significant. The exponential notation is an excellent way of expressing significant figures in very large or very small measurements. For example, Avogadro's constant is expressed as 6.022×10^{23} mol^{-1} and Planck's constant as 6.62×10^{-34} Js.

Lastly, it may be emphasized that **exact integral numbers,** such as number of pencils in a dozen or number of grams in a kilogram do not have any uncertainty associated with them and as such *these numbers have an infinite number of significant figures*.

CALCULATIONS WITH SIGNIFICANT FIGURES

During quantitative studies, scientists have to do calculations with numbers used for various measured physical quantities. These numbers generally have different number of significant digits depending upon the accuracy with which a particular measurement is made. While carrying out calculations with these numbers, the rule used is that *the accuracy of the final result is limited to the least accurate measurement*. In other words, *the final result cannot be more accurate than the least accurate number involved in the calculation*.

In order to understand this let us suppose that in a simple experiment two samples A and B are weighed by balances of different accuracy. Let the weight of sample A be 2.3 g and that of sample B be 14 g. The first weighing has 2 as the certain digit and 3 as the doubtful digit. The second weighing has 1 as the certain digit and 4 as the doubtful digit. If the samples A and B are mixed, the final weight would be expressed as 2.3 + 14 = 16 and not as 16.3. In 16.3 the second digit is certain whereas it is not so in case of the weight of sample B. Thus, if the result is expressed as 16.3, the final result would be more accurate than one of the measurements involved in the calculation which is not possible.

SOLVED EXAMPLES

Example 2.1. *Express the following numbers to four significant figures:*

(*i*) *5.607892* (*ii*) *32.392800*
(*iii*) *1.78986×10^3* (*iv*) *0.007837.*

Solution. (*i*) The fifth digit 8 is greater than 5, therefore the result will be expressed as 5.608.

(*ii*) It will be expressed as 32.39. The digit 2 is dropped and since it is less than 5 the figure is not rounded off to next number.

(*iii*) The fifth digit is greater than 5, therefore the result is expressed as 1.790.

(*iv*) The four digits to be retained are 7, 8, 3 and 7 therefore the number is expressed as 7.837×10^{-3}. The exponential term does not add to the significant figures.

Example 2.2. *Express the following numbers to three significant figures:*

(*i*) *6.022×10^{23}* (*ii*) *5.359*
(*iii*) *0.04597* (*iv*) *34.216.*

Solution. (*i*) The last digit to be retained is 2 and the digit to be dropped is 2 which is less than 5. The result will be expressed as **6.02×10^{23}.**

(*ii*) In this case the last digit to be retained is 5 and the digit to be dropped is 9, which is greater than 5. Hence, the last digit to be retained is increased by one. The number will be written as **5.36.**

(*iii*) The three digits to be retained in this case are 4, 5 and 9. The digit 7 is to be dropped which is greater than 5. Hence, the last digit to be retained will therefore be increased by one. The number will be rounded off to three significant digits as **0.0460.**

(*iv*) In this case the digits 1 and 6 will be dropped. Since, the digit following the last digit to be retained is 1, the last digit will be kept unchanged and the number is written as **34.2.**

2.3 DEMONSTRATE THE CORRECT HANDLING AND USE OF LABORATORY EQUIPMENT

The proper use of equipment for the measurement of various physical properties is very important. The correct handling and use of the laboratory equipment helps to easily and speedily use the apparatus to get desired precision and accuracy. One must be familiar with the following laboratory apparatus (Figs. 2.12 and 2.13).

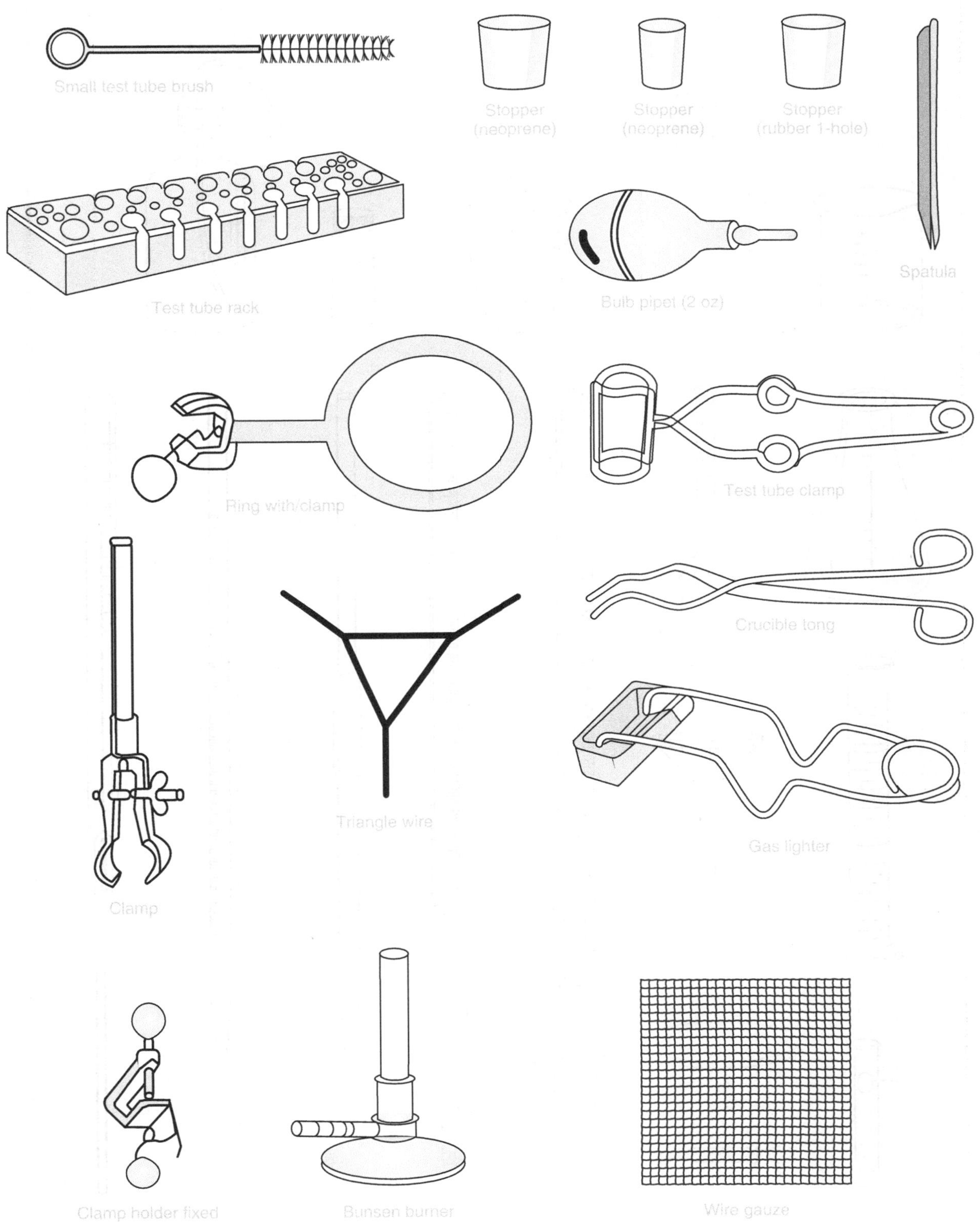
Small test tube brush
Stopper (neoprene)
Stopper (neoprene)
Stopper (rubber 1-hole)
Spatula
Test tube rack
Bulb pipet (2 oz)
Ring with/clamp
Test tube clamp
Crucible tong
Clamp
Triangle wire
Gas lighter
Clamp holder fixed
Bunsen burner
Wire gauze

Fig. 2.12

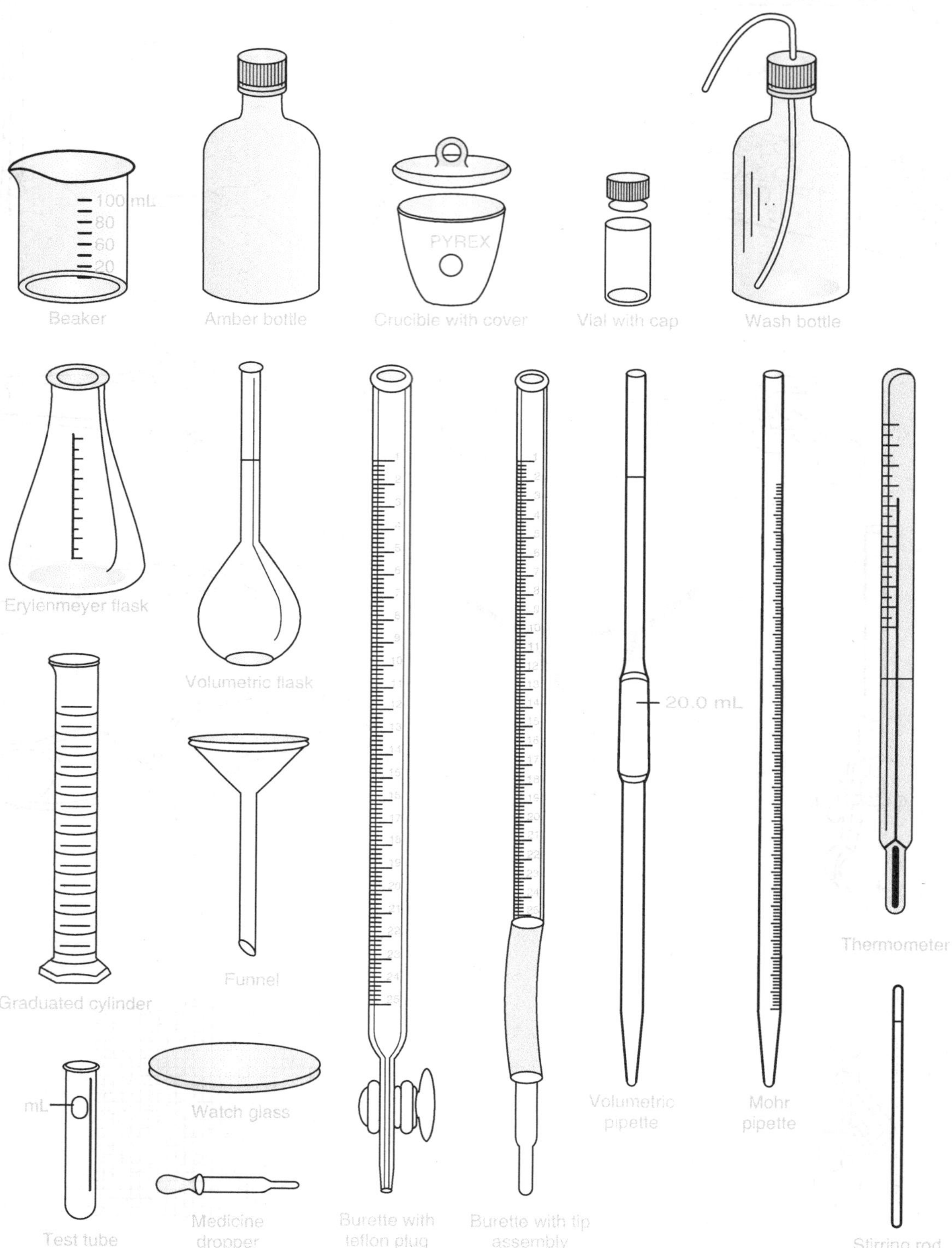
100 mL
80
60
20
Beaker
Amber bottle
PYREX
Crucible with cover
Vial with cap
Wash bottle
Erylenmeyer flask
Volumetric flask
Graduated cylinder
Funnel
20.0 mL
Volumetric pipette
Mohr pipette
Thermometer
mL
Test tube
Watch glass
Medicine dropper
Burette with teflon plug
Burette with tip assembly
Stirring rod

Fig. 2.13

Some experimental skills involved in the proper handling of some apparatus is diagrammatically explained below:

- Properly shaking the contents in a test tube.

(*i*)

Fig. 2.14

- Correctly heating a substance in a test tube holding it in a test tube holder. Note that the tube with solution is held slightly slantingly with the tube pointing away from you.

(*ii*)

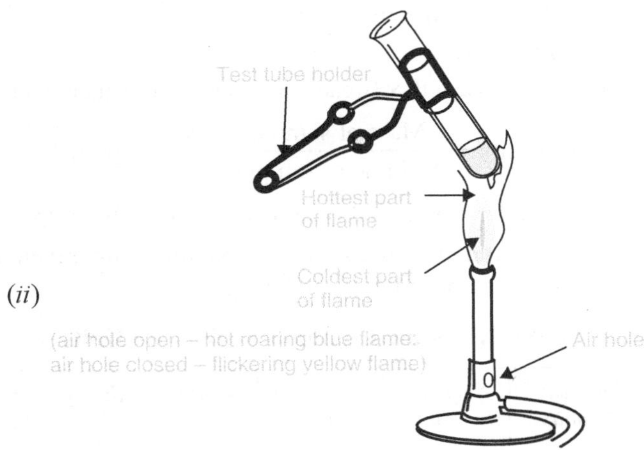

Fig. 2.15

- Correct way of adding solute and then stirring it with a rod to prepare a solution.

(*iii*)

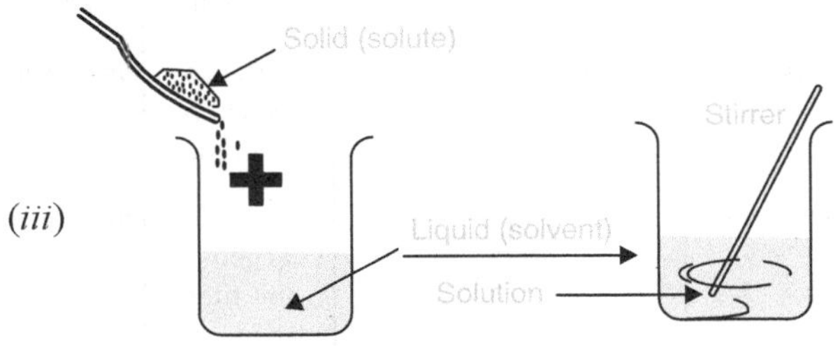

Fig. 2.16

- Correct handling of the equipment leads to fast filteration.

(*iv*)

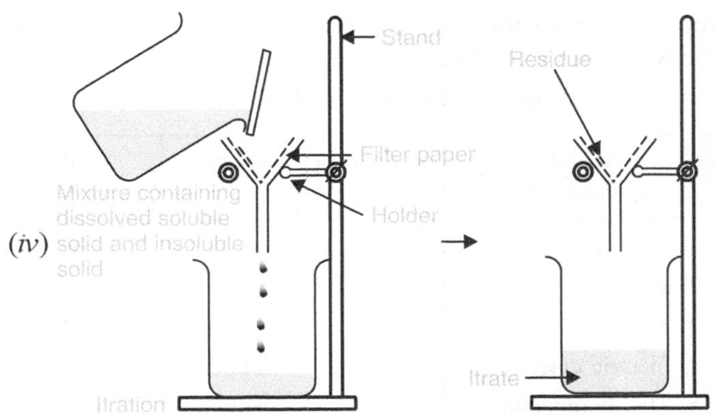

Fig. 2.17

- Proper and correct handling of a pipette and burette in use.

(*v*)

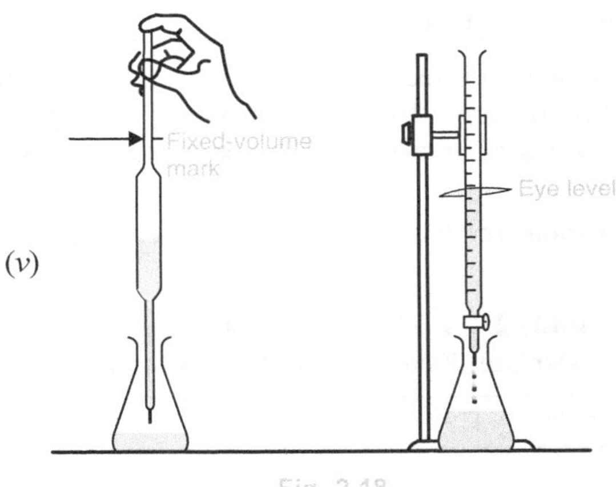

Fig. 2.18

2.4 BASE AND DERIVED UNIT OF MEASUREMENT [INTERNATIONAL SYSTEM OF UNITS (SI)]

The International System of Units (in French *Le System International d'Unités*—abbreviated as SI) was established by the 11th General Conference on Weights and Measures (CGPM from *Conference Generale des Poids at Measures*). The CGPM is an inter governmental treaty organization created by a diplomatic treaty known as Meter Convention which was signed in Paris in 1875. The metric system was found to be more convenient as it was based on the decimal system. The fundamental units of the **metric system** are, **gram** *for mass*, **the metre** *for length* and **the litre** *for the volume* of fluids.

In 1960, the International Committee of Weights and Measures recommended the use of an **International System of Units,** abbreviated as **SI Units.** SI units are largely a modification of the metric system. The seven basic units in the SI system are given in Table 2.1.

Table 2.1. The Seven Basic SI Units

Physical Quantity	*Abbreviation*	*Name of Unit*	*Symbol*
1. Length	*l*	metre	m
2. Mass	*m*	kilogram	kg
3. Time	*t*	second	s
4. Electric current	*I*	ampere	A
5. Thermodynamic temperature	*T*	kelvin	K
6. Luminous intensity	I_v	candela	cd
7. Amount of substance	*n*	mole	mol

The SI units of some of the physical quantities are either too small or too large. Two change the order of magnitude, these are expressed by using prefixes before the name of the base units.

The various prefixes used for this purpose are listed in Table 2.2.

Table 2.2. Prefixes for Expressing the Decimal Fractions in the SI System

Multiple	*Prefix*	*Symbol*
10^{24}	yotta	Y
10^{21}	zetta	Z
10^{18}	exa	E
10^{15}	peta	P
10^{12}	tera	T
10^{9}	giga	G
10^{6}	mega	M
10^{3}	kilo	k
10^{2}	hecto	h
10	deka	da
10^{-1}	deci	d
10^{-2}	centi	c
10^{-3}	milli	m
10^{-6}	micro	μ
10^{-9}	nano	n
10^{-12}	pico	p
10^{-15}	femto	f
10^{-18}	atto	a
10^{-21}	zepto	z
10^{-24}	yocto	y

The idea of using prefixes is illustrated by an example below:

A hundredth of a metre, corresponding to 10^{-2}, is a *centimetre* for which the symbol *cm* is used. Similarly for 10^3 metre we can use the term *kilometre*, symbolized as *km*.

DERIVED UNITS

In addition to the seven basic units given in Table 2.1, there are many derived units. These units are obtained by a combination of basic units. For deriving these units, we can multiply or divide the symbols for units as if they are algebric quantities. For example,

(*i*) The volume of a box is determined by multiplying length, breadth and height *i.e.*,

Volume = Length × Breadth × Height

if the units of length are *m*,

Volume = m × m × m = m^3

Similarly,

(*ii*) Area = Length × Breadth = m × m = m^2 (square metre)

(*iii*) Density = $\dfrac{\text{Mass of sample}}{\text{Volume of sample}} = \dfrac{\text{kg}}{\text{m}^3} = \text{kg m}^{-3}$

In a similar manner other units can also be built up.

Some commonly used SI derived units are given in Table 2.3 for the purpose of reading reference.

Table 2.3. Some Common SI Derived Units

Physical Quantity	*Name of Unit*	*Symbol for Unit*	*Definition of Quantity Unit*	*Expression in Terms of SI Basic Units*
Area	—	—	Length squared	m^2
Volume	—	—	Length cubed	m^3
Density	—	—	Mass per unit volume	kg/m^3 or $kg\ m^{-3}$
Speed	—	—	Distance travelled per unit time	m/s or ms^{-1}
Acceleration	—	—	Speed changed per unit time	m/s^2 or ms^{-2}
Force	newton	N	Mass times acceleration	kg ms^{-2}

Pressure	pascal	Pa	Force per unit area	kg m^{-1} s^{-2} or Nm^{-2}
Energy (work, heat)	joule	J	Force times distance travel	kg m^2 s^{-2}
Power	watt	W	Energy per second	kg m^2 s^{-3} or J s^{-1}
Frequency	hertz	Hz	Cycles per second	s^{-1}
Electric charge	coulomb	C	Ampere times second	A s
Electric potential difference	volt	V	Energy per unit charge	JA^{-1} s^{-1} or kg m^2 s^{-3} A

Some Important Facts About SI Units

- In chemistry *Angstrom*, abbreviated as Å is commonly used as a unit of *length* because it is almost of the size of an atom. However, it does not belong to the international system of units. In SI units we are supposed to use *nanometres* **(nm)** or *picometres* **(pm).**
- By convention the unit named after a scientist is written starting with a small letter and not with a capital letter. For example, unit for pressure is written as **pascal** and not as **Pascal.** Likewise unit of heat or work is written as joule and not as Joule.
- Abbreviations of units do not have a plural ending. For example it is incorrect to write **20 cms.** It should be **20 cm.**
- Prefixes are used with the basic units *e.g.,* kilometre is equal to 1000 m (metre (m) is the basic unit)
- Until 1964, a *litre* was defined as the volume of 1 kg of water at 4°C. Accordingly millilitre was a little larger than cubic centimetre. In 1964, a *litre* was redefined as exactly equal to 1000 cubic centimetres, thus, removing the confusion.

2.5 SCIENTIFIC METHOD

The scientific method is a logical and rational order of steps by which scientists come to conclusions about the world around them. **The scientific method helps to organize thoughts and procedures** so that scientists can be confident in the answers they find. Scientists use observations, hypotheses and deductions to make these conclusions, just like you will use the scientific method in your science fair project. **The steps of the scientific method are:**

Observation/Research
Hypothesis
Prediction
Experimentation
Conclusion

OBSERVATION

This step could also be called "research". It is the first stage in understanding the problem you have chosen. For this stage of the scientific method, it is important to use as many sources as you can find. The more information you have, the better the design of your experiment is going to be.

HYPOTHESIS

The next stage of the scientific method is known as the "hypothesis". This word basically means "a possible solution to a problem, based on knowledge and research". The hypothesis is a simple statement that defines what you think the outcome of your experiment will be.

PREDICTION

The hypothesis is your general statement of how you think the scientific phenomenon in question works. Your prediction lets you get specific as to how will you demonstrate that your hypothesis is true. The experiment that you will design is done to test the prediction. An **important thing to remember** during this stage of the scientific method is that once you develop a hypothesis and a prediction, you shouldn't change it, even if the results of your experiment show that you were wrong. An incorrect prediction doesn't mean that you "failed". It just means that the experiment brought some new facts to light that maybe you hadn't thought about before.

EXPERIMENTATION

This is the part of the scientific method that tests your hypothesis. An experiment is a tool that you design to find out if your ideas about your topic are right or wrong. The experiment is the most important part of the scientific method.

CONCLUSION

The final step in the scientific method is the conclusion. This is a summary of the experiment's results, and how those results match up to your hypothesis.

There are two options for your conclusions: Based on your results, either you can reject the hypothesis, or you cannot reject the hypothesis. A hypothesis connot be proved with a single experiment, because there is a chance that you made an error somewhere along the way. Hence, on the basis of the results of many experiments, what you can say is that your results support the original hypothesis.

If your original hypothesis didn't match up with the final results of your experiment, don't change the hypothesis. Instead, try to explain what might have been wrong with your original hypothesis. What information did you not have originally that caused you to be wrong in your prediction?

What are the reasons why the hypothesis and experimental results didn't match up? The flow sheet for the scientific method is shown in Fig. 2.19.

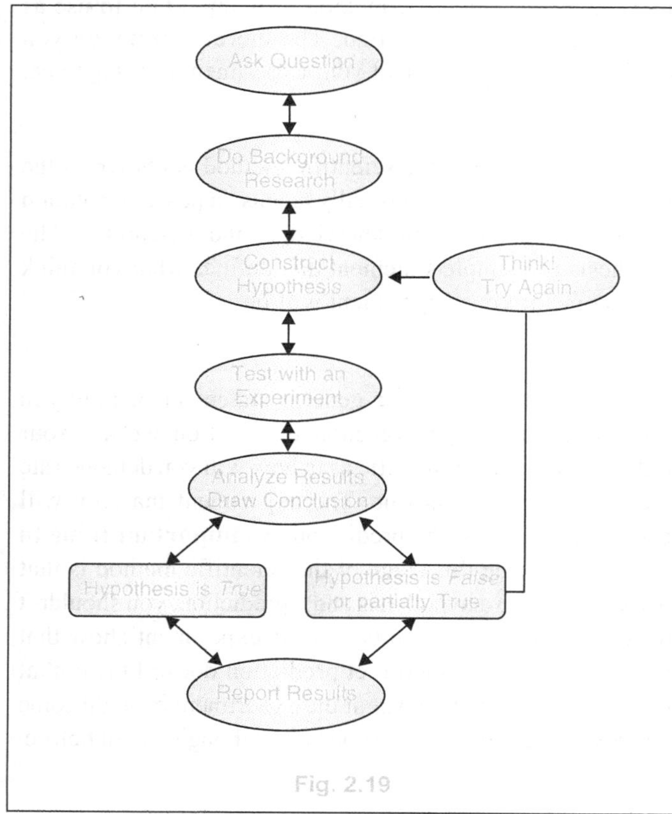

Fig. 2.19

EVALUATE YOURSELF

I. Objective Type Questions

Select the most appropriate choice from the options given as (a), (b), (c) and (d) after each question:

1. Two students A and B report the mass of the same substance as 12.0 g and 12.00 g, respectively. Which of the following statements is correct?
   (*a*) Both are equally accurate
   (*b*) A is more accurate than B
   (*c*) B is more accurate than A
   (*d*) Both are scientifically inaccurate.
2. Which of the following is incorrect about SI units?
   (*a*) Density in kg m^{-3} (*b*) Force in Newton's
   (*c*) Pressure in Pascals (*d*) Amount of substance in mol L^{-1}.
3. Picometer (pm) and nanometer (nm) are correctly related as
   (*a*) 1 pm = 10 nm (*b*) 1 nm = 10 pm
   (*c*) 1 pm = 100 nm (*d*) 1 nm = 1000 pm.
4. kg m^{-1} s^{-2} is the unit of
   (*a*) momentum (*b*) velocity
   (*c*) pressure (*d*) acceleration.
5. The prefix atto (a) stands for
   (*a*) 10^{-9} (*b*) 10^{-12}
   (*c*) 10^{-18} (*d*) 10^{-15}.
6. Acceleration is the increase in velocity of a body per unit time. The correct unit for it is:
   (*a*) N kg^{-1} (*b*) m^2 s^{-2}
   (*c*) Jm^{-1} (*d*) kg m^{-3}.
7. Which of the following is not an SI unit?
   (*a*) kelvin (*b*) metre (*c*) ampere (*d*) yard.
8. A sample of compound contains 20.45% of chlorine. The chemical analysis of the sample by two methods A and B gave a mean value of 20.60% and 20.54% respectively. The percentage accuracy of
   (*a*) A is less than B hence B is more accurate
   (*b*) A is greater than B but B is more accurate
   (*c*) A is less than B hence A is more accurate
   (*d*) Both the results are accurate.
9. Which of the following reported values is more precise?
   (*a*) 20.45 ± 0.15 (*b*) 20.60 ± 0.05
   (*c*) 20.20 ± 0.10 (*d*) 20.50 ± 0.02.
10. Four students A, B, C and D reported the percentage of Aluminium in a sample as

Student A	*Student B*	*Student C*	*Student D*
20.50	20.75	20.20	20.20
20.48	20.65	20.25	20.40
20.52	20.40	20.15	20.00
20.50	20.28	20.20	20.60

On the basis of these results it can be concluded that
(*a*) A has good precision and good accuracy
(*b*) B has good precision but low accuracy
(*c*) C has good precision and good accuracy
(*d*) D has poor precision but good accuracy.

II. Fill in the Blanks

11. Complete the following sentences by supplying the appropriate words:
   (*i*) A systematic error which makes all the measurements wrong by a certain amount is known as ....
   (*ii*) Two numbers are reported a 2.7182820135423 and 2.718281828. Among these numbers the first is more .... and second is more ....

(*iii*) The mass of the moon is reported as 7.3×10^{22} kg. In a calculation it is found to be 6.9×10^{22} kg, the accuracy of the calculation is ....

(*iv*) A femto (f) is .... whereas a peta (P) is ....

(*v*) Force is expressed in terms of the unit newton (N). Its expression in terms of SI basic units will be ....

(*vi*) If joule is the unit of energy then the unit of power is ....

(*vii*) There are .... basic SI units.

(*viii*) A block of substance weighs 48 μg. Its mass in SI units will be ....

III. Discussion Questions

12. Match the following prefixes with their multiples

Prefixes	*Multiple*
(*i*) *yocto*	10^{24}
(*ii*) *atto*	10^{21}
(*iii*) *zetta*	10^{-24}
(*iv*) *yotta*	10^{18}

13. Convert each of the following:
    (*i*) 456 nm to metres (*ii*) 45.6 mL to microlitres
    (*iii*) 10 mg to kilograms (*iv*) 50 km/h to metres per second.
14. Density is usually reported as grams per cubic centimetre. Convert this to the SI unit of density. Indicate clearly the conversion factor used.
15. Convert the following into metre
    (*i*) 7 nm (diameter of a small virus)
    (*ii*) 40 Em (thickness of milky way galaxy)
    (*iii*) 1.4 Gm (diameter of the sun)
    (*iv*) 41 Pm (distance of the nearest star).
16. Convert into kilogram
    (*i*) 0.91×10^{-27} (mass of electron)
    (*ii*) 1 fg (mass of human DNA molecule)
    (*iii*) 500 Mg (mass of Jumbo jet)
    (*iv*) 3.34×10^{-24} g (mass of hydrogen molecule).
17. Define precision and accuracy.
18. Give the unit of the final answer obtained after each of the following operations:
    (*i*) A density in gram per cubic centimetre is divided by a mass in gram.
    (*ii*) A volume in cubic centimetres is divided by an area in square metres.
    (*iii*) A mass in kilograms is multiplied by velocity in metres per second and divided by energy in joules.
19. What is the number of significant figures in each of the following?
    (*i*) 2.56×10^3 (*ii*) 256 (*iii*) 5000
    (*iv*) 0.00256 (*v*) 0.0320 (*vi*) 13.560.
20. A block of wood having the dimensions 10 cm × 5 cm × 20 cm weighs 1.5 kg. What is the density of wood, expressed in grams per cubic centimetre?
21. Express the following SI base units using a power of 10 notation
    (*i*) 1.35 mm (*ii*) 1 day (*iii*) 6.45 mL
    (*iv*) 48 μg (*v*) 0.0426 in.
22. Vanadium metal is added to steel to impart strength. The density of Vanadium is 5.96 g/cm^3. Express it in SI unit (kg/m^3).
23. Can you make a measurement that's very precise, but not very accurate?
24. Your partner uses a ruler to measure the length of a pencil and states that the length is 20.35232403 cm. What is your response to the given value in terms of precision?
25. Outline the steps of a scientific method.
26. The mass of the moon is 7.3×10^{23} kg. If a student reports the mass as 6.9×10^{22} kg, what is the accuracy of his calculation?

Answers

I. Objective Type Questions

1. (*c*) **2.** (*d*) **3.** (*d*)
4. (*c*) **5.** (*c*) **6.** (*a*)
7. (*d*) **8.** (*b*) **9.** (*d*)
10. (*a*).

II. Fill in the Blanks

11. (*i*) bias (*ii*) precise, accurate
(*iii*) 0.4×10^{22} kg (*iv*) 10^{-15}, 10^{15}
(*v*) kg ms^{-2} (*vi*) watt
(*vii*) Seven (*viii*) 4.8×10^{-8} kg.

III. Discussion Questions

12. yocto is 10^{-24}; atto is 10^{-18}; zetta is 10^{21} and yotta is 10^{24}.
13. (*i*) 4.65×10^{-7} m (*ii*) 45.6×10^3 micro litres
(*iii*) 10×10^{-6} kg (*iv*) 13.9 m/s.
14. g cm^{-3} = 10^3 kg m^{-3}, $\dfrac{10^3 \text{ kg m}^{-3}}{\text{g cm}^{-3}} = 1$.
15. (*i*) 7×10^{-9} m (*ii*) 40×10^{18} m $= 4 \times 10^{19}$ m
(*iii*) 1.4×10^9 m (*iv*) 41×10^{15} m $= 4.1 \times 10^{16}$ m.
16. (*i*) 9.1×10^{-31} kg (*ii*) 1×10^{-15} g $= 1 \times 10^{-18}$ kg
(*iii*) 500×10^6 g $= 5 \times 10^5$ kg
(*iv*) 3.34×10^{-4} g $= 3.34 \times 10^{-7}$ kg.

18. (*i*) cm^{-3} (*ii*) cm (*iii*) no unit

19. (*i*) Three (*ii*) Three (*iii*) One

(*iv*) Three (*v*) Three (*vi*) Five

20. 1.5 g cm^{-3}

21. (*i*) 1 m = 10^3 mm, therefore 1.35 mm = 1.35×10^{-3} m

(*ii*) 1 day has 24 *h i.e.*, 24 × 60 min *i.e.*,

24 × 60 × 60 sec

Therefore 1 day = 24 × 60 × 60

= 86400 s = 8.64×10^4 s

(*iii*) $(1\ m)^3$ = (100 cm), therefore 1 mL = $10^{-6}\ m^3$

Hence, 6.45 mL = $6.45 \times 10^{-6}\ m^3$

(*iv*) 1 μ = 10^{-6}, therefore

48 μg = 48×10^{-6} g = 4.8×10^{-5} g

(*v*) 1 inch = 2.54×10^{-2} m, therefore

0.0426 inches = $0.0426 \times 2.54 \times 10^{-2}$ m

= 1.082×10^{-3} m.

22. 5.96 g cm^{-3} = $\dfrac{5.96 \times 10^{-3}\ kg}{10^{-6}\ m^3}$ = 5960 kg m^{-3}.

23. Yes. Suppose you measured the voltage of a 9-volt battery and get an answer of 17.1453 volts. You must have used a device that allowed 4 decimal places of precision, but your answer was wrong. It was inaccurate by over 8 volts.

SECTION-1

UNIT-3

BASIC SAFETY LABORATORY PRACTICES

Specific Objectives

After studying this unit, the student should be able to:

- read and follow rules and instructions in the laboratory
- explain what hazard symbols are and relate their importance to the handling and use of laboratory chemicals and equipment
- handle some minor laboratory accidents and give first aid
- outline the need to have personal protective equipment in the chemistry laboratory
- explain why some chemicals should not be stored alphabetically
- quench small fire in the laboratory.

Laboratory work is subject to hazards of many kinds. A person working in the laboratory should be aware of all these hazards. Most accidents occur due to ignorance and carelessness of students. You and your laboratory partners must always practice laboratory safety. Therefore, study laboratory safety rules and follow them strictly.

3.1 RULES AND INSTRUCTIONS IN THE LABORATORY

In the laboratory, improper handling of equipment and materials can cause injury and accidents. However, a laboratory can be a safe place to work if you are careful. You and your partners must always practice laboratory safety. In the laboratory the safety rules listed below must be followed. Before beginning any laboratory work, read these rules and follow them carefully.

1. Carefully follow all the written and oral instructions. Perform only those experiments which are assigned to you by your teacher. Unauthorized experiments are forbidden.
2. Do not work alone in the laboratory.
3. Tell your teacher, if you have any medical problems that might be affected by laboratory work, such as asthma, allergies, etc.
4. Never drink, eat or smoke in the laboratory because chemicals may enter through the mouth or lungs.
5. Keep work areas clean and tidy at all times. Only laboratory manuals and notebooks should be brought to the work area.
6. Be aware of your neighbours activities; you may be the victim of their mistakes. If necessary report your concern to the instructor.
7. Record data in ink as you perform the experiment.
8. Do not engage in horseplay.

9. Always use prescribed instruments such as tongs, test-tube holder, etc.
10. Conduct any experiment involving poisonous vapour in a ***fume cupboard***.
11. Dispose off waste materials as instructed by your teacher.
12. Clean and wipe all the work surfaces at the end of class. Wash your hands thoroughly.
13. Know the location and proper use of emergency equipment, such as the ***first aid kit, fire extinguisher, fire shower*** and ***fire blanket.***
14. Keep all combustible materials away from open flames.
15. Never put your face near the mouth of the container that holds chemicals. Never smell any chemical directly. When testing for odors, use a wafting motion to direct the odors to your nose.
16. Wear appropriate clothing for working in the laboratory. Do not wear loose garments. Roll up or secure long sleeves. Wear laboratory ***apron*** or ***coat*** to protect clothing.
17. Tie back or cover long hair especially in the vicinity of an open flame. Hair may ignite near an open flame.
18. Never wear open shoes or sandals in the laboratory.
19. Wear ***goggles, safety glasses*** or ***eye-shields*** at all times during an investigation.
20. Avoid wearing contact lenses in the laboratory.
21. Whenever your skin (hands, arms, face, ...) comes into contact with laboratory chemicals, wash it quickly and thoroughly with soap and warm water. Use the ***eye-wash fountain*** to flush chemicals from the eyes or face. Do *not* rub the affected area, especially the face or eyes, with your hands before washing.
22. If a chemical is spilled over a large part of the body, use the ***safety shower*** and flood the affected area for about 15 minutes. Remove all contaminated clothing. Get medical attention.
23. Wash your hands and arms before leaving the laboratory.
24. Report any accident or injury to your class teacher.

Handling Chemicals

1. Read carefully labels on chemical bottles before removing any chemical. Take only as much chemical as you need.
2. To avoid contamination, do not return any unused chemicals to stock bottles.
3. Avoid touching chemicals with your hands. If chemicals do come in contact with your hands, wash them immediately.
4. When mixing an acid and water, always add the acid to the water.
5. Wear ***latex gloves*** when handling concentrated acids and bases.

3.2 COMMON HAZARD SYMBOLS AND THEIR MEANING

Hazard means risk of exposure to some danger. In the laboratory a person may come across many hazardous chemicals. Hazardous chemicals or equipments are labelled with hazard symbols. Be aware of these symbols and follow the precautions that you must take while dealing with such substance. Some common hazard symbols are given below:

Flammable

This type of symbol is present on containers containing flammable chemicals. For example, benzene, ether, petrol, etc. Such chemicals should be kept away from open flames because they would catch fire if brought near a flame.

Explosive

This type of symbol is present on containers containing explosives. For example, sulphur, phosphorus, etc. Such containers should not be subjected to any pressure or heat.

Corrosive

This symbol is present on containers containing corrosive chemicals. Precautions should be taken while handling such chemicals. Such chemicals should not be brought in contact with the skin or clothes.

Poison

This type of symbol indicates danger. It is present on bottles containing poisonous or toxic chemicals.

Some more hazard symbols are:

Oxidizer

Low Level Hazard

Severe Chronic Hazard

Environmental Hazard

3.3 LABORATORY EMERGENCIES AND FIRST AID

Accidents do not often happen in well-equipped chemistry laboratories if students understand safe laboratory procedures and are careful to follow them. When an occasional accident does occur, it is likely to be a minor one.

1. Bleeding from a Minor Cut

Most cuts that occur in the chemistry laboratory are minor. Allow the blood to flow for a few seconds, flush the wound thoroughly with water for minor cuts, apply pressure to the wound with a clean absorbent cloth. Apply antiseptic and bandage the wound to prevent contamination. Take the victim to the doctor.

2. Thermal Burns

In the case of a burn, apply cold water and/or ice immediately to the burned area until the pain subsides. Wrap the burned area to protect it from infection. It is best to avoid oils and ointments in first aid treatment. Go to the doctor.

3. Acid or Base Spilled on the Skin

Remove all clothing that has the chemical on it and flush the skin with water for at least 15 minutes. Take the victim to the doctor.

4. Chemical Splashes

Safety showers and ***Eye Washes*** are located in each of the laboratories. It has been conclusively proven that immediate washing of the skin with a generous amount of water is the most effective first aid treatment for chemical burns. Pulling the ring hanging from a safety shower will cause the rapid discharge of water. All of the research and teaching laboratories are also equipped with *eye wash fountains* in case a chemical splashes in the eyes. Go to the doctor.

5. Chemicals in the Mouth

Many chemicals are poisonous to varying degrees. Any chemical taken into the mouth should be spat out and the mouth rinsed thoroughly with water. Tell the victim NOT to swallow the water. Note the name of the chemical and go to the doctor immediately.

If the victim swallows a chemical, note the name of the chemical and notify the doctor immediately. Do NOT give the victim anything to drink.

6. Clothing or Hair on Fire

If your clothing catches fire, drop to the floor and roll around to extinguish the flames. If you are helping another person whose clothing is on fire, smother the flames by rolling the person on the floor, in a fire blanket, or in a heavy coat. For hair fires, use a fire blanket to smother the flames. Go to the doctor immediately.

7. Breathing Smoke, or Chemical Fumes

Inhalation of smoke or chemical fumes is unlikely if all experiments that give off smoke or noxious gases are conducted in a ***fume cupboard***.

If smoke or toxic gases such as chlorine, hydrogen sulphide are present in the laboratory, all persons—even those who do not feel ill-should leave the laboratory immediately. Make sure the room is thoroughly ventilated before anyone goes back in.

8. Shock

People who are suffering from any severe injury may be in a state of shock. A person in shock is usually pale and faint. The person may be sweating, with cold, moist skin and a weak, rapid pulse.

Shock is a serious medical condition. Do NOT allow a person in shock to walk anywhere even to the doctor. Call for emergency help immediately. While emergency help is being summoned, loosen any tight fitting clothing and keep the person comfortable.

3.4 PERSONAL PROTECTIVE EQUIPMENT

Some equipment used for personal protection are given below:

1. ***Safety glasses, eye shields*** or ***chemical goggles*** are used to protect the eyes from coming into contact with chemicals while working in a chemistry laboratory.

2. ***Safety showers*** and ***eye wash stations*** are located in chemical laboratories for first aid treatment of chemical burns and chemical splashes in the eyes. When the ring hanging from a safety shower is pulled, water is discharged rapidly.
3. ***Corrosive resistant gloves*** such as ***latex gloves*** are used while handling corrosive chemicals such as concentrated acids and concentrated alkalies.
4. ***Apron*** or ***lab coat*** protects the clothes and body parts from chemicals while working in laboratory.
5. ***Fume cupboard*** should be used to carry out reactions involving toxic fumes of vapours. ***Gas masks*** are also used while dealing with toxic gases.
6. ***Face shield.*** While carrying out reactions where an explosion may occur face shield should be used to protect the face from chemicals and glass pieces.

3.5 SAFE STORAGE OF CHEMICALS

The basic rules for the safe storage of chemicals in the laboratory are:

1. Segregate all the incompatible chemicals for proper storage of chemicals for hazard class codes.
   - In other words, store like chemicals together and away from other groups of chemicals that might cause reactions if mixed.
2. Do not simply store chemicals in alphabetical order. When chemicals are stored alphabetically there is a great possibility of incompatible substances coming into contact.
   - Flammable materials should be stored in an approved, flammable materials storage cabinet or room.
   - Non-compatible chemicals shall be stored separately.
   - Liquids should be stored in unbreakable or double-contained packaging.
   - Store acids in a dedicated place.
   - Store severe poisons in a dedicated poison cabinet.
3. Chemicals should not be stored on the floor.
4. Liquid chemicals should not be stored above eye level.
5. All chemicals should be labelled and dated. Look for unusual conditions in chemical storage areas, such as:
   - improper storage of chemicals, leaking or deteriorating containers, spilled chemicals, temperature extremes (too hot or cold in storage area).
   - lack of or low lighting levels, blocked exits.
6. Never store a chemical in standard refrigerator.
7. Chemical storage under, over or near a sink should be avoided since many chemicals are affected by moisture. Some chemicals may cause explosion on coming into contact with water.

Problems Associated with Alphabetical Chemical Storage—Some examples:

- Aluminium metal + Ammonium nitrate
  – *likely to explode*
- Acetic anhydride + Acetaldehyde
  – *can cause violent explosion*
- Lead perchlorate + Methanol
  – *explosive mixture*
- Sodium cyanide + Sulphuric acid
  – *on reaction, deadly poisonous gas HCN is released.*
- Silver oxide + Sulphur
  – *explosive mixture.*

3.6 PUTTING OFF FIRES

Small fires in the laboratory may be extinguished using the ***fire extinguisher*** provided in the laboratory. For this purpose learn how to use the fire extinguisher.

A ***fire blanket*** can also be used to smother small fires on a bench or other flat surface. In order to extinguish the fire, the fire is covered with a fire blanket. The fire gets extinguished due to lack of air.

Dry sand can also be used to put out small fires. For this purpose buckets of dry sand may be provided in the laboratory.

EVALUATE YOURSELF

I. Discussion Questions

1. Draw four hazard symbols and explain what they mean.
2. List any four personal protective equipment. Explain how each of these protects you from accidents.
3. What are the basic rules for safe storage of chemicals?
4. Explain why chemicals should not be stored simply in alphabetical order.
5. Name two devices which can be used to extinguish small fires in the laboratory.
6. What precautions should be taken while handling chemicals in the laboratory?

SHS 2

SECTION 2

ATOMIC STRUCTURE

General Objectives

After studying this section, the student should be able to:

1. recognize that atomic models are used to explain atoms and help us understand the interaction of elements and compounds observed on a macroscopic scale.
2. recognize discoveries from Dalton (atomic theory), Thomson (the electron), Rutherford (the nucleus) and Bohr (planetary model of atom) and understand how these discoveries lead to the modern atomic theory.
3. understand how periodic properties like atomic radii, ionization energy and electron affinity change with atomic number and principal quantum number.
4. show understanding of both IUPAC and the North American ways of grouping elements on the periodic table.

In this section, we shall study

Unit 4: Particulate Nature of Matter

Unit 5: Structure of Atom

Unit 6: Periodicity

SECTION-2

UNIT-4

PARTICULATE NATURE OF MATTER

Specific Objective

After studying this unit, the student should be able to:

- describe the characteristics and nature of matter.

4.1 INTRODUCTION

In our daily life, we come across many objects, the knowledge about which can be gained by one or more of our senses like sight, touch, hearing, taste and smelling. These objects possess mass, occupy space and may have different shapes, sizes and colours. All these objects constitute matter. **Matter** may thus, be defined as *anything that occupies space, possesses mass, offers resistance and can be felt by one or more of our senses*. Some examples of matter are, water, air, metals, plants, animals, etc. Thus, matter has countless forms. Matter can be classified into different categories depending upon its physical or chemical nature. Matter is categorized as a gas, a liquid or a solid on the basis of physical state. Air is gas, water is liquid whereas sand is solid. Gases and liquids are fluids but solids are rigid.

On the basis of chemical nature matter may be classified as an element, a compound or a mixture. Elements and compounds are pure substances whereas mixtures contain two or more pure substances.

In this Unit, we shall study the classification of matter on the basis of its physical properties.

4.2 PARTICULATE NATURE OF MATTER

- **Matter is made up of small particles**

The particle nature of matter can be demonstrated in activity 4.1:

ACTIVITY 4.1 | **To demonstrate particle nature of matter**

1. Take about 50 cm^3 water in a 100 cm^3 beaker.
2. Mark the level of water.
3. Add some sugar to the beaker and stir with the help of a glass rod.
4. Observe the change in water level.

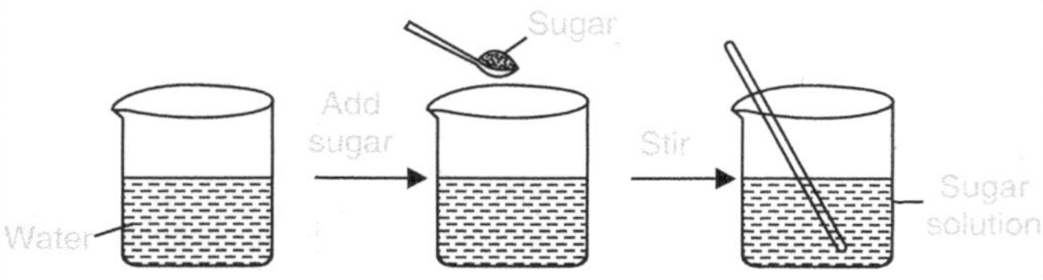

Fig. 4.1. Dissolution of sugar in water. In solution particles of sugar are present in the spaces between particles of water

It is observed that the crystals of sugar disappear. The level of water remains unchanged. These observations can be explained by assuming that matter is made up of small particles. On dissolution, the particles of sugar get distributed into the spaces between particles of water.

- **The constituent particles of matter are extremely small in size**

The following activity demonstrates that the constituent particles of matter are very small.

ACTIVITY 4.2 | **To demonstrate that the particles of matter are very small**

1. Take a 250 cm^3 beaker and put 100 cm^3 water into it.
2. Now add 2–3 crystals of $KMnO_4$ and stir with a glass rod in order to dissolve the crystals.
3. Take 10 cm^3 of this solution and add to 100 cm^3 of water taken in another beaker.
4. Take 10 cm^3 of this diluted solution and put into 100 cm^3 of water taken in still another beaker.
5. Repeat this process 10 times. Observe the colour of the solution in the last beaker.

   It is observed that the water in the last beaker is still coloured but the intensity of colour becomes light. It indicates that $KMnO_4$ crystal contains millions of tiny particles, some of which are still present even in the last beaker after so much dilution.

- **There are spaces between particles of matter**

  In activity 4.1 we observed that when sugar is dissolved in water, the volume of the liquid remains unchanged. During dissolution, the particles of sugar get into the spaces between the particles of water. As a result, they get evenly distributed and there is no noticeable change in volume. Similarly, when $KMnO_4$ is dissolved in water, its particles get evenly distributed throughout the bulk of water. This is indicated by uniform colour of the solution. This indicates that there are spaces between particles of matter. The particles of $KMnO_4$ get uniformly distributed in the spaces between water molecules.

- **The particles of matter are continuously moving**

  The motion of particles of matter can be demonstrated by the following practical activities:

ACTIVITY 4.3 | **To demonstrate motion of particles of matter**

Place a bottle containing concentrated aqueous solution of ammonia in a corner of the room. Remove the stopper.

What do you observe?

It is observed that ammoniacal smell can be sensed sitting at a distance.

It demonstrates that the particles of ammonia are moving. Due to this motion they are able to reach the observer.

Similarly, if an incense stick is lighted and placed in one corner of a room, its pleasant smell spreads in the whole room quickly. It demonstrates that the particles of matter possess motion. A burning incense stick produces some gases (vapour) having pleasant smell. The particles of these gases due to motion spread in the entire room and their presence can be felt by sensing the smell.

ACTIVITY 4.4 | **To demonstrate motion of particles in water and ink**

1. Take a 250 cm^3 beaker and put about 100 cm^3 of water into it.
2. Put a drop of blue ink to the water taken in the beaker.

   What do you observe?

   It is observed that the blue ink gets evenly distributed in the water.

   This demonstrates that the particles of water and ink possess motion.

   Due to the motion of the particles, the particles of the two liquids are able to mix with each other.

ACTIVITY 4.5	**To demonstrate that the kinetic energy of particles increases with increase in temperature**

1. Take two beakers. To one beaker add 100 cm^3 of cold water and to the other beaker add 100 cm^3 of hot water.
2. Now add a crystal of potassium permanganate to both beakers.

   What do you observe?

   It is observed that the purple colour of potassium permanganate starts spreading and after sometime the entire solution becomes purple. The rate of mixing is faster in case of hot water. This experiment demonstrates that the particles of matter possess motion and that the kinetic energy of the particles increases with increase in temperature.

The above activities demonstrate that when two different forms of matter are brought in contact they intermix spontaneously. This intermixing is possible due to the motion of the particles of matter and also due to the spaces between them. The intermixing takes place due to movement of particles of one form into the spaces between the particles of the other form of matter. *This spontaneous intermixing of particles of two different types of matter is called* ***diffusion.*** The rate of diffusion becomes faster with an increase in temperature because at higher temperature, the particles have more energy and hence move faster.

- **Particles of matter attract each other**

  There are forces of attraction between particles of matter. The evidence for forces of attraction in gases is obtained from the fact that they can be liquified by applying pressure.

The ***important characteristics of particles of matter*** are summarized below:

1. The particles of matter are extremely small in size.
2. The particles of matter have spaces between them.
3. The particles of matter are continuously moving.
4. The particles of matter attract each other.

The constituent particles of matter may be ***atoms, molecules*** or ***ions.*** Some examples are given below in tabular form:

Constituent particles	*Examples*
1. Atoms	Argon, neon, helium, diamond
2. Molecules	Sucrose, glucose, urea, methane, carbon(IV) oxide
3. Ions	Sodium chloride, magnesium oxide, zinc sulphide

4.3 STATES OF MATTER

Matter can be classified into three categories depending upon its physical state, namely: *solids, liquids* and *gases.* These states of matter arise due to variations in the characteristics of the particles of matter.

PROPERTIES OF SOLIDS

(*i*) Matter in a solid state possesses *a definite volume, a definite shape, distinct boundaries* and *a definite mass.*

(*ii*) Solids are *rigid* and *almost incompressible.*

(*iii*) Solids may break under force but it is difficult to change their shape.

(*iv*) Solids generally possess *high densities.*

(*v*) Solids *do not exhibit diffusion.*

Some common **examples** are: table, chair, common salt, silver, ice, diamond, etc.

PROPERTIES OF LIQUIDS

(*i*) Matter in a liquid state possesses *a definite volume, a definite mass,* but *no definite shape.*

(*ii*) Liquids are also *almost incompressible* but are *not rigid.* In fact, they can flow and acquire the shape of the container in which they are kept.

(*iii*) Liquids can *undergo diffusion.*

(*iv*) Liquids also have *high densities* but less than that of solids.

Some **examples** are: milk, water, alcohol, petrol, kerosene, fruit juices, etc.

PROPERTIES OF GASES

(*i*) Matter in a gaseous state *has neither a definite volume nor definite shape but it has a definite mass.* It acquires the shape and volume of the container.

(*ii*) *Gases are highly compressible.* For example, natural gas in compressed form is used as fuel (Compressed Natural Gas—**CNG**) in internal combustion engines. Oxygen supplied to hospitals in cylinders is also in a compressed form. Due to its high compressibility large volumes of gas can be compressed into a small cylinder and transported easily.

(*iii*) *The gases exhibit the property of diffusing very fast into other gases.*

(*iv*) *Gases exert pressure on the walls of the container in which they are stored.*

(*v*) *Gases have very low densities.*

Some common **examples** of gases are: air, hydrogen, carbon(IV) oxide, hydrogen sulphide, ammonia, oxygen, nitrogen, etc.

A comparison of the characteristic properties of solids, liquids and gases are given in Table 4.1.

Table 4.1. Comparison of the Characteristic Properties of Solids, Liquids and Gases

Property	*Solids*	*Liquids*	*Gases*
1. Shape	Definite	Take the shape of the container, but do not necessarily occupy all of it.	Take the shape of the container by occupying the whole of the space available to them.
2. Volume	Definite	Definite	Take the volume of the container.
3. Compressibility	Almost nil	Almost nil	Very large.
4. Fluidity or Rigidity	Rigid	Fluid	Fluid
5. Density	Large	Large	Very small.
6. Diffusion	Generally do not diffuse	Diffuse slowly	Diffuse rapidly.
7. Free Surfaces	Any number of free surfaces	Only one free surface	No free surface.

Why Solids, Liquids and Gases Exhibit Different Properties?

The properties of matter in the three states of matter are different because the characteristics of the particles vary in the three states of matter.

Now let us understand how the characteristics of particles vary in the three states of matter.

- **In solids,** the interparticle spaces are small. They have smaller amounts of energy than the same particles in the liquid and gaseous states. Consequently, the particles in the solid state cannot overcome the strong forces of attraction which are holding them together. In solids, particles can only vibrate about fixed positions. Thus, particles in a solid have vibrational and rotational motion but no translational motion. Because of smaller interparticle spaces, solids are almost incompressible and rigid due to absence of translational motion.
- **In liquids,** interparticle spaces are somewhat larger than in solids and the particles have larger amounts of energy. The particles in liquids can overcome the interparticle forces between each other to some extent and hence can move freely. However, the intermolecular forces in liquids are strong enough to keep the particles within the bulk. The particles in liquid state possess vibrational, rotational and translational motion.

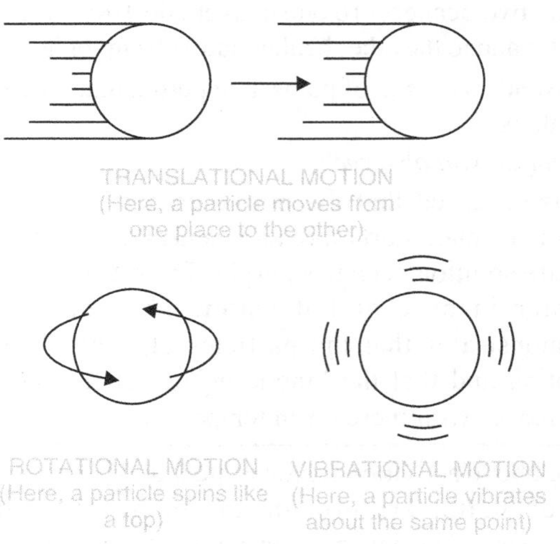

Fig. 4.2. Different types of motion.

- **In gases,** the interparticle spaces are very large and the particle possess much larger amounts of energy than those in solids and liquids. The gas particles have sufficient energy to overcome the interparticle attractive forces almost completely. As a result the gas particles move rapidly and randomly into any space available to them. Thus, a gas fills completely the vessel in which it is kept. That is why gases have neither definite shape nor definite volume. Since particles in gaseous state are free to move, they collide with one another and also against the walls of the container. The **pressure of the gas** is due to collisions of molecules against the walls of the container.

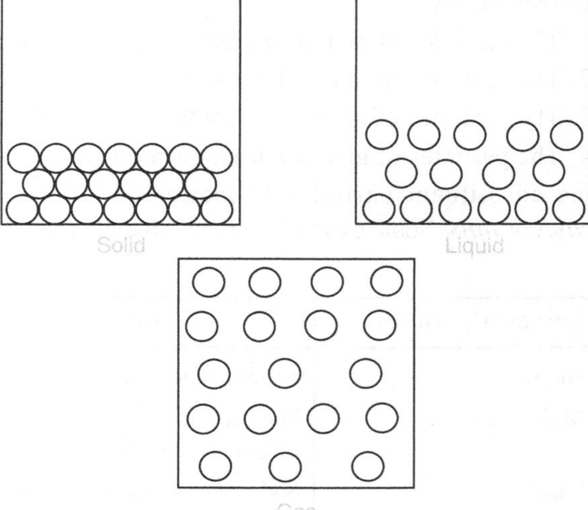

Fig. 4.3. Packing of particles in the three states of matter.

- ***Solid and liquid states** are known as **condensed states of matter** due to smaller interparticle spaces and negligible compressibility.*
- ***Liquids and gases** are known as **fluids** because of their ability to flow and take the shape of a container.*

PLASMA STATE—The Fourth State of Matter

Matter in this state is in the form of ionized gas. It consists of a neutral mixture of positive ions and unbound electrons. Matter exists in this state at temperatures in the range 10000°C to 15000°C. Matter in the sun and stars exists in the plasma state. It is estimated that 99% of matter in the universe exists in the plasma state. Neon in neon lights is also in the plasma state.

SUMMARY

- **Matter** is anything that occupies space and has mass.
- Matter can be classified as solids, liquids and gases on the basis of its physical state.
- Matter is made up of extremely small particles.
- There are spaces between particles of matter.
- The particles of matter are continuously moving.
- The particles of matter attract each other.
- The spaces between particles are minimum in the solid state and maximum in gases.
- The kinetic energy of particles is minimum in the solid state and maximum in the gaseous state.
- The force of attraction between particles is maximum in the solid state and negligible in the gaseous state.
- Liquids and gases exhibit diffusion because their particles possess translatory motion and larger interparticle spaces.
- Solid and liquid states are known as **condensed states** of matter due to smaller interparticle spaces and very little compressibility.
- Liquids and gases are known as **fluids** because of their ability to flow and take the shape of a container.

EVALUATE YOURSELF

I. Objective Type Questions

Select the most appropriate choice from the options given as (a), (b), (c) and (d) after each question:

1. Which of the following is not an example of matter?
   (*a*) Air (*b*) Almonds
   (*c*) Cold-drink (*d*) Love.
2. Which of the following has the strongest interparticle forces?
   (*a*) Nitrogen (*b*) Water
   (*c*) Iron (*d*) Neon.
3. Which of the following has atoms as the constituent particles?
   (*a*) Dry ice (*b*) Argon
   (*c*) Glucose (*d*) Potassium chloride.

II. Fill in the Blanks

4. Complete the following sentences by supplying appropriate words:
   (*i*) The particles in ...... state do not possess translatory motion.
   (*ii*) ...... and ...... states of matter are known as fluid states of matter.
   (*iii*) Particles in ...... state possess maximum kinetic energy.
   (*iv*) Kinetic energy of particles of matter ...... with increase in temperature.

III. Discussion Questions

5. What are the characteristics of the particles of matter?
6. Which out of iron and chalk has stronger interparticle forces?
7. Give reasons for the following observations:
   We can get the smell of perfume sitting from several metres away.
8. A diver is able to cut through water in a swimming pool. Which property of matter does this observation show?
9. Describe an activity to demonstrate that matter consists of particles and that the particles are of extremely small size.
10. What are the characteristics of matter in a solid state?
11. Explain why:
    (*i*) Solids do not undergo diffusion whereas liquids and gases undergo diffusion readily.
    (*ii*) Gases are highly compressible.
12. Give reasons for the following:
    (*i*) A gas fills completely the vessel in which it is kept.
    (*ii*) A gas exerts pressure on the walls of a container.
13. Explain why solid and liquid states are known as condensed states of matter.
14. Give reasons:
    (*i*) Sponge is a solid yet we are able to compress it.
    (*ii*) Sugar when kept in jars of different shapes it takes the shape of the jar yet we call it a solid.

Answers

I. Objective Type Questions

1. (*d*) **2.** (*c*) **3.** (*b*).

II. Fill in the Blanks

4. (*i*) solid (*ii*) Gaseous, liquid
(*iii*) gaseous (*iv*) increases.

SECTION-2

UNIT-5

STRUCTURE OF ATOM

Specific Objectives

After studying this unit, the student should be able to:

- explain Dalton's atomic theory
- describe the various experiments that were carried out to reveal the structure of the atom
- write detailed electron configuration (*s, p, d*) for atoms of the first thirty elements
- describe an isotope
- describe the operations of the mass spectrometer.

5.1 INTRODUCTION

In the fifth century B.C. the Greek Philosopher Democritus proposed that all matter consists of very small indivisible particles called *atomos* (meaning uncuttable or indivisible). These earlier ideas were mere speculations and did not have any experimental basis. It was in 1808 that an English scientist, John Dalton formulated a precise definition of the indivisible building blocks of matter that are known as atoms. By the end of the 19th century it was discovered that atoms consist of sub-atomic particles such as electrons, protons and neutrons. The **protons** are positively charged particles and are present in the nucleus of the atom. The **electrons** are negatively charged particles and are present in the extra-nuclear part of the atom. The **neutrons** are neutral particles and are present in the nucleus along with the protons. Most of the mass of the atom is concentrated in the **nucleus.**

During the past few years it has been found experimentally that some of the sub-atomic particles can be further split into two elementary particles, **quarks** and **leptons.**

In this unit, we shall discuss the various experiments which led to the discovery of the various sub-atomic particles and the evolution of the model for the internal structure of atom.

5.2 DALTON'S ATOMIC THEORY

On the basis of the laws of chemical combination John Dalton, an English school teacher in Manchester, proposed that the behaviour of matter could be explained using an atomic theory. He published his work about atomic theory in 1808.

The main points of Dalton's atomic theory are:

1. Matter is composed of extremely small indivisible particles called atoms.
2. An element consists of only one type of atoms. A sample of an element that is large enough to see, contains a very large number of atoms, but all the atoms of a particular

element have identical properties such as mass, shape, colour, density, chemical properties, etc.

3. Atoms of one element differ from atoms of all other elements in mass and chemical properties.
4. Atom is the smallest particle that takes part in chemical reactions.
5. Atoms of different elements may combine with each other in a fixed, simple whole number ratio to form compound atoms or molecules.
6. Atoms can neither be created nor destroyed during chemical reactions.

Dalton suggested symbols for the atoms of different elements as shown in Fig. 5.1.

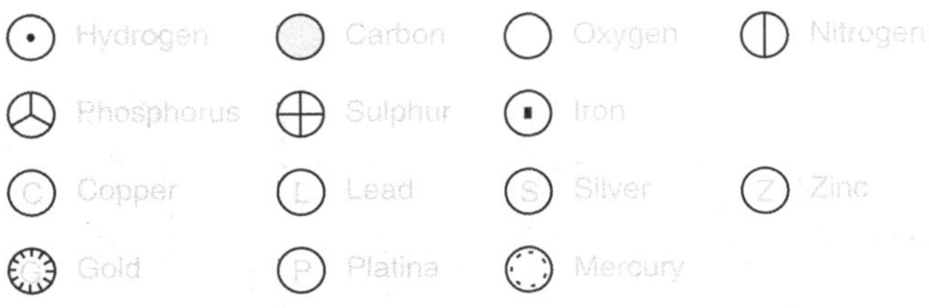

Fig. 5.1. Symbols for some elements as proposed by Dalton.

He was the first scientist to use symbols for elements in a quantitative sense. When he used a symbol for an element he meant a definite quantity of that element, that is, one atom of the element.

John Dalton (1766–1844), the English chemist was born in a poor weaver's family. He started his career as a school teacher and later became a school principal. He presented the atomic theory of matter in 1808.

LIMITATIONS OF DALTON'S ATOMIC THEORY

Dalton's atomic theory was the first successful attempt which gave us some idea about the inner structure of matter. It could very well explain the laws of chemical combination by mass. The main failures of the Dalton's atomic theory are:

1. It failed to explain how atoms of different elements differ from each other, *i.e.*, it did not tell anything about the internal structure of the atom.
2. It could not explain how and why atoms of different elements combine with each other to form compound-atoms or molecules.

Dalton's atomic theory held the ground for about a century and no changes were made to it. However, towards the end of the 19th century and in the beginning of the 20th century new discoveries were made by Sir J.J. Thomson, Neils Bohr, Chadwick and others which revealed the inner structure of the atom. In the light of these findings Dalton's atomic theory was suitably modified.

5.3 DISCOVERY OF THE ELECTRON—CATHODE RAYS

Most of the information about electrons is obtained from the study of cathode rays, which were discovered during experiments with gas discharge tubes. A discharge tube is a long glass tube, such as that shown in Fig. 5.2. This tube is fitted with metal electrodes on either end across which high voltage can be applied. The electrode which is connected to the negative terminal of the power source is called **cathode** while the electrode which is connected to the positive terminal is called **anode.** The tube is also connected to a vacuum pump for controlling the pressure of gas inside the discharge tube.

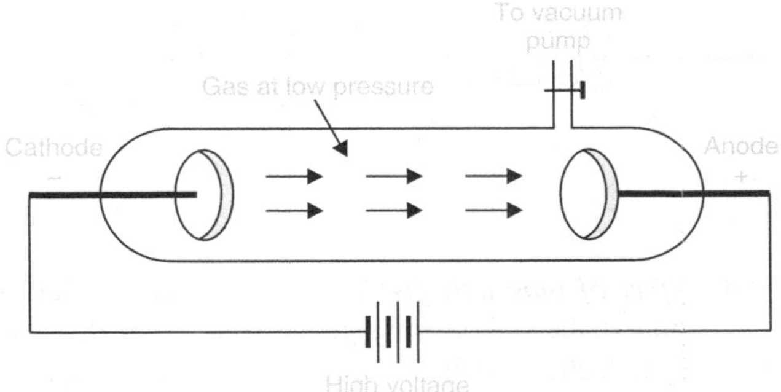

Fig. 5.2. Cathode rays.

When the gas pressure inside the discharge tube is one atmosphere, no electric current flows through the tube. If the gas pressure is reduced to about 10^{-2} atmospheres and a potential difference of about 10000 volts is applied to the electrodes, an electric current flows and at the same time light is emitted by the gas. The colour of the light depends on the nature of the gas taken in the discharge tube. The colour of light is magenta red, if the tube is filled with normal air in the beginning. The neon lights (reddish orange) used in advertising signs operate in this way. As the gas pressure in the discharge tube is reduced further, a dark space appears in the vicinity of the cathode and alternate light and dark bands can be seen between the two electrodes. If the gas pressure is reduced to 10^{-6} atmospheres, the emission of light ceases, instead the end of the glass tube opposite the cathode glows (fluoresces) with a *faint greenish light.* Further investigations revealed that the fluorescence was caused due to the bombardment of the walls of the tube by rays emanating from the cathode. These rays were called **cathode rays.** These rays were found to consist of negatively charged material particles, called **electrons.**

PROPERTIES OF CATHODE RAYS

The cathode rays possess the following properties:

1. ***Cathode rays travel in straight lines.*** An object placed in the path of cathod rays casts a sharp shadow. It shows that cathode rays travel in straight lines.
2. ***Cathode rays consist of material particles.*** This was indicated by the fact that a light paddle wheel placed in the path of cathode rays starts rotating (Fig. 5.3).
3. ***Effect of electric field.*** When electric field is applied to a stream of cathode rays, they get deflected towards the positive plate (Fig. 5.3). It showed that cathode rays themselves are negatively charged.

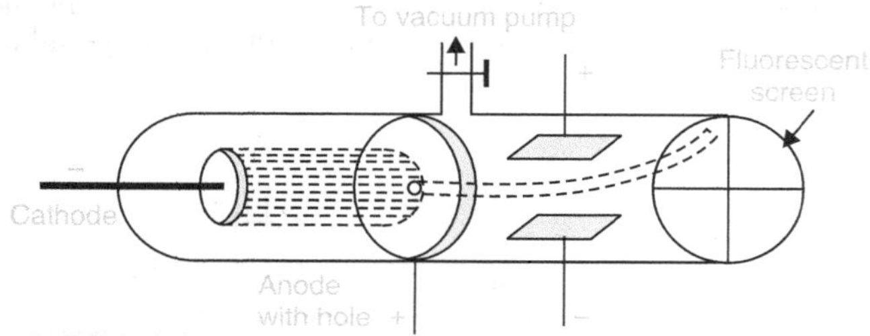

Fig. 5.3. Effect of electric field on cathode rays.

4. ***Effect of magnetic field.*** When a magnetic field is applied, the cathode rays get deflected. The direction of deflection again indicates that cathode rays are negatively charged.
5. On striking against the walls of the discharge tube ***cathode rays produce faint greenish fluorescence.***

The above mentioned properties of cathode rays indicate that the cathode rays consist of a fast-moving stream of negatively charged material particles. These particles were named **electrons.**

Do You Know?

The name **'electron'** was suggested by G. J. Stoney, in 1874, for the tiny negative particles which made up an electric current.

DETERMINATION OF CHARGE TO MASS RATIO (e/m) OF ELECTRONS

Sir J. J. Thomson studied the deflection of cathode rays under the simultaneous application of electric and magnetic fields, applied perpendicular to each other (Fig. 5.4). It is possible to adjust the two fields so that the cathode rays strike the fluorescent screen at the same position as they do when neither field is applied. From the magnitudes of electric field and magnetic field thus applied it is possible to calculate charge (e) to mass (m) ratio of the electrons. ***The value of e/m was found to be 1.76×10^8 coulombs/g or 1.76×10^{11} coulombs/kg.*** *The e/m ratio for the particles in the cathode rays was found to be same irrespective of the nature of the cathode or the nature of the gas taken in the discharge tube. It showed that the electrons are a universal constituent of matter.*

It may be mentioned here that the charge per unit mass (e/m), for a charged particle, is known as the **specific charge** of the particle.

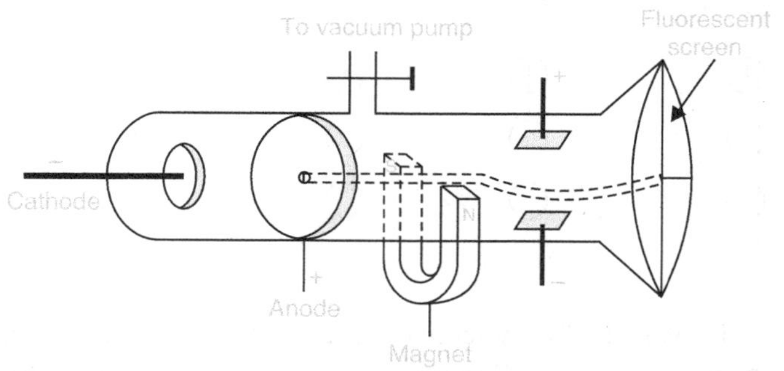

Fig. 5.4. Cathode rays deflected through magnetic and electric fields.

Sir J.J. Thomson (1856–1940), a British physicist, discovered the electron in 1897. He won the Nobel prize for physics in 1906.

CHARGE ON THE ELECTRON

Sir J. J. Thomson's experiment led to the precise determination of the e/m ratio for an electron. The charge (e) on an electron was determined by Robert Milikan in 1909 by the '***oil drop***' experiment.

The charge on the electron is found to be 1.60×10^{-19} coulombs. This is taken as one unit negative charge.

MASS OF THE ELECTRON

By combining the e/m ratio and charge (e) of the electron, it is possible to calculate the mass of an electron.

Thomson's experiment: $e/m = 1.76 \times 10^{11}$ coulombs/kg

Milikan's experiment: $e = 1.60 \times 10^{-19}$ coulombs/electron

Mass of the electron, m

$$= \frac{e}{e/m} = \frac{1.60 \times 10^{-19} \text{ coulombs/electron}}{1.76 \times 10^{11} \text{ coulombs/kg}}$$

$= 9.1 \times 10^{-31}$ kg/electron.

The mass of the electron is very small and is approximately 1/1840 times the mass of an atom of hydrogen.

In general, an electron is represented as e^-.

> *An **electron** is a fundamental particle of an atom carrying one unit negative charge and having mass nearly equal to 1/1840th the mass of an atom of hydrogen.*

5.4 ANODE RAYS OR CANAL RAYS

E. Goldstein, a German scientist, in 1886, discovered the existence of a new type of rays in the discharge tube. He used a perforated cathode (Fig. 5.5) in the discharge tube. The cathode divided the discharge tube in two chambers. On passing the electric discharge at low pressure he observed a new type of rays streaming behind the cathode. The path of these rays became visible due to the glow of the residual gas. These rays also produced fluorescent glow on striking the walls of the tube behind the cathode. These rays were named **anode rays** or **canal rays.** These rays were named canal rays because they passed through '*holes*' or '*canals*' in the cathode.

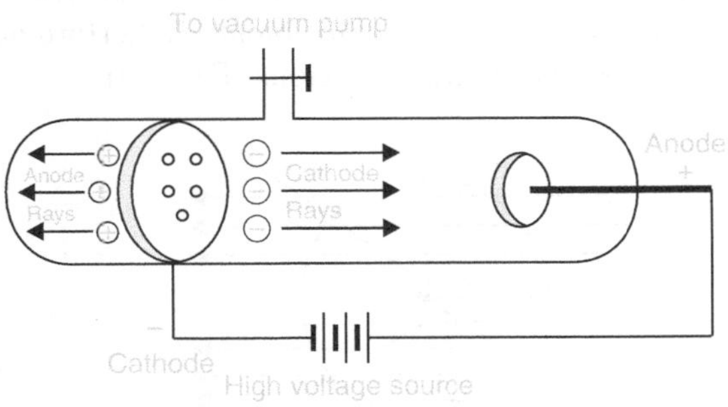

Fig. 5.5. Canal rays.

Further investigations of these rays showed that they consist of positively charged material particles.

Some of the **characteristic properties of anode rays** are:

1. Anode rays consist of material particles.
2. Anode rays are deflected by electric field towards the negatively charged plate. This indicates that *they are positively charged.*
3. The charge to mass ratio of the particles in the anode rays was determined by **W. Wien** by using Thomson's technique. Charge to mass ratio of the particles in the anode rays depends upon nature of the gas taken in the discharged tube.

DISCOVERY OF THE PROTON

As already mentioned, the charge to mass ratio of the particles in the anode rays was found to depend upon the nature of the gas in the discharge tube. It was observed that the e/m ratio was maximum when hydrogen gas was taken in the discharge tube. This indicated that the positive ions formed from hydrogen are the lightest. These lightest positively charged particles were named **protons.** The charge and mass of the proton were determined in the same manner as the one discussed in the case of the electron. Charge to mass ratio for protons was found to be 9.58×10^7 C/kg. Charge on proton is opposite but equal in magnitude to the charge on the electron *i.e.*, 1.60×10^{-19} C. From these two observations the mass of a proton works out to be 1.67×10^{-27} kg. It is practically the same as the mass of a hydrogen atom and is about 1837 times the mass of an electron.

In general, a proton is represented as p^+.

> *A **proton** is a fundamental particle of an atom carrying one unit positive charge and having mass nearly equal to the mass of an atom of hydrogen.*

5.5 THOMSON'S MODEL OF THE ATOM

Once it was known that atoms contain electrons and protons, the next problem was to know how these are arranged in an atom. **Thomson** was the first to propose a detailed model of the atom. He proposed that an atom consists of a uniform sphere of positive electricity in which the electrons are distributed more or less uniformly. The negative and the positive charge are equal in magnitude. Thus, the atom as a whole is electrically neutral. This model of the atom is known as the ***"Plum pudding model"*** because of its similarity to a christmas pudding with currants in it (Fig. 5.6).

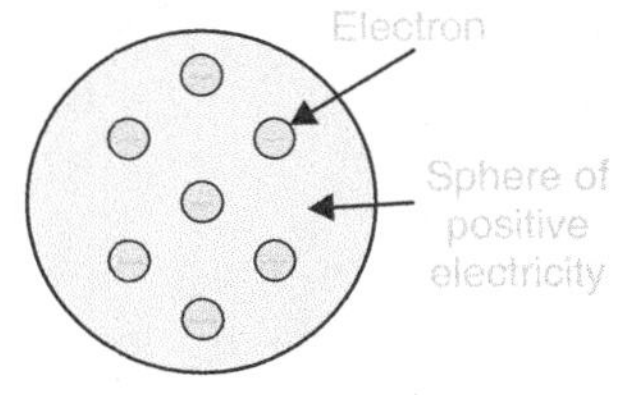

Fig. 5.6. Thomson's model of atom.

5.6 RUTHERFORD'S MODEL OF THE ATOM

Rutherford in 1911 performed an experiment which led to the downfall of Thomson's model. Most of the experimental work was carried out by Geiger and Marsden, two of Rutherford's students at Manchester University. Geiger and Marsden built the apparatus as shown in Fig. 5.7. The experiment involved the bombardment of a thin sheet of gold (thickness 4×10^{-5} cm) by α-particles. The source of α-particles was a piece of radium placed in a lead block. A narrow hole in the lead block allowed the α-particles to travel only in one direction through the evacuated vessel towards a fluorescent zinc sulphide screen. When an α-particle hits the fluorescent screen a tiny flash of light is produced which is observed with the help of a microscope. A thin gold foil was placed between the slit and the screen. The screen and the microscope were rotated to detect scattering of α-particles.

Ernest Rutherford (1871–1937) was born in New Zealand. In 1908 he was awarded the Nobel prize for chemistry for his work on radioactivity. In 1911, he discovered that α-particles could be deflected by thin metal foils.

Rutherford observed that:

1. *Most of the α-particles (nearly 99%) passed through the gold foil undeflected.*
2. *Some of the α-particles were deflected by small angles.*

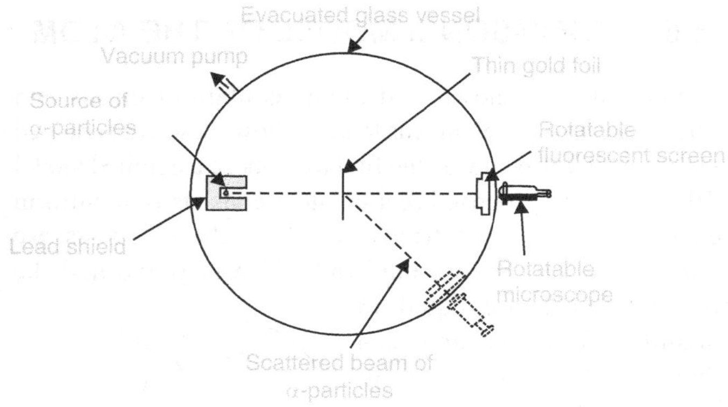

Fig. 5.7. The apparatus used by Geiger and Marsden to investigate the scattering of α-particles by a thin metal foil.

3. *Very few particles (1 in about 10^4) were either deflected by very large angles or were actually reflected back along their path.*

These observations led to the downfall of Thomson model. According to the Thomson model the positive charge is uniformly distributed in the volume of the atom. Therefore, it is so diffused that the fast moving α-particles (He^{2+}) should pass through this weak electric field largely undeflected or may suffer slight deflection.

In order to explain the observations of his scattering experiment, Rutherford assumed that the solid gold foil consists of layers of individual atoms which are touching each other so that there is hardly any empty space between them. As such the α-particles striking the gold foil must pass through the atoms. Rutherford explained his observations as follows:

1. Since most of the α-particles pass through the foil undeflected, it indicates that *most of the space in an atom is empty* (Fig. 5.8).

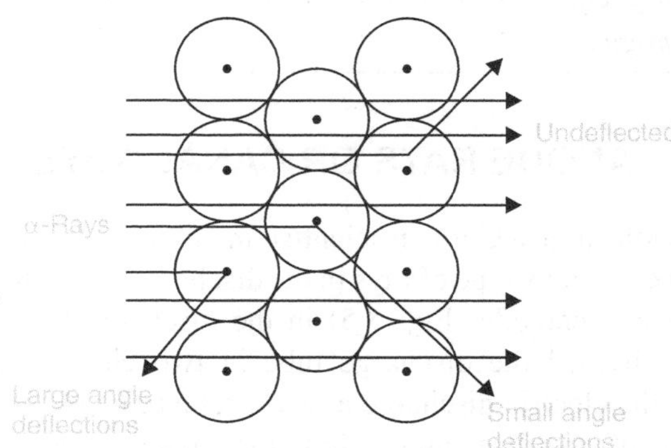

Fig. 5.8. Scattering of α-particles by gold atoms.

2. α-particles being positively charged and having considerable mass could be deflected only by some heavy, positively charged centre. The small angle of deflection of α-particles indicates the presence of a *heavy positive centre in the atom*. Rutherford named this positive centre as **nucleus** (Fig. 5.9).

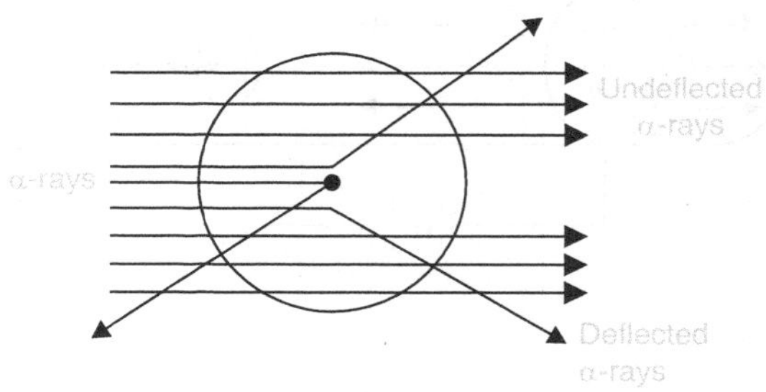

Fig. 5.9

3. α-particles which make a head-on collision with the heavy positive centre are deflected through large angles. Since the number of such α-particles is very small, *the space occupied by the heavy positive centre must be very small.*

From the data of the scattering experiment, Rutherford was able to calculate the radius of the nucleus. Rutherford calculated that the nucleus of an atom would have a radius of about 10^{-14} m. He showed that the radius of the nucleus is about 10^{-4} times the radius of the atom which is about 10^{-10} m.

Rutherford's Nuclear Model of Atom

On the basis of the scattering experiment, Rutherford put forward the ***nuclear model of atom.*** Main points of this model are:

1. *Most of the mass and all the positive charge of an atom is concentrated in a very small region called* ***nucleus.*** *Size of the nucleus is extremely small as compared with the size of the atom.* Radius of the nucleus is of the order of 10^{-14} m, whereas radius of atoms is of the order of 10^{-10} m.
2. *The nucleus is surrounded by electrons which are revolving around it at very high speeds. The electrostatic force of attraction between electrons and the nucleus is balanced by the centrifugal force acting on the revolving electrons.*
3. *Total negative charge on the electrons is equal to the total positive charge on the nucleus so that an atom on the whole is electrically neutral.*

Nuclear model of the atom can be compared with the solar system. In an atom electrons revolve around the nucleus in just the same way as the planets revolve around the sun. Due to this comparison, revolving electrons are sometimes called ***planetary electrons.***

Although Rutherford's idea of planetary electrons has now been discarded, his idea of small positive nucleus still holds true.

5.7 DISCOVERY OF THE NEUTRON

We have seen that the mass of the atom is mainly concentrated in the nucleus. The nucleus contains not only protons but also another type of particles called neutrons. The existence of neutrons in the nucleus was first predicted by **Rutherford** in 1920 to account for the difference in the mass of the atom and the total mass of protons. Neutrons were discovered experimentally by **James Chadwick** in 1932. He bombarded a thin foil of beryllium with fast moving α-particles and observed that highly penetrating rays consisting of neutral particles were produced.

These neutral particles were found to have a mass of 1.675×10^{-27} kg and were named *neutrons.* In general, a neutron is represented as '***n***'.

A ***neutron*** *is a subatomic particle carrying no charge and having a mass of* 1.675×10^{-27} *kg which is almost equal to that of a hydrogen atom.*

The mass of the atom is largely due to protons and neutrons in the nucleus of the atom. *Atoms of all elements except ordinary hydrogen contain one or more neutrons in their nuclei.*

The relative masses and relative charges of these three subatomic particles are summarized in Table 5.1.

5.8 ATOMIC NUMBER AND MASS NUMBER

ATOMIC NUMBER

The atomic number of an element is equal to the number of protons in the nucleus of its atom. The atomic number is also known as the ***proton number*** of the element. Further, in an atom, the number of protons is equal to the number of electrons. Hence, atomic number is also equal to the number of electrons in an atom of an element. Thus,

Atomic number of an element *is equal to the number of protons in the nucleus of its atom or the number of extra-nuclear electrons.*

Atomic number is denoted by the letter Z.

Atomic Number (Z) = Number of protons

= Number of electrons.

Table 5.1. The Relative Masses and Relative Charges of Three Fundamental Particles

Particle	*Relative Mass*	*Mass in kg*	*Relative Charge*	*Charge in coulombs*
Proton (p^+)	1	1.673×10^{-27} kg	+ 1 unit	$+1.602 \times 10^{-19}$ C
Neutron (n)	1	1.675×10^{-27} kg	0	0
Electron (e^-)	$\frac{1}{1840}$	9.110×10^{-31} kg	– 1 unit	-1.602×10^{-19} C

For example, an atom of magnesium contains 12 protons, therefore, its atomic number is 12.

All the atoms of a particular element contain the same number of protons in their nuclei. Therefore, all the atoms of an element have the same atomic number. Atoms of different elements contain different number of protons and hence no two elements can have the same atomic number. Thus, *each element has its characteristic atomic number which can be used to identify the element.*

For example, when we say an element has atomic number 13, we are referring to aluminium. No other element has atomic number 13.

MASS NUMBER

It has already been stated that the mass of an atom is mainly concentrated in the nucleus. In the nucleus, there are protons and neutrons. From this it follows that the mass of an atom is mainly due to protons and neutrons. Protons and neutrons are collectively called ***nucleons*** as they are present in the nucleus of an atom.

The total number of protons and neutrons in the nucleus is called ***mass number of the atom.***

It is generally represented by the letter A.

Mass Number (A) = Number of protons (Z)
+ Number of neutrons (N)
= Number of nucleons

The number of protons in the nucleus is equal to the atomic number, Z while *the number of neutrons in the nucleus is sometimes called* ***neutron number, N.***

Mass number gives the approximate atomic mass of the element. For example, the mass number of magnesium is 24 as it contains 12 protons and 12 neutrons in its nucleus. Each proton and neutron has a relative mass approximately 1 *u*. Therefore, the atomic mass of magnesium is $12\,u + 12\,u = 24\,u$.

Calculation of Number of Electrons, Protons and Neutrons

From the knowledge of atomic number and mass number of an element it is possible to calculate the number of electrons, protons and neutrons in an atom of an element. For example, the atomic number and mass number of aluminium are 13 and 27 respectively. Number of electrons, protons and neutrons in an atom of it can be calculated as:

Number of protons = Atomic number = 13
Number of electrons = Atomic number = 13
Number of neutrons = Mass number – Atomic number
= 27 – 13 = 14.

SOLVED EXAMPLES

Example 5.1. *Calculate the number of electrons, protons and neutrons in the following species:*

(i) Phosphorus atom *(ii) Phosphide ion* (P^{3-})
(iii) Magnesium ion (Mg^{2+})
Mass numbers: P = 31, Mg = 24
Atomic numbers: P = 15, Mg = 12.

Solution. *(i) Phosphorus atom* (P)

Number of electrons = Atomic number = 15
Number of protons = Atomic number = 15
Number of neutrons = Mass number – Atomic number
= 31 – 15 = 16.

(ii) Phosphide ion (P^{3-})

Phosphide ion (P^{3-}) = Phosphorus atom + 3 electrons

$$P^{3-} \equiv P + 3e^-$$

Thus, phosphide ion has same number of protons and neutrons as phosphorus atom but has three electrons more.

Number of electrons = 15 + 3 = 18
Number of protons = 15
Number of neutrons = 31 – 15 = 16.

(iii) Magnesium ion (Mg^{2+})

Mg^{2+} ion is formed by loss of two electrons by Mg atom. Therefore, it has two electrons less than the number of electrons in Mg atom.

$$Mg^{2+} \equiv Mg - 2e^-$$

Number of electrons = 12 – 2 = **10**
Number of protons = **12**
Number of neutrons = (24 – 12) = **12.**

Example 5.2. *The number of protons in the nucleus of an atom of mass number 97 is 41. What will be the number of neutrons in its isotope of mass number 99?*

Solution. The atomic number of isotopes is same. Therefore, the number of protons in both the atoms is same.

Mass number = Number of protons + Number of neutrons
∴ Number of neutrons
= Mass number – Number of protons
= 99 – 41 = **58.**

Example 5.3. *If the number of electrons in an atom is 8 and number of protons is also 8, then*

(i) What is the atomic number of the element?
(ii) What is the charge on the atom?

Solution. *(i)* Atomic number of the element
= Number of protons in the atom = 8

(*ii*) There is no charge on the atom as the number of protons is equal to the number of electrons.

Example 5.4. *The nucleus of an atom of a certain isotope contains 6 protons and 7 neutrons. What is the atomic number and the mass number of this isotope? To which element does it belong?*

Solution. We know that:

Atomic number = Number of protons

∴ Atomic number = 6

Mass number = Number of protons + Number of neutrons

= 6 + 7

= **13.**

Since atomic number of the given isotope is 6, it indicates that this isotope is of the element carbon.

5.9 ISOTOPES

All the atoms of a particular element have the same number of protons in their nuclei. However, the number of neutrons may be different. Such atoms have the same atomic number but different mass numbers and are known as **isotopes** of the element. Thus,

***Isotopes** of an element are the atoms of the element with same atomic number but different mass numbers.*

For example, hydrogen has three isotopes, protium (H), deuterium (D) and tritium (T). All the three isotopes have atomic number 1, however, their mass numbers are 1, 2 and 3 respectively. The isotopes of other elements do not have special names; they are indicated by the given mass number value on their symbol. Mass number and atomic number of an atom are generally indicated as shown below:

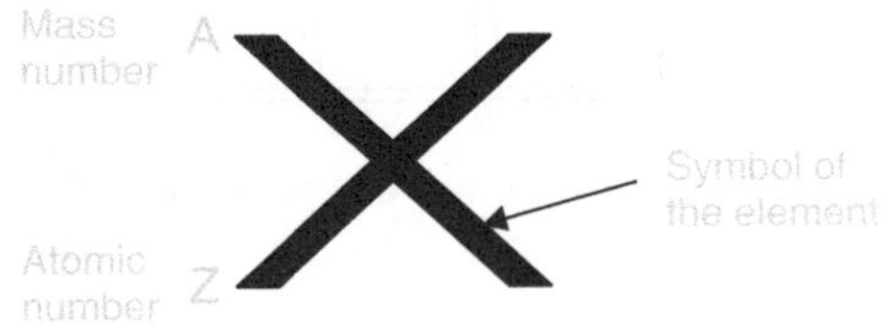

Representation of an isotope.

Thus, the three isotopes of hydrogen can be represented as

$$^{1}_{1}H,\ ^{2}_{1}H \text{ and } ^{3}_{1}H$$

It may be noted that there is a large difference between the masses of the various isotopes of hydrogen. Mass of deuterium is twice that of protium.

The two isotopes of chlorine are represented as

$$^{35}_{17}Cl \text{ and } ^{37}_{17}Cl$$

Another way to describe an isotope is to cite its elemental name and mass number. The two isotopes of chlorine may be represented as chlorine-35 and chlorine-37.

Isotopes have similar chemical properties but different physical properties.

An atom which is characterized by a specified atomic number and mass number is called a **nuclide.** For example, $^{235}_{92}U$ is a nuclide of uranium.

Add To Your Knowledge

Why Atomic Masses of Most of the Elements are Fractional?

The atomic masses of most of the elements are in fractions and not whole numbers. For example, the atomic mass of chlorine is 35.5 *u*. The reason for fractional atomic masses is that most of the elements exist in the form of two or more isotopes. These isotopes have different relative masses. In such cases the atomic mass of the element is equal to the average of the relative masses of the various isotopes of the element. While taking the average the results appear as fractions.

As an illustration, let us consider the case of chlorine. Chlorine exists in the form of two isotopes Cl-35 and Cl-37, and their relative abundance is about 75% and 25% respectively. The average atomic mass of Cl can be calculated as given below:

$$\text{Average relative mass of Cl} = \frac{35 \times 75 + 37 \times 25}{100} = 35.5\ u$$

Thus, the atomic mass of Cl is 35.5 *u*, which is a fraction.

APPLICATIONS OF ISOTOPES

Radioactive isotopes of some elements find many applications in the field of medicine, research, agriculture, etc. Some applications of these isotopes are listed below:

1. An isotope of uranium (U-235) is used ***as a fuel in nuclear reactors***.
2. An isotope of cobalt (Co-60) is used in the ***treatment of cancer*** by radiation therapy.
3. An isotope of iodine is used in the ***treatment of goitre.***
4. ***Ages of old wooden articles*** are determined by observing the radioactivity of the C-14 isotope of carbon.
5. ***Ages of rocks*** are determined by determining the relative amounts of U-238 and Pb-206 isotopes in the rock.
6. Uptake of phosphorus from fertilised soil by plants can be traced by using fertilisers containing the P-32 isotope of phosphorus.

Isobars

It may be mentioned here that atoms of different elements may have the same mass number.

The atoms of different elements with different atomic numbers, which have the same mass number are called ***isobars.***

For example, ${}^{14}_{6}C$ and ${}^{14}_{7}N$ are isobars. Similarly, ${}^{40}_{20}Ca$ and ${}^{40}_{18}Ar$ are also isobars. Isobars contain different number of electrons.

Do isobars have the same or different chemical properties?

5.10 MASS SPECTROMETER

The relative masses of isotopes and their relative abundance can be determined using a mass spectrometer. The mass spectrograph was invented by a British physicist, F.W. Aston, in 1919. It is an instrument that separates ions of different masses. The results are recorded on a photographic plate. The modern version records results electrically and is called mass spectrometer. Fig. 5.10 is a schematic diagram of a mass spectrometer.

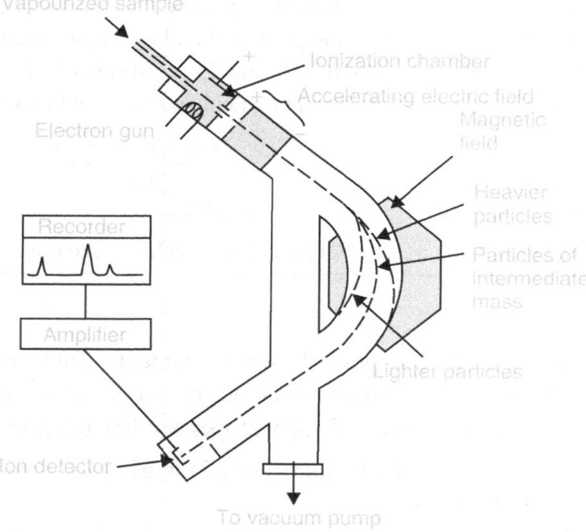

Fig. 5.10. A diagram of a mass spectrometer.

The five main steps which occur in a mass spectrometer are:

1. **Vapourization.** A sample of the element is vapourized.
2. **Ionization.** The vapour is ionized to give positive ions.
3. **Acceleration.** The positive ions are accelerated by an electric field.
4. **Deflection.** The positive ions are deflected by a magnetic field.
5. **Detection.** The ions are detected by the detector and a record is made.

The substance to be analysed is vapourized by heating and introduced into an ionization chamber. In the ionization chamber, atoms of the element are bombarded with a stream of high energy electrons. The high energy electrons knock out one or occasionally two electrons from the atoms resulting in the formation of positive ions.

$$\underset{\text{Atom}}{X} + \underset{\text{High energy electron}}{e^-} \longrightarrow \underset{\text{Positive ion}}{X^+} + e^- + e^-$$

The positive ions are accelerated by passing then through an electric field. The accelerated beam of ions passes through the magnetic field where the ions are deflected according to their mass and their charge.

For a specific accelerating electric field and deflecting magnetic field, ions of only one particular mass/charge ratio hit the detector at the end of the apparatus. Ions of smaller mass/charge ratio will be deflected too much whereas ions of greater mass/charge ratio will be deflected too little. The ion detector is linked through an amplifier to a recorder. As the strength of the magnetic field is slowly increased ions of increasing mass will be detected and a mass spectrum as shown in Fig. 5.11 is recorded.

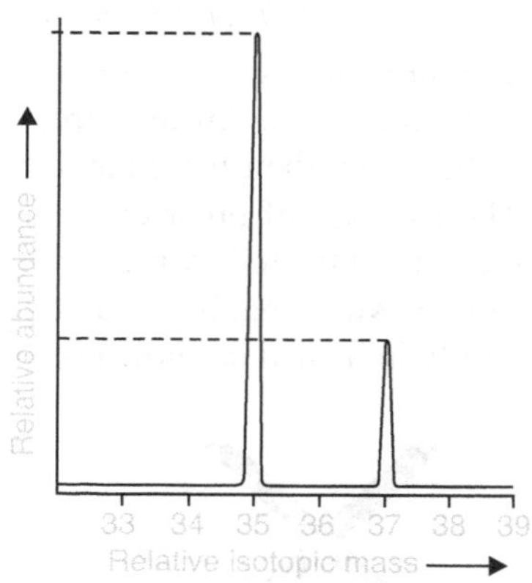

Fig. 5.11. A mass spectrum for chlorine.

The relative heights of the peaks in the mass spectrum gives in the relative abundance of different ions present. The mass spectrometer can be used to find the number of isotopes of an element, and also to determine the relative isotopic masses of various isotopes. From the knowledge of relative isotopic masses and their relative abundances, the relative atomic mass of an element can be calculated.

A naturally occurring element which has no isotope will show only one peak in a mass spectrum.

DETERMINATION OF RELATIVE MOLECULAR MASS

Mass spectrometer also helps in determining the relative molecular mass and molecular formula of compounds, especially organic compounds. When the molecules of an organic compound pass through a mass spectrometer, they undergo fragmentation. As a result, the mass spectrum of the compound contains several peaks, each corresponding to a particular fragment. However, a few molecules may pass through intact. These intact molecules give a peak showing the relative molecular mass of the compound. For example, from the mass spectrum of an unknown compound, as given below, we find that the relative mass of the heaviest particle recorded is 58. We can assume that this corresponds to the intact molecule with a single positive charge. Hence, the relative molecular mass of the compound must be 58. Knowing the empirical formula of the compound, its molecular formula can be determined.

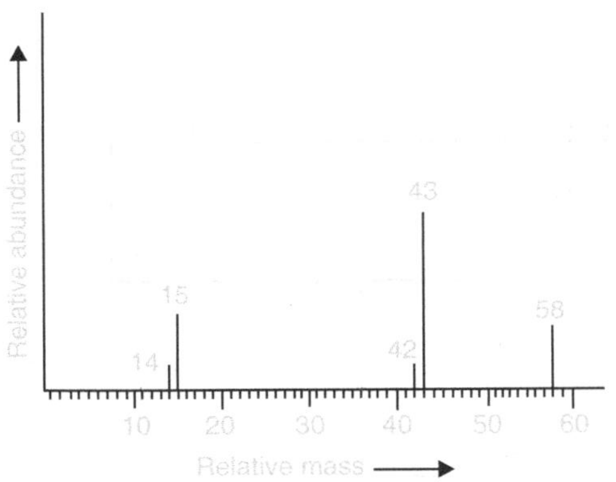

Fig. 5.12. Mass spectrum of a compound

SOLVED EXAMPLE

Example 5.5. *If bromine atom is available in the form of, say, two isotopes $^{79}_{35}Br$ (49.7%) and $^{81}_{35}Br$ (50.3%), calculate the relative atomic mass of bromine.*

Solution. % of Br-79 = 49.7% ; % of Br-81 = 50.3%

$$\text{Relative atomic mass of Br} = \frac{79 \times 49.7 + 81 \times 50.3}{100}$$

$$= \mathbf{80.0}$$

Thus, the relative atomic mass of bromine is 80.0.

5.11 NATURE OF ELECTROMAGNETIC RADIATIONS

WAVE THEORY OF RADIATIONS

Light is an important form of energy. According to the **Maxwell's wave theory,** light is transmitted in the form of electromagnetic waves. These waves are associated with oscillating electric and magnetic fields. These two fields are perpendicular to each other and also to the direction of propagation as shown in Fig. 5.13.

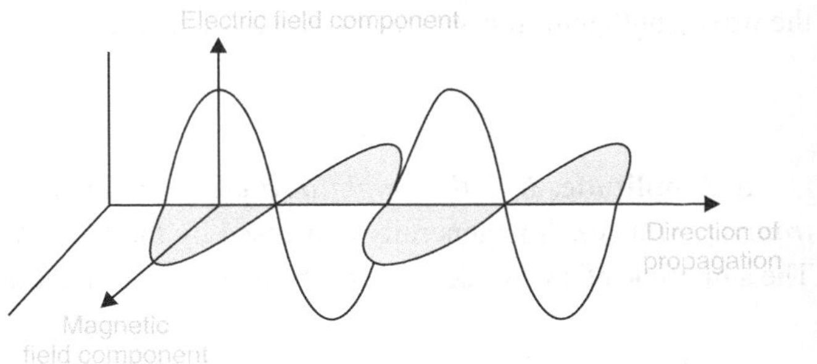

Fig. 5.13. Electromagnetic wave.

In order to characterize electromagnetic waves the following parameters are used:

1. Wavelength. *It may be defined as the distance between two neighbouring crests or troughs of a wave* as shown in Fig. 5.14. It is represented by the Greek letter λ (*lambda*) and is generally measured in Angstrom units (Å) or nanometers (nm).

$$1 \text{ Å} = 10^{-10} \text{ m}$$
$$1 \text{ nm} = 10^{-9} \text{ m}.$$

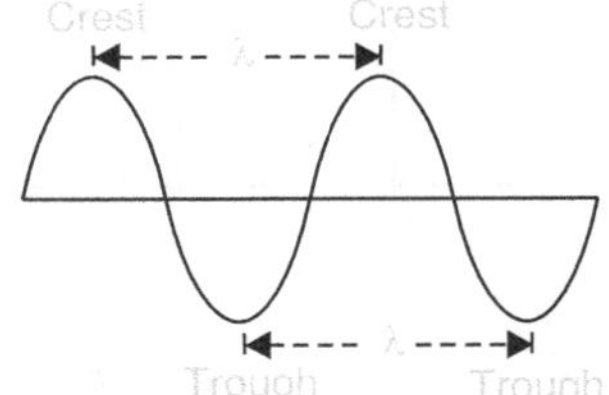

Fig. 5.14. Propagation of wave motion.

2. Frequency. *It may be defined as the number of waves which pass through a particular point in one second.* It is represented by the Greek letter ν (*nu*). Its units are:

Cycles per second (*cps*) or **Hertz** (*Hz*).

1 cps = 1 Hz

3. Velocity. *The distance travelled by a wave in one second is called the velocity of the wave.* It is denoted by the letter c. The frequency (ν) and wavelength (λ) are related to velocity (c) by the following relation:

$$c = \nu\lambda \quad \text{or} \quad \nu = \frac{c}{\lambda}$$

The velocity of all electromagnetic radiations in space or in a vacuum is the same and is equal to 3 × 10^8 m/sec.

4. Wave number. *It may be defined as the number of wavelengths per unit length. It is equal to the inverse of wavelength.* It is denoted by $\bar{\nu}$. Its unit is the reciprocal of the wavelength unit. It is generally expressed as cm^{-1} or m^{-1}.

$$\bar{\nu} = \frac{1}{\lambda} = \frac{\nu}{c}.$$

5. Amplitude. *It is the height of a crest or depth of a trough of a wave.* It is generally expressed by the letter '*a*'. The amplitude of a wave determines the intensity of radiation.

ELECTROMAGNETIC SPECTRUM

The different electromagnetic radiations have different wavelengths. The visible light in the presence of which our eye can see contains radiations having wavelengths between 400 nm–750 nm. Different colours in the visible light correspond to radiations of different wavelengths. In addition to visible light there are so many other electromagnetic radiations, such as X-rays, ultraviolet rays, infrared rays, microwaves and radiowaves.

Arrangement of all the electromagnetic radiations in increasing order of their wavelengths or decreasing order of their frequencies is called the ***electromagnetic spectrum.***

Different regions of the electromagnetic spectrum are identified by different names. Complete electromagnetic spectrum is shown in Fig. 5.15.

The various types of electromagnetic radiations have different energies and are used for different purposes as listed in Table 5.2.

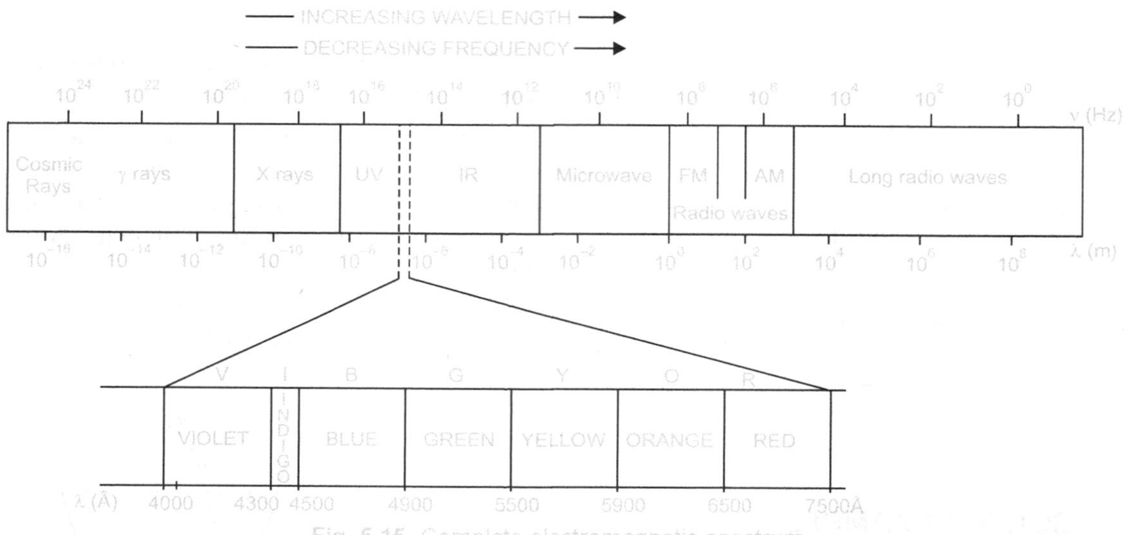

Fig. 5.15. Complete electromagnetic spectrum.

Table 5.2. Some Applications of Electromagnetic Waves

Electromagnetic wave	*Frequency (Hz)*	*Wavelength (m)*	*Uses*
Gamma rays	10^{20} to 10^{21}	10^{-12}	Cancer treatment
X-rays	10^{17} to 10^{19}	10^{-10}	Medical "pictures", materials testing
Ultraviolet	10^{15} to 10^{16}	10^{-7}	Germicidal lamps
Visible	10^{13} to 10^{14}	10^{-6}	Illumination
Infrared	10^{12} to 10^{13}	10^{-4}	Heating
Microwave	10^{9} to 10^{11}	10^{-2}	Cooking, radar
Radio frequency	10^{5} to 10^{8}	10^{2}	Signal transmission

SOLVED EXAMPLES

Example 5.6. *Yellow light emitted from a sodium lamp has a wavelength (λ) of 580 nm. Calculate frequency (ν) and wave number ($\bar{\nu}$) of the yellow light.*

Solution. Wavelength of the radiation λ = 580 nm

$$= 580 \times 10^{-9}\ \text{m} = 5.80 \times 10^{-7}\ \text{m}$$

Velocity of radiation, $c = 3 \times 10^8\ \text{ms}^{-1}$

$$c = \nu\lambda$$

Frequency $\nu = \frac{c}{\lambda} = \frac{3 \times 10^8\ \text{ms}^{-1}}{5.80 \times 10^{-7}\ \text{m}}$

$$= \mathbf{5.17 \times 10^{14}\ s^{-1}}.$$

Wave number $\bar{\nu} = \frac{1}{\lambda} = \frac{1}{5.80 \times 10^{-7}\ \text{m}}$

$$= \mathbf{1.72 \times 10^{6}\ m^{-1}}.$$

Example 5.7. *The wave number of a beam of light is 400 cm^{-1}. What is the wavelength of the light in nanometers? Also find out the frequency of the light.*

Solution. *Calculation of wavelength*

$$\bar{\nu} = 400\ \text{cm}^{-1}$$

$$\bar{\nu} = \frac{1}{\lambda}$$

or $\lambda = \frac{1}{\bar{\nu}} = \frac{1}{400}\ \text{cm} = 2.5 \times 10^{-3}\ \text{cm}$

$$= 2.5 \times 10^{-3} \times 10^{7}\ \text{nm}$$

$$= \mathbf{2.5 \times 10^{4}\ nm}.$$

Calculation of frequency

$$c = \nu\lambda$$

$$\nu = \frac{c}{\lambda} = c\bar{\nu} = 3 \times 10^{10}\ \text{cm s}^{-1} \times 400\ \text{cm}^{-1}$$

$$= \mathbf{1.2 \times 10^{13}\ s^{-1}}.$$

LIMITATIONS OF THE WAVE THEORY OF RADIATION

Though the wave theory of radiation could successfully explain the phenomena of interference and diffraction, it failed to explain the following phenomena:

(*i*) *Black-body radiation i.e.*, emission of radiation from hot bodies.

(*ii*) *Photoelectric effect i.e.*, ejection of electrons from metal surface when radiation of a suitable frequency strikes it.

5.12 PLANCK'S QUANTUM THEORY

The failure of the classical electromagnetic theory of radiation to explain the phenomenon of photoelectric effect led Max Planck (1901) to propose a new theory known as **Quantum theory of radiation.** The main points of this theory are:

1. *Radiant energy is emitted or absorbed not continuously but discontinuously in the form of small packets of energy called quanta.*

Each wave packet or **quantum** is associated with a definite amount of energy. *In case of light, the quantum of energy is often called* **photon.**

2. *The amount of energy associated with a quantum of radiation is proportional to the frequency of the radiation.*

$$E \propto \nu$$

or

$$\mathbf{E} = h\nu$$

where h is a constant known as **Planck's constant** and is equal to **6.625 × 10^{-34} joule-sec.**

The energy of a photon of light in terms of wavelength λ and velocity of light c, is given as

$$E = h\nu$$

$$E = h\frac{c}{\lambda} \qquad (\because\ c = \nu\lambda\ ;\ \nu = c/\lambda)$$

3. *A body can emit or absorb energy only in terms of the integral multiples of a quantum, i.e.,*

$$E = nh\nu$$

where $n = 1, 2, 3, 4, ...$

This means that a body can emit or absorb energy as $h\nu$, $2h\nu$, $3h\nu$, $4h\nu$, ... but it cannot emit or absorb energy in fractional values of $h\nu$ such as 1.5 $h\nu$ or 2.5 $h\nu$.

5.13 EVIDENCE FOR THE QUANTIZED ELECTRONIC STATES—ATOMIC SPECTRA

When an element is excited by some method such as *heating, passing electric current or passing electric discharge,* the atoms of the element emit electromagnetic radiations of definite frequencies. *The arrangement of these radiations in the order of increasing wavelength or decreasing frequencies* is called the **emission spectrum** of the element. Since the radiations in the spectrum are emitted due to energy changes taking place in the atoms, this spectrum is also known as **atomic spectrum.**

The instrument used for obtaining a spectrum is called *spectroscope* or *spectrograph*. In the spectroscope light is passed through a prism and the emergent light is observed. It is a well-established fact that a ray of light undergoes deviation when passed through a prism. The angle of deviation is directly proportional to the frequency of the radiation. Thus, if we pass a beam of radiations having different frequencies through a prism, these radiations undergo unequal deviation and get arranged in order of decreasing frequencies. For example, a beam of ordinary light splits up into seven colours (VIBGYOR) after passing through a prism. *This phenomenon of splitting of a beam of light into radiations of different frequencies after passing through a prism is called dispersion and the pattern of radiations obtained after dispersion is called spectrum.*

In case of dispersion of sunlight, the seven colours obtained change from violet to red without any discontinuity, which means each colour blends into the other. Such a spectrum is called **continuous spectrum.**

In a continuous spectrum, the radiations corresponding to all the wavelengths (within a certain range) are present.

On the other hand, atomic spectra of most of the elements consist of a number of coloured lines separated by dark bands. That is why atomic spectrum is also known as **line spectrum.** The various lines in the line spectrum correspond to the radiations of different wavelengths, emitted by the excited element. The lines in the line spectrum of an element are characteristic of the atoms of the element. Therefore, *atomic spectrum of an element can be used to identify the element and is sometimes called fingerprint of its atoms.*

Elements like rubidium (Rb), caesium (Cs), indium (In), scandium (Sc), etc., were discovered when their minerals were analysed by spectroscopic methods. The element helium was discovered in the sun by a study of spectrum of sunlight.

ABSORPTION SPECTRUM

When a beam of continuous light is passed through a tube containing vapours or a solution of the substance and the transmitted light is analysed with the help of a spectrometer, it is observed that the spectrum obtained contains a number of dark lines in otherwise continuous spectrum (Fig. 5.16). These dark lines appear due to the absorption of radiations of corresponding wavelengths by the substance. The dark lines in the absorption spectrum of a substance appear at the same position as the bright lines in the emission spectrum of a substance. For example, the emission spectrum of sodium has two bright lines in the yellow region of 589.0 nm and 589.6 nm whereas its absorption spectrum has two dark lines corresponding to the same wavelengths. Generally more lines are observed in the emission spectrum than in its absorption spectrum. Therefore, an absorption spectrum is taken if a simplification of the spectrum is required.

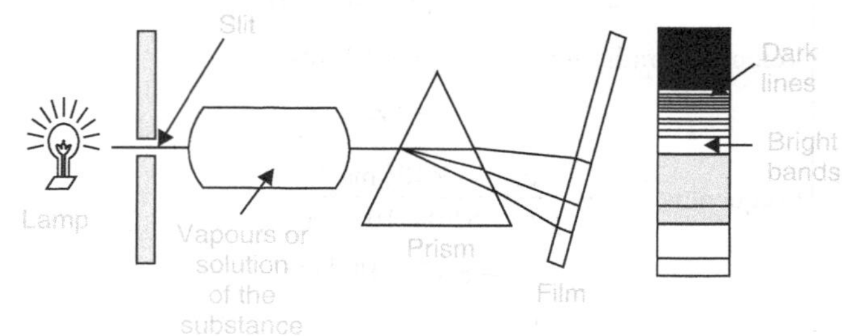

Fig. 5.16. Absorption spectrum.

The study of an emission or absorption spectra is called **spectroscopy.**

HYDROGEN SPECTRUM

The atomic spectrum of hydrogen has proved quite helpful in the development of atomic structure. It can be obtained by passing light emitted from a discharge tube containing hydrogen at low pressure, through the spectrograph as shown in Fig. 5.17.

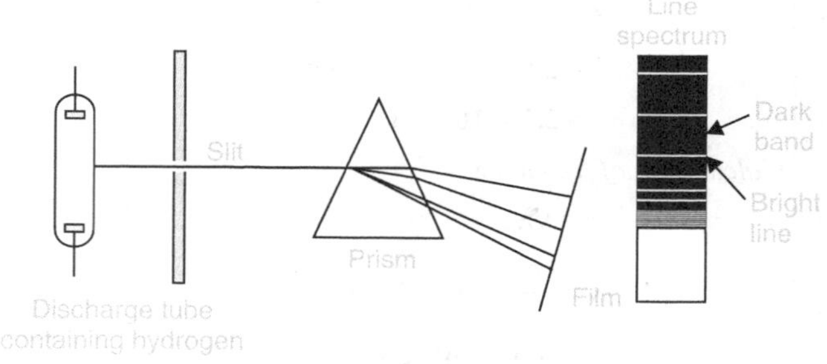

Fig. 5.17. Emission spectrum of hydrogen.

When an electric discharge is passed through hydrogen gas its molecules dissociate into hydrogen atoms. The excited hydrogen atoms, thus produced, emit electromagnetic radiations of discrete frequencies.

The spectrum obtained consists of a large number of sharp lines. Each line corresponds to a particular frequency of light emitted by the hydrogen atoms. Lines in the emission spectrum of hydrogen are present in the ultraviolet, visible and infrared regions. While the lines in the ultraviolet and visible regions can be directly obtained on a photographic plate, the lines in the infrared region are detected by other methods.

The lines in the emission spectrum of hydrogen are classified into five series as follows:

(*i*) *Lyman series* ...... Ultraviolet region

(*ii*) *Balmer series* ...... Visible region

(*iii*) *Paschen series* ...... Infrared region

(*iv*) *Brackett series* ...... Infrared region

(*v*) *Pfund series* ...... Infrared region

The complete spectrum of hydrogen is shown in Fig. 5.17(*a*)

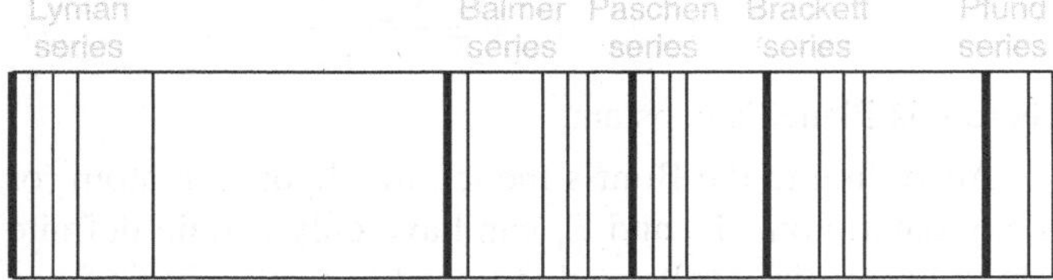

Fig. 5.17. (*a*) Atomic spectrum of hydrogen.

5.14 BOHR'S MODEL OF ATOM

In order to overcome the shortcomings of Rutherford's model, Neils Bohr (1913) proposed a new model of the atom based upon the quantum theory of radiations.

Main points of this model are:

1. *The electrons in an atom revolve around the nucleus only in certain selected circular orbits. These orbits are associated with definite energies and are called* **energy shells or energy levels.** These are numbered as 1, 2, 3, 4, ... etc., or designated as

K, L, M, N, ..., etc., shells (Fig. 5.18).

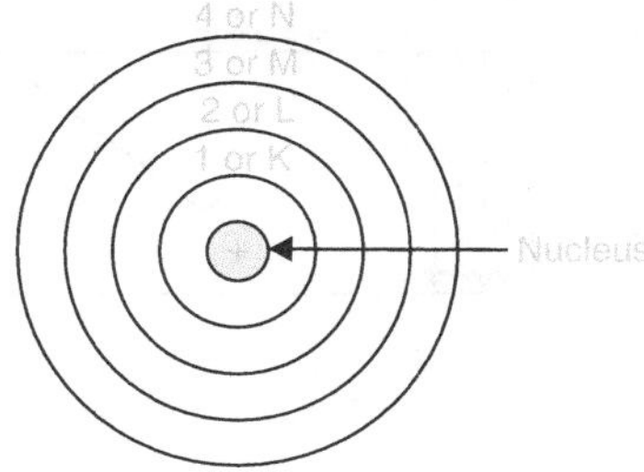

Fig. 5.18. Bohr's orbitals.

The energy of the electron is minimum in the orbit nearest to the nucleus *i.e.*, K shell. The energy of the electron increases as it moves away from the nucleus.

2. *As long as the electron remains in a particular orbit, it does not lose or gain energy. This means that the energy of an electron in a particular orbit remains constant. That is why these orbits are also called* **stationary states.**

3. *When energy from some external source is supplied to the electron, it may jump to some higher energy level by absorbing a definite amount of energy* (equal to the difference in energy between the two energy levels). *When the electron jumps back to the lower energy level it radiates the difference in energy in the form of a photon of electromagnetic radiation.*

DISTRIBUTION OF ELECTRONS IN ENERGY SHELLS

As already mentioned electrons in an atom revolve around the nucleus in well defined circular orbits or energy shells. Let us now study how electrons are distributed in the energy shells of an atom.

The arrangement of electrons in various energy levels of an atom is known as the electronic configuration of the atom. While writing the electronic configuration the following rules as given by Bohr and Bury are followed:

1. *Electrons fill up the lowest energy levels first.*

2. *The maximum number of electrons in any orbit is given by the formula $2n^2$, where n is the number of the orbit.*

For example,

For the first orbit ($n = 1$), the maximum number of electrons $= 2 \times (1)^2 = 2$

For the second orbit ($n = 2$), the maximum number of electrons $= 2 \times (2)^2 = 8$

For the third orbit ($n = 3$), the maximum number of electrons $= 2 \times (3)^2 = 18$

3. *The outermost orbit in a stable atom cannot have more than 8 electrons even if it can accommodate more electrons according to rule 2.*

4. *The penultimate energy shell, i.e., the energy shell preceeding the outermost shell, cannot have more than 18 electrons.*

Let us apply the above rules to write the electron distribution about the nucleus in some elements.

I. Case of Hydrogen Atom

Atomic number of hydrogen is 1. There is only one electron in hydrogen atom which goes into the lowest energy shell *i.e.*, K shell. Thus, the electronic configuration of hydrogen atom is:

H: K
1

Atomic diagram of hydrogen atom is shown in Fig. 5.19.

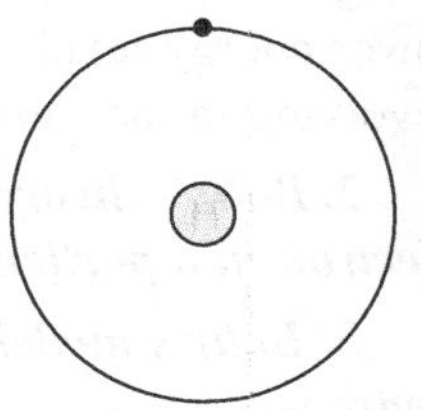

Fig. 5.19. Hydrogen atom.

Here, symbol of the element represents the nucleus and the circle drawn around the symbol represents the first energy shell. The dot in the ring represents the electron.

II. Case of Oxygen Atom

Now consider an atom of oxygen. Atomic number of oxygen is 8. Thus, there are eight electrons in an atom of oxygen. Two electrons are accommodated in the first shell (K shell) and the remaining six electrons are accommodated in second shell (L shell).

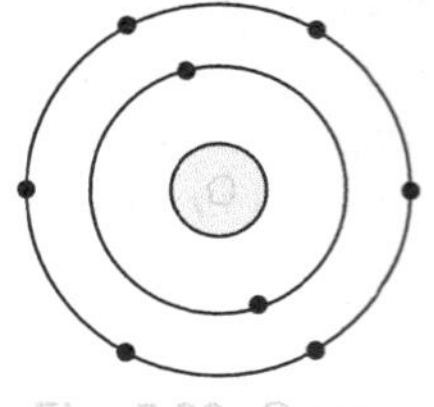

Fig. 5.20. Oxygen atom.

O: K L
2 6

The atomic diagram of oxygen is shown in Fig. 5.20.

III. Case of Calcium Atom

Atomic number of calcium is 20. An atom of calcium contains 20 electrons. 2 electrons go in the first shell (K shell), 8 electrons go to the second shell (L shell), next 8 electrons go to the third shell (M shell) and the remaining 2 electrons go to the fourth shell (N shell). Although the third shell can accommodate a maximum of 18 electrons, all the 10 electrons cannot go into it because that would make 10 electrons in the outermost shell whereas the outermost shell cannot have more than 8 electrons. After the third level acquires 8 electrons, the fourth level begins to fill. The next 2 electrons go to the fourth energy shell (N shell)

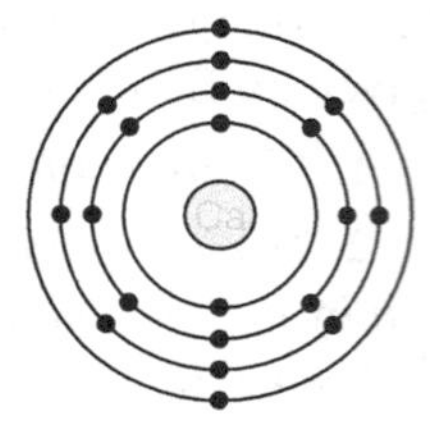

Fig. 5.21. Atomic diagram of calcium.

Ca:	K	L	M	N
	2	8	8	2

SUCCESSES OF BOHR'S MODEL

The main successes of Bohr's model are:

1. *Bohr's model could explain the stability of an atom.*

According to Bohr's model, an electron revolving in a particular orbit cannot lose energy. The electron can lose energy only if it jumps to some lower energy level. If no lower energy level is vacant then the electron will keep on revolving in the same orbit without losing energy.

2. *Bohr's theory helped in calculating the energy of an electron in a particular orbit of hydrogen.*

3. *Bohr's model could explain the atomic spectrum of hydrogen.*

According to the Bohr's model, an electron in an atom can have only certain definite energy levels. When the electron is present in the lowest possible energy level, it is said to be in the *ground state.* When energy is supplied from some external source, the electron may absorb energy and jump to some higher energy level. The electron in such a state is said to be in an *excited state*. The excited state is unstable and, therefore, the electron has a tendency to come back to the ground state. When the electron jumps back to lower energy levels, it gives out energy in the form of a quantum equal to the difference of energies between the two energy levels. If E_1 and E_2 are the energies of the lower and higher energy levels respectively, then the frequency (ν) of the radiation emitted is given by the following relation:

$$E_2 - E_1 = h\nu \quad \text{or} \quad \nu = \frac{E_2 - E_1}{h} \qquad ...(i)$$

where h is Planck's constant.

According to the Bohr's model, in a hydrogen atom (or in any other atom) E_2 and E_1 can have only certain definite values. From this, it follows that ν can have only certain fixed values. Thus, Bohr's model explains why there are certain discrete lines in the spectrum of hydrogen.

The lines which arise due to the transitions from higher energy levels to the first energy level are grouped as **Lyman series.** Similarly, the lines obtained as a result of transitions of electrons from higher energy levels to the *second, third, fourth* and *fifth* energy levels give rise to the **Balmer, Paschen, Brackett** and **Pfund series** respectively as illustrated in Fig. 5.22.

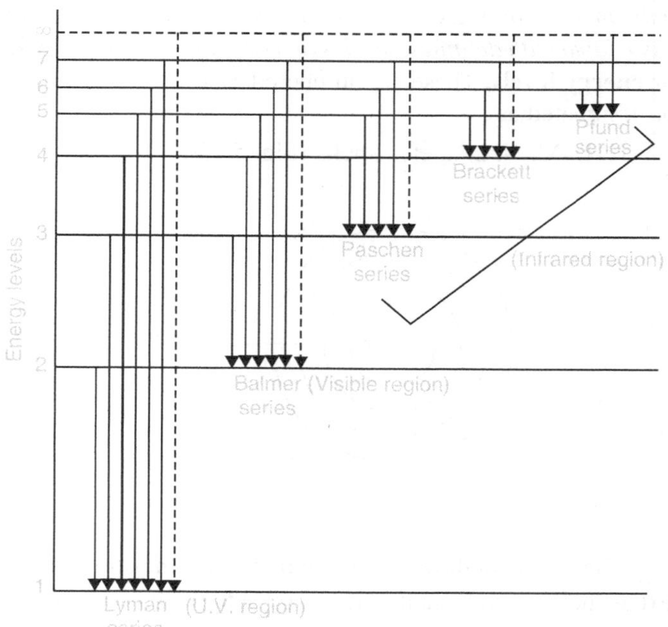

Fig. 5.22. Generation of various spectral series in the hydrogen spectrum.

LIMITATIONS OF THE BOHR'S MODEL

Bohr's theory could explain successfully the positions of the various series of lines in the hydrogen spectrum but could not explain the spectra of atoms containing more than one electron. Bohr's model could not explain even the hydrogen spectrum obtained using high resolution spectroscopes. Each spectral line, on high resolution, was found to consist of two closely spaced lines.

It was observed that in the presence of a magnetic field, *each spectral line gets split up into closely spaced lines.* This phenomenon, known as **Zeeman effect,** could not be explained by Bohr's model.

In order to overcome the limitations of the Bohr's model, attempts were made to develop a more suitable and general model for atoms. The two new concepts which led to the development of such a model were:

1. The dual behaviour of matter.
2. Heisenberg's uncertainty principle.

5.15 THE DUAL BEHAVIOUR OF MATTER

Einstein had suggested in 1905 that light has a dual nature, that is, *wave nature* as well as *particle nature.* In 1923, Louis de-Broglie, the French physicist proposed that like light, matter also has a dual character. It exhibits wave as well as particle nature. He even derived a relationship for the calculation of wavelength (λ) of the wave associated with a particle of mass m, moving with velocity v as given below:

$$\lambda = \frac{h}{mv} \quad \text{or} \quad \lambda = \frac{h}{p}$$

where h is Planck's constant and p is the momentum of the particle.

The waves associated with material particles or objects in motion are called **matter waves** or **de-Broglie waves.** These waves are distinctively different from the electromagnetic waves. The wave nature of electrons was experimentally verified by Davisson and Germer by electron diffraction experiment.

SIGNIFICANCE OF de-BROGLIE WAVES

The wave character puts some restriction on how precisely we can express the position of an electron or any other small moving particle. This is due to the reason that unlike particles, waves do not occupy a well defined position in space and are delocalised. The wave nature of matter, however, has no significance for objects of ordinary size because the wavelength of the wave associated with them is too small to be detected.

SOLVED EXAMPLES

Example 5.8. *Calculate the wavelength of a body of mass 1 mg moving with a velocity of 10 m sec^{-1}.*

Solution. We know, $\lambda = \frac{h}{mv}$

Substituting the values, $m = 1 \text{ mg} = 10^{-6}$ kg,

$$v = 10 \text{ m sec}^{-1}$$

and $$h = 6.626 \times 10^{-34} \text{ kg m}^2 \text{ s}^{-1}$$

$$\lambda = \frac{6.626 \times 10^{-34} \text{ kg m}^2 \text{ s}^{-1}}{10^{-6} \text{ kg} \times 10 \text{ ms}^{-1}}$$

$$= \mathbf{6.626 \times 10^{-29} \text{ m.}}$$

Example 5.9. *Calculate the wavelength of an electron moving with a velocity of 2.05 × 10^7 m s^{-1}.*

Solution. $$\lambda = \frac{h}{mv}$$

$$= \frac{6.626 \times 10^{-34} \text{ kg m}^2 \text{ s}^{-1}}{9.1 \times 10^{-31} \text{ kg} \times 2.05 \times 10^7 \text{ ms}^{-1}}$$

$$= \mathbf{3.55 \times 10^{-11} \text{ m.}}$$

5.16 HEISENBERG'S UNCERTAINTY PRINCIPLE

The position and the velocity of the bodies which we come across in our day-to-day life can be determined precisely at a particular moment of time. Hence, the paths or *trajectories* of such bodies can be predicted. However, in the case of small particles such as electrons it is impossible to determine simultaneously its position and velocity at a given instant with absolute certainty. Hence, it is not possible to talk of the trajectory of an electron. Werner Heisenberg (1927) put forward this fact in the form of a well-known principle, called **Heisenberg's uncertainty principle.** This principle, which is a direct consequence of the dual nature of matter and radiation, states that

It is impossible to measure simultaneously both the position and velocity (or momentum) of a microscopic particle with absolute accuracy or certainty.

Heisenberg's principle rules out the concept of definite paths or trajectories of electrons in an atom as proposed by Bohr. In order to define the trajectory of an object one must know the location and velocity of the object at various moments. Since for a subatomic particle such as an electron, it is not possible to simultaneously determine the position and velocity with absolute accuracy, it is not possible to talk of the trajectory of an electron.

Instead it is only possible to predict the probable region in a given space where we can find an electron. *Thus, Heisenberg replaced the concept of definite orbits by the concept of probability.*

5.17 QUANTUM MECHANICAL MODEL OF ATOM

The motion of all macroscopic objects such as a falling stone, moving car, orbiting satellite, etc., can be successfully described in terms of classical mechanics, based on Newton's laws of motion. These objects have predominantly particle character. However, a microscopic object such as electron has both observable wave-like and particle-like properties. The behaviour of such particles cannot be explained on the basis of classical mechanics which is based on Newton's laws of motion. In order to explain the behaviour of electrons and other microscopic particles a new branch of science called quantum mechanics was developed. ***Quantum mechanics** is a theoretical science that takes into account the dual nature of matter.* Quantum mechanics was developed independently in 1926 by Werner Heisenberg and Erwin Schrödinger.

Based on quantum mechanics a new model of the atom was developed during the 1920's. This model is known as the **quantum mechanical model.** In this model the behaviour of the electron in an atom is described by an equation known as the **Schrödinger wave equation.**

The solutions of the Schrödinger wave equation are known as wave functions (ψ). When the Schrödinger wave equation is solved for the hydrogen atom, several solutions are obtained. Out of these only certain solutions are permissible. Each permitted solution* or wave function corresponds to a definite energy state and is called **orbital.** Each energy state or orbital is characterized by a set of three quantum numbers, principal quantum number (n), azimuthal quantum number (l) and magnetic quantum number (m_l). Each orbital can be considered as a quantum mechanical analog of the electron orbits as proposed by Bohr. The electron orbitals in atoms are called *atomic orbitals* while those in a molecule are called *molecular orbitals*. Since in an atom (or a molecule) only specific or definite energy states are permitted, it implies that an electron in an atom (or a molecule) can have only certain specific values of energy. In other words, the **energy of an electron in an atom is quantised.**

*The permitted solutions of wave equations are called **eigen functions.**

SIGNIFICANCE OF ψ

A moving electron is associated with a wave and the wave function ψ gives the amplitude of electron wave. It has got no physical significance. However, the square of ψ *i.e.*, $|\psi|^2$ has a physical significance. Just like light radiations where the square of amplitude gives the intensity of light, similarly, in electron wave, $|\psi|^2$ gives the intensity of the *electron at any point*. In other words, the knowledge of $|\psi|^2$ is helpful in assessing the probability of finding an electron in a particular region.

CONCEPT OF ORBITALS

According to Heisenberg's uncertainty principle it is not possible to determine precisely the position and momentum of an electron in the atom simultaneously. Therefore, Bohr's concept of well defined orbits is ruled out. In the quantum mechanical model, we speak of probability or possibility of an electron with a particular energy being present in a certain region of space around the nucleus. The probability of finding the electron at a particular location is given by the square of the wave function ($|\psi|^2$) corresponding to that location. There are certain regions around the nucleus where the probability of finding the electron is high and there are certain regions where the probability of finding the electron is low. The probability of finding the electron does not become zero even at large distances from the nucleus, although it may become negligible. Therefore, it is not possible to draw a boundary that will enclose the region of 100% probability. However, for the sake of simplicity we draw arbitrary boundaries which enclose the regions where probability of finding the electron is maximum (about 90–95%). These regions of space around the nucleus where probability of finding the electron is maximum are called orbitals.

*An **orbital** may be defined as that region of space around the nucleus where the probability of finding an electron is maximum (90–95%).*

It is rather difficult to represent an orbital by a simple picture. Different methods are employed for this purpose. But the most common method is to represent an orbital as a charge cloud in terms of small dots, the intensity of dots being more in some regions and less in others.

In terms of charge cloud representation, the probability of finding an electron in a particular region of space is directly proportional to the density of dots in that region.

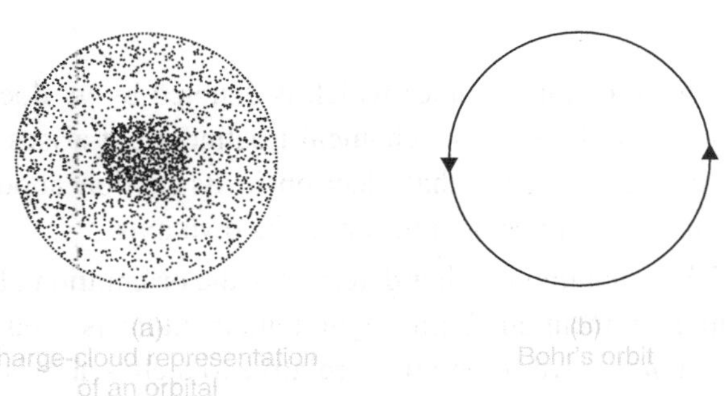

Fig. 5.23. Orbital and an orbit.

5.18 QUANTUM NUMBERS

In an atom a large number of electron orbitals are permissible. These orbitals are designated by a set of numbers known as quantum numbers. In order to specify energy, size, shape and orientation of the electron orbital three quantum numbers are required. These are, principal quantum number, azimuthal quantum number and magnetic quantum number. These quantum numbers arise as a natural consequence during the solution of the Schrödinger wave equation. In order to designate the electron, an additional quantum number called spin quantum number is needed to specify spin of the electron. Thus, each orbital in an atom is designated by a set of three quantum numbers and each electron is designated by a set of four quantum numbers. These quantum numbers are discussed below:

THE PRINCIPAL QUANTUM NUMBER (n)

This is the most important quantum number as it determines to a large extent the energy of an electron. It also determines the average distance of an electron from the nucleus. It is denoted by the letter n. This quantum number tells us in which principal energy level or shell the electron is present. It can have any whole number value such as 1, 2, 3, 4, ..., etc. The energy levels or energy shells corresponding to these numbers are designated as K, L, M, N, ..., etc. As the value of n increases, the electron gets farther away from the nucleus and its energy increases. The higher the value of n, the higher is the electronic energy. For hydrogen and hydrogen-like species, the energy and size of the orbital are determined by the principal quantum number alone.

Energy of the electron in a hydrogen atom is related to the principal quantum number by the following relation:

$$E_n = -k^2 \frac{2\pi^2 me^4}{n^2h^2} = -\frac{2.18 \times 10^{-18}}{n^2} \text{ J}$$

where, m = Mass of electron

e = Charge on electron

h = Planck's constant

E_n = Energy of the electron in nth principal shell

n = Principal quantum number used to designate the principal shell

k = Coulomb's law constant.

This relation is similar to the expression given by Bohr.

THE ORBITAL ANGULAR MOMENTUM, QUANTUM NUMBER OR AZIMUTHAL QUANTUM NUMBER (l)

This quantum number determines the angular momentum of the electron. This is denoted by l. The value of l gives the **sub-level** or **sub-shell** in which the electron is located. *It determines the three dimensional shape of the orbital* in which the electron is located. The number of sub-shells within a principal shell is determined by the value of n for that principal energy level. *Thus, l may have all possible whole number values from 0 to n—1* for each principal energy level. For a given value of n, l can have n values. The various sub-levels are designated as *s, p, d, f* depending upon the value of l, as follows:

Value of l →	0	1	2	3	4	5	6
Designation of sub-shell →	*s**	*p*	*d*	*f*	*g*	*h*	*i*

Table 5.3 shows the permissible values of l for a given principal quantum number and the corresponding sub-shell notation.

Table 5.3. Sub-shell Notations

n	l	*Sub-shell notation*
1	0	1*s*
2	0	2*s*
2	1	2*p*
3	0	3*s*
3	1	3*p*
3	2	3*d*
4	0	4*s*
4	1	4*p*
4	2	4*d*
4	3	4*f*

*The names **s**, **p**, **d** and **f** are derived from spectroscopic terms sharp, principal, diffuse and fundamental respectively.

THE MAGNETIC QUANTUM NUMBER (m_l)

This quantum number which is denoted by m_l refers to the different orientations of the electron cloud in a particular sub-shell. These different orientations are called **orbitals.** The number of orbitals in a particular sub-shell within a principal energy level is given by the number of values allowed to m_l which in turn depends on the values of l. The possible values of m_l range from $+l$ through 0 to $-l$, thus making a total of $(2l + 1)$ values. Thus, in a sub-shell, the number of orbitals is equal to $(2l + 1)$.

For l = 0 (*i.e.*, *s*-sub-shell), m_l can have only one value, $m_l = 0$. It means that **s-sub-shell** has only **one orbital.**

For l = 1 (*i.e.*, *p*-sub-shell), m_l can have three values, +1, 0 and −1. This implies that **p-sub-shell** has **three orbitals.**

For l = 2 (*i.e.*, *d*-sub-shell), m_l can have five values, +2, +1, 0, −1, −2. It means that **d-sub-shell** has **five orbitals.**

For l = 3 (*i.e.*, *f*-sub-shell), m_l can have seven values, +3, +2, +1, 0, −1, −2, −3. It means that **f-sub-shell** has **seven orbitals.**

For l = 2 (*i.e.*, *d*-sub-shell), m_l can have five values, +2, +1, 0, −1, −2. It means that **d-sub-shell** has **five orbitals.**

For l = 3 (*i.e.*, *f*-sub-shell), m_l can have seven values, + 3, +2, +1, 0, −1, −2, −3. It means that **f-sub-shell** has **seven orbitals.**

The number of orbitals in various types of sub-shells are given below in tabular form.

Sub-shell	*s*	*p*	*d*	*f*	*g*
Value of l	0	1	2	3	4
No. of orbitals ($2l$ + 1)	1	3	5	7	9

The relationship between the principal quantum number (n), angular momentum quantum number (l) and magnetic quantum number (m_l) is summed up in Table 5.4.

THE SPIN QUANTUM NUMBER (m_s)

This quantum number which is denoted by m_s does not follow from the wave mechanical treatment but arises from the spectral evidence that electron in its motion about the nucleus also rotates or spins about its own axis.

This quantum number determines the orientation of spin angular momentum. Spin angular momentum is quantised and can have two orientations relative to a chosen axis. The spin quantum number can have only two values which are $+\frac{1}{2}$ and $-\frac{1}{2}$. The $+\frac{1}{2}$ value indicates clockwise spin (generally represented by an arrow pointing upwards, *i.e.*, ↑) and the other indicates anti-clockwise spin (generally represented by an arrow pointing downwards *i.e.*, ↓).

Due to its spin, the electron behaves as a tiny magnet. The spin of the electron is responsible for most of the magnetic properties of atoms, molecules or ions. If all the electrons in an atom or molecule are paired, it behaves as a **diamagnetic substance** *i.e.*, it is weakly repelled by the magnetic field. On the other hand, if atoms or molecules of a substance have one or more unpaired or odd electrons, it behaves as a **paramagnetic substance,** *i.e.*, it is weakly attracted by the magnetic field.

5.19 SHAPES OF ORBITALS

BOUNDARY SURFACE DIAGRAMS

The shapes of the orbitals can be represented quite accurately with the help of **boundary surface diagrams**. In this representation, a boundary surface (also known as contour surface) is drawn in space for an orbital on which the value of probability density (ψ^2) is constant.

Table 5.4. Relationship among Values of n, l and m_l

Energy Level	*Principal Quantum Numbers (n)*	*Possible Values of (l)*	*Designation of Sub-shell*	*Possible Values of (m_l)*	*Number of orbitals In a given Sub-shell*	*Number of orbitals In a given Energy Level*
K	1	0	1*s*	0	1	1
L	2	0	2*s*	0	1	4
		1	2*p*	+ 1, 0, − 1	3	
M	3	0	3*s*	0	1	9
		1	3*p*	+ 1, 0, − 1	3	
		2	3*d*	+ 2, + 1, 0, − 1, − 2	5	

A boundary surface which encloses the regions of maximum probability (say 90%) best describes the shape of the orbital. It is not possible to draw a boundary surface diagram which encloses the region of 100% probability because probability density has always some value, howsoever small it may be, at any finite distance from the nucleus. The boundary surface diagrams for *s*-orbitals are spherical in shape. The size of the *s*-orbital, however, increases with increase in the value of *n*. If we observe the probability density curve for 2*s* orbital we find that the probability density function for 2*s* orbital decreases to zero, increases to maximum value and finally approaches zero with increase in distance from the nucleus. The region where the probability density function reduces to zero is called nodal surface or node. In any *ns* orbital there are $(n - 1)$ nodes. Thus, 2*s* orbital has only one node.

The boundary surface diagrams of the three 2*p* orbitals are not spherical. Each *p*-orbital consists of two lobes which are separated by a region of zero probability called node. The three 2*p* orbitals lie along *x*-, *y*- and *z*-axis respectively and are designated as $2p_x$, $2p_y$ and $2p_z$.

Like *s*-orbitals, *p*-orbitals also increase in size with the increase in the value of *n*.

The boundary surface diagrams of 1*s*, $2p_x$, $2p_y$ and $2p_z$ orbitals are shown in Fig. 5.24.

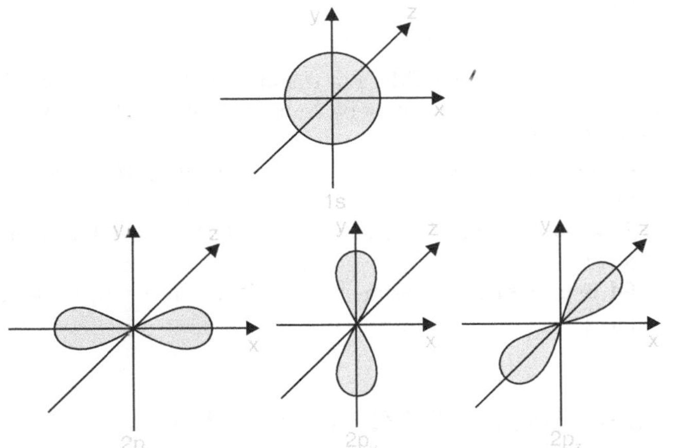

Fig. 5.24. The boundary surface diagrams of 1*s* and 2*p* orbitals.

The five *d*-orbitals are designated as d_{xy}, d_{yz}, d_{xz}, $d_{x^2-y^2}$ and d_{z^2}. The boundary surface diagrams of the five 3*d* orbitals are shown in Fig. 5.25.

Each *d*-orbital has two nodal planes or angular nodes. For example, d_{xy} orbital has two nodal planes passing through the origin and bisecting the principal axes.

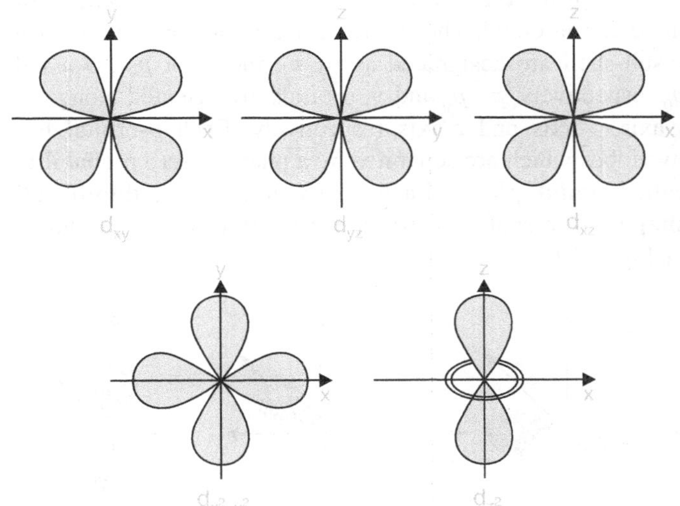

Fig. 5.25. Boundary surface diagrams of the five 3*d* orbitals.

CHARGE CLOUD PICTURE OF ORBITALS

An alternative way to describe the probability is in terms of negative charge cloud, the density of charge cloud being proportional to ψ^2 or probability. In this method, the charge cloud is represented in term of small dots. To understand this method of representation, let us imagine the electron as a very small dot. Let us further suppose that we take a very large number of photographs of the electron in a hydrogen atom at very short intervals of time on the same film. If the film were developed, the picture obtained would be similar to that shown in Fig. 5.26 (*a*). In terms of electron cloud representation the probability of finding the electron in a particular region of space is directly proportional to the density of such dots in that region.

The charge cloud pictures of 1*s*, 2*s* and 3*s* orbitals have been given in Fig. 5.26.

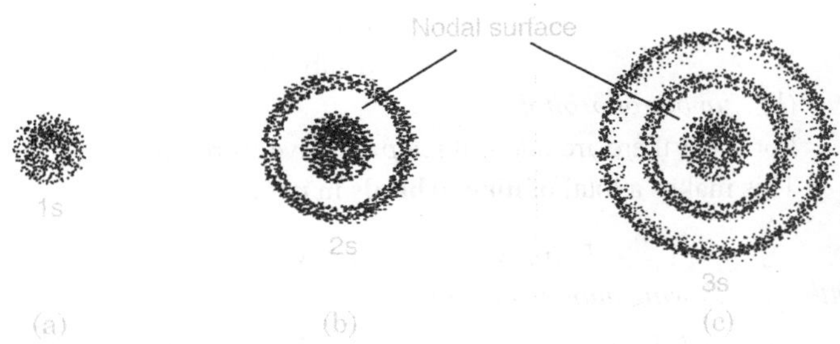

Fig. 5.26. (*a*) Charge cloud picture of 1s orbital.
(*b*) Charge cloud picture of 2s orbital.
(*c*) Charge cloud picture of 3s orbital.

For **p-orbitals** there are three possible orientations of the electron cloud. These three orientations or orbitals of a p-sub-shell are designated as p_x, p_y and p_z or p_{+1}, p_{-1} and p_0 respectively. p_x, p_y and p_z orbitals are oriented along the x-axis, y-axis and z-axis respectively. Each p-orbital has two lobes which are separated by a plane of zero probability called **nodal plane.** Each p-orbital is, thus, **dumb-bell shaped.** The spatial distributions of $2p$ orbitals are shown in Fig. 5.27.

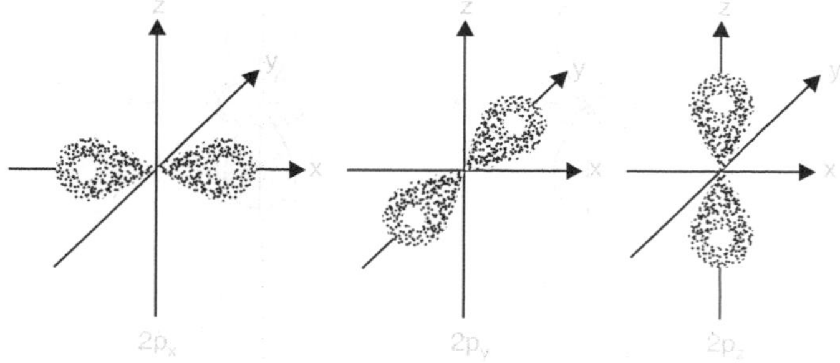

Fig. 5.27. $2p$ orbitals.

In the absence of an external electric or magnetic field, the three p-orbitals of particular energy level have the same energy. Such orbitals which have equal energy are called **degenerate orbitals.** In the presence of an external magnetic or electric field this degeneracy is lost because the three p-orbitals are oriented differently with respect to the applied field.

SOLVED EXAMPLES

Example 5.10. *(a) What sub-shells are possible in $n = 3$ energy level?*

(b) How many orbitals (of all kinds) are possible in this level?

Solution. *(a) Sub-shells in $n = 3$ energy level:*

For $n = 3$, the possible values of l are 0, 1 and 2.

The corresponding sub-shells are

$l = 0$, s sub-shell

$l = 1$, p sub-shell

$l = 2$, d sub-shell

(b) Number of orbitals:

For $n = 3$ there are one s, three p and five d orbitals.

This makes a total of **nine orbitals** in $n = 3$ level.

Example 5.11. *Using s, p, d notations describe the orbitals with the following quantum numbers:*

(a) $n = 1, l = 0$ *(b)* $n = 3, l = 1$

(c) $n = 4, l = 2$ *(d)* $n = 4, l = 3$.

Solution. *(a)* 1s *(b)* 3p *(c)* 4d *(d)* 4f.

Example 5.12. *State which of the following sets of quantum numbers would be possible and which would not be permissible for an electron in an atom.*

(i) $n = 3, l = 3, m_l = +3, m_s = +\frac{1}{2}$

(ii) $n = 3, l = 0, m_l = +1, m_s = -\frac{1}{2}$

(iii) $n = 5, l = 4, m_l = +3, m_s = 1$

(iv) $n = 0, l = 0, m_l = 0, m_s = +\frac{1}{2}$

(v) $n = 4, l = 3, m_l = -2, m_s = -\frac{1}{2}$.

Solution. *(i)* Not possible because when $n = 3$, l cannot be equal to 3. The value of l cannot be numerically equal to the value of n.

(ii) Not possible, because when $l = 0$, m_l cannot be +1. The value of m_l cannot be numerically more than the value of l.

(iii) Not possible because m_s cannot be 1. The only allowed values of m_s are $+\frac{1}{2}$ and $-\frac{1}{2}$.

(iv) Not possible, because n cannot be 0.

(v) This set is permitted because when $n = 4$, l can be 3. When $l = 3$, m_l can be – 2. m_s can have $-\frac{1}{2}$ value.

Example 5.13. *Which of the following orbitals are not possible?*

2d, 4f, 4g and 6d.

Solution. **2d is not possible.** For d sub-shell $l = 2$. When $n = 2$, l can be 0 and 1. Since l cannot be 2 for $n = 2$ hence $2d$ sub-shell is not possible.

4f is possible. For f-sub-shell $l = 3$. When $n = 4$, l can be equal to 3.

4g is not possible. For g-sub-shell $l = 4$. When $n = 4$, l cannot be equal to 4.

6d is possible. For d-sub-shell $l = 2$. When $n = 6$, l can be equal to 2.

5.20 ENERGIES OF ORBITALS

The energy of an electron in a hydrogen atom is determined solely by the principal quantum number. Therefore, **all the orbitals of a particular energy level have the same energy in hydrogen and hydrogen-like atoms.** For example, $2s$ and $2p$ orbitals have the same energy. The orbitals having equal energy are called **degenerate orbitals.**

Thus, the energy of the orbitals in hydrogen and hydrogen-like species increases as follows:

$1s < 2s = 2p < 3s = 3p = 3d < 4s = 4p = 4d = 4f$..... and so on.

In case of multielectron atoms the energy of the electron depends not only on the principal quantum number but also on the azimuthal quantum number. The order in which the energies of the orbitals increase is as follows:

$1s < 2s < 2p < 3s < 3p < 4s < 3d < 4p < 5s < 4d < 5p < 4f < 5d < 6p < 7s$

This order may be remembered using the memory aid given in Fig. 5.28. Starting from the top, the direction of the arrows gives the increasing order of energies of various orbitals in a multielectron atom.

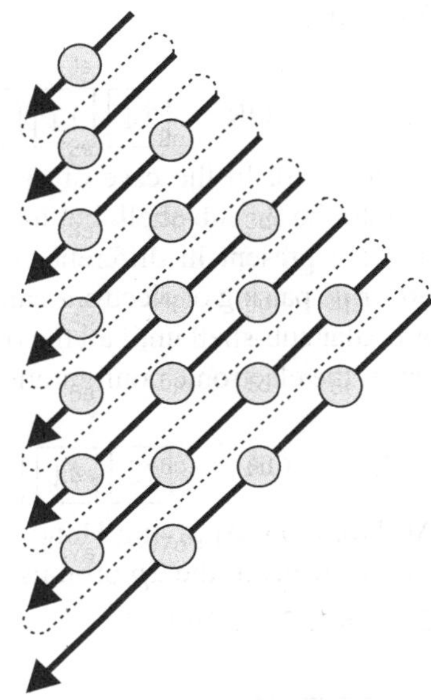

Fig. 5.28. Memory aid for remembering order of energies of various orbitals in a multi-electron atom.

The relative order of energies of various sub-shells in a multi-electron atom can be predicted with the help of the **(n + l)** rule or the **Bohr-Bury's rule.** According to this rule:

(*i*) *In neutral atoms a sub-shell with lower value of (n + l) has lower energy.*

For example, $4s$ orbital has lower energy than $3d$ orbital.

For $4s$ orbital $n = 4$ and $l = 0$.

Hence, $n + l = 4 + 0 = 4$.

For $3d$ orbital $n = 3$ and $l = 2$.

Hence, $n + l = 3 + 2 = 5$.

(*ii*) *If two sub-shells have equal value of (n + l), the sub-shell with lower value of n has lower energy.*

For example, consider $3p$ and $4s$ orbitals.

For $4s$ orbital $n = 4$ and $l = 0$.

Hence, $n + l = 4 + 0 = 4$.

For $3p$ orbital $n = 3$ and $l = 1$.

Hence, $n + l = 3 + 1 = 4$.

Here $3p$ orbital has lower energy than $4s$ because it has a lower value of n.

5.21 RULES FOR FILLING OF ORBITALS IN AN ATOM

An atom in its lowest energy state is said to be in the normal state or the *ground state*. The ground state is the most stable state for the atom. The filling of orbitals in the ground state is determined by the following ***rules:***

AUFBAU RULE

According to this rule,

the electrons are added progressively to the various orbitals in their order of increasing energies, starting with the orbital of lowest energy.

Increasing order of energies of various orbitals is:

$1s, 2s, 2p, 3s, 3p, 4s, 3d, 4p, 5s, 4d, 5p, 6s, 4f, 5d, 6p, 7s,$ ....

PAULI'S EXCLUSION PRINCIPLE

This principle was discovered by Wolfgang Pauli in 1925. This principle states that:

No two electrons in an atom can have same values for all the four quantum numbers.

A direct implication of this principle is that *it is not possible to accommodate more than two electrons in an orbital.* This can be easily understood as follows:

All the electrons in a particular orbital have same values of principal quantum number (n), azimuthal quantum number (l) and magnetic quantum number (m_l). For example, all the electrons in $3s$ orbital have $n = 3$, $l = 0$ and $m_l = 0$. Therefore, in order to have unique sets of quantum numbers they must have different values of spin quantum number (m_s). But we know m_s can have only two values $+\frac{1}{2}$ and $-\frac{1}{2}$. Hence, *in an orbital only two electrons can be accommodated, one spinning clockwise* $\left(m_s = +\frac{1}{2}\right)$ *and the other spinning anti-clockwise* $\left(m_s = -\frac{1}{2}\right)$. From the above discussion it follows that

s-sub-shell (containing only **one orbital**) can have a maximum of **2 electrons**

p-sub-shell (containing **three orbitals**) can have a maximum of **6 electrons**

d-sub-shell (containing **five orbitals**) can have a maximum of **10 electrons**

f-sub-shell (containing **seven orbitals**) can have a maximum of **14 electrons.**

HUND'S RULE OF MAXIMUM MULTIPLICITY

This rule states that

the pairing of electrons in the orbitals of a particular sub-shell (p, d or f) does not take place until all the orbitals of the sub-shell are singly occupied. Moreover, the singly occupied orbitals must have the electrons with parallel spins.

The basis of this rule is that two electrons in a particular orbital feel greater repulsion and hence while filling orbitals of equal energy pairing of electrons is avoided as long as it is possible. Moreover, the singly occupied orbitals should have electrons with parallel spin because this corresponds to a state of lower energy. This can be explained in terms of the magnetic effects of electron spin. This rule helps us in writing the ground state configurations of those atoms which have partially filled *p, d or f* sub-shells in them. The application of these rules has been illustrated in the following electronic configurations.

5.22 ELECTRONIC CONFIGURATIONS OF ATOMS

In order to represent the electron population of an orbital, the principal quantum number (*n*) is written before the orbital symbol while the number of electrons in the orbital is written as superscript near the top right hand of the orbital symbol. For example, if we have two electrons in the *s*-orbital of the first energy level then it is written as $1s^2$. Sometimes, electronic configurations are represented in a different manner by representing each orbital by a square or a circle and the electrons are represented by putting arrows in it as illustrated below:

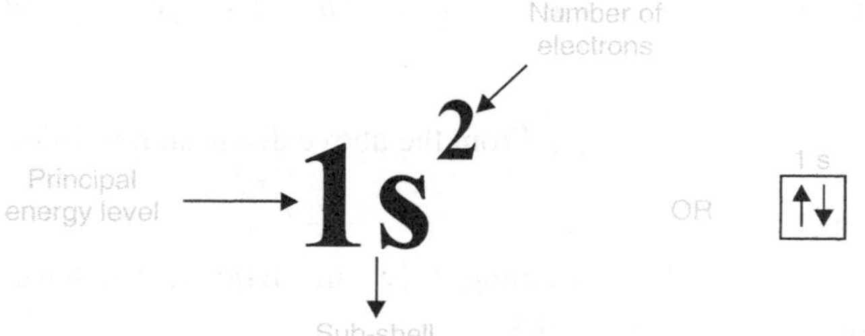

ELECTRONIC CONFIGURATIONS OF ATOMS

On the basis of the above rules and the sequence of energy levels, let us write electronic configurations for some elements.

Hydrogen (At. No. = 1). Since hydrogen has only one electron, it must go to the 1*s* orbital. The electronic configuration for hydrogen is

$_1H$: $1s^1$ OR

1s
↑

Boron (At. No. = 5). In this case, the four electrons completely fill the 1*s* and 2*s* orbitals. The fifth electron goes into one of the 2*p* orbitals.

$_5B$: $1s^2$, $2s^2$, $2p_x^{\ 1}$ OR

1s	2s	$2p_x$	$2p_y$	$2p_z$
↑↓	↑↓	↑		

Carbon (At. No. = 6). In the case of carbon, the sixth electron also goes into the 2*p* sub-shell, but the two electrons in the 2*p* sub-shell are present in different orbitals because according to Hund's rule pairing of electrons cannot take place in orbitals of a particular sub-shell until all the orbitals get one electron each. Hence, the electronic configuration of carbon is

$_6C$: $1s^2$, $2s^2$, $2p_x^{\ 1}$, $2p_y^{\ 1}$ OR

1s	2s	$2p_x$	$2p_y$	$2p_z$
↑↓	↑↓	↑	↑	

Nitrogen (At. No. = 7). Applying Hund's rule, N atom has three unpaired electrons in the 2*p* orbitals.

$_7N$: $1s^2$, $2s^2$, $2p_x^{\ 1}$, $2p_y^{\ 1}$, $2p_z^{\ 1}$

OR

1s	2s	$2p_x$	$2p_y$	$2p_z$
↑↓	↑↓	↑	↑	↑

Oxygen, Fluorine and Neon. Beginning with oxygen, the 2*p* orbitals start filling with a second electron till neon in which it is completely filled.

$_8O$: $1s^2$, $2s^2$, $2p_x^{\ 2}$, $2p_y^{\ 1}$, $2p_z^{\ 1}$

OR

1s	2s	$2p_x$	$2p_y$	$2p_z$
↑↓	↑↓	↑↓	↑	↑

$_9F$: $1s^2$, $2s^2$, $2p_x^{\ 2}$, $2p_y^{\ 2}$, $2p_z^{\ 1}$

OR

1s	2s	$2p_x$	$2p_y$	$2p_z$
↑↓	↑↓	↑↓	↑↓	↑

$_{10}Ne$: $1s^2$, $2s^2$, $2p_x^{\ 2}$, $2p_y^{\ 2}$, $2p_z^{\ 2}$

OR

1s	2s	$2p_x$	$2p_y$	$2p_z$
↑↓	↑↓	↑↓	↑↓	↑↓

From sodium (At. No. 11) to argon (At. No. 18) 3*s* and 3*p* orbitals are successively filled. After 3*p*, the 19th electron in potassium (At. No. 19) enters the 4*s* orbital instead of 3*d* because the energy of the 4*s* orbital is less than that of the 3*d*.

Electronic configurations of the first twenty elements have been given in Table 5.5.

The next ten *elements* (At. Nos. = 21 – 30) which follow calcium are *scandium* (Sc), *titanium* (Ti), *vanadium* (V), *chromium* (Cr), *manganese* (Mn), *iron* (Fe), *cobalt* (Co), *nickel* (Ni), *copper* (Cu) and *zinc* (Zn). In these elements *addition of electrons takes place in the 3d-orbitals.* All these elements are known as the **transition elements.** The electronic configuration of scandium (At. No. = 21), the first transition element, for example, is $1s^2\ 2s^2\ 2p^6\ 3s^2\ 3p^6\ 4s^2\ 3d^1$ while that of zinc (At. No. = 30), the last element of the series, is $1s^2\ 2s^2\ 2p^6\ 3s^2\ 3p^6\ 4s^2\ 3d^{10}$. Thus, while the filling of 3*d*-orbitals begins with scandium, it ends with zinc. The electronic configurations of all these 10 elements are represented in Table 5.6.

Table 5.5. Electronic Configurations of the First 20 Elements

Atomic Number	*Element*	*Orbital Electronic Configuration*			
1	Hydrogen	$1s^1$			
2	Helium	$1s^2$			
3	Lithium	$1s^2$	$2s^1$		
4	Beryllium	$1s^2$	$2s^2$		
5	Boron	$1s^2$	$2s^2\ 2p_x^{\ 1}$		
6	Carbon	$1s^2$	$2s^2\ 2p_x^{\ 1}\ 2p_y^{\ 1}$		
7	Nitrogen	$1s^2$	$2s^2\ 2p_x^{\ 1}\ 2p_y^{\ 1}\ 2p_z^{\ 1}$		
8	Oxygen	$1s^2$	$2s^2\ 2p_x^{\ 2}\ 2p_y^{\ 1}\ 2p_z^{\ 1}$		
9	Fluorine	$1s^2$	$2s^2\ 2p_x^{\ 2}\ 2p_y^{\ 2}\ 2p_z^{\ 1}$		
10	Neon	$1s^2$	$2s^2\ 2p_x^{\ 2}\ 2p_y^{\ 2}\ 2p_z^{\ 2}$		
11	Sodium	$1s^2$	$2s^2\ 2p_x^{\ 2}\ 2p_y^{\ 2}\ 2p_z^{\ 2}$	$3s^1$	
12	Magnesium	$1s^2$	$2s^2\ 2p_x^{\ 2}\ 2p_y^{\ 2}\ 2p_z^{\ 2}$	$3s^2$	
13	Aluminium	$1s^2$	$2s^2\ 2p_x^{\ 2}\ 2p_y^{\ 2}\ 2p_z^{\ 2}$	$3s^2\ 3p_x^{\ 1}$	
14	Silicon	$1s^2$	$2s^2\ 2p_x^{\ 2}\ 2p_y^{\ 2}\ 2p_z^{\ 2}$	$3s^2\ 3p_x^{\ 1}\ 3p_y^{\ 1}$	
15	Phosphorus	$1s^2$	$2s^2\ 2p_x^{\ 2}\ 2p_y^{\ 2}\ 2p_z^{\ 2}$	$3s^2\ 3p_x^{\ 1}\ 3p_y^{\ 1}\ 3p_z^{\ 1}$	
16	Sulphur	$1s^2$	$2s^2\ 2p_x^{\ 2}\ 2p_y^{\ 2}\ 2p_z^{\ 2}$	$3s^2\ 3p_x^{\ 2}\ 3p_y^{\ 1}\ 3p_z^{\ 1}$	
17	Chlorine	$1s^2$	$2s^2\ 2p_x^{\ 2}\ 2p_y^{\ 2}\ 2p_z^{\ 2}$	$3s^2\ 3p_x^{\ 2}\ 3p_y^{\ 2}\ 3p_z^{\ 1}$	
18	Argon	$1s^2$	$2s^2\ 2p_x^{\ 2}\ 2p_y^{\ 2}\ 2p_z^{\ 2}$	$3s^2\ 3p_x^{\ 2}\ 3p_y^{\ 2}\ 3p_z^{\ 2}$	
19	Potassium	$1s^2$	$2s^2\ 2p_x^{\ 2}\ 2p_y^{\ 2}\ 2p_z^{\ 2}$	$3s^2\ 3p_x^{\ 2}\ 3p_y^{\ 2}\ 3p_z^{\ 2}$	$4s^1$
20	Calcium	$1s^2$	$2s^2\ 2p_x^{\ 2}\ 2p_y^{\ 2}\ 2p_z^{\ 2}$	$3s^2\ 3p_x^{\ 2}\ 3p_y^{\ 2}\ 3p_z^{\ 2}$	$4s^2$

Table 5.6. Electronic Configurations of Transition Elements [Scandium (At. No. 21) to Zinc (At. No. 30)]

Atomic Number	*Element*	*Electronic Configuration*			
21	Scandium	$1s^2$	$2s^2\ 2p^6$	$3s^2\ 3p^6\ 3d^1$	$4s^2$
22	Titanium	$1s^2$	$2s^2\ 2p^6$	$3s^2\ 3p^6\ 3d^2$	$4s^2$
23	Vanadium	$1s^2$	$2s^2\ 2p^6$	$3s^2\ 3p^6\ 3d^3$	$4s^2$
24	**Chromium**	$\mathbf{1s^2}$	$\mathbf{2s^2\ 2p^6}$	$\mathbf{3s^2\ 3p^6\ 3d^5}$	$\mathbf{4s^1}$
25	Manganese	$1s^2$	$2s^2\ 2p^6$	$3s^2\ 3p^6\ 3d^5$	$4s^2$
26	Iron	$1s^2$	$2s^2\ 2p^6$	$3s^2\ 3p^6\ 3d^6$	$4s^2$
27	Cobalt	$1s^2$	$2s^2\ 2p^6$	$3s^2\ 3p^6\ 3d^7$	$4s^2$
28	Nickel	$1s^2$	$2s^2\ 2p^6$	$3s^2\ 3p^6\ 3d^8$	$4s^2$
29	**Copper**	$\mathbf{1s^2}$	$\mathbf{2s^2\ 2p^6}$	$\mathbf{3s^2\ 3p^6\ 3d^{10}}$	$\mathbf{4s^1}$
30	Zinc	$1s^2$	$2s^2\ 2p^6$	$3s^2\ 3p^6\ 3d^{10}$	$4s^2$

EXCEPTIONAL CONFIGURATIONS OF CHROMIUM AND COPPER

From the electronic configurations in Table 5.6, it may be noted that chromium and copper have five and ten electrons in the 3*d* orbitals rather than four and nine electrons respectively as expected. The reason is that *fully filled orbitals* and *exactly half-filled orbitals* have extra stability. Thus, the $p^3, p^6, d^5, d^{10}, f^7$ and f^{14} configurations which are either fully filled or exactly half-filled, are more stable. Therefore, to acquire more stability one of the 4*s* electron goes into the 3*d* orbitals so that the 3*d* orbitals get half-filled or completely filled in chromium and copper respectively.

Chromium

Expected Configuration: $1s^2, 2s^2\ 2p^6, 3s^2, 3p^6, 3d^4, 4s^2$

Actual Configuration: $1s^2, 2s^2, 2p^6, 3s^2, 3p^6, 3d^5, 4s^1$

Copper

Expected Configuration: $1s^2, 2s^2, 2p^6, 3s^2, 3p^6, 3d^9, 4s^2$

Actual Configuration: $1s^2, 2s^2, 2p^6, 3s^2, 3p^6, 3d^{10}, 4s^1$

Explanation of Greater Stability of Half-filled and Fully-filled Electronic Configurations

The **extra stability** of half-filled and fully-filled electronic configurations can be explained in terms of **symmetry** and **exchange energy.** The half-filled and fully-filled electronic configurations have symmetrical distribution of electrons and this symmetry leads to stability. Moreover, in such configurations electrons can exchange their positions among themselves to a maximum extent. This exchange leads to stabilization which is expressed in terms of exchange energy.

The half-filled and completely filled degenerate orbitals provide extra stability to the system due to *symmetrical arrangement* and *exchange energy* as described below:

(*i*) ***Symmetrical arrangement.*** The electronic configurations in which all the orbitals of same sub-shell are either completely-filled or exactly half-filled have relatively more symmetrical distribution of electrons and therefore, lend more stability to the system. For example, the expected configuration of chromium is $3d^4\ 4s^2$. But shifting of one electron from the 4*s* to the 3*d*-orbitals makes the configuration more symmetrical and hence relatively more stable.

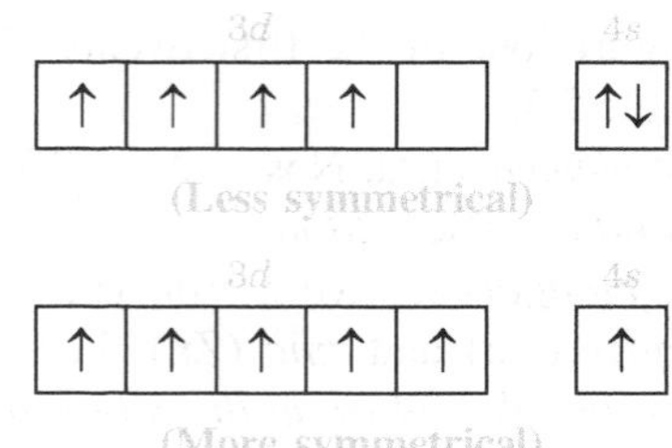

In a similar way shifting of one electron from 4*s* to 3*d* in copper also makes the configuration relatively more stable.

(*ii*) ***Stability due to exchange energy.*** The half-filled or fully-filled degenerate orbitals have more number of exchanges and consequently, have large exchange energy of stabilization. The exchange means shifting of electrons from one orbital to another within the same sub-shell. Let us compare the number of exchanges in $3d^4\ 4s^2$ and $3d^5\ 4s^1$ configurations of chromium.

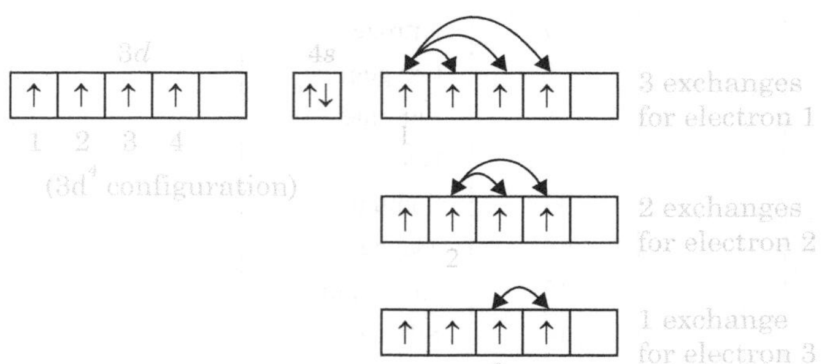

Thus, in the $\mathbf{3d^4}$ arrangement, the electron exchanges are **six** which implies that there are six possible arrangements with parallel spin in the $3d^4$ configuration.

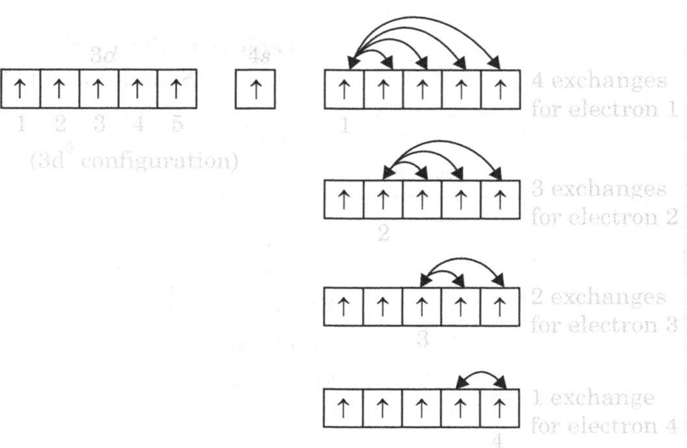

It is quite evident from the above description that the total number of electron exchanges in the $3d^5$ arrangement is larger which lends it a relatively greater stability. In a similar way it can be established that the number of exchanges in the $\mathbf{3d^{10}}$ configuration is larger than in the $\mathbf{3d^9}$ configuration which makes the $3d^{10}$ configuration relatively more stable.

ELECTRONIC CONFIGURATIONS OF IONS

Ions are formed by loss or gain of electrons by neutral atoms. A positive ion (or cation) is formed by loss of electrons whereas a negative ion (or anion) is formed by gain of electrons. The electronic configurations of ions are written just like the electronic configurations of atoms. However, it must be noted that while forming cations the electrons are lost from the shell with the highest principal quantum number. For example,

Fe : $1s^2, 2s^2, 2p^6, 3s^2, 3p^6, 3d^6, \mathbf{4s^2}$

Fe^{2+} : $1s^2, 2s^2, 2p^6, 3s^2, 3p^6, 3d^6$

Fe^{3+} : $1s^2, 2s^2, 2p^6, 3s^2, 3p^6, 3d^5$

Zn : $1s^2, 2s^2, 2p^6, 3s^2, 3p^6, 3d^{10}, \mathbf{4s^2}$

Zn^{2+} : $1s^2, 2s^2, 2p^6, 3s^2, 3p^6, 3d^{10}$

The anions are formed by adding electrons to the vacant orbital of lowest energy.

F : $1s^2, 2s^2, 2p^5$

F^- : $1s^2, 2s^2, 2p^6$

O : $1s^2, 2s^2, 2p^4$

O^{2-} : $1s^2, 2s^2, 2p^6$

SOLVED EXAMPLES

Example 5.14. *An atom has 2K, 8L, 5M electrons. Write down its electronic configuration and indicate in it*

(i) number of sub-shells

(ii) number of orbitals

(iii) number of unpaired electrons

(iv) number of electrons having $l = 1$.

Solution. In the given atom there are $(2 + 8 + 5) = 15$ electrons. The electronic configuration of the atom is

$$1s^2, 2s^2, 2p^6, 3s^2, 3p_x^{\ 1}, 3p_y^{\ 1}, 3p_z^{\ 1}$$

(*i*) Number of sub-shells is 5

(*ii*) Number of orbitals is 9

(*iii*) Number of unpaired electrons is 3

(*iv*) Number of electrons having $l = 1$ is 9.

Example 5.15. *An atom of an element contains 29 electrons and 35 neutrons. Deduce (i) the number of protons and (ii) the electronic configuration of the element.*

Solution. (*i*) Number of protons = Number of electrons = 29

(*ii*) Electronic configuration of the element is

$$1s^2, 2s^2, 2p^6, 3s^2, 3p^6, 3d^{10}, 4s^1.$$

Example 5.16. *Write the electronic configurations of Fe^{2+} and Fe^{3+} ions. Which of these has more number of unpaired electrons? Atomic no. of Fe is 26.*

Solution. Fe atom has 26 electrons. Fe^{2+} and Fe^{3+} ions are formed by removal of two and three electrons respectively from Fe atom. Their electronic configurations are:

Fe^{2+} : $1s^2, 2s^2, 2p^6, 3s^2, 3p^6, 3d^6$

Fe^{3+} : $1s^2, 2s^2, 2p^6, 3s^2, 3p^6, 3d^5$

Fe^{2+} has 4 unpaired electrons while Fe^{3+} has 5 unpaired electrons. Therefore, Fe^{3+} has more number of unpaired electrons.

WRITING THE ELECTRONIC CONFIGURATION IN CONDENSED FORM

It may be noted that configurations of atoms can also be written in condensed form by taking the configurations of noble gases as the core. The configurations of inert gases representing the core are written as $[He]^2$, $[Ne]^{10}$, $[Ar]^{18}$, $[Kr]^{36}$, $[Xe]^{54}$ and $[Rn]^{86}$. For example, the electronic configuration of scandium having atomic number 21 may be written as:

$$_{21}Sc: [Ar]^{18}, 3d^1, 4s^2.$$

SUMMARY

- **Cathode Rays.** Consist of negatively charged material particles called electrons.
- **Electrons.** Fundamental sub-atomic particles carrying a negative charge (1.602×10^{-19} coulombs) and having a mass of 9.1×10^{-31} kg. Discovered by **Sir J.J. Thomson.**
- Charge to mass (e/m) ratio for electrons is 1.76×10^8 C/g.
- **Anode Rays.** Consist of positively charged ions. Charge to mass ratio for the ions in the anode rays depends on the nature of the gas taken in the discharge tube.
- **Proton.** A sub-atomic positively charged particle, having a charge of 1.602×10^{-19} C and a mass of 1.67×10^{-27} kg. Mass of proton is nearly 1837 times the mass of an electron.
- **Alpha Particles.** He^{2+} ions or helium nuclei.
- **Rutherford's Experiment.** Led to the discovery of the nucleus. Radius of nucleus ($\sim 10^{-14}$ m) is very small as compared to the radius of an atom ($\sim 10^{-10}$ m).
- **Neutrons.** Sub-atomic neutral particles having a mass of 1.675×10^{-27} kg. Discovered by **James Chadwick.**

- **Atomic Number (Z)** = Number of protons
= Number of electrons.
- **Mass Number (A)** = Number of protons + Number of neutrons.
- **Isotopes.** Atoms of the same element having different mass numbers.
- Relative isotopic masses and relative molecular masses can be determined with the help of a mass spectrometer.
- For electromagnetic radiations, $\mathbf{c} = \nu\lambda$.
- $\bar{\nu} = \dfrac{\mathbf{1}}{\lambda} = \dfrac{\nu}{\mathbf{c}}$.
- Each element has its own **characteristic emission spectrum.**
- $E = h\nu = \dfrac{hc}{\lambda}$.
- $\nu = \dfrac{E_2 - E_1}{h}$
- **de-Broglie relationship**

$$\lambda = \frac{h}{mv} = \frac{h}{p}$$

- **Heisenberg's uncertainty principle** states that it is impossible to determine simultaneously the position and velocity of a small particle such as an electron with absolute accuracy.
- **Orbital.** Region of space around the nucleus where probability of finding the electrons is maximum.
- **Quantum Mechanics** is a theoretical science that takes into account the dual nature of matter.
- **Quantum Numbers.** A set of four numbers used to specify energy, size, shape, orientation of the electron orbital and spin of the electron.
- **Principal Quantum Number (n).** Tells us about the energy of the electron and size of the orbital.
- **Azimuthal Quantum Number (l).** Tells us about the shape of the orbital.
- **Magnetic Quantum Number (m_l).** Tells us about the orientations of the electron cloud in a sub-shell.
- **Spin Quantum Number (m_s).** Tells us about the spin of the electron.
- **For nth energy level,** the maximum number of

  sub-shells $= n$

  orbitals $= n^2$

  electrons $= 2n^2$.
- ***s*-Orbitals** are spherically symmetrical.
- ***p*-Orbitals** are dumb-bell shaped.
- The five ***d*-orbitals** are d_{xy}, d_{yz}, d_{xz}, $d_{x^2-y^2}$ and d_{z^2}.
- **Degenerate Orbitals.** The orbitals having equal energy.
- In hydrogen atom and hydrogen-like atoms all the orbitals having the same value of *n* have the same energy.
- In multi-electron atoms, for a particular energy level, a sub-shell having a greater value of *l* has greater energy.
- **Ground State.** The lowest energy state of the atom.
- **Excited State.** An atom is said to be in an excited state if some lower energy orbital is vacant and the electron is present in a higher energy orbital.
- **Electronic Configuration.** Distribution of electrons among various orbitals in an atom.
- **Aufbau Rule.** The orbitals are filled in the order of increasing energy, starting with the orbital of lowest energy.
- **Pauli's Exclusion Principle.** No two electrons in an atom can have the same set of all the four quantum numbers. From this principle only two electrons having opposite spins can be accommodated in an orbital.
- **Hund's Rule of Maximum Spin Multiplicity.** The pairing of electrons in orbitals of a particular sub-shell cannot take place until all the orbitals of the sub-shell are singly occupied. These singly occupied orbitals should have electrons with parallel spin.

EVALUATE YOURSELF

I. Objective Type Questions

Select the most appropriate choice from the options given as (a), (b), (c) and (d) after each question:

1. '*Plum pudding*' model of the atom was proposed by
   (*a*) Rutherford (*b*) Dalton
   (*c*) Bohr (*d*) J. J. Thomson.
2. An isotope of magnesium is represented as ${}^{26}_{12}\text{Mg}$. The number of nucleons in the isotope is
   (*a*) 12 (*b*) 14
   (*c*) 26 (*d*) 38.
3. Naturally occurring gold does not have an isotope. The number of peaks in the mass spectrum of naturally occurring gold would be
   (*a*) 3 (*b*) 2
   (*c*) 1 (*d*) 4.
4. The credit of discovery of the neutron goes to
   (*a*) Rutherford (*b*) Thomson
   (*c*) Goldstein (*d*) Chadwick.

5. The protons and neutrons are collectively called
(*a*) deutrons (*b*) positrons
(*c*) mesons (*d*) nucleons.

6. The ratio of the radius of the atom to the radius of the nucleus is of the order of
(*a*) 10^4 (*b*) 10^6
(*c*) 10^{-5} (*d*) 10^{-8}.

7. ${}^{13}_{6}C$ and ${}^{12}_{6}C$ differ from each other in respect of the number of
(*a*) electrons (*b*) protons
(*c*) neutrons (*d*) none of these.

8. Rutherford experiment which established the nuclear model of the atom used a beam of
(*a*) β-particles which impinged on a metal foil and got absorbed
(*b*) γ-rays which impinged on a metal foil and ejected electrons
(*c*) helium atoms, which impinged on a metal foil and got scattered
(*d*) helium nuclei which impinged on a metal foil and got scattered.

9. The charge to mass (e/m) ratio of the proton is
(*a*) 9.55×10^{-4} C/g (*b*) 9.55×10^{4} C/g
(*c*) 1.76×10^{8} C/g (*d*) 1.76×10^{-8} C/g.

10. Isotopes of an element have
(*a*) similar chemical properties but different physical properties
(*b*) similar chemical and physical properties
(*c*) similar physical properties but different chemical properties
(*d*) different chemical and physical properties.

11. For a given principal level $n = 4$, the energy of its sub-shells is in the order
(*a*) $s < p < d < f$ (*b*) $s > p > d > f$
(*c*) $s < p < f < d$ (*d*) $f < p < d < s$.

12. The electrons, identified by quantum numbers n and l, (*i*) $n = 4$, $l = 1$ (*ii*) $n = 4$, $l = 0$ (*iii*) $n = 3$, $l = 2$ (*iv*) $n = 3$, $l = 1$ can be placed in order of increasing energy, from the lowest to the highest, as
(*a*) (*iv*) < (*ii*) < (*iii*) < (*i*) (*b*) (*ii*) < (*iv*) < (*i*) < (*iii*)
(*c*) (*i*) < (*iii*) < (*ii*) < (*iv*) (*d*) (*iii*) < (*i*) < (*iv*) < (*ii*).

13. A sub-shell with $n = 6$, $l = 2$ can accommodate a maximum of
(*a*) 12 electrons (*b*) 36 electrons
(*c*) 10 electrons (*d*) 72 electrons.

14. Which of the following sets of quantum numbers is not possible?

(*a*) $n = 2, l = 1, m_l = -1, m_s = -\frac{1}{2}$

(*b*) $n = 2, l = 0, m_l = 0, m_s = +\frac{1}{2}$

(*c*) $n = 3, l = 2, m_l = -2, m_s = +\frac{1}{2}$

(*d*) $n = 3, l = 2, m_l = -3, m_s = +\frac{1}{2}$.

15. The correct configuration of ${}_{29}Cu$ is
(*a*) [Ar] $4s^1$ (*b*) [Ar] $4s^2$
(*c*) [Ar] $3d^{10}4s^1$ (*d*) [Ar] $3d^9 4s^2$.

16. The atomic orbitals are progressively filled in order of increasing energy. This principle is called
(*a*) Hund's rule (*b*) Aufbau principle
(*c*) Exclusion principle (*d*) de-Broglie rule.

17. According to the Aufbau principle, the 19th electron in an atom goes into the
(*a*) 4*s*-orbital (*b*) 3*d*-orbital
(*c*) 4*p*-orbital (*d*) 3*p*-orbital.

18. In a multi-electron atom, the energy of the electron in a particular orbital is determined by
(*a*) n only (*b*) l only
(*c*) n and l (*d*) n, l and m_l.

19. The orbital diagram in which both Pauli's exclusion principle and Hund's rule are violated is

(*a*) [⇅] [⇈ | ↿ |] (*b*) [⇅] [⇅ | ⇅ |]

(*c*) [⇅] [⇂ | ⇂ | ⇂] (*d*) [⇅] [⇅ | ⇅ | ⇂]

20. Which one of the following pairs of ions have the same electronic configuration?
(*a*) Cr^{3+}, Fe^{3+} (*b*) Fe^{3+}, Mn^{2+}
(*c*) Fe^{3+}, Co^{3+} (*d*) Sc^{3+}, Cr^{3+}.

21. The number of *d*-electrons retained in Fe^{3+} (At. no. of Fe = 26) ion is
(*a*) 6 (*b*) 3
(*c*) 4 (*d*) 5.

22. Consider the following sets of quantum numbers:

	n	l	m_l	m_s
(*i*)	3	0	0	+ 1/2
(*ii*)	2	2	1	+ 1/2
(*iii*)	4	0	– 2	– 1/2
(*iv*)	1	0	– 1	– 1/2
(*v*)	3	2	3	+ 1/2

Which of the following sets of quantum numbers is not possible?
(*a*) (*i*) and (*iii*) (*b*) (*ii*), (*iii*) and (*iv*)
(*c*) (*i*), (*ii*), (*iii*) and (*iv*) (*d*) (*ii*), (*iii*), (*iv*) and (*v*).

23. Which of the following sets of quantum numbers represents the highest energy of an atom?
(*a*) $n = 3, l = 1, m_l = 1, m_s = +1/2$
(*b*) $n = 3, l = 2, m_l = 1, m_s = +1/2$
(*c*) $n = 4, l = 0, m_l = 0, m_s = +1/2$
(*d*) $n = 3, l = 0, m_l = 0, m_s = +1/2$.

24. The electrons of the same orbitals can be distinguished by
(*a*) principal quantum number
(*b*) azimuthal quantum number

(*c*) spin quantum number

(*d*) magnetic quantum number.

25. A body of mass 10 mg is moving with a velocity of 100 m s^{-1}. The wavelength of the de-Broglie wave associated with it would be

(*a*) 6.63×10^{-37} m (*b*) 6.63×10^{-31} m

(*c*) 6.63×10^{-34} m (*d*) 6.63×10^{-35} m.

26. Which of the following sets of quantum numbers is permitted?

(*a*) $n = 4, l = 2, m_l = +3, m_s = +\frac{1}{2}$

(*b*) $n = 3, l = 3, m_l = +3, m_s = +\frac{1}{2}$

(*c*) $n = 4, l = 0, m_l = 0, m_s = +\frac{1}{2}$

(*d*) $n = 4, l = 3, m_l = +1, m_s = 0$.

27. The maximum number of electrons that can be accommodated in the fifth energy level is

(*a*) 10 (*b*) 25

(*c*) 50 (*d*) 32.

II. Fill in the Blanks

28. Complete the following sentences by supplying the appropriate words:

(*i*) The radius of atoms is of the order of ...... *m*.

(*ii*) The neutral sub-atomic particles are ....

(*iii*) In the atom represented as $^{23}_{11}\text{Na}$, the number of neutrons is ....

(*iv*) Isotopes have the same ...... number.

(*v*) A naturally occurring element which has two isotopes will show ...... peaks in its mass spectrum.

(*vi*) Alpha particle contains ...... protons.

(*vii*) *s*-orbitals are ...... in shape.

(*viii*) The orbitals having equal energy are called ...... orbitals.

(*ix*) The number of unpaired electrons in an atom of Mn (Z = 25) is ....

(*x*) Specific charge of electron is ....

III. Discussion Questions

29. Discuss Dalton's atomic theory. What are its limitations?

30. What are cathode rays? Give four characteristics of cathode rays.

31. How will you demonstrate that cathode rays carry negative charge?

32. What is the charge and charge to mass (*e*/*m*) ratio of particles in cathode rays?

33. How will you demonstrate with the help of an experiment that:

(*i*) Cathode rays travel in straight lines

(*ii*) Cathode rays consist of material particles.

34. The charge to mass ratio for an electron is 1.76×10^8 coulombs per gram. The charge on the electron is 1.602×10^{-19} coulombs. Calculate mass of the electron.

35. If an atom contains one electron and one proton, will it carry any charge or not?

36. Name the three subatomic particles of an atom.

37. On the basis of Thomson's model of an atom explain how the atom is neutral as a whole.

38. Compare the properties of the three fundamental sub-particles of an atom.

39. What were the main observations of Rutherford's scattering experiment?

40. What is a proton? How does it differ from a neutron?

41. Give experimental evidence to show that:

(*i*) Most of the space inside the atom is empty.

(*ii*) An atom contains heavy positively charge centre.

(*iii*) Size of the nucleus is very small.

(*iv*) Nucleus of an atom is positively charged.

42. What experiment established the presence of the atomic nucleus? What features of the nucleus were deduced from this experiment?

43. Describe the essential features of the model of the atom as proposed by E. Rutherford. How is it different from that proposed by J.J. Thomson?

44. Discuss the working of the mass spectrometer. How does it help us in determining the number of isotopes and relative atomic mass of an element?

45. Describe the significance of each term in the symbol $^{A}_{Z}\text{X}$.

46. What is meant by the atomic number of an element? Can two elements have the same atomic number?

47. What is meant by the term mass number of an atom? Do all atoms of an element have the same mass number?

48. Composition of the nuclei of two atomic species X and Y are given as:

	X	Y
Protons	= 6	6
Neutrons	= 6	8

Give the mass numbers of X and Y. What is the relation between the two species?

49. An isotope of an element may be represented as $^{37}_{17}\text{X}$. Answer the following:

(*i*) What is the atomic number of the element?

(*ii*) What is the number of neutrons in an atom of this isotope?

(*iii*) What is the mass number of the element?

50. Calculate the number of electrons, protons and neutrons in Cr atom and Cr^{3+} ions. The atomic number of chromium is 24 and its mass number is 52.

51. Fill in the following blanks:

S. No.	Atomic number	Mass number	Protons	Neu-trons	Elec-trons	Symbol
1.	13	—	—	14	—	—
2.	—	—	16	16	—	S^{2-}
3.	11	23	—	—	—	—

52. The mass number of the three isotopes of hydrogen, called protium, deuterium and tritium are, 1, 2 and 3 respectively. What is the basic difference among these three types of hydrogen atoms?

53. Give the number of protons, neutrons and electrons per atom in the two isotopes of chlorine, ${}^{35}_{17}Cl$ and ${}^{37}_{17}Cl$.

54. Which of the following pairs are isotopes? Give reasons:

(*i*) ${}^{14}_{7}A$, ${}^{14}_{6}B$ (*ii*) ${}^{14}_{6}X$, ${}^{12}_{6}Y$.

55. Give one similarity and one difference between a pair of isotopes.

56. The average atomic mass of a sample of an element X is 16.2 *u*. What are the percentages of isotopes ${}^{16}_{8}X$ and ${}^{18}_{8}X$ in the sample?

57. Explain why atomic masses of most of the elements are fractional.

58. Elemental boron is 20.0% B-10 and 80% B-11. Calculate the atomic mass of boron.

59. Calculate the wavelengths of an electromagnetic radiation having a frequency of 3 MHz.

60. The wavelength of the green light from a traffic signal is 522 nm. What is the frequency and wave number of this radiation?

61. Name the five series in the hydrogen spectrum.

62. How would you obtain the atomic spectrum of hydrogen?

63. Give the various postulates of the Bohr's atomic model. How does it explain the emission spectrum of hydrogen?

64. Explain the terms spectrum, electromagnetic spectrum and atomic spectrum.

65. How does Bohr's model account for only certain specific lines in the emission spectrum of hydrogen?

66. What will be the wavelength of a ball of mass 0.1 kg moving with a velocity of 10 ms^{-1}?

67. How many quantum numbers are needed to designate an orbital? Name them.

68. The principal quantum number, *n* of an atomic orbital is 5. What are the possible values of *l*?

69. What is the lowest value of *n* that allows *f* orbitals to exist?

70. Give the notation for the sub-shells denoted by the following quantum numbers:

(*i*) $n = 5, l = 2$ (*ii*) $n = 6, l = 3$

(*iii*) $n = 4, l = 0$ (*iv*) $n = 5, l = 4$.

71. Explain, giving reasons, which of the following sets of quantum numbers are *not* possible.

(*a*) $n = 0$ $l = 0$ $m_l = 0$ $m_s = +1/2$

(*b*) $n = 1$ $l = 0$ $m_l = 0$ $m_s = -1/2$

(*c*) $n = 1$ $l = 1$ $m_l = -0$ $m_s = +1/2$

(*d*) $n = 2$ $l = 1$ $m_l = 0$ $m_s = -1/2$

(*e*) $n = 3$ $l = 3$ $m_l = -3$ $m_s = +1/2$

(*f*) $n = 3$ $l = 1$ $m_l = 0$ $m_s = +1/2$.

72. How many orbitals are possible in

(*i*) 4th energy level (*ii*) 5*f*-sub-shell?

73. What is the shape of a 2*s*-orbital? Give two points of difference between 1*s* and 2*s* orbitals.

74. State and explain Heisenberg's uncertainty principle.

75. Is it possible to know the exact path of an electron in a 2*p* orbital? Name and state the principle you will rely upon to answer the question.

76. Draw the shapes of *s* and *p* orbitals.

77. Which of the following orbitals are not possible?

1*p*, 2*s*, 2*p*, 3*f*.

78. What is the maximum number of electrons that can be accommodated in

(*i*) 1st main shell

(*ii*) 3rd main shell

(*iii*) 4*d*-sub-shell?

79. State Pauli's exclusion principle. How does it put a limit on the accommodation of electrons in an orbital? Explain.

80. Describe the various rules of filling of orbitals. Illustrate with the help of suitable examples.

81. Give the electronic configuration of scandium (Z = 21) and chromium (Z = 24).

82. Give the main differences between an orbit and an orbital.

83. State Hund's rule of maximum multiplicity. How is it used in the electronic distribution of the nitrogen atom (Z = 7)?

84. What is the nuclear charge of Mn^{2+} ion (Z = 25, A = 55)? How many neutrons, protons and electrons are present in it?

Answers

I. Objective Type Questions

1. (*d*)	**2.** (*c*)	**3.** (*c*)
4. (*d*)	**5.** (*d*)	**6.** (*a*)
7. (*c*)	**8.** (*d*)	**9.** (*b*)
10. (*a*)	**11.** (*a*)	**12.** (*a*)
13. (*c*)	**14.** (*d*)	**15.** (*c*)
16. (*b*)	**17.** (*a*)	**18.** (*c*)
19. (*a*)	**20.** (*b*)	**21.** (*d*)
22. (*d*)	**23.** (*b*)	**24.** (*c*)
25. (*b*)	**26.** (*c*)	**27.** (*c*).

II. Fill in the Blanks

28. (*i*) 10^{-10} m (*ii*) neutrons (*iii*) 12
(*iv*) mass (*v*) two (*vi*) 2
(*vii*) spherical (*viii*) degenerate (*ix*) 5
(*x*) 1.76×10^{11} C/kg.

6.1 THE NEED FOR CLASSIFICATION OF ELEMENTS

Before the beginning of the eighteenth century, only a very few elements were known and it was quite easy to study and remember their individual properties. In 1800, only 31 elements were known. This number of elements grew to 63 by 1865. With the discovery of a large number of elements it became difficult to study individually the properties of these elements and their compounds. At this stage, the scientists felt the need for some simple methods to facilitate the study of the properties of various elements and their compounds. After numerous attempts the scientists were ultimately successful in arranging the elements in such a way so that similar elements were grouped together and different elements were separated.

The arrangement of elements in such a way that the similar elements fall within the same vertical group and the dissimilar elements are separated, is known as the **classification of elements.**

6.2 MENDELEEV'S PERIODIC TABLE

Attempts to find regularities among the elements led the Russian Scientist, Dmitri I. Mendeleev to put forward a scheme of classification of elements in 1869. He gave a periodic law known after his name as **Mendeleev's periodic law.** This law states that:

The properties of elements are a periodic function of their atomic weights.

It means that when the elements are arranged in order of increasing atomic weights, the elements with similar properties recur after regular intervals. On the basis of this periodic law, Mendeleev constructed a *periodic table* in such a way that the elements were arranged

Specific Objectives

After studying this unit, the student should be able to:

- relate the position of an element in the periodic table to its atomic number and electron configuration
- use the periodic table to identify metals, semi-metals, non-metals and halogens
- describe the physical properties of some representative elements
- distinguish between the terms 'group' and 'period'
- explain the periodic law
- identify trends in atomic size, ionization energy, electron affinity, electronegativity and ionic size for elements on the periodic table
- describe the periodic gradation of elements in the third period
- use the periodic table to determine the number of electrons available for bonding.

horizontally in the order of their increasing atomic weights. However, he also kept in mind the similarities in the chemical properties of the elements. *The main criterion of the judgement of similarities in the properties was the valency of the elements.* Mendeleev observed that some of the elements did not fit in with his scheme of classification if the order of atomic weights was strictly followed. He ignored the order of atomic weights and placed the elements with similar chemical properties together. For example, iodine having atomic weight 127 was placed after tellurium (atomic weight 128), together with fluorine, chlorine and bromine due to similarities in properties.

Mendeleev left certain vacant places in his table which provided a clue for the discovery of new elements. Some of the properties could be predicted with a fair accuracy. For example, both gallium and germanium were not discovered at the time when Mendeleev proposed his periodic table. Mendeleev named these elements as **Eka-Aluminium** and **Eka-Silicon** respectively. Later on, when these elements were discovered, Mendeleev's prediction proved remarkably correct. Some of the properties predicted by Mendeleev for these elements and those found experimentally are given in Table 6.1.

6.3 MODERN PERIODIC LAW

In 1913, the English Physicist Henry Moseley studied the X-ray spectra of many elements. He observed that a plot of $\sqrt{\nu}$ (where ν is the frequency of the X-rays emitted) against atomic number (Z) gave a straight line and not the plot of $\sqrt{\nu}$ against atomic weight. He proposed that atomic number is a more fundamental property of an element than its atomic weight. Therefore, the physical and chemical properties of the elements are determined by their **atomic numbers** instead of their atomic weights. This observation led to the development of the modern periodic law. The **modern periodic law** states that:

The physical and chemical properties of the elements are a periodic function of their atomic numbers.

It means that if the elements are arranged in order of increasing atomic numbers, the elements with similar properties recur after regular intervals.

PHYSICAL AND CHEMICAL PROPERTIES OF SOME ELEMENTS

Electronic configuration, physical properties and chemical properties are used to classify elements into various groups. The elements in a vertical column have largely similar properties which show regular gradation with increase in atomic number. The elements in a horizontal row show relatively sharp change in physical and chemical properties.

For example, some physical and chemical properties of alkali metals (Na, K), alkaline earth metals (Mg, Ca) and halogens (F_2, Cl_2, Br_2 and I_2) are listed below:

- Sodium and potassium are soft metals. Their hardness decreases with increase in atomic number.
- They have **low melting points** which decreases with increase in atomic number.
- They react with water liberating hydrogen gas and form soluble metal hydroxide

$$2Na\ (s) + H_2O\ (l) \longrightarrow 2NaOH\ (aq) + H_2\ (g)$$
$$2K\ (s) + H_2O\ (l) \longrightarrow 2KOH\ (aq) + H_2\ (g)$$

Table 6.1. Comparison of the Properties of Eka-aluminium and Eka-silicon as Predicted by Mendeleev with those Observed Experimentally for Gallium and Germanium

Property	*Eka-aluminium (predicted)*	*Gallium (found)*	*Eka-silicon (predicted)*	*Germanium (found)*
Atomic weight	68	70	72	72.6
Density / (g/cm^3)	5.9	5.94	5.5	5.36
Melting point / K	Low	29.78	High	1231
Formula of oxide	E_2O_3	Ga_2O_3	EO_2	GeO_2
Formula of chloride	ECl_3	$GaCl_3$	ECl_4	$GeCl_4$

Reactivity increases with increase in atomic number. The reaction of water with potassium is more vigorous than with sodium.

Caution: Reaction of alkali metals with water is very violent.

- Both Na and K have one electron in their outermost shell.
- They exhibit an oxidation state of +1
- They are strong reducing agents.
- **Magnesium** and **Calcium** have two electrons in their outermost shells.
- They are harder than alkali metals and have higher melting points.
- They are less reactive with water than alkali metals. Magnesium reacts only with boiling water.
- Mg and Ca have two electrons in the outermost shell.
- They exhibit an oxidation state of +2
- **Chlorine** and **bromine** are non-metals. They have very low melting and boiling points.
- Chlorine is a gas whereas bromine is a liquid.
- They react with water slowly in the presence of sunlight.

$$2Cl_2 + 2H_2O \longrightarrow 4HCl + O_2$$
$$2Br_2 + 2H_2O \longrightarrow 4HBr + O_2$$

Cl_2 is more reactive than Br_2.

- They have seven electrons in their outermost shell.
- They exhibit an oxidation state of −1
- They are strong oxidizing agents.

By studying the physical and chemical properties of different elements, the elements have been classified into various groups or families.

For example, alkali metals, alkaline earth metals, halogens, noble gases, etc.

Many new forms of the periodic table have been proposed in recent times with the modern periodic law as the guiding principle, but the general plan of the table remained the same as proposed by Mendeleev. ***The most commonly known periodic table is the long form of the periodic table.*** Before discussing the general plan of the long form periodic table, let us look into the basic cause of the periodic repetition of properties.

6.4 CAUSE OF SIMILARITIES IN THE PROPERTIES OF ELEMENTS IN THE SAME GROUP

It has been pointed out that if the elements are arranged in order of increasing atomic numbers, the elements with similar properties are repeated after regular intervals.

The periodic repetition of elements with similar properties after certain regular intervals when the elements are arranged in order of increasing atomic number is called periodicity.

In order to understand the cause of periodicity, let us first answer the question; *why do certain elements exhibit similar properties*?

From the knowledge of atomic structure it may be recalled that an atom has a small positively charged nucleus with electrons distributed around it. The atomic nucleus does not undergo any change during ordinary chemical reactions. Thus, it may be assumed that the physical and chemical properties of the elements must be related to the arrangement of electrons in their atoms. Since the electrons present in the inner shells do not take part in chemical combination, it must be the electrons in the outermost shell which control the properties of the atoms. Thus, if the arrangement of electrons in the outermost shell (valence shell) of the atoms is similar, their properties will also be similar. For example, the electronic configurations of alkali metals as given in Table 6.2, shows the presence of one electron in their valence shells.

Table 6.2. Electronic Configurations of Some Groups

Group 1 Elements—Alkali metals	
Element	*Electronic configuration*
Lithium (Li)	2, **1**
Sodium (Na)	2, 8, **1**
Potassium (K)	2, 8, 8, **1**
Rubidium (Rb)	2, 8, 18, 8, **1**
Group 2 Elements—Alkaline-earth metals	
Element	*Electronic configuration*
Beryllium (Be)	2, **2**
Magnesium (Mg)	2, 8, **2**
Calcium (Ca)	2, 8, 8, **2**

Group 17 Elements—Halogens	
Element	*Electronic configuration*
Fluorine (F)	2, **7**
Chlorine (Cl)	2, 8, **7**
Bromine (Br)	2, 8, 18, **7**

Group 18 Elements—Noble gases	
Element	*Electronic configuration*
Helium (He)	2,
Neon (Ne)	2, **8**
Argon (Ar)	2, 8, **8**
Krypton (Kr)	2, 8, 18, **8**

Similar behaviour of alkali metals is attributed to the similar valence shell configuration of their atoms. For example, all alkali metals react with water and liberate hydrogen gas.

$$2Li + 2H_2O \longrightarrow 2LiOH + H_2$$
$$2Na + 2H_2O \longrightarrow 2NaOH + H_2$$
$$2K + 2H_2O \longrightarrow 2KOH + H_2$$

Similarly, if we examine the electronic configurations of other elements, we shall find that there is a repetition of the similar valence shell configuration after certain regular intervals with the regular increase of atomic number. Thus, it can be concluded that *the periodic repetition of properties is due to the recurrence of similar valence shell configurations after certain regular intervals.*

Thus, all the elements in a group have similar valence shell electronic configuration. The similar chemical nature of elements in a group is due to similar electron structures of their atoms. The electron structures of the first 20 elements are given in Table 6.3.

6.5 ELECTRONIC CONFIGURATION AND THE POSITION OF ELEMENTS IN THE PERIODIC TABLE—The Long Form of Periodic Table

The long form of the periodic table is an improved form of the periodic table which is based upon the modern periodic law. The long form of the periodic table is given on page 67. Let us now, study the **structural features of the periodic table.**

DESCRIPTION OF PERIODS

A horizontal row on the periodic table is called a **period.** *A period consists of a series of elements having the same valence shell.* There are **seven periods** in all, which are numbered as 1, 2, 3, 4, 5, 6 and 7.

There is a close connection between the electronic configurations of the elements and the long form of the periodic table. As pointed out earlier in Unit 5, the principal quantum number n defines the main energy level of the electron also called *main energy shell.* Each period of the periodic table begins with the filling of a new energy shell. In fact, **the number of the period also represents the highest principal quantum number of the elements present in it.** The number of elements in each period is equal to the number of electrons which can be accommodated in the orbitals belonging to that electron shell.

The **first period** corresponds to the filling of electrons in first energy shell (*i.e.*, $n = 1$). Now this energy level has only one orbital (*i.e.*, 1*s*) and, therefore, it can accommodate two electrons. This means that there can be only **two elements** in the first period.

The **second period** starts with the electrons beginning to enter the second energy shell ($n = 2$). There are only four orbitals (one 2*s* and three 2*p* orbitals) to be filled which can accommodate eight electrons. Thus, the second period has **eight elements** in it.

Table 6.3. Electronic Configurations of the First 20 Elements in the Periodic Table

Period 1	H 1	He 2						
Period 2	Li 2, 1	Be 2, 2	B 2, 3	C 2, 4	N 2, 5	O 2, 6	F 2, 7	Ne 2, 8
Period 3	Na 2, 8, 1	Mg 2, 8, 2	Al 2, 8, 3	Si 2, 8, 4	P 2, 8, 5	S 2, 8, 6	Cl 2, 8, 7	Ar 2, 8, 8
Period 4	K 2, 8, 8, 1	Ca 2, 8, 8, 2						

Long Form of the Periodic Table

Representative Elements (s-Block) | Transition Elements (d-Block) | Representative Elements (p-Block) | Noble gases

	1	2	3	4	5	6	7	8	9	10	11	12	13	14	15	16	17	18
1.	1 H $1s^1$																	2 He $1s^2$
2.	3 Li $2s^1$	4 Be $2s^2$											5 B $2s^2 2p^1$	6 C $2s^2 2p^2$	7 N $2s^2 2p^3$	8 O $2s^2 2p^4$	9 F $2s^2 2p^5$	10 Ne $2s^2 2p^6$
3.	11 Na $3s^1$	12 Mg $3s^2$											13 Al $3s^2 3p^1$	14 Si $3s^2 3p^2$	15 P $3s^2 3p^3$	16 S $3s^2 3p^4$	17 Cl $3s^2 3p^5$	18 Ar $3s^2 3p^6$
4.	19 K $4s^1$	20 Ca $4s^2$	21 Sc $3d^1 4s^2$	22 Ti $3d^2 4s^2$	23 V $3d^3 4s^2$	24 Cr $3d^5 4s^1$	25 Mn $3d^5 4s^2$	26 Fe $3d^6 4s^2$	27 Co $3d^7 4s^2$	28 Ni $3d^8 4s^2$	29 Cu $3d^{10} 4s^1$	30 Zn $3d^{10} 4s^2$	31 Ga $4s^2 4p^1$	32 Ge $4s^2 4p^2$	33 As $4s^2 4p^3$	34 Se $4s^2 4p^4$	35 Br $4s^2 4p^5$	36 Kr $4s^2 4p^6$
5.	37 Rb $5s^1$	38 Sr $5s^2$	39 Y $4d^1 5s^2$	40 Zr $4d^2 5s^2$	41 Nb $4d^4 5s^1$	42 Mo $4d^5 5s^1$	43 Tc $4d^5 5s^2$	44 Ru $4d^7 5s^1$	45 Rh $4d^8 5s^1$	46 Pd $4d^{10}$	47 Ag $4d^{10} 5s^1$	48 Cd $4d^{10} 5s^2$	49 In $5s^2 5p^1$	50 Sn $5s^2 5p^2$	51 Sb $5s^2 5p^3$	52 Te $5s^2 5p^4$	53 I $5s^2 5p^5$	54 Xe $5s^2 5p^6$
6.	55 Cs $6s^1$	56 Ba $6s^2$	57 La* $5d^1 6s^2$	72 Hf $4f^{14} 5d^2 6s^2$	73 Ta $5d^3 6s^2$	74 W $5d^4 6s^2$	75 Re $5d^5 6s^2$	76 Os $5d^6 6s^2$	77 Ir $5d^7 6s^2$	78 Pt $5d^9 6s^1$	79 Au $5d^{10} 6s^1$	80 Hg $5d^{10} 6s^2$	81 Tl $6s^2 6p^1$	82 Pb $6s^2 6p^2$	83 Bi $6s^2 6p^3$	84 Po $6s^2 6p^4$	85 At $6s^2 6p^5$	86 Rn $6s^2 6p^6$
7.	87 Fr $7s^1$	88 Ra $7s^2$	89 Ac** $6d^1 7s^2$	104 Rf $5f^{14} 6d^2 7s^2$	105 Db $5f^{14} 6d^3 7s^2$	106 Sg $5f^{14} 6d^4 7s^2$	107 Bh $5f^{14} 6d^5 7s^2$	108 Hs $5f^{14} 6d^6 7s^2$	109 Mt $5f^{14} 6d^7 7s^2$	110 Ds $5f^{14} 6d^8 7s^2$	111 Rg $5f^{14} 6d^9 7s^2$	112 Uub $5f^{14} 6d^{10} 7s^2$	—	114 Uuq	—	116 Uuh	—	—

Inner Transition Elements (f-Block)

*Lanthanides $4f^n 5d^{0-1} 6s^2$	58 Ce $4f^1 5d^1 6s^2$	59 Pr $4f^3 5d^0 6s^2$	60 Nd $4f^4 5d^0 6s^2$	61 Pm $4f^5 5d^0 6s^2$	62 Sm $4f^6 5d^0 6s^2$	63 Eu $4f^7 5d^0 6s^2$	64 Gd $4f^7 5d^1 6s^2$	65 Tb $4f^9 5d^0 6s^2$	66 Dy $4f^{10} 5d^0 6s^2$	67 Ho $4f^{11} 5d^0 6s^2$	68 Er $4f^{12} 5d^0 6s^2$	69 Tm $4f^{13} 5d^0 6s^2$	70 Yb $4f^{14} 5d^0 6s^2$	71 Lu $4f^{14} 5d^1 6s^2$
**Actinides $5f^n 6d^{0-1} 7s^2$	90 Th $5f^0 6d^2 7s^2$	91 Pa $5f^2 6d^1 7s^2$	92 U $5f^3 6d^1 7s^2$	93 Np $5f^4 6d^1 7s^2$	94 Pu $5f^6 6d^0 7s^2$	95 Am $5f^7 6d^0 7s^2$	96 Cm $5f^7 6d^1 7s^2$	97 Bk $5f^9 6d^0 7s^2$	98 Cf $5f^{10} 6d^0 7s^2$	99 Es $5f^{11} 6d^0 7s^2$	100 Fm $5f^{12} 6d^0 7s^2$	101 Md $5f^{13} 6d^0 7s^2$	102 No $5f^{14} 6d^0 7s^2$	103 Lr $5f^{14} 6d^1 7s^2$

Fig. 6.1. Long form of periodic table.

The **third period** begins with the electrons entering the third energy shell ($n = 3$). It may be recalled that out of nine orbitals of this energy level (one *s*, three *p* and five *d*), the five 3*d* orbitals have higher energy than the 4*s* orbitals. As such only four orbitals (one 3*s* and three 3*p*) corresponding to $n = 3$ are filled before the fourth energy level begins to be formed. Hence, there are only **eight elements** in the third period.

The **fourth period** corresponds to $n = 4$. It starts with the filling of the 4*s*-orbitals. However, after the 4*s* but before the 4*p* orbitals, there are five 3*d* orbitals also to be filled. Thus, in all, nine orbitals (one 4*s*, five 3*d* and three 4*p*) have to be filled and as such there are **eighteen** elements in the fourth period. It may be noted that the filling of 3*d*-orbitals starts from Sc (Z = 21). The elements from Sc (Z = 21) to Zn (Z = 30) are called 3*d*-transition series.

Similarly, we can show that there are 18 elements in the fifth period and 32 elements in the sixth period. The seventh period is still incomplete.

DESCRIPTION OF GROUPS

A vertical column of the periodic table is called **a group.** A group consists of *a series of elements having a similar configuration of the outer energy shell*. For example, all the group 1 elements have ns^1 valence shell electronic configuration. There are eighteen vertical columns in the long form of the periodic table. According to the recommendation of the International Union of Pure and Applied Chemistry (IUPAC), these groups are numbered from **1 to 18.**

It may be noted that the elements belonging to same group are said to constitute a *family*. For example,

- The elements of group 1 are known as **alkali metals.**
- The elements of group 2 are known as **alkaline earth metals.**
- The elements of group 17 are known as **halogens.**
- The elements of group 18 are known as **noble gases.**

6.6 POSITION OF HYDROGEN IN THE PERIODIC TABLE

Hydrogen is the first element of the periodic table having atomic number 1. It has electronic configuration $1s^1$. It is difficult to give a proper place to hydrogen in the periodic table because of its resemblance with halogens as well as alkali metals. The dual behaviour of hydrogen is attributed to its electronic configuration.

Resemblance with Alkali Metals

- Electronic configuration of hydrogen resembles that of alkali metals.
- By losing one electron hydrogen forms H^+ ion and exhibits an oxidation state of + 1.
- Like alkali metals, hydrogen is a good reducing agent.

Resemblance with Halogens

- Like halogens it can attain a noble gas configuration by gaining one electron to form a hydride ion, H^-.
- It exhibits an oxidation state of –1 in metal hydrides.
- It exists as diatomic molecule, H_2.
- In most of its compounds it forms a bond by the sharing of electrons.

Due to its resemblance with both alkali metals and halogens it is placed separately as shown in Fig. 6.1.

Table 6.4. Number of Elements in Different Periods

Period	*Principal valence shell (= n)*	*Orbitals being filled up*	*Electrons to be accommodated*	*Number of elements*
First	$n = 1$	1*s*	2	2
Second	$n = 2$	2*s*, 2*p*	2 + 6	8
Third	$n = 3$	3*s*, 3*p*	2 + 6	8
Fourth	$n = 4$	4*s*, 3*d*, 4*p*	2 + 10 + 6	18
Fifth	$n = 5$	5*s*, 4*d*, 5*p*	2 + 10 + 6	18
Sixth	$n = 6$	6*s*, 4*f*, 5*d*, 6*p*	2 + 14 + 10 + 6	32
Seventh (incomplete)	$n = 7$	7*s*, 5*f*, 6*d*, 7*p*	2 + 14 + 10 + 6	32

6.7 DIVISION OF PERIODIC TABLE INTO s-, p-, d- AND f-BLOCKS ON THE BASIS OF ELECTRONIC CONFIGURATIONS

The long form of the periodic table can be divided into four main blocks. These are **s-, p-, d- and f-blocks.** The division of elements into blocks is primarily based upon their electronic configuration as shown in Fig. 6.2.

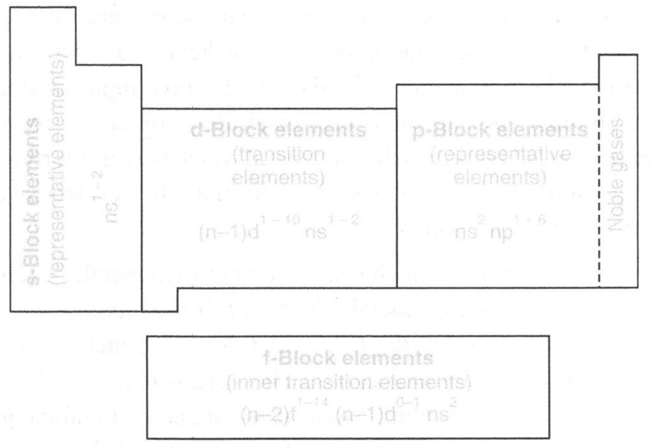

Fig. 6.2. Division of the periodic table into various blocks.

s-BLOCK ELEMENTS

The elements in which the last electron enters the s-sub-shell of their outermost energy level are called **s-block elements.** This block is situated at the extreme left of the periodic table. It contains elements of groups **1 and 2.** Their general configuration is $\mathbf{ns^{1-2}}$, where *n* represents the outermost shell. The elements of group **1** are called **alkali metals** whereas the elements of group 2 are called **alkaline earth metals.**

Some of the **general characteristics of the s-block elements are:**

(*i*) They are soft metals. Hardness decreases with increasing atomic number.

(*ii*) They have low melting and boiling points.

(*iii*) They have low ionization energies and hence are highly electropositive.

(*iv*) They are very reactive metals.

(*v*) They show oxidation states of +1 (in case of alkali metals) or +2 (in case of alkaline earth metals).

(*vi*) They are good reducing agents.

(*vii*) The compounds of *s*-block elements are predominantly ionic.

p-BLOCK ELEMENTS

The elements in which the last electron enters the p-sub-shell of their outermost energy level are called **p-block elements.** The general configuration of their outermost shell is $\mathbf{ns^2\ np^{1-6}}$. The only exception is helium ($1s^2$). Strictly, helium belongs to the *s*-block but its positioning in the *p*-block along with other group 18 elements is justified because it has a completely filled valence shell ($1s^2$) and as a result, exhibits properties characteristic of other noble gases. This block is situated at the extreme right of the periodic table and contains elements of groups **13, 14, 15, 16, 17** and **18** of the periodic table. Most of these elements are non-metals, some are metalloids and a few others are heavy elements which exhibit metallic character. The non-metallic character increases as we move from left to right across a period and metallic character increases as we go down the group. Some of the **general characteristics of p-block elements** are:

(*i*) They show variable oxidation states.

(*ii*) They form ionic as well as covalent compounds.

(*iii*) They have relatively higher values of ionization energy.

(*iv*) Most of them are non-metals.

(*v*) Most of them are highly electronegative.

(*vi*) Most of them form acidic oxides.

d-BLOCK ELEMENTS

The elements in which the last electron enters the d-sub-shell of the penultimate energy level are called **d-block elements.** Their general valence shell configuration is $\mathbf{(n-1)d^{1-10},\ ns^{1-2}}$, where *n* represents the outermost energy level. *d*-Block contains three complete rows of ten elements in each. The fourth row is incomplete. The three rows are called *first, second and third transition series.* They involve the filling of 3*d*, 4*d* and 5*d* orbitals respectively. The *d*-block contains elements of groups **3** to **12** of the periodic table. The **general characteristics of the d-block elements** are:

(*i*) They are hard, high melting metals.

(*ii*) They show variable oxidation states.

(*iii*) They form coloured complexes.

(*iv*) They form ionic as well as covalent compounds.

(*v*) Most of them exhibit paramagnetism.

(*vi*) Most of them possess catalytic properties.

(*vii*) They form alloys. For example, brass is an alloy of copper and zinc.

(*viii*) They are good conductors of heat and electricity.

f-BLOCK ELEMENTS

The elements in which the last electron enters the f-sub-level of the anti-penultimate (third to the outermost) shell are called **f-block elements.**

Their general configuration is **$(n-2)f^{1-14}$ $(n-1)$ d^{0-1}, ns^2,** where *n* represents the outermost shell. They consist of two series of elements placed at the bottom of the periodic table. The elements of the first series follow lanthanum ($_{57}La$) and are called **lanthanoids.** The elements of the second series follow actinium ($_{89}Ac$) and are called **actinoids.** Actinoid elements are radioactive. Many of them have been made only in nanogram quantities or even less by nuclear reactions. Chemistry of the actinoids is complicated and is not fully studied. The ***general characteristics of f-block elements*** are:

(*i*) They show variable oxidation states.
(*ii*) They are high melting metals.
(*iii*) They have high densities.
(*iv*) They form complexes, most of which are coloured.
(*v*) Most of the elements of the actinoid series are radioactive.

Add to Your Knowledge

1. The element hydrogen, having electronic configuration $1s^1$, can be placed in the *s*-block along with alkali metals but is placed separately at the top centre of the periodic table as shown in Fig. 6.1. This is because hydrogen, like halogens, can gain an electron to achieve a noble gas configuration and hence behave like halogens. Due to its resemblance with alkali metals as well as halogens, hydrogen is placed separately at the top of the periodic table.
2. The elements of group 18, the last column of the *p*-block, are known as ***inert gases*** or ***noble gases*** or ***aerogens.***
3. The elements of the *s* and *p*-blocks are called ***representative elements.*** They have only their outermost shell incomplete. They are also known as ***main group elements.***
4. The elements of the *d*-block are called ***d-transition elements.*** The name is derived from the fact that they represent transition (change) in character from reactive metals (elements of group 1 and 2) on one side and less active metals of groups 13 and 14 on the other side.
5. The elements of the *f*-block are called ***f-transition*** or ***inner transition elements.*** In this group of elements the outer three principal shells are incomplete.
6. The *f*-block elements after uranium are called ***trans-uranium elements.***

6.8 CLASSIFICATION OF ELEMENTS AS METALS, NON-METALS AND SEMI-METALS

In addition to the classification of elements into *s*-, *p*-, *d*- and *f*-blocks, it is possible to divide them into metals, non-metals and metalloids. More than 78% of elements are metals. *Metals are present on the left side and the centre of the periodic table.*

Metals are the elements which are malleable and ductile, possess lustre, are good conductors of heat and electricity and have high densities. Metals usually have high melting and boiling points, and are generally solids at room temperature. Mercury is the only metal which is liquid at room temperature. Gallium (303 K) and caesium (302 K) also have very low melting points.

Non-metals are much less in number than metals. There are only about 20 non-metals. *Non-metals are located at the top right hand side of the Periodic Table.* Non-metals have low melting and boiling points. They are usually solids or gases at room temperature. Non-metals are neither malleable nor ductile. They are poor conductors of heat and electricity. In a period, the non-metallic character increases as we move from left to right. In a group, the non-metallic character decreases and metallic character increases on going down a group. There is no sharp line dividing metals from non-metals. A zig-zag line separates metals from non-metals as shown in Fig. 6.3. The borderline elements such as silicon, germanium, arsenic, antimony and tellurium exhibit characteristic properties of metals as well as non-metals. These elements are called **semi-metals** or **metalloids.**

METHOD TO PREDICT THE PERIOD, GROUP AND BLOCK OF A GIVEN ELEMENT

Following steps are followed to predict the group, period and block of the elements:

1. Electronic configuration of the element is written following various rules.
2. Period of the element is represented by the principal quantum number of the valence shell.
3. Block of the element is predicted on the basis of sub-shell which receives the last electron.
4. Group is predicted from the number of electrons in the outermost or penultimate shell as follows:

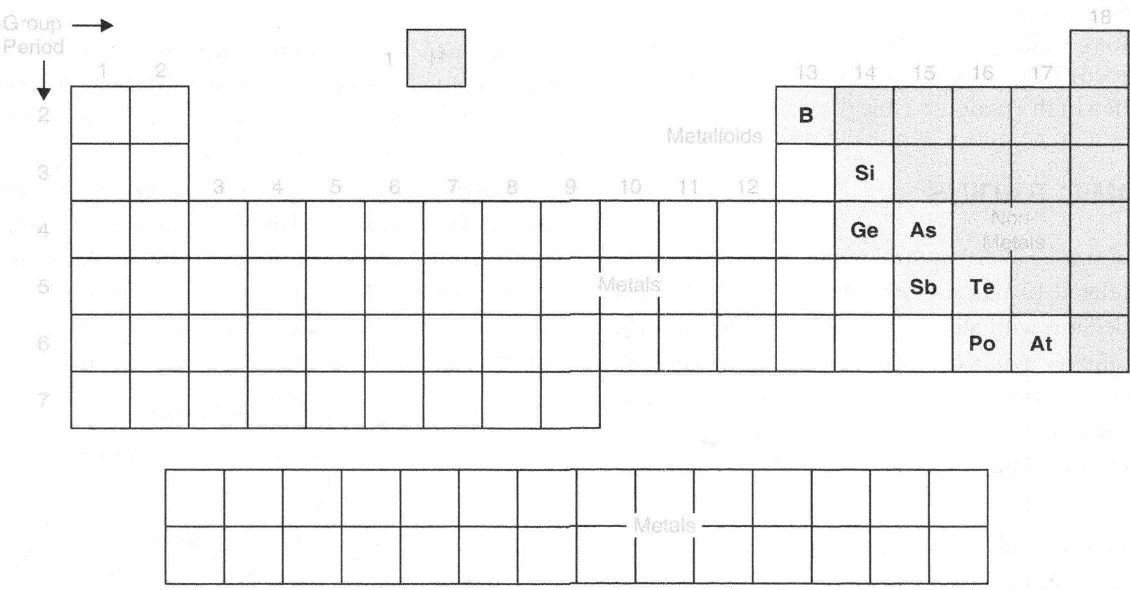

Fig. 6.3. Position of metals, non-metals and metalloids in the periodic table.

(*a*) In case the element belongs to the **s-block,** then *the group = (number of valence electrons)*

(*b*) In case the element belongs to the **p-block,** then *the group = (10 + total number of valence electrons)*

(*c*) In case the element belongs to the **d-block,** then *the group = number of electrons in (n – 1) d-subshell and n s-subshell.*

SOLVED EXAMPLES

Example 6.1. *Write the electronic configuration of the elements given below and also predict their period, group and block.*

A (At. No. 16), E (At. No. 37) and G (At. No. 30).

Solution. The electronic configurations of the elements are:

$_{16}A : 1s^2, 2s^2, 2p^6, 3s^2, 3p^4$

$_{37}E : 1s^2, 2s^2, 2p^6, 3s^2, 3p^6, 3d^{10}, 4s^2, 4p^6, 5s^1$

$_{30}G : 1s^2, 2s^2, 2p^6, 3s^2, 3p^6, 3d^{10}, 4s^2.$

As is evident from the electronic configurations:

(*i*) $_{16}$A receives last electrons in 3*p* sub-shell, therefore, its:

period = **3rd**

block = **p**

group = 10 + Number of valence electrons

= 10 + 6 = **16.**

(*ii*) $_{37}$E receives last electrons in 5*s*-orbitals. Hence its:

period = **5th**

block = **s**

group = Number of valence electrons

= **1.**

(*iii*) $_{30}$G receives last electrons in 3*d*-orbitals. Hence, its:

block = **d**

Since its valence shell is 4th. Therefore,

period = **4th**

group = Number of *n s*-electrons

+ Number of (*n* – 1) *d*-electrons

= 2 + 10 = **12.**

Example 6.2. *How would you justify that there are only 18 elements in the fifth period of the periodic table?*

Solution. The fifth period begins with the filling of the 5*s* orbital and continues until the filling of the sixth energy level (6*s*). The sub-shells which follow 5*s* are, 4*d*, 5*p*, 6*s*, ... Thus, the elements which involve filling of 5*s*, 4*d* and 5*p* sub-shells are accommodated in the fifth period. These sub-shells have nine orbitals that can accept 18 electrons in all. Hence, there are 18 elements in the fifth period.

PERIODIC TRENDS IN PHYSICAL PROPERTIES

Most of the properties of the elements such as *atomic volume, atomic size, ionization enthalpy, electron affinity and electronegativity* are directly related to the electronic configuration of the atoms. *These properties undergo periodic variation with the change in the atomic number within a period or a group.* These properties indirectly control the physical properties such as melting point, boiling point, density, etc. Let us now proceed to study the variation of some of the atomic properties in the periodic table.

6.9 ATOMIC RADIUS

The atomic size is a very important property of the atom because it is related to many other chemical and physical properties. In dealing with atomic size, the atom is assumed to be a sphere and its radius determines the size. In general, **atomic radius** is defined as ***the distance of closest approach to another identical atom.*** However, it is not possible to find precisely the radius of the atoms because of the following reasons:

1. Atom is too small to be isolated.
2. The electron cloud of an atom has no well-defined boundary.

It is for this reason that atomic size is expressed in terms of different types of radii. Some of these are discussed below:

1. COVALENT RADIUS

The approximate radii of atoms can be determined by measuring the distance between the atoms in a covalent molecule by X-ray diffraction and other spectroscopic techniques. This radius of an atom is referred to as **covalent radius.** It may be defined as *one-half of the distance between the centres of the nuclei of two similar atoms bonded by a single covalent bond.*

In case of homonuclear bonds,

$$r_{covalent} = \frac{1}{2}\ \textbf{[Internuclear distance between two bonded atoms]}$$

For example, the internuclear distance between two hydrogen atoms in H_2 molecule is 74 pm (Fig. 6.4). Therefore, the covalent radius of hydrogen atom is 37 pm.

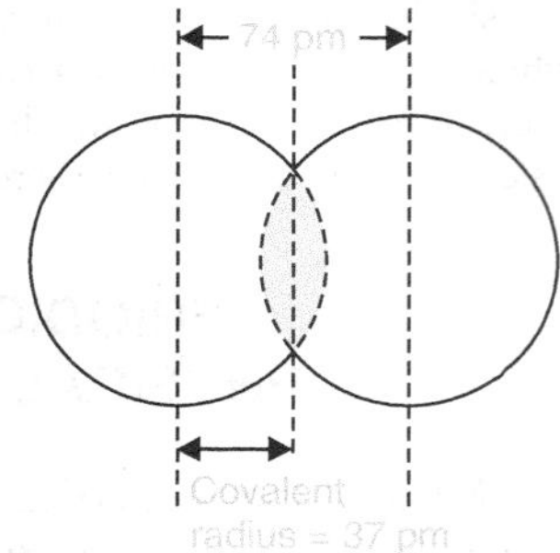

Fig. 6.4. Covalent radius of hydrogen atoms.

Similarly the internuclear distance between two chlorine atoms in chlorine molecule is 198 pm. Therefore, the covalent radius of chlorine is 198/2 = 99 pm.

2. VAN DER WAALS' RADIUS

van der Waals' radius *may be defined as half of the internuclear distance between two adjacent atoms of the same element belonging to two nearest neighbouring molecules of the same substance in the solid state.*

Covalent radius of an element is shorter than its van der Waal radius. The formation of a covalent bond involves overlapping of atomic orbitals. As a result of this, the internuclear distance between covalently bonded atoms is less than the internuclear distance between the non-bonded atoms. This has been shown in Fig. 6.5. Thus,

$$\mathbf{r_{(cov.)} < r_{(van\ der\ Waal)}}$$

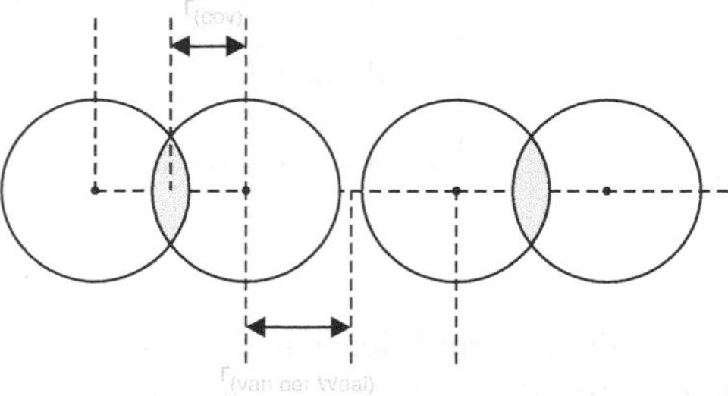

Fig. 6.5. Comparison of covalent and van der Waal radii.

3. METALLIC OR CRYSTAL RADIUS

The term is applied to metals only. A metallic lattice is considered to be consisting of closely packed atoms which are spherical in shape. ***Metallic radius*** *may be defined as half of the internuclear distance between two adjacent atoms in the metallic lattice.* It is measured in angstrom units. Internuclear distances are measured by X-ray studies. The metallic radius of sodium is found to be 186 pm while that of copper is 128 pm.

The metallic radius of an atom is always larger than its covalent radius because a metallic bond is weaker than a covalent bond. Therefore, two atoms held by a covalent bond are closer to each other. For example, the metallic radius of sodium is 186 pm, whereas its covalent radius as determined from its vapours which exist as Na_2 molecules is 154 pm. Similarly, the metallic radius of potassium is 231 pm whereas the covalent radius is 203 pm.

VARIATION OF ATOMIC RADII IN THE PERIODIC TABLE

Atomic radii usually depend upon *nuclear charge* and *number of main energy levels* of an atom. The periodic trends in atomic radii are described as follows:

Table 6.5. Atomic Radii of Elements of Second Period

Element	$_3$Li	$_4$Be	$_5$B	$_6$C	$_7$N	$_8$O	$_9$F
Nuclear Charge	+ 3	+ 4	+ 5	+ 6	+ 7	+ 8	+ 9
Configuration	$2s^1$	$2s^2$	$2s^22p^1$	$2s^22p^2$	$2s^22p^3$	$2s^22p^4$	$2s^22p^5$
Atomic Radii (pm)	152	111	88	77	75	74	72

(*a*) **Variation in a Period**

In general, *the atomic radii decrease with the increase in atomic number across a period.* For example, the atomic radii decrease from lithium to fluorine in the second period.

The decrease of atomic radii along a period can be explained on the basis of nuclear charge. In moving from left to right across the period, the nuclear charge increases progressively by one unit but the additional electron goes to the same principal shell. As a result, the electron cloud is pulled closer to the nucleus by the increased effective nuclear charge. This causes the decrease in atomic size. The atomic radii of elements of the second period are given in Table 6.5.

(*b*) **Variation in a Group**

In general, *the atomic radii increase from top to bottom within a group on the periodic table.* For example, the atomic radii increase from lithium to cesium among alkali metals and a similar trend is exhibited by halogens from fluorine to iodine as shown in Table 6.6.

Table 6.6. Variation of Atomic Radii of Group-1 and Group-17 elements

Group-1 elements (Alkali metals)		*Group-17 elements (Halogens)*	
Element	*Atomic radius*	*Element*	*Atomic radius*
Li	152	F	72
Na	186	Cl	99
K	231	Br	114
Rb	244	I	133
Cs	262	At	140

In moving down a group, the nuclear charge increases with increase in atomic number but at the same time, there is a progressive increase in the principal energy levels. The number of electrons in the outermost shell, however, remains the same. Since, the effect of additional energy level is more pronounced than the effect of increased nuclear charge, the effective nuclear charge decreases. Consequently, the distance of the outermost electron from the nucleus increases on going down the group. In other words, *the atomic size goes on increasing as we move down a group.*

In short,

Atomic radii increase down the group.
Atomic radii decrease across the period.

6.10 IONIC RADIUS

Ions are formed when the neutral atoms lose or gain electrons. A *positive ion* or *cation* is formed by the loss of one or more electrons by the neutral atom whereas a *negative ion* or *anion* is formed by the gain of one or more electrons by the atom. The term *ionic radii* refers to the size of the ions in the ionic crystals. The ionic radius of an ion may vary from crystal to crystal because of changes in the surrounding ions.

The equilibrium distance between the nuclei of the two adjacent ions can be determined by X-ray analysis of ionic crystals. Assuming ions to be spheres, the internuclear distance can be taken as the sum of the ionic radii of adjacent ions (Fig. 6.6). Knowing the ionic radius of one of the ions, the ionic radius of the other can be calculated.

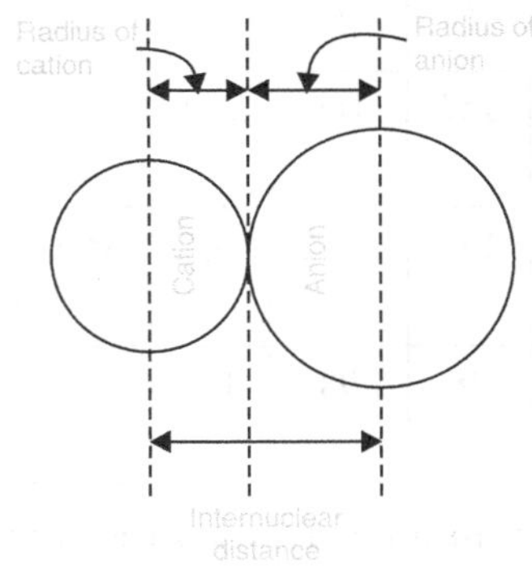

Fig. 6.6. Illustration of ionic radius.

- *The size of the cation is smaller as compared with that of the parent atom.*
- *The size of the anion is larger as compared with that of the parent atom.*

The radius of the cation is smaller than that of the parent atom.

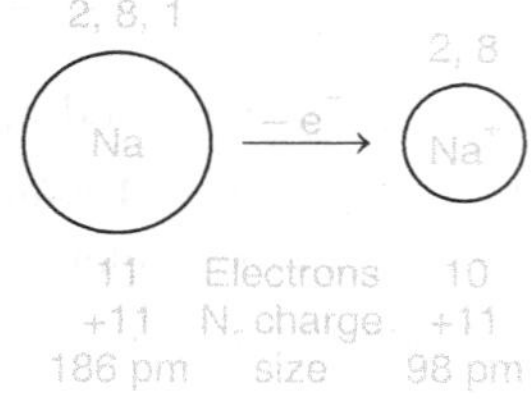

Fig. 6.7. Relative sizes of Na atom and Na^+ ion.

Cation is formed by the loss of one or more electrons from the gaseous atom. Now, in the cation the nuclear charge remains the same as that in the parent atom but the number of electrons becomes less. As a result of this, the nuclear hold on the remaining electrons increases because of the *increase in the effective nuclear charge per electron.* This causes a decrease in the size.

In many cases the formation of cation also involves the removal of the valence shell completely. For example, the formation of Na^+ ion from Na atom involves the removal of the third shell completely. This also results in the decrease in the size of the ion.

$$\underset{2,\,8,\,1}{Na} \longrightarrow \underset{2,\,8}{Na^+} + e^-$$

The comparative sizes of certain atoms and their corresponding cations are given in Table 6.7.

Table 6.7. Atomic and Ionic Radii of Some Elements

Atoms	$_{3}Li$	$_{11}Na$	$_{19}K$	$_{12}Mg$	$_{20}Ca$	$_{13}Al$
Atomic Radii (pm)	152	186	231	160	197	143
Corresponding Ions	Li^+	Na^+	K^+	Mg^{2+}	Ca^{2+}	Al^{3+}
Ionic Radii (pm)	78	98	133	78	106	57

The radius of the anion is larger than that of the parent atom. Anion is formed by the gain of one or more electrons by a gaseous atom. In the anion, the nuclear charge is the same as that in the parent atom but the number of electrons has increased. Since the same nuclear charge now acts on an increased number of electrons, *the effective nuclear charge per electron decreases in the anion.* The electron cloud is held less tightly by the nucleus. This causes an increase in the size. The relative sizes of chlorine atom and chloride ion have been shown in Fig. 6.8.

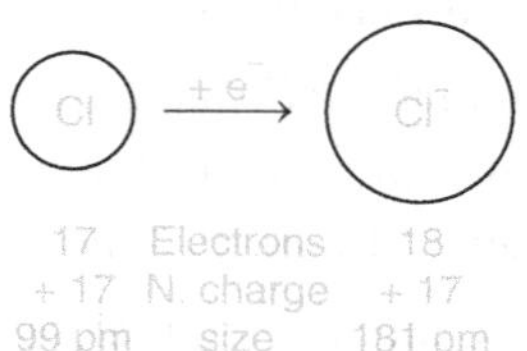

Fig. 6.8. Relative sizes of Cl atom and Cl^- ion.

The comparative sizes of some atoms and their corresponding anions are given in Table 6.8.

Table 6.8. Atomic and Ionic Radii of Some Elements

Atoms	$_{17}Cl$	$_{35}Br$	$_{53}I$	$_{8}O$	$_{7}N$
Atomic Radii (pm)	99	114	133	74	75
Corresponding Anion	Cl^-	Br^-	I^-	O^{2-}	N^{3-}
Ionic Radii (pm)	181	195	220	140	171

ISO-ELECTRONIC IONS

The ions having the same number of electrons but different magnitude of nuclear charge are called iso-electronic ions. In fact, these are the ions of different elements having the same electronic arrangement. For example, each one of sulphide (S^{2-}), chloride (Cl^-) and potassium (K^+) ion has eighteen electrons but they have different nuclear charge, +16, +17 and +19 respectively.

Variation of size among iso-electronic ions. Within the series of iso-electronic ions, as the nuclear charge increases, the attractive force between the electrons and nucleus also increases. This results in a decrease of the ionic radius. In other words, *size of the iso-electronic ions decreases with an increase in the magnitude of the nuclear charge.* For example, N^{3-}, O^{2-}, F^-, Na^+, Mg^{2+}, Al^{3+} are iso-electronic and have 10 electrons each. The sizes of these ions are in the order:

$$Mg^{2+} < Na^+ < F^- < O^{2-} < N^{3-}$$

Mg^{2+} having the highest nuclear charge (12 units) has the smallest size whereas N^{3-} ion having the smallest nuclear charge (7 units) has the largest size. Variation of size among these ions has been shown in Table 6.9.

Table 6.9. Variation of Size Among Iso-electronic Ions

Ion	N^{3-}	O^{2-}	F^-	Na^+	Mg^{2+}	Al^{3+}
Nuclear Charge	+ 7	+ 8	+ 9	+ 11	+ 12	+ 13
Electrons	10	10	10	10	10	10
Size (pm)	171	140	136	98	78	57

SOLVED EXAMPLES

Example 6.3. *The following species are isoelectronic with the noble gas neon. Arrange them in order of increasing size: Na^+, F^-, O^{2-}, Mg^{2+}, Al^{3+}.*

Solution. In Na^+, F^-, O^{2-}, Mg^{2+}, Al^{3+}, the nuclear charges are 11, 9, 8, 12 and 13 respectively. Among isoelectronic species, the greater the nuclear charge the smaller the size. Therefore, the sizes of the above ionic species are in the order:

$$\mathbf{Al^{3+} < Mg^{2+} < Na^+ < F^- < O^{2-}.}$$

Example 6.4. *Select from each group the species which has the smallest radius stating appropriate reason.*

(*i*) O, O^-, O^{2-} (*ii*) K^+, Sr^{2+}, Ar

(*iii*) Si, P, Cl.

Solution. (*i*) The species O has the smallest radius because the radius of an anion is always larger than the radius of the atom from which it is formed. O^- and O^{2-} are anions of oxygen.

(*ii*) K^+ has the smallest radius. In K^+ and Ar the outermost shell is the third whereas in Sr^{2+} it is the fourth. Out of K^+ and Ar, K^+ has a smaller size because it has a greater nuclear charge.

(*iii*) Cl has the smallest radius. Si, P and Cl belong to the same period. In a period the atomic radius decreases with an increase in atomic number due to an increase in effective nuclear charge.

Example 6.5. *Name a species that will be isoelectronic with each of the following atoms or ions:*

(*i*) Ne (*ii*) Cl^- (*iii*) Ca^{2+} (*iv*) Rb.

Solution. (*i*) Sodium ion, Na^+ (*ii*) Potassium ion, K^+

(*iii*) Sulphide ion, S^{2-} (*iv*) Sr^+

Example 6.6. *Out of Na^+ and Na which has a smaller size and why?*

Solution. Na^+ has a smaller size than Na. Na^+ is formed by the removal of one electron from Na. Therefore, Na^+ has one electron less than Na. However, Na and Na^+ have the same nuclear charge. Therefore, electrons in Na^+ are more tightly held than in Na. Moreover, the removal of one electron from Na leads to the complete removal of the third shell so that in Na^+, the outermost shell is the second. Hence, Na^+ has a smaller size than Na.

Example 6.7. *Give examples of three cations and three anions which are isoelectronic with argon.*

Solution. *Cations:* K^+, Ca^{2+}, Sc^{3+}

Anions: Cl^-, S^{2-}, P^{3-}.

6.11 IONIZATION ENERGY

It is a well-known fact that the electrons in an atom are attracted by the positively charged nucleus. In order to remove an electron from an atom, energy has to be supplied to it to overcome the attractive force. This energy is referred to as ionization energy. Thus,

Ionization energy *may be defined as the amount of energy required to remove the most loosely bound electron from the isolated gaseous atom in its ground state.*

$$\mathbf{A\,(g) + Energy\,(I.E.) \longrightarrow A^+\,(g) + e^-}$$

Ionization energy is a very important property which gives an idea about the tendency of an atom to form a gaseous positive ion.

Ionization energy is expressed in terms of *kilo Joules per mole of atoms (kJ mol^{-1})*

SUCCESSIVE IONIZATION ENERGIES

Once the first electron has been removed from the gaseous atom, it is possible to remove a second and successive electrons from positive ions one after the other. For example,

$$A\,(g) + I.E._1 \longrightarrow A^+(g) + e^-$$
$$A^+(g) + I.E._2 \longrightarrow A^{2+}(g) + e^-$$
$$A^{2+}\,(g) + I.E._3 \longrightarrow A^{3+}(g) + e^-$$

The amounts of energies required to remove loosely bound electrons from a unipositive, dipositive, tripositive ..... ions of an element in the gaseous state are called *second, third, fourth ...... ionization energies respectively.*

The second, third, fourth, etc. ionization energies are collectively known as **successive ionization energies.** It may be noted that:

$$\mathbf{I.E._3 > I.E._2 > I.E._1}$$

The variation in the values of successive ionization energies can be explained, in general, as follows:

After the removal of the first electron, the atom changes into a monopositive ion. In the ion, the number of electrons decreases but the nuclear charge remains the same as in the parent atom. As a result of this, the effective nuclear charge per electron increases. The remaining electrons are, therefore, held more tightly by the nucleus. Thus, more energy is required to remove the second electron. Hence, the value of the second ionization energy is greater than the first.

Similarly, the removal of the second electron results in the formation of a divalent positive ion and the attraction between the nucleus and the remaining electrons increases further. This accounts for the progressive increase in the values of successive ionization energies.

If the removal of the first, second or third electron also results in the removal of the valence shell, the next electron (which belongs to lower energy shell) will require very large amount of energy for its removal. For example, in sodium, the removal of the first electron leads to the removal of the third shell completely. Therefore, for sodium, $I.E._2 >> I.E._1$.

$$Na(g) - e^- \longrightarrow Na^+(g); \quad I.E._1 = 495 \text{ kJ mol}^{-1}$$
(2, 8, 1) (2, 8)

$$Na^+(g) - e^- \longrightarrow Na^{2+}(g); \quad I.E._2 = 4581 \text{ kJ mol}^{-1}$$
(2, 8) (2, 7)

The first, second and third ionization energies of some elements are given in Table 6.10.

Table 6.10. $I.E._1$, $I.E._2$ and $I.E._3$ Values of Some Elements

Element	*$I.E._1$ (kJ mol^{-1})*	*$I.E._2$ (kJ mol^{-1})*	*$I.E._3$ (kJ mol^{-1})*
Li	520	7297	11810
Be	899.5	1757	14850
B	800.6	2427	3650
C	1086.4	2352	4619
N	1402	2860	4577
O	1314	3389	5301

FACTORS ON WHICH IONIZATION ENERGY DEPENDS

The ionization energy depends upon the following factors:

1. *Size of the atom*;
2. *Magnitude of nuclear charge*;
3. *Screening effect of the inner electron*;
4. *Penetration effect of the electrons; and*
5. *Electronic configuration.*

1. Size of the Atom. The attractive force between the electron and the nucleus is inversely proportional to the distance between them. Therefore, *as the size of the atom increases, the outermost electrons are less tightly held by the nucleus. As a result, it becomes easier to remove the electron. Therefore, ionization energy decreases with increase in atomic size.*

2. Magnitude of Nuclear Charge. The attractive force between the nucleus and the electrons increases with an increase in nuclear charge provided their main energy shell remains the same. This is because, the force of attraction is directly proportional to the product of charges on the nucleus and that on the electron. Therefore, *with an increase in the nuclear charge, it becomes more difficult to remove an electron and ionization energy increases.*

3. Screening Effect of the Inner Electrons. In multi-electron atoms, the electrons present in the outermost shell do not experience the complete nuclear charge because of the repulsive interaction of the intervening electrons. Thus, the outermost electrons are *shielded* or *screened* from the nucleus by the inner electrons. This is known as **screening effect.**

If the number of electrons in the inner shells is large, the screening effect will be large. As a result, the attractive interactions between the nucleus and outermost electrons will be less. Consequently, the ionization energy will decrease. Thus, if other factors do not change, *an increase in the number of inner electrons tends to decrease the ionization energy.*

4. Penetration Effect of the Electrons. It is a well-known fact that in case of multi-electron atoms, the electrons in the *s*-orbital have the maximum probability of being found near the nucleus and this probability goes on decreasing in case of *p, d* and *f* orbitals. In other words, *s*-electrons are more penetrating towards the nucleus than *p*-electrons. The penetration power decreases in a given shell (same value of n) in the order: $s > p > d > f$.

Now, if the penetration of the electron is high, it experiences less shielding effect by the inner electrons and will be held firmly. Consequently, the ionization energy will be high. This means that *ionization energy increases with increase in the penetration power of the electrons.* Thus, *for the same shell, it is easier to remove the p-electrons in comparison to the s-electrons.*

5. Electronic Configuration. It has been noticed that certain electronic configurations are more stable than the other. *The atom having a more stable configuration has less tendency to lose the electron and consequently, has a high value of ionization energy.* For example:

(*i*) The noble gases have stable configuration (ns^2np^6). They have the highest ionization energies within their respective periods.

(*ii*) The elements like N ($1s^2$, $2s^2$, $2p_x^1$, $2p_y^1$, $2p_z^1$) and P ($1s^2$, $2s^2$, $2p^6$, $3s^2$, $3p_x^1$, $3p_y^1$, $3p_z^1$) have configurations in which orbitals belonging to the same sub-shell are exactly half-filled. Such configurations are quite stable and consequently, require more energy for the removal of an electron. Hence, their ionization energies are relatively high.

VARIATION OF IONIZATION ENERGY IN THE PERIODIC TABLE

Let us now study the variation of ionization energy across the period and along the group of representative elements on the basis of the above factors.

Variation Across the Period

In general, the *value of ionization energy increases with the increase in atomic number across the period.* This can be attributed to the fact that in moving across the period from left to right,

(*i*) nuclear charge increases regularly;

(*ii*) addition of electrons occurs in the same energy level;

(*iii*) atomic size decreases.

Thus, due to the gradual increase in nuclear charge and simultaneous decrease in atomic size, the attractive force between the nucleus and the electron cloud increases. Consequently, the electrons are more and more tightly bound to the nucleus. This results in the gradual increase in ionization energy across the period. The ionization energies of the elements of 2nd period are given in Fig. 6.9.

On carefully examining the values given in Fig. 6.9, we find some *exceptions* within the period. These can be explained on the basis of other factors governing ionization energy.

(*i*) There is an *increase* in ionization energy from Li to Be. This is due to increased nuclear charge and smaller atomic size.

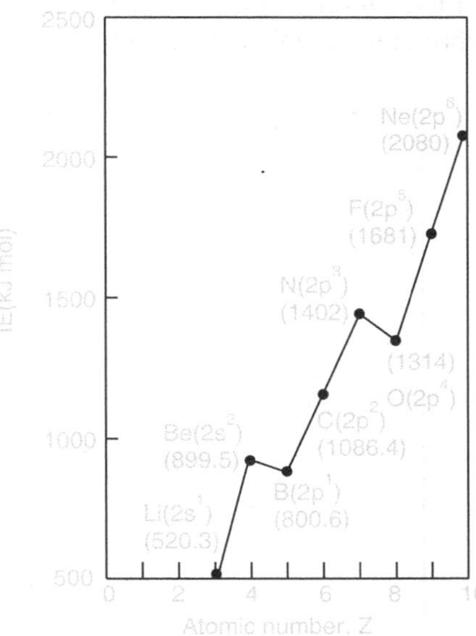

Fig. 6.9. Variation of ionization energies among elements of 2nd period.

(*ii*) There is a *decrease* in the value of ionization energy from **Be** to **B** inspite of increased nuclear charge. The effect of increased nuclear charge is cancelled by (*a*) *greater penetration of 2s electron as compared to 2p electron;* (*b*) *better shielding of 2p electrons by the inner electrons. Thus,* the 2*p* electron of boron is relatively less tightly held by its nucleus in comparison to the 2*s* electrons of Be.

(*iii*) There is a regular *increase* in ionization energy from **B** to **C** to **N.** It is again due to the gradual increase in nuclear charge and decrease in atomic size.

(*iv*) There is a slight *decrease* in ionization energy from **N** to **O.** It is attributed to the relatively stable configuration of nitrogen due to a half filled 2*p*-orbital. In the nitrogen atom, three 2*p*-electrons are present in different atomic orbitals (Hund's rule) whereas in the oxygen atom, two of the four 2*p*-electrons are present in the same 2*p*-orbital resulting in an increased electron-electron repulsion. Therefore, it is easier to remove one of the 2*p*-electron from oxygen than it is to remove one of the 2*p*-electrons from nitrogen.

(*v*) There is an expected *increase* in ionization energy from **O** to **F** to **Ne.**

Variation in a Group

The *values of ionization energies of elements decrease regularly with the increase in atomic number within a group.*

The values of ionization energies of the elements of group 1 have been represented graphically in Fig. 6.10.

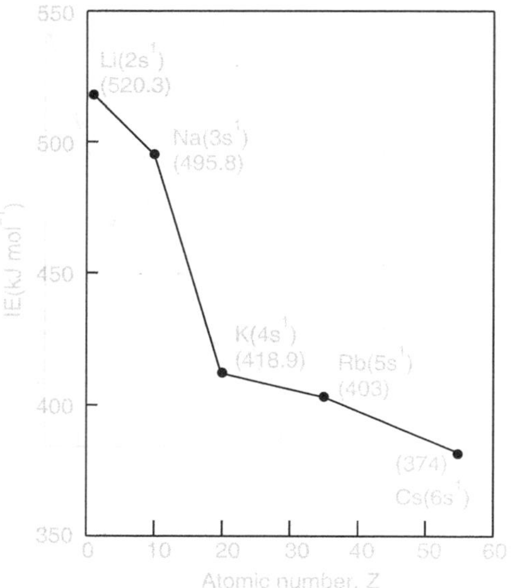

Fig. 6.10. Variation of ionization energy among elements of group 1.

The decrease in the value of ionization energy within the group can be explained on the basis of the net effect of the following factors:

As we move down the group there is:

(*i*) A gradual increase in the atomic size due to the progressive addition of new energy shells;

(*ii*) Increase in the shielding effect on the outermost electron due to an increase in the number of inner electrons.

The nuclear charge also increases but the effect of increased nuclear charge is cancelled by the *increase in atomic size* and *the shielding effect*. Consequently, the nuclear hold on the valence electron decreases gradually and ionization energy also decreases. The variation of the first ionization energy as a function of atomic number for elements with atomic numbers up to 60 have been shown in Fig. 6.11. The maxima in the curve represents noble gases which means that the ionization energies of noble gases are highest within their periods. The minima in the curve represent alkali metals which implies that ionization energies of alkali metals are the lowest within their respective periods.

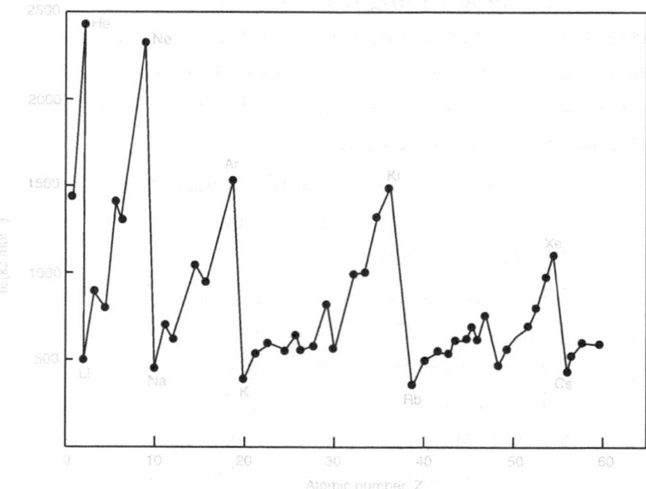

Fig. 6.11. Ionization energies of elements up to Z = 60.

SOLVED EXAMPLES

Example 6.8. *From each set, choose the atom which has the largest ionization energy and explain your answer*

(*i*) *F, O, N.* (*ii*) *Mg, P, Ar.*

(*iii*) *B, Al, Ga.*

Solution. (*i*) F has the highest ionization energy among F, O and N because it has the smallest size and highest nuclear charge. In general, ionization energy increases as we go from left to right in a period.

(*ii*) Ar (a noble gas) has the highest ionization energy among the elements Mg, P and Ar because it has a stable electronic configuration and maximum nuclear charge.

(*iii*) B, Al and Ga belong to group-13. Among these elements B has the largest ionization energy because on moving down a group, from top to bottom, the ionization energy decreases. B is the first element of group 13.

Example 6.9. *Out of Na^+ and Ne which has a higher ionization energy? Explain why.*

Solution. Na^+ has higher ionization energy than Ne. Na^+ and Ne are isoelectronic species. However, the nuclear charge in Na^+ is higher than in Ne. Hence, the electrons are more tightly held in Na^+ and it has higher ionization energy.

Example 6.10. *How do you explain that $_{31}Ga$ has a slightly higher ionization energy than $_{13}Al$, although it occupies a lower position in the group?*

Solution. $_{13}Al : 1s^2, 2s^2, 2p^6, 3s^2, 3p^1$

$_{13}Ga : 1s^2, 2s^2, 2p^6, 3s^2, 3p^6, 3d^{10}, 4s^2, 4p^1$

In Ga, the 10 electrons present in the 3*d*-sub-shell do not shield the outer electrons from the nucleus effectively. As a result, the effective nuclear charge in Ga increases. This explains why the ionization energy of Ga is slightly higher than that of $_{13}Al$.

6.12 ELECTRON AFFINITY, E_{ea}

We have already learnt that ionization energy is a measure of the tendency of the atom to form a cation. In a similar way, the tendency of a gaseous atom to form an anion is expressed in terms of *electron affinity.*

Electron affinity *may be defined as the energy change taking place when an isolated gaseous atom accepts an electron to form a monovalent gaseous anion.*

The values of electron affinity are expressed in *kilo joules per mole of atoms.* For example, the electron affinity of chlorine is – 348 kJ mol^{-1}, *i.e.*,

$$\mathbf{Cl(g) + e^- \longrightarrow Cl^- (g) \quad E_{ea} = -\ 348\ kJ\ mol^{-1}.}$$

Depending on the element, the process of adding an electron can be either exothermic or endothermic.

The magnitude of the electron affinity measures the tightness with which the atom can hold the additional electron. The large negative value of electron affinity reflects the greater tendency of an atom to accept the electron. For example, the elements of group-17 (the halogens) have very high negative electron affinities because they can attain stable noble gas electronic configuration by gaining an electron. Thus, halogens have a great tendency to accept an electron. On the other hand, the elements of group-18 (the noble gases) have large positive electron affinities because the additional electron goes to the next principal shell resulting in a very unstable electronic configuration.

FACTORS AFFECTING ELECTRON AFFINITY

Some of the important factors which affect electron affinity are discussed below:

1. Nuclear Charge. The greater the magnitude of the nuclear charge the greater the attraction for the incoming electron and the larger the negative value of electron affinity.

2. Atomic Size. The larger the size of an atom, the larger the distance between the nucleus and the additional electron and the smaller the negative value of electron affinity.

3. Electronic Configuration. The stable the electronic configuration of an atom, the lesser its tendency to accept an electron and the larger the positive value of its electron affinity. For example, the elements having completely filled sub-levels of the valence shell have relatively stable configurations and consequently, possess large positive values of electron affinity.

VARIATION OF ELECTRON AFFINITY IN THE PERIODIC TABLE

Since the experimental determination of electron affinity is not as easy as that of ionization energy, sufficient data regarding electron affinities is not available. Consequently, the varying trends of electron affinities are not well-defined. The electron affinities of some elements are given in Table 6.11.

Table 6.11. Electron Affinities of Some Elements (kJ mol^{-1})

Groups ⟶ *Periods* ↓	*1*	*2*	*13*	*14*	*15*	*16*	*17*	*18*
1	H – 73							He + 48
2	Li – 60	Be + 240				O – 141	F – 328	Ne + 116
3	Na – 53	Mg + 230				S – 200	Cl – 348	Ar + 96
4	K – 48					Se – 195	Br – 325	Kr + 96
5	Rb – 47					Te – 190	I – 295	Xe + 77
6	Cs – 46					Po – 174	At – 270	Rn + 68

Variation in a Period

On moving across the period, *the atomic size decreases* and *nuclear charge increases.* Both these factors result into a greater attraction for the incoming electron, therefore, *electron affinities tend to become more negative as we go from left to right across a period.* However, some irregularities are observed among the elements of group 2, group 15 and group 18. The electronic configurations of these elements are relatively stable and hence, these elements have positive or very low negative electron gain enthalpies.

Variation Down a Group

On moving down a group, *the atomic size* as well as *nuclear charge increases.* But the effect of increase in atomic size is much more pronounced than that of nuclear charge and thus, the additional electron feels less attraction. Consequently, *electron affinity becomes less negative on going down the group.* Let us now examine the values of electron affinity for halogens as shown in Table 6.11.

It may be noted that the electron affinity becomes less negative as we go from chlorine to bromine to iodine. However, as we move from fluorine to chlorine the electron affinity becomes more negative whereas the reverse was expected. This is because when an electron is added to the fluorine atom, it goes to the relatively compact second energy level. As a result, it experiences significant repulsion from the other electrons present in this shell. On the other hand, in the chlorine atom, the added electron goes to the third energy shell which is relatively larger. Hence, it experiences less electron-electron repulsion. Therefore, the electron affinity of fluorine is less negative as compared with chlorine. The unexpected trend is observed in case of the many other elements of the third period as their electron affinities are more negative than those of the elements of the second period.

SUCCESSIVE ELECTRON AFFINITIES

When the first electron is added to the gaseous atom, it forms a uninegative ion and the energy change during the process is called the *first electron affinity*. Now, if an electron is added to the uninegative ion, it experiences a repulsive force from the anion. As a result, energy has to be supplied to overcome the repulsive force. Thus, in order to add the second electron, energy is rather required and not released. Therefore, **the value of the second electron affinity is positive.** Similarly, the addition of the third, the fourth electrons, etc., also requires energy. Hence, the values of *successive electron affinities are positive*. For example, let us study the addition of electrons to the oxygen atom

$$O(g) + e^- \longrightarrow O^-(g) \qquad E_{ea1} = -141 \text{ kJ}$$

$$O^-(g) + e^- \longrightarrow O^{2-}(g) \qquad E_{ea2} = +780 \text{ kJ}$$

SOLVED EXAMPLE

Example 6.11. *Which of the following will have the most negative electron affinity and which the least negative?*

P, S, Cl, F.

Explain your answer.

Solution. Electron affinity generally becomes more negative across a period as we move from left to right. Therefore, among P, S and Cl the order of negative electron affinity is Cl > S > P. Within a group, electron affinity becomes less negative down a group. However, adding an electron to the 2*p*-orbital leads to greater repulsion than adding an electron to the larger 3*p*-orbital. Therefore, the electron affinity of Cl is more negative than that of F. Hence, the element with the most negative electron affinity is Cl and the one with the least negative electron affinity is P.

6.13 ELECTRONEGATIVITY

Electronegativity *may be defined as the tendency of an atom in a molecule to attract towards itself the shared pair of electrons.*

It may be mentioned here that unlike ionization energy and electron affinity, electronegativity is not a measurable quantity. However, in order to compare the electronegativity values of different elements a number of numerical scales of electronegativity of elements have been proposed. The most common and widely used scale of electronegativity is the ***Pauling scale.***

Pauling's scale of electronegativity is based on excess bond energies.

Electronegativity of fluorine, the most electronegative element, is arbitrarily taken as 4.0 on this scale.

The main factors on which the electronegativity depends are effective nuclear charge and atomic radius.

- *The greater the effective nuclear charge the greater the electronegativity.*
- *The smaller the atomic radius the greater the electronegativity.*

The electronegativity of any given element is not constant but varies depending on the element to which it is bound.

In a period electronegativity increases in moving from left to right. This is because the nuclear charge increases whereas the atomic radius decreases as we move from left to right in a period. Halogens have the highest value of electronegativity in their respective periods. The electronegativity values (on the Pauling scale) of elements of the second and third periods are listed in Table 6.12.

Table 6.12. Electronegativity Values of Elements of Second and Third Periods

Elements of Second Period	Li	Be	B	C	N	O	F
Electronegativity	1.0	1.5	2.0	2.5	3.0	3.5	4.0
Elements of Third Period	Na	Mg	Al	Si	P	S	Cl
Electronegativity	0.9	1.2	1.5	1.8	2.1	2.5	3.0

In a group electronegativity decreases on moving down the group. This is due to the effect of increased atomic radius. For example, among halogens fluorine has the highest electronegativity. In fact, fluorine has the highest value of electronegativity among all the elements.

The electronegativity values (Pauling scale) of group 1 and group 17 elements are given in Table 6.13.

Table 6.13. Electronegativity Values of Group-1 and Group-17 Elements

Group-1		*Group-17*	
Element	*Electronegativity*	*Element*	*Electronegativity*
Li	1.0	F	4.0
Na	0.9	Cl	3.5
K	0.8	Br	2.8
Rb	0.8	I	2.5
Cs	0.7	At	2.2

Relationship between Electronegativity and Non-metallic (or Metallic) Character of an Element

Non-metallic elements have strong tendency to gain electrons. Therefore, electronegativity is directly related to the non-metallic character of elements. We can also say that the electronegativity is inversely related to the metallic character of elements. Thus, the increase in electronegativities across a period is accompanied by an increase in non-metallic character (or decrease in metallic character) of elements. Similarly, the decrease in electronegativity down a group is accompanied by a decrease in the non-metallic character (or increase in metallic character) of elements.

6.14 ELECTROPOSITIVITY OR METALLIC CHARACTER

Tendency of atoms of an element to lose electrons and form positive ions is known as ***electropositivity.***

A more electropositive element has more metallic character.

Whether an element behaves as a metal or a non-metal is directly related to its ionization energy. The elements having low values of ionization energies are metals whereas elements having high values of ionization energies are non-metals. The border line elements behave as metalloids.

Variation of Metallic Character Across a Period

Metallic character decreases across a period from left to right. On the other hand non-metallic character increases with increase in atomic number across a period. For example, let us consider elements of the second and third periods.

Metals Metalloid Non-metals

Period 2	Li	Be	B	C	N	O	F	Ne
Period 3	Na	Mg	Al	Si	P	S	Cl	Ar

Metals Metalloid Non-metals

In the **second period,** lithium and beryllium are metals, boron is a metalloid while carbon, nitrogen, oxygen, fluorine and neon are non-metals.

In the **third period,** sodium, magnesium and aluminium are metals, silicon is a metalloid while phosphorus, sulphur chlorine and argon are non-metals.

Variation of Metallic Character Along a Group

On going down a group from top to bottom, the metallic character of elements increases.

In each group, the first element is the least metallic while the last element is the most metallic. For example, let us consider the elements of groups 14 and 15.

In **group 14,** the first element carbon is a non-metal, silicon and germanium are metalloids while tin and lead are metals.

	Group 14	Group 15	
Non-metals	C Carbon	N Nitrogen	Non-metals
Metalloids	Si Silicon	P Phosphorus	Non-metals
Metalloids	Ge Germanium	As Arsenic	Metalloids
Metals	Sn Tin	Sb Antimony	Metalloids
Metals	Pb Lead	Bi Bismuth	Metals

In **group 15,** the first two elements, nitrogen and phosphorus, are non-metals while arsenic and antimony are metalloids and bismuth is a metal.

It may be mentioned here that metals generally form cations by losing electrons from the outermost shell while non-metals generally form anions by accepting one or more electrons. For example, alkali metals form M^+ ions by losing one electron while alkaline earth metals form M^{2+} ions by losing two electrons from the outermost shell. The non-metallic elements of group 17 form anions (X^-) by accepting one electron.

Add To Your Knowledge

The elements of groups 1 and 2 form positive ions by losing electrons whereas elements of groups 16 and 17 have the tendency to form negative ions by gaining electrons. Hence, the compounds of elements of groups 1 and 2 with elements of group 16 and 17 are generally ionic. On the other hand, elements of the groups 13, 14, 15, 16 and 17 generally form covalent compounds with each other.

For example,

NaF, KCl, NaBr, MgO, etc., are ionic in nature. CO_2, SO_2, CS_2, PF_5, PCl_5, etc., are covalent compounds.

The periodic trends of various physical properties are given in Fig. 6.12.

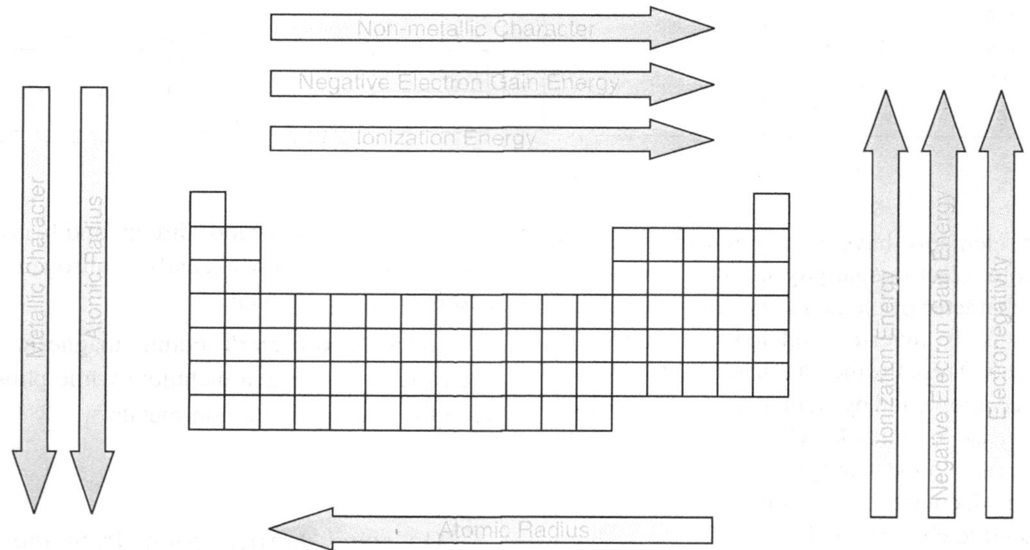

Fig. 6.12. The general periodic trends of various physical properties in the periodic table.

SOLVED EXAMPLES

Example 6.12. *Arrange the following elements the increasing order of metallic character:*

B, Al, Mg, K.

Solution. Metallic character increases on moving down the group and decreases on going across a period from left to right. Hence, the order of increasing metallic character is

B < Al < Mg < K.

Example 6.13. *Arrange the following elements in increasing order of non-metallic character:*

B, C, Si, N, F.

Solution. Non-metallic character increases across a period from left to right and decreases on moving down the group from top to bottom. Hence, the order of increasing non-metallic character is

Si < B < C < N < F.

Example 6.14. *Among the elements B, Al, C and Si*

(*i*) *Which has the highest first ionization energy?*

(*ii*) *Which has the most negative electron gain energy?*

(*iii*) *Which has the largest atomic radius?*

(*iv*) *Which has the most metallic character?*

Solution. (*i*) Carbon (C) has the highest first ionization energy.

(*ii*) Carbon (C) has the most negative electron gain energy.

(*iii*) Aluminium (Al) has the largest atomic radius.

(*iv*) Aluminium (Al) has the most metallic character.

6.15 VALENCY

The valency of an element may be defined as the combining capacity of the element.

The valency of an element may be expressed in terms of the number of electrons that an atom of the element makes available for bonding. The valency of an element is determined by the number of electrons in the outermost shell. These electrons are known as valence electrons.

In case of *representative elements*, the valency is generally equal to either the number of valence electrons or eight minus the number of valence electrons. However, the transition elements exhibit variable valency.

VARIATION OF VALENCY IN THE PERIODIC TABLE

Variation in a Period

The number of valence electrons increases from 1 to 8 on moving across a period. The valency of the elements with respect to hydrogen and chlorine increases from 1 to 4 and then decreases from 4 to zero. Table 6.14 shows the valencies of the elements of the 2nd and 3rd periods. The valencies of the elements have been shown in brackets.

Variation in a Group

On moving down a group, the number of valence electrons remains the same and, therefore, all the elements in a group exhibit the same valency. For example, all the elements of group **1** have a valency equal to **1** and those of group **2** have a valency equal to **2.**

SOLVED EXAMPLE

Example 6.15. *Predict the formulae of the stable binary compounds that would be formed by the following pairs of elements:*

(*i*) *silicon and oxygen* (*ii*) *aluminium and bromine*

(*iii*) *calcium and iodine.*

Solution. (*i*) Silicon belongs to group 14. Its valency is 4. The valency of oxygen is 2. Thus, the binary compound between silicon and oxygen would have the formula $\mathbf{SiO_2}$.

(*ii*) Aluminium belongs to group 13. Its valency is 3. The valency of bromine is 1. Thus, the binary compound between aluminium and bromine would have the formula $\mathbf{AlBr_3}$.

(*iii*) Calcium belongs to group 2. Its valency is 2. The valency of iodine is 1. Thus, the compound between calcium and iodine would have the formula $\mathbf{CaI_2}$.

6.16 PERIODIC TRENDS AND CHEMICAL REACTIVITY

We have already studied the periodic trends in the various fundamental properties such as atomic and ionic radii, ionization energy, electron affinity, electronegativity and

Table 6.14. Variation of Valencies and Type of Bonding Among Elements of Second and Third Periods

Elements of second period	Li 2, 1	Be 2, 2	B 2, 3	C 2, 4	N 2, 5	O 2, 6	F 2, 7
Valence	*1*	*2*	*3*	*4*	*3*	*2*	*1*
Examples	LiCl	$BeCl_2$	BCl_3	CCl_4	NH_3	H_2O	HF
Elements of third period	Na 2, 8, 1	Mg 2, 8, 2	Al 2, 8, 3	Si 2, 8, 4	P 2, 8, 5	S 2, 8, 6	Cl 2, 8, 7
Valence	*1*	*2*	*3*	*4*	*3, 5*	*2, 6*	*1, 7*
Examples	NaCl	$MgCl_2$	$AlCl_3$	SiH_4	PCl_3 PCl_5	H_2S SO_3	HCl Cl_2O_7
Nature of bonding	Ionic	Ionic	Covalent	Covalent	Covalent	Covalent	Covalent

valence. The periodic trends in these properties can be explained on the basis of the electronic configuration of the elements. The chemical reactivity of elements can be related to the fundamental properties of elements.

As already discussed the ionization energy is least for the element at the extreme left of the period and the electron affinity is most negative for the element at the extreme right of the period (For group-17 elements). The elements of group-18 have positive electron affinity due to their stable electronic configurations. As a result, **the chemical reactivity is maximum at the two extremes and lowest in the centre**. The extreme reactivity of group-1 elements is due to the ease with which these elements can lose an electron leading to the formation of the corresponding cation. On the other hand, the reactivity of halogens is due to the ease with which these elements can gain an electron to form the corresponding anion. Thus, elements at the extreme left exhibit strong reducing behaviour whereas the elements at the extreme right exhibit strong oxidizing behaviour.

Since the ionization energy of alkali metals decreases on moving down the group, their reactivity increases from lithium to caesium. For example, lithium reacts with water slowly, sodium reacts vigorously whereas potassium reacts almost violently. The reaction of caesium with water is explosive

$$2M + 2H_2O \longrightarrow 2MOH + H_2$$

(M = Li, Na or K)

On the other hand among halogens, the reactivity decreases on going down the group because the tendency to gain electron becomes less on descending the group.

Noble gases are almost inert due to their stable electronic configuration.

SUMMARY

- **Periodic Table.** Arrangement of elements in increasing order of atomic number such that elements with similar properties fall under the same vertical column.
- **Group.** A vertical column of elements in the periodic table.
- **Period.** A horizontal row of elements in the periodic table.
- **Long Form of Periodic Table** has 18 groups and 7 periods. It is based on the modern periodic law.
- ***s*-Block Elements.** Elements of groups 1 and 2. Their general electronic configuration is $\mathbf{ns^{1-2}}$.
- ***p*-Block Elements.** Elements of groups 13, 14, 15, 16, 17 and 18. Their general electronic configuration is $\mathbf{ns^2np^{1-6}}$.
- ***d*-Block Elements.** Elements of groups 3, 4, 5, 6, 7, 8, 9, 10, 11 and 12. Their general electronic configuration is $\mathbf{(n-1)\,d^{1-10}\,ns^{1-2}}$. $_{46}$Pd is an exception ($4d^{10}\,5s^0$).
- ***f*-Block Elements.** The two horizontal rows of elements at the bottom of the table. Also known as inner transition elements. Their general electronic configuration is $\mathbf{(n-2)\,f^{1-14}\,(n-1)\,d^{0-1}\,ns^2}$.
- **Covalent Radius.** Half of the internuclear distance between two atoms of an element held by a single covalent bond.
- **van der Waals' Radius.** Half of the internuclear distance between two nearest atoms belonging to two adjacent molecules in the solid state.
- **Metallic Radius.** Half of the internuclear distance between two nearest atoms in the metallic lattice.
- van der Waals' radius > Metallic radius > Covalent radius.
- Size of anion > Size of atom > Size of cation.
- **Isoelectronic Ions.** The ions having the same number of electrons but different nuclear charge. Examples,

  (*i*) N^{3-}, O^{2-}, F^-, Na^+, Mg^{2+}, Al^{3+}

  (*ii*) P^{3-}, S^{2-}, Cl^-, K^+, Ca^{2+}, Sc^{3+}

  Among isoelectronic ions, the greater the nuclear charge the smaller the size.
- **Ionization Energy.** The energy required to remove the outermost electron from an isolated gaseous atom of an element.
- Ionization energy increases along the period and decreases down the group.
- $\mathbf{IE_3 > IE_2 > IE_1}$
- **Electron Affinity.** The energy change taking place when an electron is added to an isolated gaseous atom of the element.
- Electron affinity becomes more negative as we move along the period and becomes less negative down the group.
- Chlorine has the highest negative electron affinity.
- **Electronegativity.** It is the tendency of an atom in a molecule to attract towards itself the shared pair of electrons.
- Fluorine is the most electronegative element whereas caesium is the least electronegative element.
- **Electropositive or metallic character** is related to the ionization energy of an element. The elements which have less value of ionization energy are more electropositive or more metallic in character.
- **Valence** of an element belonging to *s*- or *p*-block is either equal to the number of valence electrons or eight minus the number of valence electrons.
- The chemical reactivity is maximum at the two extreme ends of the periodic table and is least in the centre.
- Among alkali metals reactivity increases on descending the group while among halogens the reactivity decreases on descending the group.

EVALUATE YOURSELF

I. Objective Type Questions

Select the most appropriate choice from the options given as (a), (b), (c) and (d) after each question.

1. In the long form of the periodic table the elements are arranged according to their

(*a*) Mass number (*b*) Atomic number
(*c*) Neutron number (*d*) Atomic mass.

2. Which of the following sets of elements have a similar electronic configuration?

(*i*) H, He, Be (*ii*) Li, Be, B
(*iii*) He, Ne, Ar (*iv*) Li, Na, K
(*a*) (*iii*) (*b*) (*iii*), (*iv*)
(*c*) (*iv*) (*d*) (*ii*), (*iii*)

3. Transition metals are:

(*a*) found between groups 2 and 13
(*b*) all solids
(*c*) metals and semi-metals
(*d*) found between groups 1 and 13.

4. Hydrogen can be placed along with group 17 elements because it

(*i*) can aquire noble gas configuration by gaining one electron.
(*ii*) exhibits oxidation state of +1.
(*iii*) has one electron in the outermost shell.
(*iv*) can exhibit oxidation state of –1.
(*a*) (*ii*) and (*iii*) (*b*) (*i*) and (*iv*)
(*c*) (*i*) (*d*) (*iii*).

5. The number of elements in the 5th period of the periodic table is

(*a*) 3 (*b*) 9
(*c*) 8 (*d*) 18.

6. In the modern periodic table, the period indicates the value of:

(*a*) atomic number (*b*) atomic mass
(*c*) principal quantum number
(*d*) azimuthal quantum number.

7. The correct order of size among Cl, Cl^+ and Cl^- is

(*a*) $Cl^+ < Cl^- < Cl$ (*b*) $Cl^+ > Cl^- > Cl$
(*c*) $Cl^+ < Cl < Cl^-$ (*d*) $Cl^- < Cl < Cl^+$.

8. Considering the elements B, Al, Mg, and K, the correct order of their metallic character is

(*a*) B > Al > Mg > K (*b*) Al > Mg > B > K
(*c*) Mg > Al > K > B (*d*) K > Mg > Al > B.

9. Which of the following groups consists entirely of metals?

(*a*) 18 (*b*) 2
(*c*) 14 (*d*) 15.

10. Which of the following represents the correct order of ionic radii?

(*a*) $S^{2-} > Cl^- > K^+ > Ca^{2+}$ (*b*) $Ca^{2+} > K^+ > Cl^- > S^{2-}$
(*c*) $Cl^- > S^{2-} > K^+ > Ca^{2+}$ (*d*) $S^{2-} > Ca^{2+} > K^+ > Cl^-$.

11. The outermost electronic configuration of the most electronegative element is

(*a*) $ns^2\ np^3$ (*b*) $ns^2\ np^4$
(*c*) $ns^2\ np^5$ (*d*) $ns^2\ np^6$.

12. The first ionization energies of Na, Mg, Al and Si are in the order

(*a*) Na < Mg > Al < Si (*b*) Na > Mg > Al > Si
(*c*) Na < Mg < Al > Si (*d*) Na > Mg > Al < Si.

13. Keeping in view, the periodic law and periodic table, suggest which of the following elements should have maximum electronegative character?

(*a*) Oxygen (*b*) Nitrogen
(*c*) Fluorine (*d*) Astatine.

14. An element with atomic number 20 is placed in which period of the periodic table?

(*a*) 4 (*b*) 3
(*c*) 2 (*d*) 1.

15. Which of the following structures is associated with the biggest jump between the second and the third ionization energies?

(*a*) $1s^22s^22p^6$ (*b*) $1s^22s^22p^63s^1$
(*c*) $1s^22s^22p^63s^2$ (*d*) $1s^22s^22p^1$.

16. The statement that is ***not correct*** for periodic classification of elements is

(*a*) the properties of elements are a periodic function of their atomic numbers.
(*b*) non-metallic elements are less in number than metallic elements.
(*c*) the first ionization energies of elements along a period do not vary in a regular manner with an increase in atomic number.
(*d*) for transition elements, the *d*-subshells are filled with electrons monotonically with increase in atomic number.

17. Which one of the following statements is ***incorrect*** in relation to ionization energy?

(*a*) Ionization energy increases for each successive electron.
(*b*) The greatest increase in ionization energy is experienced on removal of an electron from a core noble gas configuration.
(*c*) End of valence electrons is marked by a big jump in ionization energy.
(*d*) Removal of electron from orbitals bearing lower *n* value is easier than from orbitals having higher *n* value.

18. The **incorrect** statement among the following is
(*a*) The first ionization energy of Al is less than the first ionization energy of Mg
(*b*) The second ionization energy of Mg is greater than the second ionization energy of Na
(*c*) The first ionization energy of Na is less than the first ionization energy of Mg
(*d*) The third ionization energy of Mg is greater than the third ionization energy of Al.

19. Which of the following ions has the highest value of ionic radius?
(*a*) Li^+ (*b*) B^{3+}
(*c*) O^{2-} (*d*) F^-

20. Which one of the following sets of ions represents a collection of isoelectronic species?
(*a*) K^+, Ca^{2+}, Sc^{3+}, Cl^- (*b*) Na^+, Ca^{2+}, Sc^{3+}, F^-
(*c*) K^+, Cl^-, Mg^{2+}, Sc^{3+} (*d*) Na^+, Mg^{2+}, Al^{3+}, Cl^-
(Atomic numbers: F= 9, Cl = 17, Na = 11, Mg = 12, Al = 13, K = 19, Ca = 20, Sc = 21).

II. Fill in the Blanks

21. Complete the following sentences by supplying appropriate words.
(*i*) The most electropositive element in the second period is ...... whereas the most electronegative element is ....
(*ii*) The ionization energy of nitrogen is ...... than that of oxygen.
(*iii*) There are ...... periods in the long form of the periodic table and ...... groups.
(*iv*) The ions Ca^{2+} and S^{2-} are isoelectronic with ...... noble gas.
(*v*) The *d*-block elements are known as ...... metals.
(*vi*) Ca^{2+} has a smaller ionic radius than K^+ because it has ....
(*vii*) Among halogens, the most negative electron affinity is of ....
(*viii*) The element with the highest value of first ionization energy is ....
(*ix*) Metallic radius is ...... than the covalent radius.
(*x*) Modern periodic law was proposed by ......
(*xi*) Metallic character ...... from top to bottom in a group.
(*xii*) Amongst the elements Na, Mg and Al the highest value of the second ionization energy is of ....

III. Discussion Questions

22. What is the modern periodic law? What is periodicity and what is its cause?

23. What is the total number of groups in the long form of the periodic table?

24. An element has atomic number 34. Deduce its period, group and block.

25. Write the general outer electronic configuration of *s*-, *p*-, *d*- and *f*-block elements.

26. Which groups constitute *d*-block of the periodic table?

27. An element has a valence shell electronic configuration of ns^2, np^3 . To which group does this element belong?

28. In what group of the periodic table is each of the following elements found?
(*i*) $1s^2, 2s^2, 2p^6, 3s^2, 3d^1, 4s^2$ (*ii*) [Xe] $4f^{14}\ 5d^4\ 6s^2$
(*iii*) [Ar] $3d^{10}\ 4s^2$ (*iv*) [Ar] $3d^{10}, 4s^2, 4p^4$.

29. Out of the metallic radius and the covalent radius of an element, which is larger and why?

30. Why is the van der Waals' radius of an element always larger than the covalent radius?

31. Out of the following ions which has the smallest ionic size?
Li^+, Na^+, K^+

32. Arrange the following ions in increasing order of their sizes.
Cl^-, P^{3-}, S^{2-}, F^-

33. Arrange the following sets of ions in decreasing order of their sizes
(*i*) Al^{3-}, Mg^{2+}, Na^+, O^{2-}, F^-
(*ii*) Na^+, Mg^{2+}, K^+.

34. Out of I and I^+ which has a larger size and why?

35. Which element has the highest ionization energy?

36. Among the alkali metals which element has the highest ionization energy?

37. Arrange each of the following sets of elements in increasing order of their ionization energies:
(*i*) O, N, S (*ii*) C, N, O
(*iii*) Li, Be, Na (*iv*) Ne, He, Ar.

38. Out of Na (Z = 11) and Mg (Z = 12), which has a higher second ionization energy and why?

39. Explain why
(*i*) Be has a higher ionization energy than B
(*ii*) O has a lower ionization energy than N and F?

40. Why halogens have highest negative electron affinities in their respective periods?

41. Why noble gases have largest positive electron affinity in their respective periods?

42. Comment on the statement that "all elements having high ionization energies also have high negative electron affinities."

43. Out of oxygen and sulphur, which has a greater negative electron affinity and why?

44. Consider the electronic configurations of the elements X, Y and Z and answer.

$$X : 1s^2, 2s^2, 2p_x^{\ 1}, 2p_y^{\ 1}$$
$$Y : 1s^2, 2s^2, 2p_x^{\ 2}, 2p_y^{\ 1}, 2p_z^{\ 1}$$
$$Z : 1s^2, 2s^2, 2p_x^{\ 1}, 2p_y^{\ 1}, 2p_z^{\ 1}$$

(*i*) Which element has highest negative electron affinity?
(*ii*) Which element has lowest negative electron affinity?

45. In each of the following sets, arrange the elements in increasing order of their negative electron affinities:

(*i*) C, N, O (*ii*) O, N, S
(*iii*) S Cl, Ar (*iv*) F, Cl, Br.

46. What is the trend of metallic character on going down from top to bottom in a group?

47. Define electronegativity. How does it vary along a period and along a group?

48. What is the basic difference between the terms electron affinity and electronegativity?

49. Why is the second ionization energy always higher than the first ionization energy?

50. A, B, C are three elements with atomic numbers, Z – 1, Z, Z + 1 respectively. B is an inert gas.

Answer the following questions:

(*i*) Predict the group of A and C.

(*ii*) Which out of the three has a positive electron gain energy and why?

(*iii*) Which of the three has least value of ionization energy?

51. The electronic configurations of some elements are given as:

(*a*) [Ne] $3s^2\ 3p^3$ (*b*) [Ne] $3s^2\ 3p^4$
(*c*) [Ne] $3s^2\ 3p^5$ (*d*) [Ne] $3s^2\ 3p^6 3d^5\ 4s^1$

(*i*) Which element will be most metallic ?

(*ii*) Which element will have the most negative electron affinity?

(*iii*) Which element belongs to the *d*-block?

(*iv*) Which element belongs to group 17?

(*v*) Out of *a*, *b* and *c* which will have the least ionization energy?

52. Account for the following:

(*i*) Be has a slightly higher value of ionization energy than that of boron (B).

(*ii*) The ionization energy of Na^+ is more than that of Ne although they have the same configuration.

(*iii*) Electron affinity of Cl is more negative than that of F.

53. Among the elements of the second period Li to Ne pick out the element

(*i*) with the highest ionization energy

(*ii*) with the highest negative electron affinity

(*iii*) with the largest covalent radius

(*iv*) that is the most reactive non-metal

(*v*) that is the most reactive metal.

54. The first (IE_1) and the second (IE_2) ionization energies (in kJ mol^{-1}) and the electron affinity (in kJ mol^{-1}) of a few elements are given as follows:

Elements	IE_1	IE_2	E_{ea}
I	520	7300	– 60
II	419	3051	– 48
III	1681	3374	– 328
IV	1008	1846	– 295
V	2372	5251	+48
VI	738	1451	– 40

Which of the above element is likely to be:

(*i*) the least reactive element.

(*ii*) the most reactive metal.

(*iii*) the most reactive non-metal.

(*iv*) the least reactive non-metal.

(*v*) the metal which can form a stable binary halide of the formulae MX_2(X = halogen).

(*vi*) the metal which can form a stable covalent halide of the formula MX (X = halogen).

55. Among the elements of the third period pick out the element

(*i*) with the highest negative electron affinity

(*ii*) with the highest positive electron affinity

(*iii*) with the highest ionization energy

(*iv*) that is the most reactive metal

(*v*) that is the most reactive non-metal.

56. How does metallic nature vary among the elements of the third period with an increase in atomic number.

Answers

I. Objective Type Questions

1. (*b*)	**2.** (*c*)	**3.** (*a*)
4. (*b*)	**5.** (*d*)	**6.** (*c*)
7. (*c*)	**8.** (*d*)	**9.** (*b*)
10. (*c*)	**11.** (*c*)	**12.** (*a*)
13. (*c*)	**14.** (*a*)	**15.** (*c*)
16. (*d*)	**17.** (*d*)	**18.** (*b*)
19. (*c*)	**20.** (*a*).	

II. Fill in the Blanks

21. (*i*) lithium (*ii*) more
(*iii*) 7, 18 (*iv*) *argon*
(*v*) transition (*vi*) higher nuclear charge
(*vii*) chlorine (*viii*) helium
(*ix*) larger (*x*) moseley
(*xi*) increases (*xii*) Na.

SHS 2

SECTION 3

CHEMICAL BONDS

General Objectives

After studying this section, the student should be able to:

1. understand that atoms combine to form molecules by sharing valence electrons via covalent or metallic bonds or by exchanging electrons to form ionic bond.
2. understand how electronegativity and ionisation energy relate to chemical bond formation.
3. predict chemical formulas based on the number of valence electrons.
4. recognize the role of modelling, evidence and theory in explaining and understanding the structure, chemical bonding and properties of ionic and molecular compounds.

In this section, we shall study

Unit 7: Interatomic Bonding

Unit 8: Intermolecular Bonding

Unit 9: Hybridization and Shapes of Molecules

SECTION-3

UNIT-7

INTERATOMIC BONDING

Specific Objectives

After studying this unit, the student should be able to:

- explain the meaning of a chemical bond
- describe how ionic bond is formed between two chemical species
- draw Lewis dot structures for simple ionic compounds
- describe how covalent characters is introduced into ionic bond
- name some binary and ternary ionic compounds from their formulae and write formulae from their names
- describe the properties of some ionic compounds
- describe how a covalent bond is formed
- draw Lewis dot structures for some covalent compounds
- describe the properties of covalent compounds
- describe how metallic bond is formed
- state and describe the properties of metals

7.1 CHEMICAL BOND

It is a well known fact that matter is made up of atoms but the atoms (other than noble gases) do not exist as independent entities under normal conditions. However, the smallest particles of matter which exist as independent entities are molecules. The molecules are nothing but a group or cluster of atoms of the same or different elements which behave as single units. For example, in hydrogen molecule there are two H atoms whereas in oxygen molecule two O atoms are held together. Water molecule contains two H atom and one O atom mutually held together by some attractive forces. In addition to this, there is another category of substances in which the constituent units are ions of opposite charges instead of atoms or molecules. Sodium chloride is a common example of this type which contains sodium (Na^+) ions and chloride ions (Cl^-) held together. *The attractive force which holds various constituents (atoms, ions, etc.) together in different chemical species is called chemical bond.* The chemical bonds can be classified into various types like *ionic bond, covalent bond, co-ordinate bond, metallic bond, hydrogen bond* and *van der Waal forces.* Since the formation of chemical compounds takes place as result of the combination of atoms of various elements in different ways, it raises many questions such as:

- *Why do atoms combine?*
- *Why a definite number of atoms are used to constitute a particular molecule, for example, hydrogen molecule is H_2 and not H_3 or H_4?*
- *Why do molecules have definite shapes?*
- *What is the nature of the force existing between the combining atoms?*

In this unit, we shall try to answer the above questions by studying the mode of formation of interatomic bonds. Various theories of valency and interpretation of the nature of chemical bonds have been related to the developments in the field of the structure of the atom, the electronic configuration of elements and the periodic table.

OCTET RULE

In the early days, the ability of the various elements to combine with one another was expressed in terms of their valency. The concept of valency was not based on any logical understanding. The developments in the field of atomic structure provided the foundations for the various theories of valency. The initial contributions in this field came from **W. Kossel** and **G. N. Lewis** in 1916. They were the first to provide some logical explanation of valency which was based on the inertness of noble gases. This view, later on, came to be known as the **octet rule.**

The rule was based on the electronic configuration of noble gases. All the noble gases (except helium), have 8 electrons in their outermost shell. Helium has 2 electrons in the outer shell. The noble gases are quite stable because they do not have any tendency to take part in chemical combination. The chemical inertness of noble gases was related to the presence of an octet of electrons in their outermost shell because other elements which were chemically reactive had less than 8 electrons in their outermost shell. These observations led Kossel and Lewis to put forward a generalisation known as the **octet rule.** The rule states that *atoms of various elements enter into chemical combination so as to attain the stable configuration of eight or two electrons in their outermost shell*. They do so either by *transference of electrons* or by *mutual sharing of electrons*. The octet rule is quite useful in explaining the normal valencies of a large number of elements. In the light of the octet rule let us now study the formation of various bonds.

LEWIS SYMBOLS

It was proposed that the electrons involved in the process of combination are outer shell electrons, therefore, these are called valence electrons. G.N. Lewis introduced the simple notations to represent the valence electrons in an atom. These notations are called **Lewis symbols** or **electron dot symbols.** According to these notations the symbol of the element represents the nucleus as well as the electrons in the inner shells. The electrons in the outer shell are represented by the dots surrounding the symbol. For example, chlorine atom has 7 electrons in the outermost shell. Its electronic configuration is 2, 8, 7. But chloride ion (Cl^-) ion has a configuration of 2, 8, 8. The Lewis symbols for chlorine atom and chloride ion are shown as follows:

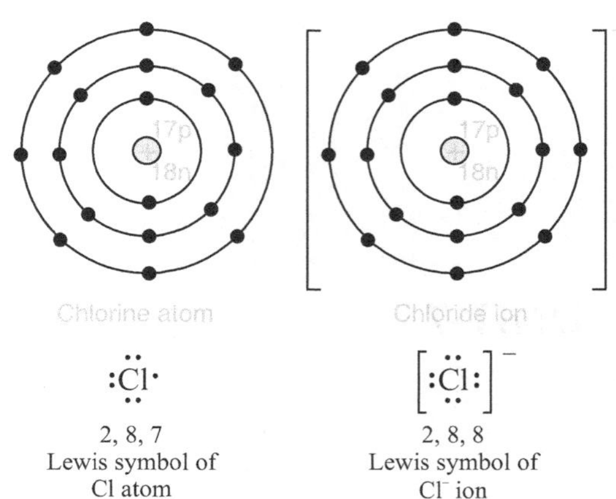

The Lewis symbols for the elements of the 2nd period are:

Li· ·Be· ·B· ·C· ·N· ·O· :F· :Ne:

Significance of Lewis Symbols

The number of dots in the Lewis symbol represents the number of valence electrons. The common valencies of the elements can be calculated from the valence electrons. The common valency of the element is either *equal to the number of dots or 8 minus the number of dots.* For example, the common valencies of Li, Be, B and C are 1, 2, 3 and 4 respectively while those of N, O, F and Ne are 8 minus the number of dots, *i.e.*, 3, 2, 1 and 0 respectively.

7.2 FORMATION OF IONIC BONDS AND IONIC COMPOUNDS

IONIC BOND OR ELECTROVALENT BOND

Ionic bond is formed by the complete transfer of one or more electrons from one atom to the other. This type of bond usually comes into existence between an electropositive (generally metal) and electronegative (generally non-metal) atom. The metal atom loses its valence electrons and changes into a **positive ion** (*cation*) while a non-metallic atom gains electrons and gets converted into a **negative ion** (*anion*). They do so in order to become stable by attaining 8 (or 2) electrons in the outermost shell. The oppositely charged ions are held together by electrostatic force of attraction. Thus, an ionic bond may be defined as *the electrostatic force of attraction which holds the positively charged (cation) and negatively charged (anion) species together.*

Let us study the formation of sodium chloride to understand the ionic bond.

Formation of Sodium Chloride (NaCl)

Sodium atom, Na (2, 8, 1) has only 1 electron in its valence shell. In order to acquire the configuration of the nearest noble gas Ne (2, 8) it has to lose its valence electron. Chlorine atom, Cl (2, 8, 7), on the other hand, has seven electrons in its valence shell. It can acquire the configuration of the nearest noble gas, Ar (2, 8, 8) by gaining one electron. Thus, sodium atom transfers its valence electron to chlorine atom resulting in the formation of sodium ion (Na^+) and chloride ion (Cl^-) respectively. The electrostatic force of attraction holds these oppositely charged ions together. Let us represent the above changes by using Lewis symbols.

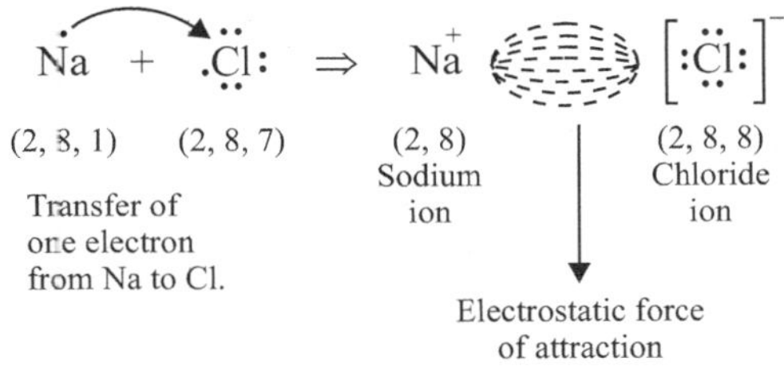

Fig. 7.1. Ionic bond formation for sodium chloride.

Formation of Calcium Oxide (CaO)

Electronic configuration of the calcium atom is 2, 8, 8, 2 whereas that of the oxygen atom is 2, 6. Calcium tends to acquire the configuration of Ar (2, 8, 8) by losing 2 valence electrons and forming a divalent cation Ca^{2+}. Oxygen atom gains 2 electrons to form oxide ion (O^{2-}) and thus, attains the configuration of Ne (2, 8, 8). The Ca^{2+} ion and O^{2-} ions are held by the electrovalent bond as shown in Fig. 7.2.

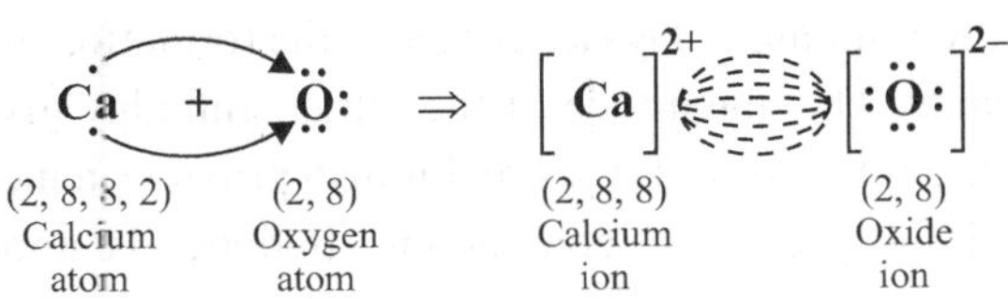

Fig. 7.2. Formation of calcium oxide (CaO).

Formation of Magnesium Chloride ($MgCl_2$)

In the formation of magnesium chloride, Mg atom (2, 8, 2) loses its two valence electrons to form divalent Mg^{2+} ion. These two electrons are gained by two chlorine atoms (one each) to form chloride ions. The oppositely charged ions are held by ionic bonds as shown in Fig. 7.3.

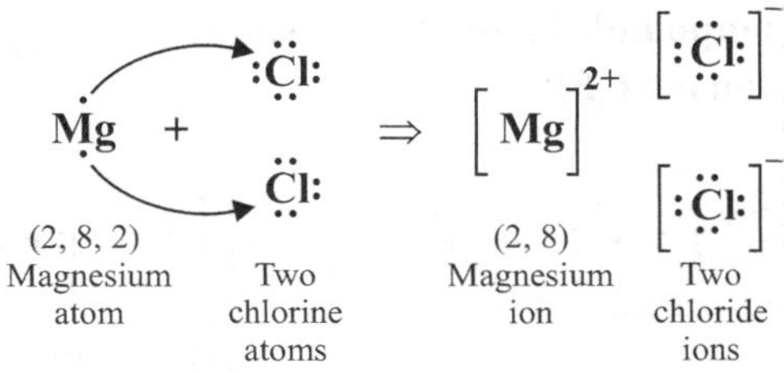

Fig. 7.3. Formation of magnesium chloride ($MgCl_2$).

It may be noted that in the ionic bond model, the ions are considered to be spherically charged particles which have a uniform field of influence around them. Hence, the ionic bond is ***non-directional*** in nature.

Electrovalency

It is the *number of electrons which an atom loses or gains while forming an ionic bond.* The atom which loses electrons is called **electropositive atom** while the atom which gains electrons is called **electronegative atom.** The positive (+) sign is put before the valency of an electropositive atom while the negative (–) sign is put before the valency of an electronegative atom. For example, electrovalency of Na is +1; Ca is +2 and that of Cl is –1. In short, **electrovalency is equal to the unit charges on the ion.**

7.3 LEWIS DOT STRUCTURES FOR SOME SIMPLE IONIC COMPOUNDS

Let us write the Lewis dot structures for some binary compounds. The compounds formed by two elements are called binary compounds.

(*i*) **Compound formed by Potassium ($_{19}K$) and Oxygen ($_8O$)**

$$\dot{K} + \dot{K} + :\ddot{O}: \Rightarrow [K]^+ [:\ddot{O}:]^{2-} [K]^+ \text{ or } K_2O$$

(2, 8, 8, 1) (2, 6) (2, 8, 8) (2, 8) (2, 8, 8)

(*ii*) **Compound formed by Calcium ($_{20}Ca$) and Fluorine ($_9F$)**

$$\dot{C}a + .\ddot{F}: + .\ddot{F}: \Rightarrow [:\ddot{F}:]^- [Ca]^+ [:\ddot{F}:]^- \text{ or } CaF_2$$

(2, 8, 8, 2) (2, 7) (2, 8) (2, 8, 8) (2, 8)

(*iii*) **Compound formed by Aluminium ($_{13}$Al) and Fluorine ($_9$F)**

$$\dot{Al}\cdot + \cdot\ddot{F}: + \ddot{F}: + \cdot\ddot{F}: \Rightarrow [:\ddot{F}:]^- [Al]^{3+} [:\ddot{F}:]^- [:\ddot{F}:]^- \text{ or } AlF_3$$

(2, 8, 3)

7.4 FACTORS WHICH INFLUENCE THE FORMATION OF IONIC BOND

In order to understand the factors which influence the formation of an ionic compound let us examine the *energy changes taking place during the formation of an ionic bond.*

The formation of an ionic bond from the gaseous atoms can be considered to proceed in three steps:

For example, let us consider the formation of the ionic compound **AX**(*s*) from atoms **A** and **X** in the gaseous state.

(*a*) *Formation of gaseous cation.*

This step involves loss of electron by the gaseous atom and is an endothermic process. The energy required for this process is expressed in terms of the ***ionisation energy***(***E_i***) which is defined as *the energy required to remove the most loosely bound electron from the gaseous atom to make it a gaseous cation.*

$$A(g) \longrightarrow A^+(g) + e^-; \quad E_i = +x \text{ kJ mol}^{-1}$$

Now the *lesser the value of ionisation energy, the greater the tendency of the atom to form a cation.* For example, alkali metals form cations more easily as compared to alkaline earth metals because of the low values of their ionisation energies.

(*b*) *Formation of gaseous anion.*

This step involves gain of electron by the gaseous atom and is an exothermic process. The energy released in this process is expressed in terms of ***electron affinity E_{ea}*** which is defined as *the energy released when a gaseous atom takes up an electron to form a monovalent gaseous anion.*

$$X(g) + e^- \longrightarrow X^-(g); \quad E_{ea} = -y \text{ kJ mol}^{-1}$$

–ve sign represents energy released.

It may be noted that this process becomes endothermic in case of the formation of a divalent or a trivalent anion because the second and successive electron affinities are positive. For example, the formation of $O^{2-}(g)$ ion from O(g) atom is endothermic.

$$O(g) + e^- \longrightarrow O^-(g); \quad E_{ea1} = -141 \text{ kJ}$$
$$O^-(g) + e^- \longrightarrow O^{2-}(g); \quad E_{ea2} = +780 \text{ kJ}$$
$$O(g) + 2e^- \longrightarrow O^{2-}(g); \quad E_{ea(net)} = +639 \text{ kJ mol}^{-1}$$

Now the *larger the value of the electron affinity the greater the tendency of an atom to form an anion.* For example, halogens have the largest tendency to form anions due to the large values of their electron affinity.

(*c*) *Packing of ions of opposite charges to form ionic solid.*

This step involves close packing of gaseous ions of opposite charges to form a crystal lattice. It is an **exothermic process** and the energy released during this process is expressed in terms of **lattice enthalpy (Δ_LH). Δ_LH** is defined as *the energy released when gaseous ions of opposite charges pack up together to form 1 mole of ionic crystal.*

$$A^+(g) + X^-(g) \longrightarrow AX(s); \quad \Delta_L H^\ominus = -z \text{ kJ}$$

The close packing of the gaseous ions of opposite charges is shown in Fig. 7.4.

Fig. 7.4. Close packing of gaseous ions of opposite charges to form solid ionic crystal.

Lattice enthalpy plays a key role in the formation of ionic compounds. The magnitude of the lattice enthalpy gives an idea of the inter-ionic forces and it also provides a measure of the stability of ionic compounds. It depends upon the following factors:

(*i*) ***Size of the ions.*** The smaller the size of the ions, the lesser the inter-nuclear distance and the greater the inter-ionic attraction. This will give a more negative value of lattice enthalpy. For example, the ionic size of K^+ (133 pm) is larger than that of Na^+ (95 pm). Consequently, the lattice enthalpy of NaCl (– 769.8 kJ mol^{-1}) is more negative than that of KCl (– 698.7 kJ mol^{-1}).

(*ii*) ***Charge on the ions.*** The larger the magnitude of the charge on the ions, the greater the attractive force and consequently, the higher the negative value of lattice enthalpy.

The lattice enthalpies of halides of alkali metals are given in Table 7.1.

Table 7.1. Lattice Enthalpies ($\Delta_L H$) of Halides of Alkali Metals (kJ mol^{-1})

M^+ / X^-	Li^+	Na^+	K^+	Rb^+	Cs^+
F^-	– 994.5	– 894.5	– 790.0	– 756.7	– 720
Cl^-	– 802.6	– 769.8	– 698.7	– 660.6	– 618.7
Br^-	– 760.8	– 714.8	– 656.2	– 631.3	– 593.6
I^-	– 710	– 669.8	– 618.7	– 597.1	– 564.5

*Now for a **stable ionic bonding,** the net effect of the above three steps should be **exothermic** i.e.,*

total energy released > total energy required.

From the above discussion we can develop the **conditions for a stable ionic bonding as follows:**

(*a*) *ionization enthalpy of cation forming element should be low;*

(*b*) *electron affinity of anion-forming element should be high , i.e., have high negative value;*

(*c*) *lattice enthalpy should also have a high negative value.*

7.5 GENERAL PROPERTIES OF IONIC COMPOUNDS

The common properties of ionic compounds have been described as follows:

1. Physical State. Ionic compounds usually exist in the form of crystalline solids. It has been shown by X-ray analysis of ionic compounds that they do not exist as independent molecules but exist in the form of ions. These ions are arranged in a well defined geometric pattern to constitute ionic crystal. Ionic crystal is a giant network of ions in three dimensions. For example, in sodium chloride crystal, there is no single unit of the Na^+ Cl^- ion pair but a large number of them arranged in a definite pattern in three dimensions. Each sodium ion (Na^+) in the crystal is surrounded by six chloride (Cl^-) ions and each chloride (Cl^-) ion is surrounded by six sodium ions. The simplest unit within the crystal which repeats itself in three dimensions to form the entire space lattice is called **unit lattice cell.** The unit lattice cell of sodium chloride is shown in Fig. 7.5. It may be noted that the pattern of arrangement of ions in the ionic crystal depends on the size of the ions and also on the magnitude of the charge on the ions.

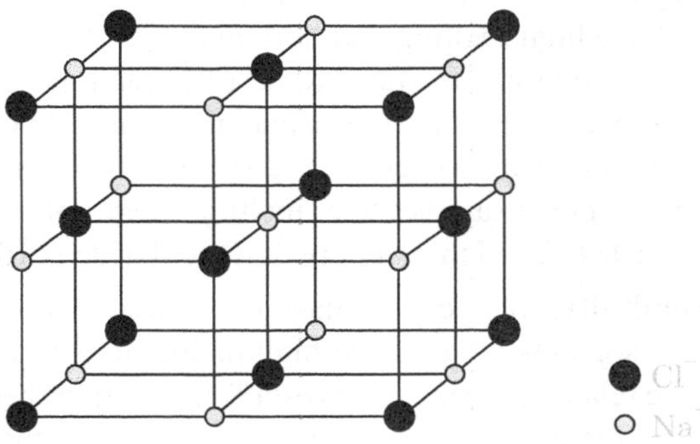

Fig. 7.5. Unit lattice cell of sodium chloride crystal.

ACTIVITY 7.1	**To build a model of a unit lattice cell of sodium chloride crystal**

1. Sodium chloride is an ionic solid. It has a giant network structure of oppositely charged ions.
   - Write the Lewis symbols of the cation and anion.
   - Which ion has a larger size? Why?
2. Take 14 coloured polystyrene balls of large size and 13 white polystyrene balls of small size. The size of the coloured ball is approximately double (1.9 times actually) the white ball.

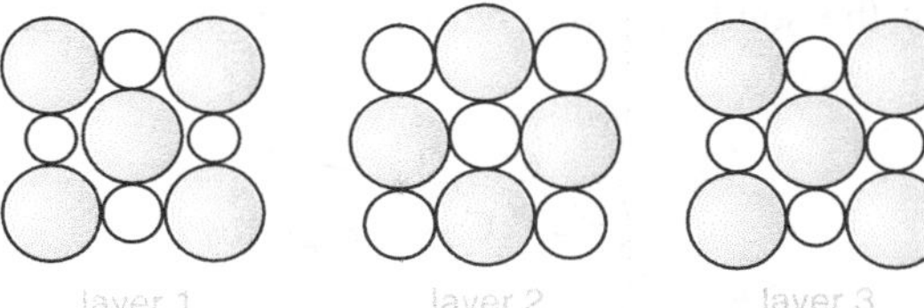

(Same as layer 1)

3. Arrange 5 coloured balls and 4 white balls with the help of tooth picks or glue as shown in layer 1. Make another similar arrangement to constitute layer 3.
4. Now make layer 2 by arranging five white balls and four coloured balls as shown.
5. Now place layer 2 just above layer 1, and layer 3 above layer 2. This will result in the formation of a unit lattice cell of sodium chloride.

Answer the following from your observations:

(*i*) Which ball represents Na^+ ion and which represents Cl^- ion?

(*ii*) How many Cl^- ions surround one Na^+ ion?

(*iii*) How many Na^+ ions touch one Cl^- ion?

(*iv*) What is the co-ordination number of Cl^- and Na^+ ions?

(*v*) How many Cl^- ions touch one Cl^- ion?

(*vi*) What is the shape of the unit lattice cell?

2. High Melting and Boiling Points. Ionic compounds generally have high melting and boiling points. It is because of the very strong electrostatic force of attraction between the oppositely charged ions. Consequently, a large amount of energy is required to overcome the strong attractive inter-ionic forces. For example, the melting point of sodium chloride is 1074 K and that of copper (II) chloride is 1085 K.

3. Solubility. Ionic compounds are soluble in water and other polar solvents which have high dielectric constants. It is due to the relatively strong electrostatic interactions between the ions and the polar solvent molecules. However, the ionic compounds are insoluble in non-polar solvents.

In order to understand these interactions let us study the dissolution of sodium chloride in water. Water is a polar solvent because H_2O molecules have oppositely charged ends with partial negative charge on the O atom and partial positive charge on the H atoms as shown in Fig. 7.6.

Fig. 7.6. Polar nature of H_2O.

When an ionic solid is dissolved in water, the cations and anions of the solid get attracted by the opposite ends of the water dipoles. The oxygen end gets attracted towards the cation while the hydrogen end gets attracted towards the anion. If the **ion-dipole** attractive forces are stronger enough to overcome the inter-ionic attraction, the ions are pulled out of the crystal lattice and they pass into solution. In the solution, the ions behave as independent entities, but they remain surrounded by the envelop of water dipoles and are called hydrated ions (Fig. 7.7). This process is referred to as **hydration of ions.**

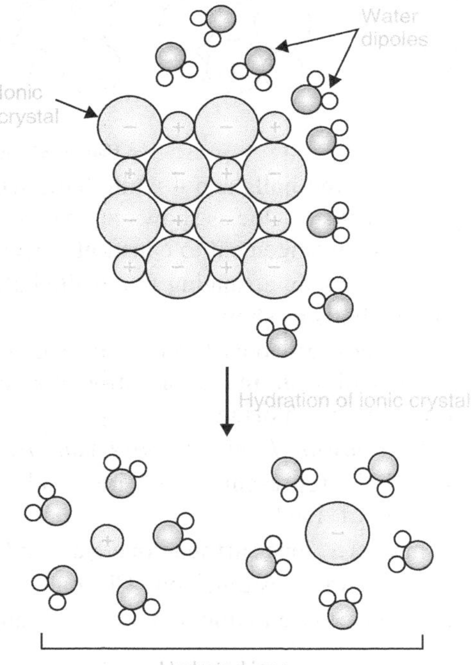

Fig. 7.7

The energy released by hydration of ions is called hydration energy. The hydration energy released compensates the lattice energy required to dismantle the crystal structure. In short, for the **dissolution of an ionic substance in water, Hydration energy > Lattice energy.**

For example, many ionic compounds such as AgCl(*s*), $PbCl_2$(*s*), $BaSO_4$(*s*), etc., do not dissolve in water because of the high value of their lattice energies.

Non-polar solvents do not have polar sites, and do not get attracted towards the ions. Therefore, ionic compounds are not soluble in non-polar solvents such as benzene, tetrachloromethane, etc.

ACTIVITY 7.2 — **To investigate the solubility of some ionic compounds**

1. Take about 1.0 g each of sodium chloride and copper (II) chloride in separate glass test tubes. Add about 5 cm^3 of water in each test tube and shake well. Observe the following:
   (*i*) Do either of the solid dissolve?
   (*ii*) Do they form clear, translucent or opaque solution?
   (*iii*) Do they have low or high solubility?
2. Repeat the experiments using the following solvents:
   (*a*) ethanol (*b*) hexane
   (*c*) tetrachloromethane (*d*) ethoxyethane (ether)

Caution: * ***The solvents a, b, d are highly inflammable.***
* ***The solvent c is non flammable but highly toxic.***

Thus these solvents must be handled with proper care.

Prepare a table as shown below and fill in the results of the solubility tests and report the conclusion.

Substances	*Water*	*Tetrachloromethane*	*Ethanol*	*Hexane*	*Ethoxyethane*
NaCl(*s*)					
$CuCl_2$(*s*)					

4. Electrical Conductivity. Ionic compounds do not conduct electricity in the solid state because the ions occupy fixed positions, and are not able to move freely. However, they are able to conduct electricity in the molten state or in their aqueous solutions because in that case their ions are free to move. Thus, *ionic compounds are bad conductors of electricity in the solid state but are good conductors of electricity in the molten or dissolved state.* Solid, molten and dissolved state of PbI_2 and NaCl are shown in Fig. 7.8.

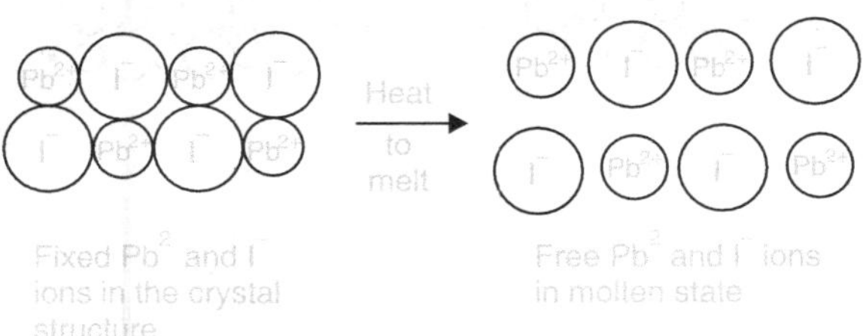

Fig. 7.8. (a) Freeing the ions by melting.

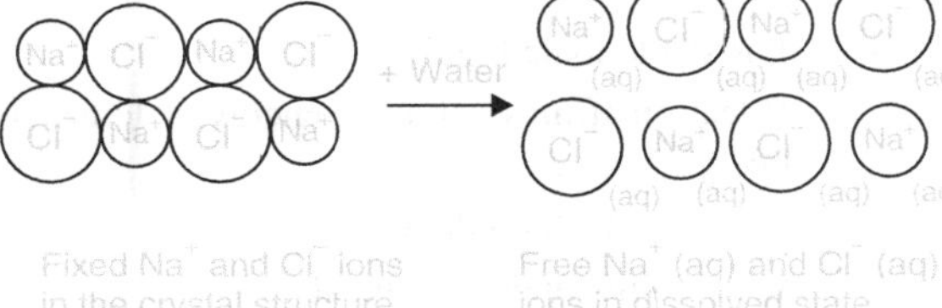

Fig. 7.8. (b) Freeing the ions by dissolution.

ACTIVITY 7.3 **To study the movement of ions in an electrical field**

1. Take a filter paper strip, soak it in potassium trioxonitrate (V) solution. Wrap the wet filter paper at its ends around the graphite electrodes (Use graphite rods of a dry cell) as shown.
2. Prepare an aqueous solution of ($CuCrO_4$) copper (II) tetraoxochromate (VI). With the help of a dropper put a drop of this aqueous solution in the centre of the filter paper.
3. Pass electrical current through the graphite rods and make these observations.

   Watch the splitting of the $CuCrO_4$ spot into blue and yellow spots. Blue spot moves towards the cathode and the yellow moves towards the anode.

   Try to answer the following:

   (*i*) What is the colour of $CuCrO_4$ solution?
   (*ii*) Blue colour of the spot is due to ......
   (*iii*) Yellow colour of the spot is due to ......
   (*iv*) Why blue colour moves towards anode?

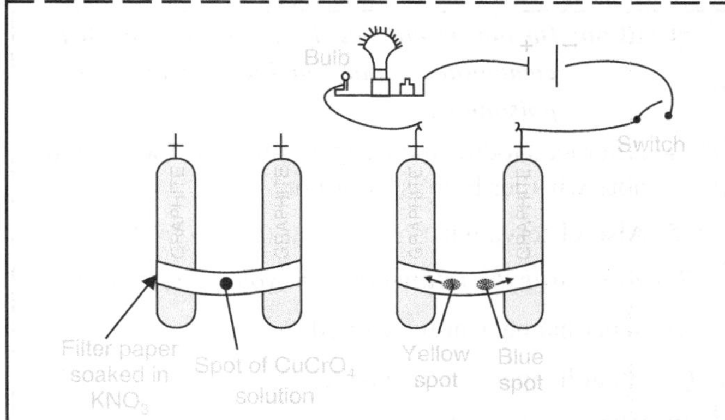

ACTIVITY 7.4 **To study the effect of electrical field on lead (II) iodide and sodium chloride**

Apparatus required:

(*i*) Four 1.5 V dry cells or 6 V power pack
(*ii*) 6 V Bulb
(*iii*) Two graphite rods (from used up dry cells)
(*iv*) Connecting wires (*v*) Silica crucible
(*vi*) Heating arrangement

A. LEAD (II) IODIDE

1. Set up the apparatus as shown in the diagram.
2. Fill the crucible half way with solid **Lead (II) iodide (PbI_2)** and press the switch. Observe whether the bulb glows or not.

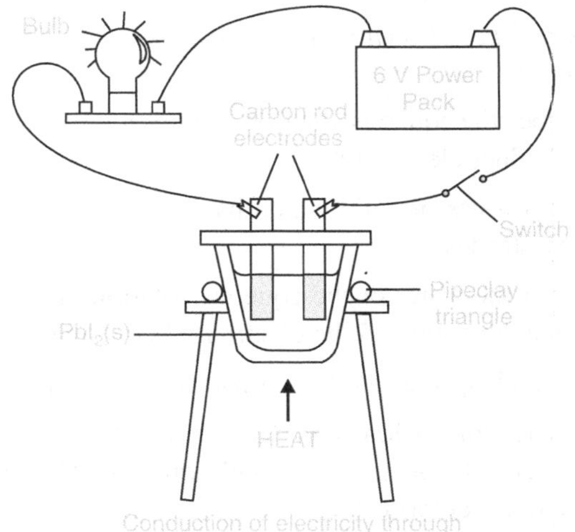

Conduction of electricity through molten lead (II) iodide

3. Heat the crucible strongly to melt the lead (II) iodide. This should only be done in a very well ventilated place or a fume cupboard.

*Caution: ***Do not breathe in the fumes because lead compounds and iodine vapours are poisonous.***

4. After lead iodide melts, again press the switch and note whether bulb is lit or not.
5. Also observe what happens at the electrodes.

Try to answer the following from your observations:

(*i*) What happens at the cathode?

(*ii*) What happens at the anode?

(*iii*) What is the colour of the gas evolved or solid formed?

(*iv*) Give the formulae of the cation and the anion of the salt used.

(*v*) Give a chemical equation for the process.

B. SODIUM CHLORIDE

1. In a new crucible add sodium chloride, press the switch. See whether the bulb glows or not.
2. Heat strongly for about two minutes. Note whether the salt melts or not.
3. Add distilled water to a beaker and insert clean electrodes into it. Now, press the switch to see whether it conducts electricity or not.
4. Cool the salt and add it to water in a beaker. Stir to dissolve the salt. Insert the electrodes in the solution so formed and press the switch. Note whether the bulb is lit or not.

Answer the following on the basis of your observations:

(*i*) Does the aqueous solution of the sodium chloride conduct electricity?

(*ii*) Is it a gas or solid which has been formed at the electrodes?

(*iii*) Is it the water or the aqueous solution of the salt which conducted the electricity? Give reasons.

(*iv*) Which species are taking part in the reaction?

(*v*) What conclusion can be drawn about the effect of electricity on the solid and aqueous solution of sodium chloride?

(*vi*) Predict whether molten sodium chloride will conduct electricity. How would you melt it?

(*vii*) Which out of lead (II) iodide and sodium chloride has a higher melting point?

5. **Ionic Crystals are Brittle.** Ionic crystals are generally brittle. It means that when a strong force or stress is applied on them they break into pieces along definite planes. This is because of the fact that stress or force causes displacement of ions along the planes and brings similarly charged ions near each other. The repulsive force between similarly charged ions causes them to fall apart along the plane as shown in Fig. 7.9.

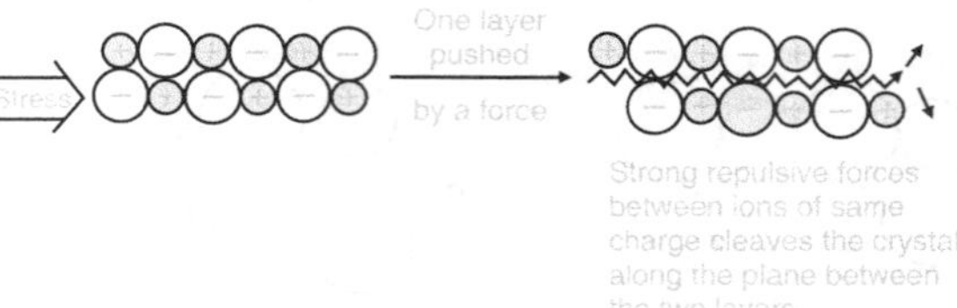

Fig. 7.9. Cleavage of ionic crystals.

- Ionic bond is formed by the transfer of electrons from electropositive atom to an electronegative atom.
- For stable ionic bonding,
  (Electron affinity + Lattice energy) > Ionization energy
- Ionic bond is non directional
- Ionic compounds in general:
  — have high melting and boiling points
  — are soluble in polar solvents and insoluble in non-polar solvents
  — are hard and brittle
  — are poor conductors of electricity in the solid state but good conductors of electricity in the dissolved or molten state.

7.6 FORMATION OF COVALENT BOND

We have studied that the ionic bond is formed between an electropositive element with low ionization energy and an electronegative element with high electron affinity. The bond results by the transference of electrons from one atom to another so that they acquire stable octets. The gain or loss of electrons cannot take place between atoms of same element or those which have similar electronegativities or high ionization energies. Such atoms combine chemically by mutual sharing of electrons so as to attain noble gas configuration.

A force which binds atoms of the same or different elements together by mutual sharing of electrons is called a covalent bond. The term ***covalent bond*** was introduced by ***Langmuir*** in 1919.

The atoms involved in bond formation *contribute equal number of electrons for sharing.* The shared electrons become a common property of both atoms and constitute a bond between them. The shared pair of electrons which is responsible for the formation of the bond is also called **bond pair of electrons.** Let us consider the formation of Cl_2 molecule.

Each chlorine atom, Cl (2, 8, 7), has seven electrons in its valence-shell and needs one more electron to acquire an octet. Both chlorine atoms contribute one electron each to share two electrons. The bond can be represented by putting a line (–) between the atoms instead of a bond pair.

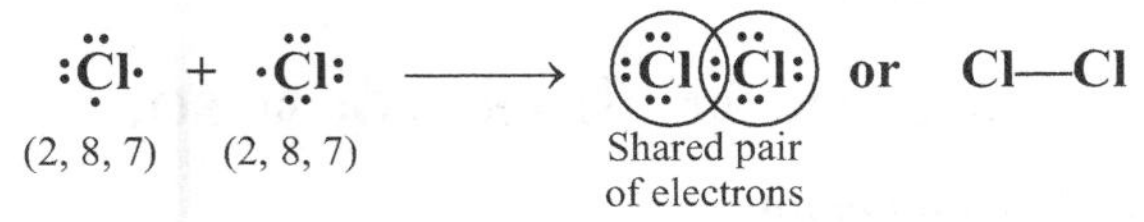

Similarly, in the formation of HCl molecule, both H and Cl atoms contribute one electron each which is shared by both. In this way H attains the configuration of helium while Cl attains the configuration of argon.

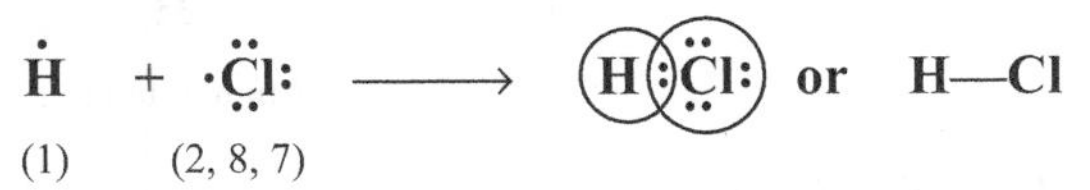

Multiple Covalent Bonds. When the atoms share one electron pair, the bond formed is called ***single covalent bond.*** However, if two electron pairs are shared by the atoms, the bond formed is called ***double covalent bond.*** Similarly, when the atoms share three electron pairs, the bond is called ***triple covalent bond.*** The double and triple covalent bonds are collectively called ***multiple covalent bonds.*** Let us now study the examples for double and triple bonds.

Formation of oxygen (O_2) molecule. Oxygen atom ($_8O$) has six electrons in its valence shell (2, 6). It has a tendency to attain the configuration of Ne (2, 8). The two oxygen atoms, therefore, contribute 2 electrons each and share four electrons, *i.e.*, 2 *bond pairs.* Thus, the oxygen atoms in oxygen molecule are held by a *double covalent bond.*

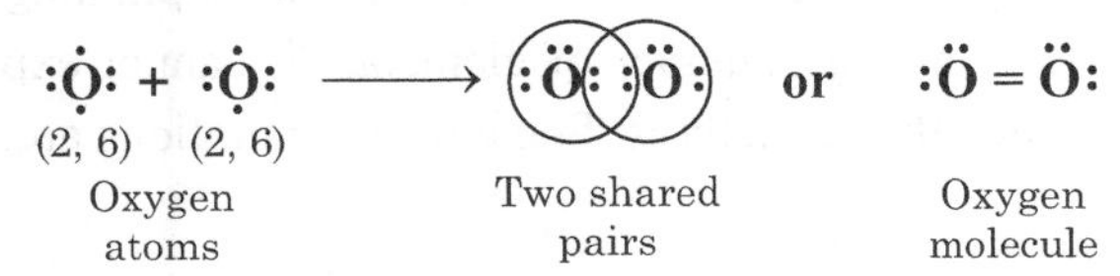

Formation of nitrogen molecule. Nitrogen atom ($_7N$) has five electrons in its valence shell (2, 5). It also tends to acquire the configuration of Ne (2, 8). The two nitrogen atoms, therefore, contribute 3 electrons each and share three pairs of electrons. Thus, the nitrogen atoms in N_2 molecule are held by a *triple covalent bond.*

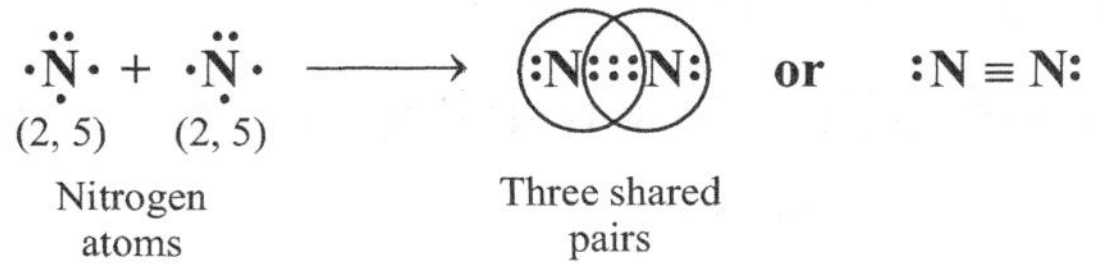

The **covalency** of an element may be defined as *the number of electrons which its atom contributes for sharing while forming a covalent bond.* For example, the covalency of chlorine is 1; oxygen is 2 and that of nitrogen is 3.

It may be noted that covalent bonded substances exist as discrete molecules. Therefore, they are also called **molecular substances.**

7.7 LEWIS DOT STRUCTURES OF SOME COVALENT MOLECULES

Let us now draw the Lewis dot structures of some covalent molecules.

(*i*) *H_2 molecule:*

(*ii*) *H_2O molecule:*

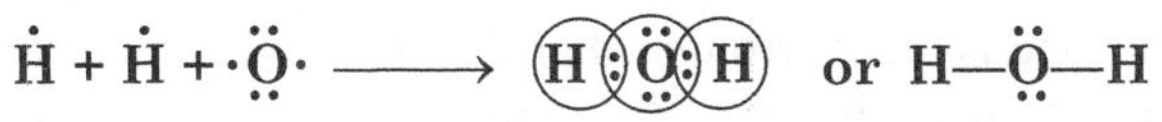

(*iii*) *Ammonia molecule:*

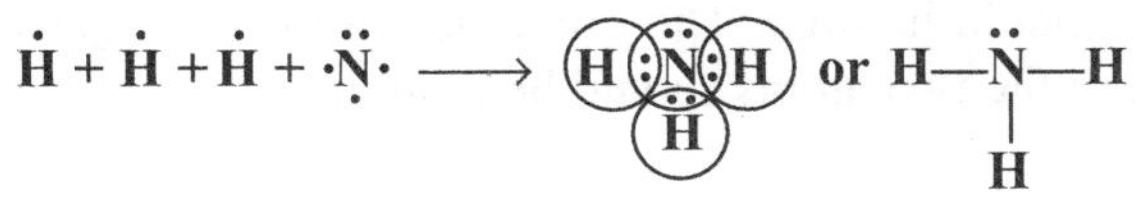

(*iv*) *Hydrogen sulphide:*

(*v*) *Methane molecule:*

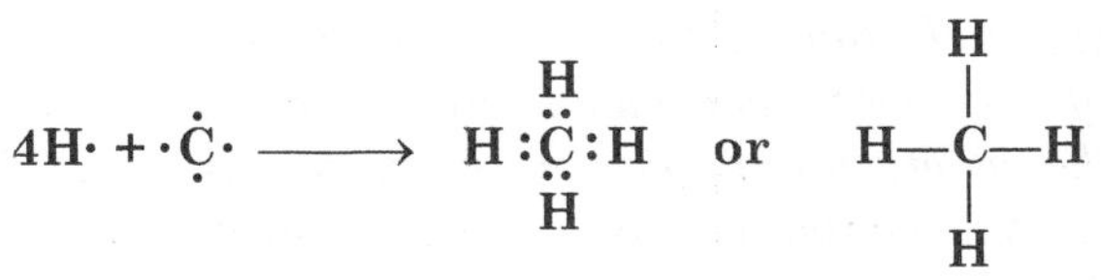

(*vi*) *Ethene molecule:*

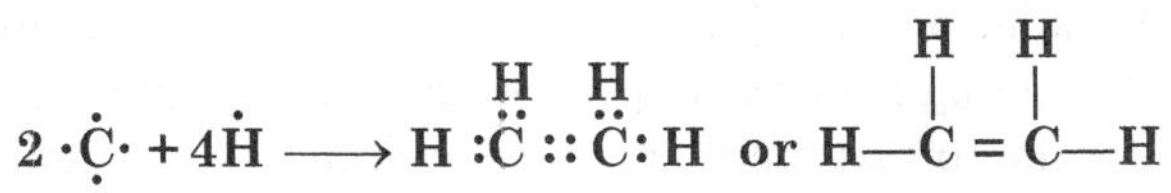

(*vii*) *Ethyne molecule:*

(*viii*) *Hydrogen cyanide molecule:*

$$\dot{H} + \cdot\dot{\underset{\cdot}{C}}\cdot + \dot{\underset{\cdot\cdot}{N}}: \longrightarrow H:C:::N: \text{ or } H—C \equiv N$$

(*ix*) *Methanol molecule:*

$$\cdot\dot{\underset{\cdot}{C}}\cdot + 4\dot{H} + \cdot\ddot{\underset{\cdot\cdot}{O}}\cdot \longrightarrow H:\overset{H}{\underset{H}{\ddot{\underset{\cdot\cdot}{C}}}}:\ddot{\underset{\cdot\cdot}{O}}:H \text{ or } H—\overset{H}{\underset{H}{C}}—\ddot{\underset{\cdot\cdot}{O}}—H$$

7.8 FACTORS WHICH INFLUENCE THE FORMATION OF COVALENT BONDS

Covalent bond is formed by the mutual sharing of electrons, and the factors which favour its formation are as follows:

1. Electron Affinity. A covalent bond is generally favoured between two atoms if both atoms have *high electron affinity.*

2. Ionization Energy. Ionization energy of both atoms participating in bonding should be *high.*

3. Atomic Size. Atomic size of the atoms forming a covalent bond should be smaller. The smaller the atomic radii of atoms, the more stronger the covalent bond. For example, *H—H* bond is more stronger than *Cl—Cl* bond which in turn is more stronger than *Br—Br* bond.

4. Electronegativity. The electronegativities of both the atoms should be *high.* The difference of electronegativities between the two atoms should be minimum.

7.9 CO-ORDINATE-COVALENT BOND OR DATIVE BOND

It is a special case of covalent bond the formation of which was postulated by ***Perkins*** (1921). It is formed *by mutual sharing of electrons between the two atoms but the shared pair of electrons is contributed only by one of the two atoms, the other atom simply participates in sharing.* The atom which donates an electron pair for sharing is called **donor** and it must have already completed its octet. On the other hand, the atom which accepts the electron pair in order to complete its octet is called **acceptor.** The bond is represented by an arrow (⟶) pointing from the donor towards the acceptor. Let us consider the formation of an ozone molecule. A molecule of oxygen contains two oxygen atoms which share four electrons and complete their octets. Now, if an atom of oxygen having six valence electrons comes close to oxygen molecule, it shares a lone pair of electrons with one of the oxygen atoms of the molecule. It can be represented as follows:

$$\underset{\text{Donor atom}}{:\ddot{O}::\ddot{O}:} + \underset{\text{Acceptor atom}}{\ddot{\underset{\cdot\cdot}{O}}:} \longrightarrow \underset{\text{Ozone molecule}}{:\ddot{\underset{\cdot\cdot}{O}}::\ddot{\underset{\cdot\cdot}{O}}:\ddot{\underset{\cdot\cdot}{O}}:} \text{ or } :\ddot{O} = \ddot{O} \rightarrow \ddot{\underset{\cdot\cdot}{O}}:$$

It is important to note that a co-ordinate bond once formed cannot be distinguished from a covalent bond.

Some more examples of molecules/polyatomic ions having co-ordinate bonds are as follows:

(*i*) *SO_2 molecule*

$$:\ddot{O}::\ddot{S}:\ \ddot{\underset{\cdot\cdot}{O}}: \quad \text{or} \quad :\ddot{O} = \ddot{S} \rightarrow \ddot{\underset{\cdot\cdot}{O}}: \quad \text{or} \quad SO_2$$

(*ii*) *Ammonium ion*

$$H:\overset{H}{\underset{H}{\ddot{N}}}: + \overset{+}{H} \longrightarrow H—\overset{H}{\underset{H}{N}}\overset{+}{\rightarrow}H \text{ or } \left[H—\overset{H}{\underset{H}{N}}—H\right]^+$$

Ammonium ion

(*iii*) *Hydronium ion*

$$H:\overset{H}{\ddot{\underset{\cdot\cdot}{O}}}: + \overset{+}{H} \longrightarrow H—\overset{H}{O}\overset{+}{\rightarrow}H \text{ or } \left[H—\overset{H}{O}—H\right]^+$$

Hydronium ion

(*iv*) *Carbon (II) oxide.* In this molecule carbon and oxygen atoms contribute two electrons to form pure covalent bonds. At the same time oxygen also act as donor atom to form a co-ordinate covalent bond.

$$\underset{(2,\,4)}{:\dot{\underset{\cdot}{C}}:} + :\ddot{O}: \longrightarrow :C:::O: \quad \text{or} \quad C \equiv O$$

LIMITATIONS OF OCTET RULE

Although the octet rule is successful in explaining the valencies of a large number of elements, it cannot explain many of the other structures. Some of its limitations are:

1. Formation of Compounds with Electron Deficient Atoms

On the basis of the octet rule, the elements of group 1, 2 and 13 should not form covalent bonds because they have less than four electrons in their valence shell. Therefore, they cannot achieve octet by the mutual sharing of electrons. But some of the elements of these groups do form covalent compounds. LiCl, BeF_2 and BCl_3 are common examples.

Li:Cl: :F:Be:F: :Cl:B:Cl: :Cl:

In LiCl, Li atom has only 2 electrons; in BeF_2, Be atom has four electrons and in BCl_3 the boron atom has only six electrons in its outer shell.

2. Formation of Super Octet Molecules Like PCl_5, SF_6 and IF_7

Violation of the octet rule is also observed in compounds like PF_5, SF_6 and IF_7 where the central atoms have more than eight electrons in their valence shell as shown below:

(10 ē around P)   (12 ē around S)   (14 ē around I)

3. Formation of Compounds of Xenon

Octet rule is primarily based upon the fact that noble gases are inert. But recent research have shown that some of the noble gases like Xe and Kr do take part in bonding and form compounds with elements like fluorine and oxygen such as XeF_2, XeF_4, $XeOF_2$, $XeOF_4$, etc.

4. Inability to Predict Energy Changes

In addition to the above exceptions, the octet rule is not able to predict energy changes during bond formation.

7.10 POLAR COVALENT BONDS

Covalent bonds between different atoms may acquire some ionic characters which can be explained on the basis of the concept of electronegativity. Electronegativity of an element is defined as *the power of its atoms to attract the bonding or shared pair of electrons towards itself.*

POLAR CHARACTER OF COVALENT BOND

When a covalent bond is formed between two similar atoms, the shared pair of electrons lies midway between the nuclei of the two atoms, because both atoms have the same attraction for the bonding electrons. The electron cloud constituting the covalent bond is symmetrically distributed around the atoms. Such a covalent bond is called a **non-polar covalent bond.**

A : A   or   A   A   or   A—A

Symmetrical electron cloud   Non-polar bond

For example, molecules like H_2, Cl_2, O_2 and N_2 contain non-polar bonds.

In case the covalent bond is formed between two dissimilar atoms, one of which has a larger value of electronegativity, the bonding pair of electrons is displaced towards the more electronegative atom. In other words, the electron cloud containing the bonding electrons gets distorted and the charge density concentrates around the more electronegative atom.

Due to the unequal distribution of electron charge density, the more electronegative atom acquires a partial negative charge (indicated as δ^-) whereas the less electronegative atom acquires a partial positive charge indicated as δ^+. Thus, *a covalent bond develops a partial ionic character as a result of the difference in electronegativities of the atoms forming the bond. Such a bond is a* **polar covalent bond** (as shown in Fig. 7.10).

A   : B   or   A   B   or $\overset{\delta^+}{A}$—$\overset{\delta^-}{B}$

Displacement of shared electrons   Distorted electron cloud   Polar covalent bond

Fig. 7.10. Polarity in covalent bond.

For example, the bond between hydrogen and chlorine atoms in HCl molecule is polar because the shared electron pair is displaced towards the chlorine atom which is more electronegative.

H :Cl: or   H⟶Cl   or $\overset{\delta^+}{H}$—$\overset{\delta^-}{Cl}$

The extent of ionic character in a covalent bond depends upon the difference in electronegativities of the two atoms forming a bond. The greater the difference in electronegativities, the greater the percentage of ionic character in a bond. For example, hydrogen fluoride is more polar than hydrogen chloride because the difference in electronegativity between H and F is higher than that between H and Cl.

	$\overset{\delta^+}{H}$—$\overset{\delta^-}{F}$	$\overset{\delta^+}{H}$—$\overset{\delta^-}{Cl}$
	2.1 4.0	2.1 3.0
Electronegativity Difference	**1.9**	**0.9**
	More ionic than H—Cl	More covalent than HF

It has been observed that a bond has 50% ionic character and 50% covalent character if the difference in electronegativities of the participating atoms is 1.7. On the other hand, the covalent character dominates if the difference in electronegativities is less than 1.7 while ionic character dominates if the difference in electronegativities is greater than 1.7.

7.11 DIPOLE MOMENT

It has been pointed out earlier that a covalent bond between two atoms acquires a partial polar character if the values of electronegativity of the two bonded atoms differ. The two charged ends of the bond behave as an *electrical dipole* just like the ends of a bar magnet has two opposite poles. The degree of polarity is measured in terms of *dipole moment.* Dipole moment is defined as *the product of the magnitude of charge on any one of the atoms and the distance between them.* Dipole moment is represented by a Greek letter 'μ'. It can be expressed mathematically, as

$$\mu = e \times d$$

where, e = charge on any one of the atoms

d = distance between the atoms.

Since the charge 'e' is of the order of 10^{-10} *e.s.u.* and 'd' is of the order of 10^{-8} cm, then μ, which is the product of 'd' and 'e' is of the order of 10^{-18} *e.s.u.-cm*. This unit is called *Debye* and is represented by D.

In SI units, dipole moment may be expressed in terms of *Coulomb-meter* abbreviated as C m.

1 D = 1 × 10^{-18} e.s.u.-cm. = 3.335 × 10^{-30} C m

For example, the dipole moment of HCl is 1.03 × 10^{-18} *e.s.u.-cm* and is expressed as 1.03 D.

Dipole moment can be determined experimentally and its value can give an idea of the polar character of a molecule. It is a *vector quantity* as it has a direction as well as magnitude. The direction of a dipole moment is usually represented by an arrow ⟼ drawn from positive end towards the negative end.

DIPOLE MOMENT AND MOLECULAR STRUCTURE

(*a*) **Diatomic molecules.** A diatomic molecule has two atoms bonded to each other by a covalent bond. In such a molecule the dipole moment of the bond also gives the dipole moment of the molecule. For example, the dipole moment of HCl molecule is the same as that of the H—Cl bond (μ = 1.03 D).

Fig. 7.11

The greater the electronegativity difference between the atoms, the higher the dipole moment of such molecules. For example, the dipole moment of HF is 1.91 D while that of HCl is 1.03 D (F is more electronegative than Cl).

(*b*) **Polyatomic molecules.** A polyatomic molecule has more than two atoms bonded by covalent bonds. In such cases the idea of a dipole can be applied to individual bonds within the molecule. The dipole moment of an individual bond in a polyatomic molecule is referred to as a **bond dipole.** The dipole moment of the molecule depends upon the orientations of the various bond dipoles.

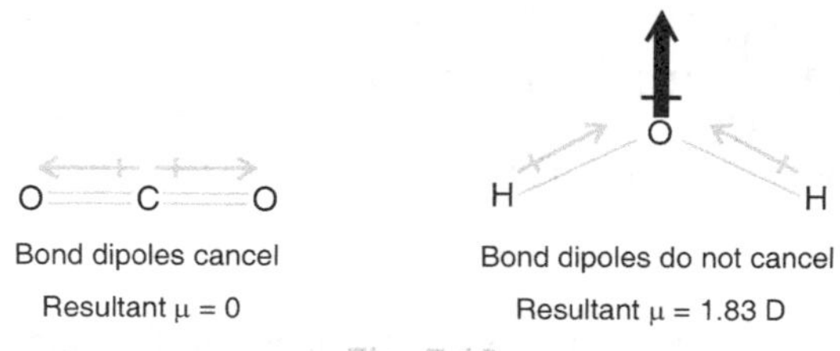

Fig. 7.12

For example, carbon dioxide (CO_2) and water (H_2O) are both triatomic molecules but the dipole moment of carbon dioxide is zero whereas that of water is 1.83 D. This can be explained on the basis of their structures.

Carbon dioxide is a *linear molecule* in which the two C = O bonds are oriented the opposite directions at an angle of 180°. The bond dipole of the C = O bond is 2.3 D but due to the linear structure of CO_2, the bond dipoles of the two C = O bonds cancel each other. Therefore, the resultant dipole moment of the molecule is zero. Hence, CO_2 is a non-polar molecule. On the other hand, water molecule has a bent structure in which the two O—H bonds are oriented at an angle 104.5°. Therefore, the bond dipoles of the two O—H bonds do not cancel each other. Hence, the resulting dipole moment of H_2O molecule is not zero but has a value equal to 1.83 D.

In a similar way, the resulting dipole moment of BF_3 molecule is zero. The bond dipoles of three B—F bonds give a net sum of zero because the resultant vector of any two is equal and opposite to the third (parallelogram law of forces) as shown in Fig. 7.13. Thus, BF_3 is a non-polar molecule.

Fig. 7.13. Dipole moment of BF_3 molecule.

On the other hand, the bond dipoles of the three N—H bonds in NH_3 molecule, do not cancel each other. Hence, the molecule has a net dipole moment of 1.49 D and is polar in nature.

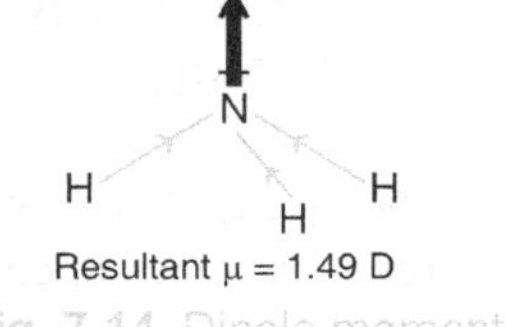

Fig. 7.14. Dipole moment of NH_3 molecule.

It may be noted that the molecules having zero dipole moment ($\mu = 0$) are called **Non-Polar Molecules.** On the other hand the molecules having dipole moment greater than zero ($\mu > 0$) are **Polar Molecules.**

The dipole moments of some selected molecules are given below in tabular form.

Type of Molecule	*Geometry/Shape*	*Examples*					
AB	Linear	HF *(1.78 D)*	HCL *(1.07 D)*	HBr *(0.79 D)*	HI *(0.38 D)*	H_2 *(0)*	F_2 *(0)*
AB_2	Linear	CO_2 *(0)*	BeF_2 *(0)*				
	Bent	H_2O *(1.83 D)*	H_2S *(0.95 D)*				
AB_3	Planar	BF_3 *(0)*	SO_3 *(0)*				
	Pyramidal	NH_3 *(1.49 D)*	NF_3 *(0.25 D)*				
AB_4	Tetrahedral	CH_4 *(0)*	CCl_4 *(0)*	$CHCl_3$ *(1.01 D)*	CH_2Cl_2 *(1.57 D)*	CH_3Cl *(1.87 D)*	

ACTIVITY 7.5 **To study the polar nature of different solvents**

Apparatus Required:

Burette, beaker, ebonite rod, fur, liquid solvents (water, tetrachloromethane, propanone, cyclohexane, ethoxyethane, etc.) whose polar nature is to be studied.

Procedure:

1. Clamp the burette and fill it with water as shown. Open the tap of the burette to allow a slow, steady and thin stream of water to flow down straight into the beaker.
2. Quickly rub ebonite rod with fur (or comb with hair) repeatedly so as to develop a static –ve charge on it.

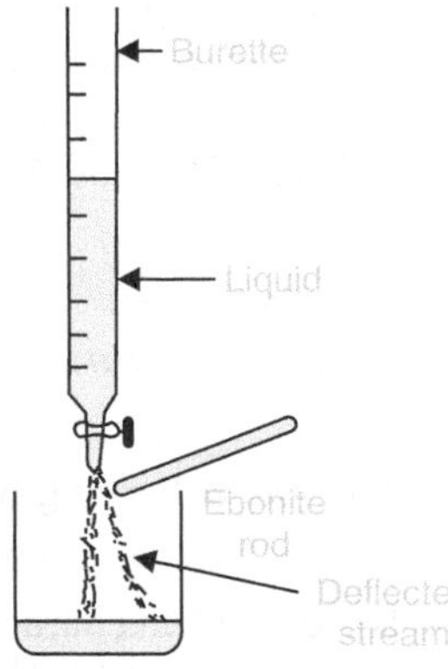

3. Hold the charged rod (or comb) near (about 1–2 cm apart) the down flowing stream and observe whether there is some deflection in the stream or not.

Try to answer the following questions from your observations:

(*i*) Sketch the structure of water and mark the positive and negative ends.

(*ii*) Which end of the water molecule was attracted to the comb or rod? Show this on your sketch.

(*iii*) Rub a warm glass or polythene rod repeatedly with tissue paper. This gives the rod a positive charge. Hold the rod 1–2 cm away from the stream of water. What do you observe? Which end of the water molecule was attracted to the rod?

Repeat the experiment using different solvents and make the following table of observations.

Caution:** ***(i)** **Use a clean burette for every solvent.
(ii)** **Some of the solvents are inflammable so handle them with care.

Liquid	*Deflection*	*No Deflection*	*Polar or Non-Polar*
1. Water			
2. Propanone			
3. Ethoxyethane			
4. Tetracholoromethane			
5. Cyclohexane			
6. Ethanol			

Note: The experiment works best in dry weather.

7.12 GENERAL PROPERTIES OF COVALENT COMPOUNDS

Compounds whose molecules consist of atoms held together by covalent bonds are called **covalent compounds.** In general, covalent compounds have the following properties:

1. **State of existence.** Atoms in covalent compounds are joined together by the mutual sharing of electrons. Covalent compounds, therefore, exist as individual molecules. The intermolecular forces in these compounds are usually weak. Therefore, they exist in the liquid or gaseous state. However, a few covalent compounds also exist in the solid state, *e.g.*, urea, sugar, etc. It may be noted that some covalent substances exist as giant network structures with very strong interatomic covalent bonds. They do not exist as simple molecules. Common examples are diamond, silica (SiO_2), graphite, etc.

2. **Non-conducting.** Covalent compounds are generally poor conductors of electricity in the fused or dissolved state due to the non-existence of ions.

3. **Low boiling and melting points.** The attractive forces between molecules of covalent compounds are generally weak. As a result, lesser amount of energy is needed to overcome these intermolecular forces. Thus, these compounds usually have low melting and boiling points.

4. **Solubility.** Covalent compounds are usually insoluble or less soluble in water and in other polar solvents. However, they are more soluble in non-polar solvents.

7.13 COVALENT CHARACTER IN IONIC BOND

Just as covalent bond develops ionic character due to the difference of electronegativities of bonded atoms, the ionic bond also develops covalent character as described below:

When two oppositely charged ions come close, the positive ion tends to attract the electron cloud of the negative ion towards itself because the electron cloud in the anion is loosely held by its nucleus. The anion no longer remains spherical but undergoes some distortion. The electron cloud of the negative ion gets polarised and the electron density is pulled in between the nuclei of the two ions. The build up of electron density in between the two nuclei is shown in Fig. 7.15. *The ionic bond does not remain 100% ionic but develops some covalent character.*

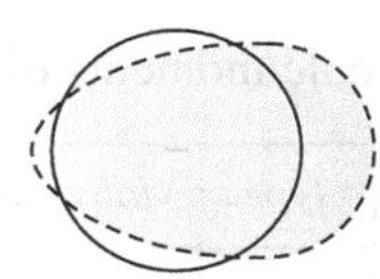
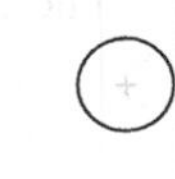

Fig. 7.15. Distortion of electron cloud of anion.

- The power of a cation to cause distortion in the electron cloud of an anion is referred to as its **polarising power.**
- The ability of an anion to undergo distortion is called its **polarisability.**
- The extent of covalent character in an ionic bond depends on the polarising power of the cation and polarisability of the anion which are decided on the basis of a set of rules called **Fajan rules.** These are described as follows:

(*i*) ***The smaller the size of a cation, the larger its polarising power.*** The polarising power of a cation depends on the electrical field used to pull electrons away from the anion. The electrical field strength depends on the charge density (q/r) *i.e.*, charge (q) to size (r) ratio of the cation. The larger the charge density of the cation, the higher its polarising power. For example, let us compare the polarising power of Na^+, Mg^{2+} and Al^{3+} ions.

	Na^+	Mg^{2+}	Al^{3+}
Charge (q):	+1	+2	+3
Size (r) (nm):	0.102	0.072	0.053
q/r:	9.8	27.8	56.6

Thus, the polarising power increases from

$$Na^+ \rightarrow Mg^{2+} \rightarrow Al^{3+}.$$

Hence covalent character increases among their respective chlorides as $NaCl < MgCl_2 < AlCl_3$.

(*ii*) *For two cations of similar size, the one with pseudo noble gas configuration* $\mathbf{ns^2np^6nd^{10}}$ *has a larger polarising power than the one with noble gas configuration* n^2p^6. For example, CuCl is more covalent than NaCl because the polarising power of Cu^+ ion which has pseudo noble gas configuration, is higher than Na^+ ion.

(*iii*) ***The larger the size of an anion, the higher its polarisability.*** The polarisability of an anion, in fact depends on the product *qr*. The larger the value of the *qr* product of an anion, the higher its polarisability. For example, the comparison of the polarisability of F^-, Cl^- and I^- ions are given as follows:

	F^-	Cl^-	I^-
Charge (q):	– 1	– 1	– 1
Size (r) (nm):	0.136	0.181	0.216
(qr) product:	0.136	0.181	0.216

As evident the *qr* product increases from F^- to I^- ion (Here, *q* is taken in magnitude only). Thus, polarisability increases from F^- to Cl^- to I^-.

Hence covalent character increases from MgF_2 to $MgCl_2$ to MgI_2.

Why Na_2CO_3 is thermally more stable than $CaCO_3$?

The stability of a metal carbonate towards heat depends upon the relative stability of the resulting metal oxide. The higher the stability of the resulting metal oxide, the lesser the stability of the metal carbonate.

A large anion is stabilised by a large cation whereas a small anion is stabilised by a small cation. Na^+ ion being larger in size stabilises CO_3^{2-} *(a larger anion)* better than O^{2-} *(a smaller anion)* ion. As a result Na_2O is less stable than Na_2CO_3. Therefore, Na_2CO_3 is stable towards heat and does not decompose. On the other hand, Ca^{2+} ion being smaller in size stabilises O^{2-} ion better than CO_3^{2-} ion. Therefore, CaO is more stable than $CaCO_3$. Hence, $CaCO_3$ easily decomposes into CaO on heating.

7.14 METALLIC BOND

Metals constitute about three-fourth of all the known elements. They have characteristic properties such as *bright lustre, high electrical* and *thermal conductivity, malleability* and *ductility* and *high tensile strength*. The attractive force which binds various metal atoms together is called **metallic bond.** The metallic bond is neither a covalent bond nor an ionic bond because neither of these bonds are able to explain the known properties of metals. For example, neither ionic nor covalent compounds conduct electricity in the solid state but metals are very good conductors of electricity. In order to explain the bonding in metals different theories have been put forward. We shall be studying here the *electron gas model* or the *electron sea model* for metallic bonding.

ELECTRON GAS MODEL OR ELECTRON SEA MODEL

This is the simplest model that explains the properties of metals. This model was proposed by Lorentz. The main features of this modal are:

1. *A metal atom is supposed to consist of two parts, valence electrons and the remaining part (the nucleus and inner shells) which is called* **kernel.**

2. *The metallic crystal consists of closely packed metal atoms in three dimensions. The kernels of metal atoms occupy fixed positions called lattice sites while the space between the kernels is occupied by valence electrons.* The arrangement of kernels and valence electrons is shown in Fig. 7.16.

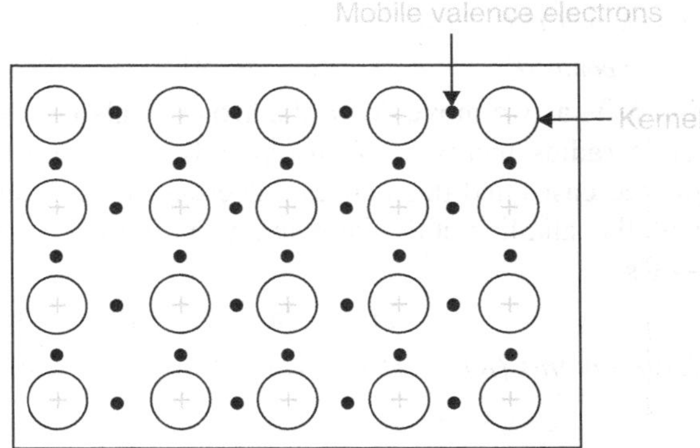

Fig. 7.16. Arrangement of metallic kernels.

3. *Due to their smaller ionisation energy, the valence electrons of metal atoms are not held by the nucleus very firmly. Therefore, they can leave the field of influence of one kernel and enter the field of influence of the other. This movement can take place through the vacant valence orbitals. Thus, the valence electrons are not localised but are* **mobile or delocalised.** The movement of electrons in a metallic crystal is just like gas molecules, and so the model is called **electron gas model.**

4. *The simultaneous force of attraction between the mobile electrons and the positive kernels is responsible for holding the metal atoms together and is known as* **metallic bond.**

The metallic bond is *non-directional* and is *weaker* than a covalent bond.

STRENGTH OF METALLIC BOND

The strength of a metallic bond depends on the magnitude of the attractive force between the positive kernels and mobile valence electrons.

The average attractive force and metal bond strength *increases* with a *decrease* in *atomic radius* and *increase* in *number* of *valence electrons.* It must be noted carefully that the existence of both factors at the same time decreases the metal character because the tendency to form a metallic crystal decreases. For example, when we move along the period from left to right, metallic character decreases. Among the elements of the 3rd period, metal character decreases from left to right. The metallic elements are only Na, Mg and Al. The strength of the metallic bond increases from Na $\rightarrow$ Al and their melting points are a reflection of that.

	Na		**Mg**		**Al**
Mobile valence electrons		$\longrightarrow$	*Increase*	$\longrightarrow$	
Atomic radius		$\longrightarrow$	*Decreases*	$\longrightarrow$	
Strength of metallic bond		$\longrightarrow$	*Increases*	$\longrightarrow$	
Melting points (K)	371		922		933

Similarly, as we move down the group of alkali metals, the atomic radius increases. Consequently, the metal bond strength decreases and this causes a decrease in the melting points of the alkali metals from top to bottom, *i.e.,* from Li $\longrightarrow$ Cs

	Li	Na	K	Rb	Cs
Melting points (K)	454	371	336	312	302

FACTORS THAT FAVOUR THE METALLIC BONDING

Metallic bonding is generally favoured by the following factors:

1. The atomic size of the element should be large.
2. Ionisation energy of the element should be low.
3. Electron affinity of the element should be low.
4. The number of valence electrons should be small (usually 1–2).
5. The number of vacant orbitals in the valence shell should be large.

EXPLANATION OF PHYSICAL PROPERTIES OF METALS

1. Metallic Lustre. When light falls on the surface of a metal, the free electrons absorb the photons of light and are set into vibrations. These vibrating electrons immediately emit energy and become a source of light. Thus, incident light appears to be reflected from the surface of the metal. Consequently, the metallic surface acquires a shining appearance which is referred to as metallic lustre.

2. Electrical Conductivity. Whenever a potential difference is applied across a metallic strip, the free mobile electrons in the metal start moving towards the positive terminal. At the same time the electrons from the negative terminal enter into the metallic crystal. Thus, the metallic crystal maintains the flow of electron from the negative to the positive terminal.

At high temperature, the metallic kernels start vibrating due to the increase in kinetic energy. This restricts the free movement of the electrons. Consequently, the resistance of metals increases with an increase in temperature.

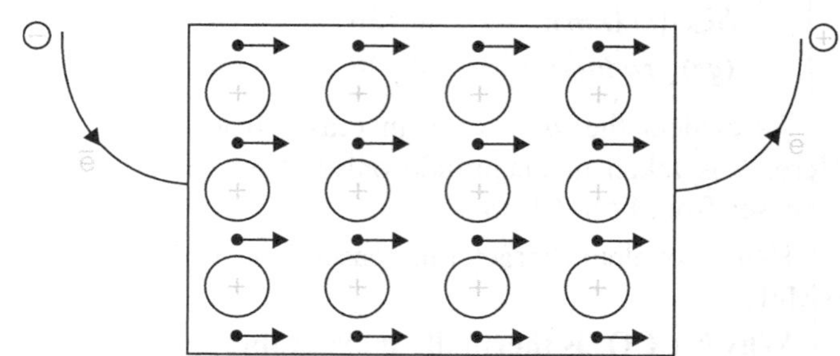

Fig. 7.17. Electrical conductivity of metals.

3. Thermal Conductivity. The conduction of heat through the metals can also be explained on the basis of the electron gas model. On heating a part of the metal, the kinetic energy of the electrons in that region increases. These energetic electrons move rapidly to the cooler parts and transfer their kinetic energy by means of collisions with other electrons. In this way, the heat travels from the hotter to the cooler parts of the metals.

4. Malleability and Ductility. Malleability is a property of metals where they can be beaten into sheets whereas ductility is a property where they can be drawn into wires. These properties are exhibited by metals *on account of the non-directional nature of a metallic bond.* Whenever any stress is applied on a metal, the position of the metallic kernels is altered without destroying the crystal. The crystal lattice gets deformed by slippage of the layers of kernels from one part to another as shown in Fig. 7.18. When one layer of kernels moves past another, the positive charges on the metal ions are shielded from each other by the valence electrons.

The electron sea model could explain the properties of metals qualitatively. However, the properties of metals can be explained more quantitatively by the molecular orbital theory which is beyond the scope of this book.

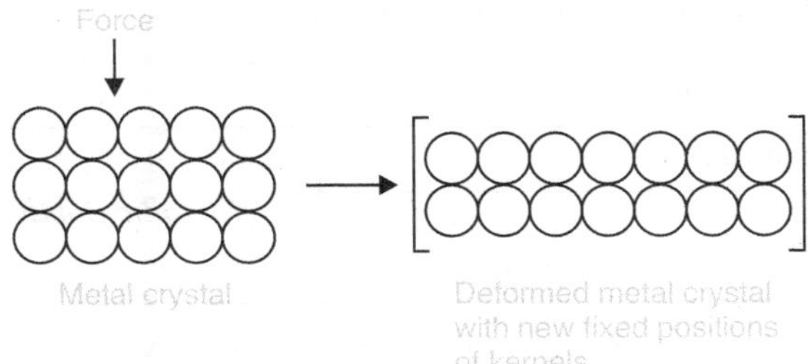

Fig. 7.18. Malleability and ductility of metals.

ACTIVITY 7.6	**To investigate the conductivity, malleability and ductility of some metals and non-metals**

Requirements: The substances in the form of small strips, blocks, rods or powder. Make the observations about each substance as directed and fill these observations in the table given at the end.

(*a*) **Nature of surface:** Observe the clean surface of each substance.

(*b*) **Thermal conductivity:** Carefully clamp each substance (metal or non-metal) which can be obtained in the form of a rod or strip, in turn. Apply a little vaseline at one end. Heat the rod or strip gently from the other end. See whether the vaseline at the other end melts or not. Try to answer what causes the vaseline to melt.

(*c*) **Conductivity of electricity:** Assemble a conductivity apparatus as shown below. Use a clean form of each substance. Observe the brightness of the lamp. Observe which substance gives the brightest light?

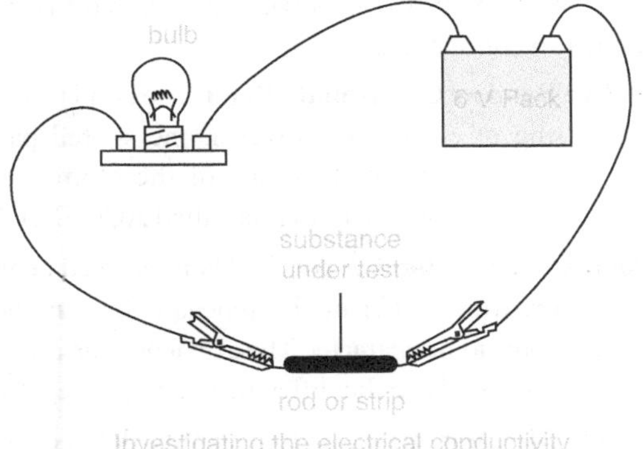

Investigating the electrical conductivity

(*d*) **Malleability and ductility**

(*i*) Put a small piece of each substance, separately on a block of iron or an anvil if available. Use a small hammer to hit it gently at first and then more strongly. Observe whether the substance break or widen in a sheet. Record which element widens most easily on hammering. What can you say about the strength of the interatomic forces in that element?

(*ii*) Fix a strip of each of the elements which do not break on hitting tightly into a vice. Hold the free end of each strip with pliers and pull as strongly as possible. Record which can be pulled apart most easily and otherwise.

Substance metal/ non-metal	*Nature of surface*	*Thermal conduc-tivity*	*Electrical conduc-tivity*	*Melleab-ility*	*Ductility*
Cu Al Fe Zn Sn					
S Charcoal graphite					
Si					

From the results of your experiment give, with brief reasons, one element that can be used to:

(*i*) conduct electricity in an electric iron;

(*ii*) transport electric current from a power plant to your town;

(*iii*) insulate an electric wire;

(*iv*) conduct electricity into a molten ionic substance;

(*v*) make earrings by moulding easily.

The general properties associated with the three primary interatomic bonds are summarized in a tabular form as follows.

General Characteristics of Substances with different Interatomic Bonding

Characteristics	*Bond Type*			
	Ionic Bond	*Covalent Bonds*		*Metallic Bonds*
		Molecular Compounds	*Giant Network Structure*	
Examples	NaCl, KCl	H_2, HCl, Cl_2, I_2	Diamond, graphite, SiO_2	Na, Mg, Al
Physical state	Solid	Gases, liquid, solids	Solids	Solids
Basic units	Cations and anions	Molecules	Atoms	Atoms; free valence electrons and positive kernels
Bonding	Ionic	Covalent within molecules, weak bonds between molecules	Covalent	Metallic
Structure	Closely packed ions of opposite charges	Molecular	Giant network	Electron gas model
Properties				
(*a*) *melting point*	High	Low	Very high	High
(*b*) *boiling point*	High	Low	Very high	High
(*c*) *hardness*	Hard, brittle	Soft in solid state	Very hard	Hard
(*d*) *electrical conductivity*	Non-conductors in solid state, conductors in molten and aqueous solution	Non-conductors except dissolved in water to form ions *e.g.*, HCl	Non-conductors except graphite	Good conductors
(*e*) *heat conductivity*	Non-conductors	Non-conductors	Non-conductors	Good conductors
(*f*) *solubility*				
(*i*) in water	Soluble	Many are insoluble	Insoluble	Insoluble
(*ii*) non-polar solvents	Insoluble	Soluble	Insoluble	Some insoluble react

SUMMARY

- **Chemical Bond.** The force of attraction which holds various chemical entities in different species.
- **Electrovalent Bond.** The attractive force between oppositely charged ions which come into existence by the transfer of electrons.
- **Electrovalency.** The number of electrons which an atom loses or gains while forming an ionic or electrovalent bond.
- **Covalent Bond.** The bond that comes into existence by the mutual sharing of electrons among the atoms participating in the bonding.
- **Valence Bond Approach of Covalent Bond.** The bond is formed by the overlapping of half-filled atomic orbitals having electrons with opposite spins.
- **Covalency.** The number of half-filled atomic orbitals which an atom provides for participation in overlapping at the time of bonding.
- **Dative Bond or Co-ordinate Bond.** The bond is formed by sharing of electrons. However, the shared pair of electrons is contributed by one of the atoms called DONOR while the other atom is called ACCEPTOR.
- **Polar Covalent Bond.** A covalent bond formed between two atoms which differ in electronegativities and hence acquire some ionic character. The extent of ionic character in a bond depends on the difference in electronegativity.
- **Dipole Moment** (μ). A vector quantity defined by the product of charge developed on any of the atom and the distance between the atoms; creating a dipole.

- **Polar and Non-Polar Molecules.** The molecules with dipole moment (μ) > 0 are called polar molecules while those with $\mu = 0$ are non-polar molecules.
- **Metallic Bonds.** The attractive forces holding metal atoms together. They arise due to an average force of attraction between metallic kernels and mobile or delocalised valence electrons.

EVALUATE YOURSELF

I. Objective Type Questions

Select the most appropriate choice from the options given as (a), (b), (c) and (d) after each question:

1. Which of the following species are constituent atoms held by non-directional bonds?
   (*a*) NH_3 (*b*) CsCl
   (*c*) NF_3 (*d*) BeF_2.
2. X and Y atoms have 2 and 6 valence electrons in their outermost shells respectively, the compound which X and Y are likely to form is:
   (*a*) XY_2 (*b*) XY
   (*c*) YX_2 (*d*) YX_3.
3. Which of the following substance is a polar covalent molecule
   (*a*) Hydrogen sulphide (*b*) Nitrogen
   (*c*) Potassium chloride (*d*) Oxygen.
4. Which of the following statement is/are correct
   I. Energy is absorbed when a chemical bond is formed
   II. Energy is released when a chemical bond is formed
   III. SF_6 is a super octet molecule
   (*a*) I and III (*b*) III and II
   (*c*) II only (*d*) III only
5. LiCl is a compound of Li and Cl_2. The compound is reported to have a partial covalent character because
   I. Lithium has sufficient non-metal character
   II. Charge density of Li^+ is very high
   III. Cl_2 has pure non-metal character
   IV. Cl^- ion has high polarizing power
   (*a*) I, II (*b*) III, IV
   (*c*) Only II (*d*) II and IV.
6. In ammonium chloride which bonds are present
   I. ionic II. covalent
   III. metallic IV. dative
   (*a*) I and II (*b*) II and III
   (*c*) I, III, IV (*d*) I, II, IV.
7. Which of the following compounds does not contain a double bond.
   (*a*) Carbon (IV) oxide (*b*) Hydrogen chloride
   (*c*) Oxygen (*d*) Sulphur (IV) oxide.
8. Which of the following species does not contain a co-ordinate covalent bond?
   I. Ozone II. Nitric acid
   III. Hydrogen peroxide IV. Calcium oxide
   (*a*) III, IV (*b*) II, III
   (*c*) IV, II (*d*) III.
9. Which pair of properties apply to ionic compounds?
   (*a*) Brittleness and solubility in benzene
   (*b*) Solubility in water and good electrical conductivity in dissolved state
   (*c*) Malleability and ductility
   (*d*) Volatility and ductility.
10. The electronegativity of Na, F, O and K are 0.9, 4.0, 3.5 and 0.8 respectively. The following is/are correct
    I. NaF is more ionic than KF
    II. O-F bond is polar
    III. Oxygen will have a partial negative charge when it forms a bond with fluorine
    IV. Fluorine and oxygen can both form non-polar diatomic molecules
    (*a*) I, II, IV (*b*) I, III, IV
    (*c*) II, IV (*d*) IV, II, III.
11. Which of the following is a non-directional bond?
    (*a*) Coordinate (*b*) Covalent
    (*c*) Metallic (*d*) None of these.
12. Which of the following elements has the largest tendency to form an electrovalent bond?
    (*a*) $1s^2$ (*b*) $1s^22s^22p^6$
    (*c*) $1s^22s^2$ (*d*) $1s^22s^22p^63s^1$.

II. Fill in the Blanks

13. Complete the following sentences by supplying appropriate words:
    (*i*) A polar covalent bond is formed between two atoms having ...... electronegativities
    (*ii*) Metals are good conductors of electricity due to the presence of ......
    (*iii*) The value of the dipole moment for a polar molecule is ......
    (*iv*) Carbon (IV) oxide has two polar bonds in its molecule but its dipole moment is ......
    (*v*) Covalency of oxygen is ...... and its electrovalency is ......
    (*vi*) Magnesium is bonded to chlorine atoms in magnesium chloride by ...... bonds

(*vii*) In sodium chloride crystal a single NaCl unit......

(*viii*) For the stability of an ionic crystal, the lattice enthalpy should be ......

(*ix*) Had H_2O been a linear molecule its dipole moment would have been ......

(*x*) Solubility of ionic crystal in water is attributed to ......

III. Discussion Questions

14. How would you differentiate between an atom and an ion? Write the Lewis symbol for an atom with atomic number 16 and its corresponding ion.
15. Write Lewis symbols of an anion with a charge of negative three and a cation with a charge of positive two.
16. What is a chemical bond? Name the various types of inter-atomic bonds.
17. Draw Lewis symbols for each of the following:

$${}_{11}Na,\ {}_{5}B,\ {}_{54}Xe,\ {}_{16}S^{2-}$$

18. What is an ionic bond? Explain its formation by taking the example of sodium and chlorine atom.
19. What energy changes take place when gaseous sodium and gaseous chlorine atoms combine to form solid sodium chloride?
20. Draw Lewis structures for the ionic compounds formed between the following pairs of element.
    (*i*) Sodium and oxygen (*ii*) Potassium and sulphur
    (*iii*) Calcium and oxygen (*iv*) Aluminium and fluorine.
21. List two important properties of ionic compounds.
22. Explain in brief why sodium chloride does not conduct electricity in the solid state but it does so in its aqueous solution?
23. Draw a well labelled unit lattice cell for sodium chloride.
24. What is a polar covalent bond? Give an example of a polar diatomic molecule. Draw its structure.
25. Distinguish between a polar covalent bond and a coordinate covalent bond.
26. Show how the co-ordinate covalent bond is formed between ammonia and the hydrogen ion? State the electronic property of the hydrogen ion which makes it possible for it to form the co-ordinate bond.
27. Three elements X, Y and Z have the Lewis symbols:

$$\dot{X} \quad \ddot{Y} \quad \cdot\ddot{Z}:$$

    (*i*) Place the elements in an appropriate group of the periodic table.
    (*ii*) Which elements are likely to form ions? What is the expected charge on the ions?
    (*iii*) Write the formulas and Lewis structures of covalent compounds formed between
    (*a*) X and Y (*b*) Y and Z.
28. For each of the following substances: ammonium chloride, hydrogen sulphide and hydrogen.
    (*i*) state their particle types (ions, atoms or molecules).
    (*ii*) state the type of forces holding the particles together.
    (*iii*) state with reasons, which is hardest in the solid state.
29. Starting from the following pairs of atoms, draw Lewis structures or other diagrams of how chemical bonds can be formed between the two. State the type of bond formed in each case.
    (*i*) Na, Cl (*ii*) O, H
    (*iii*) O, O (*iv*) H, H
    (*v*) Na, Na.
30. Hydrogen can form both ionic and covalent bonds. Give two examples of compounds which show this dual property.
31. Write the structure of the compound formed between ammonia and $AlCl_3$.
32. What is a hydronium ion? Show how it can be formed from a water molecule.
33. Draw electron dot/cross or Lewis structures for the following species. Show only the outer electrons.
    (*i*) water, H_2O (*ii*) ethane, C_2H_6
    (*iii*) sodium fluoride, NaF (*iv*) BF_3
    (*v*) CS_2 (*vi*) OH^-
    (*vii*) ozone, O_3 (*viii*) C_2H_4
    (*ix*) CH_3COOH
    Which of the species among the above does not obey the octet rule?
34. What are the limitations of the octet rule?
35. Explain the following observations:
    (*a*) Two rods, one charged positively and the other negatively, are brought near a stream of water flowing from a burette, both rods attract the stream of water.
    (*b*) Aluminium metal strips are good conductors of electricity and melt at a higher temperature than iodine crystals.
36. (*a*) Define lattice enthalpy.
    (*b*) State two factors which influence the magnitude of lattice enthalpy.
    (*c*) Which of the following salts will have the higher lattice energy: sodium chloride or magnesium chloride. Explain.
37. The two substances AB_2 and XY are solids. One is a covalent substance and the other ionic. Devise a simple experiment to show which of them is ionic, and which is covalent.
38. Explain the following observations:
    (*i*) Li^+ migrates more slowly in aqueous solution than in the molten liquid.
    (*ii*) Lithium salts are more hydrated than sodium salts.
    (*iii*) Naphthalene has a lower melting point than sodium trioxonitrate (V).

(*iv*) The iodine ion is more polarizable than the fluoride ion.

(*v*) $AlCl_3$ sublimes but NaCl does not.

39. Explain why sodium chloride is brittle but copper is malleable?

40. Given below is data for two chlorides of the elements X and Y.

Chloride	*Melting point/°C*	*Boiling point/°C*	*Solubility in 100 g of water/g*	*Solubility in 100 g benzene/g*
XCl	800	1440	38	0.07
YCl_4	– 23	77	0.08	very soluble

(*a*) What is the type of bond between
(*i*) X and Cl? (*ii*) Y and Cl?

(*b*) Which chloride is made up of molecules and why?

(*c*) Which of the elements X and Y could be a metal? Give reasons.

(*d*) Explain the differences in solubility of the chlorides in water.

41. You are given the following substances: sodium; neon; sodium bromide; diamond and trichloromethane (chloroform).
Select from the five substances:

(*i*) a volatile liquid

(*ii*) a mono-atomic substance which does not form strong chemical bonds

(*iii*) a solid with giant covalent structure

(*iv*) a solid with giant ionic structure

(*v*) a polyatomic molecular substance

(*vi*) a substance which is soft in the solid state

(*vii*) a substance which conducts electricity in the molten state but not in the solid state.

42. Explain, in terms of electronic configuration, ionization and lattice enthalpies, why the compound $NaCl_2(s)$ is not stable and cannot be formed.

43. Suggest reasons for the following:

(*a*) copper (II) tetraoxosulphate (VI) solid is brittle,

(*b*) water is a liquid and sodium chloride is a solid at room temperature,

(*c*) gold ornaments can be beaten into other shapes,

(*d*) sulphur is not an electrical or heat conductor,

(*e*) it is easier to put a *dent* in a metal object but difficult to remove it.

44. Arrange, giving reasons, chlorine, beryllium chloride and sodium chloride in order of increasing covalent character.

45. (*a*) Explain why sodium metal is a good conductor of heat but chlorine gas is not?

(*b*) Arrange, giving reasons, sodium, magnesium and potassium in order of increasing hardness.

Answers

I. Objective Type Questions

1. (*b*)	**2.** (*b*)	**3.** (*a*)
4. (*b*)	**5.** (*c*)	**6.** (*d*)
7. (*b*)	**8.** (*a*)	**9.** (*b*)
10. (*c*)	**11.** (*c*)	**12.** (*d*).

II. Fill in the Blanks

13. (*i*) different
(*ii*) mobile valence electrons
(*iii*) > 0 (*iv*) equal to zero
(*v*) 2, – 2 (*vi*) ionic
(*vii*) does not exist (*viii*) high
(*ix*) equal to zero (*x*) ion-dipole interactions.

SECTION-3

UNIT-8

INTERMOLECULAR BONDING

Specific Objectives

After studying this unit, the student should be able to:

- describe different types of intermolecular forces found in covalent compounds
- explain how intermolecular forces arise from the structural features of molecules
- describe how intermolecular forces affect the solubility and boiling points of substances
- describe the formation of hydrogen bond
- describe the existence of van der Waals' forces between covalent molecules.

8.1 INTERMOLECULAR FORCES IN COVALENT COMPOUNDS

We have studied that in covalent molecules, the constituent atoms are held by covalent bonds. These molecules may be polar or non-polar. In a covalent substance, the covalent molecules are further held by some attractive forces. For example, H_2O molecule is a descrete unit which contains two O–H covalent bonds. But a sample of water contains a large number of H_2O units held by some attractive forces. *The attractive forces operating between the molecules of a given substances (solid, liquid or gas) are called intermolecular forces.*

TYPES OF INTERMOLECULAR FORCES

The different types of intermolecular forces among covalent molecules are:

1. Hydrogen bonds

These attractive forces operate among the molecules containing O–H or N–H or F–H bonds.

2. van der Waals' forces

These attractive forces operate among polar as well as non polar molecules. They got their name from a Dutch scientist Johannes van der Waal (1837–1923) who first suggested their nature.

van der Waals' forces are further classified as:

(*i*) *Dispersion forces or London forces,*

(*ii*) *Dipole-Dipole forces and*

(*iii*) *Dipole-induced dipole forces.*

It is worth noting that the electrostatic forces between the ions of opposite charges (*interionic forces*) and those between ions and polar molecules (*ion-dipole forces*) do not fall in the category of van der Waal forces.

8.2 RELATIONSHIP BETWEEN INTERMOLECULAR FORCES AND STRUCTURAL FEATURES

The nature and magnitude/strength of intermolecular forces depends on the type of bonds and other structural features in the molecules. For example, both NH_3 and PH_3 have the same number of bonds and the same molecular shape.

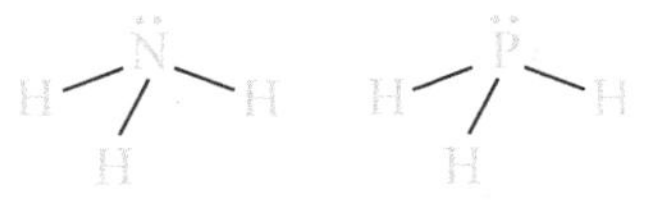

Pyramidal shape

However, there is a vast difference in their boiling points. For example, the boiling point of NH_3 is greater than that of PH_3. It means that the intermolecular forces operating in liquid ammonia and in liquid PH_3 have different strengths. Similarly, H_2S and H_2O are both triatomic molecules and possess bent-shape along with two lone pairs. But H_2S is a gas at room temperature whereas H_2O is a liquid. It also means that the intermolecular forces in $H_2O(l)$ is relatively much stronger than in H_2S

BP = 373 K BP = 213 K

Both have bent shape

The magnitude of interparticle forces also differ in H_2O, NH_3, PH_3 and H_2S because each one will have a different boiling point. Thus, we can conclude that the nature and magnitude of intermolecular forces depend upon the molecular shape and type of bonds it contains.

EFFECT OF INTERMOLECULAR FORCES ON THE PROPERTIES OF MOLECULAR SUBSTANCES

The magnitude of interparticle forces control many of the physical properties such as *melting point, boiling point, viscosity, surface tension, molar heat of vaporisation,* extent of *evaporation, etc.* In many cases they also affect the chemical properties of the substances. It may be noted that the chemical properties are mostly controlled by **intramolecular forces** which exists within the same molecule.

8.3 HYDROGEN BOND

When a H-atom is bonded to a highly electronegative atom like F, O or N (say, Z) by a covalent bond, the bond pair of electrons is displaced towards the electronegative atom. When the solitary electron of a hydrogen atom lies away from it, its nucleus gets exposed and behaves as a bare proton. Such a bare hydrogen nucleus exerts a strong electrostatic attraction on the electronegative atom of the adjacent molecule. This

$$\cdots\cdots \overset{\delta^+}{H}-\overset{\delta^-}{Z}\cdots\cdots\overset{\delta^+}{H}-\overset{\delta^-}{Z}\cdots\cdots\overset{\delta^+}{H}-\overset{\delta^-}{Z}\cdots\cdots$$

interaction between the hydrogen atom of one molecule and the electronegative atom of the other molecule is referred to as hydrogen bond. Thus, *hydrogen bond is defined as the electrostatic force of attraction which exists between the covalently bonded hydrogen atom of one molecule and the electronegative atom of the other molecule.* The hydrogen bond is represented by dotted line (.....). For example, in the case of hydrogen fluoride the hydrogen bond exists between the H atom of one molecule and the fluorine atom of another molecule as shown.

$$\cdots\cdots H-F\cdots\cdots H-F\cdots\cdots H-F\cdots\cdots$$

It may be noted that the hydrogen atom acts as a bridge between the electronegative atoms of two different molecules. To one atom it is linked through a covalent bond while to the other it is linked through a hydrogen bond. Please note that the hydrogen bond does not lie in the centre of the two atoms forming the covalent bond.

STRENGTH OF HYDROGEN BOND

The hydrogen bond is much weaker than a covalent bond and its strength ranges between 12.5 kJ mol^{-1} to 41.5 kJ mol^{-1}.

The strength of the hydrogen bond is governed by the following factors:

(*i*) *Extent of polarization within the molecule.* The molecule must contain an electronegative atom linked to the H-atom. The higher the electronegativity of the atom, the greater the polarization and the larger the strength of the hydrogen bond.

(*ii*) Size of the electronegative *atom should be small.* We know that electronegativity increases from $N \rightarrow O \rightarrow F$ and atomic radius also decreases in the same order. Therefore, H–bond strength increases as

$$H–N....H < H–O....H < H–F....H$$

(*iii*) *Lone pair* of electrons must be present on the electronegative atom which can point towards the H–atom of an adjacent or near by molecule.

(*iv*) The angle between the three atoms involved (two electronegative atoms and the central H atom) must be 180°. It may be noted that as the angle decreases, the strength of the H–bond also decreases and no H–bonding occurs when the angle falls below 140°.

EXAMPLES OF HYDROGEN BONDING

Let us study some examples of the compounds which exhibit hydrogen bonding.

1. Hydrogen fluoride. In hydrogen fluoride the hydrogen atom is bonded to a highly electronegative atom of fluorine (electronegativity = 4). It has been found that in the solid state hydrogen fluoride consists of long zig-zag chains of H-F molecules associated by H-bonds as shown in Fig. 8.1(*a*).

On heating, progressively, the length of the chain shortens, while the number of associated units reduces.

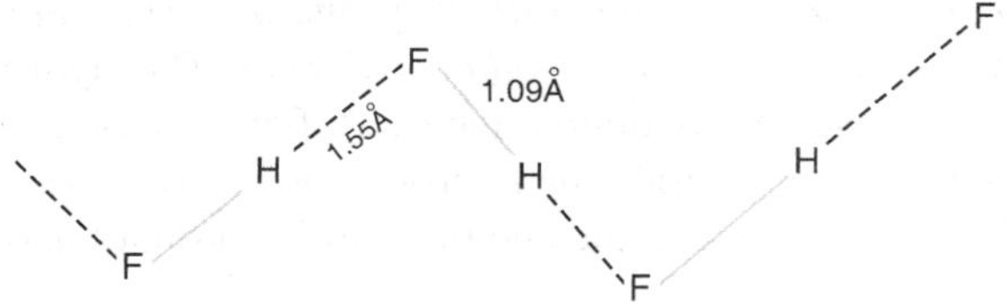

Fig. 8.1(a). Zig-zag chains of H-F molecules.

2. Water. In water molecule, oxygen atom is bonded to two hydrogen atoms. Due to its large electronegativity, the oxygen atom forms the negative centre whereas each of the hydrogen atoms acquires a partial positive charge. Each O atom can form two hydrogen bonds as shown in Fig. 8.1 (*b*).

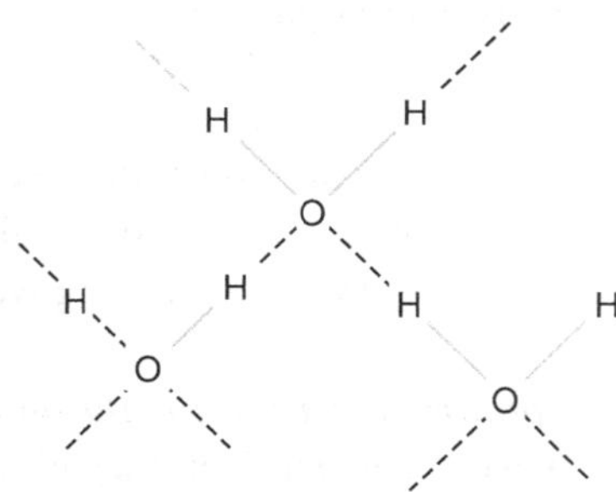

Fig. 8.1(b). H-bonds in H_2O molecules.

3. Ammonia. In ammonia molecule, nitrogen, an electronegative atom is bonded to three hydrogen atoms. The nitrogen atom forms the negative site of the molecule whereas each of three H-atoms acquires a partial positive charge. The ammonia molecules are associated by H-bonds as shown in Fig. 8.1 (*c*).

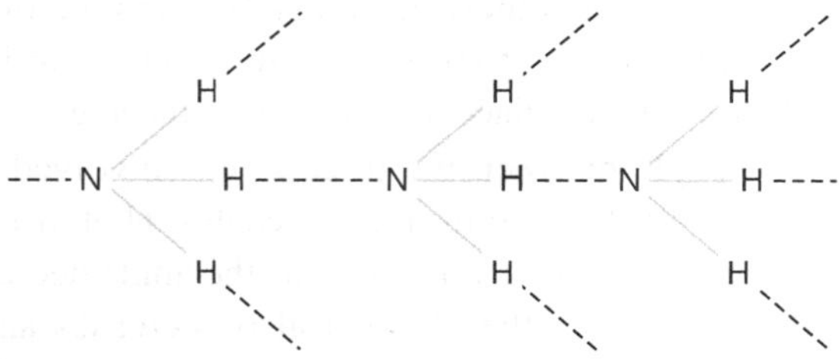

Fig. 8.1(c). H-bonds in NH_3 molecules.

4. Ethanol (C_2H_5OH) and Ethanoic acid (CH_3COOH). These molecules contain the highly electronegative oxygen atom linked to H-atom and hence, form associated molecules as shown in Fig. 8.2.

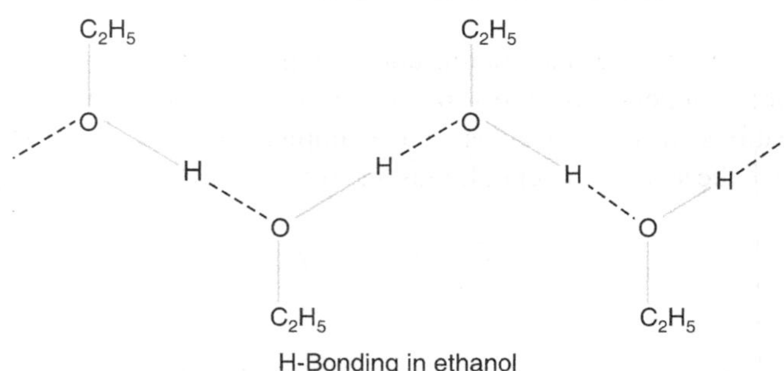

H-Bonding in ethanol

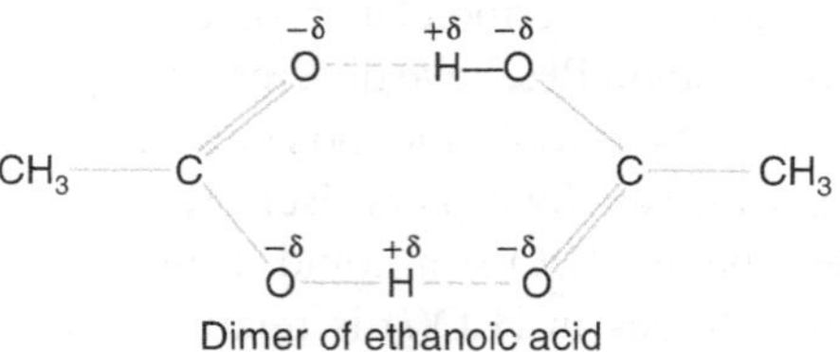

Dimer of ethanoic acid

Fig. 8.2. H–bonds in C_2H_5OH and CH_3COOH.

EFFECT OF HYDROGEN BONDING ON THE PROPERTIES

Hydrogen bond has a marked influence on the properties of various substances as discussed below:

(*i*) **Association.** The hydrogen bonds link up molecules of the same substance to form large aggregates. This is called association of molecules. For example, H-F molecules are associated with one another by hydrogen bonds. The formula of hydrogen fluoride can be written as $(HF)_n$ where the value of *n* depends on the physical state of the compound. Similarly, enthanoic acid exists as a dimer and conforms to the formula $(CH_3COOH)_2$. The association of molecules through hydrogen bonding results in the unexpected larger values of many physical properties such as *melting points, boiling points, enthalpy of vaporisation, enthalpy of fusion*, etc. For example, if we compare these physical properties among the hydrides of elements of group 16 we find that all these properties are unexpectedly higher for H_2O as compared to those of H_2S, H_2Se and H_2Te as shown in Table 8.1. This is because of the fact that H_2O molecules are associated through hydrogen bonding whereas the hydrides of other elements have van der Waal forces among their molecules.

Table 8.1. Physical Properties of Hydrides of Group 16 Elements

Hydride	*Molar mass* ($g\ mol^{-1}$)	*M.P.* (T_m) (*K*)	*B.P.* (T_b) (*K*)	ΔH_{vap} ($kJ\ mol^{-1}$)
H_2O	18	273	373	40.7
H_2S	34	188	213	18.7
H_2Se	81	207	232	19.7
H_2Te	130	225	271	23.2

(*ii*) **Melting and boiling points.** The compounds whose molecules are associated with one another by hydrogen bonds have abnormally high melting and boiling points. It is due to the fact that a large amount of energy is needed to overcome the intermolecular hydrogen bonds and to separate the molecules. Let us compare the boiling points of hydrides of elements of group 15, 16 and 17 of the periodic table.

The hydrides of the first member of each group namely; NH_3, H_2O and HF have abnormally high boiling points. For example, in the case of halogen acids the boiling point decreases with a decrease in the molecular mass, *i.e.*, from HI to HBr and from HBr to HCl. But there is a sudden rise in the value of boiling point in case of HF. The relatively high boiling point of HF is due to the presence of intermolecular hydrogen bonds in hydrogen fluoride. In the case of the other halogen acids no H-bonding occurs due to the large atomic sizes and small electronegativities of the other halogens.

A similar explanation can be used for the abnormal boiling point of H_2O among the hydrides of elements of the oxygen group and that of NH_3 among the hydrides of the elements of the nitrogen group. The relative boiling points of different hydrides of group 14, 15, 16 and 17 are shown graphically in Fig. 8.3.

It may be noted from the graph that CH_4 in group 14 does not have H–bonds as intermolecular forces. Hence, it does not follow the pattern of hydrides of elements of group 15, 16 and 17.

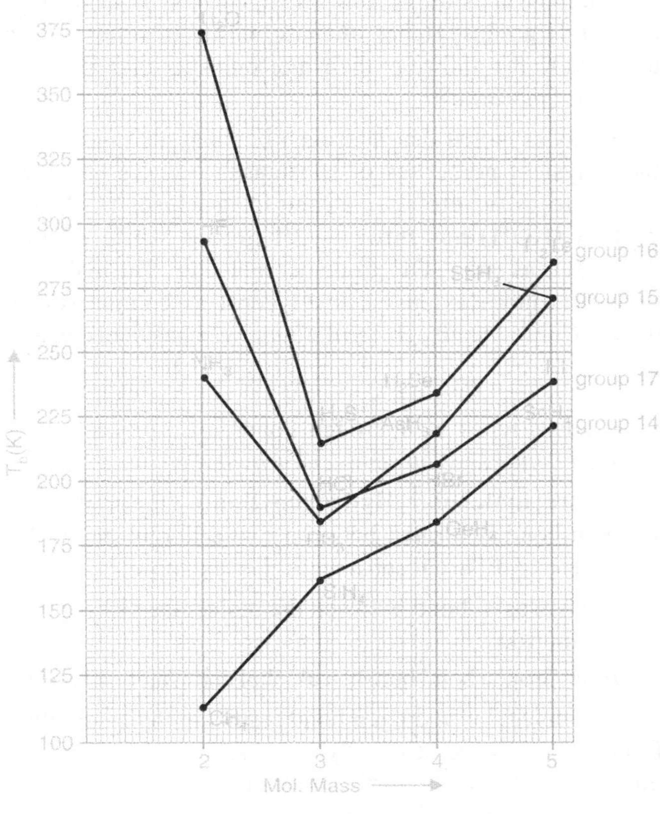

Fig. 8.3

(*iii*) **Solubility.** Hydrogen bonding is also responsible for the solubility of certain compounds in water. We are familiar with the fact that ionic compounds dissolve in water due to ion-dipole interactions. But the molecular compounds having polar groups in their molecules dissolve in water due to their tendency to form H-bonds with water molecules. For example, lower alcohols are soluble in water because their molecules can form H-bonds with water molecules.

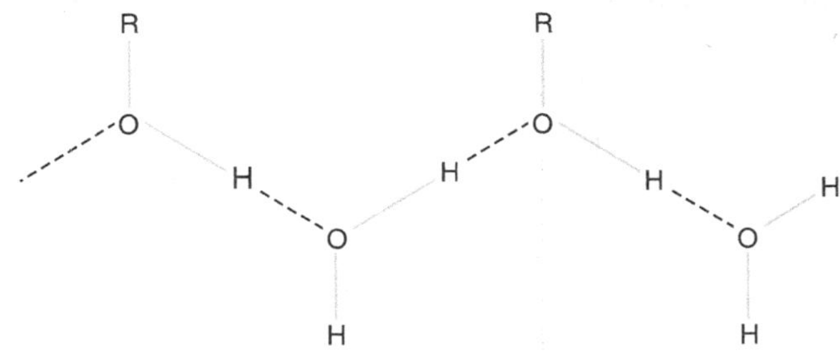

H-bonding in alcohol and water molecules.

The molecules of glucose, sugar, carboxylic acids have polar–O–H groups. Hence, their solubility in water is also attributed to the ability of their molecules to form H-bonds with water molecules.

It may be noted that hydrogen bonding also influences other physical properties of substances such as *viscosity, surface tension*, etc. For example, glycerol ($CH_2OH.CHOH.CH_2OH$) has three —OH groups while ethanol (C_2H_5OH) has one —OH group per molecule. The hydrogen bonding is more extensive in glycerol as comparison to ethanol. Consequently, glycerol is more viscous than ethanol.

TYPES OF HYDROGEN BONDS

The two types of hydrogen bonds are *intermolecular hydrogen bonds* and *intramolecular hydrogen bonds.*

1. Intermolecular Hydrogen Bond. Intermolecular hydrogen bond is formed between two different molecules of the same or different substances. For example:

(*i*) Hydrogen bond between the molecules of hydrogen fluoride.

(*ii*) Hydrogen bond between alcohol and water molecules.

Intermolecular hydrogen bond results in the association of molecules. Hence, it usually increases the melting point, boiling point, viscosity, surface tension, solubility, etc.

2. Intramolecular Hydrogen Bond. Intramolecular hydrogen bond is formed between a hydrogen atom and a highly electronegative atom (F, O or N) present in different bonds within the same molecule.

Intramolecular hydrogen bond results in the cyclisation of molecules and prevents their association. Consequently, the effect of intramolecular hydrogen bond on the physical properties is negligible.

A few examples of molecules which form intramolecular hydrogen bonds are given below:

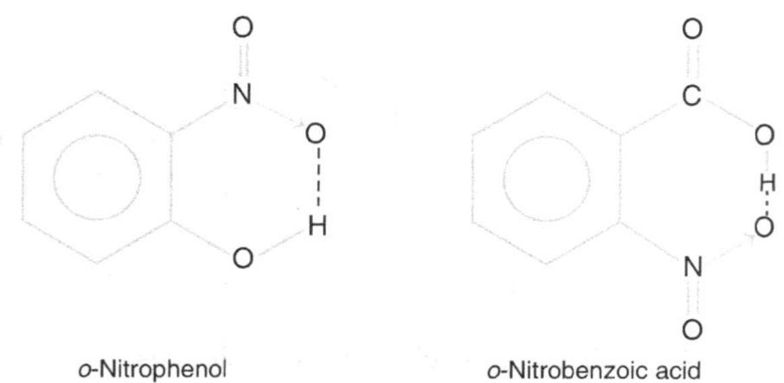

H–Bonding in Water and Ice

We have studied in the previous unit that water is a polar molecule having a dipole moment (μ) = 1.84D as shown in Fig. 8.4.

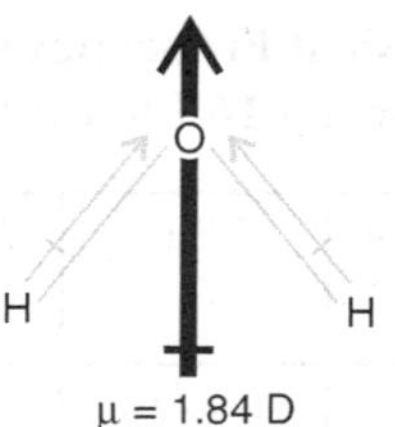

Fig. 8.4. Polar nature of H_2O.

AGGREGATION OF WATER MOLECULES IN THE LIQUID STATE

In the **gaseous state,** the individual covalent molecules in H_2O exist as such. However, in the **liquid state,** large aggregates of a varying number of H_2O units are formed because of their association through **intermolecular hydrogen bonds.**

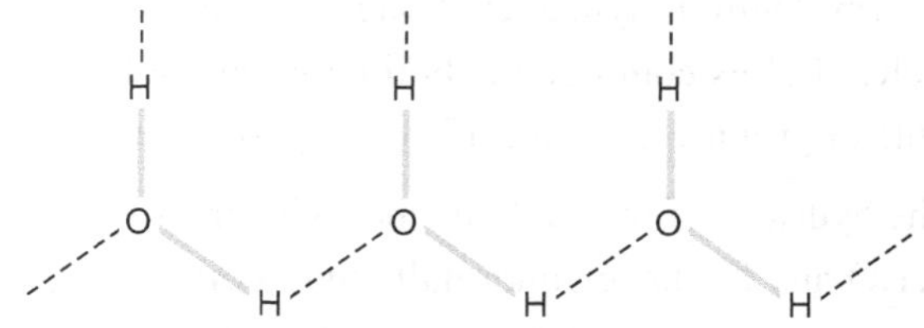

Association of H_2O molecules by H-Bonds

In fact, in liquid water, the free H_2O molecules and associated H_2O molecules exist in a state of dynamic equilibrium.

$$nH_2O \rightleftharpoons (H_2O)_n$$

The extent of association, however, depends upon the conditions of temperature and pressure. Due to intermolecular H–bonding water has an abnormally high *freezing point, boiling point, enthalpy of fusion, enthalpy of vaporisation.* It also has a high *surface tension* and *viscosity*. Due to the high surface tension water droplets are spherical in shape and surface of water bends to form a curved *meniscus* in burettes and capillary tubes.

Hydrogen Bonding in Ice

In **ice,** *the solid state of water*, each H_2O molecule is tetrahedrally surrounded by four neighbouring H_2O molecules with their oxygen atoms occupying the corners of a tetrahedron.

There are four H atoms around each O atom. Two of the four H atoms are bonded by covalent bonds (bond length 100 pm) whereas the other two are linked through hydrogen bonds (176 pm) as shown in Fig. 8.5. This gives a highly ordered three dimensional structure having large vacant spaces, Fig. 8.6 which may be compared to an open cage.

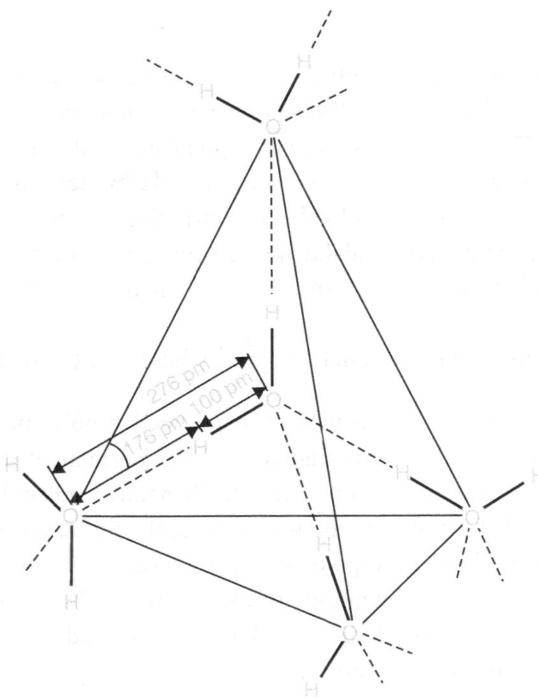

Fig. 8.5. Arrangement of H-bonds and covalent bonds around O in ice.

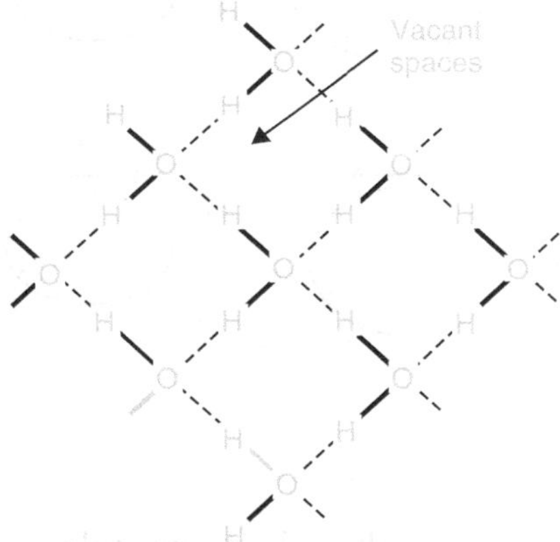

Fig. 8.6. Open cage like structure of ice.

Due to open the cage-like structure, ice has a relatively larger volume for a given mass of liquid water. Consequently, *the density of ice is less than water* and it floats over water.

As the temperature is raised beyond 273 K, the open cage like structure starts dismantling due to the cleavage of some H-bonds and the ice melts. The breaking of H-bonds causes large aggregates of H_2O molecules to move closer to each other. This results in a decrease in volume and an increase in density till 277 K when the density becomes maximum. Beyond 277 K more H-bonds cleave expansion of liquid water starts occurring due to the increased K.E. of molecules and the density starts decreasing again. However, it remains higher than ice and this makes the density of water maximum at 277 K (4°C).

This property of maximum density at 277 K is a boon for the survival of aquatic animals during winter months when the upper layer of sea water freezes.

The frozen water does not sink to the bottom but keeps floating at the surface due to its lesser density. This provides thermal insulation to the water below it.

HYDROGEN BONDS AND STRUCTURE OF MACROMOLECULES

Macromolecular structures such as proteins, carbohydrates, nucleic acids, etc., are joined together by the formation of H-bond through their active groups. Some examples are discussed as follows:

(*i*) Proteins are polymers of α-amino acids and they have $>C=O$ and $-N-H$ groups in their molecules. These groups form both *intermolecular* and *intramolecular* hydrogen bonds.

The α-*helix* structure (spiral structure) of proteins is due to *intramolecular hydrogen bonding* (Fig. 8.7.) Further, the protein chains are held together by *intermolecular hydrogen bonds* to form fibres as in hair, silk and wool, etc. It may be noted that the curly nature of hair is because of S–S cross links in addition to H-bonding. This cross linking occurs between one part of the α-helix chain over another.

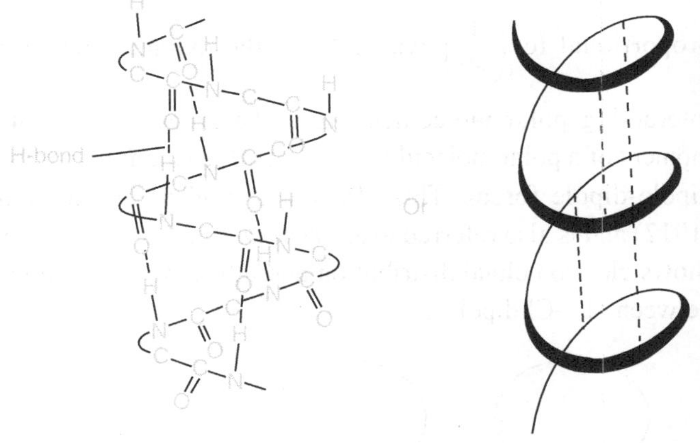

Fig. 8.7. α-helix structure of protein.

Proteins are appreciably soluble in water due to their tendency to form H-bonds with water molecules. Proteins are an important part of body tissue. Dissolved protein is absorbed into the blood and other aqueous fluids so that it can be used for growth. Proteins are an essential constituent of food and occur in meat, fish and beans.

Cellulose which forms about 50% of wood is a polymer of glucose units. The polymeric chains in cellulose are held by hydrogen bonds and it gives a fibrous nature to cellulose which can be observed in wood pulp, cotton, jute etc.

8.4 VAN DER WAALS' FORCES

It has been pointed out earlier in the beginning of this unit that the attractive forces existing between polar as well as non-polar molecules are collectively called van der Waals' forces. van der Waals' forces can be classified into the following categories:

(*i*) Dipole-Dipole forces

(*ii*) Dipole-induced dipole forces

(*iii*) Dispersion forces.

Let us proceed to discuss the nature of these attractive forces.

Dipole-Dipole Forces

Dipole-dipole forces are *attractive forces operating between the molecules possessing permanent dipole.* The ends of the dipole possess "*partial charges*" and these charges are shown by the Greek letter delta (δ). Partial charges are always less than the unit charge (1.6×10^{-19} C) because of the effect of electron sharing. The polar molecules interact with neighbouring molecules. This interaction is weak as compared to ion-ion interactions because only partial charges are involved. The interaction energy decreases with an increase in the distance between the dipoles. It is also proportional to $\left(\frac{1}{r^6}\right)$ where '*r*' is the distance between interacting polar molecules. Now, the greater the dipole moment of a polar molecule, the greater the magnitude of the dipole-dipole forces. This effect was studied by Keesom (1912) and is also referred to as *orientation effect.* Figure 8.8 shows electron cloud distribution and attractive interactions between H—Cl dipoles.

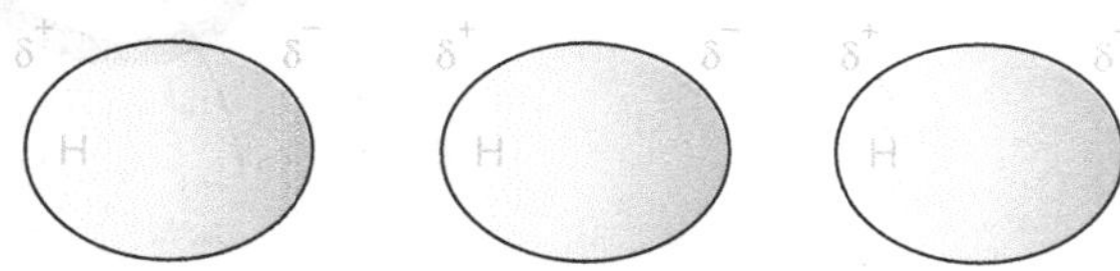

Fig. 8.8. Attractive interactions between H—Cl dipoles

The magnitude of dipole-dipole forces in different polar molecules can be predicted on the basis of the polarity. In general, for the molecules of similar molecular masses, the magnitude of the dipole-dipole forces is higher if the molecule is more polar molecules. For example, the molecular masses of PH_3 and H_2S are the same but H_2S is more polar (μ = 1.10D) than PH_3 (μ = 0.55D). Hence the boiling point of H_2S (187 K) is higher than that of PH_3 (T_b = 137 *k*).

Dipole-Induced Dipole Forces

This type of attractive forces operate *between polar molecules having $\mu > 0$ and non-polar molecules having $\mu = 0$.* The permanent dipole of a polar molecule induces a dipole on an electrically neutral molecule by deforming or polarising its electron cloud. The attractive forces between the permanent dipole and the induced dipole are called *dipole-induced dipole forces.* In this case also the energy of interaction is proportional to $\left(\frac{1}{r^6}\right)$ where *r* is the distance between the two molecules. The induced dipole moment depends upon the dipole moment of the parmanent dipole and the polarizability of the electrically neutral molecule. As the size of the molecule/atom increases, the influence of the permanent electrical dipole on it also increases, resulting in an increase in the magnitude of these forces. The existence of these forces was studied by Debye (1920) and this effect is termed as *induction effect.*

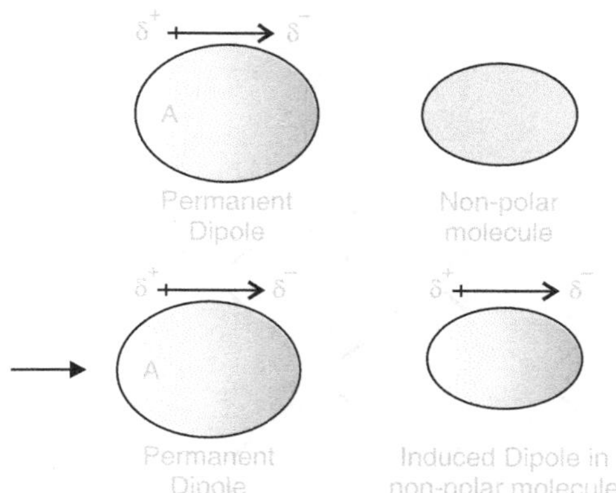

Fig. 8.9. Dipole-induced dipole forces between permanent dipole and induced dipole.

In this case also there is a cummulative effect of dispersion forces and dipole-induced dipole interactions.

Dispersion Forces or London Forces

These are the interparticle forces among monoatomic or non-polar molecules such as N_2, He, H_2, CO_2, etc. Atoms and other non-polar molecules are electrically symmetrical and have no dipole moment because their electronic charge cloud is symmetrically distributed. However, an instantaneous dipole may develop in such atoms and molecules. In order to understand this let us assume two atoms of neon 'A' and 'B' are in close vicinity of each other (Fig. 8.10 (*a*)). Rapid movement of electrons in A may cause momentary accumulation of electron density on one side, causing the charge distribution to become unsymmetrical. In other words, the charge cloud is more concentrated on one side than the other (Figs. 8.10 (*b*) and 8.10 (*c*)). This will result in the development of a temporary dipole on the atom 'A' for a very short time. This instantaneous or transient dipole distorts the electron density of the atom 'B' which is close to it. In other words, a dipole is also induced in the atom 'B'.

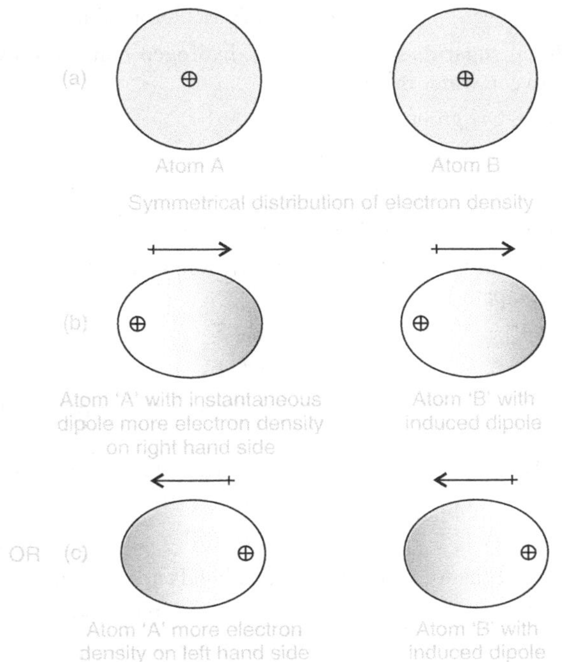

Fig. 8.10. Dispersion forces or London forces between atoms

The temporary dipoles of the atoms 'A' and 'B' attract each other. Similar temporary dipoles are also induced in non-polar molecules. The magnitude of such a force of attraction was first calculated by the German physicist *Fritz London*. For this reason, the force of attraction between two temporary dipoles is known as **London force.** Another name for this force is **dispersion force.** These forces are always attractive and the interaction energy is proportional to $\left(\frac{1}{r^6}\right)$ where r is the distance between the two interacting particles. These forces are important only at short distances (~500 pm) and their magnitude depends on the ability of the particles to undergo polarization. Dispersion forces are present among all particles and they are dependent on;

- *molecular size (Molar mass)*
- *number of electrons* and
- *surface area of molecule*

Strength of van der Waals' Forces

Strength and magnitude of van der Waals' forces depend on the following factors:

(*i*) *Molecular Size* (Molar mass). The larger the molecular size or complexity, the greater the probability of polarization or distortion and the higher the magnitude of the van der Waals' forces

(*ii*) *Number of electrons per molecule.* The larger the number of electrons per molecule, the more the chances of polarization and the higher the magnitude of the van der Waals' forces

(*iii*) *Surface area and geometry of molecule.* The larger the surface area of the molecule, the higher the chances of intermolecular interactions.

Let us now compare the boiling points of halogens. All halogens are non-polar molecules. But molecular size and number of electrons per molecule increases from F_2 to I_2. Consequently, the magnitude of the van der Waals' forces and boiling points increase in the same order.

Halogens	F_2	Cl_2	Br_2	I_2
Molecular mass	38	71	160	254
Number of electrons per molecule	18	34	70	106
Magnitude of van der Waal forces	——— INCREASE ———→			
Boiling points (K)	84.9	239	332	458

In a similar way the boiling points of noble gases increase from He to Xe due to the increase in molecular mass and number of electrons per molecule.

Noble gases	He	Ne	Ar	Kr	Xe
Molecular Mass	4	20.18	40	83.3	131.30
No. of electrons per molecule	2	10	18	36	54
Boiling point (K)	4.2	27.1	87.2	119.7	165.0

SUMMARY

- **Intermolecular Forces.** The attractive forces which hold molecules of a substance together in different states of matter. The magnitude of these forces control the physical properties such as melting points, boiling points, viscosity, surface tension, etc.
- **Hydrogen Bond.** The electrostatic force of attraction between a covalently bonded H-atom of one molecule and the electronegative atom (F or N or O) of another molecule.
- **Dipole-Dipole Interactions.** The attractive interactions among the opposite ends of polar molecules in the liquid or solid state.
- **Dipole-Induced Dipole Forces.** These are the attractive forces operating between polar molecules ($\mu > 0$) and non-polar molecules ($\mu = 0$)
- **Dispersion Forces.** The attractive forces among non-polar molecules in the solid or liquid state. They arise due to interaction between instantaneous oscillating dipoles and induced dipoles.
- **van der Waals' Forces.** Dipole-dipole forces, dipole-induced dipole forces and dispersion forces are collectively called van der Waal forces. The magnitude of van der Waal forces increases with increase in molecular size (molecular mass), number of electrons per molecule and surface area of a molecule.

EVALUATE YOURSELF

I. Objective Type Questions

Select the most appropriate choice from the options given as (a), (b), (c) and (d) after each question:

1. Interparticle forces among the non-polar molecules are
   (*a*) temporary dipole-permanent dipole
   (*b*) ion-dipole
   (*c*) temporary induced dipole-induced dipole
   (*d*) permanent dipole-permanent dipole.
2. The boiling point of HF is higher than that of NH_3 because.
   (*a*) H-bonding in H-F is more stronger
   (*b*) HF has a high bond strength
   (*c*) NH_3 is a bigger molecule
   (*d*) NH_3 is bonded to other NH_3 molecules by hydrogen bonds.
3. Proteins are soluble in water through which of the following interactions?
   (*a*) dipole–induced dipole (*b*) hydrogen bond
   (*c*) ion-hydrogen bond (*d*) dipole-dipole interactions.
4. The spiral type structure in proteins is maintained by
   (*a*) intermolecular H-bonds
   (*b*) permanent dipole
   (*c*) intramolecular hydrogen bonds
   (*d*) covalent bonds.
5. Which of the following liquids will have the lowest surface tension?
   (*a*) cyclohexane (*b*) ethanol
   (*c*) water (*d*) trichloromethane.
6. When sugar dissolves in water, hydrogen bonds are formed between water molecules and
   (*a*) —OH groups in sugar
   (*b*) C—H bonds in sugar
   (*c*) C = O groups in sugar
   (*d*) Polarised C—C bonds in sugar.
7. The number of H–bonds in which water molecule can participate is
   (*a*) 4 (*b*) 3
   (*c*) 2 (*d*) 1.
8. Which of the following substances are liquid at room temperature?
   I. Water II. Hydrogen chloride III. Ethanol IV. Iodine
   (*a*) I, III, IV (*b*) II, I, III
   (*c*) II, IV (*d*) I, III.
9. When Ethanol vapourizes which bonds are broken?
   (*a*) Covalent bonds (*b*) Hydrogen bonds
   (*c*) Dispersion forces (*d*) Ionic bonds.
10. Which among the following has lowest boiling point?
   (*a*) HCl (*b*) H–F
   (*c*) H–Br (*d*) H–I.
11. The pair of species which cannot form H-bonds?
   (*a*) Methanol-water (*b*) Ammonia-water
   (*c*) Methane-water (*d*) Water-water.
12. The intermolecular bonds responsible for the surface tension of carbon-tetrachloride is
   (*a*) Hydrogen bonds (*b*) Covalent bonds
   (*c*) Dipole-Dipole forces (*d*) Dispersion forces.

II. Fill in the Blanks

13. Complete the following sentences by supplying appropriate words:
   (*i*) The strength of a H–bond ranges between....
   (*ii*) In ice each oxygen atom is surrounded by...... H-atoms.
   (*iii*) The H–bonds around each O atom in ice is equal to ......
   (*iv*) The density of ice is less than that of water because of.... .
   (*v*) The interparticle forces among methane molecules in liquid methane are......

(*vi*) As the molecular size increases the magnitude of van der Waal forces......

III. Discussion Questions

14. State the type of bonds which are cleaved when the following process takes place.
    (*a*) melting of sodium chloride
    (*b*) vaporization of water
    (*c*) sublimation of naphthalene
    (*d*) dissolution of hydrogen iodide gas in water
    (*e*) melting of ice.
15. The following statements are either true or false. State which one(s) are true and which are false. Give reasons for the false statements and give one example which does not agree with the statement.
    (*i*) van der Waals' forces exist between molecules of every substance.
    (*ii*) Some covalent substances exist as solids.
    (*iii*) Boiling points always increase with increasing molar mass.
    (*iv*) Hydrogen bonds exist in all hydrides.
16. A solid substance, X, melts easily when heated. It is insoluble in water but dissolves well in hexane.
    Predict the following for X;
    (*i*) type of interatomic bonds,
    (*ii*) type of intermolecular bonds and how it is formed,
    (*iii*) electrical conductivity,
    (*iv*) another substance which will behave in the same way.
17. Draw labelled diagrams to show clearly the following bonds:
    (*a*) hydrogen bonds in water;
    (*b*) sugar molecules dissolved in water and held by hydrogen bonds;
    (*c*) van der Waals' forces in iodine;
    (*d*) van der Waals' forces in tetrachloromethane;
    (*e*) van der Waals' forces in iodine dissolved in tetrachloromethane.
18. What is a hydrogen bond? Describe how it is formed between two molecules of water. Sketch the bond formation.
19. (*a*) Give two properties of water arising from its hydrogen bonds.
    (*b*) State two other molecules, one large or complex, and the other small, in which hydrogen bonds exist.
20. Given hydrogen chloride, methane and water, describe for each the nature of the forces holding their molecules together.
    Arrange the given compounds according to their increasing enthalpy of vaporization, giving reasons.
21. Arrange the substances H_2, HCl and Cl_2 in order of
    (*i*) increasing polarity (*ii*) increasing boiling point.
22. (*i*) State the simplest unit (atom, ion or molecule) in oxygen, carbon dioxide and diamond.
    (*ii*) Which of the substances will be most soluble in water and why?
    (*iii*) Arrange the substances in increasing order of enthalpy of vaporization.
23. Describe briefly the nature of the forces which hold the atoms of inert gases together. How do the boiling points vary among them?
24. Explain each of the following observations in terms of intermolecular forces.
    (*i*) Iodine and chlorine are both halogens but iodine exists as a solid whereas chlorine is a gas at room temperature.
    (*ii*) H_2S has a lower melting point than water.
    (*iii*) Sugar is not an ionic substance but it dissolves in water.
25. Which of the following hydrogen bonds will be stronger:
    N—H····F or O—H····F
    Give reasons for your answer.
26. Give the names of three large, biologically important molecules in which hydrogen bonds can be found.
27. Ethanol and 1, 2, 3-propanetriol (glycerol) have the formulae CH_3CH_2OH and $CH_2OHCHOHCH_2OH$
    (*i*) State the type of intermolecular force between molecules of each substance.
    (*ii*) Explain why glycerol is more viscous than ethanol?
    (*iii*) Which of the two liquids will be more volatile? Give reasons.
    (*iv*) Which substance will form hydrogen bonds with water more easily? Explain briefly.
28. With the help of a diagram, show how ethanoic acid dissolves in water. Predict whether ethanoic acid will have a higher boiling point than ethanol or not.

Answers

I. Objective Type Questions

1. (*c*)	**2.** (*a*)	**3.** (*b*)
4. (*a*)	**5.** (*a*)	**6.** (*a*)
7. (*a*)	**8.** (*d*)	**9.** (*b*)
10. (*b*)	**11.** (*c*)	**12.** (*d*).

II. Fill in the Blanks

13. (*i*) 12.5 to 41.5 kJ mol^{-1}
(*ii*) four
(*iii*) 2
(*iv*) open cage structure
(*v*) van der Waal dispersion forces
(*vi*) also increases.

SECTION-3

UNIT-9

HYBRIDIZATION AND SHAPES OF MOLECULES

9.1 INTRODUCTION

In unit 7 we studied the formation of molecules through interatomic bonding. However, one question which still lingers in our mind at this stage is *why different molecules have different shapes.* For example, BCl_3 and PCl_3 are both tetra atomic covalent molecules, yet they have different shapes. BCl_3 is *trigonal planar* and PCl_3 has a *pyramidal* shape. In this unit, we shall try to provide answers to the above question by using the concept of **Hybridization of atomic orbitals.** Before we look into the details of the hybridization concept we have to understand how the atoms participating in bonding involve their atomic orbitals in forming the covalent bonds. We shall also study the formation of **sigma covalent bonds** and ***pi* covalent bonds.**

Specific Objectives

After studying this unit, the student should be able to:

- explain the term 'hybridization'
- describe how *sp*, sp^2, sp^3 and sp^3d^2 hybrid orbitals are formed
- describe how sigma and π bonds are formed
- illustrate the shapes of given molecular compounds.

9.2 ORBITAL CONCEPT OF COVALENT BONDING

We know, that a covalent bond between two atoms is formed by the mutual sharing of valence electrons so as to achieve an octet. According to the orbital concept, when two atoms approach each other for bonding their half-filled atomic orbitals overlap. Overlapping of atomic orbitals means that *a part of the electron cloud of each of the two half-filled atomic orbitals become merged. The probability of finding the shared electrons in the region of overlap is maximum than at other places.*

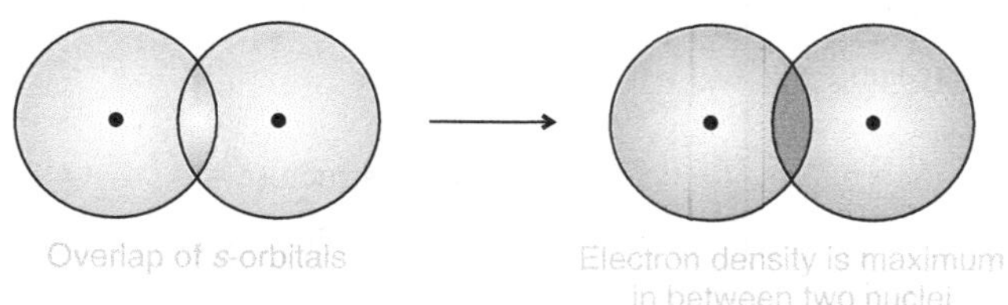

Overlap of *s*-orbitals

Electron density is maximum in between two nuclei

Let us sum up the various postulates of the orbital concept regarding the formation of covalent bonds.

- *Covalent bonds are formed by the overlapping of half-filled atomic orbitals present in the valence shell of the atoms participating in bonding.*
- *The orbitals undergoing overlapping must have electrons with opposite spins.*
- *The strength of a covalent bond depends upon the extent of overlapping. The greater the overlapping, the higher the energy released and consequently, the stronger the covalent bond.*

9.3 FORMATION OF SIGMA (σ) AND Pi (π) BONDS

Covalent bonds can be classified into two different categories depending upon the type of overlapping. These are:

(*a*) *Sigma covalent bond,*

(*b*) *Pi covalent bond.*

(*a*) Sigma (σ) Bond

This type of covalent bond is formed by the **axial overlapping** *of half-filled atomic orbitals.* The atomic orbitals overlap along their inter-nuclear axis and involve *end to end* or *head on overlap.* The electron cloud formed as a result of axial overlap is cylindrically symmetrical about the inter-nuclear axis. The electrons constituting the sigma bond are called **sigma electrons.** There can be three types of axial overlap among *s* and *p*-orbitals as discussed below:

(*i*) **s-s overlap.** It involves mutual overlap of half-filled *s*-orbitals of the atoms approaching to form a bond. The bond formed is called **s-s σ bond.**

(*ii*) **s-p overlap.** It involves mutual overlap of the half-filled *s*-orbital of one atom with a half-filled *p*-orbital of the other. The bond so formed is called **s-p σ bond.**

(*iii*) **p-p overlap.** It involves mutual overlap of the half-filled *p*-orbitals of two atoms. The bond so formed is called **p-p σ bond.**

The *s-s, s-p* and *p-p* overlaps have been shown diagrammatically in Fig. 9.1.

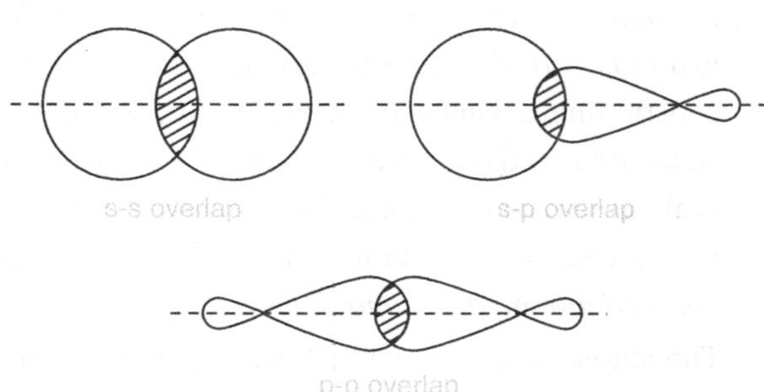

Fig. 9.1. *s-s, s-p* and *p-p* overlap of atomic orbitals.

Strength of the three types of sigma bonds. The strength of the three types of sigma bonds varies as follows:

p-p > p-s > s-s

It is because of the fact, that *p*-orbitals allow overlap to a greater extent than *s*-orbitals.

(*b*) Pi (π) Bond

This type of covalent bond is formed by the **lateral** or **sidewise overlap** *of the atomic orbitals.* The orbital overlap takes place in such a way that their axes are parallel to each other but perpendicular to the inter-nuclear axis. The *pi* bond consists of two charge clouds above and below the plane of the atoms involved in the bond formation. The electrons involved in π-bond formation are called **π-electrons.**

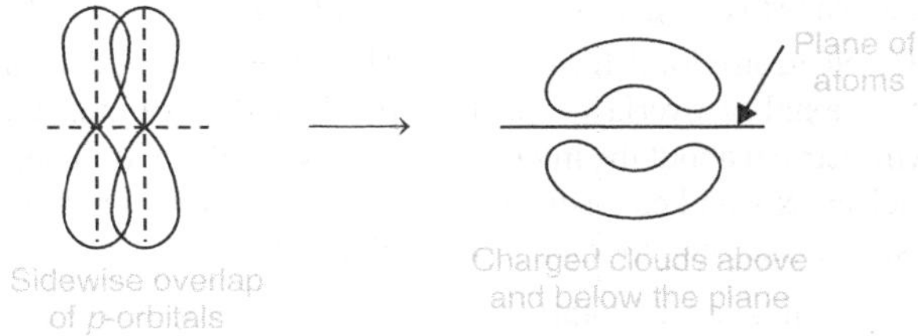

Fig. 9.2. Sidewise overlap of *p*-orbitals.

It may be noted that:

(*i*) ***Sigma bond is stronger than pi bond.*** It is because of the fact that the overlapping of atomic orbitals can take place to a greater extent during the formation of a sigma bond whereas overlapping of orbitals occurs to a smaller extent during the formation of a *pi* bond.

(*ii*) *Pi bond between two atoms is formed only in addition to a sigma bond.* It is because of the fact that the atoms constituting a single bond prefer to form a strong sigma bond rather than a weak *pi* bond. Thus, *pi* bond is always present in molecules having multiple bonds, *i.e.*, double or triple bond. In other words, *a single bond cannot be a pi bond.*

(*iii*) The **shape** of a molecule is controlled by the sigma framework (*orientations of sigma bonds*) around the central atom. *Pi* bonds are superimposed on sigma bonds hence they simply modify the dimensions of a molecule.

COMPARISON BETWEEN SIGMA AND *PI* BONDS

The various points of distinction between sigma and *pi* bonds are given in Table 9.1.

Table 9.1. Comparison of Sigma and *Pi* Bonds

Sigma Bond	*Pi Bond*
1. This bond is formed by the axial overlap of atomic orbitals.	1. The bond is formed by the sidewise overlap of atomic orbitals.
2. This bond can be formed by overlap of *s-s*, *s-p* or *p-p* orbitals.	2. It involves the overlap of *p*-orbitals only.
3. The bond is stronger because overlapping can take place to a larger extent.	3. The bond is weaker because the overlapping occurs to a smaller extent.
4. The electron cloud formed by axial overlap is symmetrical about the inter-nuclear axis and consists of a single charged cloud.	4. The electron cloud of a *pi* bond is discontinuous and consists of two charged clouds above and below the plane of atoms.
5. There can be a free rotation of atoms around the σ bond.	5. Free rotation of atoms around π bond is not possible because it will involve the breaking of the π bond.
6. The bond may be present between the two atoms either alone or along with π bond.	6. The bond is always present between the two atoms along with the sigma bond, *i.e.*, it is always superimposed on a sigma bond.
7. The shape of a molecule is determined by the sigma framework around the central atom.	7. The π bonds do not contribute to the shape, however, they modify the dimensions of the molecule.

Formation of sigma and *pi* bonds in F_2, HF and O_2 molecules are given in Fig. 9.3.

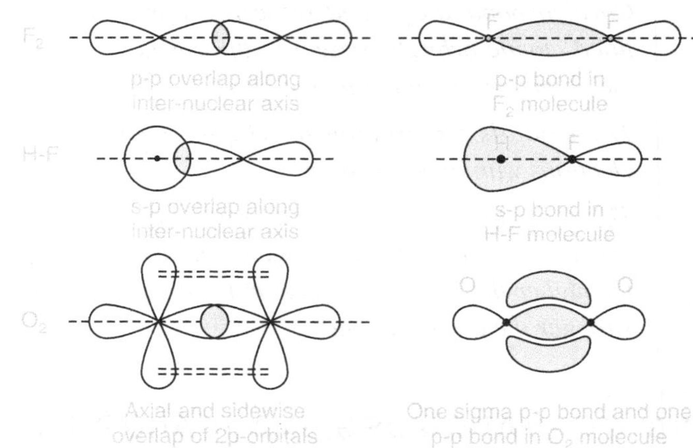

Fig. 9.3. Formation of F_2, HF and O_2 molecules.

9.4 HYBRIDIZATION OF ATOMIC ORBITALS

Hybridization is a theoretical concept which has been introduced to explain some structural properties such as the shapes of molecules or equivalency of bonds, etc., which cannot be explained by simple theories of valency.

According to this concept, valence orbitals of the atom intermix to give rise to another set of equivalent orbitals before the formation of bonds. These orbitals are called **hybrid orbitals** or **hybridized orbitals** and the phenomenon is referred to as **hybridization.** Thus, **hybridization** may be defined as *the phenomenon of intermixing of atomic orbitals of slightly different energies of the atom (by redistributing their energies) to form new set of orbitals of equivalent energies and identical shape.* See box item to understand the concept of hybridization.

The concept of hybridization can be understood by the following analogy.

If we mix two cans containing white and red paint respectively, the colour of the resulting mixture of paints is pink and at the same time we shall get only two cans of pink paint. The pink paint can be considered as a hybrid of red and white paint, because its shade is intermediate of red and white colours.

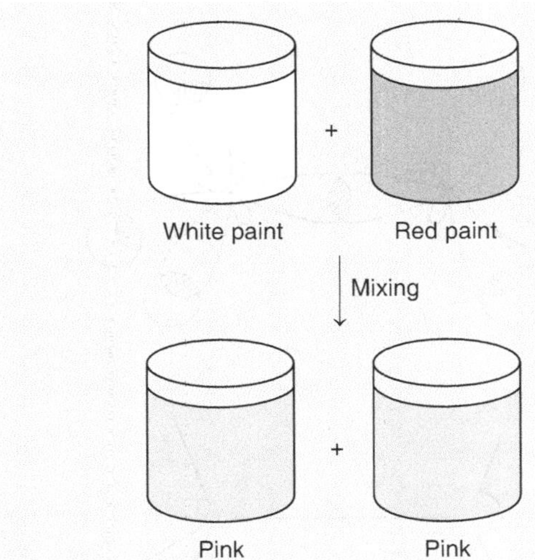

Hybridization is a familiar process in nature also. It involves the mixing of two closely related but different entities to produce another entity with intermediate characteristics. For example, a mule is a biological hybrid of a horse and a donkey.

SALIENT FEATURES OF HYBRIDIZATION

Some salient features of hybridization are:

- *The orbitals taking part in hybridization must have only a small difference of energies.*
- *The hybridized orbitals have equivalent energies and identical shapes.*
- *The number of hybrid orbitals is equal to the number of orbitals taking part in hybridization.*
- *Both half-filled as well as completely filled orbitals of valence shell can take part in hybridization. It implies that the promotion of electrons from a lower sub-shell to a higher sub-shell is not an essential condition prior to hybridization.*
- *The hybrid orbital has electron density concentrated on one side of the nucleus, i.e., it has one lobe relatively larger than the other as shown below.*

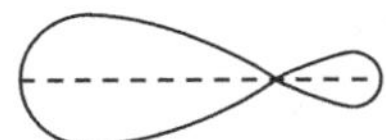

Representation of a hybrid orbital.

- *The hybrid orbitals can form stronger bonds as compared to the unhybridized or pure atomic orbitals because they can undergo more effective overlap.*
- *The hybrid orbitals are directed in space in some preferred directions so as to have minimum repulsive interactions and attain maximum stability. Thus, the type of hybridization controls the geometry of a molecule.*

9.5 TYPES OF HYBRIDIZATION

There are many different types of hybridization depending upon the type of orbitals involved in mixing such as sp^3, sp^2, sp, sp^3d, sp^3d^2, etc. Let us discuss the various types of hybridization along with some examples.

sp^3 HYBRIDIZATION

This type of hybridization involves the mixing of one orbital of the *s*-sub-level and three orbitals of the *p*-sub-level of the valence shell to form four sp^3 hybrid orbitals of equivalent energies and shape. Each sp^3 hybrid orbital has 25% *s*-character and 75% *p*-character. These hybridized orbitals tend to lie as far apart in space as possible so that the repulsive interactions between them are minimum. The four sp^3 hybrid orbitals are directed towards the four corners of a tetrahedron. The angle between the sp^3 hybrid orbitals is 109.5° (Fig. 9.4).

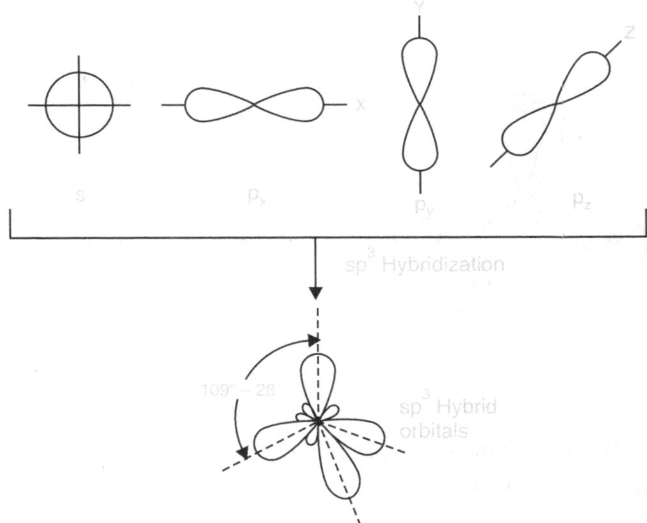

Fig. 9.4. Tetrahedral orientation of sp^3 hybrid orbitals.

sp^3 hybridization is also known as **tetrahedral hybridization.** The molecules in which the central atom which is sp^3 hybridized is linked to four other atoms directly have **tetrahedral shape.**

Some Examples of Molecules where Central Atom is sp^3 Hybridized

1. Formation of methane (CH_4). In methane the carbon atom adopts an sp^3 hybrid state as described below:

Here, one orbital of the 2*s*-sub-shell and three orbitals of the 2*p*-sub-shell of excited carbon undergo hybridization to form four sp^3 hybrid orbitals. The process involves promotion of a 2*s*-electron followed by hybridization as shown in Fig. 9.5.

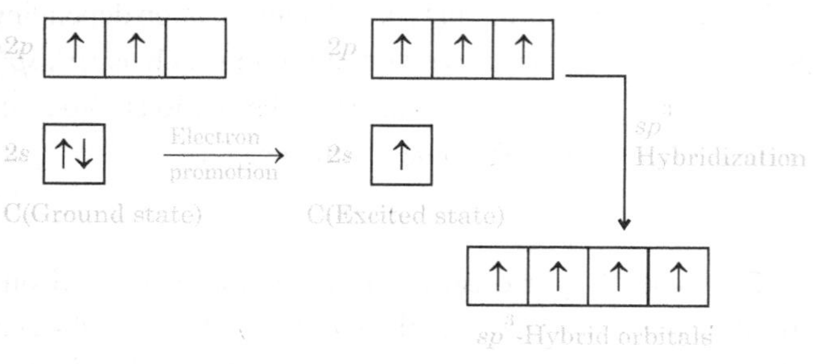

Fig. 9.5. sp^3 hybridization of carbon.

As pointed out earlier the sp^3 hybrid orbitals of carbon atom are directed towards the corners of a regular tetrahedron. Each of the sp^3 hybrid orbitals overlap axially with the half-filled 1*s*-orbital of hydrogen atom to form a sigma bond (Fig. 9.6).

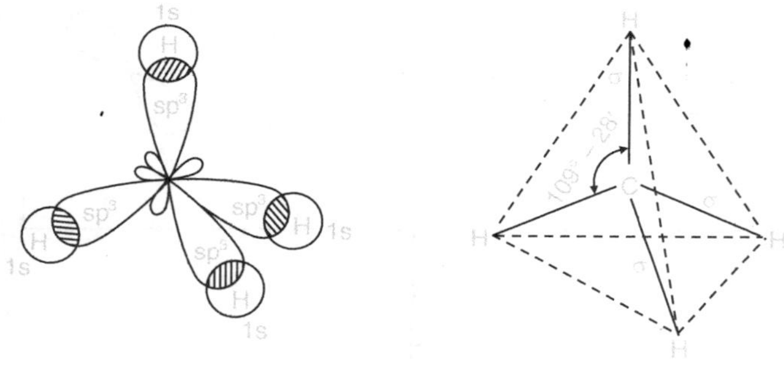

Fig. 9.6. Orbital picture of methane.

Because of the sp^3 hybridization of the carbon atom, CH_4 has a **tetrahedral shape.**

2. Formation of ethane (CH_3—CH_3). In ethane both carbon atoms adopts an sp^3 hybrid state. One of the hybrid orbitals of carbon overlaps axially with a similar orbital of the other carbon atom to form sp^3-sp^3 sigma bond. The other three hybrid orbitals of each carbon atom are used in forming sp^3-*s* sigma bonds with hydrogen atoms as described in Fig. 9.7.

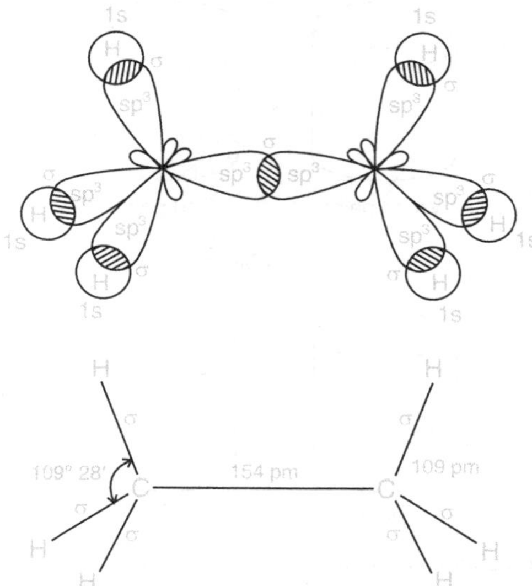

Fig. 9.7. Orbital picture of ethane.

Each C—H bond in ethane is an sp^3-*s* sigma bond with bond length **109 pm.** The C—C bond is an sp^3-sp^3 sigma bond with bond length **154 pm.**

3. Formation of ammonia (NH_3) molecule. In NH_3 the nitrogen atom adopts an sp^3-hybrid state. Three sp^3-hybrid orbitals of the N atom are used for forming sp^3-*s* σ (sigma) bonds with H atoms. The fourth sp^3-hybrid orbital carries a lone pair of electrons. The relatively larger *lone pair-bond pair* interactions causes the HNH angle to decrease from 109°.28′ to 107°. The ground state, hybrid state of N atom and orbital overlap in NH_3 are shown in Fig. 9.8.

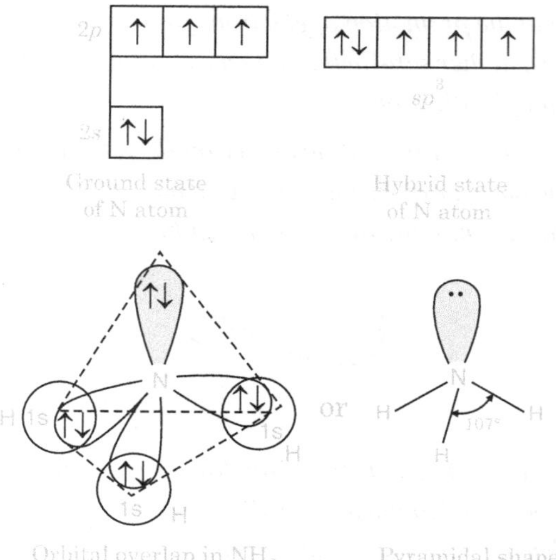

Fig. 9.8. Formation of NH_3 molecule.

4. Formation of water (H_2O) molecule. In H_2O, the oxygen atom adopts an sp^3 hybrid state. Two of the sp^3-hybrid-orbitals of oxygen contain lone pairs of electrons whereas the other two hybrid orbitals form sp^3-s σ (sigma) bonds with H atoms. The lone pair orbitals exert relatively greater repulsive interactions on bond pair-orbitals causing the HOH angle to decrease from 109°28′ to 104.5°. The hybridization of O atom along with orbital overlap in H_2O is shown in Fig. 9.9.

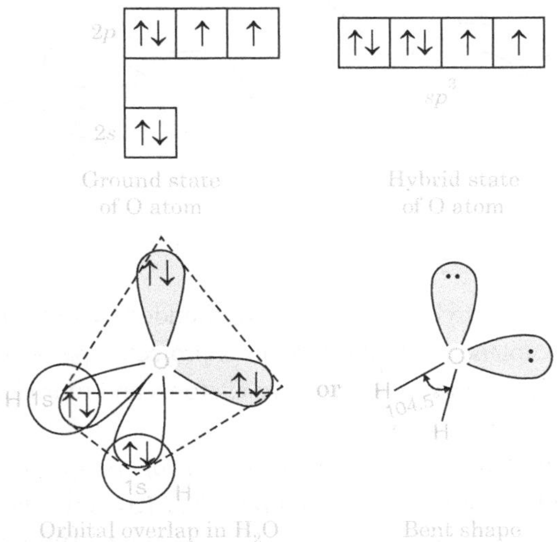

Fig. 9.9. Formation of H_2O molecule.

sp^2 HYBRIDIZATION

This type of hybridization involves the mixing of one orbital of the s-sub-level and two orbitals of the p-sub-level of the valence shell to form three sp^2 hybrid orbitals. These sp^2 hybrid orbitals lie in a plane and are directed towards the corners of an equilateral triangle (Fig. 9.10).

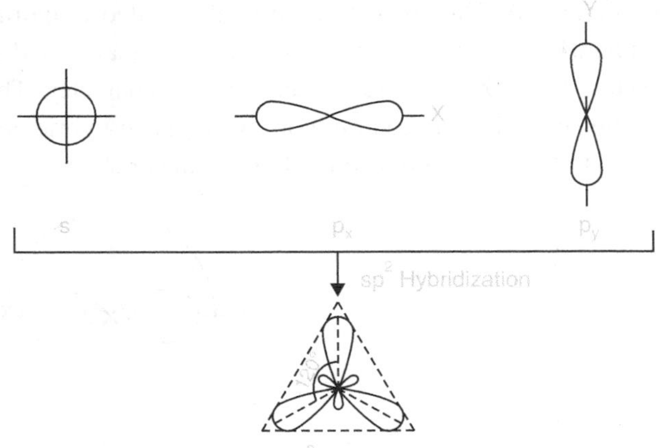

Fig. 9.10. Trigonal planar orientation of sp^2 hybrid orbitals.

Each sp^2 hybrid orbital has one-third s-character and two-third p-character. sp^2 hybridization is also called **trigonal hybridization.** The molecules in which the central atom which is sp^2 hybridized is linked to three other atoms directly have triangular planar shape.

Let us study some examples of molecules which undergo sp^2 hybridization.

1. Formation of boron trichloride (BCl_3). Boron ($_5B$) atom has a ground state configuration of $1s^2$, $2s^2$, $2p^1$. But in the excited state its configuration is $1s^2$, $2s^1$, $2p_x^1$, $2p_y^1$. One $2s$-orbital of excited boron intermixes with two $2p$-orbitals to form three sp^2 hybrid orbitals as shown in Fig. 9.11.

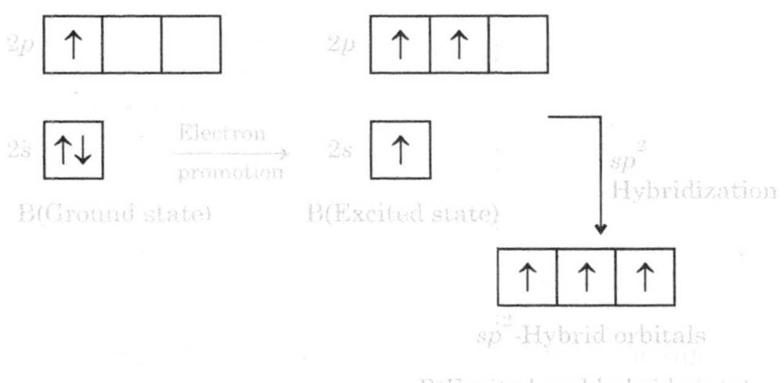

Fig. 9.11. sp^2 hybrid state of boron.

The sp^2 hybrid orbitals of boron are directed towards the corners of an equilateral triangle and lie in a plane. Each of the sp^2 hybrid orbitals of boron overlaps axially with a $3p$-half-filled orbital of chlorine atom to form three B-Cl sigma bonds as shown in Fig. 9.12.

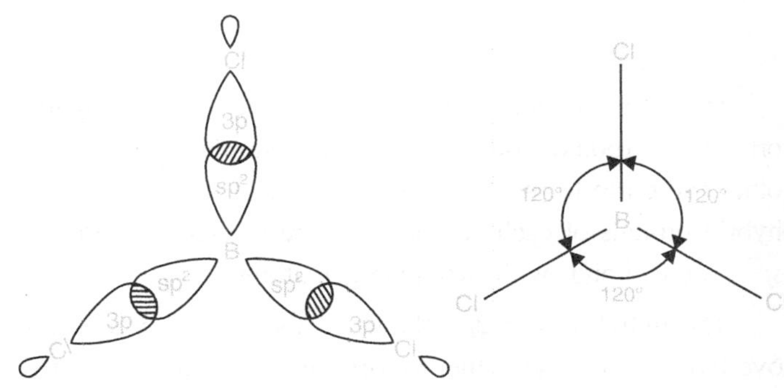

Fig. 9.12. Orbital diagram of BCl_3.

Because of the sp^2 hybridization of boron, BCl_3 molecule has a **triangular planar shape.**

2. Formation of ethylene (C_2H_4). Both carbon atoms in ethylene adopt an sp^2 hybrid state. In acquiring the sp^2 hybrid state, one 2*s*-orbital and two 2*p*-orbitals of excited carbon get hybridized to form three sp^2 hybrid orbitals. However, one orbital of the 2*p*-sub-shell of the excited carbon atom does not take part in the hybridization. The promotion of an electron and hybridization in the carbon atom is shown in Fig. 9.13.

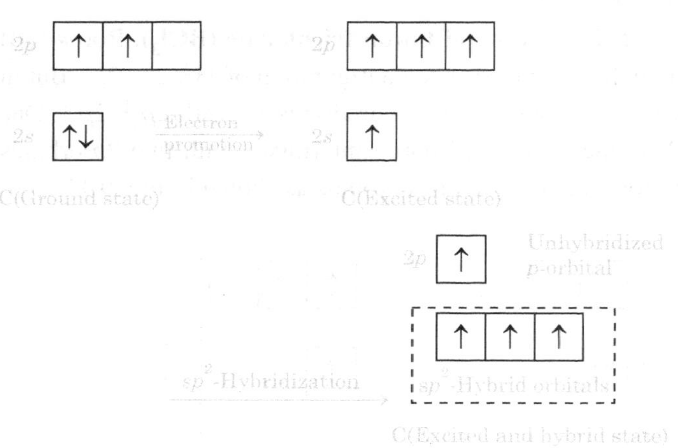

Fig. 9.13. sp^2 hybrid state of carbon.

As already indicated, the three sp^2 hybrid orbitals lie in one plane and are oriented in space at an angle of 120° to one another. The unhybridized 2*p*-orbital is perpendicular to the plane of the sp^2 hybrid orbitals as shown in Fig. 9.14.

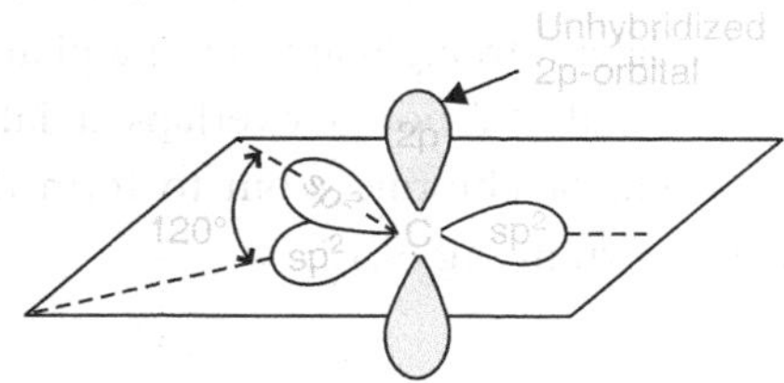

Fig. 9.14. sp^2 hybrid state of C.

In the formation of ethylene (C_2H_4), one of the sp^2 hybrid orbital of carbon overlaps axially with a similar orbital of the other carbon to form a C—C sigma bond. The other two sp^2 hybrid orbitals of each carbon atom are utilised for forming sp^2-*s* sigma bonds with two hydrogen atoms.

The unhybridized 2*p*-orbitals of the two carbon atoms overlap sidewise each other to form two π clouds distributed above and below the plane of the carbon and hydrogen atoms (Fig. 9.15).

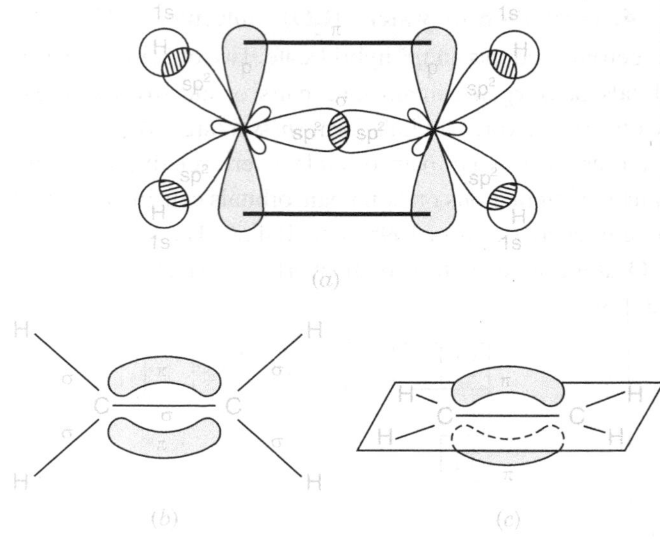

Fig. 9.15. Orbital diagram of ethylene.

Thus, in ethylene, the six atoms (bonded by sigma bonds) lie in one plane while the π bond is projected perpendicular to the plane of the six atoms (two C atoms and four H atoms) as shown in Fig. 9.15 (*c*).

In ethylene, the *C = C bond consists of one* sp^2-sp^2 sigma bond and one π bond. *Its bond length is* **134 pm.** *The C—H bond is an* sp^2-*s sigma bond with bond length* **108 pm.** *The H—C—H angle is* **117.5°** *while the H—C—C angle is* **121°.**

sp-HYBRIDIZATION

This type of hybridization involves the mixing of one orbital of the *s*-sub-level and one orbital of the *p*-sub-level of the valence shell of the atom to form two *sp*-hybridized orbitals of equivalent shapes and energies. These *sp*-hybridized orbitals are oriented in space at an angle of 180° (Fig. 9.16). This hybridization is also called **diagonal hybridization.** Each *sp*-hybrid orbital has equal *s* and *p* character, *i.e.*, 50% *s*-character and 50% *p*-character. The molecules in which the central atom which is *sp*-hybridized is linked to two other atoms directly have linear shape.

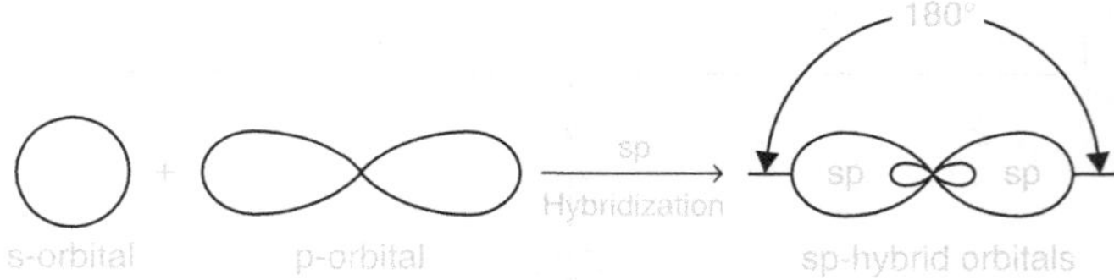

Fig. 9.16. Formation of *sp*-hybrid orbitals.

Let us study some examples of molecules with *sp*-hybridization.

1. Formation of beryllium chloride ($BeCl_2$). Beryllium ($_4Be$) atom has a ground state electron configuration of $1s^2$, $2s^2$. In the excited state one of the 2*s*-electrons is promoted to the 2*p*-orbitals. One 2*s*-orbital and one 2*p*-orbital of excited beryllium atom undergo *sp*-hybridization to form two *sp*-hybridized orbitals as described in Fig. 9.17.

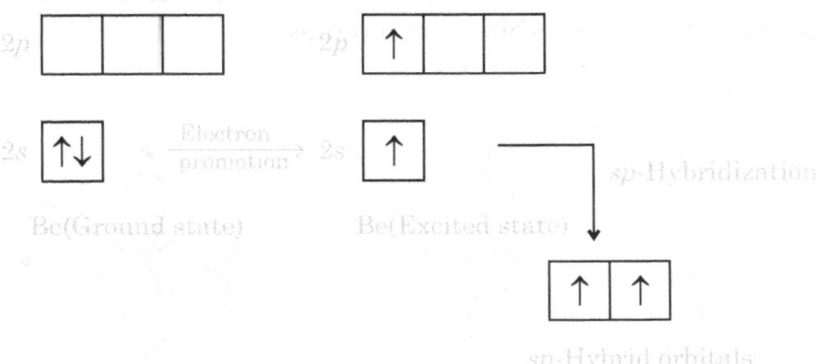

Fig. 9.17. *sp*-hybrid state of Be.

The two *sp*-hybrid orbitals are linear and oriented in opposite directions at an angle of 180°. Each of the *sp*-hybridized orbital overlaps axially with a 3*p*-half-filled orbital of chlorine atom to form two Be—Cl sigma bonds (Fig. 9.18).

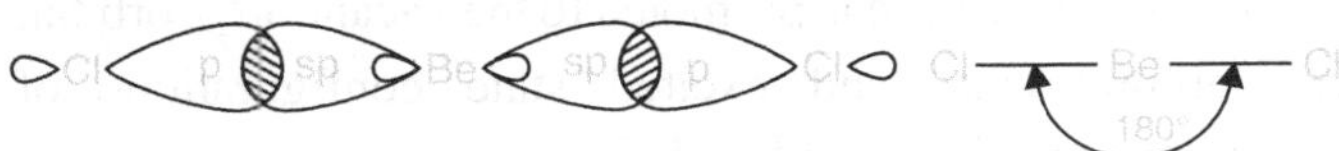

Fig. 9.18. Orbital diagram of $BeCl_2$.

Due to the *sp*-hybridised state of beryllium, $BeCl_2$ molecule has a **linear shape.**

2. Formation of ethyne (CH ≡ CH). Both carbon atoms in ethyne assume an *sp*-hybrid state. In acquiring the *sp*-hybrid state, one 2*s* orbital and one 2*p*-orbital of excited carbon ($1s^2\ 2s^1\ 2p_x^{\ 1}\ 2p_y^{\ 1}\ 2p_z^{\ 1}$) get hybridized to form two *sp*-hybridized orbitals (Fig. 9.19).

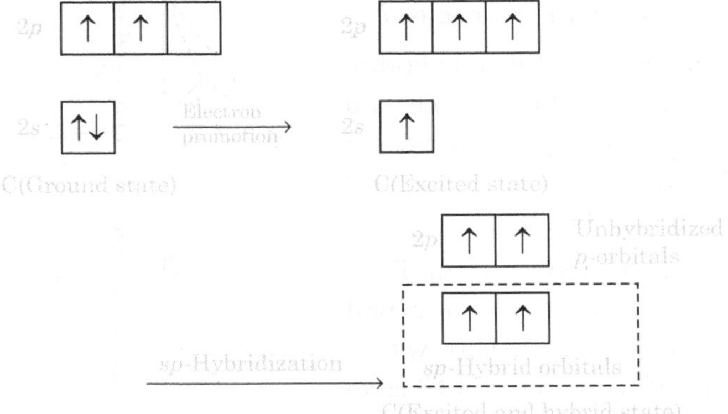

Fig. 9.19. *sp*-hybridization of carbon.

The two *sp*-hybrid orbitals of carbon are linear and are directed at an angle of 180° whereas the unhybridized *p*-orbitals are perpendicular to the *sp*-hybrid orbitals and also perpendicular to each other as shown in Fig. 9.20.

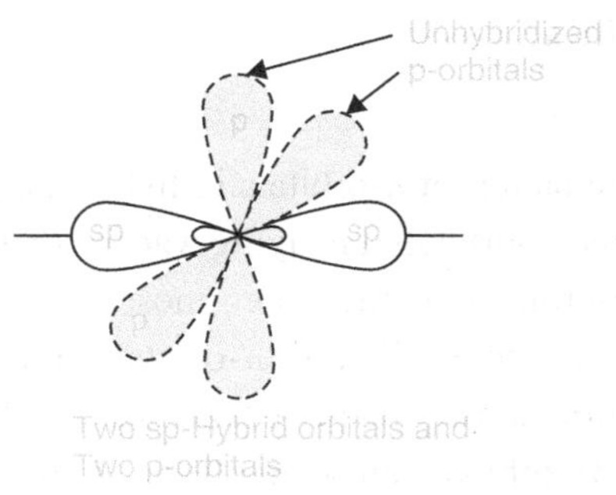

Fig. 9.20. *sp*-hybrid state of carbon.

In the formation of ethyne, carbon atom uses one of the *sp*-hybrid orbitals for overlapping with a similar orbital of the other carbon to form a C—C sigma bond. The other *sp*-hybrid orbital of each C atom overlaps axially with a 1*s*-orbital of H atom to form a C—H sigma bond. Each of the two unhybridized orbitals of both carbon atoms overlap sidewise to form two π-bonds. The electron clouds of one π-bond lie above and below the inter-nuclear axis whereas those of the other π bond lie in front and back of the inter-nuclear axis.

The overlapping of orbitals has been shown in Fig. 9.21.

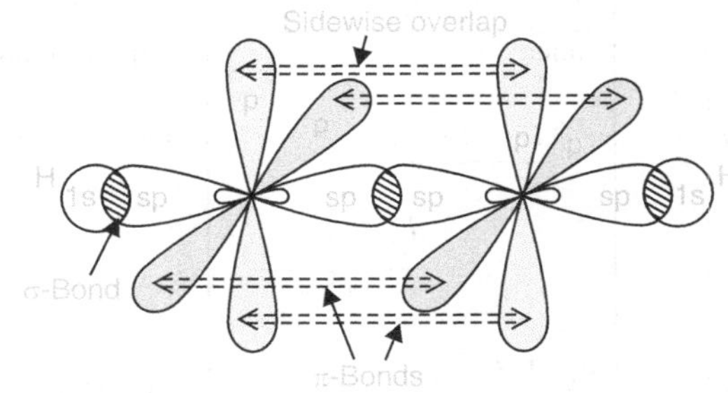

Fig. 9.21. Orbital diagram of ethyne.

The four π clouds so formed further merge into one another to form a single cylindrical electron cloud around the inter-nuclear axis representing a C—C sigma bond. It has been shown in Fig. 9.22.

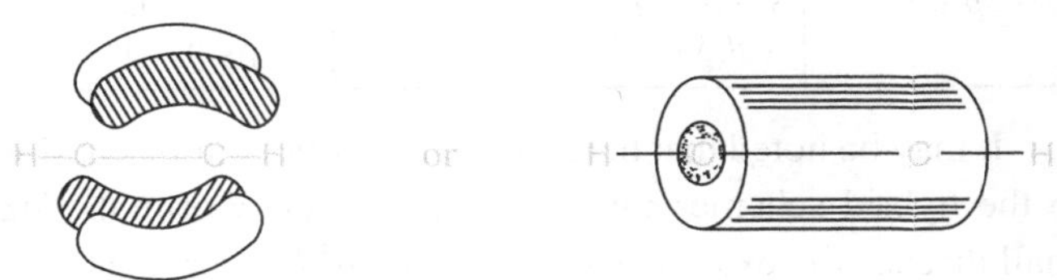

Fig. 9.22. Cylindrical π-electron cloud in ethyne.

Thus, in ethyne,

the C ≡ C bond consists of one sp-sp σ bond along with two π-bonds. The C ≡ C bond length is 120 pm. The C—H bond is an sp-s sigma bond. The H—C—C angle is 180°, i.e., the molecule is **linear.**

HYBRIDIZATION INVOLVING *d*-ORBITALS

The participation of *d*-orbitals in hybridization can take place when the *d*-orbitals are lying vacant or partly filled. The *s* and *p*-orbitals of the outer most shell can use the *d*-orbitals of the outer shell or the *d*-orbitals of the lower shell for hybridization depending upon the nature of the molecule. For example, 3*d*-orbitals can be involved in hybridization with 3*s* and 3*p*-orbitals and also with 4*s* and 4*p*-orbitals. It is because the energy of 3*d*-orbitals is comparable to 3*s* and 3*p*-orbitals and also to *s* and *p*-orbitals of the 4th shell. The important hybridization schemes of *s, p, d*-orbitals are summarised below in a tabular form:

Type of Hybridization	*Atomic Orbitals Involved*	*Spatial Orientation of Hybrid Orbitals*
sp^3d	$\underbrace{s + p_x + p_y + p_z} + \underbrace{d_{z^2}}_{\text{outer}}$	Trigonal bipyramidal
dsp^3	$\underbrace{d_{z^2}}_{\text{inner}} + \underbrace{s + p_x + p_y + p_z}$	Trigonal bipyramidal
sp^3d^2	$\underbrace{s + p_x + p_y + p_z} + \underbrace{d_{z^2} + d_{x^2-y^2}}_{\text{outer}}$	Octahedral
d^2sp^3	$\underbrace{d_{z^2} + d_{x^2-y^2}}_{\text{inner}} + \underbrace{s + p_x + p_y + p_z}$	Octahedral
dsp^2	$\underbrace{d_{x^2-y^2}}_{\text{inner}} + \underbrace{s + p_x + p_y}$	Square planar
sp^3d^3	$s + p_x + p_y + p_z + d_{xy} + d_{yz} + d_{zx}$	Pentagonal bipyramidal

It may be noted that here, we shall restrict our discussion to the hybrid schemes involving *outer shell d*-orbitals. We shall discuss the examples of sp^3d and sp^3d^2 hybridization.

sp^3d-HYBRIDIZATION

This type of hybridization involves mixing of three *p* and one *d*-orbital to form five sp^3d-hybridized orbitals which adopt trigonal bipyramidal geometry as shown in Fig. 9.23. Three of the hybrid orbitals lie in a horizontal plane at an angle of 120° to one another. These are called equatorial orbitals (marked as *e*). The other two hybrid orbitals, lie in a vertical plane at right angles to the plane of the equatorial orbitals and are called axial orbitals (marked as *a*).

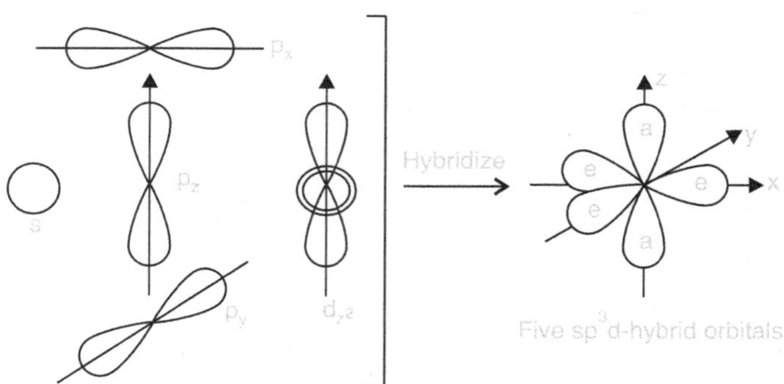

Fig. 9.23. Formation of five sp^3d hybrid orbitals.

1. Formation of PCl_5. The ground state electronic configuration of phosphorus is $1s^2\ 2s^2\ 2p^6\ 3s^2\ 3p^3$. Under conditions of bond formation the 3*s*-electrons get unpaired and one of the electron is promoted to the vacant $3d_{z^2}$ orbital. The ground state and excited state configurations of phosphorus are shown in Fig. 9.24.

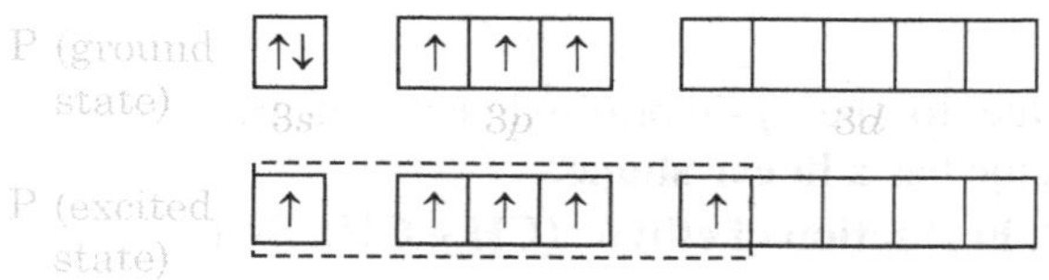

Fig. 9.24. Ground state and excited state configuration of phosphorus.

Now the five orbitals, one (3*s*), three (3*p*) and one ($3d_{z^2}$) which are half-filled, hybridize to yield a set of five sp^3d hybrid orbitals, which point towards the five corners of a trigonal bipyramid (Fig. 9.25).

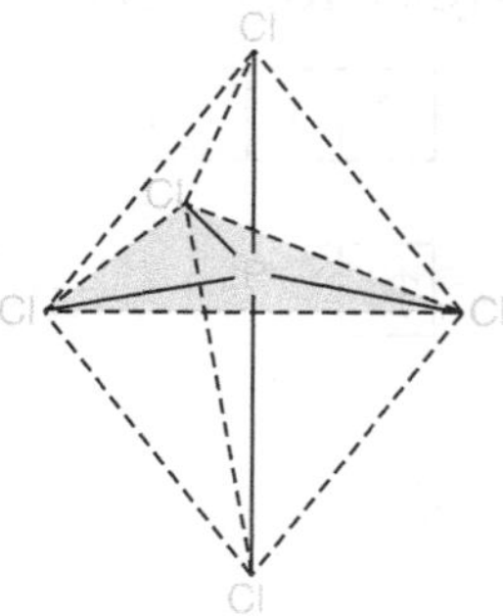

Fig. 9.25. Trigonal bipyramidal geometry of PCl_5 molecule.

In the formation of PCl_5, each of the five sp^3d hybrid orbitals overlap axially with a half-filled 3*p*-orbital of Cl atom to form five P—Cl sigma bonds. Three of the five P—Cl bonds lie in one plane at an angle of 120° to one another. These are called **equatorial bonds.** The other two P—Cl bonds are at right angles to the plane of the equatorial bonds, *i.e.*, one above and the other below the plane. These are called **axial bonds.** As the axial bond pairs suffer more repulsive interaction from the equatorial bond pairs, the **axial P—Cl bonds become slightly longer (219 pm) than the equatorial bonds (204 pm).** The unequal length of the axial and equatorial P—Cl bonds in PCl_5 has been confirmed by X-ray diffraction techniques. This causes axial bonds to become slightly weaker than the equatorial bonds; which makes PCl_5 molecule more reactive.

sp^3d^2-HYBRIDIZATION

In this type of hybridization one *s*, three *p* and two *d*-orbitals undergo intermixing to form six identical sp^3d^2 hybrid orbitals. These six orbitals are directed towards the corners of an octahedron and lie in space at an angle of 90° to one another.

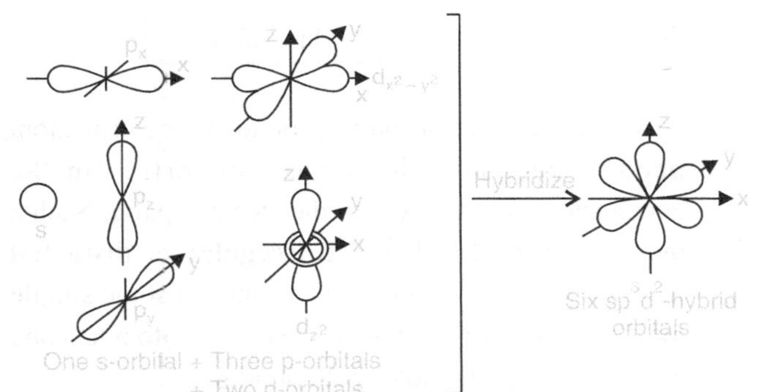

Fig. 9.26. Formation of six sp^3d^2-hybrid orbitals.

1. Formation of SF_6. The ground state outer configuration of $_{16}S$ is $3s^2 3p^4$. In the excited state the electron pairs in the 3*s* and $3p_x$ orbitals get unpaired, and one out of each pair is promoted to the vacant $3d_{z^2}$ and $3d_{x^2-y^2}$ orbitals. The ground state and excited state configurations of $_{16}S$ are given as follows:

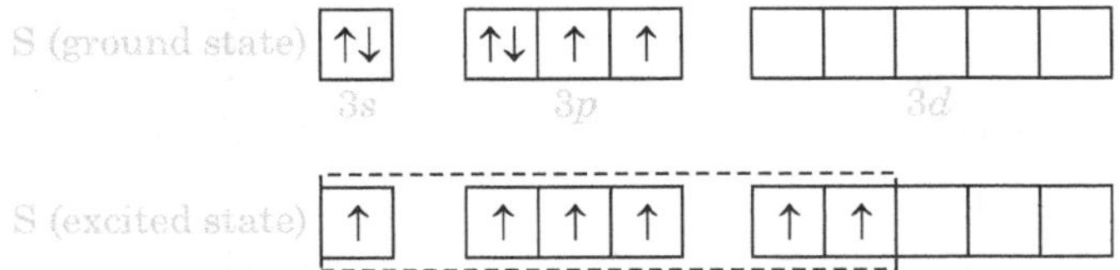

Now, six orbitals; one (3*s*), three (3*p*) and two (3*d*) orbitals which are half-filled, hybridize to form six new sp^3d^2 hybrid orbitals which are projected towards the six corners of a regular octahedron as shown in Fig. 9.27.

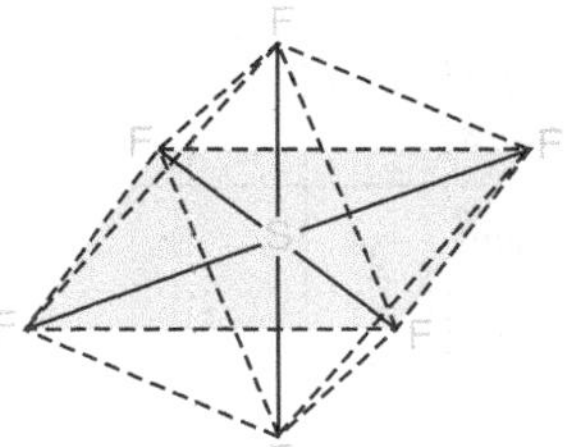

Fig. 9.27. Octahedral geometry of SF_6.

In the formation of SF_6, these six sp^3d^2 hybrid orbitals overlap with half-filled orbitals of fluorine atoms to form six S—F sigma bonds.

Because of the sp^3d^2 hybridization of sulphur, SF_6 has a **regular octahedral geometry** as shown in Fig. 9.27. All the six S—F bonds in SF_6 have the same bond length.

9.6 VSEPR MODEL AND SHAPE OF MOLECULAR COMPOUNDS

Valence Shell Electron Pair Repulsion (VSEPR) Model provides a simple method to predict the shapes of simple covalent molecules or polyatomic ions. This model was developed earlier by **Sidgwick** and **Powell** in 1940, and it was further improved by **Gillespie** and **Nyholm** in 1957. It is primarily based upon the fact that in a polyatomic molecule the direction of bonds around the central atom depends upon the total number of electron pairs (bond pairs as well as lone pairs) in its valence shell. These electron pairs place themselves as far apart as possible in space so as to have minimum repulsive interactions between them. The minimum repulsions correspond to the state of minimum energy and maximum stability of the molecule. The main points of this model are:

- *The geometry and shape of a molecule depends upon the number of electron pairs (bonded as well as non-bonded) in the valence shell of the central atom.* The geometrical arrangement of different numbers of electron pairs around the central atom A is shown in Table 9.2.
- *The electron pairs surrounding the central atom repel one another as their orbitals/electron clouds are negatively charged.*
- *In order to minimise repulsion, these electron pairs tend to occupy such positions in space where the distance between them is maximum.* In other words, if the valence shell of the central atom is taken as a sphere, the electron pairs are localised on this spherical surface at maximum distance from one another.

Table 9.2. Number of Electron Pairs and their Geometrical Arrangement

Number of electron pairs	*2*	*3*	*4*
Arrangement of electron pairs around central atom (A)	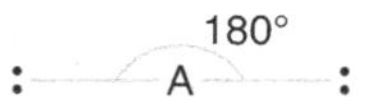 180° A Linear	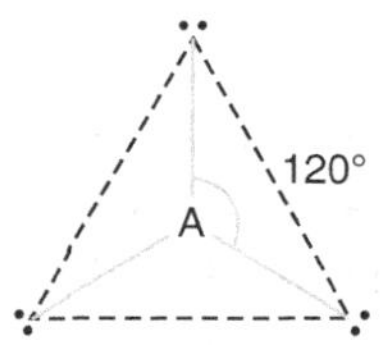 120° A Trigonal planar	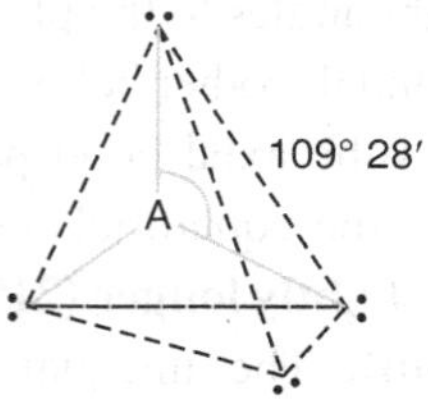 109° 28′ A Tetrahedral
Number of electron pairs	*5*	*6*	*7*
Arrangement of electron pairs around central atom (A)	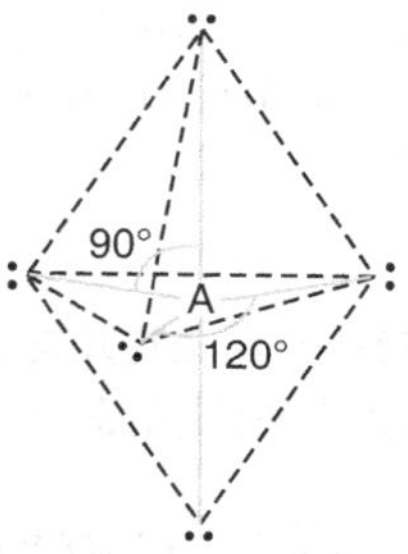 90° A 120° Trigonal Bipyramidal	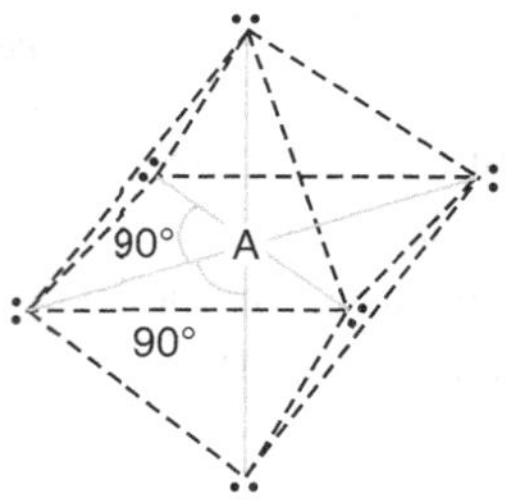 90° A 90° Octahedral	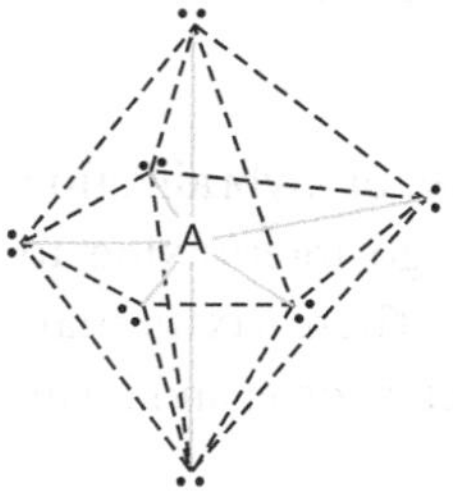 A Pentagonal Bipyramidal

- *The repulsive interactions between two lone pairs (lp) are different from those between two bond pairs (bp)* or *those between a lone pair and a bond pair. The repulsive interactions between various electron pairs decrease in the order*:

lp-lp > lp-bp > bp-bp

Nyholm and **Gillespie** postulated that a bond pair of electrons is shared by two atoms whereas a lone pair is under the influence of only the central atom. Hence, in a molecule, the electron cloud containing a lone pair is more spread out and occupy more space as compared to the electron cloud containing a bond pair. This causes relatively greater repulsive interactions between the lone pairs in comparison to lone pair-bond pair and bond pair-bond pair repulsive interactions.

- When all the electron pairs around the central atom are **bond pairs** only and the surrounding atoms are similar, the molecule is said to possess **regular geometry.** Table 9.3 gives the geometry of molecules in which the central atom does not have any lone pair.
- When **lone pairs** are also present in addition to bond pairs around the central atom, the repulsive interactions between the electron pairs around the central atom become unequal. This causes distortion in the geometrical arrangement of the electron pairs. Such a molecule is said to have an **irregular** or **distorted geometry**. Table 9.4 gives the shapes of some simple molecules and ions in which the central atom has one or more lone pairs along with bond pairs.
- It may be noted that for the purpose of applying the VSEPR theory, a **multiple bond** is treated as a **single electron pair**. For example in methanal (formaldehyde, HCHO), the bond pairs around the carbon atom are counted as three (2 bp for two C—H bonds and 1 bp for the C = O bond). Similarly the two C = O bonds in CO_2 molecule are counted as two bond pairs. The shapes of HCHO and CO_2 are shown below:

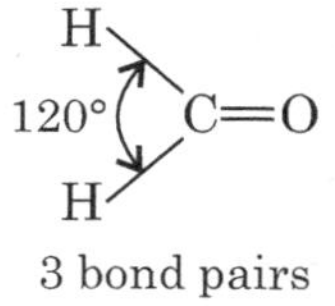

H, 120°, C=O, H
3 bond pairs
Trigonal planar

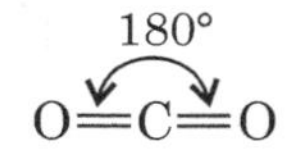

180°, O=C=O
2 bond pairs
Linear

Table 9.3. Geometry of Molecules in which the Central Atom has No Lone Pair of Electrons (here the central atom is A and the surrounding atom is B)

Number of electron pairs	*Molecular formula*	*Arrangement of electron pairs*	*Examples*
2	AB_2	***Linear***	BeF_2, $BeCl_2$, $HgCl_2$, CO_2
3	AB_3	***Trigonal planar***	BF_3, $AlCl_3$
4	AB_4	***Tetrahedral***	CH_4, SiF_4, NH_4^+
5	AB_5	***Trigonal bipyramidal*** In this arrangement, three of the bond pairs lie in the same plane at an angle of 120°. These are marked as *e* and are called **equatorial pairs.** The other two bond pairs lie at right angle to the plane of the equatorial pairs. These are called **axial pairs** and are marked as *a*. • Axial pairs in this arrangement experience relatively larger repulsive interactions. Hence *axial bonds are relatively longer than equatorial bonds.* • Any **lone pair** appearing in this arrangement tend to occupy **equatorial positions** due to the relatively lower repulsive interactions.	PF_5, PCl_5, $SbCl_5$, AsF_5
6	AB_6	***Octahedral***	SF_6, TeF_6

Table 9.4. Geometry/Shapes of Molecules containing Bond Pairs and Lone Pairs (here A is central atom, B is surrounding atom and L is lone pair)

Type of Molecule	*No. of bond pairs (bp)*	*No. of lone pairs (lp)*	*Arrangement of electron pairs (bp + lp)*	*Shape of molecule*	*Example*	*Reason/other examples*
AB_2L	2	1	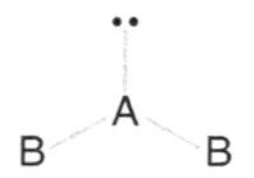 Trigonal Planar	Bent	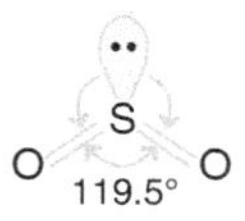 119.5° SO_2 molecule	The OSO angle around the central atom decreases from 120° to 119.5° due to larger *lp– bp* repulsions. Other examples: $\mathbf{O_3}$, $\mathbf{SnCl_2}$, $\mathbf{NO_2^-}$
AB_3L	3	1	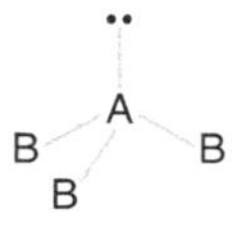 Tetrahedral	Pyramidal	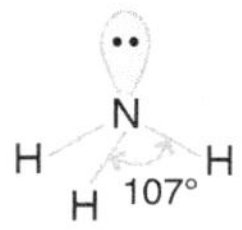 107° NH_3 molecule	The HNH angle decreases from ideal angle 109°.28′ to 107° because of relatively larger *lp-bp* repulsive interactions. Other examples: PCl_3, NCl_3, PH_3, H_3O^+
AB_2L_2	2	2	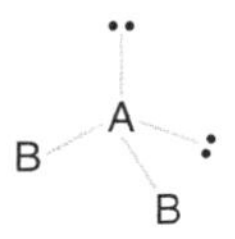 Tetrahedral	Bent	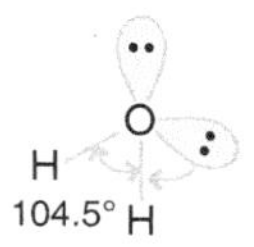 104.5° H_2O molecule	The larger *lp-bp* repulsive interactions push O—H bonds closer decreasing HOH angle from 109°.28′ to 104.5°. Other examples: H_2S, NH_2^-
AB_4L	4	1	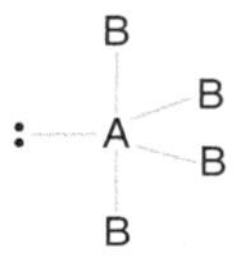 Trigonal Bipyramidal	See-Saw	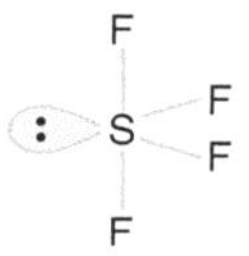 SF_4 molecule	The lone pair preferably occupies the equatorial position because the repulsive interactions at the equatorial positions are less pronounced than those at the axial position.
AB_3L_2	3	2	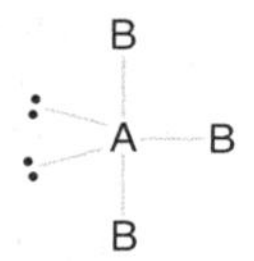 Trigonal Bipyramidal	T-Shape	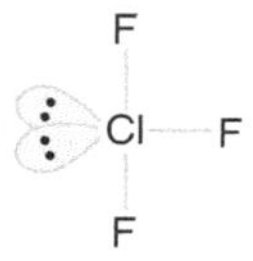 ClF_3 molecule	The two lone pairs prefer to occupy equatorial positions at an angle of 120° to each other. Other examples: BrF_3, IF_3
AB_2L_3	2	3	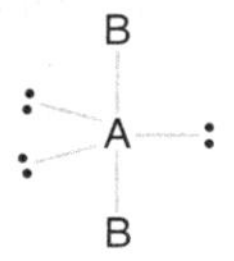 Trigonal Bipyramidal	Linear	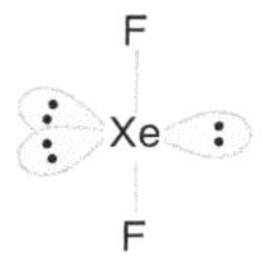 XeF_2 molecule	All the three lone pairs occupy equatorial positions at an angle of 120°. Other examples: I_3^-, Br_3^-
AB_5L	5	1	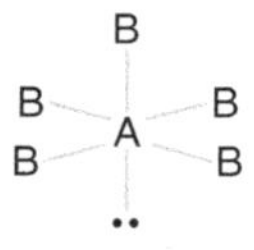 Octahedral	Square pyramidal	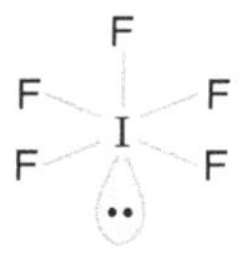 IF_5 molecule	Other examples: BrF_5
AB_4L_2	4	2	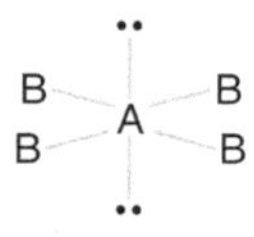 Octahedral	Square planar	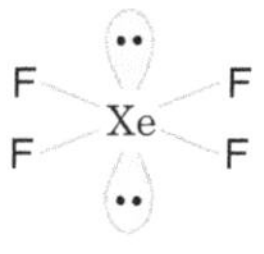 XeF_4 molecule	The two lone pairs tend to lie opposite each other at an angle of 180°.

SOLVED EXAMPLES

Example 9.1. *Describe the shapes of the following molecules on the basis of the VSEPR Model*

$BeCl_2$, BF_3, CH_4, NH_3, H_3O^+, CO_2, SF_6

Solution. The Lewis dot structures and shapes of the various species are given below in tabular form:

Species	*Lewis dot structure*	*Electron pairs around central atom*		*Arrangement of electron pairs*	*Shape*
		bp	*lp*		
$BeCl_2$	Cl:Be:Cl	2	0	Cl—Be—Cl 180°	Linear
BF_3	F:B:F with F below	3	0	F, B, F, F 120°	Trigonal planar
CH_4	H:C:H with H above and below	4	0	H, C, H, H, H 109°.28′	Tetrahedral
NH_3	H:N:H with H below, lone pair on N	3	1	N, H, H, H	Pyramidal
H_3O^+	[H:O:H]$^+$ with H below, lone pair on O	3	1	[O, H, H, H]$^+$	Pyramidal
CO_2	O::C::O	2	0	O=C=O	Linear
SF_6	S surrounded by six F	6	0	S with six F	Octahedral

Example 9.2. *Describe the bonding in methanal (formaldehyde) in terms of sigma and pi bonding.*

Solution. The electron dot structure of methanal (CH_2O) is

H:C::Ö: (with H below C) or H—C=O (with H below C)

In methanal, the carbon atom assumes an sp^2 hybrid state (see Figs. 9.13 and 9.14).

The two *sp*-hybrid orbitals of the carbon atom overlap with the 1*s* orbital of each of the two H atoms to form two C—H σ bonds. The third sp^2-hybrid orbital of the carbon atom overlaps axially with a similar orbital of oxygen atom to form an sp^2-sp^2 sigma bond. The unhybridized 2*p* orbital of carbon overlap with a similar orbital of oxygen sidewise to forms *p*π-*p*π bond. The π clouds and the sigma cloud have higher electron density around the oxygen atom due to its greater electronegativity.

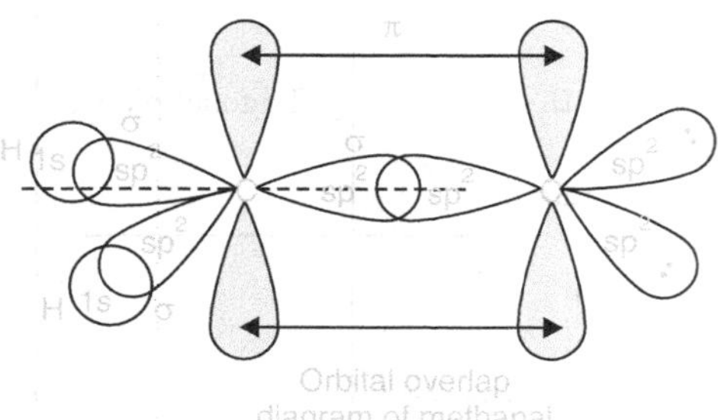

Orbital overlap diagram of methanal

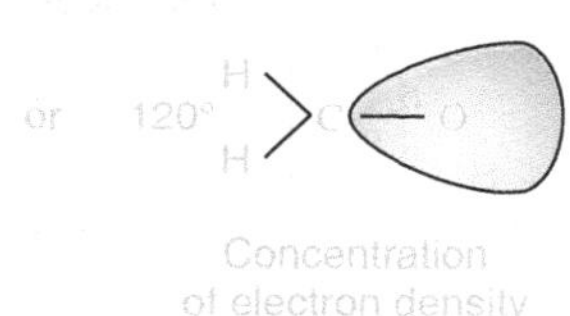

Concentration of electron density around O atom

SUMMARY

- **Covalent Bonds.** The bond formed by the overlapping of half-filled atomic orbitals present in the valence shell of the atoms participating in bonding.
- **Strength of a Bond.** The strength of a bond depends on the extent of overlap of atomic orbitals.
- **Sigma Bond.** Bond formed by axial overlap of atomic orbitals.
- **Pi Bond.** The bond formed by the sidewise overlap of atomic orbitals.
- **Hybridization.** The process of mixing or merging of orbitals (of slightly different energies) of an atom to form another set of orbitals with equivalent shape and energy.

EVALUATE YOURSELF

I. Objective Type Questions

Select the most appropriate choice from the options given as (a), (b), (c) and (d) after each question:

1. The bond formed by sidewise overlap of atomic orbitals is
   (*a*) sigma bond (*b*) coordinate bond
   (*c*) *pi* bond (*d*) ionic bond
2. A carbon (IV) oxide molecule contains how many sigma bonds?
   (*a*) 4 (*b*) 3
   (*c*) 2 (*d*) 0
3. Which type of hybridization leads to a three dimensional geometry of bonds around C atom?
   (*a*) sp^3 (*b*) *sp*
   (*c*) sp^2 (*d*) All of these
4. The hybrid state of O in ethanol molecule is
   (*a*) sp^2 (*b*) sp^3d
   (*c*) *sp* (*d*) sp^3
5. The hybrid state of S in SO_2 molecule is
   (*a*) sp^3 (*b*) sp^2
   (*c*) sp^3d (*d*) *sp*
6. The number of sigma and pi bonds in HCN molecule are respectively
   (*a*) 3, 2 (*b*) 2, 2
   (*c*) 2, 3 (*d*) 3, 3
7. Which among the following is octahedral?
   (*a*) SF_4 (*b*) SF_6
   (*c*) BF_3 (*d*) BF_4^-

II. Fill in the Blanks

8. Complete the following sentences by supplying appropriate words:
   (*i*) Tetra covalency of carbon is explained by configuration ......
   (*ii*) The correct bond angle between sp^3-hybrid orbitals is ......
   (*iii*) The hybridization assumed by boron in boron trifluoride molecule is......
   (*iv*) In carbon (IV) oxide molecule the hybridization state assumed by carbon is ......
   (*v*) The sigma and *pi* bonds in ethyne molecule are ...... respectively.

III. Discussion Questions

9. What is the orbital concept of formation of covalent bonds?
10. How are sigma and *pi* bonds formed? Are they equally stronger? Comment.

11. Give three differences between the two bonds holding oxygen atoms together in O_2 molecule.

12. What are hybrid orbitals? How do they differ from pure atomic orbitals?

13. Discuss the shape of ammonia molecule on the basis of hybridization.

14. What is the hybrid state of N in NO_3^-, NH_3 and NH_4^+?

15. Oxygen atom assumes an sp^3 hybrid state in H_2O molecule but the bond angle around O atom is not 109°.28′. Explain.

16. Using the concept of hybridization give the orbital diagram of (*i*) Ethyne (*ii*) $BeCl_2$.

17. Predict the hybrid state of sulphur in SO_4^{2-} ion and SO_3 molecule.

18. What is meant by hybridization? Give an example of a molecule in which the central atom assumes sp^3d^2 hybridization.

19. Give Lewis structures of molecules in which the carbon atoms assume (*i*) *sp*-hybridization (*ii*) sp^3-hybridization and (*iii*) sp^2-hybridization.

20. Explain the type of bonding in (*i*) methanal (HCHO) (*ii*) Ethanol (C_2H_5OH).

Answers

I. Objective Type Questions

1. (*c*) **2.** (*c*) **3.** (*a*) **4.** (*d*)
5. (*b*) **6.** (*b*) **7.** (*b*).

II. Fill in the Blanks

8. (*i*) $1s^2\ 2s^1\ 2p_x^1\ 2p_y^1\ 2p_z^1$ (*ii*) 109°.28′
(*iii*) sp^2 (*iv*) *sp* (*v*) 3 and 2.

III. Discussion Question

17. sp^3, sp^2.

SHS 2

SECTION

4

CONSERVATION OF MATTER AND STOICHIOMETRY

General Objectives

After studying this section, the student should be able to:

1. understand that balanced chemical equations indicate the quantitative relationships between reactants and products involved in chemical changes and uses stoichiometry in quantitative analysis.
2. recognise that the introduction of the Carbon-12 scale was the beginning of the quantitative measurement of the amount of substance contained in different materials.
3. recognise that the particles in the nucleus of an atom repel each other and that a strong force needs to be present to keep the nucleus.

In this section, we shall study

SECTION-4

UNIT-10

CARBON-12 SCALE

Specific Objectives

After studying this unit, the student should be able to:

- describe the carbon-12 scale of measurement of mass
- define relative atomic mass and molecular mass
- define the mole as a unit of measurement of an amount of substance
- identify molar quantities of substances.
- calculate amount of substance, number of entities and molar quantities from a given data.

10.1 CARBON-12 SCALE OF MEASUREMENT OF MASS

Measurement of the quantities of a substance forms an important part of our daily life. For example, in laboratories we measure volume and mass to find out the quantity of matter present in it. The knowledge of quantity helps us to avoid the use of an excess of a substance. In chemistry, the quantity of matter can be measured in two basic ways:

(*i*) *mass quantity* (*ii*) *amount of substance quantity*

All substances are made up of atoms, molecules or ions which together makes the total matter of the substance. As atoms are very tiny particles their absolute masses are difficult to measure. The problem was resolved by measuring the relative amounts, that is, comparing amounts. For this purpose, "*the mass of one atom of hydrogen was assumed as unity and accepted as the standard*".

Hydrogen is the lightest atom and is found to have a mass of 1.67×10^{-24} g *i.e.*, 0.000 000 000 000 000 000 000 00167 g. Similarly the mass of the heaviest atom Lawrencium (Lr) is 4.29×10^{-22} g *i.e.*, 0.000 000 000 000 000 000 000 429 g. On comparing masses we get a simpler and more convenient measurement of relative mass:

$$\frac{\text{mass of one atom of Lr}}{\text{mass of one atom of H}} = \frac{4.29 \times 10^{-22}}{1.67 \times 10^{-24}} = 257$$

It indicates that one atom of lawrencium is 257 times heavier than one atom of hydrogen.

In 1858, *oxygen atom* was adopted *as a standard* on account of the following reasons:

(*i*) It is much easier to obtain compounds of elements with oxygen than with hydrogen as oxygen is more reactive than hydrogen.

(*ii*) The atomic masses of most of the elements become approximately whole numbers but with hydrogen as the standard the atomic masses of most of the elements are fractional.

The mass of one atom of oxygen was taken as 16.0. Thus, atomic mass of an element

$$= \frac{\text{Mass of one atom of element}}{1/16^{\text{th}} \text{ part of the mass of one atom of Oxygen}}.$$

Later on even the oxygen atom as reference was rejected in favour of C-12. It is because carbon has a more stable isotope C-12 whose mass is exactly taken as 12u.

In 1961, the International Union of Chemists selected the most stable isotope of carbon (C-12 isotope) as the standard for comparison of the atomic masses of various elements. The mass of C-12 isotope was taken as 12 atomic mass units. *The scale in which the relative atomic masses of different atoms are expressed* is called **atomic mass unit scale** or **amu scale.** (In the recent years, the new symbol '*u*' (which means ***unified mass***) is recommended to be used in place of ***amu.***

The **atomic mass unit** or **unified mass** may be defined as *one-twelfth of the actual mass of an atom of carbon (carbon-12 isotope).*

The atomic mass of an element tells us as to how many times an atom of the element is heavier than one-twelfth of an atom of carbon (C-12). This can be represented as

$$\textbf{Atomic mass} = \frac{\text{Mass of an atom of the element}}{\frac{1}{12} \times \text{mass of an atom of carbon (C-12)}}$$

For example, the atomic mass of magnesium is 24 u. It means that an atom of magnesium is 24 times heavier than one-twelfth the mass of a carbon atom (C-12).

The atomic masses of some elements on the basis of the C-12 scale are given below:

Hydrogen	1.008 *u*
Oxygen	16.00 *u*
Chlorine	35.453 *u*
Magnesium	24.305 *u*
Copper	63.546 *u*
Iron	55.847 *u*
Sodium	22.989 *u*
Zinc	65.38 *u*
Silver	107.868 *u*

10.2 RELATIVE ATOMIC MASS AND MOLECULAR MASS

There are many cases where different atoms of the same element possess different relative masses. Such *atoms of the same element which have different relative masses are called* **isotopes.** In such cases the atomic mass of the element is the average of relative masses of the different isotopes of the element. For example, chlorine contains two types of atoms having relative masses 35 u and 37 u. The relative abundance of these isotopes in nature is in the ratio 3 : 1. Thus, the atomic mass of chlorine is the average of these different relative masses as described below:

$$\text{Atomic mass of chlorine} = \frac{(35 \times 3 + 37 \times 1)}{4}$$

$$= 35.5 \text{ u.}$$

Thus, the **atomic mass** of an element may be defined as *the average relative mass of an atom of the element as compared with the mass of an atom of carbon (C-12) taken as 12 u.*

Do You Know?

The use of relative masses of atoms is preferred over absolute masses because the gram is too big a unit for expressing their masses. If somehow we express the mass of an atom in gram the numerical value which we get is extremely small and hence, inconvenient to use. For example, the mass of an atom of carbon (C-12) is 1.99×10^{-23} gram.

SOLVED EXAMPLES

Example 10.1. *Nitrogen occurs in nature in the form of two isotopes with atomic masses 14 and 15 respectively. If the average atomic mass of nitrogen is 14.0067, what is the % abundance of the two isotopes?*

Solution. Let % abundance of N-14 isotope = x

Then % abundance of N-15 isotope = $(100 - x)$

The average atomic mass $= \dfrac{x \times 14 + (100 - x)\,15}{100}$

Now, $\dfrac{14x + (100 - x)\,15}{100} = 14.0067$

or $-x = 1400.67 - 1500 = -99.33$ or $x = 99.33$

Thus, % abundance of N-14 isotope = **99.33%**

% abundance of N-15 isotope = **0.67%.**

Example 10.2. *Use the data given in the table to calculate the molar mass of carbon:*

Isotope	*Relative Abundance*	*At. mass*
^{12}C	*98.892*	*12*
^{13}C	*1.108*	*13.00335*
^{14}C	2×10^{-10}	*14.00317*

Solution.

Molar mass of carbon

$$= \frac{\Sigma \text{ Isotopic mass} \times \text{Relative abundance}}{\Sigma \text{ Relative abundance}}$$

$$= 0.98892 \times (12 \text{ u}) + 0.01108 \times (13.0033 \text{ u}) + 2 \times 10^{-12} \times (14.00317 \text{ u})$$

$= \mathbf{12.011\ u.}$

GRAM ATOMIC MASS (GAM)

It may be defined as *that much quantity of the element whose mass in gram is numerically equal to its atomic mass.* Gram atomic mass is also called **one gram-atom** of the element. For example, atomic mass of magnesium (Mg) is 24 u, therefore,

1 gram-atom of Mg = gram atomic mass of Mg

= 24 g

2 gram-atom of Mg = 2 × gram atomic mass of Mg

= (2 × 24) g = 48 g.

From the above discussion, we can get the relationship between the mass of an element and its gram-atoms as:

$$\textbf{Number of gram-atom} = \frac{\textbf{Mass in gram}}{\textbf{Gram atomic mass}} = \frac{W_{(g)}}{\text{GAM}}.$$

MOLECULAR MASS

Like atoms, the molecules are also very small in size and hence, their actual masses cannot be determined by direct weighing. Therefore, masses of molecules are also determined relative to the mass of the carbon atom (C-12) taken as 12 u. These relative masses of molecules are called molecular masses.

Molecular mass of a substance (element or compound) may be defined as *the average relative mass of a molecule of the substance as compared with the mass of an atom of carbon (C-12) taken as 12 u.*

Molecular mass of a substance tells us the number of times a molecule of the substance is heavier than 1/12th of the mass of a carbon (C-12) atom. For example, the molecular mass of water (H_2O) is 18 u. It means that a molecule of water is 18 times heavier than 1/12th of the mass of a carbon atom (C-12).

The molecular mass of a substance can be obtained by adding the atomic masses of all the atoms present in a molecule of the substance. For example,

Molecular masses of NH_3

= atomic mass of N + 3 (atomic mass of H)

= 14 + 3 × 1 = 17 u.

GRAM MOLECULAR MASS (GMM)

It may be defined as *that much quantity of a substance (element or compound) whose mass in gram is numerically equal to its molecular mass.* Gram molecular mass is also called one **gram-molecule** of a substance. For example, the molecular mass of ammonia is 17 u; therefore,

1 gram-molecule of NH_3

= gram molecular mass of NH_3

= 17 g

2 gram-molecule of NH_3

= 2 × gram molecular mass of NH_3

= (2 × 17) g = 34 g.

From the above discussion it follows that

Number of gram-molecule

$$= \frac{\textbf{Mass in gram}}{\textbf{Gram atomic mass}} = \frac{W_{(g)}}{\text{GMM}}.$$

SOLVED EXAMPLES

Example 10.3. *(a) Calculate the mass of 2.5 gram-atom of calcium. Atomic mass of calcium is 40 u.*

(b) Calculate the mass of 1.5 gram-molecule of water (H_2O).

Solution. (*a*) 1 gram-atom of calcium

= gram atomic mass of calcium = 40 g

2.5 gram-atom of calcium

= 40 × 2.5 = **100 g.**

(*b*) Molecular mass of water (H_2O)

= 1 × 2 + 16 = 18 u

1 gram-molecule of H_2O

= gram molecular mass of H_2O = 18 g.

1.5 gram-molecule of H_2O

= 18 × 1.5 = **27 g.**

Example 10.4. *Calculate the number of gram-atom and gram-molecule in 25.4 mg of iodine (I_2). Atomic mass of I = 127 u.*

Solution.

$$\text{Gram-atom} = \frac{\text{Mass (g)}}{\text{GAM}} = \frac{25.4 \times 10^{-3}\text{ g}}{127\text{ g}}$$

$= \mathbf{2 \times 10^{-4}}$ **g-atom**

$$\text{Gram-molecule} = \frac{\text{Mass (g)}}{\text{GAM}} = \frac{25.4 \times 10^{-3}\text{ g}}{254\text{ g}}$$

$= \mathbf{1 \times 10^{-4}}$ **g-molecule.**

Example 10.5. *Calculate the molar mass of glucose ($C_6H_{12}O_6$) and the number of atoms of each kind in it.*

Solution. Molecular mass of glucose ($C_6H_{12}O_6$)

$= 6 \times (12.011\ u) + 12 \times (1.008\ u) + 6 \times (16.00\ u)$

$= 72.066\ u + 12.096\ u + 96.00\ u$

$= \mathbf{180.162\ u}$

Calculation of number of atoms of each kind

1 mole of glucose ($C_6H_{12}O_6$)

≡ 6 moles of carbon + 12 moles of hydrogen + 6 moles of oxygen

Hence, Atoms of carbon $= 6 \times 6.02 \times 10^{23}$

$= \mathbf{36.12 \times 10^{23}}$

Atoms of hydrogen $= 12 \times 6.02 \times 10^{23}$

$= \mathbf{72.24 \times 10^{23}}$

Atoms of oxygen $= 6 \times 6.02 \times 10^{23}$

$= \mathbf{36.12 \times 10^{23}}$

Review of particulate nature of matter
- Matter is made up of small particles.
- The constituent particles of matter are extremely small in size.
- There are spaces between particles of matter.
- The particles of matter are continuously moving.
- Particles of matter attract each other.

10.3 MOLE AS A UNIT OF MEASUREMENT OF AMOUNT OF A SUBSTANCE

In chemical investigations, we are often interested in knowing the number of atoms and molecules we are dealing with. For example, let us consider the following reaction:

$$N_2 + 3H_2 \longrightarrow 2NH_3$$

In this reaction, one molecule of nitrogen reacts with three molecules of hydrogen. So, it would be desirable to take the molecules of nitrogen and hydrogen in the ratio 1 : 3, so that the reactants are completely consumed during the reaction. But we know atoms and molecules are so small in size that it is not possible to count them individually. Moreover, the smallest sample of the substance that we may take contains a very large number of atoms or molecules. In order to overcome these difficulties, the ***concept of mole*** was introduced. According to this concept, the number of particles of a substance is related to the mass of the substance. For the sake of counting elementary particles, the number of atoms in 12 g of carbon-12 isotope is taken as standard. This number has been experimentally found to be 6.023×10^{23} and is called one mole. This number is also known as Avogadro's number or **Avogadro constant (N_A).**

The **mole*** may be defined as *"the amount of a substance that contains as many specified elementary particles (atoms, molecules, ions or electrons etc.,) as the number of atoms in 12 g of carbon-12 isotope".*

Mole is thus a unit for counting atoms, molecules, ions, etc. It stands for 6.023×10^{23} particles irrespective of their nature. *The term mole should not be confused with the amount of a substance (a physical quantity) because the amount of substance is not a number.*

SUMMARY of the various relationships of the mole has been illustrated as follows in a tabular form.

ONE MOLE			
6.02×10^{23} Atoms	6.02×10^{23} Molecules	6.02×10^{23} Formula Units	**PARTICLES**
GAM or 1 g-atom	GMM or 1 g-molecule	GFM or 1 g-Formula mass	**MASS**
22.4 L *(if gas)*	GMM/Density *(if solid or liquid)*		**VOLUME**

IMPORTANT FORMULAE ABOUT MOLE CONCEPT

The following formulae are quite helpful in solving the problems on mole concept. In these formulae N_A represents **6.02×10^{23}**

- Mass of **1 atom** of an element $= \dfrac{\mathbf{GAM}}{\mathbf{N_A}}$
- Mass of **1 molecule** of a substance $= \dfrac{\mathbf{GMM}}{\mathbf{N_A}}$
- Number of molecules in **W** g of a substance $= \dfrac{\mathbf{WN_A}}{\mathbf{GMM}}$
- Number of molecules in **V** litre of a gas at S.T.P. $= \dfrac{\mathbf{VN_A}}{22.4}$
- Number of molecules in **n** moles of a substance $= \mathbf{n} \times \mathbf{N_A}$
- 1 **u** or (1 **amu**) = **$1/N_A$ g** and is also called **1 avogram** or **1 dalton.**

*The word mole was introduced by *Wilhelm Ostwald* who derived the term from the Latin word *moles* meaning a *heap* or *pile*. A substance may be considered as a heap of atoms or molecules. In 1967, the mole was accepted as a unit which provides a simple way of reporting a *large number* (6.023×10^{23}), *i.e.*, a massive heap of atoms or molecules in a sample.

The interconversion of mole, number of molecules and mass in gram is represented diagrammatically in Fig. 10.1.

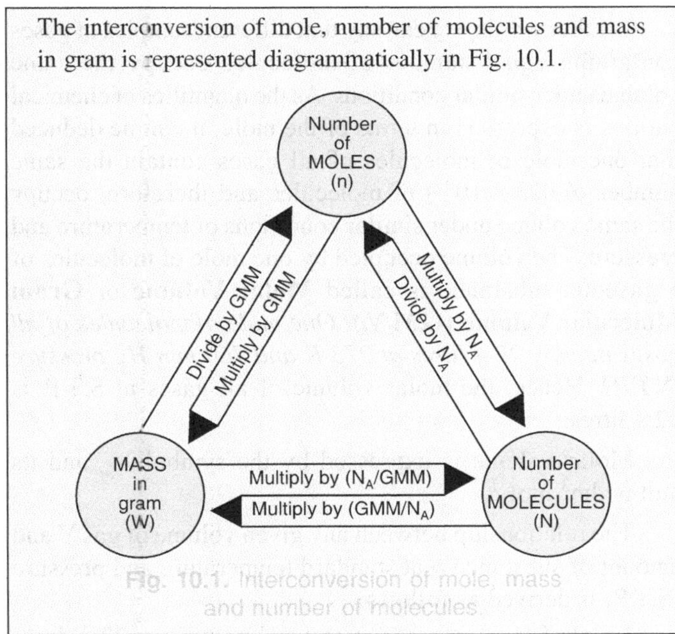

Fig. 10.1. Interconversion of mole, mass and number of molecules

10.4 MOLAR QUANTITIES OF SUBSTANCES

AVOGADRO'S CONSTANT, N_A

From measurements based on the carbon-12 scale the number of atoms in 0.012 kg, or 12.00 g, of C–12 is 6.02×10^{23}. From the definition of the mole, *one mole of every substance contains this same number of entities.*

This number is called the *Avogadro constant*. It has the symbol N_A.

- The Avogadro constant is the number of atoms in 12.00 g of ^{12}C or C–12.

  Or

- It is the number of **specified entities** in 1 mol of any substance.

The Avogadro constant has units mol^{-1} and can be quoted as $N_A = 6.02 \times 10^{23}$ entities mol^{-1}.

From the definition of Avogadro's constant:

1 mol of any substance contains N_A entities

½ mol of any substance contains ½ × N_A entities

3 mol of any substance contains 3 × N_A entities

X mole of any substance contains X × N_A entities

∴ Number of entities, N, in a given amount of substance, *n*, is given by:

$$N = n \times N_A$$

$$n = \frac{N}{N_A}$$

MOLAR MASS (M)

The mass of one mole of particles is called **molar mass.** The number of entities N_A(atoms or molecules) contained in molar mass is equal to 6.023×10^{23}. For example,

(*i*) One mol of sodium means 6.023×10^{23} Na atoms weighing 23.0 g (GAM of sodium). In brief we can write that

Molar mass of sodium = 23.0 g mol^{-1}

(*ii*) One mol of ammonia contains 6.023×10^{23} NH_3 molecules weighing 17.0 g (GMM of ammonia). Thus we can write

Molar mass of ammonia = 17.0 g mol^{-1}.

To conclude:

6.023×10^{23} atoms of oxygen weigh

= 16 gram = GAM of oxygen or 1 g atom

6.023×10^{23} atoms of nitrogen weigh

= 14 gram = GAM of nitrogen or 1 g atom

6.023×10^{23} molecules of oxygen weigh

= 32 gram = GMM of oxygen or 1 g molecule

6.023×10^{23} molecules of water weigh

= 18 gram = GMM of water or 1 g molecule.

Add to Your Knowledge

*In case of elementary substances, it is essential to specify whether we refer to **1 mol atoms** or **1 mol molecules.** If it is not mentioned then we should infer that it is the natural state of existence of that substance. For example, 1 mol of hydrogen refers to 1 mol of hydrogen molecules (H_2). However, in order to avoid this confusion the molecular states of elements are given separate names. For example, molecular states of hydrogen, nitrogen and oxygen are named as **dihydrogen** (H_2), **dinitrogen** (N_2) and **dioxygen** (O_2) respectively.*

SOLVED EXAMPLES

Example 10.6. (*i*) *Calculate the number of atoms in 16 g of Cu.*

(*ii*) *How many moles are there in 2.7 g of Al?*

(*Given: atomic mass of Cu = 64u, Al = 27u*)

Solution. (*i*) Atomic mass of Cu = 64

$$\text{Number of moles in 16 g Cu}(n) = \frac{\text{Mass in g}}{\text{Atomic mass}} = \frac{16}{64} = \frac{1}{4}$$

$$= 0.25 \text{ mol}$$

$$\therefore \quad \text{Number of atoms} = \text{mol} \times N_A = 0.25 \times 6.02 \times 10^{23}$$

$$= \mathbf{1.50 \times 10^{23}}$$

(*ii*) *Atomic mass of Al = 27*

∴ Number of moles in 2.7 g Al (n)

$$= \frac{2.7}{27} = \frac{1}{10} = \mathbf{0.1\ mol.}$$

Example 10.7. (*i*) *What is the mass of 0.600 mol of K_2CO_3?*

(*ii*) *What is the amount of substance in 5 g $CaCO_3$?*

(Relative atomic masses are: K = 39, C = 12, Ca = 40).

Solution. (*i*) Molar mass of $K_2CO_3 = (2 \times 39) + 12 + (3 \times 16)$

$= 138\ g\ mol^{-1}$

$$\text{mass of substance} = \begin{pmatrix}\text{amount of}\\ \text{substance}\end{pmatrix} \times \begin{pmatrix}\text{molar}\\ \text{mass}\end{pmatrix}$$

mass of 0.600 mol of $K_2CO_3 = 0.600\ mol \times 138\ g\ mol^{-1}$

$= \mathbf{82.8\ g}$

(*ii*) mass of $CaCO_3 = 5\ g$

molar mass of $CaCO_3 = 100\ g\ mol^{-1}$

$$\text{amount of } CaCO_3 \text{ in 5 g} = \frac{5\ g}{100\ g\ mol^{-1}}$$

$= \mathbf{0.05\ mol.}$

Example 10.8. (*i*) *How many molecules of O_2 gas are present in 16 g of oxygen gas?*

(*ii*) *What is the molar mass and relative molecular mass of a substance if the mass of 0.10 mol is 3.9 g?*

Solution. (*i*) mass of O_2 (m) = 16 g

molar mass of O_2 (M) = 32 g mol^{-1}

$$\therefore \text{ moles of oxygen, } n(O_2) = \frac{m}{M} = \frac{16\ g}{32\ g\ mol^{-1}} = 0.50\ mol$$

Now, No. of molecules (N) = moles (n) × N_A

∴ number of molecules, N(O_2)

$= 0.50\ mol \times 6.02 \times 10^{23}$ molecules mol^{-1}

$= \mathbf{3.01 \times 10^{23}}$ **molecules**

(*ii*) mass of substance = 3.9 g

amount of substance = 0.10 mol

$$\text{molar mass} = \frac{3.9\ g}{0.10\ mol}$$

$= 39\ g\ mol^{-1}$

relative molecular mass = 39.

MOLAR VOLUME (V_m)

In 1811, Amedeo Avogadro, an Italian scientist proposed as law known as Avogadro's Law.

"Avogadro's Law states that equal volumes of all gases contain equal number of molecules under similar conditions."

In other words, we can say that different samples of gases containing equal number of molecules occupy the same volume under similar conditions. As the quantities of chemical entities is expressed in terms of the mole, it can be deduced that one mole of molecules of all gases contain the same number (6.023×10^{23}) of molecules and therefore, occupy the same volume under similar conditions of temperature and pressure. The volume occupied by one mole of molecules of a gaseous substance is called **Molar Volume** or **Gram Molecular Volume (G.M.V.).** *One mole of molecules of all gases occupy 22.4 litres at 273 K and 760 mm Hg pressure (S.T.P.).* Hence, the molar volume of all gases at S.T.P. is 22.4 litres.

Molar volume is expressed by the symbol V_m and its unit is dm^3 mol^{-1}.

The relationship between any given volume of gas V and amount of substance n at standard temperature and pressure (S.T.P) is derived as follows:

1 mol of every gas occupies a volume of 1×22.4 dm^3

X mol of every gas occupies a volume of X × 22.4 dm^3

$$\therefore \quad V(dm^3) = n \times V_m$$

$$n = \frac{V(dm^3)}{V_m}.$$

SOLVED EXAMPLE

Example 10.9. *Calculate the*

(*i*) *amount of CO_2*

(*ii*) *mass of gas*

(*iii*) *number of molecules of gas in 2.8 dm^3 of CO_2 at S.T.P.*

(V_m = 22.4 dm^3 mol^{-1}, atomic mass; C = 12, O = 16)

Solution. (*i*) volume of CO_2, V, = 2.80 dm^3

molar volume of CO_2, V_m, = 22.4 dm^3 mol^{-1}

$$\text{amount of } CO_2,\ n, = \frac{V(dm^3)}{V_m}$$

$$\therefore \quad \text{amount of } CO_2,\ n, = \frac{2.80\ dm^3}{22.4\ dm^3\ mol^{-1}}$$

$= 0.125$ mol

(*ii*) molar mass of CO_2, M, = 44 g mol^{-1}

amount of CO_2, n, = 0.125 mol

But $m = n \times M$

∴ mass of $CO_2 = 0.125\ mol \times 44\ g\ mol^{-1}$

$= 5.5$ g

(*iii*) $N = n \times L$, $n = 0.125$ mol, $L = 6.02 \times 10^{23}$ mol^{-1}

∴ number of molecules of CO_2, N(CO_2)

$= 0.125\ mol \times (6.02 \times 10^{23}$ molecules mol^{-1})

$= \mathbf{7.53 \times 10^{22}}$ **molecules.**

THE FARADAY CONSTANT, F

It is the total charge in coulombs carried by 1 mol i.e., 6.02×10^{23} electrons or protons.

Accordingly, if the charge on electrons is e^-, then

$$F = N_A \times e^-$$

It is important to note that the quantity of electric charge is measured in coulombs or passage of an amount of electrons or protons. The symbol of coulombs is C.

The word Faraday (F) is no longer used as a unit of charge. It is a physical quantity and a constant. For example, the old expression 0.2 F charge should now read 0.2 mol electrons or be expressed in coulombs as 0.2×96500 C

$$\because \quad 1\ F = 96500\ C\ mol^{-1}.$$

The quantity of charge (Q) is given by the product of the electric current (I) in amperes (A), and the time (t) in seconds for which the current flows.

Mathematically, $Q = I \times t$

If $I = 1$ ampere and $t = 1$ sec then

Quantity of charge = 1 amp × 1 sec = 1 C.

SOLVED EXAMPLES

Example 10.10. *Calculate the charge, in coulombs carried by*

(*i*) *2.5 mol electrons,*

(*ii*) *3.01×10^{21} electrons,*

(*iii*) *passage of 2A of current for 90 minutes.*

Solution. (*i*) 1 mol electrons carry a charge = 96500 C

$\therefore$ 2.5 mol electrons = 2.5×96500 C

= **241250 C**

(*ii*) 6.02×10^{23} electrons carry a charge = 96500 C

$$\therefore \quad 1\ e^- = \frac{96500}{6.02 \times 10^{23}}\ C$$

$$3.01 \times 10^{21} \text{ electrons carry a charge} = \frac{96500 \times 3.01 \times 10^{23}}{6.02 \times 10^{21}}$$

$$= 96500 \times \frac{1}{2} \times 10^2\ C$$

$$= 4825000\ C$$

or $= \mathbf{4825 \times 10^3\ C}$

(*iii*) $I = 2A$, $t = 90$ min $= 90\ \text{min} \times \frac{60\ s}{\text{min}}$

= 5400 s

The quantity of charge, $Q = It$

= 2A × 5400 s

= **10800 C.**

Example 10.11. *How many electrons will flow through a conductor if a charge of 100 A is passed for 10 minutes. (1 mol electrons = 96500 C)*

Solution. Quantity of charge,

$$Q = I \times t = 100 \times 10 \times 60$$

$$= 60000\ C$$

We know that $96500\ C \equiv 6.02 \times 10^{23}\ e^-$

$$1\ C = \frac{6.02 \times 10^{23}\ e^-}{96500}$$

$$\therefore \quad 60000\ C = \frac{6.02 \times 10^{23}}{96500} \times 60000$$

$$= \mathbf{3.743 \times 10^{23}\ electrons.}$$

10.5 CALCULATION OF AMOUNT OF SUBSTANCE, NUMBER OF ENTITIES AND MOLAR QUANTITIES FROM A GIVEN DATA

Example 10.12. *Calculate the mass of the following:*

(*i*) *one atom of calcium*

(*ii*) *one molecule of sulphur dioxide (SO_2).*

Solution. (*i*) Mass of 6.023×10^{23} atoms of calcium

= gram atomic mass of calcium = 40 g

$\therefore$ Mass of 1 atom of calcium

$$= \frac{40\ g}{6.023 \times 10^{23}} = \mathbf{6.6 \times 10^{-23}\ g.}$$

(*ii*) Mass of 6.023×10^{23} molecules of SO_2

= gram molecular mass of SO_2 = 64 g

$\therefore$ Mass of 1 molecule of SO_2

$$= \frac{64\ g}{6.023 \times 10^{23}} = \mathbf{1.06 \times 10^{-22}\ g.}$$

Example 10.13. (*a*) *Calculate the number of atoms in each of the following:*

(*i*) *0.5 mole atoms of nitrogen*

(*ii*) *0.2 mole molecules of nitrogen*

(*iii*) *3.2 g of sulphur.*

(*b*) *Calculate the number of molecules in each of the following:*

(*i*) *14 g of nitrogen*

(*ii*) *3.4 g of hydrogen sulphide (H_2S).*

Solution. (*a*)

(*i*) 1 mole atoms of nitrogen

= 6.023×10^{23} atoms

$\therefore$ 0.5 mole atoms of nitrogen

= $6.023 \times 10^{23} \times 0.5$ = $\mathbf{3.01 \times 10^{23}}$ **atoms.**

(*ii*) 1 mole molecules of nitrogen

$= 6.023 \times 10^{23}$ molecules

$\therefore$ 0.2 mole molecules of nitrogen

$= 6.023 \times 10^{23} \times 0.2$

$= 1.2046 \times 10^{23}$ molecules.

1 molecule of nitrogen

= 2 atoms

1.2046×10^{23} molecules of nitrogen

$= 1.2046 \times 10^{23} \times 2 = \mathbf{2.409 \times 10^{23}}$ **atoms.**

(*iii*) 32 g sulphur contain

$= 6.023 \times 10^{23}$ atoms

$\therefore$ 3.2 g of sulphur contain

$$= \frac{6.023 \times 10^{23} \times 3.2}{32} = \mathbf{6.023 \times 10^{22}} \textbf{ atoms.}$$

(*b*) (*i*) Gram molecular mass of nitrogen

= 28 g

28 g of nitrogen contain

$= 6.023 \times 10^{23}$ molecules

$\therefore$ 14 g of nitrogen contain

$$= \frac{6.023 \times 10^{23}}{28} \times 14$$

$= \mathbf{3.01 \times 10^{23}}$ **molecules.**

(*ii*) Gram molecular mass of H_2S

= (2 + 32) g = 34 g

34 g of H_2S contain

$= 6.023 \times 10^{23}$ molecules

3.4 g of H_2S contain

$$= \frac{6.023 \times 10^{23}}{34} \times 3.4 = \mathbf{6.023 \times 10^{22}} \textbf{ molecules.}$$

Example 10.14. *How many atoms of oxygen are present in 300 g of $CaCO_3$?*

Solution. Gram formula mass of $CaCO_3$ = 100 g

Now, 1 mole of $CaCO_3$ contain

3 mole of O atoms.

or 100 g of $CaCO_3$ contain

$3 \times 6.02 \times 10^{23}$ O atoms

$\therefore$ 300 g of $CaCO_3$ contain O atoms

$$= \frac{3 \times 6.023 \times 10^{23}}{100} \times 300 = 54.207 \times 10^{23}$$

$= \mathbf{5.4207 \times 10^{24}}$ **Oxygen atoms.**

Example 10.15. *How many molecules of water of hydration are present in 252 mg of oxalic acid, ($H_2C_2O_4.2H_2O$)?*

Solution. Gram molecular mass of $H_2C_2O_4.2\,H_2O$

= 126 g

Now, the water molecules in 1 mol of oxalic acid

= 2 mol.

or the water molecules in 126 g of oxalic acid

$= 2 \times 6.023 \times 10^{23}$

$\therefore$ Water molecules in 252×10^{-3} g of oxalic acid

$$= \frac{2 \times 6.023 \times 10^{23} \times 252 \times 10^{-3}}{126} = \mathbf{2.4 \times 10^{21}}.$$

Example 10.16. *Chlorophyll, the green colouring material of plants contains 2.68% of magnesium by mass. Calculate the number of magnesium atoms in 5.00 g of this complex.*

Solution. Mass of magnesium in 5.00 g of complex

$$= \frac{2.68}{100} \times 5.00 = 0.134 \text{ g}.$$

Gram atomic mass of magnesium

= 24 g

24 g of magnesium contain

$= 6.02 \times 10^{23}$ atoms

0.134 g of magnesium would contain

$$= \frac{6.02 \times 10^{23}}{24} \times 0.134 = 3.36 \times 10^{21} \text{ atoms}$$

Therefore, 5.00 g of the given complex would contain $\mathbf{3.36 \times 10^{21}}$ **atoms of magnesium.**

Example 10.17. *Calculate the number of atoms of each type in 3.42 g of sucrose ($C_{12}H_{22}O_{11}$).*

Solution. Molecular mass of sucrose

$= (12 \times 12 + 1 \times 22 + 16 \times 11) = 342$

342 g of sucrose contain

$= 6.023 \times 10^{23}$ molecules

$\therefore$ 3.42 g of sucrose contain

$$= \frac{6.023 \times 10^{23}}{342} \times 3.42$$

$= 6.023 \times 10^{21}$ molecules.

Number of atoms of carbon in 3.42 g of sucrose

1 molecule of sucrose contains

= 12 atoms of carbon

6.023×10^{21} molecules of sucrose contain

$= 12 \times 6.023 \times 10^{21}$ atoms of carbon

$= \mathbf{72.28 \times 10^{21}}$ **atoms of carbon**

Number of atoms of hydrogen in 3.42 g of sucrose

1 molecule of sucrose contains

= 22 atoms of hydrogen

6.023×10^{21} molecules of sucrose contain

$= 22 \times 6.023 \times 10^{21}$ atoms of hydrogen

= 132.5×10^{21} atoms of hydrogen

Number of atoms of oxygen in 3.42 g of sucrose

1 molecule of sucrose contains

= 11 atoms of oxygen

6.023×10^{21} molecules of sucrose contain

$= 11 \times 6.023 \times 10^{21}$ atoms of oxygen.

= 66.25×10^{21} atoms of oxygen.

EVALUATE YOURSELF

I. Objective Type Questions

Select the most appropriate choice from the options given as (a), (b), (c) and (d) after each question:

1. Avogadro's number (N_A) represents the number of atoms in
(*a*) 12 g of C^{12} (*b*) 320 g of sulphur
(*c*) 32 g of oxygen (*d*) 12.7 g of iodine.

2. The number of moles of carbon dioxide which contain 8 g of oxygen is
(*a*) 0.50 mol (*b*) 0.20 mol
(*c*) 0.40 mol (*d*) 0.25 mol.

3. The number of atoms present in 0.5 g-atom of nitrogen is the same as the atoms in
(*a*) 12 g of carbon (*b*) 32 g of sulphur
(*c*) 8 g of the oxygen (*d*) 24 g of magnesium.

4. An atom is 10 times heavier than 1/12th the mass of a carbon atom (C-12). The mass of the atom in *u* is
(*a*) 10 (*b*) 120
(*c*) 1.2 (*d*) 12.

5. The total number of ions present in 111 g of $CaCl_2$ is
(*a*) one mole (*b*) two mole
(*c*) three mole (*d*) four mole.

6. Which of the following weighs the most?
(*a*) One g-atom of nitrogen
(*b*) One mole of water
(*c*) One mole of sodium
(*d*) One molecule of H_2SO_4.

7. Which of the following contains the same number of carbon atoms as are in 6.0 g of carbon (C-12)?
(*a*) 6.0 g ethane (*b*) 8.0 g methane
(*c*) 21.0 g propane (*d*) 28.0 g CO.

8. Which one among the following is the standard for atomic mass?
(*a*) $^{1}_{1}H$ (*b*) $^{12}_{6}C$
(*c*) $^{14}_{6}C$ (*d*) $^{16}_{8}O$.

9. Rearrange the following (I to IV) in order of increasing masses and choose the correct answer from (*a*), (*b*), (*c*) and (*d*) [Atomic masses: N = 14, O = 16, Cu = 63]
I. 1 molecule of oxygen
II. 1 atom of nitrogen
III. 1×10^{-10} g-molecule of oxygen
IV. 1×10^{-10} g-atom of copper
(*a*) II < I < III < IV (*b*) IV < III < II < I
(*c*) II < III < I < IV (*d*) III < IV < I < II.

10. The molecular masses of methane (CH_4) and oxygen (O_2) are 16 and 32 respectively. If one litre of methane at S.T.P. contains N molecules, the number of molecules in 5 L of oxygen at S.T.P. would be
(*a*) 10 N (*b*) N
(*c*) 5 N (*d*) 2.5 N.

11. 0.224 L of H_2 gas at S.T.P. is equivalent to
(*a*) 1 mol (*b*) 1 g
(*c*) 6.02×10^{22} molecules (*d*) 0.01 mol.

12. Which of the following pair of species have the same number of atoms under similar conditions?
(*a*) 1 L each of SO_2 and CO_2
(*b*) 1 L each of NH_3 and Cl_2
(*c*) 2 L each of O_3 and O_2
(*d*) 1 L of NH_3 and 2 L of SO_2.

13. 2.0 g of oxygen contains the same number of atoms as in
(*a*) 4 g of S (*b*) 7 g of nitrogen
(*c*) 0.5 g of H_2 (*d*) 12.3 g of Na.

14. 500 mL of NH_3 contains 6.00×10^{23} molecules at S.T.P. How many molecules are present in 100 mL of CO_2 at S.T.P.?
(*a*) 6×10^{23} (*b*) 1.5×10^{23}
(*c*) 1.2×10^{23} (*d*) None of these.

15. Haemoglobin contains 0.33% of iron by mass. The molecular mass of haemoglobin is approximately 67200. The number of iron atoms (at. mass 56) in one molecule of haemoglobin is
(*a*) 6 (*b*) 1
(*c*) 4 (*d*) 2.

16. How many electrons weigh one kilogram?
(*a*) 6.023×10^{23} (*b*) $\frac{10^{31}}{9.108}$
(*c*) $\frac{6.023}{9.108} \times 10^{54}$ (*d*) $\frac{1 \times 10^{8}}{9.108 \times 6.023}$.

17. A signature written in carbon pencil weighs 1 mg. What is the number of carbon atoms present in the signature?
(*a*) 6.02×10^{-20} (*b*) 0.502×10^{20}
(*c*) 5.02×10^{23} (*d*) 5.02×10^{20}.
18. Which has the maximum number of atoms?
(*a*) 24 g of C (12) (*b*) 56 g of Fe (56)
(*c*) 27 g of Al (27) (*d*) 108 g of Ag (108).
19. Number of atoms of oxygen present in 10.6 g Na_2CO_3 will be
(*a*) 6.02×10^{22} (*b*) 12.04×10^{22}
(*c*) 1.806×10^{23} (*d*) 31.80×10^{28}
20. Number of water molecules in a drop of water, if 1 mL of water has 20 drops and A is the Avogadro's number, is
(*a*) 0.5 A/18 (*b*) 0.05 A
(*c*) 0.5 A (*d*) 0.05 A/18.

II. Fill in the Blanks

21. Complete the following sentences by supplying appropriate words:
(*i*) One mole of a triatomic gas contains ...... atoms.
(*ii*) One mole of any gas at S.T.P. occupies a volume equal to ......
(*iii*) One mole of sulphuric acid contains ...... oxygen atoms.
(*iv*) Under similar conditions the ratio by volumes of gaseous reactants and gaseous products is ......
(*v*) 11.2 L of carbon dioxide at S.T.P. contains ...... oxygen atoms.
(*vi*) The number of hydrogen atoms in 80 *a.m.u.* of methane is ......
(*vii*) Equal volumes of different gases under similar conditions of temperature and pressure contain equal number of ......
(*viii*) 7 g of CO contains ...... O atoms.
(*ix*) The mass of 1×10^{22} formula units of $CuSO_4 . 5H_2O$ is ......

III. Discussion Questions

22. State the following terms:
(*a*) Atomic mass (*b*) Gram atomic mass
(*c*) Atomic mass unit.
23. What do you understand by the term formula mass? How does it differ from molecular mass?
24. Explain the following:
(*a*) Mole (*b*) V_m
(*c*) Molar mass (*d*) Unified mass.
25. Why do we use the terms *average and relative* in the definition of atomic mass?
26. Define atomic mass unit. What is its mass in terms of grams?
27. Calculate the molecular mass of the following:
(*a*) $CaCO_3$ (*b*) K_2SO_4.
28. Calculate (*i*) number of moles of carbon atoms (*ii*) number of molecules of oxygen (*iii*) number of moles of hydrogen that are present in Z moles of glucose ($C_6H_{12}O_6$).
29. Calculate the number of atoms in each of the following:
(*a*) 52 moles of He (*b*) 52*u* of He
(*c*) 52g of He.
30. Why do atomic masses of most of the elements in atomic mass units involve fractions?
31. The relative abundance of various isotopes of silicon is: Si-28 = 92.23%, Si-29 = 4.67% and Si-30 = 3.10%. What is the average atomic mass of silicon?
32. Boron has two isotopes, B-10 and B-11. The average atomic mass of boron is found to be 10.80 u. Calculate the percentage of abundance of these elements.
33. In naturally occurring neon, the fractional abundance of various isotopes is as follows: $^{20}Ne = 0.9051$; $^{21}Ne = 0.0027$; $^{22}Ne = 0.0922$. What is the atomic mass of neon?
34. Calculate the number of gram-atom and gram-molecule in 48 g of oxygen.
35. Calculate the mass of (*i*) 2 g-molecule of sulphuric acid, (H_2SO_4), (*ii*) 0.5 g-molecule of iodine, (*iii*) 1.5 g-molecule of sucrose ($C_{12}H_{22}O_{11}$).
36. What is the molar mass of a gas if 1.00 L of the gas weighs 1.50 g at 273 K and one atmosphere pressure?
37. What is the mass of one mole electrons? Mass of one electron is 9.11×10^{-31} kg.
38. What is the molar volume of water at 273 K? Density of water = 1.00 g cm^{-3}.

Answers

I. Objective Type Questions

1. (*a*)	**2.** (*d*)	**3.** (*c*)	**4.** (*a*)
5. (*c*)	**6.** (*c*)	**7.** (*b*)	**8.** (*b*)
9. (*a*)	**10.** (*c*)	**11.** (*d*)	**12.** (*a*)
13. (*a*)	**14.** (*c*)	**15.** (*c*)	**16.** (*b*)
17. (*b*)	**18.** (*a*)	**19.** (*c*)	**20.** (*d*).

II. Fill in the Blanks

21. (*i*) 18.06×10^{23} (*ii*) 22.4 L
(*iii*) $4 \times 6.02 \times 10^{23}$ (*iv*) simple whole number
(*v*) 6.02×10^{23} (*vi*) 20
(*vii*) molecules (*viii*) 1.505×10^{23}
(*ix*) 4.74 g.

SECTION-4

UNIT-11

SOLUTIONS

In our daily routine we come across a variety of substances which are not pure but instead a homogeneous mixture of two or more pure substances. Many of these homogeneous mixtures have great importance in our life. The utility of these homogeneous mixtures depend largely on their compositions. Sometimes the individual components are the same but they exhibit different properties in mixtures of different compositions. For *example*, the presence of fluoride ions ($\overline{F}$) in water at a concentration of about **1 ppm** prevents tooth decay but if its concentration in water becomes **1.5 ppm**, it becomes harmful and cause tooth to become mottled. The concentration of fluoride ion beyond this limit becomes poisonous (generally used for rat poisoning).

Specific Objectives

After studying this unit, the student should be able to:

- explain solute, solvent and solution and give examples
- prepare standard solutions from solid solutes
- demonstrate preparation of solutions from liquid solutes
- explain the terms primary standard, secondary standard and standardized solution.

11.1 SOLUTE, SOLVENT AND SOLUTION

A **solution** *is a homogeneous mixture of two or more pure substances whose composition may be altered within certain limits.* Though a solution is homogeneous in nature, it retains the properties of its constituents. The substances which make up the solution are generally called its *components.* The solution of two components is referred to as a *binary solution.* Similarly, the solutions of *three* and *four* components are called *ternary* and *quaternary solutions* respectively. In this unit, we shall limit our discussion to binary solutions only. The two components of a binary solutions are respectively called a *solvent* and a *solute.*

Solvent is that component in the solution whose physical state is the same as that of the resulting solution while the other component is called **solute.** For solutions in which both components have the same physical state, the component which is in excess is called **solvent** and the other one is called **solute.**

For example, in the case of a solution of sugar and water, sugar is the solute and water is the solvent. Each component of a binary solution may be in the solid, liquid or gaseous state. Consequently, there can be nine different types of solutions as shown in Table 11.1.

Table 11.1. Different Types of Solutions

Type of Solution	*Solvent*	*Examples*
Gaseous Solutions	*Gas*	
• Solid in gas		Sublimation of solid in gas; camphor vapours in N_2 gas
• Liquid in gas		Water vapours in air (mist)
• Gas in gas		Mixture of gases, air
Liquid Solutions	*Liquid*	
• Solid in liquid		Salt in water, sucrose in water
• Liquid in liquid		Alcohol in water, mixture of miscible liquids
• Gas in liquid		Aerated drinks, CO_2 in water
Solid Solutions	*Solid*	
• Solid in solid		Alloys; homogeneous mixture of two or more metals
• Liquid in solid		Mercury in zinc, mercury in gold
• Gas in solid		Adsorption of gases on metals, solution of hydrogen in palladium

EXPRESSING CONCENTRATION OF SOLUTIONS OF SOLIDS IN LIQUIDS

The concentration of a solution is a way of describing its composition. Qualitatively, it can be described by using the word *dilute* for solutions having very small quantity of solute and the word *concentrated* for solutions having large quantity of solute. However, this kind of description leads to confusion. Therefore, a quantitative description is more appropriate. Quantitatively, the concentration of a solution refers to ***the amount of solute present in a given quantity of solution or solvent.*** The concentration of the solution may be expressed in any of the following ways:

1. Mass Percentage (% mass)

Mass percentage may be defined as *the number of parts by mass of solute per hundred parts by mass of solution.* For example, a 5% (aqueous) solution of sugar by mass means that 100 g of solution contain 5 g of sugar.

If W_B is the mass of solute (B) and W_A is the mass of solvent (A), then

$$\textbf{Mass percentage of B} = \frac{\mathbf{W_B}}{\mathbf{W_A + W_B}} \times \mathbf{100} \quad ...(11.1)$$

2. Volume Percentage (% volume)

This mode of concentration is used for solutions where *solutes* and *solvents* are both *liquids.* Volume percentage may be defined as *the number of parts by volume of solute per hundred parts by volume of solution.* For example, a 25% solution of ethyl alcohol (by volume) means that 100 cm^3 of the solution contain 25 cm^3 of ethyl alcohol and 75 cm^3 of water.

If V_A and V_B are the volumes of the component A and B, then

$$\textbf{Volume percentage of B} = \frac{\mathbf{V_B}}{\mathbf{V_A + V_B}} \times \mathbf{100} \quad ...(11.2)$$

3. Normality (N)

Normality of a solution is defined as *the number of g-equivalents of the solute present in one litre (1 dm^3) of the solution* or *milliequivalents of solute present in one cm^3 of solution.* It is represented by N.

Mathematically,

$$\text{Normality (N)} = \frac{\text{G-equivalents of solute (B)}}{\text{Volume of solution in litre}}$$

G-equivalents of solute (B) represents the ratio of its mass in gram (W_B) to its gram-equivalent mass (GEM). Also, the ratio of the mass of solute in gram (W_B) to the volume of the solution in litre (V_L) represents the strength of the solution. Thus,

$$\textbf{Normality (N)} = \frac{\mathbf{W_B (g)}}{\mathbf{GEM_B \times V_{(L)}}}$$

or

$$\frac{\textbf{strength of sol (gdm}^{-3}\textbf{)}}{\mathbf{GEM_B}}$$

or

$$\frac{\mathbf{W_B (g) \times 1000}}{\mathbf{GEM_B \times V_{(cm^3)}}} \quad ...(11.3)$$

A solution having normality equal to unity is called a *normal solution.* Such a solution contains one gram equivalent of solute per litre of solution.

4. Molarity (M)

Molarity of a solution is defined as *the number of gram mole of the solute present in one litre of the solution* or *millimoles of solute present in one cm^3 of solution.* It is represented by M.

Mathematically,

$$\text{Molarity (M)} = \frac{\text{Moles of solute}}{\text{Volume of solution in litres}}$$

$$= \frac{\text{Mass of solute in gram}}{\text{(Molar mass of solute)} \times \text{(Volume of soln. (dm}^3\text{))}}$$

Now, moles of the solute (n_B) represents the ratio of mass of solute in gram (W_B) to its molar mass (MM_B). Thus, molarity can be represented by the expressions.

$$M = \frac{n_B}{V_{(L)}} \quad \text{or} \quad \frac{W_B(g)}{(MM)_B \times V_{(L)}}$$

$$\text{or} \quad \frac{\text{Strength (gdm}^{-3}\text{)}}{(MM)_B} \quad \text{or} \quad \frac{W_B(g) \times 1000}{(MM)_B \times V_{(cm^3)}} \quad ...(11.4)$$

A solution having molarity equal to unity is called *molar solution*. Such a solution contains one mole of solute per litre of solution. The solutions having molarity equal to *0.5 M*, *0.1 M* and *0.01 M* are called *semimolar, decimolar* and *centimolar* solutions respectively. Molarity is expressed in units of ***mol L***$^{-1}$ or ***mol dm***$^{-3}$. It may be noted that *both normality as well as molarity of a solution change with change in temperature*.

Do You Know?

Demal unit. *Demal is another unit for expressing the concentration of a solution. It is represented by* **D** *and expresses the molarity of a solution at 0°C. For example,* **2.5 D solution** *implies the presence* **of 2.5 mole of the** *solute in one litre* **of solution at 0°C.**

5. Molality (m)

Molality of a solution may be defined as *the number of gram mole of the solute present in 1000 g (1 kg) of the solvent.* It is represented by *m*.

Mathematically, Molality (*m*)

$$= \frac{\text{Mole of solute}}{\text{Mass of solvent (kg)}}$$

$$= \frac{\text{Mass of solute (g)}}{\text{(Molar mass of solute)} \times \text{(Mass of solvent (kg))}}$$

$$\text{or} \quad m = \frac{n_B}{W_A(kg)} \quad \text{or} \quad \frac{W_B(g)}{(MM)_B \times W_A\,(kg)}$$

$$\text{or} \quad \frac{W_B(g) \times 1000}{(MM)_B \times W_A\,(g)} \quad ...(11.5)$$

A solution containing one mole of solute per 1000 g of solvent has molality equal to one and is called a *molal solution*. Molality is expressed in units of moles per kilogram *(mol kg^{-1}). The molality of a solution does not change with temperature*.

SOME IMPORTANT FORMULAE

Some very significant formulae about different modes of concentrations of solution are given below for easy referencing.

1. Relationship between Molarity and Normality. The normality (N) and molarity (M) of a solution are related as follows:

$$\textit{Molarity} \times \textit{Mol. mass (solute)} = \textit{Normality} \times \textit{Eq. mass (solute)}$$

For **Acids;** *Normality = Molarity × Basicity*

For **Bases;** *Normality = Molarity × Acidity*

Acidity is the number of H^+ ions furnished by each molecule of acid in solution.

Basicity is the number of OH^- ions furnished by each molecule of base or number of H^+ ion which react with one molecule of base in solution.

2. Relationship between Molarity (M) and Mass Percentage (%)

$$\textit{Molarity} = \frac{\%\,\text{mass} \times d \times 10}{\text{Molar mass (solute)}} \; ;$$

$$\textit{Normality} = \frac{\%\,\text{mass} \times d \times 10}{\text{Eq. mass (solute)}}$$

3. Dilution Formula. If the solution of some substance is diluted by adding solvent from volume V_1 to volume V_2, then the initial molarity M_1 and the final molarity M_2 are related as

$$M_1V_1 = M_2V_2 \quad \text{Similarly,} \quad N_1V_1 = N_2V_2$$

4. Molarity of a Mixture. If V_1 ml of a solution of molarity M_1 is mixed with V_2 ml of another solution of the same substance with molarity M_2, then the molarity of the resulting mixture solution (M) can be obtained from the formula

$$M = \frac{M_1V_1 + M_2V_2}{V_1 + V_2}$$

6. Mole Fraction

Mole fraction may be defined as *the ratio of number of moles of one component to the total number of moles of all the components (solvent and solute) present in the solution.* It is denoted by the letter *x* followed by the subscript representing the component. Let us suppose that a solution contains the components A and B and suppose that W_A g of A and W_B g of B are present in it.

Moles of A $(n_A) = \dfrac{W_A}{(MM)_A}$;

Moles of B $(n_B) = \dfrac{W_B}{(MM)_B}$

$(MM)_A$ and $(MM)_B$ are the molar masses of A and B respectively

Total number of moles of A and B = $n_A + n_B$

Mole fraction of A, $(x_A) = \dfrac{\mathbf{n_A}}{\mathbf{n_A + n_B}}$

Mole fraction of B, $(x_B) = \dfrac{\mathbf{n_B}}{\mathbf{n_A + n_B}}$

Now, the sum of the mole fractions of solute and solvent in a binary solution is ***unity.***

$$x_A + x_B = \frac{n_A}{n_A + n_B} + \frac{n_B}{n_A + n_B} = 1.$$

This generalisation can be extended to solutions having more than two components by saying that *the sum of the mol fractions of all the components in a solution is always* **unity.** It may be noted that the ***mole fraction is independent of temperature.***

7. Formality

Formality of a solution may be defined as *the number of gram formula masses (moles) of the ionic solute present in one litre of the solution.* It is represented by F.

Mathematically,

$$\text{Formality (F)} = \frac{\text{Number of moles of ionic solute}}{\text{Volume of solution in litres}}$$

$$= \frac{\text{Mass of ionic solute (g)}}{(\text{GFM})_{\text{solute}} \times V_{\text{sol.}}(\text{dm}^3)}$$

Thus, $\mathbf{F = \dfrac{W_B\,(g)}{GFM_B \times V_{(L)}}}$ or $\mathbf{\dfrac{W_B\,(g) \times 1000}{GFM_B \times V_{(cm^3)}}}$ ...(11.6)

Commonly, the term formality is used to express the concentration of ionic solids which do not exist as molecules but exist as a network of ions. A solution containing one gram formula mass of solute per litre of the solution has formality equal to one and is called *formal solution.* It may be mentioned here that like normality and molarity, ***The formality of a solution changes with change in temperature.***

8. Parts Per Million (ppm)

When a solute is present in very minute amounts, its concentration is expressed in parts per million. It may be defined as *the number of parts by mass of solute per million parts by mass of the solution.* It is abbreviated as ppm.

$$\text{Parts per million (ppm)} = \frac{\text{Mass of solute}}{\text{Mass of solution}} \times 10^6.$$

$$\mathbf{ppm = \frac{W_B}{W_A + W_B} \times 10^6} \qquad ...(11.7)$$

This mode of concentration is generally used to express very low concentrations such as hardness of water or concentration of Cl_2 in public supply of potable water.

SOLVED EXAMPLES

Example 11.1. *Calculate the percentage composition in terms of the mass of a solution obtained by mixing 300 g of 25% solution of NH_4NO_3 with 400 g of a 40% solution of solute X.*

Solution. Total mass of solution $(W_A + W_B)$

$= 400 + 300 = 700$ g

Amount of solute present in 300 g of 25% solution

$$= 300 \times \frac{25}{100} = 75 \text{ g}$$

Amount of solute present in 400 g of 40% solution

$$= 400 \times \frac{40}{100} = 160 \text{ g}$$

Total mass of solute (W_B)

$= 75 + 160 = 235$ g

Percentage by mass of solute

$$= \frac{W_B \times 100}{W_A + W_B} = \frac{235 \times 100}{700} = \mathbf{33.57\%}$$

Percentage by mass of solvent

$= 100 - 33.57 =$ **66.43%.**

Example 11.2. *Calculate the molarity and normality of a solution containing 40 g of NaOH dissolved in two litre of the solution.*

Solution. Here, $W_B = 40$ g ; $(MM)_B$

$= 40$ (g mol^{-1}); $GEM_B = 40$ g

$$\therefore \quad \text{Molarity} = \frac{W_B}{(MM)_B \times V_{(L)}} = \frac{40}{40 \times 2} = \mathbf{0.5\ M}$$

$$\text{Normality} = \frac{W_B}{(GEM)_B \times V_{(L)}} = \frac{40}{40 \times 2} = \mathbf{0.5\ N.}$$

Example 11.3. *Calculate the resulting molarity of the solution that is obtained by adding 5 g of NaOH to 200 cm³ of $\frac{M}{4}$ NaOH solution (density = 1.05 g cm⁻³). The density of the resulting solution is 1.08 g cm⁻³.*

Solution. The mass of 200 cm³ of $\frac{M}{4}$ NaOH

$$= 200 \times 1.05 = 210.0 \text{ g}$$

1000 cm³ of 1 M NaOH contains NaOH = 40 g

∴ 200 cm³ of 1 M NaOH contains NaOH

$$= \frac{40}{1000} \times 200 \text{ g}$$

Hence 200 cm³ of $\frac{M}{4}$ NaOH contains NaOH

$$= \frac{40}{1000} \times \frac{200}{4} = \mathbf{2\ g}$$

Amount of NaOH added = 5 g

Total mass of NaOH in solution = 5 + 2 = 7 g

Total mass of resulting solution = 210 + 5.0 = 215 g

Now volume of 215 g of new solution

$$= \frac{215\text{ (g)}}{\text{density}} = \frac{215\text{ (g)}}{1.08\text{ (g cm}^{-3})} = 199.0 \text{ cm}^3$$

Molarity of new solution

$$= \frac{7}{199} \times \frac{1000}{40} = \mathbf{0.88\ M.}$$

Example 11.4. *4.0 g of NaOH are contained in one decilitre of solution. Calculate the following:*

(i) Mole fraction of NaOH

(ii) Molarity of NaOH

(iii) Molality of NaOH (Density of solution = 1.038 g/cm³).

Solution. Step I. *To calculate the mole fraction of NaOH*

Mass of 100 cm³ of solution = 1.038 × 100 = 103.8 g (1 decilitre = 100 cm³)

Mass of NaOH = 4.0 g

Mass of water = 103.8 – 4.0 = 99.8 g

∴ No. of moles of NaOH (n_A)

$$= \frac{4.0\text{ (g)}}{40\text{ (g mol}^{-1})} = 0.1 \text{ mole}$$

No. of moles of water (n_B)

$$= \frac{99.8\text{ (g)}}{18\text{ (g mol}^{-1})} = 5.54 \text{ mole}$$

∴ Mole fraction of NaOH,

$$X_A = \frac{n_A}{n_A + n_B} = \frac{0.1}{0.1 + 5.54} = \mathbf{0.018.}$$

Step II. *To calculate the molarity of NaOH (M)*

Molarity of the solution

$$= \frac{\text{Mass of NaOH per litre of solution}}{\text{Molar mass of NaOH}}$$

$$= \frac{40\text{ (g)}}{40\text{ (g)}} = \mathbf{1\ M.}$$

Step III. *To calculate the molality of NaOH*

Mass of NaOH in 99.8 g of H_2O = 4.0 g

Mass of NaOH in 1000 g of H_2O

$$= \frac{4.0}{99.8} \times 1000 = 40.08 \text{ g}$$

∴ Molality of the solution

$$= \frac{\text{Mass of NaOH per 1000 g of water}}{\text{Molar mass of NaOH}}$$

$$= \frac{40.08\text{ (g kg}^{-1})}{40\text{ (g mol}^{-1})}$$

$$= \mathbf{1.002\ mol\ kg^{-1}.}$$

Example 11.5. *You are provided with one litre each of 0.5 M NaOH and 0.25 M NaOH solutions. What maximum volume of 0.30 M NaOH can be obtained from these solutions without using water at all?*

Solution. In order to get a maximum volume of 0.3 M NaOH, the whole volume, *i.e.*, 1 dm³ of 0.25 M NaOH will be used.

Let the volume of 0.5 M needed be $= x$ dm³

∴ Total volume of resulting solution = $(1 + x)$ dm³

Now, molarity of the resulting mixture is given by

$$M = \frac{M_1V_1 + M_2V_2}{V_1 + V_2}$$

or

$$0.3 = \frac{0.25 \times 1 + 0.5 \times x}{(1 + x)} \quad \text{or} \quad x = 0.25 \text{ dm}^3$$

∴ Maximum volume obtained = 1 + 0.25 = **1.25 dm³.**

Example 11.6. *Out of 1 molal and 1 molar aqueous solutions which is more concentrated? Justify.*

Solution. For most aqueous solutions, 1 M solutions have higher concentrations than corresponding 1 *m* solutions.

It is so because in 1 *m* solutions, 1 mol of solute is present in 1000 g, *i.e.*, 1000 cm³ ($d \approx 1$) of water. But in 1 M solution 1 mol of solute is present in 1000 cm³ of solution which contain less than 1000 cm³ of solvent. However, in case of non-aqueous solutions, 1 M solution can be less, more or even has same concentration as that of 1 *m* solution depending on the density of solution.

11.2 PREPARING A STANDARD SOLUTION FROM SOLID SOLUTE

Let us discuss the preparation of a standard solution of sodium trioxocarbonate(IV) (say M/10) using water as the solvent.

Now sodium carbonate is a primary standard. Its molecular mass is 106.

To prepare $\frac{M}{10}$ Na_2CO_3 solution, 10.6 g of sodium carbonate should be dissolved per litre of the solution. Normally in the laboratory we are required to prepare 250 dm^3 of solution. Therefore, to prepare 250 cm^3 of $\frac{M}{10}$ Na_2CO_3 solution, $\frac{10.6}{4}$ = 2.650 g of sodium carbonate is dissolved in a small quantity of water and the solution diluted to exactly to 250 cm^3.

The **apparatus** required for the above preparation are *weight box, chemical balance, watch glass, 250 cm^3 beaker, glass rod, 250 cm^3 measuring flask, wash bottle*.

PROCEDURE

1. Take a watch glass, wash it with distilled water and then dry it.
2. Weigh the clean and dried watch glass accurately.
3. Weigh 2.650 g sodium trioxocarbonate(IV) on the watch glass accurately.
4. Transfer gently and carefully sodium carbonate from the watch glass into a clean 250 cm^3 beaker. Wash the watch glass with distilled water with the help of a wash bottle to transfer the particles sticking to it into beaker (Fig. 11.1). The volume of distilled water for this purpose should not be more than 50 dm^3.

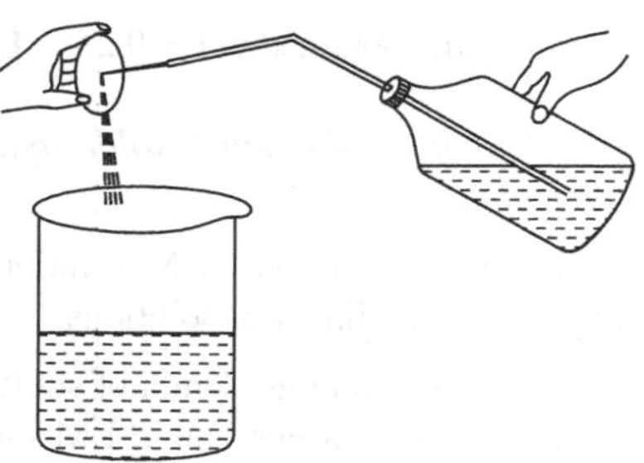

Fig. 11.1. Washing of watch glass to transfer sticking particles to beaker.

5. Dissolve sodium trioxocarbonate(IV) crystals in the beaker by gentle stirring with a clean glass rod.
6. When the sodium trioxocarbonate(IV) in the beaker is completely dissolved, transfer carefully the entire solution from the beaker into a 250 cm^3 measuring flask with the help of a funnel (Fig. 11.2).

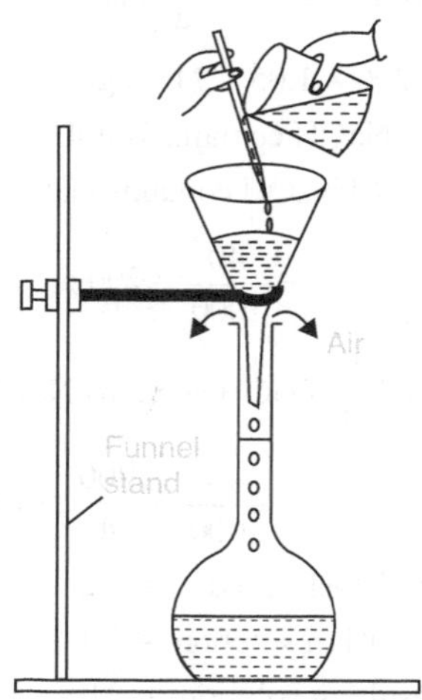

Fig. 11.2. Transferring solution to measuring flask.

7. Wash the beaker with distilled water. Transfer the washings into the measuring flask (Fig. 11.3).

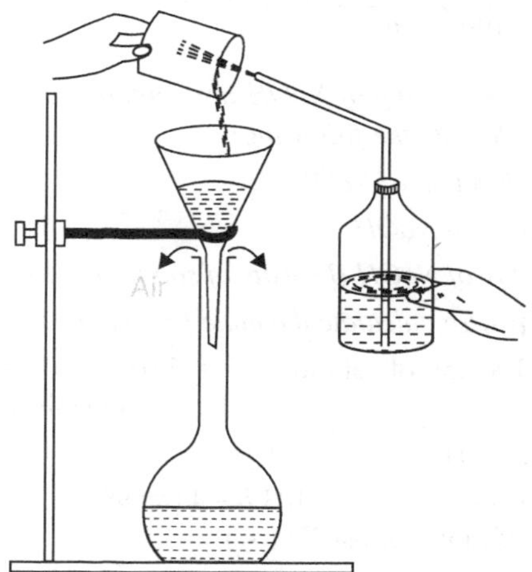

Fig. 11.3. Transferring last traces of solution to measuring flask.

8. Finally wash the funnel thoroughly with distilled water with the help of a wash bottle to transfer the solution sticking to the funnel into the measuring flask (Fig. 11.4).

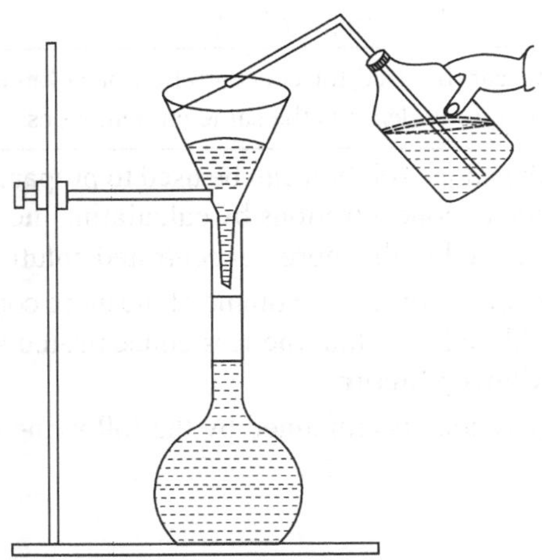

Fig. 11.4. Washing last traces of solution from funnel to the measuring flask.

9. Add enough distilled water to the measuring flask carefully upto just below the etched mark on it, with the help of a wash bottle.
10. Add the last few drops of distilled water with a pipette until the lower level of the meniscus just touches the mark on the measuring flask (Fig. 11.5).

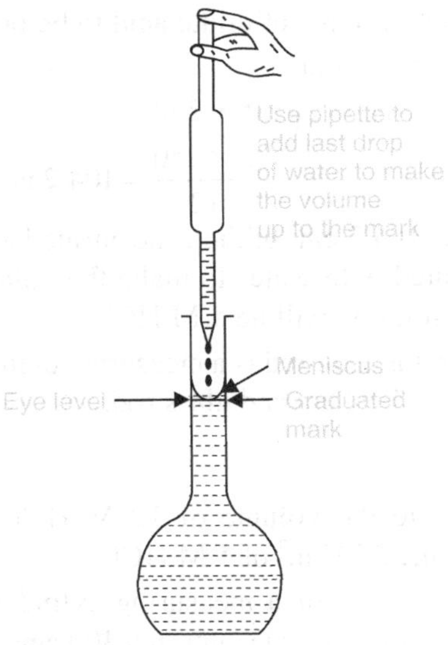

Fig. 11.5. Using pipette to add last drop of water to make the volume upto the mark.

11. Stopper the measuring flask and shake gently to make the solution uniform throughout. Label it as $\frac{M}{10}$ sodium trioxocarbonate(IV) solution.

ACTIVITY 11.1 **Preparing 1 dm³ of 0.1 M oxalic acid solution** $\begin{matrix} COOH \\ | \\ COOH \end{matrix} \cdot 2H_2O$.

The molecular mass of oxalic acid is 126. In order to prepare 0.1 M oxalic acid solution, 12.6 g of oxalic acid should be dissolved in 1 dm³ of solution. Alternatively, $\frac{12.6}{4} = 3.15$ g of oxalic acid crystals should be dissolved in water and the solution diluted to exactly 250 cm³.

Go to the laboratory and try preparing the solution in the same manner as discussed in the preparation of Na_2CO_3 solution.

ACTIVITY 11.2 **Preparing a solution of $NaNO_3$ by dissolving 0.38 g of $NaNO_3$ in 50.0 cm³ of its solution.**

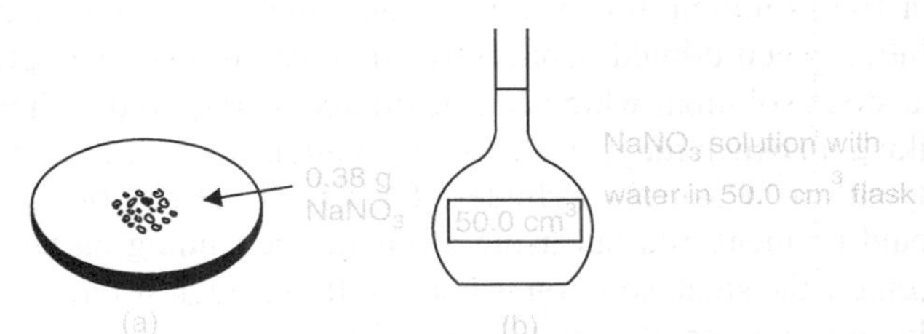

Molar mass of $NaNO_3 = 23 + 14 + 3 \times 16 = 85$

$$\text{Molarity (M) of solution} = \frac{W(g)}{\text{M. Mass}} \times \frac{1000}{V_{sol}}$$

$$= \frac{0.38}{85} \times \frac{1000}{50} = \frac{7.6}{85} = \mathbf{0.0894\ M}$$

11.3 PREPARATION OF SOLUTIONS FROM LIQUID SOLUTES

Acids like hydrochloric acid (HCl), sulphuric acid (H_2SO_4), nitric acid (HNO_3) and acetic acid (CH_3COOH) which are commonly used in the laboratory are originally available as concentrated aqueous solutions. The solutions required in the laboratory are prepared by diluting the concentrated solutions with water. Knowing the concentration of the concentrated acid we can calculate how much volume of this acid must be diluted to prepare a definite volume of dilute acid of a certain required concentration. The approximate concentrations of some acids as commercially available are given in Table 11.2.

Table 11.2. Approximate Concentrations of Some Acids

Concentrated Acid	*Approximate Normality (N)*	*Approximate Molarity (M)*
Hydrochloric acid (HCl)	12	12
Sulphuric acid (H_2SO_4)	36	18
Nitric acid (HNO_3)	16	16
Glacial acetic acid (CH_3COOH)	17	17

STOCK OR COMMERCIAL SOLUTIONS

A stock solution is a concentrated solution (standard or non-standard) whose concentration is known, which can be stored and diluted as required to give solutions of lower concentration. A stock solution should be chemically stable, so that it doesn't change with storage. Keeping suitable quantities of stock solution can save time during the term in preparing solutions, as they only need dilution. Thus 1 dm^3 of a 10M solution, for example, would make 10 dm^3 of 1M solution when diluted. Concentrated acids are one example of a stock solution, which can be diluted as required. When making up bench reagents the stock solution can be measured out with a measuring cylinder; for analytical solutions it should be measured out using a pipette, depending on how precisely the stock solution is known. If the stock solution is only approximate, then it is best to dilute without taking too much care and then standardize the final solution. The exact final concentration is then known.

DILUTION FACTOR

We often need to dilute one solution, whose concentration is known, to give another of known concentration. Although this is a simple task some students find it difficult. The basic idea to remember is that when we dilute a solution the amount of solute stays the same, although its concentration decreases.

In any solution:

moles of solute

= volume (dm^3) × concentration (mol/dm^3)

= volume (cm^3)/1000 × concentration (mol/dm^3)

Diluting with water keeps the number of moles constant *i.e.*, the product of volume and concentration is constant.

Thus:

volume (i) × concentration (i)

= volume (f) × concentration (f)

Here, i = initial; f = final

$$\therefore \quad V_i \times M_i = V_f \times M_f$$

Any unit can be used for concentration or volume in this expression, provided it is the same on both sides.

Thus, the above relation can be used to prepare solutions of various lower concentrations by calculating the volume of water to be added to the more concentrated solution.

The number of times the volume of the more concentrated solution is diluted to obtain the less concentrated solution is called the **dilution factor**.

The above idea is explained by the following example.

STEPS INVOLVED IN THE PREPARATION OF SOLUTIONS FROM LIQUID SOLUTES

Preparation of 250 cm³ of 5 M HCl from the given 12 M HCl.

To prepare 250 cm^3 of 5 M HCl from 12 M HCl, the volume of the concentrated acid required can be calculated by using the molarity equation:

$$M_1V_1 = M_2V_2$$

where, M_1 = Molarity of concentrated acid = 12 M

V_1 = Volume of concentrated acid required = ?

M_2 = Molarity of dilute acid to be prepared = 5 M

V_2 = Volume of dilute acid to be prepared = 250 cm^3.

$$\therefore \quad 12 \times V_1 = 5 \times 250$$

or

$$V_1 = \frac{5 \times 250}{12} = 104.2 \text{ cm}^3$$

Therefore, 104.2 cm^3 of the concentrated acid should be taken and diluted with water to make the volume 250 cm^3. The resulting solution will be 5 M HCl.

The apparatus required is a measuring cylinder, 250 cm^3 measuring flask, beakers and glass rod.

Procedure

1. Calculate the volume of 12 M HCl required for preparing 250 cm^3 of 5 M HCl.
2. With the help of a measuring cylinder, take about 100 cm^3 of distilled water in a 400 cm^3 beaker. (This volume of water is slightly less than the volume of water which will be required).
3. Take 104.2 cm^3 of given concentrated HCl in a measuring cylinder. Add it slowly and carefully, with stirring with a glass rod, to the water taken in the beaker. Cool the beaker under tap water from time to time.

4. After all the acid has been added, transfer the solution from the beaker to a 250 cm^3 measuring flask. Add more of water so that the total volume becomes 250 cm^3.
5. Stopper the measuring flask and shake the solution well to make it uniform. Label it as 5 M HCl.

ACTIVITY 11.3 **Prepare 250 cm^3 of 2 M H_2SO_4 (tetraoxosulphate (VI) acid) from the given 18 M H_2SO_4.**

To prepare 250 cm^3 of 2 M H_2SO_4 from 18 M H_2SO_4, the volume of the concentrated acid required can be calculated by applying the molarity equation

$M_1V_1 = M_2V_2$

where, M_1 = Molarity of concentrated acid = 18 M

V_1 = Volume of concentrated acid required = ?

M_2 = Molarity of dilute acid to be prepared = 2M

V_2 = Volume of dilute acid to be prepared = 250 cm^3

$\therefore\ 18 \times V_1 = 2 \times 250$

or $V_1 = \dfrac{2 \times 250}{18} = 27.8\ cm^3$.

Therefore, 27.8 cm^3 of the concentrated sulphuric acid should be taken and diluted with water to get 250 cm^3 of 2M H_2SO_4.

Group Activity **Preparation of solutions of known concentrations of HNO_3 acid and NH_3(aq).**

For the purpose of the preparation of above solutions.

1. The student should go to the chemistry laboratory.
2. Seek the help of the teacher for the purpose of preparation of a solution of required concentration from the stock solution.
3. Student should use the same method and technique as discussed in activity 11.3.

SOLVED EXAMPLES

Example 11.7. *(i) Calculate the concentration of a commercial solution of HCl with the following label specifications: mass per cm^3 1.18 g*

% purity 36%

M_r (HCl) 36.5

(ii) Calculate the volume of the stock solution that should be taken to dilute it to 2 dm^3 of concentration 0.80 mol per dm^3.

Solution. *(i)* $\rho(HCl) = 1.18\ g\ cm^{-3}$

$M_r\ (HCl) = 36.5$

% purity = 36%

$$\rho(\text{pure HCl}) = 1.18 \times \frac{36}{100}\ g\ cm^{-3}$$

and mass concentration in g dm^{-3}

$$\rho = \frac{1.18 \times 36 \times 1000}{100}\ g\ dm^{-3}$$

$$c = \frac{\rho}{M}$$

$$\therefore\quad c(HCl) = \frac{1.18 \times 36 \times 1000\ g\ dm^{-3}}{36.5\ g\ mol^{-1} \times 100}$$

$= 11.6\ mol\ dm^{-3}$

(ii) $c(conc) = 11.6\ mol\ dm^{-3}$

$c(dil) = 0.80\ mol\ dm^{-3}$

$V(dil) = 2\ dm^3$

$c(conc) \times V(conc) = c(dil) \times V(dil)$

$$V(conc) = \frac{c(dil) \times V(dil)}{c(conc)}$$

$$= \frac{0.8\ mol\ dm^{-3} \times 2\ dm^3}{11.6\ mol\ dm^{-3}} = 0.138\ dm^3$$

$= 138\ cm^3$.

Example 11.8 *You are given a solution of 14.8 M NH_3. How many millimetres of this solution do you require to give 100.0 cm^3 of 1.00 M NH_3 when diluted?*

Solution. Molarity of NH_3 solution (M_1) = 14.8

Volume of NH_3 required (V_1) = ?

Molarity of diluted NH_3 solution = 1.00 M

Volume of diluted NH_3 solution = 100.0 cm^3

Applying molarity equation

$$M_1V_1 = M_2V_2$$

$$14.8 \times V_1 = 1.0 \times 100$$

or
$$V_1 = \frac{1.0 \times 100}{14.8} = 6.75\ cm^3$$

6.75 cm^3 of 14.8 M NH_3 should be added to a 100.0 cm^3 flask and water added to the mark.

Example 11.9 *You are given HCl stock solution of % purity 36.5% and density 1.2 g/cm³. What is the concentration of the acid?*

Solution. % purity of HCl (stock solution) = 36.5

$$\rho(HCl) = 1.2 \text{ g cm}^{-3}$$

M. mass of HCl = 1 + 35.5 = 36.5

Mass of solution = Volume × density = 100 × 1.2 = 120 g

Volume of water = 120 – 36.5 = 83.5 cm³

$$\text{Concentration of solution (M)} = \frac{36.5}{36.5} \times \frac{1000}{83.5} = \mathbf{11.976\ M.}$$

ACTIVITY 11.4	**Practice the preparation of following solutions**

(*i*) 250 cm³ and 500 cm³ of dilute HCl solution (2.0 M and 0.1 M) from 36.5% stock solution, which has a density of 1.2 g/cm³

(*ii*) 250 cm³ and 500 cm³ of dil. HNO_3 solution (2.0 M and 0.1 M) from 63% stock solution of HNO_3, which has a density of 1.4 g/cm³.

First calculate the concentration (M) of the stock solutions and then using the molarity equation calculate the amount of these solutions required to obtain given solutions of known concentration.

11.4 PRIMARY STANDARD AND SECONDARY STANDARD AND STANDARDIZED SOLUTION

PRECISE VERSUS IMPRECISE CONCENTRATIONS

We do not always need to know the concentrations of solutions used in the laboratory very precisely *e.g.* bench reagents, test solutions, indicators etc. It is not worth spending time and effort making up solutions if the required precision is not needed.

STANDARDIZED SOLUTIONS

These are solutions used in analytical determinations. These concentrations need to be known precisely (within specified limits). They are made either by weighing out a known mass of a standard solid and making up to a specified volume (*e.g.* sodium carbonate). A standard solution is one whose concentration is known (to an appropriate number of significant figures, s.f.) and which doesn't go off with time *i.e.*, is stable.

A standard solution is prepared by dissolving a definite weight of substance (a primary standard), in a definite volume. A substance is classified as a **primary standard** if it has following characteristics:

(*i*) *It is easily available in a state of high purity.*

(*ii*) *It is neither hygroscopic nor deliquescent.*

(*iii*) *It shows high solubility in water.*

(*iv*) *It does not dissociate or decompose during storage.*

Substances whose standard solutions cannot be prepared directly are called **secondary standards.** For example, $KMnO_4$, NaOH, KOH, etc.

SUITABLE CHEMICALS FOR STANDARD SOLUTIONS

Compound	*GMM*	*Purity*	*Uses*
Sodium carbonate	105.99	99.9%	Acids
Potassium hydrogenphthalate	204.23	99.9	Bases
Potassium dichromate	294.22	99.9	Reducing agents
Potassium iodate	214.01	99.9	Sodium thiosulphate
Sodium chloride	58.45	99.9	Silver nitrate
EDTA disodium dihydrate salt	372.25	99.0	Metal salts

An alternative approach is to buy small quantities of standard solutions (ready-made or concentrated) and use these to standardize one's own solutions. Once a solution has been standardized so that we know its concentration precisely, we can use this as a 'secondary standard' to standardize other solutions.

So we can use sodium carbonate to standardize hydrochloric or sulphuric acids, which in turn can be used to standardize sodium hydroxide solutions.

Bench Reagents

By bench reagents we mean dilute acids and bases etc. that are routinely used in the laboratory for tests, simple reactions and preparations. The exact concentration is not important and they can be made with less care and precision, and do not need standardization. The concentration is usually reported as 0.5M, 1M, 2M, 4M etc., which implies it is not very precise.

EVALUATE YOURSELF

I. Objective Type Questions

Select the most appropriate choice from the options given as (a), (b), (c) and (d) after each question:

1. 10 cm^3 of which of the following solution contains the maximum number of ions assuming each solute to be fully ionized:
   (*a*) 0.2 M NaCl (*b*) 0.2 M $Cr_2(SO_4)_3$
   (*c*) 0.2 M $CaCl_2$ (*d*) 0.2 M Na_2SO_4.
2. An aqueous solution of 6.3 g of oxalic acid dihydrate is made up to 250 cm^3. The volume of 0.1 N NaOH required to completely neutralise 10 cm^3 of this solution is
   (*a*) 40 cm^3 (*b*) 20 cm^3
   (*c*) 10 cm^3 (*d*) 4 cm^3.
3. 25 cm^3 of a solution of $Ba(OH)_2$ on titration with a 0.1 M solution of HCl gave a titre value of 35 cm^3. The molarity of barium hydroxide solution is?
   (*a*) 0.07 (*b*) 0.14
   (*c*) 0.28 (*d*) 0.35.
4. Molarity of 0.2 N H_2SO_4 is
   (*a*) 0.2 (*b*) 0.4
   (*c*) 0.6 (*d*) 0.1
5. The volume of water to be added to 100 cm^3 of 0.5 N H_2SO_4 to get decinormal concentration is
   (*a*) 400 cm^3 (*b*) 450 cm^3
   (*c*) 500 cm^3 (*d*) 100 cm^3.
6. Which among the following is not affected by temperature?
   (*a*) Normality (*b*) Fermality
   (*c*) Molarity (*d*) Molality.
7. The number of cm^3 of 6M HCl which are required to prepare 150 cm^3 of solution which is 0.30 M in hydronium ions is
   (*a*) 3.0 (*b*) 7.5
   (*c*) 9.3 (*d*) 30.
8. One molar solution of sulphuric acid is equal to
   (*a*) normal solution (*b*) N/2 solution
   (*c*) 2N solution (*d*) 4N solution.
9. How many grams of H_2SO_4 should be present in 100 cm^3 of its aqueous solution to produce a solution of 0.1 N?
   (*a*) 0.49 g (*b*) 4.9 g
   (*c*) 49 g (*d*) 2.45 g.
10. How much water should be added to 200 cm^3 of 0.5 M NaOH to make it exactly 0.4 M NaOH solution
    (*a*) 250 cm^3 (*b*) 50 cm^3
    (*c*) 160 cm^3 (*d*) 100 cm^3.
11. How many moles of sodium hydroxide are present in 50 cm^3 of its 0.4 M solution?
    (*a*) 0.4 moles (*b*) 0.04 moles
    (*c*) 0.20 moles (*d*) 0.020 moles.
12. What is the concentration of a solution containing 15 g potassium hydrogen trioxocarbonate (IV) $KHCO_3$, in 3.0 dm^3 of its solution?
    (*a*) 0.01 M (*b*) 0.02 M
    (*c*) 0.04 M (*d*) 0.05 M.
13. The mass in grams of ammonia in 500 cm^3 aqueous ammonia solution of concentration 2.0 mol dm^{-3} is
    (*a*) 17 g (*b*) 1.7 g
    (*c*) 3.4 g (*d*) 34 g.
14. 100 cm^3 of 2.5 M NaOH solution is diluted to 250 cm^3. The concentration of the diluted solution is
    (*a*) 0.50 M (*b*) 0.10 M
    (*c*) 1.0 M (*d*) 2.0 M.
15. Calculate the concentration of stock trioxonitrate (V) acid which has mass density of 1.50 g cm^{-3} and has purity of 70%
    (*a*) 23.8 M (*b*) 2.38 M
    (*c*) 17.0 M (*d*) 16.7 M.
16. How many moles of chloride ions are present in 50 cm^3 of 0.50 M calcium chloride, $CaCl_2$, solution?
    (*a*) 0.10 mol (*b*) 0.02 mol
    (*c*) 0.05 mol (*d*) 0.20 mol.

II. Fill in the Blanks

17. Complete the following sentences by supplying appropriate words:
    (*i*) Molality of a solution is number of moles of the solute in each ...... kg of ......
    (*ii*) The ratio of the number of moles of solvent to the total number of moles of solute and solvent is called ......
    (*iii*) If molarity of oxalic acid solution is M/2 then its normality will be ......
    (*iv*) Aerated drinks is an example of ...... in ...... solution.
    (*v*) Substances whose standard solution cannot be prepared directly are called ....
    (*vi*) Concentrated solutions which can be diluted are known as ...... solutions.
    (*vii*) Mixing 200 cm^3 of 0.50 M NaOH with 100 cm^3 of 0.50 M NaOH will give a new solution of ...... M.
    (*viii*) The amount of solute in grams present per dm^3 of solution is known as ....
    (*ix*) Amount of water to be added to 200 cm^3 of 1 M HCl to make it exactly 0.2 M HCl is ....
    (*x*) A decimolar solution of NaOH contains ...... g NaOH per litre of the solution.

III. Discussion Questions

18. Calculate the mol fraction of ethylene glycol ($C_2H_6O_2$) in a solution containing 20% of $C_2H_6O_2$ by mass in aqueous solution.
19. Calculate the molarity of a solution containing 5.0 g of NaOH in 450 mL of solution. What additional data is required to calculate the molarity of the solution?

20. Calculate molality and the mol fraction of the solute in the following solutions:
(*i*) A sugar syrup of mass 214.2 g containing 34.2 g of sucrose
(*ii*) 5 M aqueous solution of A (Molar mass 100 g mol^{-1}) with density 1.289 g cm^{-3}.

21. What volume of decamolar aqueous solution of hydrochloric acid is required to prepare 2.0 dm^3 of 5 M HCl solution?

22. 200 cm^3 of an aqueous solution containing 8 g of KOH are mixed with 200 cm^3 of pentimolar KOH solution. What is the molarity of the resulting solution?

23. What is molarity of the following:
(*i*) Pure water if its density is 1000 kg/m^3.
(*ii*) K^+ ions in a solution obtained by mixing 250 cm^3 each of 0.2 M KCl and 0.2 M K_2SO_4.

24. Calculate the molarity of each of the following solutions (*a*) 30 g of $Co(NO_3)_2.6H_2O$ in 4.3 dm^3 of solution. (*b*) 30 cm^3 of 0.5 M H_2SO_4 diluted to 500 cm^3.

25. What maximum volume of 1.5 M HCl can be obtained from 1.0 dm^3 each of 0.5 M and 5 M HCl solutions, by mixing them without the addition of water?

26. Calculate (*a*) molality, (*b*) molarity and (*c*) mole fraction of KI if the density of 20% (mass/mass) aqueous KI is 1.202 g/cm^3.

27. (*a*) What is a standard solution?
(*b*) List six pieces of apparatus that would be used in preparing solutions of known concentration.
(*c*) Outline the main steps to be followed in preparing standard solutions.
(*d*) Calculate the concentration you would put on a label of a solution prepared by dissolving 8.0 g of potassium hydroxide, KOH, in 500 cm^3 of distilled water.
A_r (K = 39, O = 16.0, H = 1)

28. (*a*) Write down the relation between mass, *m*; molar mass, M; and amount of substance, *n*.
(*b*) Given that mass concentration, ρ, is the mass per unit volume, V, in dm^3, derive a relation between mass concentration, ρ, and amount of substance concentration, *c*.
(*c*) 9.0 g of the hydrated salt $CuSO_4.\ xH_2O$ was dissolved in 500 cm^3 of water and its concentration in mol dm^{-3} was experimentally found to be 0.0720 M. Calculate
(*i*) the molar mass of $CuSO_4.xH_2O$;
(*ii*) *x*, the water of crystallization;
(*iii*) the number of Cu^{2+} ions in 100 cm^3 of the prepared solution.
(L = 6.02×10^{23} mol^{-1})
A_r (Cu = 64, S = 32, O = 16, H = 1.0)

29. Hydrogen peroxide, H_2O_2, can be used as an antiseptic mouthwash. A doctor's prescription indicates that you should dilute five times from a given solution before use.
(*i*) Calculate the volume of water in cm^3 that you have to add to 20 cm^3 of the given H_2O_2 solution.
(*ii*) Show the number of times the concentration of the given H_2O_2 has been reduced.

30. 2 g of silver trioxonitrate (V), $AgNO_3$, is dissolved in 100 g of water. Calculate the
(*i*) molality; (*ii*) mole fraction;
(*iii*) concentration in % W/W of $AgNO_3$ in the solution. M_r ($AgNO_3$) = 170.

31. (*a*) Calculate the mass of potassium tetraoxomanganate (VII), $KMnO_4$, needed to prepare 1.0 dm^3 of a solution of concentration 0.02 mol dm^{-3} in 1.0 dm^3 standard volumetric flask.
(*b*) If 400 cm^3 of 0.02 M solution were evaporated to dryness, what mass of the potassium tetraoxomanganate (VII) solid would be left?

32. (*a*) The following label was put on a container of an injection powder:
'0.5 g of water soluble sodium thiopentone salt for injection, 5% solution'.
If a nurse were to give the injection, calculate the correct volume of sterilized, distilled water to be added to dissolve that salt for the injection. Explain why tap or river water cannot be used in place of the distilled water.
(*b*) 20 cm^3 of an eye drop of 0.5% W/V concentration is purchased in a chemist's shop. Determine the mass of the active chemical in 10 cm^3.

33. (*a*) If you were in charge of a class in which there were 25 students and each student was to be given 120 cm^3 of 0.2 M solution of sodium trioxocarbonate(IV), calculate the following:
(*i*) the total volume of solution needed by the students:
(*ii*) the total mass of sodium trioxocarbonate (IV) that you would have to weigh to prepare this solution. (Relative molecular mass of Na_2CO_3 = 106)
(*b*) In practice, would you prepare exactly the volume calculated in (*i*)? Explain your answer.

Answers

I. Objective Type Questions

1. (*b*)	**2.** (*a*)	**3.** (*a*)	**4.** (*d*)
5. (*a*)	**6.** (*d*)	**7.** (*b*)	**8.** (*b*)
9. (*a*)	**10.** (*b*)	**11.** (*d*)	**12.** (*d*)
13. (*a*)	**14.** (*c*)	**15.** (*d*)	**16.** (*c*).

II. Fill in the Blanks

17. (*i*) one, solvent (*ii*) mole fraction
(*iii*) N (*iv*) gas, liquid
(*v*) secondary standards (*vi*) stock or commercial
(*vii*) 0.50 (*viii*) strength of solution
(*ix*) 800 cm^3 (*x*) 4.

SECTION-4

UNIT-12

STOICHIOMETRY AND CHEMICAL EQUATIONS

Specific Objectives

After studying this unit, the student should be able to:

- write correct formula for named chemical compounds
- write the name of a given compound correctly
- write and balance chemical equations
- explain the laws of chemical combination
- demonstrate the principle of conservation of mass through experiment
- identify and write mole ratios for chemical species in balanced chemical equations
- determine the limiting reagents and excess reagents in a chemical reaction
- determine the formula of compounds from experimental and given data
- determine the formula of magnesium oxide through experiment.

12.1 WRITING CORRECT FORMULA FOR NAMED CHEMICAL COMPOUNDS

The elements are represented in the abbreviated form by their symbols. Similarly, a compound is represented in the abbreviated form by its chemical formula.

The expression of the composition of a substance by chemical symbols and numerical subscripts is called the ***chemical formula of the substance.***

The formula of a simple substance is obtained by writing the symbol of the element and indicating the number of atoms in a molecule of the substance by a subscript. *For example*, a molecule of hydrogen contains two atoms. Hence, hydrogen is represented by the formula H_2.

A molecule of ozone contains three atoms of oxygen. Hence, ozone is represented by the formula O_3.

In order to represent the chemical formula of a compound, one must know the elements present in the compound and the number of atoms of one element that combine with a definite number of atoms of the other element. *For example*, a molecule of water contains two hydrogen atoms and one oxygen atom. It is represented by the formula H_2O.

The formula of a compound is, thus, expressed by the symbols of the chemical elements contained in the compound and by the numerical subscripts (to the lower right of each symbol).

WRITING THE FORMULA OF A BINARY AND TERNARY MOLECULAR COMPOUND

In binary molecular compounds, the atoms of the two elements are held by covalent bonds. A covalent bond, as you know, is formed by sharing of electrons between the two atoms. The number of electrons that an atom of the element contributes for sharing is known as valency of the element. Knowing the valencies of the two elements involved in the formation of a binary molecular compound, the formula of the compound can be derived.

Step 1. *Write the constituent elements and their valencies as shown below:*

Valency	**4**	**2**
Element	**C**	**O**

Step 2. *Reduce the valency numerals to simplest whole numbers by dividing by some common factor, if any*

2	**1**
C	**O**

Step 3. *Criss-cross the reduced valency numerals and write them as subscripts at the bottom right hand side of the symbols.*

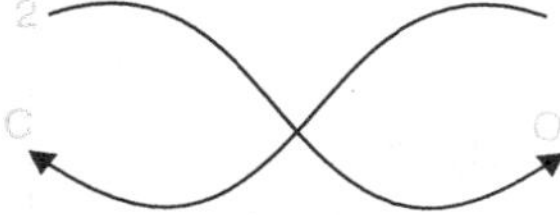

The subscript 1 is not written. Thus, the formula of the compound is CO_2. Formulae of some more compounds are given below:

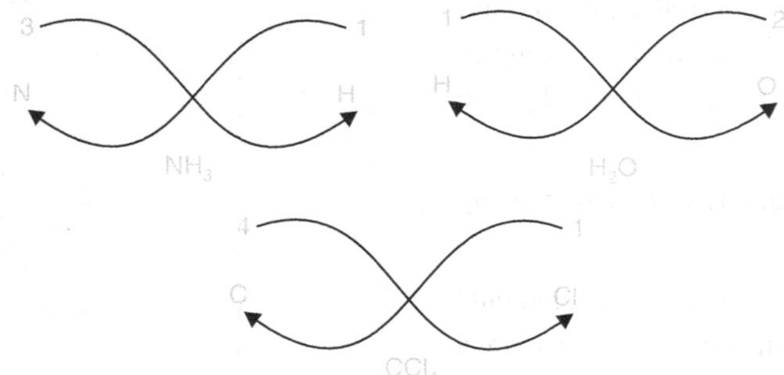

WRITING FORMULA OF SIMPLE IONS OR RADICALS OR POLYATOMIC IONS

Most of the inorganic compounds which are composed of metals and non-metals, contain oppositely charged parts called ions or radicals. The positively charged part is called **cation or electropositive radical or basic radical** whereas the negatively charged part is called **anion or electronegative radical or acidic radical.** For example, sodium chloride is made up of Na^+ ions and Cl^- ions. Similarly, nickel sulphate is made up of Ni^{2+} ions and SO_4^{2-} ions. An ion behaves as single unit in reactions. Thus,

*An **ion** is an atom or group of atoms, carrying positive or negative charge, that behaves as a single unit in reactions.*

An ion containing only one atom is known as **simple ion** namely simple cations, simple anion and oxoanion whereas an ion containing two or more atoms is known as **polyatomic ion** or **compound ion.** Thus, Na^+ is a simple ion whereas SO_4^{2-} is a compound ion.

*The charge on the ion is known as **valency of the ion.***

Simple Cations

1. The name of element is written.
2. The oxidation state is placed after the name of the element plus the word ion. The oxidation state is put in parenthesis in capital Roman numerals, with no space in between.
3. The fixed oxidation states of an element is shown.

Formulae of some common electropositive ions are given in Table 12.1.

Simple Anions

1. Anions are named by adding the suffix *-ide* for the last two or three letters in the name of the atom plus the word ion (see Table 12.2). For example, chloride ion, oxide ion, etc.
2. The ion OH^-, has the name **hydroxide**.

Oxoanions

Oxoanions are anions in which the oxygen atom is bonded to a central atom. The naming is based on the oxidation number of the central atom and the number of oxygen atoms bonded to it.

1. The suffix *-ate* replaces the last two or three letters in the name of the central atom.
2. The oxidation number of the central atom is placed in parenthesis in capital Roman numerals after its name.
3. The number of oxygen atoms is placed before the name of the central atom as dioxo, trioxo, tetraoxo, pentaoxo, hexaoxo, heptaoxo etc., respectively for 2, 3, 4, 5, 6 and 7 oxygen atoms. One oxygen atom is not given a prefix.
4. The word 'ion' is added to the name.

   The formula of some common electronegative ions is given in Table 12.2.

Table 12.1. Formulae and Valencies of Common Ions

Electropositive Ions							
Monovalent		*Bivalent*		*Trivalent*		*Tetravalent*	
Name	*Formula*	*Name*	*Formula*	*Name*	*Formula*	*Name*	*Formula*
Potassium	K^+	Barium	Ba^{2+}	Aluminium	Al^{3+}	Manganese [IV]	Mn^{4+}
Sodium	Na^+	Calcium	Ca^{2+}	Chromium	Cr^{3+}		
*Copper [I]	Cu^+	Magnesium	Mg^{2+}	Iron [III]	Fe^{3+}	Tin [IV]	Sn^{4+}
		Manganese [II]	Mn^{2+}	Gold	Au^{3+}	Lead [IV]	Pb^{4+}
						Platinum	Pt^{4+}
		Zinc	Zn^{2+}				
Silver	Ag^+	Iron [II]	Fe^{2+}				
Ammonium	NH_4^+						
Hydrogen	H^+	Nickel	Ni^{2+}				
		Cobalt	Co^{2+}				
		Tin [II]	Sn^{2+}				
		Cadmium	Cd^{2+}				
		Lead [II]	Pb^{2+}				
		Copper [II]	Cu^{2+}				
		Mercury [I]	Hg_2^{2+}				
		Mercury [II]	Hg^{2+}				
		Strontium	Sr^{2+}				

Table 12.2. The Formulae of Some Common Anions

Monovalent		*Bivalent*		*Trivalent*		*Tetravalent*	
Name	*Formula*	*Name*	*Formula*	*Name*	*Formula*	*Name*	*Formula*
Hydroxide	OH^-	Trioxocarbonate(IV) ion	CO_3^{2-}	Nitride ion	N^{3-}	Hexacyano-ferrate(II) ion	$Fe(CN)_6^{4-}$
Hydride ion	H^-	Tetraoxochromate(VI) ion	CrO_4^{2-}	Tetraoxopho-sphate(V) ion	PO_4^{3-}		
Fluoride ion	F^-	Heptaoxochromate(VI) ion	$Cr_2O_7^{2-}$	Hexacyano-ferrate(III) ion	$Fe(CN)_6^{3-}$		
Chloride ion	Cl^-	Tetraoxomanganate(VI) ion	MnO_4^{2-}	Phosphide ion	P^{3-}		
Bromide ion	Br^-						
Iodide ion	I^-	Sulphide ion	S^{2-}				
Hydrogen trioxo-carbonate(IV) ion	HCO_3^-	Trioxosulphate(IV) ion	SO_3^{2-}				
Hydrogen trioxo-sulphate(IV) ion	HSO_3^-	Tetraoxosulphate(VI) ion	SO_4^{2-}				
Hydrogen tetraoxo-sulphate(VI) ion	HSO_4^-	Oxide ion	O^{2-}				
Trioxochlorate(V) ion	ClO_3^-						
Oxochlorate(I) ion	ClO^-	Trioxosulphate(II) ion	$S_2O_3^{2-}$				
Dioxonitrate(III) ion	NO_2^-						
Trioxonitrate(V) ion	NO_3^-						
Tetraoxomanganate (VII) ion	MnO_4^-						

There are certain elements that exhibit more than one valencies in their ions (or compounds). For example, iron can exist as Fe^{2+} or Fe^{3+} in its compounds. In such cases, the name of the ion with lower valency ends with a suffix '*ous*' while that with higher valency ends with a suffix '*ic*'.

Names and formulae of some common basic ions exhibiting variable valency are listed in Table 12.3.

Table 12.3. Some Basic Ions Exhibiting Variable Valency

Name of Ions	*Formula*	*Name of Ions*	*Formula*
Cuprous or Copper(I)	Cu^{+}	Plumbous or Lead(II)	Pb^{2+}
Cupric or Copper(II)	Cu^{2+}	Plumbic or Lead(IV)	Pb^{4+}
Mercurous or Mercury(I)	Hg_2^{2+}	Stannous or Tin(II)	Sn^{2+}
Mercuric or Mercury(II)	Hg^{2+}	Stannic or Tin(IV)	Sn^{4+}
Ferrous or Iron(II)	Fe^{2+}	Aurous or Gold(I)	Au^{+}
Ferric or Iron(III)	Fe^{3+}	Auric or Gold(III)	Au^{3+}

WRITING THE FORMULA OF AN IONIC COMPOUND

The formula of an ionic compound represents the simplest whole number ratio of ions in it. The total positive charge on cations is equal to the total negative charge on anions. Knowing the formulae of the ions present in the compound, the formula of the compound can be written by the following steps:

Step 1. *Write the formulae of the ions or radicals of the compound side by side with the cation on the left hand side and the anion on the right hand side.*

Step 2. *Enclose the compound ion (if any) in a bracket.*

Step 3. *Reduce the valency numerals to a simple ratio by dividing with a common factor, if any.*

Step 4. *Criss-cross the valencies, i.e., shift the valency numerals crosswise to the lower right-hand corner of the ions. This is done to achieve electrical neutrality.*

Let us apply the above steps to write the formula of **calcium phosphate.**

Step 1. Writing the formulae of the ions.

Ca^{2+} $\quad$ PO_4^{3-}

Step 2. Enclose the compound ion phosphate in a bracket.

Ca^{2+} $\quad$ $(PO_4)^{3-}$

Step 3. Not applicable, because ratio is already simple.

Step 4. *Criss-cross the valencies.*

Thus, the formula of calcium phosphate is: $\mathbf{Ca_3(PO_4)_2}$

The correctness of the formula can be verified by checking the charge balance as given under:

Total positive charge on $3Ca^{2+}$ ions = 3 × (2) = + 6

Total negative charge $2PO_4^{3-}$ ions = 2 × (– 3) = – 6

Thus, total positive charge on cations is equal to the total negative charge on anions. Hence, the formula $Ca_3(PO_4)_2$ is correct.

On the basis of the above steps the formulae of some ionic compounds are given in Table 12.4.

Table 12.4. Writing Formula of an Ionic Compound

Compound	*Positive Radical*	*Negative Radical*	*Formula*
Potassium tetraoxo-chlorate(V)	K^1	$(ClO_3)^1$	$KClO_3$
Ammonium tetraoxo-phosphate(V)	$(NH_4)^1$	$(PO_4)^3$	$(NH_4)_3 PO_4$
Lithium nitride	Li^1	N^3	Li_3N
Sodium oxalate	Na^1	$(C_2O_4)^2$	$Na_2C_2O_4$
Silver tetraoxo-chromate(VI)	Ag^1	$(CrO_4)^2$	Ag_2CrO_4
Iron(II) tetraoxo-sulphate(VI)	Fe^2	$(SO_4)^2$	$FeSO_4$
Sodium thiosulphate or Sodium trioxosulphur sulphate(VI) or sodium trioxothiosulphate(VI)	Na^1	$(S_2O_3)^2$	$Na_2S_2O_3$
Ferric ferrocyanide or ferric(III) hexacyano-ferrate(II)	Fe^3	$[Fe(CN)_6]^4$	$Fe_4[Fe(CN)_6]_3$
Magnesium nitride	Mg^2	N^3	Mg_3N_2
Calcium chloride	Ca^2	Cl^1	$Ca(Cl)_2$
Calcium phosphide	Ca^2	P^3	Ca_3P_2
Sodium dioxozincate(II)	Na^1	$(ZnO_2)^2$	Na_2ZnO_2
Potassium tetraoxo-manganate(VI)	K^1	$(MnO_4)^2$	K_2MnO_4
Ammonium hexaoxo-μ-oxodichromate(VI)	$(NH_4)^1$	$(Cr_2O_7)^2$	$(NH_4)_2 Cr_2O_7$
Mercury(I) chloride	$(Hg_2)^2$	Cl^1	Hg_2Cl_2

SOLVED EXAMPLES

Example 12.1. *Write down the formulae of*

(*i*) *sodium oxide* (*ii*) *aluminium chloride*

(*iii*) *sodium sulphide* (*iv*) *magnesium hydroxide.*

Solution. (*i*) Sodium oxide contains Na^+ and O^{2-} ions.

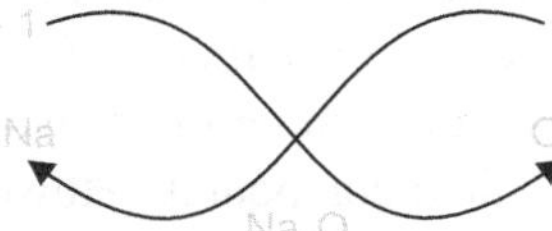

The formula of sodium oxide is $\mathbf{Na_2O}$.

(*ii*) Aluminium chloride

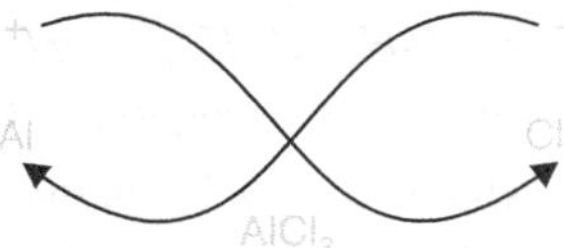

The formula of aluminium chloride is $\mathbf{AlCl_3}$.

(*iii*) Sodium sulphide

The formula of sodium sulphide is $\mathbf{Na_2S}$.

(*iv*) Magnesium hydroxide

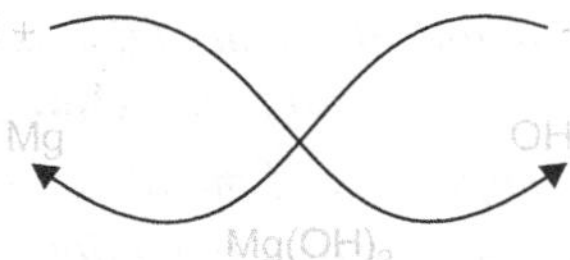

The formula of magnesium hydroxide is $\mathbf{Mg(OH)_2}$.

Example 12.2. *Write the correct formula of the following compounds:*

(*i*) *sodium trioxocarbonate(IV)*

(*ii*) *Ammonium tetraoxosulphate(VI)*

(*iii*) *trioxonitrate(V) acid*

(*iv*) *Iron(III) sulphide*

(*v*) *Hydrogen sulphide.*

Solution. The correct formula of the given compounds are:

Given Name	*Cation*	*Anion*	*Formula*
Sodium trioxocarbonate(IV)	Na^+	CO_3^{2-}	Na_2CO_3
Ammonium tetraoxosulphate(VI)	NH_4^+	SO_4^{2-}	$(NH_4)_2SO_4$
Trioxonitrate(V) acid	H^+	NO_3^-	HNO_3
Iron(III) sulphide	Fe^{3+}	S^{2-}	Fa_2S_3
Hydrogen sulphide	H^+	S^{2-}	H_2S

12.2 WRITING THE NAME OF A GIVEN COMPOUND

Compounds are made up of elements chemically combined together. The chemical formula of a compound represents this combination. The single unit can be molecules or groupings of ions. Every molecule, ion and ionic compound has a name. A system of naming chemical substances is called **chemical nomenclature.** There are various ways of naming substances. The systematic naming is the most useful. A systematic system follows a set of rules. The set of rules will give one name for one formula. For *example*, the name sulphur oxide is not a systematic name because both SO_2 and SO_3 molecules exist. The systematic name sulphur(VI) oxide, however, can only mean SO_3. The systematic naming has the added advantage of giving other information about the compound. This includes the oxidation state and number of atoms combining to form the compound.

Trivial or common names are old, traditional names. They are not necessarily based on any rules. As an example, the common name for the formula H_2SO_4 is *sulphuric acid* and the systematic name is tetraoxosulphate(VI) acid. The *West African Examination Council (WAEC)* recommends an adapted IUPAC systematic naming which will be used in this book. However, the IUPAC system still includes some trivial names which have been found to be simple and useful. For *example*, water for H_2O and ammonia for NH_3 are trivial names used in the IUPAC naming system. Some common or traditional names may be put in parentheses beside their IUPAC names.

Both inorganic and organic compounds can be named in the IUPAC system. Inorganic compounds are mainly compounds of metals and non-metals (with the exception of carbon). The naming of inorganic compounds relies on **oxidation numbers.**

Oxidation number (O.N.) of an element is defined as the residual charge which its atom has or appears to have when all other atoms from the molecule are assumed to be removed as ions by counting the shared electrons with the more electronegative atom.

For example, in hydrogen chloride molecule, chlorine is the more electronegative than hydrogen. Therefore, the shared pair is counted towards chlorine atom as shown below:

H :C̈l:

As a result of this, chlorine gets one extra electron and acquires a unit negative charge. Hence, the oxidation number of chlorine is –1. On the other hand, the hydrogen atom without an electron has a unit positive charge. Hence, the oxidation number of hydrogen in hydrogen chloride is +1.

It may be noted that *electrons shared between two similar atoms are divided equally between the sharing atoms.* For example, in chlorine molecule (Cl_2), the electron pair is equally shared between the two chlorine atoms. Therefore, one electron is counted with each chlorine atom as shown below:

:C̈l· ·C̈l:

Now, there is no net charge on each atom of chlorine. In other words, the oxidation number of chlorine in Cl_2 molecule is zero.

Thus, atoms can have *positive*, *zero* or *negative* values of oxidation numbers depending upon their state of combination. Now, since it is not always possible to remember or make out easily in a compound or ion which element is more electronegative than the other, a set of rules have been formulated to determine the oxidation number. If two or more atoms of an element are present in a molecule/ion, the oxidation number of the atom of the element will be the average of all the atoms of that element.

All elements have oxidation numbers in whatever state they occur. Some common ones are given in Table 12.5. Arabic numerals are used in the algebraic exercises to determine oxidation numbers, *e.g.*, +2, +3, +4, but capital Roman numerals, *e.g.*, I, II, IV, are used in the naming (see Table 12.1).

A combination of the knowledge of oxidation numbers, and how to indicate the number of atoms, ions or radicals, enables the correct formula of a compound to be written.

The charge on a species is written as a superscript to the right followed by its positive or negative sign. For example, Fe^{2+} and O^{2-} and not Fe^{+2} and O^{-2} since the oxidation number is not an addition or subtraction process.

RULES FOR ASSIGNING OXIDATION NUMBER TO AN ATOM

The oxidation number of various elements in their compounds can be calculated by applying the following rules. These rules have been formulated on the basis of the assumption that electrons in a covalent bond belong entirely to the more electronegative atom.

1. *The oxidation number of the element in the free or elementary state is always zero irrespective of its allotropic form.*

   For example,

   Oxidation number of helium in He = 0
   Oxidation number of chlorine in $Cl_2 = 0$
   Oxidation number of sulphur in $S_8 = 0$
   Oxidation number of phosphorus in $P_4 = 0$.

2. *The oxidation number of an element in a monoatomic ion is equal to the charge on the ion.* For example, in K^+Cl^-, the oxidation number of K is **+1** while that of Cl is **–1.** In a similar way, the oxidation number of all alkali metals is **+1** while those of alkaline earth metals is **+2** in their compounds.
3. *The oxidation number of fluorine is always* **–1** *in all its compounds.* Other halogens (Cl, Br and I) also have an oxidation number of **–1,** when they occur as halide ions in their compounds. However, in oxoacids and oxoanions they have positive oxidation numbers.
4. *Hydrogen is assigned an oxidation number of* **+1** *in all its compounds except in metal hydrides. In metal hydrides like NaH, MgH_2, CaH_2, LiH, etc., the oxidation number of hydrogen is* **–1.**
5. *Oxygen is assigned an oxidation number of* **–2** *in most of its compounds. However, in peroxides* (which contain O—O linkage) like H_2O_2, BaO_2, Na_2O_2, etc., its oxidation number is **–1.** Similarly, the exception also occurs in compounds of fluorine and oxygen like OF_2 (F—O—F) and O_2F_2 (F—O—O—F) in which the oxidation number of oxygen is **+2** and **+1** respectively
6. *In accordance with the principle of conservation of charge, the algebraic sum of the oxidation numbers of all the atoms in a molecule is* **zero.** *But in the case of polyatomic ions the sum of the oxidation numbers of all its atoms is equal to the* **charge on the ion.**
7. *In binary compounds of metals and non-metals, the metal atom has a positive oxidation number while the non-metal atom has a negative oxidation number.* For example, O.N. of K in KI is +1 but O.N. of I is –1.
8. *In binary compounds of non-metals, the more electronegative atom has a negative oxidation number, but the less electronegative atom has a positive oxidation number.* For example, O.N. of Cl in ClF_3 is positive (+3) while that in ICl is negative (–1).

REMEMBER

BRIEF SUMMARY OF RULES

Oxidation Number of:

- **Free elements = 0**
- **Fluorine = –1**
- **Simple ions = Charge on them**
- **Oxygen = –2; peroxides (– 1); F_2O (+ 2); F_2O_2 (+ 1)**
- **Hydrogen = +1; metal hydrides (–1)**
- **Sum of O.N. of atoms in molecules = 0**
- **Sum of O.N. of atoms in polyatomic ions = (Charge on them).**

Table 12.5. Oxidation Numbers in Various Species

Rule	*Formula*	*Oxidation Number*	*Notes*
1. (*i*)	H_2, H	0	
	Na	0	
	Cl, Cl_2	0	
	S_8	0	
(*ii*)	Cl^-	− 1	1. –ve charge indicates gain of electron(s) and +ve charge loss of electrons.
	O^{2-}	− 2	
	Cu^{2+}	+ 2	
	Cu^+	+ 1	
	Fe^{3+}	+ 3	
	Fe^{2+}	+ 2	2. Some elements have more than one oxidation number *e.g.*, Cu^+, Cu^{2+}.
	Na^+	+ 1	
	K^+	+ 1	
	Al^{3+}	+ 3	
	Ca^{2+}	+ 2	
	Li^+	+ 1	
	Zn^{2+}	+ 2	
(*iii*)	H_2O	H + 1, O − 2	
&	H_2O_2	H + 1, O − 1	
(*iv*)	NaH	Na + 1, H − 1	
2.	+3 −3 $Fe\,Cl_3$ +3 −1	+ 3 − 3 = 0	3. The oxidation number of each element is placed below its symbol and the total above it.
	+1 −1 $H\,Cl$ +1 −1	+ 1 − 1 = 0	4. The sum of the total oxidation numbers is added to be 0.
	+4 −4 $C\,O_2$ +4 −2	+ 4 − 4 = 0	
	+2 −2 $Cu\,O$ +2 −2	+ 2 − 2 = 0	
	+2 −2 $Cu_2\,O$ +1 −2	+ 2 − 2 = 0	
	+1 +7 −8 $K\,Mn\,O_4$ +1 +7 −2	+ 1 + 7 − 8 = 0	
	+3 +15 −18 $Al\,(N\,O_3)_3$ +3 +5 −2	+ 3 + 15 − 18 = 0	$Al(NO_3)_3 \equiv$ +3 +15 −18 $Al\,N_3\,O_9$ +3 +5 −2
3.	+6 −8 $S\,O_4^{2-}$ +6 −2	+ 6 − 8 = − 2	
	+5 −6 $N\,O_3^-$ +5 −2	+ 5 − 6 = − 1	
	−2 +1 $O\,H^-$ −2 +1	− 2 + 1 = − 1	
	−3 +4 $N\,H_4^+$ −3 +1	− 3 + 4 = + 1	
	+1 −2 $Cl\,O^-$ +1 −2	1 − 2 = − 1	

WRITING NAMES OF SIMPLE COMPOUNDS

The simplest compounds are binary compounds. A ***binary compound** is a compound that contains only two elements.* For example, NaCl is a binary compound of sodium and chlorine. CaF_2 is a binary compound of calcium and fluorine.

The following set of rules are followed while writing the correct IUPAC names:

1. The symbol of the more electronegative element is written on the right hand side while that of the less electronegative element is written on the left hand side.
2. The number of atoms of each element are indicated by subscripts on the bottom right hand side of the symbol. For example, for a compound formed by a combination of a metal with a non-metal, the symbol of the metal element is written first (left hand side) and the symbol of the non-metal element is written on the right hand side.
3. The first element (less electronegative element) is named as such while the name of the second element (more electronegative element) is written with an *-ide* ending. For example,

   KI is named as *potassium iodide*

   MgO is named as *magnesium oxide*

   NaH is named as *sodium hydride*

   $CaCl_2$ is named as *calcium chloride.*

   The names of some non-metallic elements with ***-ide*** endings are given below:

Hydrogen—Hydride–H^-	Nitrogen—Nitride–N^{3-}
Fluorine—Fluoride–F^-	Phosphorus—Phosphide–P^{3-}
Chlorine—Chloride–Cl^-	Carbon—Carbide–C^{4-}
Bromine—Bromide–Br^-	Boron—Boride–B^{3-}
Iodine—Iodide–I^-	
Oxygen—Oxide–O^{2-}	
Sulphur—Sulphide–S^{2-}	

4. While naming binary compounds of metals and non-metals, the subscript numerals are ignored. For example, $BaCl_2$ is named barium chloride and not barium dichloride.

5. In naming the binary compounds of two non-metals, the subscript numerals have to be taken into consideration and are indicated as a part of the name. *The reason for using prefixes is that the same two non-metallic elements may combine to form many compounds. For example,* phosphorus and chlorine combine to form two compounds PCl_3 and PCl_5.
   A subscript 2 is indicated by the prefix ***di***; subscript 3 by ***tri***; subscript 4 by ***tetra***, and so on. There should be no gap between the prefix and the name of the element. Some examples are given in Table 12.6.

Table 12.6. Names of Some Binary Compounds of Two Non-Metals

Formula	*Name*	*Formula*	*Name*
CO	Carbon monoxide	CCl_4	Carbon tetrachloride
CO_2	Carbon dioxide	PCl_5	Phosphorus pentachloride
NO_2	Nitrogen dioxide	SF_6	Sulphur hexafluoride
SO_2	Sulphur dioxide	N_2O_4	Dinitrogen tetroxide
SO_3	Sulphur trioxide	N_2O	Dinitrogen oxide
PCl_3	Phosphorus trichloride	N_2O_5	Dinitrogen pentoxide

- '– *o*' or '– *a*' at the end of the prefix is often dropped before another vowel. For example, tetroxide, pentoxide, etc. However, in the case of iodides '– *o*' or '–*a*' at the end of the prefix is not dropped. For example, PI_5 is named phosphorus pentaiodide and not phosphorus pentiodide.
- The prefix mono is generally not used for the first element in the formula.
- When hydrogen is the first element in the formula, no prefix is added before hydrogen irrespective of the number of hydrogen atoms in the formula. For example, H_2S is named hydrogen sulphide and not dihydrogen sulphide. Similarly, H_2Se is named hydrogen selenide.
- A few compounds of hydrogen with non-metals are named by their trivial or common names. For example, H_2O is called water and NH_3 is called ammonia.

IUPAC NAMES OF BINARY AND TERNARY COMPOUNDS (WAEC ADAPTED)

Oxoacids

The word acid replaces the word ion in the name of the oxoanion. Examples are shown in Table 12.7.

Table 12.7. Names of Some Common Inorganic Acids

Acid	*Name of Acid*
H_2SO_4	tetraoxosulphate(VI) acid
HNO_3	trioxonitrate(V) acid
HNO_2	dioxonitrate(III) acid
H_2CO_3	trioxocarbonate(IV) acid
HClO	oxochlorate(I) acid
H_3PO_4	tetraoxophosphate(V) acid

Aqueous HCl is called hydrochloric acid. It does not contain an oxoanion.

BASES AND SALTS

Bases can be divided into two groups, namely oxides and hydroxides. The salts can also be divided into two groups, namely binary salts and oxoacid salts.

1. Oxides and binary salts are named as binary compounds (see Tables 12.6 and 12.7).
2. A hydroxide is made up of a cation and the hydroxide ion, OH^-. Hydroxides are named in a similar way to binary compounds. The name of the cation is written first, followed by the oxidation number (if it varies), followed by the word hydroxide. Some examples are given in the following table.

Formula	*Common Name*	*WAEC Adapted IUPAC Name*
NaOH	Sodium hydroxide	Sodium hydroxide
$Cu(OH)_2$	Cupric hydroxide	Copper(II) hydroxide
$Fe(OH)_3$	Ferric hydroxide	Iron(III) hydroxide
$Ca(OH)_2$	Calcium hydroxide	Calcium hydroxide

3. An aqueous solution of NH_3 is a basic solution. It is preferably called **aqueous ammonia** or **ammonia solution** instead of the common name ammonium hydroxide. It is because it exists mainly as dissolved NH_3 molecules and not as ammonium NH_4^+ and hydroxide OH^- ions.

4. Oxoacid salts are named by writing the cation first followed by its oxidation number (if it varies) and the name of the oxoanion. Some examples are tabulated below:

Formula	*Common Name*	*WAEC Adapted IUPAC Name*
Na_2SO_3	Sodium sulphate	Sodium trioxosulphate(IV)
$ZnCO_3$	Zinc carbonate	Zinc trioxocarbonate(IV)
$KMnO_4$	Potassium perman-ganate	Potassium tetraoxomang-anate(VII)
NaClO	Sodium hypochlorate	Sodium oxochlorate

5. The names of salts, other than the above, are a combination of the names of cations and anions.

   For example, NaCN is named as sodium cyanide.

6. Some salts and compounds have water molecules as part of their structures.
   (*i*) The name of the hydrated *substance* is followed by the number of water molecules. *For example,* penta-(5), tetra-(4), hexa-(6), hepta-(7).
   (*ii*) The number of water molecules can also be represented as a pure number separated by a dash plus the word water. For example, $CuSO_4.5H_2O$ is named as copper(II) tetraoxosulphate(VI) pentahydrate or copper(II) tetraoxosulpahte(VI)-5-water.

   The water molecules in the substance is called ***water of crystallization***

FORMULAE AND IUPAC NAMES OF SIMPLE HYDROCARBONS

Formulae

The formulae of some simple hydrocarbons are:

Methane: CH_4

Ethane: C_2H_6 or $CH_3—CH_3$

Propane: C_3H_8 or $CH_3—CH_2—CH_3$

Butane: C_4H_{10} or $CH_3—CH_2—CH_2—CH_3$

IUPAC Names of Hydrocarbons

The IUPAC names of some simple hydrocarbons are:

1. $CH_3—\underset{}{\overset{CH_3}{\overset{|}{C}H}}—CH_3$: 2-Methyl propane
2. $CH_3\,CH_2\,CH_2\,CH_2\,CH_3$: Pentane
3. $CH_3—CH_2—CH_3$: Propane
4. $CH_3—\overset{CH_3}{\overset{|}{C}H}—\overset{CH_3}{\overset{|}{C}H}—CH_3$: 2, 3-Dimethyl butane
5. $CH_3—CH = CH_2$: Propene
6. $HC \equiv CH$: Ethyne

12.3 WRITING AND BALANCING CHEMICAL EQUATIONS

A chemical change can be represented by using symbols and formulae of various species involved in the change. Such a representation is known as a chemical equation.

Representation of a chemical change in terms of symbols and formulae of the reactants and products is known as ***chemical equation*** *of the reaction.*

For example, when zinc(II) trioxocarbonate(IV) is heated, zinc oxide and carbon dioxide are formed. This chemical change may be represented by the chemical equation given below:

$$ZnCO_3 \longrightarrow ZnO + CO_2$$

STEPS FOR WRITING CHEMICAL EQUATION

Step I: Writing Skeletal Equation

In order to write a chemical equation for a reaction, the first step is to write symbols and formulae of the various reactants and products. For example, when zinc(II) trioxocarbonate(IV) reacts with dilute hydrochloric acid it gives zinc chloride, carbon dioxide and water. This is represented by the following ***'word equation'.***

Zinc(II) trioxocarbonate(IV) + Hydrochloric acid ⟶ Zinc chloride + Carbon dioxide + water

Using symbols and formulae for the various reactants and products we get,

$$ZnCO_3 + HCl \longrightarrow ZnCl_2 + CO_2 + H_2O$$

Equations such as given above in which no attempt has been made to equalize the number of atoms of the various elements on both sides of the equation are known as **skeletal equations.** Thus:

A skeletal equation is an equation in which various reactants and products are represented by their respective formulae but no attempt is made to equalize the number of atoms of various elements on both the sides of the equation.

- Reactants are written on the left hand side while products are written on the right hand side.
- The symbols and formulae of the various reacting substances **(reactants)** are separated by a plus sign (+) between them.
- The symbols and formulae of the various substances formed **(products)** are also separated by a plus sign (+) between them.
- An **arrow** ($\longrightarrow$) is put between reactants and products.

Step II: Balancing of Chemical Equation

We know from Dalton's atomic theory that atoms can neither be created nor destroyed. Therefore, the number of atoms of the various elements on both sides of the chemical equation must be equal.

After writing the skeletal equation, the next step is to equalize the number of atoms of the various elements on both sides of the equation by multiplying various formulae by appropriate coefficients. This process is known as **balancing of chemical equations.**

In order to equalize the number of atoms of the various elements, the various species are multiplied by appropriate numbers. For example, the formation of ammonia is represented by the balanced chemical equation given below:

$$N_2 + 3H_2 \longrightarrow 2NH_3$$

An equation in which number of atoms of each element is equal on both sides of the equation is known as **balanced chemical equation.**

Sometimes the skeletal equation and the balanced equation may be identical. Some examples are given below:

$$CaCO_3 \xrightarrow{\text{Heat}} CaO + CO_2$$

$$Zn + H_2SO_4 \longrightarrow ZnSO_4 + H_2$$

Some balanced chemical equations are discussed below:

A. Synthesis (Composition)

Two or more elements or compounds may combine to form a simple or complex compound.

Examples of synthesis reactions:

1. Metal + oxygen $\longrightarrow$ metal oxide
   $2Mg(s) + O_2(g) \longrightarrow 2MgO(s)$
2. Nonmetal + oxygen $\longrightarrow$ nonmetallic oxide
   $C(s) + O_2(g) \longrightarrow CO_2(g)$
3. Metal oxide + water $\longrightarrow$ metallic hydroxide
   $MgO(s) + H_2O(l) \longrightarrow Mg(OH)_2(s)$
4. Nonmetallic oxide + water $\longrightarrow$ acid
   $CO_2(g) + H_2O(l) \longrightarrow H_2CO_3(aq)$
5. Metal + nonmetal $\longrightarrow$ salt
   $2Na(s) + Cl_2(g) \longrightarrow 2NaCl(s)$
6. A few nonmetals combine with each other.
   $2P(s) + 3Cl_2(g) \longrightarrow 2PCl_3(g)$

These two reactions must be remembered:

1. $N_2(g) + 3H_2\,(g) \longrightarrow 2NH_3(g)$
2. $NH_3(g) + H_2O(l) \longrightarrow NH_4OH(aq)$

B. Decomposition

A single compound breaks down into its component parts or simpler compounds.

Examples of decomposition reactions:

1. Metallic carbonates, when heated, form metallic oxides and $CO_2(g)$
   $CaCO_3(s) \longrightarrow CaO(s) + CO_2(g)$
2. Most metallic hydroxides, when heated, decompose into metallic oxides and water.
   $Ca(OH)_2(s) \longrightarrow CaO(s) + H_2O(g)$
3. Metallic chlorates, when heated, decompose into metallic chlorides and oxygen.
   $2KClO_3(s) \longrightarrow 2KCl(s) + 3O_2(g)$
4. Some acids, when heated, decompose into nonmetallic oxides and water.
   $H_2SO_4 \longrightarrow H_2O(l) + SO_3(g)$
5. Some oxides, when heated, decompose.
   $2HgO(s) \longrightarrow 2Hg(l) + O_2(g)$
6. Some decomposition reactions are produced by electricity.
   $2H_2O(l) \longrightarrow 2H_2(g) + O_2(g)$
   $2NaCl(l) \longrightarrow 2Na(s) + Cl_2(g)$

C. Replacement

A more active element takes the place of another element in a compound and sets the less active one free.

Examples of replacement reactions:

1. Replacement of a metal in a compound by a more active metal.
   $Fe(s) + CuSO_4(aq) \longrightarrow FeSO_4(aq) + Cu(s)$
2. Replacement of hydrogen in water by an active metal.
   $2Na(s) + 2H_2O(l) \longrightarrow 2NaOH(aq) + H_2(g)$
   $Mg(s) + H_2O(g) \longrightarrow MgO(s) + H_2(g)$
3. Replacement of hydrogen in acids by active metals.
   $Zn(s) + 2HCl(aq) \longrightarrow ZnCl_2(aq) + H_2(g)$
4. Replacement of nonmetals by more active nonmetals.
   $Cl_2(g) + 2NaBr(aq) \longrightarrow 2NaCl(aq) + Br_2(l)$

D. Ionic Reactions

These occur between ions in aqueous solution. A reaction will occur when a pair of ions come together to produce at least one of the following:

(*i*) a precipitate (*ii*) a gas

(*iii*) water or some other non-ionized substance.

Examples of ionic reactions:

1. Formation of precipitate.

$$NaCl(aq) + AgNO_3(aq) \longrightarrow NaNO_3(aq) + AgCl(s)$$
$$BaCl_2(aq) + Na_2SO_4(aq) \longrightarrow 2NaCl(aq) + BaSO_4(s)$$

2. Formation of a gas.

$$HCl(aq) + FeS(s) \longrightarrow FeCl_2(aq) + H_2S(g)$$

3. Formation of water. (If the reaction is between an acid and a base it is called a neutralization reaction.)

$$HCl(aq) + NaOH(aq) \longrightarrow NaCl(aq) + H_2O(l)$$

4. Formation of a product which decomposes.

$$CaCO_3(s) + HCl(aq) \longrightarrow CaCl_2(aq) + CO_2(g) + H_2O(l)$$

E. Combustion of Hydrocarbons

Another important type of reaction, in addition to the four types above, is that of the combustion of a hydrocarbon. When a hydrocarbon is burned with sufficient oxygen supply, the products are always carbon dioxide and water vapour. If the supply of oxygen is low or restricted, then carbon monoxide will be produced. This is why it is so dangerous to have an automobile engine running inside a closed garage or to use a charcoal grill indoors.

$$\text{Hydrocarbon } (C_xH_y) + O_2(g) \longrightarrow CO_2(g) + H_2O(g)$$
$$CH_4(g) + 2O_2(g) \longrightarrow CO_2(g) + 2H_2O(g)$$
$$2C_4H_{10}(g) + 13O_2(g) \longrightarrow 8CO_2(g) + 10H_2O(g)$$

- **Complete combustion** means a higher oxidation number is attained.
- **Incomplete combustion** means a lower oxidation number is attained.
- The phrase **"To burn"** means to add oxygen unless told otherwise.

ESSENTIALS OF A CHEMICAL EQUATION

A chemical equation must satisfy the following conditions:

1. ***It should represent a true chemical change,*** *i.e.*, it should be possible to carry out the reaction represented by the equation.
2. ***It should be balanced,*** *i.e.*, the number of atoms of each element on both sides of the chemical equation must be equal.
3. ***It should be molecular,*** *i.e.*, the elementary substances such as hydrogen, oxygen, phosphorus, etc., should be represented in their molecular form as H_2, O_2 and P_4 respectively.

If the equation is not satisfying any of the above conditions, it cannot be termed a chemical equation.

BALANCING OF SIMPLE CHEMICAL EQUATIONS

The balancing of chemical equations is based upon the **Dalton's atomic theory** and the **law of conservation of mass.** According to the Dalton's atomic theory, the atom is the smallest unit of an element that takes part in chemical reactions and that during chemical reactions atoms are neither created nor destroyed. Therefore, ***the number of atoms of each element should remain the same before and after the reaction.***

To make the number of atoms of all the elements equal on both sides in a skeletal equation is known as ***balancing.***

During balancing, the symbols and formulae of the various species given in the equation are multiplied with appropriate coefficients. A coefficient is a small whole number, like the coefficients used in algebraic equations. The simple equations can be balanced by **Hit and trial** method. The following steps may be followed while balancing the chemical equations by this method:

Step 1. Write the correct skeleton equation. The skeleton equation contains the formulae of only one molecule of each reactant and product. Once you are satisfied that the skeleton equation is correct, do not change the subscripts in any of the formulae.

Step 2. Start with the compound that has the maximum atoms or the maximum kinds of atoms and the atoms present in it are balanced first.

Step 3. Balance elements that appear only once on each side of the arrow first. Then balance elements that appear more than once on a side.

Step 4. Elementary substances are balanced last.

Step 5. If required the whole equation is multiplied by some suitable number in order to make all the coefficients whole numbers.

Balancing of Ionic Equations

A molecular equation is balanced on the basis of **mass or material balancing** by equating the number of atoms of the various elements on both sides of the chemical equation. However, in case of ionic equations we have to consider not only mass or material balancing but also **charge balancing.**

For example, consider the following equation:

$$Zn + Ag^+ \longrightarrow Zn^{2+} + Ag$$

This equation is balanced as far as mass balancing is considered. However, there is +1 charge on reactant side and +2 charge on product side. In order to balance charges Ag^+ on the reactant side and Ag on the product side are multiplied by 2. Thus, the correct balanced equation is

$$Zn + 2Ag^+ \longrightarrow Zn^{2+} + 2Ag$$

Thus, **a balanced chemical equation must satisfy mass balance as well as charge balance.**

The application of these rules is illustrated in the following solved examples.

SOLVED EXAMPLES

Example 12.3. *Aluminium reacts with oxygen on heating to form aluminium oxide. Write a balanced equation for the reaction.*

Solution. Aluminium + Oxygen $\longrightarrow$ Aluminium oxide

The skeleton equation for the reaction is

$$Al + O_2 \longrightarrow Al_2O_3$$

Start with Al_2O_3 because it contains the maximum number of atoms and most kinds of atoms. To balance aluminium atoms multiply Al by 2.

$$2Al + O_2 \longrightarrow Al_2O_3$$

In order to balance oxygen atoms, multiply O_2 by $\frac{3}{2}$ so that number of O atoms becomes three on both sides.

$$2Al + \frac{3}{2} O_2 \longrightarrow Al_2O_3$$

In order to make all the coefficients whole numbers, the whole equation is multiplied by 2.

$$2 \times \left[2Al + \frac{3}{2} O_2 \longrightarrow Al_2O_2\right]$$

$$\mathbf{4Al + 3O_2 \longrightarrow 2Al_2O_3}$$

The equation is balanced. There are 4 aluminium atoms and 6 oxygen atoms on each side.

Example 12.4. *Write a balanced equation for the following reaction:*

Methane burns in oxygen to form carbon dioxide and water.

Solution. Methane + Oxygen $\longrightarrow$ Carbon dioxide + Water

The skeleton equation for the reaction is

$$CH_4 + O_2 \longrightarrow CO_2 + H_2O$$

Since CH_4 contains the maximum number of atoms, the atoms present in it are balanced first of all. Multiply H_2O by 2 so that number of H atoms becomes equal an each side.

$$CH_4 + O_2 \longrightarrow CO_2 + 2H_2O$$

In order to balance oxygen atoms multiply O_2 by 2 so that there are 4 oxygen atoms on each side of the equation.

$$CH_4 + 2O_2 \longrightarrow CO_2 + 2H_2O$$

The equation is balanced. There is one carbon atom, 4 oxygen atoms and 4 hydrogen atoms on each side.

Example 12.5. *The equation for the oxidation of ammonia is:*

$$NH_3 + O_2 \longrightarrow N_2 + H_2O$$

Rewrite the equation in a balanced form.

Solution. The skeleton equation is

$$NH_3 + O_2 \longrightarrow N_2 + H_2O$$

NH_3 contains maximum number of atoms, so the atoms present in it are balanced first.

Multiply NH_3 by 2 in order to balance N atoms.

$$2NH_3 + O_2 \longrightarrow N_2 + H_2O$$

Multiply H_2O by 3 in order to balance H atoms.

$$2NH_3 + O_2 \longrightarrow N_2 + 3H_2O$$

Oxygen is balanced last of all. In order to balance oxygen atoms multiply O_2 by $\frac{3}{2}$.

$$2NH_3 + \frac{3}{2} O_2 \longrightarrow N_2 + 3H_2O$$

Since all the coefficients should be whole numbers, multiply the above equation by 2 in order to convert all the coefficients to whole numbers.

$$2 \times \left[2NH_3 + \frac{3}{2} O_2 \longrightarrow N_2 + 3H_2O\right]$$

$$\mathbf{4NH_3 + 3O_2 \longrightarrow 2N_2 + 6H_2O}$$

This is balanced equation. There are 4 N atoms, 12 H atoms and 6 O atoms on each side of the equation.

Example 12.6. *Balance the following equations:*

(*i*) $Na_2S_2O_3 + I_2 \longrightarrow NaI + Na_2S_4O_6$

(*ii*) $CaCl_2 + Na_2CO_3 \longrightarrow CaCO_3 + NaCl.$

Solution.

(*i*) The iodine atoms are balanced first because of the simple formula

$\therefore$ $$Na_2S_2O_3 + I_2 \longrightarrow 2NaI + Na_2S_4O_6$$

Next the sodium atoms are balanced

$$2Na_2S_2O_3 + I_2 \longrightarrow 2NaI + Na_2S_4O_6$$

The above equation is a perfect balanced equation

(*ii*) In this equation chlorine atoms are balanced

$$CaCl_2 + Na_2CO_3 \longrightarrow CaCO_3 + 2NaCl$$

The above equation is balanced.

Example 12.7. *Solid calcium phosphate and aqueous sulphuric acid solution react to give calcium sulphate, which gets precipitated. The other product is phosphoric acid, which remains in solution. Write a balanced chemical equation for the reaction and clearly indicate the states of the various reactants and products.*

Solution. Calcium phosphate + Sulphuric acid $\longrightarrow$ Calcium sulphate + Phosphoric acid

The skeleton equation for the reaction is

$$Ca_3(PO_4)_2 + H_2SO_4 \longrightarrow CaSO_4 + H_3PO_4$$

$Ca_3(PO_4)_2$ contains the maximum number of atoms hence the atoms present in it will be balanced first of all. In order to balance Ca atoms multiply $CaSO_4$ by 3 and in order to balance P atoms multiply H_3PO_4 by 2

$$Ca_3(PO_4)_2 + H_2SO_4 \longrightarrow 3CaSO_4 + 2H_3PO_4$$

Now, in order to balance S atoms multiply H_2SO_4 by 3

$$Ca_3(PO_4)_2 + 3H_2SO_4 \longrightarrow 3CaSO_4 + 2H_3PO_4$$

The H and O atoms have become equal on both sides. So, this is the balanced equation. Introducing the states of the various reactants and products, the equation becomes:

$$\mathbf{Ca_3(PO_4)_2(s) + 3H_2SO_4(aq) \longrightarrow 3CaSO_4(s) + 2H_3PO_4(aq).}$$

12.4 LAWS OF CHEMICAL COMBINATION

By studying the results of quantitative measurements of many reactions it was observed that *whenever substances react, they do so according to certain laws*. These laws are called the **laws of chemical combination.** These laws formed the basis of the Dalton's atomic theory of matter. Before we discuss this theory it would be appropriate to briefly study these laws.

LAW OF CONSERVATION OF MASS

This law was stated by the French chemist Antoine Laurent Lavoisier (1774). This law states that:

During any physical or chemical change, the total mass of the products remains equal to the total mass of the reactants.

Antoine Lavoisier (1743–1794)

Lavoisier showed that when mercuric oxide was heated it produced free mercury and oxygen. The sum of masses of mercury and oxygen was found to be equal to the mass of mercuric oxide

$$\underset{100\text{ g}}{\textit{Mercuric oxide}} \xrightarrow{\textit{Heat}} \underset{92.6\text{ g}}{\textit{Mercury}} + \underset{7.4\text{ g}}{\textit{Oxygen}}$$

Law of conservation of mass is also known as *law of indestructibility* of matter. The law is illustrated in activity 12.1.

SOLVED EXAMPLES

Example 12.8. *10.0 g of $CaCO_3$ on heating gave 4.4 g of CO_2 and 5.6 g of CaO. Show that these observations are in agreement with the law of conservation of mass.*

Solution. Mass of the reactants = 10.0 g

Mass of the products = 4.4 + 5.6 = 10.0 g

Since the mass of the reactants is equal to the mass of the products, these observations are in agreement with the law of conservation of mass.

Example 12.9. *When 4.2 g of $NaHCO_3$ is added to a solution of CH_3COOH weighing 10 g, it is observed that 2.2 g of CO_2 is released into the atmosphere. The residue is found to weigh 12.0 g. Show that these observations are in agreement with the law of conservation of mass.*

Solution. According to the law of conservation of mass the mass of reactants and products must be equal.

Thus,

mass of reactants = 4.2 + 10 = 14.2 g

and mass of products = 2.2 + 12.0 = 14.2 g

Hence, the law of conservation of mass is proved.

ACTIVITY 12.1 **Experiment to show that mass is conserved in a chemical reaction**

1. Prepare separately a 5% solution of barium chloride and a 5% solution of sodium sulphate.
2. Take about 20 cm^3 of barium chloride solution in a conical flask.
3. Take sodium sulphate solution in a small test tube. Hang the test tube in the conical flask with the help of a thread. Close the mouth of the flask with a cork (Fig. 12.1).
4. Weigh the flask along with its contents.

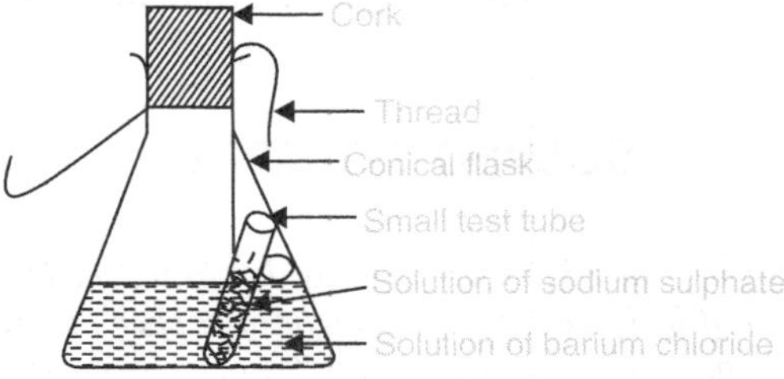

Fig. 12.1. Study of reaction between barium chloride and sodium sulphate.

5. Now tilt the flask so that the two solutions get mixed.
6. Weigh the flask again along with its contents.
   (*i*) It is observed that on mixing the two solutions a chemical reaction takes place which is indicated by the formation of a white precipitate.
   $$BaCl_2 + Na_2SO_4 \longrightarrow \underset{\text{(white ppt.)}}{BaSO_4} + 2NaCl$$
   (*ii*) The mass of the flask and its contents remains constant. Thus, during a chemical reaction mass is neither created nor destroyed.

This activity can also be carried out with the following pairs:

(*a*) Silver nitrate and sodium chloride.
(*b*) Copper sulphate and sodium carbonate.
(*c*) Sodium carbonate and calcium chloride.

LAW OF CONSTANT COMPOSITION OR DEFINITE PROPORTIONS

Joseph Louis Proust (1799)

This law deals with the composition of chemical compounds. It was discovered by the French chemist, Joseph Proust (1799). This law states that:

A chemical compound always contains the same elements combined together in the same proportion by mass.

It implies from this law that in a chemical compound the elements are present in *fixed* and *not arbitrary* ratios by mass. For example, pure water obtained from different sources such as, *river, well, spring, sea, etc.*, always contains hydrogen and oxygen combined together in the ratio of 1 : 8 by mass. Similarly, carbon dioxide can be obtained by different methods such as:

(*a*) burning of carbon,
(*b*) heating limestone, or
(*c*) the action of dilute hydrochloric acid on marble pieces.

It can be shown experimentally that different samples of carbon dioxide contain carbon and oxygen in the ratio of 3 : 8 by mass.

SOLVED EXAMPLE

Example 12.10. *1.375 g of cupric oxide was reduced by heating in a current of hydrogen and the weight of copper that remained was 1.098 g. In another experiment, 1.179 g of copper was dissolved in nitric acid and the resulting copper nitrate converted into cupric oxide by ignition. The weight of cupric oxide formed was 1.476 g. Show that these results illustrate the law of constant composition.*

Solution. *First experiment*:

Copper oxide = 1.375 g

Copper left = 1.098 g

∴ Oxygen present = 1.375 – 1.098 = 0.277 g

$$\text{Hence \% of oxygen in CuO} = \frac{0.277 \times 100}{1.375} = \mathbf{20.14.}$$

Second experiment:

Copper taken = 1.179 g

Copper oxide formed = 1.476 g

∴ Oxygen present = 1.476 – 1.179 g = 0.297 g

$$\text{Hence \% of oxygen of CuO} = \frac{0.297 \times 100}{1.476} = \mathbf{20.12.}$$

Percentage of oxygen is the same in both cases as shown above, so the law of constant composition is illustrated.

LAW OF MULTIPLE PROPORTIONS

John Dalton (1803)

This law was discovered by John Dalton (1803). This law states that:

When two elements combine with each other to form two or more compounds, the masses of one of the elements which combine with a fixed mass of the other, bear a simple whole number ratio to one another.

For example, carbon and oxygen combine with each other to form carbon monoxide (CO) and carbon dioxide (CO_2).

In carbon monoxide (CO): 12 parts by mass of carbon combine with 16 parts by mass of oxygen.

In carbon dioxide (CO_2): 12 parts by mass of carbon combine with 32 parts by mass of oxygen.

Ratio of the masses of oxygen which combine with a fixed mass of carbon (12 parts) in these compounds is 16 : 32 or 1 : 2, which is a simple whole number ratio.

Similarly, copper and oxygen combine to form two oxides, the red cuprous oxide (Cu_2O) and the black cupric oxide (CuO).

In red oxide (Cu_2O): 16 parts by mass of oxygen combine with 63.5 × 2 parts by mass of copper.

In black oxide (CuO): 16 parts by mass of oxygen combine with 63.5 parts by mass of copper.

Ratio of the masses of copper that combine with a fixed mass of oxygen (16 parts) in these oxides is 63.5 × 2 : 63.5 or 2 : 1, which is a simple whole number ratio.

SOLVED EXAMPLES

Example 12.11. *Hydrogen and oxygen are known to form two compounds. The hydrogen content in one of these is 5.93% while in the other it is 11.2%. Show that this data illustrates the law of multiple proportions.*

Solution. *In the first compound*

Hydrogen = 5.93%

Oxygen = (100 – 5.93)% = 94.07%.

In the second compound

Hydrogen = 11.2%

Oxygen = (100 – 11.2)% = 88.8%.

In the first compound the number of parts by mass of oxygen that combine with one part by mass of hydrogen $= \frac{94.07}{5.93} = 15.86$ parts.

In the second compound the number of parts by mass of oxygen that combine with one part by mass of hydrogen $= \frac{88.8}{11.2} = 7.9$ parts.

The ratio of masses of oxygen that combine with a fixed mass (1 part) by mass of hydrogen is 15.86 : 7.9 or 2 : 1.

Since this ratio is a simple whole number ratio, the given data illustrates the law of multiple proportions.

Example 12.12. *Carbon and oxygen are known to form two compounds. The carbon content in one of these is 42.9% while in the other it is 27.3%. Show that this data is in agreement with the law of multiple proportions.*

Solution. In compound I C = 42.9% ; O = 57.1%

In compound II C = 27.3% ; O = 72.7%

For a fixed mass of carbon (or parts) = 1 the ratio of oxygen in the two compounds would be

$$\text{Oxygen in, I : II} = \frac{57.1}{42.9} : \frac{72.7}{27.3} = 1.33 : 2.66 = 1 : 2$$

Since it is a simple whole number ratio, it illustrates the law of multiple proportions.

12.5 WRITING MOLE RATIO FOR CHEMICAL SPECIES IN BALANCED CHEMICAL EQUATIONS

We know that the chemical equation gives us the quantitative relationship between the reactants and the products. This quantitative information can be utilised to carry out a variety of calculations which are required many a time, to assess the economic viability of the chemical process. Calculations based on the quantitative relationship between the reactants and the products are also referred to as **stoichiometry.** The term stoichiometry is derived from the Greek words *Stoicheion* meaning *element* and *metron* meaning *measure*. Stoichiometry is, therefore, that area of chemistry and chemical technology on which determination of quantities of reactants and products of chemical reaction is based.

CALCULATIONS OF QUANTITIES OF SUBSTANCES IN CHEMICAL REACTIONS

A balanced equation indicates the mole ratio of substances involved in the reaction. Quantities of substances can be an amount of substance in mol, mass in g or kg, volume in cm^3 or dm^3, concentration in mol dm^{-3} or g dm^{-3}, or number of particles.

The mole ratio can be used to calculate unknown quantities when known quantities of reactants or products are given. Such calculations are done in industry on a large scale to find the maximum theoretical yield of product from a given mass of reactant. Such calculations have important economic implications.

In order to solve the problems based on chemical calculations the following steps, in general, are quite helpful.

(*i*) *Write the balanced chemical equation.*

(*ii*) *From the chemical equation write the molar ratio between the given and required substance.*

(*iii*) *Convert these moles into the desired parameters such as mass or volume.*

(*iv*) *Calculate the final result by applying the unitary method.*

Reading Chemical Equations

Chemical equations can be interpreted in many ways *i.e.*, number of entities, amount of substance, mass of substance, concentration, volume of substance, percentage yield of products etc. The idea is illustrated by the following example:

	$Mg(s)$	$+ 2HCl(g)$	$\rightarrow MgCl_2(s)$	$+ H_2(g)$
(*i*) **Number of entities**	1 atom Mg	2 molecules of HCl	1 formula unit of $MgCl_2$	1 molecule of H_2
(*ii*) **Amount of substance (mol)**	1 mol	2 mol	1 mol	1 mol
(*iii*) **Mass in g/mol**	56 g	2 × 36.5 g	95 g	2 g
(*iv*) **Volume at STP**	–	2 × 22.4 dm^3	–	22.4 dm^3
(*v*) **Number of particles (mol)**	N_A atoms	$2 \times N_A$ molecules	N_A formula unit	N_A molecules

The percentage yield of products can be calculated if we start with a given mass of sample.

SOLVED EXAMPLES

Example 12.13. *How many grams of oxygen (O_2) are required to completely react with 0.200 g of hydrogen (H_2) to yield water (H_2O)? Also calculate the amount of water formed. (At. mass H = 1; O = 32).*

Solution. The balanced equation for the reaction is

$2H_2$	+	O_2	$\longrightarrow$	$2H_2O$
2 mol		1 mol		2 mol
4 g		32 g		36 g

Now, 4 g of H_2 require oxygen

$= 32$ g

0.200 g of H_2 require oxygen

$$= \frac{32}{4} \times 0.200 = \mathbf{1.6\ g}$$

Again, 4 g of H_2 produce H_2O

$= 36$ g

0.200 g of H_2 produce H_2O

$$= \frac{36}{4} \times 0.200 = \mathbf{1.8\ g.}$$

Example 12.14. *What volume of oxygen at S.T.P. can be produced by 6.125 g of potassium chlorate according to the reaction $2KClO_3 \longrightarrow 2KCl + 3O_2$?*

Solution. The given chemical equation is:

$2KClO_3$	$\longrightarrow$	$2KCl$	+	$3O_2$
2 mol		2 mol		3 mol
2 × 122.5 g				3 × 22.4 L at S.T.P.

Now 245 g of $KClO_3$ produce oxygen at S.T.P.

$= 3 \times 22.4$ L

6.125 g of $KClO_3$ produce oxygen

$$= \frac{3 \times 22.4}{245} \times 6.125 = \mathbf{1.68\ L\ at\ S.T.P.}$$

Example 12.15. *How many grams of carbon are there in 83.5 g of formaldehyde, CH_2O?*

Solution.

Mol. mass of $CH_2O = 12 + 2 + 16 = 30$

Now, $CH_2O \equiv C$

Amount in g/mol 30 g 12 g

$$\therefore\ 83.5 \text{ g contains C} = \frac{12}{30} \times 83.5 = \mathbf{33.4\ g}$$

Example 12.16. *4.0 g of magnesium ribbon was burnt in excess of oxygen. What is the mass of magnesium oxide that will be formed?*

Solution.

	$2Mg(s)$ +	$O_2(g)$	$\longrightarrow 2MgO(s)$
Amount in g/mol	2 × 24 g	excess	2 × 40 g

On burning 4.0 g of magnesium the amount of MgO formed will be

$$= \frac{2 \times 40}{2 \times 24} \times 4 = \mathbf{6.67\ g}$$

12.6 DETERMINING THE LIMITING AND EXCESS REAGENTS IN A CHEMICAL REACTION

In many situations during chemical reactions, one of the reactants is taken in excess. This is to ensure that the reaction goes to completion. However, on completion of the reaction, some of the reactant, taken in excess, is left over. For example, consider the combustion of hydrogen

$$2H_2(g) + O_2(g) \longrightarrow 2H_2O(g)$$

Suppose that 2 mole of H_2 and 2 mole of O_2 are available for the reaction. It follows from the equation that only 1 mole of O_2 is required for complete combustion of 2 mole of H_2; 1 mole of O_2 will, therefore, be left over on completion of the reaction. The amount of the product obtained is determined by the amount of the reactant that is completely consumed in the reaction. This reactant is called the *limiting reagent*. Thus, the **limiting reagent** may be defined as ***the reactant which is completely consumed during the reaction.*** The reactant which is not completely consumed is referred to as **excess reactant.**

In the above example, H_2 is the limiting reagent and O_2 is the excess reactant. The amount of H_2O formed will, therefore, be determined by the amount of H_2. Since 2 mole of H_2 are taken, it will form 2 mole of H_2O on combustion.

SOLVED EXAMPLES

Example 12.17. *How much magnesium sulphide can be obtained from 2.00 g of magnesium and 2.00 g of sulphur by the reaction $Mg + S \rightarrow MgS$?*

Which is the limiting reagent?

Calculate the amount of one of the reactants which remains unreacted.

Solution. First of all each of the masses are expressed in moles:

$$2.00 \text{ g of Mg} = \frac{2.00}{24.3} = 0.0824 \text{ moles of Mg}$$

$$2.00 \text{ g of S} = \frac{2.00}{32.1} = 0.0624 \text{ moles of S}$$

From the equation, Mg + S → MgS, it follows that one mole of Mg reacts with one moles of S. We are given more moles of Mg than of S therefore, Mg is in excess and some of it will remain unreacted when the reaction is over. **S is the limiting reagent** and will control the amount of product. From the equation we note that one mole of S gives one mole of MgS, so 0.0624 mole of S will react with 0.0624 mole of Mg to form 0.0624 mole of MgS.

Molar mass of MgS = 56.4 g

∴ Mass of MgS formed

= 0.0624 × 56.4 g = **3.52 g of MgS**

Mole of Mg left unreacted

= 0.0824 – 0.0624 moles of Mg

= 0.0200 moles of Mg

Mass of Mg left unreacted

= moles of Mg × molar mass of Mg

= 0.0200 × 24.3 g of Mg = **0.486 g of Mg.**

Example 12.18. *50.0 kg of $N_2(g)$ and 10.0 kg of $H_2(g)$ are mixed to produce $NH_3(g)$. Calculate the $NH_3(g)$ formed. Identify the limiting reagent in the production of NH_3 in this situation.*

Solution. The balanced equation of the reaction is

$$N_2(g) + 3H_2(g) \rightleftharpoons 2NH_3(g)$$

1 mole   3 moles   2 moles

Calculation of number of moles:

(*i*) Number of moles of N_2

$$= 50.0 \text{ kg} \times \frac{1000 \text{ g}}{1 \text{ kg}} \times \frac{1 \text{ mol}}{28 \text{ g}}$$

$= 17.86 \times 10^2$ mol

(*ii*) Number of moles of H_2

$$= 10.00 \text{ kg} \times \frac{1000 \text{ g}}{1 \text{ kg}} \times \frac{1 \text{ mol}}{2.016 \text{ g}}$$

$= 4.96 \times 10^3$ mol

According to the balanced equation,

17.86×10^2 mol of N_2 will react with

$$= \frac{3}{1} \times 17.86 \times 10^2 = 5.36 \times 10^3 \text{ mol}$$

Since the number of moles of H_2 is 4.96×10^3 hydrogen is the limiting reagent.

PERCENTAGE COMPOSITION AND MOLECULAR FORMULA

Percentage composition of a compound refers to the amount of the various constituent elements present per 100 parts by mass of the substance. It can be calculated from the formula of the compound by using the following relation:

$$\textbf{\% (mass) of an element} = \frac{\textbf{(No. of atoms of element)} \times \textbf{(At. mass of element)} \times 100}{\textbf{Formula mass of substance}}$$

In case of minerals or ores the mass percentage of a particular constituent species in a formula can also be calculated using a similar relation:

$$\textbf{\% (mass) of constituent species} = \frac{\textbf{[Total mass of constituent species in one formula unit]} \times 100}{\textbf{Formula mass}}$$

For example, the percentage of iron (Fe) and FeO can be calculated in magnetic ore (Fe_3O_4) as follows:

Formula mass of Fe_3O_4

$= 3 \times 56 + 4 \times 16$

Atomic mass of Iron (Fe) = 56

Formula mass of FeO

$= 56 + 16 = 72$

$$\text{\% age of Fe} = \frac{3 \times 56 \times 100}{232};$$

$$\text{\% age of FeO} = \frac{72 \times 100}{232}$$

$= \textbf{72.4\%} = \textbf{31\%}$

SOLVED EXAMPLES

Example 12.19. *Ferric sulphate is a crystalline compound of iron. It is used in water and sewage treatment to help with the removal of suspended impurities. Calculate the mass percentage of iron, sulphur and oxygen in the compound.*

Solution. The formula of the compound is $Fe_2(SO_4)_3$.

The formula mass $= 2 \times 56 + 3(32 + 64)$

$= 400$ u

$$\therefore \text{ Percentage of iron} = \frac{2 \times 56 \times 100}{400} = \textbf{28\%}$$

$$\text{Percentage of sulphur} = \frac{3 \times 32 \times 100}{400} = \textbf{24\%}$$

$$\text{Percentage of oxygen} = \frac{3 \times 64 \times 100}{400} = \textbf{48\%}.$$

Example 12.20. *Calculate the mass percent of different elements present in ethyl alcohol (C_2H_5OH).*

Solution. Molar mass of ethanol

$= 2 \times 12.01 + 6 \times 1.008 + 16.00$

$= 46.068$ g

Mass percent of carbon

$$= \frac{24.02}{46.068} \times 100 = \mathbf{52.14\%}$$

Mass percent of hydrogen

$$= \frac{6.048}{46.068} \times 100 = \mathbf{13.13\%}$$

Mass percent of oxygen

$$= \frac{16.00}{46.068} \times 100 = \mathbf{34.73\%}$$

12.7 DETERMINATION OF FORMULA OF COMPOUNDS FROM EXPERIMENTAL AND GIVEN DATA

The **molecular formula** of a compound may be defined as *the formula which gives the actual number of atoms of various elements present in the molecule of the compound.* For example, the molecular formula of the compound glucose can be represented as $C_6H_{12}O_6$. A molecule of glucose contains six atoms of carbon, twelve atoms of hydrogen and six atoms of oxygen.

In order to find the molecular formula of a compound, you must first determine its empirical formula from the percentage composition.

The **empirical formula** of a compound may be defined as *the formula which gives the simplest whole number ratio of atoms of the various elements present in the molecule of the compound.* The empirical formula of the compound glucose ($C_6H_{12}O_6$), is CH_2O, which shows that C, H and O are present in the simplest ratio of 1 : 2 : 1.

Empirical formula mass of substance is equal to the sum of atomic masses of all the atoms in the empirical formula of the substance.

Relation between the Two Formulae

Molecular formula is a whole number multiple of empirical formula. Thus,

Molecular formula = Empirical formula × n

where $n = 1, 2, 3, ...$

$$n = \frac{\text{Molecular formula}}{\text{Empirical formula}} = \frac{\text{Molecular mass}}{\text{Empirical formula mass}}$$

STEPS FOR WRITING THE EMPIRICAL FORMULA

The percentage of the elements in the compound is determined by suitable methods and from the data collected, the empirical formula is determined by the following steps:

(*i*) *Divide the percentage of each element by its atomic mass*. This will give the relative number of moles of various elements present in the compound.

(*ii*) *Divide the quotients obtained in the above step by the smallest of them* so as to get a simple ratio of moles of various elements.

(*iii*) *Multiply the figures, so obtained by a suitable integer if necessary* in order to obtain a whole number ratio.

(*iv*) *Finally write down the symbols of the various elements side by side and put the above numbers as the subscripts to the lower right hand corner of each symbol.* This will represent the **empirical formula** of the compound.

STEPS FOR WRITING THE MOLECULAR FORMULA

(*i*) *Calculate the empirical formula as described above.*

(*ii*) *Find out the empirical formula mass by adding the atomic masses of all the atoms present in the empirical formula of the compound.*

(*iii*) *Divide the molecular mass (determined experimentally by some suitable method) by the empirical formula mass and find the value of n.*

(*iv*) *Multiply the empirical formula of the compound with n so as to find the* **molecular formula** *of the compound.*

The determination of the empirical formula and molecular formula of a compound involving the above steps is well illustrated by the following examples.

SOLVED EXAMPLES

Example 12.21. *Write the empirical formula of the compounds having molecular formulae:*

(*i*) C_6H_6	(*ii*) C_6H_{12}	(*iii*) H_2O_2
(*iv*) H_2O	(*v*) Na_2CO_3	(*vi*) B_2H_6
(*vii*) N_2O_4	(*viii*) H_3PO_4	(*ix*) Fe_2O_3
(*x*) C_2H_2.		

Solution. Empirical formula is the simplest whole number ratio of atoms in the molecule, therefore the empirical formula of the given compounds are:

(*i*) CH	(*ii*) CH_2	(*iii*) HO
(*iv*) H_2O	(*v*) Na_2CO_3	(*vi*) BH_3
(*vii*) NO_2	(*viii*) H_3PO_4	(*ix*) Fe_2O_3
(*x*) CH.		

Example 12.22. *A compound of nitrogen and oxygen is analyzed and a sample weighing 1.587 g is found to contain 0.483 g N and 1.104 g O. What is the empirical formula of the compound?*

Solution.

$$\underset{\text{Compound}}{\lfloor N, O \rfloor} \equiv \underset{0.483\text{ g}}{N} + \underset{1.104\text{ g}}{O}$$

$$\%\text{ age of N in compound} = \frac{0.483}{1.587} \times 100 = \mathbf{30.43}$$

$$\%\text{ age of O in compound} = \frac{1.104}{1.587} \times 100 = \mathbf{69.57}$$

Element	*% age*	*At. Mass*	*Relative No. of Moles*	*Simple Ratio*
N	30.43	14	$\frac{30.43}{14} = 2.176$	$\frac{2.176}{2.176} = 1.0$
O	69.57	16	$\frac{69.57}{16} = 4.348$	$\frac{4.348}{2.176} = 1.998 \approx 2.0$

Hence, the empirical formula = NO_2.

Example 12.23. *A compound has the following composition: Mg = 9.76%, S = 13.01%, O = 26.01%, H_2O = 51.22%. What is its empirical formula? [Mg = 24, S = 32, O = 16, H = 1]*

Solution.

Element	*Per-centage*	*At. Mass*	*Relative No. of Moles*	*Simple Ratio*	*Simplest Whole No. Ratio*
Magne-sium	9.76	24	$\frac{9.76}{24} = 0.406$	$\frac{0.406}{0.406} = 1$	1
Sulphur	13.01	32	$\frac{13.01}{32} = 0.406$	$\frac{0.406}{0.406} = 1$	1
Oxygen	26.01	16	$\frac{26.01}{16} = 1.625$	$\frac{1.625}{0.406} = 4$	4
Water	51.22	18 (mol. mass)	$\frac{51.22}{18} = 2.846$	$\frac{2.846}{0.406} = 7$	7

Hence, the empirical formula is $\mathbf{MgSO_4 \cdot 7H_2O}$.

Example 12.24. *What is the simplest formula of a compound which has the following percentage composition: Carbon 80%, Hydrogen 20%? If the molecular mass is 30, calculate its molecular formula.*

Solution. *Calculation of empirical formula:*

Element	*Per-centage*	*At. Mass*	*Relative No. of Moles*	*Simple Ratio*	*Simplest Whole No. Ratio*
C	80	12	$\frac{18}{12} = 6.66$	$\frac{6.66}{6.66} = 1$	1
H	20	1	$\frac{20}{1} = 20$	$\frac{20}{6.66} = 3$	3

∴ Empirical formula is CH_3.

Calculation of molecular formula:

Empirical formula mass

$$= 12 \times 1 + 1 \times 3 = 15$$

$$n = \frac{\text{Molecular mass}}{\text{Empirical formula mass}} = \frac{30}{15} = 2$$

Molecular formula

$$= \text{Empirical formula} \times 2$$

$$= CH_3 \times 2 = \mathbf{C_2H_6}.$$

Example 12.25. *An organic compound on analysis gave the following data: C = 57.82%, H = 3.6%, and the rest is oxygen. Its vapour density is 83. Find its empirical and molecular formula.*

Solution. *Calculation of empirical formula:*

Element	*Per-centage*	*At. Mass*	*Relative No. of Moles*	*Simple Ratio*	*Simplest Whole No. Ratio*
C	57.82	12	$\frac{57.82}{12} = 4.818$	$\frac{4.818}{2.4} \simeq 2$	4
H	3.60	1	$\frac{3.60}{1} = 3.60$	$\frac{3.6}{2.4} = 1.5$	3
O	38.58	16	$\frac{38.58}{16} = 2.40$	$\frac{2.4}{2.4} = 1$	2

∴ Empirical formula is $C_4H_3O_2$.

Calculation of molecular formula:

Empirical formula mass

$$= 12 \times 4 + 1 \times 3 + 2 \times 16 = 83$$

Molecular mass $= 2 \times \text{V.D.} = 2 \times 83 = 166$

$$n = \frac{\text{Molecular mass}}{\text{Empirical formula mass}} = \frac{166}{83} = 2$$

Molecular formula = Empirical formula × n

$= C_4H_3O_2 \times 2 = \mathbf{C_8H_6O_4}$.

12.8 DETERMINATION OF FORMULA THROUGH EXPERIMENT

Two basic ways by which the amount of each element in a given mass of a compound is found is by **synthesis** or **analysis**.

- The process of synthesis is the formation of a compound from its elements or simpler compounds.
- Analysis involves the breaking down of a compound into its elements or smaller compounds.

The knowledge of the mass of each element in a known mass of the compound is one of the keys to the molecular formula determination.

ACTIVITY 12.2 **Determining the empirical formula of magnesium oxide**

1. Weigh a clean, dry crucible with a lid on (m_1).
2. Cut a piece of magnesium ribbon about 15 cm long. Use sandpaper to remove the oxide coating.
3. Fold the ribbon loosely from one end to fit into the crucible. Put the lid on and weigh the whole crucible again (m_2).
4. Put the crucible on a pipe-clay triangle and heat it strongly until it is red-hot (Fig. 12.2). Using tongs, carefully open the lid slightly for a second or two to allow air into the crucible. Quickly put the lid back in place to make sure the product powder does not escape. Repeat this process until the reaction is complete.
5. Remove the source of heat and let the crucible cool. Add a few drops of water to the product until the pungent smell produced ceases.
6. Heat the crucible, allow it to cool and then weigh it. Repeat this process until constant mass is obtained (m_3).
7. Calculate in terms of m_1, m_2 and m_3 the mass of
   (*i*) magnesium metal;
   (*ii*) oxide of magnesium formed;
   (*iii*) oxygen that combined with the magnesium metal.
8. Calculate the empirical formula of the oxide of magnesium formed.
   (**Relative atomic mass:** Mg = 24, O = 16)

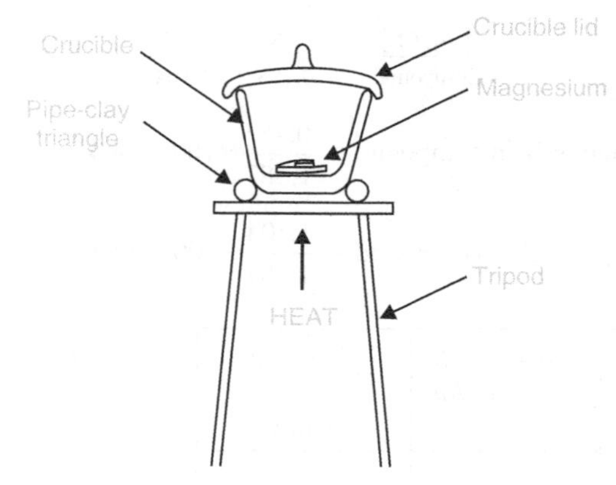

Fig. 12.2. Combustion of magnesium and oxygen.

SOLVED EXAMPLE

Example 12.26. *3.6 g of magnesium metal were reacted completely in a closed container of pure nitrogen gas to give 5 g of a compound of these two elements. Calculate the empirical formula of this compound. (Relative atomic mass, A_r (Mg = 24, N = 14))*

Solution.

Mass of magnesium (Mg) = 3.6 g
Mass of compound = 5.0 g
Mass of nitrogen (N) = (5.0 – 3.6) g = 1.4 g
Number of moles of each element

$\frac{\text{Mass of element}}{\text{Atomic mass}}$ Mg = $\frac{3.6}{24} = 0.15$; N = $\frac{1.4}{14} = 0.10$

Relative amount of element $\frac{0.15}{0.10} = 1.5$; $= \frac{0.1}{0.1} = 1$

Simple whole number ratio of elements: $1.5 \times 2 = 3$; $= 1 \times 2 = 2$

∴ Empirical formula = $\mathbf{Mg_3N_2}$

ACTIVITY 12.3 **Determining the empirical formula of hydrated magnesium tetraoxosulphate(VI), $MgSO_4 \cdot xH_2O$**

1. Weigh a clean, dry evaporating basin (m_1)
2. Add about 5 g of the hydrated salt, $MgSO_4 . xH_2O$ to the basin and weigh the whole (m_2).

3. Heat the basin gently and stir the solid at the same time to avoid spitting.
4. Continue to heat until no more change occurs. Scrape off any solid stuck to the stirrer into the basin with a clean dry spatula.

   Weigh the basin with the solid.
5. Repeat the heating and weighing until a constant mass of the solid is obtained (m_3).

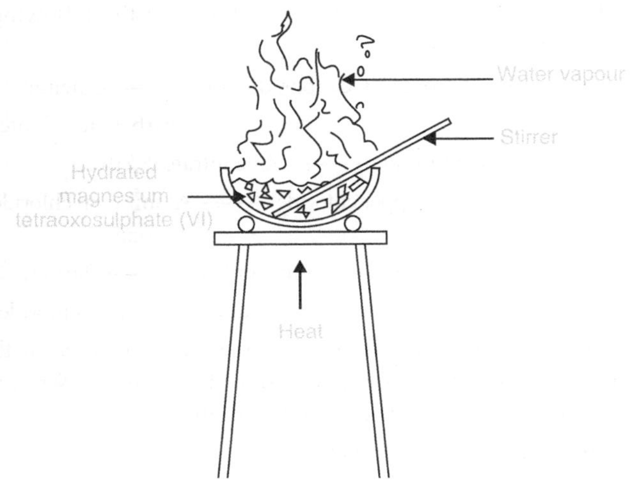

Fig. 12.3. Dehydration of hydrated magnesium tetraoxosulphate(VI).

6. (*a*) Calculate the
   (*i*) mass of the hydrated salt;
   (*ii*) mass of the dehydrated salt;
   (*iii*) mass of water lost by the hydrated salt;
   (*iv*) amount of the dehydrated salt in mols;
   (*v*) amount of water lost in mols.

   (*b*) (*i*) Calculate the empirical formula of the hydrated salt, and give the value of *x*.
   (*ii*) What is the name given to *x*?

EVALUATE YOURSELF

I. Objective Type Questions

Select the most appropriate choice from the options given as (a),(b),(c) and (d) after each question:

1. Which of the following compounds has the same empirical formula as that of glucose?

   (*a*) CH_3CHO (*b*) CH_3COOH
   (*c*) CH_3OH (*d*) C_2H_6.

2. Two elements X (atomic mass = 50) and Y (atomic mass = 16) combine to give a compound having 32% Y. The formula of the compound is

   (*a*) X_2Y_3 (*b*) XY
   (*c*) XY_2 (*d*) X_3Y_4.

3. The vapour density of a certain gas X is 4 times that of Y. The molecular mass of X is M. The molecular mass of Y is

   (*a*) 4 M (*b*) 0.5 M
   (*c*) 0.25 M (*d*) 0.33 M.

4. A symbol not only represents the name of an element but also represents

   (*a*) its atomic number (*b*) one gram-atom
   (*c*) its atomicity (*d*) reactivity.

5. The balancing of equations is based upon which of the following law?

   (*a*) Law of multiple proportions
   (*b*) Law of conservation of mass
   (*c*) Boyle's law
   (*d*) Law of reciprocal proportions.

6. Simplest formula of the compound containing nitrogen (14 g) for each 40 g of oxygen is

   (*a*) NO (*b*) N_2O_3
   (*c*) N_2O_4 (*d*) N_2O_5.

7. Analysis shows that a binary compound of X (at. mass = 10) and Y (at. mass = 20) contains 50% X. The formula of the compound is

   (*a*) XY (*b*) X_2Y
   (*c*) XY_2 (*d*) X_2Y_3.

8. Hydrogen phosphate of certain metal has a formula $MHPO_4$, the formula of its metal chloride is

   (*a*) MCl (*b*) MCl_3
   (*c*) MCl_2 (*d*) MCl_4.

9. An organic compound containing C, H, N gave the following analysis:

   C = 40%, H = 13.33%, N = 46.47%. Its empirical formula is

   (*a*) $C_2H_2N_2$ (*b*) CH_5N
   (*c*) CH_4N (*d*) C_2H_7N.

10. In the reaction, $4NH_3 + 5O_2 \longrightarrow 4NO + 6H_2O$. When one mole of ammonia and one mole of oxygen are made to react to completion, then

    (*a*) 1.0 mole of H_2O is produced
    (*b*) All the oxygen is consumed
    (*c*) 1.0 mol of NO is formed
    (*d*) All the ammonia is consumed.

11. 100 g $CaCO_3$ is treated with 1 litre of 1 N HCl. What would be the weight of CO_2 liberated after the completion of the reaction?

    (*a*) 5.5 g (*b*) 11 g
    (*c*) 22 g (*d*) 33 g
    (*e*) 44 g.

12. In the reaction, $4NH_3 + 5O_2 \longrightarrow 4NO + 6H_2O$, when one mole of ammonia and one mole of oxygen are made to react to completion, then
(*a*) 1.0 mole of H_2O is produced
(*b*) all the oxygen is consumed
(*c*) 1.5 mole of NO is formed
(*d*) all the ammonia is consumed.

13. The number of moles of $KMnO_4$, that will be needed to react with one mole of sulphite ion in acidic solution is
(*a*) 1 (*b*) 3/5
(*c*) 4/5 (*d*) 2/5.

14. In the reaction,
$$2\,Al(s) + 6\,HCl(aq) \longrightarrow 2\,Al^{3+}(aq) + 6\,Cl^-(aq) + 3\,H_2(g)$$
(*a*) 6 L HCl (*aq*) is consumed for every 3 L $H_2(g)$ produced
(*b*) 33.6 L $H_2(g)$ is produced regardless of temperature and pressure for every mole of Al that reacts
(*c*) 67.2 L $H_2(g)$ at STP is produced for every mole of Al that reacts
(*d*) 11.2 L $H_2(g)$ at STP is produced for every mole of HCl(*aq*) consumed.

II. Fill in the Blanks

15. Complete the following sentences by supplying appropriate words:
(*i*) Hydrogen and oxygen are present in water in the ratio ...... by mass.
(*ii*) A chemical compound always contains the same elements combined together in the same proportion by mass. This law is known as ....
(*iii*) The number of moles of KI required to produce 0.4 moles of K_2HgI_4 by its reaction with $HgCl_2$ in aqueous solution is ......

III. Discussion Questions

16. What do you understand by the chemical formula of a substance?

17. What is an ion? Give the formulae of the following ions:
(*i*) Nitrate (*ii*) Nitride
(*iii*) Phosphate (*iv*) Ammonium
(*v*) Manganese.

18. Write the formulae and names of the compounds formed between:
(*i*) Potassium and iodide ions
(*ii*) Sodium and sulphide ions
(*iii*) Aluminium and chloride ions.

19. Write the formulae of the compounds formed by combination of:
(*i*) Al^{3+} and Cl^- ions (*ii*) Ba^{2+} and PO_4^{3-} ions
(*iii*) Cr^{3+} and SO_4^{2-} (*iv*) Hg^{2+} and I^- ions
(*v*) Ni^{2+} and S^{2-} ions (*vi*) NH_4^+ and CO_3^{2-} ions.

20. Write the formulae of the following compounds:
(*i*) Sodium sulphate (*ii*) Ammonium phosphate
(*iii*) Calcium hydride (*iv*) Lithium nitride.

21. Write the formulae of the compounds formed by:
(*i*) Cr^{3+} and F^- (*ii*) Hg^{2+} and S^{2-}
(*iii*) Pb^{2+} and PO_4^{3-}.

22. What is a chemical equation? What is a balanced chemical equation? Why should the chemical equations be balanced?

23. Write the balanced chemical equations for the following reactions:
(*i*) Calcium hydroxide + Carbon dioxide $\longrightarrow$ Calcium carbonate + Water
(*ii*) Zinc + Silver nitrate $\longrightarrow$ Zinc nitrate + Silver.
(*iii*) Aluminium + Copper chloride $\longrightarrow$ Aluminium chloride + Copper.
(*iv*) Barium chloride + Potassium sulphate $\longrightarrow$ Barium sulphate + Potassium chloride.

24. When methyl alcohol (CH_3OH) burns in air, it reacts with oxygen from air to form carbon dioxide and water. Write a balanced chemical equation for the reaction.

25. Balance the following equations:
(*i*) $NH_3 + O_2 \longrightarrow NO + H_2O$
(*ii*) $HNO_3 \longrightarrow H_2O + NO_2 + O_2$
(*iii*) $H_2S + SO_2 \longrightarrow S + H_2O$
(*iv*) $BaCl_2 + Al_2(SO_4)_3 \longrightarrow AlCl_3 + BaSO_4$.

26. In the commercial manufacture of nitric acid, how many moles of NO_2 produces 7.33 mol of HNO_3 in the reaction;
$$3NO_2(g) + H_2O(l) \longrightarrow 2HNO_3 + NO(g).$$

27. Zinc and hydrochloric acid react according to the reaction:
$$Zn(s) + 2HCl(aq) \longrightarrow ZnCl_2(aq) + H_2(g)$$
If 0.30 mol Zn are added to hydrochloric acid containing 0.52 mol HCl, how many moles of H_2 are produced?

28. Chlorine is prepared according to the reaction:
$$4HCl(aq) + MnO_2(s) \longrightarrow 2H_2O(l) + MnCl_2(aq) + Cl_2(g)$$
How many grams of HCl react with 5.0 g of manganese dioxide? (At. mass of Mn = 55).

29. 0.6 mol of Cu_2S is roasted in excess of oxygen to yield Cu and SO_2, according to the reaction.
$$Cu_2S + O_2 \longrightarrow 2Cu + SO_2$$
Calculate the mass of Cu formed.

30. 80 g of H_2 is reacted with 80 g of O_2 to form water. Find out the mass of water obtained. Which substance is the limiting reagent?

31. A chemical compound is found to have the following composition: K = 42.5%; Fe = 15.2%; C = 19.5% and N = 22.8%. Calculate the simplest formula of the compound.

32. An organic compound on analysis is found to have the following composition: C = 41.4%; H = 5.75%; N = 16.08% and the rest is oxygen. Find out the empirical formula of the compound.

33. 1.615 g of an anhydrous salt was placed in moist air. After a few days it was found to be 2.875 g. Assuming that complete hydration has taken place, calculate the simplest formula of the hydrated salt. The composition of the anhydrous salt is Zn = 40.6%; S = 19.8%; O = 39.6%.

34. What is the empirical formula of the mineral having the composition CaO = 48%; P_2O_5 = 41.3%; $CaCl_2$ = 10.7%?

35. The percentage composition of Mohr's salt is as follows: Fe^{2+} = 14.32%; NH_4^+ = 9.20%, SO_4^{2-} = 49%; Rest being water of crystalisation.

Find the empirical formula of Mohr's salt.

36. An organic compound has the following percentage composition C = 12.36%; H = 2.13%; Br = 85%. Its V.D. is 94. Find its molecular formula.

37. A compound contains 32% carbon, 4% hydrogen and the rest oxygen. Its vapour density is 75. Calculate the empirical and molecular formula.

38. A crystalline salt on being rendered anhydrous loses 46% of its mass. The percentage composition of anhydrous salt is:

Al = 10.50%, K = 15.1%, S = 24.8%, O = 49.6%.

Find the simplest formula of the anhydrous and crystalline salt.

39. Calculate the empirical formula of a compound whose percentage composition is:

C = 21.9%, H = 4.6%, Br = 73.4%.

40. An acid of molecular mass 104 contains 34.6% carbon, 3.85% hydrogen and the rest is oxygen. Calculate the molecular formula of the acid.

41. State and explain the law of multiple proportions.

42. State and explain the law of definite proportions.

43. How is empirical formula of a compound related to its molecular formula?

44. What are the various steps in calculating the molecular formula of a compound from the percentage composition of the compound?

45. Can two different compounds have the same empirical formula? Illustrate your answer with two examples.

46. Can two different compounds have the same molecular formula? Illustrate your answer with two examples.

47. What do you understand by the terms?

(*i*) Limiting reagent (*ii*) Stoichiometry

48. Calculate the amount of carbon dioxide that could be produced when

(*i*) 1 mole of carbon is burnt in air

(*ii*) 1 mole of carbon is burnt in 16 g of dioxygen.

49. Determine the molecular formula of an oxide of iron in which the mass percent of iron and oxygen are 69.9 and 30.1 respectively. Given molar mass of oxide is 159.8 g mol^{-1} (at. mass: Fe = 55.85, O = 16.00 u).

Answers

I. Objective Type Questions

1. (*b*)	**2.** (*a*)	**3.** (*c*)
4. (*b*)	**5.** (*b*)	**6.** (*b*)
7. (*b*)	**8.** (*c*)	**9.** (*c*)
10. (*b*)	**11.** (*c*)	**12.** (*b*)
13. (*d*)	**14.** (*d*).	

II. Fill in the Blanks

15. (*i*) 1 : 8 (*ii*) law of constant composition (*iii*) 1.6.

SECTION-4

UNIT-13

NUCLEAR CHEMISTRY

It is a well known fact that atoms contain three fundamental subatomic particles namely: *electrons, protons* and *neutrons.* Protons and neutrons constitute the **nucleus** whereas electrons are present in the extranuclear region of the atom. Chemical transformations involve the changes in the valence electrons of the participating atoms while their atomic nuclei remain unaffected. In the beginning of the 20th century, another kind of transformation, which involve changes in the atomic nucleus, were observed by scientists. These transformations are called **nuclear reactions.** The nuclear reactions have become tremendous source of energy all over the world both for peaceful and destructive purposes. The study of nuclear reactions led to the development of a new branch of chemistry called *nuclear chemistry.* **Nuclear chemistry** may, therefore, be defined as *the branch of chemistry which deals with the study of the atomic nucleus and nuclear changes.*

Specific Objectives

After studying this unit, the student should be able to:

- explain some terms associated with the nucleus of an atom
- explain radioactivity
- identify the types of nuclear radiation and describe their nature
- identify stable and unstable nuclide
- explain radioactive decay
- differentiate between nuclear fission and fusion reactions
- differentiate between nuclear and chemical reactions
- describe some applications of nuclear chemistry.

13.1 NUCLEUS AND ITS COMPOSITION

NUCLEUS AND NUCLEAR PARTICLES

Nucleus is a small positively charged part of an atom at the centre where the entire mass of the atom is concentrated. The existence of nucleus was first of all proposed in 1911 on the basis of series of scattering experiments conducted by Rutherford and his co-workers. These experiments were conducted to understand the arrangement of sub-atomic particles (electrons, protons and neutrons) in an atom. Experimental observations led to the conclusion that

(*i*) the protons and neutrons are present in the nucleus.

(*ii*) the electrons are present in empty space around the nucleus.

(*iii*) the number of protons in the nucleus or electrons present outside the nucleus is known as **atomic number**.

(*iv*) *the sub-atomic particles (protons and neutrons) of the nucleus* are collectively called **nucleons.** The total number of nucleons is denoted by **A** and is called **mass number** of the nucleus

Mass Number (**A**) = *No. of protons* (**p**) + *No. of neutrons* (**n**)

Representation of nuclide

The nucleus of any atom may be represented by specifying the *atomic number* as a subscript at the left hand bottom of the atomic symbol and the mass number as a superscript at the left hand top of the symbol. For example, the nucleus of most common isotope of carbon is designated as $^{12}_{6}C$. Such symbols are referred to as **nuclides.**

Nuclear Dimensions

The radius of a nucleus is of the order of $10^{-12} - 10^{-13}$ cm as compared to that of the atom which is of the order of 10^{-8} cm. The radii of various nuclei (assuming them to be spheres) can be calculated from the following empirical relation

$$r = R_0 A^{1/3}$$

where, r is the radius of the nucleus ; A is the mass number and R_0 is a constant whose value has been found to be 1.4×10^{-13} cm.

Add to Your Knowledge

- Nuclear dimensions are expressed generally in Fermi units, **1 Fermi = 10^{-13} cm.**
- Area of cross-section of nucleus is expressed in terms of the unit called *barn* **(1 barn = 10^{-24} cm^2).**
- Radius of nucleus is about 10^5 times smaller than the atom.
- Density of nucleus is very high. It is of the order of 10^{14} g cm^{-3} (or 10^{17} kg m^{-3} or 10^8 tonnes cm^{-3}).
- Volume of the nucleus is of the order of 10^{-38} cm^3.

ISOTOPES, ISOBARS, ISOTONES

Isotopes. These are the atoms having same atomic number but different mass numbers. They contain *same number of protons but different number of nucleons.*

Some examples are ^1_1H, ^2_1H, ^3_1H are isotopes of hydrogen. Similarly, $^{215}_{84}Po$ and $^{211}_{84}Po$ are isotopes of polonium.

Isobars. These are the atoms having different atomic numbers but same mass number. They contain same *number of nucleons but different number of protons.*

Some examples are $^{40}_{20}Ca$ and $^{40}_{20}Ar$; $^{211}_{82}Pb$ and $^{211}_{83}Bi$.

Isotones. These are *the atoms having different number of nucleons* but same number of *neutrons.* Some examples are $^{30}_{14}Si$, $^{31}_{15}P$, $^{32}_{16}S$.

Do You Know?

- **Isodiaphers.** These are the species having different atomic numbers as well as mass numbers but same isotopic number (A – 2Z) or (n – Z). For example, $^{19}_{9}F$ and $^{39}_{19}K$ are isodiaphers because the isotopic number of both is 1.
- **Nuclear isomers.** Nuclear *species having the same atomic number, the same mass number but possessing different radioactive properties* are called **nuclear isomers.** The difference in the properties of nuclear isomers is due to the difference in their energy states.

13.2 RADIOACTIVITY

In 1896, Henri Becquerel was performing experiments with uranium compounds. He accidently placed a crystal of potassium uranyl sulphate [$K_2UO_2(SO_4)_2.2H_2O$] over photographic plate wrapped with usual black paper. On developing the photographic plate, he found the shadow of the crystal on the photographic plate. He repeated this phenomenon a number of times and came to a conclusion that the crystal of potassium uranyl sulphate had emitted some mysterious rays which could penetrate the black paper and had affected the photographic plate. Later on, it was found that thorium compounds also emitted similar rays.

This phenomenon of spontaneous emission of radiations by an element or its compound was given the name **radioactivity** and the **substances *which exhibit this phenomenon*** were called **radioactive substances.** Once discovered, radioactivity became a popular subject for scientific investigations.

Two very popular names associated with the study of radioactivity are *Madam Curie* and her husband *Pierre Curie* who discovered the radioactive elements, *polonium, thorium, radium* and *actinium*. Some of these elements possessed radioactive power much larger than uranium. In subsequent years a very large number of radioactive elements have been discovered.

NATURAL AND ARTIFICIAL RADIOACTIVITY

Radioactivity can be broadly classified into two categories: *natural radioactivity* and *artificial* or *induced radioactivity*.

(*i*) *If a substance emits radiations by itself,* it is said to possess **natural radioactivity.**

(*ii*) *If a substance does not possess radioactivity* but *starts emitting radiations on exposure to rays from a natural radioactive substance,* it is said to possess **induced** or **artificial radioactivity.**

DIFFERENCE BETWEEN NATURAL AND ARTIFICIAL RADIOACTIVITY

Some points of difference between natural and artificial radioactivity are as follows:

Natural Radioactivity	*Artificial Radioactivity*
1. It involves the spontaneous disintegration of an unstable nuclei with emission of α or β particles or γ-radiations giving rise to new nuclide.	1. Stable nuclei are bombarded with high energy particles to produce radioactive nuclides.
2. It cannot be controlled.	2. It can be controlled by controlling the speed of the bombarding projectiles.
3. It is shown by heavy elements *i.e.*, elements with high atomic number and mass number.	3. It can be induced even in the lighter elements.

UNITS OF RADIOACTIVITY

The SI unit of radioactivity is, *Becqueral* (**Bq**) which is defined as *one disintegration per second (dps)*. Earlier, radioactivity was expressed in terms of *curies* (***Ci***). **One curie** refers to the activity of one gram of radium, and is equal to *3.7×10^{10} disintegrations per second.*

$$\therefore \quad \mathbf{1Ci = 3.7 \times 10^{10}\ dps = 3.7 \times 10^{10}\ Bq.}$$

1 Millicurie (m Ci) = 3.7×10^7 dps and ***1 Microcurie (μ Ci) = 3.7×10^4 dps.***

A more recent unit of radioactivity is **Rutherford (Rd)**.

$$\mathbf{1\ Rd = 10^6\ dps.}$$

13.3 TYPES OF NUCLEAR RADIATIONS AND THEIR NATURE

The nature of the mysterious invisible rays emitted by radioactive substances was determined by Ernest Rutherford. He placed a sample of uranium mineral in a lead box (Fig. 13.1). The lead absorbed all other radiations except those which pass through the slit to constitute the beam. The radiations emitted by the uranium mineral could be resolved into three types by the application of electrical field.

(*a*) The rays which deflected slightly towards a –ve electrode were named as ***alpha*** **(α)** ***rays.***

(*b*) The rays which deflected towards a +ve electrode were named as ***beta*** (β) ***rays.***

(*c*) The rays which remained undeflected were named as ***gamma*** (γ) ***rays.***

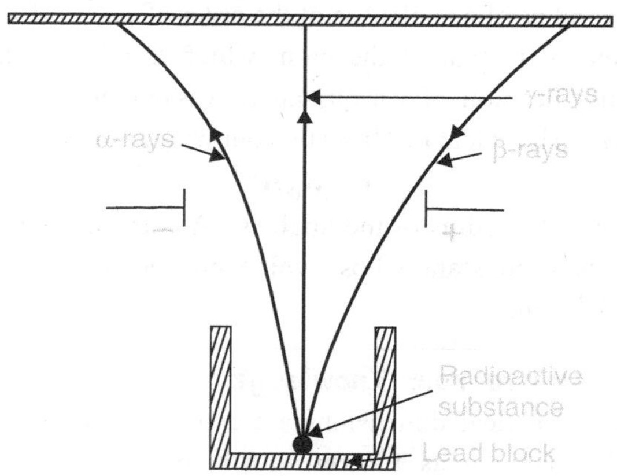

Fig. 13.1. Behaviour of α, β and γ-rays in electrical field

Later on, it was established that these three types of radiations were given off from all radioactive substances. Let us study the properties of these three types of radiations.

ALPHA (α) RAYS

(*i*) *Nature.* The study of α-rays in electrical and magnetic field has shown that these rays consist of material particles carrying 2 units of positive charge and a mass equal to 4 u. In other words, α-particles are helium nuclei [He^{2+}].

(*ii*) *Velocity.* They are shot out with a velocity about 1/10th the velocity of light.

In fact, the actual speed of α-particles depends upon the nature of the substance from which they are ejected. Energy range of α-particles is 4.9 MeV.

(*iii*) *Penetrating power.* They do not possess high penetrating power because of their large mass. They can penetrate a mica sheet, a thin metal foil or through a few centimetres air. They can penetrate through a sheet of aluminium up to a thickness of 0.02 cm.

(*iv*) *Ionizing power.* α-particles can ionize the gas through which they pass. This is because of the fact that they knock off electrons from neutral gas molecules on colliding with them resulting in the ionization of gas.

(*v*) *Effect on zinc sulphide.* Due to the high kinetic energy, they produce luminosity when they strike a zinc sulphide plate.

(*vi*) *Effect on photographic plate.* α-particles also affect the photographic plate.

BETA (β) RAYS

(*i*) *Nature.* β-rays have been found to consist of particles having the same charge and mass as electrons. Their *(e/m)* ratio is the same as that of electrons. These particles are designated as $_{-1}^{0}e$ or $_{-1}^{0}\beta$.

(*ii*) *Velocity.* β-particles move much faster in comparison to α-particles. They travel with speed almost approaching the speed of light. Energy range is 0.5 to 2 MeV.

(*iii*) *Penetrating power.* β-particles have far more penetrating power due to small mass and high velocity. They can penetrate through a sheet of aluminium having thickness up to 0.2 cm.

(*iv*) *Ionizing power.* β-rays have poor ionizing power in spite of their high speed. It is because of the fact that they possess low kinetic energy due to their extremely small mass.

(*v*) *Effect on zinc sulphide and photographic plate.* β-particles have very little effect on zinc sulphide due to their low kinetic energy. However, their photographic activity is greater than that of α-particles.

GAMMA (γ) RAYS

(*i*) *Nature.* These rays do not consist of material particles but they are electromagnetic waves having wavelength of the order of 10^{-8} to 10^{-11} cm. They are electrically neutral.

(*ii*) *Velocity.* They travel with the speed of light. Their energy range is 0.1 – 2 MeV.

(*iii*) *Penetrating power.* They have high penetrating power because of their high velocity and small wavelength. These rays can penetrate through a sheet of aluminium having thickness up to 100 cm.

(*iv*) *Ionizing power.* They have very poor ionising power because of their non-material nature.

(*v*) *Effect on zinc sulphide and photographic plate.* They produce very little effect on zinc sulphide or photographic plate.

The characteristics of α, β and γ-rays have been summed up in Table 13.1.

Table 13.1. Characteristics of α, β and γ-rays

Property	*α-rays*	*β-rays*	*γ-rays*
1. Mass	1. 6.67×10^{-27} kg or 4 amu.	1. 9.11×10^{-31} kg	1. Negligible
2. Charge	2. + 2 units	2. – 1 units	2. 0
3. Identity	3. Helium nuclei (He^{2+})	3. Electrons	3. High energy radiations
4. Velocity	4. Nearly $\frac{1}{10}$ th that of light	4. Nearly same as that of light	4. Same as that of light
5. Effect of electric and magnetic fields	5. Deflected towards negative pole	5. Deflected towards positive Pole	5. Not deflected
6. Penetrating power	6. Small	6. Large, 100 times that of α-rays	6. Very large, 10000 times that of α-rays
7. Effect on photographic plate and zinc sulphide	7. Affected more strongly than by β and γ-rays	7. Effect is less than α-rays	7. Least effect

NUCLEAR CHANGES DURING EMISSION OF α, β, γ, RADIATIONS

It has been pointed out earlier that a radioactive element emits, α, β and γ-rays. The emission of α and β-rays bring some changes in the nucleus of the atom and the new element formed as a result of these changes is called **daughter element.**

The equations representing the nuclear changes are called **nuclear chemical equations.** In writing the nuclear chemical equation, the following points must be kept in mind:

(*i*) *Sum of the mass numbers of various reactant species must be equal to the sum of the mass numbers of the product species.*

(*ii*) *Sum of the charge numbers (atomic numbers) of reactant species and that of the product species must be equal.*

Species	*Notation*
α-particle	${}^{4}_{2}He$
Proton	${}^{1}_{1}H$
Deutron	${}^{2}_{1}H$
Neutron	${}^{1}_{0}n$
Electron or β-particle	${}^{0}_{-1}e$ or β^-
Positron	${}^{0}_{+1}e$ or β^+

The notations of some of the species that participate in nuclear reactions are given below in tabular form.

Let us now proceed to discuss the changes produced by α, β and γ emission.

Emission of α-rays. When a nucleus emits an α-particle (${}^{4}_{2}He$) its *atomic number* is reduced by *two units* and its *mass number* is reduced by *four units.* The daughter element, thus, formed occupies a position in the periodic table *two places left* of the parent element. For example, emission of α-particle from uranium-238 (${}^{238}_{92}U$) produces a thorium nucleus (${}^{234}_{90}Th$). The above nuclear change can be represented as:

$${}^{238}_{92}U \longrightarrow {}^{234}_{90}Th + {}^{4}_{2}He.$$

Some more examples:

$${}^{226}_{88}Ra \longrightarrow {}^{222}_{86}Rn + {}^{4}_{2}He$$

$${}^{215}_{84}Po \longrightarrow {}^{211}_{82}Pb + {}^{4}_{2}He$$

Emission of β-rays. β-particle is merely an electron. When a β-particle is emitted, the nucleus of the new element formed possesses the same mass but the nuclear charge or atomic number increases by 1 unit. Beta particle emission is because of the result of decay of neutron into proton and electron. The daughter element formed is an isobar of the parent element.

$${}^{1}_{0}n \longrightarrow {}^{1}_{1}p + {}^{0}_{-1}e$$

Some examples are given below:

$${}^{214}_{82}Pb \longrightarrow {}^{214}_{83}Bi + {}^{0}_{-1}e$$

$${}^{234}_{90}Th \longrightarrow {}^{234}_{91}Pa + {}^{0}_{-1}e$$

Emission of γ-rays. Gamma rays are emitted due to secondary effects. With the emission of an α-or β-particle, the nucleus left behind remains in the excited state due to recoil. The excess of energy is released in the form of γ-rays. As γ-rays are electromagnetic radiations with no charge and no mass, their emission from a radioactive element does not produce a new element.

Detection of Radiations

Nuclear radiations are very useful to us, but at the same time these can also be very dangerous. The human senses cannot detect ionizing radiations. The detection of these radiations is done by the use of a Geiger-Müller counter (Fig. 13.2), the cloud chamber, spark counters and scintillation counters.

The Geiger-Müller tube

Radiations, from a radioactive source, are passed into a gas-filled tube (usually argon at low pressure) through a thin mica window. The radiation produces positive ions and electrons in the gas by ionization. The positive ions move

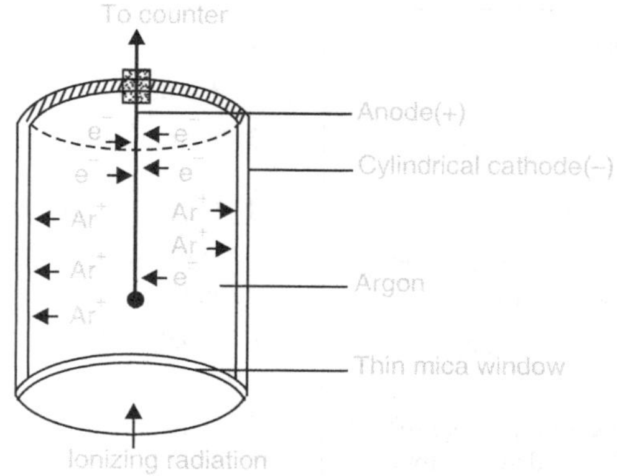

Fig. 13.2. Detection of radiation using Geiger-Müller counter.

towards the wall of the counter, which is the cathode, and electrons (negative) go to the central anode. Movement of these particles allows a pulse of electric current to pass through a conductor to the counter. The electrical pulse is amplified and counted as light flashes, clicks, or shown as meter readings. The counter does not distinguish between α, β or γ radiation.

CALCULATING NUMBER OF α AND β-PARTICLES

SOLVED EXAMPLES

Example 13.1. *$^{238}_{92}U$ emits one α-particle followed by the emission of two β-particles as shown*

$$^{238}_{92}U \xrightarrow{-\alpha} X \xrightarrow{-\beta} Y \xrightarrow{-\beta} Z$$

Label the above elements in the consecutive nuclear reactions.

Solution. Indicating the atomic numbers and mass number of the various elements, we have

$$^{238}_{92}\mathrm{U} \xrightarrow{-\alpha} {}^{234}_{90}\mathrm{X} \xrightarrow{-\beta} {}^{234}_{91}\mathrm{Y} \xrightarrow{-\beta} {}^{234}_{92}\mathbf{Z}$$

We find that the mass number of Z is 234 while the atomic number is 92. Thus, Z is an **isotope** of $^{238}_{92}\mathrm{U}$. Thus, **X is $^{238}_{90}\mathbf{Th}$; Y is $^{234}_{91}\mathbf{Pa}$ and Z is $^{234}_{92}\mathbf{U}$.**

Example 13.2. *How many α and β-particles are emitted in passing down from $^{232}_{90}Th$ to $^{208}_{82}Pb$?*

Solution. Let the number of α-particles emitted be x and β-particles be y. The net nuclear reaction can be represented as:

$$^{232}_{90}\mathrm{Th} \longrightarrow {}^{208}_{82}\mathrm{Pb} + x\,{}^{4}_{2}\mathrm{He} + y\,{}^{0}_{-1}e \qquad ...(i)$$

Balancing the mass numbers (A) on both sides of equation (*i*)

$$232 = 208 + 4x + 0 \times y = 208 + 4x$$

or $$x = \frac{232 - 208}{4} = \frac{24}{4} = 6$$

Thus, **the number of α-particles emitted = 6.**

Now balance the atomic number on both sides of equation (*i*)

$$90 = 82 + 2x + y(-1) = 82 + 2 \times 6 - y = 82 + 12 - y$$

or $$y = 82 + 12 - 90 = 4$$

Thus, **the number of β-particles emitted = 4.**

CAUSE OF RADIOACTIVITY

It has been found that the stability of a nucleus depends upon its *neutron to proton (n/p) ratio.* If we plot the number of neutrons (n) vs. number of protons (p) for nuclei of various elements as shown in Fig. 13.3, we find that:

(*i*) *light nuclei (A < 20) have n/p ratio close to unity*

(*ii*) *for heavy nuclei (A > 20) the n/p ratio increases progressively due to the dominance of number of neutrons.* It is evident that most of the stable nuclei (non-radioactive) have n/p ratio between **1–1.6.** These nuclei lie in the shaded region of the plot which is also called ***stability belt*** or ***stability zone.***

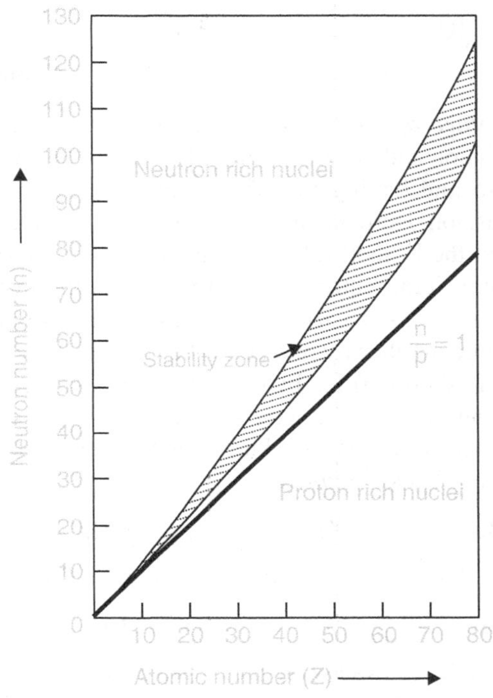

Fig. 13.3. A plot of number of neutrons (N) against the atomic number (Z) for nuclei of various elements.

The elements whose nuclei do not fall within the stability zone are said to be unstable. The unstable nuclei, whose n/p ratio is either less than 1 or greater than 1.6, disintegrate giving out α, β or γ-rays in their attempt to attain stability. The process of disintegration continues till the n/p ratio falls within the stability limit.

Add To Your Knowledge

(*i*) *n/p ratio* ***decreases,*** *in general, by the emission of β-particle*

$$^{14}_{6}\mathbf{C} \longrightarrow {}^{14}_{7}\mathbf{N} + {}^{0}_{-1}\mathbf{e}$$

(n/p = 1.33) (n/p = 1) β-particle

$$^{29}_{13}\mathbf{Al} \longrightarrow {}^{29}_{14}\mathbf{Si} + {}^{0}_{-1}\mathbf{e}$$

$(n/p = 1.23)$ $(n/p = 1.07)$ β-Particle

(*ii*) *n/p ratio* ***increases*** *either by the emission of α-particle or by the emission of a positron or by the capture of an orbital electron (K-electron).*

(*a*) α-***Decay***

$$^{238}_{92}\mathbf{U} \longrightarrow {}^{234}_{90}\mathbf{Th} + {}^{4}_{2}\mathbf{He}$$

$(n/p = 1.5)$ $(n/p = 1.6)$ α-Particle

(*b*) ***Positron emission***

$$^{11}_{6}\mathbf{C} \longrightarrow {}^{11}_{5}\mathbf{B} + {}^{0}_{+1}\mathbf{e}$$

$(n/p = 0.83)$ $(n/p = 1.2)$ Positron

(*c*) ***Capture of K-electron.*** This process is not very common and occurs in case of nuclei with low n/p ratio and having insufficient energy for positron emission. Here, the nucleus may capture an electron from the K shell (closest to nucleus) and convert a proton into neutron which remains in the nucleus ($p^+ + e^- \longrightarrow n$). This causes a decrease of the atomic number by one unit. The vacancy created in the K shell is filled by electron from a higher level giving rise to characteristic X-rays or release of nutrino. The process is known as *K-electron capture* or simply as *K-capture.*

$$^{133}_{56}\mathbf{Ba} + {}^{0}_{-1}\mathbf{e} \longrightarrow {}^{133}_{55}\mathbf{Cs} + \text{X-rays}$$

$(n/p = 1.375)$ (K-electron) $(n/p = 1.41)$

$$^{7}_{4}\mathbf{Be} + {}^{0}_{-1}\mathbf{e} \longrightarrow {}^{7}_{3}\mathbf{Li} + \nu$$

$(n/p = 0.75)$ (K-electron) $(n/p = 1.33)$ nutrino

13.4 NUCLEAR STABILITY

BINDING ENERGY OF NUCLEUS

As pointed out earlier, the particles present in the nucleus (protons and neutrons) are called **nucleons.** The sum of the individual masses of various particles in the nucleus must be equal to the nuclear mass. But this is not so in actual practice. The nuclear mass is somewhat less than the sum of the individual masses of the various nuclear particles. The difference between the actual nuclear mass and the expected nuclear mass (sum of the individual masses of nuclear particles) is referred to as **mass defect.** The *mass defect* can be converted into equivalent energy by means of the Einstein equation ($E = mc^2$). *The energy equivalent to mass-defect is responsible for holding the nucleons together and is called* **binding energy of the nucleus.**

For example, let us calculate the binding energy of the helium nucleus which contains two protons and two neutrons.

Mass of 2 free neutrons ($2 \times 1.00867\ u$) $= 2.01734\ u$

Mass of 2 free protons ($2 \times 1.00728\ u$) $= 2.01456\ u$

Sum of the masses of 2 free neutrons and 2 free protons $= 4.03190\ u$

Observed mass of 2 protons and 2 neutrons merged in a helium nucleus $= 4.00150\ u$

Mass defect (Δm) $= 0.03040\ u$

Mass defect (Δm) in gram

$$= \frac{0.03040}{6.02 \times 10^{23}}\ \text{g} \qquad \left(1\,u = \frac{1}{6.02 \times 10^{23}}\ \text{g}\right)$$

ΔE (Binding energy) $= mc^2$

$$= \frac{0.03040 \times 10^{-3}}{6.02 \times 10^{23}}\ (\text{kg}) \times (2.998 \times 10^8\ \text{ms}^{-1})^2$$

$$= 4.539 \times 10^{-12}\ \text{J} = \frac{4.539 \times 10^{-12}}{1.602 \times 10^{-19}}\ \text{eV}$$

$= 2.83 \times 10^7$ eV or $2.83 \times 10^7 \times 10^{-6}$ MeV

$=$ **28.3 MeV.**

[*MeV* represents *million electron volt* also called *mega electron volt.*]

Thus, the binding energy of the helium nucleus is 28.3 MeV. The binding energy per nuclear particle (nucleon) is a measure of the stability of the nucleus. Binding energy may also be considered as the energy required to separate the individual particles of the nucleus.

It may be noted that binding energy in MeV can be found directly by multiplying the mass defect (Δm) in u (amu) with (931.48), a conversion factor as explained below:

$$1\ u = 1.6603 \times 10^{-27}\ \text{kg}$$

$$c = 2.9979 \times 10^8\ \text{ms}^1$$

Using $E = mc^2$

$$= 1.6603 \times 10^{-27}\ \text{kg} \times (2.9979 \times 10^8\ \text{ms}^{-1})^2$$

$$= 1.4924 \times 10^{-10}\ \text{kg m}^2\ \text{s}^{-2}$$

$$= 1.4924 \times 10^{-10}\ \text{J}$$

$$= \frac{1.4924 \times 10^{-10}}{1.602 \times 10^{-19}} \text{ eV} = 931.48 \times 10^6 \text{ eV}$$

$$= 931.48 \text{ MeV}$$

To sum up,

Binding Energy (MeV) = Δm (u) × 931.48
Binding Energy (J) = Δm (u) × 1.4924 × 10^{-10}

Mass defect **Δm** in *u* or amu can be directly obtained by using the following formula

$$\Delta m = [Zm_p + (A - Z)\, m_n] - m_N$$

where, Z is the atomic number, A is the mass number, m_p mass of proton, m_n mass of neutron, m_N mass of nucleus $^{A}_{Z}X$.

Packing Fraction. The ratio of the difference of isotopic mass and mass number of the nuclide to its mass number is called packing fraction

Packing Fraction

$$= \frac{\textbf{Isotopic mass} - \textbf{Mass number}}{\textbf{Mass number}} \times 10^4.$$

Some **important facts** about packing fraction are:

- Quite often the packing fraction is multiplied by 10^4 for convenience.
- Packing fraction is negative or zero for a stable nuclei.
- Packing fraction is positive for an unstable nuclide.

NUCLEAR FORCES

The forces which hold nuclear particles together are referred to as **nuclear forces.** The exact nature of these forces is still not clear. However, a Japanese physicist **Yukawa** (1935) has provided a theoretical explanation for the origin of these forces. He postulated the presence of a new fundamental particle **meson** (π) in the nucleus of the atom. The meson is approximately 273 times heavier than an electron. The meson may be *positively charged* (π^+) or *negatively charged* (π^-) or *neutral* (π°). Yukawa further postulated that in the nucleus, *there is interconversion of proton into neutron* and *vice versa through the exchange of meson.* For example, a proton may change into a neutron by losing a positive meson (π^+) or by capture of a negative meson (π^-).

$$p \longrightarrow n + \pi^+$$

or $$p + \pi^- \longrightarrow n$$

Similarly, a neutron may change into a proton either by capture of a positive meson (π^+) or by loss of a negative meson (π^-)

$$n + \pi^+ \longrightarrow p$$

or $$n \longrightarrow p + \pi^-$$

The rapid exchange of mesons among the nucleons gives rise to the exchange forces which are responsible for holding the nucleons together. Some interesting features about nuclear forces are:

- Nuclear forces are about 10^{38} times as strong as gravitational forces and 100 times as strong as coulombic forces.
- They are short range forces and are independent of charge.
- They are not governed by the inverse square law.

Add To Your Knowledge

SIGNIFICANT FACTS ABOUT STABLE NUCLEI

A study of stable nuclei has revealed the following interesting facts.

- **Even-even nuclides.** About 60% of stable nuclei have both even number of protons and even number of neutrons. Some examples are $^{12}_{6}C$, $^{16}_{8}O$, $^{40}_{20}Ca$, etc. Almost all the remaining stable nuclei have either odd number of protons or odd number of neutrons.
- **Odd-odd nuclei.** Nuclei with odd number of protons and odd number of neutrons are least stable. The only exception being $^{2}_{1}H$, $^{6}_{3}Li$, $^{10}_{5}B$, $^{14}_{7}N$ and $^{180}_{73}Ta$.
- Nuclei with Z > 83 and A > 209 are unstable and spontaneously transform themselves into lighter ones through radioactive emissions.
- **Magic numbers.** Nuclei with number of neutrons or number of protons equal to 2, 8, 20, 50, 82 or 126 are stable. These numbers are called ***magic numbers.*** The nuclei with a magic number of protons as well as a magic number of neutrons are most stable. Some examples are $^{4}_{2}He$, $^{16}_{8}O$, $^{40}_{20}Ca$ and $^{208}_{82}Pb$.

13.5 RADIOACTIVE DECAY

The activity of a radioactive element is measured by the rate at which it changes into its daughter element. The rate of radioactive disintegration is *independent of temperature, pressure or the state of chemical combination of the element.* However, it does depend upon the nature of the element. The rate of nuclear disintegration of a radioactive substance is determined by counting the number of particles emitted per unit time by using instruments like the *Scintillation Counter,* or *Geiger-Muller Counter or Wilson cloud chamber.*

It has been experimentally observed that in all nuclear disintegration reactions the *rate of disintegration (or decay) is directly proportional to the number of atoms present in a sample at that time.* This is a characteristic feature of a first order reaction. Therefore, radioactive disintegration can be regarded as similar to a first order chemical reaction. Consider a radioactive element **A** disintegrating to give products.

$$A \longrightarrow \text{Products}$$

Let N_0 be the number of atoms in the beginning at time $t = 0$

N be the number of atoms of A at time 't'.

The rate of disintegration *i.e.,* $\frac{-dN}{dt}$ at time 't' is directly proportional to N. It can be expressed as

$$\frac{-dN}{dt} \propto N$$

or

$$\frac{-dN}{dt} = \lambda N \quad ...(13.1)$$

where λ is called *decay constant* or *disintegration constant.*

Rearrange equation (13.1) and integrate it

$$-\int \frac{dN}{N} = \int \lambda\, dt$$

or

$$-\ln N = \lambda t + C \quad ...(13.2)$$

where C is the constant of integration.

Now, when $t = 0$, number of atoms $= N_0$

$\therefore \quad -\ln N_0 = 0 + C$ or $C = -\ln N_0$

Substitute the value of C in equation (13.2),

$$-\ln N = \lambda t - \ln N_0$$

or

$$\ln N_0 - \ln N = \lambda t$$

or

$$\ln \frac{N_0}{N} = \lambda t$$

Converting natural logarithm to logarithm to base 10, we have

$$2.303 \log \frac{N_0}{N} = \lambda t \quad \text{or} \quad \log \frac{N_0}{N} = \frac{\lambda t}{2.303}$$

or

$$\boxed{\lambda = \frac{2.303}{t} \log \frac{N_0}{N}} \quad ...(13.3)$$

DECAY CONSTANT or DISINTEGRATION CONSTANT (λ)

The definition of decay constant follows from the following mathematical expressions.

(*i*) $\frac{-dN}{dt} = \lambda N$ or $\lambda dt = \frac{-dN}{N}$

If dt is 1 second, then $\lambda = \frac{-dN}{N}$

Thus, disintegration constant can be defined as ***the fraction of the total number of atoms which disintegrate in one second.***

(*ii*) The expression $\lambda t = \ln \frac{N_0}{N}$ can also be written in the exponential form as

$$\frac{N_0}{N} = e^{\lambda t} \quad \text{or} \quad \frac{N}{N_0} = e^{-\lambda t}$$

or

$$\boxed{N = N_0 e^{-\lambda t}}$$

In the above expression, if we put $t = \frac{1}{\lambda}$, we get

$$N = N_0 \cdot e^{-1} = \frac{N_0}{e} = \frac{N_0}{2.718} = 0.37\, N_0$$

Thus, the decay constant can also be defined as ***the reciprocal of the time in which the amount of radioactive substance is reduced to 37% of its initial value.***

HALF-LIFE

The time required for the decay of the radioactive element to one-half of its original amount is called **half-life.** This is usually denoted by $t_{1/2}$. Each radioactive element has its characteristic half-life. For example, the half-life for the radioactive decay of *uranium-238* is 4.51×10^9 years and that of *radium-226* is 1.62×10^3 years.

The half-life ($t_{1/2}$) does not depend upon the initial amount of radioactive element but it depends upon the value of λ. This can be proved as follows:

After time $t_{1/2}$, half of the atoms of a radioactive substance disintegrate, so that $N = \frac{N_0}{2}$.

Substituting in equation (13.3), we have

$$\log \frac{N_0}{N_0/2} = \frac{\lambda t_{1/2}}{2.303} \quad \text{or} \quad t_{1/2} = \frac{2.303}{\lambda} \times \log 2$$

or

$$\boxed{t_{1/2} = \frac{0.693}{\lambda}} \quad (\log 2 = 0.3010)$$

Since λ is a constant, *the time required for the disintegration of one-half of the original amount of a radioactive substance is independent of the initial amount of the radioactive substance.*

Add To Your Knowledge

AVERAGE LIFE PERIOD

The reciprocal of the decay constant is called average life period. It is generally denoted by τ(*tau*)

$$\tau = 1/\lambda$$

Relationship between half-life and average life period

Half-life $t_{1/2}$ is given by

$$t_{1/2} = \frac{0.693}{\lambda} = 0.693\ \tau$$

or $\tau = t_{1/2}/0.693 = 1.44\ t_{1/2}$

$\therefore$ **Average life period = 1.44 × Half-life.**

SIGNIFICANCE OF HALF-LIFE

(*i*) **Stability of nuclei.** *The value of a half-life can give some idea regarding the relative stability of radio-isotopes.* An isotope having longer half-life is relatively more stable than an isotope with shorter half-life. For example, ^{238}U has a longer half-life (4.5×10^9 years) whereas ^{214}Po has a very short half-life (1.64×10^{-8} seconds). It implies that ^{214}Po is highly unstable whereas ^{238}U is quite stable as far as radioactive decay is concerned.

(*ii*) *Half-life is independent of the initial amount of radioactive substance.*

(*iii*) *The amount of substance left after n half lives is given by*

$$\mathbf{N = N_0(1/2)^n}$$

where, N is amount after n half lives

N_0 is the initial amount of radioactive substance.

IMPORTANT RELATIONS

- $\lambda = \frac{2.303}{t} \log \frac{N_0}{N}$
- $t_{1/2} = \frac{0.693}{\lambda}$
- $N = N_0 . e^{-\lambda t}$
- *t (average life period)*

$$= \frac{1}{\lambda} = \frac{t_{1/2}}{0.693} = 1.44 \times t_{1/2}$$

- 1 C*i* = 3.7×10^{10} dps = 3.7×10^{10} Bq = 3.7×10^4 Rd
- Amount of substance left after n half lives (N) = $N_0 \times \left(\frac{1}{2}\right)^n$, where $n = \frac{t_{total}}{t_{1/2}}$

SOLVED EXAMPLES

Example 13.3. *The half-life of a certain radioactive element is 100 seconds. Calculate the disintegration constant. Also calculate the time during which 1 g of the element is reduced to 0.01 g.*

Solution. For radioactive decay, $t_{1/2} = \frac{0.693}{\lambda}$

$$\therefore \quad \lambda = \frac{0.693}{t_{1/2}} = \frac{0.693}{100} = \mathbf{6.93 \times 10^{-3}\ sec^{-1}}$$

Now, $$t = \frac{2.303}{\lambda} \log \frac{N_0}{N}$$

Here, $$\frac{N_0}{N} = \frac{1}{0.01} = 100$$

$$\therefore \quad t = \frac{2.303}{6.93 \times 10^{-3}} \log 100$$

$$= \frac{2.303}{6.93} \times 10^3 \log 100 = \frac{2.303}{6.93} \times 2$$

$$= 664.64 \text{ sec} = \mathbf{11.08\ min.}$$

Example 13.4. *2.0 g of a radioactive element is reduced to 0.25 g in 24 hrs. What is the half life of the element?*

Solution. Here, $N_0 = 2$ g; N = 0.25 g

The decrease in mass is obtained in n-half lives.

$$\therefore \quad N = N_0 \left(\frac{1}{2}\right)^n \quad \text{or} \quad 0.25 = 2\left(\frac{1}{2}\right)^n$$

or $$(2)^n = \frac{2}{0.250} = 8 \text{ hrs} = (2)^3 \text{ hrs} \quad \text{or} \quad n = 3.$$

Now, total time = 24 hr = 3 half lives

Hence half life = $\frac{24}{3}$ = **8 hrs.**

Example 13.5. *The activity of 1 gram of radium is found to be 0.5 curie. Calculate the half life of radium. Also calculate the time required for 4.0 g of radium to decay to 0.25 g (Atomic mass of Ra = 226).*

Solution. Number of atoms in 1 g of radium = $\frac{6.02 \times 10^{23}}{226}$

Activity of radium, $\frac{dN}{dt}$ = 0.5 cu

$$= 0.5 \times 3.7 \times 10^{10} \text{ dps}$$

$$= 1.85 \times 10^{10} \text{ dps}$$

We know, that $\frac{dN}{dt} = \lambda N \quad \therefore \quad \lambda = \frac{1}{N} \times \frac{dN}{dt}$

or $$\lambda = \frac{226}{6.02 \times 10^{23}} \times 1.85 \times 10^{10} = 6.945 \times 10^{-12}\ s^{-1}$$

Now, $$t_{1/2} = \frac{0.693}{\lambda} = \frac{0.693}{6.945 \times 10^{-12}} = 9.978 \times 10^{10}\ s$$

$$= \frac{9.978 \times 10^{10}}{60 \times 60 \times 24 \times 365} = \mathbf{3164\ years}$$

Now, time required for 4 g of Ra to decay to 0.25 g = 4 half lives

$$= 4 \times 3164 = \mathbf{1.2656 \times 10^4\ yrs.}$$

13.6 NUCLEAR FISSION AND FUSION REACTIONS

NUCLEAR FISSION

When a nucleus is bombarded with some sub-atomic particles such as *α-particles, neutrons, protons,* etc., these particles are either captured by the target nucleus or the nucleus disintegrates ejecting some other sub-atomic particles. The new element formed has mass either *slightly greater than or slightly smaller* than that of the parent element. In 1939, it was observed by Hahn and Strassmann that when uranium-235 is hit by slow neutrons, it splits into a number of fragments each of which has mass much smaller than that of uranium. The two main fragments formed are *barium* and *krypton.* Usually 2 or 3 neutrons are also ejected from every target uranium atom. The process is accompanied by the liberation of a large amount of energy. The reaction may be written as

$${}^{235}_{92}\mathrm{U} + {}^{1}_{0}\mathrm{n} \longrightarrow {}^{140}_{56}\mathrm{Ba} + {}^{93}_{36}\mathrm{Kr} + 3\,{}^{1}_{0}\mathrm{n}$$

The process of splitting a heavier nucleus like that of uranium-235 into a number of fragments of much smaller mass, by suitable bombardment with sub-atomic particles is called nuclear fission.

When uranium-235 is bombarded with slow neutron, uranium-236 is formed which being unstable, further breaks up in several different ways as described below:

$${}^{235}_{92}\mathrm{U} + {}^{1}_{0}n \longrightarrow [{}^{236}_{92}\mathrm{U}] \longrightarrow \begin{cases} {}^{140}_{56}\mathrm{Ba} + {}^{93}_{36}\mathrm{Kr} + 3\,{}^{1}_{0}\mathrm{n} \\ {}^{144}_{54}\mathrm{Xe} + {}^{90}_{38}\mathrm{Sr} + 2\,{}^{1}_{0}\mathrm{n} \\ {}^{144}_{55}\mathrm{Cs} + {}^{90}_{37}\mathrm{Rb} + 2\,{}^{1}_{0}\mathrm{n} \end{cases}$$

The splitting of ^{235}U has been shown in Fig. 13.4.

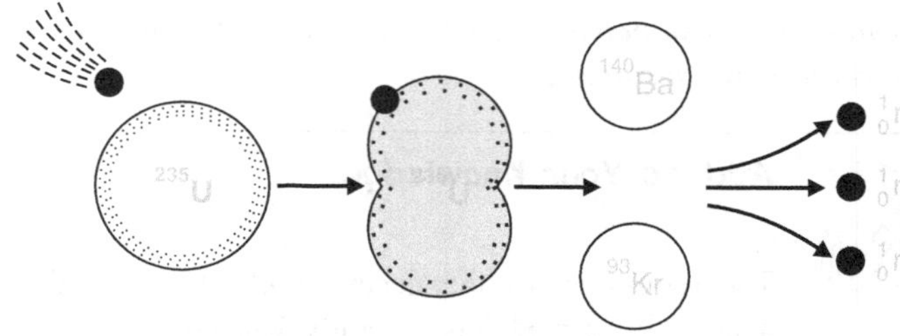

Fig. 13.4. Fission of ^{235}U.

The tremendous amount of energy released during nuclear fission is because of the loss in mass. The sum of the masses of the *fragments produced* and *neutrons released* as a result of fission is less than the sum of the masses of the *target* ^{235}U and *bombarding neutron*. The loss in mass gets converted into energy according to the Einstein equation,

$$E = mc^2$$

where, *m* is mass, *c* is the velocity of light and E is energy.

For example, let us calculate the loss of mass when ^{235}U-nuclide splits up into ^{144}Ba and ^{90}Kr along with the release of two neutrons.

$${}^{235}_{92}\mathrm{U} + {}^{1}_{0}n \longrightarrow {}^{144}_{56}\mathrm{Ba} + {}^{90}_{36}\mathrm{Kr} + 2\,{}^{1}_{0}n$$

235.118 + 1.009 = 236.127 amu; 143.881 + 89.947 + 2.018 = 235.846 amu

Thus, mass defect or mass converted into energy is given by

Mass defect $(\Delta m) = 236.127 - 235.846 = 0.281$ *amu*

But we know, *1 amu = 931.48 MeV*

∴ Energy released = 931.48 × 0.281 = **261.75 MeV.**

Hence, for every uranium atom undergoing fission, the energy released is approximately 261.75 MeV which corresponds to about 8.5×10^7 kJ per gram or 2×10^{10} kJ per mol of uranium.

We have seen that fission of U-235 nucleus results in the emission of 2 to 3 neutrons.

On the average, 2.5 neutrons per U-235 nucleus are emitted.

THE NUCLEAR CHAIN PROCESS

The neutrons emitted from the fission of the first uranium atom hit other uranium nuclei and cause their fission resulting in the release of more neutrons which further continue the fission process. In this way, *a nuclear chain reaction* sets up releasing tremendous amount of energy. A schematic view of a nuclear chain process is shown in Fig. 13.5.

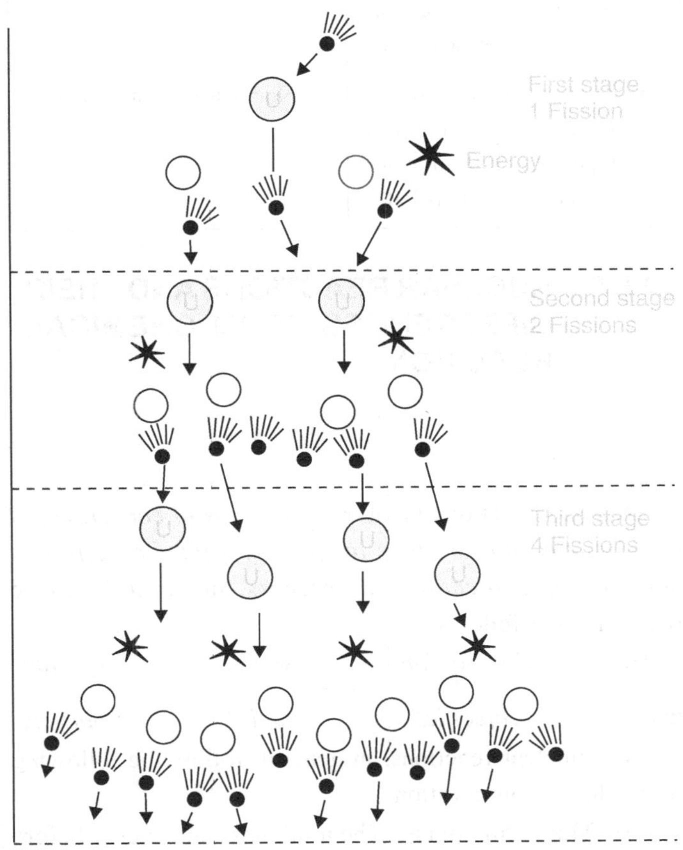

Fig. 13.5. Fission chain process.

If the chain reaction takes place completely as mentioned above, the energy released would be extremely high. However, some of the released neutrons escape from the surface unused and do not involve in the chain process. In spite of this, the net amount of energy released is quite high. It has been found that the energy available from 1 kg of uranium-235 is equivalent to that available from 2×10^4 kg of coal.

Concept of Critical mass. One of the important aspect of the chain process is that in order to continue or sustain the chain nuclear process there should be sufficient *fissionable material. The minimum amount that the fissionable material must have so as to continue the chain reaction process under the given set of conditions* is called the **critical mass.** If the mass of the material is more than the critical mass, it is referred to as **super-critical mass** whereas if the mass of the fissionable material is smaller than the critical mass, it is called **sub-critical mass.** The critical mass of *U-235* is between *1 kg to 100 kg*.

It may be noted that the naturally occurring uranium contains mostly U-238 isotope (about 99.3%) which is not fissionable with slow neutrons.

ATOMIC BOMB

The tremendous amount of energy released during the fission process if uncontrolled, can be used for destructive purposes as for example, in the **atomic bomb.** However, if the chain reaction is controlled, the energy released can be used for constructive purposes as for example, in the **nuclear reactor.** The principle of each of these devices is briefly explained as follows:

The basic principal of an atomic bomb is the nuclear chain fission reaction. It consists of several sub-critical samples of U-235 (or plutonium-239). A conventional explosive such as TNT is placed behind each sample. The explosion causes these sub-critical samples of U-235 to join together so as to constitute a **super critical mass.** A stray neutron from Ra-Be source (S) initiate the fission process which starts the chain reaction finally leading to explosion and also releasing large amount of heat energy. The rapid release of heat raises the temperature enormously and generates a very high pressure-front in the surrounding atmosphere.

The arrangement in its simple form is shown in Fig. 13.6.

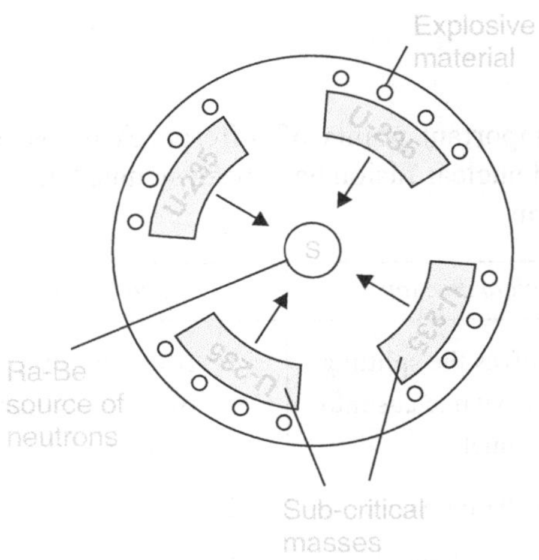

Fig. 13.6. Simple design of Atomic Bomb.

NUCLEAR FUSION

We have studied that in fission reactions, a heavy nucleus splits into smaller fragments and the difference in mass is converted into energy. Nuclear fusion is opposite to nuclear fission in the sense that it is a reaction in which two or more nuclei combine to form a heavy nuclide. However, the similar aspect of both processes is that there is liberation of large amount of energy. Thus, **nuclear fusion** may be defined as a **process in which lighter nuclei fuse together to form a heavier nucleus.** Like nuclear fission, mass defect also occurs in nuclear fusion reactions which is responsible for the tremendous amount of energy released. Some examples of nuclear fusion reactions are:

Fusion Reaction	*Mass Decrease (amu)*	*Energy Released (kJ/mol He)*
${}_1^2H + {}_1^2H \longrightarrow {}_2^4He$	0.026	23×10^8
${}_1^2H + {}_1^3H \longrightarrow {}_2^4He + {}_0^1n$	0.012	17.2×10^8
$4{}_1^1H \longrightarrow {}_2^4He + 2\,{}_{+1}^{0}e$ (Positron)	0.029	26×10^8

Nuclear fusion reactions require very high temperature, *generally* $> 10^6$ *K* in order to overcome electrostatic repulsion between the nuclei when they come together to fuse. The nuclear fusion reactions, therefore, are also called **thermonuclear reactions.**

The high temperature required for fusion to start can be made available only by a *nuclear fission process.*

DIFFERENCE BETWEEN NUCLEAR FISSION AND NUCLEAR FUSION

The important points of differences between nuclear fission and nuclear fusion have been summed up as below in tabular form.

Nuclear Fission	*Nuclear Fusion*
1. It involves the splitting of a heavy nucleus into lighter nuclei.	1. It involves the union of two or more lighter nuclei to form a heavy nucleus.
2. It is a chain process.	2. It is not a chain process.
3. High temperature is not essential.	3. Being a thermonuclear process, it requires extremely high temperature ($>10^6$ K).
4. Fission process can be carried out in a controlled manner and energy released can be used for peaceful purposes.	4. Fusion process is extremely difficult to be carried in a controlled manner.
5. There is a production of large number of radioisotopes and large amount of nuclear waste is left.	5. No nuclear waste is left.

13.7 NUCLEAR REACTIONS AND THEIR DIFFERENCES FROM CHEMICAL REACTION

NUCLEAR REACTIONS AND THEIR REPRESENTATION

Nuclear reactions involve *the process of conversion of one nucleus into another artificially with the help of a bombarding projectile.* The process, in general can be represented as follows.

$$\text{Target nuclide} + \text{Bombarding projectile} \longrightarrow \text{New nuclide} + \text{Ejected particle}$$

The nuclear reactions, in general, follow the following principles of conservation:

(*i*) **Mass and energy.** The total mass and energy before and after the nuclear change remains the same.

(*ii*) **Mass number and atomic number.** The sum total of mass numbers and that of atomic numbers before and after the nuclear change remains unaltered.

Representation of Nuclear Reactions

For the sake of brevity, the nature of a nuclear reaction is represented by writing the *bombarding projectile* and *ejected particle* within brackets. For example, the nuclear reaction represented by equation (*i*) can be termed as (α, *p*) *reaction* because here, α-particle is a bombarding projectile and proton is the ejected particle. The same can be briefly represented as ${}_7^{14}N\ (\alpha, p)\ {}_8^{17}O$.

It may be noted that out of the various bombarding particles the *neutrons* are particularly useful because they do not carry any charge. On the other hand, the *protons, deutrons* and α-*particles* are positively charged particles and are repelled by the target nuclei. Some examples of nuclear reactions involving the use of different bombarding particles are as follows:

Bombarding particle	*Examples*	*Type*
α-particle ($^{4}_{2}He$)	$^{9}_{4}Be + ^{4}_{2}He \longrightarrow ^{12}_{6}C + ^{1}_{0}n$	(α, n)
	$^{19}_{9}F + ^{4}_{2}He \longrightarrow ^{22}_{10}Ne + ^{1}_{1}H$	(α, p)
Proton ($^{1}_{1}H$)	$^{31}_{15}P + ^{1}_{1}H \longrightarrow ^{31}_{16}S + ^{1}_{0}n$	(p, n)
	$^{12}_{6}C + ^{1}_{1}H \longrightarrow ^{13}_{7}N + \gamma$	(p, r)
	$^{9}_{4}Be + ^{1}_{1}H \longrightarrow ^{8}_{4}Be + ^{2}_{1}H$	(p, α)
	$^{9}_{4}Be + ^{1}_{1}H \longrightarrow ^{6}_{3}Li + ^{4}_{2}He$	(p, α)
Neutron ($^{1}_{0}n$)	$^{27}_{13}Al + ^{1}_{0}n \longrightarrow ^{27}_{12}Mg + ^{1}_{1}H$	(n, p)
	$^{16}_{8}O + ^{1}_{0}n \longrightarrow ^{13}_{6}C + ^{4}_{2}He$	(n, α)
Deutron ($^{2}_{1}D$ or $^{2}_{1}H$)	$^{6}_{3}Li + ^{2}_{1}H \longrightarrow ^{7}_{3}Li + ^{1}_{1}H$	(d, p)
	$^{75}_{33}As + ^{2}_{1}H \longrightarrow ^{76}_{33}As + ^{1}_{1}H$	(d, p)
γ-radiation	$^{9}_{4}Be + \gamma \longrightarrow ^{8}_{4}Be + ^{1}_{0}n$	(γ, n)

Add To Your Knowledge

CATEGORIES OF NUCLEAR REACTIONS

Nuclear reactions, in general, can be classified into various types.

1. **Projectile Capture Particle Emission.** This type of reactions involve the emission of another elementary particle along with the product. Majority of nuclear reactions fall in this category.

$$^{23}_{11}Na + ^{4}_{2}He \longrightarrow ^{26}_{12}Mg + ^{1}_{1}H$$

Projectile Particle

2. **Projectile Capture Reactions.** In these reactions the bombarding particle is absorbed with or without the emission of γ-radiations.

$$^{27}_{13}Al + ^{1}_{0}n \longrightarrow ^{28}_{13}Al + \gamma$$

3. **Spallation Reactions.** In these reactions, the high speed projectiles (Energy approx. 40 MeV) chip out fragments from the heavy target nuclide.

$$^{63}_{29}Cu + ^{4}_{2}He \longrightarrow ^{37}_{17}Cl + 14\ ^{1}_{1}H + 16\ ^{1}_{0}n$$

4. **Nuclear Fission.** In these reactions a heavy nucleus is broken down into two or more medium heavy fragments along with the liberation of large amount of energy.

$$^{235}_{92}U + ^{1}_{0}n \longrightarrow ^{141}_{56}Ba + ^{92}_{36}Kr + 3\ ^{1}_{0}n + 2000\ \text{MeV}$$

5. **Fusion Reactions.** These reactions involve fusion of two light nuclei to produce a heavy nuclide along with the liberation of a large amount of energy

$$^{2}_{1}H + ^{3}_{1}H \longrightarrow ^{4}_{2}He + ^{1}_{0}n + 17.6\ \text{MeV}$$

DIFFERENCE BETWEEN NUCLEAR AND CHEMICAL REACTIONS

Some important differences between chemical reactions and nuclear reactions are given below in tabular form.

Nuclear Reactions	*Chemical Reactions*
1. No breaking or making of bonds is involved.	1. They generally involve breaking of old bonds and formation of new chemical bonds.
2. Generally proceed with tremendous release of energy.	2. They may involve evolution or absorption of energy.
3. Temperature and pressure do not affect their rates.	3. Their rates of reaction are affected by temperature and pressure conditions.
4. Such reactions are generally irreversible.	4. They can be reversible as well as irreversible.
5. They involve the conversion of one nuclide into the other.	5. They simply involve the rearrangement of atoms and do not involve any change in the nucleus.
6. Large amount of energy is released for example loss of 1u mass produces 931.48 MeV of energy.	6. In chemical reactions the amount of energy released is much less because $q_p = \Delta H = H_p - H_R$.

It must be noted here that matter is not conserved in nuclear reactions. It is because energy is produced at the expense of loss in mass during the nuclear disintegration.

13.8 APPLICATIONS OF RADIOACTIVITY AND RADIOISOTOPES

It has already been pointed out that each radioactive isotope has a specific half-life. The radiations that it emits can be detected and measured. Also the chemical properties of isotopes of a given element are identical. This makes it possible to incorporate a small amount of radioisotopes in a *system* and *trace* the fate of a particular element or a compound in a series of chemical or physical changes. Radioisotopes have large number of applications in different fields like medicine, engineering, biology, chemistry, archeology, agriculture and industry. Some of the applications of radioactivity have been discussed as follows:

AGE OF MINERALS AND ROCKS

During nuclear disintegration, the end product in a natural radioactive disintegration series is an *isotope of lead.* Each disintegration step has a definite *half-life* and hence a definite *decay constant.* Thus, by determining the amounts of parent radioactive element and the isotope of lead in a sample of rock, the age of the rock can be calculated. For example, if we wish to determine the age of a rock containing $^{238}_{92}U$ (*half life* $= 4.5 \times 10^9$ *years*), we measure the ratio of the concentrations of $^{238}_{92}U$ and its end product $^{206}_{82}Pb$. We assume that the rock did not contain any *lead isotope* initially.

Suppose, the ratio of uranium-238 and lead-206 is *unity.* It implies that half of the uranium originally present has been converted into lead isotope. The age of the rock must therefore be equal to the half-life of uranium-238 *i.e.*, 4.5×10^9 years. Pb/U ratio of most of the rocks is 1.33×10^{-2} indicating that their age is of the order of 10^8 years.

Age of the earth. The abundance ratio of the two isotopes of uranium, *i.e.*, $^{235}U : ^{238}U$ at present is 1 : 140. The half-life of ^{235}U is 7×10^8 years. Assuming that in the beginning the proportion of the two isotopes was equal, then the above data gives the age of the earth as 5×10^9 years.

SOLVED EXAMPLE

Example 13.6. *A sample of pitch blende contains 0.05 g of ^{206}Pb for every gram of ^{238}U present in it. Calculate the age of the rock ($t_{1/2}$ of $^{238}U = 4.5 \times 10^9$ years).*

Solution. $t_{1/2} = 4.5 \times 10^9$ years

$$\lambda = \frac{0.693}{4.5 \times 10^9} \text{ year}^{-1}$$

Amount of ^{238}U present at the time of analysis $(N_t) = 1$ g

Amount of ^{238}U originally present (N_0)

$$= \left[\text{Amount of } ^{238}\text{U now present}\right] + \left[\text{Amount of } ^{238}\text{U changed into } ^{206}\text{Pb}\right]$$

$$= 1\text{ g} + \frac{238}{206} \times 0.05\text{ g} = 1\text{ g} + 0.0577\text{ g}.$$

$$= 1.0577\text{ g}.$$

Let t be the time in which 1.0577 g of ^{238}U is reduced to 1 g of ^{238}U.

Now apply the relationship,

$$t = \frac{203}{\lambda} \log \frac{N_0}{N_t}$$

$$= \frac{2.303}{0.693} \times 4.5 \times 10^9 \times \log \frac{1.0577}{1}$$

$$= \mathbf{3.64 \times 10^8 \text{ years.}}$$

RADIO CARBON DATING

It is a method of determining the ages of *archeological objects* (wood, dead plants and animals). This technique was developed by *Willard Libby*. It helps in determining the date at which a particular plant or animal died. Libby was awarded a Nobel Prize for his technique.

The *principle* of this technique lies in the fact that due to bombardment by cosmic rays, nitrogen atoms present in the upper atmosphere are converted into radioactive carbon, ^{14}C according to the reaction:

$$^{14}_{7}N + ^{1}_{0}n \longrightarrow ^{14}_{6}C + ^{1}_{1}H$$

Carbon-14 is radioactive and has a half-life of 5770 years. It is oxidised in air to give radioactive carbon dioxide. Thus, the atmospheric carbon dioxide contains a small proportion of ^{14}C carbon dioxide ($^{14}CO_2$) which is assimilated by plants and animals. Living plants and animals have a definite and constant proportion of ^{12}C and ^{14}C. When a plant or animal dies, no fresh $^{14}_{6}C$ nuclei as $^{14}CO_2$ are received by the plant or animal. The ^{14}C present in these, decays according to the reaction

$$^{14}_{6}C \longrightarrow ^{14}_{7}N + ^{0}_{-1}e; t_{1/2} = 5770 \text{ years.}$$

The amount of ^{14}C in a sample can be accurately determined by counting the number of β-particles emitted per minute by one gram of the sample. By knowing the ^{14}C content and half-life, the age of the sample can be determined.

It may be noted that the *older the sample is,* the smaller the $^{14}_{6}C$ *count rate*. Ages of samples up to 50,000 years old can be determined accurately by this method.

SOLVED EXAMPLES

Example 13.7. *A sample of wood from an archeological find has 5.00 counts per minute per gram of carbon. Calculate the age of the wood sample. A freshly cut wood gives 15.3 counts per minute per gram of carbon. Half-life of ^{14}C is 5770 years.*

Solution.
$$t_{1/2} = \frac{0.693}{\lambda}$$

$$\therefore \quad \lambda = \frac{0.693}{t_{1/2}} = \frac{0.693}{5770} = 1.201 \times 10^{-4} \text{ year}^{-1}$$

Since the rate of count is directly proportional to the number of atoms.

Therefore,
$$t = \frac{2.303}{\lambda} \log \frac{N_0}{N_t}$$

$$= \frac{2.303}{1.201 \times 10^{-4}} \log \frac{15.3}{5} = 9313 \text{ years.}$$

Example 13.8. *An old chair of wood shows $^{14}_{6}C$ activity which is 80% of the activity found today. Find the age of the old chair of wood. Half-life period of ^{14}C is 5770 years.*

Solution. Suppose the amount of ^{14}C present in the wood originally (*i.e.*, same as the wood has today)

$$= x$$

The amount of ^{14}C present in the old chair of wood

$$= \frac{80}{100} x = 0.8\, x$$

Let t be time in which ^{14}C changes from x to 0.8 x

Now,
$$t_{1/2} = \frac{0.693}{\lambda}$$

$$\therefore \quad \lambda = \frac{0.693}{t_{1/2}} = \frac{0.693}{5770} = 1.201 \times 10^{-4} \text{ year}^{-1}$$

$$t = \frac{2.303}{\lambda} \log \frac{N_0}{N_t}$$

$$= \frac{2.303}{1.201 \times 10^{-4}} \log \frac{x}{0.8\, x}$$

$$= \frac{2.303}{1.201} \times 10^4 \times 0.0969 = 1.858 \times 10^3 \text{ years.}$$

TRACER TECHNIQUES

Radio isotopes are frequently used *as tracers* or *tagged atoms* in many processes in different fields like surgery, medicine, agriculture, industry and chemistry. In tracer techniques, a radioactive isotope is added to the reactants and its movement is studied by measuring radioactivity in different parts. The common examples of tracer techniques are:

(*i*) **In medical diagnosis.** For example, in order to find out if blood is circulating to a wound or not, a radioactive isotope is injected into the blood stream. After a suitable time, the blood from the wound is examined for its radioactivity. If no radioactive isotope is detected, it means that the passage of blood is hindered. The rate of circulation of blood can also be detected by this method.

Tracer technique is also used for the detection of *thyroid disorder* and *brain tumours.*

(*ii*) **In agriculture,** the uptake of phosphorus by plants is studied by mixing radioactive phosphorus with phosphatic fertilizers.

(*iii*) **In chemistry,** the use of tracer technique in chemistry is based on the fact that in chemical reactions radioactive isotopes have the same behaviour as ordinary isotopes. Some examples are:

(*a*) *Solubility of sparingly soluble salt.* The solubility of lead sulphate can be determined as follows. A lead salt containing a known amount of radioactive lead is taken and dissolved in water. Sulphuric acid is then added to the aqueous solution to precipitate lead as lead sulphate.

$$Pb^*(NO_3)_2 + H_2SO_4 \rightarrow Pb^*SO_4 + 2HNO_3$$

The precipitate is filtered. The radioactivity of the filtrate is measured. From this, the amount of lead still remaining in solution is calculated.

(*b*) *Tracer technique can be used to study the mechanisms* or *rates of chemical reactions.* For example, consider the esterification of benzoic acid

$$C_6H_5COO\boxed{H \quad + \quad HO}CH_3 \longrightarrow C_6H_5COOCH_3 + H_2O \quad ...(i)$$

$$C_6H_5CO\boxed{OH \quad + \quad H}OCH_3 \longrightarrow C_6H_5COOCH_3 + H_2O \quad ...(ii)$$

In order to decide whether the oxygen of the ester comes from the alcohol or from the acid we use radioactive alcohol containing ^{18}O isotope. The ester formed is found to contain O^{18} isotope but no radioactivity is detected in H_2O. This indicates that the reaction follows the II path and not the I. This implies that during esterification the oxygen comes from the alcohol and the reaction involves the replacement of the OH part of the acid with the OCH_3 part of the alcohol.

$$C_6H_5\text{—}C(=O)\text{—}O\text{—}H + H\ {}^*OCH_3 \rightarrow C_6H_5\text{—}C(=O)\text{—}\overset{*}{O}\text{—}CH_3 + H_2O$$

CANCER THERAPY

γ-rays emitted by the radio isotopes $\left({}^{60}_{27}Co\right)$ can be used in the treatment of cancer. The basis of such a treatment is that the radiation tend to destroy cancerous cells more easily than the normal cells. Thus, a carefully controlled beam of γ-rays of appropriate doses may be used to arrest the growth of cancerous cells. Radiophosphorus (P-32) is also used in the treatment of Leukaemia.

NEUTRON ACTIVATION ANALYSIS

This technique is gaining importance in *analytical chemistry these days.* It can help in the detection of even one part of the element per billion parts of the sample. In this technique, the sample containing very small amount of the stable isotope of the element to be investigated is bombarded with neutrons and the element of interest is converted (*i.e.,* activated) into its radioactive isotope. The radioactivity of this radioisotope is measured. Quantitative measurement of radioactivity and the knowledge of other factors such as *half-life of radioisotope, efficiency of radiation detector,* rate of *neutron bombardment,* etc., help in the calculation of the amount of the element in the sample. This technique is quite advantageous because:

(*i*) trace amounts of elements can be determined;

(*ii*) a sample can be tested without destroying it; and

(*iii*) the sample can be in any state of matter including biological material.

IN INDUSTRY

(*a*) Metal castings can be tested for cracks by putting them in baths of radioactive salts. The castings are then inspected for radioactivity to find out any penetration of salts into cracks. Absence of salt penetration indicates absence of cracks.

(*b*) Radioactive isotopes can be used to detect any leakage in underground pipes carrying oils, gas or water. To check the point of leakage a small quantity of compound of radioactive isotope is introduced at the starting place and the detector is moved along the pipe. At the point of crack or leakage, the detector will show high level of radiations.

HAZARDS OF RADIATIONS

Although nuclear radiations are quite useful to mankind, they are one of the major causes of atmospheric pollution. When living organisms are exposed to these radiations, the complex organic molecules in the body get ionized, break up and ultimately disrupt the normal functioning of the living organism. The damage caused by the radiations, however, depends upon the doses of radiations received. The radiations damage the living organisms in the following ways:

(*a*) **Pathological damage.** It is a permanent damage produced in the body which ultimately leads to death. Larger doses of radiations cause immediate death whereas smaller doses can cause the development of diseases like cancers or leukemia which are fatal.

(*b*) **Genetic damage.** This type of damage is produced when radiations affect the chromosomes of the cellular nuclei and injure the genes in reproductive cells.

One of the major problems before scientists is to check and deal with the radioactive waste formed in nuclear reactors. Any accidental leakage of the radioactive radiations can prove disastrous. One such accident occurred in 1986 in the U.S.S.R. which led to many deaths and caused injuries to several others.

Problems Associated with Nuclear Waste Disposal

From radioactive materials, very harmful radiation rays are released. When the human body is exposed to radiations, it can cause tumours and can do extreme damage to the reproductive organs. For this reason, the problems associated with radioactivity can be passed on to the victim's children as well. That is why radioactive waste produced by nuclear power plants is so dangerous.

Many scientists have argued about a long term storage for global nuclear waste. So far, continuing debates have prevented much if anything from being done about nuclear waste. Unfortunately, after being buried underground, nuclear waste can take millions of years to decay.

SUMMARY

- **Radioactivity.** Spontaneous emission of active rays by certain unstable elements.
- **Alpha-particles.** He^{2+} ion (Helium nuclei).
- **Beta-rays.** Consist of electrons.

- **Gamma-rays.** High energy electromagnetic rays.
- **N/P Ratio.** Stable nuclei have N/P ratio in the range 1 –1.6.
- **Group Displacement Law.** When a nucleus emits an alpha particle, the daughter element formed lies two places left to the parent element and when it emits a beta particle, the daughter element lies one place right to the parent element.
- Emission of a β-particle results in the formation of an **isobar** of the parent element.
- Emission of an α-particle results in the formation of an **isodiapher** of the parent nuclide.
- Emission of one α-particle and two β-particles in succession leads to the formation of an **isotope** of the parent element.
- **Decay constant,** $\lambda = \frac{2.303}{t} \log \frac{N_0}{N_t}$
- **Isosters.** The species with same number of atoms and same number of electrons *e.g.*, CO_2, N_2O.
- **Isodiaphers.** The species with the same value of isotopic number (A – 2Z).
- **Nuclear Isomers.** The nuclides with same values of A and Z but having different quantum states or energy states.
- **Half-life,** $t_{1/2} = \frac{0.693}{\lambda}$

  Half-life does not depend upon the amount of the radioactive substance.
- **Activity.** The rate of nuclear disintegration.

  $$\text{Activity} = -\frac{dN}{dt} = \lambda N.$$
- **Units of Activity:**

  1 Bacqueral (Bq) = 1 disintegration per second

  Curie (*Ci*) = 3.7×10^{10} dis/sec = 3.7×10^{10} Bq
- The first **artificial transmutation** was carried out by Rutherford.

  $${}^{14}_{7}N + {}^{4}_{2}He \longrightarrow {}^{17}_{8}O + {}^{1}_{1}H.$$

EVALUATE YOURSELF

I. Objective Type Questions

Select the most appropriate choice from the options given as (a), (b), (c) and (d) after each question:

1. Radioactive decay follows
   (*a*) first order kinetics
   (*b*) zero order kinetic
   (*c*) second order kinetics
   (*d*) third order kinetics.

2. The bombarding projectile in the following transformation is ${}^{14}_{7}N + ? \longrightarrow {}^{14}_{6}C + {}^{1}_{1}H$
   (*a*) proton (*b*) deutron
   (*c*) neutron (*d*) ${}^{0}_{-1}e$.

3. Which is the missing particle in the following nuclear reaction,

   $${}^{7}_{3}Li + {}^{2}_{1}H \longrightarrow {}^{8}_{3}Li + ?$$
   (*a*) proton (*b*) deutron
   (*c*) positron (*d*) α-particle.

4. The following nuclear transmutation:

   ${}^{23}_{11}Na + {}^{1}_{1}H \longrightarrow {}^{23}_{12}Mg + {}^{1}_{0}n$ belongs to
   (*a*) (*n*, α) type (*b*) (*p*, *n*) type
   (*c*) (α, *n*) type (*d*) (*d*, *p*) type.

5. If λ is the decay constant, then the half-life ($t_{1/2}$) for the radioactive decay is equal to
   (*a*) λ/0.693 (*b*) 0.693/λ
   (*c*) 2.303 log λ (*d*) 0.693 λ.

6. The nuclear fission reaction involves the
   (*a*) combination of two nuclei to form a single nuclide
   (*b*) splitting of single nuclide into two nuclei
   (*c*) combination of α and β-particles
   (*d*) combination of α and deutron.

7. $t_{1/2}$ of a radioactive element is 5 years. The amount of it remaining out of 10 g in 20 years will be
   (*a*) 0.625 g (*b*) 1.25 g
   (*c*) 0.3125 g (*d*) It will totally disappear.

8. The reaction ${}^{2}_{1}D + {}^{3}_{1}T \longrightarrow {}^{4}_{2}He + {}^{1}_{0}n$ is an example of
   (*a*) nuclear fission (*b*) nuclear fusion
   (*c*) artificial radioactivity (*d*) radio disintegration.

9. The radioactive decay of ${}^{88}_{35}X$ by β-emission produces an unstable nuclide which spontaneously emits a neutron. The final product is
   (*a*) ${}_{37}X^{88}$ (*b*) ${}^{89}_{35}Y$
   (*c*) ${}^{88}_{34}Z$ (*d*) ${}^{87}_{36}W$.

10. In the reaction, ${}^{239}_{93}Np \longrightarrow {}^{239}_{94}Pu + ?$ The missing particle is
    (*a*) neutron (*b*) proton
    (*c*) positron (*d*) electron.

11. If 2.0 g of an isotope has a half-life of 7 days, the half-life of 1 g of the sample will be
(*a*) 7 days (*b*) 3.5 days
(*c*) 14 days (*d*) 28 days.

12. If $\frac{3}{4}$ of a radioactive substance disintegrates in two hours, its half-life would be
(*a*) 1 hr (*b*) 45 min
(*c*) 30 min (*d*) 15 min.

13. A radioactive substance has a constant activity of 2000 disintegrations per minutes. The material is separated into two fractions one of which has initial activity of 1000 disintegrations per minute while the other fraction decays with $t_{1/2}$ = 24 years. The total activity in both samples after 48 hour of separation is
(*a*) 1500 (*b*) 1000
(*c*) 1250 (*d*) 2000.

14. $^{27}_{13}Al$ is a stable isotope $^{29}_{13}Al$ is expected to disintegrate by
(*a*) α-emission (*b*) β-emission
(*c*) positron emission (*d*) proton-emission.

15. In the nuclear reaction, $^{238}_{92}U \longrightarrow ^{206}_{82}Pb$, the number of α and β-particles emitted are
(*a*) 7α, 5β (*b*) 6α, 4β
(*c*) 4α, 3β (*d*) 8α, 6β.

16. C-14 dating method is related to the following
(*a*) C-14 fraction is same is all objects
(*b*) C-14 highly insoluble
(*c*) radio carbon-14 and C-12 are constant
(*d*) all the above.

17. The number of neutrons in a parent nucleus X which gives $^{14}_{7}N$ after two successive β-emissions would be
(*a*) 6 (*b*) 7
(*c*) 8 (*d*) 9.

18. The half-life of $^{14}_{6}C$ (with decay constant $2.31 \times 10^{-4}\ yr^{-1}$) is
(*a*) 2×10^3 yrs (*b*) 3×10^3 yrs
(*c*) 3.5×10^4 yrs (*d*) 4×10^3 yrs.

19. The number of neutrons accompanying the formation of $^{139}_{92}Xe$ and $^{94}_{38}Sr$ from the absorption of slow neutron by $^{235}_{92}U$ followed by nuclear fission is
(*a*) 0 (*b*) 2
(*c*) 3 (*d*) 1.

20. The equation $^{6}_{3}Li + ^{2}_{1}H \longrightarrow 2\,^{4}_{2}He$ represents
(*a*) spallation reaction (*b*) fission process
(*c*) fusion process (*d*) synthesis of helium.

21. Half-life of a certain radioactive substance is 45 min. The time during which 99.9% of it will disintegrate is
(*a*) $5\frac{1}{2}$ hr (*b*) $6\frac{3}{4}$ hr
(*c*) $7\frac{1}{2}$ hr (*d*) $7\frac{3}{4}$ hr.

22. The process, $^{2}_{1}H + ^{3}_{1}H \longrightarrow ^{4}_{2}He + ^{1}_{0}n$ represent the type of reaction known as
(*a*) Fusion (*b*) Chemical combination
(*c*) Autocatalytic (*d*) Fission

23. Given that radioactive species decays according to the experimental law $N = N_0 e^{-\lambda t}$, the half-life of the species is
(*a*) λ (*b*) N_0
(*c*) $\frac{\lambda}{\ln(2)}$ (*d*) $\frac{\ln(2)}{\lambda}$.

24. A radioisotope tritium ($^{3}_{1}H$) has a half-life of 12.3 years. If the initial amount of tritium is 32 mg, how many mg of it would remain after 40.2 years?
(*a*) 8 mg (*b*) 1 mg
(*c*) 2 mg (*d*) 4 mg.

II. Fill in the Blanks

25. Complete the following sentences by supplying appropriate words:

(*i*) Out of α, β and γ-rays, the one having maximum penetration power is ....

(*ii*) The number of α and β-particles emitted, when the following nuclear transformation occurs $^{232}_{92}X \longrightarrow ^{208}_{82}Y$ are .... respectively.

(*iii*) The particle having a mass of one amu and no charge is represented by the symbol ......

(*iv*) The species with different number of nucleons but same number of neutrons are called ......

(*v*) The species with the same number of protons but different number of neutrons are called ......

(*vi*) $^{27}_{13}Al + ^{4}_{2}He \longrightarrow ^{30}_{15}P +$ ......

(*vii*) $^{241}_{95}Am + ^{4}_{2}He \longrightarrow ^{244}_{97}Bk +$ ......

(*viii*) $^{96}_{42}Mo + ^{4}_{2}He \longrightarrow ^{100}_{43}Tc +$ ......

(*ix*) A radioactive nucleus decays emitting one α and two β-particles. The daughter nucleus is ...... of the parent.

(*x*) The synthetic elements prepared by artificial transmutation have atomic numbers ...... to ......

(*xi*) The half-life of a radioactive element if 87.5% of it disintegrate in 40 min. is ......

(*xii*) The radio carbon dating method was developed by ......

(*xiii*) Radioactive ^{11}C decays by ...... emission, whereas ^{14}C decays by ...... emission.

(*xiv*) $^{239}Pa + ... \longrightarrow {}^{242}_{96}Cm + {}^{1}_{0}n$.

III. Discussion Questions

26. Name the three main kinds of radiations that are emitted from radioactive elements. Compare their properties.

27. What changes are caused in atomic number and mass number of a parent atom when α- and β-particles are emitted? Give examples.

28. A nucleus does not have any electron yet a β-particle is emitted from the nucleus. How do you justify it?

29. Write the nuclear equations for the following transformations:

    (*i*) Lead-212 emits a β-particle.

    (*ii*) Uranium-238 emits an α-particle followed by emission of two β-particles.

    (*iii*) Thorium-232 decays to radium-228.

30. Define the term half-life. Is it dependent on the initial amount of radioactive substance? What is the relationship between half-life and decay constant?

31. What is meant by artificial transmutation? What kind of particles are used for carrying out such a process? Give suitable examples.

32. Why are positively charged projectiles given high energies before bombarding the atomic nuclei?

33. What is the difference between natural radioactivity and induced radioactivity?

34. Complete the following:

    (*i*) ${}^{246}_{96}Cm + {}^{12}_{6}C \longrightarrow ...... + 4\,({}^{1}_{0}n)$

    (*ii*) .... $(\alpha, 2n) \longrightarrow {}^{211}_{85}At$

    (*iii*) ${}^{96}_{42}Mo$ (........ n) $\longrightarrow {}^{97}_{43}Tc$

    (*iv*) ${}^{14}_{7}N$ (α ....) $\longrightarrow {}^{17}_{8}O$.

35. What is the difference between nuclear fission and nuclear fusion? (Give four points of difference).

36. (*a*) Explain the principle of neutron activation analysis.

    (*b*) Why is neutron activation analysis so useful in contrast to the ordinary methods of analysis?

37. Give the importance of radioisotopes in the field of surgery, chemistry and radio carbon dating.

38. Write short notes on:

    (*i*) radio carbon dating

    (*ii*) use of radio isotope in archaeology

    (*iii*) radio isotopes as tracers.

39. Complete the following nuclear equations:

    (*i*) ${}^{23}_{11}Na + {}^{2}_{1}H \longrightarrow ? + {}^{1}_{1}H$

    (*ii*) ${}^{59}_{27}Co + ? \longrightarrow {}^{56}_{25}Mn + {}^{4}_{2}He$

    (*iii*) ${}^{238}_{92}U + {}^{2}_{1}H \longrightarrow ? + {}^{0}_{-1}e$

    (*iv*) ${}^{246}_{96}Cm + {}^{13}_{6}C \longrightarrow {}^{254}_{102}No + ?$

    (*v*) ${}^{238}_{92}U + {}^{14}_{7}N \longrightarrow {}^{246}_{(...)}Es + (\ )\,{}^{1}_{0}n$

40. Write an equation to represent each of the following nuclear processes:

    (*i*) the reaction of two deutrons to produce a nucleus of ^{3}He.

    (*ii*) the production of ${}^{243}_{97}Bk$ by bombardment of ${}^{241}_{95}Am$ with α-particles.

    (*iii*) the bombardment of ^{121}Sb by α-particles to produce ${}^{124}_{53}I$ followed by its radioactive decay by positron (β^{+}) emission.

41. Which of the following nuclear reactions are accompanied by emission of neutrons:

    (*a*) ${}^{27}_{13}Al + {}^{4}_{2}He \longrightarrow {}^{30}_{15}P$

    (*b*) ${}^{12}_{6}C + {}^{1}_{1}H \longrightarrow {}^{13}_{7}N$

    (*c*) ${}^{30}_{15}P \longrightarrow {}^{30}_{14}Si + {}^{0}_{1}e$

    (*d*) ${}^{241}_{96}Am + {}^{4}_{2}He \longrightarrow {}^{244}_{97}Bk + {}^{0}_{1}e$.

42. What is the main use of the following nuclear disintegration?

    $${}^{14}_{6}C \longrightarrow {}^{14}_{7}N + {}^{0}_{-1}e.$$

43. Calculate the binding energy per nucleon in ${}^{56}_{26}Fe$. The observed mass of Fe atom is 55.9349 amu. Given that the mass of an electron, proton and neutron are 0.0005486, 1.00728 and 1.00866 amu respectively.

44. After emitting a series of α and β-particles ${}^{238}_{92}U$ changes to ${}^{206}_{82}Pb$. Calculate the number of α and β-particles emitted.

45. A radioactive element decays at such a rate that after 68 minutes only one-fourth of its original amount remains. Calculate the disintegration constant and half-life.

46. The half-life of Hg-203 is 49.6 days. How much of 0.200 mg sample of Hg-203 will remain after 6 months?

47. Half-life of a radioactive element is 100 seconds. Calculate the disintegration constant and average life period. How much time will it take for 90% to decay?

48. Starting with 1 g of radioactive sample, 0.25 g is left after 20 days. Calculate the rate constant λ for the decay of the radioisotope.

49. Predict the bombarding projectile in each of the following cases:

(*i*) $^{209}_{83}Bi + ? \longrightarrow ^{210}_{84}Po + ^{1}_{0}n$

(*ii*) $^{2}_{1}H + ? \longrightarrow ^{1}_{1}H + ^{1}_{0}n$

(*iii*) $^{82}_{34}Se + ? \longrightarrow ^{83}_{34}Se + \gamma\text{-ray}$

(*iv*) $^{27}_{13}Al + ? \longrightarrow ^{28}_{14}Si + ^{1}_{0}n$.

50. A mineral rock contains 0.257 mg of Pb-206 for each milligram of ^{238}U. Calculate the age of the rock if the half-life of ^{238}U is 4.5×10^9 years.

51. A piece of wood shows a C-14 activity, which is 70% of the activity found in fresh wood of today. Calculate the age of the piece of wood. ($t_{1/2}$ for $^{14}_{6}C$ is 5770 years).

52. The $^{14}_{6}C$ to $^{12}_{6}C$ ratio in a wooden article is 13% that of the atmosphere. Calculate the age of the wooden article. Given that half-life of ^{14}C as 5770 years.

53. A piece of preserved hair has 40% as much ^{14}C as a fresh one has today. Calculate the approximate life of the hair if $t_{1/2}$ of C-14 is 5770 years.

54. A living plant acquires a definite fraction of C-14 nuclei. If the freshly cut wood gives 16.3 counts per min. per gram and an old wooden bowl gives 9.3 counts what is the age of the wooden bowl? $t_{1/2}$ of C-14 is 5770 years.

55. What is the age of a vegetarian beverage whose tritium content is only 15% of the level in living plants? The half-life of tritium is 12.3 years.

Answers

I. Objective Type Questions

1. (*a*)	**2.** (*c*)	**3.** (*a*)	**4.** (*b*)
5. (*b*)	**6.** (*b*)	**7.** (*a*)	**8.** (*b*)
9. (*d*)	**10.** (*d*)	**11.** (*a*)	**12.** (*a*)
13. (*d*)	**14.** (*b*)	**15.** (*d*)	**16.** (*a*)
17. (*d*)	**18.** (*b*)	**19.** (*c*)	**20.** (*c*)
21. (*c*)	**22.** (*a*)	**23.** (*d*)	**24.** (*c*).

II. Fill in the Blanks

25. (*i*) γ-rays (*ii*) 6 and 4 (*iii*) $^{1}_{0}n$

(*iv*) isotones (*v*) isobars (*vi*) $^{1}_{0}n$

(*vii*) $^{1}_{0}n$ (*viii*) $^{0}_{+1}e$ (*ix*) isotope

(*x*) 93 to 104 (*xi*) 13 min-20 sec

(*xii*) Willard Libby

(*xiii*) β, α (*xiv*) $^{4}_{2}He$.

III. Discussion Questions

34. (*i*) $^{254}_{102}No$ (*ii*) $^{209}_{83}Bi$ (*iii*) $^{2}_{1}H$ (*iv*) $^{1}_{1}H$

39. (*i*) $^{24}_{11}Na$ (*ii*) $^{1}_{0}n$ (*iii*) $^{240}_{94}Pu$

(*iv*) $5\,^{1}_{0}n$ (*v*) $^{246}_{99}Es$, $6\,^{1}_{0}n$

43. 492.231 MeV ; 8.79 MeV **44.** 8α, 6β

45. $t_{1/2} = 34$ min ; $\lambda = 2.04 \times 10^{-2}$ min^{-1}

46. 0.0163 mg

47. $\lambda = 6.93 \times 10^{-3}$ sec ; $\tau = 144.3$ sec ; 332.2 sec

48. 0.0693 days^{-1}

49. (*i*) $^{2}_{1}H$, (*ii*) γ-rays, (*iii*) $^{1}_{0}n$, (*iv*) $^{2}_{1}H$

50. 1.69×10^9 years **51.** 2.97×10^3 years

52. 1.6989×10^4 years **53.** 7629.7 years

54. 4672.9 years **55.** 33.7 years.

SHS 2

SECTION 5

STATES OF MATTER

General Objectives

After studying this section, the student should be able to:

1. understand that the characteristics, nature and properties of solids and liquids depend on the type of chemical bonds that exist between the species forming the solid or liquid.
2. appreciate the domestic and industrial uses of some solids and liquids.
3. show understanding of the preparation, uses and behaviour of some gases under different conditions.

In this section, we shall study

Unit 14: Solids and Liquids

Unit 15: Kinetic Molecular Model, Gases and their Properties

SECTION-5

UNIT-14

SOLIDS AND LIQUIDS

Specific Objectives

After studying this unit, the student should be able to:

- describe the characteristics and nature of solids
- relate the properties of solids to the type of interatomic or intermolecular bonding in the solids
- outline some uses of diamond and graphite
- determine the melting point of some covalent solids
- describe the characteristics and nature of liquids
- distinguish between vapour and gas
- make a liquid boil at a temperature below its boiling point

14.1 DIFFERENT STATES OF MATTER

It is a well known fact that matter can be classified into three categories depending upon its physical state, namely: *solid, liquid* and *gaseous states.*

(*a*) **Solid state.** Matter in the solid state possesses *a definite volume, a definite shape* and *a definite mass*. Some common examples are: table, chair, common salt, silver, etc.

(*b*) **Liquid state.** Matter in the liquid state possesses *a definite volume, a definite mass,* but no *definite shape.* In fact, it acquires the shape of the container. Some common examples are: milk, water, alcohol, etc.

(*c*) **Gaseous state.** Matter in the gaseous state *neither has definite volume nor definite shape* but it has *definite mass*. It acquires the shape and volume of the container. Some common examples are: air, oxygen, hydrogen, sulphur dioxide, etc.

FOURTH AND FIFTH STATES OF MATTER

Recent research has led to the discovery of a fourth and fifth states of matter.

Fourth state called **plasma state** refers to the super heated gaseous state consisting of a mixture of electrons and positively charged ions with unusual properties. It is found at extremely high temperatures such as the interiors of the sun or stars or in intense electrical fields as in discharge tubes. Astronomers reveal that 99% of all matter in the universe is present in this plasma state.

Fifth state refers to a super cooled solid in which atoms lose their separate identity. In fact, they get condensed and behave like a single super atom. The study of this phenomenon was based upon the Bose-Einstein Condensation (BEC) concept developed in 1924.

KINETIC MODEL

Intensive investigation by scientists over the years, led to the development of a model regarding the constitution of matter. This model is known as the **dynamic particle model** or **kinetic model.** Some of its assumptions are:

1. *All matter is made up of tiny particles. However, the state of aggregation of particles is different in the three states of matter.*
2. *The particles are separated by empty spaces called* **voids.**
3. *The particles exert attractive forces on one another. However, the magnitudes of these interparticle forces differ in three states of matter.*
4. *The particles are not stationary, but they have the tendency to acquire motion.*

The basic difference between the three states of matter can be explained on the basis of the kinetic particle model.

EVIDENCE FOR MOLECULAR MOTION IN MATTER

Tiny particles of matter are not visible even with the help of a powerful microscope. However, their existence and movement particularly in liquids/gases can be indirectly noticed by Brownian movement. Brownian movement refers to the *random movement* of *specks of a solid in a liquid/or gas.*

It occurs because the particles of a liquid/or gas are in constant motion and they constantly bombard any particle of solid on surface of the liquid. Due to continuous bombardment, the particle of solid itself also acquires rapid zig-zag motion.

This observation was first made by a Scottish botanist, *Robert Brown,* (1827). He made this discovery when he was looking at some pollen grains on the surface of a liquid (water) through a microscope.

ACTIVITY 14.1	**Demonstration of Brownian movement**

1. Add some pollen grains to water, stir the suspension, then filter it.
2. Put a drop of the filtrate on a slide and examine it under a high-powered microscope.
3. Record your observations, including a sketch.
4. Attempt to explain your observation.
5. Suggest what would happen if powdered sulphur were used instead of pollen grains.

14.2 CHARACTERISTICS AND NATURE OF SOLIDS

In the **solid state,** the constituent particles are *closely packed* and the voids between them are very small. The average distance between the particles is of the order of $\mathbf{10^{-8}}$ **cm.** Due to the small voids, the particles motion is limited to *vibratory* or *oscillatory motion* only about their mean position. The magnitude of interparticle forces is quite large. Some of the *general characteristics* of solids are explained as follows:

- The definite shape and volumes of solids are attributed to the definite geometric pattern involved in the close packing of particles.
- The particles possess low thermal energy and cannot break free from mutual attraction. Consequently, solids cannot flow.
- The voids between the particles are extremely small and cannot be further decreased. Hence, solids are incompressible.
- When we heat a solid, thermal energy gradually increases which weakens the intermolecular forces and causes them to move apart. When the average distance between the particles increases beyond 10^{-7} cm, the solid melts into liquid as shown in Fig. 14.1.

The temperature at which a solid starts melting into liquid and the two states coexist in equilibrium at a standard atmospheric pressure of 100 kPa, is called ***melting point*** **(T_m)** ***of the solid.***

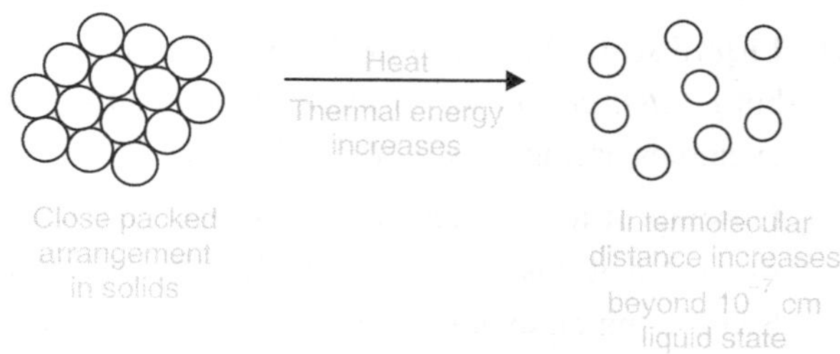

Fig. 14.1 Change of state from solid to liquid.

There are some solids, whose particles on being heated, by-pass the liquid state and change directly into the vapour phase. These solids are said to **sublime** and the process is called **sublimation.** Examples are iodine, naphthalene, solid carbon dioxide (dry ice) and ammonium chloride.

It may be noted that even though heat energy is continuously supplied to a melting solid, no temperature rise is recorded whilst melting takes place. The reason for this is that the heat energy supplied is used to overcome the cohesive forces and does not increase the kinetic energy. Overcoming the cohesive forces, however, increases the potential energy. This heat is referred to as the **latent (hidden) heat of fusion.**

14.3 CLASSIFICATION AND PROPERTIES OF SOLIDS

Solids are broadly classified into two types: *crystalline solids* and *amorphous solids*.

CRYSTALLINE SOLIDS

A **crystalline solid** *is a substance whose constituent particles possess a regular orderly arrangement.* Some examples of crystalline solids are sodium chloride (rock salt), sucrose (sugar), diamond and quartz.

AMORPHOUS SOLIDS

An **amorphous solid** *is a substance whose constituent particles do not possess a regular orderly arrangement.* Some examples of amorphous solids are glass, plastics, rubber, starch, etc. Quartz is a crystalline form of silica (SiO_2).

14.4 CLASSIFICATION OF CRYSTALLINE SOLIDS BASED ON INTERPARTICLE BINDING FORCES

Crystalline solids can be classified into different categories depending upon the type of constituent particles and the nature of attractive forces operating between them. Various categories of crystalline solids are:

1. *Molecular solids*
2. *Ionic solids*
3. *Covalent solids*
4. *Metallic solids*.

1. Molecular Solids

In these solids, the constituent particles which pack up together are molecules of the substance. These molecules may be *non-polar* (*dipole moment* = *0*) such as CO_2, I_2, CCl_4, etc. or **polar** (*dipole moment* > *0*) like H_2O, HCl, etc.

(*i*) In case of **non-polar molecules,** the attractive forces operating between the molecules are *van der Waal forces* (also called *dispersion forces*). Examples of such solids are: *dry ice* (solid CO_2), *iodine, etc*.

(*ii*) In case of polar molecules, the attractive forces operating between the molecules in the solid state are **dipole-dipole forces.** The examples of such solids are: solid SO_2, solid HCl. However, in some solids with polar molecules, the interparticle forces are **hydrogen bonds.** Some examples of such solids are *ice* (H_2O); *solid hydrogen fluoride* (HF); *solid ammonia* (NH_3), etc.

Characteristics of Molecular Solids

Some of the general characteristics of molecular solids are:

(*i*) They are generally soft.

(*ii*) Their melting points are low to moderately high. The melting points of solids with non-polar molecules are relatively low whereas solids with polar molecules have moderately high melting points.

(*iii*) They have low enthalpies of vaporization and hence are volatile.

(*iv*) They are generally bad conductors of heat and electricity.

(*v*) They have generally low density.

2. Ionic Solids

In ionic solids, the constituent particles are ions of opposite charges. Each ion is surrounded by a definite number of ions of opposite charge. The number of ions that surround a particular ion of opposite charge is called *co-ordination number* of the ion. For example, in sodium chloride crystal each sodium ion (Na^+) is surrounded by six chloride ions. Hence, co-ordination number of Na^+ is 6. At the same time each chloride ion is surrounded by six Na^+ ions. Therefore, the co-ordination number of Cl^- ion is also 6. However, in calcium fluoride crystal each Ca^{2+} ion is surrounded by eight fluoride (F^-) ions and each F^- ion is surrounded by four Ca^{2+} ions. Thus, in CaF_2 crystal, co-ordination numbers of Ca^{2+} and F^- ions are respectively 8 and 4. The interparticle forces in ionic solids are the **ionic bonds** operating between the ions of opposite charges. Some examples of ionic solids are: sodium chloride (NaCl), ceasium chloride (CsCl), zinc sulphide (ZnS), calcium fluoride (CaF_2), etc.

Characteristics of Ionic Solids

Some common characteristics of ionic solids are as follows:

(*i*) They are *hard* and *brittle*.

(*ii*) They have high enthalpies of fusion and vaporization, therefore, they possess *high melting points* and *low volatility*.

(*iii*) They are poor conductors of electricity in the solid state, however, they become good conductors of electricity in the molten state or in a dissolved state.

(*iv*) They are generally soluble in polar solvents like water.

3. Covalent or Network Solids

In these types of solids the constituent particles are atoms of the same or different elements connected to each other by a covalent bond network. For example, in diamond only carbon atoms constitute the covalent network while in carborundum, the covalent bond network is constituted by silicon and carbon atoms. Obviously, the interparticle forces operating in these solids are **covalent bonds.** These solids are also called **network solids** because the covalent bonds extend in three dimensions forming a giant interlocking structure. Some examples of covalent solids are:

Diamond, silicon carbide, silica, etc.

Characteristics of Covalent Solids

Some common characteristics of covalent solids are:

(*i*) They are very hard and brittle. Diamond is the hardest substance that occurs in nature.

(*ii*) They have very high melting points.

(*iii*) They are poor conductors of heat and electricity.

(*iv*) They have high enthalpies of fusion.

It may be mentioned here that graphite is also a network solid but it shows exceptional properties due to its typical structure. Graphite is a soft solid and is used as lubricant. It is also a good conductor of electricity and is used for making electrodes.

4. Metallic Solids

In these types of solids, the constituent particles are metal atoms. The interparticle forces in these solids are **metallic bonds.** In the metallic crystals the metal atoms occupy the fixed positions but their valence electrons are mobile. The close packed assembly of metal kernels (part of metal atom without valence electrons) remain immersed in the sea of mobile valence electrons. The attractive force between the kernels and mobile valence electrons is termed as metallic bond.

Characteristics of Metallic Solids

The common characteristics of metallic solids are as follows:

(*i*) They generally range from soft to very hard.

(*ii*) They are malleable and ductile.

(*iii*) They are good conductors of heat and electricity.

(*iv*) They possess bright lustre.

(*v*) They have high melting and boiling points.

(*vi*) They have moderate enthalpies of fusion.

The summary of classification of solids on the basis of interparticle forces is given in Table 14.1.

Table 14.1. Classification of Solids on the Basis of Binding Forces

Crystal Classification	*Constituent Particles*	*Binding Forces*	*Properties*	*Examples*
1. Molecular	Polar or non-polar molecules	van der Waals' forces (London dispersion, dipole-dipole forces, hydrogen bonds)	Fairly soft, low to moderately high melting points, poor thermal and electrical conductors	Dry ice (solid CO_2), methane (CH_4), argon, chloroform ($CHCl_3$), iodine (I_2).
2. Ionic	Positive and negative ions	Ionic bonds	Hard and brittle, high melting points, high enthalpies of fusion, poor thermal and electrical conductors	NaCl, ZnS, MgO, CaF_2, KNO_3.
3. Covalent	Atoms that are connected in covalent bond network	Covalent bonds	Very hard, very high melting points, poor thermal and electrical conductors	Diamond, quartz, silica, silicon carbide.
4. Metallic	Cations in electron cloud	Metallic bonds	Soft to very hard, low to very high melting points, excellent thermal and electrical conductors, malleable and ductile	All metallic elements, for example, Cu, Fe, Zn, Ag, Au.

14.5 DIAMOND AND GRAPHITE

STRUCTURE OF DIAMOND

In diamond, each carbon atom is sp^3 hybridized and is linked tetrahedrally to four other carbon atoms. C—C bond lengths are equal to 1.54 Å (154 pm) and angle is 109° 28′. As a result, in diamond there is a *three dimensional network of strong covalent bonds* (Fig. 14.2). This makes diamond **extremely hard** with *very high melting point* (3843 K) and *high density* 3.51 g/cm^3. Diamond is a *poor conductor of electricity* because all the valence electrons of carbon are involved in carbon-carbon sigma covalent bonds and hence are localized and not free to conduct electricity.

USES OF DIAMOND

1. Diamond is the hardest naturally occurring substance. Due to its extreme hardness, it is used for making ***tools for cutting and grinding*** other hard materials. It is also used in oil-well drills for drilling holes through the earth's rocky layers.
2. It is used in dies for the manufacture of tungsten filaments for electric light bulbs.
3. Diamond is transparent and has very high refractive index. It possesses extraordinary brilliance. Because of these properties it is used ***for making jewellery.***
4. Sharp-edged diamonds are ***used by eye surgeons to remove cataract from eyes*** with high precision.

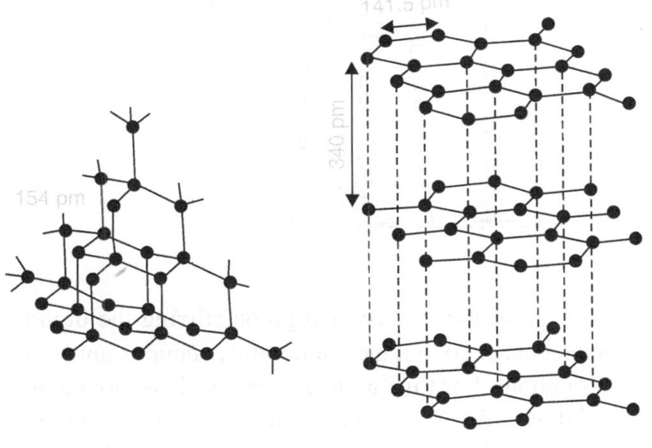

Fig. 14.2. Network structure of diamond

Fig. 14.3. Structure of graphite

5. Because of its extraordinary sensitivity to heat rays, it is used *in high precision thermometers.*
6. It is used *in protective windows for space probes* as it can keep out harmful radiations.

The value of diamond depends upon its weight and freedom from impurities. Weight of diamond is expressed in terms of *carats.* One *carat* is equal to 0.2 g or 200 mg.

STRUCTURE OF GRAPHITE

In graphite, each carbon atoms is sp^2 hybridized and is linked to three other carbon atoms directly in the same plane to form hexagonal rings. These rings constitute huge sheets or layers of atoms as shown in Fig. 14.3. The unhybridized *p*-orbitals overlap sidewise with each other to form π clouds. The C—C bond length in the rings is 142 pm which indicates the presence of π bond character. The different sheets of carbon atoms are held by weak van der Waals' forces and are separated by a distance of 340 pm. These layers can easily slide over one another. This explains the *slippery nature* of graphite and its use as *lubricant*. The *electrical conductivity* of graphite is attributed to the *mobile (delocalized)* π *electrons.* The *low density* (2.26 g/cm^3) of graphite is due to the large distance between the different layers of carbon atoms.

USES OF GRAPHITE

1. Graphite is used *as a lubricant* either as a powder or as a dispersion in oil or water. The dispersion of graphite in oil is known as *oil dag* and in water is known as *aqua dag.*
2. Mixed with clay it is *used in 'lead' pencils.*
3. Since it is a good conductor of electricity and is inert, it is used *for making carbon electrodes* in electrolytic cells and in dry cells.
4. Because of its high melting point it is used *for making graphite crucibles*. Crucibles made of graphite are not attacked by dilute acids or fused alkalies.
5. It is a component of *printers' ink.*
6. It is used as moderator in nuclear reactors.

Add to Your Knowledge

Some uses of Iodine (I_2) in daily life

- Iodine is used as an antiseptic in the form of an alcoholic solution, which is known as *tincture of iodine.*
- It is also used for the preparation of iodoform and potassium iodide.
- Iodide ion is necessary for the normal functioning of the thyroid gland. Insufficient iodide in the diet leads to *goitre* (enlargement of the thyroid gland). Hence, sodium or potassium iodide is added to table salt and this type of salt is known as *"iodized"* salt.

SOLVED EXAMPLES

Example 14.1. *Identify the type of chemical bonds in Fe, NaCl, SiO_2, I_2, diamond and graphite.*

Solution:

Substance	*Type of solid*	*Interparticle bonds*
Fe	Metal	Metallic bonds
NaCl	Ionic compound	Ionic bonds
SiO_2	Network solid	Covalent bonds
I_2	Molecular solid	Dispersion forces
Diamond	Network solid	Covalent bonds
Graphite	Network solid	Covalent bonds and dispersion forces.

Example 14.2. *Why is it that solid NaCl does not conduct electricity while solid iron does?*

Solution. Sodium Chloride is an ionic solid in which the position of the ions is fixed. Due to immobility of ions in solid NaCl it is not able to conduct electricity.

Iron, on the other hand is a metal in which positive kernels occupying fixed positions are surrounded by moving electrons. The free movement of electrons in metal crystals is responsible for the conduction of electricity.

ACTIVITY 14.2 **To determine the melting point of a covalent solid**

(*i*) Benzoic acid (Phenylmethanoic acid)
(*ii*) Ethanedioic acid (oxalic acid)
(*iii*) Ethanamide.

Apparatus required

100 cm^3 beaker, thermometer, thin walled capillary tube 8–10 cm long and 1–2 mm diameter, spatula and liquid paraffin.

Procedure

1. Powder the given sample of benzoic acid. (The sample should be dry.)
2. Take a capillary tube and seal it at one end by heating in a flame.
3. Make a heap of powdered substance on a porous plate. Push the open end of the capillary tube into the heap. Some substance will enter into it. Now tap the sealed end of the capillary tube on the porous plate gently. Fill the capillary tube up to 2–3 mm.

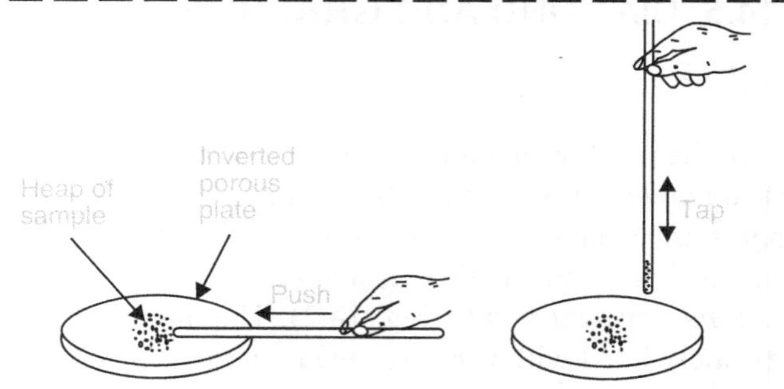

Fig. 14.4 Charging the capillary tube.

4. Attach the capillary tube to a thermometer which is immersed in a bath of liquid paraffin. The surface tension of the bath liquid is sufficient to hold the capillary tube in position.

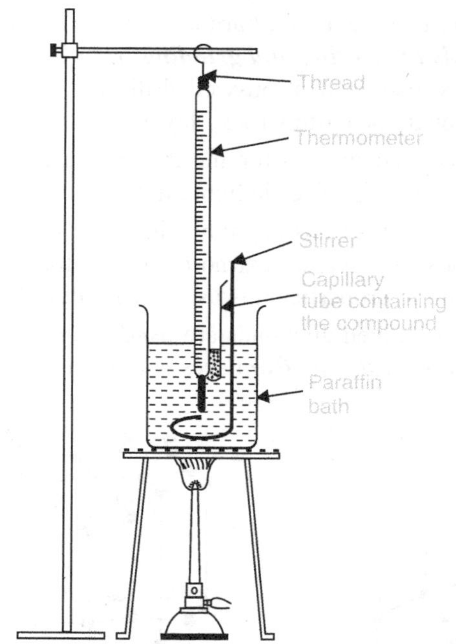

Fig. 14.5. Taking the melting point.

5. Heat the beaker slowly and go on stirring the liquid in the beaker so that the temperature remains uniform throughout. For this, a glass loop stirrer is moved up and down. When the temperature is within 15° of the melting point of the pure substance, the flame is lowered. Now the temperature is allowed to rise slowly.
6. Note the temperature when the substance starts melting. The temperature is again noted when the substance is completely melted. The average of the two readings gives the correct melting point of the substance.

7. Repeat the experiment using oxalic acid and ethanamide. Make the following table.

Substance 1	*Temperature when substance begins to melt (t_1°C)*	*Temperature when substance has melted (t_2°C)*	*Melting point ($t_1 + t_2$)/2*
Phenylmethanoic acid Ethanedioic acid Ethanamide			

14.6 CHARACTERISTICS AND NATURE OF LIQUIDS

The **liquid state** is the intermediate state between the gaseous state and the solid state. In the liquid state, particles are close together in comparison to those in gases but they are relatively loose in comparison to solids. Similarly, the interparticle forces are stronger than those in gases but relatively weaker than solids. Unlike solids, the particles in liquids do not occupy fixed positions. They possess translatory motion but it is not as vigorous as in gases. At the same time the particles cannot be separated from one another. Let us discuss some important characteristics of liquids.

SHAPE AND VOLUME

Liquid molecules are quite close but they do not occupy fixed positions. Therefore liquids do not have *definite shape*. But they do occupy *definite volume* because liquid molecules are not free to occupy all the space available to them. Liquids take up the shape of the container because the molecules are still free to move about. The slow movement of liquid molecules also give them the *tendency to flow.*

DENSITY AND COMPRESSIBILITY

In liquids, molecules are quite closely packed, therefore, their densities are much higher than the densities of gases under comparable conditions. For example, the density of water at 373 K and at 1 atmospheric pressure is 0.958 g cm^{-3}, while that of water vapour at the same temperature and pressure is equal to 0.000588 g cm^{-3}.

The space between the molecules of a liquid is reduced almost to a minimum by the intermolecular attractions, therefore, liquids are much less compressible than gases. A change in pressure has almost no effect on the densities of liquids.

DIFFUSION

Diffusion involves movement of molecules from one place to another. In liquids, since molecules can move from one place to another they show diffusion, however, they diffuse rather slowly. *The slow diffusion in liquids can be explained in terms of the smaller intermolecular spaces and restricted motion of molecules.* The diffusion in liquids can be demonstrated by releasing a drop of ink in water. At first there is a sharp boundary between the ink cloud and water, eventually the colour spreads uniformly throughout the water.

FLUIDITY AND VISCOSITY

As pointed out earlier, liquids have definite volume and also possess a tendency to flow. Therefore, they are able to transmit the same pressure from one point to another during flow. This pressure transmitting property of liquids is used in the *break system* of cars, lorries and moving parts of heavy machinary in industry. This system *of breaks* is referred to as **hydraulic system.**

Although liquids have a tendency to flow, the intermolecular forces also provide some resistance to flow. This *resistance to flow* is called **viscosity**. Now the higher the magnitude of the inter-particle forces the higher the viscosity of the liquid. Some examples of more viscous liquids are honey, glycerine, caster oil, etc.

EVAPORATION

When a liquid is placed in an open vessel, it slowly escapes into the gas phase, eventually leaving the vessel empty. This phenomenon is known as **evaporation.** It may be noted that in a given sample of liquid, the speed and kinetic energy of all the liquid molecules is not same. Some molecules move fast while others may be slow. In fact, they move with ever changing speeds. The fast moving liquid molecules having sufficient energy to overcome the intermolecular attractive forces can escape into the gas phase if such molecules happen to come near the surface.

Evaporation causes cooling. This is due to the reason that the molecules which undergo evaporation are high energy molecules, therefore the kinetic energy of the molecules, which are left behind is less. Since the remaining molecules have lower average kinetic energy, the temperature becomes lower.

The rate of evaporation increases with

(*a*) increase of temperature

(*b*) increase in surface area of the liquid and

(*c*) blowing a current of air across the surface of liquid.

CONCEPT OF VAPOUR PRESSURE

If a liquid is placed in a closed vessel, the molecules entering the gas phase cannot escape. The vapour formed by the escape of particles of the liquid in a close vessel exerts a pressure called **vapour pressure**. The molecules in the vapour phase in their random motion may strike the liquid surface and get captured. Thus, when a liquid is placed in a closed vessel, two opposing processes, *evaporation* and *condensation*, take place simultaneously. The rate of condensation increases as the number of molecules in the vapour phase increases. Eventually, the rate of condensation becomes equal to the rate of evaporation. This *state in which there is dynamic balance between evaporation and condensation is the state of equilibrium* (Fig. 14.6).

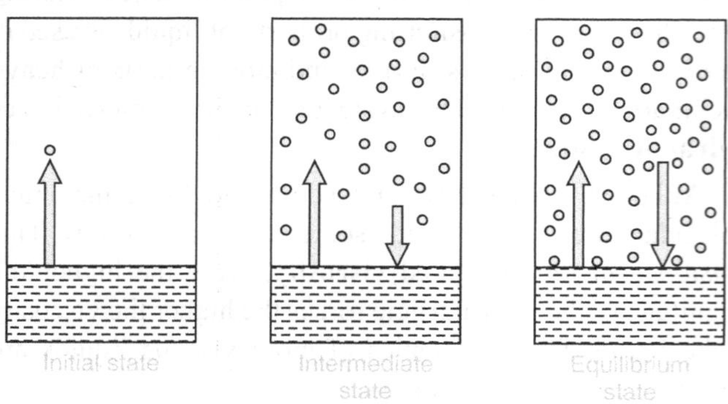

Fig. 14.6. Liquid ⇌ gas.

When equilibrium is reached, there is no further change in the number of particles in the vapour phase. The molecules in the vapour phase exert pressure in the closed vessel which is called *vapour pressure*. However, at the equilibrium point this pressure is known as *equilibrium vapour pressure* or *saturated vapour pressure*. At a given temperature, the **saturated vapour pressure** of a liquid is defined as ***the pressure exerted by the vapours of the liquid in equilibrium with the liquid at that temperature.***

The saturated vapour pressure of a liquid depends upon the following factors:

(*a*) *Nature of liquid*. More volatile liquids have higher vapour pressures.

(*b*) *Temperature*. The saturated vapour pressure of the liquid increases with increase in temperature because at higher temperature larger number of molecules escape into the vapour phase. Figure 14.7 shows temperature dependence of saturated vapour pressures of some liquids.

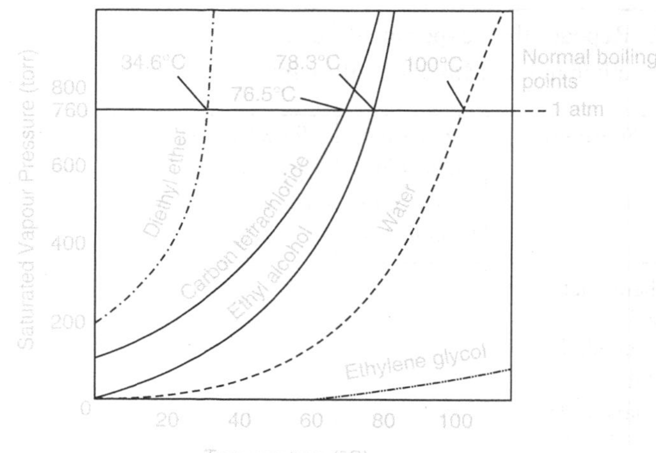

Fig. 14.7. Vapour pressure curves of some liquids.

BOILING

Boiling is a special case of evaporation, it is the rapid conversion of a liquid into vapour by means of formation of bubbles. During boiling the evaporation is not restricted to the surface only but takes place throughout the bulk of the liquid. The process of boiling can be understood as follows:

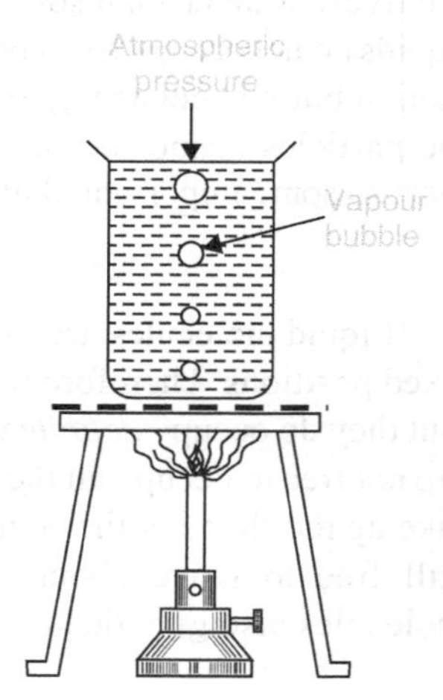

Fig. 14.8. Process of boiling

When a liquid is heated in an open vessel, its saturated vapour pressure increases. At the temperature where the equilibrium vapour pressure is lower than the atmospheric pressure, the vapours formed in the bubbles inside the bulk of the liquid remain trapped and the evaporation takes place only from the surface. As the temperature is increased, the vapour pressure increases. When the saturated vapour pressure becomes equal to the atmospheric pressure, the bubbles of vapours formed within the liquid can freely rise to the surface and can escape into the air. Evaporation taking place like this is called boiling (Fig. 14.8) and the temperature at which boiling takes place is called boiling point. Thus, *boiling point* of a liquid may be defined as *the temperature at which the saturated vapour pressure of the liquid becomes equal to the atmospheric pressure*.

From the above discussion it is clear that boiling point depends upon atmospheric pressure. The *boiling point of a liquid at one atmosphere pressure* is called ***normal boiling point.*** At normal boiling point vapour pressure of the liquid is equal to one atmosphere. The *boiling point of the liquid at one bar* (100 kPa) *pressure* is called ***standard boiling point.*** Normal boiling points can be determined from the vapour pressure-temperature curves as shown in Fig. 14.7.

Effect of Changing the External Pressure on Boiling Point

The boiling point of a liquid can be altered by changing the pressure over the liquid. The higher the pressure over the liquid, the higher the boiling point. *A liquid may be made to boil at a lower temperature by reducing the prevailing pressure.* This fact finds application in the purification of liquids, which decompose at their normal boiling points, by distillation under reduced pressure. Under reduced pressure the liquid distils over at a temperature lower than the normal boiling point. The effect of pressure on boiling point also explains why it takes longer to cook food at higher altitudes. At mountains the atmospheric pressure is lower and water boils at a lower temperature.

Difference Between Boiling and Evaporation

Both boiling and evaporation involve the conversion of liquid into vapour and appear to be similar. However, they differ in some respects. The main points of differences are given below.

Evaporation	*Boiling*
1. Evaporation occurs at the surface of a liquid.	1. It involves the formation of bubbles even below the surface within the bulk of the liquid.
2. It occurs spontaneously at all temperatures.	2. It occurs only at a specific temperature at which vapour pressure equals the imposed pressure on the liquid surface.
3. It is a slow phenomenon.	3. It is a rapid phenomenon.

14.7 DIFFERENCE BETWEEN VAPOUR AND GAS

Vapour is a name given to the gaseous state obtained by evaporation or vaporization of a substance which exists as liquid or solid at room temperature. The substance which exists as gas at room temperature is called **gas**.

A more authentic difference between gas and vapours can be deduced on the basis of the critical temperature (T_C) of the substance. *It is a temperature above which a gas cannot be liquified at all by compressing it.*

Gaseous substance above its critical temperature is called *gas*. On the other hand, *gaseous substance below its critical temperature* is called *vapour*.

For example, critical temperatures of H_2, He, etc., are much lower than room temperature. Hence they are permanent gases while the critical temperature of water is much higher than room temperature. Hence, water molecules in the gaseous state are called vapour at room temperature.

14.8 BOILING AT REDUCED PRESSURE

We have studied that the normal boiling point of a liquid is equal to the temperature at which the saturated vapour pressure of the liquid becomes equal to the atmospheric pressure. If the pressure above the liquid is reduced, then the liquid will boil at a temperature lower than the normal boiling point. This is illustrated in activity 14.3.

ACTIVITY 14.3 — **To determine the normal boiling point of water and to make it boil at a temperature below its normal boiling point.**

A. *Determining the normal boiling point of water.*

1. Set up apparatus as shown in Fig. 14.9. The conical flask should carry two hole stopper. Insert thermometer in one hole. The thermometer bulb should be just above the surface of water. The other hole is kept open so that pressure does not build up in the flask.

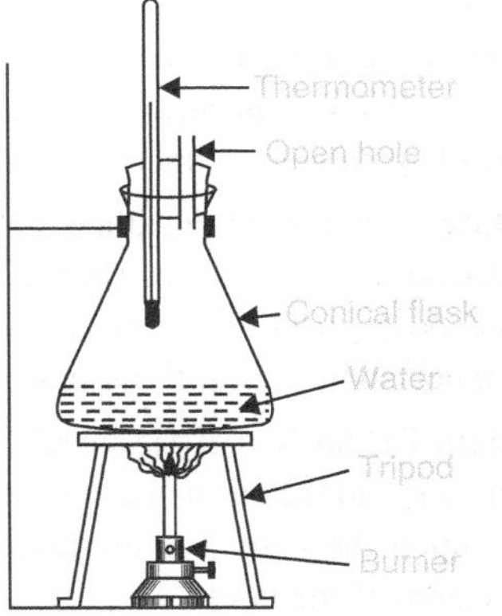

Fig. 14.9 Normal boiling of water.

2. Heat water and note the temperature at which water boils

   (*i*) Is this boiling point less or more than 100°C?

B. *Making the water boil at a temperature lower than the normal boiling point.*

1. Now cool the boiling water and connect the open hole of the stopper with hand vacuum pump (Fig. 14.10). As the vacuum pump is used to lower the pressure, the water will boil at a lower temperature. Note this temperature and compare it with the normal boiling point. Explain your observations.

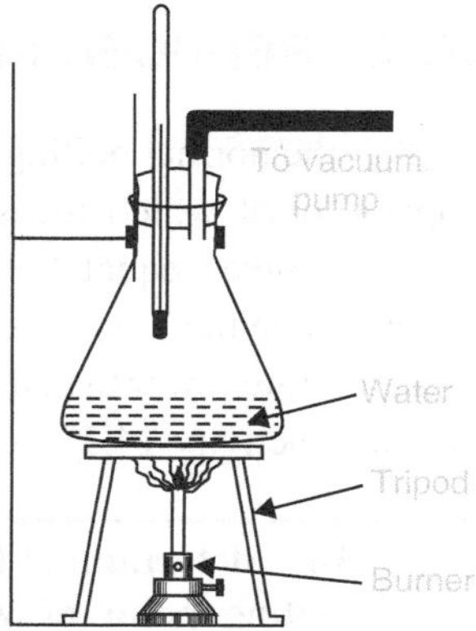

Fig. 14.10 Boiling water at reduced pressure.

SUMMARY

- **Matter** is classified into three different states; Solids, Liquids and Gases. Plasma is, however, the fourth state.
- **Solid State.** Particles in the solid state are closely packed and particle motion is restricted to only vibratory motion. Attractive forces among them are very large. Particles possess very low thermal energy.
- **Crystalline Solids** are generally classified into, metals, ionic compounds, covalent network solids and molecular solids depending upon the attractive forces among them.
- **Liquid State.** It is the intermediate state between the gaseous state and the solid state. In the liquid state the particles are loosely packed with voids and the molecular motion is much smaller than those in gases.
- **Evaporation.** The process of escape of liquid molecules from its surface into the gas phase is called evaporation. It depends upon the nature of the liquid, temperature and surface area of the liquid.
- **Saturated Vapour Pressure.** The pressure exerted by the vapours of a liquid in equilibrium with the liquid at a given temperature. Vapour pressure of a liquid depends upon temperature.
- **Boiling Point of Liquid.** The temperature at which, the saturated vapour pressure of the liquid becomes equal to the atmospheric pressure (or pressure applied on the surface of a liquid) is called boiling point of the liquid.
- **Normal Boiling Point.** Boiling point of a liquid at 1 atm pressure.
- **Standard Boiling Point.** Boiling point of a liquid at 1 bar or 100 kPa pressure.

EVALUATE YOURSELF

I. Objective Type Questions

Select the most Appropriate choice from the options given as (a), (b), (c) and (d) after each question:

1. Which of the following properties does not depend on the strength of interparticle forces in solids?
   (*a*) Enthalpy of fusion (*b*) Melting point
   (*c*) Definite shape (*d*) Definite mass.
2. Which of the following solid does not belong to the category of ionic solid?
   (*a*) Sodium chloride (*b*) Potassium trioxonitrate (V)
   (*c*) Glass (*d*) Potassium chloride.
3. When thermal energy is given to a solid
   (*a*) It ionizes (*b*) It melts
   (*c*) It gets compressed
   (*d*) Kinetic energy of particles lowers.
4. Which of the following is not a true solid?
   (*a*) Diamond (*b*) Glass
   (*c*) Sucrose (*d*) Quartz.
5. Among the following, which is not a molecular solid?
   (*a*) Ice (*b*) Dry ice
   (*c*) Solid hydrogen fluoride
   (*d*) Quartz.
6. Which of the following properties does not apply to graphite?
   (*a*) Lubricant
   (*b*) Greater thermodynamic stability than diamond
   (*c*) Hardest substance (*d*) Use in lead pencils.
7. Which liquid among the following has highest boiling point?
   (*a*) Water (*b*) Ethanol
   (*c*) Methanol (*d*) Ether.

II. Fill in the Blanks

8. Complete the following sentences by supplying appropriate words:
   (*a*) Liquids can be changed into solid by ......
   (*b*) A process in which a solid on heating changes to vapour without passing through the liquid state is called ......
   (*c*) A liquid having a higher saturated vapour pressure at a given temperature has a lower magnitude of intermolecular forces ......
   (*d*) A gaseous state of a substance which exist as liquid at room temperature is called ......
   (*e*) Water can be made to boil below 100°C at ......
   (*f*) In the liquid state the molecular interaction energy is ...... than that in solids.
   (*g*) The hybrid state of carbon atoms in diamond is ....

III. Discussion Questions

9. Use the kinetic model to explain
   (*a*) melting of a solid (*b*) Evaporation of a liquid.
10. What is the structural difference between diamond and graphite?
11. Why is graphite a good conductor of electricity while diamond is not?
12. Give three general characteristics of liquids and explain them with the help of the kinetic molecular model.
13. Point out the difference between gas and vapour.
14. What is vapour pressure of a liquid? How does saturated vapour pressure of a liquid change with temperature?
15. What is meant by boiling? Differentiate between the terms normal boiling point and standard boiling point.
16. Suggest modifications to
   (*i*) raise the boiling point of a liquid
   (*ii*) lower the boiling point of a liquid.
17. Suggest an experiment to make water boil below its boiling point.
18.

Temperature (°C)	0	20	40	60	70	80
Saturated vapour pressure of liquid (L) (kPa)	4	12	34	82	123	180
Saturated vapour pressure of water (kPa)	1	2	7	20	31	47

Using the given data answer the following:
   (*i*) Plot a graph between saturated vapour pressure *vs* temperature both for water and liquid L.
   (*ii*) What conclusion do you draw about the magnitude intermolecular forces among water and liquid L?
   (*iii*) If the atmospheric pressure where the experiment is conducted is 95 kPa, what is the boiling point of liquid L and water? (Predict it from the graph.)
19. Point out the differences between boiling and evaporation.
20. What is Brownian movement? Suggest an experiment to demonstrate it.

Answers

I. Objective Type Questions

1. (*d*) **2.** (*c*) **3.** (*b*)
4. (*b*) **5.** (*d*) **6.** (*c*)
7. (*a*).

II. Fill in the Blanks

8. (*a*) lowering temperatures
(*b*) sublimation (*c*) and high boiling points
(*d*) vapour (*e*) reduced pressure
(*f*) less (*g*) sp^3.

SECTION-5

UNIT-15

KINETIC MOLECULAR MODEL, GASES AND THEIR PROPERTIES

Specific Objectives

After studying this unit, the student should be able to:

- explain the kinetic theory of matter
- relate the speed at which different gas particles move to the masses of the particles
- demonstrate diffusion in liquids
- describe the characteristics and nature of gases
- describe the laboratory preparation of hydrogen, ammonia and carbon dioxide
- use the gas laws to explain the behaviour of gases under different conditions.

15.1 CHARACTERISTICS AND NATURE OF GASES

In the **gaseous state** the particles are very *loosely packed*. The voids between the particles are quite large to the extent that the actual volume of the particles is negligible as compared to the total volume of gas. The attractive forces between the particles are *negligibly small.* The particle motion is very rapid and they move in all the directions in straight lines. Let us explain some of the characteristics of gases.

- Because of the random motion and weak interparticle forces, the particles do not have any bonding surface. As a result they **do not have definite shape** or **definite volume**. They tend to occupy each nook and corner of the space available to them. Hence they take up the shape and volume of the container.
- They have very **low densities** because of large empty space between the molecules.
- Gases are **highly compressible.** This is because of the fact that the large voids between the gas molecules can be decreased by applying external pressure.
- Gases ***exert pressure*** on the walls of the container because of the bombardment of gas molecules against the walls of the container.
- Gases mix evenly and uniformly without any mechanical aid. This property is called diffusion. It occurs because of rapid movement of gas molecules and their tendency to enter into the voids of each other.

SOME MEASURABLE PROPERTIES OF GASES

The important measurable properties of gases are *mass, volume, temperature* and *pressure* because the state of the gas is described by the relationship between these variables.

1. Mass

The gases do possess mass. The amount of gas is generally expressed in terms of its number of moles. For a gas of molar mass (**M**), the mass in gram (**w**) is related to the number of moles (**n**) as

$$\mathbf{n} = \frac{\mathbf{w}}{\mathbf{M}}$$

2. Volume

The volume of a gas is the *space occupied by its molecules* under a given set of conditions. Since gases occupy the entire space available to them, the measurement of the gas volume simply requires the measurement of the volume of the container in which the gas is enclosed.

Units of Volume

The SI unit of volume is *Cubic metre* (m^3). Since it is too big a unit, the volume of a gas is generally expressed in terms of *litres (L)* or cm^3 or dm^3. The relationship between different units are:

$$\mathbf{1\ m^3 = 10^3\ L = 10^3\ dm^3 = 10^6\ cm^3}$$

3. Pressure

Gases exert outward force on the walls of the container in which they are enclosed. The outward force experienced by the walls is due to the bombardment of gas molecules on the walls. This *outward force per unit area of the walls* is termed as **gas pressure.**

Units of Gas Pressure

Pressure is force per unit area

$$\text{Pressure} = \frac{\text{Force}}{\text{Area}} \qquad \ldots(i)$$

By definition,

$$\text{Force} = \text{Mass} \times \text{Acceleration}$$

$$= \frac{\text{Mass} \times \text{Velocity}}{\text{time}} = \frac{\text{Mass} \times \text{Distance}}{\text{time} \times \text{time}}$$

$$= \frac{\text{kg} \times \text{m}}{\text{s} \times \text{s}}$$

Thus, the SI unit of force is $kg\ ms^{-2}$, which is called *Newton* (**N**)

$$1\ \text{N} = 1\ \text{kg ms}^{-2}$$

Putting the SI units of force (N) and area (m^2) we get SI unit of pressure as N/m^2 or $\mathbf{Nm^{-2}}$. This SI unit of pressure is called *pascal* (**Pa**) in honour of the French mathematician and physicist *Blaise Pascal*

$$1\ \text{Pa} = 1\ \text{N m}^{-2}$$

Now, 1000 Pa = 1 kilo pascal abbreviated as **kPa.**

For gases, this unit of pressure is very small. Hence pressure is generally expressed in terms of a bigger unit named as **bar.** One bar represents 100 kilo pascal. Thus

$$\mathbf{1\ bar = 100\ kPa = 10^5\ Pa}$$

It may be noted that the older unit of the pressure of gases was *atmosphere* which was abbreviated as **atm.** The relationship between these units is given as follows:

$$\mathbf{1\ atm = 1.01325\ bar = 1.01325 \times 10^5\ Pa}$$

or

$$\mathbf{1\ bar = 0.987\ atm = 10^5\ Pa}$$

4. Temperature

In general, **temperature** may be defined as the *degree of hotness*. Temperature can be measured in terms of the effect that its change produces on other measurable properties such as expansion of objects. Expansion of mercury is commonly used for the measurement of temperature. One of the common temperature-measuring device is a *thermometer*. It has a mercury column in a thin capillary tube. The length of the mercury column changes with a change in temperature.

Various Temperature Scales

- **Celsius scale of temperature (°C).** This is most common scale of temperature and was earlier known as centigrade scale. In this scale, *the melting point of ice is 0°C whereas the boiling point of water is 100°C at one atmospheric pressure*. The range between these two points has been divided into hundred equal parts. Each division corresponds to *one degree* in celsius scale. As the zero in celsius scale is arbitrarily fixed, therefore, it is possible to have temperature below the freezing point. It may appear that the celsius scale can be extended to a negative temperature indefinitely, but the experimental behaviour of gases shows that temperatures below – **273.15°C** are impossible to be attained.
- **Kelvin scale of temperature (K).** This scale of temperature is quite significant and is very useful for scientific work and can be justified by thermodynamic arguments. It is therefore, also called *thermodynamic scale of temperature. Kelvin* is also an *SI unit* of temperature. The zero point on the kelvin scale is equal to the lowest possible temperature (– 273.15°C) and is known as absolute zero (0 K). The temperature in celsius scale can be converted into kelvin by adding 273.15 (or 273 for the sake of simplicity).

$$\mathbf{K = {}^\circ C + 273.15}$$

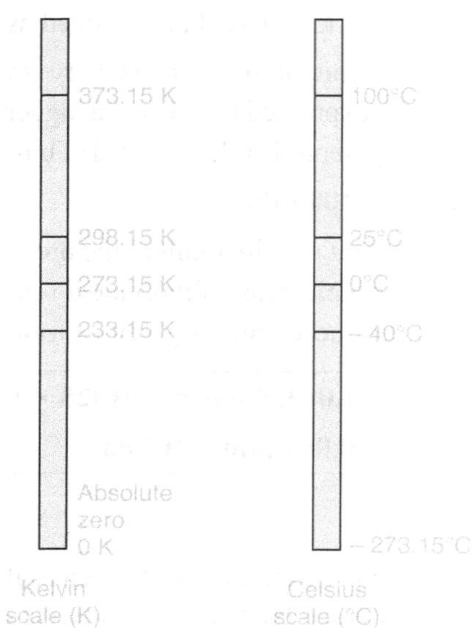

Fig. 15.1. Comparison of kelvin and celsius

5. Standard Temperature and Pressure (S.T.P.)

Since the volume of a given mass of a gas depends on both temperature and pressure, it is necessary to specify the values of P and T when the value of V is stated. In general, the comparison of the volumes of different gases is made with reference to standard temperature and pressure. The standard temperature is taken as **0°C (or 273.15 K).** The standard pressure according to the latest recommendations is taken as **1 bar (or 10^5 pascal).** The molar volume of an ideal gas at S.T.P. is **22.71098 L** mol^{-1} **(≈ 22.7 dm^3 mol^{-1}).** These conditions are quite often abbreviated as **S.T.P.** meaning *standard temperature and pressure*.

It is worth noting that previous S.T.P. conditions **(0°C or 273.15 K)** temperature and **1 atm (= 1.01325 bar)** pressure are still used in many books. At these conditions the molar volume of an ideal gas is 22.413996 dm^3 mol^{-1} (≈ 22.4 dm^3 mol^{-1}).

Conditions	*T*	*P*	*V_m (Molar Volume)*
Latest S.T.P.	273.15 K	1 bar	22.7 dm^3 mol^{-1} or 22.7 × 10^{-3} m^3 mol^{-1}
Earlier S.T.P.	273.15 K	1 atm	22.4 dm^3 mol^{-1} or 22.4. × 10^{-3} m^3 mol^{-1}

15.2 KINETIC MOLECULAR THEORY OF GASES

This theoretical model was developed by the contribution of Maxwell, Boltzmann, Clausius, etc., which is known as the *kinetic molecular theory of gases* or *microscopic model of gases.*

The **various postulates** of the theory are:

1. *All gases are made up of a very large number of extremely small particles called* **molecules.**
2. *The molecules are separated from one another by large spaces so that the actual volume occupied by the molecules is negligible as compared to the total volume of the gas.*
3. *The molecules are not at rest but possess rapid random motion. During their motion, they collide with one another and also against the walls of the container.*
4. *The pressure of the gas is due to bombardment of the gas molecules against the walls of the container.*
5. *The collisions of the molecules with each other and with the walls of the container are perfectly elastic, i.e., there is no loss or gain of kinetic energy. However, there may be redistribution of energy during such collisions.*
6. *There are no attractive or repulsive forces between the molecules of the gas. They are completely independent of each other.*
7. *At any instant, different molecules possess different velocities and hence, different energies. However, the average kinetic energy of the molecules is directly proportional to the absolute temperature.*

JUSTIFICATION FOR THE POSTULATES

There is ample evidence in favour of the postulates of the kinetic theory as described below:

(*i*) The first postulate is in accordance with the particle nature of matter.

(*ii*) The high compressibility of gases provides ample proof for the existence of large empty spaces between gas molecules. For example, at S.T.P. the empty space in 1 mole of N_2 gas has been calculated to be 99.91%.

(*iii*) Dust particles or smoke particles are found to be in a constant zig-zag movement called **Brownian movement**. This is evident due to the fact that they are being tossed about by the fast moving molecules of the gas present in air (see Activity 14.1).

(*iv*) As a result of their movement, the molecules go on colliding with the walls of the container. As a result the walls of the container experience outward force.

This force per unit area on the walls is called pressure of the gas.

(*v*) During the course of their rapid movement, the molecules of the gas collide with one another and also with the walls of the container. If these collisions were non-elastic, there would have been loss of energy and ultimately the molecules would come to rest and gases should gradually settle down, but this does not happen.

(*vi*) The indefinite expansion of gases and their tendency to occupy the available space implies that the movement of molecules occurs independent of one another. This is possible if the attractive forces between the molecules are negligible.

(*vii*) On increasing the temperature of the gas, the molecular motion becomes rapid resulting in an increase in the average kinetic energy of the gas molecules.

GAS LAWS AND BEHAVIOUR OF GASES

The extensive research on the physical properties of gases carried out by different scientists for several centuries led to the formulation of certain generalisations. These generalisations are known as **gas laws.** We shall now, study some of the gas laws in details.

15.3 BOYLE'S LAW

This law describes the ***pressure-volume*** relationship of gases at constant temperature. It was given by an Anglo-Irish scientist Robert Boyle (1662) and is known after his name as **Boyle's law**. The law states that:

The volume of a fixed mass of a gas is inversely proportional to its pressure at constant temperature.

MATHEMATICAL INTERPRETATION OF THE LAW

Mathematically, the law may be expressed as:

$$V \propto \frac{1}{P} \quad \textit{(Temperature and mass being constant)}$$

or $V = k_1 \frac{1}{P}$ where k_1 is constant of proportionality

or $PV = k_1 = \text{constant}$.

The value of the constant k_1 depends upon *amount of the gas, temperature of the gas* and the *units* in which P and V are expressed.

Thus, another statement of Boyle's law may be given as follows:

For a given amount of gas, the product of pressure and volume is constant at constant temperature.

Let V_1 be the volume of a given mass of gas having pressure P_1 at temperature T. Now if the pressure is changed to P_2 at the same temperature, let the volume change to V_2. The quantitative relationship between the four variables P_1, V_1, P_2 and V_2 is:

$$\boxed{P_1V_1 = P_2V_2} \text{ or } \boxed{\frac{P_1}{P_2} = \frac{V_2}{V_1}} \quad \textit{(T and n remain constant)}$$

ACTIVITY 15.1	**To study the effect of pressure on volume of a fixed mass of gas.**

Apparatus Required

Burette, open glass tube, rubber tube mercury.

Procedure

Set up the apparatus as shown in Fig. 15.2 and follow the following steps:

1. Open the stopcock and lower the reservoir till the Hg level almost touches the stopcock.
2. Attach the tube containing anhydrous calcium chloride to the burette and raise the reservoir again until the burette is about half-filled with dry air. Now close the stopcock and remove the calcium chloride tube.
3. Adjust the height of the reservoir until the mercury levels in the arms are equal. Now record
   (*i*) the volume (V cm^3) of the *trapped air* in the burette;
   (*ii*) the barometric, or the atmospheric pressure reading (let it be H mm Hg)
   (*iii*) the mercury levels, A and B.

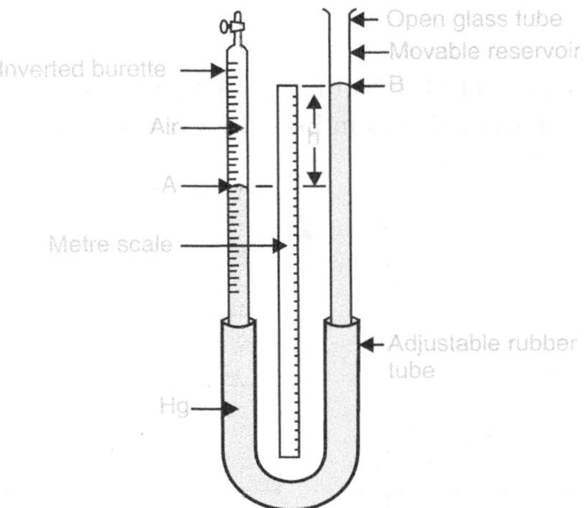

Fig. 15.2. The effect of pressure on the volume of a gas at constant temperature.

4. Now, lower the reservoir until its mercury level is 4–5 cm below that of the burette. Then repeat the readings as done in step 3.
5. Obtain similar readings at six different levels, three with the mercury level in the reservoir above and three with the mercury level below the level inside the burette.

Volume of trapped air (V cm^3)	*Difference in Height of mercury levels in A and B (h mm)*	*Pressure of the trapped air *(P mm Hg)*	*PV = K*	$\frac{1}{P}$	$\frac{1}{V}$

**Pressure of trapped air (P) is (H + h) mm Hg when level B is above A, while it is (H – h) mm Hg when level B is below A.*

6. Now plot the following graphs:
   (*a*) between P(vertical axis) *vs* V(horizontal axis)
   (*b*) between V(vertical axis) *vs* $\frac{1}{P}$ (horizontal axis)
   (*c*) between P(vertical axis) *vs* 1/V (horizontal axis)
   (*d*) between PV(vertical axis) *vs* P (horizontal axis)

   Try to answer the following:
   (*i*) What is the shape of the graph in each case?
   (*ii*) What is the necessity of attaching the calcium chloride tube?
   (*iii*) Which of the graph is not a straight line and why?
   (*iv*) What conclusion can you draw from this activity?

GRAPHICAL REPRESENTATION OF BOYLE'S LAW

The law can also be illustrated by means of *pressure-volume* curves as shown in Figs. 15.3 (*a*), 15.3 (*b*), 15.3 (*c*) and 15.3 (*d*).

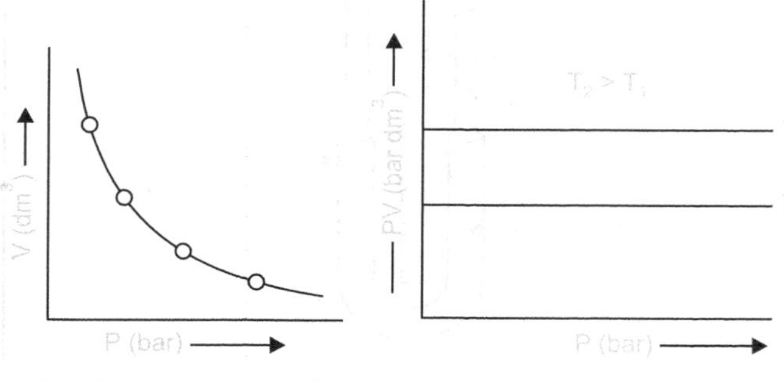

(a) Plot of V against P. (b) Plot of PV against P.

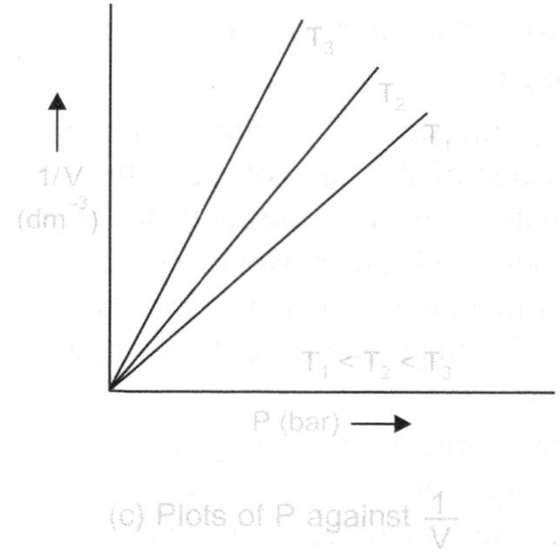

(c) Plots of P against $\frac{1}{V}$

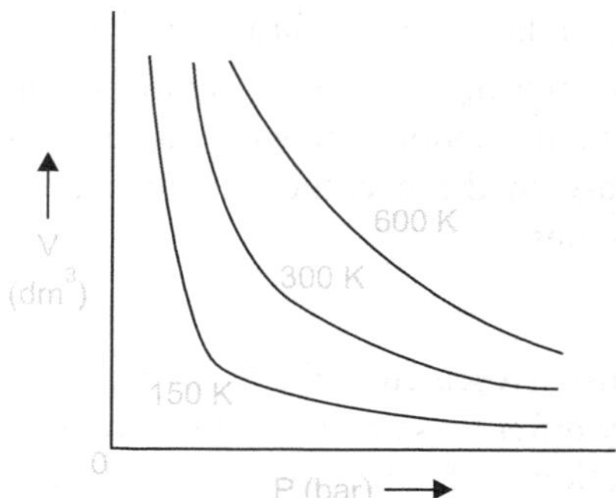

(d) P-V plots at different temperatures for fixed amount of gas.

Fig. 15.3. Different plots of pressure-volume relationship of a gas.

The plots depicting the pressure-volume behaviour at constant temperature are called **isotherms.**

KINETIC THEORY AND BOYLE'S LAW

The kinetic molecular theory assumes that the pressure exerted by a gas is due to the bombardment of its molecules on the walls of a container. Thus, the pressure depends upon the *number of molecular impacts* or *collisions per unit area of the walls per second.* If the temperature is kept constant, the average speed of the molecules $\left(\frac{1}{2}mv^2 \propto \text{T}\right)$ remains the same. Now, if the volume is increased, there will be less number of molecules colliding each unit area of the vessel in a unit time. Consequently, the pressure will decrease. On the other hand, if volume is decreased, there will be more molecules colliding each unit area of walls and hence, the pressure will increase. Thus, it can be concluded that *at a given temperature, the pressure exerted by a gas is inversely proportional to the volume.* This is **Boyle's law.**

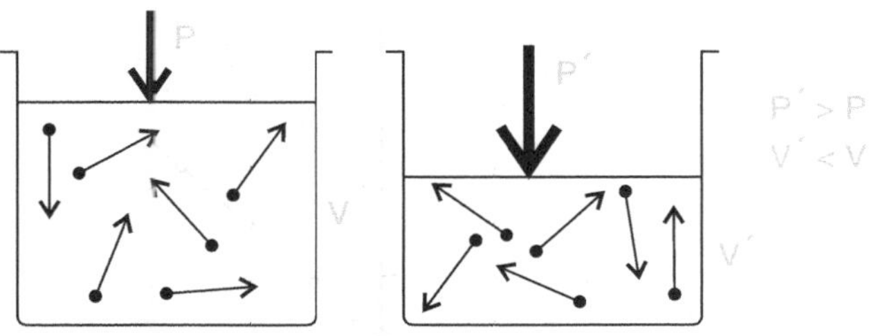

Fig. 15.4. Effect of increase in pressure

SOLVED EXAMPLES

Example 15.1. *A sample of SO_2 gas at a pressure of 1.0 atm occupies a volume of 3.0 L. What volume will it occupy when the pressure is changed to 1.5 atm?*

Solution. Here, $P_1 = 1.0$ atm $\qquad P_2 = 1.5$ atm

$V_1 = 3.0$ L $\qquad V_2 = ?$

From the Boyle's Law equation,

$$P_1V_1 = P_2V_2 \quad \text{or} \quad V_2 = \frac{P_1V_1}{P_2}$$

$$= \frac{1(\text{atm}) \times 3.0(\text{L})}{1.5(\text{atm})}$$

$$= \mathbf{2.0\ L.}$$

Example 15.2. *In a J tube partially filled with mercury the volume of the air column is 4.2 mL and the mercury level in the two limbs is the same. Some mercury is, now added to the tube so that the volume of air enclosed in the shorter limb is now 2.8 cm³. What is the difference in the levels of mercury in this situation? Atmospheric pressure is reported to be 1.0 bar.*

Solution.

Initial pressure $P_1 = 1$ bar $= 0.987$ cm^3

$= 750.12$ mm Hg

Final pressure $P_2 = (P_1 + h)$

$V_1 = 4.2$ cm^3 ; $V_2 = 2.8$ cm^3

$$P_2V_2 = P_1V_1 \quad \text{or} \quad P_2 = \frac{P_1V_1}{V_2}$$

$$= \frac{750.12 \times 4.2}{2.8} = 1125.18 \text{ mm}$$

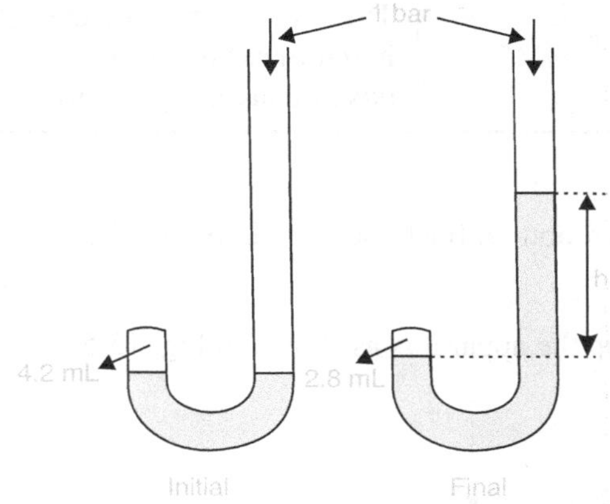

Now, $\quad 1125.18 = 750.12 + h$

or $\quad h = 1125.18 - 750.12$

$= 375.06$ mm or **37.50 cm.**

15.4 CHARLES' LAW

This law describes the ***volume-temperature*** relationship of gases at constant pressure. It was put forwarded by the French chemist *Jacques Charles* in 1787 and was further developed in 1802 by *Joseph Gay Lussac*. This law can be stated as:

The volume of a fixed mass of a gas at constant pressure is directly proportional to the temperature in Kelvin.

MATHEMATICAL INTERPRETATION

Mathematically, the law may be expressed as

$V \propto T$ or $V = k_2T$ (*Pressure and mass remain constant*)

or
$$\frac{V}{T} = k_2$$

where k_2 is a constant of proportionality whose value depends on pressure, the amount of gas and the units in which V is expressed.

Let V_1 be the volume of a certain mass of gas at temperature T_1 and at pressure P. If temperature is changed to T_2 keeping pressure constant, the volume changes to V_2. The relationship between the four variables V_1, T_1, V_2 and T_2 is:

$$\boxed{\frac{V_1}{T_1} = \frac{V_2}{T_2}} \qquad (\textit{P and n remain constant})$$

ACTIVITY 15.2 **To study the effect of an increase in temperature on the volume of gas at constant pressure.**

Apparatus Required

Round bottom flask, beaker, delivery tube.

Procedure

1. Set up the apparatus as shown in Fig. 15.5.

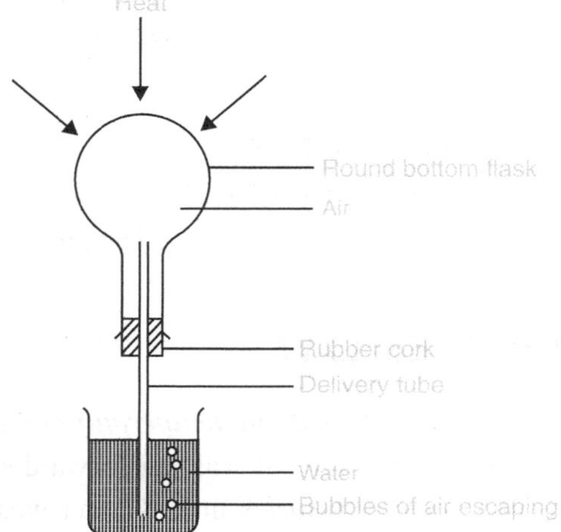

Fig. 15.5

2. Now, warm the air inside the flask gently by placing the hands on the flask. Note your observation.
3. Warm the flask gently with a low bunsen flame. Again note your observation.

   Try to explain your observations.

   Suggest a conclusion for the activity.

GRAPHICAL REPRESENTATION OF CHARLES' LAW

The law can also be illustrated by volume-temperature curves. Fig. 15.6 (*a*) gives a plot of the volume of a given mass of gas against temperature at constant pressure. It also depicts that the volume of a gas at constant pressure is a linear function of its temperature. If we plot a graph of variation of volume with temperature (expressed on the kelvin scale) at constant pressure, we get a graph as shown in Fig. 15.6 (*b*). The curve obtained by plotting volume (V) *vs* temperature at any given pressure is called **isobar.** The slopes of various isobars at different pressures are different, however, all these lines on extrapolation meet the temperature axis at 0 (K) or –273.15°C.

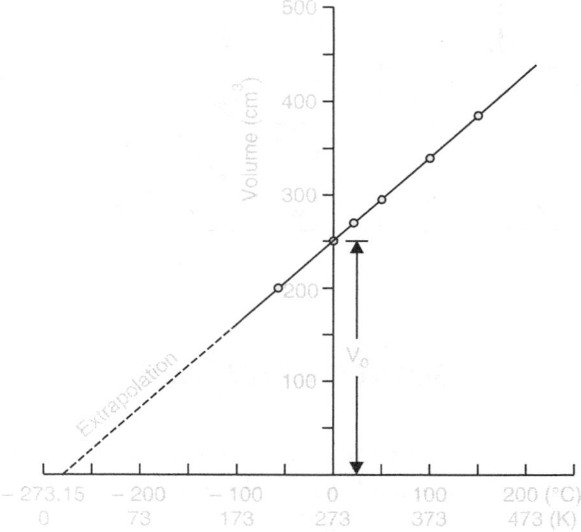

Fig. 15.6. (a) A plot of volume *vs.* temperature

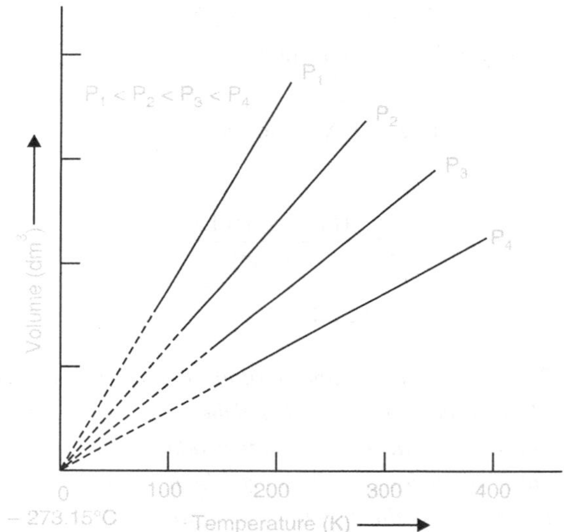

Fig. 15.6. (b) Isobars of ideal gas

KINETIC THEORY AND CHARLES' LAW

When temperature of a gas is increased, the kinetic energy of the molecules increases $\left(\frac{1}{2}mv^2 \propto T\right)$. As a result, the velocity of the molecules increases. Consequently, they collide with the walls of the container more frequently and more vigorously. Therefore, the pressure of the gas is expected to increase. However, if pressure is to be kept **constant,** the force due to *molecular collision on the walls of the container must be kept same*. This is only possible if the volume of the gas increases, *i.e.*, gas expands. On the other hand, if the temperature is decreased, the average velocity of the molecules decreases and, therefore, the pressure decreases. To keep the pressure constant, the volume of the gas must decrease so that the force due to collisions per unit area remains the same. Thus, it can be concluded that "*at a given pressure, the volume of a gas increases with a rise in temperature, and decreases with a decrease in temperature.*" Hence, volume is directly proportional to the temperature at constant pressure. This is **Charles' Law.**

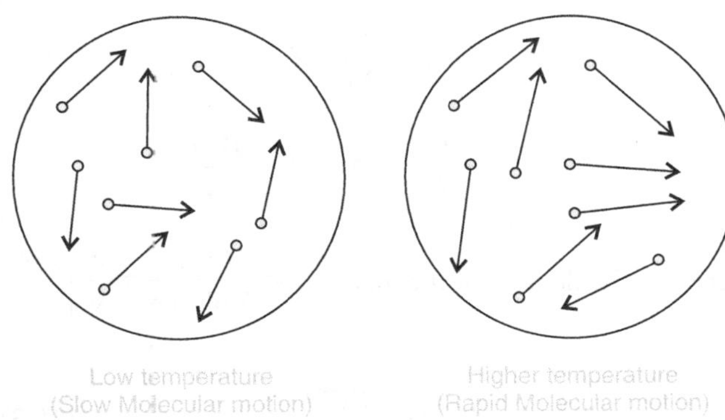

Fig. 15.7. Effect of increase of temperature.

Example 15.3 *A sample of a gas is found to occupy a volume of 900 cm³ at 27°celsius. Calculate the temperature at which it will occupy a volume of 300 cm³, provided the pressure is kept constant.*

Solution. Here, $V_1 = 900\ cm^3$, $V_2 = 300\ cm^3$

$T_1 = (27 + 273)K = 300\ K$, $T_2 = ?$

Applying Charles' law, $\frac{V_1}{T_1} = \frac{V_2}{T_2}$

$$T_2 = \frac{V_2 \times T_1}{V_1} = \frac{300\ cm^3 \times 300\ K}{900\ cm^3}$$

$$= 100\ K = 100 - 273 = -\mathbf{173°C.}$$

15.5 AVOGADRO'S LAW

This law describes the ***volume-amount*** relationship of gases at constant temperature and pressure. It was given by **Amedeo Avogadro** in 1811. It states that *equal volumes of all gases under similar conditions of temperature and pressure contain equal number of molecules.*

For example, 1 mol of all the gases contain 6.023×10^{23} molecules. At the same time 1 mol of all the gases at 273.15 K (0°C) and 100 kPa pressure occupy a volume of 22.7 L ($22.7 \times 10^{-3}\ m^3$). This means that as long as the temperature and pressure remain constant the volume of a gas is directly proportional to the number of molecules or in other words, the amount of the gas. Mathematically we can write

$$V \propto N \qquad (T\ and\ P\ are\ constant)$$

The number of molecules N is directly proportional to the number of moles (n)

$$\therefore \quad V \propto n \qquad (T\ and\ P\ are\ constant)$$

or $V = k_3 n$, k_3 is a constant of proportionality

Now, if m is the mass of the gas having molar mass equal to M, then the number of moles (n) is given as

$$n = \frac{m}{M}$$

$$\therefore \quad V = k_4 \frac{m}{M}$$

or

$$M = k_4 \frac{m}{V} \quad or \quad k_4 . d$$

where d is the density of the gas.

The above relationship implies that the *density of a gas at a given temperature and pressure is directly proportional to its molar mass.*

15.6 IDEAL GAS EQUATION

A gas that follows Boyle's law, Charles' law and Avogadro's law strictly at all conditions, is called an **ideal gas.** It is assumed that intermolecular forces are not present between the molecules of an ideal gas. Real gases follow these laws only under certain specific conditions when forces of interaction are practically negligible. In all other situations the real gases deviate from ideal behaviour.

The combination of various gas laws namely; *Boyle's law, Charles' law* and *Avogadro's law* leads to the development of the mathematical relation which relates four variables *pressure, volume, absolute temperature* and *number of moles of an ideal gas*. The equation so formulated is called **ideal gas equation.**

The ideal gas equation can be derived by combining Boyle's law, Charles' law and Avogadro's law as follows:

According to Boyle's law

$$V \propto \frac{1}{P} \quad (at\ constant\ T\ and\ n) \qquad ...(i)$$

According to Charles' law

$$V \propto T \ (at\ constant\ P\ and\ n) \qquad ...(ii)$$

According to Avogadro's law

$$V \propto n \text{ (at constant } T \text{ and } P) \quad ...(iii)$$

Combining (*i*), (*ii*) and (*iii*)

$$V \propto \frac{nT}{P}$$

or $$PV \propto nT$$

or $$\boxed{PV = nRT} \quad ...(15.1)$$

where, **R** is a constant of proportionality and is known as **universal gas constant.**

Equation (15.1) is called **ideal gas equation.** This equation is applicable to any gas under conditions where the behaviour of the gas approaches ideal behaviour. Since this equation relates four variables which are used to describe the state of any gas, it is also known as the **equation of state for an ideal gas.** It may be noted that out of the four variables, the two namely; *pressure* (P) and *temperature* (T) are **intensive** variables as they do not depend on the bulk or quantity of the gas. The other two variables, *i.e., volume* (V) and *mole* (*n*) are **extensive** variables as they depend upon the bulk or quantity of the gas.

COMBINED GAS LAW OR GENERAL GAS EQUATION

If temperature, volume and pressure of a fixed amount (say **n** mol) of gas vary from T_1, V_1 and P_1 to T_2, V_2 and P_2 respectively, then the ideal gas equation for the two states can be written as

$$P_1V_1 = nRT_1 \quad \text{or} \quad \frac{P_1V_1}{T_1} = nR \quad ...(i)$$

$$P_2V_2 = nRT_2 \quad \text{or} \quad \frac{P_2V_2}{T_2} = nR \quad ...(ii)$$

Combining (*i*) and (*ii*) we have

$$\boxed{\frac{P_1V_1}{T_1} = \frac{P_2V_2}{T_2}} \quad ...(15.2)$$

Equation (15.2) is called a **combined gas law** or a **general gas equation**. This equation is very useful in calculations involving gas laws. The knowledge of the five variables help in calculating the unknown sixth variable.

RELATIONSHIP BETWEEN DENSITY AND MOLECULAR MASS

We have studied in unit one that the number of moles (*n*) of a substance is related to the molar mass (M) as

$$n = \frac{w}{M}$$

where, w = mass in gram.

Substituting this value in the ideal gas equation we can get the relationship between density and molar mass,

$$PV = nRT = \frac{w}{M} RT$$

$$P = \frac{wRT}{MV} \quad \text{or} \quad P = \frac{\mathbf{dRT}}{\mathbf{M}}$$

where d is the density of the gas $\left[d = \frac{\text{Mass}}{\text{Volume}} = \frac{w}{V}\right]$

or $$\mathbf{P = \frac{dRT}{M}}. \quad ...(15.3)$$

NATURE AND NUMERICAL VALUES OF THE GAS CONSTANT (R)

In order to understand the significance of R, let us examine the nature of quantities in the ideal gas equation:

$$PV = nRT \quad \text{or} \quad R = \frac{PV}{nT} = \frac{\text{Pressure} \times \text{Volume}}{\text{Moles} \times \text{Temperature}}$$

$$R = \frac{\frac{\text{Force}}{\text{Area}} \times \text{Volume}}{\text{Moles} \times \text{Temperature}} = \frac{\frac{\text{Force}}{(\text{Length})^2} \times (\text{Length})^3}{\text{Moles} \times \text{Temperature}}$$

$$= \frac{\text{Force} \times \text{Length}}{\text{Moles} \times \text{Temperature}}$$

Since force × length = work energy

$$\therefore \quad R = \frac{\text{Work}}{\text{Moles} \times \text{Temperature}}$$

Thus, **R** *represents* ***work done per degree per mole.***

Since work can be expressed in different systems of units, **R** will have different numerical values in different systems.

Different numerical values of R are given in Table 15.1. It may be noted that although R can be expressed in different units, but for pressure volume calculations R must be taken in the same units as those used for pressure and volume.

Table 15.1. Values of R in Different Units

Value of R	*Units of P*	*Units of V*
0.0821 L atm K^{-1} mol^{-1}	atm	litres
0.0831 L bar K^{-1} mol^{-1}	bar	litres
8.314×10^7 ergs K^{-1} mol^{-1}	dynes cm^{-2}	cm^3
8.314 kPa dm^3 K^{-1} mol^{-1}	kPa	dm^3
8.314 J K^{-1} mol^{-1}	Pa or Nm^{-2}	cm^3
1.99 cal K^{-1} mol^{-1}	dynes cm^{-2}	cm^3

Let us now apply the **ideal gas equation** and the **combined gas laws** to solve some numerical problems. Students may note that *in many of these problems the conversion of temperature in celsius scale to kelvin scale has been done by adding 273 instead of 273.15 in order to make the calculations simple.*

Type I. Solved Examples Based on Combined Gas law: $\left(\frac{P_1V_1}{T_1} = \frac{P_2V_2}{T_2}\right)$

Example 15.4. *A sample of nitrogen gas occupies a volume of 320 cm³ at S.T.P. Calculate its volume at 66°C and 82.5 kPa pressure.*

Solution. Here, $P_1 = 100$ kPa, $P_2 = 82.5$ kPa

$V_1 = 320$ cm³, $V_2 = ?$

$T_1 = 273.15$ K

$T_2 = 66°C = (66 + 273.15)$ K $= 339.15$ K

According to the gas equation,

$$\frac{P_1V_1}{T_1} = \frac{P_2V_2}{T_2} = \frac{320 \times 339.15}{273.15 \times 0.825}$$

$$\therefore \quad V_2 = \frac{P_1V_1T_2}{T_1P_2} = \frac{100 \times 320 \times 339}{273 \times 82.5} = \mathbf{481.6\ cm^3}.$$

Example 15.5. *1.0 mol of pure dinitrogen gas at 25°C and 100 kPa pressure has a volume 24.8×10^{-3} m³. It was then put into a vessel of volume 0.025 m³, maintained at the temperature of 50°C. What is the pressure of the gas in the vessel?*

Solution. Volume of 1.0 mol of a gas at STP (V_1) $= 24.8 \times 10^{-3}$ m³

Initial conditions	*Final conditions*
$V_1 = 24.8 \times 10^{-3}$ m³	$V_2 = 0.1$ m³
$P_1 = 100$ kPa	$P_2 = ?$
$T_1 = 298.15$ K	$T_2 = 50°C = 323.15$ K

According to the general gas equation

$$\frac{P_1V_1}{T_1} = \frac{P_2V_2}{T_2}$$

or

$$P_2 = \frac{P_1V_1T_2}{T_1V_2} = \frac{100 \times 24.8 \times 10^{-3} \times 323.15}{298.15 \times 0.025}$$

$= \mathbf{107.6\ kPa.}$

Type II. Solved Examples Based on Ideal Gas Equation: (PV = nRT)

Example 15.6. *How many moles of oxygen are present in a 400 cm³ sample of a gas at a pressure of 760 mm of Hg and a temperature of 300 K. (The value of R is given to be 8.31 kPa dm³ k^{-1} mol^{-1}.)*

Solution. Since the value of R is given in SI units, it is necessary to express the pressure and volume in *kilo pascals* and *dm³* respectively.

Thus, $P = 760$ mm $= 101.3$ kPa

$V = 400$ cm³ $= 400 \times 10^{-3}$ dm³ $= 0.4$ dm³

$T = 300$ K

$n = ?$

According to the ideal gas equation.

$$PV = nRT$$

$$\therefore \quad n = \frac{PV}{RT} = \frac{101.3 \times 0.40}{8.31 \times 300}$$

$= 0.0162 = \mathbf{1.62 \times 10^{-2}\ mol.}$

Example 15.7. *Calculate the pressure of 1×10^{22} molecules of sulphur dioxide when enclosed in a vessel of 2.5 dm³ capacity at a temperature of 27°C.*

Solution. Step I. *Calculation of number of moles*

The number of moles of SO_2

$$= \frac{1 \times 10^{22}}{6.02 \times 10^{23}} = 0.166 \times 10^{-1} \text{ mol}$$

$= 1.66 \times 10^{-2}$ mol.

Step II. *Calculation of pressure*

Here, $V = 2.5$ dm³ or 2.5×10^{-3} m³, $P = ?$

$T = 27°C = 300.15$ K

$n = 1.66 \times 10^{-2}$ mol.

Applying ideal gas equation,

$$PV = nRT$$

$$P = \frac{nRT}{V}$$

$$= \frac{1.66 \times 10^{-2} \times 8.314 \times 300.15}{2.5}$$

$$= \mathbf{16.57\ kPa}$$

Example 15.8. *A balloon contains 800 cm³ of Hg at 0°C and 1 bar pressure. The balloon can attain a maximum volume of 1.5 dm³ after which it will burst. How will (i) temperature and (ii) quantity of H_2 be changed to attain the maximum volume?*

Solution. (*i*) $V_1 = 800\ cm^3 \quad T_1 = 0°C = 273\ K$

$V_2 = 1500\ cm^3 \quad T_2 = ?$

According to Charle's law

$$\frac{V_1}{T_1} = \frac{V_2}{T_2} \quad \text{or} \quad T_2 = \frac{V_2 \times T_1}{V_1} = \frac{1500 \times 273}{800} = 511.87\ K$$

or $511.87 - 273 = \mathbf{238.8°C}$

Thus, the temperature should be raised to 238.8°C.

(*ii*) **Quantity of H_2**

Calculate moles of H_2 already present

$PV = nRT$

$$\therefore \quad n = \frac{PV}{RT} = \frac{1(\text{bar}) \times 800 \times 10^{-3}\ (\text{L})}{0.083\ (\text{L-bar mol}^{-1}\ \text{K}^{-1}) \times 273\ (\text{K})}$$

$= 0.035$ mol

Calculate moles present in 1.5 L volume

$$n = \frac{PV}{RT} = \frac{1(\text{bar}) \times 1.5(\text{L})}{0.083\ (\text{L-bar mol}^{-1}\ \text{K}^{-1}) \times 273\ \text{K}}$$

$= 0.066$ mol.

Thus additional H_2 required for attaining a volume of 1.5 L

$= 0.066 - 0.035$

$= \mathbf{0.031\ mol}$

or $0.031 \times 2 = \mathbf{0.062\ g.}$

Type III. Solved Examples Involving Molar Mass and Density $\left(P = \frac{dRT}{M}\right)$

Example 15.9. *The density of a certain gaseous oxide at 1.5 bar pressure at 10°C is the same as that of dioxygen at 20°C and 4.5 bar pressure. Calculate the molar mass of the gaseous oxide.*

Solution. Density of dioxygen (O_2) at 4.5 bar pressure and 10°C

$$d_{O_2} = \frac{PM}{RT} = \frac{4.5\ (\text{bar}) \times 32\ (\text{g mol}^{-1})}{R \times 283.15\ (\text{K})}$$

$$d_{\text{oxide}} = \frac{PM}{RT} = \frac{1.5\ (\text{bar}) \times M\ (\text{g mol}^{-1})}{R \times 293.15\ (\text{K})}$$

Now, $d_{O_2} = d_{\text{oxide}}$

$$\therefore \quad \frac{4.5\ (\text{bar}) \times 32\ (\text{g mol}^{-1})}{R \times 283.15\ (\text{K})} = \frac{1.5\ (\text{bar}) \times M\ (\text{g mol}^{-1})}{R \times 293.15\ (\text{K})}$$

$$\text{or} \quad M = \frac{4.5 \times 293.15 \times 32}{283.15 \times 1.5}$$

$= \mathbf{99.39\ g\ mol^{-1}.}$

Example 15.10. *Density of a gas is found to be 5.46 g/dm³ at 27°C and 200 kPa pressure. What will be its density at S.T.P.?*

Solution.

$$d_1 = \frac{P_1M}{RT_1};\ d_2 = \frac{P_2M}{RT_2}$$

$$\text{or} \quad \frac{d_1}{d_2} = \frac{P_1T_2}{T_1P_2} \quad \text{or} \quad d_2 = \frac{d_1T_1P_2}{P_1T_2}$$

Substituting the values, we have

$$d_2 = \frac{5.46\ (\text{g dm}^{-3}) \times 300\ (\text{K}) \times 100\ (\text{kPa})}{200\ (\text{kPa}) \times 273.15\ (\text{K})}$$

$= 2.998$ or $\mathbf{3.0\ g\ dm^{-3}.}$

15.7 DALTON'S LAW OF PARTIAL PRESSURES

This law describes the relation between the pressure of the mixture of non-reacting gases enclosed in a vessel to their individual pressures. The law was given by John Dalton in 1807. It states:

At constant temperature, the pressure exerted by a mixture of two or more non-reacting gases enclosed in a definite volume, is equal to the sum of the individual pressures which each gas would exert if present alone in the same volume at the same temperature.

The individual pressures of gases are known as **partial pressures.**

If P is the total pressure of the mixture of non-reacting gases at temperature T and volume V, and P_1, P_2, P_3, ... represent the partial pressures of the gases, then

$$P = P_1 + P_2 + P_3 + ... \quad (T, V \text{ are constant})$$

The law can be **illustrated** by considering the following example:

Suppose we have three containers of capacity 1 litre each; one containing *x moles* of nitrogen, the other *y moles* of oxygen and the third having a mixture of *x moles* of nitrogen and *y moles* oxygen. All the three containers, are kept at the same temperature.

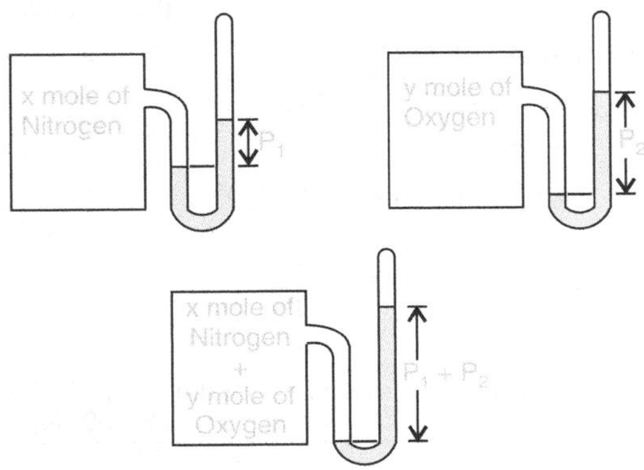

Fig. 15.8. Illustration of Dalton's law of partial pressures

Now, if the manometer, attached to first container shows a pressure P_1 and that attached to second container shows a pressure P_2, then the pressure in the third container is $P_1 + P_2$.

Utility of Dalton's law. This law is useful in calculating the pressure of gas collected by the displacement of water. The gas being collected over water also contains water vapour. The observed pressure of the moist gas is equal to the sum of the pressure of the dry gas and the pressure of the water vapour. The pressure of the water vapour is constant at a given temperature and is known as **aqueous tension** at that temperature.

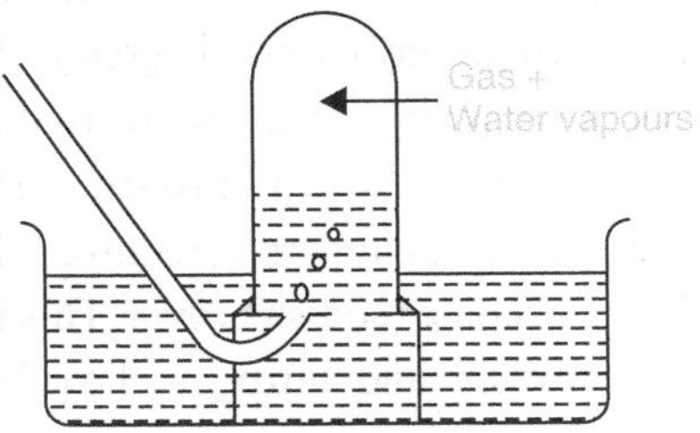

Fig. 15.9. Collection of gas over water.

Thus,

$$P_{observed} = P_{gas} + \textit{aqueous tension}$$

$$\therefore \quad P_{gas} = P_{observed} - \textit{aqueous tension}$$

Aqueous tension of water at different temperatures is given in Table 15.2.

Table 15.2. Aqueous Tension (Vapour Pressure) of Water as a Function of Temperature

Temperature (K)	*Pressure × 10^2 kPa*	*Temperature (K)*	*Pressure × 10^2 kPa*
273.15	0.0060	295.15	0.0260
283.15	0.0121	297.15	0.0295
288.15	0.0168	299.15	0.0331
291.15	0.0204	301.15	0.0372
293.15	0.0230	303.15	0.0418

Partial Pressure in Terms of Mol Fraction

Mol fraction is one of the method of expressing the concentration of a component in a mixture. Mole fraction *is the ratio of the number of mol of a component to the total number of moles*. If n_1 is the number of moles of any component present in a mixture and n is the total number of moles of all the components in a mixture, then, the mole fraction (x_1) of the component is given as

$$x_1 = \frac{n_1}{n}$$

Now consider a mixture of two gases A and B with p_A and p_B as their respective partial pressures in a vessel of volume V

$$p_A = n_A RT/V; \; p_B = n_B RT/V$$

Total pressure of the mixture

$$p_{Total} = p_A + p_B$$

or $$p_{Total} = p_A + p_B = (n_A + n_B)\, RT/V$$

Now, $$\frac{p_A}{p_{Total}} = \frac{n_A}{(n_A + n_B)} = x_A$$

or $$p_A = x_A \times p_{Total}$$

Similarly $$p_B = x_B \times p_{Total}$$

Thus, *the partial pressure of a gas in a mixture is equal to the product of its mol fraction and the total pressure of the mixture.*

SOLVED EXAMPLES

Example 15.11. *A certain quantity of a gas occupies a volume of 0.1 dm³ when collected over water at 288 K and a pressure 92 kPa. The same gas occupied a volume of 0.085 dm³ at S.T.P. in dry conditions. Calculate the aqueous tension at 288 K.*

Solution. Let the aqueous tension at 288 K be p kPa

Thus, pressure of the dry gas at 288 K

$= (92 - p)$ kPa.

Now, $P_1 = (92 - p)$ kPa $\quad P_2 = 100$ kPa.

$V_1 = 0.1\ dm^3 \quad V_2 = 0.085\ dm^3$

$T_1 = 288$ K $\quad T_2 = 273.15$ K.

According to the gas equation,

$$\frac{P_1V_1}{T_1} = \frac{P_2V_2}{T_2} \quad \text{or} \quad P_1 = \frac{P_2V_2T_1}{V_1T_2}$$

Substituting the values,

$$92 - p = \frac{100 \times 0.085 \times 288}{0.1 \times 273.15} = 89.6$$

$p =$ **2.4 kPa.**

Example 15.12. *At sea level, the composition of dry air is approximately, N_2 = 75.5% ; O_2 = 23.2% and Ar = 1.3% by mass. If the total pressure at sea level is 1 bar, what is the partial pressure of each component?*

Solution. Consider 100 g of air

Moles of $N_2 (n_{N_2}) = \frac{w}{\text{Molar mass}} = \frac{75.5}{28} = 2.7$

Moles of $O_2 (n_{O_2}) = \frac{23.2}{32} = 0.725$

Moles of Ar $(n_{Ar}) = \frac{1.3}{40} = 0.0325$

Total moles $= 2.7 + 0.725 + 0.0325 = 3.4575$

Mole fraction of $N_2 (x_{N_2}) = \frac{2.7}{3.4575} = 0.781$

Mole fraction of $O_2 (x_{O_2}) = \frac{0.725}{3.4575} = 0.209$

Mole fraction of Ar $(x_{Ar}) = \frac{0.0325}{3.4575} = 9.39 \times 10^{-3}$

$p_{N_2} = p_{total} \times x_{N_2} = 1 \times 0.781 =$ **0.781 bar**

$p_{O_2} = p_{total} \times x_{O_2} = 1 \times 0.209 =$ **0.209 bar**

$p_{Ar} = p_{total} \times x_{Ar} = 1 \times 9.39 \times 10^{-3}$

$=$ **9.39×10^{-3} bar.**

Example 15.13. *10 g of potassium trioxochlorate (V), $KClO_3$ was heated to liberate oxygen which was collected over water at 300 K and 90 kPa pressure. What is the volume of dry oxygen at S.T.P. (Molecular masses $KClO_3$ = 122.5; O_2 = 32.0, Molar volume = 22.4 dm³, saturated vapour pressure of water at 300 K = 37 kPa.)?*

Solution. The reaction involved is

$$2KClO_3(s) \longrightarrow 2KCl(s) + 3O_2(g)$$

2 mol $\qquad$ 3 mol

or $\quad 2 \times 122.5$ g $\qquad$ 3 mol

Now, mol of O_2 produced by 2×122.5 g of $KClO_3$ = 3 mol

Mol of O_2 produced by 10 g of $KClO_3 = \frac{3 \times 10}{2 \times 122.5}$

$= 0.1224$ mol.

Calculation of Vol. of O_2

$P_{\text{dry gas}} = 90 - 37$ kPa $= 53$ kPa

$T = 300$ K

$R = 8.314$ kPa dm³ mol⁻¹ K⁻¹

$$PV = nRT \quad \text{or} \quad V = \frac{nRT}{P}$$

$$\therefore \quad V_{O_2} = \frac{0.1224\ (\text{mol}) \times 8.314\ (\text{kPa dm}^3\ \text{mol}^{-1}\ \text{K}^{-1}) \times 300\ (\text{K})}{53\ \text{kPa}}$$

$= 5.76$ dm³

convert this volume at S.T.P

$$\frac{P_1V_1}{T_1} = \frac{P_2V_2}{T_2} \quad \text{or} \quad V_2 = \frac{P_1V_1 \times T_2}{T_1 \times P_2}$$

(STP)

or $\quad V_2 = \frac{53 \times 5.76 \times 273}{300 \times 100}$

$=$ **2.78 dm³.**

15.8 DIFFUSION OF GASES

It is a matter of common observation that gases intermix freely without the help of any external agency. For example, if a cylinder containing hydrogen gas is inverted over another cylinder containing reddish brown nitrogen dioxide gas and the lid separating the two cylinders is removed then both the cylinders gradually become reddish brown. This indicates the intermixing of hydrogen and nitrogen dioxide forming a uniform mixture.

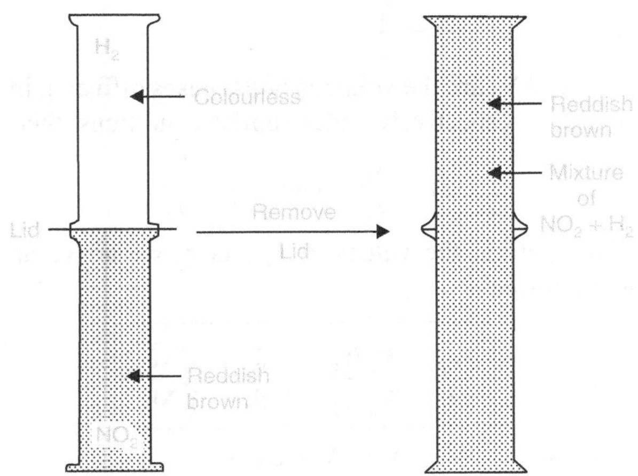

Fig. 15.10. Diffusion of gases.

This property of intermixing of gases is referred to as *diffusion*. **Diffusion** may, thus, be defined as the *process of intermixing of two or more gases irrespective of density and without the help of external agency.*

The diffusion in gases is attributed to the *rapid movement of gas molecules* and the *existence of large empty space between* the molecules. When the two gases are brought in contact with each other, the molecules of one gas move into the empty spaces of the other gas gradually and this ultimately leads to the formation of a homogeneous mixture.

EFFUSION

Effusion is a process by which a gas under pressure escapes from a vessel through a small opening or an orifice. Escape of air through a punctured tyre or football bladder and escape of perfume molecules through an atomizer are common examples of effusion. The process of effusion is always followed by diffusion.

ACTIVITY 15.3	**To study the relationship between the rate of diffusion of a gas and its molar mass by using H_2 and CO_2 gases.**

A. Diffusion of Hydrogen and Air

1. Set up the apparatus as shown in Fig. 15.11 but without the dilute hydrochloric acid. Note the levels of the *coloured water* in arms A and B of the U-tube.

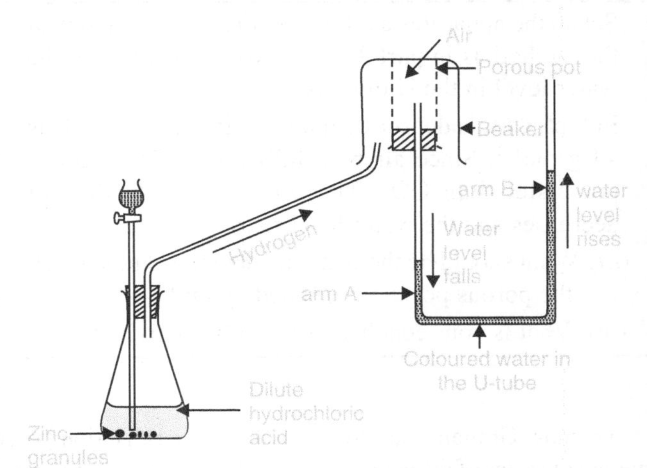

Fig. 15.11. H_2 diffuses faster than air.

2. Add some dilute hydrochloric acid to the zinc metal. Note the levels of the coloured water in the U-tube as hydrogen gas is produced.

   (*i*) Explain your observation.

3. Stop the hydrogen supply by removing the conical flask. Then remove the beaker. Note the levels of the *coloured water* again.

   (*ii*) Explain your observation.

   Molar mass of hydrogen z g mol^{-1}. Gas is lighter than air. Lowering of water level indicates that diffusion of hydrogen into the porous pot is faster than the diffusion of air out of the porous pot. The more accumulation of hydrogen increases the pressure inside the pot and the water level is pushed down.

B. Diffusion of CO_2 and Air

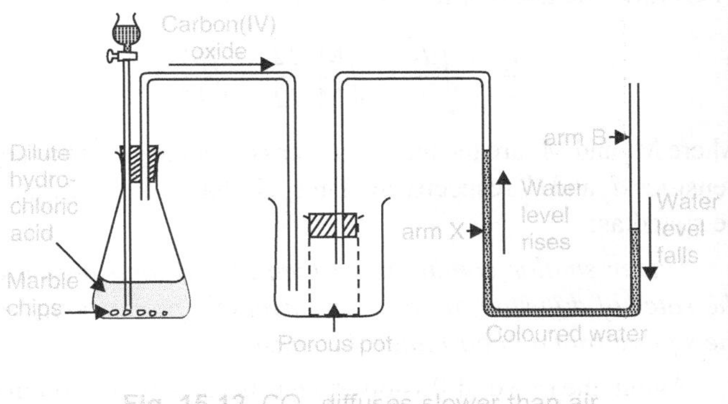

Fig. 15.12. CO_2 diffuses slower than air.

1. Set up the apparatus as shown in Fig. 15.12 and repeat the method as in part A. Here, you will find that the water level in the U tube rises.

   In fact, CO_2 is denser than air. Molar mass of CO_2 is 44 g mol^{-1}. Since air will diffuse out of the porous pot faster than CO_2. The pressure inside the pot decreases and the water level rises.

   (*iii*) What should be the water level in the U tube when the porous pot is surrounded by air?

   (*iv*) What is your conclusion about the activity?

GRAHAM'S LAW OF DIFFUSION/EFFUSION

Thomas Graham put forward a generalisation after studying the rates of diffusion of different gases, which is known after his name as *Graham's law of diffusion.* The law states:

Under similar conditions of temperature and pressure, the rates of diffusion of gases are inversely proportional to the square roots of their densities.

The law can be mathematically put as

$$r \propto \frac{1}{\sqrt{d}}$$

where r is the rate of diffusion and d is the density of the gas.

Now, if there are two gases A and B having r_1 and r_2 as their rates of diffusion, and d_1 and d_2 as their densities respectively, then

$$r_1 \propto \frac{1}{\sqrt{d_1}} \quad \text{and} \quad r_2 \propto \frac{1}{\sqrt{d_2}}$$

or

$$\frac{r_1}{r_2} = \sqrt{\frac{d_2}{d_1}} \quad \text{(at same T and P)}$$

We know that molecular mass is twice the vapour density. Therefore, the above expression may be written as

$$\frac{r_1}{r_2} = \sqrt{\frac{d_2}{d_1}} = \sqrt{\frac{M_2/2}{M_1/2}} = \sqrt{\frac{M_2}{M_1}}$$

where M_1 and M_2 are the molecular masses of the gases having densities d_1 and d_2 respectively. Thus, Graham's law may also be stated as:

Under similar conditions of temperature and pressure the rates of diffusion of gases are inversely proportional to the square root of their molecular masses.

Again, the rate of diffusion of a gas is equal to the volume of the gas which diffuses per unit time, *i.e.*,

$$\text{Rate of diffusion} = \frac{\text{Volume of the gas diffused}}{\text{Time taken for diffusion}}$$

or

$$r = \frac{V}{t}$$

If V_1 and V_2 are the volumes of the gases diffusing in the time t_1 and t_2 respectively under similar conditions, then

$$r_1 = \frac{V_1}{t_1} \quad \text{and} \quad r_2 = \frac{V_2}{t_2}$$

Thus, putting the values of r_1 and r_2 we arrive at the following formula:

$$\boxed{\frac{\mathbf{r_1}}{\mathbf{r_2}} = \frac{\mathbf{V_1/t_1}}{\mathbf{V_2/t_2}} = \sqrt{\frac{\mathbf{d_2}}{\mathbf{d_1}}} = \sqrt{\frac{\mathbf{M_2}}{\mathbf{M_1}}}}$$

Now, if $V_1 = V_2 = V$, we get

$$\frac{t_2}{t_1} = \sqrt{\frac{d_2}{d_1}} = \sqrt{\frac{M_2}{M_1}}$$

This implies that *the time taken for the diffusion of equal volumes of two gases under similar conditions of temperature and pressure is directly proportional to the square root of their densities or molecular masses.*

Similarly, if $t_1 = t_2 = t$, then $\frac{V_1}{V_2} = \sqrt{\frac{d_2}{d_1}} = \sqrt{\frac{M_2}{M_1}}$

It means that *the volumes of the two gases which diffuse in the same time under similar conditions are inversely proportional to the square roots of their densities or molecular masses.*

ACTIVITY 15.4 | **To study the diffusion of two gases**

1. Clamp a long glass tube about (1 m × 25 mm) dimensions in a horizontal position as shown in Fig. 15.13. Fix long strips of blue and red litmus papers with the help of a cellotape. Insert cotton wool plugs at both ends.
2. With the help of a medicine dropper place 5 drops of concentrated hydrochloric acid on one cotton plug.

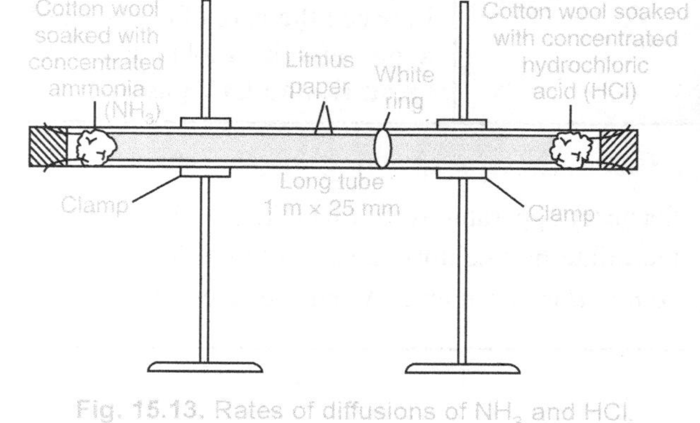

Fig. 15.13. Rates of diffusions of NH_3 and HCl.

At the same time another student must place 5 drops of concentrated ammonia on the cotton plug at the other end. As quickly as possible start the stopwatch and cork at both ends of the tube.

3. When a white ring of smoke is formed stop the stopwatch and record the time. Mark the point where the ring appears first and measure the distance of this point from both ends.

Make the following record of observations

(*a*) Initial time (t_1) when drops are introduced ......

(*b*) Final time (t_2) when the white ring appears first ......

(*c*) Time taken for diffusion 't' = $(t_2 - t_1)$

(*d*) Distance travelled by HCl (x_1) =

(*e*) Distance travelled by NH_3 (x_2) =

(*f*) Rate of diffusion of HCl (r_1) = x_1/t

(*g*) Rate of diffusion of NH_3 (r_2) = x_2/t

(*h*) Find the ratio r_2/r_1

(*i*) Compare this ratio to another ratio

$$\sqrt{\frac{M_{HCl}}{M_{NH_3}}} \text{ or } \sqrt{\frac{36.5}{17}}$$

(*j*) Verify Graham's Law.

SOLVED EXAMPLES

Let us now apply **Graham's law** to solve some **numerical problems.**

Example 15.14. *Compare the rates of diffusion of* $^{235}UF_6$ *and* $^{238}UF_6$

Solution. Let the rate of diffusion of $^{235}UF_6$ be r_1 and the rate of diffusion of $^{238}UF_6$ be r_2

Now, molecular mass of $^{235}UF_6(M_1) = 235 + 6 \times 19 = 349$

Molecular mass of $^{238}UF_6(M_2) = 238 + 6 \times 19 = 352$

According to Graham's law $\frac{r_1}{r_2} = \sqrt{\frac{M_2}{M_1}}$

$$\frac{r_1}{r_2} = \sqrt{\frac{352}{349}} = 1.004.$$

Thus, **r(^{235}UF$_6$) : r(^{238}UF$_6$) is 1.004 : 1.**

Example 15.15. *Relative densities of oxygen and carbon (IV) oxide are 16 and 22 respectively. If 25 cm^3 of carbon (IV) oxide effuse out in 75 seconds, what volume of oxygen will effuse out in 96 seconds under similar conditions.*

Solution. Here, volume of carbon (IV) oxide

$$V_{CO_2} = 25 \text{ cm}^3$$

Time taken $t_{CO_2} = 75$ seconds

Let the volume of oxygen effused

$$V_{O_2} = V \text{ cm}^3$$

Time taken $t_{O_2} = 96$ seconds

Relative density of carbon (IV) oxide, $d_{CO_2} = 22$

Relative density of oxygen $d_{O_2} = 16$ seconds

Now, $$\frac{r_{O_2}}{r_{CO_2}} = \sqrt{\frac{d_{CO_2}}{d_{O_2}}} \quad \text{or} \quad \frac{V_{O_2}}{t_{O_2}} \Big/ \frac{V_{CO_2}}{t_{CO_2}} = \sqrt{\frac{22}{16}}$$

or $$\frac{V_{O_2}}{t_{O_2}} \times \frac{t_{CO_2}}{V_{CO_2}} = \sqrt{\frac{22}{16}}$$

Substituting the values

$$\frac{V_{O_2} \times 75}{96 \times 25} = \sqrt{\frac{22}{16}} = 1.172$$

or $$V_{O_2} = \frac{1.172 \times 96 \times 25}{75} = 37.5 \text{ cm}^3$$

Thus, volume of oxygen effused = **37.5 cm^3.**

Example 15.16 *A certain gaseous fluoride of phosphorus has a formula* PF_x. *Under similar conditions fluorine diffuses 1.82 times faster than the gaseous phosphorus fluoride. Find the value of x and formula of the phosphorus fluoride. (Atomic masses, P = 31.0; F = 19.0).*

Solution. According Graham's law

$$\frac{r_{F_2}}{r(PF_x)} = \frac{1.82}{1} = \sqrt{\frac{M_{(PF_x)}}{M_{(F_2)}}}$$

$$M_{(F_2)} = 2 \times 19 = 38$$

$$\frac{1.82}{1} = \sqrt{\frac{M_{(PF_x)}}{38}} \text{ or } M_{(PF_x)} = \frac{1.82 \times 1.82 \times 38}{1}$$

$$= 125.8 \quad ...(i)$$

Molecular mass of PF_x from formula = $31 + 19x$ ...(*ii*)

Equate (*i*) and (*ii*)

$$31 + 19x = 125.8 \quad \text{or} \quad x = \frac{125.8 - 31}{19} = 4.98 \approx 5$$

∴ The formula of phosphorus fluoride = **PF_5.**

DIFFUSION IN LIQUIDS IS MUCH SLOWER THAN THAT IN GASES

We have studied in the previous unit that the process of diffusion also occurs in liquids but the rate of diffusion in liquids is very slow. The slow rate of diffusion in liquids can be observed by the following simple experiment:

1. Take water in a long, clean transparent glass jar.
2. With the help of a long pipette put 4–5 drops of concentrated blue coloured solution of *copper (II) tetraoxo-sulphate(VI)* at the bottom of the jar.

   Keep it undisturbed and observe the change.

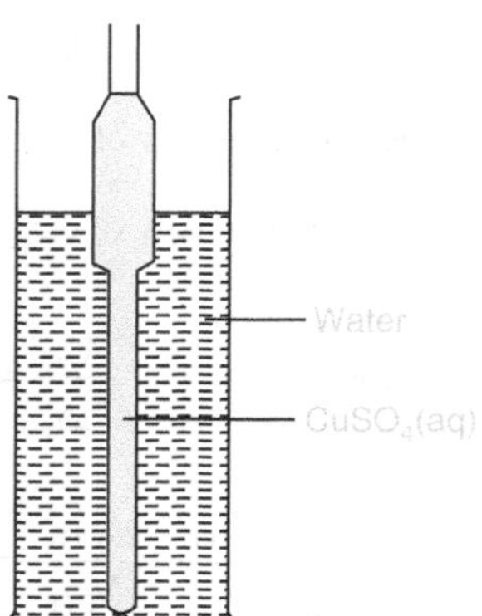

3. You will notice the slow intermixing of liquids and it will take quite a long time for the blue colour to become uniform throughout the jar.

Why Diffusion in Liquids is Slow?

The molecular motion in gases is very rapid and intermolecular voids are large. On the other hand, molecular motion in liquids is much slower and restricted. This makes the intermixing of liquids very slow.

ACTIVITY 15.5	**Experiment to Demonstrate the Diffusion of Solute Particles in Liquids**

Requirements

1M $Pb(NO_3)_2$ (*aq*); 1M KI (*aq*); Deionised water or Distilled water. 1 m long transparent tube bent at right angle at both ends.

Procedure

1. Set up the apparatus as shown in Fig. 15.14.
2. Add deionised water to fill the tube up to mark '*a*'.
3. With the help of a pipette add 1 cm^3 of 1M $Pb(NO_3)_2$ solution in end A of the tube. At the same time another student should add 1 cm^3 of KI solution in end B. Cork both ends of the tube loosely and keep it undisturbed for 1–2 hours.

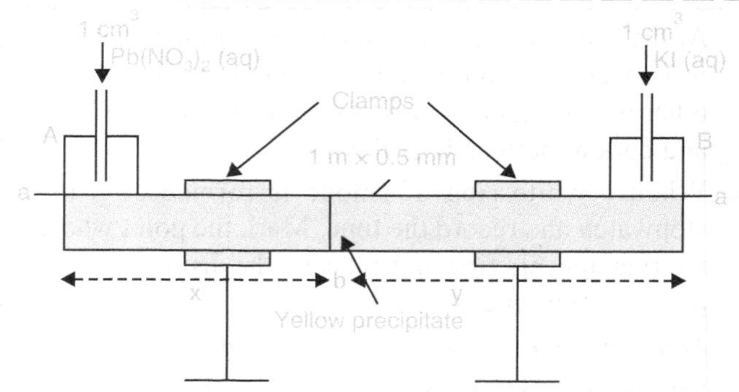

Fig. 15.14. Diffusion of solute particles through liquid.

4. Keep a watch on the tube and notice the formation of a yellow precipitate first. Mark that point as '*b*' with the help of a marker. Measure the distance of *b* from both ends

   Distance travelled by Pb^{2+} ions (*x*) = ...

   Distance travelled by I^- ions (*y*) = ...

   Why *x* and *y* are different?

   Both lead (II) trioxonitrate (V) and potassium iodide ionise as

$$Pb(NO_3)_2 \longrightarrow Pb^{2+} + 2NO_3^-$$
$$KI \longrightarrow K^+ + I^-$$

The ions diffuse through water and on meeting, the yellow precipitate of PbI_2 *is formed*

$$Pb^{2+}(aq) + 2I^-(aq) \longrightarrow \underset{\text{yellow}}{PbI_2}$$

Since Pb^{2+} *ions are heavier (Ionic mass = 207) than* I^- *ions (Ionic mass = 127). Hence, iodide ions move and diffuse at a faster rate. Consequently, the distance y must be larger than x.*

15.9 LABORATORY PREPARATION OF GASES

Let us now, study the **laboratory preparation of some gases,** namely: **hydrogen, ammonia and carbon dioxide.** The focus will primarily be on their *preparation, drying,* and *mode of collection*. The choice of the method of collection and drying depends upon some physical characteristics of these gases. In general,

- *Gases which are lighter than air, i.e., which have molecular mass less than the average molecular mass of air 28.8 can rise upward and can be collected by downward displacement of air.*
- *Gases which have molecular mass greater than 28.8 i.e., heavier than air tend to move downward and can be collected by upward displacement of air.*

- *The drying agent employed for drying of the gas should be that which does not react chemically with the gas.* The common drying agents are anhydrous calcium chloride ($CaCl_2$), phosphorus (V) oxide (P_4O_{10}), conc. tetraoxosulphate (VI) acid, (H_2SO_4), silica gel, *etc.*, because all such substances have the ability to absorb moisture.

The students are advised to carry out the preparation of these gases either in a fume cupboard or in a well ventilated room. Let us now proceed to study the preparation of these gases through activities.

ACTIVITY 15.6	**To prepare a sample of hydrogen gas and study some of its physical properties**

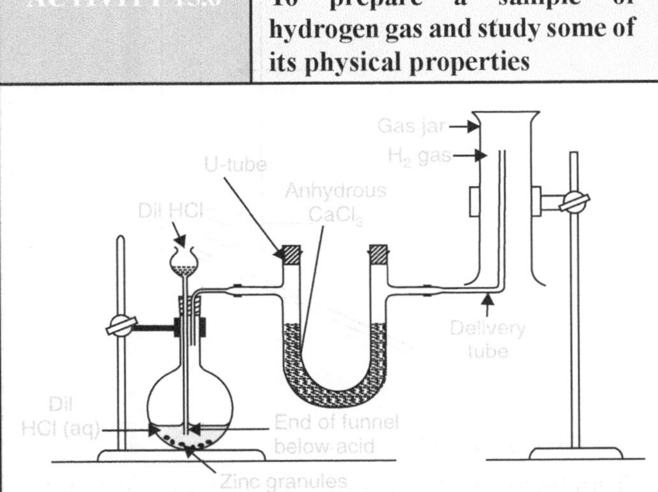

Fig. 15.15

In the laboratory hydrogen gas is prepared by the reaction of dilute hydrochloric acid and zinc.

$$Zn(s) + 2HCl\,(aq) \longrightarrow ZnCl_2\,(aq) + H_2\uparrow$$

Being much lighter (mol. mass = 2) than air (average mol. mass 28.8) it can be collected by downward displacement of air.

Procedure

1. Set up the apparatus as shown in Fig. 15.15 but without adding dilute hydrochloric acid at first.
2. Add enough dilute hydrochloric acid to cover the pieces of zinc. Add a few drops of copper (II) tetraoxosulphate (VI) solution to speed up the reaction if the zinc is pure.
3. Collect three gas jars full of the gas and cover them with glass lids to perform different experiments.
4. Observe the colour of the gas. Invert one of the jar of gas and quickly test the gas with damp blue and red litmus paper. Then immerse the inverted jar in a trough of water.
5. Insert a lighted splint into another gas jar of the gas, well away from the main apparatus. Do not hold the gas jar in your hand.
   What do you observe?
   Explain why it happens so?
6. Invert a gas jar containing air over the third jar of the gas and remove the lid. Insert a lighted splint into each gas jar after 15–20 seconds. Make sure the lighted splint is well away from the main apparatus.
   What happens?

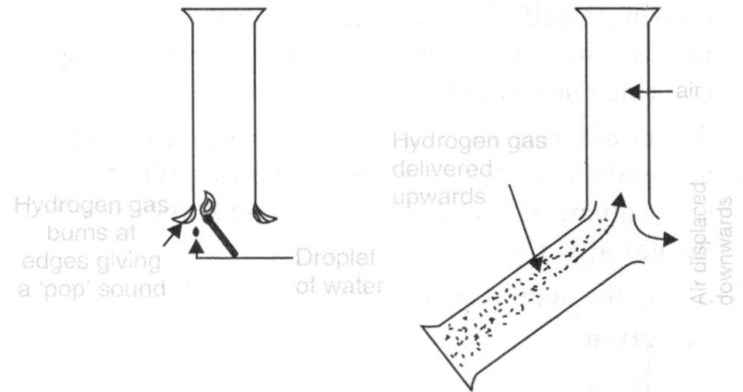

Fig. 15.16. Burning of the gas.

Fig. 15.17. Hydrogen is lighter than air

Try to answer some of the questions as given below:

(*i*) What is the colour and smell of the hydrogen gas?
(*ii*) Why was the gas passed through fused calcium chloride?
(*iii*) What other substances could be used instead of calcium chloride?
(*iv*) Why was the gas collected in an inverted gas jar?
(*v*) What happens when the gas jar is immersed in a trough of water?
(*vi*) What would happen if the hydrogen gas jar is inverted over the gas jar of air and a lighted splinter applied to each gas? Explain your answer.

Hints. (*i*) The hydrogen gas is inflammable and burns with a pop sound because it reacts rapidly with oxygen to form water

$$2H_2(g) + O_2(g) \longrightarrow 2H_2O(l)$$

(*ii*) Calcium chloride is used as drying agent
(*iii*) Silica gel, conc. tetraoxosulphate (VI) acid
(*iv*) It is much lighter than air
(*v*) Water does not rise inside the jar because the gas is not soluble in water
(*vi*) Hydrogen being lighter cannot flow downward.

ACTIVITY 15.7 **To prepare a sample of ammonia gas and study some of its physical characteristics.**

Ammonia is prepared by heating a finely powdered mixture of calcium hydroxide and ammonium chloride

$$Ca(OH)_2 + 2NH_4Cl \longrightarrow CaCl_2 + 2NH_3 + H_2O$$

The gas is lighter (molecular mass 17) than air (average molecular mass 28.8). Thus, it is collected by downward displacement of air.

Ammonia is a colourless alkaline gas with a pungent choking smell. The gas readily forms dense white fumes when it comes into contact with hydrogen chloride due to the formation of NH_4Cl.

Ammonia readily reacts with water to produce ammonium (NH_4^+) ions and hydroxide (OH^-) ions. One volume of water dissolves about 800 volumes of the gas at S.T.P.

1. Set up the apparatus as shown in Fig. 15.18.

   ***Caution.**

   - *The neck of the round bottom flask has the slope downwards as shown in Fig. 15.18. This is to prevent water which is one of the products of the reaction from trickling back into the hot part of the flask and cracking it.*
   - *The reactants being solids have to be powdered in order to provide a maximum surface area for the reaction.*

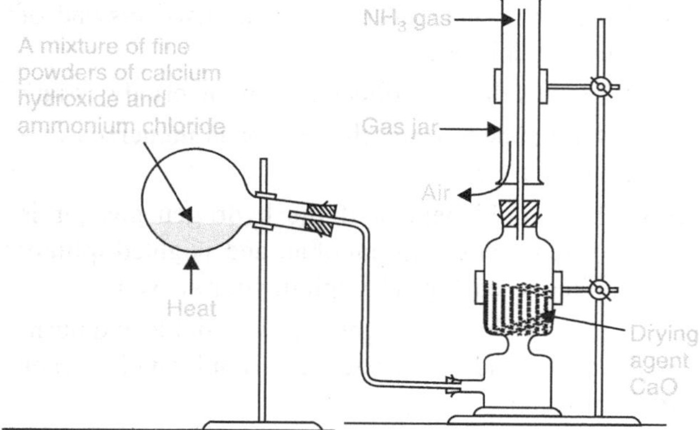

Fig. 15.18. Preparation of ammonia

2. Heat the flask gently and make three collections: two using *dry gas jars* and one with a clean and dry 500 cm^3 round bottom flask with a cork. To find out if each container is full of the gas, hold a piece of moist universal indicator paper at its mouth. The indicator paper turns blue when the container is full. Collect the gas in two jars and in one round bottom flask. Cover them with a lid.
3. Remove the cover from one of the gas jars. Observe the colour and ***very cautiously*** smell the gas.
4. Invert the second gas jar of ammonia and remove the cover. Dip a glass rod into bench hydrochloric acid and bring it immediately beneath the mouth of the gas jar (Fig. 15.19).

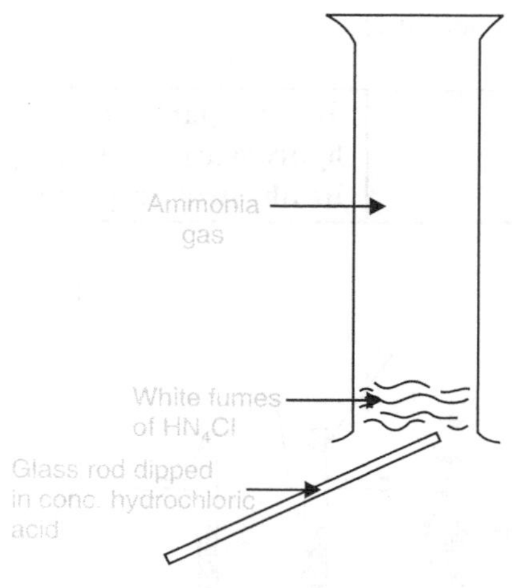

Fig. 15.19. Testing of NH_3 gas.

   (*i*) *What happens?*

5. Turn the flask upside down and uncork it. Then fit it with a cork with delivery tube and set up the apparatus as shown in Fig. 15.20.

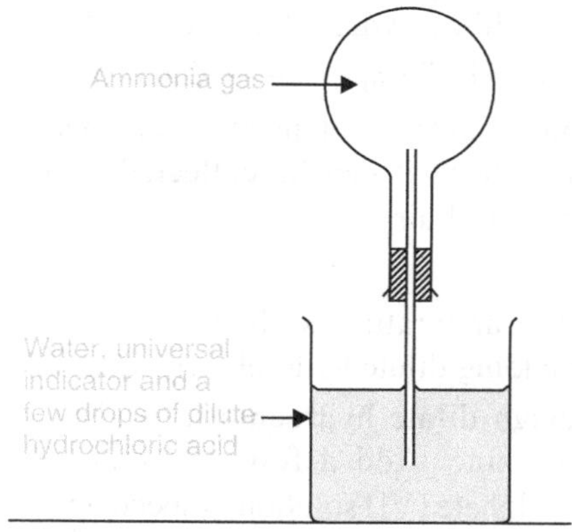

Fig. 15.20. Solubility of ammonia in water.

   (*ii*) *What happens?*

   (*iii*) *What conclusions can be drawn from this observation?*

   (*iv*) *Could the gas be collected over water?*

   (*v*) *Explain your answer.*

ACTIVITY 15.8 **To prepare a sample of carbon (IV) oxide, CO_2 gas and study some of its physical characteristics.**

In the laboratory, carbon (IV) oxide gas is prepared by the action of dilute hydrochloric acid on marble ($CaCO_3$) chips.

$$CaCO_3(s) + 2HCl\,(aq) \longrightarrow CaCl_2\,(aq) + H_2O(l) + CO_2(g)$$

The gas is colourless, odourless and is more denser than air. Hence, it is collected by upward displacement of air.

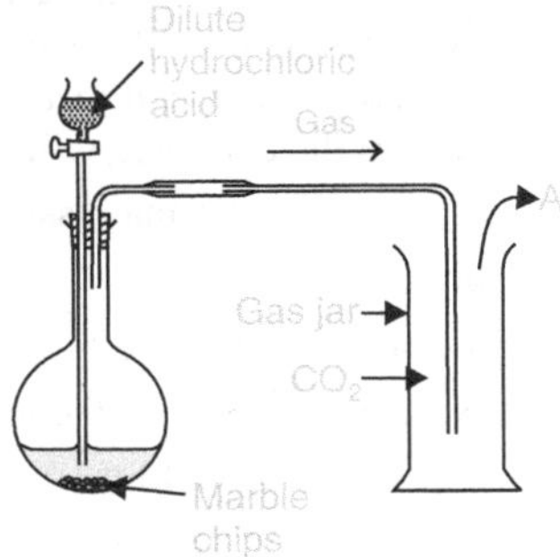

Fig. 15.21. Preparation of carbon (IV) oxide CO_2 gas.

1. Set up the apparatus as shown in Fig. 15.21.
2. Open the stopcock of the funnel to add dilute hydrochloric acid sufficient enough to cover the marble chips.
3. Collect 2 to 3 jars of the gas by upward displacement of air.
4. Note the colour of gas and its smell.
5. Dip a strip of moist blue litmus paper in one of the jar.
   (*i*) *What happens?*
6. Put a burning splinter in the second gas jar.
   (*ii*) *What happens?*
7. Invert the 3rd jar of gas and immerse it in water.
   (*iii*) *What do you observe?*

   Draw the following table of record for all the gases you have prepared.

Gas	*Colour*	*Odour*	*Effect on litmus*	*Molecular Mass*	*Method of Coll-ection*	*Solubility in water*
H_2						
NH_3						
CO_2						

SUMMARY

- In the **Gaseous State,** the particles move rapidly in all directions and have large voids.
- **Gas Laws.** The generalisations regarding the behaviour of gases.
- **Boyle's Law.** The volume of a fixed mass of a gas is inversely proportional to the pressure at constant temperature.
- **Charles' Law.** The volume of fixed mass of a gas is directly proportional to the temperature (in Kelvin scale) at constant pressure.
- **S.T.P. Conditions.** The standard temperature and pressure conditions are 0°C or 273.15 K and 1 bar or 100 kPa pressure.
- **Avogadro's Law.** Under similar conditions equal volumes of various gases contain equal numbers of molecules.

  Or

  At constant temperature and pressure the volume of the gas is directly proportional to its number of moles.
- **Dalton's Law.** The pressure of the mixture of non-reacting gases is equal to the sum of their partial pressures when enclosed in the same volume under similar conditions.
- **Graham's Law of Diffusion.** The rate of diffusion of gases (V/t) is inversely proportional to the square root of their densities or molecular masses under similar conditions.
- **Absolute Zero.** The hypothetical temperature at which all gases are supposed to occupy zero volume. Its numerical value on the celsius scale is **– 273.15°C.**
- **Vapour Density of a Gas.** It is the ratio of the mass of a certain volume of gas to the mass of the same volume of hydrogen under similar conditions.

  Molecular mass of gas = 2 × vapour density
- **Hydrogen gas** is prepared by reaction of dilute HCl on zinc granules. The gas is dried by anhydrous $CaCl_2$ or conc. tetraoxosulphate (VI) acid.
- **Ammonia** is prepared by heating a mixture of solid calcium hydroxide and ammonium chloride. The gas is dried by calcium oxide.
- **Carbon (IV) oxide** gas is prepared by the action of dilute hydrochloric acid on marble pieces.

EVALUATE YOURSELF

I. Objective Type Questions

Select the most appropriate choice from the options given as (a), (b), (c) and (d) after each question:

1. Which of following conditions does not pertain to S.T.P. conditions?
   (*a*) 273 K (*b*) 25°C
   (*c*) 1 bar (*d*) 100 kPa.
2. Which property as given below does not apply to gases?
   (*a*) High compressibility (*b*) Low density
   (*c*) Free intermixing (*d*) Low compressibility.
3. In one mole of nitrogen gas at S.T.P. the empty space between the gas molecules is
   (*a*) 80% (*b*) 99.9%
   (*c*) 100% (*d*) 50%.
4. The Boyle's law equation is PV = k (constant). The value of k does not depend upon
   (*a*) Mass of the gas (*b*) Temperature of the gas
   (*c*) Nature of gas (*d*) The product nT.
5. The volume of a fixed mass of gas at a given pressure should become zero when its temperature becomes
   (*a*) – 273.15 K (*b*) – 273.15°C
   (*c*) 0°C (*d*) 273.15 K.
6. At a given temperature and pressure the volume of a given sample of gas is directly proportional to its moles. The above statement pertains to
   (*a*) Charles' Law (*b*) Dalton's Law
   (*c*) Graham's Law (*d*) Avogadro's Law.
7. Which of the following gases are lighter than dry air?
   I. CO_2 II. Moist-air
   III. Hydrogen IV. Chlorine
   (*a*) III, II (*b*) I, III
   (*c*) II, IV (*d*) III, IV.
8. Which of the following volume-temperature plots represents the behaviour of 1 mol of an ideal gas at 100 kPa pressure?

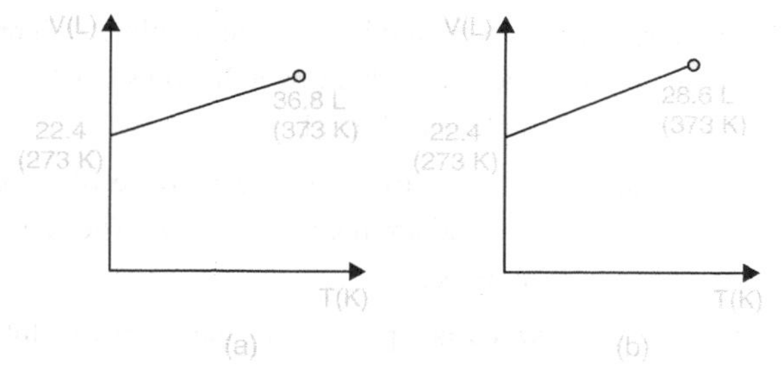

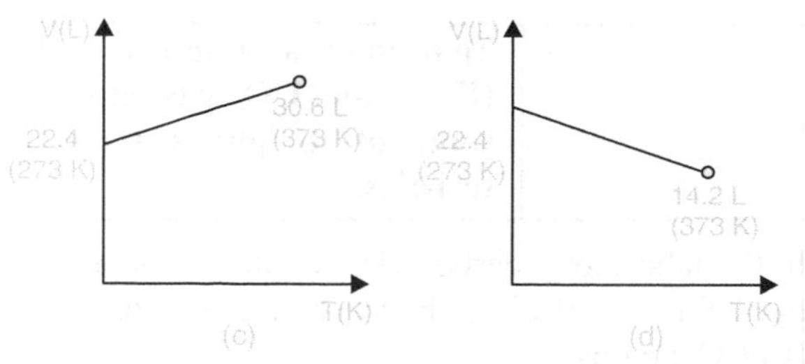

9. If one litre of dry oxygen is bubbled through water and collected at the same temperature and pressure, the volume of it would now be
   (*a*) 1 L (*b*) < 1 L
   (*c*) > 1 L (*d*) 2 L.
10. A cylinder of V litre capacity containing ammonia gas is inverted over another vessel of V litre capacity containing hydrogen chloride at the same temperature and pressure. After some time the pressure in the cylinders will
    (*a*) become double (*b*) remain same
    (*c*) drop considerably
    (*d*) become 3/2 of the original pressure.
11. Mass of 112 ml of N_2 at S.T.P. on liquefaction will be
    (*a*) 0.28 g (*b*) 0.14 g
    (*c*) 0.56 g (*d*) 0.84 g.

II. Fill in the Blanks

12. Complete the following sentences by supplying appropriate words:
    (*i*) The law describing the relationship between pressure and volume of an ideal gas at constant temperature is called ......
    (*ii*) 1 bar = ... kPa = ... atm.
    (*iii*) The volume of 1 mole of gas at 273 K and 100 kPa is ......
    (*iv*) The plot of ...... vs P at a given temperature is parallel to the pressure axis.
    (*v*) Density of a gas is ...... proportional to pressure.
    (*vi*) Vapour density of a diatomic elementary gas is 16, the atomic mass of the gas is ......
    (*vii*) Helium (mol mass = 4) diffuses ...... times faster than sulphur (IV) oxide gas.
    (*viii*) Ammonia is generally dried by using ....

III. Discussion Questions

13. How does the kinetic model explain the ability of free intermixing of gases? What is the name given to this phenomenon?
14. Give the general characteristics of gases.
15. State and explain Boyle's law. Derive the mathematical equation for Boyle's law. Draw a plot between P and 1/V (Vertical axis).
16. (*i*) Name and state the law which relates volume of a fixed mass of gas at a given pressure to the temperature?
    (*ii*) How can you explain the law by the kinetic molecular mode?

17. Suggest an experiment to compare the rates of diffusion of ammonia and hydrogen chloride. What conclusion do you get from the experiment?

18. Derive the ideal gas equation by using Boyle's law, Charles' law and Avogadro's law.

19. The following table shows that the effect of changing pressure on the volume of a sample of gas keeping the temperature constant:

Pressure (P) (Atm)	1.00	0.90	0.85	0.75	0.65	0.55	0.45	0.30	0.20
Volume (V) (L)	22.4	24.9	26.3	29.9	40.2	40.7	49.8	74.7	112

(*a*) Plot the following graph:

(*i*) P *vs*. V (*ii*) P *vs*. $\frac{1}{V}$ (*iii*) PV *vs*. P

Interpret each graph in terms of Boyle's law.

(*b*) What name is given to these graphs?

(*c*) One measurement in the table is wrong, identify it giving reasons.

20. States Dalton's law of partial pressure. Describe the utility of the law.

21. What will be the minimum pressure required to compress 500 dm^3 of air at 100 kPa bar to 200 dm^3 at 30°C?

22. Using the equation of state PV = *n*RT show that at a given temperature, the density of the gas is proportional to the gas pressure P.

23. A student forgot to add the reaction mixture to the round bottomed flask at 27°C but put it on the flame. After a lapse of time, he realised his mistake. By using a pyrometer, he found that the temperature of the flask was 477°C. What fraction of air would have been expelled out?

24. The drain cleaner, 'Drainex' contains small bits of aluminium which react with caustic soda to produce hydrogen. What volume of hydrogen at 20°C and 100 kPa will be released when 0.15 g of aluminium reacts?

The skeleton equation of the reaction involved is

$$Al + NaOH + H_2O \longrightarrow NaAlO_2 + H_2$$

25. 34.05 cm^3 of phosphorus vapours weigh 0.0625 g at 546°C and 100 kPa pressure. What is the molar mass of phosphorus?

26. Calculate the total pressure of a mixture of 0.25 mole of oxygen and 4 g of hydrogen confined in a vessel of 1 dm^3 at 27°C (R = 8.314 kPa dm^3 K^{-1} mol^{-1}).

27. Two gases A and B have densities 0.09 g dm^{-3} and 1.45 g dm^{-3} respectively. What is the ratio of their rates of diffusion under similar conditions?

28. A mixture of gases whose total pressure is 80 kPa contains 1.5 g of hydrogen, 3.2 g of oxygen and 1.4 g of nitrogen at 273 K. Calculate

(*i*) the total volume of the gaseous mixture and

(*ii*) the partial pressure of each gas.

29. A gas diffuses 1.27 times as rapidly as Cl_2 under the same conditions of temperature and pressure. Answer the following:

(*i*) Is the gas heavier or lighter than Cl_2?

(*ii*) What is the ratio of the molecular mass of the gas to that of Cl_2?

30. Read the following paragraph and answer the questions given after that:

A metal was reacted with dilute hydrochloric acid to give a colourless and odourless gas, G. Gas G was mistakenly passed over very hot copper (II) oxide before collection. The error was corrected by replacing the oxide with anhydrous calcium chloride. The gas was then collected in an inverted test tube.

(*i*) Name gas G.

(*ii*) Why was it an error to pass the gas over very hot copper (II) oxide before collection?

(*iii*) Why was the copper (II) oxide replaced with anhydrous calcium chloride?

(*iv*) Name two substances that could be used in place of the fused anhydrous calcium chloride.

(*v*) Name the actual method of collection of the gas.

(*vi*) Why is this particular method of collection used?

(*vii*) Can we collect the gas by downward displacement of water?

(*viii*) If the metal is zinc write the chemical equation for its reaction with dil. hydrochloric acid and also with dilute tetraoxosulphate (VI) acid.

31. Name the gas that could be obtained by the action of lime-stone and dilute hydrochloric acid.

(*i*) Give the chemical reaction involved

(*ii*) How can the gas be collected?

(*iii*) What will happen if the burning splinter is put in a jar of the gas?

(*iv*) Suggest one chemical test for the gas.

Answers

I. Objective Type Questions

1. (*b*) **2.** (*d*) **3.** (*b*) **4.** (*c*) **5.** (*b*) **6.** (*d*) **7.** (*a*) **8.** (*c*) **9.** (*c*) **10.** (*c*) **11.** (*b*).

II. Fill in the Blanks

12. (*i*) Charles' law (*ii*) 100, 0.987 (*iii*) 22.7 dm^3 (*iv*) PV (*v*) directly (*vi*) 16 (*vii*) four (*viii*) CaO.

III. Discussion Questions

21. 250 kPa **23.** 0.6 **24.** 202.68 dm^3

25. 124.8 g mol^{-1} **26.** 5605 kPa **27.** $r_A : r_B$ is 4 : 1

28. V = 25.53 dm^3

P_{O_2} = 8.89 kPa; P_{N_2} = 4.44 kPa

P_{H_2} = 66.67 kPa

SHS 3

SECTION 1

ENERGY AND ENERGY CHANGES

General Objectives

After studying this section, the student should be able to:

1. understand, determine, interpret and communicate energy changes in chemical reactions.
2. recognise that reorganisation of atoms in chemical reactions results in the release or absorption of heat energy.

In this section, we shall study

SECTION-1

UNIT-16

ENERGY CHANGES IN PHYSICAL AND CHEMICAL PROCESSES

Specific Objectives

After studying this unit, the student should be able to:

- explain the terms systems and surrounding
- explain the terms energy and enthalpy
- measure the enthalpy change of a reaction experimentally
- identify the type of enthalpy change associated with a given chemical process
- describe the enthalpy change associated with the burning of food and fuels.

16.1 INTRODUCTION

Energy is one of the important factors which has significant contribution to our living standards. The economic prosperity of any nation depends on its industries which in turn are linked with the availability of energy. The three primary sources of energy are: *solar energy, nuclear energy* and *chemical energy*. The main focus of our discussion in this unit is on chemical energy. It is a well known fact that almost all chemical reactions are associated with energy changes. Quite often, the energy change accompanying a chemical reaction is more significant than the reaction itself. For example, liberation of heat energy during the burning of coal is more significant to a common man than the production of carbon dioxide. The energy changes occurring during chemical reactions may not always appear as *heat energy;* but also as *electrical energy, work energy* and *radiant energy*.

Some examples are:

- Burning of *coal or cooking gas* in air.

$$C + O_2 \longrightarrow CO_2 + \textit{Heat energy}$$

$$C_3H_8 + 5O_2 \longrightarrow 3CO_2 + 4H_2O + \textit{Heat energy.}$$

- Burning of *magnesium ribbon* in oxygen.

$$2Mg + O_2 \longrightarrow 2MgO + \textit{Heat} + \textit{Light.}$$

- *Slow combustion* of *phosphorus* in dark.

$$P_4 + 5O_2 \longrightarrow 2P_2O_5 + \textit{Light energy.}$$

- Combustion of petrol in an internal combustion engine.

$$C_{10}H_{22} + 15\tfrac{1}{2}O_2 \longrightarrow 10CO_2 + 11H_2O + \textit{Mechanical energy} + \textit{Heat.}$$

- Indirect *redox reaction* occurring in a *galvanic cell.*

$$Zn(s) + CuSO_4(aq) \longrightarrow Cu(s) + ZnSO_4(aq) + \textit{Electric energy}$$

Thus, we find that chemical reactions are accompanied by energy changes appearing in different forms. These energy changes take place because during chemical reactions certain bonds are cleaved and certain new bonds are formed. Energy is consumed during cleavage of bonds while energy is released during the formation of bonds. Since the bond energy varies from one bond to another, the chemical reactions are always accompanied by absorption or release of energy.

16.2 SYSTEM, SURROUNDINGS AND SOME BASIC TERMS

In this section we shall study some basic terms/concepts which are needed in the study of chemical energetics.

SYSTEM

A system is defined as that part of the Universe which is under investigation. For example, if we are interested in investigating the neutralization of NaOH with HCl, the solutions of sodium hydroxide and hydrochloric acid constitute the system. Similarly, if we are studying the effect of temperature on the properties of water, then water will be taken as the system.

SURROUNDINGS

The part of the Universe other than the system is known as surroundings.

Thus, *Universe = System + Surroundings*

The entire universe, other than the system is not affected by the changes taking place in the system. Therefore, for all practical purposes, the surroundings are that portion of the remaining universe which can interact with the system. Generally, the region of space in the neighbourhood of the system constitutes its surroundings.

In order to keep track of the exchange of matter and energy between system and surroundings, it is necessary to think of the system as separated from the surroundings by some sort of wall which may be real or imaginary. The wall which separates the system from the surroundings is called **boundary.** For example, if a reaction mixture is taken in a beaker, the reaction mixture constitutes the system, the walls of the beaker constitute the boundary and everything else is the surroundings.

TYPES OF SYSTEM

Systems may be classified into three types on the basis of movement of matter and energy in and out of the system:

(*a*) *Open system* (*b*) *Closed system*

(*c*) *Isolated system*.

(*a*) **Open System.** *A system which can exchange mass as well as energy with the surroundings,* is called an **open system.** For example, heating of calcium carbonate in an open vessel. In this case heat is supplied to the system by the burner, while CO_2 escapes into the surroundings.

$$CaCO_3 \xrightarrow{\text{Heat}} CaO + CO_2 \uparrow$$

(*b*) **Closed System.** *A system which can exchange energy with the surroundings but not mass* is called a **closed system.** For example, calcination of $CaCO_3$ taken in a sealed bulb. On heating, $CaCO_3$ decomposes into CaO and CO_2, however, CO_2 cannot escape and remains trapped in the bulb.

(*c*) **Isolated System.** *A system which can neither exchange mass nor energy with the surroundings* is called an **isolated system.** For example, a reaction carried out in a closed, well insulated container (such as thermous flask). The open, closed and isolated systems have been shown in Fig. 16.1.

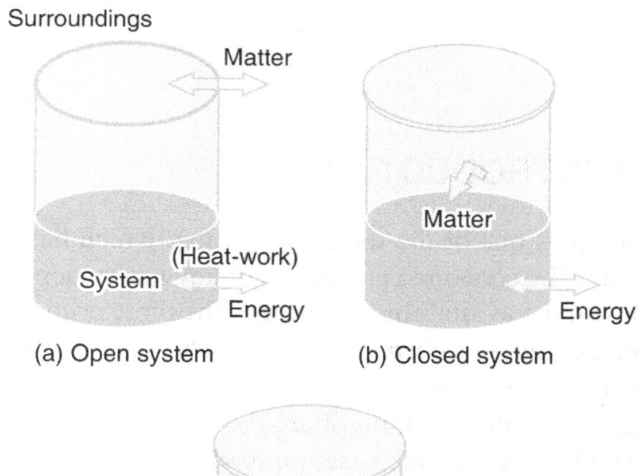

(a) Open system (b) Closed system

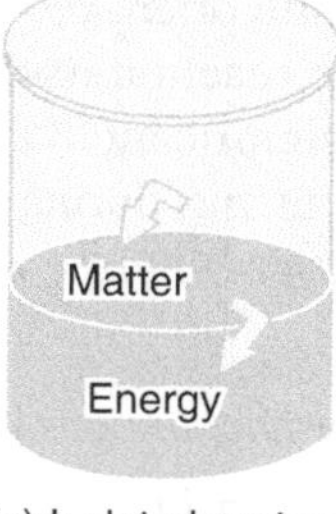

(c) Isolated system

Fig. 16.1. Open, closed and isolated systems

MACROSCOPIC SYSTEM

A system containing a large number of chemical species (atoms, ions or molecules) is called **macroscopic system.** In fact, the word macroscopic conveys the sense of appreciable quantities. *The properties of the system which arise from the collective behaviour of a large number of species* is called **macroscopic properties.** These properties are *pressure,* *temperature, volume, composition, refractive index, surface tension, viscosity, density,* etc. For example, in order to note the temperature of water we do not deal with individual molecules but we consider the molecules in bulk.

EXTENSIVE AND INTENSIVE PROPERTIES

The various physical properties of the system may be classified into two types:

(*a*) **Extensive Properties.** *The properties of the system which depend upon the quantity or size of matter present in it* are called **extensive properties.** Some common examples of these properties are *mass, volume, internal energy, enthalpy, heat capacity,* etc.

(*b*) **Intensive Properties.** *The properties of the system which are independent of the quantity or size of matter present in it* are called **intensive properties.** Some common examples are *temperature, pressure, refractive index, viscosity, specific heat, density,* etc.

It may be noted that *the ratio of two extensive properties become intensive in nature.* For example, mass and volume are extensive properties, but the ratio *mass/volume, i.e., density* is independent of the quantity of matter and is intensive. Similarly, heat capacity is extensive, but molar heat capacity is intensive. In general, if χ is any extensive property of n mol of a system, then the molar property of the system, χ_m is intensive because it refers to the property of 1 mol of the system and is independent of the quantity of matter. $\chi_m = \chi/n$.

STATE OF THE SYSTEM

State of the system refers to the *conditions of existence of a system* when *its macroscopic properties have definite values.* In order make useful calculations, it is necessary to describe the system completely before and after it undergoes any change. The state of the system can be described by specifying a certain minimum number of properties such as *pressure* **(P)**, *volume* **(V)**, *temperature* **(T)** and *amount* **(n)** or *composition.* Once these minimum number of macroscopic properties are fixed, the other properties automatically acquire definite values. Some notable features about a thermodynamic state are:

- Variation in one or more macroscopic properties brings a change in the state of the system, when other macroscopic properties attain new values. The macroscopic properties are thus, called **state variables** or **state functions.**
- **Initial state** refers to the starting state of a system in equilibrium. After interaction with surroundings (involving exchange of matter or energy or both) the system attains another equilibrium state which is referred to as a **final state** of the system.
- Thermodynamic state of the system must not be confused with *physical state* or *phase.*
- A system is said to be in a state of thermodynamic equilibrium if its macroscopic properties do not change with time.

STATE FUNCTIONS

The properties whose values depend only upon the initial and final states of the system and are independent of the manner the change is brought about, are called **state functions.** The concept of state function can be easily understood from the following analogy. If we consider '*h*' as the height between the top and bottom of a mountain, then '*h*' is independent of the path followed in reaching the top of the mountain. Here, the parameter *h* is analogous to a state function. In thermodynamics, some common state functions are *internal energy* **(U)**, *enthalpy* **(H)**, *entropy* **(S)**, *Gibb's energy* **(G)**, *pressure* **(P)**, *temperature* **(T)**, *volume* **(V)**, etc. It may be noted that two very important thermodynamic parameters namely; *heat* **(q)** and *work* **(w)** are not state functions because they are *path dependent.*

16.3 MODES OF EXCHANGE OF ENERGY

As pointed out earlier chemical reactions are accompanied by energy changes. The exchange of energy between the system and surroundings can occur in several ways. Two of the important modes of transference of energy are *heat* and *work* which have been described briefly as follows:

HEAT (q)

The transference of energy takes place as **heat** if the system (reaction mixture) and surroundings are at different temperatures. The exchange of heat occurs through thermally conducting walls which constitute the boundary between a system and its surrounding. If the system is at a higher temperature, the energy is lost to the surrounding as heat, causing a fall in the temperature of the system. The energy transfer continues till the system and the surroundings attain the same temperature. If the system is at a lower temperature than the surroundings, the energy is gained by the system from the surroundings causing a rise in the temperature of the system. The amount of heat gained or lost by the system is represented by **q.** According to international conventions,

(i) ***heat absorbed by the system is positive, i.e.,*** $q > 0$

(ii) ***heat given out by the system is negative, i.e.,*** $q < 0$.

WORK (w)

Exchange of energy between system and surrounding can occur in the form of work which may be *mechanical work, pressure volume work* or *electrical work*. For example, let us take a system containing some quantity of water in a thermos flask which does not allow any exchange of heat between system and surrounding. Let temperature of water be T_A. We now insert a set of small paddles in the water and rotate them to churn the water for sometime. This will result in an increase in the temperature of the water. This indicates an increase in the internal energy of the water by adding ***mechanical work*** to the system. The same increase in the temperature of water can be produced by adding an equal amount of ***electrical work*** with the help of an immersion rod. Let us now familiarise ourselves with ***pressure volume work.*** The exchange of energy as pressure-volume work can occur if the system consists of gaseous substance and there is a difference of pressure between the system and its surrounding. The amount of work done by the system or on the system is denoted by *w*. According to international conventions,

(i) ***work done on the system is positive, i.e.,*** $w > 0$

(ii) ***work done by the system is negative, i.e.,*** $w < 0$

The sign conventions of heat and work have been diagrammatically represented in Fig. 16.2.

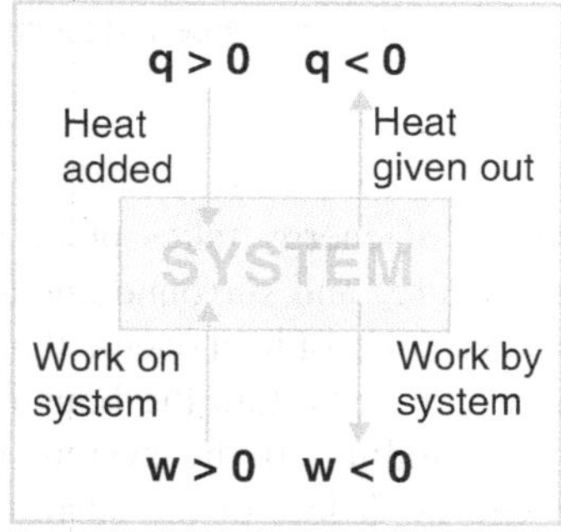

Fig. 16.2. Sign conventions of heat (*q*) and work (*w*)

Units of work. In *CGS* units, the work is expressed in *ergs* but in *SI* units, work is expressed in *joules*. The equivalence between joules and other units of work is

$$1\ J = 10^7\ ergs = 1\ Nm = 1\ kg\ m^2\ s^{-2}.$$

DIFFERENCE BETWEEN HEAT AND WORK

In order to understand the difference between heat and work, let us compare the effect of adding heat and work to a gaseous system.

(*i*) When heat is added to a gas, its molecules start moving faster in different directions causing an increase in disorder or randomness. Thus heat can be regarded as a mode which stimulates the random motion.

(*ii*) When work is done on a system by forcing the piston down, the initial effect is to force the molecules to move in the direction of movement of the piston. Thus, work can be regarded as a mode that stimulates the organised motion.

To conclude, *HEAT refers to a random form of energy whereas WORK refers to an organised form of energy.*

16.4 HEAT CAPACITY

We have learnt about the exchange of heat between a system and its surrounding. Now, we shall focus on the measurement of heat. The heat transferred to a system appears as a rise of its temperature. The increase in temperature (**ΔT**) is directly proportional to the quantity of heat (**q**) absorbed by the system. It can be put as

$$\mathbf{q \propto \Delta T \quad or \quad q = C.\Delta T} \qquad ...(16.1)$$

where **C** is the **heat capacity** of the system, the value of which depends upon the size, composition and nature of the system.

Now, if the temperature difference (**ΔT**) is unity, *i.e.,* 1° celsius or 1° kelvin then, **q = C**. Thus, *the heat capacity of a system is defined as the quantity of heat required to raise its temperature by 1°.*

Equation (16.1) reveals that a given amount of heat will raise the temperature of a system to a smaller extent if its heat capacity is large and *vice versa.*

MOLAR HEAT CAPACITY

The molar heat capacity of a substance (C_m) is the heat required to raise the temperature of 1 mol of the system by 1°. If C is the heat capacity of *n* mol of the system then its molar heat capacity, C_m is given by

$$C_m = \frac{\text{Heat capacity}}{\text{Number of moles}} = \frac{C}{n}. \qquad ...(16.2)$$

SPECIFIC HEAT CAPACITY (*c*)

This term is used more frequently for *solids* and *liquids*. We know heat capacity of a system depends upon the quantity of matter in the system. For example, a large block of aluminium has a higher heat capacity than a small piece of the same metal. ***Specific heat capacity*** *(c) is the heat capacity of 1 g of the sample. It is the quantity of heat required to raise the temperature of 1 gram of substance by 1 K (or 1°C).*

$$\text{Specific heat capacity } (c) = \frac{\text{Heat capacity}}{\text{Mass}} = \frac{C}{m}$$

Now, $$C = \frac{q}{\Delta T}$$

- Specific heat capacity of water
 $= \mathbf{4.184\ kJ\ kg^{-1}K^{-1}}$ or $\mathbf{4.184\ Jg^{-1}K^{-1}}$

$$\therefore \quad \textbf{Specific heat capacity (c)} = \frac{\mathbf{q}}{\mathbf{m\ \Delta T}}$$

or $$q = c \times m \times \Delta T \qquad ...(16.3)$$

Units of specific heat capacity are:

$$\mathbf{Jg^{-1}\ K^{-1}} \text{ or } \mathbf{Jg^{-1}\ (°C^{-1})}$$

The specific heat capacity (c) and molar heat capacity (C_m) of the substance are related as

$$\mathbf{c \times molar\ mass = C_m} \qquad ...(16.4)$$

ACTIVITY 16.1	**To determine the heat capacity (*c*) of a calorimeter system.**

Apparatus Required

Glass beaker or polythene bottle fitted with thermometer and stirrer.

Procedure

1. Take about 100 g of water (≈ 100 cm^3) in a beaker/ polythene bottle. Note the initial temperature [Let it be t_1°C (say 25°C)]
2. Now heat 100 g of water in another beaker to say 80°C.

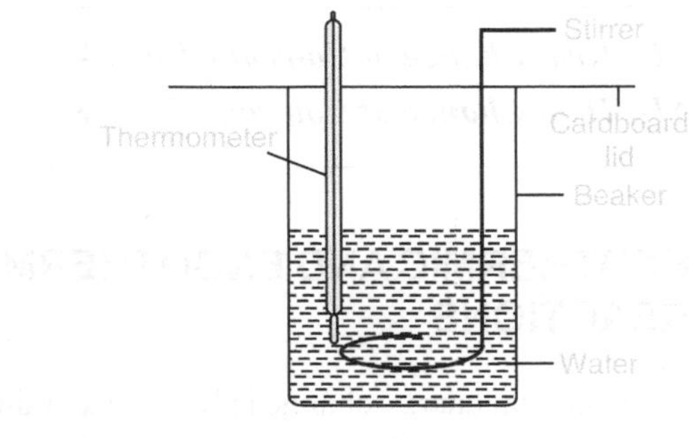

Fig. 16.3. Calorimeter system.

3. Put hot water into the cold water and stir properly. Record the temperature of the mixture and note its final temperature (t_2). Let t_2 be 52.3°C.
4. Heat lost by hot water is gained by cold water and the beaker.

Heat lost by hot water = [mass × specific heat capacity] × temperature diff.

$= m \times c \times \Delta T$

$= 100(g) \times 4.184\ (Jg^{-1}K^{-1}) \times (80–52.3)\ K$

$= 11589.68\ J$

Heat gained by cold water and beaker

$= 100(g) \times 4.184\ (Jg^{-1}K^{-1}) \times (52.3 - 25)\ K + C\ (52.3–25)$

$= 11422.32 + 27.3C$

Now, $11589.68 = 11422.32 + 27.3C$

or $27.3C = 167.36$

$$C = \frac{167.36}{27.3}\ JK^{-1}$$

$$= \mathbf{6.130\ JK^{-1}}$$

16.5 ENERGY AND ENTHALPY

INTERNAL ENERGY

Chemical changes are always accompanied by energy changes. This indicates that reactants and products must have certain amounts of energy. A fixed quantity of any substance is associated with a definite amount of energy which depends upon the chemical nature of the substance and its state of existence. This energy is called **internal energy or intrinsic energy** of the substance and is represented by **U**. Earlier, internal energy was represented by the symbol **E**. However, **U** is the latest notation as per the recommendations of the IUPAC. Internal energy of a system is the energy possessed by all its constituent molecules. The various forms of energies which contribute towards the energy of a molecule are *translational energy* (E_t), *rotational energy* (E_r), *vibrational energy* (E_v), *electronic energy* (E_e), *nuclear energies of constituent atoms* (E_n) and *potential energy due to interaction with neighbouring molecules* (E_{PE}). Internal energy is an *extensive property* and is also a *state function.* Its value depends upon the state of the substance but does not depend upon how that state is achieved. For example, CO_2 can be obtained by various methods such as by heating calcium(II) trioxocarbonate(IV) or by burning coal. However, one mole of CO_2 at S.T.P. is associated with a definite amount of internal energy which does not depend upon the source from which it is obtained.

It may be noted that the absolute value of internal energy cannot be determined because it is not possible to determine the exact values for the constituent energies such as translational, vibrational, rotational energies, etc. However, we can determine the change in internal energy (ΔU) of the system when it undergoes a change from the initial state (U_i) to the final state (U_f). For example, the change in internal energy (ΔU) of a reaction is the difference between the internal energies of the products (ΣU_P) and the reactants (ΣU_R).

$$\Delta U = \Sigma U_P - \Sigma U_R$$

Significant Features of Internal Energy (U)

- Internal energy depends upon the quantity of substance in the system and hence, it is an *extensive property.*
- Change in internal energy (ΔU) represents the *heat evolved or absorbed in a reaction at constant temperature and constant volume.*
- For processes involving the evolution of energy, $U_P < U_R$. Thus, **ΔU < 0** or *sign of* ΔU *is* **negative.**
- For processes involving absorption of energy, $U_P > U_R$. Thus, **ΔU > 0** or *sign of* ΔU *is* **positive.**

ENTHALPY

We have studied that the energy change that occurs during a reaction at *constant temperature* and *constant volume* is given by the internal energy change. However, most of the reactions in the laboratory are carried out in open beakers or test tubes, etc. In such cases, the reacting system is open to the atmosphere. Since atmospheric pressure is almost constant, such reactions may involve changes in volume. The energy change occurring during such reactions may not be equal to the change in internal energy. In order to understand this, let us assume that a chemical reaction involving gaseous substances proceeds with the evolution of heat. When the reaction is carried out at constant pressure, two possibilities arise.

(*a*) If the reaction proceeds with an increase in volume the system has to expand against the atmospheric pressure and energy is required for this purpose. The heat evolved in this case would be a little less than the heat evolved at constant volume because a part of the energy has to be utilised for expansion.

(*b*) If the reaction proceeds with a decrease in volume at constant pressure, the work is done on the system and heat evolved will be greater than the heat evolved at constant volume.

Thus, it can be concluded from the above discussion that heat changes occurring at constant pressure and constant temperature are not simply due to the changes in internal energy alone but also include energy changes due to expansion or contraction against atmospheric pressure. In order to study the heat changes of chemical reactions at constant temperature and pressure a new function, *enthalpy* is introduced. **Enthalpy** *is the total energy associated with any system which includes its internal energy and also energy due to environmental factors such as pressure-volume conditions.* This can be understood as follows:

A substance has to occupy some space in its surroundings depending upon its volume (V). It does so against the compressing influence of the atmospheric pressure (P). Due to this, the substance possesses an additional energy called PV energy.

The sum of internal energy and PV energy of any system, under a given set of conditions, is called **enthalpy.** It is denoted by H and is also called ***heat content*** of the system.

Mathematically, it may be put as

$$H = U + PV \qquad ...(16.5)$$

Some *important features* of enthalpy are:

- It is a state function and is an extensive property,
- It is also called ***heat content*** of a system,
- Its value depends upon the amount of the substance, chemical nature of the substance and conditions of temperature and pressure.

It is not possible to determine the absolute value of the enthalpy of a system because the absolute value of the internal energy (U) is not known. However, the change in enthalpy (ΔH) taking place during the process can be experimentally determined. Change in enthalpy is equal to the difference between the enthalpies of products (ΣH_P) and reactants (ΣH_R).

The change in enthalpy may be expressed as

$$\Delta H = \Sigma H_P - \Sigma H_R$$

ΔU: Heat Change at Constant T and V

ΔH: Heat Change at Constant T and P.

16.6 EXOTHERMIC AND ENDOTHERMIC REACTIONS

We have studied that energy changes always occur during chemical reactions. Depending upon the evolution or absorption of energy, the chemical reactions can be classified into two categories: *exothermic* and *endothermic.*

EXOTHERMIC REACTIONS

The chemical reactions which proceed with the evolution of heat energy are called **exothermic reactions.** The heat energy produced during the reaction is indicated by writing $+q$ or more precisely by giving the actual numerical value along with the products. In general, exothermic reactions may be represented as:

$$A + B \longrightarrow C + D + \mathbf{q} \text{ (heat energy)}$$

The heat evolved is expressed in the units of *Joules (J)* or *kilo Joules (kJ).*

Most of the reactions that we carry out in the laboratory are processes carried out at *constant temperature* and *constant pressure* where the heat evolved is equal to the enthalpy change (ΔH). In exothermic reactions, the enthalpy of products (ΣH_P) becomes less than the enthalpy of reactants (ΣH_R). Therefore, the enthalpy change for an exothermic reaction is **negative.**

$$\Delta H = \Sigma H_P - \Sigma H_R$$

As $\quad \Sigma H_P < \Sigma H_R$

$$\mathbf{\Delta H = -ve.}$$

For example, formation of one mole of water from hydrogen and oxygen gases is accompanied by the evolution of 286.0 kJ of heat. It can be expressed as:

$$H_2(g) + O_2(g) \longrightarrow H_2O(l) + 286.0 \text{ kJ}$$

or $\quad H_2(g) + O_2(g) \longrightarrow H_2O(l);\ \Delta H = -\ 286.0 \text{ kJ}$

ENDOTHERMIC REACTIONS

The chemical reactions which proceed with the absorption of heat energy are called **endothermic reactions.** The heat energy absorbed during the reaction can be indicated by writing **+q** (or the actual numerical value) with the reactants. It can be indicated by writing **–q** (or the actual numerical value) with the products. In general, an endothermic reaction can be represented as:

$$A + B + \mathbf{q} \text{ (heat)} \longrightarrow C + D$$

or $\quad A + B \longrightarrow C + D - \mathbf{q} \text{ (heat)}$

where **q** is the heat absorbed.

When an endothermic reaction is carried out at *constant temperature* and *constant pressure,* the heat absorbed increases the enthalpy of the products. In such a case the **enthalpy change** of the reaction is **positive.**

$$\Delta H = \Sigma H_P - \Sigma H_R, \text{ As } \Sigma H_P > \Sigma H_R$$

$\therefore \quad \mathbf{\Delta H = +ve}$

For example, decomposition of 2 moles of mercuric oxide at constant temperature and pressure proceeds with the absorption of 180.4 kJ of heat. It can be represented as:

$$2HgO(l) \longrightarrow Hg(l) + O_2(g) - 180.4 \text{ kJ}$$

or $\quad 2HgO(s) \longrightarrow 2Hg(l) + O_2(g);\ \Delta H = +180.4 \text{ kJ}$

The enthalpy changes for exothermic and endothermic reactions can be represented by **energy level diagrams** as shown in Fig. 16.4(*a*) and (*b*) respectively. In these diagrams, the energy levels of reactants and products are drawn on a vertical scale according to which, a substance with higher heat content is represented higher up on the scale.

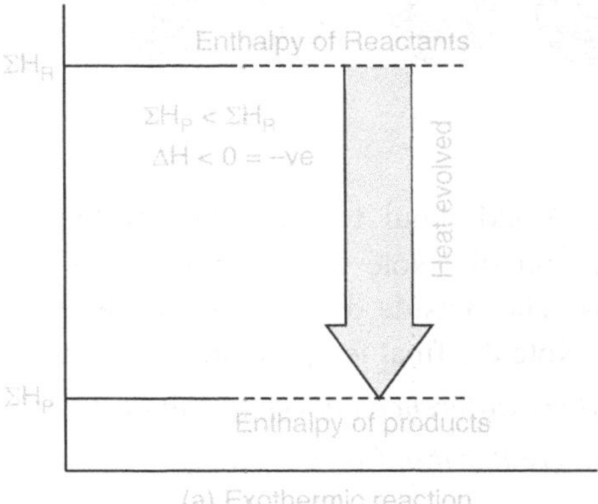

(a) Exothermic reaction

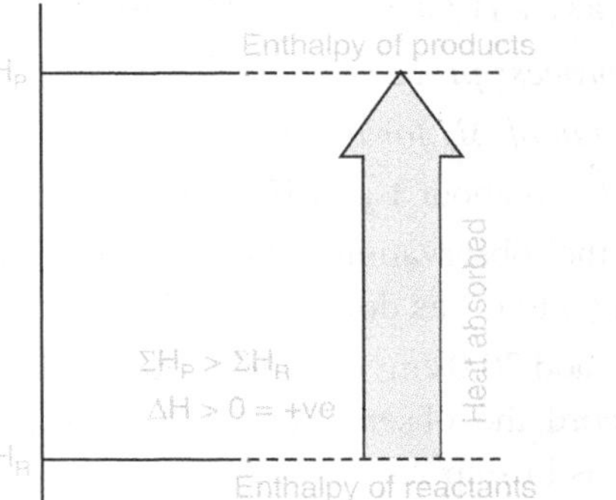

(b) Endothermic reaction

Fig. 16.4

For exothermic reactions: ΔU or $\Delta H = -ve$
For endothermic reactions: ΔU or $\Delta H = +ve$.

ACTIVITY 16.2	**To estimate the heat change involved in a chemical/physical process.**

1. Take four boiling test tubes marked A, B, C, D and mount on a stand. Put 50 cm^3 of water by using a measuring cylinder into each tube. Insert a thermometer and a small stirrer in each tube. Note the initial temperature in each tube (t_1).

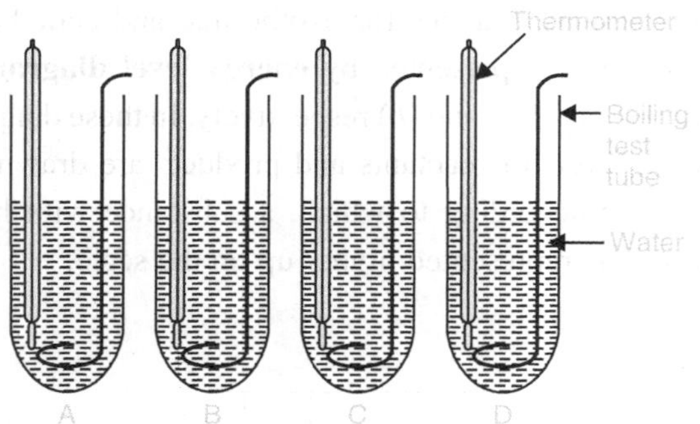

Fig. 16.5

2. In tube A add about 1 g of CaO with the help of a spatula and stir. Note whether the temperature rises or falls. The outside of the tube can be felt by the hands. Note the final temperature of the mixture (t_2).
   (*i*) *Is there difference between t_1 and t_2?*
   (*ii*) *t_2 is greater than/(or less than) t_1*
   (*iii*) *The reaction involved is*
   $$CaO(s) + H_2O(l) \longrightarrow Ca(OH)_2(aq)$$
   (*iv*) *The process proceeds with ...... of heat*
   (*v*) *The sign of ΔH for the process is ......*
3. In tube B add about 1 g of NH_4Cl
   Record the observations and provide answers to question (*i*) to (*v*) as done in step 2.
4. In tube C add 20.0 cm^3 of *conc. tetraoxosulphate (VI) acid,* record the observations and answer question (*i*) to (*v*) as in step 2.
5. In tube D add 1 g of *solid NaOH* and record same observations.

16.7 THERMOCHEMICAL EQUATIONS

A thermochemical equation may be defined as *a chemical equation which indicates the amount of heat evolved or absorbed during a chemical reaction in addition to the quantities of reactants or products.* It may be noted that in thermochemical equations the fractional coefficients are also commonly used contrary to our usual practice for balancing chemical equations. A thermochemical equation can be written in two ways:

(*a*) Heat effect can be written as one of the term along with the products.
For example,
$$2SO_2(g) + O_2(g) \longrightarrow 2SO_3(g) + 694.6 \text{ kJ}$$

(*b*) Heat evolved or absorbed can be expressed in terms of ΔH.
For example,
$$2SO_2(g) + O_2(g) \longrightarrow 2SO_3(g); \Delta H = -694.6 \text{ kJ}.$$

It is very important to mention the **physical states** *of the various reactants and products while writing thermochemical equations because a change of physical state is also accompanied by enthalpy changes.* For example:

When 1 mol of hydrogen gas reacts with $\frac{1}{2}$ mol of oxygen gas to produce 1 mol of liquid water, 286 kJ of heat is produced.

$$H_2(g) + \frac{1}{2} O_2(g) \longrightarrow H_2O(l); \quad \Delta H = -286 \text{ kJ}$$

On the other hand, if 1 mol of water vapour is produced instead of 1 mol of liquid water, the value of ΔH will be different.

$$H_2(g) + \frac{1}{2} O_2(g) \longrightarrow H_2O(g); \quad \Delta H = -249 \text{ kJ}$$

Some Important Conventions about Thermochemical Equations

1. *The coefficients of the various substances of a chemical equation represents the number of their respective moles. In thermodynamic interpretation of an equation, we never interpret the coefficients as the number of molecules. Hence, it is acceptable to write coefficients in fractions wherever necessary.*

2. *The value of ΔH in a thermochemical equation corresponds to the enthalpy change taking place when a specified number of moles of various reactants and products* (as indicated by the coefficients of the various substances in the chemical equation) *are involved in the reaction.*

3. *In the case where the coefficients in the chemical equation are multiplied or divided by some integer, the ΔH value must also be multiplied or divided by the same integer.* For example:

$$H_2(g) + \frac{1}{2} O_2(g) \longrightarrow H_2O(l); \quad \Delta H = -286 \text{ kJ}$$

If the whole equation is multiplied by 2, the ΔH for the new thermochemical equation is given as:

$$2H_2(g) + O_2(g) \longrightarrow 2H_2O(l);$$
$$\Delta H = 2 \times (-286) \text{ kJ} = -572 \text{ kJ}$$

4. *When a chemical equation is reversed, the magnitude of the ΔH remains the same, however, its sign is reversed.* For example, if ΔH is +ve for the forward reaction, it would be negative for the reverse reaction.

$$H_2(g) + \frac{1}{2} O_2(g) \longrightarrow H_2O(l); \quad \Delta H = -286 \text{ kJ}$$

$$H_2O(l) \longrightarrow H_2(g) + \frac{1}{2} O_2(g); \Delta H = +286 \text{ kJ}$$

The thermochemical equations for some exothermic and endothermic reactions are given as follows:

Exothermic reactions (ΔH = –ve):

$$H_2(g) + \frac{1}{2} O_2(g) \longrightarrow H_2O(l); \quad \Delta H = -286.0 \text{ kJ}$$

$$CH_4(g) + 2O_2(g) \longrightarrow CO_2(g) + 2H_2O(l);$$
$$\Delta H = -890.4 \text{ kJ}$$

$$N_2(g) + 3H_2(g) \longrightarrow 2NH_3(g); \quad \Delta H = -92.3 \text{ kJ}$$

Endothermic reactions (ΔH = +ve):

$$C(s) + 2S(l) \longrightarrow CS_2(l); \quad \Delta H = +92.00 \text{ kJ}$$

$$H_2(g) + I_2(g) \longrightarrow 2HI(g); \quad \Delta H = +53.6 \text{ kJ}$$

$$C(s) + H_2O(g) \longrightarrow CO(g) + H_2(g);$$
$$\Delta H = +131.2 \text{ kJ}$$

SOLVED EXAMPLE

Example 16.1. *Heat produced by burning 1 mol of liquid ethanol (C_2H_5OH) in oxygen gas to produce carbon dioxide gas and liquid water at 100 kPa is 1367 kJ mol^{-1}.*

(*a*) *Write the thermochemical equation.*

(*b*) *What mass of ethanol should be burnt to get 4 × 10^3 kJ of heat?*

Solution. (*a*) The thermochemical equation for the process is

$$C_2H_5OH(l) + 3O_2(g) \longrightarrow 2CO_2(g) + 3H_2O(l);$$
$$\Delta H = -1367 \text{ kJ mol}^{-1}$$

(*b*) Molar mass of ethanol = 46 g mol^{-1}

Ethanol required for producing 1367 kJ of heat = 1 mol or 46 g

Ethanol required for producing 4 × 10^3 kJ of heat

$$= \frac{46 \times 4 \times 10^3}{1367} = \mathbf{134.6\ g.}$$

16.8 ENTHALPY CHANGES IN SOME PHYSICAL AND CHEMICAL PROCESSES

Whenever a substance undergoes any physical or chemical transformation, energy changes also occur. In this section, we shall study the enthalpy changes involved in some physical processes like *evaporation, fusion, dissolution,* etc. We shall also become familiar with the energy changes involved in some chemical processes, like combustion, hydration, etc.

16.9 ENTHALPY CHANGES DURING PHASE TRANSFORMATIONS

It is a matter of common observation that energy is required to convert a *solid* into *liquid* or a liquid into gas. The conversion of a solid into liquid is called *melting* or *fusion* and the process of conversion of a liquid into gas is termed as *vaporisation.* These processes are collectively known as **phase transformations** or **phase changes.** The magnitude of the enthalpy change accompanying the phase transition depends largely on the strength of intermolecular forces in the substance undergoing phase transformations. Let us study the enthalpy changes accompanying the phase changes.

(a) Standard or Molar Enthalpy of Fusion ($\Delta_{fus}H^{\circ}$)

It is the enthalpy change accompanying the melting of 1 mole of a solid substance in standard state at its melting point. For example, when one mole of ice changes into water at its melting point 273 K, 6.0 kJ of heat is absorbed. It is represented as:

$$H_2O(s) \longrightarrow H_2O(l); \Delta_{fus}H^{\ominus} = +6.0 \text{ kJ mol}^{-1}$$

The value of the enthalpy of fusion gives an idea of the magnitude of interparticle forces in solids. For example, ionic solids have very strong interparticle forces. Consequently, they have high values of enthalpy of fusion. On the other hand, molecular solids have weak interparticle forces which reflect in their low values of enthalpy of fusion. The enthalpies of fusion of some substances are given in Table 16.1.

Freezing of ice is exothermic. When water freezes, attractive interactions between water molecules increase and energy is released. The freezing of ice, is therefore, an *exothermic process.* The enthalpy of freezing of ice is same in magnitude as enthalpy of fusion but opposite in sign.

$$H_2O(l) \longrightarrow H_2O(s); \quad \Delta H = -6.0 \text{ kJ mol}^{-1}$$

(b) Standard or Molar Enthalpy of Vaporisation ($\Delta_{vap}H^\ominus$)

It is the enthalpy change accompanying the conversion of one mole of a liquid in standard state into its vapours at the boiling point and standard pressure (100 kPa). For example, when one mole of water is converted into steam at 273 K, the enthalpy change accompanying the process is 40.6 kJ. It can be expressed as:

$$H_2O(l) \longrightarrow H_2O(g); \ \Delta_{vap}H^\ominus = +40.6 \text{ kJ mol}^{-1}$$

The value of the enthalpy of vaporisation gives an idea of the magnitude of the interparticle forces in liquids. The values of the enthalpy of vaporisation of some substances are given in Table 16.1.

Table 16.1. Standard Enthalpy of Fusion and Vaporisation

Substance	$T_f(K)$	$\Delta_{fus}H^\ominus$ (kJ mol^{-1})	$T_b(K)$	$\Delta_{vap}H^\ominus$ (kJ mol^{-1})
O_2	55	0.45	90	0.67
He	03	0.021	04	0.092
N_2	63.15	0.72	77.35	5.59
NH_3	195.40	5.65	239.73	23.35
HCl	159.0	1.992	188.0	16.15
CO	68.0	6.836	82.0	6.04
CH_3COCH_3	177.8	5.72	329.4	29.1
CCl_4	250.16	2.5	349.69	30.0
H_2O	273.15	6.01	373.15	40.79
NaCl	1081.0	28.8	1665.0	170.0
C_6H_6	278.65	9.83	353.25	30.8

(c) Enthalpy of Sublimation ($\Delta_{sub}H^\ominus$)

It is the enthalpy change accompanying the sublimation of one mole of a solid substance into its gaseous state at a constant temperature below its melting point at the standard pressure.

Sublimation is a direct conversion of a solid into vapour. For example, solid CO_2 (*dry ice*) sublimes at 195 K with $\Delta_{sub}H^\ominus = 25.2$ kJ mol^{-1}. Similarly $\Delta_{sub}H^\ominus$ of naphthalene and iodine are 73.0 kJ mol^{-1} and 62.4 kJ mol^{-1} respectively. The thermochemical equations are as follows:

$$CO_2(s) \longrightarrow CO_2(g); \ \Delta_{sub}H^\ominus = 25.2 \text{ kJ mol}^{-1}$$

$$C_{10}H_8(s) \longrightarrow C_{10}H_8(g); \ \Delta_{sub}H^\ominus = +73 \text{ kJ mol}^{-1}$$

$$I_2(s) \longrightarrow I_2(g); \ \Delta_{sub}H^\ominus = +62.4 \text{ kJ}$$

The process of sublimation consists of a change from the solid state into the vapour state in one step. The same change may be brought about in two steps:

(*i*) The melting of solid into its liquid and the enthalpy change is $\Delta_{fus}H$

(*ii*) The vaporisation of a liquid into its gaseous state and the enthalpy change is ΔH_{vap}

$$\mathbf{\Delta_{sub}H = \Delta_{fus}H + \Delta_{vap}H.}$$

16.10 ENTHALPY CHANGES DURING DISSOLUTION

(a) Enthalpy of Solution ($\Delta_{Sol}H$)

It is the enthalpy change accompanying the dissolution of one mole of a substance in a large excess of a solvent so that further addition of solvent does not involve any more enthalpy change. It may be noted that the enthalpy change accompanying the dissolution of 1 mole of a substance in a specific number of moles of solvent is called ***integral enthalpy of solution*** whereas the enthalpy change observed by dissolving one mole of substance in an infinite amount of solvent so that the interactions between the solute molecules are negligible, is called ***enthalpy of solution*** or more specifically, the ***enthalpy at infinite dilution.***

If water is the solvent, then the symbol *aq* (aqueous) is used to represent large or infinite dilutions. For example, the thermochemical equations for the dissolution of KCl and $CuSO_4.5H_2O$ are:

$$KCl(s) + aq \longrightarrow K^+(aq) + Cl^-(aq); \ \Delta H^\ominus = +18.6 \text{ kJ}$$

$$CuSO_4.5H_2O(s) + aq \longrightarrow Cu^{2+}(aq) + SO_4^{-2}(aq) + 5H_2O$$

$$\Delta H^\ominus = +11.7 \text{ kJ mol}^{-1}$$

Some notable features about $\Delta_{sol}H$ are:

- *$\Delta_{sol}H$ of hydrated salts like $CuSO_4.5H_2O$, $CaCl_2.6H_2O$, etc., or the salts which do not form hydrates like NaCl, KCl, NH_4Cl, etc., is generally* ***positive.***
- *$\Delta_{sol}H$ of anhydrous salts which form hydrates like $CuSO_4$, etc., is* ***negative.***

While doing thermochemical calculations, if any equation is multiplied by any integer, the word aq which represents large dilutions is not multiplied by it and is retained as such. In other words *aq* does not have quantitative significance.

(b) Dissolution of Ionic Salts

When ionic compounds dissolves in a solvent, the ions leave their ordered position in the crystal lattice and start moving freely in the solution. In the solution, the ions remain surrounded by an envelop of solvent molecules (see properties of ionic compounds in *unit-7*). This interaction between the ions and the solvent molecules is known as ***solvation.*** In case the solvent is water this process is called ***hydration.*** The dismentalling of ionic crystal, *i.e.*, breaking away the ions requires energy which is generally expressed in terms of ***lattice energy*** ($\Delta_L H$) and solvation/hydration of ions releases energy, which is generally expressed in terms of ***solvation energy or hydration energy,*** $\Delta_{Hyd}H$ in case the solvent is water. For aqueous solutions, these energy changes can be related to $\Delta_{Sol}H$ according to the energy level diagram as shown in Fig. 16.6.

Energy

$A^+(g) + B^-(g) + aq$

$AB(s) + aq$

$A^+(aq) + B^-(aq)$

$\Delta_L H^\ominus$

$\Delta_{Sol}H^\ominus$

$\Delta_{Hyd}H^\ominus$

Fig.16.6. Energy level diagram for dissolution of ionic salt AB

It is quite clear that

$$\Delta_{Sol}H = \Delta_L H + \Delta_{Hyd}H$$

The value of Δ_{Sol} H can be +ve or –ve depending upon the magnitudes of $\Delta_L H$ and $\Delta_{Hyd}H$. Dissolution of salt may not occur if the value of $\Delta_L H$ is very high.

(c) Enthalpy of Hydration of Anhydrous Salt ($\Delta_{Hyd}H$)

It is the enthalpy change accompanying the hydration of one mole of an anhydrous salt when it combines with a specific number of moles of water. For example, the enthalpy of hydration of anhydrous copper(II) tetraoxosulphate(VI) can be represented as:

$$CuSO_4(s) + 5H_2O(l) \longrightarrow CuSO_4.5H_2O(s); \quad \Delta H = -78.2 \text{ kJ}.$$

ACTIVITY 16.3 — Determining the enthalpy of solution of anhydrous and hydrated copper(II) tetraoxosulphate(VI) salts.

1. Pour 50 cm^3 of distilled water into a clean and dry beaker containing a thermometer and a stirrer.
2. Set up the apparatus as shown in Fig. 16.7. Determine the heat capacity of the beaker, C.
3. Stir the water well and record the temperature.
4. Accurately weigh out about 4.0 g of anhydrous copper(II) tetraoxosulphate(VI) powder on a watch glass and put the salt into the water and stir.

   Whilst stirring until dissolution is complete, record the temperature every 30 seconds for 5 minutes. Note the highest temperature.
5. Draw a temperature-time graph and find ΔT or the corrected highest or lowest temperature.

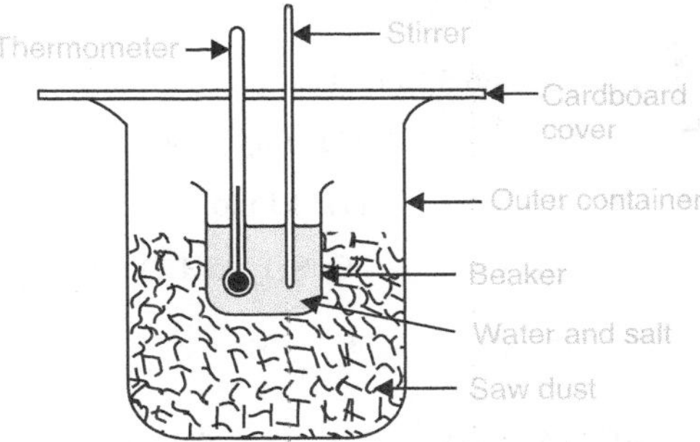

Fig. 16.7. Determination of enthalpy change of solution.

6. Clean and dry the beaker, stirrer and thermometer. Then repeat the experiment, using about 6.25 g of copper(II) tetraoxosulphate(VI) pentahydrate in place of the anhydrous salt.
7. Calculate the enthalpy of solution of each salt following sample calculations.
8. Express the result in the form of an equation and also in the form of an energy level diagram, using a graph sheet. Explain the results. Write an equation for the hydration of the anhydrous $CuSO_4(s)$.

   (*a*) **Sample calculations**

   Volume of water 50 cm^3 ≈ 50 g

   Mass of $CuSO_4(s) = w$ g

   Initial temperature of water = t_1

   Final temperature of solution = t_2

Change in temperature = $\Delta T = (t_2 - t_1)$

Molar mass of salt = 159.5 g mol^{-1}

Heat change = $50 \times 4.184 \times \Delta T + C\Delta T$

= x (Suppose)

$$\Delta_{sol}\ CuSO_4(s) = \frac{x \times 159.5}{w}\ \text{kJ mol}^{-1} \quad ...(1)$$

In a similar way find the enthalpy of dissolution of $CuSO_4.5H_2O(s)$.

(*b*) Use the results in 7 and 8 to calculate the enthalpy change of hydration of $CuSO_4(s)$ to $CuSO_4.5H_2O(s)$.

Theoretical values of the two experiments

$$\Delta_{sol}H\ [CuSO_4(s)] = -\ 66.5\ \text{kJ mol}^{-1}$$

$$\Delta_{sol}\ H\ [CuSO_4.5H_2O(s)] = +\ 11.3\ \text{kJ}$$

The energy level diagram for the two experiments is as shown in Fig. 16.8.

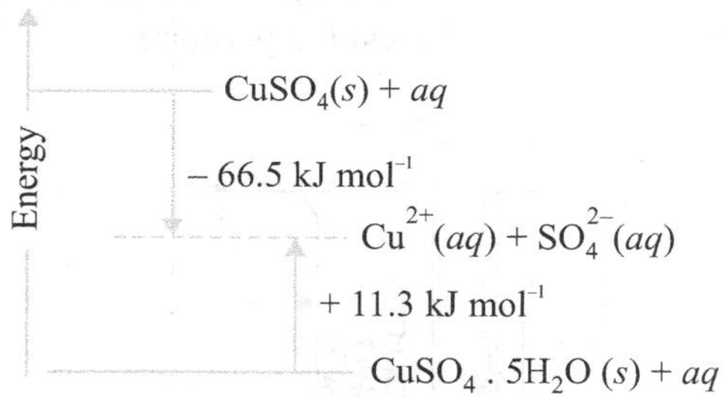

Fig. 16.8. Energy level diagram for the dissolution of $CuSO_4(s)$ and $CuSO_4.5H_2O(s)$.

$$\Delta H_{Hyd.}\ \text{of}\ CuSO_4(s) \longrightarrow CuSO_4.5H_2O$$

$$= -\ 66.5 - 11.3$$

$$= \mathbf{-\ 77.8\ kJ\ mol^{-1}}$$

16.11 ENTHALPY CHANGE DURING COMBUSTION

Combustion is a chemical process involving the burning of a substance in excess of air or oxygen. Energy is always released when a substance undergoes combustion. Some familiar examples of combustion are:

- burning of coal, wood.
- burning of cooking gas, petrol, gasoline.
- burning of food in our body, etc.

Enthalpy of combustion. *It is the enthalpy change accompanying the combustion of one mole of a substance in excess of oxygen or air.*

It is denoted by $\Delta_c H$.

Since combustion reactions are always exothermic, the enthalpy of combustion is always negative. Some examples are as follows:

$$C(s) + O_2(g) \longrightarrow CO_2(g);\ \Delta_c H = -\ 393.5\ \text{kJ mol}^{-1}$$

$$C_3H_8(s) + 5O_2(g) \longrightarrow 3CO_2(g) + 4H_2O(l);$$

$$\Delta_c H = -\ 1222\ \text{kJ mol}^{-1}$$

16.12 ENERGY LEVEL DIAGRAMS AND PROGRESS OF REACTION

We have studied the energy level diagrams for exothermic and endothermic reactions drawn on a vertical scale. But enthalpy change of a reaction does not tell the whole story about the reaction process. Many reactions which are energetically favourable, *i.e.*, have –ve value of $\Delta_r H^\ominus$ do not occur unless and until they are given some amount of energy. For example, the combustion of carbon is exothermic reaction.

$$C(s) + O_2(g) \longrightarrow CO_2(g);\ \Delta H = -\ 393.5\ \text{kJ mol}^{-1}$$

But this reaction does not occur on its own inspite of the abundant availability of carbon and oxygen. For this reaction to occur we have to add some heat energy to the carbon and oxygen mixture in the form of burning. This means that the reactants do not directly change into products but instead, there exists some barrier of energy between the reactants and products. In fact, some additional energy is required by the reactants to commence the cleavage of old bonds.

CONCEPT OF ACTIVATION ENERGY

Let us consider the exothermic reaction

$$A_2 + B_2 \longrightarrow 2AB;\ \Delta_r H = -ve$$

The simple energy level diagram is represented in Fig. 16.9. The reaction involves the breaking of bonds between A – A and B – B molecules and the formation of A – B bonds. The net energy released is more than that required. Therefore, $\Delta_r H$ *i.e.*, enthalpy of reaction = –ve. Now in going from A_2 and B_2 to A – B, the reactants form a complex structure of high energy which represents partial elongation of A – A and B – B bonds and commencement of the formation of A – B bonds.

The partial breaking of bonds in reactants and making of new bonds have been shown as follows by broken lines.

$$\begin{bmatrix} A & \cdots & B \\ \vdots & & \vdots \\ A & \cdots & B \end{bmatrix}$$

This arrangement is more energetic than the reactants as well as products and is called activated complex.

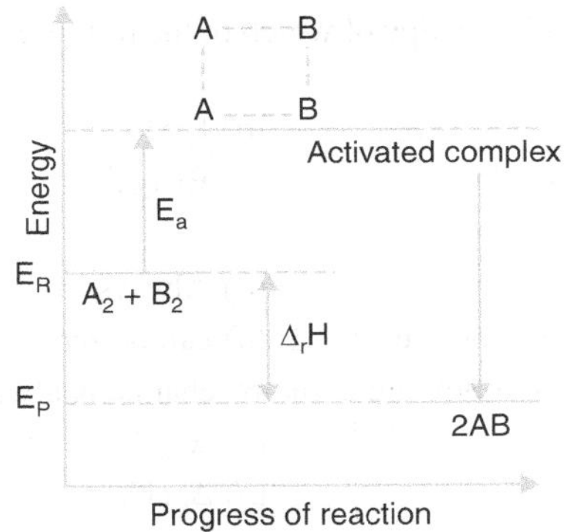

Fig. 16.9(*a*). Energy level diagram for reaction progress.

The additional energy required by the reacting species over and above their average energy is called **activation energy** *represented by* E_a. The values of E_a is different for different reactions.

It can be noted that the *higher* the value of E_a for a reaction, the *slower* will be the chemical process.

Function of Catalyst

A catalyst provides an alternative path of lower activation energy to the reactants. Since activation energy gets lowered, more number of reactant molecules can change into the products per unit time. Hence the reaction rate is accelerated.

The energy level diagram for a catalysed and uncatalysed reaction is shown in Fig. 16.9(*b*).

It can be easily understood from the diagram shown in Fig. 16.9(*b*) that a catalyst does not change the individual energy of the reactants and products. Hence, the value of $\Delta_r H$ for uncatalysed and catalysed reaction remains the same.

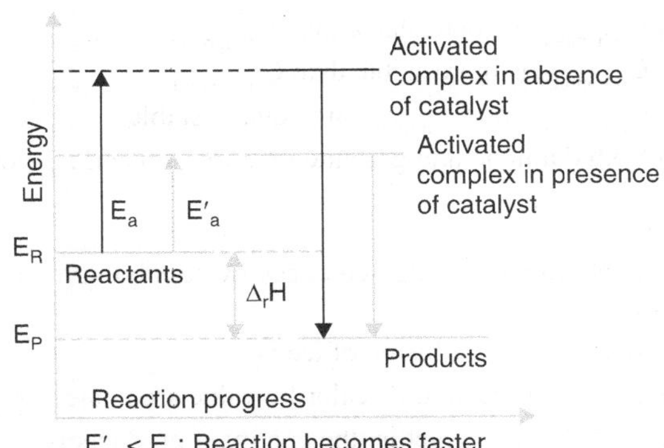

Fig. 16.9(*b*). Energy level diagram for uncatalysed and catalysed reaction

16.13 ENERGY IN FOODS AND FUELS

COMBUSTION OF FOOD IN OUR BODIES

Our body requires energy for performing various functions. This energy mainly comes from carbohydrates and fats that we take as food. Carbohydrates are decomposed into glucose or its derivatives in the stomach. Glucose is soluble in the blood. It is transported by blood cells to various parts of the body where it reacts with oxygen in a series of steps producing CO_2, H_2O and energy.

$$C_6H_{12}O_6(s) + 6O_2(g) \longrightarrow 6CO_2(g) + 6H_2O(g); \quad \Delta H = -2900 \text{ kJ}$$

This combustion of glucose, in the human body, takes place at the body temperature because of the catalytic action of enzymes.

EFFICIENCY OF FOODS AND FUELS

Just as the major source of energy in our bodies is the combustion of foods, in a similar way, the major source of energy for industries is the combustion of fuels. The efficiency of foods and fuels is compared in terms of their *calorific values*. The **calorific value** of food or fuel is defined as *the amount of heat produced when one gram of the substance (food and fuel) is completely burnt.* The calorific values of some common foods and fuels are given below:

Food	*Calorific Value* ($kJ\ g^{-1}$)	*Fuel*	*Calorific Value* ($kJ\ g^{-1}$)
Curd	2.5	*Wood*	17
Milk	3.2	*Charcoal*	33
Egg	7.3	*Kerosene*	48
Meat	12.0	*Fuel oil*	45
Honey	13.3	*Butane (LPG)*	55
Ghee	37.6	*Hydrogen*	150

The normal human needs 3000 kcal (≈ 14450 kJ) of energy per day. The food consisting of carbohydrates, fats, oils, vitamins, proteins, mineral salts which together provides the required calories is termed **balanced diet.**

SUMMARY

- **Chemical Energetics.** It is a branch of science which deals with the energy changes associated with chemical reactions.
- **System.** Part of the universe selected for investigations.
- **Surroundings.** Part of the universe other than the system.
- **State of System.** The conditions of existence of a system when its macroscopic properties have definite values.

- **State Functions.** The thermodynamic quantities which depend only on initial and final state of a system.
- **Heat (q).** It is a random form of energy. It is one of the mode of transference of energy between system and surroundings across a boundary.
- **Work (w).** It is an organised form of energy. It is another mode of transference of energy between system and surroundings across a boundary.
- **Internal Energy (U).** The energy associated with a system at given conditions of temperature and pressure.
- **Internal Energy Change (ΔU).** It is the measure of heat change occurring during the process at constant temperature and constant volume.
- **Enthalpy (H).** It is sum total of internal energy and PV-energy of a system at given conditions of temperature and pressure. It is also called heat content of a system

  ($H = U + PV$).
- **Enthalpy Change (ΔH).** It is a measure of the heat change taking place during the process at constant temperature and constant pressure.
- **Enthalpy of Solution ($\Delta_{sol.} H^\ominus$).** It is the enthalpy change taking place when 1 mole of a solute is dissolved in a large excess of solvent so that on further dilution no enthalpy occurs.
- **Standard Enthalpy of Fusion ($\Delta_{fus.} H^\ominus$).** It is the enthalpy change taking place during the fusion of one mole of solid at its melting point.
- **Standard Enthalpy of Vaporisation ($\Delta_{vap.} H^\ominus$).** It is the enthalpy change taking place during the vaporisation of one mole of liquid at its boiling point.
- **Activation Energy.** The additional energy required by the reacting molecules over and above their average energy to form products.
- **Catalyst.** A foreign substance which accelerates the reaction rate by providing an alternative path of lower activation energy to the reactants.

EVALUATE YOURSELF

I. Objective Type Questions

Select the most appropriate choice from the options given as (a), (b), (c) and (d) after each question:

1. A test tube containing dilute HCl and zinc pieces represents
   (*a*) closed system (*b*) isolated system
   (*c*) open system (*d*) microscopic system.
2. The standard enthalpy of which of the following substance is zero
   (*a*) $I_2(l)$ (*b*) $I_2(s)$
   (*c*) c (diamond) (*d*) $Br_2(g)$.
3. $\Delta_{vap}H^\ominus$ is
   (*a*) Always +ve (*b*) Always –ve
   (*c*) can be +ve or –ve (*d*) can be zero.
4. Which of the following enthalpy change is always negative?
   (*a*) $\Delta_f H^\ominus$ (*b*) $\Delta_{sol}H^\ominus$
   (*c*) $\Delta H^\ominus_{fusion}$ (*d*) $\Delta_C H^\ominus$.
5. $\Delta_C H^\ominus$ of carbon is the same as
   (*a*) $\Delta_{vap}H$ of $CO_2(s)$ (*b*) $\Delta_f H$ of $CO_2(g)$
   (*c*) $\Delta_C H$ of CO (*d*) $\Delta H_{atomisation}$ of CO_2.
6. For the two reactions given below:

$$H_2(g) + \frac{1}{2} O_2(g) \longrightarrow H_2O(l) + x \text{ kJ mol}^{-1}$$

$$H_2(g) + \frac{1}{2} O_2(g) \longrightarrow H_2O(g) + y \text{ kJ mol}^{-1}$$

   (*a*) $x = y$ (*b*) $x < y$
   (*c*) $x > y$ (*d*) $x + y =$ zero.
7. Which of the following reaction represent Δ_{Hyd} H of anhydrous salt?
   (*a*) $BaCl_2(s) + 2H_2O(l) \longrightarrow BaCl_2.2H_2O(s)$
   (*b*) $CuSO_4(s) + 5H_2O(l) + aq \longrightarrow Cu^{2+} (aq) + SO_4^{2-} (aq) + 5H_2O(l)$
   (*c*) $CuSO_4(s) + aq \longrightarrow Cu^{2+}(aq) + SO_4^{2-} (aq)$
   (*d*) None of these.
8. In an endothermic reaction, the value of ΔH is always
   (*a*) = 0 (*b*) > 0
   (*c*) < 0 (*d*) Constant.
9. The value of $\Delta H_{transition}$ of Graphite $\longrightarrow$ Diamond is 1.9 kJ mol^{-1}. This implies that
   (*a*) $C_{(Diamond)}$ is more stable than $C_{(Graphite)}$
   (*b*) $C_{(Graphite)}$ is more stable than $C_{(Diamond)}$
   (*c*) $C_{(Diamond)}$ and $C_{(Graphite)}$ are equally stable
   (*d*) Both diamond and graphite give CO_2 on combustion.

II. Fill in the Blanks

10. Complete the following sentences by supplying appropriate words:
   (*i*) The enthalpy of fusion of ice is ......
   (*ii*) $\Delta_C H^\ominus$ of a fuel and its calorific value are related as ......
   (*iii*) In exothermic reaction the energy of products is ...... than that of reactant.
   (*iv*) Slow combustion of phosphorus releases energy in the form of ......

(*v*) Specific heat capacity (c) and molar heat capacity (C_m) of a substance are related as ......

(*vi*) ΔH represents heat change at ...... temperature and ......

(*vii*) Energy is always ...... when water freezes into ice.

(*viii*) $\Delta_f H^\ominus$ of $C_2H_4(g)$ = 52.26 kJ mol^{-1}. The thermochemical equation that it represents is ......

III. Discussion Questions

11. Define the terms system and surrounding. Give one example each of open, closed and isolated systems.
12. What are the unit of heat, work and standard enthalpy of a substance?
13. What are standard state conditions? Why is it necessary to define the standard state condition?
14. Identify from the following: Open, closed or nearly isolated systems
    (*i*) the earth (*ii*) human beings
    (*iii*) can of tomato soup
    (*iv*) ice cube tray filled with water
    (*v*) satellite in orbit (*vi*) coffee in a thermos flask
    (*vii*) helium filled balloon.
15. (*i*) Will the process $N^{2-}(g) + e^- \longrightarrow N^{3-}(g)$ be exothermic or endothermic?
    (*ii*) $\Delta_f H^\ominus$ of freon $CHClF_2$ is – 480.0 kJ mol^{-1}. Write a thermochemical equation for this notation.
16. How are enthalpy of vaporisation and enthalpy of fusion related to enthalpy of sublimation?
17. What are exothermic and endothermic reactions? Give examples and also draw an energy level diagram.
18. Explain what information is conveyed by
    (*i*) $\Delta_{sol}H^\ominus$ of NaCl(s) = + 5 kJ mol^{-1}
    (*ii*) Describe briefly how would you determine experimentally the value of $\Delta_{sol}H^\ominus$
    (*iii*) Draw the energy level diagram for the process.
19. $CuSO_4(s)$ and $CuSO_4.5H_2O(s)$ were separately dissolved in water and values of their $\Delta_{sol}H^\ominus$ were found to be $-x$ kJ mol^{-1}. and $+y$ kJ mol^{-1}. Draw the energy level diagram for the two and from the diagram determine the ΔH for the process
    $$CuSO_4(s) + 5H_2O(l) \longrightarrow CuSO_4.5H_2O(s)$$
20. What do you understand by the terms enthalpy of combustion and calorific value of fuel?

Answers

I. Objective Type Questions

1. (*c*)	**2.** (*b*)	**3.** (*c*)
4. (*d*)	**5.** (*b*)	**6.** (*c*)
7. (*a*)	**8.** (*b*)	**9.** (*b*).

II. Fill in the Blanks

10. (*i*) > 0 (*ii*) calorific value = $\dfrac{\Delta_c H^\circ}{\text{molar mass}}$

(*iii*) less (*iv*) light

(*v*) $c \times$ molar mass = C_m (*vi*) constant, pressure

(*vii*) released

(*viii*) 2C (graphite) + $2H_2(g) \longrightarrow C_2H_4(g)$.

SECTION-1

UNIT-17

ENERGY CYCLES AND BOND ENTHALPIES

Specific Objectives

After studying this unit, the student should be able to:

- define and use the terms standard state and standard enthalpy change of formation ($\Delta H_f^\ominus$), combustion ($\Delta H_c^\ominus$) and neutralization
- measure the enthalpy change associated with chemical reactions
- determine the heat of combustion of a given fuel
- explain Hess's law of constant heat summation and its application in the development of the Born-Haber cycle
- determine the enthalpy of neutralization for a given acid-base reaction
- explain bond energy and bond dissociation energy
- use Hess's law to determine the heat of reaction for reactions whose energy cannot be easily measured directly.

In the previous unit, we learnt the concept of energy, enthalpy and also the enthalpy changes which take place during physical and chemical processes. In this unit, we shall mainly focus on the standard state conditions and standard enthalpy changes of some important chemical processes along with other aspects related to them.

17.1 ENTHALPY OF A REACTION OR HEAT OF A REACTION

The **enthalpy of a reaction** is defined as the *enthalpy change accompanying the chemical reaction when the molar quantities of reactants and products are the same as indicated in the chemical equation.* It is also known as *heat of reaction.* It is represented by $\Delta_r H$.

FACTORS ON WHICH ENTHALPY OF A REACTION ($\Delta_r H$) DEPENDS

The enthalpy of a reaction depends on the following factors:

1. Physical state of reactants and products. As pointed out earlier the change in physical state also involve heat changes. Thus, the value of ΔH does depend upon the physical state of the reactants and products.

2. Quantities of reactants and products. The amount of heat evolved or absorbed depends upon the amounts of reactants and products. For example, combustion of 12 g (1 mol) of carbon produces 393.5 kJ of heat, while combustion of 1.2 g of carbon produces 39.35 kJ of heat.

3. Allotropic modifications. For different allotropic forms of substances, different amount of heat is evolved or absorbed.

$$C\,(\text{graphite}) + O_2(g) \longrightarrow CO_2(g); \quad \Delta H = -393.5 \text{ kJ}$$

$$C\,(\text{diamond}) + O_2(g) \longrightarrow CO_2; \quad \Delta H = -395.4 \text{ kJ.}$$

4. Temperature. The value of ΔH also depends upon the temperature at which reactants and products are considered. For example, for a reaction

$$H_2(g) + Cl_2(g) \longrightarrow 2HCl$$

the value of ΔH at 298 K and that at 348 K are –184.6 kJ and –184.8 kJ respectively.

The variation of ΔH with temperature is given by Kirchhoff's equation as

$$[\Delta H_2 - \Delta H_1] = (T_2 - T_1)[\Sigma C_p(P) - \Sigma C_p(R)]$$

$\Sigma C_p(P)$ and $\Sigma C_p(R)$ being heat capacities of products and reactants respectively at constant pressure.

5. Conditions of constant pressure or constant volume. The heat of reaction also depends on the conditions of constant volume or constant pressure at a given temperature.

17.2 STANDARD STATE AND STANDARD ENTHALPY OF REACTION ($\Delta_r H^\ominus$)

ΔH of a reaction varies with temperature and pressure conditions, therefore, for the sake of comparison, the values of ΔH for different reactions are expressed at their standard state conditions. The **standard conditions** usually chosen are *298 K temperature*, and *1 bar or 100 kPa* pressure; and *1 mol dm⁻³ concentration for a solution.* ***A standard state (also called reference state) of a substance is its most stable state at 100 kPa pressure and 298 K.***

According to IUPAC recommendations, the enthalpy change under standard state conditions is called **standard enthalpy of the reaction** and is denoted by $\Delta_r H^\ominus$. The superscript (⊖) represents standard state. It is worthwhile to mention here, *that unless otherwise mentioned, the ΔH values are the standard enthalpy changes of reactions.*

Note. Earlier, the notation used for standard enthalpy of reaction was $\Delta H_r°$. This is still used even now in a large number of books.

Let us now, discuss the standard enthalpies of some chemical processes such as ***formation, combustion*** and ***neutralisation.***

MOLAR ENTHALPY OF FORMATION ($\Delta_f H$)

It is the enthalpy change accompanying the formation of one mole of a compound from its constituent elements. It is generally denoted by $\Delta_f H$. For example, enthalpy of formation of carbon dioxide (CO_2) and that of ethyl alcohol (C_2H_5OH) are – 393.5 and – 277.0 kJ mol⁻¹ respectively. These are expressed as follows:

$$C(s) + O_2(g) \longrightarrow CO_2(g); \quad \Delta H = -393.5 \text{ kJ}$$

$$2C(s) + 3H_2(g) + \tfrac{1}{2} O_2(g) \longrightarrow C_2H_5OH(l); \quad \Delta H = -277.0 \text{ kJ.}$$

Standard Molar Enthalpy of Formation ($\Delta_f H^\ominus$)

When all the species of a chemical reaction are in their standard states, the enthalpy of formation is called **standard molar enthalpy of formation.** The standard molar enthalpy of formation is denoted by $\Delta_f H^\ominus$. Thus, **standard molar enthalpy of formation** of a compound is defined as *the enthalpy change accompanying the formation of one mole of a compound from its constituent elements, all the substances being in their standard states (1 bar or 100 kPa pressure and 298 K).*

It may be noted that:

(*i*) *Standard molar enthalpy of formation of a substance is also called its standard enthalpy and can also be denoted as* $\mathbf{H}^\ominus$.

(*ii*) *By convention, the standard molar enthalpies of free elements are taken to be zero.*

(*iii*) *Standard state of elements refers to the pure element in its most stable form at 298 K and 100 kPa pressure.* For example, the standard state of *carbon, oxygen, sulphur, mercury, bromine, and iodine* are: ***C (graphite), $O_2(g)$, S_8 (rhombic), $Br_2(l)$*** and ***$I_2(s)$*** respectively.

The values of the standard molar enthalpies of formation of some substances at 298 K are given in Table 17.1.

Table 17.1. Standard Molar Enthalpies of Formation of Some Substances at 298 K

Substance	$\Delta_f H^\ominus$ *(kJ mol⁻¹)*	*Substances*	$\Delta_f H^\ominus$ *(kJ mol⁻¹)*
$Al_2O_3(s)$	– 1675.7	$HI(g)$	+ 26.48
$BaCO_3(s)$	– 1216.3	$KCl(s)$	– 436.74
$Br_2(l)$	0	$KBr(s)$	– 393.8
$Br_2(g)$	30.91	$MgO(s)$	– 601.70
$CaCO_3(s)$	– 1206.92	$Mg(OH)_2(s)$	– 924.54
$CaO(s)$	– 635.09	$NaF(s)$	– 573.65
$CH_4(g)$	– 74.81	$NaCl(s)$	– 411.15

$C_2H_4(g)$	52.26	$NaBr(s)$	– 361.06
$CH_3OH(l)$	– 238.66	$NaI(s)$	– 287.78
$C_2H_5OH(l)$	– 277.69	$NH_3(g)$	– 46.11
$C_6H_6(l)$	49.03	$NO(g)$	+ 90.25
$CO(g)$	– 110.525	$NO_2(g)$	+ 33.18
$CO_2(g)$	– 393.51	$PCl_3(l)$	– 319.70
$C_2H_6(g)$	– 84.68	$PCl_5(s)$	– 443.5
$Cl_2(g)$	0	$SiO_2(s)$ (quartz)	– 910.94
C_3H_8	– 103.2	$SnCl_2(s)$	– 325.1
n-$C_4H_{10}(g)$	– 888.0	$SnCl_4(l)$	– 511.3
$HgS(s)$ red	– 58.2	$SO_2(g)$	– 296.83
$H_2(g)$	0	$SO_3(g)$	– 395.72
$H_2O(g)$	– 241.82	$SiH_4(g)$	34
$H_2O(l)$	– 285.83	$SiCl_4(g)$	– 657.0
$HF(g)$	– 271.1	$C(g)$	+ 715.0
$HCl(g)$	– 92.30	$Cl(g)$	+ 121.0
$HBr(g)$	– 36.40	$H(g)$	+ 217.7

Importance of standard molar enthalpies of formation. The knowledge of standard molar enthalpies of various substances can help us to calculate standard enthalpy change of any reaction. Standard enthalpy change of a reaction is equal to the difference of the standard molar enthalpies of all the products and standard enthalpies of all the reactants.

$$\begin{bmatrix}\textbf{Standard}\\ \textbf{enthalpy}\\ \textbf{of reaction}\end{bmatrix} = \begin{bmatrix}\textbf{Sum of standard}\\ \textbf{molar enthalpies}\\ \textbf{of formation}\\ \textbf{of products}\end{bmatrix} - \begin{bmatrix}\textbf{Sum of standard}\\ \textbf{molar enthalpies}\\ \textbf{of formation}\\ \textbf{of reactants}\end{bmatrix}$$

or $\Delta_r H^\ominus = \Sigma \Delta_f H^\ominus$ (Products) – $\Sigma \Delta_f H^\ominus$ (Reactants)

For example, consider a hypothetical reaction

$$aA + bB \longrightarrow lL + mM$$

$$\Delta_r H^\ominus = \Sigma\, \Delta_f H^\ominus\, (Products) - \Sigma\, \Delta_f H^\ominus\, (Reactants)$$
$$= [l\Delta_f H^\ominus (L) + m\Delta_f H^\ominus (M)] - [a\Delta_f H^\ominus (A) + b\Delta_f H^\ominus (B)].$$

SOLVED EXAMPLES

Example 17.1. *Calculate the standard enthalpy of formation of $C_2H_4(g)$ from the following thermochemical equation:*

$$C_2H_4(g) + 3O_2(g) \longrightarrow 2CO_2(g) + 2H_2O(g); \quad \Delta H^\ominus = -1323\ kJ.$$

Given that $\Delta_f H^\ominus$ of $CO_2(g)$ and $H_2O(g)$ are – 393.5 and – 249 kJ mol^{-1} respectively.

Solution. We know

$$\Delta_r H^\ominus = \Sigma\, \Delta_f H^\ominus\, (Products) - \Sigma\, \Delta_f H^\ominus\, (Reactants)$$

Substituting the given values

$$-1323 = [2(-393.5) + 2 \times (-249.0)] - [\Delta_f H^\ominus (C_2H_4) + 0]$$
$$= [-787 - 498] - \Delta_f H^\ominus (C_2H_4)$$
$$= -1285 - \Delta_f H^\ominus (C_2H_4)$$

or $\Delta_f H^\ominus (C_2H_4) = 1323 - 1285$

$= +$ **38 kJ mol^{-1}.**

Example 17.2. *The enthalpy of combustion of ethanoic acid is given as follows:*

$$CH_3COOH(l) + 2O_2(g) \longrightarrow 2CO_2(g) + 2H_2O(l); \quad \Delta H = -867.0\ kJ\ mol^{-1}$$

If values of $\Delta_f H^\ominus$ of $CO_2(g)$ and $H_2O(l)$ are – 393.5 kJ mol^{-1} and – 286.0 kJ mol^{-1} respectively, calculate the enthalpy of formation of ethanoic acid.

Solution. $CH_3COOH(l) + 2O_2(g) \longrightarrow 2CO_2(g) + 2H_2O(l)$; $\Delta H = -867.0$ kJ mol^{-1}

$$\Delta_r H^\ominus = \Sigma\Delta_f H^\ominus\, (Products) - \Sigma\Delta_f H^\ominus\, (Reactants)$$
$$= 2\,\Delta_f H^\ominus (H_2O) + 2\Delta_f H^\ominus (CO_2) - \Delta_f H^\ominus (CH_3COOH) + 2\Delta_f H^\ominus (O_2)$$

or $-867.0 = 2 \times (-286.0) + 2 \times (-393.5) - \Delta_f H^\ominus (CH_3COOH) + 2(0)$

or $\Delta_f H^\ominus (CH_3COOH) = 867.0 - 572 - 787$

$=$ **– 492.0 kJ mol^{-1}.**

STANDARD ENTHALPY OF COMBUSTION ($\Delta_c H^\ominus$)

It is the enthalpy change accompanying the complete combustion of one mole of a substance in excess of oxygen or air. Some examples of combustion reactions are given as follows:

(*i*) Combustion of carbon

$$C(s) + O_2(g) \longrightarrow CO_2(g); \quad \Delta H^\ominus = -393.5\ kJ\ mol^{-1}$$

(*ii*) Combustion of methane (CH_4)

$$CH_4(g) + 2O_2(g) \longrightarrow CO_2(g) + 2H_2O(l); \quad \Delta H^\ominus = -891 \text{ kJ mol}^{-1}$$

(*iii*) Combustion of butane (C_4H_{10})

$$C_4H_{10}(g) + \frac{13}{2} O_2(g) \longrightarrow 4CO_2(g) + 5H_2O(l);$$

$$\Delta H^\ominus = -2658 \text{ kJ mol}^{-1}$$

Combustion reactions are always accompanied by the evolution of heat, therefore, the value of $\Delta_c H^\ominus$ is always **negative.**

Standard enthalpies of combustion of some substances are given in Table 17.2.

Table 17.2. Standard Enthalpies of Combustion

Substance	$\Delta_c H^\ominus$ (*kJ mol*$^{-1}$)
$CH_4(g)$	– 890
$C_2H_6(g)$	– 1560
$C_2H_4(g)$	– 1411
$C_2H_2(g)$	– 1300
$C_6H_6(l)$	– 3268
$CH_3OH(l)$	– 726.1
$C_2H_5OH(l)$	– 1368.0
$CH_3COOH(l)$	– 875
$(NH_2)_2CO(s)$ (*urea*)	– 632
$C_6H_{12}O_6(s)$ (*glucose*)	– 2902
$C_{12}H_{22}O_{11}$ (*s*) (*sucrose*)	– 5645

ACTIVITY 17.1 To determine the enthalpy of combustion of ethanol.

Requirements

H-shaped calorimeter fitted with a copper spiral, thermometer and stirrer, Spirit lamps, ethanol.

Procedure

Part I:

1. Set up the apparatus as shown in Fig. 17.1. Add water to it and attach the other end of the copper spiral to the suction pump. Note the constant temperature of water (t_1°C).
2. Fill the spirit lamp to about 2/3 level with ethanol, and weigh it with its cover on it (W_1 g). Keep this cover on until it is time to ignite.

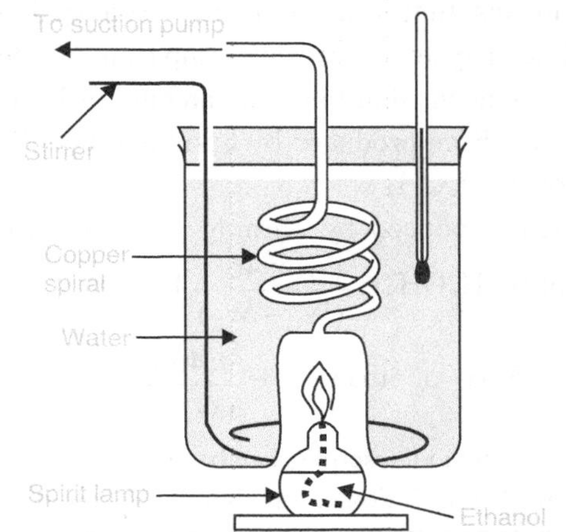

Fig. 17.1. An apparatus for measuring the heat of combustion of alcohol.

3. Start the suction pump to draw the slow current of air through the copper spiral.
4. Uncover the lamp and ignite it. Stir the water in the calorimeter gently. When a temperature rise of about 10–20°C has been achieved, stop heating and again cover the lamp. Note the final temperature (t_2°C).
5. Weigh the spirit lamp again (W_2 g).

Part II:

6. To find the *heat capacity* of the calorimeter is ordinarily not simple because the heat capacity of water, copper and glass is different. This problem can be tackled in a most simple way as follows.
7. Let the calorimeter system be cooled to the same initial temperature (t_1°C).
8. Now take another spirit lamp, filled about 2/3, with methanol (whose $\Delta_c H^\ominus$ is already known) and weigh it (W_3 g) along with its cover on. Uncover it and ignite it.
9. Keep heating along with stirring till the same temperature (t_2°C) is attained. Stop heating. Cover the lamp again and take its weight (W_4 g).

Sample Calculations

From **experiment**—Part (II) with methanol

$\Delta_C H^\ominus$ for methanol = – 726.3 kJ mol^{-1} (*already known*)

Mass of methanol burnt = $(W_3 - W_4)$ g

Heat produced by burning

1 mol (= 32 g) of CH_3OH (*l*) = 726.3 kJ

∴ Heat produced by burning

$$(W_3 - W_4) \text{ g of } CH_3OH(l) = \frac{726.3 \times (W_3 - W_4)}{32}$$

$$= x \text{ kJ (say)}$$

This means that x kJ of heat is required by the calorimeter system to raise its temp from t_1°C to t_2°C. This also means that the same amount of heat (x kJ) must have been produced by burning $(W_1 - W_2)$g of ethanol (C_2H_5OH)

$\therefore$ Heat produced by burning 1 mol (= 46 g) of ethanol (C_2H_5OH) $= \dfrac{x \times 46}{(W_1 - W_2)}$ kJ

$$\therefore \quad \Delta_C H^{\ominus} \text{ of ethanol} = -\frac{46 \times x}{(W_1 - W_2)} \text{ kJ mol}^{-1}$$

Note: The method described can be used to determine the enthalpy of combustion of any flammable liquid or fuel.

17.3 STANDARD ENTHALPY OF NEUTRALIZATION (Δ_nH)

It is the enthalpy change accompanying the complete neutralization of an acid by a base or vice versa involving combination of 1 mol of H^+ ions (from acid) and 1 mol of OH^- ions (from base) to form 1 mol of $H_2O(l)$ in dilute aqueous solutions. For example, the enthalpy change accompanying the neutralization of NaOH and HCl is represented as:

$$HCl(aq) + NaOH(aq) \longrightarrow NaCl(aq) + H_2O(l); \quad \Delta_n H = -57.1 \text{ kJ mol}^{-1}$$

Similarly,

$$HCl\,(aq) + \frac{1}{2}\,Ca(OH)_2\,(aq) \longrightarrow \frac{1}{2}\,CaCl_2\,(aq) + H_2O(l)$$

$$\Delta_n H^{\ominus} = -57.2 \text{ kJ mol}^{-1}.$$

Some acid-base reactions and their standard enthalpy of neutralization are given in Table 17.3.

Table 17.3. Standard Enthalpies of Neutralization of Some Acids and Bases

Reactants	*Strength*		$\Delta_n H^{\ominus}$
	Acid	*Base*	*(kJ mol^{-1})*
$HCl(aq) + NaOH\,(aq)$	strong	strong	– 57.1
$HNO_3\,(aq) + NaOH\,(aq)$	strong	strong	– 57.1
$CH_3COOH\,(aq) + NaOH\,(aq)$	weak	strong	– 56.1
$HCl\,(aq) + NH_3\,(aq)$	strong	weak	– 53.4
$CH_3COOH\,(aq) + NH_3\,(aq)$	weak	weak	– 50.4

The enthalpy of neutralization of all strong acids and strong bases is always constant, i.e., –57.1 kJ. The explanation to this generalization can be provided on the basis of the Arrhenius theory of ionization. The strong acids and strong bases are almost completely ionized in dilute aqueous solutions. The neutralization of strong acid and strong bases simply involve the combination of H^+ ions (from acid) and OH^- ions (from base) to form water molecules. It has been found experimentally that when 1 mole of water is formed by the neutralization of 1 mole of H^+ ions and 1 mol of OH^- ions in aqueous solutions, 57.1 kJ of energy is released.

$$H^+(aq) + OH^-(aq) \longrightarrow H_2O(l); \Delta H = -57.1 \text{ kJ}.$$

Now, the net process in the neutralization of HCl(aq) and NaOH(aq) can be represented as:

$$H^+(aq) + Cl^-(aq) + Na^+(aq) + OH^-(aq) \longrightarrow H_2O(l) + Na^+\,(aq) + Cl^-\,(aq)$$

or cancelling out the common spectator ions on both sides of arrow, we get

$$\mathbf{H^+(aq) + OH^-(aq) \longrightarrow H_2O(l)}$$

In a similar way, the neutralization of $HNO_3(aq)$ and KOH(aq) can be represented as

$$H^+(aq) + NO_3^-(aq) + K^+(aq) + OH^-(aq) \longrightarrow H_2O(l) + K^+(aq) + NO_3^-(aq)$$

or $$\mathbf{H^+(aq) + OH^-(aq) \longrightarrow H_2O(l)}$$

Thus, during neutralization of all strong acids and strong bases the same reaction takes place. Hence *the value of the enthalpy of neutralization of strong acid or strong base is constant.*

However, if either acid or base is weak, then its ionization is not complete in solution. Therefore, a part of the energy liberated during the combination of H^+ ions and OH^- ions is utilised for the ionization of the weak acid (or base). Consequently, *the value of the enthalpy of neutralization of weak acid/strong base or strong acid/weak base is numerically less than 57.1 kJ.* For example, the neutralization of acetic acid and sodium hydroxide can be represented as

$$\underset{\text{(weak acid)}}{CH_3\,COOH} \rightleftharpoons CH_3COO^- + H^+; \quad \Delta H_{\text{ionization}} = +1.0 \text{ kJ}$$

$$\underset{\text{(strong base)}}{NaOH} \longrightarrow Na^+ + OH^-$$

$$H^+(aq) + OH^-(aq) \longrightarrow H_2O\,(l)\,; \Delta H = -57.1 \text{ kJ}$$

Net reaction:

$$CH_3COOH + Na^+ + OH^- \rightleftharpoons CH_3COO^- + Na^+ + H_2O; \Delta H = -56.1 \text{ kJ}$$

or $$CH_3COOH + OH^- \rightleftharpoons CH_3COO^- + H_2O; \quad \Delta H = -56.1 \text{ kJ}.$$

Thus, the enthalpy of neutralization of acetic acid and sodium hydroxide is –56.1 kJ. Similarly, the enthalpy of neutralization of ammonium hydroxide (weak base) and hydrochloric acid (strong acid) is –51.5 kJ.

ACTIVITY 17.2 **To determine the enthalpy of neutralization for the reaction between HCl and NaOH.**

Requirments

100 cm^3 1 M HCl solution, 100 cm^3 1M NaOH solution, three 250 cm^3 beakers, Polystyrene cup with a lid fitted with stirrer and thermometer.

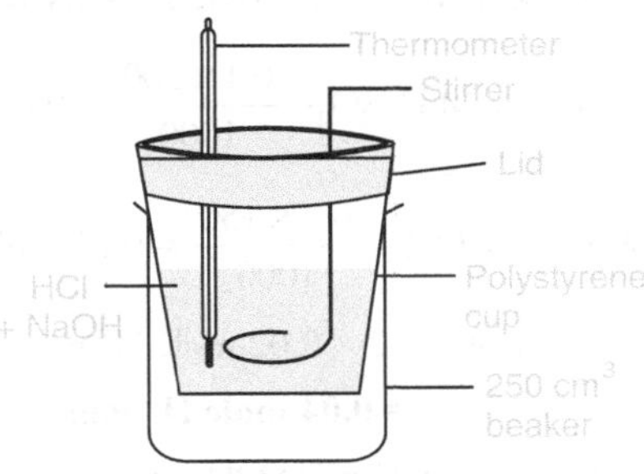

Fig. 17.2. Determining $\Delta_n H$ of HCl and NaOH

Procedure

1. Set up a clean and thoroughly dried apparatus as shown in Fig. 17.2.
2. Pour 100 cm^3 of 1M HCl into the polystyrene cup and note its temperature every one minute till constant (t_1°C).
3. At the same time note the temperature of 100 cm^3 of 1M NaOH. (It should be the same as t_1°C.) If it is less than t_1, then rotate a test tube with hot water in the solution till the temperature becomes t_1°C. If it is more than t_1, then rotate a test tube with cold water in the solution till it becomes t_1°C.
4. Open the lid and quickly pour 100 cm^3 of 1M NaOH and stir gently. Record the temperature after every 20 seconds from the moment of mixing till at least 5 minutes.
5. Plot a graph of temperature (t_2) of the mixture vs time as shown in Fig. 17.3.
6. Find Δt as shown in Fig. 17.3 and make calculations by assuming density of solutions as 1g cm^{-3} and specific heat capacity of calorimeter system as 4.3 Jg^{-1} K^{-1} and specific heat capacity of water as 4.183 Jg^{-1} K^{-1}. (For calculations follow the solved example 17.3.)

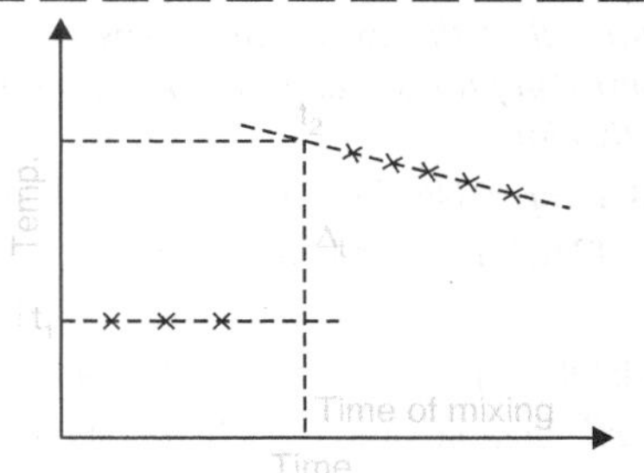

Fig. 17.3. Temperature-time graph.

7. Clean the apparatus, dry it and use it to repeat the experiments with each of the following pairs:
   (*a*) 100 cm^3 0.5 M H_2SO_4 + 100 cm^3 1M NaOH solutions.
   (*b*) 100 cm^3 1M Ethanedioic acid (CH_3COOH) + 100 cm^3 1M NaOH solution
   (*c*) 100 cm^3 1M HCl + 100 cm^3 1M NH_4OH solution.

   Try to answer the following

   (*i*) *Write the values of:*

	Experimental Values	*Theoretical Values*
$\Delta_n H^\ominus$ *for HCl and NaOH*	......	......
$\Delta_n H^\ominus$ *for H_2SO_4 and NaOH*	......	......
$\Delta_n H^\ominus$ *for CH_3COOH and NaOH*	......	......
$\Delta_n H^\ominus$ *for HCl and NH_4OH*	......	......

Account for the differences in theoretical and experimental values.

(*ii*) *What conclusion can you draw about the enthalpy of neutralization of*
strong acid – strong base,
weak acid – strong base,
strong acid – weak base?

(*iii*) *Can you compute the enthalpy of ionization of a weak acid or weak base?*

SOLVED EXAMPLES

Example 17.3. *100 cm^3 of 1M hydrochloric acid, HCl (aq), was carefully added to 100 cm^3 of 1M sodium hydroxide solution, NaOH (aq), in a well-insulated beaker. The heat capacity of the beaker is known to be 180 J K^{-1}. The initial temperature of the acid and alkali was 25°C. On adding the acid, the temperature of the mixture rose to 30.6°C. (The specific heat capacity of water, c, is 4.18 kJ kg^{-1} K^{-1}.)*

(*i*) *Write a balanced equation for the reaction.*

(*ii*) *Calculate the amount of water formed in the reaction.*

(iii) Assuming that the experiment was carried out under standard conditions, calculate the standard enthalpy of neutralization.

Solution. (*i*) $HCl\ (aq) + NaOH\ (aq) \longrightarrow NaCl\ (aq) + H_2O(l)$

(*ii*) Moles of HCl used = Molarity × $V_{(L)}$

$= 1 \times 100 \times 10^{-3} = 0.1$ mol

Moles of NaOH = $1 \times 100 \times 10^{-3} = 0.1$ mol

Equation reveals mole ratio of HCl, NaOH, H_2O as 1 : 1 : 1

∴ Moles of water formed = **0.1** mol or 0.1 × 18 = **1.8 g**

(*iii*) Heat evolved by neutralization of 0.1 mol of HCl is equal to the heat gained by the reaction mixture (200 cm^3) and the beaker.

If heat evolved is x, then $x = mc\Delta T + C\Delta T$

Here, m = mass of mixture

$= 200\ cm^{-3} \times 1g\ cm^{-3} = 200$ g

c = Specific heat capacity of water

$= 4.184\ JK^{-1}\ g^{-1}$

C = Heat capacity of beaker = 180 JK^{-1}

$\Delta T = 30.6 - 25 = 5.6$ °C or 4.4 K

$= 200\ (g) \times 4.184\ (Jg^{-1}\ k^{-1}) \times 5.6\ (K) + 180\ (JK^{-1}) \times 5.6\ (K)$

$= 4686.08\ (J) + 1008\ (J) = 5694.08$ J

Heat produced by 1 mol of HCl = $\dfrac{5694.08\ (J)}{0.1}$ or 56940.8 J

or = 56.941 kJ.

∴ $\Delta_n H^\ominus = -$ **56.941 kJ mol^{-1}.**

Example 17.4. *What would be the heat evolved when:*

(*a*) *0.5 mole of HCl is neutralized by 0.5 mole of NaOH in aqueous solutions.*

(*b*) *An aqueous solution containing 0.5 mole of HNO_3 is mixed with an aqueous solution containing 0.3 mole of NaOH.*

(*c*) *100 cm^3 of 0.2 M HCl solution is mixed with 200 cm^3 of 0.3 M KOH.*

(*d*) *200 cm^3 of 0.1 M H_2SO_4 is mixed with 150 cm^3 of 0.2 M KOH.*

Solution. (*a*) 0.5 mole of HCl ≡ 0.5 mole of H^+ ions

0.5 mole of NaOH ≡ 0.5 mole of OH^- ions

$$\underset{0.5\ \text{mole}}{H^+(aq)} + \underset{0.5\ \text{mole}}{OH^-(aq)} \longrightarrow \underset{0.5\ \text{mole}}{H_2O(l)}$$

When 1 mole of water is formed by the combination of H^+ and OH^- ions in aqueous solutions, 57.1 kJ of heat is produced.

∴ Heat produced during the formation of 0.5 mole of H_2O would be

$= 57.1 \times 0.5 =$ **28.5 kJ.**

(*b*) 0.5 mole of HNO_3 ≡ 0.5 mole of H^+ ions

0.3 mole of NaOH ≡ 0.3 mole of OH^- ions

0.3 mole OH^- ions would react with 0.3 mole H^+ ions to give 0.3 mole H_2O.

∴ The heat liberated = 57.1 × 0.3 = **17.1 kJ.**

(*c*) 100 cm^3 of 0.2 M HCl contain

$\dfrac{0.2 \times 100}{1000} = 0.02$ mole HCl

200 cm^3 of 0.3 M KOH contain $\dfrac{0.3 \times 200}{1000} = 0.06$ mole KOH

0.02 mole HCl ≡ 0.02 mole H^+ ions

0.06 mole KOH ≡ 0.06 mole OH^- ions

0.02 mole H^+ ions would react with 0.02 mole of OH^- ions to form 0.02 mole of H_2O.

∴ The heat liberated = 57.1 × 0.02 kJ = **1.1 kJ.**

(*d*) 200 cm^3 of 0.1 M H_2SO_4= $\dfrac{0.1 \times 200}{1000} = 0.02$ mole H_2SO_4

150 cm^3 of 0.2 M KOH = $\dfrac{0.2 \times 150}{1000} = 0.03$ mole KOH

0.02 mole H_2SO_4 ≡ 2 × 0.02 mole H^+ ions

= **0.04 mole H^+ ions**

0.03 mole of KOH ≡ 0.03 mole OH^- ions

0.03 mole OH^- ions would react with 0.03 mole of H^+ ions to form 0.03 mol of H_2O.

∴ Heat liberated = 57.1 × 0.03 = **1.7 kJ.**

17.4 HESS'S LAW OF CONSTANT HEAT SUMMATION

On the basis of numerous thermochemical observations, the Russian Chemist, *Germain Henry Hess* gave one of the most important generalisation of thermochemistry in 1840 which is known after his name as **Hess's Law.** This generalisation is primarily based on the fact that enthalpy is a path independent thermodynamic function. In other words, the enthalpy change in a chemical reaction is the same whether it occurs in one step or in more than one step. This may be stated in the form of Hess's law as follows. *The standard enthalpy of reaction is the algebraic sum of the standard enthalpies of reactions into which the overall reaction may be splitted or divided.*

Consider a process involving the conversion of reactant A into B in one step (path I). The enthalpy change of a process is represented by ΔH. Now suppose the process is carried out in two steps involving a change from A to C and C to B according to path II, (Fig. 17.4). Let ΔH_1 and ΔH_2 be enthalpy changes from A to C and C to B respectively. Then, according to Hess's law:

$$\Delta H = \Delta H_1 + \Delta H_2$$

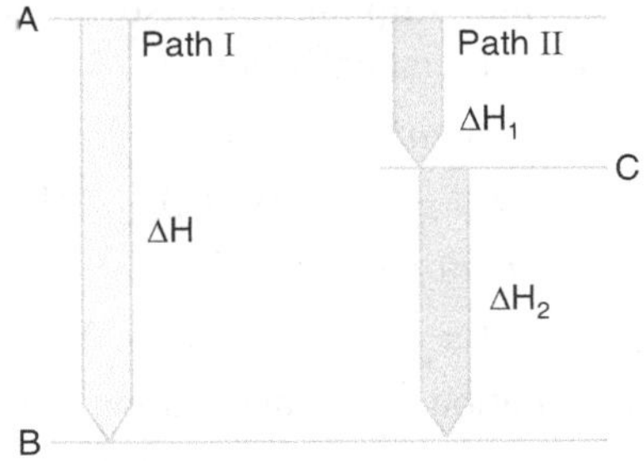

Fig. 17.4. Illustration of Hess's law.

Hess's law is simply a corollary of the law of conservation of energy. It implies that the enthalpy change of a reaction depends only on the states of the reactants and products and not on the manner by which the change is brought about. If ΔH were not independent of the manner in which the corresponding change was brought about, it would have been possible to create or destroy energy, by taking a system from A to B by one path and then returning it to state A by a second path.

ILLUSTRATION OF HESS'S LAW

Let us consider the formation of carbon dioxide from carbon and oxygen. There are two ways by which the change can be brought about.

(*a*) *Conversion of carbon to carbon dioxide*

$$C(s) + O_2(g) \longrightarrow CO_2(g);\ \Delta H = -393.5 \text{ kJ}$$

(*b*) *Conversion of carbon to carbon monoxide and subsequent oxidation of carbon monoxide to carbon dioxide.*

$$C(s) + \frac{1}{2} O_2(g) \longrightarrow CO(g);\quad \Delta H_1 = -110.5 \text{ kJ}$$

$$CO(g) + \frac{1}{2} O_2(g) \longrightarrow CO_2(g);\ \Delta H_2 = -283.0 \text{ kJ}$$

The two paths by which carbon can be converted into carbon dioxide have been shown in Fig. 17.5.

As is clear, $\Delta H = \Delta H_1 + \Delta H_2$

$$-393.5 \text{ kJ} = -110.5 + (-283.0) \text{ kJ}$$

$$= -393.5 \text{ kJ.}$$

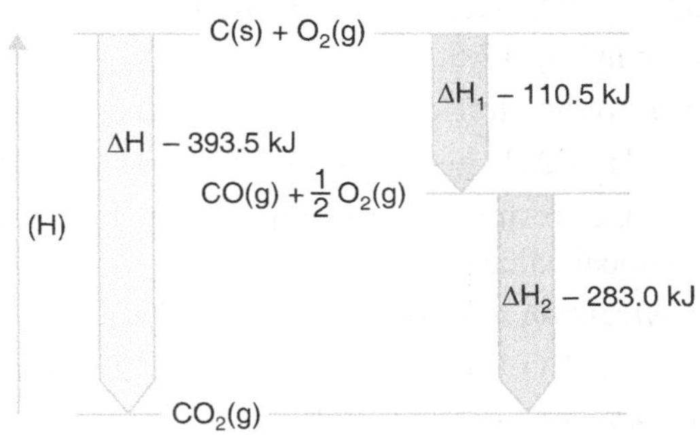

Fig. 17.5. Hess's law cycle for CO_2.

APPLICATIONS OF HESS'S LAW

The practical utility of Hess's law lies in the fact that it allows us to carry out thermochemical calculations to predict the enthalpies of different reactions whose direct measurement is not possible. The thermochemical equations like algebraic equations can be added or substracted, multiplied or divided by any numerical factor. Some of the important applications of the Hess's law are discussed below:

1. Determination of Enthalpy of Formation

There are large number of compounds such as methane, benzene, carbon monoxide, ethane, etc., which cannot be directly synthesized from their elements. Hence, their enthalpy of formation cannot be determined by the calorimetric method. In case of such substances the enthalpy of formation can be determined by an indirect method based upon Hess's law. The unknown reaction is made one segment of the closed cycle of reactions. The enthalpy changes of the reactions which represent other segments of the cycle can be determined experimentally. Then by applying Hess's law the desired value of ΔH can be calculated.

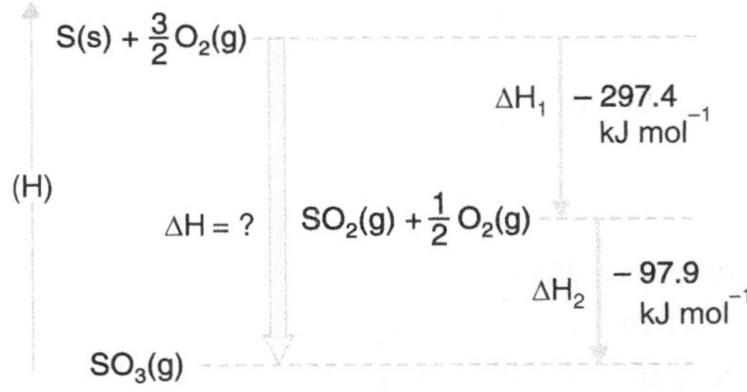

Fig. 17.6. Hess's law cycle for SO_3.

For example, let us study the calculation of the enthalpy of formation of SO_3.

The enthalpy of formation of SO_2 (ΔH_1) and the enthalpy of combustion of SO_2 (ΔH_2) have been found to be –297.4 kJ

mol^{-1} and – 97.9 kJ mol^{-1} respectively. These changes have been shown in Fig. 17.6.

Now according to Hess's law,

$$\Delta H = \Delta H_1 + \Delta H_2 = -297.4 + (-97.9) = \mathbf{-395.3\ kJ.}$$

The same result can be arrived at by means of thermochemical calculations

(*i*) *Conversion of sulphur to sulphur dioxide*

$$S + O_2 \longrightarrow SO_2; \quad \Delta H_1 = -297.4 \text{ kJ}$$

(*ii*) *Combustion of sulphur dioxide to* SO_3

$$SO_2 + \frac{1}{2} O_2 \longrightarrow SO_3; \quad \Delta H_2 = -97.9 \text{ kJ}$$

Add equations (*i*) and (*ii*),

$$S + \frac{3}{2} O_2 \longrightarrow SO_3;$$

$$\Delta H = \Delta H_1 + \Delta H_2 = -297.4 - 97.9 = \mathbf{-395.3\ kJ.}$$

See solved example for further illustration.

2. Determination of Enthalpy of Transition

Transition implies the conversion of one allotropic form of a substance to another. For example, change of graphite to diamond, red phosphorus to yellow phosphorus and rhombic sulphur to monoclinic sulphur, etc. Such reactions are very slow and the enthalpy change accompanying them cannot be measured directly. However, Hess's law is quite helpful in determining the enthalpy of transition.

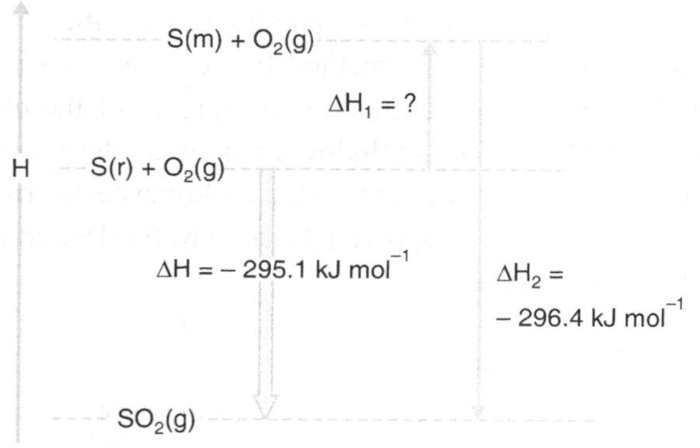

Fig. 17.7. Hess's law cycle for $S_{(R)}$ to SO_2.

For example, let us calculate the enthalpy of the process $S_R \longrightarrow S_M$ from the enthalpies of combustion of monoclinic sulphur (S_M) and rhombic sulphur (S_R) which have been found to be – 296.4 and – 295.1 kJ mol^{-1} respectively. Now Hess's law cycle is constituted as $S_{(R)} \longrightarrow SO_2$ (path I) and $S_{(R)} \longrightarrow S_{(M)} \longrightarrow SO_2$ (path II). These changes have been shown in Fig. 17.7.

Now according to Hess's law:

$$\Delta H = \Delta H_1 + \Delta H_2$$

or $$\Delta H_1 = \Delta H - \Delta H_2 = -295.1 - (-296.4) = \mathbf{1.3\ kJ}$$

The enthalpy of the allotropic transformation of rhombic sulphur to monoclinic sulphur can also be calculated as follows by thermochemical calculations:

1. *Combustion of rhombic sulphur*

$$S_{(R)} + O_2(g) \longrightarrow SO_2(g); \quad \Delta H_1 = -295.1 \text{ kJ}$$

2. *Combustion of monoclinic sulphur*

$$S_{(M)} + O_2(g) \longrightarrow SO_2(g); \quad \Delta H_2 = -296.4 \text{ kJ}$$

Subtracting equation (2) from equation (1)

$$S_{(R)} \longrightarrow S_{(M)}; \quad \Delta H = \Delta H_1 - \Delta H_2$$
$$= -295.1 - (-296.4) = \mathbf{+1.3\ kJ.}$$

3. Determination of Enthalpy of Hydration of Anhydrous Salt

Hydration means the conversion of an anhydrous salt into its hydrate by combining with a specific number of moles of water. For example, one mole of anhydrous copper sulphate combines with five moles of water to form one mole of hydrated copper sulphate.

$$\underset{\text{Anhydrous salt}}{CuSO_4(s)} + 5H_2O(l) \longrightarrow \underset{\text{Hydrated salt}}{CuSO_4 \cdot 5H_2O(s)}$$

Direct measurement of enthalpy of hydration is not possible. However, use of Hess's law makes it possible as described below:

The enthalpies of dissolution of hydrated copper sulphate and that of anhydrous copper sulphate have been experimentally found to be +11.70 kJ mol^{-1} and – 66.5 kJ mol^{-1} respectively. From these change the Hess's law cycle is constituted according to Fig. 17.8.

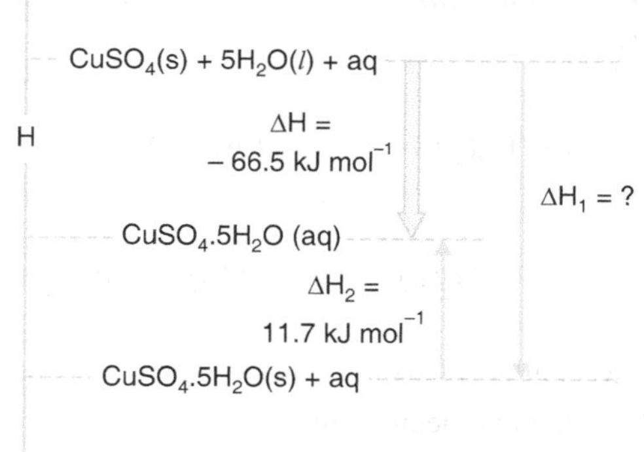

Fig. 17.8. Hess's law cycle for enthalpy of hydration of $CuSO_4$.

According to Hess's law,

$$\Delta H = \Delta H_1 + \Delta H_2$$

or $$\Delta H_1 = \Delta H - \Delta H_2 = -66.5 \text{ kJ} - 11.70 \text{ kJ}$$

$$= \mathbf{-78.20 \text{ kJ.}}$$

4. Determination of Standard Heat of Reaction

Hess's law cycle can also be used for the determination of the standard heat of a reaction from the knowledge of standard heats of formation of various reactants and products. The Hess's law cycle for elements in their standard states, reactants and products can be written as shown in Fig. 17.9.

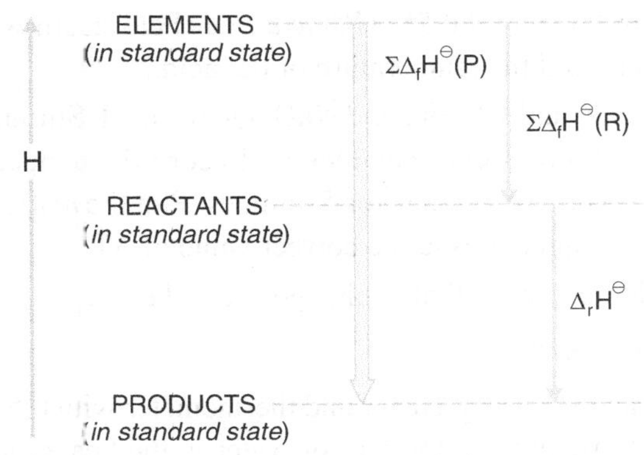

Fig. 17.9. Hess's law cycle for enthalpy of reaction.

$\Sigma\Delta_f H^{\ominus}$ (R) = Sum of the standard enthalpy of formation of reactants

$\Sigma\Delta_f H^{\ominus}$ (P) = Sum of the standard enthalpy of formation of the products

$\Delta_r H^{\ominus}$ = Standard enthalpy change of the reaction

Now according to Hess's law,

$$\Sigma\Delta_f H^{\ominus} (P) = \Sigma\Delta_f H^{\ominus} (R) + \Delta_r H^{\ominus}$$

or $$\boxed{\mathbf{\Delta_r H^{\ominus} = \Sigma\Delta_f H^{\ominus} (P) - \Sigma\Delta_f H^{\ominus} (R)}}$$

SOLVED EXAMPLES

Example 17.5. *Given that $\Delta_C H^{\ominus}$ for the combustion of propane C_3H_8 (g), H_2(g) and C(s) are respectively – 2220.1 kJ mol^{-1}, – 286.0 kJ mol^{-1} and – 393.5 kJ mol^{-1}. Find the value of $\Delta_f H^{\ominus}$ of C_3H_8(g). Also represent the enthalpy changes by the Hess's law cycle.*

Solution. Let us write all the thermochemical equations

$$3C(s) + 4H_2(g) \longrightarrow C_3H_8(g); \ \Delta_f H^{\ominus} = ? \text{ Required}$$

Given:

(*i*) $$C(s) + O_2(g) \longrightarrow CO_2(g); \ \Delta H_1^{\ominus} = -393.5 \text{ kJ}$$

(*ii*) $$H_2(g) + \frac{1}{2} O_2(g) \longrightarrow H_2O(l); \ \Delta H_2^{\ominus} = -286 \text{ kJ}$$

(*iii*) $$C_3H_8(g) + 5O_2(g) \longrightarrow 4H_2O(l) + 3CO_2(g); \quad \Delta H_3^{\ominus} = -2220.1 \text{ kJ}$$

In order to get the required equation multiply equation (*i*) by 3; equation (*ii*) by 4 and add these. Then subtract equation (*iii*) from the resulting equation,

$$3(i) + 4(ii) - (iii)$$

$$\begin{bmatrix} 3C(s) + 3O_2(g) \\ + 4H_2(g) + 2O_2(g) \\ - C_3H_8 - 5O_2(g) \end{bmatrix} \longrightarrow \begin{bmatrix} 3CO_2(g) \\ + 4H_2O(l) \\ - 4H_2O(l) - 3CO_2(g) \end{bmatrix};$$

$$\Delta_f H^{\ominus} = 3\Delta H_1^{\ominus} + 4\Delta H_2^{\ominus} - \Delta H_3^{\ominus}$$

or $$3C(s) + 4H_2(g) - C_3H_8(g) \longrightarrow 0;$$

$$\Delta_f H^{\ominus} = 3(-393.5) + 4(-286) - (-2220.1)$$

or $$3C(s) + 4H_2(g) \longrightarrow C_3H_8(g); \ \Delta_f H^{\ominus} = -102.9 \text{ kJ mol}^{-1}$$

These enthalpy changes are represented by a Hess's law cycle as follows.

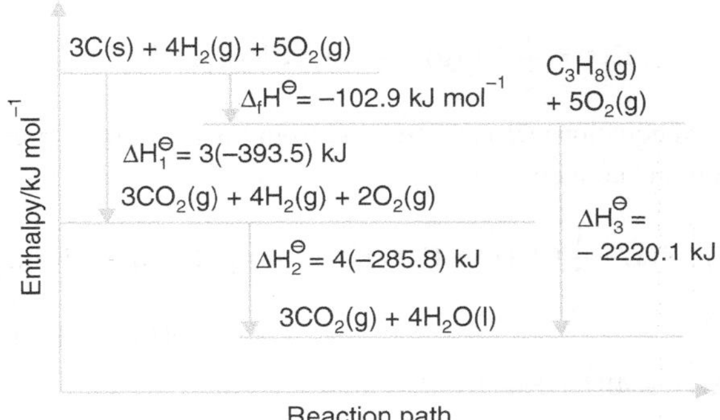

Fig. 17.10. Enthalpy diagram for the standard enthalpy of formation of propane.

Example 17.6. *Calculate the enthalpy of combustion of glucose from the following data:*

(*i*) $C_{(graphite)} + O_2(g) \longrightarrow CO_2(g)$; $\Delta H = -395.0$ kJ

(*ii*) $H_2(g) + \frac{1}{2} O_2(g) \longrightarrow H_2O(l)$; $\Delta H = -269.4$ kJ

(*iii*) $6C_{(graphite)} + 6H_2(g) + 3O_2(g) \longrightarrow C_6H_{12}O_6(s)$; $\Delta H = -1169.8$ kJ.

Solution. The required equation is:

$$C_6H_{12}O_6(s) + 6O_2(g) \longrightarrow 6CO_2(g) + 6H_2O(l)$$

Multiply equations (*i*) and (*ii*) by 6 and add them up:

(*iv*) $$6C_{(graphite)} + 6H_2(g) + 9O_2(g) \longrightarrow 6CO_2(g) + 6H_2O(l); \ \Delta H = -3986.4 \text{ kJ}$$

Subtracting equation (*iii*) from equation (*iv*), we get

$$C_6H_{12}O_6(s) + 6O_2(g) \longrightarrow 6CO_2(g) + 6H_2O(l); \quad \Delta H = -2816.6 \text{ kJ}$$

$\therefore$ Enthalpy of combustion of glucose = **– 2816.6 kJ.**

Example 17.7. *Calculate the enthalpy of formation of KCl from the following data:*

(i) $KOH(aq) + HCl(aq) \longrightarrow KCl(aq) + H_2O(l)$; $\Delta H = -57.3$ kJ

(ii) $H_2(g) + \frac{1}{2} O_2(g) \longrightarrow H_2O(l)$; $\Delta H = -286.2$ kJ

(iii) $\frac{1}{2} H_2(g) + \frac{1}{2} Cl_2(g) + aq \longrightarrow HCl(aq)$; $\Delta H = -164.4$ kJ

(iv) $K(s) + \frac{1}{2} O_2(g) + \frac{1}{2} H_2(g) + aq \longrightarrow KOH(aq)$; $\Delta H = -487.4$ kJ

(v) $KCl(s) + aq \longrightarrow KCl(aq)$; $\Delta H = +18.4$ kJ

Solution. Our object is to determine the enthalpy change of the following reaction:

$$K(s) + \frac{1}{2} Cl_2(g) \longrightarrow KCl(s);\ \Delta_f H = ?$$

Add equations (*iii*) and (*iv*) and from the resulting expression and subtract equation (*v*)

$$(vi)\ \left[K(s) + \frac{1}{2} Cl_2(g) + H_2(g)\right] \longrightarrow [KCl(aq) + HCl(aq) + KOH(aq) - KCl(aq)]$$

$\Delta H = -487.4 + (-164.4) - 18.4 = -670.2$

Now add equation (*i*) to equation (*vi*) and then subtract equation (*ii*) from the resulting expression.

$$K(s) + \frac{1}{2} Cl_2(g) \longrightarrow KCl(s);$$

$$\Delta H = -670.2 - 57.3 - (-286.2)$$
$$= \mathbf{-441.3\ kJ.}$$

Example 17.8. *Enthalpies of a solution of $BaCl_2(s)$ and $BaCl_2.2H_2O(s)$ are – 20.6 kJ mol^{-1} and 8.8 kJ mol^{-1} respectively. Calculate the $\Delta_{Hyd}H$ of $BaCl_2$ to $BaCl_2.2H_2O$.*

Solution. We require the enthalpy change of the following reaction:

$BaCl_2(s) + 2H_2O(l) \longrightarrow BaCl_2.2H_2O(s)$; $\Delta H = ?$

Thermochemical equations for the given changes are:

(*i*) $BaCl_2(s) + aq + 2H_2O(l) \longrightarrow BaCl_2.2H_2O(aq)$; $\Delta H = -20.6$ kJ

(*ii*) $BaCl_2.2H_2O(s) + aq \longrightarrow BaCl_2.2H_2O(aq)$; $\Delta H = +8.8$ kJ

Subtract equation (*ii*) from equation (*i*) and rearrange

$BaCl_2(s) + 2H_2O(l) \longrightarrow BaCl_2.2H_2O(s)$;

$\Delta H = -20.6 - 8.8 = \mathbf{-29.4\ kJ.}$

ACTIVITY 17.3	**To demonstrate the Hess's Law of constant heat summation.**

Requirments

A beaker engulfed in saw dust or wood shavings or a polythene calorimeter whose heat capacity is either known or has been found fitted with a thermometer and a stirrer, solid NaOH, 500 cm^3 of 0.5 M HCl.

Procedure

Experiment 1:

1. Pour 250 cm^3 of 0.5M HCl in a clean beaker, stir well and record the temperature of the acid.
2. Quickly add 10 g of solid NaOH to the acid. Stir until the solid dissolves completely. Record the temperature every 30 seconds for 5 minutes. Plot the *temp.* vs *time* graph and find the correct value of ΔT.
3. Calculate the enthalpy change. Let it be ΔH_1.

Experiment 2:

1. Clean the beaker, stirrer and thermometer with 0.5 M HCl. Again take 250 cm^3 of water in the beaker and record its temperature.
2. Now add 10.0 g of sodium hydroxide to it, stir well until the solid is completely dissolved. In the mean time record the temperature every 30 second for 5 min. Draw the temperature time graph and find the correct value of ΔT. Calculate the enthalpy change. Let it be ΔH_2.
3. Now quickly add 250 cm^3 0.5 M Hydrochloric acid (whose temperature has been maintained the same as that of the NaOH sol.). Stir and again record the temperature after every 30 seconds for 5 min. Find the correct ΔT from the temperature-time graph. Calculate the enthalpy change. Let it be ΔH_3.

(*i*) *Write the names associated with three enthalpy changes ΔH_1, ΔH_2, ΔH_3.*

(*ii*) *Write thermochemical equations for each process.*

(*iii*) *How are ΔH_1, ΔH_2 and ΔH_3 related?*

(*iv*) *Do your findings obey Hess's Law?*

Hint: The process in reference to ΔH_1 pertains to *dissolution + neutralisation;* that for ΔH_2 pertains to *dissolution* and the one for ΔH_3 pertains to *neutralisation.*

ACTIVITY 17.4 **To determine the enthalpy of combustion of the process, $Mg(s) + O_2(g) \longrightarrow MgO(s)$ using Hess's Law.**

Requirements

A polythene bottle or glass beaker whose heat capacity is either known or has been experimentally determined (C). 1.0 dm³ 1M HCl, 1.0 g each of Mg(s) and MgO(s).

Procedure

Part I:

1. Clean the beaker and set up the apparatus as shown in Fig. 17.11.
2. Pour 250 cm³ of 1M HCl solution. Record its temperature (t_1).
3. Quickly add 1.0 g of MgO, stir well and record the temperature after every 30 seconds till MgO is completely dissolved.
4. Find ΔT_1 from the temperature-time graph and calculate ΔH_1.

$$\Delta H_1 = -\frac{(1 + 250)\,\Delta T_1 + C\Delta T_1}{1} \times 56 \text{ J mol}^{-1}$$

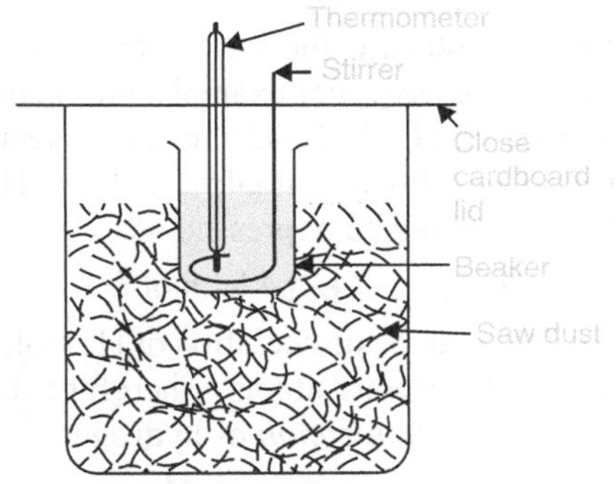

Fig. 17.11

Part II:

5. Clean the beaker, again take 250 cm³ of 1M HCl. Repeat the experiment by adding 1.0 g of Mg. (*Take care H_2 will be evolved and no flame should be brought near the apparatus*). Find ΔT_2 in a similar way and calculate ΔH_2.

$$\Delta H_2 = -\frac{(1 + 250)\,\Delta T_2 + C\Delta T_2}{1} \times 24 \text{ J mol}^{-1}.$$

(*i*) Now, ΔH_1 pertains to the chemical process

$$MgO(s) + 2HCl\,(aq) \longrightarrow MgCl_2\,(aq) + H_2O\,(l);\quad \Delta H_1 = ...$$

(*ii*) ΔH_2 pertains to the chemical process

$$Mg(s) + 2HCl\,(aq) \longrightarrow MgCl_2\,(aq) + H_2(g);\quad \Delta H_2 = ...$$

(*iii*) Use ΔH_3 for the process

$$H_2(g) + \frac{1}{2}O_2(g) \longrightarrow H_2O(l);$$

$$\Delta H_3 = -286 \text{ kJ mol}^{-1} \text{ (From tables)}$$

Required equation is $Mg(s) + \frac{1}{2}O_2(g) \longrightarrow MgO(s)$

Add equations (*ii*) and (*iii*) and from the resulting equation subtract equation (*i*)

Thus, $\Delta_c H$ for combustion of Mg = $\Delta H_2 + \Delta H_3 - \Delta H_1$.

17.5 BORN HABER CYCLE AND LATTICE ENTHALPY

LATTICE ENTHALPY

The Lattice enthalpy of an ionic compound is the enthalpy change which occurs when one mole of the ionic compound in the solid state dissociates into its gaseous ions.

$$Na^+Cl^-(s) \longrightarrow Na^+(g) + Cl^-(g);\quad \Delta_L H^\ominus = +788 \text{ kJ mol}^{-1}$$

Lattice enthalpy may also be defined as the energy released when one mole of ionic crystal is formed by the close packing of the constituent ions in the gaseous state.

$$Na^+(g) + Cl^-(g) \longrightarrow NaCl(s);\quad \Delta_L H^\ominus = -788 \text{ kJ mol}^{-1}.$$

The magnitude of lattice enthalpy gives an idea of the stability of the ionic crystal. It is not possible to determine lattice enthalpies of ionic compounds directly by experimental techniques. However, we can use indirect method by constructing an enthalpy diagram called ***Born Haber Cycle.*** Born Haber cycle is a simplified method which was developed in 1919 by *Max Born* and *Fritz Haber* to correlate lattice energies of ionic crystals to other thermodynamic data. The development of the Born Haber cycle was primarily based on the Hess's law. Let us consider the energy changes during the formation of sodium chloride crystal from *metallic sodium* and *chlorine gas* to calculate the lattice enthalpy of NaCl(*s*). The net enthalpy of formation of NaCl, $\Delta_f H^\ominus = -411.2$ kJ mol⁻¹.

$$\mathbf{Na(s) + \frac{1}{2}\,Cl_2(g) \longrightarrow NaCl(s);\quad \Delta_f H^\ominus = -411.2 \text{ kJ mol}^{-1}}$$

The overall process can be imagined to occur in following steps:

(*i*) *Sublimation of metallic sodium*

$$Na(s) \longrightarrow Na(g);\ \Delta_{sub}H^{\ominus} = 108.4 \text{ kJ mol}^{-1}$$

(*ii*) *Ionisation of sodium atoms*

$$Na(g) \longrightarrow Na^+(g) + e^-;\ \Delta_i H^{\ominus} = 495.6 \text{ kJ mol}^{-1}$$

$\Delta_{ie}H$ is ionisation enthalpy of sodium.

(*iii*) *Atomisation of Cl_2*. This step involves the dissociation of $Cl_2(g)$ into $Cl(g)$ atoms. The reaction enthalpy is half of the *bond dissociation enthalpy* of chlorine.

$$\frac{1}{2}Cl_2(g) \longrightarrow Cl(g)\ ;\ \Delta_a H^{\ominus} = \frac{1}{2}\Delta H^{\ominus}_{Cl-Cl}$$

$$= 121 \text{ kJ mol}^{-1}.$$

(*iv*) *Conversion of Cl(g) to $Cl^-(g)$.*

$$Cl(g) + e^- \longrightarrow Cl^-(g);\ \Delta_e H^{\ominus} = -348.6 \text{ kJ mol}^{-1}$$

$\Delta_{ea}H$ is electron affinity of chlorine.

(*v*) Combination of $Na^+(g)$ and $Cl^-(g)$ ions to form 1 mole of NaCl(*s*). The energy released here is called lattice enthalpy ($\Delta_L H^{\ominus}$). The sequence of steps (*i*) – (*v*) is shown in Fig. 17.12 and is known as Born Haber cycle. The sum of the enthalpy changes round a cycle is zero.

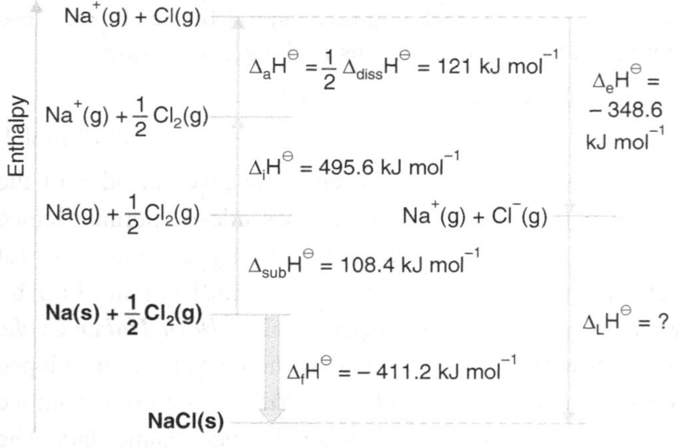

Fig. 17.12. Born Haber cycle for lattice enthalpy of NaCl.

The various changes in the Born Haber Cycle for sodium chloride are shown in Fig 17.13.

Applying Hess's law we get

$$\Delta_f H^{\ominus} = \Delta_{sub}H^{\ominus} + \Delta_i H^{\ominus} + \frac{1}{2}\Delta_{diss}H^{\ominus} + \Delta_e H^{\ominus} + \Delta_L H^{\ominus}$$

or $$\Delta_L H^{\ominus} = -411.2 - 108.4 - 121 - 495.6 + 348.6$$

$$= -787.6 \text{ kJ}.$$

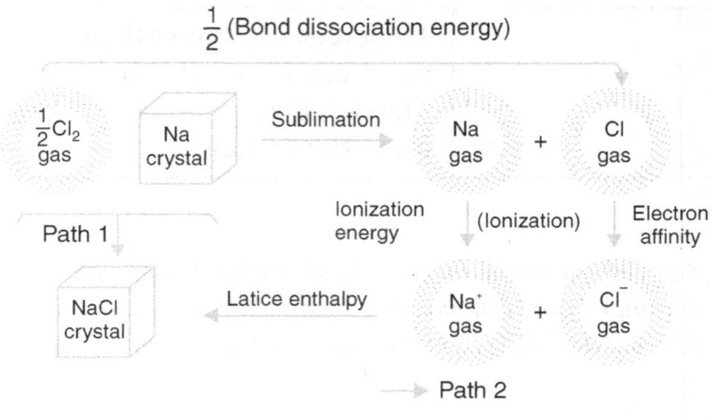

Fig. 17.13. Born Haber cycle for sodium chloride

Let us now proceed to solve some numerical problems based upon *thermochemical calculations.*

17.6 BOND ENERGY OR BOND ENTHALPY ($\Delta H^{\ominus}_{bond}$)

It is a well known fact that the process of formation of a chemical bond is accompanied by the release of energy. Conversely, the energy has to be supplied for the breaking of a bond. *The energy required to break a particular bond in a gaseous molecule* is referred to as **bond dissociation energy.** It is a definite quantity and is expressed in kJ mol^{-1}.

In a **diatomic molecule**, the *bond dissociation* energy is the same as the *bond energy*. For example, the energy required to break one mol of H—H bonds in the gaseous state is 435.4 kJ. Therefore, bond enthalpy of H—H bond is 435.4 kJ mol^{-1}. This may be expressed as:

$$\Delta H^{\ominus}_{H-H} = 435.4 \text{ kJ mol}^{-1}$$

It is quite clear that the dissociation of H_2 molecule leads to the formation of gaseous atoms. Therefore, this is also known as *enthalpy of atomization* of hydrogen.

The **enthalpy of atomization ($\Delta_a H^{\ominus}$)** *may be defined as the enthalpy change accompanying the dissociation of a one mole of a substance completely into its atoms in the gaseous state.* For example, enthalpy of atomization of CH_4 is represented as

$$CH_4(g) \longrightarrow C(g) + 4H(g);\ \Delta_a H^{\ominus} = 1665 \text{ kJ mol}^{-1}.$$

Similarly, the enthalpy of atomization of Na(*s*) is 108.4 kJ mol^{-1} which also represents the enthalpy of sublimation of Na(*s*).

However, in a **polyatomic molecule** having more than one similar bonds the bond dissociation enthalpy is not the same for successive bonds. Therefore, in such a case the bond energy is not equal to the bond dissociation energy. For example, in water, H_2O, the bond dissociation energies of the two O—H bonds differ from one another as described as follows:

$$HO—H(g) \longrightarrow H(g) + OH(g); \Delta H = 498 \text{ kJ}$$

$$O—H(g) \longrightarrow H(g) + O(g); \quad \Delta H = 430 \text{ kJ}.$$

In such a case, therefore, the *bond energy* is expressed as the average of the *bond dissociation* energies of various similar bonds. For example, the ***average bond energy*** of the O—H bond is expressed as

$$\Delta H^{\ominus}_{O—H} = \frac{498 + 430}{2} = 464 \text{ kJ mol}^{-1}.$$

Thus, **mean bond energy,** in general, may be defined as the *average amount of energy required to break one mole bond of that type in gaseous molecules.*

CALCULATION OF BOND ENERGIES

The thermochemical data makes it possible to calculate the bond energies of different bonds. As an illustration the bond enthalpy of the C—H bond in methane can be obtained if ΔH of the following chemical reaction is known

$$CH_4(g) \longrightarrow C(g) + 4H(g); \Delta H = ?$$

The bond enthalpy of the C—H bond is one fourth of the ΔH value for the above reaction. The value of ΔH can be calculated from the type of data available.

(*a*) *Direct use of the Hess's law if $\Delta_f H^{\ominus}$ values of H(g), C(g) and $CH_4(g)$ are given.*

$\Delta_f H^{\ominus}$ values of H(g); C(g) and $CH_4(g)$ are 218.0, 717.0 and 750 kJ mol^{-1} respectively.

Now, $\Delta_r H^{\ominus} = \Sigma \Delta_f H^{\ominus}$ (*Products*) $- \Sigma \Delta_f H^{\ominus}$ (*Reactants*)

$$= 4\Delta_f H^{\ominus} H(g) + \Delta_f H^{\ominus} C(g) - \Delta H^{\ominus}_f CH_4(g)$$

$$= 4(218) + 717 - (-75.0) = 1664 \text{ kJ}$$

$$\therefore \Delta H^{\ominus}_{C—H} = \frac{1664}{4} = 416 \textbf{ kJ mol}^{-1}.$$

(*b*) *Use of thermochemical calculations if the data has values other than $\Delta_f H$.*

Let the given data be:

$\Delta_a H^{\ominus}$ of $H_2(g) = +436$ kJ mol^{-1}

$\Delta_{Sub} H^{\ominus}$ of carbon $= +717$ kJ mol^{-1}

$\Delta_c H^{\ominus}$ of Carbon, Hydrogen and Methane

are – 394 kJ, – 286 kJ and – 891 kJ respectively.

Let us now write thermochemical equations for each of the above.

(*i*) *Enthalpy of combustion of methane*

$$CH_4(g) + 2O_2(g) \longrightarrow CO_2(g) + 2H_2O(l); \quad \Delta H_1 = -891 \text{ kJ}$$

(*ii*) *Enthalpy of combustion of carbon*

$$C(s) + O_2(g) \longrightarrow CO_2(g); \quad \Delta H_2 = -394 \text{ kJ}$$

(*iii*) *Enthalpy of combustion of hydrogen*

$$H_2(g) + \frac{1}{2} O_2(g) \longrightarrow H_2O(l); \quad \Delta H_3 = -286 \text{ kJ}$$

(*iv*) *Enthalpy of sublimation of carbon*

$$C(s) \longrightarrow C(g); \quad \Delta H_4 = +717 \text{ kJ}$$

(*v*) *Enthalpy of dissociation of hydrogen molecule*

$$H_2(g) \longrightarrow 2H(g); \quad \Delta H_5 = +436 \text{ kJ}$$

From the above equations the enthalpy change for the reaction

$$CH_4(g) \longrightarrow C(g) + 4H(g)$$

can be calculated as follows:

Multiply equation (*v*) and also equation (*iii*) by 2. Now add *equation (i), equation (iv)* and *twice of equation (v)*. From the resulting expression subtract *equation (ii)* and *twice of equation (iii)*. The resulting value of the enthalpy change comes out to be

$$\Delta H = \Delta H_1 - \Delta H_2 - (2 \times \Delta H_3) + \Delta H_4 + (2 \times \Delta H_5)$$

$$= -891 - (-394) - (-2 \times 286) + (717) + (2 \times 436)$$

$$= +1664 \text{ kJ}$$

This represents the energy required for the cleavage of **four** C—H bonds.

Therefore, $\Delta H_{C—H} = \frac{1664}{4} = \mathbf{416\ kJ\ mol^{-1}}$.

The bond enthalpies of various bonds are given Table 17.4.

Table 17.4. Bond Enthalpies of Some Bonds

Bond	*Bond enthalpy* (kJ mol^{-1})*	*Bond*	*Average bond enthalpy** (kJ mol^{-1})*
H—H	435.4	C = C	619
H—F	565	C ≡ C	812
H—Cl	431	O—H	463
H—Br	364	N—H	389
H—I	297	C—H	413
F—F	155	C—Cl	328
Cl—Cl	242	C—O	335
Br—Br	190	C = O	707
I—I	149	C—N	293
O = O	494	C = N	616
N ≡ N	941	C ≡ N	879
		C—Br	275.6
		O—O	138
		N—N	159
		N = N	418
		C—C	347

* These bond enthalpies are the dissociation energies of diatomic molecules that have only one bond, they are, therefore, exact values.

** These bond enthalpies are obtained from molecules that contain more than one bond, therefore, they are average values.

USE OF BOND ENTHALPY DATA

The bond enthalpy data given in Table 17.2 is quite useful in determining the standard enthalpies of reactions. The chemical reaction involves the cleavage of old bonds in reacting species and formation of new bonds to give product molecules. In general,

$$\begin{bmatrix}\textbf{Enthalpy}\\ \textbf{of reaction}\end{bmatrix} = \begin{bmatrix}\textbf{Sum of bond enthalpies}\\ \textbf{of reactants}\end{bmatrix} - \begin{bmatrix}\textbf{Sum of bond enthalpies}\\ \textbf{of products}\end{bmatrix}$$

or $\Delta_r H^\ominus = \Sigma$ **Bond Enthalpies**$_{(reactants)}$ $- \Sigma$ **Bond Enthalpies**$_{(Products)}$

Thus, the values given in bond enthalpy data can help us in:

(*i*) *calculating the standard enthalpy of reactions.*

(*ii*) *calculation of bond enthalpies of some specific bond in a molecule.*

(*iii*) *comparing the strength of various bonds.*

Some examples are discussed as follows:

SOLVED EXAMPLES

Example 17.9. *Calculate the enthalpy change (ΔH) of the following reaction:*

$2C_2H_2(g) + 5O_2(g) \longrightarrow 4CO_2(g) + 2H_2O(g)$ given average bond enthalpies of various bonds, i.e., C—H, C ≡ C, O = O, C = O, O—H as 414, 814, 499, 724 and 640 kJ mol^{-1} respectively.

Solution. The equation representing the various substances involved in the chemical reaction is:

$$2(H—C \equiv C - H) + 5(O = O) \longrightarrow 4(O = C = O) + 2(H—O—H)$$

The reaction involves the breaking of 4 C—H bonds, 2C ≡ C bonds, 5 O = O bonds and formation of 8 C = O bonds and 4 O—H bonds.

$$\Delta H = \begin{bmatrix}\text{Total energy required}\\ \text{to break the bonds}\end{bmatrix} - \begin{bmatrix}\text{Energy given out in}\\ \text{forming the bonds}\end{bmatrix}$$

$$= [4\,\Delta H_{C—H} + 2\,\Delta H_{C\equiv C} + 5\,\Delta H_{O=O}] - [8\,\Delta H_{C=O} + 4\,\Delta H_{O—H}]$$

$$= (4 \times 414 + 2 \times 810 + 5 \times 499) - (8 \times 724 + 4 \times 460)$$

$$= 5771 - 7632 = -\mathbf{1861\ kJ.}$$

Example 17.10. *Calculate the bond enthalpy of the Cl—Cl bond from the following data: $CH_4(g) + Cl_2(g) \longrightarrow CH_3Cl(g) + HCl(g)$; ΔH = – 100.3 kJ. Also the bond enthalpies of C—H, C—Cl, H—Cl bonds are 413, 326 and 431 kJ mol^{-1} respectively.*

Solution. The given reaction is

$$CH_4(g) + Cl—Cl(g) \longrightarrow CH_3—Cl(g) + HCl(g); \quad \Delta H = -100.3\ kJ$$

As is evident from the equation, the reaction involves the cleavage of one C—H bond, one Cl—Cl bond and formation of one C—Cl bond and one H—Cl bond.

$$\Delta H = [\text{Total energy required}] - [\text{Total energy released}]$$

$$-100.3 = [\Delta H_{C—H} + \Delta H_{Cl—Cl}] - [\Delta H_{C—Cl} + \Delta H_{HCl}]$$

$$= 413 + x - 326 - 431 = x - 344$$

or $x = 344 - 100.3 = \mathbf{243.7\ kJ\ mol^{-1}}$.

SUMMARY

- **Enthalpy of Reaction.** The enthalpy change accompanying a chemical reaction in which the number of moles of reactants consumed and those of products formed are the same as the stoichiometric coefficients.
- **Standard Enthalpy of Formation ($\Delta_f H^\ominus$).** It is the enthalpy change accompanying the formation of 1 mole of a substance from its constituent elements in their standard state. $\Delta_f H^\ominus$ can be > 0 or < 0.
- **Standard Enthalpy of Combustion ($\Delta_c H^\ominus$).** It is the enthalpy change occurring during the combustion of one mole of a substance in excess of oxygen. $\Delta_c H^\ominus$ is always less than zero.
- **Enthalpy of Neutralization ($\Delta_n H$).** It is the enthalpy change taking place during neutralisation involving combination of 1 mol of H^+ ions from an acid and 1 mol of OH^- ions from a base in dilute aqueous solutions.
- **Enthalpy of Hydration ($\Delta_{hyd.} H^\ominus$).** It is the enthalpy change occurring during the hydration of one mole of an anhydrous salt by combining with a specific number of moles of water.
- **Hess's Law.** The enthalpy change in a chemical or physical process is the same whether the process is carried out in one step or in several steps.
- **Bond Enthalpy ($\Delta H^\ominus_{bond}$).** The average amount of energy required to break one mole of bonds of a particular type in gaseous molecules.
- **Enthalpy of Atomisation ($\Delta_a H^\ominus$).** It is the enthalpy change accompanying the dissociation of one mole of a substance into gaseous atoms.

EVALUATE YOURSELF

I. Objective Type Questions

Select the most appropriate choice from the options given as (a), (b), (c) and (d) after each question:

1. The enthalpy of formation of gaseous N_2O and NO at 298 K are 82 and 90 kJ mol^{-1}. The enthalpy of the reaction

$N_2O(g) + \frac{1}{2} O_2 \longrightarrow 2NO(g)$ is

(*a*) – 8 kJ (*b*) 98 kJ
(*c*) – 74 kJ (*d*) + 8 kJ.

2. The enthalpy of neutralization of HCl by NaOH under certain conditions is – 55.9 kJ and that of HCN by NaOH is – 12.1 kJ. The enthalpy of ionization of HCN is

(*a*) – 68.0 kJ mol^{-1} (*b*) – 43.8 kJ mol^{-1}
(*c*) 68.0 kJ mol^{-1} (*d*) 43.8 kJ mol^{-1}.

3. In which of the following reactions does the heat of neutralization has the highest numerical value?

(*a*) NH_4OH and HCl
(*b*) NaOH and HCl
(*c*) CH_3COOH and NaOH
(*d*) NH_4OH and CH_3COOH.

4. The enthalpy of vaporisation of liquid water using the data below is

$H_2(g) + \frac{1}{2} O_2(g) \longrightarrow H_2O(l)$; $\Delta H = -285.77$ kJ

$H_2(g) + \frac{1}{2} O_2(g) \longrightarrow H_2O(g)$; $\Delta H = -241.84$ kJ

(*a*) + 43.93 kJ mol^{-1} (*b*) – 43.93 kJ mol^{-1}
(*c*) + 527.61 kJ mol^{-1} (*d*) – 527.61 kJ mol^{-1}.

5. Which compound is most stable on the basis of the $\Delta_f H^\ominus$ values given in parenthesis?

(*a*) $C_3H_8(g)$ [– 103.8 kJ mol^{-1}]
(*b*) $C_4H_{12}(g)$ [– 124.7 kJ mol^{-1}]
(*c*) *n*-hexane [– 167.2 kJ mol^{-1}]
(*d*) *n*-octane [– 208.4 kJ mol^{-1}].

6. Energy required to dissociate 4 g of $H_2(g)$ into free gaseous atoms is *x* kJ. The value of $\Delta H_{H—H}$ will be

(*a*) *x* kJ mol^{-1} (*b*) *x*/2 kJ mol^{-1}
(*c*) *x*/3 kJ mol^{-1} (*d*) 0.25*x* kJ mol^{-1}.

7. The bond dissociation enthalpies of four bonds C—A, C—B, C—D and C—E are respectively 240, 382, 276 and 486 kJ mol^{-1}. The atom with the smallest size is

(*a*) A (*b*) B
(*c*) E (*d*) D.

8. If the bond dissociation energies of XY, X_2, Y_2 (all diatomic molecules) are in the ratio 1 : 1 : 0.5 and Δ_f H of XY is – 200 kJ mol^{-1}.

The bond dissociation energy of X_2 will be

(*a*) 400 kJ mol^{-1} (*b*) 300 kJ mol^{-1}
(*c*) 200 kJ mol^{-1} (*d*) 100 kJ mol^{-1}.

9. Absolute enthalpy of neutralization of the reaction

$MgO(s) + 2HCl(aq) \longrightarrow MgCl_2(aq) + H_2O(l)$ will be

(*a*) – 57.33 kJ mol^{-1}
(*b*) more negative than – 57.33 kJ mol^{-1}
(*c*) less negative than – 57.33 kJ mol^{-1}
(*d*) 57.33 kJ.

10. The enthalpies of combustion of ethene, C_2H_4, and butane, C_4H_8, are:

$C_2H_4 + 3O_2 \longrightarrow 2CO_2 + 2H_2O$; ΔH_1

$C_4H_8 + 6O_2 \longrightarrow 4CO_2 + 4H_2O$; ΔH_2

Find the value, ΔH, for the enthalpy of the reaction of C_2H_4 to form C_4H_8 given by the equation:

$2C_2H_4 \longrightarrow C_4H_8$

(*a*) $\Delta H = \Delta H_1 - \Delta H_2$ (*b*) $\Delta H = \Delta H_1 + \Delta H_2$
(*c*) $\Delta H = 2\Delta H_1 - \Delta H_2$ (*d*) $\Delta H = 2\Delta H_1 + \Delta H_2$.

II. Fill in the Blanks

11. Complete the following sentences by supplying appropriate words:

(*i*) If $\Delta H_{H–H}$ is 436 kJ mol, the value of $\Delta_f H^\ominus$ of H(*g*) is ....
(*ii*) Heat change during the process $BaCl_2(s) + 2H_2O(l) \longrightarrow BaCl_2.2H_2O(s)$ is called ....
(*iii*) The value of ΔH for the process $2F(g) \longrightarrow F_2(g)$ is .....
(*iv*) $\Delta_n H^\ominus$ of a strong acid-strong base is always ......
(*v*) The energy required to break off 1 mol of solid crystal into its constituent gaseous ions is called ....
(*vi*) $\Delta_f H^\ominus$ of a substance can be ...... zero.
(*vii*) Combustion reaction is always ......

III. Discussion Questions

12. What is activation energy? What is its role in the progress of a reaction?

13. Explain the function of a catalyst with the help of energy level diagrams for catalysed and uncatalysed reaction.

14. Calculate the lattice energy for the change

$Li^+(g) + Cl^-(g) \longrightarrow LiCl(s)$ using Born Haber Cycle.

Given that

$\Delta_{sub}H^\ominus$ of Li = 160.67 kJ mol^{-1}
$\Delta_{diss}H^\ominus$ of Cl_2 = 244.34 kJ mol^{-1}
$\Delta_{ie}H^\ominus$ of Li(*g*) = 520.07 kJ mol^{-1}
$\Delta_{eg}H^\ominus$ of Cl(*g*) = – 365.26 kJ mol^{-1}
$\Delta_f H^\ominus$ of LiCl(*s*) = – 401.66 kJ mol^{-1}.

15. Calculate bond enthalpy of the C = C bond in propene ($CH_3 - CH = CH_2$). Given that $\Delta_f H^\ominus$ of propene is – 87.78 kJ mol^{-1}. Bond enthalpies of C—C, C—H and H—H bonds are 347, 413, 435.4 kJ mol^{-1} respectively. Also $C(s) \longrightarrow C(g)$; $\Delta H = 715$ kJ mol^{-1}.

16. Define enthalpy of neutralization. Why is the enthalpy of neutralisation of a strong acid-strong base always constant?

17. What is meant by the term lattice enthalpy? How does it differ from enthalpy of atomisation?

18. Use the Born Haber Cycle for calcium chloride to illustrate various energy changes.

19. When 100 cm^3 of 0.5 M sodium hydroxide solution is neutralized with 100 cm^3 of 0.5 M hydrochloric acid, the temperature changes from 29.0°C to 31.5°C.

(*a*) Calculate (*i*) the amount of heat given out in the reaction in kilojoules (sp. heat capacity of water is 4.184 Joules); (*ii*) $\Delta_n H$. Assume the product mixture has density, $\rho = 1$ g cm^{-3}.

(*b*) Draw an energy level diagram for the reaction.

20. Using methane (CH_4) as example, distinguish between the terms bond enthalpy and bond dissociation enthalpy.

21. Calculate the enthalpy of formation of ethanoic acid if its enthalpy of combustion is – 867 kJ mol^{-1}. The $\Delta_f H°$ for $H_2O(l)$ and $CO_2(g)$ are – 286 kJ mol^{-1} and – 393.5 kJ mol^{-1} respectively.

22. Distinguish between the term *standard enthalpy of formation* of a substance and *standard enthalpy of combustion* of a substance.

23. Calculate the standard enthalpy of formation of methane (CH_4) from the following data:

Reaction	ΔH_c^θ *kJ mol*$^{-1}$
$CH_4 + 2O_2(g) \longrightarrow CO_2(g) + 2H_2O(l)$	– 890.4
$C(graphite) + O_2(g) \longrightarrow CO_2(g)$	– 393.5
$2H_2(g) + O_2(g) \longrightarrow 2H_2O(l)$	– 571.7

24. What is meant by standard enthalpy of formation? Give its importance.

25. Explain these terms giving examples

(*i*) standard enthalpy of combustion

(*ii*) standard enthalpy of neutralization.

26. Suggest an experiment to determine the enthalpy of combustion of ethanol.

Answers

I. Objective Type Questions

1. (*b*)	**2.** (*d*)	**3.** (*b*)
4. (*a*)	**5.** (*d*)	**6.** (*b*)
7. (*c*)	**8.** (*a*)	**9.** (*c*)
10. (*c*).		

II. Fill in the Blanks

11. (*i*) 218 kJ mol^{-1} (*ii*) enthalpy of hydration of $BaCl_2$ (*iii*) negative (*iv*) constant (*v*) lattice enthalpy (*vi*) less than or greater than (*vii*) exothermic.

III. Discussion Questions

14. – 961.48 kJ mol^{-1} **15.** 713.98 kJ mol^{-1}

21. – 492.0 kJ mol^{-1}.

SHS 3

SECTION

2

INORGANIC CHEMISTRY

General Objectives

After studying this section, the student should be able to:

1. understand the physical and chemical properties and uses of compounds formed by the elements on the periodic table.
2. recognize the colour production, catalytic properties and many applications of transition elements in industries.
3. use inorganic chemistry to explain the many processes that go on in our environment.

In this section, we shall study

SECTION-2

UNIT-18

PERIODIC CHEMISTRY

Specific Objectives

After studying this unit, the student should be able to:

- describe the reactions between acids and metals, their oxides and trioxocarbonate(IV)s
- describe the acidic properties of the oxides of non-metals.
- describe the patterns in physical and chemical properties of the chlorides of period 3 metals
- account for the differences in thermal stability of the trioxocarbonate(IV) and trioxocarbonate(V) of some metals
- demonstrate the thermal stabilities of some trioxocarbonate(IV) in the laboratory
- describe the uses of Silicon.
- explain the physical properties of the halogens (Group 17 elements)
- relate the electron configurations of the halogens to their chemical properties
- describe the reactions of the halogens with water and alkalis
- explain why there are differences in the acid strengths of hydrogen halides.

18.1 REACTIONS BETWEEN ACIDS AND METALS, THEIR OXIDES AND TRIOXOCARBONATE(IV)S

REACTION OF METALS WITH ACIDS

Metals are known to react with acids resulting in the evolution of hydrogen gas. Only the metals which lie above the electrochemical or activity series can evolve hydrogen gas. The reactions of Na, K, Ca, Mg and Al with HCl are given below:

$$2Na(s) + 2HCl(aq) \longrightarrow 2NaCl(s) + H_2(g)\uparrow$$
$$2K(s) + 2HCl(aq) \longrightarrow 2KCl(s) + H_2(g)\uparrow$$
$$Mg(s) + 2HCl(aq) \longrightarrow MgCl_2(s) + H_2(g)$$
$$Ca(s) + 2HCl(aq) \longrightarrow CaCl_2(s) + H_2(g)$$
$$2Al(s) + 6HCl \longrightarrow 2AlCl_3 + 3H_2(g)$$

The rate of evolution of H_2 gas is different in all the reactions because metals have different reactivity with HCl

REACTION OF METAL OXIDES WITH ACIDS

Metal oxides react with acids to produce salts and water

$$Na_2O(s) + 2\,HCl(aq) \longrightarrow 2NaCl(s) + H_2O(l)$$
$$K_2O(s) + 2HCl(aq) \longrightarrow 2KCl(s) + H_2O(l)$$
$$CaO(s) + 2HCl(aq) \longrightarrow CaCl_2(s) + H_2O(l)$$
$$MgO(s) + 2HCl(aq) \longrightarrow MgCl_2(s) + H_2O(l)$$
$$Al_2O_3(s) + 6HCl(aq) \longrightarrow 2AlCl_3 + 3H_2O(l)$$

REACTIONS OF TRIOXOCARBONATE(IV)S WITH ACID

The metal carbonates react vigorously with acids like HCl to give strong effervescence along with the evolution of carbon dioxide gas.

$$Na_2CO_3(s) + 2HCl(aq) \longrightarrow 2NaCl(s) + H_2O(l) + CO_2(g)$$

$$K_2CO_3(s) + 2HCl(aq) \longrightarrow 2KCl(s) + H_2O(l) + CO_2(g)$$
$$CaCO_3(s) + 2HCl(aq) \longrightarrow CaCl_2(s) + H_2O(l) + CO_2(g)$$
$$MgCO_3(s) + 2HCl(aq) \longrightarrow MgCl_2(s) + H_2O(l) + CO_2(g)$$
$$Al_2CO_3(s) + 6HCl(aq) \longrightarrow 2AlCl_3(s) + 3H_2O(l) + CO_2(g)$$

NATURE OF AQUEOUS SOLUTIONS OF OXIDES AND TRIOXOCARBONATE(IV)S

All the metals (Na, K, Ca, Mg) combine with oxygen to form metal oxides which are soluble in water. The aqueous solution of metal oxides are basic in character. The basic nature is supported by the fact that on dipping litmus paper in the solution of oxides, its colour changes.

The colour of litmus changes from red to blue.

The aqueous solution of the oxide of Al_2O_3 is amphoteric in character and will not change the colour of litmus paper.

The aqueous solution of trioxocarbonates of metals (Na, K, Mg) are weakly basic in character and will change red litmus blue.

18.2 ACIDIC PROPERTIES OF THE OXIDES OF NON-METALS

Across a period the number of protons in the nucleus increases, as do the number of electrons. This also happens down a Group; but the key difference is that across a period the shell of electrons is not completed until the noble gas at the end of the period is reached. As we go across a period the efficiency of screening is not so great because the shell of electrons is not completed. As a result of this the effective nuclear pull increases and the electrons are held more tightly.

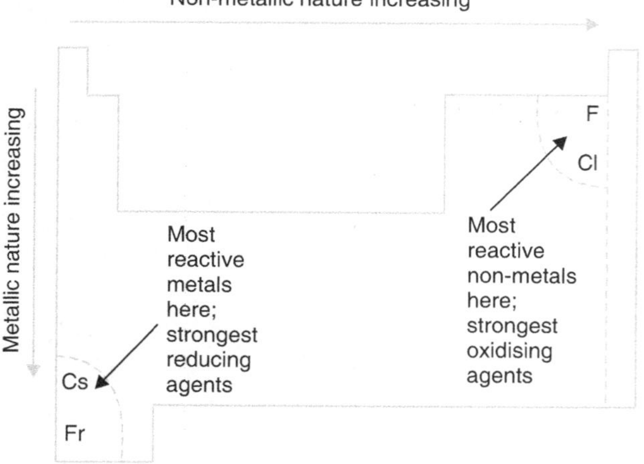

Fig. 18.1. How metallic and non-metallic behaviour change in the Periodic Table

As a consequence, the elements become harder to ionise, and they also tend to attract electrons towards them. Thus we see a change from the metals (which give positive ions) on the left of a period to the non-metals on the right. With this change there is a change in the nature of the hydrides, oxides and chlorides of the elements. These changes are summarised in Fig. 18.1.

OXIDES OF CARBON

Carbon on heating in air or oxygen produces two oxides namely; carbon monoxide and carbon dioxide

$$2C(g) + O_2(g) \longrightarrow 2CO(g)$$
$$C(s) + O_2(g) \longrightarrow CO_2(g)$$

Carbon monoxide is formed by incomplete combustion of carbon or carbon containing compounds.

Carbon dioxide is obtained when carbon is heated in excess supply of oxygen or air.

Solid CO_2 is called *dry ice* and readily sublimes.

CO is *neutral* in character whereas CO_2 is *acidic*.

Selected physical properties of CO and CO_2 are given in Table 18.1. The structure of CO and CO_2 is explained below:

Table 18.1. Selected Properties of CO and CO_2

Property	*CO*	*CO_2*
Melting point/K	68	–
Boiling point/K	82	195 (sublimes)
$\Delta_f H^\ominus$ (298 K)/kJ mol^{-1}	– 110.5	– 393.5
$\Delta_f G^\ominus$ (298 K)/kJ mol^{-1}	– 137	– 394
Bond energy/kJ mol^{-1}	1075	806
C—O bond distance/pm	112.8	116.0
Dipole moment/D	0.11	0

OXIDES OF NITROGEN AND PHOSPHORUS

Some common oxides of nitrogen and phosphorus are given below:

Oxides of Nitrogen and Phosphorus

Oxidation State of the Element	*N*	*P*
+5	N_2O_5	P_4O_{10}
+4	$\begin{bmatrix} N_2O_4 \\ NO_2 \end{bmatrix}$	P_4O_8
+3	N_2O_3	P_4O_6
+2	NO	—
+1	N_2O	—

Nitrogen forms six different oxides by combination with oxygen. The oxidation state of nitrogen in these oxides ranges from +1 to +5. The molecular formula of the various oxides of nitrogen along with the oxidation number and some physical characteristics of nitrogen are given in Tables 18.2 and 18.3 respectively.

Table 18.2. Oxides of Nitrogen

Formula and Name	*Oxidation Number of N*	*Physical Appearance*	*Method of Preparation*
N_2O Nitrous oxide (Laughing gas)	+ 1	Colourless gas (neutral)	$NH_4NO_3 \xrightarrow{\text{Heat}} N_2O + 2H_2O$
NO Nitric oxide	+ 2	Colourless gas (neutral)	$4NH_3 + 5O_2 \longrightarrow 4NO + 6H_2O$
N_2O_3 Dinitrogentrioxide	+ 3	Blue solid (acidic)	$NO(g) + NO_2(g) \xrightarrow{-20°C} N_2O_3$
N_2O_4 Dinitrogen tetroxide	+ 4	Colourless liquid/solid (acidic)	$2NO_2 \underset{}{\overset{0°C}{\rightleftharpoons}} N_2O_4$
NO_2 Nitrogen dioxide	+ 4	Brown gas acidic	$2Pb(NO_3)_2 \xrightarrow{\text{Heat}} 2PbO + 4NO_2 + O_2$
N_2O_5 Dinitrogen pentoxide	+ 5	Colourless gas (acidic)	$4HNO_3 + P_4O_{10} \longrightarrow 2N_2O_5 + 4HPO_3$ (Meta phosphoric acid)

Table 18.3. Selected data for the Oxides of Nitrogen

Formula	N_2O	NO	N_2O_3	NO_2	N_2O_4	N_2O_5
Name	Dinitrogen monoxide‡	Nitrogen monoxide‡	Dinitrogen trioxide	Nitrogen dioxide	Dinitrogen tetraoxide	Dinitrogen pentaoxide
Melting point/K	182	109	173	–	262	303
Boiling point/K	185	121	277 dec.	–	294	305 sublimes
Physical appearance	Colourless gas	Colourless gas	Blue solid or liquid	Brown gas	Colourless solid or liquid, but see text	Colourless solid, stable below 273 K
$\Delta_f H°(298\ K)/kJ\ mol^{-1}$	82.1(*g*)	90.2(*g*)	50.3(*l*) 83.7(*g*)	33.2(*g*)	– 19.5(*l*) 9.2(*g*)	– 43.1(*s*)
Dipole moment of gasphase molecule/D	0.16	0.16	—	0.315	—	—
Magnetic properties	Diamagnetic	Paramagnetic	Diamagnetic	Paramagnetic	Diamagnetic	Diamagnetic

‡N_2O and NO are commonly called nitrous oxide and nitric oxide, respectively.

NATURE OF OXIDES

All the oxides of nitrogen (except NO and N_2O) and phosphorus are strongly acidic.

Reason. *The change in character from acidic to basic can be explained on the basis of the size of atoms. As the size of the nitrogen atom is small and it has a strong positive field, it interacts with water more strongly pulling the electron pair between the O—H bond and thus helps in the release of H^+ ions. However, this tendency diminishes with an increase in size and a decrease in acidic character or conversely an increase in basic character.*

OXIDES OF SULPHUR

Sulphur combines with oxygen to form monoxide, dioxide and trioxide

$$S + O_2 \longrightarrow SO_2(g)$$

$$2S + 3O_2 \xrightarrow[\text{excess}]{} 2SO_3(g)$$

Both oxides of sulphur are *acidic* in character.

OXIDES OF HALOGENS

Halogens do not combine with oxygen directly but their oxides with oxygen can be prepared indirectly. For *example,* oxygen difluoride, OF_2 is prepared by the action of fluorine on 2% sodium hydroxide solution.

$$2F_2 + 2NaOH \longrightarrow 2NaF + OF_2 + H_2O$$

Similarly, dichlorine oxide, Cl_2O is obtained indirectly by passing chlorine over precipitated mercuric oxide.

$$2Cl_2 + 2HgO \xrightarrow{575\ K} HgO.HgCl_2 + Cl_2O$$

Add To Your Knowledge

The compounds of oxygen with fluorine are called fluorides because fluorine is more electronegative than oxygen.

Some oxides of halogens are given below in Table 18.4.

Table 18.4. Oxides of Halogens

Oxidation State	*Fluorine*	*Chlorine*	*Bromine*	*Iodine*
+1	OF_2	Cl_2O	Br_2O	—
+2	F_2O_2			
+3	—	Cl_2O_3		
+4	—	ClO_2	BrO_2	I_2O_4
+5	—	—	—	I_2O_5
+6	—	Cl_2O_6	—	—
+7	—	Cl_2O_7	—	—

All the oxides are *powerful oxidising agents* and decompose explosively when subjected to mechanical shock or heat.

18.3 PHYSICAL AND CHEMICAL PROPERTIES OF SOME PERIOD 3 (Na → Cl) COMPOUNDS

HYDRIDES

The properties of the hydrides are summarized in Table 18.5. The hydrides of the less electronegative elements, such as lithium hydride and sodium hydride, are ionic. The remaining hydrides become progressively more covalent in nature as we move across the period. In water, most hydrides react to form either alkaline hydroxide solutions or acidic solutions. The remainder do not react with water but produce a neutral solution.

Table 18.5. Properties of Some Period 3 Hydrides

Hydrides	NaH	MgH_2	AlH_3	SiH_4
Bonding/Units	ionic/ions	ionic/covalent chain	covalent (polymer)	covalent (molecules)
Physical state (appearance)	*s* (white)	*s* (white)	*s* (white)	*g* (colourless)
$T_m/K(°C)$	1073 (800)	533 (260)	—	92 (– 181)
$T_b/K(°C)$	—	—	—	161 (– 112)
Acid/alkali nature in water	alkaline solution	alkaline solution	alkaline solution	neutral solution

Reactions of Hydrides with Air

Hydrides do not react with air at room temperature. However, a few react with oxygen in the air to form their oxides and water when heated to high temperatures. Examples are shown in the equations which follow.

$$CH_4(g) + O_2(g) \xrightarrow{\text{heat}} CO_2(g) + 2H_2O(g)$$

$$4NH_3(g) + 5O_2(g) \xrightarrow{\text{catalyst}} 4NO(g) + 6H_2O(g)$$

$$SiH_4(g) + 2O_2(g) \xrightarrow{\text{heat}} SiO_2(s) + 2H_2O(g)$$

B_2H_6 is unstable and very reactive. It burns with a green flame.

Carbon forms many other stable hydrides—hydrocarbons such as alkanes, alkenes, alkynes and aromatic hydrocarbons which burn in air. Silicon has only alkane analogues, stable up to Si_6H_{14}. These are not as stable as the hydrocarbons and hence more reactive in air.

Reactions of Hydrides with Water

All the ionic and many of the covalent hydrides react with water to different extents. The ionic hydrides give hydroxides and hydrogen. Reactions of ionic hydrides are shown by their equations as:

$$LiH(s) + H_2O(l) \longrightarrow LiOH(aq) + H_2(g)$$

$$NaH(s) + H_2O(l) \longrightarrow NaOH(aq) + H_2(g)$$

$$MgH_2(s) + 2H_2O(l) \longrightarrow Mg(OH)_2(s) + 2H_2(g)$$

Covalent hydrides such as aluminium hydride, boron hydride, ammonia and hydrogen fluoride react with water to give *basic* and *acidic* solutions. The reactions are as follows:

$$2AlH_3(s) + 3H_2O(l) \longrightarrow 2Al(OH)_3(s) + 2H_2(g)$$

Aluminium hydride

$$B_2H_6(g) + 6H_2O(l) \longrightarrow 2H_3BO_3(aq) + 6H_2(g)$$

diborane

$$NH_3(g) + H_2O(l) \rightleftharpoons NH_4^+(aq) + OH^-(aq)$$

ammonia

$$HF(g) + H_2O(l) \rightleftharpoons H_3O^+(aq) + F^-(aq)$$

Hydrofluoric acid

Methane does not react with water at room temperature. Silicon hydride does not react with water at room temperature except in the presence of an alkali.

$$SiH_4(g) + H_2O(l) + 2OH^-(aq) \longrightarrow SiO_3^{2-}(aq) + 4H_2(g)$$

OXIDES

The oxides of the third period of elements are:

Na_2O	MgO	Al_2O_3	SiO_2	P_4O_{10}	SO_3	Cl_2O_7
				P_4O_6	SO_2	Cl_2O

Those oxides in the top row are known as the ***highest oxides*** of the various elements. These are the oxides where the period 3 elements are in their highest oxidation states. In these oxides, all the outer electrons in the period 3 element are being involved in the bonding *i.e.*, from one with sodium, to all seven of chlorine's outer electrons.

The Structures

The trend in structure is from the metallic oxides containing giant structures of ions on the left of the period via a giant covalent oxide (silicon dioxide) in the middle to molecular oxides on the right. Sodium, magnesium and aluminium oxides consist of ***giant structures containing metal ions and oxide ions***. Magnesium oxide has a structure just like sodium chloride. The other two have more complicated arrangements of ions beyond the scope of the syllabuses at this level.

Melting and Boiling Points

The giant structures (the metal oxides and silicon dioxide) will have high melting and boiling points because a lot of energy is needed to break the strong bonds (ionic or covalent) operating in three dimensions.

The oxides of phosphorus, sulphur and chlorine consist of individual molecules *i.e.*, some small and simple; others polymeric.

The attractive forces between these molecules are van der Waals dispersion and dipole-dipole interactions. These vary depending on the size, shape and polarity of the various molecules but will always be much weaker than the ionic or covalent bonds in a giant structure. These oxides tend to be gases, liquids or low melting solids.

Electrical Conductivity

None of these oxides has any free or mobile electrons. That means that none of them will conduct electricity when they are solid.

The ionic oxides can, however, undergo ***electrolysis*** when they are molten. They can conduct electricity because of the movement of the ions in the molten state.

Electrolysis of molten sodium oxide depends on whether it actually melts or sublimes or decomposes under ordinary circumstances. If it sublimes, you won't get any liquid to electrolyse! Magnesium and aluminium oxides have melting points far too high to be able to electrolyse them in a simple lab.

Silicon Dioxide (Silicon(IV) Oxide)

The Structure

The electronegativity of the elements increases as you go across the period, and by the time you get to silicon, there isn't enough electronegativity difference between the silicon and the oxygen to form an ionic bond. Silicon dioxide is a ***giant covalent structure***.

There are three different crystal forms of SiO_2. Crystalline silicon has the same structure as diamond. To turn it into silicon dioxide, all you need to do is to modify the silicon structure by including some oxygen atoms (see Fig. 18.2).

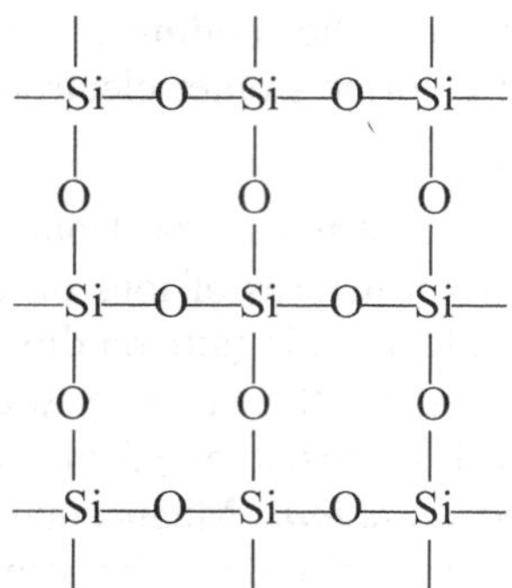

Fig. 18.2. Crystal structure of silica

Notice that each silicon atom is bridged to its neighbours by an oxygen atom. Do not forget that this is just a tiny part of a giant structure extending in all 3 dimensions.

Melting and Boiling Points

Silicon dioxide has a high melting point around 1700°C. Very strong silicon-oxygen covalent bonds have to be broken throughout the structure before melting occurs. Silicon dioxide boils at 2230°C.

In simple words silicon dioxide has a giant structure, and the melting and boiling points are high.

Electrical Conductivity

Silicon dioxide does not have any mobile electrons or ions—so it does not conduct electricity either as a solid or a liquid.

The Molecular Oxides

Phosphorus, sulphur and chlorine all form oxides which consist of molecules. Some of these molecules are fairly simple while others are polymeric. Melting and boiling points of these oxides will be much lower than those of the metal oxides or silicon dioxide. The intermolecular forces holding one molecule to its neighbours will be van der Waals dispersion forces or dipole-dipole interactions. The strength of these will vary depending on the size of the molecules.

None of these oxides conducts electricity either as solids or as liquids. None of them contains ions or free electrons.

The Phosphorus Oxides

Phosphorus has two common oxides, phosphorus(III) oxide, P_4O_6, and phosphorus(V) oxide, P_4O_{10}.

Phosphorus(III) oxide

Phosphorus(III) oxide is a white solid, melting at 24°C and boiling at 173°C.

a P_4O_6 molecule

The phosphorus uses only three of its outer electrons (the 3 unpaired *p* electrons) to form bonds with the oxygen atoms.

Phosphorus(V) oxide

Phosphorus(V) oxide is also a white solid, subliming (turning straight from solid to vapour) at 300°C. In this case, the phosphorus uses all five of its outer electrons in bonding.

Solid phosphorus(V) oxide exists in several different forms, some of them are polymeric.

The structure of P_4O_{10} in a simple molecular form is given below:

a P_4O_{10} molecule

The Sulphur Oxides

Sulphur has two common oxides, sulphur dioxide (sulphur(IV) oxide), SO_2, and sulphur trioxide (sulphur(VI) oxide), SO_3.

Sulphur dioxide

Sulphur dioxide is a colourless gas at room temperature with an easily recognised choking smell. It consits of simple SO_2 molecules.

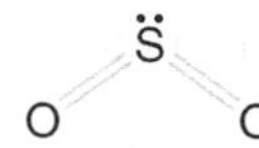

The sulphur atom uses 4 of its outer electrons to form the double bonds with the oxygen, leaving the other two as a lone pair on the sulphur. The bent shape of SO_2 is due to this lone pair.

Sulphur trioxide

Pure sulphur trioxide is a white solid with a low melting and boiling point. It reacts very rapidly with water vapour in the air to form sulphuric acid. That means that if you make some in the laboratory, you tend to see it as a white sludge which fumes dramatically in moist air (forming a fog of sulphuric acid droplets).

Gaseous sulphur trioxide consists of simple SO_3 molecules in which all six of the sulphur's outer electrons are involved in the bonding.

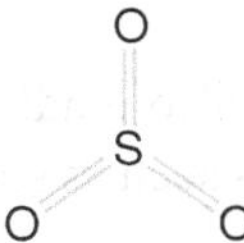

There are various forms of solid sulphur trioxide. The simplest one is a trimer, S_3O_9, where three SO_3 molecules are joined up and arranged in a ring.

There are also other polymeric forms in which the SO_3 molecules join together in long chains as shown below:

The Chlorine Oxides

Chlorine forms several oxides. Here we are just looking at two of them (*i*) chlorine(I) oxide, Cl_2O, and (*ii*) chlorine(VII) oxide, Cl_2O_7.

Chlorine(I) oxide

Chlorine(I) oxide is a yellowish-red gas at room temperature. It consists of simple small molecules.

There's nothing surprising about this molecule and its physical properties are just what one would expect for a molecule of this size.

Chlorine(VII) oxide

In chlorine(VII) oxide, the chlorine uses all of its seven outer electrons in bonds with oxygen. This produces a much bigger molecule. Its melting point and boiling point is higher than chlorine(I) oxide.

Chlorine(VII) oxide is a colourless oily liquid at room temperature.

The structures of oxides of chlorine are given below. The shape is tetrahedral around both chlorines, and V-shaped around the central oxygen.

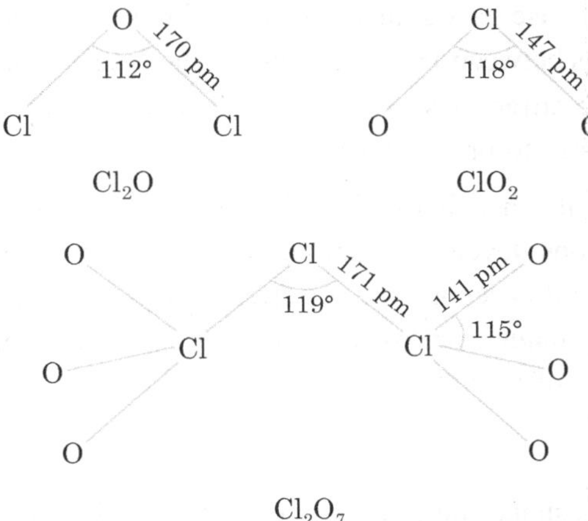

Acid-Base Behaviour of the Period 3 Oxides

The trend in acid-base behaviour is summarized below:

- On moving from left to right among the oxides we observe that metal oxides, form strongly basic oxides on the left-hand side to strongly acidic ones on the right, via an amphoteric oxide (aluminium oxide) in the middle. An *amphoteric oxide* is one which shows both acidic and basic properties.

The oxides with the highest oxidation state of its central element are known to be acidic in character. The acidic character decreases with decrease in oxidation state of the atom.

CHEMISTRY OF THE INDIVIDUAL OXIDES

Sodium Oxide

Sodium oxide is a simple strongly basic oxide. It is basic because it contains the oxide ion, O^{2-}, which is a very strong base with a high tendency to combine with hydrogen ions.

Reaction with water

Sodium oxide reacts exothermically with cold water to produce sodium hydroxide solution. Depending on its concentration, this will have a pH around 14.

$$Na_2O + H_2O \longrightarrow 2NaOH$$

Reaction with acids

As a strong base, sodium oxide also reacts with acids. For example, it would react with dilute hydrochloric acid to produce sodium chloride solution.

$$Na_2O + 2HCl \longrightarrow 2NaCl + H_2O$$

Magnesium Oxide

Magnesium oxide is a simple basic oxide, because it also contains oxide ions. However, it is not as strongly basic as sodium oxide because the oxide ions are not so free.

In the case of sodium oxide, the solid is held together by attractions between 1+ and 2– ions. In the case of magnesium oxide, the attractions are between 2+ and 2– ions. It takes more energy to break these.

Even if other factors like the energy released when the positive ions develop attractions with water in the solution are taken into account, the net effect of this is that reactions involving magnesium oxide will always be less exothermic than those of sodium oxide.

Reaction with water

If you shake some white magnesium oxide powder with water, nothing seems to happen, it does not look as if it reacts. However, if you test the pH of the liquid, you find that it is somewhere around pH 9-showing that it is slightly alkaline.

There must have been some slight reaction with the water to produce hydroxide ions in solution. Some magnesium hydroxide is formed in the reaction, but this is almost insoluble and so not many hydroxide ions actually get into solution.

$$MgO(s) + H_2O(l) \longrightarrow Mg(OH)_2(aq)$$

Reaction with acids

Magnesium oxide reacts with warm dilute hydrochloric acid to give magnesium chloride solution.

$$MgO(s) + 2HCl(aq) \longrightarrow MgCl_2\,(aq) + H_2O(l)$$

Aluminium Oxide

Describing the properties of aluminium oxide can be confusing because it exists in a number of different forms. One of those forms is very unreactive. It is known chemically as alpha-Al_2O_3 and is produced at high temperatures.

Aluminium oxide is ***amphoteric***. It has reactions as both a base and an acid.

Reaction with water

Aluminium oxide does not react in a simple way with water like sodium oxide and magnesium oxide do, and does not dissolve in it. Although it contains oxide ions, they are held too strongly in the solid lattice to react with water.

Reaction with acids

Aluminium oxide contains oxide ions and so reacts with acids in the same way as sodium or magnesium oxides. Aluminium oxide reacts with hot dilute hydrochloric acid to give aluminium chloride.

$$Al_2O_3 + 6HCl \longrightarrow 2AlCl_3 + 3H_2O$$

The above reaction supports the basic character of aluminium.

Reaction with bases

Aluminium oxide is also acidic in its nature. It shows this by reacting with bases such as sodium hydroxide solution and forms aluminate.

With hot, concentrated sodium hydroxide solution, aluminium oxide reacts to give a colourless solution of sodium tetrahydroxoaluminate.

$$Al_2O_3 + 2NaOH + 3H_2O \longrightarrow 2NaAl(OH)_4$$

Silicon Dioxide (Silicon(IV) Oxide)

The bonds between Si and oxygen are covalent in character due to the small electronegativity difference between the two.

Silicon dioxide has no basic properties *i.e.*, it does not react with acids. Instead, it is very weakly acidic, and reacts with strong bases.

Reaction with water

Silicon dioxide does not react with water, because of the difficulty of breaking up the giant covalent structure.

Reaction with bases

Silicon dioxide reacts with sodium hydroxide solution, but only if it is hot and concentrated. A colourless solution of sodium silicate is formed.

$$SiO_2 + 2NaOH \longrightarrow Na_2SiO_3 + H_2O$$

You may also be familiar with one of the reactions happening in the Blast Furnace extraction of iron in which calcium oxide (from the limestone which is one of the raw materials) reacts with silicon dioxide to produce a liquid slag, calcium silicate. This is also an example of acidic silicon dioxide reacting with a base.

$$SiO_2 + CaO \longrightarrow CaSiO_3$$

The Phosphorus Oxides

Phosphorus forms two oxides, phosphorus(III) oxide, P_4O_6, and phosphorus(V) oxide, P_4O_{10}.

Phosphorus(III) oxide

Phosphorus(III) oxide reacts with *cold* water to give a solution of the weak acid, H_3PO_3 known as phosphorus acid, orthophosphorus acid or phosphoric acid. Its reaction with hot water is much more complicated.

$$P_4O_6 + 6H_2O \longrightarrow 4H_3PO_3$$

The pure un-ionised acid has the structure:

$$\begin{array}{c} O \\ \| \\ HO-P-OH \\ | \\ H \end{array}$$

Phosphorus acid is a weak acid. Phosphorus acid has a pK_a of 2.00 which makes it stronger than common organic acids like ethanoic acid ($pK_a = 4.76$). It is pretty unlikely that you would ever react phosphorus(III) oxide directly with a base, but you might need to know what happens if you react the phosphorus acid with a base.

In phosphorus acid, the two hydrogen atoms in the –OH groups are acidic, but the other one is not. This means that you can get two possible reactions with, sodium hydroxide solution depending on the proportions used.

$$NaOH + H_3PO_3 \longrightarrow NaH_2PO_3 + H_2O$$

$$2NaOH + H_3PO_3 \longrightarrow Na_2HPO_3 + 2H_2O$$

In the first case, only one of the acidic hydrogens has reacted with the hydroxide ions from the base. In the second case (using twice as much sodium hydroxide), both have reacted.

Phosphorus(V) Oxide

Phosphorus(V) oxide reacts violently with water to give a solution containing a mixture of acids, the nature of which depends on the conditions. We usually consider one of these, tetraoxophosphate(V) acid H_3PO_4. It is also known as orthophosphoric acid.

$$P_4O_{10} + 6\,H_2O \longrightarrow 4H_3PO_4$$

The pure un-ionised acid has the structure:

$$\begin{array}{c} O \\ \| \\ HO-P-OH \\ | \\ OH \end{array}$$

Tetraoxophosphate(V) acid is also a weak acid with a pK_a of 2.15. That makes it *fractionally* weaker than phosphorus acid. Solutions of both of these acids of concentrations around 1 mol dm^{-3} will have a pH of about 1.

If we look at the structure, there are three-OH groups, and each of these has an acidic hydrogen atom. Hence, the reaction with sodium hydroxide occurs in three stages, with one after another of these hydrogens reacting with the hydroxide ions.

$$NaOH + H_3PO_4 \longrightarrow NaH_2PO_4 + H_2O$$

$$2NaOH + H_3PO_4 \longrightarrow Na_2HPO_4 + 2H_2O$$

$$3NaOH + H_3PO_4 \longrightarrow Na_3PO_4 + 3H_2O$$

Sulphur Dioxide

Sulphur dioxide is fairly soluble in water, reacting with it to give a solution of sulphurous acid (trioxosulphate(IV) acid), H_2SO_3. This only exists in solution, and any attempt to isolate it causes sulphur dioxide to be given off again.

$$SO_2 + H_2O \longrightarrow H_2SO_3$$

The un-ionised acid has the structure:

$$\begin{array}{c} O \\ \| \\ HO-S-OH \end{array}$$

Trioxosulphate(IV) acid is also a weak acid with a pK_a of around 1.8—very slightly stronger than the two phosphorus containing acids above. A reasonably concentrated solution of sulphurous acid will again have a pH of about 1. Sulphur dioxide will also react directly with bases such as sodium hydroxide solution. If sulphur dioxide is bubbled through sodium hydroxide solution, sodium sulphite solution is formed first followed by sodium hydrogen sulphite solution when the sulphur dioxide is in excess.

$$SO_2 + 2NaOH \longrightarrow Na_2SO_3 + H_2O$$

$$Na_2SO_3 + H_2O + SO_2 \longrightarrow 2NaHSO_3$$

Another important reaction of sulphur dioxide is with the base calcium oxide to form calcium sulphite (calcium sulphate(IV)). This is at the heart of one of the methods of removing sulphur dioxide from flue gases in power stations.

$$CaO + SO_2 \longrightarrow CaSO_3$$

Sulphur Trioxide

Sulphur trioxide reacts violently with water to produce a fog of concentrated sulphuric acid droplets.

$$SO_3 + H_2O \longrightarrow H_2SO_4$$

Sulphur trioxide itself will also react directly with bases to form sulphates. For example, it will react with calcium oxide to form calcium sulphate. This is just like the reaction with sulphur dioxide described above.

$$CaO + SO_3 \longrightarrow CaSO_4$$

Pure un-ionised tetraoxosulphate(VI) acid has the structure:

```
       O
       ||
HO — S — OH
       ||
       O
```

Tetraoxosulphate(VI) acid is a strong acid, and solutions will typically have pH's of around 0.

The acid reacts with water to give a hydroxonium ion (a hydrogen ion in solution) and a hydrogen sulphate ion. This reaction is virtually 100% complete.

$$H_2SO_4(aq) + H_2O(l) \longrightarrow H_3O^+(aq) + HSO_4^-(aq)$$

The second hydrogen is more difficult to remove. In fact the hydrogen sulphate ion is a relatively weak acid.

$$HSO_4^-(aq) + H_2O(l) \rightleftharpoons H_3O^+(aq) + SO_4^{2-}(aq)$$

It should be noted that sulphuric acid is a stronger acid than sulphurous acid.

When a hydrogen ion is lost from one of the –OH groups, the negative charge left on the oxygen is spread out (delocalised) over the ion by interacting with the doubly-bonded oxygens. The higher the delocalisation, the greater the stability of the resulting ion and hence the higher the acidic character.

In tetraoxosulphate(VI) acid the negative charge is distributed on four oxygen atoms while in H_2SO_3 it is on three. Hence sulphuric acid is a stronger acid.

It reacts with sodium hydroxide solution to form sodium sulphate solution, in which both of the acidic hydrogens react with hydroxide ions.

$$2NaOH + H_2SO_4 \longrightarrow Na_2SO_4 + 2H_2O$$

The Chlorine Oxides

Chlorine forms several oxides. The important among these are chlorine(VII) oxide, Cl_2O_7, and chlorine(I) oxide, Cl_2O. Chlorine(VII) oxide is also known as dichlorine heptoxide, and chlorine(I) oxide as dichlorine monoxide.

Chlorine(VII) oxide

Chlorine(VII) oxide is the highest oxide of chlorine. The chlorine is in its maximum oxidation state of +7.

Chlorine(VII) oxide reacts with water to give a very strong acid, tetraoxochlorate(VII) acid is also known as perchloric acid. The pH of solutions will be around 0.

$$Cl_2O_7 + H_2O \longrightarrow 2HClO_4$$

Un-ionised tetraoxochlorate(VII) acid has the structure:

```
        O
        ||
HO — Cl=O
        ||
        O
```

The acid when it reacts with water, loses H^+ ion to form the tetraoxochlorate(VII) ion (perchlorate ion) since the negative charge can be delocalised over every oxygen atom in the ion. That makes it very stable, and it means that tetraoxochlorate(VII) acid is a very strong acid.

Tetraoxochlorate(VII) acid reacts with sodium hydroxide solution to form a solution of sodium tetraoxochlorate(VII).

$$NaOH + HClO_4 \longrightarrow NaClO_4 + H_2O$$

Chlorine(VII) oxide can also react with sodium hydroxide solution to give the same product.

$$2NaOH + Cl_2O_7 \longrightarrow 2NaClO_4 + H_2O$$

Chlorine(I) oxide

Chlorine(I) oxide is far less acidic than chlorine(VII) oxide. It reacts with water to some extent to give tetraoxochlorate(I) acid, HOCl, also known as hypochlorous acid.

$$Cl_2O + H_2O \rightleftharpoons 2HOCl$$

The structure of tetraoxochlorate(I) acid is exactly as shown by its formula, HOCl. It has no doubly-bonded oxygens, and no way of delocalising the charge over the negative ion formed by loss of the hydrogen. This means that the negative ion formed is not very stable, and readily reclaims its hydrogen to revert to the acid. Tetraoxochlorate(I) acid is a very weak acid ($pK_a = 7.43$).

Tetraoxochlorate(I) acid reacts with sodium hydroxide solution to give a solution of sodium tetraoxochlorate(I) (sodium hypochlorite).

$$NaOH + HOCl \longrightarrow NaOCl + H_2O$$

Chlorine(I) oxide also reacts directly with sodium hydroxide to give the same product.

$$2NaOH + Cl_2O \longrightarrow 2NaOCl + H_2O$$

CHLORIDES

The chlorides of the third period of elements are:

NaCl	$MgCl_2$	$AlCl_3$	$SiCl_4$	PCl_5	S_2Cl_2
				PCl_3	

The Structures

Sodium chloride and magnesium chloride are ionic and consists of giant ionic lattices at room temperature. Aluminium chloride and phosphorus(V) chloride are tricky! They change their structure from ionic to covalent when the solid turns to a liquid or vapour. There is much more about this later on this page.

The others are simple covalent molecules.

Melting and Boiling Points

Sodium and magnesium chlorides are solids with high melting and boiling points because of the large amount of heat which is needed to break the strong ionic attractions.

The rest are liquids or low melting point solids. Leaving aside the aluminium chloride and phosphorus(V) chloride cases where the situation is quite complicated, the attractions in the others will be much weaker intermolecular forces such as van der Waals dispersion forces. These vary depending on the size and shape of the molecule, but will always be far weaker than ionic bonds.

Electrical Conductivity

Sodium and magnesium chlorides are ionic and so will undergo ***electrolysis*** when they are molten. Electricity is carried by the movement of the ions and their discharge at the electrodes.

In the aluminium chloride and phosphorus(V) chloride cases, the solid does not conduct electricity because the ions are not free to move. In the liquid state also, they do not conduct due to its covalent nature.

The rest of the chlorides do not conduct electricity either in the solid or molten state because they do not have any ions or any mobile electrons.

Reactions with Water

The simple ionic chlorides (sodium and magnesium chloride) dissolve in water. The other chlorides all react with water in a variety of ways described to form different products. The reaction with water is known as ***hydrolysis***.

CHEMISTRY OF INDIVIDUAL CHLORIDES

Sodium Chloride, NaCl

Sodium chloride is a simple ionic compound consisting of a giant array of sodium and chloride ions.

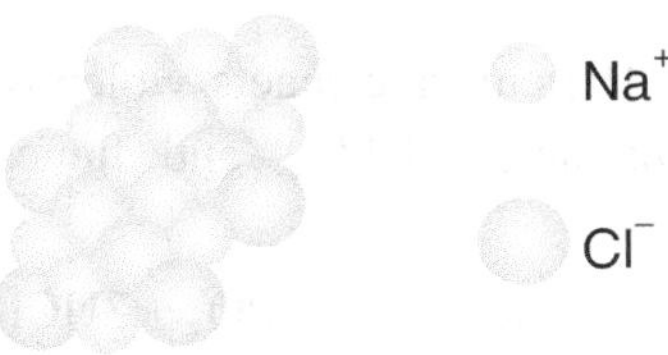

A simplified 3-D structure of NaCl is shown below:

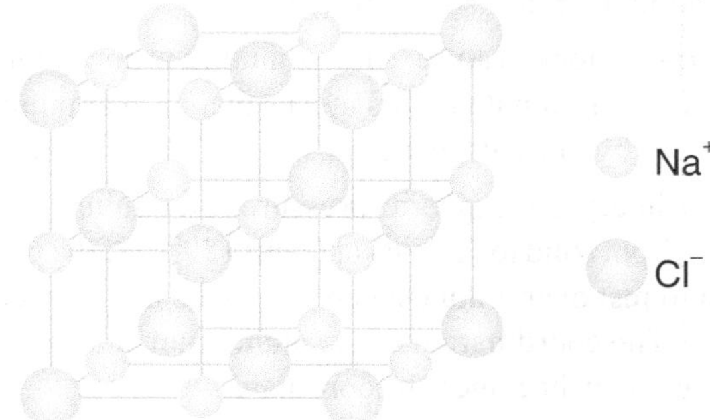

The strong attractions between the positive and negative ions need a lot of heat energy to break, and therefore sodium chloride has high ***melting and boiling points.***

It does not conduct electricity in the solid state because it does not have any mobile electrons and the ions are not free to move. However, when it melts it undergoes electrolysis.

Sodium chloride dissolves in water to give a neutral solution.

Magnesium Chloride, $MgCl_2$

Magnesium chloride is also ionic, but with a more complicated arrangement of the ions to allow for having twice as many chloride ions as magnesium ions. Again, lots of heat energy is needed to overcome the attractions between the ions, and so the melting and boiling points are again high. Solid magnesium chloride is a non-conductor of electricity because the ions are not free to move. However, it undergoes electrolysis when the ions become free on melting.

Magnesium chloride dissolves in water to give a faintly acidic solution (pH = approximately 6). When magnesium ions are broken off the solid lattice and go into solution, there is enough attraction between the Mg^{2+} ions and the water molecules to get coordinate (dative covalent) bonds formed between the magnesium ions and lone pairs on surrounding water molecules.

In the aqueous solution, $[Mg(H_2O)_6]^{2+}$, are formed.

$$MgCl_2(s) + 6H_2O(l) \longrightarrow [Mg(H_2O)_6]^{2+}(aq) + 2Cl^-(aq)$$

The aqueous solution of $MgCl_2$ is only weakly acidic.

$$[Mg(H_2O)_6]^{2+}(aq) \rightleftharpoons [Mg(H_2O)_5(OH)]^+(aq) + H^+(aq)$$

Hydrogen ions in solution are hydroxonium ions. If you use this form, it is essential to include the state symbols.

Aluminium Chloride, $AlCl_3$

We know that electronegativity increases as we go across the period. As a result there is not enough electronegativity difference between aluminium and chlorine. Hence, an ionic bond is not formed.

At room temperature, the aluminium in aluminium chloride is 6-coordinated. This means that each aluminium is surrounded by six chlorine atoms.

At ordinary atmospheric pressure, aluminium chloride sublimes (from solid to vapour) at about 180°C. If the pressure is raised to just over 2 atmospheres, it melts at a temperature of 192°C. The coordination of the aluminium changes at this temperatures. It becomes 4-coordinate.

The structure of Al_2Cl_6 is shown as:

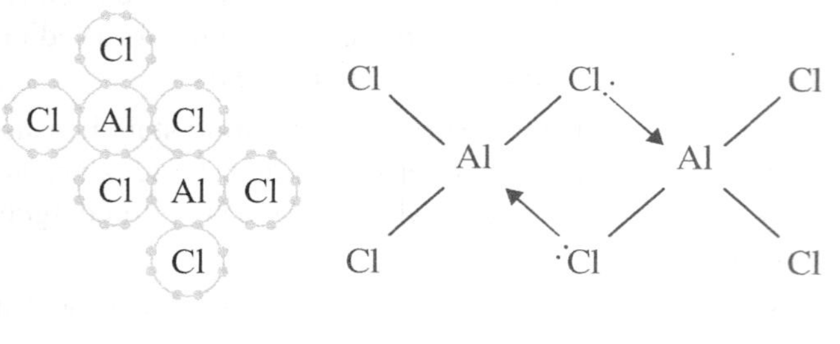

Lewis dot structure

There is an equilibrium between these dimers and simple $AlCl_3$ molecules. As the temperature increases further, the position of equilibrium shifts more and more towards the right.

$$Al_2Cl_6 \rightleftharpoons 2AlCl_3$$

Solid aluminium chloride does not conduct electricity at room temperature because the ions are not free to move. Molten aluminium chloride (only possible at increased pressures) does not conduct electricity because there are no ions any more. The reaction of aluminium chloride with water is typical. If water is dropped onto solid aluminium chloride, a violent reaction occurs producing clouds of steamy fumes of hydrogen chloride gas. If solid aluminium chloride is added to an excess of water, it still splutters, but instead of hydrogen chloride gas being given off, we get an acidic solution. A solution of aluminium chloride at ordinary concentrations (around 1 mole dm^{-3},) will have a pH around 2–3.

Solid aluminium chloride reacts with water rather than just dissolving in it. In the first instance, hexaaquaaluminium ions are formed together with chloride ions.

$$AlCl_3(s) + 6H_2O(l) \longrightarrow [Al(H_2O)_6]^{3+}(aq) + 3Cl^-(aq)$$

This is very similar to the magnesium chloride equation given above and the only real difference is the charge on the ion. The extra charge pulls electrons from the water molecules quite strongly towards the aluminium. That makes the hydrogens more positive and easier to remove from the ion. In other words, this ion is much more acidic than in the corresponding case of magnesium.

$$[Al(H_2O)_6]^{3+} + H_2O \rightleftharpoons [Al(H_2O)_5(OH)]^{2+} + H_3O^+$$

These equilibria lie to the right, and so the solution formed is more acidic as there are more hydroxonium ions in it.

Violent evolution of HCl happens because due to the heat produced in the reaction hydrogen ions and chloride ions in the mixture combine together as hydrogen chloride molecules and are given off as a gas. With a large excess of water, the temperature never gets high enough for that to happen and the ions just stay in solution.

Silicon Tetrachloride, $SiCl_4$

Silicon tetrachloride is a simple covalent chloride. There is not enough electronegativity difference between the silicon and the chlorine for the two to form ionic bonds.

Silicon tetrachloride is a colourless liquid at room temperature which fumes in moist air. The only attractions between the molecules are van der Waals dispersion forces.

It fumes in moist air because it reacts with water in the air to produce hydrogen chloride. If water is added to silicon tetrachloride, there is a violent reaction producing silicon dioxide and fumes of hydrogen chloride. In a large excess of water, the hydrogen chloride will, of course, dissolve to give a strongly acidic solution containing hydrochloric acid.

It does not conduct electricity because of the lack of ions or mobile electrons.

$$SiCl_4 + 2H_2O \longrightarrow SiO_2 + 4HCl$$

The Phosphorus Chlorides

There are two phosphorus chlorides *i.e.*, phosphorus(III) chloride, PCl_3, and phosphorus(V) chloride, PCl_5.

Phosphorus(III) Chloride (Phosphorus Trichloride), PCl_3

This is another simple covalent chloride *i.e.*, again a fuming liquid at room temperature.

It is a liquid because there are only van der Waals dispersion forces and dipole-dipole attractions between the molecules.

Phosphorus(III) chloride reacts violently with water to form phosphorus acid, H_3PO_3, and fumes of hydrogen chloride.

$$PCl_3 + 3H_2O \longrightarrow H_3PO_3 + 3HCl$$

It does not conduct electricity because of the lack of ions or mobile electrons.

Phosphorus(V) Chloride (Phosphorus pentachloride), PCl_5

Phosphorus(V) chloride is a white solid which sublimes at 163°C. The higher the temperature goes above that, the more the phosphorus(V) chloride dissociates (splits up reversibly) to give phosphorus(III) chloride and chlorine.

$$PCl_5 \rightleftharpoons PCl_3 + Cl_2$$

Solid phosphorus(V) chloride contains ions which is why it is a solid at room temperature. The formation of the ions involves two molecules of PCl_5.

A chloride ion transfers from one of the original molecules to the other, leaving a positive ion, $[PCl_4^+]$, and a negative ion, $[PCl_6]^-$.

$$2PCl_5 \longrightarrow [PCl_4]^+ + [PCl_6]^-$$

At 163°C, the phosphorus(V) chloride converts to a simple molecular form containing PCl_5 molecules.

Phosphorus(V) chloride has a violent reaction with water producing fumes of hydrogen chloride. As with the other covalent chlorides, if there is enough water present, these will dissolve to give a solution containing hydrochloric acid. The reaction happens in two stages. In the first, with cold water, phosphorus oxychloride, $POCl_3$, is produced along with HCl.

$$PCl_5 + H_2O \longrightarrow POCl_3 + 2HCl$$

If the water is boiling, the phosphorus(V) chloride reacts further to give phosphoric(V) acid and more HCl. Phosphoric(V) acid is also known as phosphoric acid or orthophosphoric acid.

$$POCl_3 + 3H_2O \longrightarrow H_3PO_4 + 3HCl$$

The overall equation in boiling water is just a combination of these:

$$PCl_5 + 4H_2O \longrightarrow H_3PO_4 + 5HCl$$

Solid phosphorus(V) chloride does not conduct electricity because the ions are not free to move.

Disulphur Dichloride, S_2Cl_2

Disulphur dichloride is just one of three sulphur chlorides formed when chlorine reacts with hot sulphur. Disulphur dichloride is a simple covalent liquid. It is orange in colour and has typical smell.

The shape of the molecule is difficult to draw convincingly. There is no plane of symmetry in the molecule and this means that it will have an overall permanent dipole. The liquid will have van der Waals dispersion forces and dipole-dipole attractions.

Disulphur dichloride reacts slowly with water to produce a complex mixture of things including hydrochloric acid, sulphur, hydrogen sulphide and various sulphur containing acids and anions (negative ions). There is no way that you can write a single equation for this.

There are no ions in disulphur dichloride and no mobile electrons so it never conducts electricity.

HYDROXIDES

The word "hydroxide" means to include anything which contains either a hydroxide ion or an —OH group covalently bound to the element in question.

Both react with acids to form salts. For example, with dilute hydrochloric acid, you get colourless solutions of sodium chloride or magnesium chloride.

$$NaOH + HCl \longrightarrow NaCl + H_2O$$

$$Mg(OH)_2 + 2HCl \longrightarrow MgCl_2 + 2H_2O$$

Aluminium Hydroxide

Aluminium hydroxide is amphoteric.

Like sodium or magnesium hydroxides, it will react with acids. With dilute hydrochloric acid, a colourless solution of aluminium chloride is formed. It shows the basic nature of it.

$$Al(OH)_3 + 3HCl \longrightarrow AlCl_3 + 3H_2O$$

But aluminium hydroxide also has an acidic nature. It will react with sodium hydroxide solution to give a colourless solution of sodium tetrahydroxoaluminate.

$$Al(OH)_3 + NaOH \longrightarrow NaAl(OH)_4$$

The other hydroxides of period 3 elements are:

```
     OH                 O
     |                  ||
HO—Si—OH           HO—P—OH
     |                  |
     OH                 OH
Orthosilicic acid   Phosphoric acid

     O                  O
     ||                 ||
HO—S—OH            HO—Cl=O
     ||                 ||
     O                  O
Sulphuric acid      Perchloric acid
```

None of these contains hydroxide ions. In each case the –OH group is covalently bound to the period 3 element, and in each case it is possible for the hydrogens on these –OH groups to be removed by a base. In other words, all of these compounds are acidic.

But they vary considerably in strength:

- Orthosilicic acid is very weak indeed.
- Tetraoxophosphate(V) acid is a weak acid although somewhat stronger than simple organic acids like ethanoic acid.
- Tetraoxosulphate(VI) acid and tetraoxochlorate(VII) acids are both very strong acids.

The main factor in determining the strength of an acid is how stable the anion (the negative ion) is once the hydrogen has been removed. This in turn depends on how much the negative charge can be spread around the rest of the ion.

If the negative charge stays entirely on the oxygen atom left behind from the —OH group, it will be very attractive to hydrogen ions. The lost hydrogen ion will be easily recaptured and the acid will be weak.

On the other hand, if the charge can be spread out (delocalised) over the whole of the ion, it will be so "dilute" that it won't attract the hydrogen back very easily. The acid will then be strong.

Wherever possible, the negative charge is delocalised by interacting with doubly-bonded oxygens.

For example, in tetraoxochlorate(VII) acid, the ion produced is the tetraoxochlorate(VII) ion (also known as the perchlorate ion), ClO_4^-.

The structure of the ion is:

$$^{-}O{-}\overset{\overset{\displaystyle O}{\|}}{\underset{\underset{\displaystyle O}{\|}}{Cl}}{=}O \quad \text{or} \quad (ClO_4)^-$$

The negative charge is delocalised over the whole ion, and all four chlorine-oxygen bonds are identical as shown below.

$$\left[O \cdots \overset{\overset{\displaystyle O}{\vdots}}{\underset{\underset{\displaystyle O}{\vdots}}{Cl}} \cdots O \right]^-$$

When tetraoxosulphate(VI) acid loses a hydrogen ion to form the hydrogensulphate ion, HSO_4^-, the charge can be spread over three oxygens (the original one with the negative charge, and the two sulphur-oxygen double bonds. That is still an effective delocalisation, and tetraoxosulphate(VI) acid is almost as strong as perchloric acid.

Tetraoxophosphate(V) acid is much weaker than tetraoxosulphate(VI) acid because it only has one phosphorus-oxygen double bond which it can use to help delocalise the charge on the ion formed by losing one hydrogen ion so the charge on that ion is delocalised less effectively.

In orthosilicic acid, there are no silicon-oxygen double bonds to delocalise the charge. This means the ion formed by loss of a hydrogen ion is not at all stable, and easily recovers its hydrogen.

18.4 THERMAL STABILITIES OF TRIOXOCARBONATE(IV) AND TRIOXONITRATE(V) OF SOME METALS

The trioxocarbonate(IV) of Group-1 metals are known to be relatively more stable to heat as compared to those of group-2 metals which are less stable.

CARBONATES AND HYDROGEN CARBONATES

The trioxocarbonate(IV) of alkali metals except lithium trioxocarbonate(IV) are stable to heat. The trioxocarbonate(IV) of group-2 metals *i.e.*, magnesium and calcium and that of lithium decompose on heating, forming an oxide along with the evolution of carbon dioxide. For example,

$$Li_2CO_3 \xrightarrow{\Delta} Li_2O + CO_2$$

$$MgCO_3 \xrightarrow{\Delta} MgO + CO_2$$

$$Na_2CO_3 \xrightarrow{\Delta} \text{no effect.}$$

Reason. *The stability of carbonate towards heat depends upon the relative stability of the resulting metal oxide. The higher the stability of the resulting metal oxide, the lesser the stability of the carbonate towards heat and vice versa. Now the stability of the resulting metal oxides decreases down the group due to a decrease in lattice enthalpy, (because of bigger size) therefore the stability of carbonates towards heat increases.*

Li_2CO_3 decomposes on heating because it gives lithium oxide. The small O^{2-} anion is very strongly attracted to Li^+ ion resulting in an ionic compound with high lattice energy making it is more stable than Li_2CO_3. The greater the stability of the resulting oxide, the more its tendency of formation and the lower the thermal stability of the carbonate.

In period 3, magnesium trioxocarbonate(IV) decomposes more readily than sodium trioxocarbonate(IV). It is because the resulting MgO is more stable than $MgCO_3$ because of the strong ionic bond between Mg^{2+} and O^{2-} ions. Thus its lattice energy is high and hence the stability. The stabilities of carbonates of group-2 metals increase on moving down the group. For example, $BeCO_3$ decomposes at 373 K, $MgCO_3$ at 813 K, $CaCO_3$ at 1173 K, $SrCO_3$ at 1563 K and $BeCO_3$ at 1633 K.

Lithium and group-2 metals do not form solid hydrogencarbonates, although they exist in solution. On heating these solutions, the hydrogencarbonates decompose to form carbonates and CO_2 gas is liberated. The solid hydrogencarbonates of alkali metals decompose between 375 and 575 K.

$$2NaHCO_3(s) \longrightarrow Na_2CO_3(s) + H_2O(g) + CO_2(g)$$

NITRATES

Trioxonitrate(V) of alkali metals (Na, K, etc.), except $LiNO_3$, decompose on strong heating forming nitrites and oxygen. For example,

$$2KNO_3(s) \xrightarrow{\text{heat}} 2KNO_2(s) + O_2(g)$$

Trioxonitrate(V) of Mg and Ca metals and $LiNO_3$ decompose on heating to form oxides, nitrogen dioxide and oxygen.

$$2LiNO_3(s) \xrightarrow{\Delta} Li_2O(s) + 2NO_2(g) + O_2(g)$$

$$2Ca(NO_3)_2(s) \xrightarrow{\Delta} 2CaO(s) + 4NO_2(g) + O_2(g)$$

Thermal stabilities of trioxonitrate(V) of group-1 *i.e.,* Li, Na, K, Rb, Cs, etc., and group-2 metals Be, Mg, Ca, Sr, Ba etc., increase on moving down the group from top to bottom.

18.5 DEMONSTRATION OF THERMAL STABILITIES OF SOME TRIOXOCARBONATE(IV) IN THE LABORATORY

The thermal stabilities of some of the trioxocarbonate(IV)s of metals like lithium, sodium and copper can be assessed by performing simple experiments in the laboratory.

Requirements

The experiment requires, three test tubes, gas holder, lime water.

Procedure

Take three test tubes and to each of it add about 1g each of Li_2CO_3, Na_2CO_3 and $CuCO_3$. To the mouth of the test tubes attach a gas detector containing lime water. The simple experimental set up is shown below:

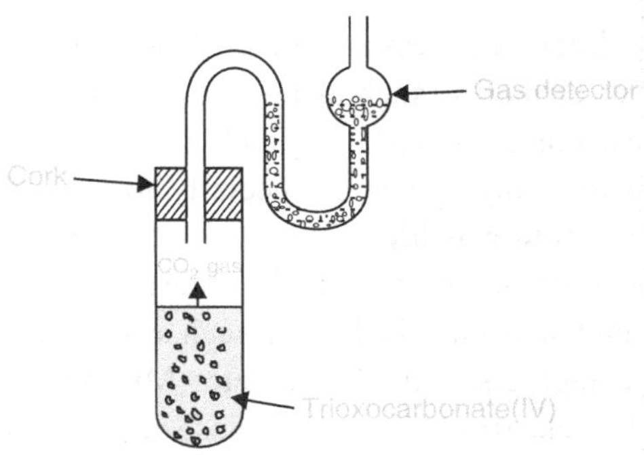

Fig. 18.3

Observations

The trioxocarbonate(IV)s of lithium and copper will decompose on heating and will turn lime water milky.

$$Li_2CO_3(s) \xrightarrow{\Delta} Li_2O(s) + CO_2(g)$$

$$CuCO_3(s) \xrightarrow{\Delta} CuO(s) + CO_2(g)$$

The trioxocarbonate of copper decomposes at a much slower rate than that of lithium.

Na_2CO_3 will not decompose to produce carbon dioxide gas.

18.6 SILICON AND ITS USES

Silicon is the most important element used in semi-conductors such as transistors and diodes. Without a supply of pure silicon it would be impossible to make the range of high quality computers, calculators, telephones, radios, etc., upon which modern industrial societies have come to rely. Crude silicon can be obtained by reducing silica (sand) with coke:

$$SiO_2(s) + 2C(s) \longrightarrow Si(s) + 2CO(g)$$

The product is contaminated by silicon carbide, SiC_2. This carbide is also known as ***carborundum.*** It is extremely hard. The bits of steel drills are often impregnated with carborundum in order to reduce the wear on them.

Much better quality silicon is produced by reducing silicon tetrachloride with hydrogen:

$$SiCl_4(l) + 2H_2(g) \longrightarrow Si(s) + 4HCl(g)$$

IMPORTANT USES OF SILICON

1. Silicon is one of the most abundant of elements and an essential ingredient of a large number of minerals that make up the Earth's surface. Sand, clay and the stuff we call 'earth' is a mixture of silicon and oxygen.
2. Pure silicon is a semiconductor. When small amounts of other elements (such as boron) are added to silicon, its resistance is reduced and it becomes a good conductor. Tiny pieces of specially treated silicon are the basis of **microchips**. Transistors and other electronic components are built into the microchip. Microchips are found in electronic devices such as radios, television sets, computers, calculators, etc. Microchips contain complete electronic circuits in these devices. Computers contain many microchips.

   The **main advantage** of microchips is that they occupy little space and act as effective conductors in many different ways. The microchip industry is one of the fastest growing industries worldwide. Microchips have revolutionized our way of life.
3. Silicon and oxygen make giant three dimensional structures called ***silicates.*** Silicates are also used as the basis of ceramics. Ceramics are much better electrical and thermal insulators than metals, and have greater rigidity, hardness and temperature stability than organic polymers. Following a great deal of research, new ceramic materials have been developed that contain, *for example*, borides, carbides and nitrides. They have found an increasing number of uses, *for example,* glasses for covering solar panels, parts of turbines and internal combustion engines, and refractory brick linings for high temperature furnaces.
4. Silicon is also used in photoelectric cells in solar panels to convert solar energy to electrical energy for use in space-craft, artificial satellites, electronic watches and pocket size solar calculators, domestic heating, lighting and the provision of other sources of energy.
5. Silicon is widely distributed as silica or sand (SiO_2) and tetraoxosilicates(IV), SiO_4^{4-}. Glass is made from tetraoxosilicate(IV) materials.

STRUCTURE OF CO_2 AND SiO_2

There is a marked difference in the structure and properties of CO_2 and other dioxides. CO_2 is a gas at room temperature. On the other hand, silicon dioxide or silica is a solid at room temperature. CO_2 is a monomeric, linear, non-polar molecule in which carbon forms *pπ-pπ* bonds with oxygen.

It may be considered to be a resonance hybrid of the following structures.

$$[:\ddot{O}::C::\ddot{O}: \leftrightarrow \overset{-}{:\ddot{O}}:C:::\overset{+}{O}: \leftrightarrow :\overset{+}{O}:::C:\overset{-}{\ddot{O}:}]$$

Silicon is unable to form *pπ-pπ* bond with oxygen atom due to its relatively large size. Thus, it satisfies its tetra covalency with four oxygen atoms and constitutes a three dimensional network structure in which each silicon atom forms four Si—O bonds which are tetrahedrally disposed.

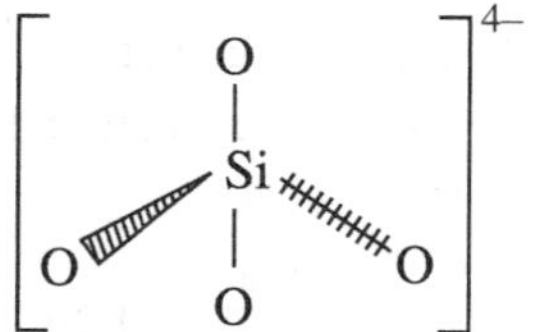

Fig. 18.4. Structure of silicate.

In turn each oxygen atom of SiO_2 is shared by two silicon atoms. Three crystalline modifications of SiO_2 are **quartz, cristobalite** and **tridymite.** In addition to these, several amorphous forms of silica such as *silica gel* and *fumed silica* are known.

Since Si—O bonds are very strong (368 kJ mol^{-1}), silica is relatively inert and has a very high m.p.

COMPARISON OF PROPERTIES CO_2 AND SiO_2

CO_2	SiO_2
1. It exists as gas.	1. It is a solid.
2. It is a linear, monomeric non-polar molecule.	2. It is a three dimensional network solid.
3. Appreciably soluble in water.	3. Insoluble in water.
4. On reduction with coke gives carbon monoxide $CO_2 + C \longrightarrow 2CO$	4. On reduction with coke in electrical furnace gives silicon carbide $SiO_2 + 3C \rightarrow SiC + 2CO$.

18.7 GROUP-17 ELEMENTS—THE HALOGEN FAMILY

The non-metallic elements *Fluorine* (F), *Chlorine* (Cl), *Bromine* (Br), *Iodine* (I) and *Astatine* (At) are grouped together to form the group 17 elements of the Periodic Table. Their salts are present in sea-water and hence they are collectively known as ***halogens*** (Greek, *Halos* means *sea salt producers*). The group collectively is called *halogen family*. All halogens precede the noble gases in the periodic table because their atoms contain one electron less than the electronic configuration of the inert gas. They have very strong tendency to acquire stable inert gas configuration by accepting one electron. For this reason halogens exhibit non-metallic behaviour. Astatine the last member of the halogen family is a radioactive element with a very short life. All other halogens are abundant in nature in the combined state.

The sources of halogens along with their relative abundance are given in Table 18.6.

PHYSICAL PROPERTIES OF HALOGENS (GROUP 17 ELEMENTS)

There is a regular gradation in the physical as well as chemical properties of halogens. These are:

1. Physical State. The halogens are all diatomic and exist as F_2, Cl_2, Br_2 and I_2. The intermolecular forces are very weak in halogens. The nature of forces is van der Waals' and their magnitude increases down the group.

Thus, F_2 and Cl_2 are gases, bromine is a volatile liquid and iodine is a volatile solid.

2. Colour. Halogens are coloured.

The colour of different halogens are as follows:

Halogen	*Fluorine*	*Chlorine*	*Bromine*	*Iodine*
Colour	Pale yellow	Greenish yellow	Reddish brown	Dark violet

Reason. *The colour of the halogen is due to absorption of visible light by their molecules resulting in the excitation of outer electron to higher energy levels. Fluorine being smaller in size absorbs violet light (of high energy) for excitation and appears pale yellow while iodine absorbs yellow light (of less energy) for excitation and therefore appears violet black. In between, the colour of chlorine is yellowish-green and of bromine is reddish-brown or orange. Thus, the colour deepens down the group.*

3. Melting and Boiling Points. Melting and boiling points increase with increase in atomic number. The heat of fusion as well as the heat of vapourisation also increase with an increase in atomic number. This indicates that the strength of intermolecular forces of attraction between the molecules increases with the increase in atomic number.

4. Interatomic Bond Energy. The interatomic bond energy of group 17 elements shows a periodic variation. In general, the bond dissociation energy decreases down the group as atomic radius increases. With the increase in size of atoms, the shared electron cloud in the covalent bond experiences less force of attraction hence a weaker bond forms. The energy needed to break the weaker covalent bond therefore decreases. Fluorine shows an unexpected (anomalous) bond dissociation energy (atomization energy, E_{at}) due to its very small size. The electrons in the shells of the two atoms forming the bond are packed near each atom because of their small size, and repel each other strongly. This reduces the covalent bond attraction between the positive nuclei and the shared electrons. The bond is weakened and the energy needed to break it is greatly reduced compared with that of the other halogens.

Table 18.6. Sources and Relative Abundance of Halogens

Halogen	*Abundance % Earth's Crust*	*Main Sources*
Fluorine	0.07	Fluorspar (CaF_2); fluorapatite [$3Ca_3(PO_4)_2CaF_2$]
Chlorine	0.14	Sea-water (2.8%), salt wells, beds ($NaCl$, KCl, $MgCl_2$, $CaCl_2$)
Bromine	2.5×10^{-4}	Sea-water, salt lakes ($NaBr$, KBr, $MgBr_2$)
Iodine	8×10^{-5}	Sea-weeds (I^- 0.5% by mass); Chile salt peter (0.02% $NaIO_3$)

Some of the properties of group 17 elements are given in Table 18.7.

Table 18.7. Atomic and Physical Properties of Halogens

Property	*F*	*Cl*	*Br*	*I*	*At*[a]
Atomic number	9	17	35	53	85
Atomic mass/g mol^{-1}	19.00	35.45	79.90	126.90	210
Electronic configuration	$[He]2s^22p^5$	$[Ne]3s^23p^5$	$[Ar]3d^{10}4s^24p^5$	$[Kr]4d^{10}5s^25p^5$	$[Xe]4f^{14}5d^{10}6s^26p^5$
Covalent radius/pm	64	99	114	133	–
Ionic radius X^-/pm	133	184	196	220	–
Ionisation enthalpy/kJ mol^{-1}	1680	1256	1142	1008	–
Electron gain enthalpy/kJ mol^{-1}	– 333	– 349	– 325	– 296	–
Electronegativity[b]	4	3.2	3.0	2.7	2.2
$\Delta_{Hyd}H(X^-)$/kJ mol^{-1}	515	381	347	305	–
	F_2	*Cl_2*	*Br_2*	*I_2*	–
Melting point/K	54.4	172.0	265.8	386.6	–
Boiling point/K	84.9	239.0	332.5	458.2	–
Density/g cm^{-3}	1.5(85)[c]	1.66(203)[c]	3.19(273)[c]	4.94(293)[d]	–
Distance X–X/pm	143	199	228	266	–
Bond dissociation enthalpy/(kJ mol^{-1})	158.8	242.2	192.8	151.1	–
$E^⦵/V^e$	2.87	1.36	1.09	0.54	–

[a]Radioactive; [b]Pauling scale; [c]For the liquid at temperatures (K) given in parentheses; [d]solid; [e]The half-cell reaction is $X_2(g) + 2e^- \rightarrow 2X^- (aq)$.

Table 18.8. Variation of Atomization Energy

Halogen	*Covalent radius/nm*	*E_{at}/kJ mol^{-1}*
F_2	0.071	79
Cl_2	0.099	121.7
Br_2	0.114	111.9
I_2	0.133	106.8

5. Atomic and Ionic Radii. Atomic and ionic radii are small and increase regularly down the group from fluorine to iodine because new electronic shells are added.

6. Ionization Energies. Ionization energies of all the halogens are very high. Therefore, they have a less tendency to lose electron. However, this tendency increases down the group because the distance of valence shell from nucleus increases. Iodine is capable of forming stable compounds in which it exists as I^+ ion.

7. Non-metallic Character. All the halogens have very high values of ionization energies and exhibit non-metallic character. The non-metallic character, however, decreases down the group. Iodine shows some distinct metallic properties *e.g.*, it possesses metallic lustre and forms positive ions like I^+, I^{3+} etc.

8. Electron Gain Enthalpy. Since halogens have strong tendency to accept electrons, they have high values of electron gain enthalpy. On moving down the group electron gain enthalpy values decrease because size of the atom increases. However, it has been observed that fluorine has less electron affinity than chlorine. *The low value of electron affinity for fluorine is due to the small size of the fluorine atom. In a compact 2p subshell of fluorine the added electron does not feel much attraction resulting in low value of electron gain enthalpy.*

9. Nature of Bonds with Other Elements. The halogens combine with metals to form predominantly ionic halides. The ionic character of these halides decreases while covalent character increases on going down the halogen group. This is because an increase in size of the halide ion increases its polarizability and this imparts covalent character to the compound. For example, AlF_3 is predominantly ionic whereas $AlCl_3$ and $AlBr_3$ are predominantly covalent.

With metals in higher oxidation states or with non-metals, halogens form covalent compounds which are polar in character. Polarity of these compounds decreases as we move down the group.

Due to the similarity in chemical nature the halides readily combine with each other to form polyhalide ions.

18.8 ELECTRONIC CONFIGURATION AND VARIABLE OXIDATION STATES OF HALOGENS

ELECTRONIC CONFIGURATION

The general electronic configuration of halogens is **ns^2np^5**. The electronic configurations of halogens are given in Table 18.9.

All the halogens are very reactive but amongst them fluorine is the most reactive. As we move down the group, reactivity decreases. This is due to the decrease in electronegativity and increase in bond dissociation energy. In general, a halogen of low atomic number oxidises halide ions of higher atomic number *i.e.*, F_2 displaces Cl_2, Br_2 and I_2 from their salts, Cl_2 displaces Br_2 and I_2 whereas Br_2 displaces iodine from its salts, *e.g.*,

$$F_2 + 2X^- \longrightarrow 2F^- + X_2 \qquad (X = Cl, Br, I)$$

$$F_2 + 2KX \longrightarrow 2KF + X_2 \qquad (X = Cl, Br, I)$$

$$Cl_2 + 2X^- \longrightarrow 2Cl^- + X_2 \qquad (X = Br, I)$$

$$Br_2 + 2I^- \longrightarrow 2Br^- + I_2$$

Halogens exist in the elemental form as diatomic molecules having characteristic colours. Fluorine exhibits only a negative oxidation state of –1. Other halogens in addition to –1 oxidation state exhibit formal positive oxidation states.

OXIDATION STATES

All the elements of the halogen family have a tendency to acquire noble gas electronic configuration by either accepting an electron resulting in the formation of monovalent anion or by sharing one electron with those of other elements. Thus, they show an oxidation state of –1 or +1 depending on whether the element combining with the halogens is less electronegative or more electronegative than the halogen. Since fluorine is the most electronegative element it always has a –1 oxidation state and never has a positive oxidation state. Also higher oxidation states are not possible in fluorine as it does not have vacant *d*-orbitals.

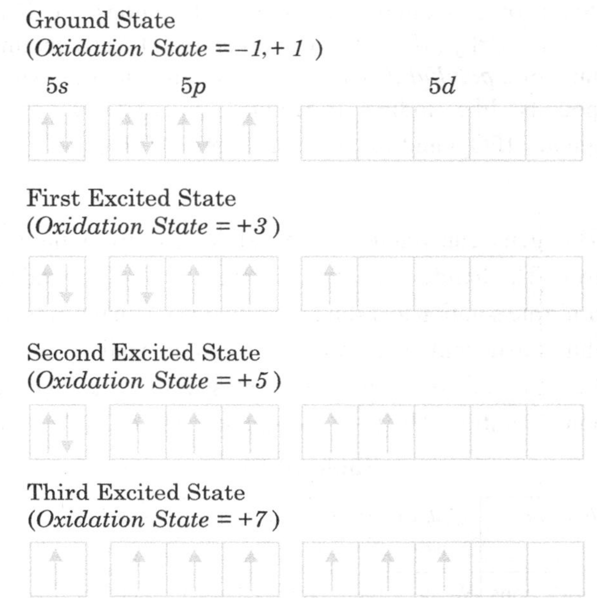

Other elements of this group show oxidation states of + 1, + 3, + 5 and + 7 in addition to the – 1 state which is most common. Higher oxidation states of these elements are due to the presence of vacant *d*-orbitals. The outer *s* or *p* electrons can easily be excited to these vacant *d*-orbitals as shown below in the case of iodine.

All the halogens have high electron affinity and high electronegativity. Therefore, these show similarity in the chemical nature.

Table 18.9. Electronic Configuration of Halogens

Elements	*At. No.*	*Electronic Configuration*	*Abundance in Earth's Crust (ppm)*
Fluorine (F)	9	[He] $2s^2\ 2p^5$	800
Chlorine (Cl)	17	[Ne] $3s^2\ 3p^5$	480
Bromine (Br)	35	[Ar] $3d^{10}\ 4s^2\ 4p^5$	2
Iodine (I)	53	[Kr] $4d^{10}\ 5s^2\ 5p^5$	0.3
Astatine (At)	85	[Xe] $4f^{14}\ 5d^{10}\ 6s^2\ 6p^5$	—

POLYHALIDE IONS

In additions to neutral interhalogen compounds, polyhalide ions containing more than two halogen atoms have also been prepared. These are formed by a combination between the halide ion with another halogen atom or with some compound of the halogen. For example, I_3^- (triodide/ion is formed by a combination of I_2 molecule and iodide in (I^-).

$$I^-(aq) + I_2 \rightleftharpoons I^-_3(aq)$$

Some other examples of polyhalide anions are ICl_2^-, IBr_2^-, ICl_4^-, BrF_4^-, IF_6^-, etc. Besides anions, I_3^+ is a common example of a ***polyhalide cation.*** This ion is found to exist in compounds like iodine triacetate, $I(CH_3COO)_3$ iodine phosphate (IPO_4) and iodine perchlorate, $I(ClO_4)_3$.

HALIDE IONS

Halogens can readily form halide ions by gaining an electron. The halide ions can be identified easily by adding silver nitrate solution to a solution of halide ion in the presence of dilute nitric acid. The halide ions can be easily identified on the basis of colour of precipitates or by reacting with ammonia solution. The tests are summarized in Table 18.10.

Table 18.10

Halide ion	*Colour of precipitate with silver nitrate*	*Solubility in ammonia solution*
Chloride ions	White precipitate $Ag^+(aq) + Cl^-(aq) \longrightarrow AgCl(s)$	Precipitates dissolve in ammonia solution
Bromide ions	Pale yellow precipitates $Ag^+(aq) + Br^-(aq) \longrightarrow AgBr(s)$	Precipitates partly dissolve
Iodide ions	Yellow precipitate $Ag^+(aq) + I^-(aq) \longrightarrow AgI(s)$	Precipitates do not dissolve.

18.9 REACTIONS OF HALOGENS WITH WATER AND ALKALIES

All the halogens are very reactive but amongst them fluorine is the most reactive. As we move down the group, reactivity decreases. This is due to the decrease in electronegativity and increase in bond dissociation energy. In general, a halogen of low atomic number oxidises halide ions of higher atomic number *i.e.*, F_2 displaces Cl_2, Br_2 and I_2 from their salts, Cl_2 displaces Br_2 and I_2 whereas Br_2 displaces iodine from its salts, *e.g.*,

$$F_2 + 2X^- \longrightarrow 2F^- + X_2 \qquad (X = Cl, Br, I)$$

$$F_2 + 2KX \longrightarrow 2KF + X_2 \qquad (X = Cl, Br, I)$$

$$Cl_2 + 2X^- \longrightarrow 2Cl^- + X_2 \qquad (X = Br, I)$$

$$Br_2 + 2I^- \longrightarrow 2Br^- + I_2$$

1. Reaction with water. Both fluorine and chlorine oxidise water. Fluorine gives a mixture of oxygen and trioxygen (O_3).

$$2F_2(g) + 2H_2O(l) \longrightarrow O_2(g) + 4HF(aq)$$

Chlorine, bromine and iodine dissolve and react with water to form an aqueous mixture in which the halogens are in equilibrium with reduced and oxidized species. The halogens are considered dissolved because some of the molecules remain hydrated and some react to give new products, forming an equilibrium.

$$X_2(g) + H_2O(l) \rightleftharpoons HX(aq) + HXO(aq)$$

Chlorine is not such a powerful oxidising agent. It does not release oxygen. Its reaction with water produces a greenish-yellow solution which contains hydrochloric and chloric (I) acid *i.e.*, hypochlorous acid. It is also known as chlorine water.

$$Cl_2(g) + H_2O(l) \longrightarrow \underbrace{H^+(aq) + Cl^-(aq)}_{\text{hydrochloric acid}} + \underbrace{H^+(aq) + ClO^-(aq)}_{\text{Chloric (I) acid}}$$

The chlorate (I) ions, ClO^-, are responsible for bleaching action. It is because the ClO^- ion quickly loses dioxygen if exposed to sunlight and bring about an oxidation process.

$$2ClO^-(aq) \longrightarrow 2Cl^-(aq) + O_2(g)$$

Bromine interacts slightly with water to give an acidic solution. This is prepared by adding a few drops of liquid bromine to water and shaking. A light yellow solution is produced.

Iodine crystals have very strong van der Waals' intermolecular forces which are difficult to overcome. It is only sparingly soluble in water. When iodine is shaken strongly with water a light yellow-brown solution is formed. Iodine, however, dissolves very well in aqueous potassium iodide solution due to the formation of polyhalide ion.

2. Reaction of the halogens with alkalis. All the halogens react with the alkalis potassium hydroxide or sodium hydroxide. However, different products are formed under different conditions of temperature and concentration of the alkali. Fluorine reacts with cold and hot alkali solution as follows:

$$2F_2(g) + 2NaOH(aq) \xrightarrow[\text{cold, dilute}]{} 2NaF(aq) + F_2O(g) + H_2O(l)$$

$$2F_2(g) + 4NaOH(aq) \xrightarrow[\text{hot, concentrated}]{} 4NaF(aq) + O_2(g) + 2H_2O(l)$$

Chlorine, bromine and iodine react with alkali in essentially the same way with only slight differences of products.

$$Cl_2(g) + 2NaOH(aq) \xrightarrow[\text{cold, dilute}]{} NaCl(aq) + NaClO(aq) + H_2O(l)$$

$$3Cl_2(g) + 6NaOH(aq) \xrightarrow[\text{hot, concentrated}]{} 5NaCl(aq) + NaClO_3(aq) + 3H_2O(l)$$

As pointed out earlier, the sodium oxochlorate (I) solution is a more stable bleaching agent than chlorine water which decomposes more easily.

NaClO is commercially prepared by the electrolysis of brine and used as bleach in the home. It bleaches by oxidation in the same way as chlorine water does because it contains the ClO^- ion which is responsible for the bleaching by production of oxygen.

Bromine reacts with alkalis as follows:

$$Br_2(l) + 2KOH(aq) \xrightarrow[\text{cold, dilute}]{} KBr(aq) + KBrO(aq) + H_2O(l)$$

On standing or warming BrO_3^- is formed:

$$3KBrO(aq) \longrightarrow KBrO_3(aq) + 2KBr(aq)$$

Iodine reacts in alkali solution to form the iodide and trioxoiodate(V) (IO_3^-) ions directly without heating. Any oxoiodate (I) (IO^-) formed is quickly oxidised to trioxoiodate(V) which is more stable.

$$3I_2(s) + 6KOH(aq) \longrightarrow 5KI(aq) + KIO_3(aq) + 3H_2O(l)$$

A closer look at the equations of the reactions of the halogens with water and the alkalis shows that the halogen atom is oxidized and reduced.

18.10 ACID STRENGTH OF HYDROGEN HALIDES

Halogens combine with hydrogen to form volatile halides of the formula HX. Some characteristics of these hydrides are given below:

(*i*) *Physical Nature*. Except HF, all other hydrides *viz.*, HCl, HBr and HI are gases. HF is a liquid because of *intermolecular hydrogen bonding.*

—H—F....H—F....H—F....

The bond between hydrogen and halogen is covalent in all the cases.

(*ii*) *Acidic Strength.* All the hydrogen halides act as acids in their aqueous solutions. The acidic strength varies in the order

$$HF < HCl < HBr < HI$$

Reason. *All the halogen acids ionise to give H^+ ion and halide ion, X^-.*

$$HX \longrightarrow H^+ + X^- \; ; \text{(where } X^- = F^-, Cl^-, Br^-, I^-)$$

The above order of acidic character can be explained in terms of the strength of H—X bonds, which is in the order H—I < H—Br < H—Cl < H—F. Since the H—I bond is weakest, ***HI is the strongest acid.*** *On the other hand H—F bond is strongest, hence it is the weakest acid among all the halogen acids.*

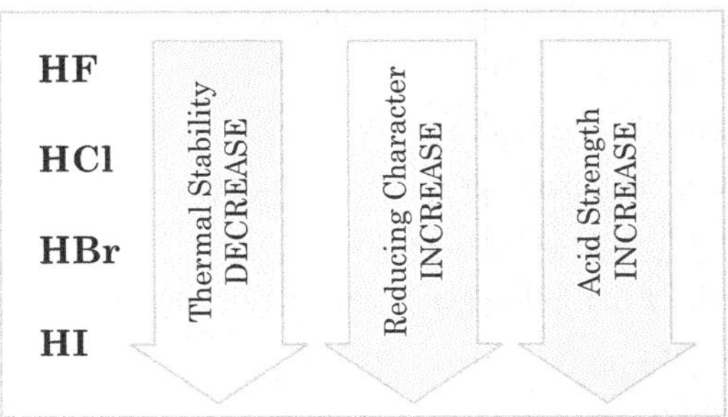

(*iii*) *Reducing Character.* The reducing character of the hydrides increases down the group. It is because the strength of the H—X bond decreases from HF to HI. The lower the strength of the bond, the greater the ease with which it can be cleaved and the greater the reducing character.

Some physical properties of hydrogen halides are summarised in Table 18.11.

Table 18.11. Properties of Hydrogen Halides

Properties	*HF*	*HCl*	*HBr*	*HI*
Melting point/K	190	159	185	222
Boiling point/K	293	189	206	238
Bond length(H—X)/pm	91.7	127.4	141.4	160.9
$\Delta_{diss} H^{\ominus}$/kJ mol^{-1}	574	432	363	295
pK_a	3.2	– 7.0	– 9.5	– 10.0

EVALUATE YOURSELF

I. Objective Type Questions

Select the most appropriate choice from the options given as (a), (b), (c) and (d) after each question:

1. The correct order of acidic strength is
   (*a*) $Cl_2O_7 > SO_2 > P_4O_{10}$
   (*b*) $CO_2 > N_2O_4 > SO_3$
   (*c*) $Na_2O > MgO > Al_2O_3$
   (*d*) $K_2O > CaO > MgO$.
2. Which of the following oxides will produce dioxygen on treatment with water?
   (*a*) $KClO_3$ (*b*) Na_2O_2
   (*c*) CaO (*d*) SO_3.
3. Which of the following set contains oxide in the sequence, basic, amphoteric and acidic respectively?
   (*a*) SO_2, P_2O_5, CO (*b*) BaO, Al_2O_3, SiO_2
   (*c*) CaO, SiO_2, Al_2O_3 (*d*) Al_2O_3, CO_2, CO.
4. Which of the following set contains only amphoteric oxides?
   (*a*) ZnO, PbO, SnO, Al_2O_3
   (*b*) ZnO, BaO, SO_2, CO
   (*c*) Co, PbO, BaO_2, SO_2
   (*d*) P_2O_5, N_2O_5, Cl_2O_7, BaO.
5. Which of the following is called laughing gas?
   (*a*) N_2O (*b*) No
   (*c*) N_2O_5 (*d*) N_2O_3.
6. The greater reactivity of fluorine is due to
   (*a*) lower energy of F—F bond
   (*b*) high energy of F—F bond
   (*c*) F_2 is a gas at room temperature
   (*d*) none of these is correct.
7. Halogens are all coloured
   (*a*) due to absorption of U-V light
   (*b*) due to absorption of I.R. light
   (*c*) due to absorption of visible light
   (*d*) none of the above is correct.
8. The halogens are chemically similar because
   (*a*) for each element, the atomicity is 2
   (*b*) their outermost electrons have the same principal quantum number
   (*c*) their atoms can lose 7 electrons to form ions with the electronic structure of an inert gas
   (*d*) they each have atoms which need only one electron to fill completely as outermost *p*-sub-shell.
9. Which of the following contains a maximum number of oxo groups in its structure?
   (*a*) H_2SO_4 (*b*) H_2SO_3
   (*c*) H_3PO_3 (*d*) H_3PO_4.
10. Among the halogen acid, the one which is weakest is
   (*a*) HF (*b*) HCl
   (*c*) HBr (*d*) HI.
11. In the reaction $I_2 + I^- \longrightarrow I_3^-$, the lewis acid is
   (*a*) I_2 (*b*) I^-
   (*c*) I_3^- (*d*) none.
12. Which of the following hydrogen halide is most volatile?
   (*a*) HF (*b*) HCl
   (*c*) HBr (*d*) HI.
13. Order of boiling point is
   (*a*) HF > HI > HBr > HCl
   (*b*) HF > HBr > HI > HCl
   (*c*) HCl > HBr > HI > HF
   (*d*) HCl > HI > HBr > HF.
14. The correct order of acidity of halogenic acids is
   (*a*) HF < HCl < HBr < HI
   (*b*) HI < HBr < HCl < HF
   (*c*) HF < HCl < HBr < HI < HCl
   (*d*) HF < HBr < HI < HCl.
15. The correct order of electron affinity is
   (*a*) Cl > F > Br > I (*b*) F > Cl > Br > I
   (*c*) F > Br > Cl > I (*d*) Cl > Br > F > I.
16. Which of the following orders is **not** in accordance with the property stated against it?
   (*a*) $F_2 > Cl_2\ Br_2 > I_2$: Bond dissociation energy
   (*b*) $F_2 > Cl_2 > Br_2 > I_2$: Oxidising power
   (*c*) HI > HBr > HCl > HF : Acidic property in water
   (*d*) $F_2 > Cl_2 > Br_2 > I_2$: Electronegativity.

II. Fill in the Blanks

17. Complete the following sentences by supplying appropriate words:
   (*i*) SiO_2 is ...... oxide.
   (*ii*) Laughing gas is obtained by heating of ....
   (*iii*) I_2 dissolves in KI(*aq*) to form ....
   (*iv*) Most common oxidation state of phosphorus element is ....
   (*v*) The formula of laughing gas is ....
   (*vi*) The basicity of H_3PO_4 is ....
   (*vii*) In the reaction; $I_2 + I^- \longrightarrow I_3^-$ the Lewis acid is ....
   (*viii*) The solubility of iodine in water is greatly increased by adding iodide ions because of the formation of ....
   (*ix*) The most volatile halogen acid is ....

III. Discussion Questions

18. Why fluorine shows an oxidation state of –1 only while other halogens show oxidation states of +1, +3, +5 and +7 as well.
19. F atom is more electronegative than I atom, yet HF has a lower acid strength than HI. Explain.
20. Iodine is more soluble in KI than in water, why?
21. Explain the following:
    (*i*) Why halogens are highly reactive?
    (*ii*) Why do halogens possess a colour?
22. Give chemical equations for reaction of sodium, potassium and aluminium with hydrochloric acid.
23. How does the acidic properties of oxides of non-metals change
    (*i*) on moving across the period?
    (*ii*) on moving down the group?
24. How many different oxides are formed by?
    (*i*) Nitrogen (*ii*) Phosphorus
    Write their formulas.
25. Discuss (*i*) the melting and boiling points and (*ii*) the structure of oxides of sodium and silicon.
26. Explain how magnesium oxide and Aluminium oxide react with water. Write chemical equations also.
27. Explain why Aluminium chloride has a dimeric structure? What happens when it is heated (*i*) at 180°C, 1 atmospheric pressure (*ii*) at 190°C, 2 atmospheric pressure?
28. Explain what type of solutions will be obtained (neutral, acidic, basic) when the following chlorides are dissolved in water
    (*i*) NaCl (*ii*) $AlCl_3$ (*iii*) PCl_3
29. Explain why PCl_5 exists as a solid?
30. Classify the hydroxides of the following elements as acidic, basic, neutral or amphoteric.
    (*i*) Mg (*ii*) Al
    (*iii*) P (*iv*) S
31. Explain why Na_2CO_3 is stable to heat while Li_2CO_3 decomposes on heating.
32. Describe an experiment to distinguish between chloride, bromide and iodide solutions. Give chemical equations also.
33. Give important uses of silicon.
34. Explain the physical states of halogen, giving reasons for its state of existence.
35. How does halogen react with
    (*i*) water (*ii*) caustic soda solution?

Answers

I. Objective Type Questions

1. (*a*)	**2.** (*b*)	**3.** (*b*)
4. (*a*)	**5.** (*a*)	**6.** (*a*)
7. (*c*)	**8.** (*d*)	**9.** (*a*)
10. (*a*)	**11.** (*a*)	**12.** (*b*)
13. (*a*)	**14.** (*a*)	**15.** (*a*)
16. (*a*).		

II. Fill in the Blanks

17. (*i*) Amphoteric (*ii*) Ammonium nitrate
(*iii*) KI_3 (*iv*) +5
(*v*) N_2O (*vi*) 3
(*vii*) I_2 (*viii*) I_3^- complex ion
(*ix*) HCl.

SECTION-2

UNIT-19

TRANSITION CHEMISTRY

Specific Objectives

After studying this unit, the student should be able to:

- write detail electron configuration of the first row transition elements
- State and describe the properties of transition elements
- demonstrate through experiment, the catalytic properties of transition elements and their compounds
- describe the type of bonding of complex compunds
- name complex compounds
- draw the shapes of complex compounds
- outline the similarities and differences between transition metals and representative (main group) metals.

19.1 INTRODUCTION

In the Periodic Table, the elements are classified into four blocks namely; *s, p, d* and *f* after the type of the atomic orbital which receives the last electron. The *d*-block elements are found *from the fourth period and onwards.* The *d*-block contains the elements of the groups 3-12. These elements possess properties that are transitional between the *s*- and *p*-block elements. In other words, *they represent a transition or change in properties from the s-block to the p-block and are, therefore, called transition elements.*

The ***transition elements*** *may be defined as the elements whose atoms or simple ions contain partially filled d-orbitals.*

This definition of transition elements does not cover *zinc, cadmium* and *mercury* because they have completely filled *d*-subshells in their atoms and common oxidation states. As a result Zn, Cd, Hg do not exhibit the characteristic properties of transition elements. However, they are studied along with other *d*-block elements.

All the elements belonging to this section are metals. Most of the transition metals have immense industrial importance. For example, the precious metals, gold, platinum, palladium, silver, etc., and commonly used metals such as iron, copper, mercury, etc., all belong to the transition elements. In this unit, we shall discuss the general characteristics of the transition elements.

d-Block

Transition Elements

s-Block Groups 1–2	3	4	5	6	7	8	9	10	11	12	p-Block Groups 13–18
	21 Sc	22 Ti	23 V	24 Cr	25 Mn	26 Fe	27 Co	28 Ni	29 Cu	30 Zn	
	39 Y	40 Zr	41 Nb	42 Mo	43 Tc	44 Ru	45 Rh	46 Pd	47 Ag	48 Cd	
	57 La	72 Hf	73 Ta	74 W	75 Re	76 Os	77 Ir	78 Pt	79 Au	80 Hg	
	89 Ac	104 Rf	105 Db	106 Sg	107 Bh	108 Hs	109 Mt	110 Ds	111 Rg	112 Uub	

19.2 ELECTRONIC CONFIGURATIONS

In the transition elements, the *d*-orbitals of the penultimate shell are progressively filled. There are mainly three transition series. The **first transition series** involves the filling of 3*d*-orbitals. It starts from *scandium* (Z = 21) and goes up to *zinc* (Z = 30).

The **second transition series** involves the filling of 4*d*-orbitals and includes 10 elements from *yttrium* (Z = 39) to *cadmium* (Z = 48).

The **third transition series** involves filling of 5*d*-orbitals. The first element of this series is *lanthanum* (Z = 57). It is followed by fourteen elements called lanthanoids which involve the filling of 4*f*-orbitals. The next nine elements from *hafnium* (Z = 72) to *mercury* (Z = 80) belong to the third transition series.

The general electronic configuration of the transition elements is $(n-1)\, d^{1-10}\, ns^{1-2}$. The electronic configurations of the first row transition elements are given in Table 19.1.

Table 19.1. Electronic Configurations of Elements of the First Transition Series ($_{21}$Sc to $_{30}$Zn)

Atomic Number	*Element*	*Electronic Configuration*
21	Sc	[Ar] $3d^1 4s^2$
22	Ti	[Ar] $3d^2 4s^2$
23	V	[Ar] $3d^3 4s^2$
24	Cr	**[Ar] $3d^5 4s^1$**
25	Mn	[Ar] $3d^5 4s^2$
26	Fe	[Ar] $3d^6 4s^2$
27	Co	[Ar] $3d^7 4s^2$
28	Ni	[Ar] $3d^8 4s^2$
29	Cu	**[Ar] $3d^{10} 4s^1$**
30	Zn	[Ar] $3d^{10} 4s^2$

It may be noted that the electronic configurations of zinc, cadmium and mercury can be represented by the general formula $(n-1)\, d^{10}\, ns^2$. The *d*-subshell in these elements is completely filled in the ground state as well as in their common oxidation states. Because of this reason, *the elements Zn, Cd and Hg are not regarded as transition elements.*

Scandium (Sc) also does not exhibit the typical properties of *d*-block elements because in Sc(III) which is the only oxidation state of scandium, it has no electron in the *d*-subshell.

$$Sc^{3+} : 1s^2, 2s^2\, 2p^6, 3s^2\, 3p^6$$

SOLVED EXAMPLES

Example 19.1. *Write down the electronic configuration of:*

(*i*) Cr^{3+} (*ii*) Cu^+

(*iii*) Co^{2+} (*iv*) Mn^{2+}

Solution. (*i*) Cr^{3+}: $1s^2, 2s^2 2p^6, 3s^2 3p^6\, 3d^3$

(*ii*) Cu^+: $1s^2, 2s^2 2p^6, 3s^2\, 3p^6\, 3d^{10}$

(*iii*) Co^{2+}: $1s^2, 2s^2 2p^6, 3s^2 3p^6 3d^7$

(*iv*) Mn^{2+}: $1s^2, 2s^2 2p^6, 3s^2 3p^6 3d^5$.

Example 19.2. *Explain why zinc, cadmium and mercury are not regarded as transition elements?*

Solution. The electronic configurations of zinc, cadmium and mercury are represented by the general formula $(n-1)\, d^{10}\, ns^2$. The *d*-orbitals in these elements are completely filled in their ground states as well as in their common oxidation states. Hence, these elements do not exhibit characteristic properties of transition elements. Therefore, they are not regarded as transition elements.

Example 19.3. *Why are Mn^{2+} compounds more stable than Fe^{2+} towards oxidation to their + 3 state?*

Solution. The electronic configuration of Mn^{2+} is $3d^5$, which is stable due to the half filled subshell. Hence, Mn^{2+} is not easily oxidized to Mn^{3+}. On the other hand, Fe^{2+} has electronic configuration $3d^6$. After losing one electron it changes to Fe^{3+} which has a stable $3d^5$ electronic configuration. Hence, Fe^{2+} is relatively easily oxidized to Fe^{3+}

$$\underset{3d^5\,(\text{stable})}{Mn^{2+}} \longrightarrow \underset{3d^4}{Mn^{3+} + e^-}$$

$$\underset{3d^6}{Fe^{2+}} \longrightarrow \underset{3d^5\,(\text{stable})}{Fe^{3+} + e^-}$$

19.3 PROPERTIES OF TRANSITION ELEMENTS

Transition elements having partly filled *d*-orbitals exhibit several interesting properties. For example, they exhibit variable oxidation states, form coloured complexes with different anions and neutral molecules and show paramagnetic behaviour. Transition metals and their compounds also possess catalytic properties. Some of the important properties of transition metals are discussed below:

METALLIC CHARACTER

All transition elements are metallic in nature. Transition elements exhibit typical metallic properties such as lustre, high tensile strength, ductility, malleability and high thermal and electrical conductivity. They show a gradual decrease in electropositive character in moving from left to right. The metallic bonds in transition metals are quite strong. This is due to the greater effective nuclear charge and the large number of valence electrons. ***Due to the presence of strong metallic bonds, the transition metals are hard, possess high densities and high energies of atomisation.*** Atomisation energies of the first transition series are represented graphically in Fig. 19.1.

It may be observed that atomisation energies exhibit the maxima at about the middle of the series. It indicates that interatomic interactions become stronger with increase in half-filled *d*-orbitals.

Most of the transition elements have densities higher than 5 g cm^{-3}, the highest being that of iridium (22.6 g cm^{-3}). Scandium has the least density among transition metals.

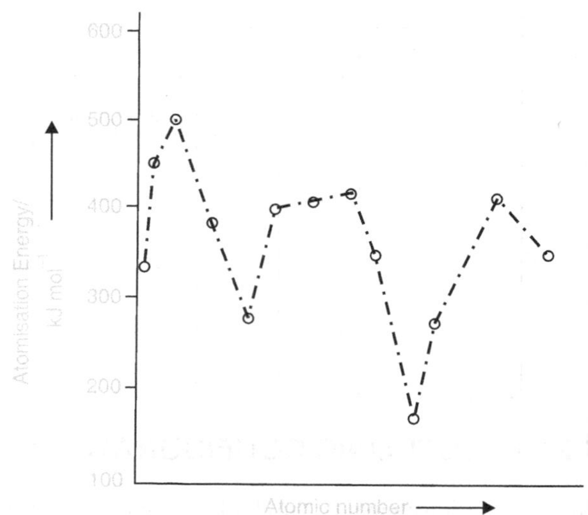

Fig. 19.1. Trends in atomisation energies of first row transition elements.

MELTING AND BOILING POINTS

The melting and boiling points of transition elements are generally very high. *This is due to the strong metallic bond and the presence of half-filled d-orbitals in them. Due to these half-filled orbitals, some covalent bonds also exist between atoms of transition elements.* ***Because of stronger interatomic bonding, transition elements have high melting and boiling points.***

In moving along the period from left to right, the melting points of these metals first increase to a maximum and then decrease regularly towards the end of the period. In any row the melting points of these metals rise to a maximum at d^5.

IONIZATION ENERGIES

The ionization energies of transition elements are higher than those of *s*-block elements but lower than *p*-block elements. In a particular transition series, ionization energy increases gradually as we move from left to right. However, the relative difference of ionization energy values of any two consecutive *d*-block elements of a particular period is much smaller than those of *s*- and *p*-block elements.

Reason. *The increase in ionization energy is primarily due to increase in nuclear charge. As the transition elements involve the gradual filling of (n – 1) d-orbitals, the effect of increase in nuclear charge is partly cancelled by the increase*

Table 19.2. Some Physical Properties of the First Row Transition Elements

Physical Properties	*Scandium* *Sc*	*Titanium* *Ti*	*Vanadium* *V*	*Chromium* *Cr*	*Manganese* *Mn*	*Iron* *Fe*	*Cobalt* *Co*	*Nickel* *Ni*	*Copper* *Cu*	*Zinc* *Zn*
1. Atomic (metallic) radius/pm	161	145	131	129	137	124	125	125	128	133
2. First ionization energy/kJ mol^{-1}	631	656	650	653	717	762	758	736	745	906
3. Melting point/K	1813	1943	2173	2173	1523	1813	1763	1723	1356	692
4. Density/g cm^{-3} at 293 K (°C)	3.0	4.50	5.80	7.19	7.43	7.86	8.90	8.90	8.96	7.14
5. Common oxidation states	3	2, 3, 4	2, 3, 4, 5	2, 3, 6	2, 3, 4, 6, 7	2, 3	2, 3	2	1, 2	2
6. Standard redox potentials/V										
$M^{2+} + 2e^- \rightarrow M$	–	– 1.63	– 1.18	– 0.91	– 1.18	– 0.44	– 0.28	– 0.25	+ 0.34	– 0.76
$M^{3+} + 3e^- \rightarrow M$	– 2.08	– 1.21	– 0.85	– 0.74	– 0.28	– 0.04	+ 0.04	–	–	–

in the screening effect. Consequently, the increase in ionization energy along the period of d-block elements is very small.

The ionization energies of 3*d*-transition series are given in Table 19.2 and graphically represented in Fig. 19.2.

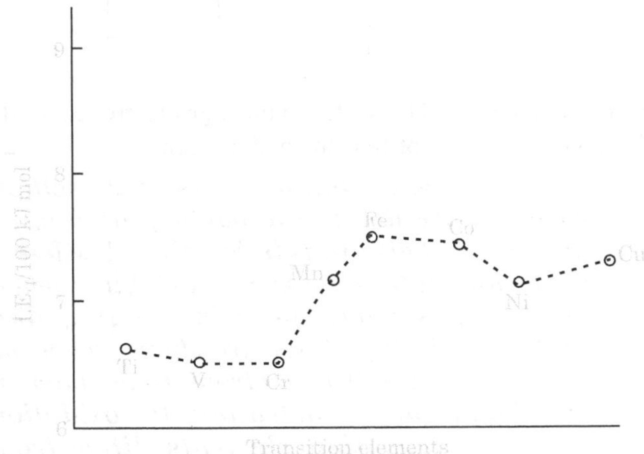

Fig. 19.2. Ionization energies of first transition series.

Significance of Ionization Energy Values

The magnitudes of ionization energies give some indication of the energy required to raise the metal to a particular oxidation state. From the knowledge of values of ionization energies of the metals it is possible to rationalize the relative stabilities of the various oxidation states. For example, the sums of the first two and first four ionization energies of nickel and platinum are given in Table 19.3.

Table 19.3. Ionization Energies of Nickel and Platinum

Element	*Sum of First Two Ionization Energies (kJ mol⁻¹)*	*Sum of First Four Ionization Energies (kJ mol⁻¹)*
Nickel	2.49×10^3	11.29×10^3
Platinum	2.66×10^3	8.36×10^3

Since the sum of the first two ionization energies is less for nickel, Ni(II) compounds are thermodynamically more stable than Pt(II) compounds. On the other hand, Pt(IV) compounds are more stable than Ni(IV) compounds because the sum of the first four ionization energies is less for platinum. K_2PtCl_6 is well known compound of platinum with +4 oxidation state. The corresponding nickel compound does not exist.

In addition to ionization energy, the other factors that determine the stability of a particular state are the atomisation energy of the metal and the lattice energy or the solvation energy.

ATOMIC AND IONIC RADII

The atomic and ionic radii of transition elements are smaller than those of *s*-block elements and larger than those of *p*-block elements. Among the elements of a particular transition series, as the atomic number increases, the atomic radii first decrease till the middle, become almost constant and then increase towards the end of the period. For example, the atomic radii of the first transition series decrease from Sc to Cr, remain almost constant till Cu and then increase towards the end.

Table 19.4. Atomisation Energies, Ionization Energies and Standard Electrode Potentials for Elements of the First Transition Series

Element	*Sc*	*Ti*	*V*	*Cr*	*Mn*	*Fe*	*Co*	*Ni*	*Cu*	*Zn*
Atomic number	*21*	*22*	*23*	*24*	*25*	*26*	*27*	*28*	*29*	*30*
Atomisation energy/kJ mol^{-1}										
	326	473	515	397	281	416	425	430	339	126
Ionization energy/kJ mol^{-1}										
I.E.$_1$	631	656	650	653	717	762	758	736	745	906
I.E.$_2$	1235	1309	1414	1592	1509	1561	1644	1752	1958	1734
I.E.$_3$	2393	2657	2833	2990	3260	2962	3243	3402	3556	3829
Standard electrode potential $E^{\ominus}/V$										
M^{2+}/M	—	− 1.63	− 1.18	− 0.90	− 1.18	− 0.44	−0.28	− 0.25	+ 0.34	− 0.76
M^{3+}/M^{2+}	—	− 0.37	− 0.26	− 0.41	+ 1.57	+ 0.77	+ 1.97	—	—	—

Table 19.5. Atomic and Ionic Radii of Elements of First Transition Series

Element	Sc	Ti	V	Cr	Mn	Fe	Co	Ni	Cu	Zn
Atomic Radii (pm) (metallic)	161	145	131	129	137	124	125	125	128	133
Ionic Radii (pm)										
M^{2+}	—	90	88	84	80	76	74	72	72	74
M^{3+}	81	76	74	69	66	64	63	—	—	—

Reason. *The decrease in size in the beginning is attributed to the increase in nuclear charge. However, the increased nuclear charge is partly cancelled by the increased screening effect of electrons in the d-orbitals of the penultimate shell. When the increased nuclear charge and increased screening effect balance each other, the atomic radii become almost constant. Increase in atomic radii towards the end may be attributed to the electron-electron repulsions. In fact, the pairing of electrons in the d-orbitals occurs after the d^5 configuration. The repulsive interactions between the paired electrons in the d-orbitals become very dominant towards the end of the period and cause the expansion of electron cloud which results in increased atomic size. The ionic radii also follow a similar trend.*

OXIDATION STATES

All transition metals exhibit a great variety of oxidation states. The oxidation states of the first row transition elements are listed in Table 19.6. The less common and unstable oxidation states are given in parentheses. The stability of a particular oxidation state depends upon the nature of the element with which the transition metal forms the compound. The highest oxidation states are found in compounds of fluorine and oxygen. This is due to the high electronegativity values and small size of fluorine and oxygen.

The variable oxidation states of transition elements are due to the participation of ns and (n – 1) d-electrons in bonding. The lower oxidation state is generally, exhibited when ns-electrons participate in bonding and higher oxidation states are shown when ns as well as (n – 1) d-electrons take part in bonding. It may be noted that the oxidation states of transition elements differ from each other by unity whereas the oxidation states of non-transition elements generally differ by two.

Some noteworthy features of the oxidation states of the transition elements are:

1. *In each group, the highest oxidation state increases with increase in atomic number, reaches a maximum in the middle and then starts decreasing.* For example, for the first transition series the maximum oxidation state is shown by manganese.

Table 19.6. Different Oxidation States of Transition Metals

Sc	Ti	V	Cr	Mn	Fe	Co	Ni	Cu	Zn
	(+ 2)	+ 2	+ 2	+ 2	+ 2	+ 2	+ 2	+ 1	+ 2
+ 3	+ 3	+ 3	+ 3	(+ 3)	+ 3	+ 3	(+ 3)	+ 2	
	+ 4	+ 4	(+ 4)	+ 4	(+ 4)	(+ 4)	(+ 4)		
		+ 5	(+ 5)	(+ 6)	(+ 6)				
			+ 6	+ 7					

Note. Less common and unstable oxidation states are given in parentheses.

2. The elements which exhibit the maximum number of oxidation states occur either in or near the middle of the series. For example, in the first transition series manganese exhibits a maximum number of oxidation states (+ 2 to + 7).
3. The elements in the beginning of the series exhibit fewer oxidation states because they have small number of electrons which they can lose or contribute for sharing. The elements at the end of the series exhibit fewer oxidation states because they have too many *d*-electrons and hence have fewer vacant *d*-orbitals which can be involved in bonding.
4. *For the elements of the first transition series* (except scandium) + 2 *oxidation state is the most common oxidation state.* This oxidation state arises due to the loss of the 4*s*-electrons.
5. *The transition elements in the lower oxidation states (+ 2 and + 3) generally form ionic bonds. In the higher oxidation states, the bonds formed are essentially covalent.* For example, in tetraoxochromate(VI) ion (CrO_4^{2-}), the bonds formed between chromium and oxygen are covalent.
6. *Some transition metals also show oxidation state of zero in their compounds.* $[Ni(CO)_4]$ and $[Fe(CO)_5]$ are common examples.

DISPROPORTIONATION OF AN OXIDATION STATE

Reactions in which a single substance undergoes change to produce products, one of which is in higher oxidation state and the other in lower oxidation state are called **disproportionation reactions.**

Disproportionation reaction takes place when a particular oxidation state of the element becomes less stable than the other oxidation state, one lower and one higher. A part of the substance is oxidized and a part is reduced during disproportionation. For example, Mn^{3+} ion undergoes disproportionation in aqueous solution.

$$\overset{III}{2Mn^{3+}(aq)} + 2H_2O(l) \longrightarrow \overset{IV}{MnO_2(s)} + \overset{II}{Mn^{2+}(aq)} + 4H^+(aq)$$

Similarly, Mn(VI) and Cr(V) undergo disproportionation reactions in acidic solutions.

$$\overset{VI}{3MnO_4^{2-}} + 4H^+ \longrightarrow \overset{VII}{2MnO_4^-} + \overset{IV}{MnO_2} + 2H_2O$$

$$\overset{V}{3CrO_4^{3-}} + 8H^+ \longrightarrow \overset{VI}{2CrO_4^{2-}} + \overset{III}{Cr^{3+}} + 4H_2O.$$

COMPLEX FORMATION

In contrast to the representative elements, transition elements form many coordination complexes. Their tendency to form complexes is attributed to the following reasons:

1. *Small size and high charge density of the ions of transition metals.*
2. *Presence of vacant orbitals of appropriate energy which can accept lone pairs of electrons donated by other groups (ligands).*

Some examples of coordination complexes are:

(*i*) $[Ag(NH_3)_2]$ Cl (*ii*) $K_4[Fe(CN)_6]$

(*iii*) $[Cu(NH_3)_4]^{2+}$

Coordination complexes have been discussed in detail in *Section 19.4.*

COLOURED COMPLEXES

The compounds of transition elements are usually coloured both in the solid state and in aqueous solution. The colour of these complexes is due to absorption of some radiation from visible light, which is used in promoting an electron from one of the *d*-orbitals to another. This can be explained as:

The d-orbitals in the transition elements do not have the same energy in their complexes. Under the influence of the ligands attached, the d-orbitals split into two sets of orbitals having slightly different energies. In the transition elements, which have partly filled d-orbitals, the transition of electron can take place from one of the lower d-orbitals to some higher d-orbital within the same subshell. The energy required for this transition falls in the visible region. So when white light falls on these complexes they absorb a particular colour from the radiation for the promotion of electron and the remaining colours are emitted. The colour of the complex is due to this emitted radiation. For example, copper(II) salts are bluish green due to the absorption of red light. Ti^{3+} salts appear purple due to the absorption of yellow light.

The energy difference between the two sets of *d*-orbitals in the central atom of the complex depends on the nature of the ligands and the structure of the complex ion. As a result different complexes of the same metal ion, with different ligands, may have different colours. For example, $[CoCl_4]^{2-}$ is blue in colour whereas $[Co(H_2O)_6]^{2+}$ is pink

$[Fe(H_2O)_6]^{2+}$ is green in colour whereas $[Fe(CN)_6]^{4-}$ is yellow.

Zn^{2+} and Ti^{4+} salts are white because they do not absorb any radiation in the visible region. In these compounds, *d-d* transitions are not possible because in Zn^{2+} all the *d*-orbitals are fully filled whereas in Ti^{4+} all the *d*-orbitals are vacant. The colours of some transition metal ions in aqueous solutions are given in Table 19.7.

Table 19.7. The Colours of Some Transition Metal Ions in Aqueous Solution

Metal Ion	*Configuration*	*Colour*
Sc^{3+}	$3d^0$	Colourless
Ti^{4+}	$3d^0$	Colourless
Ti^{3+}	$3d^1$	Purple
V^{4+}	$3d^1$	Blue
V^{3+}	$3d^2$	Green
Cr^{3+}	$3d^3$	Violet
Mn^{3+}	$3d^4$	Violet
Mn^{2+}	$3d^5$	Pink
Fe^{3+}	$3d^5$	Yellow
Fe^{2+}	$3d^6$	Green
Co^{2+}	$3d^7$	Pink
Ni^{2+}	$3d^8$	Green
Cu^{2+}	$3d^9$	Blue
Zn^{2+}	$3d^{10}$	Colourless
Cd^{2+}	$4d^{10}$	Colourless

PARAMAGNETISM

The substances, which contain some species (atoms, ions or molecules) with unpaired electrons in their orbitals, behave as *paramagnetic* substances. Such substances are weakly attracted by a magnetic field. On the other hand, the substances whose constituent particles do not contain any unpaired electrons are repelled by a magnetic field and are called *diamagnetic.*

The transition metal ions generally contain one or more unpaired electrons in them and hence their complexes are generally paramagnetic. The paramagnetic character increases with an increase in number of unpaired electrons.

CATALYTIC PROPERTIES

Many transition metals and their compounds are known to act as catalysts. For example, finely divided iron acts as catalyst in the manufacture of ammonia by the **Haber Process**, vanadium pentoxide (V_2O_5) or platinum act as catalyst for the oxidation of SO_2 to SO_3 in the **Contact Process**, ferrous sulphate and hydrogen peroxide (Fenton's reagent) are used for the oxidation of alcohols to aldehydes.

The catalytic activity of transition metals is attributed to the following reasons:

1. *Because of their variable oxidation states, transition metals sometimes form unstable intermediate compounds and provide a new path with lower activation energy for a reaction.*

For example, V_2O_5 catalyses the oxidation of SO_2 to SO_3.

$$2SO_2 + O_2 \xrightarrow{V_2O_5} 2SO_3$$

The catalytic action of V_2O_5 can be understood as follows:

During the conversion of SO_2 to SO_3, V_2O_5 adsorbs SO_2 molecule on its surface and gives oxygen to it to form SO_3 and V_2O_4. V_2O_4 then reacts with oxygen to form V_2O_5.

$$\mathbf{V_2O_5 + SO_2 \longrightarrow SO_3 + \underset{\text{Divanadium tetroxide}}{V_2O_4}}$$

$$\mathbf{V_2O_4 + \tfrac{1}{2}O_2 \longrightarrow V_2O_5}$$

Similarly, iron(III) catalyses the reaction between iodide and persulphate ions.

$$2I^- + S_2O_8^{2-} \xrightarrow{Fe(III)} I_2 + 2SO_4^{2-}$$

The catalytic action of iron(III) in this reaction is explained as follows:

$$\mathbf{2Fe^{3+} + 2I^- \longrightarrow 2Fe^{2+} + I_2}$$

$$\mathbf{2Fe^{2+} + S_2O_8^{2-} \longrightarrow 2Fe^{3+} + 2SO_4^{2-}}$$

2. *In some cases, transition metals provide a suitable surface for a reaction to take place. The reactants are adsorbed on the surface of the catalyst where the reaction occurs.*

Adsorption results in increased concentration of reactants at the surface and also weakens the bonds between the atoms in the reactant molecules.

Catalytic Properties of Transition Elements and their Compounds

Some first row metals and their compounds used as catalysts are given below:

Catalyst	*Process Catalysed*
$TiCl_4$	Used as the Ziegler-Natta catalyst in the **polymerization of ethene** and propene
V_2O_5	Used as catalyst during conversion of SO_2 to SO_3 in the **Contact process** for the manufacture of H_2SO_4. $2SO_2 + O_2 \rightleftharpoons 2SO_3$
MnO_2	Used as catalyst to decompose $KClO_3$ to produce oxygen gas. $2KClO_3 \longrightarrow 2KCl + 3O_2$
Fe	Iron in the presence of a promotor act as catalyst in the **Haber process** for the manufacture of ammonia $N_2 + 3H_2 \rightleftharpoons 2NH_3$
$FeCl_3$	Used as catalyst in the production of CCl_4 from CS_2 and Cl_2
$Co_2(CO)_8$	**Oxo process** for the conversion of alkenes to alkanals and alkanones
Ni	**Hydrogenation** of vegetable oils
$CuCl_2$	Used as catalyst in the manufacture of chlorine from HCl (**Deacon process**).

The catalytic activity of transition metal compounds can be demonstrated by the following activity.

ACTIVITY 19.1 **Demonstration of catalytic properties of transition elements and their compounds**

- Take a 250 cm^3 conical flask and add 25 cm^3 of 0.1 M sodium thiosulphate solution to it.
- Now add 25 cm^3 of 0.1 M iron(III) trioxonitrate(V) solution to the flask and start the stopwatch.
- Shake the contents of the conical flask and place it on the tile with a cross mark as shown in Fig. 19.3.

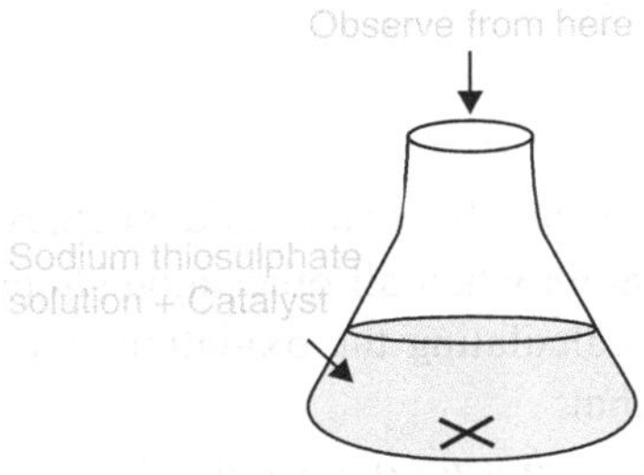

Fig. 19.3

- Go on observing from top downwards in the flask and note down the time from the stopwatch when the cross mark just becomes invisible.
- Repeat the experiment with various catalysts and complete the observation table given below:

Catalyst	No catalyst	$NiSO_4$	$CuSO_4$	$FeSO_4$	$CoSO_4$
Time (s)	—	—	—	—	—

Conclude which catalyst is most effective.

SOLVED EXAMPLES

Example 19.4. *Explain why*

(*i*) *Sc^{3+} salts are diamagnetic?*

(*ii*) *Ti^{3+} salts are coloured whereas Ti^{4+} salts are white?*

Solution. (*i*) Sc^{3+} : $1s^2, 2s^2, 2p^6, 3s^2, 3p^6$

In Sc^{3+}, all the electrons are paired. Hence, Sc^{3+} salts are diamagnetic.

(*ii*) Ti^{3+} has one electron in the 3*d*-subshell whereas Ti^{4+} has no electron in the 3*d*-subshell. Therefore, in Ti^{3+} ($3d^1$) *d-d* transitions are possible whereas in Ti^{4+} ($3d^0$) *d-d* transitions are not possible. Hence, Ti^{3+} salts are coloured whereas Ti^{4+} salts are white.

Example 19.5. *Explain why*

(*i*) *Transitions metals have higher density than alkaline earth metals?*

(*ii*) *Transition metals are less electropositive than alkaline earth metals?*

(*iii*) *Transition elements exhibit higher energies of atomisation?*

Solution. (*i*) Transition metals have smaller atomic size and stronger interparticle attractive forces (metallic bond) than alkaline-earth metals and hence have higher density than alkaline earth metals.

(*ii*) Transition metals due to their smaller atomic size and greater effective nuclear charge, have higher ionization energy than alkaline earth metals and hence are less electropositive.

(*iii*) Transition elements have very strong interatomic bonds due to their small size and also due to the presence of a large number of unpaired electrons in their atoms. As a result, they have very high energies of atomisation.

Example 19.6. *Why is the highest oxidation state of a metal exhibited in its oxide or fluoride only?*

Solution. Due to its very high electronegativity and small size oxygen and fluorine can oxidize a metal to its highest oxidation state.

Example 19.7. *Explain why*

Hydrated copper(II) tetraoxosulphate(VI) is blue whereas anhydrous copper(II) tetraoxosulphate(VI) is white?

Solution. In the presence of ligands (water molecules), the *d*-orbitals of Cu(II) split into two sets having slightly different energies. Hence, *d-d* transitions become possible in hydrated copper(II) tetraoxosulphate(VI) and it acquires a blue colour due to absorption of a part of visible light. On the other hand, in the case of anhydrous copper(II) tetraoxosulphate(VI) all the *d*-orbitals have the same energy in the Cu(II) ion and hence *d-d* transitions are not possible. So, the entire white light is reflected and it appears white.

Example 19.8. *Name a transition element which does not exhibit variable oxidation states.*

Solution. Scandium (Z = 21) does not exhibit a variable oxidation states. It exhibits only +3 oxidation state.

Example 19.9. *Which element of the first transition series exhibits the largest number of oxidation states?*

Solution. Manganese (Z = 25) exhibits the maximum number of oxidation states (from +2 to +7) among the elements of the first transition series.

Example 19.10. *The $E^\ominus$ (M^{2+}/M) value for copper is positive (+0.34V). What is possibily the reason for this?*

Solution. The positive value of $E^\ominus(M^{2+}/M)$ for copper is due to its high energy of atomization and high values of ionization energies.

Example 19.11. *How would you account for the following:*

Of the d^4 species, Cr^{2+} is strongly reducing while manganese (III) is strongly oxidizing.

Solution. For chromium the + III state is more stable as compared to the +II state. Therefore, Cr^{2+} readily changes into Cr^{3+} and behaves as a strong reducing agent. On the other hand, for manganese, the + II state is more stable than the + III state. Hence, Mn^{3+} readily changes into Mn^{2+} ($3d^5$) by gaining an electron and behaves as a strong oxidising agent.

19.4 BONDING IN COMPLEX COMPOUNDS

*A **coordination complex** contains a central metal atom or cation which is surrounded by a suitable number of anions or neutral molecules (called ligands).*

Ligands are bonded to the central metal atom or ion through coordinate bonds.

A complex may be cationic, anionic or a neutral species.

A complex retains its identity in solution as well as in the solid state. The number of electron pairs accepted by the metal atom in a complex is called the **coordination number** of the metal atom.

A complex may be represented by a general formula

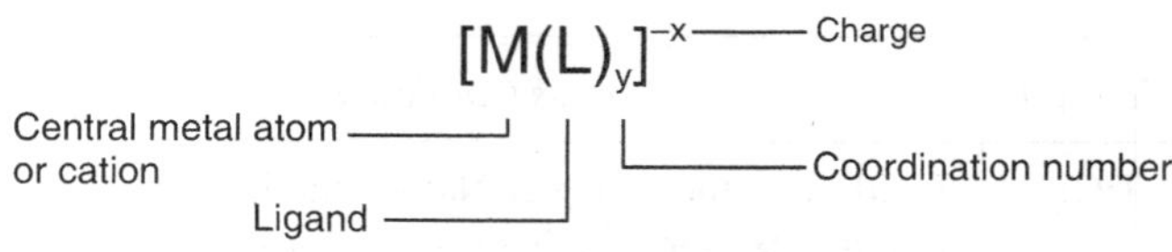

- The **central metal atom or cation** is an electron-pair acceptor. It must have vacant orbitals which can accommodate lone-pairs of electrons donated by ligands.
- A **ligand** is an electron-pair donor. It is a molecule or anion bonded to the central metal atom of a complex ion by means of a **coordinate or dative bond.** Some common ligands are CO, NH_3, H_2O, Cl^-, CN^-.
- A ligand which is attached to the central atom only through one atom is called **monodentate ligand.** A ligand which is attached to the central atom through more than one atom is called **multidentate ligand.** For example, ethylene diamine is a bidentate ligand.

$$\begin{array}{l} CH_2-\ddot{N}H_2 \\ | \\ CH_2-\ddot{N}H_2 \end{array}$$

Ethylenediamine

- Charges on complex ions are delocalized over the whole ion. Positive charge gives rise to a **cationic complex**. Negative charge gives rise to an **anionic complex**. Some complexes carry no charge and hence are **neutral**.
- The **coordination number** is the number of ligands attached to the central atom. It can also be regarded as *the number of coordinate bonds formed between the ligands and the central atom.*

Some examples of coordination complexes are

$[Cu(NH_3)_4]^{2+}$, $[Ag(NH_3)_2]^+$, $[Fe(CO)_5]$, $[Co(H_2O)_6]^{2+}$, $[Ni(CO)_4]$, $[Ni(CN)_4]^{2-}$, $[Fe(CN)_6]^{3-}$, $[CoCl_4]^{2-}$.

OXIDATION NUMBER OR OXIDATION STATE OF THE CENTRAL METAL ION

The oxidation number is the residual charge that appears on the central atom when all other atoms/ions are removed from it. While calculating the oxidation state of an atom it must be noted that:

(*a*) *the sum of charges of the constituents is equal to zero if the complex is neutral.*

(*b*) *for an anionic or cationic complex, the sum of charges of the constituents is equal to the charge on the complex.*

The above points are illustrated by taking the following examples:

(*i*) In the complex ion $[Co(NH_3)_2Cl_4]^{2-}$ the charge on the complex is – 2. The oxidation number of the central metal ion may be calculated as:

$$\begin{array}{ccc} [Co & (NH_3)_2 & Cl_4]^{2-} \\ \downarrow & \downarrow & \downarrow \\ x & 2(0) & 4(-1) \end{array}$$

$$x + 2(0) + 4(-1) = -2$$

$$x = +2$$

(*ii*) In the complex, $[Fe(CN)_6]^{4-}$, the charge on the complex ion is – 4. The oxidation number of the central metal may be calculated as:

$$\begin{array}{cc} [Fe & (CN)_6]^{4-} \\ \downarrow & \downarrow \\ x & 6(-1) \end{array}$$

$$x + 6(-1) = -4$$

$$x = +2$$

SOLVED EXAMPLE

Example 19.12. *What is the oxidation state of the metal in each of the following complexes:*

(*i*) $[Fe(CN)_6]^{3-}$ (*ii*) $[Cu(NH_3)_4]^{2+}$

(*iii*) $[Cr(CO)_6]$ (*iv*) $[Co(H_2O)_6]\,Cl_2$

(*v*) $[Ni(CN)_4]^{2-}$.

Solution. (*i*) $[Fe(CN)_6]^{3-}$

Let the oxidation number of Fe be x

$$x + 6(-1) = -3$$
$$x = +3$$

(*ii*) $[Cu(NH_3)_4]^{2+}$

$$x + 4(0) = +2$$
$$x = +2$$

(*iii*) $[Cr(CO)_6]$

$$x + 6(0) = 0$$
$$x = 0$$

(*iv*) $[Co(H_2O)_6]\,Cl_2$

$$x + 6(0) + 2(-1) = 0$$
$$x = +2$$

(*v*) $[Ni\,(CN)_4]^{2-}$

$$x + 4(-1) = -2$$
$$x - 4 = -2$$
$$x = +2.$$

19.5 NOMENCLATURE OF COMPLEX COMPOUNDS

1. *If the compound is ionic, the name of the cation is mentioned first followed by the name of the anion.*

   For the *non-ionic complexes the name of complex is written as one word.*

2. *In naming the complex, the names of the ligands are written first followed by the name of the central ion. The oxidation number of the central metal atom is expressed by Roman numeral in parentheses just after the name of the central metal atom.*

3. **Naming of ligands.**

(*a*) *For the ligands carrying a negative charge the name has a characteristic ending "***o***".* A few example are:

Cl^- – chloro

F^- – fluoro

I^- – iodo

CN^- – cyano

NO_3^- – trioxonitrato(V)

SO_4^{2-} – tetraoxosulphato(VI)

(*b*) *For the ligands carrying a positive charge, the name of the ligand has a characteristic ending of "***ium***".* For example,

NO_2^+ – *nitronium*

(*c*) *For neutral ligands no characteristic ending is used.* For example,

The following examples are, however, exceptions:

H_2O – aqua, NO – nitrosyl,

NH_3 – ammine, CO – carbonyl.

4. Order of naming ligands. According to *old conventions,* the names of more than one type of ligands surrounding the central atom are written in the order, (*i*) *Negative* (*ii*) *Neutral* (*iii*) *Positive,* without separating them with a hyphen.

According to the latest **IUPAC conventions** ***the names of ligands surrounding the central metal atom are written in alphabetical order of preference irrespective of whether they are negative or neutral.***

In the present text the **new** *IUPAC system has been followed.*

5. The number of each kind of ligands are specified by the prefixes *di, tri, tetra,* etc. If the ligand is complex, such as ethylenediamine (en), and is repeated two, three or four times then the words *bis, tris, tetrakis*, etc., are used, followed by the name of the ligand in parentheses.

6. Ending of name of the central atom. *Names of complex cations and neutral molecules have no distinguishing termination but in the case of anionic complexes the suffix* "**ate**" *is attached to the name of the central atom.* For example,

$K_4[Ni(CN)_4]$ is named as

Potassium tetracyanonickelate(0).

7. The oxidation state of the central metal atom or ion is designated by a Roman numeral (I, II, III, etc.) in the bracket at the end of the name of the complex without a gap between the two.

Some **examples** illustrating the above rules are given below:

1. $K_2[PtCl_6]$
   potassium hexachloroplatinate(IV)
2. $Co(NH_3)_5Cl]\ Cl_2$
   pentamminechlorocobalt(III) chloride
3. $[Ni(NH_3)_4\ Cl_2]$
   tetraamminedichloronickel(II)
4. $[Co(NH_3)_6]^{3+}$
   hexaamminecobalt(III) ion
5. $K_4[Fe(CN)_6]$
   potassium hexacyanoferrate(II)
6. $[Co(en)_3]\ Cl_3$
   tris (ethylenediamine) cobalt(III) chloride
7. $Ni(CO)_4$
   tetracarbonylnickel (0)
8. $K_4[Cr(CN)_6]$
   potassium hexacyanochromate(II)
9. $[CrCl_2(H_2O)_4]^+$
   tetraaquadichlorochromium(III) ion
10. $[Cu\ (NH_3)_4]^{2+}$
    tetraamminecopper(II) ion

19.6 SHAPES OF COMPLEX IONS

- A complex with coordination number 4 may be **tetrahedral** or **square planar** in shape, depending upon the orbitals the metal atom or ion uses for bonding.
- A complex with coordination number 6 is likely to be **octahedral** in shape.
- A complex with coordination number 5 may be **trigonal bipyramidal** or **square pyramidal.** Some shapes of complex ions are shown in Fig. 19.4

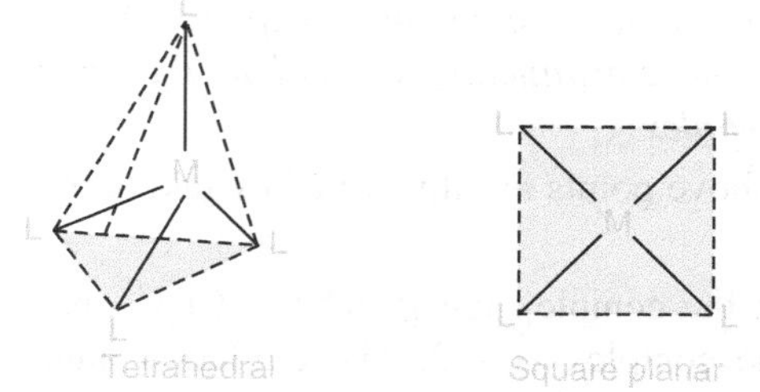

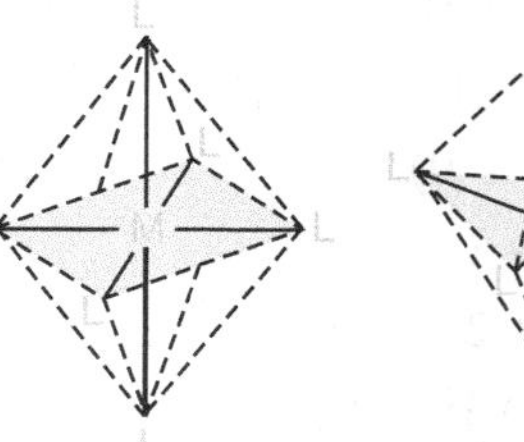

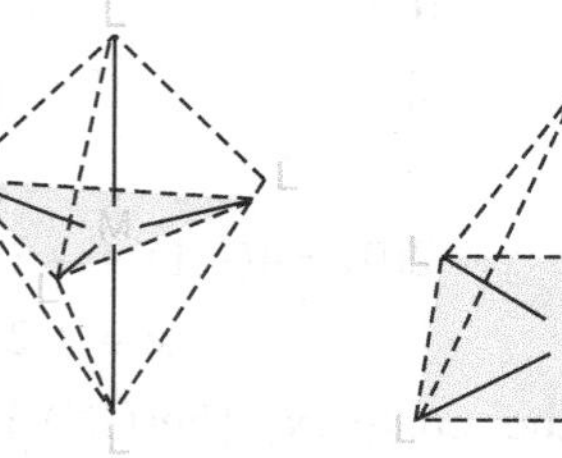

Fig. 19.4. Shapes of Complexes

SOLVED EXAMPLE

Example 19.13. *Draw the shapes of the following complexes:*

(*i*) $[Fe(CN)_6]^{3-}$ (*ii*) $[Cu(NH_3)_4]^{2+}$

(*iii*) $[Ag(NH_3)_2]^+$ (*iv*) $[Cu(CN)_4]^{2-}$.

Solution. (*i*) $Fe(CN)_6]^{3-}$ is octahedral in shape.

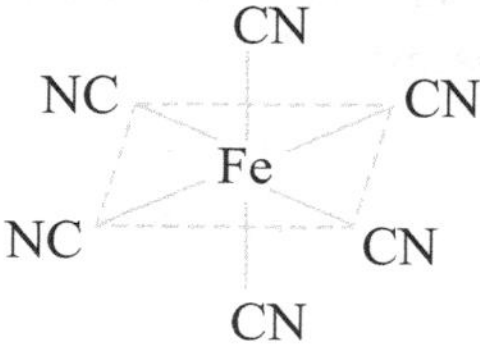

(*ii*) $[Cu(NH_3)_4]^{2+}$ is square planar.

H_3N NH_3
Cu
H_3N NH_3

(*iii*) $[Ag(NH_3)_2]^+$ is linear in shape.

$$H_3N \rightarrow Ag \leftarrow NH_3$$

(*iv*) $[Cu(CN)_4]^{2-}$ is tetrahedral in shape.

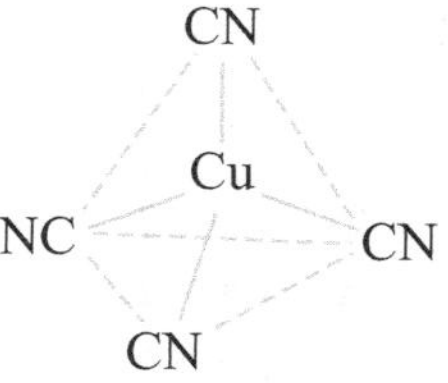

19.7 SIMILARITIES AND DIFFERENCES BETWEEN TRANSITION METALS AND REPRESENTATIVE METALS

- Transition metals have stronger interparticle forces and hence are harder than representative metals. They have higher densities and higher melting points.
- Transition metals exhibit variable oxidation states whereas representative metals rarely show variable oxidation states.
- Transition metals form a large number of coordination complexes. Most of these complexes are coloured and paramagnetic. Representative metals have less tendency to form complexes. These complexes are generally white.
- Transition metals are less reactive than representative metals. This is due to their less electropositive nature.
- A comparison of some chemical properties of transition metals and some representative metals is given in Table 19.8.

Table 19.8. Comparison of Some Reactions of Elements of First Transition Series with that of *s*-Block Elements

Metal	*Reaction with acids*	*Reaction with water*	*Reaction with air*
Ti	$2Ti(s) + \underset{\text{hot and conc.}}{6HCl(aq)} \longrightarrow 2TiCl_3(aq) + 3H_2(g)$ $\underset{\text{hot and conc.}}{Ti(s) + HNO_3(aq)} \longrightarrow \underset{\text{hydrated titanium(IV) oxide}}{TiO_2 . xH_2O}$ No reaction with HCl and HNO_3 in cold. $2Ti(s) + \underset{\text{(from dil } H_2SO_4)}{6H^+(aq)} \longrightarrow 2Ti^{3+}(aq) + 3H_2(aq)$	$Ti(s) + 2H_2O(g) \xrightarrow[\text{heating}]{\text{strong}} TiO_2(s) + 2H_2(g)$	$Ti(s) \xrightarrow[\text{heating in air}]{\text{strong}} TiO_2(s) + TiN(s)$
V	No reaction with HCl at room temperature. Dissolves in hot, concentrated HNO_3, H_2SO_4 and aqua regia.	No reaction at room temperature	No reaction at room temperature.
Cr	$Cr(s) + 2HCl(aq) \xrightarrow[\text{slow}]{\text{very}} CrCl_2(aq) + H_2(g)$ It also dissolves in cold H_2SO_4. It is rendered passive by the oxide formed when it reacts with conc. HNO_3 or aqua regia.	$\underset{\text{red-hot}}{2Cr(s)} + 3H_2O(g) \longrightarrow Cr_2O_3(s) + 3H_2(g)$	Metal forms green coloured Cr_2O_3 on strong heating
Mn	$Mn(s) + 2HCl(aq) \longrightarrow MnCl_2(aq) + H_2(g)$ It reacts readily with dil. H_2SO_4 also.	It reacts slowly with water liberating H_2	Reacts at high temperature forming Mn_3O_4 and Mn_3N_2
Fe	$Fe(s) + \underset{(HCl \text{ or } H_2SO_4)}{2H^+(aq)} \longrightarrow Fe^{2+}(aq) + H_2(g)$ $2Fe(s) + \underset{\text{(From dil.} HNO_3)}{6H^+(aq)} \longrightarrow 2Fe^{3+} + 3H_2$ $2Fe(s) + \underset{\text{conc.}}{6H_2SO_4(aq)} \longrightarrow Fe_2(SO_4)_3(aq) + 6H_2O(l) + 3SO_2(g)$ It is also attacked by dil. HNO_3 Conc. HNO_3 renders it passive.	$3Fe(s) + \underset{\text{steam}}{4H_2O(g)} \rightleftharpoons Fe_3O_4(s) + 4H_2(g)$	It is slowly attacked by moist air forming hydrated reddish brown iron (III) oxide, $Fe_2O_3.xH_2O$ or iron rust.
Co	$Co(s) + 2HCl(aq) \xrightarrow[\text{slow}]{\text{very}} CoCl_2(aq) + H_2(g)$ It reacts with dil. H_2SO_4 at room temperature also. Conc. HNO_3 renders it passive.	No reaction at room temperature	No reaction at room temperature. On strong heating gives Co_3O_4

Ni	$Ni(s) + 2H^+(aq) \longrightarrow Ni^{2+}(aq) + H_2(g)$ From dil HCl or dil HNO_3 Conc. HNO_3 renders it passive	No reaction	No reaction
Cu	$3Cu(s) + 8HNO_3(aq) \longrightarrow 3Cu(NO_3)_2(aq) + 4H_2O(l) + 2NO(g)$ dilute $Cu(s) + 4HNO_3(aq) \longrightarrow Cu(NO_3)_2(aq) + 2H_2O(l) + 2NO_2(g)$ conc. No reaction with dil. HCl or H_2SO_4 $Cu(s) + 2H_2SO_4(aq) \xrightarrow{\text{warm}} CuSO_4(aq) + 2H_2O(l) + SO_2(g)$ conc.	No reaction with water	No reaction in pure air. In moist air it gradually becomes coated with $CuCO_3$ and $Cu(OH)_2$ which is green and is called verdigris.
Zn	$Zn(s) + 2H^+(aq) \longrightarrow Zn_2\ (aq) + H_2(g)$ Very slow due to a protective layer of oxide. $Zn(s) + 2H_2SO_4(aq) \longrightarrow ZnSO_4(aq) + 2H_2O(l) + SO_2(g)$ (hot and conc.)	No reaction with cold water $Zn(s) + H_2O(g) \longrightarrow ZnO(s) + H_2(g)$	On prolonged exposure zinc gradually becomes coated with a layer of zinc oxide and basic zinc carbonate which prevents further reaction.
s-block elements (Li, Na, Be, Mg)	Lithium and sodium react vigorously with acids $Mg(s) + 2H^+(aq) \longrightarrow Mg^{2+}(aq) + H_2(g)$ Beryllium reacts only slowly.	$2Li(s)\ 2H_2O(l) \xrightarrow[\text{temperature}]{\text{room}} 2LiOH(aq) + H_2\ (g)$ Less vigorous reaction than that of sodium. $2Na(s) + H_2O(l) \xrightarrow[\text{temperature}]{\text{room}} 2NaOH(aq) + H_2(g)$ Beryllium hardly reacts with cold water or steam. $Mg(s) + H_2O(g) \xrightarrow[\text{fast}]{\text{very}} MgO(s) + H_2(g)$ Very slow reaction with cold water.	Lithium and sodium readily tranish when they come into contact with atmospheric oxygen. Tarnishing of beryllium and magnesium is comparatively slow. The transition metals tarnish far less than the *s*-block metals in air.

SUMMARY

- **Transition Elements.** Elements having incomplete *d*-orbitals in their atoms or in their simple ions.
- All transition elements are metals.
- Transition elements have high melting and boiling points due to stronger interatomic bonding.
- The atomic and ionic radii of transition elements are smaller than those of *s*-block elements but larger than those of *p*-block elements.
- **Variable Oxidation States.** Transition elements exhibit variable oxidation states due to participation of $(n - 1)$ *d*-electrons in bonding.
- Transition metals and their compounds act as catalysts in many reactions. This is due to their variable oxidation states and ability to adsorb strongly.
- **Coloured Complexes.** Transition elements form coloured complexes. Colour of the complexes is due to absorption of light energy due to *d-d* transitions in partially filled *d*-orbitals.
- **Paramagnetic Nature.** Most of the compounds of transition elements are paramagnetic due to the presence of unpaired electrons in them.
- **Zinc, Cadmium, Mercury** do not show characteristic properties of transition elements because they do not contain partially filled *d*-orbitals in their atoms or common oxidation states.
- A coordination complex may be represented as $[M(L)_y]^{x-}$
- A coordination complex may be neutral or charged.
- In a coordination complex, ligands are bonded to the central metal ion through coordinate or dative bonds.

EVALUATE YOURSELF

I. Objective Type Questions

Select the most appropriate choice from the options given as (a), (b), (c) and (d) after each question:

1. The general outer electronic configuration of transition elements is
   (*a*) $(n-1)\,d^{1-10}\,ns^1$ (*b*) $(n-1)\,d^{10}\,ns^2$
   (*c*) $(n-1)\,d^{1-10}\,ns^{1-2}$ (*d*) $(n-1)\,d^5\,ns^1$.
2. Which one of the following does not show variable oxidation states?
   (*a*) Iron (*b*) Copper
   (*c*) Zinc (*d*) Manganese.
3. Which of the following is not a condition for complex salt formation?
   (*a*) Small size (*b*) Higher nuclear charge
   (*c*) Availability of vacant *d*-orbitals
   (*d*) Variable oxidation states.
4. The outer electronic configuration of chromium is
   (*a*) $4s^1\,3d^5$ (*b*) $4s^2\,3d^4$
   (*c*) $4s^0\,3d^6$ (*d*) $4s^2\,3d^5$.
5. Which ion gives coloured solution?
   (*a*) Cu^+ (*b*) Zn^{2+}
   (*c*) Ag^+ (*d*) Fe^{2+}.
6. Which one of the following metals is used as a catalyst in the Haber's process?
   (*a*) Tungsten (*b*) Molybdenum
   (*c*) Chromium (*d*) Iron.
7. Which of the following is likely to form white salts?
   (*a*) Cu^{2+} (*b*) Sc^{3+}
   (*c*) Ti^{3+} (*d*) Fe^{3+}.
8. Which of the following has the maximum number of unpaired electrons?
   (*a*) Fe^{2+} (*b*) Cr^{3+}
   (*c*) Fe^{3+} (*d*) Co^{2+}.
9. In the first transition series, the element with the highest melting point is
   (*a*) Mn (*b*) Fe
   (*c*) Cr (*d*) Cu.
10. Ziegler-Natta catalyst is $TiCl_4$ dissolved in
    (*a*) terithylaluminium (*b*) ether
    (*c*) water (*d*) ammonia.
11. How many ions are produced from $[Co(NH_3)_5Cl]Cl_2$ in solution
    (*a*) 3 (*b*) 4
    (*c*) 2 (*d*) 5.
12. The oxidation number of cobalt in $K[Co(CO)_4]$ is
    (*a*) + 1 (*b*) + 3
    (*c*) − 1 (*d*) − 3.

II. Fill in the Blanks

13. Complete the following sentences by supplying appropriate words:
    (*i*) V_2O_5 is used as catalyst in ...... process for the manufacture of H_2SO_4.
    (*ii*) The only oxidation state exhibited by scandium is ....
    (*iii*) A ligand which is attached to the central atom through two atoms is called ...... ligand.
    (*iv*) All zinc salts are ...... in colour.
    (*v*) IUPAC name of the complex $[Ni(CN)_4]^{2-}$ is ....
    (*vi*) Oxidation number of Fe in $[Fe(CN)_6]^{4-}$ is ....

III. Discussion Questions

14. Name the first and the last element in the first transition series.
15. Name the *d*-block elements which do not have partially filled *d*-orbitals in their atoms or in their simple ions.
16. On what ground can you say that scandium (Z = 21) is a transition element but zinc (Z = 30) is not.
17. What is the general electronic configuration of transition elements? Write down the electronic configuration of the following ions:
    (*i*) Mn^{2+} (*ii*) Fe^{3+}
    (*iii*) Cr^{3+} (*iv*) Ni^{2+}.
18. Which three elements belonging to the *d*-block are usually not considered as transition elements?
19. Why are the *d*-block elements called transition elements?
20. Explain giving reason:
    (*i*) Transition metals and many of their compounds show paramagnetic behaviour
    (*ii*) The atomisation energies of the transition metals are high.
21. Explain why:
    (*i*) The transition metals generally form coloured compounds.
    (*ii*) Transition metals and their many compounds act as good catalyst.
22. Describe briefly the following properties of transition metals
    (*i*) metallic character
    (*ii*) complex formation.
23. How is it that several transition metals act as catalysts? Give two examples of reactions catalysed by them.
24. Why are Ni(II) compounds thermodynamically more stable than Pt(II) compounds while Pt(IV) compounds are relatively more stable than Ni(IV) compounds?
25. Account for the following giving reasons
    (*i*) Zn^{2+} salts are white while Cu^{2+} salts are blue

(*ii*) Zn, Cd and Hg are normally not considered transition metals

(*iii*) Transition elements have very high heats of atomisation, high densities and high melting points.

26. Explain giving reason:

(*i*) Copper(I) is diamagnetic whereas copper(II) is paramagnetic. Explain.

(*ii*) Mn(II) ion shows maximum paramagnetic character among the bivalent ions of the first transition series.

27. Decide giving reason which one of the following pairs exhibits the property indicated.

(*i*) Sc^{3+} or Cr^{3+} exhibits paramagnetism

(*ii*) V or Mn exhibits more number of oxidation states.

(Atomic numbers: Sc = 21, Cr = 24, V = 23, Mn = 25)

28. Account for the following:

Of the d^4 species Cr(II) is strongly reducing but Mn(III) is strongly oxidizing.

(Atomic numbers: Cr = 24, Mn = 25, Co = 27)

29. Give reasons for the following features of transition metal chemistry:

(*i*) Most of the transition metal ions are coloured in solution

(*ii*) Transition metals are well known to form complex compounds.

30. Of the ions Co^{2+}, Sc^{3+} and Cr^{3+} which ones will give coloured aqueous solutions and how will each of them respond to a magnetic field and why? (Atomic numbers: Co = 27, Sc = 21, Cr = 24)

31. What is a ligand? Give two examples each of negative and neutral ligands.

32. What is a coordination complex? What is the nature of bonding in coordination complexes? Give two examples of coordiantion complexes of the first transition series.

33. What is the oxidation number of iron in the following complexes?

(*i*) $[Fe(CO)_5]$ (*ii*) $[Fe(C_2O_4)_3]^{3-}$

(*iii*) $K_3[Fe(CN)_6]$.

34. Compare the reactions of chromium and sodium with

(*i*) air (*ii*) water.

35. Compare the following properties of transition metals with *s*-block metals:

(*i*) hardness (*ii*) density

(*iii*) tendency to form complexes.

Answers

I. Objective Type Questions

1. (*c*)	**2.** (*c*)	**3.** (*d*)	**4.** (*a*)
5. (*d*)	**6.** (*d*)	**7.** (*b*)	**8.** (*c*)
9. (*c*)	**10.** (*a*)	**11.** (*a*)	**12.** (*c*).

II. Fill in the Blanks

13. (*i*) Contact (*ii*) +3

(*iii*) bidentate (*iv*) white

(*v*) tetracyanonickelate(II) ion

(*vi*) +2.

SHS 3

SECTION 3

CHEMICAL KINETICS AND EQUILIBRIUM

General Objectives

After studying this section, the student should be able to:

1. understand that chemical reactions proceed at different rates depending on the conditions set for the reaction to occur.
2. design simple experiments to determine the effects of the various factors on rate of reactions and systems in equilibrium.
3. show awareness of the application of the principles of rate of reaction and equilibrium systems in industrial processes.
4. recognize that there is a balance of opposing reactions in chemical equilibrium systems and that chemical changes eventually attain equilibrium.
5. perform simple calculations based on rate and equilibrium laws.

In this section, we shall study

SECTION-3

UNIT-20

RATE OF REACTIONS

Specific Objectives

After studying this unit, the student should be able to:

- explain rate of reaction
- monitor the speed of a chemical reaction using a simple experiment
- describe the factors that affect the rate of a chemical reaction
- demonstrate experiment to show how changes in temperature affect the rate of a reaction
- analyse and interpret simple graphs on rates of reactions
- describe the collision theory of reaction rate
- Relate the collision theory to reaction rate
- identify the role of activation energy in chemical reactions
- deduce the rate law from a given data
- draw and analyze graphical representation for zero, first and 2nd order reactions
- deduce half-life from first order reaction
- describe the effect of temperature and catalyst on the rate constant
- deduce the rate law from an experiment
- identify the rate determining step of a multi-step reaction.

Chemistry, by its very nature, is concerned with change. Substances with well defined properties are converted by chemical reactions into other substances with different properties. For any chemical reaction, chemists try to find out

(*a*) *the feasibility of a chemical reaction*

(*b*) *the extent to which a reaction will proceed*

(*c*) *the speed of a reaction i.e., time taken for a reaction to reach equilibrium.*

It is important to know the rate and the factors controlling the rate of a chemical reaction. For example,

(*i*) Which parameters determine how rapidly food gets spoiled?

(*ii*) How do you design a rapidly setting material for dental filling?

(*iii*) What controls the rate at which fuel burns in an auto engine?

During the course of our studies we come across a wide variety of chemical reactions having different rates. In general, the various reactions can be categorised into three types depending upon the reaction rates.

1. Very Fast Reactions (Type 1). This type consists of reactions which occur almost instantaneously. For example, the precipitation of AgCl occurs instantaneously by addition of an aqueous solution of sodium chloride to an aqueous solution of silver nitrate.

$$AgNO_3(aq) + NaCl(aq) \longrightarrow AgCl(s) + NaNO_3(aq)$$

The rates of such reactions (ionic reactions) cannot be determined easily.

2. Very Slow Reactions (Type 2). This type comprises reactions which occur at a very slow rate. These reactions may require months or even years for their completion. For example, rusting of iron.

The rates of such reactions are hardly of any physical importance.

3. Moderately Slow Reactions. This type refers to reactions in between the very fast (*type 1*) and very slow (*type 2*) reactions. These reactions proceed at moderate speeds which can be easily measured. Inversion of cane sugar and hydrolysis of starch are common examples of this type of reactions.

$$\underset{\text{Cane sugar}}{C_{12}H_{22}O_{11}} + H_2O \longrightarrow \underset{\text{Glucose}}{C_6H_{12}O_6} + \underset{\text{Fructose}}{C_6H_{12}O_6}$$

$$2(C_6H_{10}O_5)_n + nH_2O \longrightarrow \underset{\text{Maltose}}{nC_{12}H_{22}O_{11}}$$

20.1 RATE OF A REACTION

The rate of a chemical reaction is the speed or velocity with which a reaction takes place. It can be expressed *qualitatively* as well as *quantitatively*.

(*a*) **Qualitative Rate.** The qualitative rate is based on certain visual parameters like disappearance of reactants, colour change, effervescence, etc. The rate expressed as such provides arbitrary conclusions about the rate of a reaction.

(*b*) **Quantitative Rate.** The quantitative rate of a chemical reaction provides us with much better information about the rate data and the accurate rate of a reaction.

Quantitatively, the ***rate of a chemical reaction*** may be expressed in any of the following ways :

(*i*) *The rate of decrease in the concentration of any one of the reactant,* or

(*ii*) *The rate of increase in the concentration of any one of the product.*

For example, for a hypothetical reaction, assuming that the volume of the system remains constant.

$$R \longrightarrow P$$

One mole of the reactant R produces one mole of the product P. If $[R]_1$ and $[P]_1$ are the concentrations of R and P respectively at time t_1 and $[R]_2$ and $[P]_2$ are their concentrations at time t_2 then,

$$\Delta t = t_2 - t_1$$

$$\Delta[R] = [R]_2 - [R]_1$$

$$\Delta[P] = [P]_2 - [P]_1$$

The square brackets in the above expressions are used to express molar concentration.

Rate of disappearance of R

$$= \frac{\text{Decrease in concentration of R}}{\text{Time taken}} = -\frac{\Delta[R]}{\Delta t}$$

Rate of appearance of P

$$= \frac{\text{Increase in concentration of P}}{\text{Time taken}} = +\frac{\Delta[P]}{\Delta t}$$

While writing the expressions for rate we use a negative sign in the expression employed for expressing rate in terms of reactants.

SIGNIFICANCE OF NEGATIVE SIGN

The negative sign in the first expression does not mean that the rate is negative, rather it indicates the decrease in the concentration of the reactant. As we know with the passage of time, the concentrations of reactants decrease. Therefore, the change in concentration, $\Delta[R]$ = (Final conc.– Initial conc.) will have a negative value. In other words since the rate of a reaction is a positive quantity, we can get a positive rate by putting a –ve sign in the rate expression.

Thus, *Rate of reaction* = *Rate of disappearance of R*
= *Rate of appearance of P*

$$= -\frac{\Delta[R]}{\Delta t} = +\frac{\Delta[P]}{\Delta t}$$

PHYSICAL CHANGES WHICH INDICATE PROGRESS OF A REACTION

Experimental determination of reaction rates depends upon the nature of the reaction under investigation. Rates can be conveniently studied by measuring a suitable physical property of the system as a whole or of one of the constituent of the reaction. Some common examples are given below:

1. Disappearance of Reactants and Appearance of Products. The visual disappearance of reactants or appearance of products indicates progress of a reaction. For example, in the reaction of zinc granules with HCl, the granules disappear after sometime and the reaction stops.

2. Colorimetry or Colour Change. Here increase or decrease in the intensity of colour is observed at regular intervals

$$\underset{\text{(Colourless)}}{Cl_2 + Br^-} \longrightarrow \underset{\text{(Coloured)}}{2Cl^- + Br_2}$$

$$\underset{\text{(Coloured)}}{I_2 + 2S_2O_3^{2-}} \longrightarrow \underset{\text{(Colourless)}}{S_4O_6^{2-} + 2I^-}$$

3. Rotation of the Plane of Polarised Light. It is observed that some molecules rotate the plane of polarised light. The amount of rotation can be measured using a polarimeter. The reaction of sucrose with water is an example of this category. In the reaction, surcose changes into a mixture of glucose and fructose. This brings about a change in the amount of rotation. The rotation can be measured at various time intervals. A graph of angle of rotation plotted against time allows the rate of the reaction to be determined.

$$\underset{\substack{\text{Sucrose} \\ (\textit{dextrorotatory})}}{C_{12}H_{22}O_{11}} + H_2O \xrightarrow{H^+} \underbrace{\underset{\textit{Glucose}}{C_6H_{12}O_6} + \underset{\textit{Fructose}}{C_6H_{12}O}}_{(\textit{Levorotatory})}$$

4. Volume Changes. We have already seen how to measure volume changes when gases are given off in a reaction. Some reactions involving liquids also show volume changes. An example is the reaction between 2-methylpropene and water in the presence of H_2SO_4:

$$(CH_3)_2C{=}CH_2 + H_2O \longrightarrow (CH_3)_3COH$$

In the reaction, two molecules are replaced by one and there is a small decrease in volume. The reaction can be followed using a dilatometer.

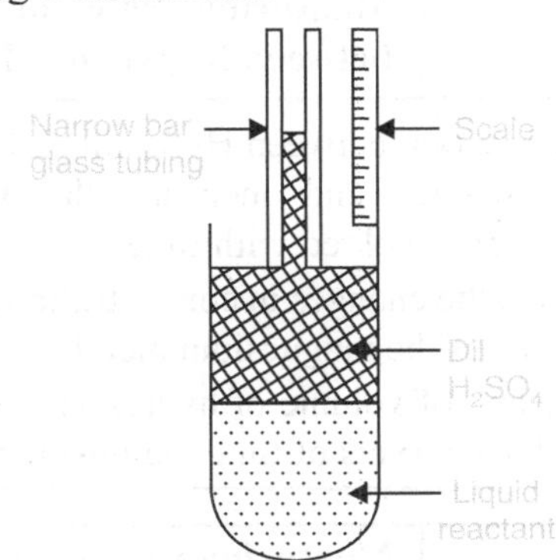

Fig. 20.1. A sample dilatometer.

5. pH Measurement. In the hydrolysis of ethyl acetate, the gradual increase in the concentration of acetic acid is measured by measuring the pH of the solution

$$CH_3COOC_2H_5 + H_2O \xrightarrow{H^+} C_2H_5OH + CH_3COOH$$

6. Pressure Changes. Reactions between gases are often followed by measuring changes in pressure rather than volume. (It is easier to keep the volume of the apparatus constant and measure pressure changes). Examples of reactions that have been followed using pressure changes are:

$$2NO(g) + O_2(g) \longrightarrow 2NO_2(g)$$

$$2N_2O(g) \longrightarrow 2N_2(g) + O_2(g)$$

Notice that in both cases there is a change in the number of moles of gas. Gradual increase or decrease of pressure is noticed in gaseous system.

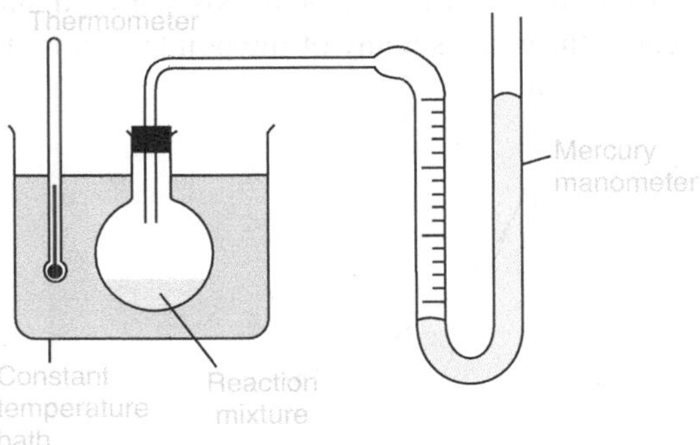

Fig. 20.2. Measuring the pressure of a gaseous reaction component.

7. Change in Mass. The rate of a reaction can also be determined by change in mass during the reaction. For example, when a reaction is carried out between $CaCO_3$ and HCl, carbon dioxide gas is evolved causing a loss in mass. This indicates progress of the reaction.

8. Change in Temperature. Changing temperature can also help in predicting progress of reaction. For example, the colour of $KMnO_4$ in its reaction with oxalic acid discharges very slowly at room temperature but when the solution is warmed *i.e.*, temperature is increased, it decolourises faster.

20.2 MONITORING SPEED OF A CHEMICAL REACTION USING A SIMPLE EXPERIMENT

To measure the rate of a chemical reaction, we must find out the change in the number of particles of a reactant that is disappearing or that of a product which is being made. We must also record the time taken for the change to take place.

Let us discuss a simple example. Suppose we want to measure the rate at which hydrogen peroxide solution decomposes in the presence of a catalyst. The equation for the reaction is

$$2H_2O_2(aq) \longrightarrow 2H_2O(l) + O_2(g)$$

There are three different types of molecules in the reaction, and hence,

$$\text{rate of disappearance of } H_2O_2 = \frac{\text{number of molecules of } H_2O_2 \text{ lost}}{\text{time taken}}$$

or $$\text{rate of appearance of } H_2O = \frac{\text{number of molecules of } H_2O \text{ produced}}{\text{time taken}}$$

or $$\text{rate of appearance of } O_2 = \frac{\text{number of molecules of } O_2 \text{ produced}}{\text{time taken}}$$

We cannot count the actual number of molecules of hydrogen peroxide, water, or oxygen. Therefore, we measure the concentrations of substances which are directly related to the number of molecules.

At this stage we should decide which measurements would be the easiest to make. In the present example, the measurement of oxygen in terms of volume of oxygen which collected can be easily done so it is the third equation that we should use. For a gas, the concentration is proportional to its volume at constant pressure and temperature, or its pressure at a constant volume and temperature. Thus we can put,

$$\text{rate of appearance of } O_2 = \frac{\text{volume of } O_2 \text{ produced}}{\text{time taken}}.$$

The sample set of apparatus that can be used to measure the volume of oxygen collected every 10 s or so is given in Fig. 20.3

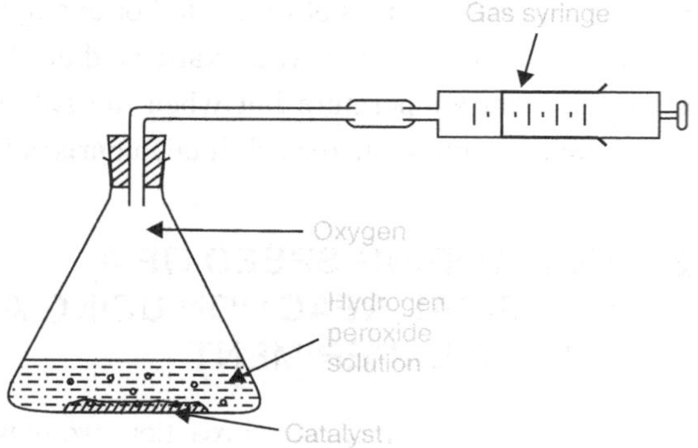

Fig. 20.3. Apparatus for measuring rate of decomposition of H_2O_2

On plotting the volume of oxygen that is given off at regular intervals of time we get a plot as shown in Fig. 20.4.

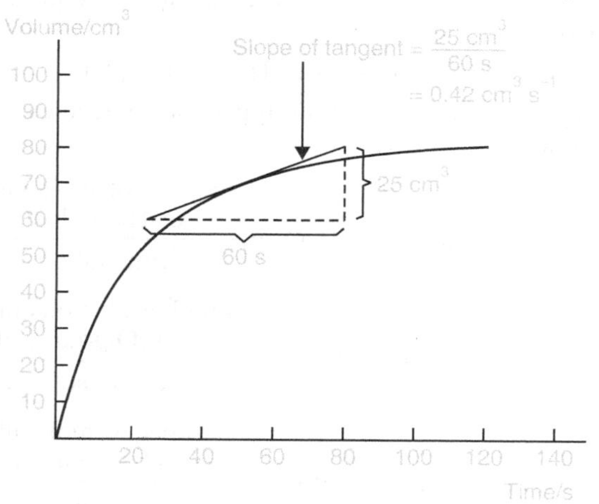

Fig. 20.4. Plot of volume of oxygen collected vs time.

An alternative way of producing this graph is to connect the pen of a chart recorder to the plunger of the gas syringe. The pen will then produce the graph automatically.

On analysing the graph it is clear that, the oxygen gas is given off very quickly at the start. This is where the curve is the steepest. As time goes on, the curve becomes less steep, until at about 120 s it levels off. At this stage the reaction has stopped. We can measure the rate at which oxygen is evolved at any time by finding out how steep the curve is. Say, let us measure the rate at time 60 s. The slope is found by drawing a tangent to the curve at time 60 s. This involves a certain amount of estimation if you do it by hand and eye. (There are more accurate methods, which make use of advanced mathematics and computers.) We have,

$$\text{slope of line} = 0.42 \text{ cm}^3 \text{ s}^{-1}$$

i.e., rate of appearance of oxygen is 0.42 cm^3 s^{-1} at time 60 s.

If we measured the slope at time zero, we would have measured the *initial rate* of the reaction.

ACTIVITY 20.1 Measuring rate of reaction between Mg(s) and HCl

(*i*) Use the reaction between HCl acid and Mg(*s*) in a similar manner and measure the volume of hydrogen gas evolved with time.

(*ii*) Determine the changes in concentration of the acid and nature of the magnesium metal.

(*iii*) Plot a graph of volume of hydrogen gas collected (at regular intervals of time) against time.

ACTIVITY 20.2 Measuring rate of a reaction

1. Set up the apparatus as shown in Fig. 20.3 but *without* the marble chips at first.
2. Add calcium trioxocarbonate (IV) (marble chips). Plug the cotton wool in position immediately. Read the mass of the flask and its contents and start a stop-clock. Record the mass at one-minute intervals until the reaction is over.

$$CaCO_3(s) + 2HCl(aq) \longrightarrow CaCl_2(aq) + H_2O(l) + CO_2(g)$$

3. Use a table like the one shown below to record your results.

Time/min	0	1	2	3	4	5	6	7	8	9	10
Mass/g											
Loss in mass/g											

Plot a graph with time on the horizontal axis and loss in mass on the vertical axis. Draw a smooth curve through as many of the points as possible.

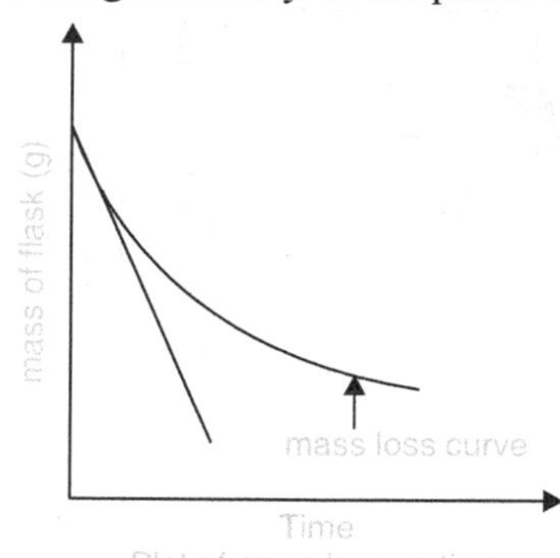

Plot of mass loss vs time

Note that in this experiment change in mass is proportional to change in concentration.

After recording the experimental data try to explore

- What interpretation can be made from the *slope* of this graph at any point?
- Explain why the graph is:
  (*a*) steep at the start of the reaction?
  (*b*) less steep in the middle of the reaction?
  (*c*) horizontal at the end of the reaction?
  At what time does the reaction stop?
- What is
  (*a*) the average reaction rate of the reaction?
  (*b*) the reaction rate at 2 minutes (calculate)?

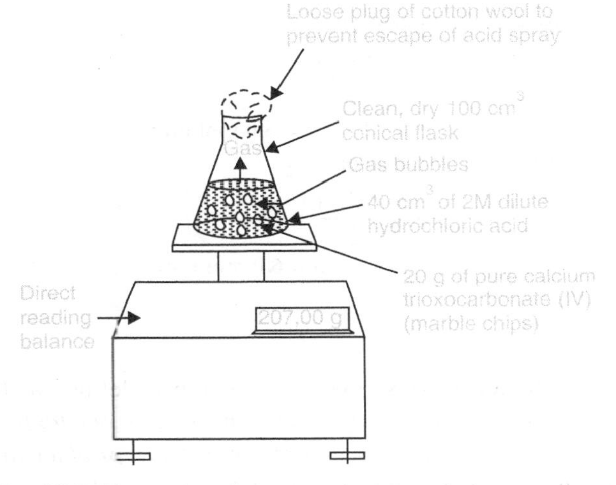

Fig. 20.5. Measuring decreases in mass during reaction of calcium trioxocarbonate(IV) with dilute acid.

WRITING EXPRESSIONS FOR REACTION RATES

Expression 1. The expressions for the rates of gaseous reaction,

$$NO_2(g) + CO(g) \longrightarrow CO_2(g) + NO(g)$$

are given as:

$$\text{Rate of reaction} = \begin{bmatrix}\text{rate of}\\ \text{disappearance}\\ \text{of } NO_2\end{bmatrix}$$

$$= \begin{bmatrix}\text{rate of}\\ \text{disappearance}\\ \text{of CO}\end{bmatrix} = \begin{bmatrix}\text{rate of}\\ \text{appearance}\\ \text{of } CO_2\end{bmatrix} = \begin{bmatrix}\text{rate of}\\ \text{appearance}\\ \text{of NO}\end{bmatrix}$$

$$= -\frac{\Delta[NO_2]}{\Delta t} = -\frac{\Delta[CO]}{\Delta t}$$

$$= \frac{\Delta[CO_2]}{\Delta t} = \frac{\Delta[NO]}{\Delta t}$$

Expression 2. Consider another gaseous reaction between hydrogen and iodine to give hydro-iodic acid.

$$H_2(g) + I_2(g) \longrightarrow 2HI(g)$$

The rate of this reaction can also be given in terms of rate of disappearance of one of the reactants H_2 or I_2 or in terms of the rate of formation of HI.

In the present case, for every mole of H_2 or I_2 reacting we get two moles of HI. Thus, the rate of formation of HI will be twice the rate of disappearance of H_2 or I_2. Now in order to avoid the difference in the two rates, *i.e.*, to get a **unique value** of the reaction rate irrespective of the species selected, we divide the rate of change of concentration by the number of moles of reactant or product present in the reaction. Thus, we have

$$\text{Rate of reaction, } r = -\frac{\Delta[H_2]}{\Delta t} = -\frac{\Delta[I_2]}{\Delta t} = \frac{1}{2}\frac{\Delta[HI]}{\Delta t}$$

The rate given above is the **average rate** of a reaction.

It is calculated by dividing the total change in concentration of any one of the reactant or product by the total time taken to do so.

Average rate

$$= \frac{\textit{Change in concentration in a given time}}{\textit{Time taken}}$$

i.e.,
$$= \frac{\Delta x}{\Delta t} \text{ or } = -\frac{\Delta R}{\Delta t} \text{ or } = +\frac{\Delta P}{\Delta t}$$

In general, for a hypothetical reaction,

$$aA + bB \longrightarrow cC + dD$$

$$\text{Rate of reaction} = -\frac{1}{a}\frac{\Delta[A]}{\Delta t} = -\frac{1}{b}\frac{\Delta[B]}{\Delta t} = \frac{1}{c}\frac{\Delta[C]}{\Delta t}$$

$$= \frac{1}{d}\frac{\Delta[D]}{\Delta t}$$

UNITS OF RATE OF REACTION

The units of the rate of a reaction are concentration time^{-1}. As concentration of a substance is expressed in *mol dm*$^{-3}$ and the time is expressed in *seconds* or *minutes* or *hours* the units for reaction rate, therefore, are **moles dm^{-3} sec^{-1}** or **moles dm^{-3} min^{-1}** or **moles dm^{-3} hr^{-1}.**

For gaseous reactions, if the concentrations of reactants and products are given in terms of partial pressures, then the units of the rate of reaction will be **atm min^{-1}** or **atm sec^{-1}** or **atm hr^{-1}** (Pressure is expressed in atmosphere).

SOLVED EXAMPLES

Example 20.1. *Describe the relationship between rate of reaction, rate of disappearance of X, Y and rate of appearance of X_2Y_3 in the reaction, $2X + 3Y \longrightarrow X_2Y_3$.*

Solution. Rate of reaction $= -\frac{\Delta[X]}{2\Delta t} = -\frac{\Delta[Y]}{3\Delta t} = \frac{\Delta[X_2Y_3]}{\Delta t}$.

Example 20.2. *A gaseous hypothetical chemical reaction $2A \rightleftharpoons 4B + C$ is carried out in a closed vessel. The concentration of B is found to increase by 5×10^{-3} mol dm^{-3} in 10 seconds. Calculate*

(*i*) *Rate of appearance of B*

(*ii*) *Rate of reaction and*

(*iii*) *Rate of disappearance of A.*

Solution. (*i*) Rate of appearance of B

$$= \frac{\text{Increase in conc. of B}}{\text{Time taken}}$$

$$= \frac{5 \times 10^{-3} (\text{mol dm}^{-3})}{10\,(\text{sec})} = \mathbf{5 \times 10^{-4}\ mol\ dm^{-3}\ s^{-1}}$$

(*ii*) Rate of reaction $= \frac{1}{4} \times$ Rate of appearance of B

$$= \frac{1}{4} \times 5 \times 10^{-4} = \mathbf{1.25 \times 10^{-4}\ mol\ dm^{-3}\ s^{-1}}$$

(*iii*) Rate of disappearance of A

$$= \frac{1}{2} \times \text{Rate of disappearance of B}$$

$$= \frac{1}{2} \times 5 \times 10^{-4}\ \text{mol dm}^{-3}\ \text{s}^{-1}$$

$$= \mathbf{2.5 \times 10^{-4}\ mol\ dm^{-3}\ s^{-1}}.$$

Example 20.3. *The following reaction was carried out in water:*

$$Cl_2 + 2I^- \longrightarrow 2Cl^- + I_2$$

The initial concentration of I^- was 0.50 mol dm^{-3} and the concentration after 10 minutes was 0.46 mol dm^{-3}. Calculate the rate of disappearance of I^- and rate of appearance of Iodine.

Solution. $\Delta[I^-] = 0.46 - 0.50 = -0.04$ mol dm^{-3}

$\Delta t = 10.0 - 0.0 = 10$ min

Rate of disappearance of $I^- = -\frac{\Delta[I^-]}{\Delta t} = \frac{-(-0.04)}{10}$

$= \mathbf{0.004\ mol\ dm^{-3}\ min^{-1}}$.

Rate of appearance of iodine

$$= \frac{1}{2}\ (\text{Rate of disappearance of I}^-)$$

$$= \frac{1}{2} \times 0.004 = \mathbf{0.002\ mol\ dm^{-}\ min^{-1}}.$$

Example 20.4. *If the rate of formation of oxygen gas is 48 g mm^{-1}*

(*i*) *What is the rate of disappearance of N_2O_5?*

(*ii*) *What is the rate of appearance of NO_2?*

Solution. The rate of a reaction is always expressed in terms of mol dm^{-3}, therefore, rate of formation of oxygen gas,

$$O_2 = \frac{48}{32} = \mathbf{1.5\ mol\ mm^{-1}}$$

(*i*) Rate of disappearance of N_2O_5,

$$-\frac{1}{2}\frac{\Delta[N_2O_5]}{\Delta t} = \frac{\Delta[O_2]}{\Delta t}$$

$$= 2 \times 1.5 = \mathbf{3.0\ mol\ mm^{-1}}$$

(*ii*) Rate of appearance of NO_2,

$$\frac{1}{4}\frac{\Delta[NO_2]}{\Delta t} = \frac{\Delta[O_2]}{\Delta t}$$

$$= 4 \times 1.5 = \mathbf{6.0\ mol\ mm^{-1}}.$$

INSTANTANEOUS RATE AND ITS DETERMINATION

The **instantaneous rate** of a reaction is defined as the *decrease in concentration of any one of the reactants or increase in concentration of any one of the product at a particular instant of time and at a given temperature.*

Mathematically, instantaneous rate may be expressed by the expression $\frac{dx}{dt}$ or $-\frac{d[R]}{dt}$ or $+\frac{d[P]}{dt}$ instead of $\frac{\Delta x}{\Delta t}$ or $-\frac{\Delta[R]}{\Delta t}$ or $+\frac{\Delta[P]}{\Delta t}$. Here, *dt* is an infinitesimally small time interval and *dx* is the change in concentration of any of the species within the time interval *dt*. Thus, the instantaneous rate is given by

$$r_{inst} = -\frac{\Delta[R]}{\Delta t} = \frac{\Delta[P]}{\Delta t}$$

In differential calculus, when Δt becomes small and approaches zero, the ratio $\frac{\Delta x}{\Delta t}$ (average rate) may be replaced by the derivative $\frac{dx}{dt}$ which means rate of change of *x* with time '*t*'. It may be put as

$$\lim_{\Delta t \to 0} \frac{\Delta x}{\Delta t} = \frac{dx}{dt}.$$

As $\Delta t \to 0$,

$$r_{\text{inst}} = -\frac{d[R]}{dt} = \frac{d[P]}{dt}$$

To **Sum up**, the expression $\frac{\Delta x}{\Delta t}$ or $-\frac{\Delta[R]}{\Delta t}$ or $\frac{\Delta[P]}{\Delta t}$ becomes $\frac{dx}{dt}$ or $-\frac{d[R]}{dt}$ or $\frac{d[P]}{dt}$ as Δt approaches zero.

CALCULATION OF INSTANTANEOUS RATE

The instantaneous rate can also be calculated from the plot of concentration vs time. Suppose the rate of reaction is to be determined at time interval '*t*' (marked in the figure). A straight line is then drawn from this point to cut the curve at point P. The slope of line gives the instantaneous rate of the reaction. The calculation of both average as well as instantaneous rate is shown in Figs. 20.6 and 20.7.

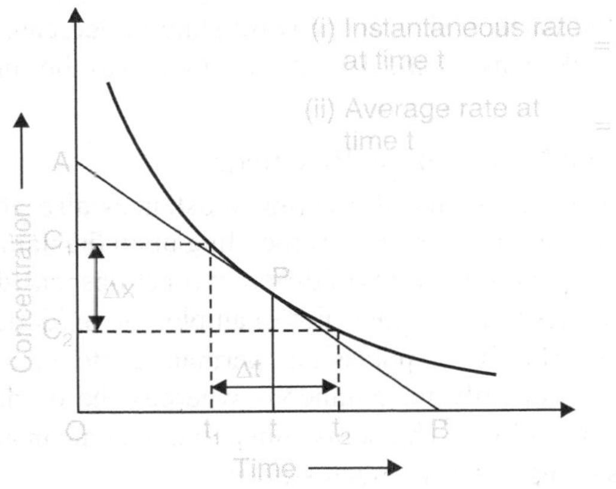

Fig. 20.6

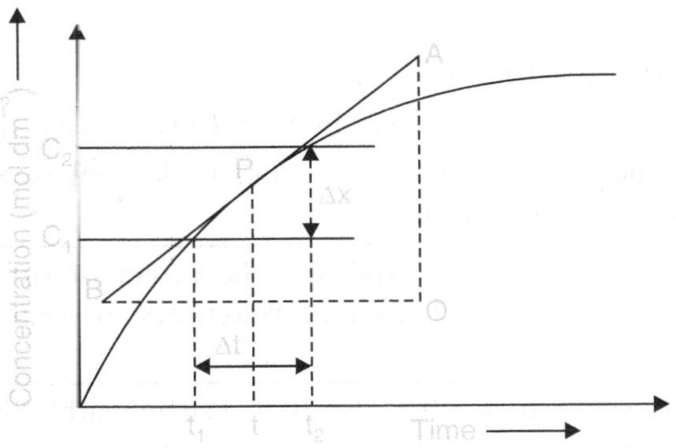

(i) Instantaneous rate at time t

$$r_{inst} = \frac{OA}{OB} = \text{slope}$$

(ii) Average rate at time t

$$r_{av} = \frac{\Delta x}{\Delta t} = \frac{C_2 - C_1}{t_2 - t_1}$$

Fig. 20.7

TO SUM UP

(*i*) Rate of reaction $= \frac{\text{Change in conc.}}{\text{Time interval}}$.

(*ii*) The rate can be expressed in term of any reactant or product whichever is convenient.

(*iii*) The rate expression carries a negative sign when written in terms of concentration of reactants.

(*iv*) To get unique rate of reaction, the rate expressions are divided by stoichiometric coefficients present in the balanced equation.

(*v*) Instant rate of reaction = slope of tangent to the curve between concentration versus time at a given instant.

20.3 FACTORS INFLUENCING RATE OF A REACTION

The rate of reactions is influenced by the following factors:

(*i*) *Concentration of the reactants,*

(*ii*) *Temperature of the reactants,*

(*iii*) *Particle size and nature of the reacting substances,*

(*iv*) *Presence of catalyst, and*

(*v*) *Exposure to radiations.*

1. Dependence of Rate on Concentration

When a chemical reaction occurs, the reactants are converted to products. It is observed that with the passage of time the concentrations of reactants decrease while those of products increase. It is graphically shown in Fig. 20.8.

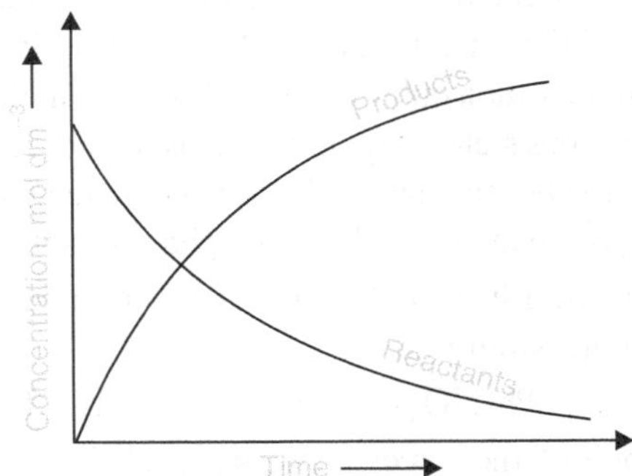

Fig. 20.8. Time dependence of the concentrations of reactants and products in a reaction.

Now, if we assume that other factors are constant then the rate of a chemical reaction decreases with a decrease in the concentration of the reactants. For example, *we find that a piece of wood burns at a much faster rate in oxygen than in air. It is because of the higher concentration of O_2 in the former.*

Cato Guldberg and Peter Waage proposed a qualitative relationship between the rates of reactions and the concentration of the reacting species. This generalisation is known as **"Law of Mass Action".** It states that:

At a given temperature, the rate of a chemical reaction is directly proportional to the product of the molar concentrations of the reacting species with each concentration term raised to the power of the numerical coefficient of that species in the chemical equation.

Thus, for a hypothetical reaction;

$$A + B \longrightarrow \text{Products}$$

The rate according to the law of mass action is given as

Rate of reaction, $r \propto [A][B] = k[A][B]$

where [A] and [B] are the molar concentrations of the reactants A and B respectively and k is a constant of proportionality.

2. Particle Size and Nature of Reactants

It is commonly observed that the nature of the reacting substances has a marked influence on the reaction rates. The effect of the nature of reactants can be described in terms of the following factors:

(*a*) *Physical state of reactants*

(*b*) *Surface area and particle size of reactants*

(*c*) *Chemical nature of reactants.*

(*a*) Physical State of the Reactants

We know that a chemical reaction takes place as a result of the collisions between the reacting particles. In order to collide with one another, the reacting particles must come close and intermix. But the intermixing of reacting particles cannot take place easily if they are in the solid state. This is due to the restriction in their molecular motion. However, in the liquid as well as the gaseous state, the intermixing is possible and chances of their collision become larger. Thus, the reaction will be faster if the reactants are mixed in the liquid phase or solution phase.

This can be illustrated with the reaction between lead nitrate and potassium iodide.

$$Pb(NO_3)_2 + KI \longrightarrow PbI_2 + KNO_3$$

(*i*) In one of the experiment, solid $Pb(NO_3)_2$ is mixed with solid KI in a china dish with a spatula.

(*ii*) In another experiment, an aqueous solution of $Pb(NO_3)_2$ is mixed with an aqueous solution of KI in a beaker.

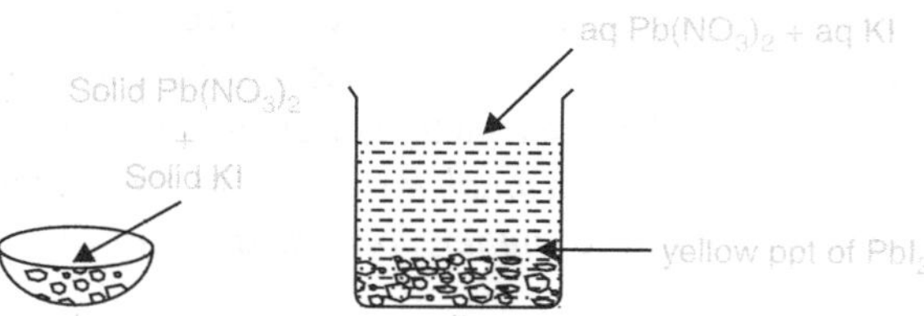

It is observed that the rate of formation of yellow coloured PbI_2 is much faster in the second experiment because the probability of encounters between the reacting particles is much higher in the liquid phase than in the solid phase.

(*b*) Surface Area and Particle Size of Reactants

Particle size and surface area play an important role in rates of **heterogeneous reactions** particularly when one of the reacting species is in the solid phase. The smaller the size of the reacting particles present in the solid phase, the larger the surface area. As a result, the probability of the reacting species coming in contact with each other is higher. For example,

- combustion of a piece of coal in air is relatively slower, whereas combustion of coal dust in air is much faster.
- dissolution of sugar crystals in water is slower but dissolution of powdered sugar is rapid.
- Thus, we find that the rate of the process is increased, if the particle size in the solid state is decreased. In fact, decrease in the particle size increases the surface area.

(*c*) Chemical Nature of the Reactants

The chemical nature of reacting substances also affects the reaction rates significantly. Depending upon the nature of the reacting species, the rates of different reactions may differ very widely from one another. For example, the oxidation of ferrous ions (Fe^{+2}) by potassium permanganate in acidic medium is practically instantaneous whereas the oxidation of oxalate ions ($C_2O_4^{-2}$) by potassium permanganate in acidic solution is comparatively much slower:

$$MnO_4^- + 8H^+ + 5Fe^{+2} \longrightarrow Mn^{+2} + 5Fe^{+3} + 4H_2O \quad (\textit{fast})$$

$$2MnO_4^- + 16H^+ + 5C_2O_4^{-2} \longrightarrow 2Mn^{+2} + 10CO_2 + 8H_2O \quad (\textit{slow})$$

In these two reactions, everything is identical except the nature of reducing agents.

ACTIVITY 20.3	**Studying the effect of surface area of reactants on reaction rate**

1. Repeat activity 20.2. Use 20 g of small marble chips and then powdered marble in place of the large ones.
2. Compare the rate curve, average rate and reaction rate at 2 minutes with that of activity 20.2.
3. Sketch graphs for the consumption of each size of marble chips (large, small and powder) against time on the same graph sheet.
4. A reactant with small particle size therefore has a faster reaction rate. For example, fine iron filings react

faster with dilute acid solution than an equal mass of coarse iron filings (Fig. 20.4). Other solids such as wood and metals behave in a similar way.

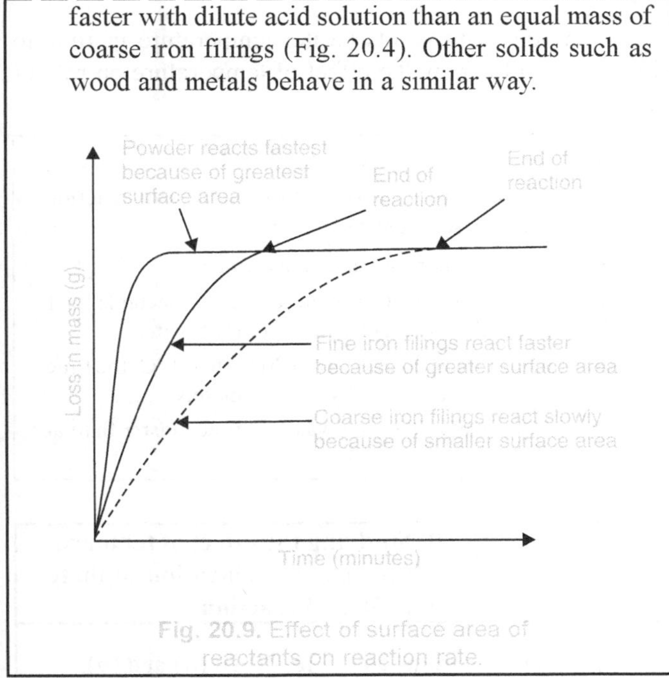

Fig. 20.9. Effect of surface area of reactants on reaction rate.

3. The Effect of Pressure

The change in pressure also affects the rate of a reaction.

- **If one or more of the reactants is a gas** then **increasing pressure** will effectively increase the concentration of the reacting molecules and **speed up the reaction**.
- Changing the pressure *i.e.*, lower or higher, results in lesser or greater concentration and so slower or faster reaction, all because of the **increased chance of a 'fruitful' collision**.

4. The Effect of Stirring

In doing rate experiments with a solid and a solution as reactants *e.g.*, marble chips-acid solution or a solid catalyst like manganese(IV) oxide catalysing the decomposition of hydrogen peroxide solution, it is **sometimes forgotten that stirring the mixture is an important rate factor**.

- If the reacting mixture is not stirred 'evenly', the reactant concentration in solution becomes much less near the solid, which tends to settle at the bottom of the flask.
- Therefore, at the bottom of the flask the reaction prematurely slows down distorting the overall rate measurement and making the results uneven and inaccurate. The 'unevenness' of the results is even more evident by giving the reaction mixture the 'odd stir'! You get jumps in the graph!!!

5. The Effect of Radiations

Reactions carried out in the presence of light are known as photochemical reactions. The presence/absence of radiations greatly affects the rate of a reaction. Some examples are given below to illustrate the idea.

- Halogenation of methane can readily occur in the presence of uv radiations. The reaction does not occur in the dark. The radiations supply sufficient energy to break the bonds of Cl_2 molecule giving free radicals which start the reaction.

Photosynthesis in green plants occurs in the presence of sunlight. The green chlorophyll molecules absorb the photon energy packets of sunlight and initiate the chemical changes summarised below:

$$\text{water} + \text{carbon dioxide} \longrightarrow \text{glucose} + \text{oxygen}$$

$$\mathbf{6H_2O(\mathit{l}) + 6CO_2(\mathit{g}) \longrightarrow C_6H_{12}O_6(\mathit{aq}) + 6O_2(\mathit{g})}$$

6. Effect of Temperature

The rates of almost all reactions increase with an increase in temperature. For example,

- In the decomposition of N_2O_5, the time taken for half of the original amount of material to decompose is 12 min at 50°C, 5 hours at 25°C and 10 days at 0°C.
- In a mixture of $KMnO_4$ and Oxalic acid ($H_2C_2O_4$), potassium permanganate gets decolourised faster at higher temperature than at lower temperature.

In most of the cases it has been found that the rate of the reaction becomes almost double for every 10° rise in temperature. This is also expressed in terms of the **Temperature coefficient** which is the ratio of the rate constants of the reaction at two temperatures differing by 10°. The two temperatures generally selected are 298 K and 308 K. Thus,

$$\textit{Temperature coefficient} = \frac{\text{Rate constant at 308 K}}{\text{Rate constant at 298 K}}.$$

The rate constants for the decomposition of N_2O_5 at different temperatures is given in Table 20.1. From the table, it is seen that the rate constant at 273 K is 7.87×10^{-7} and at 298 K is 3.56×10^{-5}. This shows that for a 25° rise in temperature, the rate constant increases 43 times.

Table 20.1. Rate Constants for the Decomposition of N_2O_5 at Different Temperatures

Temperature (K)	*k (sec^{-1})*
273	7.87×10^{-7}
298	3.56×10^{-5}
308	13.5×10^{-5}
318	4.98×10^{-4}
328	15.0×10^{-4}
338	4.87×10^{-3}

20.4 EXPERIMENT TO SHOW HOW CHANGES IN TEMPERATURE AFFECT THE RATE OF A REACTION

Students, do you think that temperature will have an effect on how fast or slow a reaction might take place? In order for a chemical reaction to occur, the molecules, which are reactants, must physically come into contact with one another. Anything that increases the frequency of these encounters will increase the rate at which products are formed. Your hypothesis can be that the rate of a chemical reaction will be increased by raising the temperature of the reactants.

Requirement

Six clear cups, a measuring cup, a thermometer (–20°C to 110°C), three original formula effervescent Alka-Seltzer® tablets, a stopwatch, a mortar and pestle, a source of hot water, ice cubes, and a graph paper.

Procedure

Fill a clear glass with exactly 50 cm^3 of hot water. Use the thermometer to take the temperature and record it on a data sheet. Drop 1 Alka-Seltzer® tablet into the water. Measure the time required for the tablet to fully dissolve. Be prepared to start and stop on time. The reaction will take less than 15 seconds. Record the time. Repeat this experiment using water at room temperature.

For the cold water test, the procedure is a little different. Fill a clear glass with 25 cm^3 of water and add enough ice to adjust the level to 50 cm^3. Stir the ice water for about 15 seconds so the temperature will come to equilibrium. Use the thermometer to take the temperature and record it on your data sheet. (Leave the ice cubes in the water!) Drop 1 Alka-Seltzer tablet into the water. Measure and record the time required for the reaction to be completed.

Analysis

Graph your data points (water temperature *vs* time to fully dissolve) to show the effect of temperature on rate of reaction.

So now can you answer these questions?

As the temperature increases, does the rate of reaction increases, decreases, or stays the same?

How about these more advanced questions?

1. At a temperature of 10 °C, it would take ...... seconds for 1 Alka-Seltzer tablet to react with 50 cm^3 of water.
2. If the temperature is doubled from 20 °C to 40 °C, the time for the rate of reaction ...... by approximately ....
3. Using hot tap water, the rate was ...... times faster than at 0 °C.

ACTIVITY 20.4 — Studying the effect of temperature on the trioxothiosulphate (VI)/acid reaction

1. Set up the apparatus as in Fig. 20.10 (*a*) and (*b*).
2. Cool the solutions to 273 K (0 °C).
3. Pour the solutions of dilute hydrochloric acid and sodium thiosulphate into the 100 cm^3 beaker and start a stop-clock at the same time. Carefully stir the mixture with a thermometer and record both the time taken for the cross to disappear and the average temperature during the reaction.
4. Rinse the 100 cm^3 beaker and thermometer very well. Measure the same volume of reactants into test tubes at room temperature. Repeat instruction 3.
5. Repeat the whole of instruction 4 four times at 10 K (10 °C) intervals from room temperature. Tabulate your results as temperature in K versus time in minutes.
6. Plot a graph of temperature on the horizontal axis against time on the vertical axis.
   At what temperature does the cross take (*a*) the shortest time and (*b*) the longest time to disappear?
7. Plot a graph of temperature on the horizontal axis against 1/time on the vertical axis.
   What shape is this graph?
   Is the reaction rate proportional to time or proportional to 1/time? Explain your answer.
8. Draw a conclusion about the relationships between the time taken for the precipitation of sulphur and temperature, and the average reaction rate and temperature.

9. *Explain your conclusion.*
Using your graph, how long will it take to form a sulphur precipitate if a trioxothiosulphate(VI) acid reaction takes place at (*a*) 296 K (*b*) 330 K?

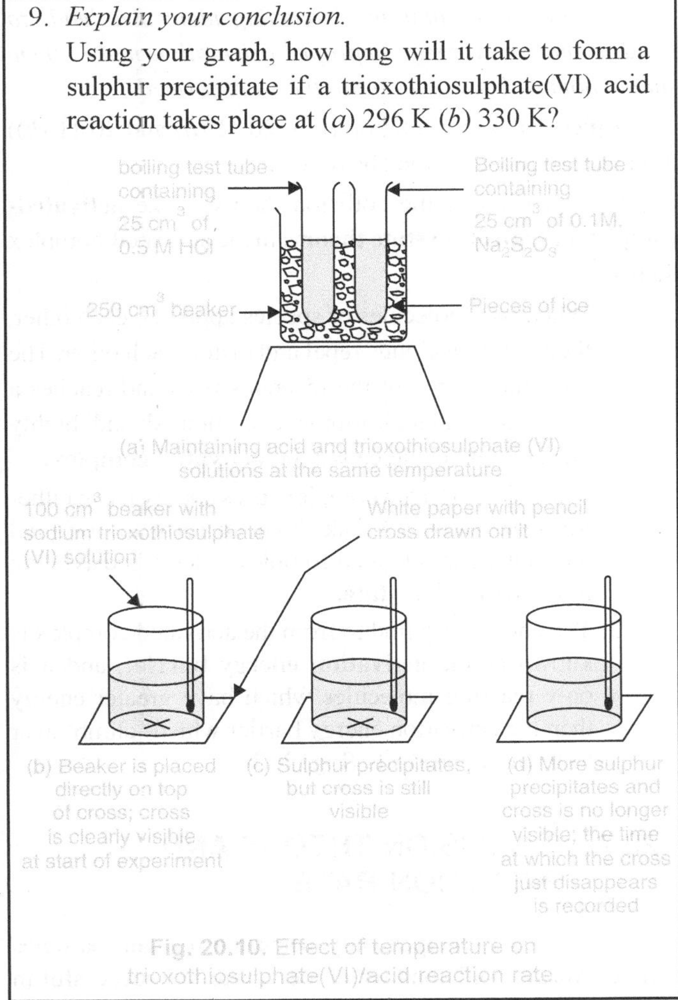

Fig. 20.10. Effect of temperature on trioxothiosulphate(VI)/acid reaction rate.

20.5 ANALYSIS AND INTERPRETATION OF SIMPLE GRAPHS ON RATES OF REACTION

Graph 1 shows the decrease in the amount of a solid reactant with time. The graph is curved, becoming less steep because the reactants are being used up, so the speed decreases. In the first few minutes the graph will (*i*) decline less steeply for larger 'lumps' and (*ii*) decline more steeply with a fine powder *i.e.*, (*i*) less surface area gives slower reaction and (*ii*) more surface area a faster reaction.

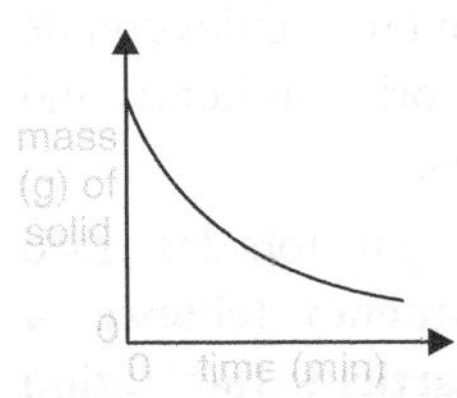

Graph 2 shows the increase in the amount of a solid product with time. The graph tends towards a maximum amount possible when all the solid reactant is used up and the graph becomes horizontal. This means the speed has become zero as the reaction has stopped.

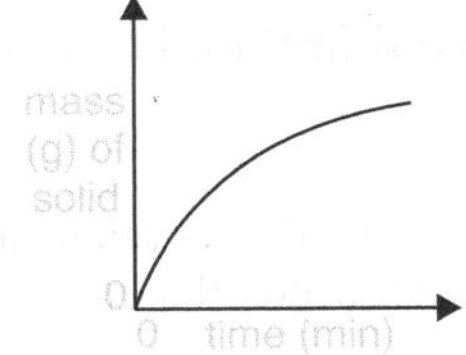

Graph 3 shows the decrease in reaction time with increase in temperature as the reaction speeds up. The reaction time can represent how long it takes to form a fixed amount of gas in the first few minutes of a metal/carbonate-acid reaction, or the time it takes for so much sulphur to form in a sodium thiosulphate-hydrochloric acid reaction.

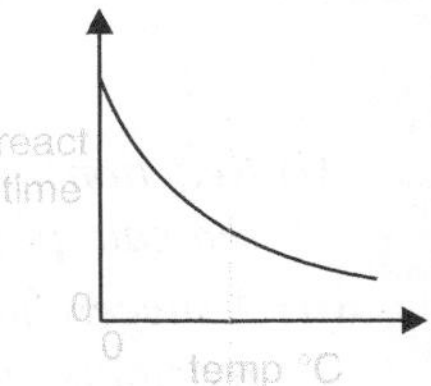

Graph 4 shows the increase in speed of a reaction with increase in temperature as the particles have more and more kinetic energy. The rate of reaction is proportional to $1/t$, where t is the reaction time.

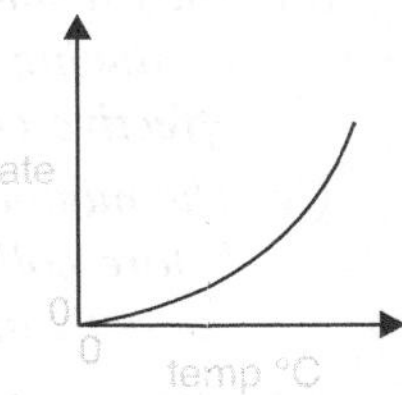

Graph 5 shows the increase in the amount of a gas formed in a reaction with time. Again, the graph becomes horizontal as the reaction stops when one of the reactants is all used up.

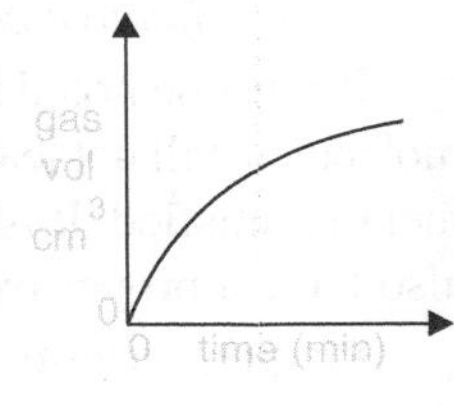

Graph 6 shows the effect of increasing concentration, which decreases the reaction time, as the speed increases because the greater the concentration the greater the chances of a fruitful collision.

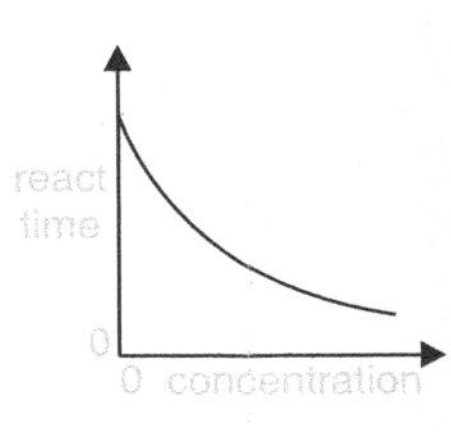

Graph 7 shows that the rate/speed of a reaction is often proportional to the concentration of one particular reactant. This is due to the chance of a fruitful collision forming products being proportional to the concentration.

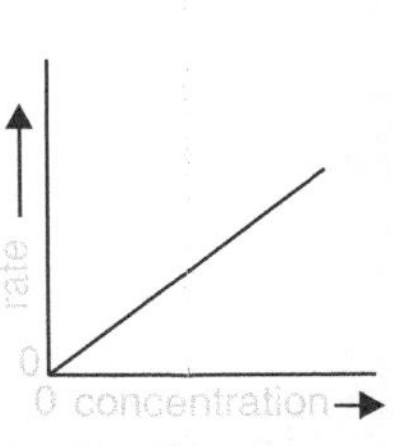

20.6 COLLISION THEORY OF REACTION RATES

Collision theory, was developed by Max Trautz and William Lewis in 1916–18 and, provides a greater insight into the energetic and mechanistic aspects of reactions and their rates. According to this theory, the reactant molecules are assumed to be hard spheres and a reaction can occur when molecules collide with each other. **The number of collisions per second per unit volume of the reaction**

mixture is known as the collision frequency (Z). The more frequent the collisions, the faster the reaction.

POSTULATES OF THE COLLISION THEORY

(*i*) *Reactions occur due to approach and collisions of reactant particles (atoms, molecules or ions)*

(*ii*) *A successful collision, the one required for changing reactants to products, can occur between two molecules only if they possess a certain minimum amount of energy in excess of the normal energy of molecules.*

(*iii*) *Only a small fraction of collisions are successful in producing a reaction. These collisions are called* ***effective collisions****.*

(*iv*) *The minimum energy which molecules must possess before collision should be equal to or greater than the activation energy.*

(*v*) *The rate of reaction is proportional to the frequency of effective collisions per second.*

It must be noted here that the collisions between reactant molecules will not lead to a reaction even if the energy requirement is satisfied. It is because the colliding molecules should also have a proper orientation.

Proper Orientation:

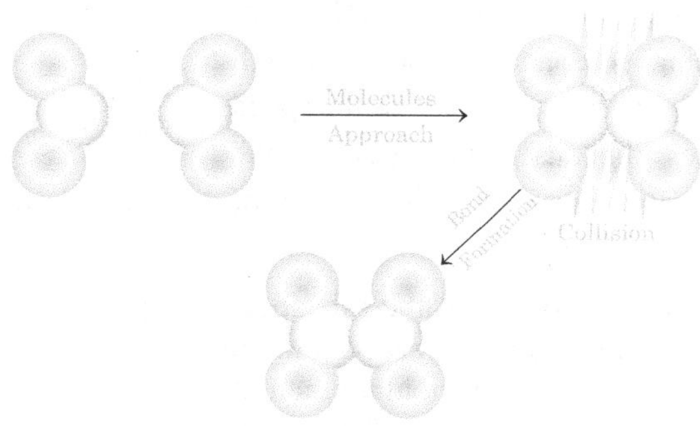

Fig. 20.11 (a)

Improper Orientation:

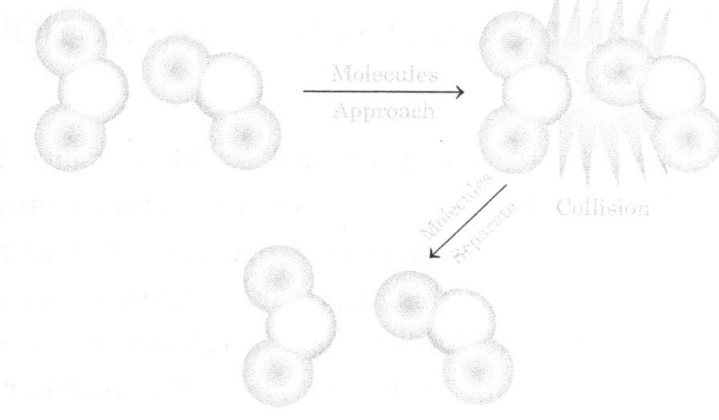

Fig. 20.11 (b)

The proper orientation of reacting molecules leads to bond formation whereas improper orientation makes them simply bounce back and no products are formed.

A pictorial representation (Fig. 20.11 (*a*) and 20.11 (*b*)) is shown, to illustrate the above idea.

An extension of the collision theory is the **activated-complex or transition state theory.** In the activated-complex theory:

1. When two molecules or species approach each other, their electron clouds repel and distort each other. The potential energy of the reactants rises and reaches a maximum value, forming a combined and highly unstable molecule called the **activated complex.**
2. When the activated complex loses energy it can either form products or break down to form the original reactants. The activated complex is described as being in the **transition state.**
3. The energy required to from the activated complex is known as the **activation energy barrier,** and it is only reacting molecules which have greater energy than the activation energy barrier which 'climb' over to form the products (Fig. 20.11).

20.7 COLLISION THEORY AND REACTION RATE

Reactions can happen only when the reactant particles collide. All the collisions which occur are not successful in forming product molecules. The reason is that *particles have a wide range of kinetic energy but only a small fraction of particles have enough kinetic energy* to break bonds and bring about chemical change. The minimum kinetic energy required for reaction is known as the **activation energy**.

The small number of high kinetic energy collisions between particles which do produce a chemical change are called **'fruitful collisions'**. Here the reactant molecules collide with enough kinetic energy to break the original bonds and form new bonds in the product molecules.

It must be noted that the rate-controlling factors described below are to do with the **collision frequency (chance of collision) or the energy of reacting particles (activation energy)** which can be summed up as the **'chance of a fruitful collision'** leading to product formation.

THE EFFECT OF CONCENTRATION

If the concentration of any reactant in a solution is increased, the rate of the reaction is increased. Increasing the concentration, increases the probability of collision between reacting particles because there are more of them in the same volume. This increases the chance of a fruitful collision to form products.

For example, increasing the concentration of acid molecules in a reaction of zinc with HCl, increases the frequency at which they hit the surface of the zinc chips to dissolve them faster.

THE EFFECT OF TEMPERATURE

When gases or liquids are heated the particles gain kinetic energy and move faster (see diagrams below.) The increased speed increases the chance (frequency) of collision between reacting molecules and the rate increases.

It must be noted that this is not the main reason for the increased reaction speed, so be **careful** in your theory explanations if investigating the effect of temperature.

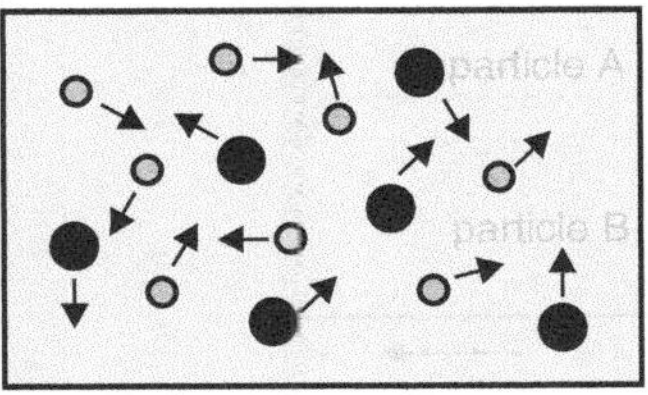

low temperature

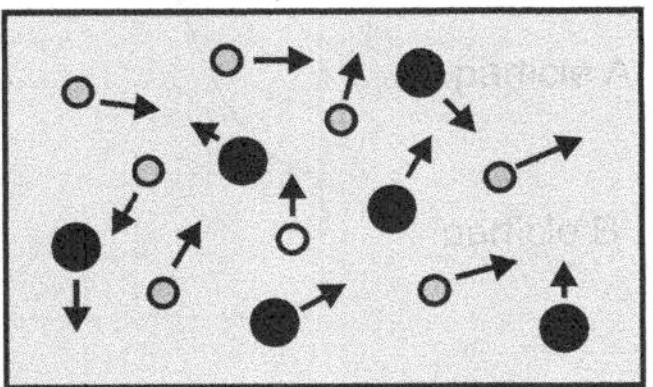

high temperature

THE EFFECT OF SURFACE AREA-PARTICLE SIZE OF A SOLID REACTANT

If a solid reactant or a solid catalyst is broken down into smaller pieces the rate of reaction increases.

The speed increase happens because smaller pieces of the same mass of solid have a greater surface area compared to larger pieces of the solid.

Therefore, there is more chance that a reactant particle will hit the solid surface and react.

20.8 ROLE OF ACTIVATION ENERGY IN CHEMICAL REACTIONS

A reaction can occur when molecules of reactants collide with each other to form an unstable intermediate (Fig. 20.12). The intermediate exists for a very short time and then breaks up to form product molecules. The energy required to form this intermediate, called **activated complex (C),** is known as **activation energy (E_a)**. The distribution of the kinetic energy of different fractions (N_E/N_T) of molecules is shown in Fig. 20.12.

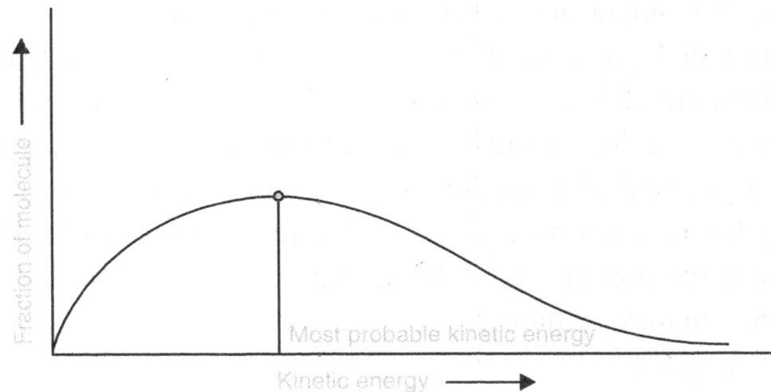

Fig. 20.12. Distribution curve showing energies among gaseous molecules.

Only those collisions resulting in the formation of products possess energy equal to or more than the minimum energy called *threshold energy.* Collisions of the molecules possessing energy less than the threshold energy do not form products. It means that between reactants and products there is an energy barrier which must be crossed before the reactants are converted into products. The energy required for crossing this energy barrier is supplied by the kinetic energy of the molecules.

***"The minimum extra energy over and above the average potential energy of the reactants which must be supplied to the reactants to enable them to cross over the energy barrier between reactants and products is called* Activation energy".**

Thus,

Activation energy = (Threshold energy) – (Average energy of the reactants)

or $$E_a = E_T - E_R$$

The idea of activation energy and the energy barrier involved in a reaction is given in Fig. 20.13.

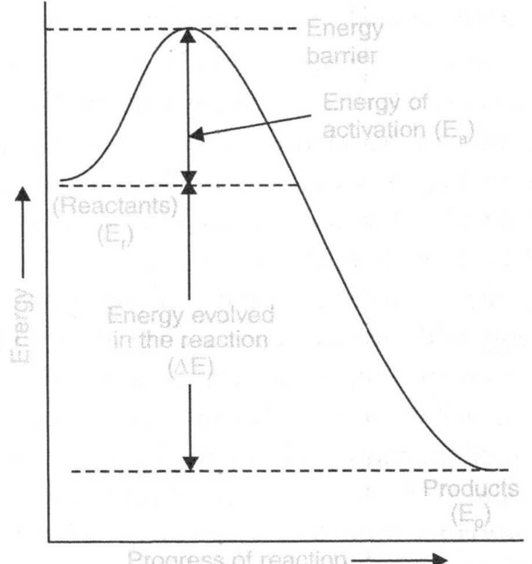

Fig. 20.13. Illustration of activation energy and energy barrier involved in a reaction.

It is important to note that each reaction has a definite value of E_a and this decides the fraction of total collisions which are effective. Obviously, if the activation energy for a reaction is low, large number of molecules can have this energy and the fraction of effective collision, *f*, will be large. Such a reaction proceeds at high rate. On the other hand, if the activation energy is high, then *f* will be small and the reaction may be quite slow.

In short,

For fast reactions; *activation energies are low.*

For slow reactions; *activation energies are high.*

For example, for the reaction between NO and O_2, E_a is low and the reaction is fast whereas for the reaction between CO and O_2, E_a is high and the reaction is slow.

$$2NO + O_2 \longrightarrow 2NO_2 \qquad (Fast)$$

$$2CO + O_2 \longrightarrow 2CO_2 \qquad (Slow)$$

The two theories can be successfully used to explain how the factors of surface area, concentration or pressure of gases, temperature and catalysts affect reaction rates.

In the case of surface area, concentration and pressure of gases at constant temperature, any increase in any individual value is equivalent to an increase in the number of molecules colliding per second. An increase in the frequency of effective collisions results in an increase in reaction rate.

ACTIVATED COMPLEX OR TRANSITION STATE THEORY

It has been pointed out earlier that during a chemical reaction certain bonds are broken and new bonds are formed. The breaking of bonds requires energy whereas the formation of bonds results in the release of energy. For example, in the reaction of hydrogen with iodine to form hydrogen iodine, when a molecule of hydrogen approaches that of iodine, H—H and I—I bonds start breaking and H—I bonds start forming. In the beginning, breaking of bonds predominates and the energy of the system starts increasing till it reaches a maximum (corresponding to the threshold energy). After this, the energy starts decreasing because the process of bond formation predominates and finally leads to the product hydrogen iodide. ***The arrangement of atoms corresponding to energy maxima** (threshold energy)* is called **transition state or activated complex.** In the transition state, the system has partial reactant character and partial product character as shown in Fig. 20.14 and Fig. 20.15.

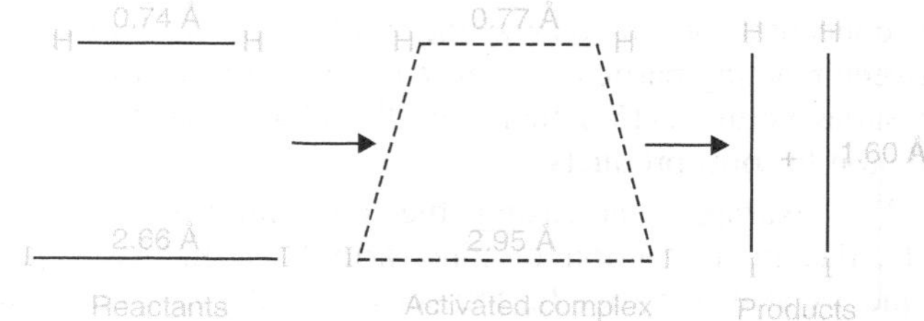

Fig. 20.14. Formation of activated complex during the reaction of H_2 and I_2 from HI.

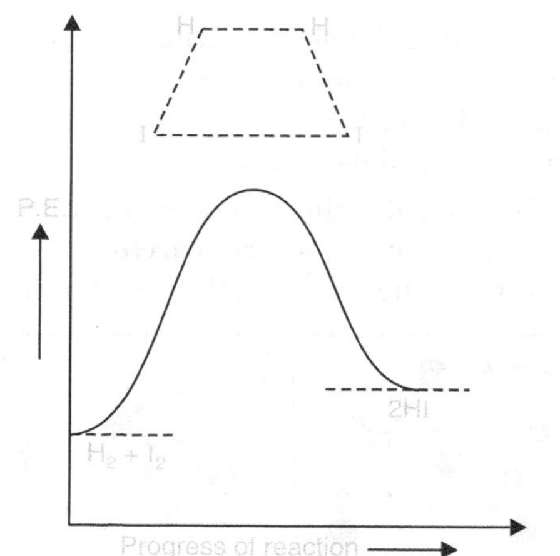

Fig. 20.15. Transition state or activated complex.

The difference between the energy of the transition state and the energy of the reactants is equal to the activation energy.

$$E_{\text{transition state}} - E_{\text{reactants}} = E_{\text{activation}}$$

20.9 EFFECT OF TEMPERATURE AND CATALYST ON THE RATE CONSTANT

TEMPERATURE

Let us now consider the effect of an increase in temperature on the number of effective collisions.

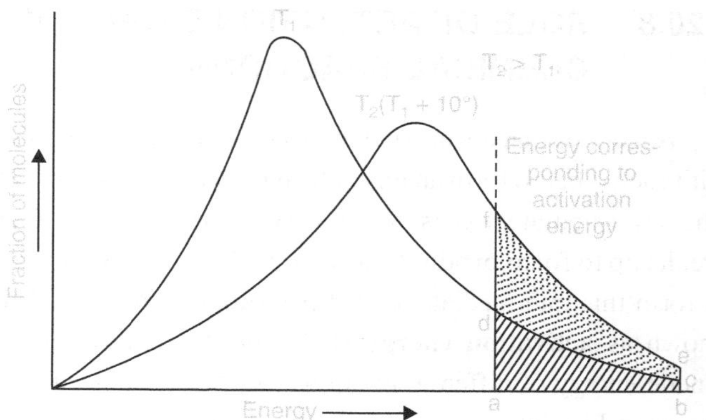

Fig. 20.16. Energy distribution at different temperatures.

On the basis of probability Fig. 20.16 is drawn to give the energy distribution curves at temperatures T_1 and T_2 (where $T_2 = T_1 + 10°$). Now as we know, the rise in temperature increases the kinetic energy of molecules ($\because$ K.E. $\propto$ T). So we should expect the energy distribution curve to be flattened and shift towards a higher energy region. A close examination of the curves clearly reveals that the fraction of molecules possessing higher kinetic energy, *i.e.*, energy greater than the threshold energy, as indicated by the shaded portion becomes almost double and therefore *the rate of reaction almost doubles for a 10° rise of temperature.* Thus, an increase in the rate of reaction with an increase in temperature is mainly due to an increase in the number of collisions which are energetically effective.

Temperature Dependence on Rate Constant

Arrhenius developed a mathematical relation between *temperature* and the *rate constant* on the basis of observations from a large number of experiments. This temperature dependence of the rate constant is expressed algebraically as,

$$k = Ae^{-E_a/RT} \qquad ...(20.1)$$

Equation (20.1) is called the ***Arrhenius equation.*** Here A is a pre exponential factor called the *frequency factor,* E_a is the *energy of activation* and T is the temperature in kelvin. The term $e^{-E_a/RT}$ in the above equation is also called the **Boltzmann Factor.** Both A and E_a are characteristic of the reaction. Another form of the equation which is more useful for calculations is obtained by taking the logarithm of Eqn. (20.1), thus,

$$\log \frac{k_2}{k_1} = \frac{E_a}{2.303R}\left[\frac{1}{T_1} - \frac{1}{T_2}\right] \qquad ...(20.2)$$

Here k_2 and k_1 are the rate constants at temperatures T_1 and T_2 respectively, E_a is the activation energy and R is the universal gas constant.

Add to Your Knowledge

Arrhenius Equation—A Justification for Reaction Rate Doubling

Let us now see how the rate constant of a reaction varies when the temperature is increased by 10° from 300 to 310 K. In general, for many of the reactions the activation energy (E_a) is about 55 kJ mol^{-1} near room temperature. Hence,

$$\log \frac{k_2}{k_1} = \frac{55{,}000}{2.303 \times 8.314}\left(\frac{310 - 300}{300 \times 310}\right) = 0.30887$$

or $$\frac{k_2}{k_1} = \text{antilog } 0.30887 = 2.03 \simeq 2.0$$

Thus, it is quite clear that for majority of reactions the rate constant doubles for a 10° rise in temperature near room temperature.

EFFECT OF CATALYST ON RATE OF REACTION

A catalyst is a substance that alters the rate of a chemical reaction without itself being chemically changed permanently. Never state things like "it doesn't react, just speeds it up". It must take part in the reaction and it must change chemically, albeit on a temporary basis. A catalyst provides a different 'pathway' or mechanism that makes the bond breaking processes (or other electronic changes in the reactants) occur more readily. In general,

- *A catalyst speeds up a reaction, but it must be involved 'chemically', however temporarity, in some way, and is continually changed and reformed as the reaction proceeds.*
- *Catalysts work by providing an alternative reaction pathway of lower activation energy.*

Thus, the function of a catalyst is to lower the activation energy. In simple words, the greater the decrease in activation energy caused by the catalyst, the higher the reaction rate. In the presence of a catalyst, the reaction follows a path of lower activation energy. Under this condition, a large number of reacting molecules are able to cross-over the energy barrier and the rate of reaction increases. The energy profile diagram for catalysed and uncatalysed reactions are as shown in the Fig. 20.17. The dotted curve represents the progress of a catalysed reaction and the solid curve represents an uncatalysed reaction.

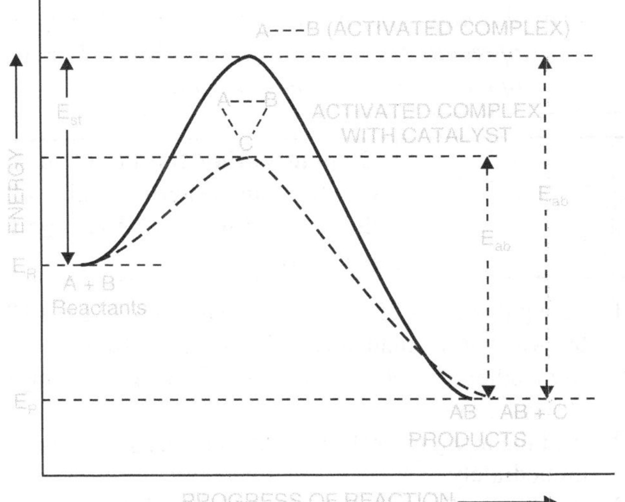

Fig. 20.17. Potential energy curves for catalysed and uncatalysed reactions.

For a general reaction of the type A + B ⟶ AB the course of uncatalysed and catalysed reaction may be represented as:

(*a*) *Uncatalysed reaction:*

$$A + B \longrightarrow \underset{\text{Activated complex}}{[A......B]} \longrightarrow AB$$

(*b*) *Catalysed reaction:*

$$\underset{\text{Reactants}}{[A + B]} + \underset{\text{Catalyst}}{C} \longrightarrow \underset{\text{Activated complex}}{\begin{bmatrix} A - - - - B \\ \diagdown \quad \diagup \\ C \end{bmatrix}} \longrightarrow \underset{\text{Product}}{AB} + \underset{\text{Catalyst}}{C}$$

Though the catalyst increases the rate of the reaction, it does not affect the state of equilibrium in the case of reversible reactinons. It is because the activation energy for the forward reaction and the backward reaction is reduced to the same extent.

Add to Your Knowledge

CHARACTERISTICS OF A CATALYST

(*i*) *A catalyst is highly specific in its action. This means that a catalyst may increase the rate of one reaction but be completely ineffective in another reaction.*

(*ii*) *A catalyst may undergo intermediate physical change and may even form temporary chemical bonds with the reactants but is recovered unchanged in mass and composition at the end of the reaction.*

(*iii*) *A catalyst never starts a reaction which has no tendency to proceed.*

(*iv*) *A catalyst does not alter the individual energies of reactants and products, therefore ΔH or ΔU of the reaction remains unaltered.*

ACTIVITY 20.5 **Demonstrating the effect of a catalyst on reaction rate of decomposition of hydrogen peroxide**

1. Set up three sets of apparatus as shown in Fig. 20.18 but *without* the manganese (IV) oxide at first. Remember to set the volume of the gas syringe at 0 cm^3.
2. Add the manganese(IV) oxide and start a stop-clock immediately.
3. Read the volume of gas collected in each gas syringe at minute intervals until each reaction is complete. Identify the gas and write a balanced chemical equation for the reaction. Which gas syringe contains more of the gas at any one time?
   Give an explanation for this observation.
   Describe the appearance of the oxide before and after the reaction.
4. Tabulate the results. On the same axes plot graphs of volume of gas (cm^3) on the vertical axis and time (minutes) on the horizontal axis for two different sets of readings.
   Which sets of readings were used and why?
5. Draw a conclusion about the reaction rate and
   (*a*) the presence of a catalyst
   (*b*) different amounts of the same catalyst
   (*c*) the absence of a catalyst.
6. Repeat steps 1–5 but replace 1 g and 3 g of manganese (IV) oxide with 5 cm^3 and 15 cm^3 of dilute 1 M HCl respectively.
7. Draw a conclusion about the reaction rate and the presence of
   (*a*) manganese(IV) oxide and
   (*b*) dilute 1 M HCl.

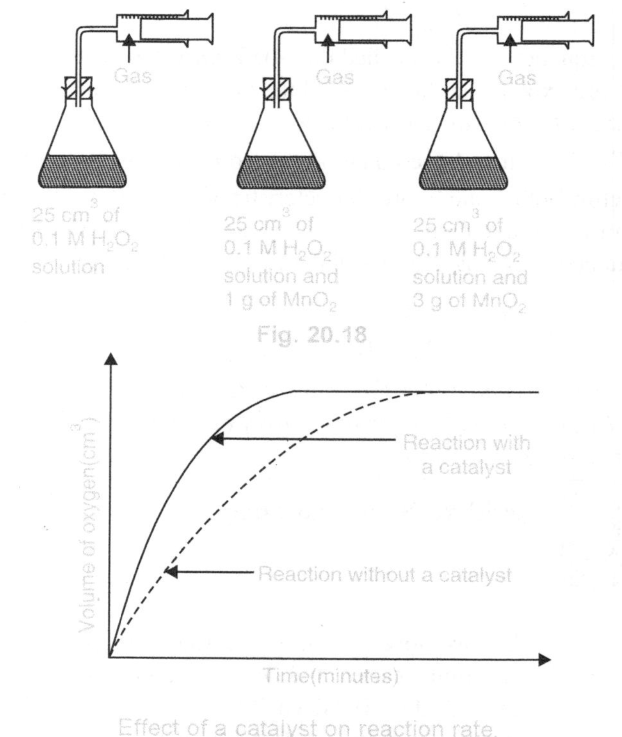

Fig. 20.18

Effect of a catalyst on reaction rate.

20.10 DEDUCING RATE LAW FROM A GIVEN DATA

According to the **Collision Theory** a chemical reaction takes place due to collisions between the particles of the reactants. *The number of reacting species (atoms, ions or molecules) which must collide simultaneously in order to bring about a chemical reaction is called* **molecularity** *of a reaction*. The molecularity of a reaction can be 1, 2 or 3. For **example,**

(*i*) The decomposition of ammonium nitrite is a **unimolecular** reaction.

$$NH_4NO_2 \longrightarrow N_2 + 2H_2O$$

(*ii*) The reaction involving simultaneous collision between *two species* is a **bimolecular** reaction. For **example,** dissociation of HI is a bimolecular reaction.

$$2HI(g) \longrightarrow H_2(g) + I_2(g)$$

(*iii*) In the same way, the reaction between NO and O_2 is a **trimolecular** reaction because it involves collisions among three species

$$2NO + O_2 \longrightarrow 2NO_2$$

In all the above chemical reactions molecularity is simply the sum of molecules of the different reactants as represented by the balanced chemical equation. Such reactions are known as **elementary reactions.**

For most reactions, the molecularity does not exceed three. It is because the *chances of simultaneous collisions between three or more particles* are rare.

A large number of chemical reactions show balanced equations with four or more species/molecules/ions participating in the reaction. Some common examples are:

(*i*) $N_2 + 3H_2 \longrightarrow 2NH_3$

(*ii*) $4HBr + O_2 \longrightarrow 2H_2O + 2Br_2$

(*iii*) $2MnO_4^{-1} + 5C_2O_4^{2-} + 16H^+ \longrightarrow 2Mn^{+2} + 10CO_2 + 8H_2O$

On the basis of balanced equations, the molecularity of above reactions are 4, 5 and 23 respectively.

Since a molecularity greater than three is not possible, the above reactions does not involve the simultaneous collision of all the reacting species in a single step. In fact, such chemical reactions proceed through a sequence of steps. Each step is an elementary step and involves the simultaneous collision of two or three species only. Such ***chemical reactions which proceed through more than one step are termed complex reactions.***

The detailed description of the various steps of a complex chemical reaction is called the **mechanism of a reaction.** For **example,** the reaction,

$$4HBr + O_2 \longrightarrow 2H_2O + 2Br_2$$

occurs by the following steps:

Step I. $HBr + O_2 \xrightarrow{slow} HOOBr$

Step II. $HOOBr + HBr \xrightarrow{fast} 2HOBr$

Step III. $[HOBr + HBr \xrightarrow{fast} H_2O + Br_2] \times 2$

$$4HBr + O_2 \longrightarrow 2H_2O + 2Br_2$$

The different steps of the given reaction are written based upon experimental evidence like *detection of the presence of some shortlived intermediates, etc.* All the above steps are known to proceed at different rates.

RATE LAW OR RATE EQUATION

The determination of the rate law expression of a complex reaction is not an easy task. Its determination requires,

(*i*) *Information about the number of moles of reactants consumed during the reaction.*

(*ii*) *A knowledge of the intermediates produced during the reaction and how much they accumulate during the early stages of the reaction.*

(*iii*) *The rate data to be supplemented by different techniques so that certain elementary steps are verified to the maximum.*

In complex reactions a rate expression written on the basis of the overall balanced equation has no significance at all. It merely represents a theoretical expression. The true rate expression for such complex reactions can be evaluated on the basis of the experimental data only. For example, the reaction between NO_2 and F_2 to yield nitryl fluoride (NO_2F), *i.e.*,

$$2NO_2(g) + F_2(g) \longrightarrow 2NO_2F(g)$$

Experimentally, it has been observed that the rate of this reaction is proportional to the product of a single concentration term of NO_2 and F_2. Thus, the experimental rate of the reaction or the actual rate of the reaction is given as

$$\text{rate} = k[NO_2]\,[F_2].$$

A mathematical expression that gives the true rate of a reaction in terms of the concentration of the reactants, which actually influence the rate, is called **Rate Law.**

For a general reaction,

$$aA + bB \longrightarrow cC + dD$$

where *a*, *b*, *c* and *d* are the stoichiometric coefficients of reactants and products, the rate expression for the reaction may be

$$\text{Rate} \propto [A]^x\,[B]^y = k\,[A]^x\,[B]^y$$

k is known as the **rate constant** or **velocity constant** or **specific reaction rate.** If the concentration of all reacting species is taken as unity then,

$$\text{rate} = k \qquad [\because \ [A] = [B] = 1 \text{ mol dm}^{-3}]$$

Hence, the **rate constant** is defined as *the rate of a reaction when the concentration of each reactant is taken as unity.* It is for this reason that the rate constant is also called **specific reaction rate.**

CHARACTERISTICS OF RATE CONSTANT

(*i*) The value of a rate constant gives an idea about the speed of a reaction, *i.e.*, the greater the value of a rate constant the faster the reaction. In simple words it is a measure of the rate of a reaction.

(*ii*) Each reaction has a definite value of a rate constant at a particular temperature.

(*iii*) The value of a rate constant depends on the temperature *i.e.*, changes with change in temperature.

(*iv*) The value of a rate constant is independent of the concentration of the reacting species.

(*v*) The units of a rate constant depend on the order of a reaction.

DIFFERENCES BETWEEN RATE AND RATE CONSTANT

Rate of Reaction	*Rate Constant of Reaction*
1. It is the speed at which the reactants are converted into products at any moment of time.	1. It is a constant of proportionality in the rate law expression.
2. It depends upon the concentration of reactants at any moment of time.	2. It refers to the rate of a reaction at a specific point where the concentration of every reacting species is unity.
3. It generally decreases with the progress of a reaction.	3. It is constant and does not depend on the progress of the reaction.

RATE CONTROLLING STEP

As pointed out, already a complex reaction proceeds through more than one step. Let us now examine which of these steps can be used to write the rate expression. Out of various steps of a reaction, the **slowest step** will decide the rate of the overall reaction because the reaction cannot take place faster than the slowest step. Thus, the ***slowest step of a complex reaction* is called the Rate Controlling Step** or **Rate Determining Step.** A few examples are discussed to illustrate the above point(s) more clearly:

Reactions Involving Slow Steps

(*i*) **Thermal decomposition of nitrous oxide**

$$2N_2O \longrightarrow 2N_2 + O_2$$

On the basis of experimental data it has been is observed that the rate of this reaction depends on the concentration of N_2O raised to the power one. A possible mechanism of the reaction may be proposed as:

Step I. $N_2O \xrightarrow{\text{slow step}} N_2 + O$

Step II. $N_2O + O \xrightarrow{\text{fast step}} N_2 + O_2$

$$2N_2O \longrightarrow 2N_2 + O_2$$

... *Overall reaction*

Thus, the rate law expression for the reaction can be written as

$$\text{Rate} = k[N_2O].$$

The above postulated mechanism is consistent with the rate law expression.

*It must be **noted** here that there is no way to verify the absolute validity of the proposed mechanism. All we can do is to postulate a mechanism based on experimental data and intuitive power. If further investigations reveal information that is inconsistent with a postulated mechanism, it has to be modified accordingly so as to explain all the observations.*

(*ii*) **Thermal decomposition of dinitrogen pentaoxide**

$$2N_2O_5 \longrightarrow 4NO_2 + O_2.$$

The rate of the above reaction has been found to be dependent on the concentration of N_2O_5 raised to the power one. Thus, the following mechanism can be proposed for this reaction.

Step I. $N_2O_5 \xrightarrow{\text{slow}} NO_2 + NO_3$

Step II. $N_2O_5 + NO_3 \xrightarrow{\text{fast}} 3NO_2 + O_2$

The above reaction is a unimolecular reaction and the rate law expression will be

$$\text{Rate} = k\,[N_2O_5]$$

(*iii*) **Reaction between nitric oxide and hydrogen**

$$2NO + 2H_2 \longrightarrow N_2 + 2H_2O$$

Experimentally, it has been observed that the rate of this reaction depends upon the concentration of NO raised to the power two and the concentration of H_2 raised to the power one. Consistent with this data the logical mechanism for the reaction may be written as:

Step I. $2NO + H_2 \xrightarrow{\text{slow}} N_2 + H_2O_2$

Step II. $H_2O_2 + H_2 \xrightarrow{\text{fast}} 2H_2O$

The rate law expression for this reaction can be written as

$$\text{Rate} = k\,[NO]^2\,[H_2]$$

SOLVED EXAMPLES

Example 20.5. *For a reaction,* $2NO_2 + F_2 \longrightarrow 2NO_2F$ *the experimental rate law is* $r = k[NO_2]\,[F_2]$*. Propose the mechanism of the reaction.*

Solution. The rate law is $r = k[NO_2]\,[F_2]$

Since the rate law is proportional to $[NO_2]$ and $[F_2]$, it implies that only one molecule of NO_2 and one molecule of F_2 are involved in the slow step. Thus, the various steps are:

Step I. $NO_2 + F_2 \longrightarrow NO_2F + F$ ...*(slow)*

Step II. $NO_2 + F \longrightarrow NO_2F$ ...*(fast)*

$2NO_2 + F_2 \longrightarrow 2NO_2F$ ...*(Overall reaction)*

Example 20.6. *The reaction,* $2NO + Br_2 \longrightarrow 2NOBr$ *proceeds by the following mechanism:*

$$(i)\quad NO + Br_2 \underset{fast}{\overset{step\ I}{\rightleftharpoons}} NOBr_2$$

$$(ii)\quad NOBr_2 + NO \xrightarrow[slow]{step\ II} 2NOBr$$

Suggest a rate law consistent with the mechanism.

Solution. The slowest step is the rate determining step. Therefore the rate law is

$$r = k\,[NOBr_2]\,[NO]$$

Now since $NOBr_2$ is an intermediate its concentration can be calculated from step I as follows:

$$K = \frac{[NOBr_2]}{[NO]\,[Br_2]} \qquad [K = \text{Eq. constant}]$$

or $$[NOBr_2] = K\,[NO]\,[Br_2]$$

Substituting this value in the above equation

$$r = k\,.\,K\,[NO]\,[Br_2]\,[NO]$$

or $$\text{rate} = k'\,[NO]^2\,[Br_2] \qquad [\because\ k\,.\,K = \text{constant } k']$$

$\therefore$ Rate law is rate $= k'\,[NO]^2\,[Br_2]$.

Example 20.7. *The reaction,* $NO_2 + CO \longrightarrow CO_2 + NO$ *is experimentally found to have a rate given as* $r = k[NO_2]^2$*. Suggest a mechanism consistent with the data.*

Solution. The possible mechanism can be

Step I. $NO_2 + NO_2 \xrightarrow{slow} NO + NO_3$

Step II. $NO_3 + CO \xrightarrow{fast} CO_2 + NO_2$.

Example 20.8. *Reaction between NO and* O_2 *to give* NO_2 *takes place by the following mechanism:*

$$NO(g) + O_2(g) \xrightarrow{Slow} NO_3$$

$$NO_3(g) + NO(g) \xrightarrow{Fast} 2NO_2$$

Write the rate expression for the reaction.

Solution. The rate of a complex reaction is given by the slowest step in the reaction

$$\therefore \quad \frac{-d[NO]}{dt} = k[NO][O_2].$$

20.11 ZERO, FIRST AND SECOND ORDER REACTIONS

Order of a reaction is an important parameter for every chemical reaction. It is always determined experimentally and cannot be written from the balanced chemical equation. It refers to the number of reacting particles whose concentration terms determine the reaction rate.

It may be defined as, ***"the sum of powers or exponents to which the concentration terms are raised in the rate law expression."***

For example, for a hypothetical reaction,

$$aA + bB \longrightarrow \text{Products}$$

if the rate law expression for this reactions is,

$$\text{Rate} = k[A]^m\,[B]^n$$

The **order** of the above reaction is equal to **(m + n)**. These powers or exponents, *i.e.*, *m* and *n* have no relation to the stoichiometric coefficients *a* and *b* of the balanced chemical equation. Order of the reaction with respect to A is *m* and that with respect to B is *n*. If the sum of the powers is equal to *one*, the reaction is called **first order reaction.** If the sum of the powers is **two or three,** the reaction is a **second order or third order** reaction respectively. *The order of a reaction can also be **zero** or a **fraction.***

For illustration, a few examples are given below:

(*i*) Nitrous oxide (N_2O) decomposes as:

$$2N_2O(g) \longrightarrow 2N_2(g) + O_2(g)$$

The rate law expression for this reaction is

$$\text{Rate} = k\,[N_2O];\ \text{Order} = 1$$

Therefore, it is a ***first order*** reaction.

(*ii*) For the reaction, $H_2(g) + I_2(g) \longrightarrow 2HI(g)$; the rate law is

$$\text{Rate} = k[H_2][I_2]; \quad \text{Order} = 1 + 1 = 2$$

So, it is a ***second order*** reaction.

(*iii*) The reaction between nitric oxide and oxygen is an example of a ***third order*** reaction.

$$2NO(g) + O_2(g) \longrightarrow 2NO_2(g);$$

The rate law is

$$\text{Rate} = k[NO]^2[O_2]; \text{ Order} = 2 + 1 = 3.$$

(*iv*) The reaction between hydrogen and bromine to form hydrogen bromide has a ***fractional order***

$$H_2(g) + Br_2(g) \longrightarrow 2HBr(g);$$

The rate law is

$$\text{Rate} = k\,[H_2][Br_2]^{1/2}; \quad \text{Order} = 1 + \frac{1}{2} = \frac{3}{2}.$$

(*v*) Decomposition of ammonia over platinum or gold is a ***zero order reaction***

$$2NH_3 \xrightarrow{\text{Platinum}} N_2 + 3H_2$$

$$\text{Rate} = k\,[NH_3]^0; \quad \text{order = zero}$$

In the above decomposition reaction, the metal acts a catalyst.

UNITS OF RATE CONSTANT FOR REACTIONS OF DIFFERENT ORDERS

The rate constant *k* has different units for reactions having different orders.

(*i*) For *zero order reactions,*

$$\text{Rate} = \frac{dx}{dt} = k[A]^0$$

$$\frac{\text{mol dm}^{-3}}{\text{sec}} = k\,.\,1 \quad \text{or} \quad k = \textbf{mol dm}^{-3}\,\textbf{s}^{-1}.$$

Thus, *the units of the rate constant for a* **zero order reaction** *is*: **mol dm^{-3} s^{-1}.**

(*ii*) For *first order reactions,*

$$\text{Rate} = \frac{dx}{dt} = k\,[A]$$

$$\frac{\text{mol dm}^{-3}}{\text{sec}} = k\,.\,\text{mol dm}^{-3} \quad \text{or} \quad k = \frac{1}{\text{sec}} = \text{sec}^{-1}.$$

∴ *The units of the rate constant for a* **first order reaction** *is*: **sec^{-1}.**

(*iii*) For *second order reactions,*

$$\text{Rate} = \frac{dx}{dt} = k[A]^2$$

$$\frac{\text{mol dm}^{-3}}{\text{sec}} = k\,(\text{mol dm}^{-3})^2$$

or $$k = \frac{1}{\text{mol dm}^{-3}\text{ sec}} = \text{dm}^3\text{ mol}^{-1}\text{ sec}^{-1}.$$

∴ *The units of the rate constant for a* **second order reaction** *is*: **dm^3 mol^{-1} sec^{-1}.**

(*iv*) For *third order reactions,*

$$\text{Rate} = \frac{dx}{dt} = k\,[A]^3$$

$$\frac{\text{mol dm}^{-3}}{\text{sec}} = k\,(\text{mol dm}^{-3})^3$$

or $$k = \frac{1}{\text{sec mol}^2\,.\,\text{dm}^{-6}} = \text{dm}^6\text{ mol}^{-2}\text{ sec}^{-1}.$$

∴ *The units of the rate constant for a* **third order reaction** *is*: **dm^6 mol^{-2} sec^{-1}.**

SOLVED EXAMPLES

Example 20.9. *(a) Draw a schematic graph showing how the rate of a first order reaction changes with changes in the concentration of the reactant.*

(b) Rate of a reaction is given by the equation:

$$Rate = k[A]^2\,[B]^2$$

What are the units for the rate and rate constant of this reaction?

Solution. (*a*) Rate of a first order reaction is directly proportional to the molar concentration of the reactant *i.e.*, rate ∝ conc. So the graph is a straight line as shown in the figure given below:

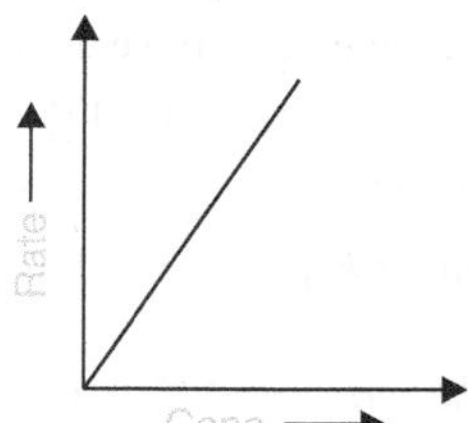

(*b*) Unit of rate = mol dm^{-3} sec^{-1}

$$\left(\because \text{ rate} = \frac{\text{Change in conc.}}{\text{time}}\right)$$

Unit of rate constant,

$$k = \frac{\text{rate}}{[A]^2[B]} = \frac{\text{mol dm}^{-3}\text{ sec}^{-1}}{(\text{mol dm}^{-3})^2(\text{mol dm}^{-3})}$$

$$= \text{mol}^{-2}\text{ dm}^6\text{ sec}^{-1}$$

In terms of molarity, M, these can also be written as

Unit of rate of reaction = M s^{-1} (here M = mol dm^{-3})

Unit of rate constant = M^{-2} s^{-1}

Example 20.10. *Identify the reaction order from each of the following rate constant:*

(*i*) $k = 3.1 \times 10^{-4}\ s^{-1}$

(*ii*) $k = 4.2 \times 10^{-5}\ dm^3\ mol^{-1}\ s^{-1}$

(*iii*) $k = 6.5 \times 10^{-4}\ mol\ dm^{-3}\ s^{-1}$.

Solution. On the basis of units of rate constant the order of reactions are

(*i*) 1st order (*ii*) 2nd order

(*iii*) zero order.

INTEGRATED RATE EQUATIONS

The integrated rate equations are different for reactions of different reaction orders. Let us illustrate this method by applying it to a **zero order** and a **first order reaction.**

1. ZERO ORDER REACTION

A *reaction is zero order if its rate is independent of the concentration of the reactants* or a zero order reaction means that the rate of reaction is proportional to the concentration of reactants raised to the power zero. For a zero order reaction of the form

$$R \longrightarrow P$$

the rate of reaction, $$\text{Rate} = -\frac{d[R]}{dt} = k[R]^0 \quad ...(20.3)$$

As any quantity raised to the power zero is unity

$$\therefore \quad \text{Rate} = -\frac{d[R]}{dt} = k \times 1$$

or $$d[R] = -\,kdt$$

Integrating both sides, we get

$$[R] = -\,kt + I \quad ...(20.4)$$

where, I is the constant of integration

At $t = 0$, $[R] = [R]_0$, where $[R]_0$ is the initial concentration of the reactant.

Substituting this value in equation (20.4), we get

$$[R] = -\,kt + [R]_0 \quad \text{or} \quad kt = [R]_0 - [R]$$

or $$k = \frac{1}{t}\left[[R]_0 - [R]\right] \quad ...(20.5)$$

SIGNIFICANCE OF INTEGRATED RATE EQUATION

(*i*) All zero order reactions obey the equation.

(*ii*) The value of k can be evaluated if $[R]_0$ and [R] at time t are known.

On plotting a graph of [R] vs t we get a straight line as shown in the figure below. The slope of the line is given as

$$\text{slope} = -\,k$$

The intercept on the concentration axis = $[R]_0$

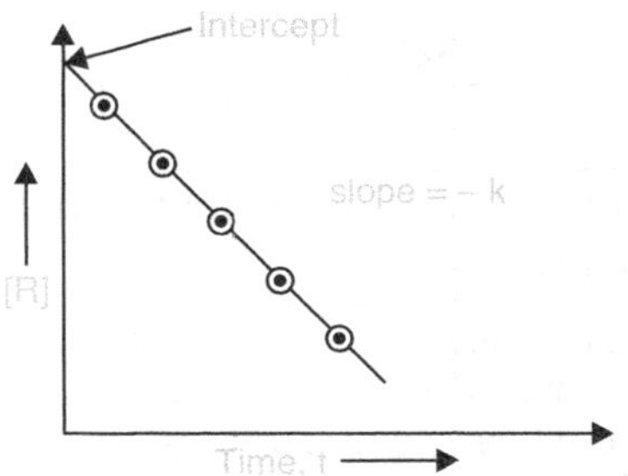

2. FIRST ORDER REACTION

Let us assume a simple hypothetical first order reaction represented as,

$$R \longrightarrow P$$

If the initial concentration of R is $[R]_0$, k is the rate constant and [R] is conc. at time t then the differential form of this first order reaction will be

$$\frac{-\,d[R]}{dt} = k[R]$$

Rearranging the equation, we get

$$\frac{-\,d[R]}{[R]} = kdt \quad ...(20.6)$$

On integrating equation (20.6),

$$\int \frac{-\,d[R]}{[R]} = \int kdt$$

we get, $$-\ln [R] = k\,t + \text{constant} \quad ...(20.7)$$

$$\left[\text{Integral of } \frac{d[R]}{[R]} \text{ is } \ln [R]\right]$$

To get the value of the constant k let us put the value of $t = 0$ in Eqn. (20.7), then

$$-\ln [R]_0 = k \times 0 + \text{constant}$$

or $$\text{constant} = -\ln [R]_0 \qquad [\because \text{ When } t = 0 \text{ then } [R] = [R]_0 \text{ i.e., initial conc.}]$$

Substituting this value in Eqn. (20.7), we get

$$-\ln [R] = k\,t - \ln [R]_0$$

or $$k\,t = \ln [R]_0 - \ln [R]$$

$$k\,t = \ln \frac{[R]_0}{[R]} \qquad ...(20.8)$$

Changing it to common log $k\,t = 2.303 \log \frac{[R]_0}{[R]}$

or

$$\mathbf{k = \frac{2.303}{t} \log \frac{[R]_0}{[R]}} \qquad ...(20.9)$$

Equation (20.9) is the **Integrated Rate Equation** for a first order reaction.

Physical significance of k. It represents the fraction of the reactant decomposed per unit time of constant concentration. For example, if k for a given reaction is 0.00674 s^{-1}, it means that 0.00674 × 100 = 0.674% of substance would dissociate per second provided the initial concentration is kept constant.

SIGNIFICANCE OF INTEGRATED RATE EQUATION

The integrated rate equation is also used to evaluate the value of k by a **graphical method.** Equation (20.9) can be written as

$$\frac{kt}{2.303} = \log \frac{[R]_0}{[R]} \quad \text{or} \quad \frac{kt}{2.303} = \log [R]_0 - \log [R]$$

or

$$\log [R] = \frac{-k}{2.303} \cdot t + \log [R]_0$$

If we plot log [R] vs time or $\log \frac{[R]_0}{[R]}$ vs time the following graphs would be obtained.

In both plots *i.e.*, Fig. 20.19 (*a*) and 20.19 (*b*) we get straight lines whose slope would be given as

$$\text{slope} = -\frac{k}{2.303} \quad \text{(Fig. 20.19 }(a));$$

$$\text{slope} = \frac{k}{2.303} \quad \text{(Fig. 20.19 }(b))$$

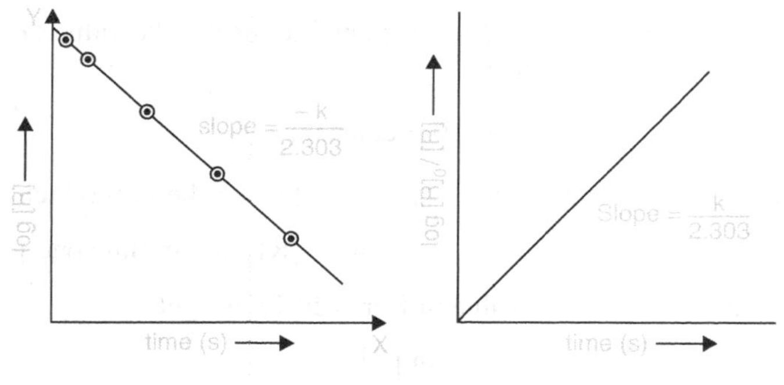

Fig. 20.19 (a) Fig. 20.19 (b)

From the value of the slope, the value of k can be calculated. The intercept made on the Y-axis would be log $[R]_0$.

All natural and artificial radioactive decay of nuclei takes place by first order reaction and follow first order kinetics.

20.12 HALF LIFE OF A FIRST ORDER REACTION

Half life is defined as ***the time during which the concentration of the reactants is reduced to half of the initial concentration or it is the time required for the completion of half of the reaction.*** It is denoted by $t_{1/2}$.

For a first order reaction, we know that

$$k = \frac{2.303}{t} \log \frac{[A]_0}{[A]} \quad \text{or} \quad t = \frac{2.303}{k} \log \frac{[A]_0}{[A]}$$

When $t = t_{1/2}$ then $[A] = \frac{1}{2}[A]_0$,

therefore, $t_{1/2} = \frac{2.303}{k} \log \frac{[A]_0}{\frac{1}{2}[A]_0} = \frac{2.303}{k} \log 2$

$$\mathbf{t_{1/2} = \frac{0.693}{k}} \qquad [\because \log 2 = 0.3010]$$

It is quite clear from the above expression that the ***half life period or half change time for a first order reaction does not depend upon the initial concentration of the reactants.*** *Similarly, the time required to reduce the concentration of the reactant to any fraction of the initial concentration for a first order reaction is also independent of the initial concentration.*

Add to Your Knowledge

TIME REQUIRED TO COMPLETE DIFFERENT FRACTIONS OF A FIRST ORDER REACTION

In addition to the calculation of $t_{1/2}$, the time required to complete different fractions of a first order reaction can also be calculated.

For example, the time required to complete one-third of the reaction will be given as

$$t_{1/3} = \frac{2.303}{k} \log \frac{a}{a - a/3}$$

$$= \frac{2.303}{k} \log \frac{3}{2} = \frac{0.406}{k}.$$

Similarly, the expressions for different fractions of time to complete are given as

(*i*) $t_{3/4}$ or $t_{75\%} = \dfrac{1.386}{k}$;

(*ii*) $t_{0.6}$ or $t_{60\%} = \dfrac{0.9165}{k}$;

(*iii*) $t_{0.9}$ or $t_{90\%} = \dfrac{2.303}{k}$;

(*iv*) $t_{0.99}$ or $t_{99.9\%} = \dfrac{6.909}{k}$.

3. SECOND ORDER REACTION

$$2A \longrightarrow \text{Products}$$

For the above reaction the rate,

$$r = \frac{-d[A]}{dt} = k[A]^2$$

$$r = -k[A]^2 \equiv y = mx^2$$

A graph of rate, r versus $[A]^2$ will give a curve as shown in Fig. 20.20.

Integrating the second-order rate equation:

$$\int \frac{d[A]}{dt} = -k\int [A]^2, \int \frac{d[A]}{[A]^2}$$

$$= \int [A]^{-2}\, d[A] = -k\int dt,$$

$$[-A]^{-1} + [A]_0^{-1} = -kt \quad \text{or} \quad [A_t]^{-1} = kt + [A]_0^{-1}$$

$$\equiv y = mx + c$$

If a graph of the reciprocal of concentration, $[A]_t^{-1}$ versus time, t, is drawn and a straight line is obtained, it indicates a second-order reaction. The gradient of such a curve is k, the rate constant. This is shown in Fig. 20.20. When evaluated, the half-life

$$t_{1/2} = \frac{1}{k[A]_0}$$

indicating that the half-life of a second-order reaction depends inversely on the initial concentration $[A]_0$, and the rate constant, k.

Example of second-order reactions are:

$CH_3Cl + OH^- \longrightarrow CH_3OH + Cl^-$

$r = k[CH_3Cl]\,[OH^-]$

$H_2(g) + I_2(g) \longrightarrow 2HI(g)$

$r = k[I_2]\,[H_2]$

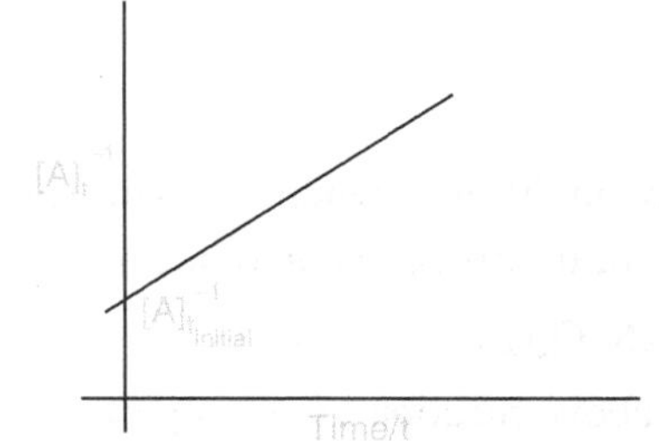

Fig. 20.20. Graph of $[A]_t^{-1}$ versus time for a second-order reaction.

20.13 EXPERIMENTAL DETERMINATION OF THE ORDER OF A REACTION

The following methods are employed to determine the *rate law, the rate constant and the order of a reaction*.

1. GRAPHICAL METHOD

This method is used to determine the rate law of a reaction which involves only *one reactant species*. The various steps involved are:

(*i*) The concentrations of the reacting substance is determined at different time intervals by some suitable method.

(*ii*) A graph is plotted between concentration and time.

(*iii*) For the plot of *concentration* vs *time*, the instantaneous rates corresponding to different concentrations are determined by drawing tangents to the curve and subsequently calculating their slopes as discussed earlier (instantaneous rate).

(*iv*) Different graphs are now plotted between:

(*a*) ***reaction rate*** vs ***concentration*** or

(*b*) ***reaction rate*** vs ***(concentration)***^{2} or in general,

(*c*) ***reaction rate*** vs ***(concentration)***^{n}, where n = 1, 2, 3 and so on.

If a straight line is obtained in the first case, it means the rate is directly proportional to concentration of the reactant. This in turn means that the rate law is given as,

$$\text{rate} = k\,[\text{Reactant}]$$

and therefore its **order is *one.***

Similarly, if a straight line is obtained in the second case, then

$$\text{rate} = k\,[\text{Reactant}]^2$$

and its **order is *two.***

EXPERIMENTAL DETERMINATION OF ORDER BY GRAPHICAL METHOD

Let us explain the above method, in details, by using the decomposition of dinitrogen pentaoxide as an example.

$$2N_2O_5(g) \longrightarrow 4NO_2(g) + O_2(g)$$

Since the above reaction involves gaseous reactants and products, it is convenient to observe the changes in pressure accompanying the reaction. From the measured values of total pressure, the partial pressure of N_2O_5 is first calculated after which the concentration in moles per litre is determined. The molar concentrations of N_2O_5 at different time intervals so obtained are given in Table 20.2.

Table 20.2. Concentration-Time Data for Decomposition of N_2O_5(g)

Time (min)	*$[N_2O_5]$ (mol dm^{-3})*
0	1.72×10^{-2}
10	1.13×10^{-2}
20	0.84×10^{-2}
30	0.62×10^{-2}
40	0.46×10^{-2}
50	0.35×10^{-2}
60	0.26×10^{-2}
70	0.19×10^{-2}
80	0.14×10^{-2}

Thereafter, the rate of reaction (Table 20.3) at different time intervals is obtained by calculating the slope of the tangent of the curve in Fig. 20.21. (*obtained by plot of $[N_2O_5]$ as a function of time*).

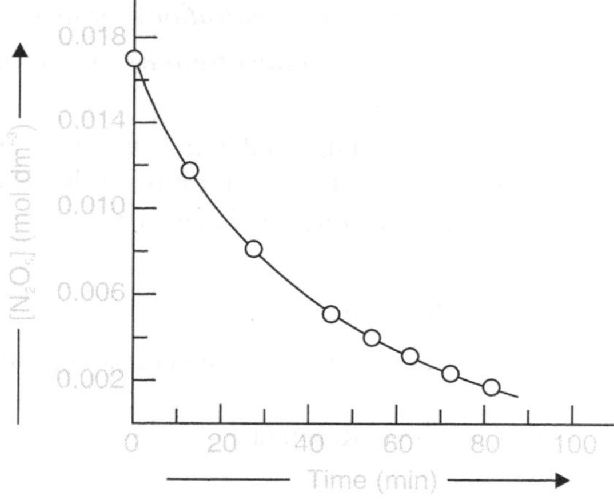

Fig. 20.21. Concentration of N_2O_5 as a function of time.

Table 20.3. Rate Data for Decomposition of N_2O_5(g)

$[N_2O_5]$ (mol dm^{-3})	*Rate (mol dm^{-3} min^{-1})*
1.13×10^{-2}	34×10^{-5}
0.84×10^{-2}	25×10^{-5}
0.62×10^{-2}	18×10^{-5}
0.46×10^{-2}	13×10^{-5}
0.35×10^{-2}	10×10^{-5}
0.26×10^{-2}	8×10^{-5}
0.19×10^{-2}	6×10^{-5}
0.14×10^{-2}	4×10^{-5}

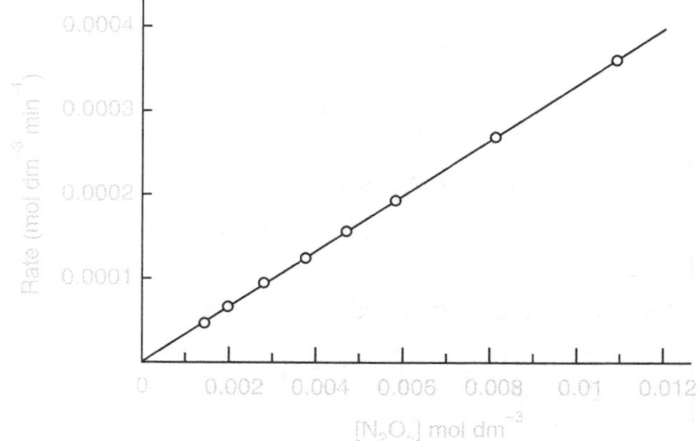

Fig. 20.22. Rate of decomposition of N_2O_5 as a function of $[N_2O_5]$.

Now two graphs are drawn by plotting

(*i*) *rate* against *concentration* of N_2O_5 *i.e.*, $[N_2O_5]$ as shown in Fig. 20.22 and

(*ii*) rate against the square of the concentration of N_2O_5 *i.e.*, $[N_2O_5]^2$ as shown in Fig. 20.23.

We find that a straight line is obtained in the first case which means that

$$\text{rate} \propto [N_2O_5]$$

and thus, the *rate law expression* is given as

$$\text{rate} = k\,[N_2O_5]$$

Also, the value of the *rate constant k*, evaluated from the rate data (Table 20.2) is,

$$k = \frac{\text{rate}}{[N_2O_5]} = 3.0 \times 10^{-2}\ \text{min}^{-1}$$

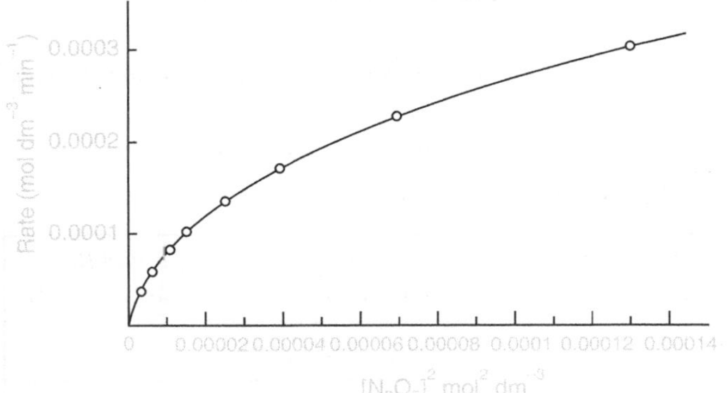

Fig. 20.23. Rate of decomposition of N_2O_5 as a function of $[N_2O_5]^2$.

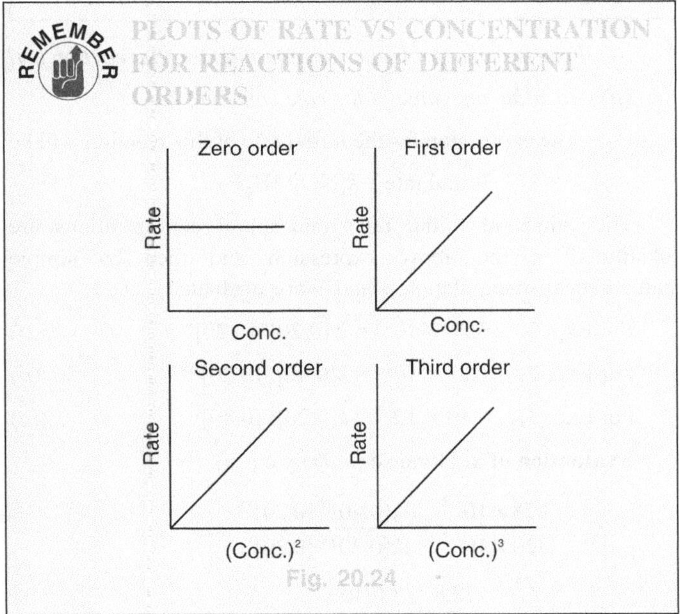

Fig. 20.24

SOLVED EXAMPLES

Example 20.11. *For the decomposition of nitrogen (V) oxide* $2N_2O_5 \longrightarrow 4NO_2 + O_2$, *following a first order kinetics,*

(*i*) *Calculate the rate constant for a 0.04 M solution, if the instantaneous rate is* 1.4×10^{-6} *mol dm*$^{-3}$ *s*$^{-1}$.

(*ii*) *Also calculate the rate of reaction when the concentration of* N_2O_5 *is 1.20 M.*

(*iii*) *What concentration of* N_2O_5 *would give a rate of* 2.45×10^{-5} *mol dm*$^{-3}$ *s*$^{-1}$?

Solution. (*i*) The given reaction is of first order, therefore,

$$\text{rate} = k\,[N_2O_5]$$

or

$$k = \frac{\text{rate}}{[N_2O_5]} = \frac{1.4 \times 10^{-6}}{0.04} \qquad [\text{M = Conc. in mol dm}^{-3}]$$

$$= \mathbf{3.5 \times 10^{-5}\ s^{-1}.}$$

(*ii*) Now if the concentration of N_2O_5 is 1.20 M, then

$$\text{rate} = k\,[N_2O_5]$$

$$= 3.5 \times 10^{-5} \times 1.20$$

$$= \mathbf{4.2 \times 10^{-5}\ mol\ dm^{-3}\ s^{-1}.}$$

(*iii*) To obtain the concentration of N_2O_5 when the rate is 2.45×10^{-5} mol dm^{-3} *s*$^{-1}$, we note that

$$[N_2O_5] = \frac{\text{Rate}}{k} = \frac{2.45 \times 10^{-5}}{3.5 \times 10^{-5}}$$

$$= \mathbf{0.7\ mol\ dm^{-3}\ or\ 0.7\ M.}$$

Example 20.12. *The rate of formation of a dimer in a second order dimerisation reaction is* 9.5×10^{-5} *mol dm*$^{-3}$ *s*$^{-1}$ *at 0.01 mol dm*$^{-3}$ *monomer concentration. Calculate the rate constant.*

Solution. If the reactant monomer is A then the reaction is

$$2A \longrightarrow (A)_2 \ \text{(dimerisation)}$$

Since it is a second order reaction

$$\therefore \qquad \text{Rate} = k[A]^2$$

$$\text{rate} = -\frac{1}{2}\frac{d[A]}{dt} \quad \text{or} \quad \frac{d[A_2]}{dt}$$

$$= 9.5 \times 10^{-5}\ \text{mol}^{-1}\ \text{s}^{-1}$$

$$\therefore \qquad 9.5 \times 10^{-5} = k(0.01)^2$$

$$k = \frac{9.5 \times 10^{-5}}{10^{-4}} = \mathbf{0.95\ dm^3\ mol^{-1}\ s^{-1}.}$$

Example 20.13. *The experimental data for the decomposition of* N_2O_5 *in the gaseous phase at 220 K is as follows:*

$$2N_2O_5(g) \longrightarrow 4NO_2(g) + O_2(g)$$

Time (sec)	*Conc. of N_2O_5 (mol dm^{-3})*
0	*1.63×10^{-2}*
400	*1.36×10^{-2}*
800	*1.14×10^{-2}*
1200	*0.95×10^{-2}*

Calculate the rates of reaction at the times given above. Also find

(*i*) *Rate constant* (*ii*) *Order of the reaction.*

Solution.

Time *(sec)*	$[N_2O_5]$ *(mol dm^{-3})*	*Rate* $= \frac{1}{2}\frac{\Delta[N_2O_5]}{\Delta t}$ *(mol dm^{-3} s^{-1})*	$k = \frac{Rate}{[N_2O_5]}$ *(s^{-1})*
0	1.63×10^{-2}		
400	1.36×10^{-2}	$\frac{1}{2}\frac{(1.63-1.36)\times 10^{-2}}{400}$ $= 3.38 \times 10^{-6}$	2.48×10^{-4}
800	1.14×10^{-2}	$\frac{1}{2}\frac{(1.36-1.14)\times 10^{-2}}{800-400}$ $= 2.75 \times 10^{-6}$	2.46×10^{-4}
1200	0.95×10^{-2}	$\frac{1}{2}\frac{(1.14-0.95)\times 10^{-2}}{1200-800}$ $= 2.38 \times 10^{-6}$	2.49×10^{-4}

The value of k is nearly the same.

∴ The rate expression is, **Rate = k[N$_2$O$_5$]**

Order of reaction = 1.

2. INITIAL RATE METHOD

The graphical method described above becomes, complicated if the reaction involves *more than one reactant.* For such cases, a simple alternative method, *i.e., Initial Rate Method* is employed. The method involves the determination of the order of the different reactants separately.

In order to determine the order of a particular reactant, a series of experiments are performed in which the concentration of that particular reactant is varied whereas the concentrations of other reactants are kept constant. In each experiment the *initial reaction rate* is determined from the plot of *conc. vs time*.

Initial reaction rate refers to the *rate at the beginning of the reaction. It may also be taken as the rate over the first feasible and smallest possible time interval.* From the data obtained, the order with respect to a particular reactant is calculated. *(See example 20.14 to 20.17.)*

The same procedure is then repeated by devising another set of experiments of the same reaction in which the concentration of the second reactant is varied whereas those of the others are kept constant. Then, by a similar method as discussed above, the initial rates for the different experimental runs and hence the order is determined.

The results are then put together give the rate law expression.

SOLVED EXAMPLES

Example 20.14. *For the reaction between nitric oxide (NO) and hydrogen (H_2),*

$$2NO + 2H_2 \longrightarrow N_2 + 2H_2O,$$

the following rate data were obtained:

Experiment	*Initial Concentrations (mol dm^{-3})*		*Initial Rate (mol dm^{-3} s^{-1})*
	[NO]	*[H$_2$]*	
1	*0.20*	*0.20*	*3.20×10^{-3}*
2	*0.40*	*0.20*	*1.28×10^{-2}*
3	*0.20*	*0.40*	*6.40×10^{-3}*

(*i*) *Write the most probable rate equation for the reaction and*

(*ii*) *Calculate the value of the rate constant, k.*

Solution. The expression for the initial rate of this reaction will be

$$\text{Initial rate} = k[\text{NO}]^x\,[\text{H}_2]^y.$$

The values of initial rates and initial concentrations are substituted in the above expression and then by simple mathematical manipulation x and y are evaluated.

For Exp. 1, $3.20 \times 10^{-3} = k\,[0.20]^x\,[0.20]^y$ ...(*i*)

For Exp. 2, $1.28 \times 10^{-2} = k[0.40]^x\,[0.20]^y$ ...(*ii*)

For Exp. 3, $6.40 \times 10^{-3} = k[0.20]^x\,[0.40]^y$ ...(*iii*)

Evaluation of x. Divide eq. (*ii*) by eq. (*i*)

$$\frac{1.28\times 10^{-2}}{320\times 10^{-3}} = \frac{k[0.40]^x\,[0.20]^y}{k[0.20]^x\,[0.20]^y}$$

$$\frac{12.8}{3.2} = \left(\frac{0.40}{0.20}\right)^x$$

$$4.0 = (2.0)^x$$

$$(2.0)^2 = (2.0)^x$$

∴ $x = \mathbf{2}.$

Evaluation of y. Divide eq. (*iii*) by eq. (*i*),

$$\frac{6.40\times 10^{-3}}{3.20\times 10^{-3}} = \frac{k[0.20]^x\,[0.40]^y}{k[0.20]^x\,[0.20]^y}$$

$$(2.0)^1 = (2.0)^y$$

∴ $y = \mathbf{1}$

(*i*) Thus, the **Rate equation** is $r = k[\text{NO}]^2[\text{H}_2]$.

(*ii*) The value of the Rate constant

$$k = \frac{\text{rate}}{[\text{NO}]^2[\text{H}_2]} = \frac{3.20\times 10^{-3}}{0.20\times 0.20\times 0.20} = \mathbf{0.4\ mol^{-2}\ dm^{6}\ s^{-1}}.$$

Example 20.15. *The experimental data for the reaction*

$$2A + B_2 \longrightarrow 2AB \text{ is}$$

Experiment	*[A] (mol dm^{-3})*	*[B$_2$] (mol dm^{-3})*	*Initial Rate*
1	0.50	0.50	1.6×10^{-4}
2	0.50	1.00	3.2×10^{-4}
3	1.00	1.00	3.2×10^{-4}

Write the rate law equation.

Solution. Let the rate law be

$$r = k\,[A]^x[B_2]^y$$

Now, $r_1 = 1.6 \times 10^{-4} = k[0.50]^x[0.50]^y$ ...(*i*)

$r_2 = 3.2 \times 10^{-4} = k[0.50]^x\,[1.00]^y$ ...(*ii*)

$r_3 = 3.2 \times 10^{-4} = k[1.00]^x\,[1.00]^y$ ...(*iii*)

Divide (*ii*) by (*i*),

$$\frac{r_2}{r_1} = \frac{k(0.50)^x\,(1.00)^y}{k(0.50)^x\,(0.50)^y} = \frac{3.2 \times 10^{-4}}{1.6 \times 10^{-4}}$$

$$\left(\frac{1.00}{0.50}\right)^y = 2 \quad \text{or} \quad 2^y = 2 \quad \text{or} \quad y = 1$$

Now divide (*iii*) by (*ii*),

$$\frac{r_3}{r_2} = \frac{k(1.00)^x\,(1.00)^y}{k(0.50)^x\,(0.50)^y} = \frac{3.2 \times 10^{-4}}{3.2 \times 10^{-4}}$$

$$\left(\frac{1.00}{0.50}\right)^x = 1 \Rightarrow 2^x = 1 \Rightarrow 2^x = 2^0 \text{ and } x = 0$$

$\therefore$ Rate law is $\mathbf{r = k[A]^0\,[B_2]^1}$.

Example 20.16. *For the thermal decomposition of acetaldehyde*

$$CH_3CHO(g) \longrightarrow CH_4(g) + CO(g),$$

the following rate data was obtained.

Experiment	*Initial pressure (torr.)*	*Initial rate of increase in total pressure (torr.)*
1	300	0.61 (r_1)
2	200	0.27 (r_2)

Predict the order of the reaction.

Solution. For the given reaction,

Rate = $k\,[CH_3CHO]^n$, where n = order of reaction

or $\log \text{Rate} = \log k + n \log [CH_3CHO]$

For the two rates given above,

$$\log r_1 = \log k + n \log [CH_3CHO]_1$$
$$\log r_2 = \log k + n \log [CH_3CHO]_2$$

$$\therefore \quad \log \frac{r_1}{r_2} = n \log \frac{[CH_3CHO]_1}{[CH_3CHO]_2}$$

Hence, the order of reaction,

$$n = \frac{\log r_1 / r_2}{\log [CH_3 / CHO]_1 / [CH_3CHO]_2}$$

$$= \frac{\log (0.67 / 0.27)}{\log (300 / 200)} = \frac{\log 2.26}{\log 1.50} = \frac{0.354}{0.176} = \mathbf{2.0.}$$

Example 20.17. *A reaction between A and B is represented as A + B → C. The kinetic data of the reaction is*

S. No.	*Initial Conc. (M)*		*Duration of (h) Experiment*	*Final Conc. (M) [A]*
	[A]$_0$	*[B]$_0$*		
1.	0.100	1.0	0.50	0.0975
2.	0.100	2.0	0.50	0.0900
3.	0.050	1.0	2.00	0.0450

What is the order w.r.t. each reactant and what is the value of rate constant?

Solution. From the above data we shall calculate the rates of different the experiments first *i.e.*,

$$\text{rate} = \frac{\text{change in conc.}}{\text{duration}}$$

1. $\dfrac{0.100 - 0.0975}{0.5} = 0.005 \text{ Mh}^{-1}$

2. $\dfrac{0.100 - 0.0900}{0.5} = 0.020 \text{ Mh}^{-1}$

3. $\dfrac{0.050 - 0.0450}{2.0} = 0.0025 \text{ Mh}^{-1}$

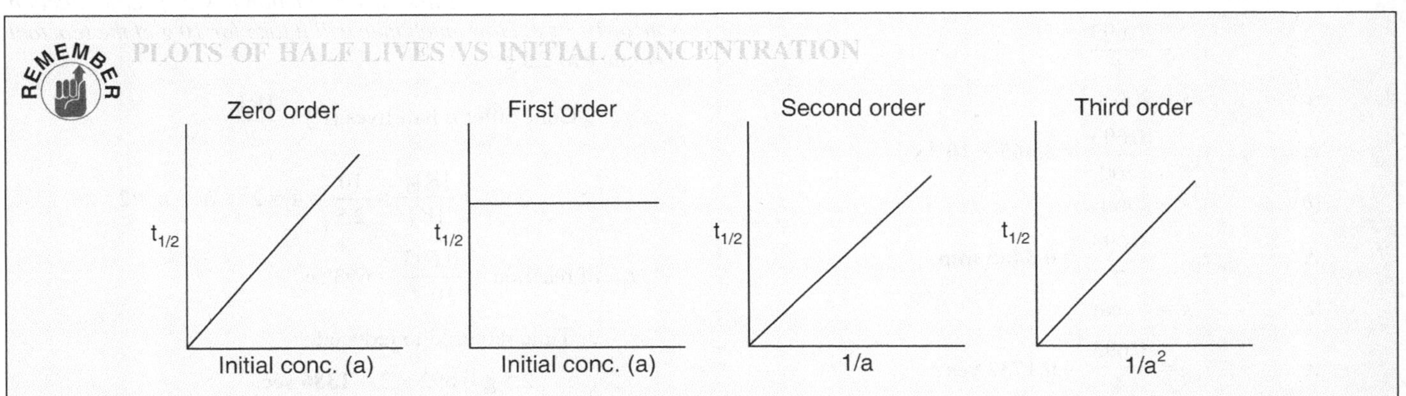

Thus, the rate expressions become

$$0.005 = k[0.1]^x[1.0]^y \quad ...(i)$$
$$0.020 = k[0.1]^x[2.0]^y \quad ...(ii)$$
$$0.0025 = k[0.05]^x[1.0]^y \quad ...(iii)$$

Dividing (*ii*) by (*i*),

$$\frac{0.020}{0.005} = \frac{k[0.1]^x[2.0]^y}{k[0.1]^x[1.0]^y}$$

$$\Rightarrow \quad 4 = 2^y \Rightarrow 2^2 = 2^y \Rightarrow \mathbf{y = 2}$$

Similarly, dividing (*i*) by (*iii*),

$$\frac{0.005}{0.0025} = \frac{k[0.1]^x[1.0]^2}{k[0.05]^x[1.0]^2} \quad \text{or} \quad 2 = 2^x \quad \text{or} \quad \mathbf{x = 1}$$

$$k = \frac{0.005}{[0.1]^1[1.0]^2} = \mathbf{5 \times 10^{-2}\ M^{-2}\ h^{-1}}.$$

SOLVED EXAMPLES

Example 20.18. *A reaction is 50% complete in 10 minutes. It is allowed to proceed another 5 minutes. By what percentage would the reaction be completed at the end of 15 minutes if it follows zero order kinetics?*

Solution. Say the initial conc. $[R]_0 = 100$ mol dm^{-3}

At $t_{1/2}$, half of this is consumed,

$$\therefore \quad k = \frac{[R]}{t} = \frac{50}{10} = 5 \text{ mol dm}^{-3} \text{ min}^{-1}$$

Conc. after 15 minutes $= [[R]_0 - [R]]$

$$= [[R]_0 - k \times t] \qquad \left[\because\ k = \frac{[R]}{t}\right]$$

$$= [100 - 5 \times 15] = 25 \text{ mol dm}^{-3}$$

$\therefore$ 75% of the reaction would be complete after 15 minutes.

Example 20.19. *Calculate the half-life of a first order reaction from their rate constants given below:*

(*a*) *200 s^{-1}* (*b*) *2 min^{-1}*

(*c*) *5 year^{-1}.*

Solution. For the first order reaction,

$$t_{1/2} = \frac{0.693}{k}$$

(*a*) $k = 200$ s^{-1}

$$\therefore \quad t_{1/2} = \frac{0.693}{200} = \mathbf{3.465 \times 10^{-3}\ s}$$

(*b*) $k = 2$ min^{-1}

$$\therefore \quad t_{1/2} = \frac{0.693}{2} = \mathbf{0.3465\ min}$$

(*c*) $k = 4$ year^{-1}

$$\therefore \quad t_{1/2} = \frac{0.693}{4} = \mathbf{0.1732\ year.}$$

Example 20.20. *The half-life of a reaction of first order is 100 sec. Calculate its rate constant.*

Solution. For a first order reaction, $t_{1/2} = \frac{0.693}{k}$

$$\therefore \quad k = \frac{0.693}{t_{1/2}}$$

Now, $t_{1/2} = 100$ sec

$$\therefore \quad k = \frac{0.693}{10^{-2}} = \mathbf{6.93 \times 10^{-3}\ sec^{-1}.}$$

Example 20.21. *A first order reaction has a rate constant of 10^{-2} sec^{-1}. Calculate the half-life for this reaction.*

Solution. For a first order reaction, $t_{1/2} = \frac{0.693}{k}$

$$k = 10^{-2} \text{ sec}^{-1}$$

$$t_{1/2} = \frac{0.693}{10^{-2}} = \mathbf{69.3\ sec.}$$

Example 20.22. *The $t_{1/2}$ of a first order reaction is 60 minutes. What percentage will be left after 240 minutes?*

Solution. Number of half lives $= \frac{240}{60} = 4$

$$\therefore \quad \text{Amount left [A]} = \frac{[A]_0}{2^4} = 0.0625 \text{ of } [A]_0 = \mathbf{6.25\%}.$$

Example 20.23. *The rate constant of a zero order reaction is 0.25 Mh^{-1}. What will be the initial concentration of the reactant if after 30 minutes its concentration is 0.075 M?*

Solution. For a zero reaction, $k = \frac{1}{t}\,[[R]_0 - [R]]$

$$\text{Thus,} \quad 0.25 = \frac{1}{0.5}\,[[R]_0 - 0.075] \quad [30 \text{ minutes} = 0.5 \text{ h}]$$

$$0.25 \times 0.5 = [R]_0 - 0.075$$

$$[R]_0 = 0.125 + 0.075 = \mathbf{0.2\ M.}$$

Example 20.24. *A first order reaction has a specific reaction rate of 10^{-3} s^{-1}. How much time will it take for 10 g of the reactant to reduce to 2.5 g?*

Solution. Conc. after *n* half lives $[R] = \frac{[R]_0}{2^n}$

$$\therefore \quad 2^n = \frac{[R]_0}{[R]} = \frac{10}{2.5} = 4 = 2^2 \quad \text{or} \quad n = 2$$

$$t_{1/2} \text{ of reaction} = \frac{0.693}{10^{-3}} = 693 \text{ sec}$$

$\therefore$ Time required to reduce to

2.5 g = 693 × 2 = **1386 sec.**

Example 20.25. *Show that in a first order reaction, the time required for the completion of 99.9% is 10 times the half-life ($t_{1/2}$) of the reaction.*

Solution. When reaction is completed 99.9%,

$$[R]_n = [R]_0 - 0.999\,[R]_0$$

$$k = \frac{2.303}{t}\log\frac{[R]_0}{[R]}$$

$$= \frac{2.303}{t}\log\frac{[R]_0}{[R]_0 - 0.999\,[R]_0}$$

$$= \frac{2.303}{t}\log 10^3$$

$$t = 6.909/k$$

For half-life of the reaction

$$t_{1/2} = 0.693/k$$

$$\frac{t}{t_{1/2}} = \frac{6.909}{k} \times \frac{k}{0.693} = 10.$$

SUMMARY

- **Chemical Kinetics.** The branch of chemistry which deals with the study of reaction rates and their mechanism.
- **Rate of Reaction.** The rate of change of concentration of any of the reactant or product with time at any particular time.
- **Average Rate.** The rate of reaction measured over a long time interval is called average rate. It is given as $\Delta x/\Delta t$.
- **Instantaneous Rate.** It is the rate of a reaction at a given instant of time *i.e.*, $\frac{\Delta x}{\Delta t}$ (average rate) becomes $\frac{dx}{dt}$ when Δt approaches zero.
- **Rate Constant (k).** It is the rate of the reaction when the concentration of each of the reacting species is unity. It is also called *velocity constant* or *specific reaction rate* of the reaction.
- **Rate Law.** The mathematical expression based on experimental fact, which describes the reaction rate in terms of the concentrations of reacting species. It cannot be written from the balanced equation.
- **Molecularity.** The number of reacting particles which collides simultaneously to bring about the chemical change. It is a theoretical concept.
- **Order of Reaction.** The sum of the exponents of the concentration terms in the experimental rate law of a reaction. It can be zero, 1, 2, 3 or any fractional value.
- **Zero Order Reaction.** Reactions in which the rate of reaction does not change with the concentration of reactants, *i.e.*, $r = k\,[A]_0 = k$
- **First Order Reaction.** A reaction in which the rate of reaction is directly proportional to the conc. of reacting substance, *i.e.*, $r = k[A]$

  or $kt = 2.303 \log [A]_0/[A]$
- **Initial Rate.** The rate at the beginning of a reaction when the conc. have not changed appreciably.
- **Elementary Process.** One step reaction is known as elementary process.
- **Rate Determing Step.** The slowest step in a reaction mechanism is the rate determining step.
- **Half-life of a Reaction ($t_{1/2}$).** The time taken for the concentration of reactants to be reduced to half of their initial concentration.
- **Activation Energy (E_a).** The additional energy required by reacting species over and above their average PE to enable them to cross the energy barrier between reactants and products.
- **Activated Complex.** The highly energetic arrangement of atoms formed during the course of a reaction which corresponds to the peak of the curve in the energy profile diagram for the progress of the reaction.
- **Effective Collisions.** The collisions responsible for changing the reactant molecules into product molecules.
- **Mechanism of Reaction.** The sequence of elementary steps leading to an overall stoichiometry of a reaction.
- **Threshold Energy.** The minimum energy that a reacting species must possess in order to undergo effective collisions.
- **Collision Frequency (z).** It is total number of collisions which all molecules undergo in a unit volume in a unit time.
- **Initial Rate.** The rate at the first feasible time interval or the rate at the beginning of a reaction.
- **Temperature Coefficient.** It is ratio of a rate constant at 308 K and 298 K.

EVALUATE YOURSELF

I. Objective Type Questions

Select the most appropriate choice from the options given as (a), (b), (c) and (d) after each question:

1. For a chemical reaction, $x + 2y \longrightarrow z$, if the rate of appearance of z is 0.50 moles per litre per hour, then the rate of disappearance of y is

   (*a*) 0.5 mol dm^{-3} hr^{-1} (*b*) 1.0 mol dm^{-3} hr^{-1}

   (*c*) 0.25 mol dm^{-3} hr^{-1} (*d*) cannot be predicted.

2. For the reaction, $NO_2(g) + CO(g) \longrightarrow NO(g) + CO_2(g)$, the correct expression for the rate of the reaction is

(*a*) rate = $\frac{-d[NO_2]}{dt}$ (*b*) rate = $\frac{-d[CO_2]}{dt}$

(*c*) rate = $\frac{d[NO_2]-d[CO]}{dt}$ (*d*) rate = $\frac{d[CO_2]}{dt}$.

3. The reactions of high molecularity are rare because
   (*a*) many body collisions have a low probability
   (*b*) many body collisions are not favoured energetically
   (*c*) activation energy of many body collisions is very large
   (*d*) very high concentration is required for such reactions.

4. For a chemical reaction A $\longrightarrow$ B, it is found that the rate of the reaction quadruples when the concentration of A is doubled. The rate expression for the reaction is, rate = $k[A]^n$ where the value of *n* is
   (*a*) 1 (*b*) 2
   (*c*) 0 (*d*) 3.

5. On increasing the temperature of the reacting system by 10^0 the rate of reaction almost becomes double. The most appropriate reason for this is that
   (*a*) activation energy decreases by increase in temperature
   (*b*) the fraction of molecules having threshold energy increases
   (*c*) collision frequency increases
   (*d*) the value of the threshold energy decreases.

6. A zero order reaction $A \longrightarrow$ *Products,* has rate constant 10^{-2} mol dm^{-3} s^{-1}. If a process is started with 10 moles of A in a one litre vessel, the number of *moles* of reactant after 10 minutes will be
   (*a*) 10 (*b*) 5
   (*c*) 6 (*d*) 4.

7. A hypothetical reaction $2p + q \longrightarrow s + r$ has rate constant as 2.0×10^{-3} mol dm^{-3} s^{-1}. The order of the reaction is
   (*a*) unpredictable (*b*) zero
   (*c*) one (*d*) two.

8. The rate law for the chemical reaction, $2NO_2Cl \longrightarrow 2NO_2 + Cl_2$ is $r = k[NO_2Cl]$. Which of the following is the rate controlling step?
   (*a*) $NO_2 + Cl \longrightarrow NO_2Cl$
   (*b*) $NO_2Cl + Cl \longrightarrow NO_2 + Cl_2$
   (*c*) $NO_2Cl \longrightarrow NO_2Cl$
   (*d*) $2NO_2Cl \longrightarrow 2NO_2 + Cl_2$.

9. If a graph is plotted between ln *k* and 1/T for a first order reaction, the slope of the straight line obtained is given by
   (*a*) $-E_a/R$ (*b*) $-E_a/2.303\ R$
   (*c*) $-\frac{2.303}{E_aR}$ (*d*) $-\frac{E_a}{2.303}$.

10. For the reaction, $2x + y \longrightarrow z$, the following kinetic data was obtained:

$[x]$ (*mol dm^{-3}*)	$[y]$ (*mol dm^{-3}*)	$\frac{d[z]}{dt}$ (*mol dm^{-3} s^{-1}*)
4.0	1.0	1.0
4.0	4.0	2.0
1.0	4.0	1.0

The rate expression is:
   (*a*) $k[x][y]^{1/2}$ (*b*) $k[x]^2[y]$
   (*c*) $k[x]^{1/2}[y]^{1/2}$ (*d*) $k[x]^{1/2}[y]$.

11. For a hypothetical reaction,
$$2A + 2B \longrightarrow X + Y,$$
the kinetic data revealed the rate expression, rate = $k[A]^2[B]$. If the reaction proceeds by the following sequence of steps:
Step I $A + B \longrightarrow C$
Step II $A + C \longrightarrow D$
Step III $B + D \longrightarrow X + Y$
then the rate determining step is
   (*a*) Step I (*b*) Step II
   (*c*) Step III
   (*d*) cannot be predicted as data is insufficient.

12. The mechanism of $2O_3 \longrightarrow 3O_2$ is given as
$$O_3 \rightleftharpoons O_2 + O \text{ (Fast)}$$
$$O + O_3 \rightleftharpoons 2O_2 \text{ (Slow)}$$
Which of the following represents the rate law?
   (*a*) $r = k[O_3]^2$ (*b*) $r = k[O_3]^2[O_2]^{-1}$
   (*c*) $r = k[O_3][O_2]$ (*d*) Unpredictable.

13. For a chemical reaction, A $\rightarrow$ B, it is found that the rate of reaction doubles when the concentration of A is increased four times. The order of the reaction with respect to A is
   (*a*) 2 (*b*) 1
   (*c*) Zero (*d*) 1/2.

14. The activation energy of a chemical reaction can be determined by
   (*a*) evaluating rate constants at two different temperatures
   (*b*) changing concentrations of reactants
   (*c*) evaluating rate constant at standard temperature
   (*d*) evaluating velocities of reaction at two different temperatures.

15. The rate constant for the reaction,
$$2N_2O_5 \longrightarrow 4NO_2 + O_2,$$
is 3.0×10^{-5} s^{-1}. If the rate is 2.4×10^{-5} mol dm^{-3} s^{-1} then the concentration of N_2O_5 (in mol dm^{-3}) is
   (*a*) 1.4 (*b*) 1.2
   (*c*) 0.04 (*d*) 0.8.

16. The reaction A $\longrightarrow$ B follows first order kinetics. The time taken for 0.8 mole of A to produce 0.6 mole of B is 1 hour. What is the time taken for conversion of 0.9 mole of A to produces 0.675 moles of B?

(*a*) 1 hour (*b*) 0.5 hour
(*c*) 0.25 hour (*d*) 2 hours.

17. The rate law of a reaction between the substances A and B is given by rate = $k[A]^n[B]^m$. On doubling the concentration of A and halving the conc. of B, the ratio of the new rate to the earlier rate of reaction will be

(*a*) $\frac{1}{2^{m+n}}$ (*b*) $m+n$
(*c*) $n-m$ (*d*) $2^{(n-m)}$.

18. In a first order reaction the concentration of reactant decreases from 800 mol/dm^3 to 50 mol/dm^3 in 2×10^4 sec. The rate constant of reaction in sec^{-1} is

(*a*) 2×10^4 (*b*) 3.45×10^{-5}
(*c*) 1.386×10^{-4} (*d*) 2×10^{-4}.

19. The reaction, X $\longrightarrow$ Product follows first order kinetics. In 40 minutes the concentration of X changes from 0.1 M to 0.025 M. The rate of reaction when the concentration of X is 0.01 M is

(*a*) 1.73×10^{-4} M min^{-1} (*b*) 3.47×10^{-5} M min^{-1}
(*c*) 3.47×10^{-4} M min^{-1} (*d*) 1.73×10^{-5} M min^{-1}.

20. Consider the reaction

$$N_2(g) + 3H_2(g) \longrightarrow 2NH_3(g)$$

The equality relationship between $\frac{d[NH_3]}{dt}$ and $-\frac{d[H_2]}{dt}$ is

(*a*) $\frac{d[NH_3]}{dt} = -\frac{d[H_2]}{dt}$

(*b*) $\frac{d[NH_3]}{dt} = -\frac{1}{3}\frac{d[H_2]}{dt}$

(*c*) $+\frac{d[NH_3]}{dt} = -\frac{2}{3}\frac{d[H_2]}{dt}$

(*d*) $+\frac{d[NH_3]}{dt} = -\frac{3}{2}\frac{d[H_2]}{dt}$.

21. For a first order reaction, $A \longrightarrow$ *products*, if k is the rate constant then the half life of the reaction is equal to

(*a*) $\frac{0.693}{k[A]}$ (*b*) $\frac{1}{k}\frac{1}{[A]}$

(*c*) $\frac{1}{k}\ln\frac{[A]}{[A]_0}$ (*d*) $\frac{\ln 2}{k}$.

22. A first order reaction has a half life of 34.65 seconds. Its rate constant is

(*a*) 2×10^{-2} s^{-1} (*b*) 4×10^{-2} s^{-1}
(*c*) 20 s^{-1} (*d*) 2×10^{-2} s^{-1}.

23. For a certain first order reaction, X $\longrightarrow$ Y + Z, the value of $t_{1/2}$ is 10 minutes. The concentration of X is reduced to 10% of the original concentration in

(*a*) 10 min (*b*) 90 min
(*c*) 33 min (*d*) 70 min.

24. The first order rate constant for the decomposition of N_2O_5 is 6.2×10^{-4} sec^{-1}. The $t_{1/2}$ of the decomposition is

(*a*) 1117.7 sec (*b*) 111.7 sec
(*c*) 228.4 sec (*d*) 168.9 sec.

II. Fill in the Blanks

25. Complete the following sentences by supplying appropriate words:

(*i*) Hydrolysis of ethyl acetate in an acidic solution is an example of .... order.

(*ii*) If the activation energy of the reaction is low, it proceeds at ...... rate.

(*iii*) The reaction taking place under the influence of visible light is called .... reaction.

(*iv*) In a multi-step reaction, the ...... step determines the rate of the reaction.

(*v*) The sum of powers to which the concentration terms are raised in the rate law is called .... of the reaction.

(*vi*) The order of a reaction with respect to B, whose rate law is, rate = $k[A]^{3/2}[B]^{1/2}$ is ......

(*vii*) The units of rate of gaseous reaction are ......

(*viii*) Temperature coefficient = ...

(*ix*) For a zero order reaction A $\longrightarrow$ Products, the plot of [A] *vs* time has a slope = ...

(*x*) For the reaction $xA + yB \longrightarrow lL + mM$, the average rate of disappearance of B = ...

III. Discussion Questions

26. What is meant by the terms average and instantaneous rates of a reaction? How are they expressed?

27. How can average and instantaneous rates of a reaction be determined from the plot of concentration *vs* time?

28. Give expressions for the rate of reaction

$$2X + 3Y \longrightarrow X_2Y_3$$

in terms of rates of disappearance of the reactants and rates of appearance of the products.

29. Why can't rate of a reaction be obtained by dividing the total amount of reactant consumed with time taken?

30. Express the rate of the reaction

$$3H_2(g) + N_2(g) \longrightarrow 2NH_3$$

in terms disappearance of reactants and appearance of the product.

31. Distinguish explicitly between the rate of a reaction and the rate constant of a reaction.

32. What are the factors on which the rate constant of a reaction depends?

33. For each of the following reactions state the order w.r.t. to each reactant and also the overall order:

(*i*) $3NO(g) \longrightarrow N_2O(g) + NO_2(g)$; $rate = k[NO]^2$

(*ii*) $H_2O_2 + 3I^- + 2H^+ \longrightarrow 2H_2O + I_3^-(aq)$;

$rate = k[H_2O_2][I^-]$

(*iii*) $CH_3CHO(g) \longrightarrow CH_4(g) + CO(g)$;

$rate = k[CH_3HO]^{3/2}$

(*iv*) $CHCl_3(g) + Cl_2(g) \longrightarrow CCl_4(g) + HCl(g)$;

$rate\ k = [CHCl_3][Cl_2]^{1/2}$

(*v*) $C_2H_5Cl(g) \longrightarrow C_2H_4(g) + HCl(g)$;

$rate = k[C_2H_5Cl]$

What are the dimensions of the rate constant in each case?

34. The reaction, $A + 2B \longrightarrow C + D$, obeys the rate equation; Rate $= k[A]^x[B]^y$. What would be the order of the reaction?

35. The first order decomposition of SO_2Cl_2 into SO_2 and Cl_2 presumably proceeds by a two step mechanism

(*i*) write a reasonable mechanism for the reaction

(*ii*) which step is the rate determining step?

36. Deduce the rate law expressions for the conversion of H_2 and I_2 to HI at 400°C, corresponding to each of the following mechanisms:

(*i*) $H_2 + I_2 \longrightarrow 2HI$ (one step)

(*ii*) $2I + H_2 \longrightarrow 2HI$ (slow)

(*iii*) $I_2 \longrightarrow 2I$; $I + H_2 \rightleftharpoons IH_2$;

$IH_2 + I \longrightarrow 2HI$ (slow).

37. Nitric oxide, NO, reacts with oxygen to produce nitrogen dioxide:

$$2NO(g) + O_2(g) \longrightarrow 2NO_2(g)$$

What is the predicted rate law, if the mechanism is

$$NO + O_2 \overset{k}{\rightleftharpoons} NO_3\ (fast)$$

$$NO_3 + NO \xrightarrow{k_1} NO_2 + NO_2\ (slow)?$$

38. The following mechanism has been proposed for the reaction

$A + C \longrightarrow D + F$ (*slow*)

$F + B \longrightarrow E$ (*fast*)

Write down the rate expression and overall reaction.

39. For the reaction $2NO_2 + F_2 \longrightarrow 2NO_2F$, the rate law is given by

$$r = k[NO_2][F_2]$$

Suggest a suitable mechanism based upon the above rate law indicating the rate determining step. Had the reaction been a single step reaction what would have been its rate law?

40. The rate expression for the chemical reaction

$2NO_2F \longrightarrow 2NO_2 + F_2$ is

$$r = k[NO_2F]$$

(*i*) propose a suitable mechanism for the reaction

(*ii*) what should be the units of the rate constant?

41. Give the units of rate constant for 1st order, 2nd order, 3rd order and *n*th order reaction.

42. Derive an expression for the rate constant of a first order reaction. Show that log [A] is a linear function of time.

43. What would be the probable plot of conc. *vs* time for a zero order reaction?

44. The initial rate of the reaction

$$A + B_2 \longrightarrow C_2 + D,$$

is doubled if the initial concentration of A is doubled, but is quadrupled if the initial concentration of B_2 is doubled:

(*i*) what is the order with respect to each of the reactants?

(*ii*) what is the overall order of the reaction?

(*iii*) could this possibly be a single step reaction? Explain.

45. The following experimental data was collected for the reaction

$$Cl_2(g) + 2NO(g) \longrightarrow 2NOCl(g)$$

Trial	*Initial Conc.* $[Cl_2]$	*Initial Conc.* $[NO]$	*Initial Rate* $(mol\ dm^{-3}\ s^{-1})$
1	0.010	0.010	1.2×10^{-4}
2	0.010	0.030	10.8×10^{-4}
3	0.020	0.030	21.6×10^{-4}

Deduce the rate equation for the reaction.

46. Consider the reaction, $2NO + Cl_2 \longrightarrow 2NOCl$, and the following experimental data. Write the most probable rate equation for the reaction.

Experiment	*[NO]* *(in mol* dm^{-3}*)*	$[Cl_2]$ *(in mol* dm^{-3}*)*	*Rate,* $(mol\ dm^{-3}\ s^{-1})$
1	0.380	0.380	5.0×10^{-3}
2	0.760	0.760	4.0×10^{-2}
3	0.380	0.760	1.0×10^{-2}

47. The rate data obtained for a hypothetical reaction,

$$A + B \longrightarrow 2C \text{ is}$$

[A] (mol dm^{-3})	[B] (mol dm^{-3})	$\frac{d[C]}{dt}$ (mol dm^{-3} min^{-1})
1.0	1.0	3.0×10^{-2}
1.5	1.0	4.5×10^{-2}
1.5	0.816	3.0×10^{-2}

Based on above data determine the rate law expression and also calculate the value of rate constant *k*.

48. A first order reaction has a specific reaction rate of 2.31×10^{-3} sec^{-1}. Calculate the half-life of the reaction.

49. The half-life for a first order reaction is 100 sec. How long will it take for two-third of the reaction to be completed?

50. The decomposition of H_2O_2 is a first order reaction. Calculate the half life and the rate constant for the decomposition assuming the fraction decomposed at the end of 50 minutes is 0.75.

51. A certain substance A is mixed with an equal amount of B. At the end of one hour 75% of a A has reacted. How much of A will be left unreacted at the end of 2 hours if the reaction is of first order with respect to A and independent of B?

52. A reaction is of first order w.r.t. reactant A and with a rate constant of 6 min^{-1}. If we start with [A] = 0.5 mol dm^{-3}, when would [A] reach the value 0.05 mol dm^{-3}?

53. A first order reaction is 75% complete in 60 minutes. Find $t_{0.5}$ of the reaction.

54. A first order reaction is 15% complete in 20 min. How long will it take for it to be 60% complete?

55. What is activation energy? How is the rate constant of a reaction related to activation energy?

56. What is activated complex? How does it differ from:
(*i*) ordinary molecule
(*ii*) reaction intermediate?

57. Express the rate of reaction,

$$2N_2O_5 \longrightarrow 2NO_2 + O_2,$$

using concentration changes of N_2O_5 and O_2.

58. Nitric oxide, NO reacts with oxygen to produce nitrogen dioxide:

$$2NO(g) + O_2(g) \longrightarrow 2NO_2(g)$$

The rate law for this reaction is

$$\text{rate} = k\,[NO]^2\,[O_2]$$

Propose a mechanism for the above reaction.

59. The reaction $SO_2Cl_2 \longrightarrow SO_2 + Cl_2$ is a first order reaction with half-life 3.15×10^4 s at 320°C. What percentage of SO_2Cl_2 would be decomposed on heating at 320°C for 90 minutes?

60. A reaction is first order in A and second order in B
(*i*) Write the differential rate equation.
(*ii*) How is the rate affected if the concentration of B is tripled?
(*iii*) How is the rate affected when the concentration of both A and B are doubled?

61. Following reaction takes place in one step,

$$2NO(g) + O_2(g) \rightleftharpoons 2NO_2(g)$$

How will the rate of the above reaction change if the volume of the reaction vessel is diminished to one-third of its original volume ? Will there be any change in the order of the reaction with the reduced volume?

Answers

I. Objective Type Questions

1. (*b*)	**2.** (*a*)	**3.** (*a*)
4. (*b*)	**5.** (*b*)	**6.** (*d*)
7. (*b*)	**8.** (*c*)	**9.** (*a*)
10. (*c*)	**11.** (*b*)	**12.** (*b*)
13. (*b*)	**14.** (*a*)	**15.** (*d*)
16. (*a*)	**17.** (*d*)	**18.** (*c*)
19. (*c*)	**20.** (*c*)	**21.** (*d*)
22. (*d*)	**23.** (*c*)	**24.** (*a*).

II. Fill in the Blanks

25. (*i*) pseudo, first (*ii*) faster
(*iii*) photochemical (*iv*) slowest
(*v*) order (*vi*) 0.5
(*vii*) atm sec^{-1} (*viii*) k_{308}/k_{298}
(*ix*) *k* (*x*) $-\Delta[B]/\Delta t$.

III. Discussion Questions

28. Rate of reaction $= -\frac{1}{2}\frac{d[X]}{dt} = -\frac{1}{3}\frac{d[Y]}{dt} = \frac{d[X_2Y_3]}{dt}$

29. It is because the rate calculated in this manner would be constant throughout the entire span of the reaction which is contrary to the actual observation.

30. $-\frac{1}{3}\frac{d[H_2]}{dt} = \frac{-d[N_2]}{dt} = +\frac{1}{2}\frac{d[NH_3]}{dt}$.

33. (*i*) Order w.r.t. NO and overall order = 2;
k = dm^3 mol^{-1} s^{-1}
(*ii*) Order w.r.t. H_2O_2 = 1, order w.r.t. I^- = 1; order w.r.t. H^+ = 0; overall order = 2
k = dm^3 mol^{-1} s^{-1}

(*iii*) Order w.r.t. CH_3CHO and overall order = 1.5

$k = dm^{3/2}\ m^{-1/2}\ s^{-1}$

(*iv*) Order w.r.t. $CHCl_3 = 1$; order w.r.t. $Cl_2 = 0.5$; overall order = 1.5

$k = dm^{3/2}\ mol^{-1/2}\ s^{-1}$

(*v*) Order w.r.t. $C_2H_5Cl = 1$; overall order = 1

$k = s^{-1}$

34. order of reaction $= x + y$

35. $SO_2Cl_2 \xrightarrow{\text{slow}} SOCl_2 + O;$

$SOCl_2 + O \xrightarrow{\text{fast}} SO_2 + Cl_2.$

36. (*i*) Rate $= k\,[H_2][I_2]$; (*ii*) Rate $= k\,[H_2][I_2]$; (*iii*) Rate $= k\,[H_2][I_2]$

37. The rate low is given by slowest step *i.e.*,

$r = k_1[NO_3]\,[NO]$

As NO_3 is obtained from the fast step, therefore its conc. is calculated from step 1

$k = \dfrac{[NO_3]}{[NO]\,[O_2]}$. Here k is equilibrium constant

or $[NO_3] = k[NO]\,[O_2]$. Substituting this value in rate equation, we get

or $r = k_1 \times k\,[NO]\,[O_2]\,[NO] = k'[NO]^2[O_2]$.

38. $r = k[A]\,[C]$, overall reaction is $A + B + C \longrightarrow D + E$

39. $NO_2 + F_2 \xrightarrow{\text{slow}} NO_2F + F\,;\ NO_2 + F \xrightarrow{\text{fast}} NO_2F$

If the reaction is elementary *i.e.* single step then

$r = k[NO_2]^2[F_2]$.

40. $NO_2F \xrightarrow{\text{slow}} NO_2 + F\,;\ NO_2F + F \xrightarrow{\text{fast}} NO_2 + F_2$

Since the reaction is of first order, therefore units of

$k = s^{-1}$.

42. The integrated equation of first order reaction is

$$k = \frac{2.303}{t}\log\frac{[A]_0}{[A]}$$

or $$t = \frac{2.303}{k}\log\,[A_0] - \frac{2.303}{k}\log\,[A]$$

As it is a straight line equation, therefore log [A] is a linear function of time.

44. 1st order w.r.t. A and 2nd order w.r.t. B_2

45. $r = k[Cl_2]\,[NO]^2$

46. Rate $= k[NO]^2[Cl_2]$.

SECTION-3

UNIT-21

EQUILIBRIUM

Specific Objectives

After studying this unit, the student should be able to:

- explain reversible and irreversible reactions
- use experiment to demonstrate reversible reaction
- explain that equilibrium is established when forward and reverse reactions are proceeding at the same rate
- describe how Le-Chatelier's principle can be used to predict the effect of changes in concentration, temperature and pressure on an equilibrium reaction
- identify the correct equilibrium constant expression and use it in computation
- establish equilibrium for a chemical reaction from an experiment.

21.1 INTRODUCTION

It is generally observed that many of the chemical reactions do not proceed to completion when they are carried out in a closed container. This implies that the reactants are not completely converted into products. Instead, after some time the *concentration* of the reactants do not undergo further decrease and the reaction appears to have stopped. This state of a system in which no further net change occurs is called a **state of equilibrium.**

Equilibrium can be attained for both physical and chemical processes. The equilibrium achieved in physical processes such as the dissolution of salt, evaporation of water, etc., is called **physical equilibrium,** whereas the equilibrium achieved in chemical processes such as the decomposition of calcium carbonate or the reaction between hydrogen and iodine is called **chemical equilibrium.**

The equilibrium situation raises many interesting questions such as:

- What is the equilibrium state and how is it recognised?
- What is the molecular behaviour when a state of equilibrium is reached?
- What are the factors which alter the state of equilibrium?

In this unit, we shall seek answers to the above questions.

21.2 REVERSIBLE AND IRREVERSIBLE REACTIONS

When a piece of sodium is dropped into water, a violent reaction occurs resulting in the formation of sodium hydroxide and hydrogen gas.

$$2Na(s) + 2H_2O(l) \longrightarrow 2NaOH(aq) + H_2(g)$$

However, it is not possible to carry out the reverse reaction under any known experimental conditions, *i.e.*, reduction of aqueous sodium hydroxide by hydrogen to form sodium and water cannot be achieved.

The reactions in which the products do not react to give back reactants are called irreversible reactions.

Now let us consider the reaction between metallic iron and steam.

$$3Fe(s) + 4H_2O(g) \longrightarrow Fe_3O_4(s) + 4H_2(g)$$

If this reaction is carried out in an open vessel, then the hydrogen gas escapes and the reaction proceeds to completion provided sufficient amount of steam is supplied. The reverse of this reaction, *i.e.*, reduction of iron oxide (Fe_3O_4) to iron by hydrogen gas can also be achieved rather easily in the laboratory.

$$Fe_3O_4(s) + 4H_2(g) \longrightarrow 3Fe(s) + 4H_2O(g)$$

To carry out this reaction hydrogen gas is passed over heated iron oxide. Again, if this reaction is carried out in an open vessel the whole of Fe_3O_4 is reduced to iron because one of the products, *i.e.*, water escapes as steam.

The reactions in which products can react to give back reactants are called reversible reactions.

A reversible reaction is represented by putting two arrows pointing in opposite directions between the formulae of the reactants and the products as shown below:

$$3Fe(s) + 4H_2O(g) \rightleftharpoons Fe_3O_4(s) + 4H_2(g)$$

Some **more examples** of reversible reactions are

$$CaCO_3(s) \rightleftharpoons CaO(s) + CO_2(g)$$

$$H_2(g) + I_2(g) \rightleftharpoons 2HI(g)$$

$$N_2O_4(g) \rightleftharpoons 2NO_2(g)$$

$$N_2(g) + 3H_2(g) \rightleftharpoons 2NH_3(g)$$

21.3 KINETIC REVERSIBLE REACTION DEMONSTRATION

The reversible nature of the reaction,

$$\underset{\text{Yellow}}{2CrO_4^{2-}}(aq) + 2H^+(aq) \rightleftharpoons \underset{\text{Orange}}{Cr_2O_7^{2-}}(aq) + H_2O(l)$$

is demonstrated in the following Activity.

CrO_4^{2-} ion is yellow in colour whereas $Cr_2O_7^{2-}$ ion is orange in colour.

ACTIVITY 21.1 To demonstrate the kinetic reversibility of a reaction

- Take 2 cm^3 of 0.05 M solution of K_2CrO_4 in a test tube
- Add 2 cm^3 of 0.1 M solution of H_2SO_4 to the above solution.
  The colour of the solution changes from yellow to orange due to formation of $Cr_2O_7^{2-}$ ions. This indicates that the forward reaction has taken place.
- Now add 2 cm^3 of 0.2 M solution of NaOH to the above reaction mixture.
  The colour again becomes yellow due to formation of CrO_4^{2-} ions. This indicates that a reverse reaction has taken place.

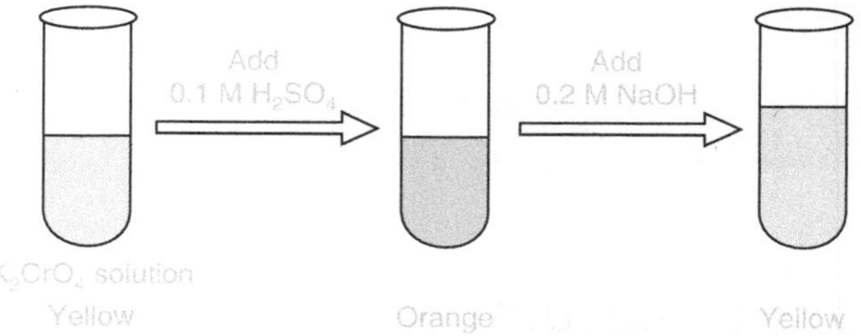

This experiment demonstrates the kinetic reversibility of a reaction.

21.4 EQUILIBRIUM IN REVERSIBLE REACTIONS

When a reversible reaction is carried out in a closed vessel it does not go to completion. In a reversible reaction both the forward as well as the backward reaction are feasible. When such a reaction is carried out in a closed vessel both reactions occur simultaneously in such a way that before the whole of the reactants are converted into products a state of equilibrium is established. In the state of equilibrium, the reaction appears to have stopped even though the reactants are still present. Therefore, reversible reactions do not proceed to completion. On the other hand, irreversible reactions are feasible only in one direction and hence, go to completion.

Let us illustrate the state of equilibrium in some reversible processes and see how it is recognised.

DECOMPOSITION OF CALCIUM CARBONATE

Calcium carbonate when heated to 800°C in a closed evacuated vessel, starts decomposing to yield calcium oxide and carbon dioxide gas. Carbon dioxide builds up the pressure within the vessel which can be recorded on a manometer, as shown in Fig. 21.1. The pressure goes on increasing as the reaction proceeds and *finally becomes constant* and remains so *as long as the temperature remains constant*. It appears as if the reaction has come to a stop although $CaCO_3$ is still present. This indicates that the system has attained an equilibrium state.

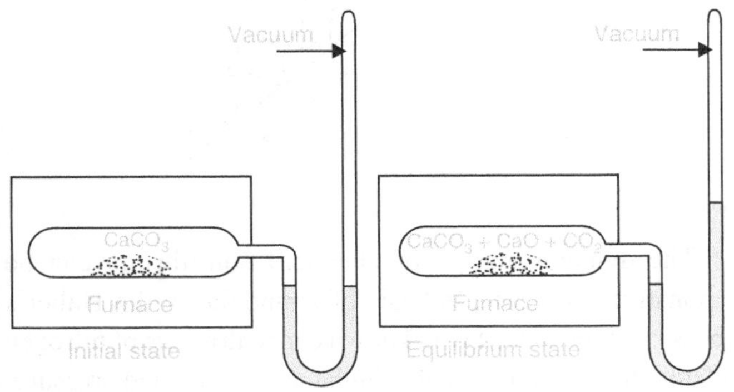

Fig. 21.1. Attainment of equilibrium in the thermal decomposition of $CaCO_3$.

REACTION BETWEEN HYDROGEN AND IODINE

When hydrogen and iodine are heated in a closed vessel at 444°C, they react to form hydrogen iodide. The colour of the reaction mixture is deep violet in the beginning due to the presence of iodine. As the reaction progresses the intensity of colour decreases due to the formation of hydrogen iodide which is colourless.

After some time, it is observed that the *intensity of colour becomes constant* indicating that the reaction has stopped although hydrogen and iodine are still present. Actually, the system acquires a state of equilibrium.

The attainment of equilibrium in a reaction is recognised by constancy of *certain observable properties*. For example, in the case of the dissociation of calcium carbonate, we recognised the attainment of equilibrium by observing that the pressure of carbon dioxide had become constant. In this case, *attainment of equilibrium has been recognised by observing constancy of pressure*. In the case of the reaction between hydrogen and iodine the attainment of equilibrium is recognised by observing the *constancy of intensity of colour* of the reaction mixture.

Sometimes, the concentration of one of the reactants or products is measured to recognise the state of equilibrium. Thus, *state of equilibrium is the state in which the measurable properties of the system do not undergo any noticeable change under a particular set of conditions.*

21.5 DYNAMIC NATURE OF CHEMICAL EQUILIBRIUM

We have seen that all the observable properties of the system become constant at equilibrium. It may lead us to think that the reaction stops altogether at equilibrium and that equilibrium is static in nature. But this is not true. Actually, at equilibrium the rate of the forward reaction becomes equal to the rate of the backward reaction so that there is no net change in the concentration of various species. In other words, we can say that the *equilibrium state is a dynamic balance between the forward and the backward reaction.* This can be illustrated by considering the reaction between hydrogen and iodine to form hydrogen iodide.

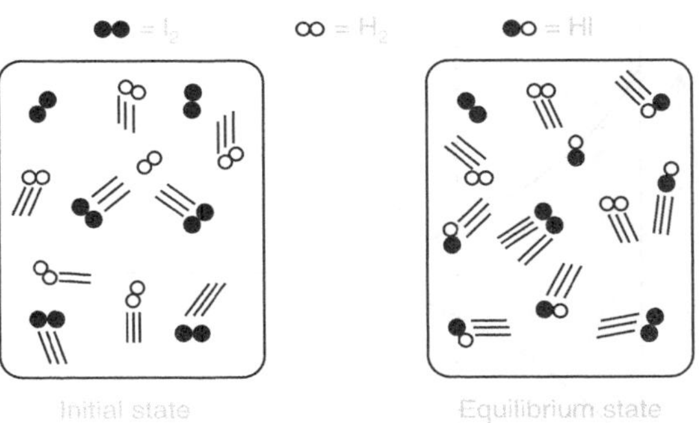

Fig. 21.2. Reaction between H_2 and I_2 in a closed vessel.

When hydrogen and iodine are taken in a closed vessel (Fig. 21.2) maintained at 717 K, hydrogen molecules combine with iodine molecules to form hydrogen iodide.

$$H_2(g) + I_2(g) \longrightarrow 2HI(g)$$

The molecules of HI formed, also begin to dissociate to form H_2 and I_2.

$$2HI(g) \longrightarrow H_2(g) + I_2(g)$$

As the reaction progresses, the concentrations of H_2 and I_2 decrease and hence, the rate of the forward reaction slows down. On the other hand, the concentration of hydrogen iodide increases and therefore the rate of the backward reaction increases. A stage is reached when the rate of the backward reaction becomes equal to the rate of the forward reaction and the system attains equilibrium. The variation of rates of forward and backward reactions have been shown in Fig. 21.3. After this no change in concentration occurs provided the temperature of the reaction mixture is kept constant. Thus, at equilibrium the reaction *does not stop* but the system acquires constant observable properties because of the equal rates of the forward and backward reactions. Thus, ***equilibrium is dynamic in nature.***

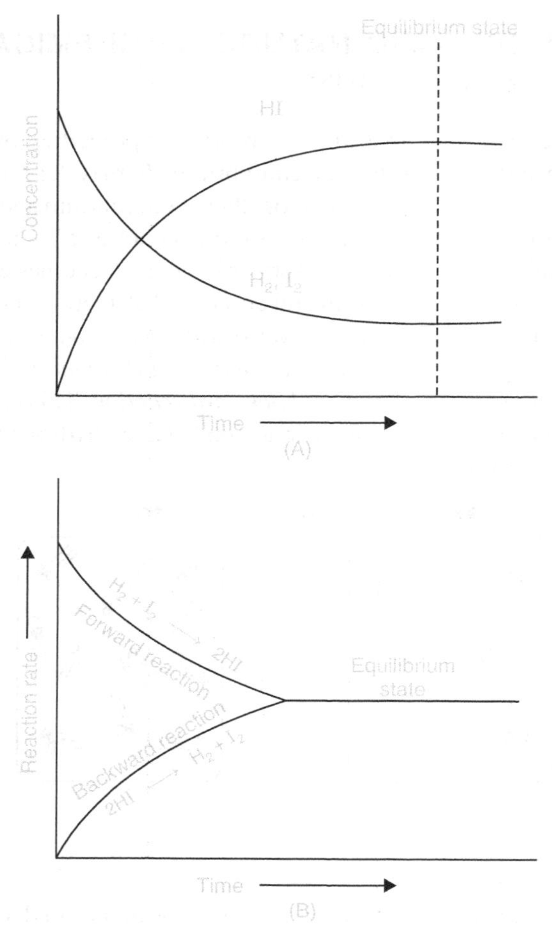

Fig. 21.3. Attainment of equilibrium state
(A) Variation of concentrations with time
(B) Variation of rate of forward and backward reactions with time.

The difference between static equilibrium and dynamic equilibrium can be understood by the following analogy:

Consider two children sitting on a see-saw. At the balance point no movement of the children or the see-saw occurs. This is analogous to **static equilibrium**.

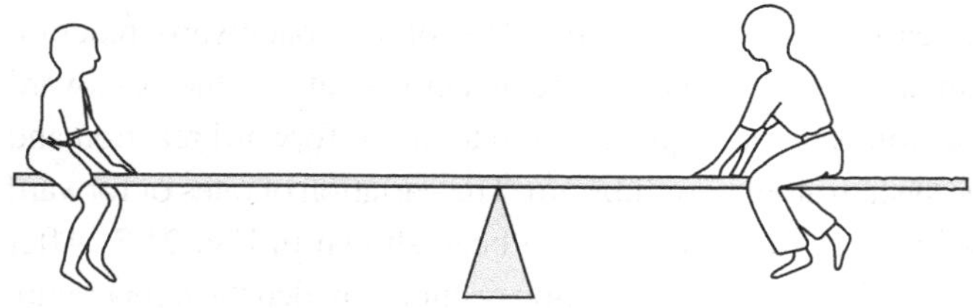

Fig. 21.4. Static equilibrium.

Now consider a child is ascending an escalator at the same rate as the escalator is descending. At this point, the child and the escalator are moving at the same rate in opposite directions and there is no net change in the position of the child. This is analogous to **dynamic equilibrium**.

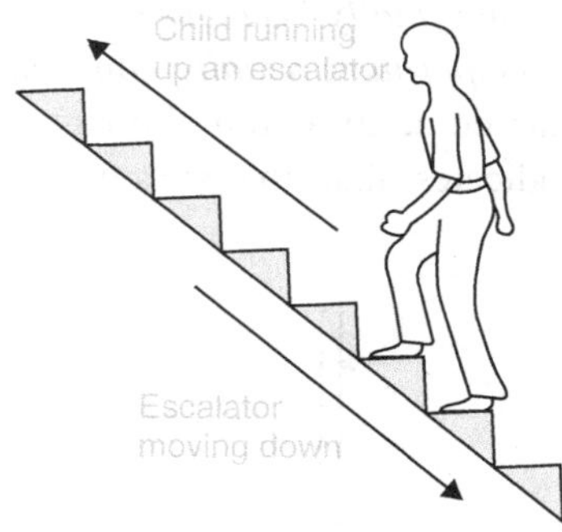

Fig. 21.5. Dynamic equilibrium.

The dynamic nature of chemical equilibrium can be demonstrated by the synthesis of ammonia by the Haber's process. Haber started with known concentrations of nitrogen and dihydrogen maintained at high temperature and pressure. He determined the concentrations of ammonia formed and, nitrogen and hydrogen left unreacted at regular intervals. He observed that after a certain time the composition of the reaction mixture became constant (Fig. 21.6) even though some of the reactants were still present. This constancy in composition indicates that the reaction has attained equilibrium.

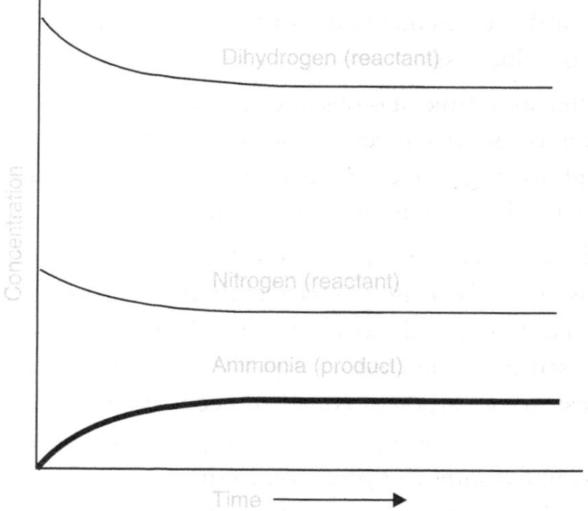

Fig. 21.6. Variation of concentrations of N_2, H_2 and NH_3 with time starting with 1 M N_2 and 3 M H_2 for the reaction $N_2(g) + 3H_2(g) \rightleftharpoons 2NH_3(g)$.

In order to demonstrate the dynamic nature of equilibrium, synthesis of ammonia was carried out with exactly the same starting conditions of partial pressures and temperature but using D_2 (deuterium) in place of H_2. The reaction was allowed to attain equilibrium. At equilibrium, the composition of the mixture was found to be the same as in the case of the reaction between N_2 and H_2, the only difference being that D_2 and ND_3 were present in place of H_2 and NH_3. The two reaction mixtures, the one containing H_2, N_2 and NH_3, and the other containing D_2, N_2 and ND_3 were mixed together and left for some time. It was observed that the concentration of ammonia remains constant with the passage of time, indicating that the equilibrium remains undisturbed. However, when the reaction mixture was analyzed with the help of a mass spectrometer it was observed that all deuterium containing forms of ammonia (NH_3, NH_2D, NHD_2 and ND_3) and hydrogen (H_2, HD and D_2) were present. It indicated that the forward as well as backward reactions are continuing even at equilibrium and that the equilibrium is dynamic in nature. If the reaction had stopped at equilibrium then there would have been no mixing of isotopes in this way. The formation of products such as HD, NH_2D, NHD_2 demonstrated that the equilibrium is dynamic in nature.

OTHER CHARACTERISTICS OF CHEMICAL EQUILIBRIUM

It has already been discussed that chemical equilibrium is dynamic in nature. In addition to this, other characteristics of chemical equilibrium are:

1. *The observable properties of the system such as pressure, colour, concentration, etc., become constant at equilibrium and remain unchanged thereafter.* For example, for the reaction between hydrogen and iodine to form hydrogen iodide, the colour of the reaction mixture becomes constant at equilibrium.

2. *The equilibrium can be approached from either direction.* This can be proven with the help of a very simple experiment involving dinitrogen tetraoxide, N_2O_4 (colourless gas) and nitrogen dioxide, NO_2 (a reddish brown gas) at equilibrium.

$$\underset{\text{Colourless}}{N_2O_4(g)} \rightleftharpoons \underset{\text{Reddish brown}}{2NO_2(g)}$$

It is observed that at ordinary temperature, say 298 K, two gases exist as an equilibrium mixture having pale brown colour. At very low temperature (273 K), N_2O_4 is stable and exists as pure N_2O_4 (colourless gas). When the temperature is increased, it decomposes to NO_2. At 373 K it decomposes almost completely to NO_2 which is dark brown in colour.

Let us take two identical bulbs A and B. Fill both of them with nitrogen dioxide and seal them. Place the bulb A in an ice bath and the bulb B in boiling water as shown in Fig. 21.7.

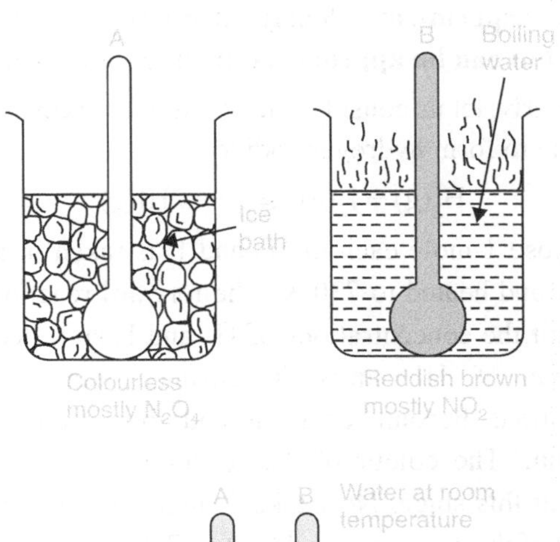

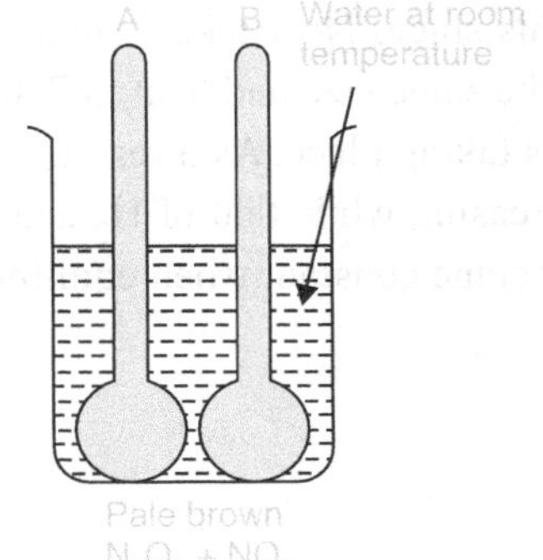

Fig. 21.7. Study of the equilibrium, $N_2O_4(g) \rightleftharpoons 2NO_2(g)$.

The gas in bulb A becomes almost colourless and consists of mostly N_2O_4 molecules. On the other hand, the gas in bulb B becomes reddish brown and consists of mainly NO_2 molecules.

Now transfer both bulbs into a bath maintained at 298 K and observe the change in colour. The colour of the gas in bulb A starts changing into brown indicating the gradual conversion of N_2O_4 into NO_2.

$$\underset{\text{Colourless}}{N_2O_4(g)} \longrightarrow \underset{\text{Reddish brown}}{2NO_2(g)} \qquad \text{(Bulb A)}$$

On the other hand, the reddish brown colour in bulb B begins to fade and becomes pale brown indicating the gradual formation of N_2O_4. This may be represented as:

$$\underset{\text{Reddish brown}}{2NO_2(g)} \longrightarrow \underset{\text{Colourless}}{N_2O_4(g)} \qquad \text{(Bulb B)}$$

After some time, when both the bulbs attain the temperature of the bath, the colour in the two bulbs becomes identical and no further change in colour occurs. This constancy of colour indicates that equilibrium has been attained in both cases and both contain a mixture of NO_2 and N_2O_4 of the same composition.

This experiment clearly demonstrates that the equilibrium can be approached from either direction.

Similarly, let us consider the reaction between hydrogen and iodine to form hydrogen iodide.

$$H_2(g) + I_2(g) \rightleftharpoons 2HI(g)$$

Suppose 1 mole each of H_2 and I_2 is taken in a closed container and heated to 730 K. The forward reaction starts. As a result the concentrations of H_2 and I_2 start decreasing while that of HI starts increasing. Finally, all the concentrations become constant and the reaction attains equilibrium. The colour of the reaction mixture becomes constant at this stage. Now take 2 moles of HI in a closed container of the same size and heat to 730 K. The backward reaction starts taking place. As a result, the concentration of HI starts decreasing while that of H_2 and I_2 start increasing and finally become constant when equilibrium is attained.

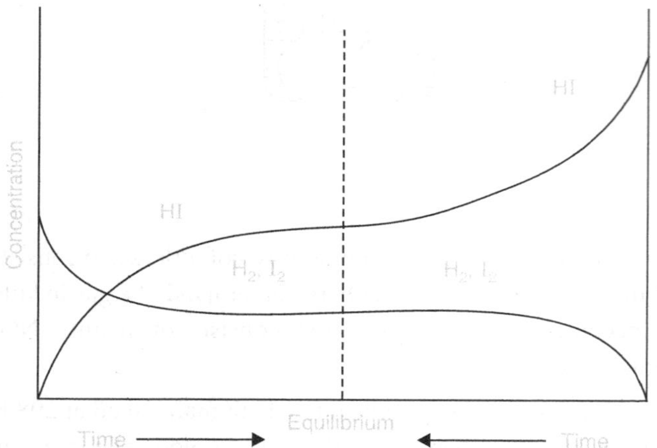

Fig. 21.8. Chemical equilibrium can be attained from either direction.

If we compare the intensity of the purple colour in the two cases we find that it is same. It indicates that the same equilibrium is attained in both cases, in one case starting with H_2 and I_2 and in the other case starting with HI. It shows that equilibrium can be attained from either direction.

3. *The equilibrium can be attained only if the system is closed.* For the establishment of equilibrium, it is necessary that the system should be a close one. If the system is not closed, the products may escape from the container and, the backward reaction may not take place.

4. *A catalyst does not alter the equilibrium point.* In a reversible reaction, a catalyst increases the rate of the forward as well as the backward reaction to the same extent. Therefore, the equilibrium point is not altered. However, the equilibrium is attained earlier.

21.6 LAW OF MASS ACTION AND EQUILIBRIUM CONSTANT

LAW OF MASS ACTION

On the basis of observations of many equilibrium reactions, two Norwegian chemists **Cato Guldberg** and **Peter Waage** (1864) suggested a qualitative relationship between rates of reaction and the concentration of the reacting species. The relationship is known as ***Law of Mass Action.*** This law states that:

At constant temperature the rate of a chemical reaction is directly proportional to the product of the molar concentrations of the reacting species with each concentration term raised to a power equal to the stoichiometric co-efficient of that species in the chemical equation.

For any general reaction,

$$aA + bB \longrightarrow \text{Products}$$

The law of mass action may be written as

$$\text{Rate} = k\,[A]^a[B]^b$$

where [A] and [B] are the molar concentrations of the reactants A and B respectively, k is a constant of proportionality called rate constant.

APPLICATION OF LAW OF MASS ACTION TO CHEMICAL EQUILIBRIUM

By applying the law of mass action to a reversible reaction in equilibrium it is possible to derive a simple mathematical expression referred to as the **law of chemical equilibrium.**

Let us consider a simple reversible reaction,

$$aA + bB \rightleftharpoons cC + dD$$

in which an equilibrium exists between the reactants A and B and the products C and D. According to the law of mass action:

$$\textit{Rate of forward reaction} = k_f\,[A]^a\,[B]^b$$

where $\boldsymbol{k_f}$ is the rate constant for the forward reaction. [A] and [B] are the molar concentrations of reactants A and B respectively.

Similarly, the rate of the backward reaction

$$= k_b[C]^c[D]^d$$

where k_b is the rate constant for the backward reaction and [C] and [D] are the molar concentrations of products C and D respectively.

At equilibrium the rates of two opposing reactions become equal. Therefore, at equilibrium,

Rate of forward reaction = Rate of backward reaction

$$\therefore \quad k_f[A]^a[B]^b = k_b[C]^c[D]^d$$

or

$$\frac{k_f}{k_b} = \frac{[C]^c[D]^d}{[A]^a[B]^b}$$

Since k_f and k_b are constants, the ratio k_f/k_b is also constant and is represented by K_c.

$$K_c = \frac{k_f}{k_b} = \frac{[C]^c[D]^d}{[A]^a[B]^b} \quad ...(21.1)$$

Thus, the expression of equilibrium constant can be written as:

$$\boxed{K_c = \frac{[C]^c[D]^d}{[A]^a[B]^b}}$$

The constant K_c is called **equilibrium constant.** The subscript '*c*' indicates that K_c is expressed in terms of concentrations (mol dm^{-3}) of reactants and products. The concentrations in expression (21.1) are the equilibrium concentrations.

The equilibrium constant, *K_c, may be defined as the ratio of product of the equilibrium concentrations of the products to that of the reactants with each concentration term raised to a power equal to the stoichiometric co-efficient of the substance in the balanced chemical equation.*

Law of chemical equilibrium may be stated as:

At a given temperature, the ratio of the product of the equilibrium concentrations of products to that of reactants with each concentration term raised to a power equal to the respective stoichiometric coefficients in the balanced chemical equations.

The concentration ratio, $\frac{[C]^c[D]^d}{[A]^a[B]^b}$, is called **concentration quotient** and is denoted by Q_c. According to the law of chemical equilibrium,

$$Q_c = K_c \text{ at equilibrium.}$$

EQUILIBRIUM CONSTANT IN GASEOUS SYSTEMS

If an equilibrium involves gaseous species, then the concentrations in the concentration quotient may be replaced by partial pressures because at a given temperature the partial pressure of a gaseous component is proportional to its concentration. If the above mentioned reaction has all the gaseous species, then

$$Q_p = \frac{(p_C)^c(p_D)^d}{(p_A)^a(p_B)^b} = K_p$$

The equilibrium constant, K_p, defined in terms of partial pressures is not the same as the equilibrium constant K_c, defined in terms of concentrations.

RELATIONSHIP BETWEEN K_p AND K_c

For an ideal gas, PV = *n*RT

$$P = \frac{n}{V}RT = CRT$$

$$\therefore \quad p_A = C_ART$$

$$p_B = C_BRT$$

$$p_C = C_CRT$$

$$p_D = C_DRT$$

where C_A, C_B, C_C and C_D are the molar concentrations of A, B, C and D respectively.

$$K_p = \frac{(C_CRT)^c(C_DRT)^d}{(C_ART)^a(C_BRT)^b}$$

$$= \frac{(C_C)^c(C_D)^d}{(C_A)^a(C_B)^b} \cdot (RT)^{(c+d)-(a+b)}$$

$$\mathbf{K_p = K_c(RT)^{\Delta n}}$$

where

$$\Delta n = \begin{pmatrix}\text{Sum of the exponents in} \\ \text{the numerator of } Q_c\end{pmatrix} - \begin{pmatrix}\text{Sum of the exponents in} \\ \text{the denominator of } Q_c\end{pmatrix}$$

For reactions such as $H_2(g) + I_2(g) \rightleftharpoons 2HI(g)$ and $N_2(g) + O_2(g) \rightleftharpoons 2NO(g)$ for which $\Delta n = 0$, the value of $K_p = K_c$.

It is necessary to be cautious with the units of R while calculating the value of K_p.

- If partial pressures are expressed in *atm*, the value of R should be 0.0821 atm $dm^3 K^{-1} mol^{-1}$.
- If partial pressures are expressed in *kPa*, the value of R should be 8.314 kPa $dm^3 K^{-1} mol^{-1}$.
- If partial pressures are expressed in *bar*, the value of R should be 0.083 bar $dm^3 K^{-1} mol^{-1}$.

SOLVED EXAMPLES

Example 21.1. *For the reaction, $N_2(g) + 3H_2(g) \rightleftharpoons 2NH_3(g)$, the value of K_p is 3.6×10^{-2} bar^{-2} at 500 K. Calculate the value of K_c for the reaction at the same temperature.*

Solution. $N_2(g) + 3H_2(g) \rightleftharpoons 2NH_3(g)$

$$\Delta n = 2 - (1 + 3) = -2$$

$$K_p = K_c(RT)^{\Delta n}$$

$$K_c = \frac{K_p}{(RT)^{\Delta n}} = \frac{3.6 \times 10^{-2}}{(0.0831 \times 500)^{-2}}$$

$$= 3.6 \times 10^{-2} \times (0.0831 \times 500)^2 = \mathbf{62.15}.$$

Example 21.2. *2.0 moles of PCl_5 were introduced in a 2 dm^3 flask and heated at 625 K to establish equilibrium when 60% of PCl_5 was dissociated into PCl_3 and Cl_2. Find the value of the equilibrium constant.*

Solution. The equilibrium reaction involved is:

$$PCl_5 \rightleftharpoons PCl_3 + Cl_2$$

$$K = \frac{[PCl_3]\,[Cl_2]}{[PCl_5]}$$

Initial moles of PCl_5 taken = 2.0 mol

%age dissociation = 60%

∴ Number of moles dissociated $= \frac{60}{100} \times 2.0 = 1.2$

∴ Number of moles left undissociated at equilibrium will be

$$= 2.0 - 1.2 = 0.8$$

∴ Number of moles of PCl_3 formed by the dissociation of 1.2 moles of PCl_5 = 1.2.

Similarly, number of moles of Cl_2 formed by the dissociation of 1.2 moles of PCl_5 = 1.2.

∴ $$[PCl_5] = \frac{0.8}{2} = 0.4 \text{ mol dm}^{-3}$$

$$[PCl_3] = \frac{1.2}{2} = 0.6 \text{ mol dm}^{-3}$$

$$[Cl_2] = \frac{1.2}{2} = 0.6 \text{ mol dm}^{-3}$$

$$K = \frac{[PCl_3]\,[Cl_2]}{[PCl_5]} = \frac{0.6 \times 0.6}{0.4} = \mathbf{0.9}.$$

Example 21.3. *The following concentrations were obtained for the formation of NH_3 from N_2 and H_2 at equilibrium at 500 K. $[N_2] = 1.5 \times 10^{-2}$ M, $[H_2] = 3.0 \times 10^{-2}$ M and $[NH_3] = 1.2 \times 10^{-2}$ M. Calculate equilibrium constant.*

Solution. The equilibrium constant for the reaction, $N_2(g) + 3H_2(g) \rightleftharpoons 2NH_3(g)$ can be written as,

$$K_c = \frac{[NH_3]^2}{[N_2]\,[H_2]^3}$$

$$= \frac{(1.2 \times 10^{-2})^2}{(1.5 \times 10^{-2})(3.0 \times 10^{-2})^3}$$

$$= 0.106 \times 10^4 = \mathbf{1.06 \times 10^3}.$$

HOMOGENEOUS AND HETEROGENEOUS EQUILIBRIA

The equilibrium in which all the substances are present in the same phase is known as homogeneous equilibrium.

For example,

$$N_2(g) + 3H_2(g) \rightleftharpoons 2NH_3(g)$$

$$2SO_2(g) + O_2(g) \rightleftharpoons 2SO_3(g)$$

$$CH_3COOH(l) + C_2H_5OH(l) \rightleftharpoons CH_3COOC_2H_5(l) + H_2O(l)$$

The equilibrium in which the substances involved are present in different phases is called heterogeneous equilibrium.

For example,

$$CaCO_3(s) \rightleftharpoons CaO(s) + CO_2(g)$$

$$H_2O(l) \rightleftharpoons H_2O(g)$$

$$AgCl(s) \rightleftharpoons Ag^+(aq) + Cl^-(aq)$$

$$Fe(OH)_3(s) \rightleftharpoons Fe^{3+}(aq) + 3OH^-(aq)$$

EQUILIBRIUM EXPRESSIONS OF SOME REACTIONS

While writing equilibrium expressions certain conventions are followed. Let us illustrate these conventions with the help of some examples.

1. In case some solid is involved in the equilibrium, *its concentration remains constant* no matter how much of it is present. *Therefore, by convention, the concentrations of all solids are taken as unity, i.e.*, [Solid] = 1. For example,

$$CaCO_3(s) \rightleftharpoons CaO(s) + CO_2(g)$$

$$K = \frac{[CaO(s)]\,[CO_2(g)]}{[CaCO_3(s)]}$$

$$= [CO_2(g)] \quad \because \quad [CaO(s)] = [CaCO_3(s)] = 1$$

or $$K_p = p_{CO_2}.$$

The above expression shows that at a particular temperature, there is a constant pressure of CO_2 in equilibrium with $CaO(s)$ and $CaCO_3(s)$.

It may be emphasized here that in heterogeneous equilibria pure solids or liquids must be present for the equilibrium to exist, but their concentrations or partial pressures do not appear in the expression of the equilibrium constant.

2. In case some pure liquid is in equilibrium with some gas or gases, the concentration of the pure liquid is taken as unity. For example,

$$H_2O(l) \rightleftharpoons H_2O(g)$$

$$K = \frac{[H_2O(g)]}{[H_2O(l)]}$$

$$= [H_2O(g)] \qquad \because \quad [H_2O(l)] = 1$$

or $$K_p = p_{H_2O}$$

3. For equilibria in an aqueous medium where H_2O is also involved in the equilibrium, its concentration will not change appreciably because it is present in a large quantity ($\approx$ 55.5 mol dm^{-3}) and hence, $[H_2O]$ is taken as constant. For example,

$$NH_3(aq) + H_2O(l) \rightleftharpoons NH_4^+(aq) + OH^-(aq)$$

$$K = \frac{[NH_4^+ (aq)]\,[OH^- (aq)]}{[NH_3(aq)][H_2O(l)]}$$

or $$K \times [H_2O] = \frac{[NH_4^+ (aq)]\,[OH^- (aq)]}{[NH_3(aq)]}$$

or $$K' = \frac{[NH_4^+ (aq)]\,[OH^- (aq)]}{[NH_3(aq)]}$$

By convention, the concentration of a solvent is taken as constant.

STEPS FOR WRITING EQUILIBRIUM CONSTANT EXPRESSION FOR A REACTION

Step I. Write an equation for the reaction.

Step II. Write the product of equilibrium concentrations of the products in the numerator. Omit pure solids, pure liquids and the solvents in dilute solutions.

Step III. Write the product of equilibrium concentrations of the reactants in the denominator. Omit pure solids, pure liquids and the solvents in dilute solutions.

Step IV. Raise each concentration term to the power of the stoichiometric co-efficient of the species in the equation.

SOLVED EXAMPLES

Example 21.4. *Classify each of the following equilibria as homogeneous or heterogeneous.*

(*i*) $CaCO_3(s) + H_2O(l) + CO_2(g) \rightleftharpoons Ca^{2+}(aq) + 2HCO_3^-(aq)$

(*ii*) $CaCO_3(s) \rightleftharpoons CaO(s) + CO_2(g)$

(*iii*) $NH_3(aq) + H_2O(aq) \rightleftharpoons NH_4^+(aq) + OH^-(aq)$

(*iv*) $(NH_4)_2CO_3(s) \rightleftharpoons 2NH_3(g) + CO_2(g) + H_2O(g)$.

Solution. (*i*) Heterogeneous. (*ii*) Heterogeneous. (*iii*) Homogeneous. (*iv*) Heterogeneous.

Example 21.5. *Write the equilibrium constant expressions for the following reactions:*

(*a*) $BaCO_3(s) \rightleftharpoons BaO(s) + CO_2(g)$

(*b*) $AgBr(s) \rightleftharpoons Ag^+ (aq) + Br^-(aq)$

(*c*) $CH_3COCH_3(l) \rightleftharpoons CH_3COCH_3(g)$

(*d*) $Ni(s) + 4CO(g) \rightleftharpoons Ni(CO)_4(g)$.

Solution. (*a*) $$K_c = \frac{[BaO(s)][CO_2(g)]}{[BaCO_3(s)]}$$

$$K_c = [CO_2(g)]$$

$\because$ $[BaO(s)] = [BaCO_3(s)]$ = Constant

$$K_p = (p_{CO_2})$$

(*b*) $$K_c = \frac{[Ag^+(aq)][Br^-(aq)]}{[AgBr(s)]}$$

$$K_c = [Ag^+(aq)][Br^-(aq)]$$

$\because$ $[AgBr(s)]$ = Constant

(*c*) $$K_c = \frac{[CH_3COCH_3(g)]}{[CH_3COCH_3(l)]}$$

$$K_c = [CH_3COCH_3(g)].$$

(*d*) $$K_c = \frac{[Ni(CO)_4(g)]}{[CO(g)]^4\,[Ni(s)]}$$

$$K_c = \frac{[Ni(CO)_4(g)]}{[CO(g)]^4}$$

$\because$ $[Ni(s)]$ = Constant

CHARACTERISTICS OF EQUILIBRIUM CONSTANT (K)

Some important characteristics of equilibrium constant are as follows:

1. *The equilibrium constant has a definite value for every chemical reaction at a particular temperature. The value of an equilibrium constant is independent of the initial concentrations of the reacting species.*

For example, the equilibrium constant for the reaction,

$$Fe^{3+}(aq) + SCN^-(aq) \rightleftharpoons FeSCN^{2+}(aq),$$

at 298 K is

$$K = \frac{[FeSCN^{2+}]}{[Fe^{3+}][SCN^-]} = 138.0$$

Whatever may be the initial concentrations of the reactants, Fe^{3+} and SCN^{-1} ions, the value of K comes out to be 138.0 at 298 K.

2. *The value of equilibrium constant, K changes with change in temperature.*

However, it may be noted that the value of K may increase or decrease with increase in temperature depending upon whether the reaction is endothermic or exothermic in nature respectively. The values of K_p at different temperatures for an endothermic reaction and for an exothermic reaction are given in Table 21.1.

Table 21.1. Variation of K_p with Temperature for Some Reactions

Reactions	*Temperature (K)*	K_p
Endothermic reaction		
$N_2O_4(g) \rightleftharpoons 2NO_2(g)$	298	0.98
	400	47.9
	500	1700
Exothermic reaction		
$N_2(g) + 3H_2(g) \rightleftharpoons 2NH_3(g)$	298	6.8×10^5
	400	41
	500	3.6×10^{-2}

3. *For a reversible reaction, the equilibrium constant for the backward reaction is the inverse of the equilibrium constant for the forward reaction.*

For example, the equilibrium constant, K, for the reaction between hydrogen and iodine at 717 K is 48.

$$H_2(g) + I_2(g) \rightleftharpoons 2HI(g);\ K = \frac{[HI]^2}{[H_2][I_2]} = 48$$

The equilibrium constant for the decomposition of hydrogen iodide at the same temperature is the inverse of the above equilibrium constant.

$$2HI(g) \rightleftharpoons H_2(g) + I_2(g)$$

$$K_1 = \frac{[H_2][I_2]}{[HI]^2} = \frac{1}{K} = \frac{1}{48}$$

4. *The equilibrium constant is independent of the presence of a catalyst.*

5. *If equilibrium constant is expressed in terms of concentration, it has different units for different reactions.*

This can be illustrated by the following examples:

(*a*) If the number of moles of the product is the same as the number of moles of the reactants, K has no units (*i.e.*, it is dimensionless). Thus, in the reaction between hydrogen and iodine to form hydrogen iodide, K has no units as illustrated below:

$$H_2(g) + I_2(g) \rightleftharpoons 2HI(g)$$

$$K = \frac{[HI]^2}{[H_2][I_2]} = \frac{(\text{mol dm}^{-3})^2}{(\text{mol dm}^{-3})(\text{mol dm}^{-3})}$$

(*b*) If the number of moles of the products is *not the same as* that of the reactants, K will have units that depend upon the change in the number of moles. Thus, for the reaction between nitrogen and hydrogen to form ammonia, K would have the units $dm^6\ mol^{-2}$ as illustrated below:

$$N_2(g) + 3H_2(g) \rightleftharpoons 2NH_3(g)$$

$$K = \frac{[NH_3]^2}{[N_2][H_2]^3} = \frac{(\text{mol dm}^{-3})^2}{(\text{mol dm}^{-3})(\text{mol dm}^{-3})^3}$$

$$= \frac{1}{(\text{mol dm}^{-3})^2}$$

Thus, K will have the units $dm^6\ mol^{-2}$.

21.7 APPLICATIONS OF EQUILIBRIUM CONSTANT

The knowledge of equilibrium constant helps us in

- predicting the extent of a reaction.
- predicting the direction in which the net reaction is taking place.
- calculation of equilibrium concentrations and equilibrium pressures.

PREDICTION OF EXTENT OF REACTION

The magnitude of the equilibrium constant tells us about the extent to which the reactants are converted into products before equilibrium is attained.

The larger the value of K, the greater the extent to which reactants are converted into products. On the other hand, if the value of K is small it indicates that equilibrium is mainly in favour of reactants and that only a small fraction of the reactants are converted into products.

We can make the following generalisations concerning the composition of equilibrium mixtures:

- If $K_c > 10^3$, products predominate over reactants. If K_c is very large, the reaction proceeds nearly to completion.
- If $K_c < 10^{-3}$, reactants predominate over products. If K_c is very small, the reaction hardly proceeds.

- If K_c is in the range of 10^{-3} to 10^3, appreciable concentrations of both reactants and products are present.

For example, consider the following reaction:

$$2CO(g) + O_2(g) \rightleftharpoons 2CO_2(g)$$
$$K = 2.2 \times 10^{22} \text{ at } 1000 \text{ K}$$

K for this reaction is given by the expression:

$$K = \frac{[CO_2]^2}{[CO]^2\,[O_2]}$$

The large value of K in this case indicates that at equilibrium, the concentration of CO_2 is quite high. In other words, this indicates that the reaction between CO and O_2 to form CO_2 is almost complete under the given conditions.

Let us now consider another reaction:

$$2H_2O(g) \rightleftharpoons 2H_2(g) + O_2(g)$$
$$K = 4.1 \times 10^{-48} \text{ at } 500 \text{ K}$$

The expression for equilibrium constant of this reaction is

$$K = \frac{[H_2]^2[O_2]}{[H_2O]^2}$$

Since the value of K in this case is very small, it indicates that at equilibrium, the concentrations of H_2 and O_2 are very small. In other words, under the given conditions only a small fraction of H_2O dissociates into H_2 and O_2. Now let us consider some reactions in which the value of K_c is neither too small nor too large. In such reactions there are appreciable concentrations of both reactants and products at equilibrium. Two such equilibria are given as follows:

$$H_2(g) + I_2(g) \rightleftharpoons 2HI(g)$$
$$K_c = 57 \text{ at } 700 \text{ K}$$
$$N_2O_4(g) \rightleftharpoons 2NO_2(g)$$
$$K_c = 4.64 \times 10^{-3} \text{ at } 298 \text{ K}$$

The relationship between K_c and the extent of reaction is illustrated in Fig. 21.9.

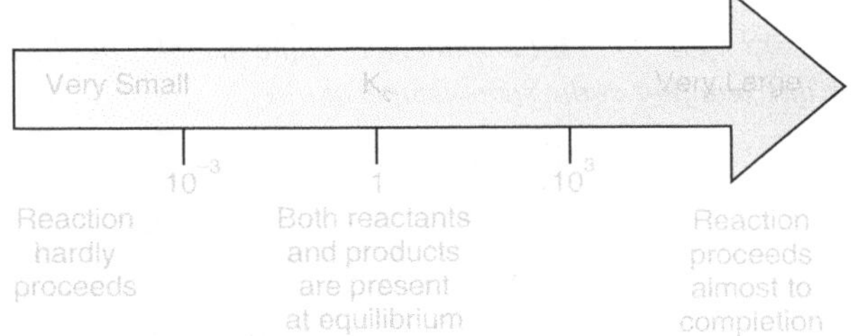

Fig. 21.9. Relationship between K_c and the extent of reaction.

PREDICTING THE DIRECTION OF THE NET REACTION

From the knowledge of equilibrium constant, it is possible to predict the direction of the net reaction at given concentrations or partial pressures of reactants and products.

From the given concentrations, the concentration quotient Q_c for the reaction is calculated. By comparing the value of Q_c with the equilibrium constant, K_c the direction of the reaction can be predicted.

- If $Q_c > K_c$, the net reaction is taking place in the backward direction *i.e.*, direction of reactants.
- If $Q_c < K_c$, the net reaction is taking place in the forward direction *i.e.*, direction of products.
- If $Q_c = K_c$, the reaction is at equilibrium *i.e.*, no net reaction is taking place.

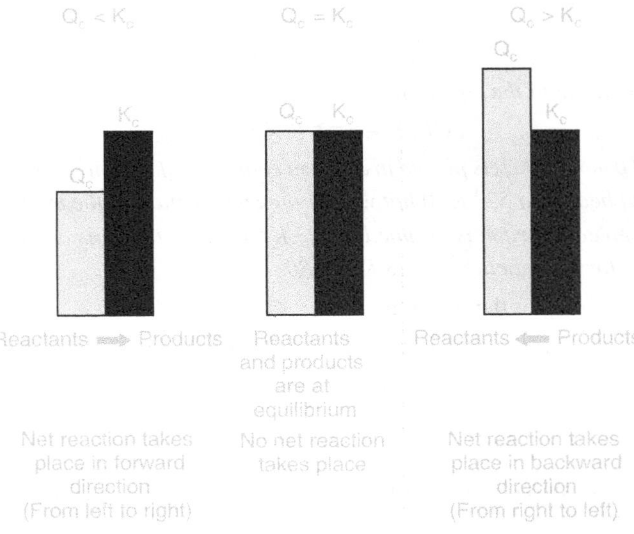

Fig. 21.10. Relation between Q_c, K_c and the direction of net reaction.

CALCULATION OF EQUILIBRIUM CONCENTRATIONS AND EQUILIBRIUM PRESSURES

If the equilibrium concentrations of various reactants and products are known in a reaction, the equilibrium constant can be calculated. On the other hand, if the equilibrium constant is known, then the equilibrium concentration can be calculated.

STEPS FOR CALCULATING EQUILIBRIUM CONCENTRATIONS AND EQUILIBRIUM PRESSURES

Step I.	Write a chemical equation for the equilibrium.
Step II.	Write the equilibrium constant expression for the reaction.
Step III.	Express all unknown concentrations or partial pressures in terms of a single variable x.
Step IV.	Substitute the equilibrium concentrations or partial pressures (in terms of x) in the equilibrium constant expression.
Step V.	Solve the equation for x.
Step VI.	Substitute the value obtained for x in the expressions in step III to calculate equilibrium concentrations or equilibrium partial pressures.

SOLVED EXAMPLES

Example 21.6. *Phosphorus pentachloride decomposes according to the equation*

$$PCl_5(g) \rightleftharpoons PCl_3(g) + Cl_2(g)$$

1.00 mol of PCl_5 is placed in a closed container of 5.0 dm³ capacity and heated to 523 K. What will be the composition of the mixture when equilibrium is attained at 523 K? The equilibrium constant, K_c, for the reaction at 523 K is 1.80.

Solution. For the reaction,

$$PCl_5(g) \rightleftharpoons PCl_3(g) + Cl_2(g)$$

$$K_c = \frac{[PCl_3][Cl_2]}{[PCl_5]}$$

The initial concentration of $PCl_5 = \frac{1.00}{5.0} = 0.20 \text{ mol dm}^{-3}$.

Suppose, when equilibrium is reached the concentration of PCl_5 decreases by x mol dm^{-3}.

	PCl_5	$\rightleftharpoons$ PCl_3	+ Cl_2
Initial conc. (mol dm^{-3})	0.20	0	0
Equilibrium conc. (mol dm^{-3})	$(0.20 - x)$	x	x

Substituting the equilibrium concentrations in the expression for equilibrium constant, we get

$$K_c = \frac{[PCl_3][Cl_2]}{[PCl_5]}$$

$$1.80 = \frac{(x)(x)}{(0.2 - x)}$$

$$x^2 + 1.80x - 0.36 = 0$$

$$x = \frac{-1.80 \pm \sqrt{(1.80)^2 - 4(1)(-0.36)}}{2}$$

$$= -1.98 \text{ and } 0.18$$

Since concentration must be positive, 0.18 is selected as the appropriate value of x.

$$\therefore \quad [PCl_5] = 0.20 - 0.18 = \mathbf{0.02 \text{ mol dm}^{-3}}$$

$$[PCl_3] = x = \mathbf{0.18 \text{ mol dm}^{-3}}$$

$$[Cl_2] = x = \mathbf{0.18 \text{ mol dm}^{-3}}.$$

Example 21.7. *The equilibrium constant at 298 K for the reaction*

$$Cu(s) + 2Ag^+(aq) \rightleftharpoons Cu^{2+}(aq) + 2Ag(s)$$

is 2.0×10^{15}. In a solution in which copper has displaced some silver ions from solution, the concentration of Cu^{2+} ions is 1.8×10^{-2} mol dm^{-3} and the concentration of Ag^+ ions is 3.0×10^{-9} mol dm^{-3}. Is this system at equilibrium?

Solution. The given reaction in a state of equilibrium may be represented as

$$Cu(s) + 2Ag^+(aq) \rightleftharpoons Cu^{2+}(aq) + 2Ag(s)$$

According to the Law of Chemical Equilibrium,

$$K_c = \frac{[Cu^{2+}(aq)][Ag(s)]^2}{[Cu(s)][Ag^+(aq)]^2} = \frac{[Cu^{2+}(aq)]}{[Ag^+(aq)]^2}$$

($\because$ Molar conc. of solids = 1)

Now, $[Cu^{2+}(aq)] = 1.8 \times 10^{-2} \text{ mol dm}^{-3}$

$[Ag^+(aq)] = 3.0 \times 10^{-9} \text{ mol dm}^{-3}$

Substituting the values, let us calculate the concentration quotient Q_c.

$$Q_c = \frac{1.8 \times 10^{-2}}{3.0 \times 10^{-9} \times 3.0 \times 10^{-9}} = \frac{1.8}{9.0} \times 10^{16}$$

$$= \mathbf{2 \times 10^{15}}.$$

The value of Q is equal to the value of the equilibrium constant. This means the reaction is in equilibrium.

Example 21.8. *At 700 K, the equilibrium constant K_p, for the reaction*

$$2SO_3(g) \rightleftharpoons 2SO_2(g) + O_2(g)$$

is 1.8×10^{-3} k Pa. What is the numerical value in moles per litre of K_c for this reaction at the same temperature?

Solution.

$$K_p = K_c (RT)^{\Delta n}$$

$$\Delta n = 3 - 2 = 1$$

$$T = 700 \text{ K}$$

$$R = 8.31 \text{ kPa dm}^3 \text{ mol}^{-1} \text{ K}^{-1}$$

$$K_p = 1.8 \times 10^{-3} \text{ kPa}$$

$$1.8 \times 10^{-3} \text{ (kPa)} = K_c \times 8.31 \text{ (kPa dm}^3 \text{ mol}^{-1} \text{ K}^{-1}) \times 700 \text{ (K)}$$

$$K_c = \frac{1.8 \times 10^{-3}}{8.31 \times 700} \text{ mol dm}^{-3} = 0.03094 \times 10^{-5}$$

$$= \mathbf{3.09 \times 10^{-7} \text{ mol dm}^{-3}}.$$

Example 21.9. *For the reaction, $CH_4(g) + 2H_2S(g) \rightleftharpoons CS_2(g) + 4H_2(g)$, at 1173 K, the magnitude of the equilibrium constant, K_c is 3.6. For each of the following composition, decide whether the reaction mixture is at equilibrium. If it is not, decide to which direction the reaction should go.*

(*i*) *$[CH_4]$ = 1.07 M, $[H_2S]$ = 1.20 M, $[CS_2]$ = 0.90 M, $[H_2]$ = 1.78 M*

(*ii*) *$[CH_4]$ = 1.45 M, $[H_2S]$ = 1.29 M, $[CS_2]$ = 1.25 M, $[H_2]$ = 1.75 M*

Solution. (*i*) $$Q_c = \frac{[CS_2][H_2]^4}{[CH_4][H_2S]^2}$$

$$Q_c = \frac{(0.90)(1.78)^4}{(1.07)(1.20)^2} = 5.86$$

Since Q_c is greater than K_c (3.6), the net reaction will go in the ***backward direction.***

(*ii*) $$Q_c = \frac{[CS_2][H_2]^4}{[CH_4][H_2S]^2} = \frac{(1.25)(1.75)^4}{(1.45)(1.29)^2} = \mathbf{4.86}$$

Q_c is greater than K_c, therefore, the net reaction will go in the ***backward direction.***

Example 21.10. *50.0 g of $CaCO_3$ is heated to 1073 K in a 5 dm³ vessel. What per cent of the $CaCO_3$ would decompose at equilibrium? K_p for the reaction,*

$CaCO_3(s) \rightleftharpoons CaO(s) + CO_2(g)$, is 1.15 atm at 1073 K.

Solution. $$CaCO_3(s) \rightleftharpoons CaO(s) + CO_2(g)$$

$$K_p = p_{CO_2} = 1.15 \text{ atm}$$

$$PV = nRT$$

$$n_{CO_2} = \frac{p_{CO_2} \cdot V}{RT} = \frac{1.15 \times 5}{0.082 \times 1073} = 0.065 \text{ mol}$$

1 mole of CO_2 is obtained by the decomposition of 1 mole $CaCO_3$.

Therefore, moles of $CaCO_3$ decomposed is equal to the moles of CO_2 = 0.065 mol

Moles of $CaCO_3$ initially present $= \frac{50}{100} = 0.5$ mol

[Mol. mass of $CaCO_3$ = 100]

Per cent of $CaCO_3$ decomposed

$$= \frac{0.065}{0.5} \times 100 = \mathbf{13\%}.$$

21.8 FACTORS AFFECTING EQUILIBRIA—Le-Chatelier's Principle

Le-Chatelier, a French Chemist, made a generalization to explain the effect of changes in concentration, temperature or pressure on the state of a system in equilibrium. The generalization is known as **Le-Chatelier's Principle.** It states that:

If a system in equilibrium is subjected to a change of concentration, pressure or temperature, the equilibrium shifts in a direction that tends to undo the effect of the change.

Let us consider the effect of these variables in the light of Le-Chatelier's Principle, using a few well known equilibria.

EFFECT OF CHANGE OF CONCENTRATION

According to Le-Chatelier's Principle, when the concentration of one of the substances in a system at equilibrium is increased, then the equilibrium will shift so as to use up the substance added. Suppose at equilibrium one of the reactants is added, the equilibrium will shift in the direction that consumes the reactants, *i.e.*, the forward direction. In other words, more of the reactants would be converted into products. On the other hand, if one of the products is added, the equilibrium will shift in the backward direction to consume the products.

In order to understand the effect of change of concentration on equilibrium, let us consider the reaction,

$$H_2(g) + I_2(g) \rightleftharpoons 2HI(g)$$

If H_2 is added to the reaction mixture at equilibrium, then the equilibrium of the reaction is disturbed. In order to restore it, the reaction proceeds in a direction wherein H_2 is consumed, *i.e.*, more of H_2 and I_2 react to form HI and finally the equilibrium is re-established. Thus, addition of H_2 shifts the equilibrium in the forward direction. In the new equilibrium mixture, the concentration of H_2 would be less than it was after addition of H_2 but more than what it was in the original equilibrium (Fig. 21.11).

The same point can be understood in terms of the reaction quotient, Q_c.

$$Q_c = [HI]^2/[H_2][I_2]$$

If the concentration of H_2 is increased, the value of Q_c decreases and it is no longer equal to K_c. In order to attain new equilibrium, the reaction proceeds in the forward direction resulting in a decrease in the concentrations of H_2

and I_2 and an increase in the concentration of HI. Finally, a new equilibrium is attained.

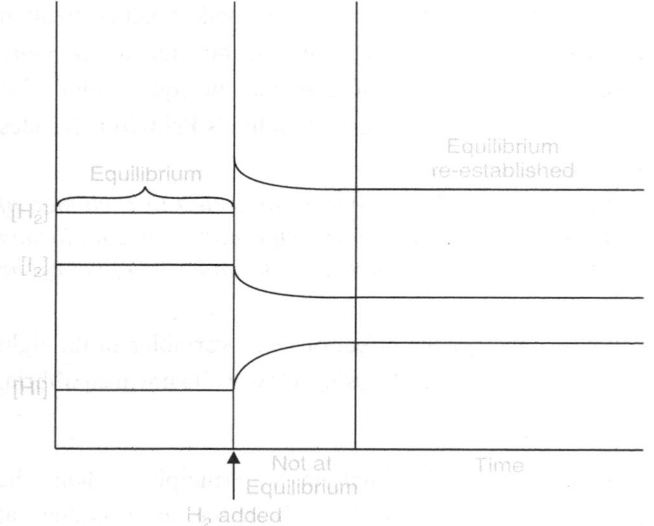

Fig. 21.11. Effect of addition of H_2 on change of concentration for the reactants and products in the reaction, $H_2(g) + I_2(g) \rightleftharpoons 2HI(g)$.

Similarly, we can say that the removal of a product also favours the forward reaction and increases the concentration of the products. For example, in the large scale production of CaO, by calcination of lime stone ($CaCO_3$), CO_2 is continuously removed from the kiln to shift the reaction in the forward direction. Continuous removal of one of the products maintains Q_c at a value less than K_c and thus the reaction keeps on proceeding in the forward direction.

The effect of concentration on equilibrium can be demonstrated with the help of the following reaction:

$$\underset{\text{(yellow)}}{Fe^{3+}(aq)} + \underset{\text{(colourless)}}{SCN^-(aq)} \rightleftharpoons \underset{\text{(deep red)}}{[FeSCN]^{2+}(aq)}$$

$$K_c = \frac{[Fe(SCN)^{2+}(aq)]}{[Fe^{3+}(aq)][SCN^-(aq)]}$$

A reddish colour appears on adding 2–3 drops of 0.002 M potassium thiocynate solution to 5 mL of 0.2 M iron(III) trioxonitrate(V) solution due to the formation of $[Fe(SCN)]^{2+}$. The intensity of the red colour becomes constant when equilibrium is attained. This equilibrium can be shifted in either the forward or backward direction depending on the choice of adding a reactant or a product.

Suppose, some iron(III) salt is added to this equilibrium. The colour of the solution deepens immediately showing that there is an increase in the concentration of the coloured complex ion $[FeSCN]^{2+}$. This change is in accordance with Le-Chatelier's principle. Addition of more of Fe^{3+} ions has resulted in increasing the concentration of the complex $[FeSCN]^{2+}$ ions. Similarly, if some potassium thiocyanate is added at equilibrium, the colour deepens again due to the formation of more of the $[FeSCN]^{2+}$ ions.

The above equilibrium can be shifted in the backward direction by adding reagents that remove Fe^{3+} or SCN^- ions. For example, oxalic acid ($H_2C_2O_4$), reacts with Fe^{3+} ions to form the stable complex ion $[Fe(C_2O_4)_3]^{3-}$, thus decreasing the concentration of Fe^{3+} ions. In accordance with the Le-Chatelier's principle, the decrease in the concentration of Fe^{3+} ions leads to a decrease in the concentration of $[FeSCN]^{2+}$ ions which is indicated by the colour of the solution becoming lighter.

ACTIVITY 21.2 | **Investigating Effect of Change of Concentration on Equilibrium State**

The effect of concentration on equilibrium state can be demonstrated by the study of the following equilibrium which exists in a solution of bromine in water.

$$\underbrace{Br_2(aq) + H_2O(l)}_{\textit{Reddish brown}} \rightleftharpoons \underbrace{2H^+(aq) + Br^-(aq) + BrO^-(aq)}_{\textit{Colourless}}$$

- Take bromine water in a test tube. In this tube the colour of the solution would be reddish brown.
- Now add dilute sodium hydroxide solution to the above solution.

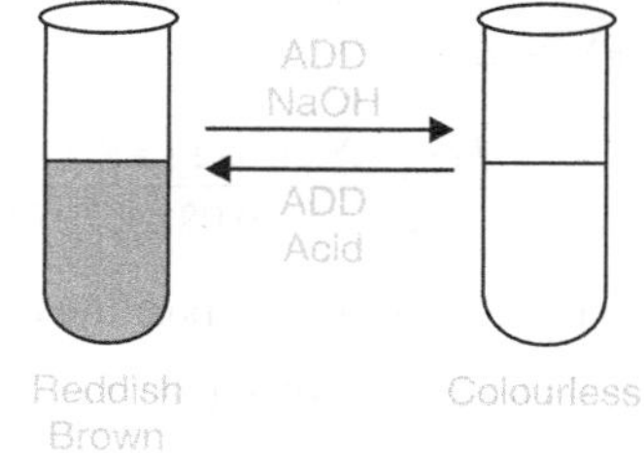

Fig. 21.12. Effect of concentration on equilibrium state.

The solution becomes colourless indicating that the equilibrium has shifted to the right. Addition of OH^- ions removes H^+ ions to form water,

$$H^+(aq) + OH^-(aq) \longrightarrow H_2O(l)$$

In order to undo the effect of the change, the equilibrium shifts in the forward direction producing a colourless species.

- Now add dilute H_2SO_4 solution.

The reddish brown colour is restored indicating that the equilibrium has shifted to the left. The acid increases the concentration of H^+ ions and shifts the equilibrium in backward direction. Due to formation of Br_2, the reddish brown colour is restored.

EFFECT OF CHANGE IN PRESSURE

The change in pressure has effect only on those equilibria which involve gaseous substances and proceed with a change in the number of moles of the gases. According to Le-Chatelier's principle, increase of external pressure should affect the equilibrium in such a way as to reduce the pressure. This implies that the equilibrium will shift in the direction which has smaller number of moles of the gaseous substances. This can be easily understood from the following equilibrium representing the formation of ammonia.

$$N_2(g) + 3H_2(g) \rightleftharpoons 2NH_3(g)$$

On increasing the pressure, the volume occupied by the system will decrease. As such there will be greater number of moles per unit volume. The effect of this change can be counteracted if the equilibrium shifts in the direction involving a decrease in the number of moles. This can happen only if nitrogen and hydrogen combine to form ammonia. Thus, increase in pressure in this case will favour the forward reaction.

This can also be understood in terms of the reaction quotient Q_c. Suppose the pressure of the reaction mixture is doubled and the volume is reduced to half, the partial pressures and concentrations of the various species in the reaction mixture would become double. We obtain Q_c by replacing each equilibrium concentration by double its value.

$$K_c = \frac{[NH_3]^2}{[N_2][H_2]^3} = \frac{(a)^2}{(b)(c)^3} = \frac{a^2}{bc^3}$$

$$Q_c = \frac{[NH_3]^2}{[N_2][H_2]^3} = \frac{(2a)^2}{(2b)(2c)^3} = \frac{a^2}{4bc^3}.$$

As the value of Q_c is less than K_c, the reaction proceeds in the forward direction.

Now, let us consider some reactions which do not involve any change in the number of moles of gaseous species:

$$H_2(g) + I_2(g) \rightleftharpoons 2HI(g)$$

$$N_2(g) + O_2(g) \rightleftharpoons 2NO(g)$$

$$CO(g) + H_2O(g) \rightleftharpoons CO_2(g) + H_2(g)$$

In such reactions the pressure does not have any effect on equilibrium.

Let us explain the effect of pressure on the solubility of gases in liquid solvents on the basis of Le-Chatelier's principle. As an example, let us study the equilibrium involving dissolution of CO_2 in water. The equilibrium may be represented as:

$$CO_2(g) \rightleftharpoons CO_2(aq)$$

On increasing the pressure of CO_2, the equilibrium shifts in a direction which results in a decrease in the pressure of CO_2. The pressure of CO_2 will be lowered only if $CO_2(g)$ dissolves in water to form $CO_2(aq)$. Thus, the *solubility of a gas in some liquid is directly proportional to the pressure of the gas in equilibrium with the solution.*

EFFECT OF TEMPERATURE

A chemical equilibrium involves two opposing reactions, one favouring the products and the other favouring the reactants. If one of the reactions is exothermic the other must be endothermic. For example, let us consider the equilibrium representing the formation of ammonia:

$$N_2(g) + 3H_2(g) \underset{\text{Endothermic}}{\overset{\text{Exothermic}}{\rightleftharpoons}} 2NH_3(g); \quad \Delta H = -\mathbf{93.6\ kJ.}$$

In this equilibrium, the forward reaction is exothermic while the backward reaction is endothermic. Now, if the temperature is increased, the equilibrium will shift in the direction of the endothermic reaction, which tends to undo the effect of added heat. Since the backward reaction is endothermic, the equilibrium shifts in favour of the backward reaction. In other words, it will result in lesser amount of ammonia. On the other hand, if temperature is decreased, the equilibrium will shift towards the exothermic reaction. So low temperature favours the formation of ammonia.

ACTIVITY 21.3	**Investigating Effect of Temperature on Equilibrium**

The effect of temperature on equilibrium position can be demonstrated by the study of the following equilibrium:

$$\underset{\textit{Colourless}}{N_2O_4} \underset{\text{Exothermic}}{\overset{\text{Endothermic}}{\rightleftharpoons}} \underset{\textit{Dark brown}}{2NO_2}$$

- Prepare NO_2 gas by heating conc. HNO_3 and copper chips.

- Collect the gas in a test tube and close the mouth of the test tube with a cork.

  The colour of the gas in the test tube is reddish brown. It is actually an equilibrium mixture of NO_2 and N_2O_4.
- Place the test tube in a beaker containing ice. Observe the change in the colour of the gas in the test tube.
- Now place the test tube in a beaker containing hot water. Again observe the change in colour of the gas.

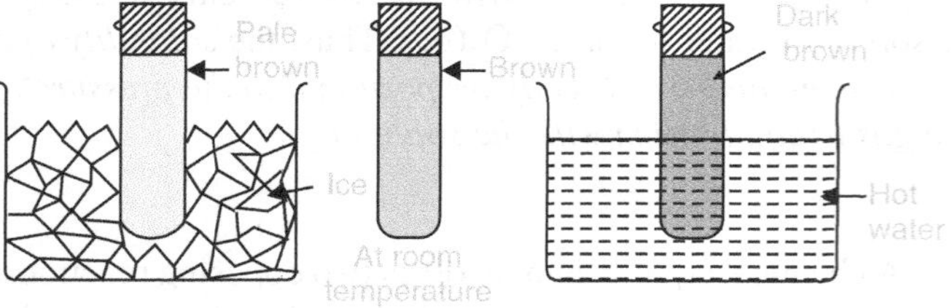

Fig. 21.13. Effect of temperature on equilibrium in the reaction, $N_2O_4 \rightleftharpoons 2NO_2$.

When the tube is placed in beaker containing ice it is observed that the colour of the gas mixture lightens indicating that the equilibrium has shifted towards the left (*exothermic direction*). When the tube is placed in a beaker containing hot water, the gas mixture becomes darker in colour, indicating that the equilibrium has shifted towards the right (*endothermic direction*).

The temperature dependence of the K_c depends on whether the reaction is exothermic or endothermic.

- The equilibrium constant for an exothermic reaction decreases as the temperature increases.
- The equilibrium constant for an endothermic reaction increases as the temperature increases.

EFFECT OF CATALYST

The presence of a catalyst does not disturb the state of equilibrium because it increases the rate of the forward as well as the backward reaction to the same extent. It simply hastens the attainment of equilibrium by providing a new pathway having a low activation energy. Catalyst does not disturb the equilibrium composition of the reaction mixture.

Certain reactions such as the formation of ammonia from nitrogen and hydrogen cannot be carried out at high temperatures because high temperature favours the backward reaction. So such reactions give poor yields at high temperatures and therefore, have to be carried out at low temperature. But at low temperature the rate of reaction becomes very slow and it takes a very long time to attain equilibrium. In order to increase the rates of such reactions, generally, catalysts are used so that equilibrium is attained early. For example, in case of formation of ammonia, a catalyst consisting of iron, is used to catalyse the reaction.

SOLVED EXAMPLES

Example 21.11. *Which of the following reactions will get affected by an increase in pressure? Also mention, whether the change will cause the reaction to go in the right or left direction?*

(*i*) $CH_4(g) + 2S_2(g) \rightleftharpoons CS_2(g) + 2H_2S(g)$

(*ii*) $CO_2(g) + C(s) \rightleftharpoons 2CO(g)$

(*iii*) $4NH_3(g) + 5O_2(g) \rightleftharpoons 4NO(g) + 6H_2O(g)$

(*iv*) $C_2H_4(g) + H_2(g) \rightleftharpoons C_2H_6(g)$.

Solution. (*i*) In this case n_p is equal to n_r

This reaction will not be affected by an increase in pressure.

(*ii*) $CO_2(g) + C(s) \rightleftharpoons 2CO(g)$

In this reaction, moles of gases on product side ($n_p = 2$) is more than that on reactant side ($n_r = 1$). This reaction ***will be affected*** by an increase in pressure. Increase in pressure shifts the equilibrium in the direction where there is less no. of moles of gases. In this reaction, an increase in pressure will cause the reaction to go in the ***left direction.***

(*iii*) $4NH_3(g) + 5O_2(g) \rightleftharpoons 4NO(g) + 6H_2O(g)$

The reaction ***would be affected*** by an increase in pressure because n_p is different from n_r.

Increase in pressure shifts the equilibrium in the ***left direction*** because n_r is less than n_p.

(*iv*) $C_2H_4(g) + H_2(g) \rightleftharpoons C_2H_6(g)$

The reaction ***would be affected*** by an increase in pressure. Increase in pressure will shift the equilibrium ***towards the right*** because n_p is less than n_r.

Example 21.12. *In the reaction,*

$$CO(g) + 2H_2(g) \rightleftharpoons CH_3OH(g); \Delta_r H^\ominus = -92.0 \text{ kJ mol}^{-1}$$

Concentrations of hydrogen, carbon monoxide and methanol become constant at equilibrium. What will happen if

(*i*) *volume of the reaction vessel in which reactants and products are contained is suddenly reduced to half?*

(*ii*) *the partial pressure of hydrogen is suddenly doubled?*

Solution. For the equilibrium,

$$CO(g) + 2H_2(g) \rightleftharpoons CH_3OH(g)$$

$$K_c = \frac{[CH_3OH]}{[CO][H_2]^2};\ K_p = \frac{p_{CH_3OH}}{p_{CO} \times p_{H_2}^{\ 2}}$$

(*i*) When the volume of the vessel is suddenly reduced to half, the partial pressures of the various species become double. Therefore,

$$Q_p = \frac{2p_{CH_3OH}}{2p_{CO} \times (2p_{H_2})^2} = \frac{1}{4}K_p$$

Since Q_p is less than K_p, the equilibrium shifts in the forward direction, producing more CH_3OH.

(*ii*) When the partial pressure of hydrogen is suddenly doubled, Q_p changes and is no longer equal to K_p

$$Q_p = \frac{p_{CH_3OH}}{p_{CO} \times (2p_{H_2})^2} = \frac{1}{4}K_p$$

Since Q_p has become less than K_p, the equilibrium shifts in favour of products, producing more CH_3OH.

SUMMARY

- **Reversible Reaction.** A chemical reaction which can take place in the forward as well as backward direction.
- **Irreversible Reaction.** A chemical reaction which cannot take place in the backward direction.
- **Equilibrium State.** A state of a system when its observable properties do not change under a given set of conditions. At equilibrium, the forward and backward reactions proceed at the same rate.
- **Physical Equilibrium.** Equilibrium achieved in physical processes.
- **Chemical Equilibrium.** Equilibrium achieved in chemical processes.
- **Homogeneous Equilibrium.** Equilibrium in which reactants and products are in the same phase.
- **Heterogeneous Equilibrium.** Equilibrium in which the various substances involved have different phases.
- **Law of Mass Action.** A theoretical concept which states that the rate of reaction is directly proportional to the product of the molar concentrations of reactants with each concentration term raised to a power which equals the stoichiometric co-efficient of that species in the balanced equation.
- **Equilibrium Constant (K).** For an equilibrium:

$$aA + bB \rightleftharpoons cC + dD$$

$$K = \frac{[C]^c[D]^d}{[A]^a[B]^b};$$

$$K_p = \frac{(p_C)^c(p_D)^d}{(p_A)^a(p_B)^b};$$

$$K_p = K_c(RT)^{\Delta n}$$

In using these expressions, the concentrations or partial pressures of solids are taken to be UNITY.

- **Applications of Equilibrium Constant**
  — To predict the extent of a reaction on the basis of its magnitude.
  — To predict the direction of a reaction.
  — To calculate equilibrium concentrations.
- If $K_c > 10^3$, products predominate over reactants, and the reaction proceeds nearly to completion.
- If $K_c < 10^{-3}$, reactants predominate over products, and the reaction proceeds rarely.
- If K_c is in the range 10^3 to 10^{-3} appriciable concentrations of both the reactants and products are present.
- **Reaction Quotient (Q_c).** The ratio of the product of molar concentrations of the products to the product of molar concentrations of the reactants with each concentration term raised to a power which is equal to the stoichiometric co-efficient of that species is called reaction quotient.

  At equilibrium, $Q_c = K_c$.

  At concentrations other than equilibrium, $Q_c \neq K_c$.

  If $Q_c > K_c$, the reaction will proceed in the direction of the reactants (reverse reaction).

  If $Q_c < K_c$, the reaction will proceed in the direction of the products (forward reaction).

  If $Q_c = K_c$, the reaction mixture is already at equilibrium.
- **Le-Chatelier's Principle.** When an equilibrium is subjected to any kind of stress (change in concentration, temperature or pressure) it shifts in a direction so as to undo the effect of the stress.
- The equilibrium constant for an exothermic reaction decreases as the temperature increases.
- The equilibrium constant for an endothermic reaction increases as the temperature increases.

EVALUATE YOURSELF

I. Objective Type Questions

Select the most appropriate choice from the options given as (a), (b), (c) and (d) after each question:

1. In a chemical reaction equilibrium is established when
   (*a*) opposing reaction ceases
   (*b*) concentrations of reactants and products are equal
   (*c*) rate of the opposing reaction is same as that of the forward reaction
   (*d*) forward reaction ceases.
2. The value of K for the reaction,

$$H_2 + I_2 \rightleftharpoons 2HI$$

is 49. The value of K for the reaction,

$$2HI \rightleftharpoons H_2 + I_2 \text{ will be}$$

   (*a*) 49 (*b*) 1/49
   (*c*) 7 (*d*) 1/7.

3. The units of equilibrium constant for the reaction

$$N_2 + 3H_2 \rightleftharpoons 2NH_3 + \text{heat},$$

will be

(*a*) $dm^6\ mol^{-2}$ (*b*) $mol\ dm^{-3}$

(*c*) $mol^2\ dm^{-6}$ (*d*) $dm^3\ mol^{-1}$.

4. One mole of SO_3 was placed in a 1 dm^3 vessel at a certain temperature. When equilibrium was established in the reaction,

$$2SO_3 \rightleftharpoons 2SO_2 + O_2$$

the vessel was found to contain 0.4 moles of SO_3. The value of the equilibrium constant is

(*a*) 0.13 (*b*) 0.36

(*c*) 0.68 (*d*) 0.45.

5. For the reaction equilibrium

$$N_2O_4(g) \rightleftharpoons 2\ NO_2(g)$$

the concentrations of N_2O_4 and NO_2 at equilibrium are 4.8×10^{-2} and 1.2×10^{-2} mol dm^{-3} respectively. The value of K_c for the reaction is

(*a*) 3×10^3 mol dm^{-3} (*b*) 3.3×10^2 mol dm^{-3}

(*c*) 3×10^{-1} mol dm^{-3} (*d*) 3×10^{-3} mol dm^{-3}.

6. The reaction quotient (Q) for the reaction

$$N_2(g) + 3H_2(g) \rightleftharpoons 2\ NH_3(g)$$

is given by $Q = \frac{[NH_3]^2}{[N_2][H_2]^3}$. The reaction will proceed from right to left if

(*a*) $Q < K_c$ (*b*) $Q > K_c$

(*c*) $Q = 0$ (*d*) $Q = K_c$

where K_c is the equilibrium constant.

7. In what manner will increase of pressure affect the following equilibrium?

$$C(s) + H_2O(g) \rightleftharpoons CO(g) + H_2(g)$$

(*a*) Shift in the forward direction

(*b*) Shift in the reverse direction

(*c*) Increase in the yield of hydrogen

(*d*) No effect.

8. The equilibrium constant for the reaction,

$$N_2(g) + O_2(g) \rightleftharpoons 2NO(g)$$

is 4×10^{-4} at 2000 K. In the presence of a catalyst, the equilibrium is attained ten times faster. Therefore, the equilibrium constant, in presence of a catalyst, at 2000 K is

(*a*) 40×10^{-4} (*b*) 4×10^{-4}

(*c*) 4×10^{-3}

(*d*) difficult to compute without more data.

9. For an equilibrium; $N_2O_4(g) \rightleftharpoons 2NO_2(g)$

the total pressure at equilibrium is P and the degree of dissociation of N_2O_4 is x.

Which one of the following is the partial pressure of NO_2?

(*a*) $\frac{2x}{(1+x)}$ (*b*) $2 \times P$

(*c*) $\frac{2}{(1-x)}P$ (*d*) $\frac{2}{3} \times P$.

10. Which of the following statements is correct about the equilibrium constant?

(*a*) Its value increases by increase in temperature

(*b*) Its value decreases by decrease in temperature

(*c*) Its value may increase or decrease with increase in temperature

(*d*) Its value is constant at all temperatures.

11. If the gas mixture for the following reaction,

$N_2O_4(g) \rightleftharpoons 2NO_2(g)$ is compressed, then

(*a*) products are favoured (*b*) reactants are favoured

(*c*) no change

(*d*) products become equal to reactants.

12. In the lime kiln, the reversible reaction,

$CaCO_3 \rightleftharpoons CaO + CO_2$ proceeds to completion because

(*a*) reaction is carried at high temperature

(*b*) CO_2 escapes out

(*c*) CaO is removed

(*d*) of low temperature.

13. Which of the following equilibria will have the same values for K_p and K_c?

(*a*) $N_2O_4(g) \rightleftharpoons 2NO_2(g)$

(*b*) $PCl_5(g) \rightleftharpoons PCl_3(g) + Cl_2(g)$

(*c*) $A(s) + 2B(g) \rightleftharpoons 2C(g)$

(*d*) $N_2(g) + 3H_2(g) \rightleftharpoons 2NH_3(g)$.

14. Which of the following is a characteristic of a reversible reaction?

(*a*) It never proceeds to completion in a closed container.

(*b*) It proceeds only in the forward direction.

(*c*) Number of moles of reactants and products are equal.

(*d*) It can be influenced by a catalyst.

15. Of the following which change will shift the reaction towards the product?

$$I_2(g) \rightleftharpoons 2I(g) \quad \Delta H_r^\ominus\ (298\ K) = +150\ kJ$$

(*a*) Increase in concentration of I.

(*b*) Decrease in concentration of I_2.

(*c*) Increase in temperature.

(*d*) Increase in total pressure.

II. Fill in the Blanks

16. Complete the following sentences by supplying the appropriate words:

(*i*) An equilibrium can be achieved only in ...... reactions.

(*ii*) For the reaction,

$$N_2 + O_2 \rightleftharpoons 2NO;\ \Delta H = 180\ kJ,$$

an increase in temperature shifts the equilibrium in ...... direction.

(*iii*) In a certain reaction, $A + B \rightleftharpoons C + D$, the value of K is 300 but $Q_c = 100$, the net reaction proceeds towards ....

(*iv*) For a certain equilibrium reaction $K_p = K_c$. It indicates the equilibrium is ...... pressure variation.

(*v*) For endothermic reactions, the value of K ...... with increase in temperature.

III. Discussion Questions

17. What are reversible and irreversible reactions? Explain with the help of some examples.
18. What do you understand by the state of equilibrium? How is it recognised?
19. What do you understand by physical equilibrium and chemical equilibrium?
20. How would you show that equilibrium is dynamic in nature?
21. Give an experiment to show that equilibrium can be approached from either side.
22. What is the difference between steady state and equilibrium state as applied to chemical systems?
23. A reaction $A \rightleftharpoons B$ is allowed to proceed in a closed vessel till it reaches a stage when no more change in the concentration of A or B is observed. What is this state and what can be said about the rates of the forward and reverse reactions at this time?
24. A process, $X \rightleftharpoons Y$, attains equilibrium at T. Can we increase the amount of the product by addition of a catalyst?
25. What type of chemical reactions attain a state of equilibrium when carried in a closed vessel?
26. What is meant by equilibrium constant? What are its characteristics?
27. How does the value of the equilibrium constant predict the extent of a reaction?
28. Write the expression for the equilibrium constant, K_c for each of the following reactions:

    (*i*) $2NOCl(g) \rightleftharpoons 2NO(g) + Cl_2(g)$

    (*ii*) $2Cu(NO_3)_2(s) \rightleftharpoons 2CuO(s) + 4NO_2(g) + O_2(g)$

    (*iii*) $Fe^{3+}(aq) + 3OH^-(aq) \rightleftharpoons Fe(OH)_3(s)$

    (*iv*) $I_2(s) + 5F_2(g) \rightleftharpoons 2IF_5(g)$.

29. Classify each of the following equilibria as homogeneous or heterogeneous:

    (*i*) $CaCO_3(s) \rightleftharpoons CaO(s) + CO_2(g)$

    (*ii*) $2SO_2(g) + O_2(g) \rightleftharpoons 2SO_3(g)$

    (*iii*) $2Mg(s) + O_2(g) \rightleftharpoons 2MgO(g)$

    (*iv*) $2N_2O(g) \rightleftharpoons 2N_2(g) + O_2(g)$

    (*v*) $CH_3COOC_2H_5(aq) + H_2O(l) \rightleftharpoons CH_3COOH(aq) + C_2H_5OH(aq)$.

30. The equilibrium constant for the reaction

    $A_2(g) + B_2(g) \rightleftharpoons 2AB(g)$ is 64 at 600 K.

    What is the equilibrium constant for

    $$2AB(g) \rightleftharpoons A_2(g) + B_2(g)$$

    at the same temperature?

31. Ammonium hydrogen sulphide dissociates according to the equation:

    $$NH_4HS(s) \rightleftharpoons NH_3(g) + H_2S(g)$$

    When solid NH_4HS is taken in an evacuated flask at a certain temperature, it is observed that the pressure of the mixture is 600 mm Hg when equilibrium is established. Calculate the value of the equilibrium constant K_p for the reaction.

32. Consider the reaction $N_2O_4(g) \rightleftharpoons 2NO_2(g)$, at 373 K. Suppose at a particular moment, the concentrations of N_2O_4 and NO_2 are 1 mol dm^{-3} each. In which direction will the reaction proceed to establish equilibrium? K for the reaction is 0.36.
33. PCl_5 decomposes into PCl_3 and Cl_2. It is found that if we put 3 moles of PCl_5 in a 2 dm^3 flask at 600 K, 40% of it decomposes. Calculate equilibrium constant for the decomposition reaction.
34. One mole of acetic acid and one mole of methyl alcohol were mixed at 298 K. At equilibrium 0.333 mole of acetic acid was found to be unreacted. Calculate the equilibrium constant for the reaction,

    $$CH_3COOH + CH_3OH \rightleftharpoons CH_3COOCH_3 + H_2O$$

35. 2 moles N_2O_4 were introduced in a flask of 10 dm^3 capacity at 350 K. At equilibrium 50% of N_2O_4 was found to be dissociated into NO_2. Find the equilibrium constant for the dissociation of N_2O_4 under the given conditions.
36. The value of K_c for the reaction $2A \rightleftharpoons B + C$ is 2×10^{-3}. At a given time, the composition of the reaction mixture is $[A] = [B] = [C] = 3 \times 10^{-4}$ M. In which direction will the reaction proceed?
37. A mixture containing equal amounts of methane and steam is passed over a catalyst at 1000 K. After the establishment of equilibrium, it is found that the concentrations of the various gases in mol dm^{-3} are:

    $$[CO] = 0.10,\ [H_2] = 0.30,\ [CH_4] = [H_2O] = 0.90$$

    Calculate K for the reaction

    $$CH_4(g) + H_2O(g) \rightleftharpoons CO(g) + 3H_2(g).$$

38. In an experiment, 2 moles of HI are taken in an evacuated 5.0 dm^3 container. At 144°C, the equilibrium constant is equal to 0.0156 for the gaseous reaction:

    $$2HI(g) \rightleftharpoons H_2(g) + I_2(g)$$

    Calculate the equilibrium concentrations of HI, H_2 and I_2.

39. State Le-Chatelier's principle. Discuss the effect of pressure and temperature on the following reactions:

$$N_2 + 3H_2 \rightleftharpoons 2NH_3 + 93.6 \text{ kJ}$$
$$N_2 + O_2 \rightleftharpoons 2NO - 180 \text{ kJ.}$$

40. Bromine water (a dilute solution of bromine in water) is brown and weakly acidic because of the equilibrium denoted by the following equation:

$$Br_2(aq) + 2H_2O(l) \rightleftharpoons HBrO(aq) + H_3O^+(l) + Br^-(aq)$$

In solution, $Br_2(aq)$ is brown, $Br^-(aq)$ is colourless and HBrO (hypobromic acid—a weak acid) is colourless. When sodium hydroxide is added to the solution, the solution becomes colourless but the colour returns when hydrochloric acid is added. Explain this observation.

41. The value of the equilibrium constant of a reaction A + B $\rightleftharpoons$ P + Q is 2.0×10^{-2} at 25°C and it is 2.0×10^{-1} at 50°C. Predict whether the formation of P and Q should be exothermic or endothermic?

Answers

I. Objective Type Questions

1. (*c*)	**2.** (*b*)	**3.** (*a*)	**4.** (*c*)
5. (*d*)	**6.** (*b*)	**7.** (*b*)	**8** (*b*)
9. (*a*)	**10.** (*c*)	**11.** (*b*)	**12.** (*b*)
13. (*c*)	**14.** (*a*)	**15.** (*c*).	

II. Fill in the Blanks

16. (*i*) reversible

(*ii*) forward

(*iii*) forward

(*iv*) independent of

(*v*) increases.

SHS 3

SECTION 4

ACIDS AND BASES

General Objectives

After studying this section, the student should be able to:

1. understand the behaviour of acid and base solutions based on their physical and chemical properties.
2. explore acidic and basic solutions qualitatively and quantitatively.
3. describe acids and bases and classify substances as acids and bases based on Arrhenius, Bronsted-Lowry and Lewis concepts.
4. show knowledge of how to test for anions and cations.
5. use the principles and practice of titration to solve analytical problems.

In this section, we shall study

Unit 22: The Concepts and Properties of Acids and Bases

Unit 23: Acid-Base Titrations

Unit 24: Classification of Acids and Bases

Unit 25: Concept of pH and pOH

Unit 26: Buffer Solutions

Unit 27: Acid-Base Indicators

Unit 28: Solubility of Substances

Unit 29: Salts and Chemicals from Salt.

SECTION-4

UNIT-22

THE CONCEPTS AND PROPERTIES OF ACIDS AND BASES

Specific Objectives

After studying this unit, the student should be able to:

- outline the characteristic properties of acids and bases in aqueous solutions
- describe Arrhenius, Bronsted-Lowry and Lewis concepts of acids and bases
- state the physical and chemical properties of acids and bases.

22.1 INTRODUCTION

Certain substances, when dissolved in water, conduct electricity. Such substances are called **electrolytes**. The various substances which act as electrolytes can be classified into three types: **acids, bases** and **salts.** The properties of aqueous solutions of acids and bases are listed below:

Acids		*Properties of an Acidic Solution*
Hydrochloric acid:	HCl	1. Tastes sour.
Tetraoxosulphate(VI) acid:	H_2SO_4	2. Reacts with metals such as zinc, magnesium, etc., liberating hydrogen gas.
Trioxonitrate(V) acid:	HNO_3	
Ethanoic acid:	CH_3COOH	3. Changes the colour of litmus from blue to red.
Tetraoxophosphate(V) acid:	H_3PO_4	4. Conducts electricity.
Bases		*Properties of a Basic Solution*
Sodium hydroxide:	NaOH	1. Tastes bitter, feels slippery.
Potassium hydroxide:	KOH	2. Reacts with an acidic solution to destroy or neutralize the properties that are characteristic of an acid.
Ammonium hydroxide:	NH_4OH	
Calcium hydroxide:	$Ca(OH)_2$	3. Changes the colour of litmus from red to blue.
Barium hydroxide:	$Ba(OH)_2$	4. Conducts electricity.

A salt is formed by the reaction of an acid with a base. The reaction is known as **neutralization.**

Acid + Base ⟶ Salt + Water

$HCl\ (aq) + NaOH\ (aq) \longrightarrow NaCl\ (aq) + H_2O\ (l)$

In this unit, you will study the important properties of acids and bases. You will also learn about important concepts of acids and bases.

22.2 PROPERTIES OF ACIDS

1. Acids have a **sour taste.**
2. Strong acids such as tetraoxosulphate(VI) acid, hydrochloric acid, trioxonitrate(V) acid, etc., are **highly corrosive in nature** and can burn the skin. Therefore, these acids should be handled with care and should not be brought in contact with skin or clothes.
3. Acids are readily soluble in water. Dissolution of acids in water is generally an **exothermic** process.

ACTIVITY 22.1	**Investigating the Exothermic Nature of Dissolution of Acids in Water**

- Take about 50 cm^3 of water in a 100 cm^3 beaker.
- Add 10–15 drops of concentrated H_2SO_4 to it. During addition, stir the solution with the help of a glass rod.
- Touch the beaker from outside to feel the change in temperature.

  (You may use a thermometer to note the change in temperature.)

  What do you observe?

An acid should be diluted by adding acid to water and not water to the acid. If water is added to the acid, the heat produced may cause the mixture to splash out and cause burns. The glass container may also break due to excessive local heating.

4. Acids **conduct electricity** in their aqueous solutions.

ACTIVITY 22.2	**Testing Conductivity of an Acidic Solution**

- Take 1M solutions of hydrochloric acid (HCl), tetraoxosulphate(VI) acid (H_2SO_4), alcohol (C_2H_5OH), glucose ($C_6H_{12}O_6$), etc.
- Put hydrochloric acid solution in a beaker and place two graphite rods or metal rods in it.
- Connect the graphite rods to a 6 volt battery through a bulb and a switch as shown in Fig. 22.1.
- Switch on the current.

  What do you observe?

  It is observed that the bulb starts glowing, indicating that the solution of hydrochloric acid conducts electricity.
- Repeat the experiment with the remaining three solutions also. Observe whether the bulb glows or not. Also observe the brightness with which the bulb glows.

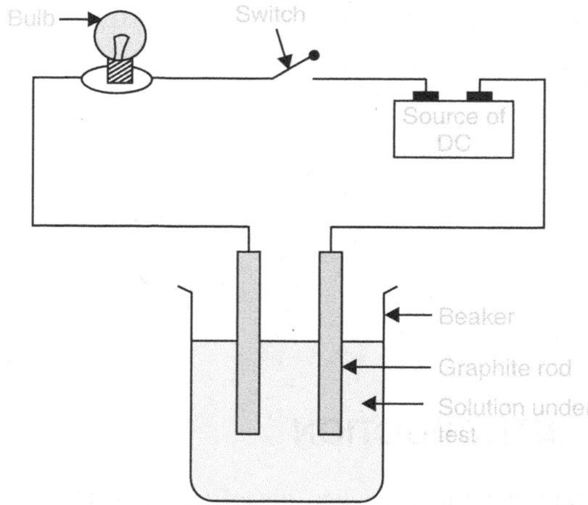

Fig. 22.1. Testing the conductivity of a solution.

What do you observe?

It is observed that the bulb starts glowing when solutions of acids are taken in the beaker. However, the bulb does not glow when glucose and ethyl alcohol solutions are taken. This indicates that the aqueous solutions of glucose and alcohol do not conduct electricity.

Acids on dissociation produce ions (H^+ ions and anions corresponding to the acid) which are responsible for conduction of electricity.

Glucose and ethyl alcohol do not produce ions in aqueous solution. Hence, they do not conduct electricity.

5. **Acids impart specific colours to the acid-base indicators**

   Acid-base indicators are the chemical compounds which have different colours in acidic and basic solutions. They can be used to test whether a given solution is acidic or basic. Some common acid-base indicators are given in Table 22.1

Table 22.1. Common Acid-Base Indicators

Indicator	*Colour in Acidic Medium*	*Colour in Basic Medium*
Litmus solution	Red	Blue
Methyl orange	Pink	Yellow
Phenolphthalein	Colourless	Pink
Methyl red	Red	Yellow

Do You know?

Litmus is a water soluble purple dye extracted from certain *lichens* (a class of plants). It turns red under acidic conditions and blue under alkaline conditions. When the solution is neither acidic nor basic its colour is purple. It is used as a rough indicator of acidity or alkalinity both in solution form and as litmus paper. Some other natural materials such as red cabbage leaves, turmeric, coloured petals of some flowers like *petunias* and *geraniums* can also act as acid-base indicators. Turmeric turns reddish-brown in alkaline medium.

ACTIVITY 22.3 Study of Acid-Base Indicators

- Collect the dilute solutions of hydrochloric acid, tetraoxosulphate(VI) acid, trioxonitrate(V) acid, acetic acid, sodium hydroxide, potassium hydroxide and ammonium hydroxide.
- Take four test tubes and add about 1 cm^3 dilute hydrochloric acid solution.
- Now add 1–2 drops of different indicators (red litmus solution, blue litmus solution, phenolphthalein and methyl orange) to different test tubes and note the change in colour (if any).

  Wash the test tubes with water and repeat the above steps with other solutions. Note the observations in a tabular form as shown below:

Sample solution	*Red litmus solution*	*Blue litmus solution*	*Phenol-phthalein solution*	*Methyl orange solution*
1. Hydrochloric acid				
2. Tetraoxosul-phate(VI) acid				
3. Trioxonitrate(V) acid				
4. Acetic acid				
5. Sodium hydroxide				
6. Potassium hydroxide				
7. Ammonium hydroxide				

6. Reaction of Acids with Active Metals

Metals lying above hydrogen in the activity series react with dilute acid solutions and displace hydrogen from them. The metal combines with the remaining part of the acid to form salt.

Metal + Acid ⟶ Salt + Hydrogen gas

$$Zn + H_2SO_4 \longrightarrow ZnSO_4 + H_2$$
$$Fe + H_2SO_4 \longrightarrow FeSO_4 + H_2$$
$$Mg + 2HCl \longrightarrow MgCl_2 + H_2$$
$$2Al + 3H_2SO_4 \longrightarrow Al_2(SO_4)_3 + 3H_2$$

Hydrogen is a highly combustible gas.

ACTIVITY 22.4 Study of Reaction of Metals with Acids

- Arrange the apparatus as shown in Fig. 22.2.

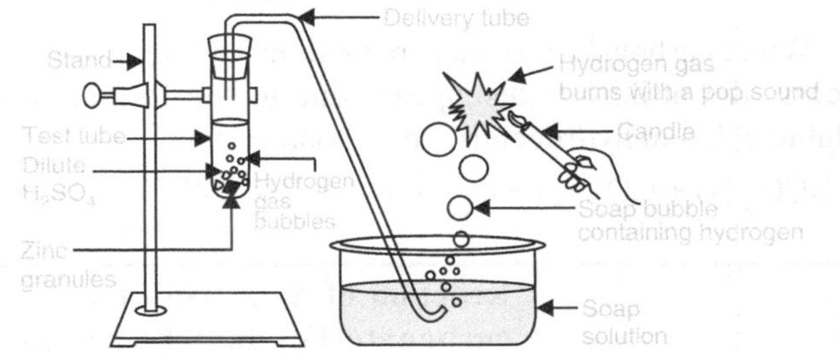

Fig. 22.2. Reaction of zinc granules with dilute sulphuric acid.

- In the test tube take about 5 cm^3 of dilute H_2SO_4 solution and a few zinc granules.

  What do you observe?

  It is observed that zinc granules react with dilute H_2SO_4 and bubbles of gas evolve from the surface of the zinc granules. The gas evolved is hydrogen.
- Pass the gas through a soap solution taken in a tub.

  What do you observe?

  It is observed that bubbles start rising from the soap solution.
- Bring a burning candle near a gas filled soap bubble.

  What do you observe?

  It is observed that the gas in the bubble burns with a pop sound, thus, indicating that the gas in the bubble is hydrogen gas.
- Repeat the activity with hydrochloric acid and acetic acid.

7. Reaction of Acids with Metal trioxocarbonate(IV) and hydrogentrioxocarbonate(IV)

Metal trioxocarbonate(IV) and hydrogen trioxocarbonate(IV) react with acid to form the corresponding salt and liberate CO_2 gas with brisk effervescence.

Metal trioxocarbonate(IV) + Acid $\longrightarrow$ Salt + Carbon(IV) oxide + Water

Metal hydrogentrioxocarbonate(IV) + Acid $\longrightarrow$ Salt + Carbon(IV) oxide + Water

$$Na_2CO_3\,(s) + 2HCl\,(aq) \longrightarrow 2NaCl\,(aq) + CO_2\,(g) + H_2O\,(l)$$

$$NaHCO_3\,(s) + 2HCl\,(aq) \longrightarrow NaCl\,(aq) + CO_2\,(g) + H_2O\,(l)$$

$$CaCO_3\,(s) + 2HCl\,(aq) \longrightarrow CaCl_2\,(aq) + CO_2\,(g) + H_2O\,(l)$$

Carbon(IV) oxide is a colourless, odourless gas. When the gas is passed through lime water, milkiness appears due to the formation of calcium carbonate.

$$Ca(OH)_2\,(aq) + CO_2\,(g) \longrightarrow \underset{\text{White ppt}}{CaCO_3\,(s)} + H_2O\,(l)$$

When carbon(IV) oxide is passed through lime water in excess, the milkiness disappears due to the formation of soluble calcium hydrogentrioxocarbonate(IV).

$$CaCO_3\,(s) + CO_2\,(g) + H_2O\,(l) \longrightarrow \underset{\text{(Soluble in water)}}{Ca(HCO_3)_2\,(aq)}$$

ACTIVITY 22.5 **Reaction of Acids with trioxocarbonate(IV) and hydrogen trioxocarbonate(IV)**

- Take about 1 g of sodium trioxocarbonate(IV) (Na_2CO_3) in a test tube.
- Add about 2 cm^3 of dilute hydrochloric acid.

  A colourless, odourless gas evolves with brisk effervescence.
- Pass the gas through freshly prepared lime water (Fig. 22.3).

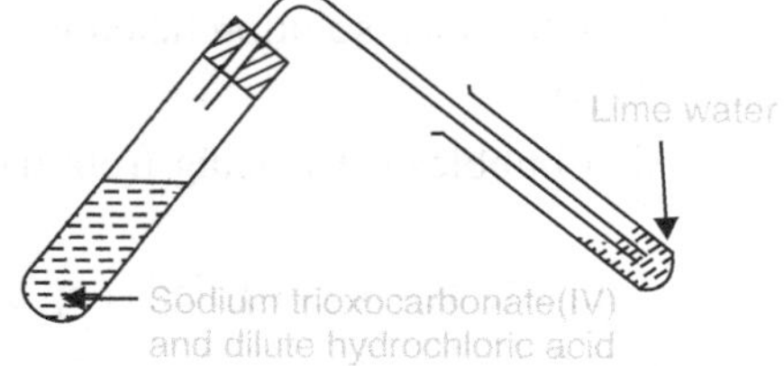

Fig. 22.3. Reaction of dilute hydrochloric acid with sodium trioxocarbonate(IV)

What do you observe?

This activity can also be performed with sodium hydrogencarbonate and calcium carbonate.

8. Reaction of Acids with Bases. Acids react with bases to from salt and water.

Acid + Base $\longrightarrow$ Salt + Water

The reaction between acids and bases to form salt and water is known as ***neutralization reaction.*** For example,

$$HCl\,(aq) + NaOH\,(aq) \longrightarrow NaCl\,(aq) + H_2O\,(l)$$

$$H_2SO_4\,(aq) + 2NaOH\,(aq) \longrightarrow Na_2SO_4\,(aq) + 2H_2O\,(l)$$

$$HNO_3\,(aq) + KOH\,(aq) \longrightarrow KNO_3\,(aq) + H_2O\,(l).$$

ACTIVITY 22.6 **Study of Neutralization Reaction**

- Take about 2 cm^3 of dilute NaOH solution in a test tube.
- Add 2–3 drops of phenolphthalein solution.

  What is the colour of the solution?

  The colour of the solution becomes dark pink.
- Add dilute HCl solution to the pink solution drop by drop.

  What do you observe?

  It is observed that when sufficient quantity of HCl solution has been added, the pink colour of the solution disappears.
- Now add a few drops of NaOH solution and again observe if there is any change in colour.

  It is observed that the pink colour reappears.

Phenolphthalein has a pink colour in basic medium. When sufficiently large quantity of HCl solution is added to the NaOH solution, the medium becomes acidic due to the presence of excess acid. In acidic medium, the phenolphthalein is colourless and hence the solution becomes colourless. When we again add NaOH solution, it neutralizes the excess acid and the solution again becomes alkaline. Therefore, the pink colour reappears.

Some Practical Applications of Neutralization

1. A person suffering from hyperacidity is advised to take antacid tablets or antacid suspension. Antacid preparations contain magnesium hydroxide as the active component which neutralizes the excess acid present in the stomach.

$$Mg(OH)_2 + 2HCl \longrightarrow MgCl_2 + 2H_2O$$

2. In acidic soils, slaked lime is added to reduce acidity.

$$\underset{\substack{\text{Slaked lime}\\\text{(Base)}}}{Ca(OH)_2} + \underset{\substack{\text{Tetraoxosulphate(VI) acid}\\\text{(Acid)}}}{H_2SO_4} \longrightarrow CaSO_4 + 2H_2O$$

3. The sting of ants and bees contains methanoic acid. It is neutralized by rubbing soap or dilute ammonia solution.
4. The sting of yellow wasps contains an alkali. It is neutralized by rubbing dilute ethanoic acid (*vinegar*).

Table 22.2. Acids Present in Some Natural Substances

Natural substance	*Acid*
Citrus fruits (Oranges, lemons, etc.)	Citric acid
Apples	Malic acid
Sour Milk (Curds)	Lactic acid
Vinegar	Ethanoic acid
Tomatoes	Oxalic acid
Rancid butter (spoiled butter)	Butanoic acid
Urine	Uric acid
Gastric juice	Hydrochloric acid
Tamarind, grapes	Tartaric acid
Fats	Stearic acid
Sting of bees and ants	Methanoic acid (Formic acid)
Nettle plant sting	Methanoic acid

SOLVED EXAMPLES

Example 22.1. *Why does distilled water not conduct electricity whereas rainwater does?*

Solution. Distilled water contains only a negligible concentration of ions due to slight ionization of water. Hence, it does not conduct electricity.

On the other hand, rainwater contains many acidic impurities which are formed by dissolution of gases present in the atmosphere.

$$SO_2 + H_2O \longrightarrow H_2SO_3$$
$$CO_2 + H_2O \longrightarrow H_2CO_3$$

These acids dissociate in rainwater to furnish an appreciable concentration of ions. Due to presence of these ions rainwater conducts electricity.

Example 22.2. *Compounds like alcohols and glucose also contain hydrogen but are not categorised as acids. Describe an activity to prove it.*

Solution. Compound like alcohols and glucose do not undergo dissociation in water to yeild H^+ (*aq*) ions and hence they do not exhibit the characteristic properties of acids. The aqueous solutions of alcohols and glucose do not conduct electricity (Activity 22.2) which shows that their aqueous solutions do not contain ions.

22.3 PROPERTIES OF BASES

1. Bases have a **bitter taste** and are **slippery to touch.**
2. Strong bases such as caustic soda (NaOH) and caustic potash are corrosive in nature and can harm the skin. These bases are highly soluble in water and the dissolution process is **exothermic**.

ACTIVITY 22.7 — Dissolution of Bases in Water is an Exothermic Process

- Take about 20 cm^3 of water in a 100 cm^3 beaker.
- Add 4–5 pellets of sodium hydroxide to it and stir the contents of the beaker with the help of a glass rod.
- Touch the bottom of the beaker from outside to feel any change in temperature.

What do you observe?

It is observed that sodium hydroxide readily dissolves in water. As a result of dissolution the solution becomes warm indicating that the dissolution of sodium hydroxide in water is an exothermic process.

The above activity can be performed with potassium hydroxide also.

3. Alkalies are good conductors of electricity.

ACTIVITY 22.8 — Aqueous Solutions of Bases Conduct Electricity

- Take aqueous solutions of sodium hydroxide (NaOH) and potassium hydroxide (KOH).
- Set the apparatus as shown in Fig. 22.1 and put NaOH solution in the beaker.
- Switch on the current.

What do you observe?

It is observed that the bulb starts glowing, indicating that an aqueous solution of NaOH conducts electricity.

NaOH gives Na^+ and OH^- ions in the solution which are responsible for conduction of electricity.

- Repeat the activity with KOH solution.

KOH solution also conducts electricity.

From this, we can conclude that aqueous solutions of all bases are good conductors of electricity.

4. Indicator Properties

Aqueous solutions of bases impart specific colours to acid-base indicators.

- They turn red litmus blue.
- They impart pink colour to phenolphthalein.

- They impart yellow colour to methyl orange.
- They impart yellow colour to methyl red.

Do You Know?

The bases such as sodium hydroxide, potassium hydroxide, barium hydroxide, etc., which are soluble in water are known as **alkalies.** It may be noted that all the bases are not alkalies. Alkalies are highly corrosive and hence should not be brought in contact with skin.

5. Reaction with Acids

Bases react with acids to form salt and water

$$\textbf{Base + Acid} \longrightarrow \textbf{Salt + Water}$$

$$KOH + HCl \longrightarrow KCl + H_2O$$

$$2NaOH + H_2SO_4 \longrightarrow Na_2SO_4 + H_2O$$

The reaction is known as **neutralization.**

6. Reaction with Metals

Certain metals such as zinc, aluminium, tin and lead react with strong alkali solutions, on heating, and hydrogen gas is evolved.

$$2NaOH + Zn \longrightarrow \underset{\text{Sodium zincate}}{Na_2ZnO_2} + H_2$$

7. Reaction with Non-metallic Oxides

Non-metallic oxides are generally acidic in nature. They react with bases to form salt and water.

$$\textbf{Base + Non-metallic oxide} \longrightarrow \textbf{Salt + Water}$$

For example,

$$Ca(OH)_2 + CO_2 \longrightarrow CaCO_3 + H_2O$$

$$2KOH + CO_2 \longrightarrow K_2CO_3 + H_2O$$

$$2NaOH + SO_2 \longrightarrow Na_2SO_3 + H_2O$$

Concentrated solution of caustic potash is used for absorbing carbon(IV) oxide gas.

8. Reaction with Metallic Oxides

Certain metal oxides can act as bases as well as acids. These are known as **amphoteric oxides.** For example, zinc oxide (ZnO), lead(II) oxide (PbO), aluminium oxide (Al_2O_3).

These oxides on reaction with hot alkali solutions form complex salts and dissolve.

$$ZnO + 2NaOH \longrightarrow \underset{\text{Sodium dioxozincate(II)}}{Na_2ZnO_2} + H_2O$$

$$PbO + 2NaOH \longrightarrow \underset{\text{Sodium dioxoplumbate (II)}}{Na_2PbO_2} + H_2O$$

$$Al_2O_3 + 2NaOH \longrightarrow \underset{\text{Sodium dioxoaluminate (III)}}{2NaAlO_2} + H_2O$$

In these reactions metal oxides are acting as acids.

Add To Your Knowledge

Certain compounds such as ammonia (NH_3), sodium trioxocarbonate(IV) (Na_2CO_3) and sodium hydrogen-trioxocarbonate(IV) ($NaHCO_3$) which are themselves not hydroxides behave as bases because they can produce hydroxyl ions in solution by interacting with water molecules.

$$NH_3 + H_2O \longrightarrow NH_4^+ + OH^-.$$

Now, let us learn about the various concepts of acids and bases:

22.4 ARRHENIUS CONCEPT OF ACIDS AND BASES

According to the Arrhenius concept,

An acid is a substance which can furnish hydrogen ions in its aqueous solution.

A base is a substance which can furnish hydroxyl ions in its aqueous solution.

For example, substances such as HNO_3, HCl, CH_3COOH are acids whereas substances such as NaOH, KOH, NH_4OH are bases, according to this concept.

$$H_2SO_4 \longrightarrow 2H^+ + SO_4^{2-}$$

$$HNO_3(aq) \longrightarrow H^+(aq) + NO_3^-(aq)$$

$$HCl(aq) \longrightarrow H^+(aq) + Cl^-(aq)$$

$$CH_3COOH(aq) \rightleftharpoons H^+(aq) + CH_3COO^-(aq)$$

$$H_2CO_3 \rightleftharpoons 2H^+ + CO_3^{2-}$$

$$NaOH(aq) \longrightarrow Na^+(aq) + OH^-(aq)$$

$$KOH(aq) \longrightarrow K^+(aq) + OH^-(aq)$$

$$NH_4OH(aq) \rightleftharpoons NH_4^+(aq) + OH^-(aq)$$

STRONG AND WEAK ACIDS AND BASES

Acids such as HCl, H_2SO_4 and HNO_3, which are almost completely ionized in aqueous solutions are termed **strong acids** whereas acids such as ethanoic acid (CH_3COOH) and trioxocarbonate(IV) acid (H_2CO_3) which are weakly ionized are called **weak acids.**

Similarly, bases which are almost completely ionized in aqueous solution are called **strong bases,** for example, NaOH and KOH. The bases such as NH_4OH are only slightly ionized and are called **weak bases.**

You will learn more about strength of acids and bases in Unit 24.

According to the Arrhenius theory, neutralization of acids and bases is basically a reaction between H^+ and OH^- ions in solutions.

$$H^+(aq) + OH^-(aq) \rightleftharpoons H_2O(l)$$

Nature of Hydrogen Ion and Hydroxyl Ion in Aqueous Solution

Hydrogen atom contains one proton and one electron. H^+ ion is formed by loss of this electron. Therefore, H^+ ion is simply a proton. Charge density of this unshielded proton is very high. Therefore, it is not likely to exist independently as H^+ ion. In an aqueous solution H^+ ion is considered to be present in a hydrated form in combination with a water molecule as H_3O^+.

$$H^+ + H_2O \longrightarrow H_3O^+$$

H_3O^+ ion is called **hydronium ion.**

In aqueous solution hydronium ion is further hydrated to give species such as $H_5O_2^+$, $H_7O_3^+$ and $H_9O_4^+$. Similarly, hydroxyl ion is hydrated to give species such as $H_3O_2^-$, H_5O_3, etc.

22.5 BRONSTED-LOWRY CONCEPT OF ACIDS AND BASES

In 1923, a Danish Chemist J.H. Bronsted and an English Chemist T.M. Lowry independently proposed new definitions for acids and bases. They proposed that:

An ***acid*** *is a substance that can donate a proton.*

A ***base*** *is a substance that can accept a proton.*

These definitions are more general than the Arrhenius definitions because according to these definitions even ions can behave as acids or bases. Moreover, these definitions are not restricted to reactions taking place in aqueous solution only.

In order to understand this concept of acids and bases, lets take some specific examples.

$$\underset{\text{Acid}}{HCl(aq)} + \underset{\text{Base}}{H_2O(l)} \rightleftharpoons H_3O^+(aq) + Cl^-(aq)$$

$$\underset{\text{Acid}}{H_2O(l)} + \underset{\text{Base}}{NH_3(aq)} \rightleftharpoons NH_4^+(aq) + OH^-(aq)$$

$$\underset{\text{Acid}}{H_2O(l)} + \underset{\text{Base}}{CO_3^{2-}(aq)} \rightleftharpoons HCO_3^-(aq) + OH^-(aq)$$

From the above equations, it is obvious that acid base reactions according to the Bronsted-Lowry concept involve transfer of proton from the acid to a base. *A substance can act as an acid only if another substance capable of accepting a proton, is present.*

CONJUGATE ACID-BASE PAIRS

An acid after losing a proton becomes a base whereas a base after accepting the proton becomes an acid. For example, let us consider the reaction between water and ammonia as represented by the following equilibrium:

Adds proton

$$\underset{\text{Acid}_1}{H_2O(l)} + \underset{\text{Base}_2}{NH_3(aq)} \rightleftharpoons \underset{\text{Acid}_2}{NH_4^+(aq)} + \underset{\text{Base}_1}{OH^-(aq)}$$

Loses proton

In the *forward reaction*, water donates a proton to ammonia (base) and acts as acid. In the *reverse reaction*, NH_4^+ ions donate a proton to the OH^- ions (base) and act as acid. *A base formed by the loss of a proton from an acid* is called a **conjugate base** of the acid whereas *an acid formed by gain of a proton from the base* is called a **conjugate acid** of the base. In the above example, $OH^-$is the conjugate base of H_2O and NH_4^+ is conjugate acid of NH_3. *Acid-base pairs such as H_2O/OH^- and NH_4^+/NH_3 which are formed by loss or gain of a proton are called conjugate acid-base pairs.*

A strong acid would have a large tendency to donate a proton. Thus, *the conjugate base of a strong acid would be a weak base.* Similarly, *the conjugate base of a weak acid would be a strong base.*

Add To Your Knowledge

It is quite interesting to note that both the Arrhenius as well as the Bronsted concepts require acid to be a source of protons. Hence, **all Arrhenius acids are also Bronsted acids.** However, there is a difference in the definition of bases. Arrhenius theory requires base to be a source of OH^- ions in aqueous medium but Bronsted theory requires base to be a proton acceptor. Hence, **Arrhenius bases may not be Bronsted bases.** For example, NaOH is a base according to the Arrhenius theory because it gives OH^- ions in aqueous solution but NaOH does not accept a proton as such. Hence, it may not be classified as a base according to the Bronsted theory.

Some more conjugated acid-base pairs have been given in the following equations:

$$\textbf{Acid}_1 + \textbf{Base}_2 \rightleftharpoons \textbf{Acid}_2 + \textbf{Base}_1$$

(1) $CH_3COOH(aq) + H_2O(l) \rightleftharpoons H_3O^+(aq) + CH_3COO^-(aq)$

(2) $H_2O(l) + CO_3^{2-}(aq) \rightleftharpoons HCO_3^-(aq) + OH^-(aq)$

(3) $HCO_3^-(aq) + NH_3(aq) \rightleftharpoons NH_4^+(aq) + CO_3^{2-}(aq)$

(4) $HCl(aq) + HCO_3^-(aq) \rightleftharpoons H_2CO_3(aq) + Cl^-(aq)$

It may be noticed that in equation (1) H_2O is behaving as a base whereas in equation (2) it is behaving as an acid. Similarly, HCO_3^- ion in equation (3) acts as an acid and in equation (4) it acts as a base. *Such substances which can act as acids as well as bases are called* ***amphoteric substances.***

SOLVED EXAMPLES

Example 22.3. *Write the conjugate bases for the following Bronsted acids:*

$$HCl,\ HNO_3,\ HSO_4^-,\ H_2SO_4 \text{ and } H_2S.$$

Solution. The conjugate base is formed by loss of a proton by the acid. Thus, the conjugate bases of the given acids are: Cl^-, NO_3^-, SO_4^{2-}, HSO_4^- and HS^- respectively.

Example 22.4. *Write the conjugate acids for the following Bronsted bases:*

$$NH_2^-,\ H_2O,\ NH_3,\ C_2H_5OH,\ {}^-OH \text{ and } HCOO^-.$$

Solution. The conjugate acid is formed by gain of a proton by the base. Thus, the conjugate acids of the given bases are: NH_3, H_3O^+, NH_4^+, $C_2H_5OH_2^+$, H_2O and HCOOH respectively.

Example 22.5. *An amphoteric species can act as an acid as well as a base. Therefore, an amphoteric species has a conjugate acid as well as a conjugate base. Write conjugate acids as well as bases for the following amphoteric species:*

$$NH_3,\ C_2H_5OH,\ HSO_4^-,\ HS^-,\ HCO_3^- \text{ and } H_2O.$$

Solution.

Species	*Conjugate Acid*	*Conjugate Base*
NH_3	NH_4^+	NH_2^-
C_2H_5OH	$C_2H_5OH_2^+$	$C_2H_5O^-$
HSO_4^-	H_2SO_4	SO_4^{2-}
HS^-	H_2S	S^{2-}
HCO_3^-	H_2CO_3	CO_3^{2-}
H_2O	H_3O^+	OH^-

22.6 LEWIS CONCEPTS OF ACIDS AND BASES

Although the Bronsted-Lowry theory was more general than the Arrhenius theory of acids and bases, it failed to explain the acid base reactions which do not involve the transfer of a proton. For example, it fails to explain how acidic oxides such as anhydrous CO_2, SO_2, SO_3, etc., can neutralize basic oxides such as CaO, BaO, etc., even in the absence of solvent.

G.N. Lewis (1923) proposed broader and more general definitions of acids and bases, which do not require the presence of protons to explain the acid base behaviour. According to the **Lewis concept.**

*An **acid** is a substance which can accept a pair of electrons.*

*A **base** is a substance which can donate a pair of electrons.*

Acid-base reactions according to this concept involve *donation of electron pair by a base to an acid to form a co-ordinate bond.* Lewis bases can be neutral molecules such as $\ddot{N}H_3$, $CH_3—\ddot{O}H$, $H_2\ddot{O}$, etc., having one or more unshared pairs of electrons, or anions such as CN^-, OH^-, Cl^-, etc.

Lewis acids are the species having vacant orbitals in the valence shell of one of its atoms. The following species can act as Lewis acids.

(a) Molecules having an atom with incomplete octet. For example, BF_3 and $AlCl_3$

$$H_3N: \;+\; BF_3 \longrightarrow H_3N:BF_3$$

LEWIS BASE LEWIS ACID

$$:\ddot{Cl}^-: \;+\; AlCl_3 \longrightarrow [:\ddot{Cl}:AlCl_3]^-$$

LEWIS BASE LEWIS ACID

(b) Simple cations. For example, H^+, Ag^+, etc.

$$H_2\ddot{O}: \;+\; H^+ \longrightarrow [H_3O]^+$$

LEWIS BASE LEWIS ACID

$$2\ddot{N}H_3 + Ag^+ \longrightarrow [H_3N : Ag : NH_3]^+$$

(c) Molecules in which the central atom has vacant orbitals and may acquire more than an octet of valence electrons. For example, SiF_4.

$$SiF_4 \;+\; 2:\ddot{F}:^- \longrightarrow [SiF_6]^{2-}$$

LEWIS ACID LEWIS BASE

(d) Molecules containing multiple bonds. For example, CO_2, SO_2, etc.

$$[Ca]^{2+}[:\ddot{O}:]^{2-} + O{=}C{=}O \longrightarrow [Ca]^{2+}[CO_3]^{2-}$$

LEWIS BASE LEWIS ACID

Add To Your Knowledge

It may be noted that ***all Bronsted bases are also Lewis bases but all Bronsted acids are not Lewis acids.*** Lewis bases generally contain one or more lone pairs of electrons and, therefore, they can also accept a proton (Bronsted base). Thus, all Lewis bases are also Bronsted bases. On the other hand, Bronsted acids are those which can give a proton, for example, HCl, H_2SO_4. But they may not be capable of accepting a pair of electrons. Hence, all Bronsted acids are not Lewis acids.

LIMITATIONS OF THE LEWIS CONCEPT

Though the Lewis concept of acids and bases is more general than the Arrhenius as well as Bronsted concepts, it has several drawbacks as discussed below:

1. It is too general and includes all the co-ordination compounds and co-ordination reactions.
2. It does not help to assign the relative strengths of acids and bases.
3. It does not explain the behaviour of protonic acids such as HCl, H_2SO_4, etc., which do not form co-ordinate bond with bases which is the primary requirement of the Lewis theory.
4. The catalytic activity of many acids is due to $H^+(aq)$ ion. Since a Lewis acid need not contain hydrogen, many Lewis acids will not possess catalytic properties.
5. Normally, the formation of co-ordination compounds is slow. Therefore, acid-base reactions should also be slow, but in actual practice, acid-base reactions are extremely fast.

SOLVED EXAMPLE

Example 22.6. *Classify the following species into Lewis acids and Lewis bases and show how these act as such:*

(*i*) ^-OH (*ii*) F^- (*iii*) H^+

(*iv*) BF_3 (*v*) $^-NH_2$.

Solution. (*i*) ^-OH ion is a **Lewis base** as it has lone pairs of electrons on oxygen which it can donate.

(*ii*) F^- ion has four lone pairs of electrons. Hence, it can act as a **Lewis base** by donating any one of these lone pairs.

(*iii*) H^+ acts as a **Lewis acid** as it has a vacant orbital and hence can accept a pair of electrons from Lewis bases such as ^-OH, NH_3, etc.

(*iv*) BF_3 acts as a **Lewis acid**. In BF_3 there are only six electrons in the valence shell of boron. It can accept a pair of electrons from bases such as NH_3 to complete its octet.

(*v*) NH_2^- is a Lewis base. It can donate a pair of electrons present on the nitrogen atom.

SUMMARY

- Aqueous solutions of acids and bases conduct electricity.
- Acids are sour in taste.
- Acids and bases impart specific colours to acid-base indicators.
- Acids on reaction with active metals liberate hydrogen gas.
- Acids on reaction with metal trioxocarbonate(IV) and metal hydrogen trioxocarbonate(IV) liberate CO_2 gas with brisk effervescence.
- The reaction between acids and bases to form salt and water is called **neutralization**.
- Bases are bitter in taste and slippery to touch.
- According to the **Arrhenius Concept** an acid is a substance which can furnish hydrogen ions in aqueous solution whereas a base is a substance which can furnish hydroxyl ions in aqueous solution.
- Acids such as **HCl, HNO_3 and H_2SO_4** which are completely ionized in aqueous solution are called **strong acids** whereas acids such as **CH_3COOH, H_2CO_3**, etc. which are weakly ionized in solution are called **weak acids.**
- Bases such as **NaOH, KOH, $Ba(OH)_2$** which are completely ionized in aqueous solution are known as **strong bases** whereas bases such as **NH_4OH, $Ca(OH)_2$**, etc. which are weakly ionized are called **weak bases.**
- According to the **Bronsted and Lowry** concept an acid is a substance that can donate a proton whereas a base is a substance that can accept a proton.
- Acid-base pairs such as H_2O/OH^- or CH_3COOH/CH_3COO^- which differ by a proton are called **conjugate acid base pairs.**
- Conjugate base of a strong acid is a weak base where as the conjugate base of a weak acid is a strong base.
- According to the **Lewis Concept** an acid is a substance which can accept a pair of electrons and base is a substance which can donate a pair of electrons.

EVALUATE YOURSELF

I. Objective Type Questions

Select the most appropriate choice from the options given as (a), (b), (c) and (d) after each question:

1. Which of the following compounds is a weak acid?

(*a*) HNO_3 (*b*) H_2SO_4

(*c*) NH_4OH (*d*) CH_3COOH

2. Which of the following compounds is a strong base?

(*a*) $Ca(OH)_2$ (*b*) $Ba(OH)_2$

(*c*) NH_4OH (*d*) $B(OH)_3$

3. The colour of methyl orange in basic medium is

(*a*) Red (*b*) Orange

(*c*) Yellow (*d*) Pink

4. Which of the following compounds is not a lewis base?

(*a*) H_2O (*b*) NH_3

(*c*) CH_3OH (*d*) BF_3

5. Which of the following pairs does not represent a conjugate acid base pair?

(*a*) H_2O/OH^- (*b*) H_3O^+/H_2O

(*c*) H_2SO_4/SO_4^{2-} (*d*) HCO_3^-/CO_3^{2-}

II. Fill in the Blanks

6. Complete the following sentences by supplying appropriate words:

(*i*) Acids are .... in taste.

(*ii*) Acids on reaction with active metals liberate .... gas.

(*iii*) Alkalies are ...... in touch.

(*iv*) The conjugate base of H_2O is ....

(*v*) The formula of hydronium ion is ....

(*vi*) According to the Bronsted-Lowry concept an acid is a .... donor.

III. Discussion Questions

7. Give any three characteristic properties of acids.

8. What is the colour of the following indicators in acidic medium?

(*i*) Litmus solution (*ii*) Methyl orange.

9. (*i*) Why does an aqueous solution of an acid conduct electricity?

(*ii*) Why do HCl, HNO_3, etc., show acidic character in aqueous solutions while solutions of compounds like C_2H_5OH and glucose do not show acidic character?

10. What is the reaction of active metals with acids?

11. Write word equations and then balanced equations for the reaction taking place when:

(*i*) dilute tetraoxosulphate(VI) acid reacts with iron filings

(*ii*) dilute hydrochloric acid reacts with sodium trioxocarbonate(IV)

(*iii*) dilute hydrochloric acid reacts with copper(II) oxide.

12. What is neutralization reaction? Give two practical applications of neutralization.

13. A metal compound A reacts with dilute hydrochloric acid to produce effervescence. The gas evolved extinguishes a burning candle. Write a balanced chemical equation for the reaction if one of the compounds formed is calcium chloride.

14. Which gas is usually liberated when an acid reacts with a metal? Illustrate with an example. How will you test for the presence of this gas?

15. Write any three chemical properties of acids.

16. Why should curd and sour substances not be kept in brass and copper vessels?

17. While diluting an acid, why is it recommended that the acid should be added to water and not water to the acid?

18. What is neutralization reaction? Give two examples.

19. Give any three characteristic properties of bases.

20. What is the colour of following indicators in acidic medium?

(*i*) Phenolphthalein (*ii*) Litmus.

21. Give examples of two amphoteric oxides.

22. Why does dry HCl gas not change the colour of the dry litmus paper.

23. What are strong acids and weak acids? Give two examples of each.

24. What are strong bases and weak bases? Give two examples of each.

25. Comment on the statement '*All Arrhenius acids are also Bronsted acids.*'

26. Why do HCl, HNO_3, etc., show acidic character in aqueous solutions while solutions of compounds like C_2H_5OH and glucose do not show acidic character?

27. How is the concentration of hydronium ions (H_3O^+) affected when a solution of an acid is diluted?

28. How is the concentration of hydroxyl ions (OH^-) affected when a solution of sodium hydroxide is diluted?

29. Which out of the following pairs is acid according to Arrhenius concept?

(*i*) HCl(*g*) and HCl(*aq*)

(*ii*) CH_3COOH(*l*) and CH_3COOH(*aq*).

30. Can we consider the solution of ammonia in diethyl ether, an Arrhenius base?

31. Which theory justifies water as amphoteric substance? Write chemical equations to show its amphoteric nature.

32. What is meant by the conjugate acid base pair? Find the conjugate acid/base for the following species:

HNO_2, CN^-, $HClO_4$, F^-, OH^-, CO_3^{2-} and S^{2-}.

33. Write the conjugate acids of the following species:
(*i*) SO_4^{2-} (*ii*) RNH_2
(*iii*) NH_2^- (*iv*) $C_2H_5OC_2H_5$
(*v*) F^-.

34. Write the conjugate acid and conjugate base of the following amphoteric species:
(*i*) HCO_3^- (*ii*) H_2O
(*iii*) HS^- (*iv*) NH_3.

35. In a reaction:
$C_2H_5NH_2 + HI \rightleftharpoons (C_2H_5NH_3)^+I^-$, which reactant would be considered a Bronsted acid?

36. Classify the following into acids and bases according to the Lewis concept:

SO_3, CaO, OH^-, BF_3, $R\ddot{N}H_2$, S^{2-}, Ag^+, H^+

37. Diethyl ether $C_2H_5OC_2H_5$ acts as a Lewis base. Which of its atoms acts as electron donor?

38. In the reaction, $SnCl_4 + 2Cl^- \longrightarrow [SnCl_6]^{2-}$
which is Lewis acid and which one is base?

Answers

I. Objective Type Questions

1. (*d*)	**2.** (*b*)	**3.** (*c*)
4. (*d*)	**5.** (*c*).	

II. Fill in the Blanks

6. (*i*) sour	(*ii*) hydrogen	(*iii*) slippery
(*iv*) OH^-	(*v*) H_3O^+	(*vi*) proton.

SECTION-4

UNIT-23

ACID-BASE TITRATIONS

Specific Objective

After studying this unit, the student should be able to:

- determine the quantity of an analyte in solution using titration.

23.1 INTRODUCTION

In volumetric analysis, the quantities of the constituents present in the given unknown solution are determined by measuring the volumes of the solutions taking part in the given chemical reaction. The main process of this analysis is called **titration** which means the determination of the volume of a reagent required to bring a definite reaction to completion. When the substance involved in titration is an acid or a base the process is called **acid-base titration.**

During volumetric analysis we have to prepare solutions of known concentrations. In this unit, we shall study how to prepare solutions of substances with known concentrations. We shall also learn about acid-base titration.

23.2 SOME IMPORTANT TERMS

1. Standard Solution

A solution whose concentration is known is called a **standard solution.** Concentration of a solution is generally expressed as normality or molarity.

2. Normality

Normality *of a solution is defined as the number of gram-equivalents of solute in 1 dm³ of solution.* It is denoted by N. Mathematically, it may be expressed as:

$$\text{Normality} = \frac{\text{Number of gram-equivalents of solute}}{\text{Volume of solution (in dm}^3\text{ or litres)}}$$

∴ Number of gram equivalents of solute

= Normality × Volume of solution (in dm^3).

A solution containing one gram-equivalent of solute per dm^3 of solution is called **normal solution.**

Equivalent mass. Equivalent mass of a substance can be obtained from its molecular mass.

Equivalent mass of an acid is defined as *the mass of the acid which can furnish one mole of hydrogen ions.*

Equivalent mass of an acid

$$= \frac{\text{Molecular mass of acid}}{\text{Number of available hydrogen ions per molecule}}$$

$$= \frac{\text{Molecular mass of acid}}{\text{Basicity of acid}}$$

Similarly, **equivalent mass of a base** is defined as *the mass of the base which can furnish one mole of hydroxyl ions.* It may also be defined as *the mass of the base which is completely neutralized by one equivalent of the acid.*

Equivalent mass of a base

$$= \frac{\text{Molecular mass of base}}{\text{Number of hydroxyl ions per molecule}}$$

$$= \frac{\text{Molecular mass of base}}{\text{Acidity of base}}$$

$$\text{Equivalent mass of a salt} = \frac{\text{Molecular mass of salt}}{\text{Total valency of metal ion}}$$

Equivalent mass of some compounds (acids, bases and salts) are given in Table 23.1.

3. Molarity

Molarity *of a solution may be defined as the number of gram moles of solute per dm^3 or per litre of the solution.* It is denoted by M. Mathematically, it may be expressed as:

$$\text{Molarity} = \frac{\text{Gram moles of solute}}{\text{Volume of solution (in dm}^3)}$$

$$= \frac{\text{Mass of solute (in grams) per dm}^3 \text{ of solution}}{\text{Gram-molecular mass of the solute}}$$

∴ Gram moles of solute

= Molarity × Volume of solution (in dm^3)

A solution containing one gram mole of solute per dm^3 of solution is called **molar solution.**

23.3 PREPARING A STANDARD SOLUTION

A standard solution is prepared by dissolving a definite weight of a substance (a primary standard), in a definite volume. A substance is classified as a **primary standard** if it has the following characteristics:

(*i*) *It is easily available in a state of high purity.*

(*ii*) *It is neither hygroscopic nor deliquescent.*

(*iii*) *It shows high solubility in water.*

(*iv*) *It does not dissociate or decompose during storage.*

Substances whose standard solutions cannot be prepared directly are called **secondary standards.** For example, $KMnO_4$, NaOH, KOH, etc. The solutions of such substances are standardized by titrating against a solution of some appropriate primary standard.

Table 23.1. Equivalent Masses of Some Compounds

Name of Compound	*Formula*	*Molecular Mass*	*Equivalent Mass*
1. Hydrochloric acid	HCl	36.5	36.5
2. Tetraoxosulphate(VI) acid	H_2SO_4	98	49
3. Ethanoic acid	CH_3COOH	60	60
4. Ethanedioic acid	COOH | COOH . $2H_2O$	126	63
5. Sodium hydroxide	NaOH	40	40
6. Sodium trioxocarbonate(IV)	Na_2CO_3	106	53
7. Sodium hydrogentrioxocarbonate(IV)	$NaHCO_3$	84	84

ACTIVITY 23.1 **Preparation of 0.1 M sodium trioxocarbonate(IV) solution**

Sodium carbonate is a primary standard. Its molecular mass is 106.

To prepare 0.1 M Na_2CO_3 solution, 10.6 g of sodium trioxocarbonate(IV) should be dissolved per dm^3 of the solution. Normally in the laboratory we are required to prepare 250 cm^3 of solution. Therefore, to prepare 250 cm^3 of 0.1 M Na_2CO_3 solution, $\frac{10.6}{4} = 2.650$ g of Na_2CO_3 are dissolved in lesser quantity of water and the solution diluted to exactly 250 cm^3.

Procedure

1. Take a watch glass, wash it with distilled water and then dry it.
2. Weigh the clean and dried watch glass accurately and record its weight in the notebook.
3. Weigh 2.650 g sodium trioxocarbonate (IV) on the watch glass accurately and record this weight in the notebook.
4. Transfer gently and carefully the sodium trioxocarbonate(IV) from the watch glass into a clean 400 cm^3 beaker. Wash the watch glass with distilled water with the help of a wash bottle to transfer the particles sticking to it into the beaker (Fig. 23.1). The volume of distilled water for this purpose should not be more than 50 cm^3.

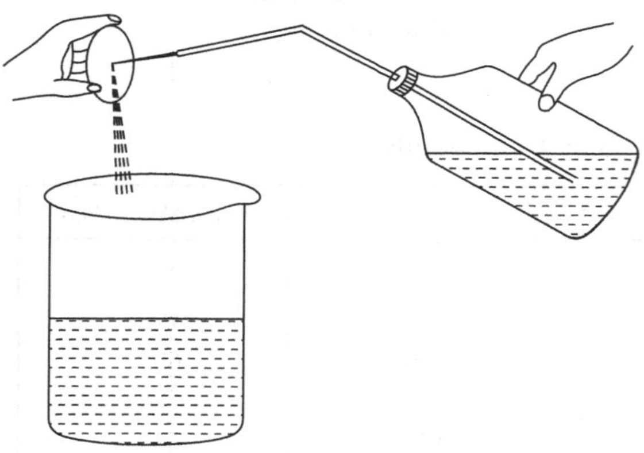

Fig. 23.1. Washing of watch glass to transfer sticking particles to beaker.

5. Dissolve sodium trioxocarbonate(IV) crystals in the beaker by gentle stirring with a clean glass rod.
6. When the sodium trioxocarbonate(IV) in the beaker is completely dissolved, transfer carefully the entire solution from the beaker into a 250 cm^3 measuring flask with the help of a funnel (Fig. 23.2).

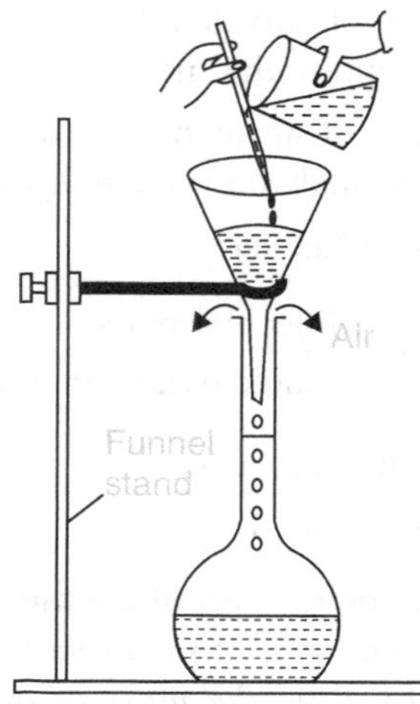

Fig. 23.2. Transferring solution to measuring flask.

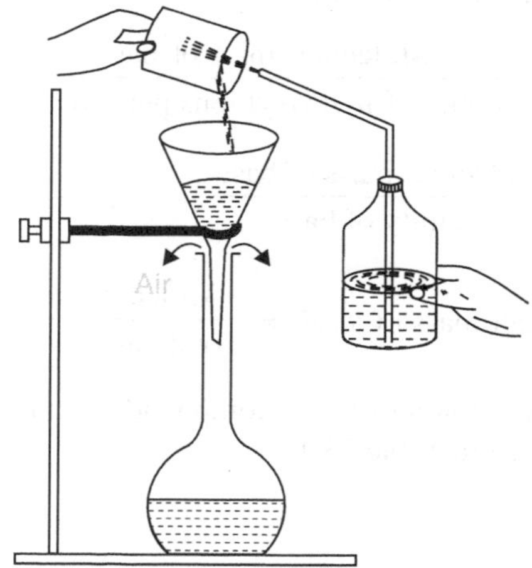

Fig. 23.3. Transferring last traces of solution to measuring flask.

7. Wash the beaker with distilled water. Transfer the washings into the measuring flask (Fig. 23.3).
8. Finally wash the funnel thoroughly with distilled water with the help of a wash bottle to transfer the solution sticking to the funnel into the measuring flask (Fig. 23.4).
9. Add enough distilled water to the measuring flask carefully up to just below the etched mark on it, with the help of wash bottle.

10. Add the last few drops of distilled water with a pipette until the lower level of the meniscus just touches the mark on the measuring flask (Fig. 23.5).
11. Stopper the measuring flask and shake gently to make the solution uniform throughout. Label it as 0.1 M sodium trioxocarbonate(IV) solution.

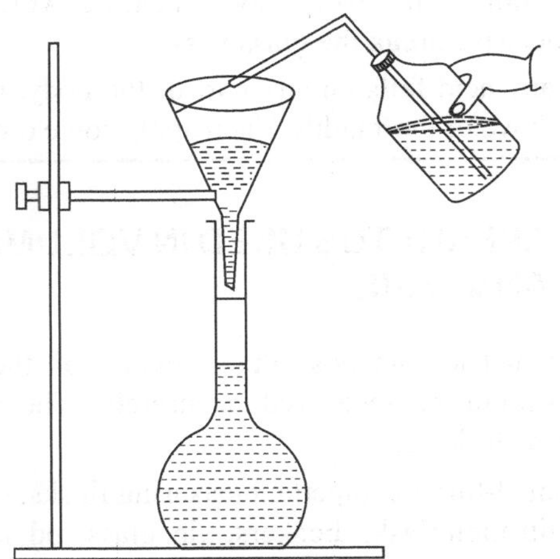

Fig. 23.4. Washing last traces of solution from funnel to the measuring flask.

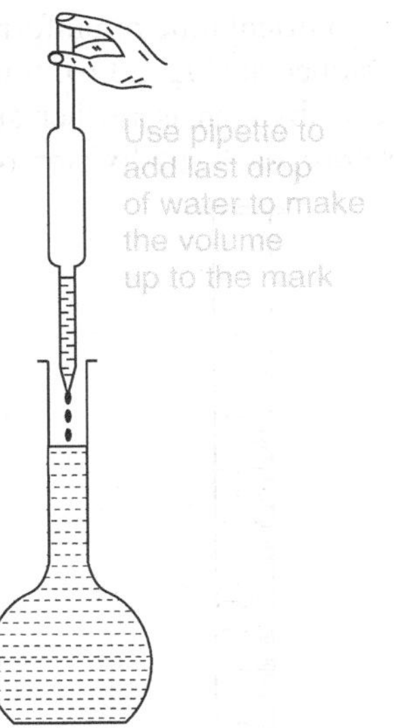

Fig. 23.5. Using pipette to add last drop of water to make the volume up to the mark

23.4 PREPARATION OF AN ACID SOLUTION OF DESIRED MOLARITY FROM A CONCENTRATED ONE

Acids like hydrochloric acid (HCl), tetraoxosulphate(VI) acid (H_2SO_4), trioxonitrate(V) acid (HNO_3) and acetic acid (CH_3COOH) which are commonly used in the laboratory are originally available as concentrated aqueous solutions. The solutions required in the laboratory are prepared by diluting the concentrated solutions with water. Knowing the concentration of the concentrated acid we can calculate how much volume of this acid must be diluted to prepare a definite volume of dilute acid of certain required concentration. The approximate concentrations of some acids as commercially available are given in Table 23.2.

Table 23.2. Approximate Concentrations of Some Acids

Concentrated Acid	*Approximate Molarity*
Hydrochloric acid (HCl)	12 M
Tetraoxosulphate(VI) acid (H_2SO_4)	18 M
Trioxonitrate(V) acid (HNO_3)	16 M
Glacial acetic acid (CH_3COOH)	17 M

Knowing the molarity of a concentrated acid, makes it possible to calculate the volume of this acid required for preparing a definite volume of a dilute solution of known molarity. This can be done with the help of the molarity equation:

$$M_1V_1 = M_2V_2$$

where, M_1 is the molarity of concentrated acid,
V_1 is the volume of concentrated acid,
M_2 is the molarity of dilute acid,
V_2 is the volume of dilute acid.

For example, let us calculate the volume of concentrated tetraoxosulphate(VI) acid (molarity = 18 M) required to prepare 250 cm³ of 2 M acid.

According to the molarity equation:

$$M_1V_1 = M_2V_2$$

Now, molarity of concentrated acid (M_1)
= 18 M

volume of concentrated acid required (V_1) = ?

molarity of dilute acid to be prepared (M_2)
= 2 M

volume of dilute acid to be prepared (V_2)
= 250 cm³.

Substituting these values in the molarity equation, we get

$$18 \times V_1 = 2 \times 250$$

$$V_1 = \frac{2 \times 250}{18} = 27.8 \text{ cm}^3$$

Therefore, 27.8 cm³ of 18 M H_2SO_4 must be diluted with water to make the volume 250 cm³. The resulting solution will be 2 M H_2SO_4.

ACTIVITY 23.2 **Preparation of 250 cm^3 of 1 M HCl from the given 12 M HCl**

To prepare 250 cm^3 of 1 M HCl from 12 M HCl, the volume of concentrated acid required can be calculated by using the equation:

$$M_1V_1 = M_2V_2$$

where,

M_1 = Molarity of concentrated acid = 12 M

V_1 = Volume of concentrated acid required = ?

M_2 = Molarity of dilute acid to be prepared = 1 M

V_2 = Volume of dilute acid to be prepared = 250 cm^3.

$\therefore \quad 12 \times V_1 = 1 \times 250$

or $\quad V_1 = \dfrac{1 \times 250}{12} = 21\ cm^3$

Therefore 21 cm^3 of concentrated acid should be taken and diluted with water to make the volume 250 cm^3. The resulting solution will be 1 M HCl.

Procedure

1. With the help of a measuring cylinder, take about 100 cm^3 of distilled water in a 400 cm^3 beaker. (This volume of water is less than the volume of water which will be required).
2. Now take 21 cm^3 of given concentrated HCl in a measuring cylinder. Add it slowly and carefully, with stirring with a glass rod, to the water taken in the beaker. Cool the beaker under tap water from time to time. Wash the cylinder with a little water and add the washings to the beaker.
3. Transfer the solution from the beaker to a 250 cm^3 measuring flask. Add more of water so that the total volume becomes 250 cm^3.
4. Stopper the measuring flask and shake the solution well to make it uniform. **Label it as 1 M HCl.**

PRECAUTIONS

1. While diluting acid with water, always **add acid to water and not the water to acid.**
2. Pour the concentrated acid into water very slowly.
3. When H_2SO_4 is diluted, dilution is done in instalments and the mixture is cooled under tap water after each addition. This is done because dilution of H_2SO_4 is exothermic and a large amount of heat is liberated during its dilution.
4. Concentrated acids are highly corrosive and therefore, should be handled with great care.
5. Measure the concentrated acid in a small measuring cylinder. If pipette is to be used, make use of a pipette filler. **Do not suck it.**
6. Go on stirring the solution with a glass rod when you are adding the concentrated acid to water. This ensures uniform rise of temperature throughout the solution. This avoids local heating which may otherwise break the glass vessel.
7. If any acid falls on any part of the body, wash it with water thoroughly. Then apply some ointment.

23.5 APPARATUS USED IN VOLUMETRIC ANALYSIS

In volumetric analysis, the volumes of the various solutions should be measured accurately. The apparatus required is as follows:

Graduated-burette, pipette, measuring flasks, measuring cylinders, titration flasks, beakers, tile, glass-rod, funnel and wash bottle.

BURETTE

It is a long, cylindrical tube of uniform bore fused at the lower end with a stopcock (Fig. 23.6). It is graduated in cm^3 from 0 to 50. Each division is further subdivided into ten equal parts. Therefore, each subdivision reads 0.1 cm^3.

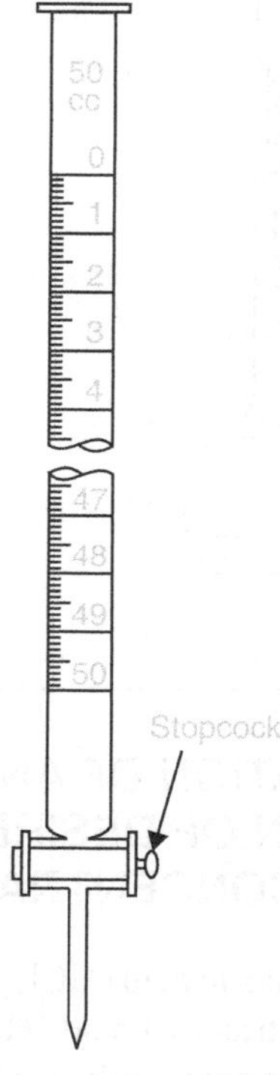

Fig. 23.6. Burette

Before a burette is filled with the solution, it is thoroughly washed with water so that no greasy matter is sticking inside or outside the burette. Take a small volume of solution (to be taken in it as titrant) in the burette and hold in a horizontal position as shown in Fig. 23.8. Rotate the burette so as to wet the inner walls of the burette. Reject this solution through the stopcock. This process is known as **rinsing.** Then the burette is filled through a funnel inserted in the top Fig. 23.7. The funnel must then be taken out after filling the burette.

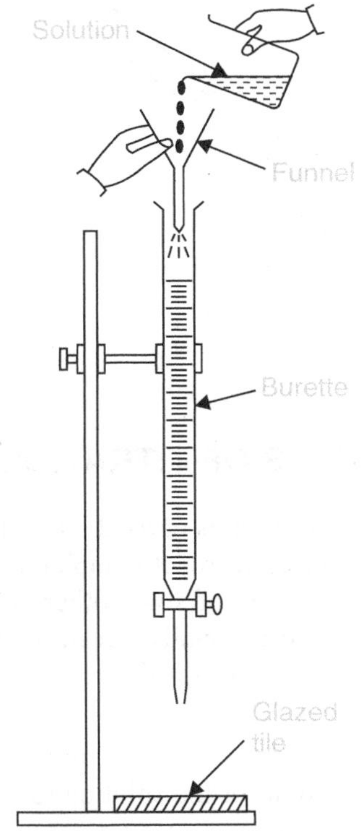

Fig. 23.7. Filling the burette.

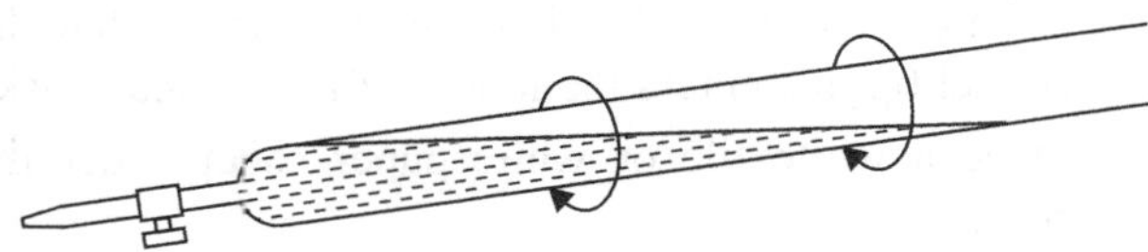

Fig. 23.8. Rinsing the burette.

Care must be taken that no air bubbles remain in the narrow bottom tip of the burette. To remove this air, the stopcock is opened and the liquid is allowed to run out rapidly into the beaker or flask.

Burette reading forms the most important aspect of the experiment, therefore, a burette should be read very carefully, after removing parallax.

To read the burette, hold behind the level of the liquid and in contact with the burette a piece of white paper to illuminate the surface of the liquid. This paper, called **antiparallax card,** eliminates errors in reading due to parallax. In order to prepare an antiparallax card take a rectangular piece of paper and fold it half way. Give two cuts as shown in Fig. 23.9. Open the fold and mount it on the burette.

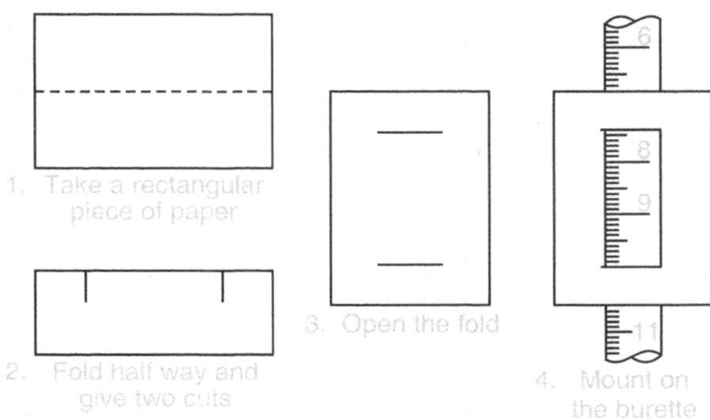

Fig. 23.9. Making and mounting an antiparallax card.

It is to be remembered that in the case of colourless solutions the lower meniscus is read while in the case of coloured solutions, the level is read from the upper meniscus. This is due to the reason that in the case of coloured solutions the lower meniscus is not visible clearly.

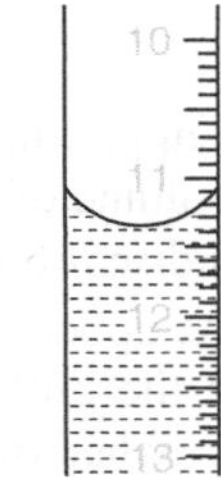

Fig. 23.10. Lower meniscus.

Placing your eye exactly in front of the meniscus (Fig. 23.11) of the solution take initial reading and then final reading at the end. Find out the difference between two readings to obtain the volume of solution consumed.

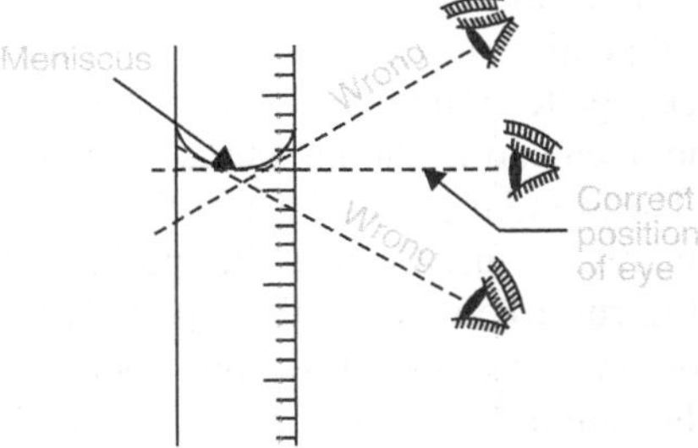

Fig. 23.11. Correct way of reading burette.

Precautions

1. See that stopcock does not leak.
2. Remove the funnel immediately after filling the burette.
3. Do not allow any air bubble to remain inside the burette.
4. Always use antiparallax card and place the eye exactly in the level of meniscus.
5. Let no drops of solution be hanging at the tip of the burette at the end point.

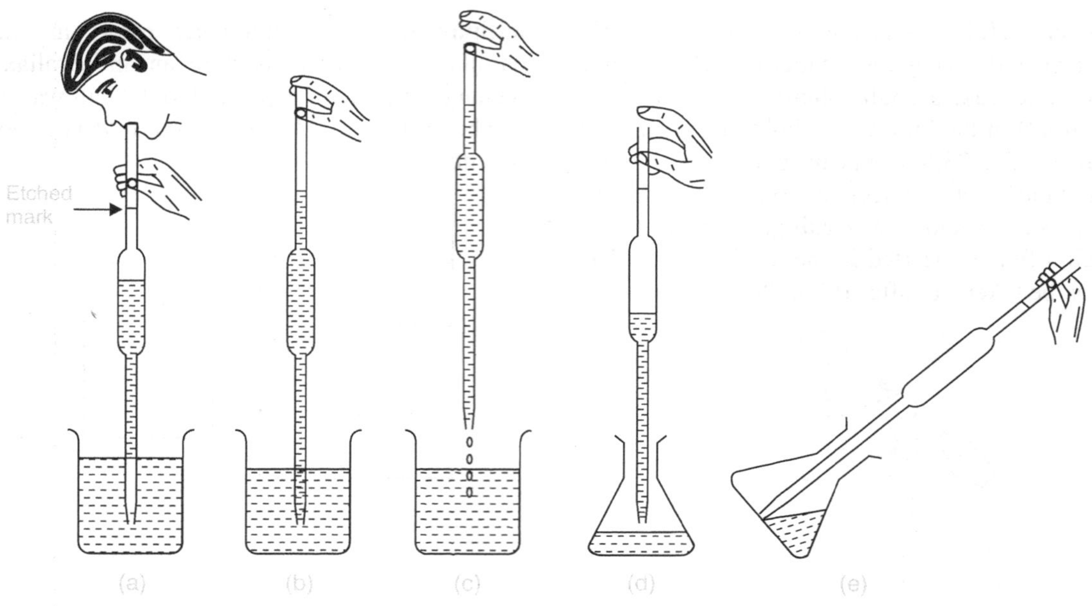

Fig. 23.12. Use of pipette

PIPETTE

This apparatus is used for accurate measurements of definite volume of solution. It consists of a long narrow tube with cylindrical bulb in the middle and a jet at its lower end.

On the upper part of the stem, there is an etched circular mark. On the bulb is marked the volume which the pipette can deliver when filled up to the circular mark [Fig. 23.12 (*a*)]. Pipettes of different capacities such as 1, 2, 5, 10, 20, 25, 50 and 100 cm^3 are available.

Before a pipette is filled with the solution, it is washed thoroughly and rinsed. The upper part of the pipette is then held by the thumb and middle finger of the right hand, the lower end is dipped into the liquid and the solution is sucked into the pipette until the liquid level is about 2 cm above the mark. The open end of the pipette is then closed with the index finger. The liquid is allowed to run slowly until the lower edge of the meniscus just touches the mark. The solution is then allowed to run freely out of the pipette [Fig. 23.12 (*d*)]. When no more of the liquid flows out, touch the tip of the pipette with the bottom of the flask [Fig. 23.12 (*e*)]. Some liquid will still remain in the pipette. Do not remove it by blowing because the pipette is calibrated taking this liquid into account.

Precautions

1. Never close the pipette with the thumb.
2. Keep the lower end always dipping in the liquid while sucking the liquid.
3. Never pipette out hot solutions or corrosive solutions.
4. Do not blow out the last drop of the solution from the jet end.

23.6 PROCESS OF TITRATION

The process of titration is carried out to find the volume of one solution required to react completely with a certain known volume of solution of some other substance. In the acid-base titration acid is usually put in the burette and the alkali in the flask. The process of titration is carried out as detailed below:

1. Wash the burette with water and then rinse it with the acid solution (or the solution to be used).
2. With the help of a funnel fill the burette with the acid solution. Flush out air bubbles from the jet of the burette. Adjust the level of the burette so that the nozzel tip just enters the mouth of the conical flask.

   Place a white tile or white paper sheet under the conical flask.
3. Rinse the pipette with alkali solution and pipette out 25 cm^3 of the alkali solution into a washed conical flask.
4. Add 1–2 drops of indicator to the solution in the conical flask and note the colour.
5. Carefully note the initial reading of the burette and record it.
6. With the neck of the flask in one hand, open the stopcock with the other hand, as shown in Fig. 23.13. Add the acid solution to the flask in small volumes at anyone time. Swirl the conical flask during and after each addition of acid solution.

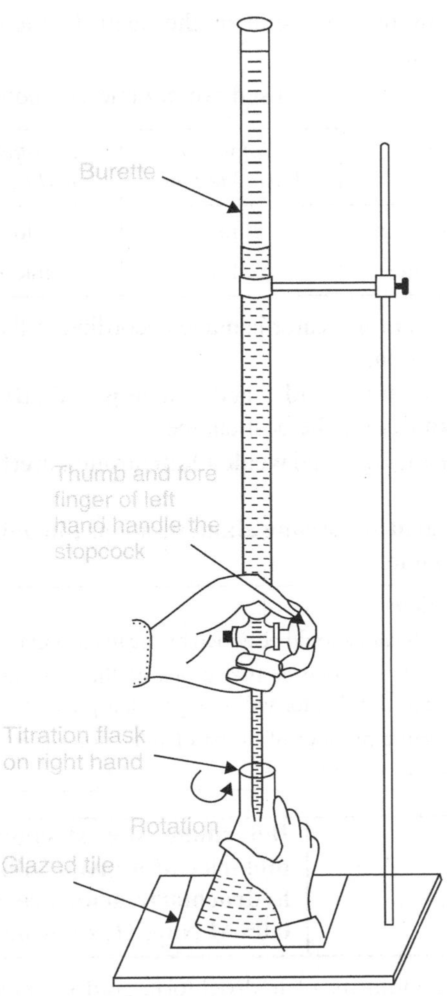

Fig. 23.13. Correct way of handling a burette.

7. Continue the addition of small volumes of the acid solution until the colour of the indicator just changes.
8. Record the final burette reading to two decimal places, carefully estimating the second decimal place.
9. The solution from the titration flask is thrown away and the titration flask is washed thoroughly first by keeping it under tap water and then with a little of distilled water. *Do not rinse the titration flask.*
10. Repeat the titration using another 25 cm^3 of alkali solution in the washed conical flask. Note the initial reading of the burette. Release acid into the alkali until you are about 1 cm^3 away from the end point. Now add the acid dropwise with swirling until the last single drop brings about a permanent change in the colour of indicator. Read the burette. Note down the final burette reading.
11. Repeat the step 10 until you get three values for the volume of acid used which differ by not more than 0.20 cm^3.
12. Average any two or more **consistent titre values to** find the volume of acid used for neutralization of 25 cm^3 of the alkali solution.

- The solution of known concentration is called **titrant.**
- The solution being titrated is called titrand or analyte.
- The fixed volume or portion of solution delivered into a conical flask is called **aliquot.**
- The volume of solution from the burette which reacts with the fixed volume of the other solution to the end point (equivalence point) is called **titre** or **titre value.**

23.7 RECORDING AND TREATMENT OF TITRATION DATA

The data from a typical titration is recorded as given below. Here, titration of sodium hydroxide solution versus 0.1 M HCl is taken as an example for illustration

Solution in burette: 0.1 M HCl
Solution in flask: NaOH (*aq*)
Volume of NaOH (*aq*): 25 cm^3
Indicator: Phenolphthalein
End point: Pink to colourless
Chemical equation:

$$NaOH\ (aq) + HCl\ (aq) \longrightarrow NaCl\ (aq) + H_2O\ (l)$$

OBSERVATIONS

Initial reading of the burette	*Final reading of the burette*	*Volume of the acid used (cm^3)*
1.00	20.50	19.50
3.50	22.65	19.15
5.00	24.10	19.10
4.00	23.10	19.10
4.00	23.05	19.05

The last four titre values are consistent. The average of the consistent titre values

$$= \frac{19.15 + 19.10 + 19.10 + 19.05}{4}$$

$$= 19.10\ cm^3$$

This volume of acid is used for further calculations.

23.8 LAW OF EQUIVALENTS

According to this law, the number of equivalents of the substance to be titrated is equal to the number of equivalents of the titrant used.

Derivation of the normality equation. Consider an acid-alkali neutralization reaction. Let V_1 cm^3 of an acid solution of N_1 normality require V_2 cm^3 of base of N_2 normality.

We know that 1000 cm^3 of 1 N acid solution contains acid = 1 gram equivalent.

V_1 cm^3 of 1 N acid contains acid $= \frac{1}{1000} \times V_1$ gram equivalents. Thus, number of gram equivalents of acid in V_1 cm^3 of N_1 acid solution $= \frac{V_1N_1}{1000}$.

Similarly, number of gram equivalents of base in V_2 cm^3 of its N_2 solution $= \frac{V_2N_2}{1000}$.

By the law of equivalents, at the end point,

$$\frac{V_1N_1}{1000} = \frac{V_2N_2}{1000}.$$

$$\boxed{N_1V_1 = N_2V_2}$$

It is known as the normality equation. If three factors (V_1, V_2, N_1) are known, N_2 can be calculated by using the above formula.

In terms of molarities we can proceed as

$$\frac{[\text{Molarity } (M_1) \times \text{Volume } (V_1)] \text{ of Acid}}{[\text{Molarity } (M_2) \times \text{Volume } (V_2)] \text{ of Base}} = \frac{\text{No. of moles of acid in the balanced equation}}{\text{No. of moles of base in the balanced equation}}$$

For a reaction between HCl and Na_2CO_3

$$\underset{\text{1 mole}}{Na_2CO_3} + \underset{\text{2 moles}}{2HCl} \longrightarrow 2NaCl + CO_2 + H_2O$$

Thus, $\frac{(\text{Molarity} \times \text{Volume}) \text{ of HCl}}{(\text{Molarity} \times \text{Volume}) \text{ of } Na_2CO_3} = \frac{2}{1}$

The mole ratio of the reactants up to the end point is equal to the mole ratio in the balanced chemical equation.

23.9 ACID-BASE TITRATIONS

In acid-base titrations the amount of the substance is determined by titrating it against a standard solution of acid or base. The chemical reaction involved in acid-base titration is called *neutralization* reaction.

To know the end point or the neutralization point, an indicator is used.

Common indicators used are organic compounds.

Indicator	*Colour in Basic Medium*	*Colour in Acidic Medium*
Phenolphthalein	Pink	Colourless
Methyl orange	Yellow	Pink or red

Selection of indicator is made according to the solutions used for titration.

For **strong acid** and **strong alkali** titration **phenolphthalein** is the best choice.

For **strong acid** and **weak alkali** titration **methyl orange** suits the best.

Weak acid and **strong alkali** titrations **phenolphthalein** is most suitable.

Precautions

1. An indicator should be added by means of a glass dropper.
2. Always use one or two drops of the indicator. Large amount of indicator will not give sharp end point.
3. The same number of drops of indicator should be used for each titration.

ACTIVITY 23.3 — **Determine the strength and molarity of a given solution of hydrochloric acid. Provided is 0.05 M Na_2CO_3 solution.**

1. The molarity of hydrochloric acid is determined by titrating it against the standard solution of Na_2CO_3 using methyl orange as indicator.
2. Strength of the acid is determined by multiplying its molarity with its molecular mass which is 36.5.

Chemical Equation

$$Na_2CO_3 + 2HCl \longrightarrow 2NaCl + CO_2 + H_2O$$

Indicator. Methyl orange.

End Point. Yellow to pink (*Acid in burette*).

Procedure

1. Take a burette and wash it with water.
2. Rinse the burette with the given solution of hydrochloric acid and fill it with it.
3. Rinse the pipette with the given Na_2CO_3 solution and pipette out 25 cm^3 of this solution in a washed titration flask.
4. Add 2–3 drops of methyl orange indicator to the titration flask and place it just below the nozzle of the burette over a white glazed tile.

5. Note down the initial reading of the burette and run the acid solution slowly and dropwise to the titration flask till the colour of the solution changes from *yellow* to *light pink*.
6. Note the final reading and find the volume of hydrochloric acid used.
7. Repeat the procedure to take a set of at least three consistent readings.

Observations

Molarity of Na_2CO_3 solution = 0.05 M

Volume of Na_2CO_3 solution taken in each titration

= 25 cm^3.

Average consistent volume = x cm^3 (say)

Calculations

Volume of HCl solution used for neutralizing 25 cm^3 of 0.05 M Na_2CO_3 solution = x cm^3

$$\frac{M_1V_1 \text{ (of acid)}}{M_2V_2 \text{ (of base)}} = \frac{\text{Stoichiometric coefficient of acid}}{\text{Stoichiometric coefficient of base}}$$

$$\frac{M_1 \times x}{0.05 \times 25} = \frac{2}{1}$$

$$M_1 = \frac{2 \times 25 \times 0.05}{x} = \frac{2.5}{x}$$

Molecular mass of HCl = 36.5.

$$\therefore \text{ Strength of hydrochloric acid} = \frac{2.5}{x} \times 36.5 \text{ g dm}^{-3}.$$

Result

The strength of the given solution of hydrochloric acid is $\frac{91.25}{x}$ g dm^{-3}.

ACTIVITY 23.4 **Determine the percentage purity of the given sample of caustic soda, 5.0 g of which are dissolved per dm^3. Ask for your requirement.**

Requirement: 0.1 M HNO_3.

Chemical Equation

$$NaOH + HNO_3 \longrightarrow NaNO_3 + H_2O.$$

Indicator. Phenolphthalein.

End Point. Pink to colourless (*Acid solution in the burette*).

Procedure

1. Rinse and fill the burette with the given standard HNO_3 solution.
2. Rinse the pipette with the given caustic soda solution and pipette out 25 cm^3 of this solution in the titration flask.
3. Add 1–2 drops of phenolphthalein to the titration flask. The solution becomes pink in colour.
4. Place the flask on the glazed tile under the burette and note the initial reading of the burette.
5. Run the acid solution in the titration flask slowly and dropwise until the end point is reached. At the end point the pink colour of the solution disappears.
6. Note the final reading and find out the volume of 0.1 M HNO_3 used for titration.
7. Repeat the procedure 4–5 times to get a set of atleast three consistent readings.

Observations

Molarity of given HNO_3 = 0.1 M

Volume of NaOH taken for each titration = 25.0 cm^3.

Average consistent volume = x cm^3 (say).

Calculations

1. *Calculation of molarity of NaOH solution.*

According to the equation, one mole of sodium hydroxide reacts with one mole of nitric acid.

$$= \frac{V_{NaOH} \times M_{NaOH}}{V_{HNO_3} \times M_{HNO_3}} = \frac{1}{1}$$

Molarity of standard HNO_3 solution = 0.1 M

$$\frac{25 \times M_{NaOH}}{x \times 0.1} = \frac{1}{1}$$

$$M_{NaOH} = \frac{x \times 0.1}{25} = \frac{x}{250}.$$

2. *Calculation of the strength of NaOH in g dm^{-3}.*

Strength (g dm^{-3}) = Molarity × Molecular mass

$$= \frac{x}{250} \times 40 = \frac{4x}{25} \text{ g dm}^{-3}.$$

3. *To calculate the percentage purity.*

5.0 g of impure sample of caustic soda contains

$$= \frac{4x}{25} \text{ g of NaOH}$$

$\therefore$ 100 g of it will contain caustic soda

$$= \frac{4x}{25 \times 5} \times 100 \text{ g of NaOH}$$

$\therefore$ Percentage purity = 3.2 x%

ACTIVITY 23.5 **Determine the percentage composition of a mixture of KOH and KCl, 8.0 gms of which are dissolved per dm^3 of the given solution. Ask for what you require.**

Requirement

Standard solution of hydrochloric acid (say 0.1 M HCl).

Theory

In this titration, molarity of KOH in the mixture solution can be determined by titrating it against a standard solution of hydrochloric acid. **KCl does not react with HCl.** This molarity is multiplied by the molecular mass of KOH to get the strength of the KOH solution. Comparing it with the total strength, the percentage composition can be calculated.

Chemical Equations

$$KOH + HCl \longrightarrow KCl + H_2O$$

$$KCl + HCl \longrightarrow \text{No reaction.}$$

Indicator. Phenolphthalein.

End Point. Pink to colourless. (*Acid in burette*)

Observations

Solution taken in the burette = 0.1 M HCl

Mixture solution taken for each titration = 25 cm^3.

Suppose average consistent volume = x cm^3.

Calculations

1. *Calculation of molarity of KOH*

According to the equation, one mole of KOH reacts with one mole of HCl

$$\frac{M_{KOH} \times V_{KOH}}{M_{HCl} \times V_{HCl}} = \frac{\text{Number of moles of KOH in balanced chemical equation}}{\text{Number of moles of HCl in balanced chemical equation}}$$

Molarity of standard HCl solution = 0.1 M

$$\frac{M_{KOH} \times 25}{0.1 \times x} = \frac{1}{1}$$

$$M_{KOH} = \frac{x \times 0.1}{25} = \frac{x}{250}$$

2. *Calculation of strength of KOH*

Strength = Molarity × Molecular mass

$$= \frac{x}{250} \times 56 \text{ g dm}^{-3}$$

$$= y \text{ g dm}^{-3} \text{ (say)}$$

3. *Calculation of percentage composition*

Percentage of KOH $= \frac{y}{8} \times 100 = A\%$ (say)

Percentage of KCl $= (100 - A)\%$.

ACTIVITY 23.6 **Determine volumetrically, the value of 'n' i.e., the number of water molecules of crystallization in washing soda, $Na_2CO_3.nH_2O$, 10 g of which have been dissolved per dm^3 of the given solution. Provided is 0.05 M H_2SO_4.**

Chemical Equation

$$Na_2CO_3.nH_2O + H_2SO_4 \longrightarrow Na_2SO_4 + (n+1)H_2O + CO_2$$

Indicator. Methyl orange.

End Point. Yellow to light pink. (*Acid in burette*)

Procedure

1. Rinse and fill the burette with the 0.05 M H_2SO_4 solution.
2. Rinse the pipette with the sodium carbonate solution and pipette out 25 cm^3 of this solution in the titration flask.
3. Add 2–3 drops of methyl orange indicator to the titration flask. The colour of the solution becomes yellow.
4. Note the initial reading of the burette and run the acid solution slowly into the titration flask till the solution becomes light pink.
5. Note the final reading of the burette and find out the volume of acid used.
6. Repeat the procedure 4–5 times to get a set of at least three consistent readings.

Observations

Molarity of given $H_2SO_4 = 0.05$ M

Volume of alkali solution taken for each titration

$$= 25 \text{ cm}^3$$

Suppose average consistent volume of $H_2SO_4 = x$ cm^3

Calculations

The molarity of standard $H_2SO_4 = 0.05$ M

Since in the balanced equation one moles of H_2SO_4 react with one mole of $Na_2CO_3.nH_2O$.

$$\therefore \quad \frac{M_{H_2SO_4} \times V_{H_2SO_4}}{M_{Na_2CO_3} \times V_{Na_2CO_3}} = \frac{1}{1};$$

$$\frac{0.05 \times x}{M_{Na_2CO_3} \times 25} = \frac{1}{1}$$

$$\therefore \quad M_{Na_2CO_3} = \frac{1 \times x \times 0.05}{1 \times 25} = \frac{x}{500}$$

$$M_{Na_2CO_3} = \frac{\text{Strength (gdm}^{-3})}{\text{Molecular mass of } Na_2CO_3.nH_2O}$$

Molecular mass of $Na_2CO_3.nH_2O$

$$= \frac{\text{Strength (g dm}^{-3})}{M_{Na_2CO_3}} = \frac{10}{x/500}$$

But molecular mass of hydrated sodium carbonate

$$= 106 + 18\,n$$

$$106 + 18\,n = \frac{10}{x/400}$$

Knowing the value of x the value of n can be calculated.

ACTIVITY 23.7	**You are provided with the solution of a hydrogen trioxo carbonate(IV) of a monovalent alkali metal (M) with strength equal to 10.0 gms per litre. Find out the atomic mass of the metal (M). Ask for your requirement.**

Requirement

Standard solution of HCl (say 0.1 M HCl).

Theory

The formula of metal hydrogentrioxocarbonate (IV) is $MHCO_3$.

This will react with HCl according to the equation

$$MHCO_3 + HCl \longrightarrow MCl + CO_2 + H_2O.$$

Molecular mass of $MHCO_3$ = (M + 61).

Knowing the molarity and strength (given), the molecular mass can be calculated

$$\text{Molecular Mass} = \frac{\text{Strength}}{\text{Molarity}}$$

Molecular mass = M + 61

$$M = \text{Molecular mass} - 61.$$

Chemical Equation

$$MHCO_3 + HCl \longrightarrow MCl + CO_2 + H_2O.$$

Indicator. Methyl orange.

End point. Yellow to light pink. (*Acid in burette*).

Procedure

1. Rinse and fill the burette with the standard HCl solution.
2. Rinse the pipette with the solution of $MHCO_3$ and pipette out 25 cm^3 of this solution in a titration flask.
3. Add 2–3 drops of methyl orange indicator.
4. Note the initial reading of the burette and run the acid solution into the titration flask slowly and with swirling until the end point (colour change from yellow to light pink) is reached.
5. Note the final reading of the burette and find out the volume of 0.1 M HCl used.
6. Repeat the procedure 4–5 times to get a set of at least three consistent readings.

Observations

Molarity of given HCl = 0.1 M.

Volume of alkali hydrogentrioxocarbonate(IV) solution taken for each titration = 25.0 cm^3.

Suppose average consistent volume of HCl = x cm^3.

Calculations

From the balanced chemical equation, it is clear that one mole of HCl reacts with one mole of $MHCO_3$.

$$\therefore \quad \frac{M_{HCl} \times V_{HCl}}{M_{MHCO_3} \times V_{MHCO_3}} = \frac{1}{1}$$

$$\frac{0.1 \times x}{M_{MHCO_3} \times 25.0} = \frac{1}{1}$$

M_{MHCO_3} (Molarity of $MHCO_3$ solution)

$$= \frac{x \times 0.1}{25.0} = \frac{x}{250}$$

Since, Molarity $= \frac{\text{Strength (g dm}^{-3})}{\text{Molecular mass}}$

Molecular mass of $MHCO_3 = \frac{10}{x/250} = X$ (say)

But molecular mass of $MHCO_3$

= Atomic mass of alkali metal + 61

∴ Atomic mass of alkali metal

= Molecular mass of $MHCO_3 - 61$

$= X - 61.$

SOLVED EXAMPLES

Example 23.1. *Solution A is 0.1 M hydrochloric acid. Solution B is contains 10.0 g of a mixture of sodium chloride and sodium hydrogentrioxocarbonate(IV) per dm³ of solution. On titration of 25 cm³ portions of solution B, the average titre value was found to be 24.50 cm³.*

Calculate:

(i) the percentage of sodium hydrogentrioxocarbonate(IV) in the mixture.

(ii) the mass of sodium chloride in the mixture per dm³.

Solution. Conc. of HCl = 0.1 M

Volume of HCl = 24.50 cm³

Volume of mixture solution = 25.0 cm³

Out of NaCl and $NaHCO_3$ only $NaHCO_3$ will react with HCl

$$NaHCO_3 + HCl \longrightarrow NaCl + CO_2 + H_2O$$

$$\frac{M_1V_1(\text{of HCl})}{M_2V_2\ (\text{of } NaHCO_3)} = \frac{\text{Moles of HCl}}{\text{Moles of } NaHCO_3}$$

$$\frac{0.1 \times 24.50}{M_2 \times 25} = \frac{1}{1}$$

$$M_2 = \frac{24.50 \times 0.1}{25} = 0.098 \text{ M}$$

∴ Conc. of $NaHCO_3$ = 0.098 mol dm^{-3}

Molar mass of $NaHCO_3$ = 23 + 1 + 12 + 48 = 84 g mol^{-1}

Mass of $NaHCO_3$ per dm³ = 0.098 × 84 = 8.23 g dm^{-3}

% of $NaHCO_3$ in the mixture $= \frac{8.23}{10} \times 100 = 82.3\%$

Mass of NaCl in the mixture per dm³ = (10 – 8.23) g dm^{-3}

= 1.77 g dm^{-3}.

23.10 BACK TITRATION

This method is used to determine the quantity of a substance which cannot be analysed by direct titration. It involves the following steps:

1. A known but excess quantity of a suitable reagent is reacted with the substance to be analysed.
2. A part of the reagent reacts with the substance and the amount of the excess reagent is determined by titration.
3. The difference between the amounts of the reagent before and after the reaction gives the amount of the reagent that reacted with the substance.
4. Knowing the amount of the reagent used and the chemical equation for the reaction between the substance and the reagent, it is possible to calculate the amount of the substance.

The amount of nitrogen or ammonia in an ammonium salt can be calculated by the back titration method.

Example 23.2. *A is 0.1 M KOH solution.*

B is a solution obtained after 0.500 g of ammonium chloride has been treated with 1 dm³ of 0.1 M KOH solution and heated to completely expel all the ammonia present in the ammonium chloride.

C is 0.10 M HCl.

25 cm³ of B containing excess KOH reacted with 19.5 cm³ of C in a back titration.

Calculate the % mass of ammonia in the given ammonium salt.

Solution. In the beginning KOH solution contains 0.10 mol of KOH per dm³.

Amount of KOH per dm³ left after reaction with ammonium salt can be obtained from the back titration results.

The volume of 0.1 M HCl that reacted with 25 cm³ of excess KOH

= 19.50 cm³

Volume of KOH = 25 cm³

$$KOH\ (aq) + HCl\ (aq) \longrightarrow KCl\ (aq) + H_2O\ (l)$$

$$\frac{M_1V_1\ (\text{of KOH})}{M_2V_2\ (\text{of HCl})} = \frac{1}{1}$$

$$M_1 = \frac{M_2V_2}{V_1} = \frac{0.1 \times 19.5}{25} = 0.078 \text{ mol dm}^{-3}$$

M_1 is the conc. of KOH after reaction with ammonium salt

Amount of KOH before reaction = 0.10 mol dm^{-3}

Amount of KOH after reaction = 0.078 mol dm^{-3}

Amount of KOH which reacted with NH_4Cl

$= (0.10 - 0.078)$

$= 0.022$ mol

In order to obtain the amount of NH_3 evolved, the mole ratio between KOH and NH_3 is used.

$$NH_4^+ (aq) + OH^- (aq) \longrightarrow NH_3 (g) + H_2O (l)$$

From the chemical equation it is clear that when one mol of KOH is consumed, one mol of NH_3 is produced.

$\therefore$ Moles of NH_3 = Moles of KOH reacted = 0.022 mol.

Mass of NH_3 = 0.022 mol × 17 g mol^{-1}

= 0.374 g

$$\% \ NH_3 \text{ in the given ammonium chloride} = \frac{0.374 \text{ g}}{0.500 \text{ g}} \times 100$$

$= 74.8\%$

23.11 DOUBLE INDICATOR TITRATIONS

In this section, you will learn about double indicator titrations involving CO_3^{2-} and HCO_3^- mixture solutions.

Both CO_3^{2-} and HCO_3^- behave as base and react with acids. The CO_3^{2-} reacts with acids in two stages.

1. $CO_3^{2-} (aq) + H^+ (aq) \longrightarrow HCO_3^- (aq)$
2. $HCO_3^- (aq) + H^+ (aq) \longrightarrow H_2O (l) + CO_2(g)$

In the first stage CO_3^{2-} changes into HCO_3^- and in the second stage HCO_3^- changes into CO_2 and H_2O. The first stage represents half neutralization of CO_3^{2-}. If the same acid with the same concentration is used, then the same volume of acid would be needed to provide an equal amount.

Phenolphthalein is used to locate the end point of stage 1 because there is a sharp pH change, at pH 8.3, within the pH range of phenolphthalein at that stage. Methyl orange is used to locate the end point of stage 2 since there is a sharp pH change, at pH 5 which is near its pH range.

In a solution of a mixture of CO_3^{2-} and HCO_3^- the first titration using phenolphthalein as indicator represents the half neutralization of CO_3^{2-} whereas the second stage of the titration represents the reaction between the other half of the CO_3^{2-} *(aq)*, *i.e.*, HCO_3^- *(aq)* from CO_3^{2-} *(aq)* and the HCO_3^- *(aq)* in the real mixture.

The end point of the second stage is indicated by methyl orange.

In a mixture of CO_3^{2-} *(aq)* and HCO_3^- *(aq)* at the phenolphthalein end point there is a colour change of red to colourless. If the volume of the acid is V_1, this volume reacted with half of the CO_3^{2-} to give HCO_3^-.

Thus, volume of acid required for complete neutralization of CO_3^{2-} in the mixture = $2 \times V_1$

On addition of methyl orange to the same reaction mixture the second end point can be located when it turns from yellow to pink. If V_2 is the volume of acid used from the start of the reaction then V_2 is the volume of acid required for complete neutralization of CO_3^{2-} and HCO_3^-. Volume of acid required for complete neutralization of $HCO_3^- = V_2 - 2V_1$.

If V_3 is the volume of the acid used after the addition of methyl orange indicator, then V_3 is the volume of the acid required for half neutralization of CO_3^{2-} and complete neutralization of HCO_3^-.

Volume of acid required for complete neutralization of HCO_3^- in that case is equal to $V_3 - V_1$.

Knowing the volumes of acid required for complete neutralization of CO_3^{2-} and HCO_3^- in the mixture, their individual concentrations can be calculated.

Alternatively, this experiment can be carried out by titrating the CO_3^{2-} and HCO_3^- mixture separately using phenolphthalein and methyl orange indicators.

SOLVED EXAMPLE

Example 23.3. *25 cm^3 of a mixture of Na_2CO_3 and $NaHCO_3$ was titrated with 0.1 M HCl using phenolphthalein as indicator. The volume of acid that reacted to the end point is 9.60 cm^3.*

A separate 25 cm^3 of the mixture was titrated using methyl orange as indicator. The volume of 0.1 M HCl that reacted to the end point was found to be 23.80 cm^3.

(a) What are the colour changes at the end points using
(i) phenolphthalein? (ii) methyl orange?

(b) Calculate the concentration in mol dm^{-3} and g dm^{-3} of
(i) Na_2CO_3 (ii) $NaHCO_3$

(c) What is the percentage composition of the mixture?

Solution.

(a) *(i)* With phenolphthalein: pink $\longrightarrow$ colourless

(ii) With methyl orange: yellow $\longrightarrow$ pink

(b) At the phenolphthalein end point only half of Na_2CO_3 is neutralized.

$\therefore$ Volume of 0.10 M HCl required for complete neutralization of $Na_2CO_3 = 2 \times 9.60 = 19.20$ cm^3

At the methyl orange end point Na_2CO_3 as well as $NaHCO_3$ are completely neutralized.

$\therefore$ Volume of 0.1 M HCl required for complete neutralization of Na_2CO_3 and $NaHCO_3$ = 23.80 cm^3

Volume of HCl required for complete neutralization of $NaHCO_3$

$$= 23.80 - 19.20 = 4.60\ cm^3$$

The chemical equations for the reaction between Na_2CO_3 and HCl, and $NaHCO_3$ and HCl are:

$$Na_2CO_3\,(aq) + 2HCl\,(aq) \longrightarrow 2NaCl\,(aq) + H_2O\,(l) + CO_2\,(g)$$

$$NaHCO_3(aq) + HCl\,(aq) \longrightarrow NaCl\,(aq) + H_2O\,(l) + CO_2\,(g)$$

$$\frac{M_1V_1(\text{of } Na_2CO_3)}{M_2V_2\,(\text{of HCl})} = \frac{1}{2}$$

$$\frac{M_1 \times 25.0}{0.10 \times 19.20} = \frac{1}{2}$$

$$M_1 = \frac{0.1 \times 19.20}{2 \times 25.0} = 0.0384 \text{ mol dm}^{-3}$$

$\therefore$ Conc. of $Na_2\,CO_3$ = **0.0384 mol dm^{-3}**

Molar mass $Na_2\,CO_3$ = 23 × 2 + 12 + 16 × 3 = 106 g mol^{-1}

Conc. of Na_2CO_3 in g dm^{-3} = Molarity × Molar mass

= 0.0384 × 106 g dm^{-3}

= 4.07 g dm^{-3}

$$\frac{M_1'V_1'(\text{of } NaHCO_3)}{M_2'V_2'\,(\text{of HCl})} = \frac{1}{1}$$

$$\frac{M_1' \times 25.0}{0.10 \times 4.6} = 1$$

$$M_1' = \frac{0.10 \times 4.6}{25.0} = 0.0184 \text{ mol dm}^{-3}$$

$\therefore$ Conc. of $NaHCO_3$ = 0.0184 mol dm^{-3}

Molar mass of $NaHCO_3$ = 23 + 1 + 12 + 16 × 3 = 84 g mol^{-1}

Conc. of $NaHCO_3$ in g dm^{-3} = Molarity × Molar mass

= 0.0184 × 84 g dm^{-3} = **1.55 g dm^{-3}.**

(*c*) $\%$ of $Na_2CO_3 = \dfrac{4.07}{(4.07 + 1.55)} \times 100 =$ **72.4 %**

$\%$ of $NaHCO_3 = \dfrac{1.55}{(4.07 + 1.55)} \times 100 =$ **27.6 %.**

SUMMARY

- **Standard Solution.** A solution whose concentration is known.
- **Molarity** of a solution may be defined as the number of moles of solute per dm^3 of solution 1.0 M solution contains 1.0 mole of solute in 1 dm^3 of solution.
- **Primary Standard** is a substance whose standard solution can be prepared by dissolving it directly in water.
- **For dilution.**

$$\underset{\text{Original Solution}}{M_1V_1} = \underset{\text{Diluted Solution}}{M_2V_2}$$

- **Titration.** If according to a chemical equation *a* moles of acid react with *b* moles of alkali then

$$\frac{M_1V_1\,(\text{of acid})}{M_2V_2\,(\text{of alkali})} = \frac{a}{b}$$

- For strong acid-strong base titrations phenolphthalein is used as indicator.

  For strong acid-weak base titrations methyl orange is used as indicator.

  For weak acid-strong alkali titrations phenolphthalein is used as indicator.
- **Phenolphthalein** is pink in alkaline medium and colourless in acidic medium.

  Methyl orange is yellow in alkaline medium and pink in acidic medium.
- Ammonium salts on heating with alkali solutions decompose to liberate ammonia gas.

$$NaOH + NH_4Cl \longrightarrow NaCl + NH_3\uparrow + H_2O$$

or $$NH_4^+ + OH^- \longrightarrow NH_3\uparrow + H_2O$$

- Neutralization of CO_3^{2-} and HCO_3^- with acids, using methyl orange as indicator, gives the volume of acid required for complete neutralization.
- Neutralization of CO_3^{2-} with acids, using phenolphthalein as indicator, gives the volume of acid required for half neutralization of CO_3^{2-}.

EVALUATE YOURSELF

I. Objective Type Questions

Select the most appropriate choice from the options given as (a), (b), (c) and (d) after each question:

1. Which of the following will introduce error in the titre value?

   (*a*) Rinsing the titration flask with distilled water.

   (*b*) Rinsing the burette with distilled water.

   (*c*) Rinsing the pipette with the solution to be pipetted.

   (*d*) Flushing out air bubbles from the nozzle of burette before titration starts.

II. Fill in the Blanks

2. Complete the following sentences by supplying appropriate words:
   (*i*) A solution whose concentration is known is called ....
   (*ii*) A solution which contains 1 mole of solute per 1000 cm^3 of solution is called ....
   (*iii*) Colour of phenolphthalein in alkaline solution is ....
   (*iv*) The solution taken in the conical flask during titration is called ....
   (*v*) In titration of strong acid and weak base, the indicator used is ....

III. Discussion Questions

3. 25.0 cm^3 of 0.10 M sodium trioxocarbonate(IV) solution will require how much volume of 0.20 M solution of hydrochloric acid.
4. What volume of 10 M HCl should be diluted with water to prepare 1000 cm^3 of 0.5 M HCl?
5. What mass of Na_2CO_3 should be dissolved in water to prepare 250 cm^3 of its 0.05 M solution?
6. A solution is prepared by dissolving 6.0 g of a mixture of KOH and KCl per 1000 cm^3 of solution. 25 cm^3 of this solution, on titration, with 0.1 M HCl required 20.00 cm^3 of the acid for complete neutralization. What is the percentage composition of the mixture?
7. Explain the following terms:
   (*i*) Standard solution (*ii*) Titrant
   (*iii*) Titrate or titrand.
8. A solution is prepared by dissolving 6.5 g of an impure sample of sodium trioxocarbonate(IV) per 1000 cm^3 of solution. Determine the percentage purity of the sample if 25 cm^3 of this solution required 22.50 cm^3 of 0.1 M HCl for complete reaction.
   What is the indicator used in the titration? What is the end point?
9. A solution contains a mixture of KOH and $KHCO_3$. Explain how you would determine the percentage composition of the mixture. You are provided with 0.10 M solution of HNO_3.
10. You are given a solution containing a mixture of Na_2CO_3 and NaOH. Explain how you would determine the percentage composition of the mixture. Provided is 0.05 M H_2SO_4.

Answers

I. Objective Type Question

1. (*b*)

II. Fill in the Blanks

2. (*i*) standard solution (*ii*) molar solution
   (*iii*) pink (*iv*) titrant
   (*v*) methyl orange.

SECTION-4

UNIT-24

CLASSIFICATION OF ACIDS AND BASES

Specific Objectives

After studying this unit, the student should be able to:

- classify acids and bases into strong and weak
- explain the conduction of strong and weak electrolytes

24.1 STRENGTH OF ACIDS AND BASES

When an acid is dissolved in water, it dissociates into ions and produces hydronium ions. The characteristic properties of acids are due to hydronium ions.

Dissociation of some acids in aqueous solution, to form H_3O^+ ions is shown as follows:

$$HCl + H_2O \longrightarrow H_3O^+ + Cl^-$$
$$H_2SO_4 + 2H_2O \longrightarrow 2H_3O^+ + SO_4^{2-}$$
$$HNO_3 + H_2O \longrightarrow H_3O^+ + NO_3^-$$

STRONG AND WEAK ACIDS

When an acid such as hydrochloric acid (HCl) dissolves in water, almost all of it dissociates. Such acids are known as **strong acids.**

On the other hand, when acetic acid dissolves in water only a small fraction of the molecules ionizes. Such acids are called a **weak acids.**

$$HCl + H_2O \longrightarrow H_3O^+ + Cl^- \qquad \textbf{Strong acid}$$
$$CH_3COOH + H_2O \rightleftharpoons H_3O^+ + CH_3COO^- \qquad \textbf{Weak acid}$$

Dissociation and Ionization

- When a polar covalent compound such as HCl, H_2SO_4, HNO_3, etc. is dissolved in water, it splits into ions. This process is called **ionization**.
- When an ionic compound such as NaCl, KCl, etc. is dissolved in water, the ions, which are held strongly in the solid state, become free to move in solution. This process is called **dissociation**.

In the present context both the terms are used in the same sense

Some common strong acids and weak acids are listed in Table 24.1

Table 24.1. Strong Acids and Weak Acids

Strong Acids			*Weak Acids*		
1. Hydrochloric acid	:	HCl	1. Ethanoic acid (Acetic acid)	:	CH_3COOH
2. Trioxonitrate(V) acid	:	HNO_3	2. Tetraoxophosphate(V) acid	:	H_3PO_4
3. Tetraoxosulphate(VI) acid	:	H_2SO_4	3. Trioxoborate(III) acid	:	H_3BO_3
4. Hydrobromic acid	:	HBr	4. Trioxocarbonate(IV) acid	:	H_2CO_3

The words strong and weak used in this way do not refer to the concentration of the acid but refer to the degree of dissociation of the acid.

ACTIVITY 24.1 Acids Exhibit Acidic Properties only in Aqueous Solutions

- Take a pinch (about 1 g) of solid sodium chloride (NaCl) in a dry test tube fitted with a delivery tube (Fig. 24.1).

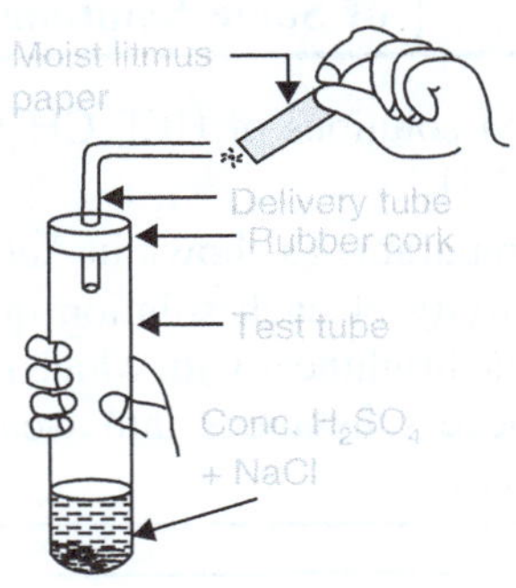

Fig. 24.1. Preparation of HCl gas.

- Add about 1 mL concentrated tetraoxosulphate(VI) acid to the test tube with the help of a dropper.

  What do you observe?

  It is observed that a colourless gas having pungent smell, evolves. The gas produced is HCl gas.

$$NaCl + H_2SO_4 \longrightarrow NaHSO_4 + HCl\uparrow$$

- Test the gas evolved first with a dry blue litmus paper and then with a moist blue litmus paper.

  What do you observe?

  It is observed that moist blue litmus turns red. It indicates that HCl exhibits acidic properties in the presence of water. This activity demonstrates that H^+ ions are produced only in the presence of water and not in dry gas.

CONCENTRATION OF AN ACID

Concentration of an acid may be expressed in terms of **moles of the acid per litre or per dm³ of solution.** This is known as **molarity** and is denoted by M. An acid solution having concentration 4 M contains 4 moles of the acid per dm^3 or per litre of solution.

A 4M solution of H_2SO_4 is a dilute acid solution while a 10M solution of H_2SO_4 is a concentrated acid. Concentration of an acid may also be expressed as **mass per cent.** A 20% solution of H_2SO_4 means that 100 g of the given solution contains 20 g H_2SO_4.

It must be clear to you that the strength of an acid is determined by the extent of dissociation of the acid and hydrogen ion concentration whereas concentration of the acid is determined by relative amounts of acid and water.

When a solution of some acid is mixed with a solution of some base, the H^+ (*aq*) ions from the acid combines with the OH^-(*aq*) from the base to form undissociated water (H_2O) molecules. Thus, during neutralization of an acid with a base the following reaction takes place:

$$\underset{\text{From acid}}{H^+(aq)} + \underset{\text{From base}}{OH^-(aq)} \longrightarrow H_2O(l)$$

STRONG AND WEAK BASES

A base when dissolved in water undergoes dissociation to produce hydroxyl ions (OH^-).

$$NaOH(s) \xrightarrow{H_2O} Na^+(aq) + OH^-(aq)$$

$$Ca(OH)_2(s) \xrightarrow{H_2O} Ca^{2+}(aq) + 2OH^-(aq)$$

The characteristic properties of bases are due to OH^- (*aq*) ions.

A base such as NaOH or KOH which is completely dissociated in aqueous solution is called a ***strong base.*** Hydroxides of group-1 metals are strong bases. $Ba(OH)_2$ is also a strong base as it is completely ionized in aqueous solution. *On the other hand, a base which is weakly dissociated in aqueous solution is called a* ***weak base****.*

For example, $Mg(OH)_2$ and NH_3 are weak bases.

Strengths of acid and bases can be compared quantitatively in terms of their ionization constants K_a and K_b.

This has been discussed in detail in Unit 25.

SOLVED EXAMPLES

Example 24.1. *Equal lengths of magnesium ribbons are taken in test tubes A and B. Hydrochloric acid (HCl) is added to test tube A while acetic acid (CH_3COOH) is added to test tube B. In which test tube will the fizzing occur more vigorously and why?*

Solution. Fizzing will occur more vigorously in **test tube A.**

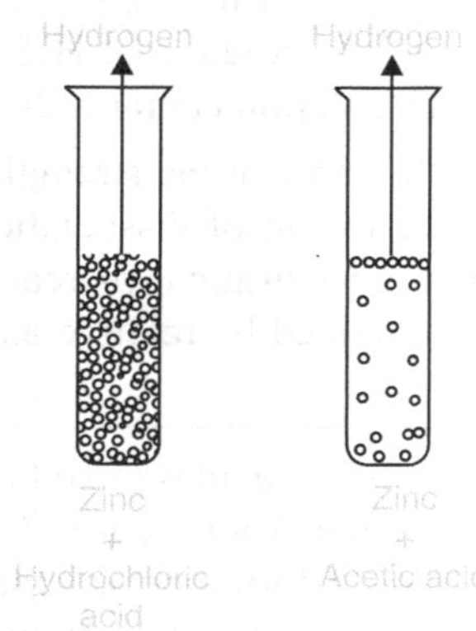

Fig. 24.2

Hydrochloric acid is a strong acid. It dissociates almost completely in aqueous solution producing a higher concentration of $H^+(aq)$ ions. Hence, it reacts more vigorously with magnesium. On the other hand, acetic acid is a weak acid and hence produces less concentration of $H^+(aq)$ ions in solution. Hence, less fizzing occurs in this case.

Example 24.2. *Why do acids not show acidic behaviour in the absence of water?*

Solution. Acidic behaviour of acids is due to the presence of hydrogen ions. An acid, when dissolved in water, dissociates to produce $H^+(aq)$ ions and hence exhibits the acidic behaviour. In the absence of water, the acids do not contain hydrogen ions and hence do not exhibit acidic behaviour.

24.2 CONDUCTIVITY OF SOLUTIONS

Strong acids and strong bases are almost completely ionized in aqueous solutions. Therefore, their solutions contain large numbers of ions and hence, they behave as **strong electrolytes.** They are very good conductors of electricity. On the other hand weak acids and weak bases are weakly ionized in aqueous solution. Therefore, they produce less ions in solution and hence behave as **weak electrolytes** and are poor conductors of electricity.

The solutions of strong acids such as HCl, H_2SO_4 and HNO_3, and strong bases such as NaOH and KOH behave as good conductors of electricity.

On the other hand, aqueous solutions of weak acids such as CH_3COOH and H_3PO_4, and weak bases such as $Ca(OH)_2$, $Mg(OH)_2$ and NH_3 behave as poor conductors of electricity.

The aqueous solutions of non-electrolytes such as glucose, sucrose and ethanol are bad conductors of electricity.

ACTIVITY 24.2 **Investigation of Conductivity of Some Solutions**

- Prepare 1 M solutions of HCl, CH_3COOH, KOH, NaOH and NH_3.
- Using the apparatus as shown in Activity 22.2 find the conductivity of each solution qualitatively by observing the brightness with which the bulb glows. Classify these substances into weak and strong, acids and bases.

SUMMARY

- Acids and bases can be classified as strong or weak depending upon their degree of ionization.
- An acid or a base is said to be strong if it undergoes complete dissociation in water.
- An acid or a base is said to be weak if it undergoes only slight dissociation in water.
- A strong acid or a base produces higher concentration of ions in solution and hence behaves as a better conductor than a weak acid or a weak base.
- HCl, H_2SO_4, HNO_3 are some common examples of strong acids.
- NaOH, KOH and $Ba(OH)_2$ are strong bases.
- CH_3COOH, H_2CO_3, HCOOH and H_3PO_4 are some weak acids.
- NH_4OH, $Ca(OH)_2$ and $Mg(OH)_2$ are weak bases.

EVALUATE YOURSELF

I. Objective Type Questions

Select the most appropriate choice from the options given as (a), (b), (c) and (d) after each question:

1. Which of the following is a strong acid?
   (*a*) CH_3COOH (*b*) H_3PO_4
   (*c*) HNO_3 (*d*) NaOH
2. Which of the following is a weak base?
   (*a*) $Ba(OH)_2$ (*b*) NaOH
   (*c*) KOH (*d*) Aqueous NH_3
3. Which of the following is a weak electrolyte?
   (*a*) KOH (*b*) H_2SO_4
   (*c*) HCOOH (*d*) HNO_3

II. Discussion Questions

4. What is a strong acid? Give two examples.
5. What is a strong base? Give three examples of strong bases.
6. Explain why $Ca(OH)_2$ is considered a weak base.
7. Describe an activity to demonstrate that hydrochloric acid is a strong acid whereas ethanoic acid is a weak acid.
8. How does conductivity behaviour of an aqueous solution of ammonium hydroxide differ from that of potassium hydroxide?
9. Give two examples of weak acids and two examples of weak bases.
10. How does a concentrated acid solution differ from a strong acid solution?
11. What is molarity of a solution? How would you prepare a 0.1 M solution of NaOH?

Answers

I. Objective Type Questions

1. (*a*) **2.** (*d*) **3.** (*c*).

II. Discussion Questions

11. Dissolve 4 g of NaOH per dm^3 of solution to prepare 0.1 M solution of NaOH.

SECTION-4

UNIT-25

CONCEPT OF pH AND pOH

Specific Objectives

After studying this unit, the student should be able to:

- distinguish between solutions that are acidic, neutral or basic using the pH scale
- explain pK_a and pK_b of weak acids and bases.

25.1 IONIZATION OF ACIDS AND BASES

Acids like perchloric acid ($HClO_4$), hydrochloric acid (HCl), hydrobromic acid (HBr), hydroiodic acid (HI), trioxonitrate(V) acid (HNO_3) and tetraoxosulphate(VI) acid (H_2SO_4) are termed ***strong*** because they are almost completely dissociated into their constituent ions in an aqueous medium, thereby acting as strong proton (H^+) donors. Similarly, strong bases like sodium hydroxide (NaOH), potassium hydroxide (KOH), caesium hydroxide (CsOH) and barium hydroxide $Ba(OH)_2$ almost completely dissociate into ions in an aqueous medium giving hydroxyl ions, OH^-. According to the Arrhenius concept they are strong bases as they are able to completely dissociate and produce OH^- ions in the medium.

The strength of acids or bases is experimentally measured by determining its ionization constants or dissociation constants.

PARTIAL IONIZATION OF WEAK ACIDS AND BASES

Strong acids ionize almost completely into hydrogen ion and the corresponding anion in water and, therefore, the molar concentration of hydrogen ions in the solution is the same as that of acid itself. But weak acids are not completely ionized. The relative strengths of weak acids can be compared in terms of their ionization constants. For example, the ionization equilibrium of an acid HA may be represented as:

$$HA(aq) + H_2O(l) \rightleftharpoons H_3O^+(aq) + A^-(aq)$$

Applying the law of chemical equilibrium,

$$K = \frac{[H_3O^+][A^-]}{[HA][H_2O]}$$

Since the concentration of water is very large and remains almost constant in solution, it can be combined with K to give another constant K_a.

$$K[H_2O] = \frac{[H_3O^+][A^-]}{[HA]}$$

$$K_a = \frac{[H_3O^+][A^-]}{[HA]}$$

Here K_a is called **the ionization constant of an acid.**

The value of the ionization constant gives an idea of the relative strength of the acid. *The larger the value of the K_a, the higher the concentration of H_3O^+, and the stronger the acid.* If dissociation constants of two acids are known, their relative strengths can be predicted. For example,

$$K_{CH_3COOH} = \frac{[H_3O^+][CH_3COO^-]}{[CH_3COOH]}$$
$$= 1.8 \times 10^{-5} \text{ at } 298 \text{ K}$$

$$K_{HF} = \frac{[H_3O^+][F^-]}{[HF]}$$
$$= 6.7 \times 10^{-4} \text{ at } 298 \text{ K}$$

Since $K_{HF} > K_{CH_3COOH}$, HF is a stronger acid than CH_3COOH.

> **$-\log K_a = pK_a$**
>
> The smaller the value of the pK_a, the greater the strength of the acid.

The value of dissociation constants of some of the monoprotic acids have been given in Table 25.1.

Table 25.1. The Ionization Constants of Some Common Weak Acids at 298 K

Acid	*Ionization Constant, K_a*
Chloroethanoic acid, $ClCH_2COOH$	1.4×10^{-3}
Nitrous acid, HNO_2	4.5×10^{-4}
Hydrofluoric acid, HF	3.5×10^{-4}
Formic acid, HCOOH	1.8×10^{-4}
Benzoic acid, C_6H_5COOH	6.5×10^{-5}
Ethanoic acid, CH_3COOH	1.8×10^{-5}
Hypochlorous acid, HOCl	3.0×10^{-8}
Ammonium ion, NH_4^+	5.6×10^{-10}
Hydrocyanic acid, HCN	4.9×10^{-10}
Phenol, C_6H_5OH	1.3×10^{-10}
Water, H_2O	2.0×10^{-16}

Calculation of $[H_3O^+]$ and Degree of Ionization

From the knowledge of K_a it is possible to calculate the hydronium ion concentration and degree of ionization of a weak acid. As an example, let us take the case of acetic acid. The following equation represents the ionization of acetic acid in aqueous solution:

$$CH_3COOH(aq) + H_2O(l) \rightleftharpoons H_3O^+(aq) + CH_3COO^-(aq)$$

$$K_a = \frac{[H_3O^+][CH_3COO^-]}{[CH_3COOH]}$$

Suppose C moles of CH_3COOH are dissolved per litre of solution and let α be the degree of ionization of CH_3COOH, then at equilibrium the concentration of the various species would be as follows:

$$[CH_3COOH] = C(1 - \alpha)$$
$$[H_3O^+] = C\alpha$$
$$[CH_3COO^-] = C\alpha$$

Therefore, $K_a = \dfrac{(C\alpha)(C\alpha)}{C(1-\alpha)} = \dfrac{C\alpha^2}{(1-\alpha)}$

Since for weak acid α is very small as compared to 1, α in the denominator can be neglected. The expression of K_a then becomes

$$K_a = C\alpha^2$$

Knowing the value of K_a it is possible to calculate the degree of ionization of the acid at any particular concentration C.

$$\alpha = \sqrt{\frac{K_a}{C}}$$

From the degree of ionization, the hydronium ion concentration can be calculated as:

$$[H_3O^+] = C\alpha$$

Ionization Constants of Bases in Water

The **relative strengths of bases** are also compared in terms of their dissociation or ionization constants. The ionization constant $\mathbf{K_b}$ for a weak base B can be represented as follows:

$$B(aq) + H_2O(l) \rightleftharpoons BH^+(aq) + OH^-(aq)$$

$$K_b = \frac{[BH^+][OH^-]}{[B]}$$

The smaller the value of the ionization constant of a base, the weaker the base.

If C is the molar concentration of base and α is its degree of dissociation, then by similar calculation (as in case of acids) we can derive the relations,

$$\alpha = \sqrt{\frac{K_b}{C}}$$

and $[OH^-] = C\alpha$.

> $-\log K_b = pK_b$
> The smaller the value of the pK_b, the greater the strength of the base.

The values of ionization constants for some weak bases are given in Table 25.2.

Table 25.2. The Ionization Constants of Some Bases at 298 K

Base	*Ionization Constant, K_b*
Ammonia, NH_3	1.8×10^{-5}
Methylamine, CH_3NH_2	4.6×10^{-4}
Dimethylamine, $(CH_3)_2NH$	5.4×10^{-4}
Trimethylamine, $(CH_3)_3N$	6.5×10^{-5}
Aniline, $C_6H_5NH_2$	4.3×10^{-10}
Pyridine, C_5H_5N	1.8×10^{-9}

Add To Your Knowledge

The acids, such as HCl, CH_3COOH, HCN, etc., which have only one ionizable hydrogen are known as **monoprotic acids.**

$$HCl(aq) \longrightarrow H^+(aq) + Cl^-(aq)$$

$$CH_3COOH(aq) \rightleftharpoons CH_3COO^-(aq) + H^+(aq)$$

On the other hand, there are many acids, such as tetraoxophosphate(V) acid (H_3PO_4), tetraoxosulphate(VI) acid (H_2SO_4), oxalic acid ($H_2C_2O_4$), which have more than one ionizable hydrogen. Such acids are known as **polyprotic acids.** Thus:

The acids which have more than one ionizable hydrogen atom are known as polyprotic acids.

Polyprotic acids are also known as **polybasic acids.**

Common Ion Effect in the Ionization of Weak Acids and Bases

The degree of ionization of a weak acid or a weak base is suppressed (reduced) by common ion effect.

An acid on ionization gives hydrogen ion and the corresponding anion. A strong acid is completely ionized while a weak acid is only partially ionized. In case of weak acids the ions formed are in equilibrium with the unionized molecules of the acid. The fraction of the total number of molecules of the acid dissolved that ionizes is known as the degree of ionization of the acid.

The degree of ionization of the acid decreases if some strong electrolyte that can provide any of the product ions is added to the solution of the acid. For example, consider the ionization of acetic acid.

$$CH_3COOH(aq) \rightleftharpoons \boxed{CH_3COO^-(aq)} + H^+(aq)$$

$$CH_3COONa(aq) \longrightarrow \boxed{CH_3COO^-(aq)} + Na^+(aq)$$

Common ion

If sodium acetate is added to the above solution, the concentration of acetate ions increases and the above equilibrium shifts in the backward direction according to Le-Chatelier's principle. Thus, addition of acetate ions reduces the degree of ionization of acetic acid. Similarly, if H^+ ions are provided from some source, such as dilute hydrochloric acid, the degree of ionization of the weak acid is suppressed.

Similarly, the degree of ionization of a weak base is suppressed by common ion effect. A base on ionization gives hydroxyl ions (OH^-) and the corresponding cation (M^+). Addition of some strong electrolyte which provides either of the ions (OH^- or M^+) reduces the degree of ionization of the weak base in accordance with Le-Chatelier's principle. For example, the degree of ionization of ammonia is suppressed by addition of ammonium chloride or sodium hydroxide.

$$NH_3(aq) + H_2O(l) \rightleftharpoons \boxed{NH_4^+(aq)} + OH^-(aq)$$

$$NH_4Cl(aq) \longrightarrow \boxed{NH_4^+(aq)} + Cl^-(aq)$$

Common ion

$$NH_3(aq) + H_2O(l) \rightleftharpoons NH_4^+(aq) + \boxed{OH^-(aq)}$$

$$NaOH(aq) \longrightarrow Na^+(aq) + \boxed{OH^-(aq)}$$

Common ion

SOLVED EXAMPLES

Example 25.1. *Calculate the degree of ionization of 0.01 M solution of HCN, K_a of HCN is 4.8×10^{-10}. Also calculate the hydronium ion concentration.*

Solution. The ionization of HCN may be represented as,

$$HCN(aq) + H_2O(l) \rightleftharpoons CN^-(aq) + H_3O^+(aq)$$

If the degree of ionization of HCN is α then the equilibrium concentration of various species are

$$[HCN] = C(1 - \alpha)$$

$$[CN^-] = C\alpha$$

$$[H_3O^+] = C\alpha$$

where C = concentration of HCN.

Applying the law of chemical equilibrium

$$K_a = \frac{[CN^-][H_3O^+]}{[HCN]} = \frac{(C\alpha)(C\alpha)}{C(1-\alpha)} = \frac{C\alpha^2}{(1-\alpha)}$$

Since α is very small as compared with unity, $1 - \alpha$ in the denominator may be taken as 1.

$$K_a = C\alpha^2$$

$$\alpha = \sqrt{\frac{K_a}{C}} = \sqrt{\frac{4.8\times10^{-10}}{0.01}} = \mathbf{2.2 \times 10^{-4}}.$$

$$[H_3O^+] = C\alpha = 0.01 \times 2.2 \times 10^{-4} = \mathbf{2.2 \times 10^{-6}\ mol\ dm^{-3}}.$$

Example 25.2. *Calculate the degree of ionization of 0.02 M acetic acid if its $K_a = 1.8 \times 10^{-5}$. What would be the degree of ionization if the solution also contains 0.01 M sodium acetate?*

Solution. $CH_3COOH(aq) \rightleftharpoons CH_3COO^-(aq) + H^+(aq)$

The degree of ionization of this weak acid can be calculated by the approximate relation:

$$\alpha = \sqrt{\frac{K_a}{C}}$$

$$= \sqrt{\frac{1.8\times10^{-5}}{0.02}} = 0.03 = \mathbf{3 \times 10^{-2}}.$$

Now let us calculate the degree of ionization when the solution also contains 0.01 M sodium acetate. Sodium acetate being a strong electrolyte would be completely ionized in solution. Let x mol L^{-1} of acetic acid be ionized.

$$\underset{(0.02 - x)M}{CH_3COOH(aq)} \rightleftharpoons \underset{x\ M}{CH_3COO^-(aq)} + \underset{x\ M}{H^+(aq)}$$

$$CH_3COONa(aq) \longrightarrow \underset{0.01\ M}{CH_3COO^-(aq)} + \underset{0.01\ M}{Na^+(aq)}$$

$$[H^+] = x \text{ mol } L^{-1}$$

$$[CH_3COO^-] = (x + 0.01) \text{ mol } L^{-1} \approx 0.01 \text{ mol } L^{-1}$$

[$\because$ x is very small as compared to 0.01]

$$[CH_3COOH] = (0.02 - x) \text{ mol } L^{-1} \approx 0.02 \text{ mol } L^{-1}$$

$$K_a = \frac{[H^+][CH_3COO^-]}{[CH_3COOH]}$$

$$1.8 \times 10^{-5} = \frac{(x)(0.01)}{(0.02)}$$

$$x = 9 \times 10^{-6}$$

Degree of ionization,

$$\alpha = \frac{x}{0.02} = \frac{9\times10^{-6}}{0.02} = \mathbf{4.5 \times 10^{-4}}.$$

Thus, it may noted that the degree of ionization of acetic acid has decreased from 3×10^{-2} to 4.5×10^{-4} due to the presence of sodium acetate.

Example 25.3. *Calculate the degree of ionization of pyridine (C_5H_5N) in its 0.1 M solution, K_b for pyridine is 1.5×10^{-9}. What would be the degree of ionization if the solution is also 0.1 M in NaOH?*

Solution. Since the value of K_b is very less, the degree of ionization of pyridine (a weak base) can be calculated by the approximate formula

$$\alpha = \sqrt{\frac{K_b}{C}} = \sqrt{\frac{1.5\times10^{-9}}{0.1}} = 1.22 \times 10^{-4}$$

Now let us calculate the degree of ionization when the solution also contains 0.1 M NaOH. Suppose x mol L^{-1} of pyridine are ionized. NaOH being strong base would be completely ionized.

$$\underset{(0.1 - x)\ M}{C_5H_5N(aq)} + H_2O(l) \rightleftharpoons \underset{x\ M}{C_5H_5\overset{+}{N}H(aq)} + \underset{x\ M}{OH^-(aq)}$$

$$NaOH(aq) \longrightarrow \underset{0.1\ M}{Na^+(aq)} + \underset{0.1\ M}{OH^-(aq)}$$

$$[C_5H_5N] = (0.1 - x)M \approx 0.1\ M$$

[$\because$ x is very small as compared to 0.1]

$$[C_5H_5\overset{+}{N}H] = x\ M$$

$$[OH^-] = (x + 0.1)M \approx 0.1\ M$$

$$K_b = \frac{[C_5H_5\overset{+}{N}H][OH^-]}{[C_5H_5N]}$$

$$1.5 \times 10^{-9} = \frac{(x)(0.1)}{0.1}$$

$$\therefore \quad x = 1.5 \times 10^{-19}$$

The degree of ionization, $\alpha = \dfrac{x}{0.1}$

$$= \frac{1.5\times10^{-9}}{0.1} = \mathbf{1.5 \times 10^{-8}}.$$

This is much smaller than the degree of ionization of pyridine in the absence of NaOH. Thus, presence of NaOH decreases the degree of ionization of pyridine.

25.2 THE IONIZATION CONSTANT OF WATER AND ITS IONIC PRODUCT

The conductivity measurements of water indicate that water is a weak electrolyte. Even in pure state water is weakly ionized to give $H^+(aq)$ and $OH^-(aq)$ ions as shown below:

$$H_2O(l) \rightleftharpoons H^+(aq) + OH^-(aq)$$

The ionization constant of water may be written as

$$K_{a(H_2O)} = \frac{[H^+][OH^-]}{[H_2O]}$$

Since the concentration of water is very high ($\approx$ 55.55 mol L^{-1}) and only a very small fraction of it undergoes ionization, $[H_2O]$ may be taken as constant and may be combined with $K_{a(H_2O)}$ to have another constant K_w.

$$K_{a(H_2O)}[H_2O] = [H^+][OH^-]$$

$$K_w = [H^+][OH^-]$$

The constant K_w is called **the ionic product constant of water.** The concentration of H^+ in pure water has been found experimentally to be 1.0×10^{-7} mol dm^{-3} at 298 K. Moreover, in pure water the concentration of H^+ is equal to the concentration of OH^- ions because the dissociation of water produces equal numbers of H^+ and OH^- ions. Therefore,

$$[H^+] = [OH^-] = 1 \times 10^{-7} \text{ mol dm}^{-3}.$$

$$K_w = [H_3O^+][OH^-]$$

or

$$K_w = (1 \times 10^{-7} \text{ mol dm}^{-3})(1 \times 10^{-7} \text{ mol dm}^{-3})$$

$$= 1 \times 10^{-14} \text{ mol}^2 \text{ dm}^{-6}$$

The ***degree of dissociation*** of water

$$= \frac{10^{-7} \text{ mol dm}^{-3}}{55.55 \text{ mol dm}^{-3}} = 1.8 \times 10^{-9}$$

Since with an increase in temperature the dissociation of water increases, the value of K_w also increases as the temperature is increased. However, at all temperatures $[H^+]$ remains equal to $[OH^-]$ in pure water. The values of K_w at different temperatures have been given in Table 25.3.

Table 25.3. Ionic Product of Water at Different Temperatures

Temperature (K)	*Ionic Product, K_w (mol^2dm^{-6})*
273	0.113×10^{-14}
283	0.292×10^{-14}
293	0.687×10^{-14}
298	1.008×10^{-14}
313	2.917×10^{-14}
323	5.474×10^{-14}
373	$(56) \times 10^{-14}$

IONIZATION CONSTANT OF WATER

Water can behave as a weak acid as it ionizes according to the equation:

$$H_2O(l) \rightleftharpoons H^+(aq) + OH^-(aq)$$

The ionization constant of water may be written as

$$K_{a(H_2O)} = \frac{[H^+][OH^-]}{[H_2O]}$$

As already discussed $[H^+]$ and $[OH^-]$ is 10^{-7} M at 298 K and $[H_2O] = 55.55$ M, therefore,

$$K_{a(H_2O)} = \frac{(10^{-7})(10^{-7})}{55.55} = \mathbf{1.8 \times 10^{-16}}$$

It may be noted that the ionization constant of water, $K_{a(H_2O)}$ is different from the ionic product of water, K_w.

$$\boxed{K_w = K_{a(H_2O)} \times 55.55}$$

RELATION BETWEEN IONIZATION CONSTANT OF A BASE AND IONIZATION CONSTANT OF ITS CONJUGATE ACID

Ionization of a base B may be expressed as:

$$B + H_2O \longrightarrow BH^+ + OH^-$$

$$K_b = \frac{[BH^+][OH^-]}{[B]} \quad ...(i)$$

Ionization of BH^+, the conjugate acid of B may be expressed as:

$$BH^+ + H_2O \longrightarrow B + H_3O^+$$

$$K_a = \frac{[B][H_3O^+]}{[BH^+]} \quad ...(ii)$$

Multiplying equation (*i*) by (*ii*), we get

$$K_b \times K_a = \frac{[BH^+][OH^-]}{[B]} \times \frac{[B][H_3O^+]}{[BH^+]}$$

$$K_b \times K_a = [OH^-][H_3O^+]$$

$$\boxed{K_b \times K_a = K_w}$$

Concentration of H_3O^+ and OH^- Ions in Aqueous Solutions of Acids and Bases

We have already seen that $[H_3O^+]$ is equal to $[OH^-]$ in pure water. But, on addition of some acid or base to water these concentrations no longer remain equal. However, the value of the *ionic product of water* (K_w) at a particular temperature always remains constant irrespective of whether the water is pure or some acid or base has been added to it. For example, if an acid (say HCl) is added to water, the concentration of hydronium ion becomes quite high. Consequently, the dissociation equilibrium of water shifts in the reverse direction (according to the Le-Chatelier's principle), *i.e.*, H_3O^+ ions would combine with OH^- ions to form undissociated water molecule, so that the value of K_w in the solution may remain the same as that in pure water. Thus, the addition of an acid to water decreases the $[OH^-]$ according to the relation,

$$[OH^-] = \frac{K_w}{[H_3O^+]}$$

Similarly, the addition of a base such as NaOH increases the $[OH^-]$ and decreases the $[H_3O^+]$ according to the relation,

$$[H_3O^+] = \frac{K_w}{[OH^-]}$$

From the above discussion, it is clear that H_3O^+ ions and OH^- ions are always present in an aqueous solution whether it is acidic or basic. However, the relative concentrations of these ions vary in different solutions. In general,

in neutral solutions $[H_3O^+] = [OH^-]$

in acidic solutions $[H_3O^+] > [OH^-]$

in basic solutions $[H_3O^+] < [OH^-]$

The concentrations of H_3O^+ ions and OH^- ions in different types of acidic and basic solutions have been calculated in the following examples:

SOLVED EXAMPLES

Example 25.4. *Calculate the concentrations of H_3O^+ ions and OH^- ions in (a) 0.01 M solution of HCl (b) 0.01 M solution of NaOH at 298 K, assuming that HCl and NaOH are completely ionized under the given conditions.*

Solution. (*a*) $HCl(aq) + H_2O(l) \longrightarrow H_3O^+(aq) + Cl^-(aq)$

Since HCl is completely ionized

$\therefore$ $[H_3O^+] = [HCl] = \mathbf{0.01\ M}$ or $\mathbf{1 \times 10^{-2}\ mol\ dm^{-3}}$

$$[OH^-] = \frac{[K_w]}{[H_3O^+]}$$

$$= \frac{1\times10^{-14}}{1\times10^{-2}} = \mathbf{1 \times 10^{-12}\ mol\ dm^{-3}}$$

(*b*) $NaOH(aq) \longrightarrow Na^+(aq) + OH^-(aq)$

Since NaOH is completely ionized

$[OH^-] = [NaOH] = \mathbf{0.01\ M}$ or $\mathbf{1 \times 10^{-2}\ mol\ dm^{-3}}$

$$[H_3O^+] = \frac{K_w}{[OH^-]} = \frac{1\times10^{-14}}{1\times10^{-2}}$$

$$= \mathbf{1 \times 10^{-12}\ mol\ dm^{-3}}.$$

Example 25.5. *Calculate the concentration of H_3O^+ ions in 0.005 M solution of $Ba(OH)_2$ at 298 K, assuming that $Ba(OH)_2$ is completely ionized under the given conditions.*

Solution. $Ba(OH)_2(aq) \longrightarrow Ba^{2+}(aq) + 2OH^-(aq)$

Since $Ba(OH)_2$ is completely ionized and on ionization its one molecule gives two OH^- ions.

$\therefore$ $[OH^-] = 2[Ba(OH)_2]$

$= 2 \times 0.005 = .01\ mol\ dm^{-3}$

$= 1 \times 10^{-2}\ mol\ dm^{-3}$

$$[H_3O^+] = \frac{K_w}{[OH^-]} = \frac{1\times10^{-14}}{1\times10^{-2}}$$

$$= \mathbf{1 \times 10^{-12}\ mol\ dm^{-3}}.$$

25.3 EXPRESSING HYDRONIUM ION CONCENTRATION—pH SCALE

We have seen that the concept of the ionic product of water enables us to classify solutions as acidic, basic, or neutral by specifying the H_3O^+ ion concentration. For expressing the H_3O^+ ion concentration a logarithmic scale was devised by P.L. Sorensen (1909). This scale is called the **pH scale.** ***The pH of a solution may be defined as the negative logarithm of the hydronium ion concentration in moles per litre.***

Thus, mathematically, pH may be expressed as

$$pH = -\log [H_3O^+]$$

From the above relation, the concentration of $[H_3O^+]$ can be written as

$$[H_3O^+] = 1 \times 10^{-pH}$$

This enables us to give an alternative definition of pH as the *negative power to which 10 must be raised in order to express the hydronium ion concentration of a solution.*

Add To Your Knowledge

The symbol pH is derived from *Potenz*, the Danish word for power, pH refers to the potency of hydronium ion in solution. The potency of hydroxyl ions in solution may also be expressed in terms of pOH.

For pure water or neutral solutions, at 298 K, $[H_3O^+]$ is equal to 1×10^{-7} mol dm^{-3}. Therefore, pH of such solutions is given as:

$$pH = -\log [H_3O^+] = -\log [1 \times 10^{-7}] = 7.$$

For acidic solutions, $[H_3O^+]$ concentration is more than 1×10^{-7} mol dm^{-3}. Therefore, the pH of an acidic solution is **less than 7.** For basic solutions, the pH value is **greater than 7.** The range of pH is from 0 to 14. The solutions having pH between 0 and 2 are strongly acidic, those with pH between 2 to 4 are moderately acidic while others having pH between 4 to 7 are weakly acidic. Similarly, solutions having pH values between 7 to 10 are weakly basic, those having pH 10 to 12 are moderately basic whereas others which have pH between 12 to 14 are strongly basic.

The complete range of $[H_3O^+]$ and pH has been illustrated in Fig. 25.1.

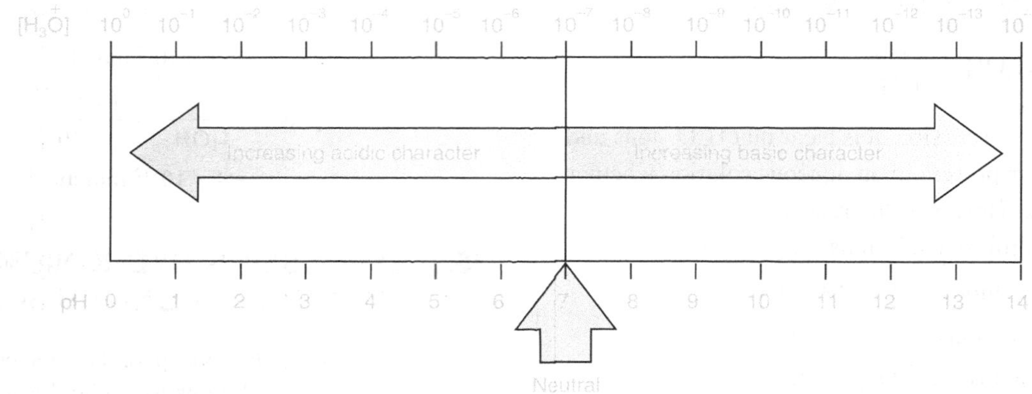

Fig. 25.1. Range of pH and $[H_3O^+]$.

The approximate pH of a solution can be determined with the help of **pH papers.** pH papers have different colours in solutions of different pH. A pH paper can determine pH of a solution with an accuracy of about 0.5. However, for accurate measurement of pH (up to accuracy of 0.001 units) **pH metres** are used.

The approximate pH of some common substances are given in Table 25.4.

Table 25.4. The pH of Some Common Substances

Name of the Fluid	*pH*	*Name of the Fluid*	*pH*
1 M NaOH solution	~ 14	Black coffee	5.0
0.1 M NaOH solution	13	Tomato juice	~ 4.2
Lime water	10.5	Soft drinks and vinegar	~ 3.0
Milk of magnesia	10	Orange juice	~ 2.6
Egg white, sea water	7.8	Lemon juice	~ 2.2
Human blood	7.4	Gastric juice	~ 1.2
Tears	7.4	1 M HCl solution	~ 0
Milk	6.8		
Human saliva	6.4		

- **pOH of the solution**

  Like pH, pOH may be defined as the negative logarithm of the hydroxyl ion concentration of a solution

  $pOH = -\log [OH^-]$.

- **Relationship between pH and pOH of a solution**

$$K_w = [H_3O^+][OH^-] = 1 \times 10^{-14}, \text{ at } 298 \text{ K}$$

Taking logarithm,

$$\log K_w = \log [H_3O^+] + \log [OH^-] = -14$$

or $-\log K_w = -\log [H_3O^+] + (-\log [OH^-]) = 14$

or $\boxed{pK_w = pH + pOH = 14}$ at 298 K

- As the temperature increases K_w increases and pK_w decreases. Consequently, the pH and the pOH of the solution also decreases.
- pH of neutral water is < 7 at high temperatures. For example, the pH of neutral water at 310 K, is 6.98.
- It is more appropriate to express the strengths of acids and bases in terms of their pK_a or pK_b values

$$pK_a = -\log K_a \; ; \; pK_b = -\log K_b$$

- The lower the value of the pK_a, the stronger the acid.
- The lower the value of the pK_b, the stronger the base.

***Acid-base indicators** or **pH indicators** are substances which undergo change in colour with pH.* These indicators can be used to find whether the solution is acidic or alkaline. Some common acid base indicators are: **litmus, phenolphthalein, methyl orange** and **methyl red.**

The colour of an indicator undergoes an observable change over a range of pH. Therefore, these indicators cannot be used to find the exact pH of a solution.

25.4 UNIVERSAL INDICATOR

A universal indicator is prepared by mixing a number of common indicators together so that the mixture obtained can pass through a series of colour changes over a much wider pH range. For example, one such mixture which may show various colours at different pH is as given in Table 25.5

Table 25.5. Colours of Universal Indicator at Different pH Values

pH	*Colour*
3.0	Red
5.0	Orange red
5.5	Orange
6.0	Orange-yellow
7.0–7.5	Greenish-yellow
8.0	Green
9.5	Blue
10.0	Violet

Such mixtures are commonly known as **universal indicators.** Universal indicators are available commercially as solutions and as test papers. A pH paper is a strip of paper which is prepared by dipping the strip in the solutions of different indicators and then drying them.

pH paper can be used to find the approximate pH of any solution. The pH paper is dipped in a given sample of the solution, the colour developed on the paper is compared with the colour chart and the approximate pH of the solution predicted.

25.5 EXPERIMENTAL MEASUREMENT OF pH OF A GIVEN SOLUTION

The practical measurement of the pH of a solution can be done by the use of a universal indicator solution, a pH paper or a pH meter.

Using indicator solution:

1. Take about 5 cm^3 of the given solution in a test tube.
2. Add two or three drops of the universal indicator to the solution. Shake well for the colour to become uniform.
3. Match the colour of the solution with the standard colour chart provided on the universal indicator bottle. The pH of the given solution is the pH of the colour with which it matches best.

Using pH or universal indicator paper:

1. Dip the strip of pH paper into the given solution or add few drops of the given solution to the pH paper. Observe the colour on the pH paper.
2. Compare the shade of the colour formed with the various colours given on the standard chart.
3. The pH of the given solution is the same as the pH corresponding to the colour on the chart.

Using pH meter:

When using the pH meter, dip the electrode into the given solution. The pH is registered on the scale.

ACTIVITY 25.1 Finding pH of Some Solutions

1. Prepare about 0.1 M solutions of HCl, H_2SO_4, CH_3COOH, NaOH, Na_2CO_3, $NaHCO_3$, NH_4Cl, NH_4OH (aqueous ammonia).
2. Take about 5 cm^3 of HCl solution in a test tube and add 2-3 drops of a universal indicator solution.
3. Note the colour developed and find the pH of the solution by comparing the colour developed with the colours on the standard chart provided with the universal indicator solution.
4. Record your observations and repeat the steps 2 and 3 with other solutions.

Solution	*Colour produced With universal indicator*	*Approximate pH*
0.1 M HCl	......	......
0.1 M NaOH	......	......
0.1 M Na_2CO_3		
⋮		

25.6 IMPORTANCE OF pH IN OUR DAILY LIFE

pH AND PLANTS

Plants require a specific pH of the soil for their proper growth. Plants cannot grow in too alkaline or too acidic soils. Therefore, for proper cultivation of crops it is important to have knowledge of the pH of the soil. If the soil is too acidic then its pH can be raised by adding lime (or slaked lime) which neutralizes the excess acid in the soil.

$$\underset{\text{Lime (Base)}}{CaO} + \underset{\text{Tetraoxosulphate(VI) acid}}{H_2SO_4} \longrightarrow CaSO_4 + H_2O$$

Similarly, if the soil is too alkaline then its pH can be lowered by adding gypsum or some other substance which can neutralize the excess alkali present in the soil.

ACTIVITY 25.2	Study of pH of the Soil

- Take about 5 g soil in a 100 cm^3 beaker and add 10 cm^3 of water to it.
- Mix the soil with water thoroughly by stirring with the help of a glass rod.
- Filter the contents and collect the filtrate in a beaker.
- Dip a strip of universal indicator paper in the filtrate and determine its pH by comparing the colour of the indicator paper with the reference chart.
- Repeat the activity by taking soil from different places and find out the ideal soil pH for the growth of plants.

pH IN OUR DIGESTIVE SYSTEM

The medium in our stomach is highly acidic and has pH around 1.2. Our stomach produces hydrochloric acid which helps in the digestion of food. Sometimes the stomach produces too much hydrochloric acid. Excess acid in the stomach causes pain and irritation. It can also cause ulcers in the digestive system. The condition of excess acid in the stomach is referred to as **hyperacidity.** In order to treat hyperacidity, a patient is advised to take antacid tablets or antacid suspensions. The antacid preparations contain some mild base such as magnesium hydroxide as the active component which neutralizes the excess acid in the stomach and thus controls the pH in the stomach.

$$Mg(OH)_2 + 2HCl \longrightarrow MgCl_2 + 2H_2O$$

pH AND TOOTH DECAY

Tooth enamel is the hardest substance in our body. It is made up of calcium tetraoxophosphate(V). It does not dissolve in water and mildly acidic solutions. However, it gets corroded slowly when the pH in the mouth is below 5.5. When we eat food, some food particles remain sticking to the teeth. Bacteria present in the mouth produce acid by the degradation of sugar and food particles present in the mouth. This results in lowering of the pH in the mouth which promotes tooth decay.

In order to prevent tooth decay one should avoid eating sugary foods. The mouth should be cleaned after eating food. Cleaning the teeth using toothpastes also helps in preventing tooth decay. Toothpastes are generally basic, therefore, they neutralize the excess acid in the mouth and thus prevent tooth decay.

The knowledge of pH also helps us in maintaining the proper acidity in many products such as *shampoos, medicines,* etc. pH of *rain water* can be used to detect the extent of certain pollutant gases in the atmosphere.

SOLVED EXAMPLES

Example 25.6. *The pH values of 1 M aqueous solutions of three acids X, Y and Z are 2.0, 1.0 and 3.0 respectively. Arrange these acids in order of increasing acid strength.*

Solution. The acid having minimum pH has the highest concentration of H^+ ions and hence is the strongest acid. Thus, acid Y is the strongest acid. The acid Z, having highest pH, is the weakest acid. The order of acid strength is $\mathbf{Z < X < Y}$.

Example 25.7. *pH of a solution changes from 3 to 4. What is the change in concentration of H^+ ions in the solution? How many times do you expect a change in hydrogen ion concentration?*

Solution.

$$pH = -\log [H^+]$$

When $pH = 3, \quad 3 = -\log [H^+]$

$$\log [H^+] = -3$$
$$[H^+] = 1 \times 10^{-3} \text{ mol L}^{-1}$$

When $pH = 4, \quad 4 = -\log [H^+]$

$$\log [H^+] = -4$$
$$[H^+] = 1 \times 10^{-4} \text{ mol L}^{-1}$$

Thus, with increase in pH of the solution from 3 to 4, the concentration of H^+ ions decreases from 1×10^{-3} mol L^{-1} to 1×10^{-4} mol L^{-1}.

$$\frac{[H^+]_{pH=4}}{[H^+]_{pH=3}} = \frac{1 \times 10^{-4} \text{ mol L}^{-1}}{1 \times 10^{-3} \text{ mol L}^{-1}} = \frac{1}{10}$$

Thus, when the pH of a solution increases by 1 unit, the concentration of H^+ ions becomes one-tenth.

Example 25.8. *Five solutions A, B, C, D and E when tested with universal indicator showed pH as 4, 1, 11, 7 and 9, respectively. Which solution is:*

(i) neutral? *(ii) strongly alkaline?*
(iii) strongly acidic? *(iv) weakly acidic?*
(v) weakly alkaline?
(vi) arrange the solutions in increasing order of hydrogen-ion concentration.

Solution. (*i*) The solution **D** having pH = 7 is **neutral.**

(*ii*) The solution **C** having very high value of pH (pH = 11) is **strongly alkaline.**

(*iii*) The solution **B** having very low value of pH (pH = 1) is strongly acidic.

(*iv*) The solution **A** having a moderate value of pH which is less than 7 (pH = 4) would be **weakly acidic.**

(*v*) The solution **E** having a moderate value of pH which is greater than 7 (pH = 9) is **weakly alkaline.**

(*vi*) A solution having smaller value of pH has greater hydrogen-ion concentration.

Hence, the solutions in increasing order of hydrogen ion concentration are:

$$\mathbf{C < E < D < A < B.}$$

Example 25.9. *Fresh milk has a pH of 6. How do you think the pH will change as it turns into curd? Explain your answer.*

Solution. As the milk changes into curd it becomes more sour in taste indicating that the hydrogen-ion concentration is increasing. This is due to the formation of lactic acid. Therefore, the pH of milk decreases as it changes into curd.

Example 25.10. *Calculate the pH value of*

(*a*) *0.003 M HCl* (*b*) *0.005 M NaOH*

(*c*) *0.001 M* $Ba(OH)_2$.

Assuming that HCl, NaOH and $Ba(OH)_2$ *are completely ionized.*

Solution. (*a*) $HCl(aq) + H_2O(l) \longrightarrow H_3O^+(aq) + Cl^-(aq)$

Since HCl is completely ionized,

$\therefore \quad [H_3O^+] = [HCl] = 0.003\text{ M} = 3 \times 10^{-3}\text{ mol L}^{-1}$

Now $\quad pH = -\log [H_3O^+] = -\log [3 \times 10^{-3}]$

$= 3 - \log 3 = 3 - 0.477 = \mathbf{2.52}$

(*b*) $\quad NaOH(aq) \longrightarrow Na^+(aq) + OH^-(aq)$

Because NaOH is completely ionized,

$\therefore \quad [OH^-] = [NaOH]$

or $\quad [OH^-] = 0.005\text{ M} = 5 \times 10^{-3}\text{ mol L}^{-1}$

Hence $[H_3O^+] = \dfrac{K_w}{[OH^-]} = \dfrac{1 \times 10^{-14}}{5 \times 10^{-3}}$

$= 2 \times 10^{-12}\text{ mol L}^{-1}$

$\therefore \quad pH = -\log [H_3O^+]$

$= -\log 2 \times 10^{-12} = 12 - \log 2$

$= 12 - 0.301 = \mathbf{11.7}$

(*c*) $\quad Ba(OH)_2(aq) \longrightarrow Ba^{2+}(aq) + 2OH^-(aq)$

Since $Ba(OH)_2$ is completely ionized and one mole of $Ba(OH)_2$ furnishes 2 moles of $OH^-(aq)$ ions,

$\therefore \quad [OH^-] = 2[Ba(OH)_2] = 2 \times 0.001\text{ M}$

$= 2 \times 10^{-3}\text{ mol L}^{-1}$

$[H_3O^+] = \dfrac{K_w}{[OH^-]} = \dfrac{1 \times 10^{-14}}{2 \times 10^{-3}}$

$= 5 \times 10^{-12}\text{ mol L}^{-1}$

$\therefore \quad pH = -\log (5 \times 10^{-12})$

$= -(\overline{12}.699) = -(-11.301) = \mathbf{11.301.}$

Example 25.11. *How many grams of NaOH must be dissolved in one litre of a solution to have a pH = 12?*

Solution. $\quad pH = -\log [H_3O^+] = 12$

$\log [H_3O^+] = -12$

$\therefore \quad [H_3O^+] = 1 \times 10^{-12}\text{ mol L}^{-1}$

We know that $K_w = [H_3O^+][OH^-]$

$\therefore \quad [OH^-] = \dfrac{K_w}{[H_3O^+]}$

$= \dfrac{1 \times 10^{-14}}{1 \times 10^{-12}} = 1 \times 10^{-2}\text{ mol L}^{-1}$

Since NaOH is a strong base, it is completely ionized.

$\therefore \quad [NaOH] = [OH^-] = 1 \times 10^{-2}\text{ mol L}^{-1}$

Molecular mass of NaOH = 40

$\therefore$ Amount of NaOH in grams per litre

$= 1 \times 10^{-2} \times 40 = \mathbf{0.4\ g.}$

Example 25.12. *Dissociation constant of a weak acid HA is* 1×10^{-7}. *Find the pH of its 0.1 M solution.*

Solution. Dissociation of this acid in water may be represented as

$$HA(aq) + H_2O(l) \rightleftharpoons H_3O^+(aq) + A^-(aq)$$

$$K_{HA} = \frac{[H_3O^+][A^-]}{[HA]}$$

If x moles of HA dissociate at equilibrium then the equilibrium concentrations of the various species are

$$[HA] = 0.1 - x, \quad [H_3O^+] = x, \quad [A^-] = x$$

$$K_a = \frac{x \times x}{0.1 - x} = \frac{x^2}{0.1 - x}$$

$$1 \times 10^{-7} = \frac{x^2}{(0.1 - x)}$$

Since x is very small hence, $0.1 - x$ may be taken as 0.1 in denominator

$$1 \times 10^{-7} = \frac{x^2}{0.1}$$

$$x^2 = 1 \times 10^{-7} \times 0.1 = 1 \times 10^{-8}$$

$$x = \sqrt{1 \times 10^{-8}} = 1 \times 10^{-4}\text{ mol L}^{-1}$$

Hence,

$$[H_3O^+] = 1 \times 10^{-4} \text{ mol L}^{-1}$$

$$pH = -\log [H_3O^+] = -\log (1 \times 10^{-4})$$

$$= -(-4) = \mathbf{4.}$$

Example 25.13. *Calculate the $[H_3O^+]$ of a solution having a pH of 10.6.*

Solution. $pH = -\log [H_3O^+]$

or $\log [H_3O^+] = -pH = -10.6$

Since mantissas are always positive in logarithmic tables, add -1 to the characteristic and $+1$ to the mantissa so that the mantissa becomes positive.

$\therefore$ $\log [H_3O^+] = (-10 - 1) + (-0.6 + 1)$

$$= (-11) + (+0.4) = \overline{11}.4$$

$\therefore$ $[H_3O^+] = \text{Antilog } \overline{11}.4 = \mathbf{2.5 \times 10^{-11} \text{ mol L}^{-1}}.$

SUMMARY

- Strength of acids or bases is measured in terms of their respective dissociation constants K_a (or pK_a) and K_b (or pK_b).

$$pK_a = -\log K_a$$
$$pK_b = -\log K_b$$

- Larger value of K_a or lower value of pK_a corresponds to greater strength of acid. Similarly larger value of K_b or lower value of pK_b corresponds to a more stronger base.
- Ionic product of water, $K_w = 1 \times 10^{-14}$ mol² dm⁻⁶
- For conjugate acid-base pairs

$$K_a \times K_b = K_w$$
$$pK_a + pK_b = pK_w = 14$$

- For a monobasic acid HA, if *c* is the molar concentration of HA in solution, α is the degree of dissociation, then

$$[H_3O^+] = C\alpha = \sqrt{K_a \times c}$$

- Similarly, for a monoacid base,

$$[OH^-] = C\alpha = \sqrt{K_b \times c}$$

- **pH of Solution.** Potency of H^+ in solution. It is the negative logarithm of the H_3O^+ ion concentration in solution.
- **pOH of Solution.** It is the negative logarithm of the OH^- ion concentration of the solution.
- For any solution at 25°C

$$pK_w = pH + pOH = 14$$
$$pH = -\log [H_3O^+];$$
$$pOH = -\log [OH^-];\ pK_w = -\log K_w$$

- **K_w**, the ionic product $[H^+][OH^-] = 1 \times 10^{-14}$ at 25° C. It increases with an increase in temperature and decreases with a decrease in temperature.

EVALUATE YOURSELF

I. Objective Type Questions

Select the most appropriate choice from the options given as (a), (b), (c) and (d) after each question:

1. A certain sample of beer has a pH of 10. The H_3O^+ ion concentration of the beer is
   (*a*) 10^{10} M (*b*) 10^{-2} M
   (*c*) 10^{-4} M (*d*) 10^{-10} M.
2. At 80°C, pure distilled water has a H_3O^+ concentration equal to 1×10^{-6} mol L⁻¹. The value of K_w at this temperature will be
   (*a*) 1×10^{-8} (*b*) 1×10^{-12}
   (*c*) 1×10^{-14} (*d*) 10^{-6}.
3. $Ba(OH)_2$ is a strong base. The pH of its 0.005 M solution at 298 K would be
   (*a*) 11.31 (*b*) 11.7
   (*c*) 12 (*d*) None of these.
4. $[OH^-]$ in a solution is 1 mol L⁻¹. The pH of the solution is
   (*a*) 1 (*b*) 0
   (*c*) 14 (*d*) 10^{-14}.
5. The value of K_w in an acidic aqueous solution at 298 K is
   (*a*) $> 10^{-14}$ (*b*) $< 10^{-14}$
   (*c*) 10^{14} (*d*) 10^{-14}.

II. Fill in the Blanks

6. Complete the following sentences by supplying appropriate words:
   (*i*) pH of 0.1 M HCl is ......
   (*ii*) pH of 0.1 M NaOH is ......
   (*iii*) The ionization constant (K_a) and degree of dissociation (α) of a weak electrolyte are related as ......
   (*iv*) K_a for a weak acid HA is 1×10^{-5}, K_b for A^- would be ......
   (*v*) A solution is said to be basic if $[H_3O^+]$ is ...... than $\sqrt{K_w}$.

III. Discussion Questions

7. What is ionization constant, K_a of an acid? What is the relationship between the value of K_a and acid strength?
8. The ionization constant of HF is 3.2×10^{-4}. Calculate the degree of ionization of HF and hydronium ion concentration in a 0.02 M solution of HF.
9. The dissociation constant of acetic acid is 1.8×10^{-5}. Calculate the $[H_3O^+]$ in its 0.05 M solution.
10. Calculate degree of ionization of a weak acid HA in its 0.1 M solution. Dissociation constant of HA is 9×10^{-5}.
11. What is the ionic product of water? What is its value at 298 K? How does its value change with an increase in temperature?
12. What do you understand by the pH of a solution? What is the pH of a neutral solution?

13. What is the concentration of H_3O^+ ions in pure water at 298 K?
14. An aqueous solution has a pH = 4. What can you say about its acid-base character?
15. What effect does the concentration of $H^+(aq)$ have on the acidic nature of a solution?
16. What is the normal range of pH? What is the pH of a 0.1 M solution of a base whose $K_b = 1 \times 10^{-5}$?
17. How can a farmer raise the pH of soil in his fields?
18. You have two solutions, A and B. The pH of solution A is 6 and the pH of solution B is 8. Which solution has more hydrogen ion concentration? Which of this is acidic and which one is basic?
19. Do basic solutions also have $H^+(aq)$ ions? If yes, then why are these basic?
20. What is pH scale? How does it tell us whether a solution is acidic or basic?
21. What is meant by pH of a solution? What is pH of neutral solution? What happens to pH and pOH of water on increasing the temperature?
22. The concentration of hydrogen ion in a sample of soft drink is 3.8×10^{-3} M. What is its pH?
23. Calculate the dissociation constant of an acid, 0.1 M solution of which has pH equal to 5.
24. The value of K_w at a certain temperature is 9.55×10^{-14}. Calculate the pH of water at this temperature.
25. Find the pH value of 0.1 M acetic acid, its degree of ionization is 0.013 under the given conditions.
26. What will be the pH of a 0.16 M solution of acetic acid? $K_{CH_3COOH} = 1.8 \times 10^{-5}$.
27. The pH of 0.004 M hydrazine solution is 9.7. Calculate its ionization constant K_b and pK_b.

Answers

I. Objective Type Questions

1. (*d*) **2.** (*b*) **3.** (*c*)
4. (*c*) **5.** (*d*).

II. Fill in the Blanks

6. (*i*) 1 (*ii*) 13 (*iii*) $\alpha = \sqrt{\frac{K}{C}}$
(*iv*) 1×10^{-9} (*v*) less.

III. Discussion Questions

8. $\alpha = 0.12$, $[H_3O^+] = 2.4 \times 10^{-3}$ M
9. $[H_3O^+] = 9.48 \times 10^{-4}$ M
10. $\alpha = 3 \times 10^{-2}$
16. pH = 11
22. pH = 2.42
23. $K_a = 1 \times 10^{-9}$
24. pH = 6.51
25. pH = 2.89
26. pH = 2.77
27. $K_b = 8.97 \times 10^{-7}$, $pK_b = 6.05$.

SECTION-4

UNIT-26

BUFFER SOLUTIONS

Specific Objective

After studying this unit, the student should be able to:

- describe a buffer solution in terms of its composition and behaviour.

26.1 BUFFER SOLUTIONS

Generally, the pH of a solution changes on addition of small amounts of acids or bases. But if the solution contains a weak acid and its conjugate base, or a weak base and its conjugate acid, such a solution can resist changes in pH and is called a buffer solution.

A buffer solution is a solution which can resist changes in pH on addition of small amount of acid or base. The ability of a buffer solution to resist changes in pH on addition of acid or base is called buffer action.

Depending upon the pH values, buffer solutions are divided into two classes. If the pH of the buffer solution is *less than 7*, it is called an **acid buffer** and if it is *more than 7*, it is called a **basic buffer.**

WEAK ACID BUFFERS

Buffer solutions with pH range from 4 to 7 can be prepared from weak acids and their salts. For example, acetic acid and sodium acetate are commonly used for this purpose.

Now let us understand how an acid *buffer* such as a solution of acetic acid (weakly dissociated) and sodium acetate (largely dissociated) resists changes in pH. This solution contains a large amount of sodium and acetate ions and a large amount of undissociated acetic acid molecules along with a small amount of hydronium ions.

$$CH_3COOH(aq) + H_2O(l) \rightleftharpoons CH_3COO^-(aq) + H_3O^+(aq)$$

$$CH_3COONa(aq) \longrightarrow CH_3COO^-(aq) + Na^+(aq)$$

Suppose, a few drops of HCl(*aq*) are added to this buffer solution. This would provide H_3O^+ ions. These H_3O^+ ions would combine with CH_3COO^- ions to form weakly ionized acetic acid molecules as shown below:

$$\underset{\text{From buffer}}{CH_3COO^-(aq)} + \underset{\text{From HCl}}{H_3O^+(aq)} \rightleftharpoons CH_3COOH(aq) + H_2O(l)$$

Since the additional H_3O^+ ions are neutralised by CH_3COO^- ions in the solution, there will be no change in its pH value. On the other hand, if a few drops of NaOH are added to the buffer solution, it would provide OH^- ions. These OH^- ions will combine with H_3O^+ ions present in the buffer solution to form unionized water molecules. This would result in the greater ionization of acetic acid in order to restore the concentration of H_3O^+ ions to its original value.

$$\underset{\text{From buffer}}{H_3O^+(aq)} + \underset{\text{From NaOH}}{OH^-(aq)} \rightleftharpoons 2H_2O(l)$$

Therefore, the pH of the solution remains unchanged.

WEAK BASE BUFFERS

Buffer solutions with pH values between 7 and 10 can be prepared from weak bases and their salts with strong acids. For example, a solution containing NH_3 and NH_4Cl acts as a buffer.

Now let us understand how a solution containing ammonia and ammonium chloride acts as a buffer and resists changes in pH.

In case of a *basic buffer* solution containing equimolar quantities of ammonia and ammonium chloride, there is a large concentration of ammonium ions, chloride ions, ammonia and small amount of OH^- ions.

$$NH_3(aq) + H_2O\ (l) \rightleftharpoons NH_4^+(aq) + OH^-(aq)$$

$$NH_4Cl(aq) \longrightarrow NH_4^+(aq) + Cl^-(aq)$$

When a few drops of HCl(*aq*) are added, the additional H_3O^+ ions are neutralised by the OH^- ions present in the buffer.

$$\underset{\text{From buffer}}{OH^-(aq)} + \underset{\text{From acid}}{H_3O^+(aq)} \rightleftharpoons 2H_2O(l)$$

As some of the OH^- ions from NH_4OH combine with H_3O^+ ions from the acid, it would result in the greater ionization of aqueous ammonia to restore the OH^- ion concentration.

On the other hand, when a few drops of NaOH solution are added to the buffer solution, it would give OH^- ions. The additional OH^- ions will combine with NH_4^+ ions to form aqueous ammonia.

$$\underset{\text{From buffer}}{NH_4^+(aq)} + \underset{\text{From NaOH}}{OH^-(aq)} \rightleftharpoons NH_3(aq) + H_2O\ (l)$$

If we observe the titration curves (Fig. 27.2) we find that in certain regions (flat portions of the titration curve) the pH of the solution is insensitive to small changes in the concentration of acid or base. These portions of the curve are known as **buffer regions.**

Some common buffer systems are given in Table 26.1

Table 26.1. Some Common Buffer Systems

Acid	*Conjugate base*	pK_a	*Approx. pH range*
H_3PO_4	$H_2PO_4^-$	2.12	1–3
HCOOH	$HCOO^-$	3.74	3–5
CH_3COOH	CH_3COO^-	4.75	4–6
NH_4^+	NH_3	9.25	8–10
HCO_3^-	CO_3^{2-}	10.25	9–11

ACTIVITY 26.1 **Demonstrating the buffer action of CH_3COOH/ CH_3COONa**

1. Take 20 cm^3 of 0.1 M CH_3COOH and 20 cm^3 of 0.1 M CH_3COONa solution in a beaker. Stir well with a glass rod.
2. Find out the pH of the above solution with a universal indicator paper or pH meter.

   What is the pH of this mixture?

   What is the general name of the mixture prepared and why is it given this name?
3. Take 5 cm^3 of the mixture prepared in step 1 into a clean test tube and add 1 cm^3 of 0.1 M HCl to it.

   What is the pH of the resultant solution?
4. In another test tube take 5 cm^3 of the mixture prepared in step 1 and add 1 cm^3 of 0.1 M NaOH solution.

   What is the pH of the resultant solution?

   Is the solution prepared in step 1 behaving as a buffer solution? If yes why?
5. Repeat steps 3 and 4 with distilled water in place of the mixture solution.

   Does water behave as a buffer?

   Repeat the above Activity with NH_4Cl/NH_3 buffer using 0.1 M solutions.

Proteins present in the blood and other body fluids maintain the pH of the blood and other body fluids by buffer action. Protein molecules have a zwitter ion structure. They contain $-NH_3^+$ and $-COO^-$ groups at the ends. The reserve acidity of protein is due to $-NH_3^+$. It neutralizes small amounts of excessive bases. Similarly, the reserve basicity of protein is due to $-COO^-$. It absorbs small amounts of excessive acids and thus maintains the pH nearly constant.

STRONG ACID BUFFERS

A strong acid such as trioxonitrate(V) acid or hydrochloric acid can act as a buffer with a low pH. Strong acids are completely ionized in aqueous solution and there the concentration of hydrogen ions is high. The addition of a small amount of acid or base to the acid will have negligible effect on the pH of the solution.

For example, if 1.0 cm^3 of 0.1 M hydrochloric acid is added to 100 cm^3 of 0.01 M trioxonitrate(V) acid, the pH of the solution changes from 2.00 to 1.96, which is a negligible change. On the other hand if 1 cm^3 of 0.1 M hydrochloric acid is added to 100 cm^3 of pure water, the pH changes from 7.00 to 4.00. Thus, pure water does not act as a buffer.

STRONG BASE BUFFERS

A strong base such as NaOH, KOH etc., can act as a buffer with a high pH. The addition of a small amount of acid or base has negligible effect on the pH of a solution of such bases. For example, when 1 cm^3 of 0.1 M HCl is added to 100 cm^3 of 0.01 M NaOH solution, the pH changes from 12.00 to 11.96, which is negligible change.

26.2 pH OF A BUFFER SOLUTION

pH of a buffer solution is calculated by applying **Henderson-Hasselbalch equation.** In order to derive this equation let us consider an acid buffer consisting of a weak acid HA and its salt NaA

$$HA + H_2O \rightleftharpoons H_3O^+ + A^- \quad \text{...Weakly dissociated}$$

$$NaA \longrightarrow Na^+ + A^- \quad \text{...Completely dissociated}$$

$$K_a = \frac{[H_3O^+][A^-]}{[HA]}$$

Since HA is weakly dissociated, the conc. of HA in solution can be assumed to be equal to the initial conc. of the acid. Since NaA is completely dissociated, conc. of A^- can be taken to be equal to the conc. of NaA (salt)

$$K_a = \frac{[H_3O^+][\text{Salt}]}{[\text{Acid}]}$$

$$[H_3O^+] = \frac{K_a\,[\text{Acid}]}{[\text{Salt}]}$$

$$-\log[H_3O^+] = -\log K_a - \log\frac{[\text{Acid}]}{[\text{Salt}]}$$

$$\mathbf{pH = pK_a + \log\frac{[Salt]}{[Acid]}}$$

Similarly, for basic buffers,

$$\mathbf{pOH = pK_b + \log\frac{[Salt]}{[Base]}}$$

It may be noted that the pH of a buffer solution does not change with dilution because the ratio of the concentration of the salt to the concentration of the acid or base does not change with dilution.

26.3 BUFFER CAPACITY

It is observed that a buffer solution resists the change in pH to a certain limit only. The addition of excess of strong acid or strong base to a buffer solution leads to a considerable change in pH. The capacity of a buffer solution in maintaining a constant pH upon the addition of acid or base is known as buffer capacity. The buffer capacity of a buffer solution is defined as the ratio of the number of moles of an acid or a base added to the buffer solution and change in pH, *i.e.*,

$$\text{Buffer capacity} = \frac{\text{No. of moles of acid or base added}}{\Delta pH}$$

A buffer solution has maximum buffer capacity when the ratio $\frac{[\text{Salt}]}{[\text{Acid}]}$ or $\frac{[\text{Salt}]}{[\text{Base}]} = 1$.

In such a solution the pH = pK_a for acidic buffers and the pOH = pK_b for basic buffers. A solution can act as a buffer only if the ratio of the conc. of salt to acid or base is between 0.1 and 1.0. Thus, the ***pH of an acid buffer can have range from*** $pK_a - 1$ ***to*** $pK_a + 1$. Similarly, a basic buffer can act as buffer in the pOH range from $pK_b - 1$ to $pK_b + 1$.

For example, CH_3COOH/CH_3COONa buffer has a pH range 3.76 to 5.76 because pK_a of CH_3COOH is 4.76.

26.4 USES OF BUFFER SOLUTIONS IN EVERYDAY LIFE

Buffer solutions play an important role in many industrial processes such as *electroplating, food processing, manufacture of medicines, dyes and photographic materials, etc.*

Many biological fluids such as blood, urine, etc., have definite pH which is maintained by the buffer action of many substances. For example, blood pH is maintained between 7.35 and 7.45 by a buffer system whose principal acid-base pair is H_2CO_3/HCO_3^-. Besides protein buffers also regulate the pH of the blood.

The pH of tears is maintained at 7.4 by protein buffers.

The pH of beer is maintained between 4 and 5 by the buffer action of dissolved CO_2 and proteins.

SOLVED EXAMPLES

Example 26.1. *Calculate the pH of a solution which is 0.1 M in HA and 0.5 M in NaA. K_a for HA is 1.8×10^{-6}.*

Solution. $$HA + H_2O \rightleftharpoons H_3O^+ + A^-$$

$$NaA \longrightarrow Na^+ + A^-$$

Suppose x moles of HA dissociate per litre of solution

$$[H_3O^+] = x$$

$$[A^-] = x + 0.5 \approx 0.5 \text{ M}$$

$$[HA] = 0.1 - x \approx 0.1 \text{ M}$$

$$K_a = \frac{[H_3O^+][A^-]}{[HA]}$$

$$1.8 \times 10^{-6} = \frac{(x)(0.5)}{0.1}$$

$$x = 3.6 \times 10^{-7}$$

$$[H_3O^+] = 3.6 \times 10^{-7} \text{ M}$$

$$pH = -\log [H_3O^+] = -\log (3.6 \times 10^{-7})$$

$$= 7 - \log 3.6 = \mathbf{6.44.}$$

Alternatively, pH can be calculated by applying the Henderson's equation

$$pH = pK_a + \log \frac{[Salt]}{[Acid]}$$

$$= -\log (1.8 \times 10^{-6}) + \log \left(\frac{0.5}{0.1}\right)$$

$$= 5.744 + 0.6990 = \mathbf{6.44.}$$

Example 26.2. *What should be the ratio of the concentrations of formate ion and formic acid in a buffer solution so that its pH should be 4? Around what pH will this buffer have maximum buffer capacity? K_a for formic acid is 1.8×10^{-4}.*

Solution. We know,

$$pH = pK_a + \log \frac{[Salt]}{[Acid]}$$

$$4 = -\log (1.8 \times 10^{-4}) + \log \frac{[Formate]}{[Formic\ acid]}$$

$$4 = 3.74 + \log \frac{[Formate]}{[Formic\ acid]}$$

$$\log \frac{[Formate]}{[Formic\ acid]} = 4 - 3.74 = 0.26$$

$$\frac{[Formate]}{[Formic\ acid]} = \mathbf{1.8.}$$

The buffer capacity of this solution would be maximum near the pK_a of the acid.

∴ For maximum buffer capacity

$$pH = pK_a = -\log K_a$$

$$= -\log (1.8 \times 10^{-4}) = \mathbf{3.74.}$$

Example 26.3. *Calculate the pH of a solution obtained by mixing 100 cm³ of 0.10 M CH_3COOH and 100 cm³ of 0.05 M NaOH solution. K_a for CH_3COOH is 1.8×10^{-5}.*

Solution. $$NaOH + CH_3COOH \longrightarrow CH_3COONa + H_2O$$

100 cm³ of 0.10 M CH_3COOH contains

$$= 100 \times 0.1 = 10 \text{ m mol } CH_3COOH$$

100 cm³ of 0.05 M NaOH contains

$$= 100 \times 0.05 = 5 \text{ m mol of NaOH}$$

5 m mol NaOH react with 5 m mol CH_3COOH to form 5 m mol of CH_3COONa. 5 m mol of CH_3COOH remain unreacted

After mixing

$$[CH_3COOH] = \frac{5}{200} \text{ M} = 2.5 \times 10^{-2} \text{ M}$$

$$[CH_3COO^-] = \frac{5}{200} \text{ M} = 2.5 \times 10^{-2} \text{ M}$$

$$pH = pK_a + \log \frac{[CH_3COO^-]}{[CH_3COOH]}$$

$$pH = pK_a + \log \frac{(2.5 \times 10^{-2})}{(2.5 \times 10^{-2})}$$

$$pH = pK_a = -\log (1.8 \times 10^{-5}) = \mathbf{4.74.}$$

SUMMARY

- **Buffer Solution.** A solution which resists the change in its pH value on addition of small amount of acid or a base.
- A buffer solution is generally a mixture of a weak acid and its conjugate base or a weak base and its conjugate acid.
- A mixture of a weak acid and its conjugate base behaves as an **acid buffer.**
- A mixture of a weak base and its conjugate acid behaves as a **basic buffer.**
- The proteins present in body fluids of living bodies helps in maintaining pH through buffer action.
- pH of a buffer solution is given by the Henderson-Hasselbalch equation.

$$pH = pK_a + \log \frac{[Salt]}{[Acid]} \quad \text{... for acid buffers}$$

$$pOH = pK_b + \log \frac{[Salt]}{[Base]} \quad \text{...for basic buffers}$$

$pH = pK_a$ when [Salt] = [Acid]

$pOH = pK_b$ when [Salt] = [Base]

EVALUATE YOURSELF

I. Objective Type Questions

Select the most appropriate choice from the options given as (a,) (b), (c), and (d) after each question:

1. Which of the following pairs of substances would not form a buffer when mixed?
   (*a*) $CH_3COONa–CH_3COOH$
   (*b*) $NH_4Cl–NH_3$
   (*c*) $(NH_4)_2SO_4–H_2SO_4$
   (*d*) HCOOH–HCOONa.
2. Which buffer solution out of the following will have a pH > 7?
   (*a*) $CH_3COOH + CH_3COONa$
   (*b*) HCOOH + HCOOK
   (*c*) CH_3COONH_4
   (*d*) $NH_4OH + NH_4Cl$.
3. In a mixture of a weak acid and its salt, the ratio of the concentration of the acid to the salt is increased by 10 fold. The pH of the solution
   (*a*) decreases by one (*b*) decreases by one-tenth
   (*c*) increase by one (*d*) increases by ten fold.
4. An equimolar solution of CH_3COOH and CH_3COONa has a pH of 6. The value of K_a for acetic acid is
   (*a*) 10^6 (*b*) 1×10^{-6}
   (*c*) 2×10^{-6} (*d*) cannot be predicted.
5. A weak acid HA has $K_a = 10^{-6}$. What would be the molar ratio of this acid and its salt with strong base so that the pH of the buffer solution is 5?
   (*a*) 1 (*b*) 2
   (*c*) 10 (*d*) 1/10.
6. The pK_a of HCN is 9.30. The pH of a solution prepared by mixing 2.5 moles of KCN and 2.5 moles of HCN in water and making up to a total volume to 500 ml is
   (*a*) 9.30 (*b*) 7.30
   (*c*) 10.30 (*d*) 8.30.

II. Discussion Questions

7. What is a buffer solution? Give any two examples of a buffer solution.
8. What is an acid buffer? Explain the buffer action of an acid buffer.
9. Explain how a mixture of NH_4Cl and aqueous ammonia act as a buffer.
10. How does a protein molecule act as a buffer? Give examples of two natural buffers.
11. What is buffer capacity?
12. What is the importance of buffer solutions?
13. A solution is 0.1 M in NH_4Cl and 0.2 M in NH_3. What is the pH of this solution? K_b for NH_3 is 1.8×10^{-5}.

Answers

I. Objective Type Questions

1. (*c*)	**2.** (*d*)	**3.** (*a*)
4. (*b*)	**5.** (*c*)	**6.** (*a*).

II. Discussion Questions

13. 9.66.

SECTION-4

UNIT-27

ACID-BASE INDICATORS

Specific Objectives

After studying this unit, the student should be able to:

- describe qualitatively how acid-base indicators work
- draw graphs for acid-base titrations.

27.1 ACID-BASE INDICATORS

Acid-base indicators are substances which change colour according to the hydrogen ion concentration (or pH) of the solution to which they are added. They are usually weak acids or weak bases. A typical indicator is a weak organic acid that has different colour from its conjugate base. Let us consider an indicator which is a weak acid (HIn). The equilibrium between the indicator and its conjugate base (In^-) may be represented as:

$$\underset{\substack{\text{Acid} \\ \text{(Colour A)}}}{HIn(aq)} + H_2O(l) \rightleftharpoons H_3O^+(aq) + \underset{\substack{\text{Conjugate base} \\ \text{(Colour B)}}}{In^-(aq)}$$

At low pH, the concentration of H_3O^+ is high so the equilibrium is pushed to the left and the equilibrium solution has colour A. At high pH the concentration of H_3O^+ is low, the equilibrium is pushed to the right and the equilibrium solution has colour B.

We can write equilibrium constant expressions for indicator equilibria. For a weak acid indicator at equilibrium we can write

$$K_{In} = \frac{[H_3O^+][In^-]}{[HIn]}$$

K_{In} is known as **indicator dissociation constant.** The colour of the indicator changes from colour A to colour B at a particular point, known as the **end point of the indicator**. At this point,

$$[HIn] = [In^-]$$

$$K_{In} = [H_3O^+]$$

$$-\log K_{In} = -\log [H_3O^+]$$

$$pK_{In} = pH$$

The pH of the solution at the end point is equal to the pK_{In}. When the pH of the solution is equal to the pK_{In} half the indicator is in the acid form and half is in the form of its conjugate base. The colour of the solution at the end point is intermediate between the colours A and B. This is known as the **midpoint colour.** For example, the indicator methyl orange is red in acidic medium and yellow in alkaline medium. Its midpoint colour is orange.

27.2 THE RANGE OF AN INDICATOR

The range of an indicator is the pH range over which it changes colour

At low pH, a weak acid indicator is mainly present as HIn (in acid form) and the colour of this form predominates. As the pH increases, the intensity of colour A starts decreasing while that of B starts increasing. The observable change from colour A to colour B takes place over a range of pH and this range of pH is known as the range of an indicator. An indicator is most effective if it undergoes a distinct change in colour over a narrow range of pH. *For most of the indicators, the range is within ±1 of the pK_{In} value.* Since indicators have quite different values of pK_{In}, the pH ranges of various indicators also vary widely as given in Table 27.1.

It may be noted that the end point of each indicator is in the centre of the pH range.

ACTIVITY 27.1 Investigation of Neutralization Reaction Using an Indicator

- Take a clean test tube and add 5.0 cm^3 of 0.1 M NaOH solution from a burette.
- Add 2–3 drops of methyl orange indicator. Note the colour of the solution in the test tube. Label this test tube **A**.
- In another test tube take 5.0 cm^3 of 0.1 M HCl solution and add 2–3 drops of methyl orange indicator. Note the colour of the solution in the test tube. Label this tube **B**.
- In another test tube take 2.5 cm^3 of 0.1 M NaOH and 2–3 drops of methyl orange indicator. Now add 2.5 cm^3 of 0.1 M HCl to this solution. Note colour of the resultant solution. Label this test tube as **C**.

Answer the following:

(*a*) What is the colour of the solution in test tube A?

(*b*) What is the colour of the solution in test tube B?

(*c*) What is the colour of the solution in test tube C?

(*d*) Name the reaction that has taken place in test tube C. Give an equation for the reaction.

Litmus contains several substances and changes colour over a broader range from about pH 5 to pH 8.

A universal indicator is a mixture of indicators which gives a gradual change from one colour to the other over a wide range of pH. The approximate pH of the solution can be determined from the colour obtained when a few drops of universal indicator are added to the solution. The colour of the solution is then compared with that given on the standard chart. The corresponding pH is the pH of the solution.

Certain dyes from flowers and vegetables can be extracted with ethanol and used as acid-base indicators. For example, turmeric gives brown colour in alkaline medium. Dye from red cabbage also serves as acid-base indicator. In order to extract dye from some flower, the flower is crushed and the dye extracted with ethanol or warm ethanol.

Table 27.1. Indicators and their pH Ranges

Indicator	*Colours*		*pK_{In}*	*pH Range*
	In acidic medium	*In basic medium*		
Thymol blue (first change)	Red	Yellow	1.5	1.2 – 2.8
Methyl orange	Red	Yellow	3.7	3.2 – 4.4
Bromocresol green	Yellow	Blue	4.7	3.8 – 5.4
Methyl red	Red	Yellow	5.1	4.8 – 6.0
Bromothymol blue	Yellow	Blue	7.0	6.0 – 7.6
Phenol red	Yellow	Red	7.9	6.3 – 8.4
Thymol blue (second change)	Yellow	Blue	8.9	8.0 – 9.6
Phenolphthalein	Colourless	Pink	9.4	8.2 – 10.0

Deep red Red Orange Yellow Green Blue Violet Deep violet
0 1 2 3 4 5 6 7 8 9 10 11 12 13 14
pH

Fig. 27.1. Colour of universal indicator at different pH values.

27.3 ACID-BASE TITRATIONS AND THEIR GRAPHS

During titration we determine the volume of one solution which just reacts with a fixed volume of an other solution.

An acid-base titration is a procedure used in quantitative chemical analysis to determine the concentration of either an acid or a base. Generally, an acid solution of known concentration is taken in a burette and is added to an alkaline solution of unknown concentration taken in a conical flask. *The solution that is titrated (the solution with fixed volume) is known as titrand, analyte or titrate while the solution with which titration is performed is called titrant.* The titrate is taken in a conical flask while the titrant is taken in a burette.

The **end point** or the **equivalence point** reaches when a stoichiometric amount of acid has been added to the alkali solution. At this point all the alkali has been neutralised and the solution contains salt and water only.

From the end point we can find the volume of the base (V_b) having molarity M_b required to neutralize a certain volume of the acid (V_a) of molarity M_a. Now, number of moles in V cm^3 of a solution of molarity M is given by M × V/1000. For the titrations involving monobasic acid and monoacid base, one mole of base is required to completely neutralize one mole of the acid.

Therefore, $$\frac{M_aV_a}{1000} = \frac{M_bV_b}{1000} \quad \text{or} \quad \mathbf{M_aV_a = M_bV_b}$$

For titration involving polyacid bases and polybasic acids the above equation takes the form

$$n_aM_aV_a = n_bM_bV_b$$

where n_a is the basicity of the acid while n_b is the acidity of the base.

Thus, for the titration of tetraoxosulphate(VI) acid with sodium hydroxide, the equation applicable is

$$\mathbf{2M_aV_a = M_bV_b}$$

or $$\frac{M_aV_a}{M_bV_b} = \frac{1}{2}.$$

Similarly, for the titration of barium hydroxide (a diacid base) against hydrochloric acid (a monobasic acid) the equation becomes

$$\mathbf{M_aV_a = 2M_bV_b}$$

or $$\frac{M_aV_a}{M_bV_b} = \frac{2}{1}$$

The end point is indicated by the colour change of the indicator or a sudden rise or fall in pH. The pH of the reaction mixture changes during the course of titration.

A graph of the pH versus the volume (V) of the titrating solution (titrant) is known as titration curve.

With the help of titration curves we can choose the proper indicator for a particular titration.

Let us assume that a titration is carried out by adding alkali solution to an acid solution. If the variation of pH is plotted against the volume of alkali solution added, four different types of titration curves are obtained depending upon whether the acid and base are strong or weak (Fig. 27.2).

(A)

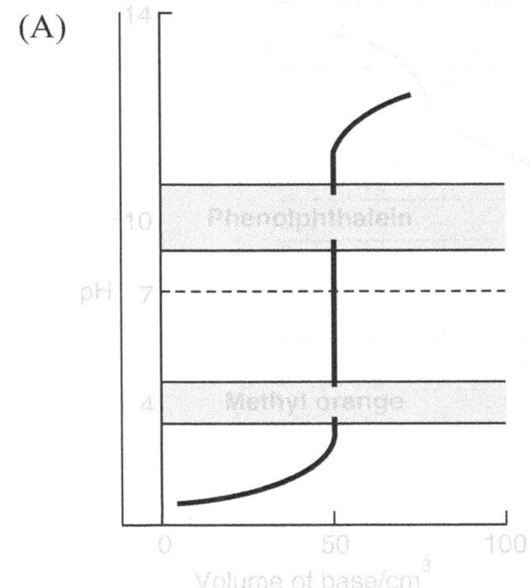

(B)

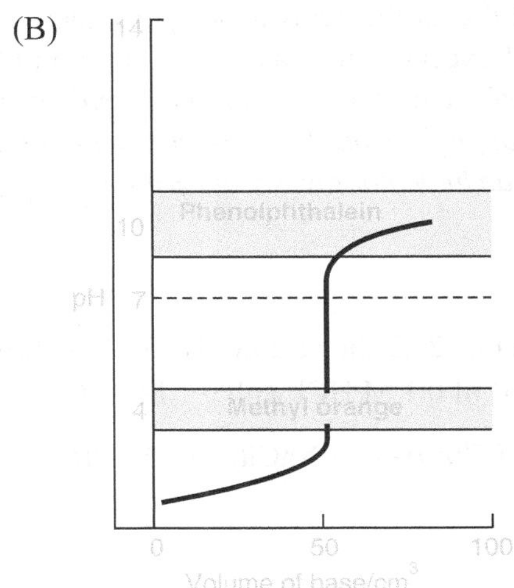

(C)

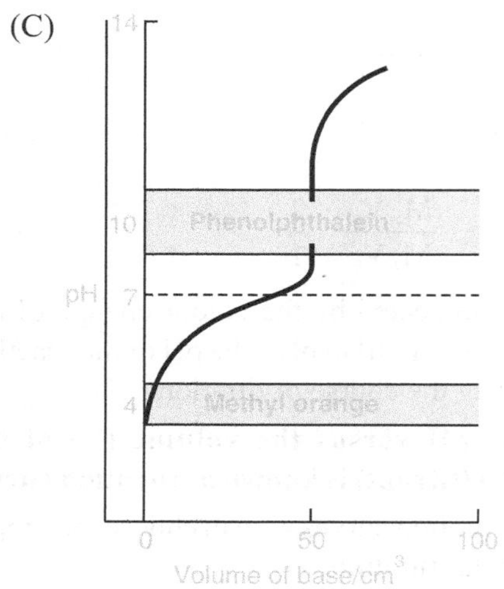

(D)

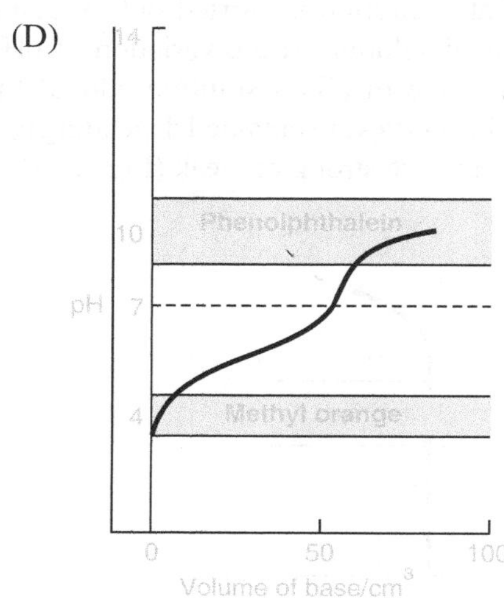

Fig. 27.2. Titration curves: (A) strong base with strong acid; (B) weak base with strong acid; (C) strong base with weak acid; (D) weak base with weak acid.

It may be noted that at the equivalence point there is a sharp increase in pH except in the case of the titration of a weak base with a weak acid. *A proper acid-base indicator is the one whose pH range falls on the vertical portion of the titration curve. Such an indicator gives a sharp colour change at the end point.*

(*i*) Titration of a Strong Acid against a Strong Alkali

The graph (A) in Fig. 27.2 shows how pH changes during the titration of 50 cm^3 of 0.1 M HCl with 0.1 M NaOH.

$$NaOH(aq) + HCl(aq) \longrightarrow NaCl(aq) + H_2O(l)$$

The pH of the 0.1 M solution of HCl in the beginning would be 1. As the alkali is added, the pH changes slowly in the beginning. However, at the equivalence point pH changes rapidly from about 3.5 to 10. It can be shown by simple calculations that the pH of the solution is 3.7 when 49.8 cm^3 of 0.1 M NaOH solution have been added. The pH suddenly changes to 10 after the addition of 50.1 cm^3 of the NaOH solution. Thus, any indicator having a pH range between 3.5 and 10 will identify the equivalence point. This means that anyone of ***phenolphthalein, methyl orange*** or ***bromothymol blue*** could be used as an indicator.

(*ii*) Titration of a Strong Acid against a Weak Alkali

The graph (B) in Fig. 27.2 shows how pH changes during titration of 50 cm^3 of 0.1 M HCl with 0.1 M NH_3.

$$HCl(aq) + NH_4OH(aq) \longrightarrow NH_4Cl(aq) + H_2O(l)$$

In this case the pH changes rapidly from 3.5 to 7.0 at the equivalence point. ***Methyl orange, methyl red and bromocresol green*** are suitable indicators for this type of titration. Phenolphthalein is unsuitable because its pH range lies outside the vertical portion of the curve.

(*iii*) Titration of a Weak Acid against a Strong Base

The graph (C) in Fig. 27.2 shows how pH changes during titration of 50 cm^3 of 0.1 M CH_3COOH with 0.1 M NaOH.

$$CH_3COOH(aq) + NaOH(aq) \longrightarrow CH_3COONa(aq) + H_2O(l)$$

The vertical portion of this titration curve lies between pH 7 and pH 10.6. ***Phenolphthalein*** is a suitable indicator for this titration. Methyl orange is not suitable for this titration because its pH range (3.2 – 4.4) is outside the sharp pH change (7 – 10) at the end point.

(*iv*) Titration of a Weak Acid against a Weak Base

The graph (D) in Fig. 27.2 represents the titration curve obtained for titration of 50 cm^3 of 0.1 M CH_3COOH with 0.1 M NH_3.

$$CH_3COOH(aq) + NH_4OH(aq) \longrightarrow CH_3COONH_4(aq) + H_2O(l)$$

For this type of titration there is no sharp increase in pH at the equivalence point. No indicator is suitable for this type of titration.

SUMMARY

- **Acid-base indicators** are substances which change colour with pH. They are generally weak acids of which either the undissociated molecule or the dissociated anion, or both are coloured.
- The equilibrium of a weak acid indicator may be represented as

$$\underset{\text{(Colour A)}}{HIn(aq)} + H_2O(l) \rightleftharpoons H_3O^+(aq) + \underset{\text{(Colour B)}}{In^-(aq)}$$

$$K_{In} = \frac{[H_3O^+][In^-]}{[HIn]}$$

- At the end point pH = pK_{In}.
- **The range of an indicator** is the pH range over which it changes colour.
  Range of an indicator = $pK_{In} \pm 1$.
- A graph of the pH versus the volume (V) of titrant is known as **titration curve.**
- A proper acid-base indicator is the one whose pH range falls on the vertical portion of the titration curve.
- For **strong acid-strong base** titrations phenolphthalein or methyl orange are suitable indicators.
  For **strong acid-weak alkali** titrations methyl orange or methyl red are suitable indicators.
  For **weak acid-strong base** titrations phenolphthalein is a suitable indicator.

EVALUATE YOURSELF

I. Objective Type Questions

Select the most appropriate choice from the options given as (a), (b), (c) and (d) after each question:

1. Which of the following indicators can be used for the titration of NaOH solution against CH_3COOH?
   (*a*) Methyl orange (*b*) Methyl red
   (*c*) Phenolphthalein (*d*) Bromocresol green.
2. What is the colour of methyl orange in alkaline medium?
   (*a*) Yellow (*b*) Orange
   (*c*) Red (*d*) Green.

II. Discussion Questions

3. What is an acid-base indicator? What is the midpoint of an indicator?
4. Explain why acid-base indicators have different colours in acidic and basic solutions.
5. What is the colour of the universal indicator in
   (*i*) strongly acidic solution
   (*ii*) strongly basic solution
   (*iii*) neutral solution?
6. Draw the titration curve for the titration of strong acid against strong alkali. What is the suitable indicator for this titration?
7. What is the colour of
   (*i*) phenolphthalein in acidic medium
   (*ii*) methyl red in basic medium?
8. Explain why:
   (*i*) methyl orange is not a suitable indicator for the titration of a weak acid against a strong alkali?
   (*ii*) phenolphthalein is not a suitable indicator for the titration of a strong acid against a weak alkali?
9. Explain the following terms:
   (*i*) titrate (*ii*) titrant (*iii*) end point

Answers

I. Objective Type Questions

1. (*c*) **2.** (*a*).

SECTION-4

UNIT-28

SOLUBILITY OF SUBSTANCES

Specific Objectives

After studying this unit, the student should be able to:

- explain the term solubility
- describe factors that affect solubility of substances
- determine the solubility and solubility product of sparingly soluble substances
- describe an experiment to determine the solubility product constant for $Ca(OH)_2$
- describe the precipitation of sparingly soluble substances.

There are many substances which dissolve in water quite appreciably while some substances either dissolve to a smaller extent or do not dissolve at all. Similarly, when we mix the aqueous solutions of $AgNO_3$ and KCl we get the formation of a white precipitate of AgCl. But we do not get any precipitation on mixing aqueous solutions of $AgNO_3$ and K_2SO_4.

How can we predict the extent of dissolution of a substance or formation of a precipitate on mixing aqueous solutions of the two substances? The answers to these basic questions comes from **solubility** and the concept of **solubility product.** The main focus in this unit is to study the *solubility of substances, concept of solubility product* and its applications to predict the precipitation of ions from their aqueous solutions.

28.1 SOLUBILITY

In general, solubility of a substance is its maximum amount that can dissolve in a specified amount of solvent under a given set of conditions. Solubility depends on the nature of the solute and solvent as well as the conditions of temperature and pressure. Let us study the effect of these factors on solutions of solids in liquids.

SOLUBILITY OF SOLIDS IN LIQUIDS

When a solid solute is added to a solvent, dissolution of solute occurs. The solute particles go into solution and acquire random movement. The concentration of the solution gradually increases as more and more solute dissolves. The dissolved solute particles also collide with solid solute and some of them separate out of solution and get deposited there. This process is called **crystallisation**. Soon a stage is reached when the two opposite processes; *dissolution* and *crystallisation* occur at the same rate and a state of dynamic equilibrium is established.

$$\text{Solid solute} + \text{Solvent} \rightleftharpoons \text{Solution}$$

At this stage, the concentration of the solution becomes constant at the given conditions because the number of solute particles going into solution will be equal to number of solute particles separating out of solution. *Such a solution at which no more solute can be dissolved at the given conditions of temperature and pressure* is called a **saturated solution.**

- The solution which can dissolve more of solute at a given temperature is called an **unsaturated solution.**
- *The amount of solute that can be dissolved in 100 g of a solvent at a given temperature* is called its **solubility.**
- Another common way of expressing solubility of a substance at a given temperature is *the maximum quantity of solute in moles (or gram) that can dissolve in a solvent to form 1 dm³ of solution.*
- The concentration of a solute, its ions or molecules in a saturated solution is constant at a given temperature.
- The units commonly used for expressing solubility of a substance are **mol dm^{-3}** or **g dm^{-3}.**

SOLUBILITY RULES

These are empirical generalisations about the solubility of ionic compounds in water. These generalisations have been made from the experimental observations regarding the solubility of ionic compounds.

These rules are very useful in predicting the formation of precipitate particularly in qualitative analysis. Of course the more logical explanation about the precipitation comes from the concept of solubility product. The solubility rules about the ionic compounds have been summed up in Table 28.1. They refer to the solubility of ionic compounds in water at room temperature.

Table 28.1. Solubility Rules for Ionic Compounds

- All trioxonitrate(V), NO_3^- salts are soluble.
- All salts of Na^+, K^+ and NH_4^+ ions are soluble.
- All tetraoxosulphate(VI), SO_4^{2-} salts are soluble except those of Ca^{2+}, Pb^{2+}, Ba^{2+} and Hg^+ ions.
- All chlorides (Cl^-), bromides (Br^-), iodides (I^-) are soluble except those of Pb^{2+} and Ag^+ ions.
- All trioxosulphates(IV), SO_3^{2-} salts are insoluble except those of Na^+, K^+ and NH_4^+ ions.
- All sulphides (S^{2-}) are insoluble except those of Na^+, K^+, NH_4^+ ions. CaS is sparingly soluble. Al_2S_3 hydrolyses to form Al $(OH)_3$.
- All trioxocarbonates (IV), CO_3^{2-} are insoluble except those of NH_4^+, Na^+ and K^+ ions.
- All oxides, O^{2-} and hydroxides, (OH^-) are insoluble except those of Na^+, K^+, Ba^{2+}. $Ca(OH)_2$ is, however, sparingly soluble.

Utility of Solubility Rules

The usefulness of the rule is clear from the following example. If a solution of an unknown salt forms a precipitate with an aqueous solution of hydrochloric acid (containing Cl^- ions), then the precipitate could be any of the insoluble chlorides *i.e.*, Hg_2Cl_2 or $PbCl_2$ or AgCl. This implies that the cation of the unknown salt must be any of Hg_2^{2+} or Pb^{2+} or Ag^+ ions.

ACTIVITY 28.1	**To measure the solubility of potassium tetraoxosulphate(VI) at room temperature**

1. Note the room temperature by suspending a thermometer in water.
2. Weigh about 10.0 g of K_2SO_4 and put it in a 250 cm³ beaker. Add 50 cm³ of water and stir well for about 10 minutes. Some salt will be left undissolved.
3. Heat the contents to dissolve the rest of the salt.
4. Stop heating. Allow to cool with occasional stirring so that crystals are formed.
5. Take a clean dry evaporating dish and find its mass (*a* g).
6. When the saturated solution is cooled to room temperature, filter it into another clean beaker.
7. Pipette out 10 cm³ of filtrate into the weighed evaporating dish.
8. Find the mass of the dish and solution, (*b* g)
9. Now evaporate the water from the saturated solution on a water bath.
10. Find the constant mass of the evaporating dish and the salt left after evaporation (*c* g)

Sample Calculations

Mass of water = $(b-c)$ g

Mass of K_2SO_4 = $(c-a)$ g

Solubility of K_2SO_4 per 100 g of water

$$= \frac{(c-a)\times 100}{(b-c)} \text{ g}$$

(*i*) Calculate the solubility in g dm^{-3} ....

(*ii*) Calculate the solubility in mol dm^{-3} ....

28.2 FACTORS THAT AFFECT SOLUBILITY OF SUBSTANCES

Solubility of solid solutes in liquid depends on the following factors:

(*i*) **Nature of solute.** Dissolution of solid solutes in liquids can be summed up in a phrase **"like dissolves like".**

This means that, polar solutes dissolve in polar solvents and non-polar solutes dissolve in non-polar solvents. For example, ionic substances such as NaCl, KCl, KNO_3, etc., have larger solubilities in polar solvents like water but they have poor solubilities in non-polar solvents like CCl_4, CS_2, etc. On the other hand, non-polar solids such as I_2, S_8, etc., are more soluble in non-polar solvents like CCl_4, CS_2 but they are almost insoluble in polar solvents like water.

(*ii*) **Temperature.** Saturated solution represents equilibrium between undissolved solute and dissolved solute

$$\text{Undissolved solute} + \text{Solvent} \rightleftharpoons \text{Solution}; \Delta H_{Sol} = \pm x.$$

If the value of $\Delta H_{Sol} < 0$; *i.e.*, the solution process is **exothermic,** then by Le-Chatelier's Principle, the solubility of such a solute will **decrease** with a *rise in temperature.* On the other hand, if $\Delta H_{Sol} > 0$; *i.e.*, the solution process is **endothermic**, then, solubility of such a solute will *increase* with a *rise in temperature.*

Measurement of Solubility at Different Temperatures

The solubility can be measured at different temperatures by repeating Activity 28.1 using saturated solutions.

An alternative method of finding the solubility of a substance is to prepare an unsaturated solution of known concentration at higher temperature. The solution is then cooled and the temperature at which crystals begin to form in the solution is recorded. This gives the solubility of the substance at that temperature. This method is used in Activity 28.2.

SOLUBILITY CURVES

The graphs obtained by plotting the solubility of a solid *vs* temperature are referred to as **solubility curves.** The various ionic substances can be divided into three categories on the basis of effect of temperature on solubility.

(*i*) ***Solids whose solubility increase continuously and gradually with a rise in temperature.*** This happens in case of solids whose ΔH_{Sol} is > 0. Some examples are NaCl, KCl, KNO_3, $NaNO_3$ and hydrated salts.

(*ii*) ***Solids whose solubility decrease gradually with a rise in temperature.*** This happens in case of solids whose $\Delta H_{Sol} < 0$. Li_2SO_4 and anhydrous salts like $CuSO_4$, Na_2SO_4 are common examples of this category.

(*iii*) ***Solids whose solubility does not increase or decrease regularly or continuously.*** This happens if a substance undergoes a change from one polymeric form to another at a particular temperature. For example, the dissolution of $Na_2SO_4.10H_2O$ is endothermic but the dissolution of Na_2SO_4 is exothermic. The conversion of $Na_2SO_4.10H_2O$ to Na_2SO_4 occurs at 34°C. Thus, the solubility of $Na_2SO_4.10H_2O$ increases with a rise in temperature till 34°C and thereafter it starts decreasing. The temperature at which this reversal of solubility occurs is often called **transition temperature.** The solubility curves of some substances are shown in Fig. 28.1.

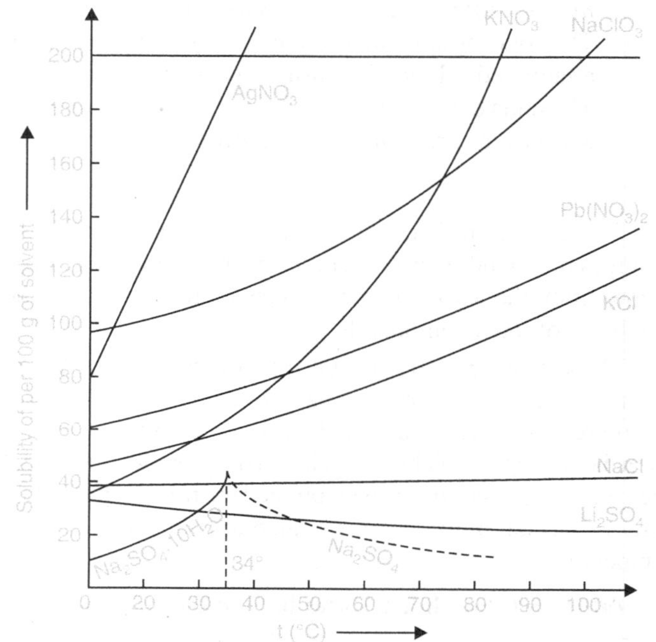

Fig. 28.1. Solubility curves of some substances.

INFORMATIONS CONVEYED BY SOLUBILITY CURVE

In order to understand the information that we can obtain from the solubility curve, consider the solubility for the dissolution of A as shown in Fig. 28.2.

The curve shows

- Solubility of A increases with increase in temperature.
- Any point on the curve (say *y*) represents a saturated solution.
- The corresponding point *a* represents the solubility at temperature t_1.
- Increasing temperature from $t_1 \rightarrow t_2$ till point *X* means that the solution becomes unsaturated.

- Decreasing temperature from $t_1 \rightarrow t_3$ results in a decrease in solubility to b mol dm^{-3}, and an amount of solute equal to $(a - b)$ moles separate out as crystals.
- Since the solubility increases with a rise in temperature, it means $\Delta H_{sol} > 0$ *i.e.*, the solution process is *endothermic*.

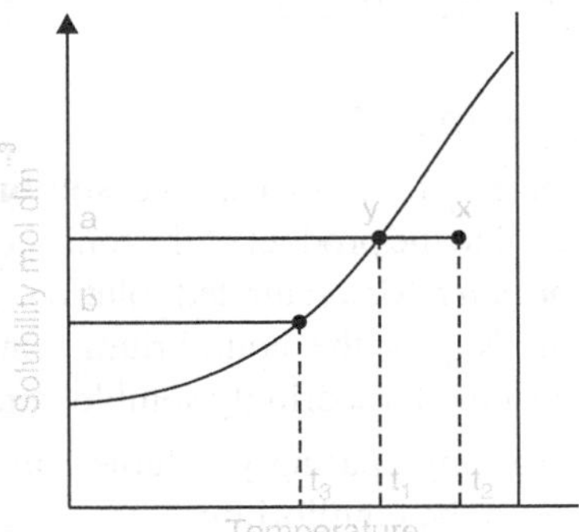

Fig. 28.2. Solubility curve of A.

ACTIVITY 28.2	**To find the solubility of copper(II) tetraoxosulphate(VI) pentahydrate at different temperatures and draw the solubility curve**

1. Weigh accurately 10 g of hydrated salt copper(II) tetraoxosulphate(VI)pentahydrate, $CuSO_4.5H_2O$ into a 100 cm^3 boiling tube.
2. From the burette, add 15.0 cm^3 of distilled water.
3. Heat and stir at the same time until all the solid dissolves (Fig. 28.3).

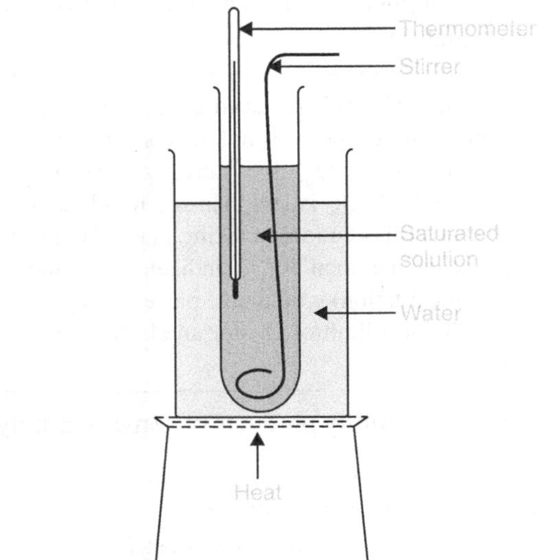

Fig. 28.3. Experimental determination of solubility at different temperatures.

Stop heating and remove the boiling tube from the hot water.

4. Continue to stir and allow to cool.

   Watch for the temperature at which crystals start to form and record this temperature in the table.

5. Add 5.0 cm^3 of distilled water from the burette to the crystals formed in step 4. Repeat steps 3 and 4. Continue adding 5.0 cm^3 of distilled water and repeating steps 3 and 4 until the total volume of solution is 40 cm^3.

   Record all temperatures of crystallization for each set in the table given in step 6.

6. Calculate the solubility of the saturated solution in each experiment and complete the table.

Mass of solute (g)	*Volume of water V (cm^3)*	*Temperature of crystallization*	*Solubility*	
			g dm^{-3} $\frac{10 \times 10^3}{V}$	*mol dm^{-3}* $\frac{10 \times 10^3}{V \times 249.5}$
10.0	15			
10.0	20			
10.0	25			
10.0	30			

7. From your results plot a solubility curve for the salt by placing the temperature on the horizontal axis and solubility on the vertical axis.

   Answer the following from your activity.

   (*i*) Deduce from the curve:

   (*a*) the solubility at 50°C; and 25°C

   (*b*) whether or not, the solution is saturated at 35°C when the total volume of water added to 10 g of salt is 35 cm^3.

   (*ii*) Why a burette is used in the measurement of the volume of water instead of a measuring cylinder.

   (*iii*) The solubility of $CuSO_4.5H_2O$ ...... with increase in temperature

   (*iv*) The dissolution of the salt is ...... (exothermic/ endothermic).

28.3 SOLUBILITY AND SOLUBILITY PRODUCT OF SPARINGLY SOLUBLE IONIC COMPOUNDS

SOLUBILITY EQUILIBRIA

We have studied in section 28.2 that some ionic solids are highly soluble in water whereas others are almost insoluble. The solubility of an ionic solid in water depends mainly on the *lattice enthalpy* of the salt and the *hydration enthalpy* of the ions in solution.

The energy required to overcome the attractive forces between ions is referred to as **lattice enthalpy** whereas the energy released during interaction between ions and the solvent molecules is referred to as **solvation enthalpy** or **hydration energy.** Thus, for a dissolution process, the lattice enthalpy is always positive while solvation enthalpy is always negative. For a salt to dissolve, its solvation enthalpy must be greater than its lattice enthalpy.

Ionic solids, in general, can be broadly classified into three categories on the basis of their solubility in water,

(*i*) *The solids having solubility greater than 0.1 M are classified as* **soluble solids.**

(*ii*) *The solids having solubility between 0.01 M and 0.1 M are classified as* **slightly soluble solids.**

(*iii*) *The ionic solids having solubility less than 0.01 M are termed as* **sparingly soluble solids.** The salts such as $BaSO_4$, AgCl, $AgCrO_4$ etc., belong to this category of salts.

28.4 SOLUBILITY PRODUCT CONSTANT

When a sparingly soluble salt is dissolved in water, then even in a saturated solution, the concentration of the salt is very low. So whatever little of electrolyte goes into solution, undergoes complete dissociation (due to the low concentration). Therefore, in saturated solutions of such electrolytes, the solid electrolyte is in equilibrium with the ions as represented below:

$$\textit{Undissolved electrolyte} \rightleftharpoons \textit{Ions in solution.}$$

For example, in an aqueous solution of silver chloride, the following equilibrium exists:

$$AgCl(s) \xrightleftharpoons{\text{Water}} Ag^+(aq) + Cl^-(aq)$$

Applying the law of chemical equilibrium,

$$K = \frac{[Ag^+][Cl^-]}{[AgCl]}$$

The concentration of the undissolved salt is constant (say k) at a particular temperature regardless of the amount of solid silver chloride present. (*This is true for any solid that its concentration is independent of the amount of solid*). We, therefore, have

$$K = \frac{[Ag^+][Cl^-]}{k}$$

$$K.k = K_{sp} = [Ag^+]\,[Cl^-]$$

The constant K_{sp} is known as the **solubility product constant.** It is equal to the product of the ionic concentrations, termed as *ionic product,* for a saturated solution. The solubility product constant, K_{sp}, is the equilibrium constant for the solubility equilibrium of a sparingly soluble ionic compound.

In general, for any sparingly soluble salt A_xB_y, which dissociates to set up the equilibrium

$$A_xB_y \rightleftharpoons xA^{y+} + yB^{x-} \qquad ...(28.1)$$

the solubility product constant may be expressed as:

$$K_{sp} = [A^{y+}]^x\,[B^{x-}]^y \qquad ...(28.2)$$

where A^{y+} and B^{x-} denote the positive and negative ions respectively and x and y represent the number of these ions in the formula of the electrolyte. Thus,

The solubility product of a sparingly soluble salt at a given temperature may be defined as the product of the concentrations of its ions in the saturated solution, with each concentration term raised to the power equal to the number of times the ion occurs in the equation representing the dissociation of the electrolyte.

Equation (28.2) is applicable for a saturated solution *i.e.,* under equilibrium situation.

> The term K_{sp} in equation (28.2) is replaced by Q_{sp} when the concentration of one or more species is not the concentration under equilibrium. The value of Q_{sp} gives the direction of the process in equation (28.1). If Q_{sp} is less than K_{sp} it indicates that the forward process (dissolution) is taking place. On the other hand if Q_{sp} is greater than K_{sp} it indicates that the backward process (precipitation) would take place. It may be noted that K_{sp} is the highest limit of Q_{sp} for an electrolyte at a given temperature.

The expressions of solubility products of some sparingly soluble salts are given below:

ZnS : $ZnS \rightleftharpoons Zn^{2+} + S^{2-}$

$K_{sp} = [Zn^{2+}][S^{2-}]$

$PbCl_2$: $PbCl_2 \rightleftharpoons Pb^{2+} + 2Cl^-$

$K_{sp} = [Pb^{2+}][Cl^-]^2$

$Mg(OH)_2$: $Mg(OH)_2 \rightleftharpoons Mg^{2+} + 2OH^-$

$$K_{sp} = [Mg^{2+}][OH^-]^2$$

$Ca_3(PO_4)_2$: $Ca_3(PO_4)_2 \rightleftharpoons 3Ca^{2+} + 2PO_4^{3-}$

$$K_{sp} = [Ca^{2+}]^3[PO_4^{3-}]^2$$

CALCULATION OF SOLUBILITY PRODUCT CONSTANT

Knowing the solubility of a salt, its solubility product constant (K_{sp}) can be calculated. For example, consider the salt AB. Suppose at a particular temperature its solubility is S mol L^{-1}. S moles of salts on ionization will give S moles of A^+ and S moles of B^-.

$$AB(s) \rightleftharpoons A^+(aq) + B^-(aq)$$

$$K_{sp} = [A^+][B^-] = [S][S] = S^2$$

Knowing the solubility S, K_{sp} can be calculated. The solubility products of some sparingly soluble salts are given in Table 28.2.

Table 28.2. Solubility Product Constants of Some Sparingly Soluble Salts at 298 K

Name	*Formula*	K_{sp}
Barium hydroxide	$Ba(OH)_2$	5×10^{-3}
Barium ethanedioate	BaC_2O_4	1.6×10^{-7}
Barium(II) tetraoxosulphate(VI)	$BaSO_4$	1.1×10^{-10}
Barium(II) trioxocarbonate(IV)	$BaCO_3$	5.1×10^{-9}
Cadmium(II) trioxocarbonate(IV)	$CdCO_3$	5.2×10^{-12}
Cadmium hydroxide	$Cd(OH)_2$	2.5×10^{-14}
Cadmium sulphide	CdS	8×10^{-28}
Calcium(II) trioxocarbonate(IV)	$CaCO_3$	2.8×10^{-9}
Calcium(II) tetraoxochromate(IV)	$CaCrO_4$	7.1×10^{-4}
Calcium fluoride	CaF_2	3.9×10^{-11}
Calcium hydroxide	$Ca(OH)_2$	5.5×10^{-6}
Calcium phosphate	$Ca_3(PO_4)_2$	2.0×10^{-29}
Calcium(II) tetraoxosulphate(IV)	$CaSO_4$	9.1×10^{-6}
Chromium(III) hydroxide	$Cr(OH)_3$	6.3×10^{-31}
Cobalt(II) trioxocarbonate(IV)	$CoCO_3$	1.4×10^{-13}
Cobalt(II) hydroxide	$Co(OH)_2$	1.6×10^{-15}
Cobalt(III) hydroxide	$Co(OH)_3$	1.6×10^{-44}
Cobalt(II) sulphide	CoS	5×10^{-22}
Copper(I) bromide	$CuBr$	5.3×10^{-9}
Copper(I) chloride	$CuCl$	1.2×10^{-6}
Copper(II) trioxocarbonate(IV)	$CuCO_3$	1.4×10^{-10}
Copper(II) tetraoxochromate(VI)	$CuCrO_4$	3.6×10^{-6}
Copper(II) hydroxide	$Cu(OH)_2$	2.2×10^{-20}
Copper(II) tetraoxophosphate(V)	$Cu_3(PO_4)_2$	1.3×10^{-37}
Copper(II) sulphide	CuS	6×10^{-37}
Gold(I) chloride	$AuCl$	2.0×10^{-13}
Gold(III) chloride	$AuCl_3$	3.2×10^{-25}
Iron(II) trioxocarbonate(IV)	$FeCO_3$	3.2×10^{-11}
Iron(II) hydroxide	$Fe(OH)_2$	8.0×10^{-16}
Iron(II) sulphide	FeS	6×10^{-19}
Iron(III) hydroxide	$Fe(OH)_3$	4×10^{-38}
Lead(II) trioxocarbonate(IV)	$PbCO_3$	7.4×10^{-14}
Lead chloride	$PbCl_2$	1.6×10^{-5}
Lead fluoride	PbF_2	2.7×10^{-8}
Lead hydroxide	$Pb(OH)_2$	1.2×10^{-15}
Lead(II) tetraoxosulphate(VI)	$PbSO_4$	1.6×10^{-8}
Lead sulphide	PbS	3×10^{-28}
Magnesium hydroxide	$Mg(OH)_2$	1.8×10^{-11}
Manganese hydroxide	$Mn(OH)_2$	1.9×10^{-13}
Manganese(II) sulphide	MnS	3×10^{-14}
Mercury(I) chloride	Hg_2Cl_2	1.3×10^{-18}
Mercury(II) hydroxide	$Hg(OH)_2$	3.0×10^{-26}
Mercury(II) sulphide	HgS	2×10^{-53}
Nickel hydroxide	$Ni(OH)_2$	1.6×10^{-14}
Nickel sulphide	NiS	3×10^{-20}
Silver bromide	$AgBr$	5.0×10^{-13}
Silver(I) trioxocarbonate(IV)	Ag_2CO_3	8.1×10^{-12}
Silver chloride	$AgCl$	1.8×10^{-10}
Silver(I) tetraoxochromate(VI)	Ag_2CrO_4	1.1×10^{-12}
Silver iodide	AgI	8.3×10^{-17}
Silver(I) tetraoxosulphate(VI)	Ag_2SO_4	1.4×10^{-5}
Silver sulphide	Ag_2S	6×10^{-51}
Strontium(II) trioxocarbonate(IV)	$SrCO_3$	1.1×10^{-10}
Tin(II) hydroxide	$Sn(OH)_2$	1.4×10^{-28}
Tin(II) sulphide	SnS	1×10^{-26}
Zinc(II) trioxocarbonate(IV)	$ZnCO_3$	1.4×10^{-11}
Zinc hydroxide	$Zn(OH)_2$	1.2×10^{-17}
Zinc(II) ethanedioate	ZnC_2O_4	2.7×10^{-8}
Zinc sulphide	ZnS	2×10^{-25}

SOLVED EXAMPLES

Example 28.1. *The solubility of $PbCl_2$ at 298 K is 2×10^{-2} mol dm^{-3}. Find the K_{sp} at this temperature.*

Solution. The solubility equilibrium of $PbCl_2$ may be represented as:

$$PbCl_2(s) \rightleftharpoons Pb^{2+}(aq) + 2Cl^-(aq)$$

The solubility of $PbCl_2 = 2 \times 10^{-2}$ mol dm^{-3}. $PbCl_2$ is a sparingly soluble salt, therefore, it is completely ionized in solution. One mole of $PbCl_2$ on dissociation gives one mole of Pb^{2+} ions and 2 moles of Cl^- ions. Hence,

$$[Pb^{2+}] = [PbCl_2] = 2 \times 10^{-2} \text{ mol dm}^{-3}$$

$[Cl^-] = 2[PbCl_2] = 2(2 \times 10^{-2})$ mol dm^{-3}

$K_{sp} = [Pb^{2+}][Cl^-]^2$

$= (2 \times 10^{-2})(2 \times 2 \times 10^{-2})^2 = \mathbf{3.2 \times 10^{-5}}$.

Example 28.2. *One litre of a solution saturated at 25°C with CaC_2O_4 (Calcium (II) ethanedioate) is evaporated to dryness giving 0.0061 g of CaC_2O_4 residue. Calculate the solubility product constant for this salt at 25°C.*

Solution. The solubility equilibrium of CaC_2O_4 is

$$CaC_2O_4\,(s) \rightleftharpoons Ca^{2+}\,(aq) + C_2O_4^{2-}\,(aq)$$

Concentration of CaC_2O_4 in saturated solution = 0.0061 *gdm*$^{-3}$

$$\therefore\ [CaC_2O_4] = \frac{0.0061}{\text{mol. mass }(CaC_2O_4)} \text{ mol dm}^{-3}$$

$$= \frac{0.0061}{128} \text{ mol dm}^{-3}$$

$$= 4.76 \times 10^{-5} \text{ mol dm}^{-3}$$

Now, $[Ca^{2+}] = [C_2O_4^{2-}] = [CaC_2O_4] = 4.76 \times 10^{-5}$ mol dm^{-3}

$K_{sp} = [Ca^{2+}]\,[SO_4^{2-}]$

$= (4.76 \times 10^{-5})\,(4.76 \times 10^{-5})$

$= \mathbf{2.265 \times 10^{-9}}$.

28.5 COMMON ION EFFECT ON SOLUBILITY OF IONIC SALTS

In accordance with Le-Chatelier's Principle if we increase the concentration of one of the ions in equilibrium with the solid salt, it should combine with the ion of its opposite charge and some of the salt will be precipitated. Precipitation will continue till Q_{sp} becomes equal to K_{sp}. Similarly, if the concentration of one of the ions is decreased, Q_{sp} becomes less than K_{sp}. Some of the salt dissolves to increase the concentrations of both the ions so that again Q_{sp} becomes equal to K_{sp}. The common ion effect finds application in the purification of sodium chloride and in the precipitation of soap. Pure sodium chloride is precipitated by passing HCl gas through a saturated solution of impure sodium chloride

$$NaCl(s) \rightleftharpoons Na^+(aq) + Cl^-(aq) \qquad ...(28.3)$$

$$HCl + H_2O \rightleftharpoons H_3O^+(aq) + Cl^-(aq)$$

Due to the increase in concentration of Cl^- ions, the equilibrium in equation (28.3) shifts in the backward direction resulting in the precipitation of pure sodium chloride. The impurities present in the sodium chloride remain in the solution.

CALCULATION OF SOLUBILITY

If the solubility product of a sparingly soluble salt at a particular temperature is known, its solubility at that temperature can be calculated.

Expressions correlating K_{sp} and solubility (mol L^{-1}) of some common types of salts are listed below:

Salt Type	*Relation between K_{sp} and s*	*Examples*
AB	$K_{sp} = (s)(s) = s^2$	$AlPO_4$, AgCl, $BaSO_4$, ZnS
AB_2	$K_{sp} = (s)(2s)^2 = 4s^3$	$PbCl_2$, $HgCl_2$
A_2B	$K_{sp} = (2s)^2(s) = 4s^3$	Ag_2CrO_4, $Ag_2C_2O_4$, Ag_2SO_4
AB_3	$K_{sp} = (s)(3s)^3 = 27s^4$	$Fe(OH)_3$, $Al(OH)_3$, $Cr(OH)_3$
A_3B_2	$K_{sp} = (3s)^3(2s)^2 = 108s^5$	$Ca_3(PO_4)_2$, $Zn_3(PO_4)_2$
A_3B_4	$K_{sp} = (3s)^3(4s)^4 = 6912s^7$	$Zr_3(PO_4)_4$

SOLVED EXAMPLES

Example 28.3. *The solubility product of Ag_2CrO_4 at 298 K is 4×10^{-12}. Find its solubility at this temperature.*

Solution. The solubility equilibrium of Ag_2CrO_4 is

$$Ag_2CrO_4(s) \rightleftharpoons 2Ag^+(aq) + CrO_4^{2-}(aq)$$

Suppose the solubility of Ag_2CrO_4 is x mol dm^{-3}.

This would give $2x$ mol dm^{-3} of Ag^+ ions and x mol dm^{-3} of CrO_4^{2-} ions on dissociation.

$\therefore$ $[Ag^+] = 2x$ mol dm^{-3}

$[CrO_4^{2-}] = x$ mol dm^{-3}

$K_{sp} = [Ag^+]^2\,[CrO_4^{2-}] = (2x)^2(x) = 4x^3$

But $K_{sp} = 4 \times 10^{-12}$

$\therefore$ $4x^3 = 4 \times 10^{-12}$ or $x^3 = 10^{-12}$

$x = 1 \times 10^{-4}$ mol dm^{-3}.

Therefore, solubility of Ag_2CrO_4 = $\mathbf{1 \times 10^{-4}}$ **mol dm^{-3}**.

Example 28.4. *The solubility product of AgBr at a certain temperature is 2.5×10^{-13}. Find the solubility of AgBr in gram per dm^3 at this temperature.*

Solution. The solubility equilibrium of AgBr is

$$AgBr(s) \rightleftharpoons Ag^+(aq) + Br^-(aq)$$

Suppose the solubility of AgBr is x mol dm^{-3}. This would give x mol dm^{-3} of Ag^+ ions and x mol dm^{-3} of Br^- ions on dissociation.

$[Ag^+] = x$ mol dm^{-3}

$[Br^-] = x$ mol dm^{-3}

Solubility product expressed for AgBr is

$K_{sp} = [Ag^+][Br^-] = [x][x] = x^2$

But $K_{sp} = 2.5 \times 10^{-13}$

Hence, $x^2 = 2.5 \times 10^{-13}$

or $x = \sqrt{2.5 \times 10^{-13}} = 5 \times 10^{-7}$ mol dm^{-3}

Molecular mass of AgBr = 108 + 80 = 188

Solubility of AgBr in g dm^{-3} = $5 \times 10^{-7} \times 188$

= $\mathbf{9.4 \times 10^{-5}}$ **g dm^{-3}**.

28.6 EXPERIMENTAL DETERMINATION OF THE SOLUBILITY PRODUCT CONSTANT OF $Ca(OH)_2$

Solubility product constant of a sparingly soluble salt can be experimentally determined at room temperature by investigating its solubility. For example, the experimental determination of the solubility product constant of calcium(II) hydroxide, $Ca(OH)_2$ is given in Activity 28.3.

ACTIVITY 28.3 — **To find the solubility and solubility product constant (K_{sp}) for $Ca(OH)_2$ in water**

Requirements

Pure $Ca(OH)_2$, Two titration/conical flasks, 0.05 M Hydrochloric acid, Burette and Filtration set.

Procedure

1. Record the room temperature by putting a thermometer in water.
2. Weigh about 10.0 g of $Ca(OH)_2$ solid and put it in 500 cm^3 of distilled water taken in a conical flask. Stir, shake and keep the contents overnight with proper cover on it or shake at least for 1 hour and filter the solution to collect filterate in a clean beaker. Cover it.
3. Rinse and fill burette with a 0.05 M Hydrochloric acid solution.
4. Pipette out 25 cm^3 of the filterate into a titration flask, add a drop of phenolphthalein indicator and titrate it against 0.05 M HCl solution till the end point (pink to colourless) is reached. Repeate the titration and record the titre value (V_1 cm^3)

Sample Calculations

The chemical equation is

$Ca(OH)_2\,(aq) + 2HCl \longrightarrow CaCl_2\,(aq) + 2H_2O$

Let molarity of $Ca(OH)_2$ be M_2,

Volume of $Ca(OH)_2$ taken (V_2) = 25 cm^3

Molarity of HCl solution $M_1 = 0.05$

Volume of HCl used $V_1 = ...$

According to the Molarity equation $\frac{M_1V_1}{2} = \frac{M_2V_2}{1}$

or $M_2 = \frac{M_1V_1}{V_2} = \frac{0.05 \times V_1}{25} = x$ (suppose)

$\therefore$ Concentration of $Ca(OH)_2 = x$ mol dm^{-3}

The equilibrium existing in saturated solution

$Ca(OH)_2\,(s) \rightleftharpoons Ca^{2+}\,(aq) + 2\bar{O}H\,(aq)$

$[Ca^{2+}\,(aq)] = x$ mol dm^{-3}; $[\bar{O}H\,(aq)] = 2x$ mol dm^{-3}

$K_{sp} = [Ca^{2+}]\,[\bar{O}H]^2 = x \times (2x)^2 = \mathbf{4x^3}$

solubility of $Ca(OH)_2 = \mathbf{x}$ **mol dm^{-3}**.

28.7 CRYSTALLISATION AND RECRYSTALLISATION

We have studied in section 28.1 that in a saturated solution of an ionic solid, an equilibrium exists between the undissolved solute and its ions in solution.

$$AB\,(s) \rightleftharpoons A^+\,(aq) + B^-\,(aq)$$

The forward process is called **dissolution** and the backward process is called **crystallisation**. Both these processes take place at the same rate at equilibrium. We have also studied that the maximum amount of a solute which dissolves in a given volume of solvent changes with temperature. This means that the saturated solutions formed at different temperatures contain different amounts of solute. When a saturated solution is cooled, crystallisation takes place. Sometimes when the cooling is done carefully (*without stirring*, or without seeding for crystallisation to start), the resulting solution contains more solute than it can normally hold and is known as a **supersaturated solution.** A supersaturated solution is not a true saturated solution. It quickly forms crystals when stirred or when a small crystal of solute for seeding is added. Crystallisation can also be affected by decreasing the volume of a solution through evaporation of solvent.

TEMPERATURE AT WHICH CRYSTALLISATION STARTS

Consider the solubility curve of solute A, whose solubility increases with increase in temperature (Fig. 28.4).

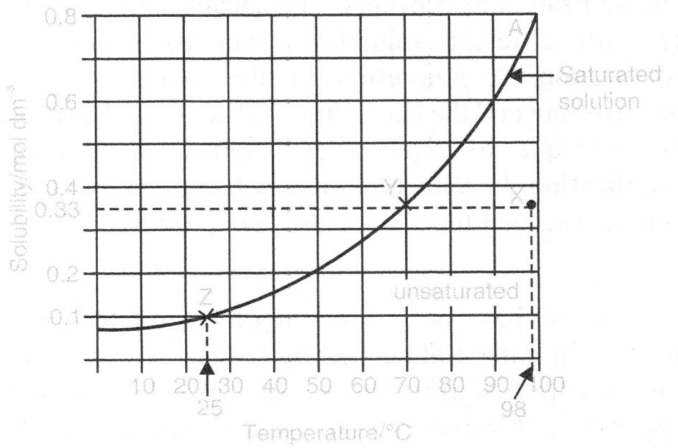

Fig. 28.4. Solubility curve for the substance A

Point X represents an unsaturated solution at about 98°C. On cooling to about 70°C, at point Y, a saturated solution consisting of about 0.33 mol of A in 1 dm^3 of solution is formed. On further cooling, crystals begin to form and the concentration of the solution follows the curve. At point Z, the saturated solution contains about 0.10 mol of A; approximately 0.27 mol (0.37 – 0.10) of substance A would have crystallised out at about 25°C.

SEPARATION OF TWO SALTS OF DIFFERENT SOLUBILITIES

Consider the solubility curves of two solutes A and B having different solubilities, as shown in Fig. 28.5.

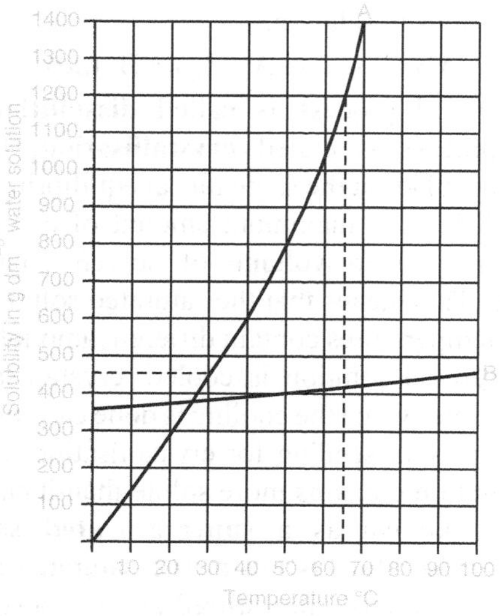

Fig. 28.5. Solubility curves for A and B.

1 dm^3 solution containing 350 g of solute B and 1100 g of solute B at 70°C was allowed to a cool. At 63°C crystals of A start to crystallize out. At a room temperature of about 30°C, the solubility of A is approximately 450 g dm^{-3}. Therefore (1100 – 450) = 650 g of A will have crystallized out.

At about 28°C solute B is an unsaturated solution. Thus, no crystallization of B occurs. This means a mixture of two salts with different solubilities can be separated by crystallization. A proportion of the more soluble salt crystallizes out and the less soluble salt stays in the solution. This technique of separation is known as **fractional crystallization.** In some cases it may be necessary to repeat the process several times to obtain pure products.

RECRYSTALLISATION

So far, we have been discussing the solubility of ionic substances in water. Solutes readily crystallize from solution when there is a significant decrease in solubility at low temperatures. However, a solute can also be recrystallized from solvents other then water provided its *solubility is high at high temperatures and low at low temperatures.* The concentration of any soluble impurity in the solute will decrease in each recrystallization process.

To sum up,

- Recrystallisation is a technique used to purify various substances.
- The substance to be purified must fulfil the following requirements.

(*a*) *It should be present in larger amount in the mixture.*

(*b*) *It should have low solubility (i.e., almost sparingly soluble) at low temperature and high solubility at high temperature.*

(*c*) *It should neither decompose, nor react with solvent at higher temperature.*

Recrystallization techniques can be used to purify both inorganic as well organic solutes.

ENERGY CHANGES DURING CRYSTALLISATION

Crystallisation is the reverse of dissolution. For a crystallisation process the *solute-solvent* interactions have to be overcome. For this process, the energy equivalent to *hydration energy* (ΔH_{Hyd}) is required to be added to the system. The release of energy occurs during solute-solute interactions. Quantitatively this release of energy is expressed in terms of lattice energy (ΔH_L). The net energy change during crystallisation ΔH_{Cry} is the difference of energy released and that which is required

For *dissolution* process

$\Delta H_{sol} = \Delta H_{Hyd} + \Delta H_L$

For crystallisation process

$\Delta H_{Cry} = -\Delta H_{sol}$

$= -(\Delta H_{Hyd} + \Delta H_L)$

Crystallisation, as well as recrystallisation is possible if the solubility of a solute is high at a higher temperature and low at a lower temperature. This means ΔH_{sol} for such a process is +ve. Consequently, ΔH_{Cry} should be –ve. In other words, crystallisation is governed by –ve sign of enthalpy change.

ACTIVITY 28.4	**To purify a given sample of copper(II) tetraoxosulphate(VI)**

Apparatus Required

Beakers, evaporating dish, glass rod and filtration set.

Procedure

1. Take 150–200 cm^3 of water in a beaker and add the given sample of $CuSO_4.5H_2O$ to it. Add 1–2 cm^3 of dilute tetraoxosulphate(VI) acid and stir to dissolve the salt.
2. Filter the solution to remove any insoluble or suspended impurities. Collect the filterate in an evaporating dish.

3. Heat the filterate with constant stirring with a glass rod. Do not let the solution boil.

   Go on evaporating till the solution becomes saturated at high temperature. This point is called *crystallisation point*. To identify this crystallisation point take out the glass rod and blow it, if a solid crust is formed, it means the saturation point is reached.

4. Stop heating and let the solution cool gradually, you can add a very small crystal of $CuSO_4.5H_2O$ for seeding.

5. When the crystals are formed decant the mother liquor and dry the crystals between pads of filter paper.

Try to answer the following questions:

(*i*) Why H_2SO_4 is added while making the solution?

(*ii*) What is crystallisation point?

(*iii*) What is the colour of $CuSO_4$. $5H_2O$ crystals?

(*iv*) What happens when the crystals are heated?

ACTIVITY 28.5 **To recrystallise N-phenylmethanoic acid (Benzoic acid) contaminated with charcoal**

Benzoic acid has a very low solubility in cold water, but it is appreciably soluble in hot water. Hence distilled water is an excellent solvent for the recrystallisation of benzoic acid. Various steps are as under:

1. Put about 5 g of the impure N-phenylmethanoic acid into a small beaker. Add about 100 cm^3 of distilled water. Stir well and heat gently. Continue to heat till all the white solid has dissolved.

2. Filter the hot solution quickly into a flask using a short stem funnel with fluted filter paper (for rapid filtration).

3. Cool the filtrate in the flask under running water. Record what happens in the flask when it is cooled. Observe the crystalline nature of any solid formed.

4. Filter the crystals using a Büchner funnel (Fig. 28.6). Wash the crystals on filter paper in the Büchner funnel with some distilled water from a wash bottle.

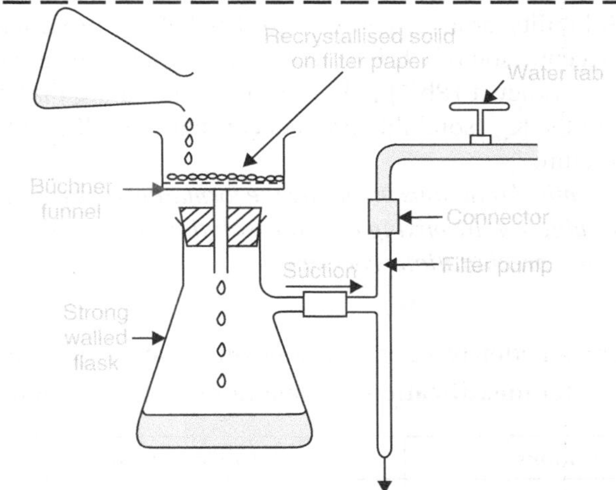

Fig. 28.6. Filtration by suction using a Büchner funnel.

5. Remove the filter paper with the crystals on it carefully. Cover it with another filter paper and leave to dry in an oven if needed quickly.

Try to answer the following questions:

(*i*) What is the colour of the cold mixture and the hot solution?

(*ii*) Why is it necessary to heat the cold suspension of mixture and water?

(*iii*) What is the colour of the recrystallised solid?

(*iv*) What is the principle of the Büchner funnel. How is it different from ordinary filtration?

(*v*) State one method to judge the purity of the recrystallised product.

(*vi*) What is the colour of the residue on the fluted filter paper?

28.8 PRECIPITATION OF INSOLUBLE SALTS

Solubility rules (section 28.1) and the concept of solubility product help us to predict the precipitation of insoluble salts from a solution. For precipitation of a sparingly soluble salt from a solution, the ***ionic product of the salt must be made to exceed the value of its solubility product constant at that temperature.*** For example, suppose we have an aqueous solution of lead (II) salt (say Pb $(NO_3)_2$) and we want to precipitate Pb^{2+} ions from the solution. Now solubility rules provides us the clue that lead (II) chloride, $PbCl_2$ is insoluble. So the addition of any solution containing Cl^- ions such as HCl (*aq*) or NaCl (*aq*) will cause the precipitation of Pb^{2+} ions.

Solubility product constant further helps us to control the concentration of Cl^- ions in such a way that the value of the ionic product $[Pb^{2+}]$ $[Cl^{-2}]$ must become more than the value of the K_{sp} (solubility product constant) of $PdCl_2$ at that temperature.

Chemical reactions involving the formation of precipitate of insoluble salt by mixing aqueous solutions of soluble salts are called precipitation reactions.

Use of Precipitation Reactions in Qualitative Analysis

Precipitation reactions are quite often used in qualitative analysis for **identification** of some cations as well anions in aqueous solutions. The reagent used for the precipitation of an insoluble salt of an ion is called a **precipitating agent.** Let us study the precipitation reactions used for the identification of some cations and anions.

Precipitation Reactions of Some Cations

The cation, its precipitating agent and the precipitation reactions are given in tabular form as follows:

Cations	*Precipitating reagents*	*Precipitation reaction*
1. Lead (II) ion **$Pb^{2+}(aq)$**	(*i*) Dilute HCl	$Pb^{2+}(aq) + 2HCl(aq) \longrightarrow PbCl_2(s) + 2H^+(aq)$ White ppt.
	(*ii*) Potassium iodide solution KI (*aq*)	$Pb^{2+}(aq) + 2KI(aq) \longrightarrow PbI_2(s) + 2K^+(aq)$ Yellow ppt.
	(*iii*) Potassium tetraoxochromate (VI) solution $K_2CrO_4(aq)$	$Pb^{2+}(aq) + K_2Cr_2O_7(aq) \longrightarrow PbCrO_4(s) + 2K^+(aq)$ Yellow ppt.
2. Calcium (II) ion **$Ca^{2+}(aq)$**	(*i*) Ammonium trioxocarbonate (IV) solution $(NH_4)_2CO_3(aq)$	$Ca^{2+}(aq) + (NH_4)_2CO_3(aq) \longrightarrow CaCO_3(s) + 2\overset{+}{NH_4}(aq)$ White ppt.
	(*ii*) Ammonium ethanedioate solution $(NH_4)_2C_2O_4(aq)$	$Ca^{2+}(aq) + (NH_4)_2C_2O_4(aq) \longrightarrow CaC_2O_4(s) + 2\overset{+}{NH_4}(aq)$ White ppt.
3. Zinc (II) ion **Zn^{2+}**	(*i*) Hydrogen sulphide solution $H_2S(aq)$	$Zn^{2+}(aq) + H_2S(aq) \longrightarrow ZnS(s) + 2H^+(aq)$ Black ppt.
	(*ii*) Potassium hexacyanoferrate(II) solution $K_4[Fe(CN)_6]$	$2Zn^{2+}(aq) + K_4[Fe(CN)_6](aq) \longrightarrow Zn_2[Fe(CN)_6](s) + 4K^+$ Bluish White ppt.
4. Aluminium (III) ion **$Al^{3+}(aq)$**	(*i*) Ammonium hydroxide solution $NH_4OH(aq)$	$Al^{3+}(aq) + NH_4OH(aq) \longrightarrow Al(OH)_3(s) + 3NH_4^+$ Geletenous White ppt.
	(*ii*) Sodium hydroxide solution NaOH (*aq*) Addition of a few drops give white precipitate which redissolve on adding more NaOH solution.	$Al^+(aq) + NaOH(aq) \xrightarrow{\text{Few drops}} Al(OH)_3(s) + 3Na^+a(aq)$ White ppt. $\underset{\text{ppt.}}{Al(OH)_3(s)} + NaOH(aq) \longrightarrow \underset{\text{Soluble}}{NaAlO_2(aq) + 2H_2O(l)}$
5. Copper (II) ion **Cu^{2+}**	(*i*) $Na_2S(aq)$ or $H_2S(aq)$	$Cu^{2+}(aq) + Na_2S(aq) \longrightarrow CuS(s) + 2Na^+$ Black ppt.
	(*ii*) Ammonium hydroxide solution $NH_4OH(aq)$	$Cu^{2+}(aq) + {}^4NH_4OH(aq) \longrightarrow [Cu(NH_3)_4]OH(aq) + {}^4H_2O(l)$ Tetraamine copper (II) hydroxide (Deepblue)

6. Iron (II) ion **Fe^{2+}** *(aq)*	(*i*) Ammonium hydroxide or sodium hydroxide solution NH_4OH *(aq)* or NaOH *(aq)*	$Fe^{2+}(aq) + NH_4OH(aq) \longrightarrow Fe(OH)_2(s) + 2N\overset{+}{H}_4(aq)$ Light green ppt.
7. Iron (III) ion **Fe^{3+}** *(aq)*	(*i*) Ammonium hydroxide or sodium hydroxide solution NH_4OH *(aq)* or NaOH *(aq)*	$Fe^{3+}(aq) + NH_4OH(aq) \longrightarrow Fe(OH)_3(s) + 3N\overset{+}{H}_4(aq)$ Reddish brown ppt.

Precipitation Reactions of Some Anions

Anion	*Precipitating agent*	*Precipitation Reaction*
1. Chloride ion **Cl^-** *(aq)*	(*i*) Silver trioxonitrate(V) solution $AgNO_3$ *(aq)*	$Cl^-(aq) + AgNO_3(aq) \longrightarrow AgCl(s) + NO_3^-(aq)$ White ppt.
	(*ii*) Lead(II) trioxonitrate(V) solution $Pb(NO_3)_2$ *(aq)*	$2Cl^-(aq) + Pb(NO_3)_2(aq) \longrightarrow PbCl_2(s) + 2NO_3^-(aq)$ White ppt.
2. Tetraoxosulphate (VI) ion **SO_4^{2-}** *(aq)*	Barium(II) chloride solution $BaCl_2$ *(aq)*	$SO_4^{2-}(aq) + BaCl_2(aq) \longrightarrow BaSO_4(s) + 2Cl^-(aq)$ White ppt.
3. Trioxosulphate (IV) ion **SO_3^{2-}** *(aq)*	Barium chloride solution $BaCl_2$ *(aq)* White precipitate soluble in dil. HCl solution.	$SO_3^{2-}(aq) + BaCl_2(aq) \longrightarrow BaSO_3(s) + 2Cl^-$ White ppt. $BaSO_3(s) + 2HCl \longrightarrow BaCl_2(aq) + SO_2(g) + H_2O$ Soluble
4. Trioxocarbonate (IV) ion **CO_3^{2-}** *(aq)*	Magnesium chloride solution $MgCl_2$ *(aq)*	$CO_3^{2-}(aq) + MgCl_2(aq) \longrightarrow MgCO_3(s) + 2Cl^-(aq)$ White ppt.

28.9 IDENTIFICATION OF CATIONS AND ANIONS IN SOLUTION

In section 28.6, we have studied the precipitation reactions of some cations and anions. The knowledge of these precipitation reactions can be utilised to identify the unknown ions in the aqueous solution of their salts. The common chemical reagents used are dil. HCl, NaOH *(aq)*, NH_4OH *(aq)* or NH_3 *(aq)*, $BaCl_2$ *(aq)* and $AgNO_3$ *(aq)*. For the identification of ions in solution, the unknown solution is treated with these precipitating reagents. The *precipitate formed* or any *gas evolved* is observed and proper inference or conclusions can be drawn. This identification of the ions in solution forms a part of **Qualitative analysis.** The identification procedure in the form of tests is given as follows in a tabular form.

IDENTIFICATION OF IONS

I. Identification of Cations

Reagents / *Ions*	*Dilute NaOH in drops then ..... in excess*	*NH_4OH in drops then in excess*	*H_2S (aq)*	*Additional Tests/Observations*
1. **Pb^{2+} *(aq)***	White, chalky precipitate of $Pb(OH)_2$ forms on addition of drops, precipitate dissolves in excess to form the soluble complex ion $[Pb(OH)_4]^{2-}$.	White precipitate of $Pb(OH)_2$ forms, precipitate does not dissolve in excess.	Black precipitate of PbS forms.	(*i*) Aq HCl or any solution containing Cl^- ions is added. *Observation*: White precipitate of $PbCl_2$ forms which dissolves on heating and recrystallises on cooling. (*ii*) Dil H_2SO_4 or any solution containing SO_4^{2-} ions is added. *Observation*: White precipitate of $PbSO_4$ forms. (*iii*) ***Confirmatory test***: K_2CrO_4 is added *Observation*: Yellow precipitate forms. (*iv*) ***Confirmatory test***: Aqueous KI solution is added in drops *Observation*: Bright yellow precipitate of PbI_2 forms, dissolves on heating and recrystallises on cooling.

2. **Ca^{2+} (*aq*)**	White, chalky precipitate, $Ca(OH)_2$ forms. It does not dissolve in excess. **Note:** with low concentration of either ion, precipitate may not be formed.	Clear solution forms, no precipitate forms, no gas evolved.	No precipitate appear.	(*i*) Dilute H_2SO_4 or any solution containing SO_4^{2-} ion is added. *Observation*: White precipitate of $CaSO_4$ is formed. No precipitate in very dilute solutions. (*ii*) ***Confirmatory test*:** Aqueous NH_3 is added followed by ammonium ethanedioate solution. *Observation*: White precipitate of calcium ethanedioate CaC_2O_4 forms.
3. **Zn^{2+} (*aq*)**	White gelatinous precipitate of $Zn(OH)_2$ forms, precipitate dissolves in excess to give soluble complex ion solution of $[Zn(OH)_4]^{2-}$.	White gelatinous precipitate of $Zn(OH)_2$ forms precipitate, dissolves in excess to give soluble complex ion $[Zn(NH_3)_4]^{2+}$.	White precipitate of ZnS forms in neutral or alkaline solutions.	(*i*) ***Confirmatory test*:** Aqueous NH_4Cl is added, followed by aq NH_3 and then H_2S gas passed through. *Observation*: White precipitate of ZnS forms. (*ii*) ***Confirmatory test*:** 2 drops aq NH_3 is added to form $Zn(OH)_2$ (do not add excess to dissolve). The mixture is then, heated to evaporate the liquid leaving a gelatinous precipitate in the crucible or test tube. The residue is heated strongly or directly. *Observation:* Yellow solid of ZnO is formed when hot and white when cold.
4. **Al^{3+} (*aq*)**	White gelatinous precipitate of $Al(OH)_3$ forms, precipitate dissolves in excess, to form soluble complex solution of $[Al(OH)_4]^-$.	White gelatinous precipitate of $Al(OH)_3$ forms, precipitate does not dissolve in excess.	No precipitate forms but white $Al(OH)_3$ forms with $(NH_4)_2S$ solution.	***Confirmatory test*:** Two drops litmus solution is added (test solution if not acidic add dil HCl and then with aq NH_3 until just alkaline). *Observation*: Blue 'precipitate' forms by litmus adsorbing $Al(OH)_3$ precipitate in aq NH_3 leaving solution almost colourless. These test is called **lake test**.
5. **Cu^2 (*aq*)**	Light blue precipitate of $Cu(OH)_2$ forms; precipitate does not dissolve in excess and turns to black CuO on heating.	Light blue precipitate of $Cu(OH)_2$ forms on addition of drops. Deep blue solution of soluble $[Cu(NH_3)_4]^{2+}$ forms on addition of excess and shaking.	Black precipitate of CuS formed.	(*i*) ***Confirmatory test*:** Aqueous $K_4[Fe(CN)_6]$, potassium hexacyanoferrate(II) is added in drops. *Observation:* Reddish brown gelatinous precipitate of $Cu_2[Fe(CN)_6]$ forms. (*ii*) Formation of deep blue complex formed by the addition of excess ammonia solution can also be used as a confirmatory test.
6. **Fe^{2+} (*aq*)**	Pale green gelatinous precipitate of $Fe(OH)_2$ forms, precipitate does not dissolve in excess, but turns brown on the surface of mixture and walls of test tube on standing.	Pale green precipitate of $Fe(OH)_2$ forms, precipitate does not dissolve in excess but turns brown on standing .	Black precipitate of FeS forms in alkaline or neutral solution.	(*i*) Aqueous acidified (H_2SO_4) $KMnO_4$,potassium tetraoxomanganate(VII) solution is added. *Observation*: Purple MnO_4^- changes to colourless (decolorized), Mn^{2+}. (*ii*) Few drops aq $K_4[Fe(CN)_6]$ potassium hexacyanoferrat(II) is added. *Observation*: White precipitate forms, turns dark blue on standing. (*iii*) ***Confirmatory test*:** Few drops of aq $K_3[Fe(CN)_6]$ potassium hexacyanoferrate(III) solution is added. *Observation*: Dark blue precipitate of $KFe[Fe(CN)_6]$ forms.

7. Fe^{3+} **(*aq*)**	Red brown or rust gelatinous precipitate of $Fe(OH)_3$ forms, precipitate insoluble in excess.	Rust or red brown precipitate of $Fe(OH)_3$ forms; precipitate is insoluble in excess.	Yellow precipitate of sulphur particles is formed.	(*i*) Aqueous $K_3[Fe(CN)_6$, potassium hexacyanoferrate(III) is added in drops. *Observation*: Reddish brown solution forms. (*ii*) ***Confirmatory test***: Aqueous KSCN or NH_4SCN (thiocyanate) is added. *Observation*: Deep 'blood' red solution of $Fe(SCN)_3$ forms. (*iii*) ***Confirmatory test***: Aqueous $K_4[Fe(CN)_6]$, potassium hexacyanoferrate (II) solution is added. *Observation*: Dark blue precipitate of $Fe_4[Fe(CN)_6]_3$ forms.

II. Identification of Anions

1. Cl^- **(*aq*)**	Clear solution, no gas, no precipitate forms.	Frothy effervescence of fuming, pungent smelling (choking) HCl gas forms. Gas gives white fumes with aq. NH_3	(*i*) Few drops aq Pb $(NO_3)_2$ or any solution containing Pb^{2+} ion is added *Observation*: White precipitate of $PbCl_2$ forms, precipitate dissolves on heating and recrystallises on cooling. (*ii*) Aqueous $AgNO_3$ is added, followed by excess dil HNO_3. *Observation*: White precipitate of AgCl insoluble in excess dil HNO_3; precipitate dissolves in excess aq NH_3 to form soluble complex $[Ag(NH_3)_4]^+$, white precipitate also darkens on standing in light.
2. SO_3^{2-} **(*aq*)**	In the cold or on heating, colourless, pungent smelling SO_2 gas forms. The gas turns $K_2Cr_2O_7$ solution green.	Same as with dil acid.	(*i*) SO_2 generated with dilute acid is bubbled through delivery tube into aqueous acidified (H_2SO_4) $KMnO_4$. *Observation*: Purple aqueous $KMnO_4$ decolourized, no precipitate of S forms. (*ii*) ***Confirmatory test***: SO_2 gas is bubbled through delivery tube into acidified $K_2Cr_2O_7$ or over filter paper soaked in $K_2Cr_2O_7$. *Observation*: *Yellow* or *orange* $K_2Cr_2O_7$ solution forms *green* Cr^{3+} ion. (*iii*) Aqueous $BaCl_2$ or Ba $(NO_3)_2$ is added, followed by excess dil HCl or dil HNO_3. *Observation*: White precipitate, soluble in dil HCl or dil HNO_3.
3. SO_4^{2-} **(*aq*)**	Clear solution, no precipitate, no gas forms.	Clear solution, no gas, no precipitate forms.	(*i*) $Pb(NO_3)_2$ solution or any solution which contains Pb^{2+} ions is added *Observation*: White precipitate of $PbSO_4$ forms. (*ii*) ***Confirmatory test***: Aqueous $BaCl_2$ or $Ba(NO_3)_2$ is added followed by excess dil HCl or HNO_3 respectively. *Observation*: White precipitate of $BaSO_4$ forms, the precipitate is insoluble in dil. HNO_3 or HCl. (See same test with SO_3^{2-}, precipitate dissolves in the case of SO_3^{2-}).

4. CO_3^{2-} **(s, aq)**	Effervescence of colourless odourless CO_2 gas forms; gas turns limewater milky.	Vigorous effervescence of colourless odourless CO_2 gas turns limewater milky.	***Confirmatory test***: Add $MgCl_2$ solution to aqueous solution of salt. *Observation*: A white precipitate of $MgCO_3$ is formed.
5. NO_3^- **(aq)**	Clear solution, no gas, no precipitate forms.	Colourless, pungent smelling HNO_3 vapour forms. Vapour turns blue litmus red and fumes in air. Pale brown vapour of NO_2 forms on further heating.	***Confirmatory test***: *Brown ring test*—freshly prepared aq $FeSO_4$ is added, followed by concentrated H_2SO_4 carefully down the sides of test tube in tilted position to form a denser lower layer. *Observation*: Brown ring or solution forms at the junction of the two layers. **Note:** If Pb^{2+} and Ca^{2+} are present a white precipitate of their insoluble tetraoxosulphate(VI) also forms.

ACTIVITY 28.6 **Identification of cations in solution**

Three solutions were prepared by dissolving of the salts, $Pb(NO_3)_2$ or $Zn(NO_3)_2$ or $Al_2(SO_4)_3$. The solutions are randomly marked as X, Y, Z.

Identify the cation present in the solution X, Y, Z by using the following solutions:

1. Dilute NaOH solution in drops and then in excess.
2. Dilute NH_3 *(aq)*.
3. Dilute HCl, heat and cool.

(*i*) Perform the experiments given in the 1st column starting with solution X and fill up the Observation and Inference.

(*ii*) Repeat the same process for solution **Y** and solution. **Z.**

Experiment	*Observations*	*Inference*
1. Dilute aq NaOH was added to A in drops and in excess.	....	....
2. Dilute aq NH_3 was added to A in drops and in excess.	....	....
3. Dilute aq HCl was added. The mixture was heated and cooled.	....	....

Conclusion:

(*i*) cation present in solution X ....

(*ii*) cation present in solution Y ....

(*iii*) cation present in solution Z ....

SUMMARY

- **Solubility.** It is the maximum amount of solute in moles or gram that can dissolve in a solvent to form 1 dm^3 of solution at a given temperature.
- **Solubility of solids** in a given liquid depends upon the nature of solute and also on temperature.
- **Solubility curve.** It is the graph representing the variation of solubility of solutes in water with temperature.
- **Solubility product constant (K_{sp}).** It is the highest limit of ionic product Q_{sp} of an electrolyte in solution at a given temperature. At the saturation point $Q_{sp} = K_{sp}$.
- **Crystallisation.** It is a process of getting a solid crystalline substance from its aqueous solution. It is affected by concentrating the solution to the saturation point at a higher temperature and then subsequent cooling.
- **Recrystallisation.** It is technique of purification of organic or inorganic compounds by using dissolution, filtration and crystallisation processes.

EVALUATE YOURSELF

I. Objective Type Questions

Select the most appropriate choice from the options given as (a), (b), (c) and (d) after each question:

1. Which of the following substance will dissolve in water at room temperature?

(*a*) lead(II) trioxocarbonate(IV)

(*b*) lead(II) tetraoxosulphate (VI)

(*c*) lead(II) trioxonitrate(V)

(*d*) lead(II) chloride.

2. On mixing aqueous solutions of Na_2CO_3 and $Ba(NO_3)_2$ the following products will be formed:

(*a*) precipitate of $BaCO_3$ only

(*b*) precipitate of $BaCO_3$ and aqueous solution of $NaNO_3$

(*c*) precipitates of $NaNO_3$ and $BaCO_3$

(*d*) precipitate of $Ba(NO_3)_2$ only.

3. Fractional crystallization is a technique based on differences in:

(*a*) solubilities (*b*) melting and boiling points

(*c*) acidity or basicity (*d*) all the above.

4. At room temperature water will dissolve appreciably.

(*a*) Benzoic acid

(*b*) Strontium trioxocarbonate(IV)

(*c*) Aluminium(III) hydroxide

(*d*) Zinc (II) sulphide.

5. Which of the following reagent will give a yellow precipitate with a solution containing lead (II) ions?

(*a*) Na_2CO_3 (*aq*) (*b*) NaOH (*aq*)

(*c*) KNO_3 (*aq*) (*d*) KI (*aq*).

6. A solution of barium(II) chloride was added to an aqueous solution of silver (I)tetraoxosulphate(VI) to form a white precipitate. The precipitate consists of:

(*a*) $BaCO_3$ only (*b*) AgCl only

(*c*) Both $BaCO_3$ and AgCl

(*d*) Neither of the two.

7. A suitable solvent for the recrystallization of a solute by solvent X implies that the solute:

(*a*) dissolves in the solvent when hot and crystallizes out at room temperature

(*b*) does not dissolve in the hot solution of X

(*c*) reacts chemically with the solvent

(*d*) decomposes in the solvent when heated.

8. The solubility of sodium chloride at room temperature is given as X g cm^{-3}. The information needed to convert it to its concentration in *mol* dm^{-3} is the:

(*a*) Relative atomic mass of sodium

(*b*) Molar volume of sodium chloride

(*c*) Molar mass of sodium chloride

(*d*) Density of sodium chloride.

9. A saturated solution has the following properties:

I. it contains more solute than it can dissolve

II. it will not dissolve any more solute at a given temperature

III. its dissolved ions are in equilibrium with any undissolved solid at a given temperature

IV. it will form crystals when it is cooled if it contains a solute whose solubility increases with increase in temperature.

(*a*) I, II, III (*b*) I only

(*c*) II, III, IV (*d*) II and III.

10. When aq Na_2S is added to solution X, the formation of a black precipitate indicates that X contains:

(*a*) Pb^{2+} (*b*) Pb^{2+}, Cu^{2+}

(*c*) Pb^{2+}, Cu^{2+}, Fe^{2+} (*d*) Pb^{2+}, Cu^{2+}, Ca^{2+}.

11. Sodium hydroxide solution was added in drops and then in excess to solution Y. A light blue precipitate insoluble in excess sodium hydroxide was produced. Solution Y, therefore, contains:

(*a*) iron (II) ions (*b*) iron (II) ions

(*c*) copper (II) ions (*d*) calcium ions.

12. The solubilities of a salt A in water at 273 K and 313 K are 0.8 and 1.60 mol dm^{-3} respectively. This means that the salt A:

I. will form saturated solutions containing less than 0.80 mol dm^{-3} of solute when it is cooled below 273 K

II. increases its solubility when temperature increases

III. will deposit 0.08 mol of solute when 1 dm^3 of saturated solution, without any solid present, is cooled from 313 K to 273 K

IV. will form a saturated solution containing more than 1.6 mol dm^{-3} of solute when it is heated above 313 K

(*a*) I (*b*) I, II

(*c*) I, II, III (*d*) I, II, III, IV.

II. Fill in the Blanks

13. Complete the following sentences by supplying appropriate words:

(*i*) In a saturated solution the molar concentration of solute is ....

(*ii*) Solubility product constant .... with the variation of temperature.

(*iii*) Amount of solute that can be dissolved in 100 g of solvent at a given temperature is called ....

(*iv*) According to solubility rules tetraoxosulphate(VI) of Ba^{2+} ions is .... in water.

(*v*) Sulphide of Ag^+ ion is .... while that of Na^+ is .... in water.

(*vi*) In a saturated solution Q_{sp} is .... K_{sp}.

(*vii*) The expression of K_{sp} for barium(II) tetraoxophosphate (V) is ....

(*viii*) ΔH_{sol} and ΔH_{cry} for an electrolyte are related as ....

III. Discussion Questions

14. What do you understand by the **term** solubility? How is solubility of a solid in water expressed?

15. Use solubility rules to answer the following:

(*i*) The insoluble salt of Ag^+ ion is ....

(*ii*) Name the cations which form insoluble tetraoxosulphate (VI) salts.

(*iii*) Write the formula of at least two sulphides which are insoluble.

(*iv*) Write the name and formulas of two hydroxides which are soluble and other two which are insoluble.

16. Which ions could be present in a solution which gives white precipitate with barium(II) chloride?

17. How can you measure the solubility of copper(II) tetraoxosulphate(VI) pentahydrate at room temperature?

18. Discuss in brief the effect of temperature on the solubility of solids in water.

19. Draw a solubility curve for two solutes A and B in water. ΔH_{sol} of A is + 3.09 kJ mol^{-1} whereas that of B is – 4.92 kJ mol^{-1}.

20. Comment on the statement that the solubility of a solid in water always increases with rise in temperature.

21. What is a solubility curve, what information does it convey?

22. (*a*) What do you understand by solubility product constant?

(*b*) Solubility of an electrolyte AB_2 is S mol dm^{-3}, write the expression of its K_{sp} in terms of S.

23. (*a*) What is the importance of solubility product?

(*b*) Calculate the solubility of silver (I) tetraoxochromate(VI) if the value of its K_{sp} is 1.1×10^{-12}.

24. What do you understand by the term recrystallization? What is the basic principle involved in recrystallization?

25. Explain crystallization. How lattice energy and hydration energy affect crystallization?

26. What are precipitation reactions? How are they useful in the identification of the ions in solution?

27. Suggest an experiment to find the solubility product constant of $Ca(OH)_2$ in a laboratory.

Unknown solutions labelled **A, B, C** and **D**, are aqueous solutions of $Pb(NO_3)_2$, $Zn(NO_3)_2$, Na_2CO_3 and HCl, but not necessarily arranged in that same order.

Small volumes each of A, B, C and D were added to each other solution. The observations were recorded in the table that follows:

Solution	*A*	*B*	*C*	*D*
A	—	Effervescence of a colourless, odourless gas	White precipitate forms	Clear solution forms
B	Effervescence of a colourless, odourless gas	—	White precipitate forms	White precipitate forms
C	White precipitate forms	White precipitate forms	—	Clear solution forms
D	Clear solution forms	White precipitate forms	Clear solution forms	—

Study the observations carefully and deduce the identifies of the solutions A, B, C and D. Write equations to show the observations made on any one identified solution.

Answers

I. Objective Type Questions

1. (*c*)	**2.** (*b*)	**3.** (*a*)
4. (*b*)	**5.** (*d*)	**6.** (*c*)
7. (*a*)	**8.** (*b*)	**9.** (*c*)
10. (*b*)	**11.** (*a*)	**12.** (*b*).

II. Fill in the Blanks

13. (*i*) constant (*ii*) does not change

(*iii*) solubility (*iv*) insoluble

(*v*) insoluble, soluble

(*vi*) equal to

(*vii*) $[Ba^{2+}]^3 [PO_4^{3-}]^2$ (*viii*) $\Delta H_{cry} = -\Delta H_{sol}$.

SECTION-4

UNIT-29

SALTS AND CHEMICALS FROM SALT

Specific Objectives

After studying this unit, the student should be able to:

- explain the meaning of salt
- state and explain how salt forms acidic, alkaline and neutral aqueous solutions
- describe the laboratory and industrial production of salt
- describe the process of obtaining chemicals from brine (sea water).

29.1 INTRODUCTION

Salts are the ionic compounds which are formed by neutralization reaction between acids and bases. Soluble salts, when dissolved in water, undergo dissociation to form cations and anions. Due to presence of these free ions, aqueous solutions of salts are good conductors of electricity. Hence, salts also behave as **electrolytes.**

Salts can also conduct electricity in their molten state.

When salts are dissolved in water sometimes their ions interact with water molecules producing acidic or basic solutions. This is known as **hydrolysis** of salts.

The most common example of salts is the common salt, sodium chloride (NaCl). Common salt has been used in food and as a preservative since ancient times. Sodium chloride is also used in the preparation of many important compounds such as Na_2CO_3, $NaHCO_3$ and NaOH.

29.2 SALTS

*A **salts** is a compound formed by replacement of either all or a part of the ionizable hydrogen atom(s) of an acid by some other cation.*

A salt is formed by neutralization of an acid by a base.

$$\textbf{Acid + Base} \longrightarrow \textbf{Salt + Water}$$

For example, sodium chloride (NaCl) is formed by neutralization of hydrochloric acid with sodium hydroxide.

$$\underset{\text{Hydrochloric acid}}{HCl(aq)} + \underset{\text{Sodium hydroxide}}{NaOH(aq)} \longrightarrow \underset{\text{Sodium chloride}}{NaCl(aq)} + \underset{\text{Water}}{H_2O(l)}$$

Similarly, potassium nitrate (KNO_3) is formed by neutralization of nitric acid with potassium hydroxide.

$$\underset{\text{Nitric acid}}{HNO_3(aq)} + \underset{\text{Potassium hydroxide}}{KOH(aq)} \longrightarrow \underset{\text{Potassium nitrate}}{KNO_3(aq)} + \underset{\text{Water}}{H_2O(l)}$$

A salt is an electrovalent compound. When dissolved in water it dissociates to yield cations other than hydrogen ions (H^+ ion) and anions other than hydroxyl ions (OH^-). Aqueous solutions of salts, therefore, conduct electricity.

TYPES OF SALTS

1. Normal Salts

Normal salts *are the salts formed by complete replacement of all the ionizable hydrogen atoms of an acid by metallic or ammonium ions.*

Some **examples** of normal salts are:

$$NaCl, KNO_3, Na_2CO_3, Na_2SO_4, (NH_4)_3PO_4$$

$$\mathbf{HCl} + NaOH \longrightarrow \mathbf{Na}Cl + H_2O$$

$$\mathbf{H_2}SO_4 + 2NaOH \longrightarrow \mathbf{Na_2}SO_4 + 2H_2O$$

2. Acid Salts

Acid salts *are the salts formed by partial replacement of ionizable hydrogen atoms of a dibasic or tribasic acid by metal or ammonium ions.*

For example,

$$NaHSO_4, NaHCO_3, KH_2PO_4, K_2HPO_4$$

Because there is still some replaceable hydrogen atoms present in this type of salts, these salts behave as an acid as well as a salt and hence are known as **acid salts.**

Acid salts ionize in water to yield hydronium ions.

$$NaHSO_4 + H_2O \longrightarrow Na^+ + H_3O^+ + SO_4^{2-}$$

An acid salt is formed by incomplete neutralization of a dibasic or a tribasic acid. For example,

$$H_2SO_4 + NaOH \longrightarrow \underset{\text{Sodium hydrogen tetraoxosulphate(VI)}}{NaHSO_4} + H_2O$$

Add To Your Knowledge

Some acid salts can behave as acids as well as bases. For example, sodium hydrogen trioxocarbonate(IV). An aqueous solution of $NaHCO_3$ turn red litmus blue due to formation of hydroxyl ions by hydrolysis of HCO_3^-.

$$HCO_3^-(aq) + H_2O(l) \rightleftharpoons H_2CO_3(aq) + OH^-(aq)$$

Due to formation of excess OH^- ions, the aqueous solution of $NaHCO_3$ behaves as a base.

On reaction with sodium hydroxide solution, sodium hydrogen trioxocarbonate(IV) is converted into sodium trioxocarbonate(IV). In this reaction, it behaves as an acid

$$NaHCO_3(aq) + NaOH(aq) \rightleftharpoons Na_2CO_3(aq) + H_2O(l)$$

3. Basic Salts

Basic salts *are the salts formed by the partial replacement of hydroxyl groups of a diacid or a triacid base by some other anion.*

For example, Basic lead chloride, $Pb(OH)Cl$, basic magnesium bromide, $Mg(OH)Br$, basic zinc chloride, $Zn(OH)Cl$.

A basic salt is formed by the partial neutralization of a diacid or a triacid base

$$Mg(OH)_2 + HBr \longrightarrow \mathbf{Mg(OH)Br} + H_2O$$

$$Pb(OH)_2 + HCl \longrightarrow Pb(OH)Cl + H_2O$$

A basic salt can be converted into normal salt by reacting with more acid.

4. Double Salts

A double salt *is obtained by mixing two simple salts in stoichiometric proportion and then crystallizing them together.*

For example, Mohr's salt, potash alum.

A double salt when dissolved exhibits the properties of the constituent salts in solution.

Mohr's salt is formed by mixing iron(II) tetraoxosulphate(VI) and ammonium tetraoxosulphate(VI). Its formula is:

$$\mathbf{FeSO_4(NH_4)_2SO_4.6H_2O}$$

Potash alum is obtained by mixing potassium tetraoxosulphate(VI) and aluminium tetraoxosulphate(VI) in the molar ratio 1 : 1. Its formula is: $\mathbf{K_2SO_4.Al_2(SO_4)_3.24H_2O}$.

5. Complex Salts

A complex salt *is formed by combination of two or more simple salts. The complex salt formed does not show the tests of all the ions present in the constituent salts.*

A complex salt yields at least one complex ion on dissociation in aqueous solution.

For example,

$K[Ag(CN)_2]$, $K_4[Fe(CN)_6]$, $Na[Al(OH)_4]$, $[Cu(NH_3)_4]SO_4$.

29.3 HYDROLYSIS OF SALTS

The aqueous solutions of salts may be neutral, acidic or basic in nature depending upon the type of salt.

The aqueous solutions of **salts of strong acid and strong base** are **neutral** and have a pH value of 7.

For example, aqueous solutions of sodium chloride, potassium trioxonitrate(V) and sodium tetraoxosulphate(VI), potassium chloride.

The aqueous solutions of **salts of strong acid and weak base** are **acidic** and have pH values less than 7.

For example, aqueous solutions of zinc tetraoxo-sulphate (VI), calcium chloride, copper(II) tetraoxosulphate (VI) and ammonium chloride.

The aqueous solutions of **salts of weak acid and strong base** are **basic** and have pH values greater than 7.

For example, aqueous solutions of sodium acetate, sodium trioxocarbonate(IV), potassium trioxocarbonate(IV).

The aqueous solutions of **salts of weak acid and weak base** are **almost neutral** and have pH values close to 7.

For example, aqueous solutions of ammonium tetraoxo-phosate(V), ammonium acetate, zinc trioxocarbonate (IV), zinc acetate, etc.

ACTIVITY 29.1 Study of pH of Some Salt Solutions

- Collect samples of the following salts:
  Sodium chloride (NaCl), sodium acetate (CH_3COONa), potassium trioxonitrate(V) (KNO_3).
  Zinc tetraoxosulphate(VI) ($ZnSO_4$), copper(II) tetraoxosulphate(VI) ($CuSO_4$), sodium trioxocarbonate(IV) (Na_2CO_3).
  Potassium trioxocarbonate(IV) (K_2CO_3), sodium hydrogencarbonate(IV) ($NaHCO_3$).
  Zinc acetate ($(CH_3COO)_2Zn$) and ammonium acetate (CH_3COONH_4).
- Try to dissolve these salts in distilled water one by one and note the observations.
- Check the effect of these salt solutions on litmus paper (blue as well as red).
- Find the pH of these solutions using a universal indicator solution or pH paper.
- Record the observations in a tabular form as given below:

S.No.	*Salt*	*Effect on Litmus Paper*	*pH of Solution (Using pH paper)*
1.	Sodium chloride	no effect	7
2.	....	....	....
3.	....	....	....

The acidic or basic behaviour of aqueous solutions of salts can be explained on the basis of hydrolysis.

***Hydrolysis** may be defined as the process in which water reacts with salt to form an acid and a base.*

$$\text{Salt} + \text{Water} \rightleftharpoons \text{Acid} + \text{Base}$$

1. Hydrolysis of Salt of a Weak Acid and a Strong Base

This type of salt includes salts of sodium and potassium except halides, trioxonitrate(V) and tetraoxosulphate(VI). For example, sodium trioxocarbonate(IV), potassium cyanide, sodium acetate, etc. This type of salt produces ***alkaline solutions*** on hydrolysis. Let us discuss the hydrolysis of sodium acetate a salt of this type.

$$CH_3COONa + H_2O \rightleftharpoons \underset{\text{Strong base}}{NaOH} + \underset{\text{Weak acid}}{CH_3COOH}$$

In solution, CH_3COONa and the strong base NaOH undergo complete dissociation whereas the acid, CH_3COOH being a weak acid remains almost undissociated. Therefore, we can write,

$$Na^+ + CH_3COO^- + H_2O \rightleftharpoons Na^+ + OH^- + CH_3COOH$$

$$CH_3COO^- + H_2O \rightleftharpoons OH^- + CH_3COOH$$

Due to formation of excess OH^- ions, the solution of sodium acetate behaves as a alkaline solution and has pH greater than 7. Such salts undergo **anionic hydrolysis.**

2. Hydrolysis of Salt of a Strong Acid and a Weak Base. This type of salt includes halides, trioxonitrate(V) and tetraoxosulphate(VI) of all metals except that of sodium and potassium. For example, zinc trioxonitrate(V), copper tetraoxo-sulphate(VI), aluminium tetraoxosulphate(VI), iron(III) chloride, etc. This type of salt produces ***acidic solutions*** on hydrolysis. Let us discuss the hydrolysis of a salt iron(III) chloride, $FeCl_3$ of this type

$$FeCl_3 + 3H_2O \rightleftharpoons \underset{\text{Weak base}}{Fe(OH)_3} + \underset{\text{Strong acid}}{3HCl}$$

In solution $FeCl_3$ and the strong acid HCl undergo complete dissociation whereas the weak base $Fe(OH)_3$ remains almost undissociated. Therefore, we can write,

$$Fe^{3+}(aq) + 3Cl^-(aq) + 3H_2O(l) \rightleftharpoons Fe(OH)_3(aq) + 3H^+(aq) + 3Cl^-(aq)$$

$$Fe^{3+}(aq) + 3H_2O(l) \rightleftharpoons Fe(OH)_3(aq) + 3H^+(aq)$$

Due to the presence of excess H^+ ions or H_3O^+ ions, the aqueous solution of such salts are acidic in nature and have pH less than 7. Such salts are said to undergo cationic hydrolysis. Similarly, an aqueous solution of ammonium chloride is acidic in nature due to cationic hydrolysis.

3. Hydrolysis of Salt of a Weak Acid and a Weak Base. This type of salt includes all the salts except halides, trioxonitrate(V) and tetraoxosulphate(IV) of all the metals except sodium and potassium. For example, zinc acetate, aluminium tetraoxophosphate(V), calcium trioxocarbonate(IV), ammonium tetraoxophasphate, etc. This type of salt produces *weak acids* and *weak bases* on hydrolysis. Let us discuss the hydrolysis of zinc acetate, $Zn(CH_3COO)_2$, a salt of this type

$$Zn(CH_3COO)_2(aq) + 2H_2O(l) \rightleftharpoons \underset{\text{Weak base}}{Zn(OH)_2(aq)} + \underset{\text{Weak acid}}{2CH_3COOH(aq)}$$

or $$Zn^{2+}(aq) + 2CH_3COO^-(aq) + 2H_2O(l) \rightleftharpoons Zn(OH)_2(aq) + 2CH_3COOH(aq)$$

Due to the formation of a weak acid and a weak base on hydrolysis, solutions of such salts are **almost neutral** and have pH nearly equal to 7.

The actual value of the pH of the solution would depend on the K_a and K_b values of the weak acid and weak base formed by hydrolysis.

In case $K_a = K_b$, the solution would be neutral and pH = 7.

In case $K_a > K_b$, the solution would be slightly acidic and the pH would be less than 7. On the other hand if $K_a < K_b$, the solution would be slightly alkaline and the pH would be greater than 7.

4. Hydrolysis of Salt of a Strong Acid and a Strong Base. This type of salt includes halides, trioxonitrate (V) and tetraoxosulphate (IV) of sodium and potassium. For example, NaCl, KCl, KNO_3, Na_2SO_4, etc. ***Salts of this type are not hydrolysed*** and hence their aqueous solutions are ***neutral.*** For example, consider the solution of sodium chloride.

$$NaCl(aq) + H_2O(l) \rightleftharpoons \underset{\text{Strong base}}{NaOH(aq)} + \underset{\text{Strong acid}}{HCl(aq)}$$

NaCl, NaOH and HCl are completely dissociated.

$$\therefore Na^+(aq) + Cl^-(aq) + H_2O(l) \rightleftharpoons Na^+(aq) + OH^-(aq) + H^+(aq) + Cl^-(aq)$$

or $$H_2O(l) \rightleftharpoons H^+(aq) + OH^-(aq)$$

or $$2H_2O(l) \rightleftharpoons H_3O^+(aq) + OH^-(aq)$$

Do You Know?

Hydrolysis is the reaction between salt and water to form acid and base.

For example,

$$CH_3COONa(aq) + H_2O(l) \rightleftharpoons CH_3COOH(aq) + NaOH(aq)$$

Hydration is the process in which an anhydrous salt combines with a specific number of moles of water to form hydrated salt. For example,

$$CuSO_4(s) + 5H_2O(l) \longrightarrow CuSO_4.5H_2O(s)$$

29.4 LABORATORY AND INDUSTRIAL PRODUCTION OF SALTS

Salts can be prepared by a veriety of methods.

LABORATORY PRODUCTION

The general methods of preparation of soluble salts and insoluble salts are different. Before a method is selected for the preparation of a salt, we must know whether the salt is soluble in water or it is insoluble.

1. By the action of acids on metals. For example,

$$Zn(s) + H_2SO_4(aq) \longrightarrow ZnSO_4(aq) + H_2(g)$$

This method is not suitable for preparing salts of sodium, potassium and calcium because these metals react violently with acids.

2. By the action of an acid on an insoluble oxide or hydroxide. For example,

$$ZnO(s) + H_2SO_4(aq) \longrightarrow ZnSO_4(aq) + H_2O(l)$$

3. By the action of an acid on a soluble hydroxide (alkali)–By titration. For example,

$$2NaOH(aq) + H_2SO_4(aq) \longrightarrow Na_2SO_4(aq) + 2H_2O(l)$$

$$NaOH(aq) + H_2SO_4(aq) \longrightarrow NaHSO_4(aq) + H_2O(l)$$

4. By the action of an acid on a trioxocarbonate(IV). For example,

$$CaCO_3(s) + 2HCl(aq) \longrightarrow CaCl_2(aq) + CO_2(g) + H_2O(l)$$

$$ZnCO_3(s) + H_2SO_4(aq) \longrightarrow ZnSO_4(aq) + CO_2(g) + H_2O(l)$$

5. By double decomposition reaction. This method is suitable for the preparation of insoluble salts. For example

$$NaCl(aq) + AgNO_3(aq) \longrightarrow AgCl\downarrow + NaNO_3(aq)$$

$$BaCl_2(aq) + H_2SO_4(aq) \longrightarrow BaSO_4\downarrow + 2HCl(aq)$$

6. By direct combination of elements. This method is suitable for preparing anhydrous salts. For example,

$$2Na + Cl_2 \longrightarrow 2NaCl$$

$$Fe + S \longrightarrow FeS$$

$$2Fe + 3Cl_2 \longrightarrow 2FeCl_3$$

Now let us study the laboratory preparation of some specific salts.

ACTIVITY 29.2	**Preparation of Zinc Tetraoxosulphate(VI), $ZnSO_4.7H_2O$ (White Vitriol)**

Zinc tetraoxosulphate(VI) is a water soluble salt. It can be prepared in the laboratory by the action of dilute H_2SO_4 on zinc granules.

$$Zn + \text{dil. } H_2SO_4 \longrightarrow ZnSO_4 + H_2\uparrow$$

1. Take about 40 cm^3 of 4 M H_2SO_4 in a 250 cm^3 beaker and add about 10 g of zinc granules to the acid. Zinc reacts with dilute H_2SO_4 and hydrogen gas is evolved.
2. When the evolution of gas ceases, it indicates that the reaction is over. Filter the solution to remove unreacted zinc.
3. The solution thus obtained is heated in a china dish and concentrated to crystallisation point.
4. Cool the solution. On cooling, needle shaped crystals of white vitriol ($ZnSO_4.7H_2O$) are obtained. The crystals are separated and dried between the folds of a paper.

ACTIVITY 29.3	**Preparation of Copper(II) Tetraoxosulphate (VI), $CuSO_4.5H_2O$ (Blue Vitriol)**

Copper(II) tetraoxosulphate(VI) can be prepared by the reaction of dilute H_2SO_4 with copper(II) oxide

$$CuO + H_2SO_4 \longrightarrow CuSO_4 + H_2O$$

1. Take about 50 cm^3 of dilute H_2SO_4 in a 250 cm^3 beaker and warm it gently.
2. Add black copper(II) oxide to the above acid in small portions. Initially, Copper(II) oxide reacts with the acid and dissolves. When no more of the oxide dissolves, the solution is filtered to remove unreacted copper(II) oxide.
3. The bluish filtrate, thus obtained, is concentrated by heating in a china dish. When the hot saturated solution is obtained, the solution is allowed to cool undisturbed.
4. On cooling, blue crystals of blue vitriol ($CuSO_4.5H_2O$) are formed. The crystals are separated and dried between the folds of a filter paper.

INDUSTRIAL PRODUCTION OF COMMON SALT

Salt is currently produced on large scale by *evaporation of sea water* or brine from other sources, such as brine wells and salt lakes, and by *mining of rock salt.* The mineral form is called *halite.*

In order to extract salt from sea water, sea water is pumped into shallow pans or pits. As the wind and sun begin to evaporate the water, the salt content is increased and the water is moved by gravity through a sequence of pans of steadily increasing concentration. When saturation is reached deposition begins in the crystallising pans.

For use within the chemical industry the salt is directly used as such. For agricultural purposes and human consumption the salt has to undergo further purification.

Table salt sold for consumption is not pure sodium chloride. In order to reduce the incidence of simple goiter trace amounts of iodine in the form of sodium iodide, potassium iodide or potassium trioxoiodate(V) is added. Iodized table salt contains 0.006% to 0.01% potassium iodide. In order to make it flow more freely magnesium carbonate may be added.

Do You Know?

Iodine deficiency in humans leads to thyroid gland problems such as goiter. The daily recommended requirement of iodine in our diet is about 150 microgram.

Uses of Common Salt (Sodium Chloride)

1. Sodium chloride is used as a food preservative to preserve food items such as pickles, fish etc. This use is based on its hygroscopic properties.
2. It is an essential constituent of our food. It is used as common salt or table salt for domestic purposes.
3. It is used in the manufacture of sodium, caustic soda, chlorine, hydrogen, washing soda, etc.
4. It is used for salting out soap during manufacture of soap.
5. In cold countries, large quantities of salt are used to clear highways of ice during winter.
6. It is used in freezing mixture.

SOLVED EXAMPLE

Example 29.1. *Assuming iodized table salt contains 0.006% KI by mass, how much iodized salt would a person consume each day in order to get the recommended 150 microgram of iodine?*

Solution. Suppose the mass of iodized salt needed to meet the daily recommended dose of iodine = x g

$$\text{Mass of KI in it} = \frac{x \times 0.006}{100} \text{ g}$$

$$= x \times 6 \times 10^{-5} \text{ g}$$

$$\text{Molar mass of KI} = (39 + 127) \text{ g mol}^{-1}$$

$$= 166 \text{ g mol}^{-1}$$

166 g of KI contains iodine = 127 g

$\therefore \quad x \times 6 \times 10^{-5}$ g of KI would supply iodine

$$= \frac{x \times 6 \times 10^{-5} \times 127}{166} \text{ g}$$

But the daily requirement of iodine

$$= 150 \text{ microgram} = 150 \times 10^{-6} \text{ g}$$

$$\therefore \quad \frac{x \times 6 \times 10^{-5} \times 127}{166} \text{ g} = 150 \times 10^{-6} \text{ g}$$

$$x = \frac{150 \times 10^{-6} \times 166}{6 \times 10^{-5} \times 127} = \mathbf{3.27 \text{ g.}}$$

29.5 SOURCES AND CHEMICALS FROM SODIUM CHLORIDE

Sodium chloride (NaCl) is known as common salt. It is present in sea water along with many other salts. Sodium chloride is separated from these salts. It also exists as solid salt, in the form of rocks, in several parts of the world. The large crystals of sodium chloride obtained from rocks are generally brown in colour due to impurities present in them. This is known as **Rock salt.** Rock salt is mined just like any other mineral. Common salt is an important component of our food. It is also used for the extraction of sodium metal.

Both inland and oceanic sources of salt exist in Ghana. Ghana has a favourable coast line and has a potential for production of salt by evaporation of sea water. The production of salt by solar evaporation is the main method used for production of salt in Ghana. In Ghana, production of salt from sea water is concentrated mainly in Central, Volta and Greater Accra Region where climatic conditions are more favourable.

Areas such as the Keta Lagoon, the Songor Lagoon and the Amisa Lagoon are the dominant mining zones.

Common salt is an important raw material for many materials of daily use. Some of these compounds are: Caustic soda (NaOH), Washing soda ($Na_2CO_3.10H_2O$) and baking soda ($NaHCO_3$).

Now let us study the preparation and uses of some of these compounds.

SODIUM HYDROXIDE, CAUSTIC SODA (NaOH)

Preparation

Sodium hydroxide is prepared by passing electricity through an aqueous solution of sodium chloride (also known as brine) in a specially designed cell. As a result of electrolysis chlorine gas is given off at the anode and hydrogen gas at the cathode. Sodium hydroxide solution is formed near the cathode.

$$\underset{\text{Brine}}{2NaCl(aq)} + 2H_2O(l) \xrightarrow{\text{Electricity}} \underset{\substack{\text{Caustic soda} \\ \textbf{(Near cathode)}}}{2NaOH(aq)} + \underset{\substack{\text{Chlorine} \\ \textbf{(At anode)}}}{Cl_2(g)} + \underset{\substack{\text{Hydrogen} \\ \textbf{(At cathode)}}}{H_2(g)}$$

This process is known as **chlor-alkali process** due to the formation of chlorine and sodium hydroxide (an alkali) as the products.

Properties of Sodium Hydroxide

1. Sodium hydroxide is a white, transluscent solid. It melts at 591 K.
2. It is readily soluble in water to give a strongly alkaline solution.
3. Crystals of sodium hydroxide are deliquescent.
4. On prolonged exposure, CO_2 is absorbed resulting in the formation at the surface, a layer of Na_2CO_3.
5. It is highly soluble in water and the solution has a very corrosive action on animal tissue.

All the three products formed in chlor-alkali process are used for variety of purposes.

Uses of Sodium Hydroxide

1. It is used in the **manufacture of paper, soap** and **artificial silk.**
2. It is used in **petroleum refining.**
3. It is used for **mercerizing cotton.**
4. It is used for the **preparation of sodium metal** and many **salts of sodium.**
5. It is used for the **purification of bauxite.**
6. It is an important **laboratory reagent.**

Uses of Chlorine

1. Chlorine is used for **disinfection of drinking water** and swimming pools.
2. It is used in the **manufacture of bleaching agent** ($CaOCl_2$).
3. It is used in the **preparation** of many **compounds** such as $NaClO_3$, $KClO_3$, NaClO, HCl, CCl_4, $CHCl_3$ and vinyl chloride (Chloroethene). Vinyl chloride on polymerization yields polyvinylchloride or **PVC.**
4. It is used in the preparation of insecticides and pesticides.

Uses of Hydrogen

1. Hydrogen is used for the **manufacture of** important compounds such as **ammonia (NH_3), hydrogen chloride (HCl), methanol (CH_3OH),** etc.

2. It is used for **hydrogenation of oils.**
3. Hydrogen is used for filling **balloons for weather, search light**, etc.
4. Hydrogen is also used **as a fuel.**

WASHING SODA ($Na_2CO_3.10H_2O$)

Preparation

Washing soda is prepared from sodium chloride by the **Ammonia-soda process** or **Solvay process.** In this process, CO_2 gas is bubbled through a brine solution saturated with ammonia. It results in the formation of sodium hydrogen-trioxocarbonate(IV).

$$NaCl + H_2O + CO_2 + NH_3 \longrightarrow \underset{\text{Ammonium chloride}}{NH_4Cl} + \underset{\text{Sodium hydrogen-trioxocarbonate (IV)}}{NaHCO_3}$$

$NaHCO_3$ so formed precipitates out in the presence of excess of sodium chloride. It is filtered off and then ignited to get sodium trioxocarbonate(IV) (Na_2CO_3).

$$\underset{\text{Sodium hydrogen-trioxo carbonateIV}}{2NaHCO_3} \longrightarrow \underset{\text{Sodium trioxo-carbonate(IV)}}{Na_2CO_3} + CO_2 + H_2O$$

Anhydrous sodium trioxocarbonate(IV) thus formed is called **soda ash.** When soda ash is dissolved in water and subjected to crystallisation, the crystals that separate out are of sodium trioxocarbonate(IV) decahydrate ($Na_2CO_3.10H_2O$) which is also known as **washing soda.**

$$\underset{\text{Soda ash}}{Na_2CO_3} + 10H_2O \longrightarrow \underset{\text{Washing soda}}{Na_2CO_3 . 10H_2O}$$

Uses of Sodium Trioxocarbonate(IV)

1. Large quantities of sodium trioxocarbonate(IV) are used in the manufacture of glass, borax, soap and caustic soda.
2. It is used in the paper, paints and textile industries.
3. It is used for softening hard water. It removes temporary as well as permanent hardness.
4. It is used for washing purposes in the laundry.
5. As an important laboratory reagent both in qualitative and quantitative analysis.

BAKING SODA, SODIUM HYDROGENTRIOXO-CARBONATE(IV) ($NaHCO_3$)

Preparation

Sodium hydrogentrioxocarbonate(IV) is obtained as a primary product in the Solvay's process for the manufacture of Na_2CO_3.

$$NaCl + NH_3 + CO_2 + H_2O \longrightarrow NH_4Cl + NaHCO_3$$

It can also be prepared by passing CO_2 through a solution of Na_2CO_3 whereby $NaHCO_3$ being less soluble crystallizes out.

$$Na_2CO_3 + CO_2 + H_2O \longrightarrow 2NaHCO_3$$

Uses of Sodium Hydrogentrioxocarbonate(IV)

1. It is used as a component of **baking powder.** In addition to sodium hydrogentrioxocarbonate(IV), baking powder contains mild edible acid such as tartaric acid or some other similar acidic compound. When mixed with water and heated, sodium hydrogentrioxocarbonate(IV) reacts with the acidic component of the baking powder producing CO_2 which causes bread or cake to swell and become light. Tartaric acid present in baking powder neutralizes sodium trioxocarbonate(IV) which would otherwise impart bitter taste to the food item.
2. It is used in **soda-acid fire extinguishers.**

   Soda-acid fire extinguishers contain a solution of $NaHCO_3$ and H_2SO_4. These two chemicals are brought into contact by pressing a knob or by inverting the extinguisher. CO_2 is produced which forces a stream of effervescing liquid on the fire. CO_2 surrounds the combustible substance and cuts off the supply of air. Thus, the fire gets extinguished.
3. It is used in **medicines.** It acts as **mild antiseptic** for skin infections. It is also present as an ingredient in **antacids.** Being alkaline it neutralizes excess acid in the stomach.

Do You Know?

EFFLORESCENCE

Certain hydrated crystalline salts when exposed to the atmosphere lose their water of crystallisation spontaneously and change into amorphous powder. *The spontaneous loss of water of crystallisation, wholly or partly, when crystals with water of crystallisation are exposed to air is called efflorescence and the substances exhibiting efflorescence are called efflorescent substances.*

For example,

Washing soda ($Na_2CO_3 . 10 H_2O$),
Glauber's salt ($Na_2SO_4 . 10 H_2O$),
Blue vitriol ($CuSO_4 . 5 H_2O$) and
Epsom salt ($MgSO_4 . 7 H_2O$).

DELIQUESCENCE

Certain crystalline substances when exposed to the atmosphere absorb moisture and change into solution.

The absorption of moisture from air by crystals to form a solution is called deliquescence.

Sodium hydroxide, potassium hydroxide, calcium chloride etc., are deliquescent substances.

HYGROSCOPIC SUBSTANCES

Certain substances absorb water from the atmosphere without undergoing change in physical state. Such substances are known as **hygroscopic substances.**

Anhydrous sodium trioxocarbonate(IV), anhydrous copper tetraoxosulphate(VI), concentrated tetraoxosulphate(VI) acid are examples of hygroscopic substances.

SUMMARY

- **Salt** is a compound formed by the replacement of either all or a part of the ionizable hydrogen atoms of an acid by some other cation.
- A salt is an ionic compound.
- **Types of Salts:**
  - (*i*) Normal salts: $NaCl$, Na_2CO_3
  - (*ii*) Acid salts: $NaHCO_3$, $NaHSO_4$
  - (*iii*) Basic salts: $Mg(OH)Br$
  - (*iv*) Double salts: $FeSO_4.(NH_4)_2SO_4.6H_2O$
  - (*v*) Complex salts: $K[Ag(CN)_2]$
- Salts on hydrolysis may yield acidic, basic or neutral solutions.
- **Hydrolysis** is the interaction between water and salt to form an acid and a base.
- **Chlor-alkali process** involves electrolysis of an aqueous solution of sodium chloride. Its products are NaOH, Cl_2 and H_2.
- Common salt is obtained from sea-water and from rocks.
- Iodized table salt contains potassium iodide or potassium iodate.

EVALUATE YOURSELF

I. Objective Type Questions

Select the most appropriate choice from the options given as (a), (b), (c) and (d) after each question:

1. Which of the following salts is an acid salt?
   (*a*) $NaHCO_3$ (*b*) Na_2CO_3
   (*c*) $CaCl_2$ (*d*) NaCl.
2. Which of the following salts when dissolved in water would produce a solution having a pH less than 7?
   (*a*) KCl (*b*) CH_3COONa
   (*c*) NH_4Cl (*d*) Na_2SO_4.
3. Which of the following compounds is not a product of the chlor-alkali process?
   (*a*) NaOH (*b*) Cl_2
   (*c*) H_2 (*d*) Na_2CO_3.

II. Fill in the Blanks

4. Complete the following sentences by supplying appropriate words:
   (*i*) Salts of strong acids and .... do not undergo hydrolysis.
   (*ii*) Na_2CO_3 solution is .... in nature.
   (*iii*) pH of NH_4Cl solution is .... than 7.
   (*iv*) $K_4[Fe(CN)_6]$ is a/an .... salt.
   (*v*) $FeSO_4.(NH_4)_2SO_4.6H_2O$ is a/an .... salt.

III. Discussion Questions

5. What is a salt? How is it formed?
6. What is a normal salt and an acid salt? Give two examples of each type.
7. What is the pH of the following salt solutions as compared to that of pure water?
   (*i*) NaCl (*ii*) $Zn(NO_3)_2$
   (*iii*) CH_3COOK (*iv*) $CaCl_2$.
8. Give an example of an acid salt whose aqueous solution is basic in nature. Explain how this salt can behave as an acid and as a base.
9. What are double salts and complex salts? Give two examples of each type.
10. Define hydrolysis of salts. Explain why an aqueous solution of Na_2CO_3 is alkaline while that of NH_4Cl is acidic.
11. The colour of phenolphthalein is pink in an aqueous solution of a salt. The salt can be
    (*a*) NH_4Cl (*b*) NaCl
    (*c*) Na_2CO_3 (*d*) KBr.
12. Give any two methods for the preparation of salts in the laboratory.
13. How is common salt prepared on an industrial scale? What are the uses of sodium chloride?
14. What is the chemical formula of rock salt? Give names and formulae of two compounds of daily use which are prepared starting from rock salt.
15. What is chlor-alkali process?
16. What are the different products formed when an aqueous solution of sodium chloride is electrolysed?
17. Give any four uses of caustic soda.

18. Give any two uses of
    (*i*) hydrogen (*ii*) chlorine.
19. Name any three chemicals which are prepared from sodium chloride.
20. What are the uses of
    (*i*) washing soda (*ii*) baking soda?
21. Name a sodium compound which is used for softening of hard water.
22. What is the chemical name of washing soda?
23. Write the chemical name and chemical formula of baking soda.
24. Write the constituents of baking powder. How does it cause the cakes to rise and become light?
25. How is washing soda converted into baking soda?
26. How is baking soda differ from baking powder?
27. Write the chemical name and formula of baking soda and mention any one of its uses.
28. What will happen if the solution of sodium hydrogen-trioxocarbonate(IV) is heated? Give the equation of the reaction involved.
29. A baker found that the cake prepared by him is hard and small in size. Which ingredient has he forgotten to add that would have made the cake fluffy. Give reasons.
30. Name a sodium compound which on heating liberates carbon dioxide gas.

Answers

I. Objective Type Questions

1. (*a*) **2.** (*c*) **3.** (*d*).

II. Fill in the Blanks

4. (*i*) strong bases
(*ii*) basic
(*iii*) less
(*iv*) Complex (*v*) double.

SHS 3

SECTION

5

REDOX REACTIONS AND ELECTROCHEMISTRY

General Objectives

After studying this section, the student should be able to:

1. understand the nature of oxidation-reduction reactions and apply its principles to electrochemical cells.
2. show awareness of corrosion as an oxidation-reduction process and its economic cost.

In this section, we shall study

Unit-30: Oxidation-reduction Processes and Oxidising-reducing Agents

Unit-31: Balancing Redox Reactions

Unit-32: Redox Titrations

Unit-33: Electrochemical Cells

Unit-34: Electrolytic Cells

Unit-35: Corrosion of Metals

SECTION-5

UNIT-30

OXIDATION-REDUCTION PROCESSES AND OXIDISING-REDUCING AGENTS

Specific Objectives

After studying this unit, the student should be able to:

- describe oxidation and reduction processes
- describe the types of redox reactions
- describe half reactions
- describe an experiment to illustrate reactivity of metals
- perform an experiment to illustrate the reactivity of halogens
- describe oxidizing and reducing agents.

In our daily life we come across processes like rusting of iron, fading of the colour of clothes, burning of combustible substances such as cooking gas, wood, coal, etc. All these processes fall in the category of **redox reactions.** A large number of industrial processes like, electroplating, extraction of metals like aluminium and sodium, bleaching of wood pulp, manufacture of caustic soda, etc., are also based upon redox reactions. Redox reactions also form the basis of electrochemical and electrolytic cells. In the present unit we shall understand the meaning of oxidation, reduction, oxidising agents and reducing agents.

30.1 CLASSICAL CONCEPT OF OXIDATION AND REDUCTION

According to the classical concept the following definitions were proposed to explain the process of oxidation and reduction.

Oxidation: It is a process of the *chemical addition of oxygen or any electronegative radical or the removal of hydrogen or any electropositive radical.*

Reduction: It is a process of the *chemical addition of hydrogen or any electropositive radical or the removal of oxygen or any electronegative radical.*

Some examples of oxidation and reduction reactions are given below:

(*i*) *Reaction of PbO and carbon*

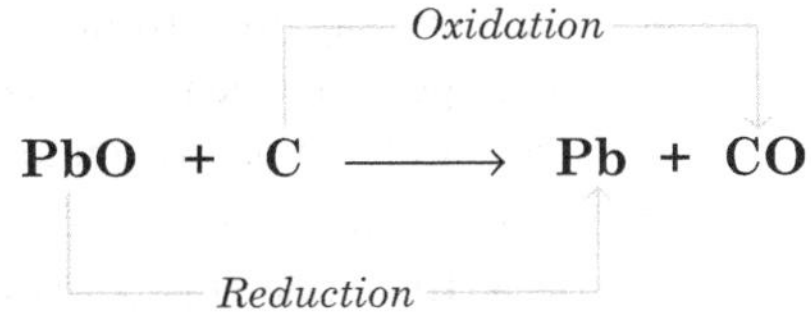

Here, oxygen is being removed from lead oxide (PbO) and is being added to carbon (C). Thus, PbO is *reduced* while C is *oxidised.*

(*ii*) *Reaction of H_2S and Cl_2*

Oxidation

$$Cl_2 + H_2S \longrightarrow 2HCl + S$$

Reduction

Here, hydrogen is being removed from hydrogen sulphide (H_2S) and is being added to chlorine (Cl_2). Thus, H_2S is *oxidised* and Cl_2 is *reduced.*

(*iii*) *Reaction between Mg and F_2*

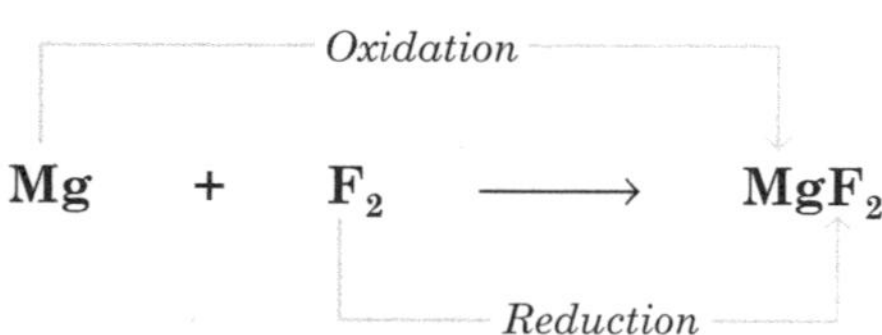

Here, electronegative radicals fluoride ion ($\bar{F}$) is added to magnesium while electropositive radical Mg^{2+} is added to fluorine. Hence, Mg is *oxidised* and F_2 is *reduced.*

Looking at all the reactions given above we observe that oxidation and reduction occurs simultaneously, hence the word **"redox"** was coined for this class of chemical reactions.

30.2 OXIDATION-REDUCTION IN TERMS OF ELECTRON TRANSFER

In order to understand the electronic concept of oxidation and reduction let us study the reaction of magnesium with oxygen. When magnesium is burnt in oxygen it gets oxidised to magnesium oxide (MgO). In the formation of magnesium oxide, two electrons from the magnesium atom are transferred to the oxygen atom.

$$\dot{\dot{M}}g + \ddot{\underset{\cdot\cdot}{O}} : \longrightarrow Mg^{2+} [: \ddot{\underset{\cdot\cdot}{O}} :]^{2-}$$

The process of transference of electrons is described as a redox process. Let us define oxidation and reduction in terms of electrons.

Oxidation is a process in which an atom or a group of atoms taking part in a chemical reaction loses one or more electrons. The loss of electrons results in the increase of positive charge or decrease of negative charge of the species. For example,

$$2Cl^- \longrightarrow Cl_2 + 2e^-$$

Chloride ion → Chlorine molecule

(*Decrease of negative charge*)

$$Fe^{2+} \longrightarrow Fe^{3+} + e^-$$

Iron (II) → Iron (III)

(*Increase of positive charge*)

$$Cu \longrightarrow Cu^{2+} + 2e^-$$

Copper → Copper (II)

(*Increase of positive charge*)

The species which undergo the loss of electrons during the reactions are called **reducing agents** or **reductants.** Cl^-, Fe^{2+} and Cu are reducing agents in the above examples.

Reduction is a process in which an atom or a group of atoms taking part in a chemical reaction gains one or more electrons. The gain of electrons results in the decrease of positive charge or increase of negative charge of the species. For example,

$$Ag^+ + e^- \longrightarrow Ag$$

(*Decrease of positive charge*)

$$Fe^{3+} + e^- \longrightarrow Fe^{2+}$$

Iron (III) → Iron (II)

(*Decrease of positive charge*)

$$Br_2 + 2e^- \longrightarrow 2Br^-$$

Bromine molecule → Bromide ions

(*Increase of negative charge*)

The species which undergo gain of electrons during the reactions are called **oxidising agents** or **oxidants.** In the above examples, Ag^+, Fe^{3+} ions, Br_2 molecule are oxidising agents.

SOLVED EXAMPLES

Example 30.1. *In the reactions given below, identify the species undergoing oxidation and reduction:*

(*i*) $H_2S(g) + Cl_2(g) \longrightarrow 2HCl(g) + S(s)$

(*ii*) $3Fe_3O_4(s) + 8Al(s) \longrightarrow 9Fe(s) + 4Al_2O_3(s)$

(*iii*) $2Na(s) + H_2(g) \longrightarrow 2NaH(s)$.

Solution.

(*i*) H_2S is oxidised because either we consider the addition of the more electronegative element chlorine to hydrogen or we may say that from it the electropositive element hydrogen has been removed. Chlorine is reduced due to the addition of hydrogen to it.

(*ii*) Aluminium is oxidised because oxygen is added to it. Iron (III) oxide (Fe_3O_4) is reduced because from Fe_3O_4, oxygen has been removed.

(*iii*) On the basis of the concept of electronegativity only we may infer that sodium is oxidised and hydrogen is reduced.

Example 30.2. *Identify the oxidising agent and reducing agent in the following reactions:*

(*a*) $Zn(s) + \frac{1}{2} O_2(g) \longrightarrow ZnO(s)$

(*b*) $Zn(s) + 2H^+ (aq) \longrightarrow Zn^{2+} (aq) + H_2 (g)$

(*c*) $10H^+ (aq) + 4Zn(s) + NO_3^- (aq)$

$\longrightarrow 4Zn^{+2}(aq) + NH_4^+ (aq) + 3H_2O(l)$.

Solution. (*a*) Zinc loses electrons and hence is a reducing agent while oxygen is an oxidising agent.

(*b*) Zinc is a reducing agent while hydrogen is an oxidising agent.

(*c*) Zinc is a reducing agent while nitrogen in nitrate ion is reduced and hence is an oxidising agent.

SIMULTANEOUS OCCURRENCE OF OXIDATION AND REDUCTION

Since oxidation involves *loss* of electrons and reduction involves *gain* of electrons, it is evident that if one substance loses electrons, another substance at the same time must gain electrons because electrons cannot be the products in any chemical change. This means that in any process, oxidation can occur only if reduction is also taking place side by side and *vice versa*. Thus, neither oxidation, nor reduction can occur alone. Both processes are complementary and so the *give* and *take* proceed simultaneously. That is why chemical reactions involving reduction-oxidation are called **redox reactions.** In fact, during a redox reaction there is a ***transfer of electrons from the reducing agent to the oxidising agent*** as shown below:

$$\overset{\quad e^- \quad}{\text{A} + \text{X}} \longrightarrow \bar{\text{A}} + \overset{+}{\text{X}}$$

A (Oxidising agent) + X (Reducing agent) ⟶ $\bar{A}$ (Reduced product) + $\overset{+}{X}$ (Oxidised product)

For example, consider a reaction between zinc and copper ions

Oxidised $-2\bar{e}$

$$Zn(s) + Cu^{2+}(aq) \longrightarrow Zn^{2+}(aq) + Cu(s)$$

Zn(s) *(Reducing agent)*; $Cu^{2+}(aq)$ *(Oxidising agent)*

Reduced $+2\bar{e}$

In this reaction zinc atoms lose electrons and are oxidised to zinc ions (Zn^{2+}) whereas cupric ions (Cu^{2+}) gain electrons and are reduced to copper atoms. Thus, cupric ions act as an *oxidising agent* and zinc atoms act as a *reducing agent.* In fact, the oxidising agent gets reduced while reducing agent gets oxidised during a redox reactions.

Oxidation: *Process involving loss of one or more electrons*

Reduction: *Process involving gain of one or more electrons*

Oxidising agents: *Species which gain one or more electrons*

Reducing agents: *Species which lose one or more electrons*

30.3 OXIDATION NUMBER OR OXIDATION STATE

In many covalent reactions such as the reaction between H_2 and Cl_2:

$$H_2(g) + Cl_2(g) \longrightarrow 2HCl(g)$$

the loss and gain of electrons cannot be easily explained. In order to explain the transfer of electrons in either of the species in a more convenient way, the concept of oxidation number has been introduced.

Oxidation number (O.N.) of an element is defined as *the residual charge which its atom has or appears to have when all other atoms from the molecule are assumed to be removed as ions by counting the shared electrons with the more electronegative atom.*

For example, in hydrogen chloride molecule, chlorine is more electronegative than hydrogen. Therefore, the shared pair is counted towards chlorine atom as shown below:

H :C̈l:

As a result of this, chlorine gets one extra electron and acquires a unit negative charge. Hence, the oxidation number of chlorine is –1. On the other hand, the hydrogen atom without electron has a unit positive charge. Hence, the oxidation number of hydrogen in hydrogen chloride is +1.

It may be noted that *electrons shared between two similar atoms are divided equally between the sharing atoms.* Hence in molecules like H_2, Cl_2, Br_2 the oxidation number of each element is zero.

RULES FOR ASSIGNING OXIDATION NUMBER TO AN ATOM

These rules have been formulated on the basis of the assumption that electrons in a covalent bond belong entirely to the more electronegative atom.

1. *The oxidation number of an element in the free or elementary state is always zero irrespective of its allotropic form.*

   For example,

   Oxidation number of helium in $He = 0$

   Oxidation number of chlorine in $Cl_2 = 0$

   Oxidation number of sulphur in $S_8 = 0$

   Oxidation number of phosphorus in $P_4 = 0$.
2. *The oxidation number of an element in a monoatomic ion is equal to the charge on the ion.* For example, in K^+Cl^-, the oxidation number of K is **+1** while that of Cl is **–1.** In a similar way, the oxidation number of all the alkali metals is **+1** while those of alkaline earth metals is **+2** in their compounds.
3. *The oxidation number of fluorine is always* **–1** *in all its compounds.* Other halogens (Cl, Br and I) also have an oxidation number of **– 1,** when they occur as halide ions in their compounds. However, in oxoacids and oxoanions they have positive oxidation numbers.
4. *Hydrogen is assigned the oxidation number* **+1** *in all its compounds except in metal hydrides. In metal hydrides like NaH, MgH_2, CaH_2, LiH, etc., the oxidation number of hydrogen is* **–1.**
5. *Oxygen is assigned the oxidation number* **–2** *in most of its compounds, however, in peroxides* (which contain O—O linkage) like H_2O_2, BaO_2, Na_2O_2, etc., its oxidation number is **–1.** Similarly, the exception also occurs in compounds of fluorine and oxygen like OF_2 (F—O—F) and O_2F_2 (F—O—O—F) in which the oxidation number of oxygen is **+2** and **+1** respectively.
6. *In accordance with the principle of conservation of charge, the algebraic sum of the oxidation numbers of all the atoms in a molecule is* **zero.** *But in the case of polyatomic ions the sum of oxidation numbers of all its atoms is equal to the* **charge on the ion.**
7. *In binary compounds of metal and non-metals, the metal atom has a positive oxidation number while the non-metal atom has a negative oxidation number.* For example, the O.N. of K in KI is +1 but O.N. of I is –1.
8. *In binary compounds of non-metals, the more electronegative atom has a negative oxidation number, but the less electronegative atom has a positive oxidation number.* For example, the O.N. of Cl in ClF_3 is positive (+3) while that in ICl is negative (–1).

BRIEF SUMMARY OF RULES

Oxidation Number of:

- **Free elements = 0**
- **Fluorine = – 1**
- **Simple ions = Charge on them**
- **Oxygen = – 2 ; peroxides (– 1) ; F_2O (+ 2) ; F_2O_2 (+ 1)**
- **Hydrogen = + 1 ; metal hydrides (– 1)**
- **Sum of O.N. of atoms in molecules = 0**
- **Sum of O.N. of atoms in polyatomic ions = (Charge on ion).**

By applying the above rules, we can find the oxidation number of the desired element in a molecule or in an ion. It may be noted that if a molecule contains two or more atoms of the same element, then the oxidation number calculated by these rules is an *average oxidation number*.

Let us apply the above rules to calculate the oxidation number of some elements.

SOLVED EXAMPLE

Example 30.3. *Calculate the oxidation number of the underlined elements in the following species:*

$\underline{C}O_2$, $\underline{Cr}_2O_7^{2-}$, $\underline{Pb}_3O_4$, $\underline{C}H_2Cl_2$, $\underline{P}O_4^{3-}$, $\underline{S}_2O_3^{2-}$,

$H\underline{N}O_3$, $K_2\underline{Mn}O_4$, $\underline{N}H_4^+$.

Solution. (*i*) **C in CO_2 :**

Let oxidation number of C be x

O.N. of each O atom $= -2$

Sum of O.N. of all atoms $= x + 2(-2) = x - 4$

As it is a neutral molecule, the sum must be equal to zero.

$\therefore \quad x - 4 = 0 \quad \text{or} \quad x = +4.$

(*ii*) **Cr in $Cr_2O_7^{2-}$:**

Let oxidation number of Cr = x

O.N. of each O atom $= -2$

Sum of O.N. of all atoms $= 2x + 7(-2) = 2x - 14$

Sum of O.N. must be equal to the charge on the ion.

Thus, $2x - 14 = -2$

$$x = \frac{2+14}{2} = \mathbf{+6}.$$

(*iii*) **Pb in Pb_3O_4:**

Let O.N. of Pb be x

O.N. of each O atom $= -2$

Sum of O.N. of all atoms $= 3x + 4(-2) = 3x - 8$

The sum of O.N. must be equal to zero

$\therefore \quad 3x - 8 = 0$ or $x = \frac{8}{3} = \mathbf{+2\frac{2}{3}}$.

(*iv*) **C in CH_2Cl_2:**

Let O.N. of C be x

O.N. of each H atom $= +1$

O.N. of each Cl atom $= -1$

Sum of O.N. of all atoms

$= x + 2(+1) + 2(-1) = x + 2 - 2 = x$

In molecule, the sum must be zero $x = \mathbf{0}$.

(*v*) **P in PO_4^{3-}:**

Let O.N. of P be x

O.N. of each O atom $= -2$

Sum of O.N. of all atoms $= x + 4(-2) = x - 8$

The sum must be equal to the total charge, *i.e.*, -3

$\therefore \quad x - 8 = -3$

$x = -3 + 8 = \mathbf{+5}.$

(*vi*) **S in $S_2O_3^{2-}$:**

Let O.N. of S be x

O.N. of each O atom $= -2$

Sum of O.N. of each atoms

$= 2x + 3(-2) = 2x - 6$

Sum must be equal to the charge, *i.e.*, -2

$\therefore \quad 2x - 6 = -2$

$$x = \frac{-2+6}{2} = \mathbf{+2}.$$

(*vii*) **N in HNO_3:**

Let O.N. of N be x

O.N. of each O atom $= -2$

O.N. of each H atom $= +1$

Sum of O.N. of various atoms

$= x + 1 + 3(-2) = x + 1 - 6 = x - 5$

In molecule, sum must be equal to zero

$\therefore \quad x - 5 = 0$ or $x = \mathbf{+5}.$

(*viii*) **Mn in K_2MnO_4:**

Let O.N. of Mn be x

O.N. of each O atom $= -2$

O.N. of each K atom $= +1$

Sum of O.N. of all atoms

$= 2 + x + 4(-2)$

$= 2 + x - 8 = x - 6$

Sum must be equal to zero

$\therefore \quad x - 6 = 0$

$x = \mathbf{+6}.$

(*ix*) **N in NH_4^+:**

Let O.N. of N be x

O.N. of each H atom $= +1$

Sum of O.N. of all atoms

$= x + 4(+1) = x + 4$

Sum must be equal to the change *i.e.*, $+1$

$\therefore \quad x + 4 = +1$

$x = \mathbf{-3}.$

30.4 OXIDATION AND REDUCTION IN TERMS OF OXIDATION NUMBER

After having discussed the concept of oxidation number, let us now define oxidation and reduction in terms of oxidation number. **Oxidation** is defined as *a chemical process in which the oxidation number of an element increases.* On the other hand, **reduction** is defined as *a chemical process in which oxidation number of an element decreases.* For example, let us consider the reaction between hydrogen sulphide and bromine to give hydrogen bromide and sulphur.

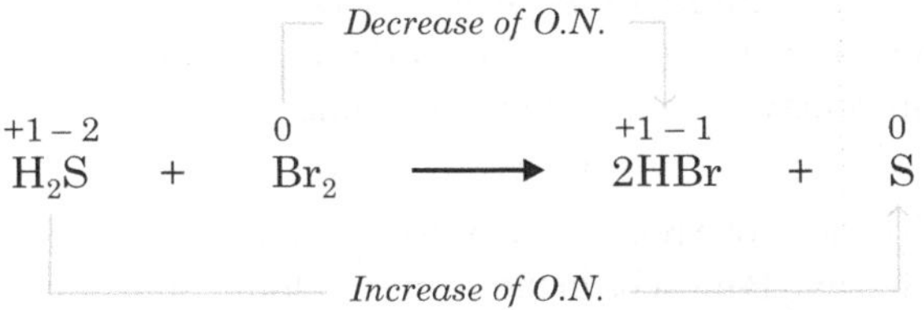

In the above example, the oxidation number of bromine decreases from 0 to -1, thus, it is reduced. The oxidation number of S increases from -2 to 0. Hence, H_2S is oxidised.

Let us now define oxidising and reducing agents in the light of the concept of oxidation number:

Oxidising agent *is a substance which undergoes a* ***decrease in the oxidation number*** *of one or more of its elements.*

Reducing agent *is a substance which undergoes an* ***increase in the oxidation number*** *of one or more of its elements.*

In the above example, H_2S is a reducing agent while Br_2 is an oxidising agent.

OXIDATION HALF AND REDUCTION HALF REACTIONS

Every redox reaction can be split up into two **half reactions.** The part of the redox reaction which represents loss of electrons or increase in oxidation number, is called **oxidation half reaction** while the other part which represents gain of electrons or decrease in oxidation number is called **reduction half reaction.** Some examples are given below:

(*i*) The reaction: $\mathbf{Zn + Cu^{2+} \longrightarrow Zn^{2+} + Cu}$, can be split up into two half equations as

$$Zn \longrightarrow Zn^{2+} + 2\bar{e}$$

(*Oxidation half reaction*)

$$Cu^{2+} + 2\bar{e} \longrightarrow Cu$$

(*Reduction half reaction*)

(*ii*) The reaction: $\mathbf{Sn^{2+} + 2Hg^{2+} \longrightarrow Sn^{4+} + Hg_2}$, can be split up into half reactions as

$$Sn^{2+} \longrightarrow Sn^{4+} + 2\bar{e}$$

(*Oxidation half reaction*)

$$2Hg^{2+} + 2\bar{e} \longrightarrow Hg_2$$

(*Reduction half reaction*)

30.5 TYPES OF REDOX REACTIONS

Some important types of redox reactions are described as follows:

1. Combination Reactions

These are reactions in which two species (atoms or molecule) combine to form a single species. A combination reactions may be denoted in the manner:

$$A + B \longrightarrow C$$

For the above reaction to be a redox reaction, either A or B or both A and B must be in the elemental form. Some important examples of this category are:

$$\overset{0}{C}(s) + \overset{0}{O_2}(g) \longrightarrow \overset{+4\,-2}{CO_2}(g)$$

$$3\overset{0}{Mg}(s) + \overset{0}{N_2}(g) \longrightarrow \overset{+2}{Mg_3}\overset{-3}{N_2}(s)$$

2. Decomposition Reactions

Decomposition reactions are just the reverse of combination reactions. *A decomposition reaction involves the breakdown of a compound into two or more components, one of which must at least be in the elemental state.* Some examples of this class of reactions are:

$$2\overset{+1\,-2}{H_2O}(l) \xrightarrow{\text{Heat}} 2\overset{0}{H_2}(g) + \overset{0}{O_2}(g)$$

$$2\overset{+1\,-1}{NaH}(s) \xrightarrow{\text{Heat}} 2\overset{0}{Na}(s) + \overset{0}{H_2}(g)$$

$$2\overset{+1+5-2}{KClO_3}(s) \xrightarrow{\text{Heat}} 2\overset{+1-1}{KCl}(s) + 3\overset{0}{O_2}(g)$$

3. Combustion Reactions

Combustion reaction is a special category of combination reaction in which one of the element is oxygen. Some examples are:

$$\overset{+2-2}{CO}(g) + \overset{0}{O_2}(g) \longrightarrow 2\overset{+4-2}{CO_2}(g)$$

$$\overset{-4\,+1}{CH_4} + 2\overset{0}{O_2} \longrightarrow \overset{+4}{C}\overset{-2}{O_2} + 2\overset{+1}{H_2}\overset{-2}{O}$$

$$\overset{0}{S} + \overset{0}{O_2} \longrightarrow \overset{+4-2}{SO_2}$$

4. Displacement Reactions

In these reactions, an atom or ion in a compound is replaced by an atom or ion of some other element. In general, it is represented by the equation,

$$X + YZ \longrightarrow XZ + Y$$

Here, from the compound YZ, the atom Y has been displaced by another atom X.

Types of displacement reactions. Displacement reactions are of the following two types:

(*a*) *Metal-displacement reactions*

(*b*) *Non-metal displacement reactions*

(*a*) **Metal-displacement reactions.** In these reactions, a metal in the compound is displaced by some other metal in the elemental state. For example,

$$\overset{+2}{Cu}\overset{+6}{S}\overset{-2}{O_4}(aq) + \overset{0}{Zn}(s) \longrightarrow \overset{0}{Cu}(s) + \overset{+2}{Zn}\overset{+6}{S}\overset{-2}{O_4}$$

$$\overset{+4}{Ti}\overset{-1}{Cl_4}(s) + 2\overset{0}{Mg}(s) \xrightarrow{\Delta} \overset{0}{Ti}(s) + 2\overset{+2}{Mg}\overset{-1}{Cl_2}(s)$$

$$\overset{+3}{Cr_2}\overset{-2}{O_3}(s) + 2\overset{0}{Al}(s) \xrightarrow{\Delta} \overset{+3}{Al_2}\overset{-2}{O_3}(s) + 2\overset{0}{Cr}(s)$$

Here, the metal causing displacement is a better reducing agent than the metal undergoing displacement.

(*b*) **Non-metal displacement reactions.** *In these reactions, a metal or a non-metal displaces another non-metal from its compound.* In most of these reactions, the non-metal getting displaced is *hydrogen*. However, there are some reactions which involve the displacement of *oxygen* or *halogens*. Let us study some reactions in which *hydrogen is being displaced.* Depending upon the capability of the reducing metal or non-metal, the following cases arise:

(*i*) *All alkali metals and some alkaline earth metals (Ca, Sr and Ba) which are very good reducing agents displace hydrogen from cold water.*

$$2\overset{0}{Na}(s) + 2\overset{+1}{H_2}\overset{-2}{O}(l) \longrightarrow 2\overset{+1}{Na}\overset{-2}{O}\overset{+1}{H}(aq) + \overset{0}{H_2}(g)$$

$$\overset{0}{Ca}(s) + 2\overset{+1}{H_2}\overset{-2}{O}(l) \longrightarrow \overset{2+}{Ca}(\overset{+2}{O}\overset{+1}{H})_2(aq) + \overset{0}{H_2}(g)$$

(*ii*) *Less active metals such as magnesium and iron react with steam to produce hydrogen gas*:

$$\overset{0}{Mg}(s) + 2\overset{+1}{H_2}\overset{-2}{O}(g) \longrightarrow \overset{+2}{Mg}(\overset{-2}{O}\overset{+1}{H})_2(aq) + \overset{0}{H_2}(g)$$

$$3Fe(s) + 4\overset{+1}{H_2}\overset{-2}{O}(g) \longrightarrow \overset{+2,\,+3}{Fe_3}\overset{-2}{O_4}(s) + 3\overset{0}{H_2}(g)$$

(*iii*) *Many metals, including those which do not react with cold water, are capable of displacing hydrogen from acids.* For example,

$$\overset{0}{Zn}(s) + 2\overset{+1}{H}\overset{-1}{Cl}(aq) \longrightarrow \overset{+2}{Zn}\overset{-1}{Cl_2}(aq) + \overset{0}{H_2}(g)$$

$$\overset{0}{Fe}(s) + 2\overset{+1}{H}\overset{-1}{Cl}(aq) \longrightarrow \overset{+2}{Fe}\overset{-1}{Cl_2}(aq) + \overset{0}{H_2}(g)$$

Metals like cadmium and tin which do not react with steam also react with acids to displace dihydrogen gas.

$$Cd(s) + 2\overset{+1}{H}\overset{-1}{Cl}(aq) \longrightarrow \overset{+2}{Cd}\overset{-1}{Cl_2}(aq) + \overset{0}{H_2}(g)$$

(*iv*) Very less reactive metals such as silver (Ag) and gold (Au) which may occur in the native state do not react even with dilute hydrochloric acid.

Reactivity of Metals

From the above discussion, it follows that the rate of evolution of H_2 by metals from water and aqueous acids can be used to determine the order of reactivity of metals. For example, sodium (Na) reacts with water at the fastest rate, magnesium (Mg) reacts slowly, iron (Fe) reacts at the slowest rate while silver (Ag) and gold (Au) do not react at all.

The *reactivity* of metals is given in the form of an **activity series** in Table 30.1.

Reactivity of Non-Metals

Like metals, an activity series also exists for **non-metals.** Non-metals have a tendency to accept electrons, therefore, *this reactivity depends upon their* ***oxidising power.*** For example, among ***halogens*** *the oxidising power decreases as we move down group 17 from fluorine to iodine.* Thus, fluorine (F_2) is the strongest oxidising agent. It displaces Cl_2, Br_2 and I_2 from the solution of chloride, bromide and iodide ions respectively. In fact, F_2 is so reactive that it even displaces oxygen from water.

$$\overset{0}{F_2}(g) + 2\overset{+2}{H_2}\overset{-2}{O}(l) \longrightarrow 4\overset{+1}{H}\overset{-1}{F}(aq) + \overset{0}{O_2}(g)$$

On the other hand, chlorine can displace bromine from bromide ions and iodine from iodide ions.

Their corresponding ionic equations are:

$$\overset{0}{Cl_2}(g) + 2\overset{-1}{Br^-}(aq) \longrightarrow 2\overset{-1}{Cl^-}(aq) + \overset{0}{Br_2}(l)$$

$$Cl_2(g) + 2\overset{-1}{I^-}(aq) \longrightarrow \overset{-1}{Cl} + \overset{0}{I_2}(s)$$

Similarly, bromine can displace iodine from iodide ions.

$$Br_2(l) + 2\overset{-1}{I^-}(aq) \longrightarrow 2\overset{-1}{Br^-}(aq) + \overset{0}{I_2}(s)$$

Thus, in general, the reactivity of halogens decreases as

$$\mathbf{F_2 > Cl_2 > Br_2 > I_2.}$$

It may be noted that halogens can also be displaced by oxidation of their corresponding halide ions using suitable chemical oxidising agents.

$$2X^- \xrightarrow{\text{Oxidising agent}} X_2 + 2e^-$$

Although a number of oxidising agents such as $KMnO_4$, $K_2Cr_2O_7$, MnO_2, etc., are available to oxidise Cl^-, Br^- and I^- ions to form Cl_2, Br_2 and I_2 respectively, no oxidising agent is available to oxidise F^- ions to F_2 because F_2 itself is the strongest oxidising agent. Therefore, the only way to prepare F_2 is to oxidise F^- ions electrolytically.

5. Disproportionation Reactions

A reaction in which the same species is simultaneously oxidised as well as reduced is called a **disproportionation reaction.** For such redox reactions to occur, the reacting species must contain an element which has at least three oxidation states. The element in the reacting species is present in the intermediate oxidation state while the higher and lower oxidation states are available for reduction and oxidation to occur.

Some examples, of disproportionation reactions are:

Reduction

$$(i)\quad 2\overset{+}{Cu}(aq) \longrightarrow \overset{0}{Cu}(s) + \overset{2+}{Cu}(aq)$$

Copper (I)

Oxidation

Reduction

$$(ii)\quad 3\overset{0}{Cl_2} + 6\overset{-}{OH} \longrightarrow 5\overset{-}{Cl} + \overset{+5\,-}{ClO_3} + 3H_2O$$

Oxidation

ACTIVITY 30.1 **To study and compare the reactivity of some metals (Zn, Al, Fe and Cu) towards displacement of hydrogen from acid.**

The chemical reaction involved is

$$M(s) + xHCl\ (aq) \longrightarrow MCl_x\ (aq) + \tfrac{x}{2}H_2(g)$$

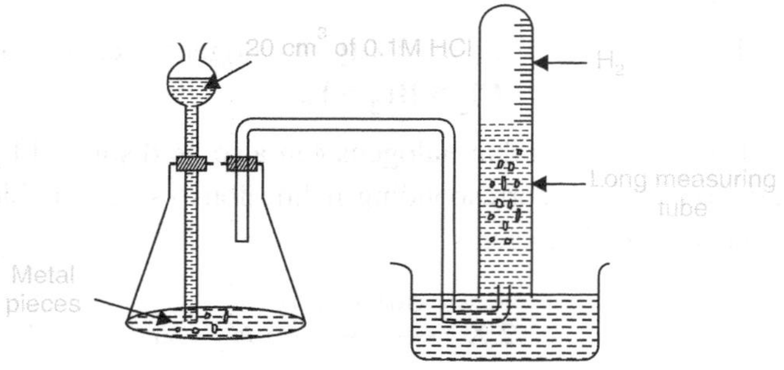

Fig. 30.1. Reaction of metals with HCl.

Requirements

Zinc granules, iron nails or iron filings, aluminium chips, copper scrap about 5 g each. 1.0 dm³ of 0.1 M HCl, conical flasks fitted with two holes, long graduated tube, trough, stop watch etc.

(I) Reaction with Zinc

Procedure

1. Set up the apparatus as shown in Fig. 30.1 (without adding HCl). Clean zinc pieces and put them in the clean and dry flask. Place an inverted graduated tube full of water over the other end of delivery tube. Now, add about 20 cm³ of hydrochloric acid and start the *stop watch*. Observe the rate of bubbling and collect hydrogen for 2 minutes by downward displacement of water.
2. Note the volume of hydrogen after exactly 2 min. and disconnect the delivery tube. Let the volume of H_2 collected be V_1 cm³.
3. Now *repeat* the whole experiment *one after the other* with flasks containing

   (*a*) Aluminium (*b*) Iron (*c*) Cu

   Record the volume of hydrogen in 2 min. in each case make the following table:

Metal used	*Volume of 0.1 M HCl taken*	*Volume of Hydrogen collected in 2 min (V)*	*Rate of evolution of hydrogen* $= \frac{volume}{time}$
Zn	20 cm³	V_1	
Al	20 cm³	V_2	
Fe	20 cm³	V_3	
Cu	20 cm³	V_4	

Rate of evolution of hydrogen reflects the reactivity of metals as ....

General **reactivity of metals** is given below in Table 30.1.

Table 30.1. Activity Series (electromotive series) of Some Metals.

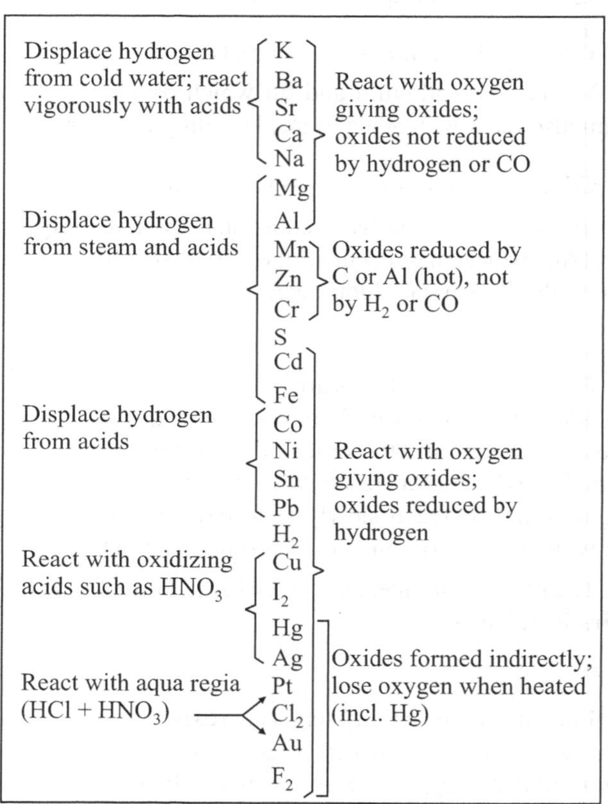

Reactivity of Halogens

Halogens are very good oxidising agents due to their tendency to gain electrons.

$$X_2 + 2e^- \longrightarrow 2X^-.$$

As discussed earlier in section 30.5, the reactivity of halogens as oxidising agents decreases as $Cl_2 > Br_2 > I_2$.

For example,

- **Cl_2 can oxidise Br^- as well as I^- ions**

$$Cl_2\,(aq) + 2Br^-\,(aq) \longrightarrow Br_2\,(aq) + 2Cl^-$$
$$Cl_2\,(aq) + 2I^-\,(aq) \longrightarrow I_2\,(aq) + 2Cl^-$$

- **Br_2 can oxidise only I^- ions but not Cl^- ions**

$$Br_2\,(aq) + I^-\,(aq) \longrightarrow I_2\,(aq) + 2Br^-$$
$$Br_2\,(aq) + Cl^-\,(aq) \longrightarrow \text{No reaction}$$

- **I_2 can neither oxidise Br^- nor Cl^- ions.**

ACTIVITY 30.2	**To investigate the relative reactivity of Halogens.**

Requirements

Aqueous solutions of NaCl, NaBr, NaI, $AgNO_3$, NH_3, etc. Beakers, Test tubes, $CHCl_3$ (*l*).

Procedure

1. Put given solutions of NaCl, NaBr, NaI in beakers A, B, C.

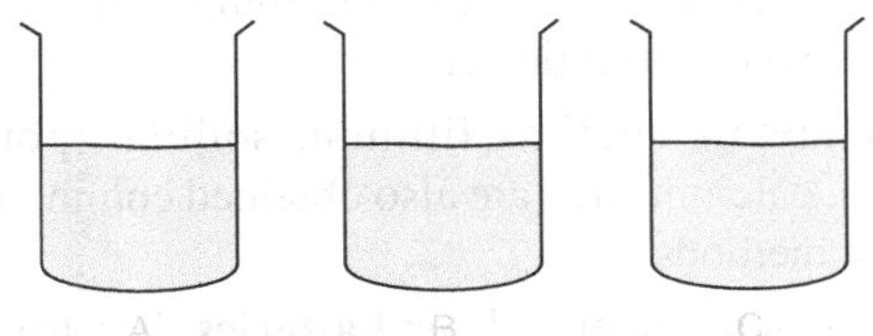

2. Take a small part of the solutions A, B, C (about 1 cm³) in test tube *a*, *b*, *c* respectively.

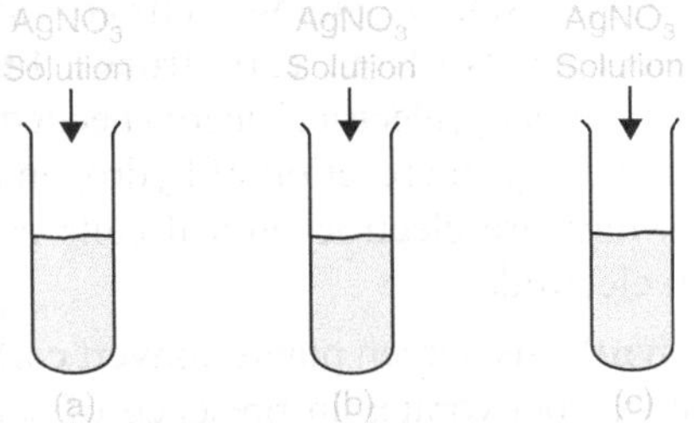

3. Add about 1 cm³ of $AgNO_3$ solution to each tube.
   - What happens ....
   - Note the colour of the precipitate in *a* ......, *b* ......, *c* ......
   - **White precipitate** soluble in NH_4OH indicate Cl^- ion,

   Dull yellow precipitate indicate Br^- ions,

   Bright yellow precipitate insoluble in NH_4OH indicate I^- ion.
4. Based on the colour of the precipitate mark *NaCl* solution as **A**. *NaBr* solution as **B** and *NaI* solution as **C**.
5. Take a small part about 2 cm³ of NaBr and NaI solutions in tubes (*e*) and (*f*) respectively. Add freshly prepared Cl_2 water to each and note the developments of any colour.
   - Change of colour in '*e*' ....
   - Change of colour in '*f*' ..... .

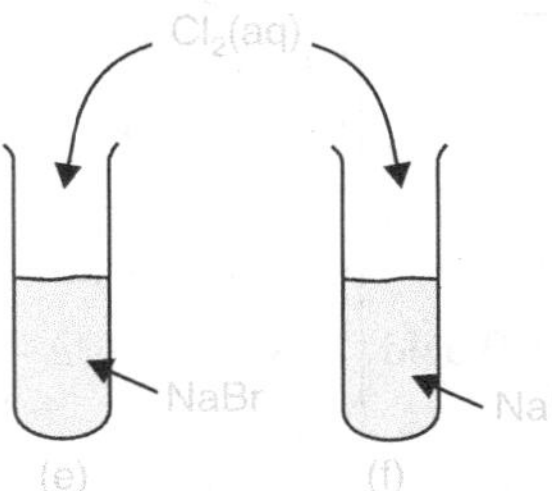

Formation of reddish colour indicates liberation of Br_2

Formation of violet colour indicates liberation of I_2.

6. Add CCl_4 or $CHCl_3$ about 1 cm³ to each tube '*e*' and '*f*' and shake vigorously. Allow it to settle. Note the colour of the $CHCl_3$ layer in each.

   Is there are yellow colour confirms Br_2

   Is there purple colour confirms I_2
   - Write the chemical reactions occurring between I^- ion and Cl_2, and also between Br^- ions and Cl_2.
7. Now, take NaCl and NaI solutions in tubes '*g*', '*h*' and add to them aqueous Br_2.

   Any reaction between Cl^- ions and Br_2?

   Any reaction between I^- ions and Br_2?
8. Again take NaCl and NaI solutions in tubes (*i*), (*j*) respectively. Add to each tube about 1 cm³ of aqueous I_2 solution.

   Notice, if there is any reaction between Cl^- ions and I_2 and also between Br^- ions and I_2

   Write your conclusions on the relative reactivity of halogens.

30.6 FUNCTIONS OF OXIDIZING AND REDUCING AGENTS

The main function of oxidising agents is to bring about the oxidation of a desired material. Similarly, reducing agents are required to bring about the reduction of a desired material in the laboratory. The choice of the oxidising agent/reducing agent to be used depends upon the nature of the material to be oxidised or reduced. The reduction half reactions of some important oxidising agents are given in Table 30.2. Similarly, the oxidation half reactions of some important reducing agents are given in Table 30.3.

Table 30.2. Reduction Half Equations for Some Oxidising Agents

Chlorine	$Cl_2(g) + 2e^- \longrightarrow 2Cl^-$
Bromine	$Br_2(g) + 2e^- \longrightarrow 2Br^-$
Iodine	$I_2(g) + 2e^- \longrightarrow 2I^-$
Oxygen	$O_2(g) + 4e^- \longrightarrow 2O^{2-}(s)$
Tetraoxosulphate (VI) acid	$H_2SO_4(l) + 2e^- \longrightarrow SO_2(g) + 2H_2O + SO_4^{2-}(aq)$
Trioxonitrate (V) ion	$NO_3^-(aq) + 4H^+ + 3e^- \longrightarrow NO(g) + 2H_2O(l)$
Tetraoxomanganate (VII)/OH^-	$MnO_4^- + 2H_2O + 3e^- \longrightarrow MnO_2(s) + 4OH^-(aq)$
Tetraoxomanganate (VII)/H^+	$MnO_4^- + 8H^+ + 5e^- \longrightarrow Mn^{2+}(aq) + 4H_2O(l)$
Oxochlorate (I)/H^+	$ClO^-(aq) + 2H^+ + 2e^- \longrightarrow Cl^-(aq) + H_2O(l)$
Trioxochlorate (V) ion/H^+	$ClO_3^-(aq) + 4H^+ + 4e^- \longrightarrow Cl^-(aq) + H_2O(l)$
Manganese (IV) oxide	$MnO_2(s) + 4H^+ + 4e^- \longrightarrow Mn^{2+}(aq) + 2H_2O(l)$
Hexaoxo-μ-oxochromate (VI)/H^+	$Cr_2O_7^{2-}(aq) + 14H^+ + 6e^- \longrightarrow 2Cr^{3+}(aq) + 7H_2O(l)$
Hydrogenperoxide/H^+	$H_2O_2(aq) + 2H^+ + 2e^- \longrightarrow 2H_2O(l)$
Trioxoiodate (V)/H^+ Conc.	$IO_3^-(aq) + 6H^+ + 6e^- \longrightarrow I^- + 3H_2O$
Trioxoiodate (V) ion/H^+ dil	$IO_3^-(aq) + 12H^+ + 10e^- \longrightarrow I_2 + 6H_2O$
Metals (Fe^{2+}, Cu^{2+}, Ag^+)	$M^{n+}(aq) + ne^- \longrightarrow M$
Iron (III) ions	$Fe^{3+} + 3e^- \longrightarrow Fe(s)$ $Fe^{3+} + e^- \longrightarrow Fe^{2+}(aq)$

Table 30.3. Oxidation Half Reactions of Some Reducing Agents

Carbon	$C(s) + O^{2-}(s) \longrightarrow CO(g) + 2e^-$
Carbon (II) oxide	$CO(g) + O^{2-}(s) \longrightarrow CO_2(g) + 2e^-$
Hydrogen	$H_2(g) + O^{2-}(s) \longrightarrow H_2O(l) + 2e^-$
Sulphide ion	$S^{2-}(aq) \longrightarrow S(s) + 2e^-$
Tin (II) ions/H^+	$Sn^{2+}(aq) \longrightarrow Sn^{4+}(aq) + 2e^-$
Trioxothiosulphate (II) ion	$S_2O_3^{2-}(aq) \longrightarrow S_4O_6^{2-}(aq) + 2e^-$
Trioxosulphate (IV)	$SO_3^{2-}(aq) \longrightarrow SO_4^{2-}(aq) + 2e^-$
Iodide/(H^+)	$2I^-(aq) \longrightarrow I_2(aq) + 2e^-$
Iron (II) salts	$Fe^{2+}(aq) \longrightarrow Fe^{3+}(aq) + e^-$
Ethanediates	$C_2O_4^{2-}(aq) \longrightarrow 2CO_2(g) + 2e^-$
Hydrogen peroxide	$H_2O_2(aq) \longrightarrow O_2(g) + 2H^+(aq) + 2e^-$
Metal atoms M (Mg, Zn , Na)	$M(s) \longrightarrow M^{n+} + ne^-$

30.7 APPLICATIONS OF REDOX REACTIONS

Oxidation-reduction reactions form the basis of many applications of chemistry in industry and in our daily life. Some important applications are as follows:

1. Extraction of metals. By using a suitable reducing agent, metal oxides can be reduced to metals. For example, Fe_2O_3 is reduced to iron in the blast furnace using coke as the reducing agent.

$$Fe_2O_3(s) + 3C(s) \longrightarrow 2Fe(s) + 3CO(g)$$

Similarly, Al_2O_3 is reduced to aluminium by cathodic reduction in an electrolytic cell.

Other metals such as lithium, sodium, potassium, magnesium, calcium, etc., are also obtained commercially by electrolytic methods.

2. Electrochemical cells or batteries. Electrochemical cells or batteries based on redox reactions are widely used in our day-to-day life to run a number of small and big gadgets and equipment. For example, storage cells are used to supply all the electrical needs of our cars, trucks, buses, trains, aeroplanes, etc. Similarly, electrical energy needed in the space capsule is obtained by the reaction of hydrogen and oxygen in fuel cells which are electrochemical cells using oxygen and hydrogen electrodes.

3. Photosynthesis. Green plants convert carbon dioxide and water into carbohydrates in presence of sunlight. This reaction is called **photosynthesis** and is sensitized by chlorophyll.

$$6\,CO_2(g) + 6\,H_2O(l) \xrightarrow[\text{Chlorophyll}]{\text{Sunlight}} \underset{\text{Carbohydrate}}{C_6H_{12}O_6(aq)} + 6\,O_2(g)$$

During this reaction, CO_2 is reduced to carbohydrates while water is oxidised to oxygen. The energy needed for the reaction is provided by sunlight.

This reaction is a source of food for plants and animals. It also maintains a constant supply of 21% of O_2 by volume in the atmosphere needed for combustion of fuels and breathing of all the living creatures.

4. Supply of energy. The energy required for our daily needs is obtained by oxidation of fuels. For example, oxidation of fuels such as wood, gas, kerosene, petrol, etc. produces a large amount of energy which we need for various purposes in our daily life.

$$\text{Fuels (wood, petrol, kerosene, gas)} + O_2 \longrightarrow CO_2 + H_2O + \text{Other products} + \text{Energy}$$

Human body also needs energy for proper functioning. This is obtained by the oxidation of glucose in our body to CO_2 and water.

$$\underset{\text{Glucose}}{C_6H_{12}O_6\,(aq)} + 6\,O_2\,(g) \longrightarrow 6\,CO_2\,(g) + 6\,H_2O\,(l) + \text{Energy}$$

5. Production of chemicals. Many chemicals of our daily needs such as caustic soda, chlorine, fluorine, etc., are produced by electrolysis which is based on redox reactions.

6. Quantitative analysis. Redox reactions are very useful in quantitative analysis via redox titrations. These titrations involve the reactions between oxidising and reducing agents and help in estimating the amount of unknown substances in solutions.

SUMMARY

- **Oxidation.** A process involving an increase in oxidation number by the loss of electrons.
- **Reduction.** A process involving a decrease in oxidation number by gain of electrons.
- **Oxidising Agent.** A substance which involves a decrease in oxidation number of one or more of its elements.
- **Reducing Agent.** A substance which involves an increase in oxidation number of one or more of its elements.
- **Oxidation Number.** A residual charge which an atom of an element appears to have when other atoms from the molecule are removed as ions by counting the electrons with the more electronegative element.
- **Categories of Redox Reactions.** Redox reactions can be divided into different categories such as, combination reactions, decomposition reactions, displacement reactions and disproportionation reactions.

EVALUATE YOURSELF

I. Objective Type Questions

Select the most appropriate choice from the options given as (a), (b), (c) and (d) after each question:

1. In the reaction, $2Ag + 2H_2SO_4 \longrightarrow Ag_2SO_4 + SO_2 + 2H_2O$, H_2SO_4 acts as
   (*a*) Dehydrating agent
   (*b*) Reducing agent as well as acid
   (*c*) Acid as well as catalyst
   (*d*) Oxidising agent as well as acid.
2. Which of the following reactions does not involve oxidation or reduction?
   (*a*) $BaO_2 + H_2SO_4 \longrightarrow BaSO_4 + H_2O_2$
   (*b*) $BaO + \frac{1}{2} O_2 \longrightarrow BaO_2$
   (*c*) $2KMnO_4 \longrightarrow K_2MnO_4 + MnO_2 + O_2$
   (*d*) $H_2O_2 \longrightarrow H_2O + \frac{1}{2} O_2$.
3. The oxidation number of Cl in Cl_2O_7 is
   (*a*) + 7 (*b*) + 5
   (*c*) + 3 (*d*) – 7.
4. The oxidation number of carbon in CH_2Cl_2 is
   (*a*) 0 (*b*) 2
   (*c*) 3 (*d*) 5.
5. CrO_5 has structure as shown

   O
   O ╲ ‖ ╱ O
   | Cr |
   O ╱ ╲ O

   The oxidation number of chromium in the above compound is
   (*a*) 4 (*b*) 5
   (*c*) 6 (*d*) 10
6. The elements undergoing reduction and oxidation during the reaction,
   $$(NH_4)_2Cr_2O_7 \longrightarrow N_2 + Cr_2O_3 + 4H_2O$$
   are respectively
   (*a*) N and Cr (*b*) Cr and N
   (*c*) N and O (*d*) Cr and O.
7. In which of the following species does valency and oxidation numbers have different numerical values?
   (*a*) CO_2 (*b*) CH_4
   (*c*) $CHCl_3$ (*d*) CCl_4.
8. In $H_2AsO_4^-$ ion the oxidation number of hydrogen, arsenic, and oxygen are respectively
   (*a*) – 1, + 3, – 2 (*b*) + 1, + 3, – 2
   (*c*) + 1, + 5, – 2 (*d*) + 1, + 4, + 1.
9. The oxidation number of Mn is maximum in
   (*a*) MnO_2 (*b*) K_2MnO_4
   (*c*) Mn_3O_4 (*d*) $KMnO_4$.
10. Oxidation number of copper in CuC_2O_4 is
    (*a*) + 2 (*b*) + 1
    (*c*) + 3 (*d*) 8/3.

II. Fill in the Blanks

11. Complete the following sentences by supplying the appropriate words:
   (*i*) Oxidation and reduction are .... processes.
   (*ii*) Oxidation numbers of sulphur in S_8 and S_2F_2 are respectively ....
   (*iii*) The category of reaction $CO(g) + \frac{1}{2}O_2(g) \longrightarrow CO_2(g)$ is ....
   (*iv*) The oxidation number of oxygen in NO_2 is .... while in KO_2 it is ....
   (*v*) If a species undergo reduction one or more of its atom must involve .... of oxidation number.
   (*vi*) The combustion of magnesium in oxygen represents .... reaction.

III. Discussion Questions

12. Write the oxidation half reaction for the following reactions.
   (*i*) $A(s) + B^{2+}(aq) \longrightarrow A^{2+}(aq) + B(s)$
   (*ii*) $2H_2S + SO_2 \longrightarrow 2H_2O + 3S$
   (*iii*) $I_2 + 10HNO_3 \longrightarrow 2HIO_3 + 10NO_2 + 4H_2O$
13. Pick up the substance oxidised and reduced in the following reactions:
   (*i*) $X(s) + H_2O(g) \longrightarrow XO(s) + H_2(g)$
   (*ii*) $H_2O_2 + O_3 \longrightarrow H_2O + 2O_2$
14. Explain redox reactions in terms of electron transfer.
15. Chlorine has a variable oxidation state in its compounds.
   Explain it by finding the oxidation number of Cl in the following.
   $KClO_4$, Cl_2O_6, CH_2Cl_2, $KOCl$, CH_3Cl_2, ClO_3
16. State whether the underlined element in each of the following equations is oxidized, reduced or neither oxidized nor reduced. In each case, give an explanation for your choice.
   (*i*) $2\underline{Cu}SO_4(aq) + 4KCN(aq) \longrightarrow 2\underline{Cu}CN(s) + K_2SO_4(aq)$
   (*ii*) $\underline{Sn}Cl_2(aq) + HgCl_2(aq) \longrightarrow Hg(l) + SnCl_4(aq)$
   (*iii*) $4\underline{N}H_3(g) + 3O_2(g) \longrightarrow 2\underline{N}_2(g) + 6H_2O(g)$
   (*iv*) $3C(s) + 2\underline{Fe}_2O_3 \longrightarrow 3CO_2(g) + 4\underline{Fe}(s)$
17. Name the various categories of redox reactions and give one example for each them.
18. Give one example for each of the following:
   (*i*) A metal displacing another one from its salt solution.
   (*ii*) A metal causing displacement of hydrogen.
   (*iii*) A non-metal displacing another non-metal from its solution.

Answers

I. Objective Type Questions

1. (*d*)	**2.** (*a*)	**3.** (*a*)
4. (*a*)	**5.** (*c*)	**6.** (*b*)
7. (*c*)	**8.** (*c*)	**9.** (*d*)
10. (*a*).		

II. Fill in the Blanks

11. (*i*) complementary
(*ii*) 0 and + 1
(*iii*) combustion (*iv*) –2 and –1
(*v*) decrease in O.N (*vi*) redox.

SECTION-5

UNIT-31

BALANCING REDOX REACTIONS

Specific Objective

After studying this unit, the student should be able to:

- explain the steps involved in balancing redox equations.

We are familiar with the balancing of chemical equations by the inspection method. However, the inspection method may not be useful for balancing redox equations because in these equations, we have to keep in mind the *conservation of charge* as well as *conservation of mass*. Redox equations are, therefore, balanced by using the concept of half equations and following certain set of rules. One of the methods used for balancing redox reactions is called the **ion-electron** method.

31.1 BALANCING BY ION ELECTRON OR HALF REACTION METHOD

We know that during redox reactions there is a change in the oxidation number of the elements due to the transfer of electrons. The **basic principle** involved in balancing a redox equation is that the *number of electrons lost during oxidation is equal to the number of electrons gained during reduction.*

STEPS INVOLVED IN BALANCING REDOX EQUATIONS

The various steps involved in balancing redox equations using the ion-electron method are as follows:

1. Indicate the oxidation number of each atom involved in the reaction. Identify the elements which undergo a change in oxidation number.
2. Divide the skeleton redox equation into two half reactions; oxidation half and reduction half. In each half reaction balance the atoms which undergo the change in oxidation number.
3. In order to make up for the difference in O.N. add electrons to the left hand side or the right hand side of the arrow in each half reaction.
4. Balance **oxygen** atoms by the addition of the appropriate number of H_2O molecules to the side which is falling short of O atoms in each half reaction.

5. This step is meant for **only ionic equations**. It involves the balancing of H atoms in each half reaction as follows:
   (*i*) ***For acidic medium.*** Add the correct number of H^+ ions to the side falling short of H atoms.
   (*ii*) ***For basic medium.*** Add the correct number of H_2O molecules to the side falling short of H atoms and equal number of OH^- ions to the other side.
6. Equalise the number of electrons lost and gained by multiplying the half reactions with a suitable integer and add them to get the final equation.
   The application of the various steps described above has been illustrated as follows by balancing the redox equation representing the reaction between iodine and nitric acid.

$$HNO_3 + I_2 \longrightarrow HIO_3 + NO_2 + H_2O.$$

Step 1. *Indication of oxidation numbers of each atom.*

$$\overset{+1}{H}\overset{+5}{N}\overset{-2}{O_3} + \overset{0}{I_2} \longrightarrow \overset{+1}{H}\overset{+5}{I}\overset{-2}{O_3} + \overset{+4}{N}\overset{-2}{O_2} + \overset{+1}{H_2}\overset{-2}{O}.$$

Thus, only nitrogen and iodine undergo change in oxidation number.

Step 2. *Division into two half reactions and balancing the atoms undergoing change in O.N.*

$$\overset{0}{I_2} \longrightarrow 2H\overset{+5}{I}O_3 \quad (\textit{Oxidation half reaction})$$

$$H\overset{+5}{N}O_3 \longrightarrow \overset{+4}{N}O_2 \quad (\textit{Reduction half reaction})$$

Step 3. *Addition of electrons to make up the difference in O.N.*

$$\overset{0}{I_2} \longrightarrow 2H\overset{+5}{I}O_3 + 10e^-$$

(Each I atom loses $5e^-$ therefore, two iodine atoms would lose $10e^-$)

$$H\overset{+5}{N}O_3 + e^- \longrightarrow \overset{+4}{N}O_2$$

(Each N atom gains 1 electron).

Step 4. *Balancing of O by adding the correct number of H_2O molecules to the side falling short of O atoms.*

$$I_2 + 6H_2O \longrightarrow 2HIO_3 + 10e^-$$
$$HNO_3 + e^- \longrightarrow NO_2 + H_2O.$$

Step 5. *Not required because the equation is not ionic.*

Step 6. *To equalise the electrons multiply reduction half reaction by 10 and add the two.*

$$I_2 + 6H_2O \longrightarrow 2HIO_3 + 10e^-$$
$$[HNO_3 + e^- \longrightarrow NO_2 + H_2O] \times 10$$

$$I_2 + 10HNO_3 \longrightarrow 2HIO_3 + 10NO_2 + 4H_2O$$

Let us now proceed to balance some ionic equations in acidic and basic mediums.

SOLVED EXAMPLES

Example 31.1. *Balance the following redox equations by the half reaction method:*

(*i*) $Cr_2O_7^{2-} + Fe^{2+} \longrightarrow Cr^{3+} + Fe^{3+} + H_2O$ *in acidic medium*

(*ii*) $Cr(OH)_3 + IO_3^- \longrightarrow I^- + CrO_4^{2-}$ *in basic medium.*

Solution. **(*i*)** $(Cr_2O_7)^{2-} + Fe^{2+} \longrightarrow Cr^{3+} + Fe^{3+} + H_2O$

Step 1. *Indication of O.N. of each atom*

$$(\overset{+6}{Cr_2}\overset{-2}{O_7})^{2-} + Fe^{2+} \longrightarrow Cr^{3+} + Fe^{3+} + \overset{+1}{H_2}\overset{-2}{O}$$

Thus, Cr in $Cr_2O_7^{2-}$ and Fe change their oxidation numbers.

Step 2. *Writing the oxidation and reduction half reactions and equalising the atoms changing their oxidation numbers.*

$$(\overset{+6}{Cr_2}\overset{-2}{O_7})^{2-} \longrightarrow 2Cr^{3+} \quad (\textit{Reduction half})$$

$$Fe^{+2} \longrightarrow Fe^{3+} \quad (\textit{Oxidation half})$$

Step 3. *Addition of e^- to make up the difference in O.N.*

$$(\overset{+6}{Cr_2}O_7)^{2-} + 6e^- \longrightarrow 2Cr^{3+}$$

(Each Cr atom gains $3e^-$. Thus, 2Cr atom will gain $6e^-$)

$$Fe^{2+} \longrightarrow Fe^{3+} + e^-.$$

Step 4. *Balance 'O' by adding equal number of H_2O molecules to the side which is deficient in O atoms.*

$$Cr_2O_7^{2-} + 6e^- \longrightarrow 2Cr^{3+} + 7H_2O.$$
$$Fe^{2+} \longrightarrow Fe^{3+} + e^-.$$

Step 5. *Balance H by adding H^+ ions to the side which is deficient in H atoms.*

$$Cr_2O_7^{2-} + 6e^- + 14H^+ \longrightarrow 2Cr^{3+} + 7H_2O$$
$$Fe^{2+} \longrightarrow Fe^{3+} + e^-.$$

Step 6. *Multiply oxidation half reaction by 6 to equalise the electrons lost and gained and add the two.*

$$Cr_2O_7^{2-} + 6e^- + 14H^+ \longrightarrow 2Cr^{3+} + 7H_2O$$
$$[Fe^{2+} \longrightarrow Fe^{3+} + e^-] \times 6$$

$$6Fe^{2+} + Cr_2O_7^{2-} + 14H^+ \longrightarrow 2Cr^{3+} + 6Fe^{3+} + 7H_2O$$

It may be remembered that *in a balanced equation the number of atoms and also the electrical charges must be* **equal** *on both sides of the arrow.*

(*ii*) $Cr(OH)_3 + IO_3^- \longrightarrow I^- + CrO_4^{2-}$ (*in basic medium*)

Step 1. *Indication of oxidation number of each element.*

$$\overset{+3}{Cr}(\overset{-2}{O}\overset{+1}{H})_3 + (\overset{+5}{I}\overset{-2}{O_3})^- \longrightarrow I^- + (\overset{+6}{Cr}\overset{-2}{O_4})^{2-}$$

Thus, we find that Cr in $Cr(OH)_3$ and iodine in IO_3^- undergo change in oxidation number.

Step 2. *Writing oxidation and reduction half reactions.*

$$\overset{+3}{Cr}(OH)_3 \longrightarrow (\overset{+6}{Cr}O_4)^{2-} \quad \textit{(Oxidation half)}$$

$$(\overset{+5}{I}O_3)^- \longrightarrow I^- \quad \textit{(Reduction half)}$$

Step 3. *Addition of e^- to make up the difference in O.N.*

$$\overset{+3}{Cr}(OH)_3 \longrightarrow (\overset{+6}{Cr}O_4)^{2-} + 3e^-$$

$$(\overset{+5}{I}O_3)^- + 6e^- \longrightarrow I^-$$

Step 4. *Balance O atoms by adding H_2O molecules to the side deficient in 'O' atoms.*

$$Cr(OH)_3 + H_2O \longrightarrow CrO_4^{2-} + 3e^-$$

$$IO_3^- + 6e^- \longrightarrow I^- + 3H_2O$$

Step 5. *Balance H atoms. Since the medium is basic, add the correct number of H_2O molecules to the side falling short of H atoms and an equal number of OH^- ions to the other side.*

$$Cr(OH)_3 + H_2O + 5OH^- \longrightarrow CrO_4^{2-} + 3e^- + 5H_2O$$

$$IO_3^- + 6e^- + 6H_2O \longrightarrow I^- + 3H_2O + 6OH^-$$

Step 6. *Equalise the electrons lost and gained by multiplying the oxidation half reaction by 2.*

$$[Cr(OH)_3 + H_2O + 5OH^- \longrightarrow CrO_4^{-2} + 3e^- + 5H_2O] \times 2$$

$$IO_3^- + 6e^- + 6H_2O \longrightarrow I^- + 3H_2O + 6OH^-$$

$$2Cr(OH)_3 + 4OH^- + IO_3^- \longrightarrow 2CrO_4^{2-} + 5H_2O + I^-$$

Example 31.2. *Write the skeleton equation for each of the following process and balance them by the ion electron method:*

(*i*) *Tetraoxomanganate (VII) ion oxidises ethanedioate ions in acidic medium to carbon dioxide and gets reduced itself to Mn^{2+} ions.*

(*ii*) *Bromine and hydrogen peroxide react to give trioxobromate (V) ions and water.*

(*iii*) *Zinc reacts with nitric acid to form zinc (II) ions, ammonium ions and water.*

Solution. **(*i*)** The skeleton equation for the process is:

$$MnO_4^- + C_2O_4^{2-} + H^+ \longrightarrow Mn^{2+} + CO_2 + H_2O.$$

Step 1. *Indication of oxidation number*

$$(\overset{+7}{Mn}\overset{-2}{O_4})^{2-} + (\overset{+3}{C_2}\overset{-2}{O_4})^{2-} \longrightarrow Mn^{2+} + \overset{+4}{C}\overset{-2}{O_2} + \overset{+1}{H_2}\overset{-2}{O}$$

Step 2. *Writing oxidation and reduction half reactions*

$$(\overset{+3}{C_2}O_4)^{2-} \longrightarrow 2\overset{+4}{C}O_2 \quad \textit{(Oxidation half reaction)}$$

$$(\overset{+7}{Mn}O_4)^- \longrightarrow Mn^{2+} \quad \textit{(Reduction half reaction)}$$

Step 3. *Adding electrons to make up the difference in O.N.*

$$(\overset{+3}{C_2}O_4)^{2-} \longrightarrow 2\overset{+4}{C}O_2 + 2e^-$$

(Each carbon atom loses one electron)

$$(\overset{+7}{Mn}O_4)^- + 5e^- \longrightarrow Mn^{2+}$$

(Each Mn atom gains $5e^-$)

Step 4. *Balancing of O atom by adding H_2O molecules*

$$C_2O_4^{2-} \longrightarrow 2CO_2 + 2e^-$$

$$MnO_4^- + 5e^- \longrightarrow Mn^{2+} + 4H_2O$$

Step 5. *Balancing of H atoms by adding H^+ ions*

$$C_2O_4^{2-} \longrightarrow 2CO_2 + 2e^-$$

$$MnO_4^- + 5e^- + 8H^+ \longrightarrow Mn^{2+} + 4H_2O$$

Step 6. *Multiply the oxidation half reaction by 2 and the reduction half reaction by 5 to equalise the electrons lost and gained and add the two half reactions.*

$$[C_2O_4^{2-} \longrightarrow 2CO_2 + 2e^-] \times 5$$

$$[MnO_4^- + 5e^- + 8H^+ \longrightarrow Mn^{+2} + 4H_2O] \times 2$$

$$2MnO_4^- + 5C_2O_4^{2-} + 16H^+ \longrightarrow 10CO_2 + 2Mn^{2+} + 8H_2O$$

(*ii*) The skeleton equation for the given process is:

$$Br_2 + H_2O_2 \longrightarrow BrO_3^- + H_2O \text{ in acidic medium}$$

Step 1. *Indicate the oxidation number of each atom.*

$$\overset{0}{Br_2} + \overset{+1}{H_2}\overset{-1}{O_2} \longrightarrow (\overset{+5}{Br}\overset{-2}{O_3})^- + \overset{+1}{H_2}\overset{-2}{O}$$

Thus, Br and O change their oxidation numbers.

Step 2. *Write the oxidation and reduction half reactions*

$$\overset{0}{Br_2} \longrightarrow 2(\overset{+5}{Br}O_3)^- \quad \textit{(Oxidation half reaction)}$$

$$H_2\overset{-1}{O_2} \longrightarrow 2H_2\overset{-2}{O} \quad \textit{(Reduction half reaction)}$$

Step 3. *Addition of electrons to make up for the difference in O.N.*

$$\overset{0}{Br_2} \longrightarrow 2(\overset{+5}{Br}O_3)^- + 10e^-$$

(Each Br atom loses $5e^-$)

$$H_2\overset{-1}{O_2} + 2e^- \longrightarrow 2H_2O^{-2} \quad \text{(Each O atom gains } 1e^-\text{)}$$

Step 4. *Balance 'O' atoms by adding H_2O molecules.*

$$Br_2 + 6H_2O \longrightarrow 2BrO_3^- + 10e^-$$

$$H_2O_2 + 2e^- \longrightarrow 2H_2O.$$

Step 5. *Balance H atoms by adding H^+ ions.*

$$Br_2 + 6H_2O \longrightarrow 2BrO_3^- + 10e^- + 12H^+$$

$$H_2O_2 + 2e^- + 2H^+ \longrightarrow 2H_2O.$$

Step 6. *Equalise the electrons by multiplying the reduction half by 5 and add the two half reactions.*

$$Br_2 + 6H_2O \longrightarrow 2BrO_3^- + 10e^- + 12H^+$$
$$[H_2O_2 + 2e^- + 2H^+ \longrightarrow 2H_2O] \times 5$$
$$\overline{Br_2 + 5H_2O_2 \longrightarrow 2BrO_3^- + 4H_2O + 2H^+}$$

(iii) The skeleton equation for the process is

$$Zn + HNO_3 \longrightarrow Zn^{2+} + NH_4^+ + H_2O$$

or $$Zn + NO_3^- + H^+ \longrightarrow Zn^{2+} + NH_4^+ + H_2O$$

Step 1. *Writing oxidation number of each atom.*

$$\overset{0}{Zn} + (\overset{+5}{N}\overset{-2}{O_3})^- + H^+ \longrightarrow \overset{2+}{Zn} + (\overset{-3}{N}\overset{+1}{H_4})^+ + \overset{+1}{H_2}\overset{-2}{O}$$

Thus, Zn and N atoms change their oxidation states.

Step 2. *Write the oxidation and reduction half reactions.*

$$Zn^0 \longrightarrow Zn^{2+} \qquad \textit{(Oxidation half reaction)}$$
$$(\overset{+5}{N}O_3)^- \longrightarrow (\overset{-3}{N}H_4)^+ \qquad \textit{(Reduction half reaction)}$$

Step 3. *Addition of electrons to make up for difference in O.N.*

$$Zn^0 \longrightarrow Zn^{2+} + 2e^-$$
$$(\overset{+5}{N}O_3)^- + 8e^- \longrightarrow (\overset{-3}{N}H_4)^+$$

Step 4. *Balance O atoms by adding H_2O molecules.*

$$Zn \longrightarrow Zn^{2+} + 2e^-$$
$$(NO_3)^- + 8e^- \longrightarrow (NH_4)^+ + 3H_2O.$$

Step 5. *Balance H atoms by the addition of H^+ ions.*

$$Zn \longrightarrow Zn^{2+} + 2e^-$$
$$NO_3^- + 8e^- + 10H^+ \longrightarrow NH_4^+ + 3H_2O.$$

Step 6. *Multiply the oxidation half reaction by 4 and add the two half reactions.*

$$[Zn \longrightarrow Zn^{2+} + 2e^-] \times 4$$
$$NO_3^- + 8e^- + 10H^+ \longrightarrow NH_4^+ + 3H_2O$$
$$\overline{4Zn + NO_3^- + 10H^+ \longrightarrow 4Zn^{2+} + NH_4^+ + 3H_2O}$$

Example 31.3. *Balance the following redox equations:*

(i) $Cl_2 + OH^- \longrightarrow Cl^- + ClO_3^- + H_2O$

(ii) $FeS_2 + O_2 \longrightarrow Fe_2O_3 + SO_2$.

Solution. ***(i)*** $Cl_2 + OH^- \longrightarrow ClO_3^- + Cl^- + H_2O.$

Step 1. *Indicate the oxidation number of each atom.*

$$\overset{0}{Cl_2} + (\overset{-2}{O}\overset{+1}{H})^- \longrightarrow Cl^{-1} + (\overset{+5}{Cl}\overset{-2}{O_3})^- + \overset{+1}{H_2}\overset{-2}{O}.$$

Thus, chlorine is the only element which undergoes the change in oxidation number. It decreases its oxidation number from 0 to –1 and also increases its oxidation number from 0 to +5.

Step 2. *Write the oxidation and reduction half reactions:*

$$\overset{0}{Cl_2} \longrightarrow 2(\overset{+5}{Cl}O_3)^- \qquad \textit{(Oxidation half reaction)}$$
$$\overset{0}{Cl_2} \longrightarrow 2Cl^- \qquad \textit{(Reduction half reaction)}$$

Step 3. *Add electrons to make up for the difference in O.N.*

$$\overset{0}{Cl_2} \longrightarrow 2(\overset{+5}{Cl}O_3)^- + 10e^-$$

Each Cl atom loses 5 electrons

$$\overset{0}{Cl_2} + 2e^- \longrightarrow 2Cl^{-1}$$

Each Cl atom gains one electron.

Step 4. *Balance O atoms by adding H_2O molecules.*

$$Cl_2 + 6H_2O \longrightarrow 2(ClO_3)^- + 10e^-$$
$$Cl_2 + 2e^- \longrightarrow 2Cl^-$$

Step 5. *Since the medium is basic, balance H atoms by adding H_2O molecules to the side falling short of H atoms and add an equal number of OH^- ions to the other side.*

$$Cl_2 + 6H_2O + 12OH^- \longrightarrow 2ClO_3^- + 10e^- + 12H_2O$$
$$Cl_2 + 2e^- \longrightarrow 2Cl^-$$

Step 6. *Multiply the oxidation half reaction by 5 and add the two half reactions.*

$$Cl_2 + 6H_2O + 12OH^- \longrightarrow 2ClO_3^- + 10e^- + 12H_2O$$
$$(Cl_2 + 2e^- \longrightarrow 2Cl^-) \times 5$$
$$\overline{Cl_2 + 5Cl_2 + 12OH^- \longrightarrow 2ClO_3^- + 10Cl^- + 6H_2O}$$

or $$6Cl_2 + 12OH^- \longrightarrow 2ClO_3^- + 10Cl^- + 6H_2O$$

or $$\mathbf{3Cl_2 + 6OH^- \longrightarrow ClO_3^- + 5Cl^- + 3H_2O}$$

(ii) $$FeS_2 + O_2 \longrightarrow Fe_2O_3 + SO_2$$

Step 1. *Indicate the oxidation number of each atom*

$$\overset{+2}{Fe}\overset{-1}{S_2} + \overset{0}{O_2} \longrightarrow \overset{+3}{Fe_2}\overset{-2}{O_3} + \overset{+4}{S}\overset{-2}{O_2}$$

(It may be noted that FeS_2 is made of Fe^{2+} and $(S—S)^{2-}$ ion. In $(S—S)^{2-}$ ion, each S atom has O.N. equal to – 1).

It is a typical reaction in which every element Fe, S and O change their oxidation numbers.

Step 2. *Write the oxidation and reduction half reactions.*

$2\overset{+2}{Fe}\overset{-1}{S_2}$ — *Oxidation* →

$\frac{11}{2}\overset{0}{O_2}$ — *Reduction* →

$\overset{+3}{Fe_2}\overset{-2}{O_3} + 4\overset{+4}{S}\overset{-2}{O_2}$

Step 3. *Indicate the electron lost and gained.*

$\overset{+2\ -1}{2Fe\ S_2}$ $\xrightarrow{-22e^-}$ $\overset{+3\ -2}{Fe_2\ O_3} + \overset{+4\ -2}{S\ O_2}$

$\frac{11}{2}\overset{0}{O_2}$ $\xrightarrow{-22e^-}$

Each Fe atom loses one electron and each S atom loses 5 electrons. The total electrons lost by 2Fe and 4S atoms are 22. Similarly, each O atom gains 2 electrons, therefore, the total electrons gained by 11 oxygen atoms are 22. Since electrons lost and those gained are equal, the final equation can be written as

$$2FeS_2 + \frac{11}{2}O_2 \longrightarrow Fe_2O_3 + 4SO_2$$

or multiply throughout by 2

$$4FeS_2 + 11O_2 \longrightarrow 2Fe_2O_3 + 8SO_2.$$

SPECTATOR IONS

Consider the chemical reaction between zinc and copper(II) tetraoxosulphate(V) solution.

$$Zn(s) + CuSO_4(aq) \longrightarrow ZnSO_4(aq) + Cu(s)$$

Writing the ionic form of each species we have

$$Zn(s) + Cu^{2+}(aq) + SO_4^{2-}(aq) \longrightarrow Zn^{2+}(aq) + SO_4^{2-}(aq) + Cu(s)$$

Now, it is evident that Cu atoms and Zn^{2+} ions undergo oxidation and reduction respectively and are active ions. SO_4^{2-} ion, on the other hand, do not participate in the reaction and remain as such. They are inactive towards the reaction and are called **spectator ions.** In a similar way K^+ ions are spectator ions in the following reaction.

$$2K^+I^-(aq) + Cl_2(g) \longrightarrow 2K^+(aq) + Cl^-(aq) + I_2(g)$$

SUMMARY

Balancing of a redox equation is based on the principle of conservation of mass as well as charge. The method involves the splitting of the redox reaction into oxidation half and reduction half reactions. The two half reactions are balanced by following a set of rules and then combined to get the final balanced redox equation.

EVALUATE YOURSELF

I. Objective Type Questions

Select the most appropriate choice from the options given as (a), (b), (c) and (d) after each of the question:

1. Which of the following species involve the transfer of five electrons during the process?

(*a*) $MnO_4^{2-} \rightarrow MnO_4^-$ (*b*) $MnO_4^- \rightarrow Mn^{2+}$

(*c*) $Cr_2O_7^{2-} \rightarrow CrO_4^{2-}$ (*d*) $MnO_4^- \rightarrow MnO_2$.

2. $I_2 + 2S_2O_3^{2-} \longrightarrow 2I^- + S_4O_6^{x-}$ the value of x is

(*a*) 4 (*b*) 1

(*c*) 2.5 (*d*) 2.

3. In the equation $xBr_2 + yOH^- \longrightarrow zBrO_3^- + wBr^- + xH_2O$ the value of x and w are respectively

(*a*) 3, 5 (*b*) 2, 6

(*c*) 3, 1 (*d*) 3, 3.

4. One mole of N_2H_4 loses 10 electrons to form a new compound Y. Assuming all the N appear in the new compound, the oxidation number of N in the new compound is

(*a*) – 1 (*b*) + 10

(*c*) + 5 (*d*) + 3.

5. In the redox equation $aSO_2 + bH_2S \longrightarrow cS + dH_2O$, the value of c is

(*a*) 3 (*b*) 2

(*c*) 1 (*d*) 4.

II. Fill in the Blanks

6. Complete the following sentences by supplying the appropriate words:

(*i*) KI gets oxidised in alkaline medium to IO_3^- ion the number of electron lost per mole of species is ......

(*ii*) During reduction of HNO_3 to NH_3 the number of electrons gained by each N atom is ......

(*iii*) Iron (II) ethanedioate FeC_2O_4 gets oxidised by $KMnO_4$ in acidic medium to iron III ion and CO_2 molecules. The number of electrons lost per molecule is ......

(*iv*) In going from $Cr_2O_7^{2-}$ to Cr^{3+} ion the species act as ......

(*v*) When MnO_4^- changes to MnO_2 the number of electrons ...... is equal to ......

III. Discussion Questions

7. The Mn^{3+} ion is unstable in solution and undergoes disproportionation to give Mn^{2+}, MnO_2, and H^+ ion. Write a balanced ionic equation for the reaction.

8. Hydrogen peroxide reacts with ozone (O_3) to produce water and oxygen gas. Write the balanced equation and indicate the oxidising and reducing agent.

9. Write skeleton equations for the following chemical reactions and balance them by the oxidation number method.

(*i*) Bromide ions reduce tetraoxomanganate (VII) ions in basic medium to form manganese (IV) oxide and trioxobromate (V) ions.

(*ii*) Arsenic (III) sulphide react with trioxonitrate (V) ions in acidic medium to form nitric oxide (NO) and tetraoxoarsenate (V) ions and sulphur.

10. Balance the following equations in acidic medium by the ion electron method and identify the oxidants and the reductants:

(*i*) $MnO_4^-(aq) + C_2H_2O_4(aq) \longrightarrow Mn^{2+}(aq) + CO_2(g) + H_2O(l)$

(*ii*) $H_2S(aq) + Cl_2(g) \longrightarrow S(s) + Cl^-(aq)$

(*iii*) $MnO_4^-(aq) + C_2H_5OH(aq) \longrightarrow Mn^{2+}(aq) + CH_3COOH(aq)$

11. Balance the following equations in basic medium by the ion-electron method and identify the reductants and the oxidants.

(*i*) $P(s) + OH^-(aq) \longrightarrow PH_3(g) + H_2PO_2^-(aq)$

(*ii*) $N_2H_4(g) + ClO_3^-(aq) \longrightarrow NO(g) + Cl^-(g)$

(*iii*) $Cl_2O_7(g) + H_2O_2(aq) \longrightarrow ClO_2^-(aq) + O_2(g)$.

12. Balance the following redox reactions by the ion-electron method:

(*a*) $MnO_4^-(aq) + I^-(aq) \longrightarrow MnO_2(s) + I_2(s)$ *(in basic medium)*

(*b*) $MnO_4^-(aq) + SO_2(g) \longrightarrow Mn^{2+}(aq) + HSO_4^-(aq)$ *(in acidic solution)*

(*c*) $H_2O_2(aq) + Fe^{2+}(aq) \longrightarrow Fe^{3+}(aq) + H_2O(l)$ *(in acidic solution)*

(*d*) $Cr_2O_7^{2-} + SO_2(g) \longrightarrow Cr^{2-}(aq) + SO_4^{2-}(aq)$ *(in acidic solution)*

Answers

I. Objective Type Questions

1. (*b*) **2.** (*d*) **3.** (*a*)
4. (*d*) **5.** (*a*).

II. Fill in the Blanks

6. (*i*) 6 mol (*ii*) $8e^-$
(*iii*) $5e^-$ (*iv*) oxidising agent
(*v*) gained, 3.

III. Discussion Questions

7. $Mn^{+3} + 2H_2O \longrightarrow MnO_2 + Mn^{+2} + 4H^+$.

8. $H_2O_2 + O_3 \longrightarrow H_2O + 2O_2$
Oxidant; O_3; Reductant: H_2O_2.

9. (*i*) $2MnO_4^- + Br^- + H_2O \longrightarrow 2MnO_2 + BrO_3^- + 2OH^-$
(*ii*) $3As_2S_3 + 10NO_3^- + 4H_2O \longrightarrow 6AsO_4^{3-} + 9S + 8H^+$.

10. (*i*) $2MnO_4^- + 5C_2H_2O_4 + 6H^+ \longrightarrow 2Mn^{+2} + 10CO_2 + 8H_2O$
(*ii*) $H_2S + Cl_2 \longrightarrow S + 2Cl^- + 2H^+$
(*iii*) $4MnO_4^{-1} + 5C_2H_5OH + 12H^+ \longrightarrow 4Mn^{+2} + 5CH_3COOH + 11H_2O$.

11. (*i*) $4P + 3OH^- + 3H_2O \longrightarrow PH_3 + 3H_2PO_2^-$
(*ii*) $3N_2H_4 + 4ClO_3^- \longrightarrow 4Cl^- + 6H_2O + 6NO$
(*iii*) $Cl_2O_7 + 4H_2O_2 + 2OH^- \longrightarrow 2ClO_2^- + 5H_2O + 4O_2$.

12. (*a*) $2MnO_4^- + 6I^- + 4H_2O \longrightarrow 2MnO_2 + 3F_2 + 8OH^-$
(*b*) $5SO_2 + 2MnO_4^- + 2H_2O + H^+ \longrightarrow 5HSO_4^- + 2Mn^{+2}$
(*c*) $2Fe^{+2} + H_2O_2 + 2H^+ \longrightarrow 2Fe^{+3} + 2H_2O$
(*d*) $3SO_2 + Cr_2O_7^{2-} + 2H^+ \longrightarrow 2Cr^{+3} + 3SO_4^{2-} + H_2O$.

SECTION-5

UNIT-32

REDOX TITRATIONS

Specific Objectives

After studying this unit, the student should be able to:

- describe and explain the processes involved in carrying out redox titrations
- describe an experiment to determine the end point of a redox titration.

32.1 REDOX REACTIONS AS THE BASIS FOR TITRATION

One importance of redox reactions in solutions is their frequent use in volumetric analysis. Volumetric analysis involves the estimation of a substance by accurate measurement of the volumes of solutions undergoing a chemical change. Here, a definite mass of the substance to be estimated is dissolved in water to get some exact volume of the solution. A known volume of this solution is allowed to react completely with some standard solution of an oxidising or reducing agent. *The process by which two solutions are allowed to react is called* **titration.** The final result is calculated from the volumes of the two reacting solutions. In this section, our main focus will be on *redox titrations* and the *calculations based on the molar relationship between the oxidising and reducing agents.*

In acid-base systems we came across a method of making one solution react with another, using glassware (*pipette* and *burette*) and indicators to establish the exact end point. The objective in such exercise is to find out something previously unknown about one of the solutions.

In a similar way, redox titrations can be used to determine the exact amount of an oxidising or reducing agent in a given solution by titrating it against a standard solution* of a suitable reducing agent or oxidising agent. However, acid-base and redox titrations differ from each other in the choice of indicators.

Choice of indicators. The main difference between the two types of titrations is that the indicators used in acid-base titrations to detect the changes in pH are not useful for redox titrations. This problem can be solved in the following three ways:

1. One solution to the problem of use of indicator is to choose one of the reagents, which is itself intensely coloured, as indicator, *e.g.*, permanganate ion, MnO_4^-. Here, *$KMnO_4$ or MnO_4^- acts as a* **self indicator**.

*A solution whose molar concentration is known is called a standard solution.

In potassium tetraoxomanganate(VII) ($KMnO_4$) titrations, when the whole of the reductant (Fe^{2+} or $C_2O_4^{2-}$) is oxidised, the end point is reached. At the end point, the violet colour of MnO_4^- solution disappears and a lasting (permanent) tinge of pink colour appears. This colour change is so sensitive that it can easily be detected even at very low concentrations (10^{-6} mol dm^{-3} or 10^{-6} mol L^{-1}) of MnO_4^-.

2. If there is no dramatic auto-colour change (as with $KMnO_4$ titrations), there are indicators which are immediately oxidised after the last drop of reactant has reacted producing a dramatic colour change. For example, $Cr_2O_7^{2-}$ (or $K_2Cr_2O_7$) is not a self indicator, therefore, **diphenylamine** is used as an indicator in $K_2Cr_2O_7$ titrations. It produces an intense blue colour at the end point.
3. Another indicator which has been used in redox reactions is **starch** which produces intense blue colour with molecular iodine. The use of this indicator is, however, limited to only those reagents which can either oxidise I^- ions such as Cu^{2+} ions or reduce I_2 such as thiosulphate ion ($S_2O_3^{2-}$) ions.

$$2Cu^{2+}(aq) + 4I^-(aq) \longrightarrow Cu_2I_2(s) + I_2(s)$$

$$I_2(s) + 2S_2O_3^{2-}(aq) \longrightarrow 2I^-(aq) + S_4O_6^{2-}(aq)$$

Thus, when I^- ions are oxidised to molecular I_2, an intense *blue colour appears at the end point*. However, when I_2 is reduced to I^- ions, the *intense blue colour disappears at the end point*.

It may be noted here that although I_2 is insoluble in water, it can be easily dissolved in KI solution where it exists as KI_3.

32.2 PROCESSES INVOLVED IN CARRYING OUT REDOX TITRATIONS

Depending upon the nature of the oxidising agent, these are divided into the following categories:

1. Potassium Tetraoxomanganate(VII) Titrations

In these titrations, reducing agents like iron (II) ions $(Fe)^{2+}$ from $FeSO_4$, Mohr's salt $[(NH_4)_2SO_4 \cdot FeSO_4 \cdot 6H_2O]$, oxalic acid $(COOH)_2$, or ethanedioate ($C_2O_4^{2-}$) ions from ethanedioic acid etc., are directly titrated against $KMnO_4$ as the oxidising agent in acidic medium. For example,

(*i*) *Oxidation of iron (II) salts:*

$$5Fe^{2+} + MnO_4^- + 8H^+ \longrightarrow 5Fe^{3+} + Mn^{2+} + 4\,H_2O$$

(*ii*) *Oxidation of ethanedioate ion:*

$$5C_2O_4^{2-} + 2MnO_4^- + 16H^+ \longrightarrow 2\,Mn^{2+} + 10\,CO_2 + 8H_2O$$

It may be noted that in titrations involving ethanedioate (oxalate) ions, the flask containing the $C_2O_4^{2-}$ ions in acidic medium is heated to about 60°C. This is to provide activation energy to the reactants. However, as the titration progresses, the manganese (II) ions (Mn^{2+}) formed during the reaction catalyse the reaction.

It may be noted that in neutral or basic conditions a brown precipitate is formed which interfere with the end point. The equation involved is

$$MnO_4^-(aq) + 5Fe^{2+}(aq) + \overline{O}H(aq) \longrightarrow \underset{\text{Brown ppt.}}{MnO_2(s)} + 5Fe^{3+}(aq) + 4H_2O(l)$$

2. Potassium Dichromate Titrations

In these titrations, the above listed reducing agents are directly titrated against $K_2Cr_2O_7$ as the oxidising agent in acidic medium. For example,

(*i*) *Oxidation of iron (II) salts:*

$$6Fe^{2+} + Cr_2O_7^{2-} + 14H^+ \longrightarrow 2Cr^{3+} + 6Fe^{3+} + 7H_2O$$

3. Iodimetric Titrations

These titrations involve the direct use of *iodine as the oxidising agent* (in neutral or slightly acidic medium) using starch as an indicator. The various reducing agents used in these titrations are *thiosulphates, sulphites, arsenites and antimonites*.

$$I_2 + \underset{\text{Thiosulphate ion}}{S_2O_3^{2-}} \longrightarrow 2I^- + \underset{\text{Tetrathionate ion}}{S_4O_6^{2-}}$$

$$I_2 + \underset{\text{Sulphite ion}}{SO_3^{2-}} + H_2O \longrightarrow 2I^- + \underset{\text{Sulphate ion}}{SO_4^{2-}} + 2H^+$$

$$I_2 + \underset{\text{Arsenite ion}}{AsO_3^{3-}} + H_2O \longrightarrow 2I^- + \underset{\text{Arsenate ion}}{AsO_4^{3-}} + 2H^+$$

$$I_2 + \underset{\text{Antimonite ion}}{SbO_3^{3-}} + H_2O \longrightarrow 2I^- + \underset{\text{Antimonate ion}}{SbO_4^{3-}} + 2H^+$$

4. Iodometric Titrations

These titrations are carried out in two steps. In the first step, oxidising agents such as $KMnO_4$, $K_2Cr_2O_7$, $CuSO_4$, etc. are treated with an excess of KI when I_2 is liberated quickly and quantitatively. For example,

$$2MnO_4^- + 16H^+ + 10I^- \longrightarrow 2Mn^{2+} + 5I_2 + 8H_2O$$

$$Cr_2O_7^{2-} + 14H^+ + 6I^- \longrightarrow 2Cr^{3+} + 3I_2 + 7H_2O$$

$$2Cu^{2+} + 4I^- \longrightarrow Cu_2I_2 + I_2$$

In the second step, the liberated iodine is titrated against a standard solution of sodium thiosulphate using starch as an indicator. *All such titrations in which iodine liberated from potassium iodide with the help of an oxidising agent is titrated against a standard solution of sodium thiosulphate are called* **iodometric titrations.**

32.3 STOICHIOMETRY OF REDOX REACTIONS IN SOLUTIONS

Stoichiometry *of redox reactions means calculation of the quantities of the oxidising and reducing agents and their products in oxidation-reduction reactions.*

Stoichiometric calculations in *redox titrations* are carried out by *using the* **molarity equation**. In both methods, we need a balanced equation for the redox reaction. The stoichiometric coefficients in the balanced equation tell us the ratio by moles in which the reactants combine and the products form. In the titration method, the molarity equation, *i.e.*,

$$\frac{M_1V_1}{n_1} = \frac{M_2V_2}{n_2}$$

is used where M_1 and M_2 are the molarities of the reductant and oxidant or vice-versa, V_1 and V_2 are their volumes and n_1 and n_2 represent their stoichiometric coefficients. This equation can be used to calculate any unknown quantity if the other three quantities are known.

32.4 EXPERIMENTS ON REDOX REACTIONS

Let us familiarise ourselves with the method of performing experiments on redox titrations between the following pairs:

- MnO_4^- versus Fe^{2+} and $C_2O_4^{2-}$ ions.
- I_2/KI versus $S_2O_3^{2-}$ ions.

1. Titrations of MnO_4^- versus Fe^{2+} and $C_2O_4^{2-}$

The chemical reaction involved in these titrations are:

(*i*) $\underset{\text{Pink}}{MnO_4^-} + 5Fe^{2+} + 8H^+ \longrightarrow \underset{\text{Colourless}}{Mn^{2+}} + 5Fe^{3+} + 4H_2O$

(*ii*) $\underset{\text{Pink}}{2MnO_4^-} + 5C_2O_4^{2-} + 16H^+ \longrightarrow \underset{\text{Colourless}}{2Mn^{2+}} + 10CO_2 + 8H_2O$

Some important points are:

- $KMnO_4$ solution is taken in the burette.
- Solution of Fe^{2+} salt or $C_2O_4^{2-}$ salts are taken in the titration flask.
- Sufficient dilute H_2SO_4 is added to the titration flask to provide an acidic medium.
- End point is from *colourless solution to light pink* colour.
- In case of $C_2O_4^{2-}$ salt the solution in the titration flask should be heated to about 60°C before the addition of the $KMnO_4$ solution to provide sufficient activation energy. (Follow Activity 32.1.)

ACTIVITY 32.1 **To standardize a given solution of potassium tetraoxomanganate(VII) with 0.05M $Na_2C_2O_4$ solution and use this solution to estimate the strength of $FeSO_4.7H_2O$ in a given solution of iron(II) tetraoxosulphate(VI).**

Apparatus required:

A burette, pipette, titration flask and beakers.

Part I. To standardize potassium tetraoxomanganate(VII) solution

1. Rinse and fill the burette with $KMnO_4$ solution.
2. Rinse the pipette with 0.05 M $Na_2C_2O_4$ solution and pipette 10 cm^3 of $Na_2C_2O_4$ solution into three thoroughly cleaned titration flasks '*a*', '*b*' and '*c*'. Add 10 cm^3 of dil tetraoxosulphate(VI) acid to each flask.
3. Heat each mixture '*a*', '*b*', '*c*' to about 60°C (just unbearable to your skin) and titrate each one against the $KMnO_4$ solution from the burette till a permanent light pink colour is formed. Record the titre value in each case as follows:

Flasks	*Initial burette reading*	*Final burette reading*	*Volume of $KMnO_4$ used*
a			
b			
c			

Let the titre value in this case be = 13.3 cm^3 (supposed)

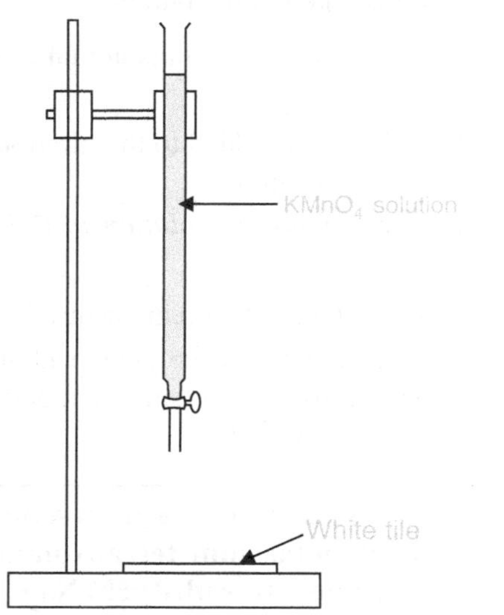

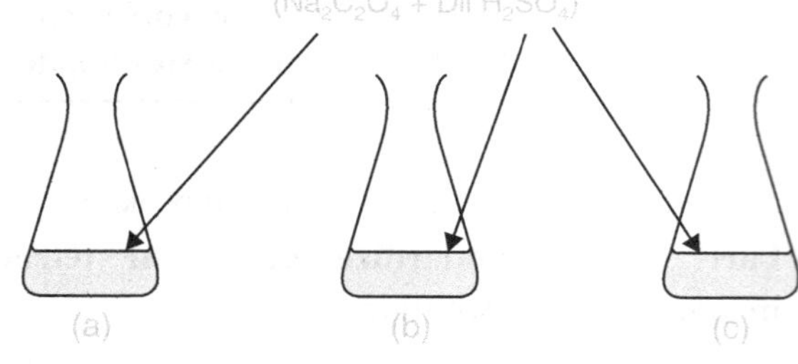

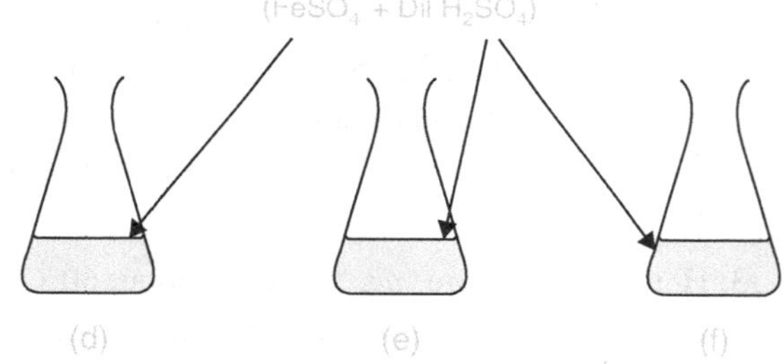

Fig. 32.1 Titration of $KMnO_4$ *vs*
I. $Na_2C_2O_4$ solution. II. $FeSO_4$ solution

Part II. To estimate the strength of $FeSO_4.7H_2O$

1. Fill the burette again
2. Now, clean and rinse the pipette with iron (II) tetraoxosulphate(VI) solution. Pipette 10 cm^3 of $FeSO_4$ solution and put it in cleaned titration flasks *d, e, f* respectively. Add 10 cm^3 of dilute tetraoxosulphate(VI) acid to each titration flask.
3. Titrate each of the three mixtures, *d, e, f* one after the other by dropping $KMnO_4$ solution from the burette till a permanent light pink colour is obtained at the end point.
4. Make the record in a similar way. Let us assume the titre value in this case is *15.2 cm^3*.

Calculations involved

Let us calculate the results by using the above supposed titre values

Part I. The chemical equation involved is

$$2MnO_4^- + 5C_2O_4^{2-} + 16H^+ \longrightarrow 2Mn^{2+} + 10CO_2 + 8H_2O$$

Volume of ethanedioate solution taken (V_2) = 10 cm^3
Molarity of ethanedioate solution (M_2) = 0.05 M
Volume of $KMnO_4$ solution used (V_1) = 13.3 cm^3 (supposed)
Molarity of $KMnO_4$ solution (M_1) = ?
According to molarity equation

$$\frac{M_1V_1}{2} = \frac{M_2V_2}{5}$$

$$\text{or} \quad M_1 = \frac{M_2V_2 \times 2}{V_1} = \frac{0.05 \times 10 \times 2}{13.3} = 0.075$$

Concentration of $KMnO_4$ solution = **0.075 mol dm^{-3}**

Part II. The chemical equation involved is

$$MnO_4^- + 5Fe^{2+} + 8H^+ \longrightarrow Mn^{2+} + 5Fe^{3+} + 4H_2O$$

Volume of iron (II) salt taken (V_3) = 10 cm^3
Molarity of iron (II) salt solution (M_3) = ?
Volume of $KMnO_4$ solution used (V_1) = 15.2 cm^3 (supposed)
Molarity of $KMnO_4$ solution (M_1) = 0.075 M
According to Molarity equation

$$\frac{M_1V_1}{1} = \frac{M_3V_3}{5}$$

$$\text{or } M_3 = \frac{M_1V_1 \times 5}{V_2} = \frac{0.075 \times 15.2 \times 5}{10} = 0.57 \text{ mol dm}^{-3}$$

Strength of $FeSO_4.7H_2O$ = 0.57 × molecular mass of $FeSO_4.7H_2O$
= 0.57 × 278 = **158.46 g dm^{-3}**

2. Titration of I_2/KI versus $S_2O_3^{2-}$ salt

The chermical reaction involved in this titration is

$$\underset{\text{Violet}}{I_2(aq) + S_2O_3^{2-}(aq)} \longrightarrow \underset{\text{Colourless}}{I^-(aq) + S_4O_6^{2-}(aq)}$$

In this titration,

- I_2/KI solution is taken in the burette.
- Aqueous solution of $S_2O_3^{2-}$ salts is taken in the titration flask.

- Indicator used is freshly prepared starch solution.
- Indicator should be added towards the end of the titration when the solution acquires a lime yellow colour.
- The end point is the appearance of a blue colour.

Follow these steps in **Activity 32.2.**

ACTIVITY 32.2 **To standardize a given solution of iodine volumetrically using 0.05M aqueous sodium trioxosulphur sulphate(VI) pentahydrate ($Na_2S_2O_3.5H_2O$).**

Requirement

Burette, 20 cm^3 pipette, Freshly prepared starch solution, Beakers, Titration flask, etc.

Procedure

1. Clean the volumetric apparatus
2. Rinse and fill the burette with $Na_2S_2O_3$ solution
3. Rinse the pipette with iodine solution. Pipette out 20 cm^3 of iodine solution into the titration flask
4. Start dropping $Na_2S_2O_3$ solution from the burette till a lime yellow colour is obtained. At this stage add starch solution and again drop sodium thiosulphate solution till blue colour just disappears. Note the final burette reading.
5. Repeat the experiment to get three constant readings. Let the constant titre value be 18.2 cm^3 (suppose). Calculate the result as follows.

The chemical equation involved is

$$I_2 + 2S_2O_3^{2-} \longrightarrow 2I^- + S_4O_6^{2-}$$

Volume of iodine solution (V_2)

= 18.2 cm^3 (supposed)

Volume of thiosulphate solution taken (V_1)

= 20 cm^3

Molarity of thiosulphate solution (M_1) = 0.05 M

Molarity of iodine solution (M_2) = ?

According to the molarity equation

$$\frac{M_1V_1}{2} = \frac{M_2V_2}{1} \quad \text{or} \quad M_2 = \frac{M_1V_1}{2\,V_2}$$

$$= \frac{0.05 \times 20}{18.2} = \mathbf{0.055\ M}$$

SOLVED EXAMPLES

Example 32.1. *Determine the volume of M/8 $KMnO_4$ solution required to react completely with 25.0 cm^3 of M/4 $FeSO_4$ solution in acidic medium.*

Solution. The balanced ionic equation for the reaction is

$$MnO_4^- + 5Fe^{2+} + 8H^+ \longrightarrow Mn^{2+} + 5Fe^{3+} + 4H_2O$$

From the balanced equation, it is evident that

1 mole of $KMnO_4$ ≡ 5 moles of $FeSO_4$

Applying the molarity equation to the balanced redox equation,

$$\frac{M_1V_1}{n_1}(KMnO_4) = \frac{M_2V_2}{n_2}(FeSO_4)$$

or $$\frac{1 \times V_1}{8 \times 1} = \frac{1}{4} \times \frac{25}{5}$$

or $$V_1 = \frac{1 \times 25 \times 8}{4 \times 5} = 10.0\ cm^3$$

Thus, the volume of M/8 $KMnO_4$ solution required = **10.0 cm^3.**

Example 32.2. *Calculate the concentration of sodium trioxosulphur sulphate (VI) pentahydrate ($Na_2S_2O_3.5H_2O$) solution in g dm^{-3} if 10 cm^3 of this solution decolourised 15 cm^3 of M/40 iodine solution.*

Solution. The balanced equation for the redox reaction is

$$2S_2O_3^{2-} + I_2 \longrightarrow 2I^- + S_4O_6^{2-}$$

From the balanced equation, it is evident that

2 moles of $Na_2S_2O_3$ ≡ 1 mole of I_2

Applying the molarity equation, we have

$$\frac{M_1V_1}{n_1}(Na_2S_2O_3) = \frac{M_2V_2}{n_2}(I_2)$$

$$\frac{M_1 \times 10}{2} = \frac{15 \times 1}{1 \times 40}$$

or $$M_1 = \frac{15 \times 1 \times 2}{10 \times 40} = \frac{3}{40}$$

Thus, the molarity of the $Na_2S_2O_3$ solution = 3/40 M

Mol. mass of $Na_2S_2O_3.5H_2O$ = 248 g mol^{-1}

∴ Concentration of $Na_2S_2O_3.5H_2O = \frac{248 \times 3}{40}$

= **18.6 g dm^{-3}.**

SUMMARY

- **Redox Titrations.** These are the volumetric titrations based on redox reactions. The choice of indicators in these titrations depends upon the nature of the redox reactions. The various categories of redox titrations are potassium tetraoxomanganate(VII) titrations, potassium dichromate titrations, iodometric titrations and iodimetric titrations. The stochiometric calculations in these titrations are based on the molarity relation

$$\frac{M_1V_1}{n_1} = \frac{M_2V_2}{n_2}.$$

EVALUATE YOURSELF

I. Objective Type Questions

Select the most appropriate choice from the options given as (a), (b), (c) and (d) after each question:

1. In $KMnO_4$ titrations, the indicator used is
 (*a*) H_2SO_4 (*b*) phenolphthalein
 (*c*) starch (*d*) $KMnO_4$ is self indicator.
2. Which of the following reducing agent is not used in iodimetric titrations?
 (*a*) $S_2O_3^{2-}$ (*b*) Fe^{3+}
 (*c*) AsO_3^{3-} (*d*) SO_3^{2-}.
3. The indicator used in iodine titrations is
 (*a*) Starch (*b*) $FeSO_4$ solution
 (*c*) $Na_2S_2O_3$ solution (*d*) I_2 acts as self indicator.
4. Which is the most appropriate conditions for $KMnO_4$ titrations?
 (*a*) Acidic medium (*b*) Basic medium
 (*c*) Neutral medium (*d*) Both (*b*) and (*c*).

II. Fill in the Blanks

5. Complete the following sentences by supplying the appropriate words:
 (*i*) In acidic medium MnO_4^- ion gets reduced to ....
 (*ii*) In iodine titration the starch indicator is generally added ....
 (*iii*) The disappearance of the pink colour of MnO_4^- during titrations is due to the formation of ....
 (*iv*) The colour of $K_2Cr_2O_7$ solution is ....
 (*v*) Ethanedioate ions get oxidised to ...... on treatment with acidified $KMnO_4$ solution.

III. Discussion Questions

6. What do you understand by redox titrations. In what respect do they differ from acid base titration?
7. During $KMnO_4$ titrations, acidic conditions are generally made by H_2SO_4. Why can't we use HCl or HNO_3 acids?
8. Write balanced chemical equations for the chemical reaction involved in titration of acidified solution of $KMnO_4$ with:
 (*i*) ethanedioate ion
 (*ii*) iron (II) ion.
9. Why is starch indicator added towards the end of a titration?
10. 2.48 g of $Na_2S_2O_3.xH_2O$ was dissolved per litre of the given solution. 20 cm^3 of this solution on titration required 10 cm^3 of 0.01 M iodine solution. Find out the value of x.
11. What is the percentage of ethanedioate ion in a given sample of ethanedioate salt, 3.0 g of which has been dissolved per litre in a given solution 'A'? 10.0 cm^3 of solution A requires 8 cm^3 of 0.01 M $KMnO_4$ solution for complete oxidation in acidic medium.

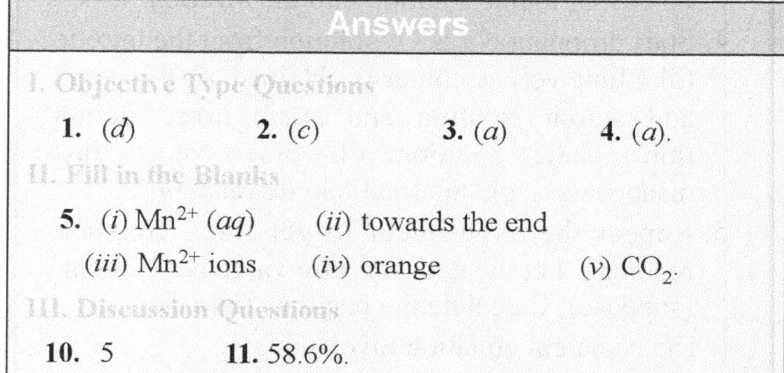

Answers

I. Objective Type Questions

1. (*d*) **2.** (*c*) **3.** (*a*) **4.** (*a*).

II. Fill in the Blanks

5. (*i*) Mn^{2+} (*aq*) (*ii*) towards the end (*iii*) Mn^{2+} ions (*iv*) orange (*v*) CO_2.

III. Discussion Questions

10. 5 **11.** 58.6%.

SECTION-5

UNIT-33

ELECTROCHEMICAL CELLS

Specific Objectives

After studying this unit, the student should be able to:

- describe the interconversion of chemical energy and electrical energy in redox reactions
- describe and explain the functions of a simple electrochemical cell
- describe and explain the function of the standard electrode potential in redox reactions
- describe how standard electrode potentials can be used to produce the electrochemical series
- explain some applications of electrochemical cells.

33.1 ELECTROCHEMISTRY

It is a well known fact that energy manifests itself in different forms which are interconvertible into one another. Among the different forms of energy, electrical energy plays a very significant role in our daily lives. Many chemical transformations and industrial processes are based on electrical energy and its relationship with chemical energy. *There are a large number of spontaneous redox reactions which form the basis of production of electrical energy.* The device in which such chemical processes are carried out is called ***electrochemical cell*** or ***galvanic cell.*** For example, Daniell cell is based on the following redox reaction:

$$Cu^{2+}(aq) + Zn(s) \longrightarrow Cu(s) + Zn^{2+} + \textit{Electrical Energy}$$

At the same time many of the ***non-spontaneous redox reactions*** can be made to occur *by **the use of electrical energy.*** Some examples are:

$$2H_2O(l) \xrightarrow{\text{Electricity}} 2H_2(g) + O_2(g)$$

$$2Al_2O_3(l) \xrightarrow{\text{Electricity}} 4Al + 3O_2$$

The process associated with such transformation is called ***Electrolysis*** and the device used for electrolysis is called ***electrolytic cell.***

*The branch of chemistry which deals with the study of the relationship between electrical energy and chemical energy and the interconversion of one form of energy into another is called **electrochemistry.*** Electrochemistry is a very vast and interdisciplinary branch of chemistry which finds tremendous applications. Its study is very important as it helps in the development of new technologies which are economical and environment-friendly. In this unit, we shall focus on the study of some aspects of electrochemical cells, batteries and fuel cells, etc.

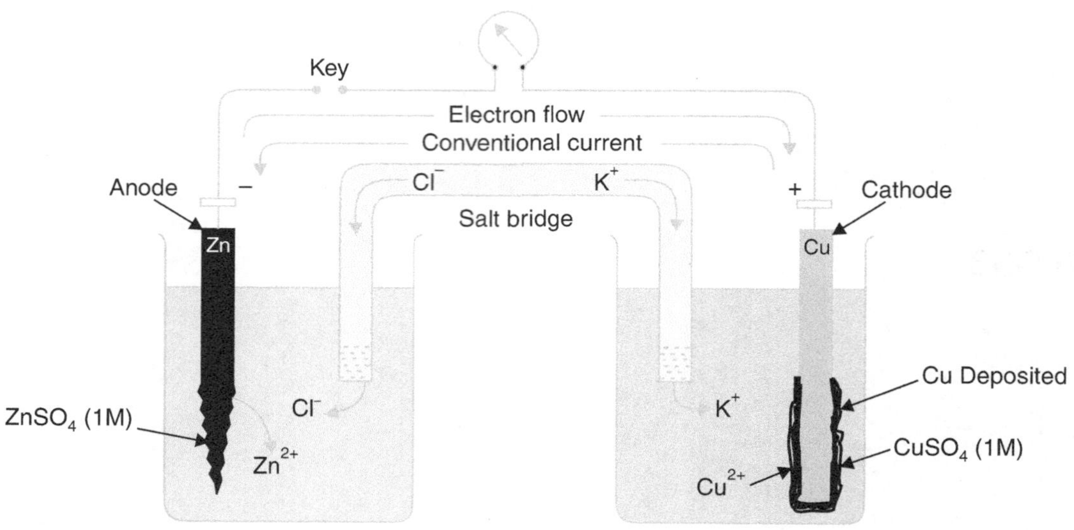

Fig. 33.1. Electrochemical cell.

33.2 ELECTROCHEMICAL CELLS

The device in which chemical energy is converted into electrical energy is called **galvanic cell** or **electrochemical cell** or **voltaic cell.** In a galvanic cell, a redox reaction is carried out in an *indirect manner* and the decrease in free energy during the chemical process appears as electrical energy. An indirect redox reaction is such that reduction and oxidation processes are carried out in separate vessels. In order to understand this phenomenon, let us consider the redox reaction between Zn and $CuSO_4$ solution.

$$Zn\,(s) + Cu^{2+}\,(aq) \longrightarrow Zn^{2+}\,(aq) + Cu\,(s)$$

The overall reaction is split up into two half reactions, $Zn\,(s) \longrightarrow Zn^{2+}\,(aq)$ and $Cu^{2+}\,(aq) + 2e^- \longrightarrow Cu\,(s)$.

Half Reactions and Redox Couples

The half reactions are used to form redox couples as described below.

In its simple form, a *zinc strip* is dipped in $ZnSO_4$ solution and a *copper strip* is dipped in $CuSO_4$ solution in separate beakers. *An arrangement involving contact between oxidised and reduced form of the substance such as Zn/Zn^{2+}, or Cu/Cu^{2+} at the interface is called* **redox couple.** The two metallic strips which act as electrodes are connected by conducting wires through a voltmeter. The two solutions are joined by an inverted U-tube known as **salt bridge.** The U-tube is filled with a solution of some electrolyte such as KCl, KNO_3 or NH_4Cl to which *gelatin* or *agar-agar* has been added to convert it into a semi-solid paste. A schematic diagram of this cell has been shown in Fig. 33.1.

The deflection in voltmeter indicates that there is a potential difference between the two electrodes. It has been found that the conventional current flows through the outer circuit from the copper to the zinc strip. It implies that the electron flow occurs from the zinc to the copper strip.

Let us now understand the working of the cell.

(*i*) Zinc undergoes oxidation to form zinc ions.

$$Zn(s) \longrightarrow Zn^{2+}(aq) + 2e^-\ (Oxidation)$$

(*ii*) The electrons liberated during oxidation flow through the connecting wires to the copper strip.

(*iii*) Copper ions move towards the copper strip, pick up the electrons, and gets reduced to copper atoms which are deposited at the copper strip.

$$Cu^{2+}(aq) + 2e^- \longrightarrow Cu(s)\ (Reduction)$$

The *overall cell reaction* is a combination of the oxidation half and reduction half reactions.

$$Zn(s) + Cu^{2+}\,(aq) \longrightarrow Cu(s) + Zn^{2+}\,(aq)$$

The electrode at which *oxidation occurs* is the **anode** and that at which *reduction occurs* is the **cathode.** In the above cell, the zinc strip is the anode and the copper strip is the cathode. Due to the oxidation process occurring at the anode, it becomes a source of electrons and acquires a negative charge in the cell. Similarly, due to the reduction process occurring at the cathode, it acquires a positive charge and becomes a receiver of electrons. Thus, in the electrochemical cell, the **anode electrode** acts as the **negative terminal** and the **cathode electrode** acts as the **positive terminal.**

Salt Bridge and its Functions

Salt Bridge is a U-shaped tube containing a semi-solid paste of some inert electrolyte like KCl, KNO_3, NH_4Cl, etc., in *agar-agar* or *gelatine*. An inert electrolyte is one which:

(*a*) does not react chemically with the solution in either of the compartment.

(*b*) does not interfere with the net cell reaction.

Function of a Salt Bridge. In the electrochemical cell a salt bridge serves two very important functions:

(*i*) *It allows the flow of current by completing the circuit.*

(*ii*) *It maintains electrical neutrality.*

The transfer of electrons from the *anode* to the *cathode* leads to a net positive charge around the anode due to an increase in the concentration of cations and a net negative charge around the cathode due to an excess of anions in solution. The positive charge around the anode will prevent electrons from flowing out from it and the negative charge around the cathode will prevent the inflow of electrons into it. The reaction would thus stop and no current will flow. The salt bridge comes to aid and restores the electroneutrality of the solutions in the two compartments. It contains a concentrated solution of an *inert electrolyte* the ions of which are not involved in electrochemical reactions. The anions of the electrolyte in the salt bridge migrate to the anode compartment and the cations to the cathode compartment. Thus, the salt bridge prevents the accumulation of charges and maintains the flow of current. In the electrochemical cell, the salt bridge can be replaced by a *porous partition* which allows the migration of ions but does not allow mixing of the two solutions.

REPRESENTATION OF REDOX COUPLE AND GALVANIC CELL

Galvanic cell is a combination of two redox couples, namely; *oxidation couple* or *oxidation half cell* and *reduction couple* or *reduction half cell*. If M represents the symbol of the element and M^{n+} represents its cation (*i.e.*, its *oxidised state*) in solution, then

Oxidation half cell is represented as **$M/M^{n+}(c)$**

Reduction half cell is represented as **$M^{n+}(c)/M$.**

In both the notations *c* refers to the molar concentration of the ions in solution. Conventionally, a cell is represented by writing the *cathode* on the *right hand side* and *anode* on the *left hand side*. Two vertical lines are put between the two half cells to indicate a salt bridge. Sometimes the formula of the electrolyte used in the salt bridge is also written below the vertical lines.

For example, zinc-copper sulphate cell is represented as follows:

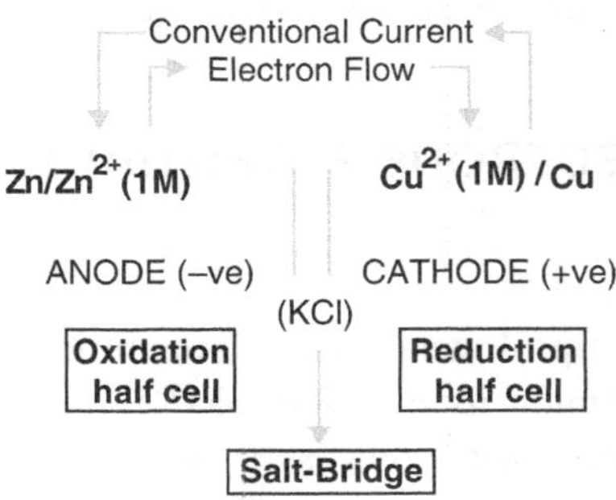

In order to avoid confusion about the nomenclature of the electrodes and the process occurring on them remember the word LOAN. Here,

L (*left*); O (*Oxidation*); A (*anode*); N (*Negative*)

This implies that the electrode represented on the left hand side involves oxidation, acts as anode and is the –ve terminal of the electrochemical cell. Once this is known, the nomenclature and process on the other electrode can be guessed.

VARIOUS TYPES OF REDOX COUPLES OR ELECTRODES

Some important types of electrodes which are frequently used in electrochemical cells are described as follows:

1. Metal-metal Ion Electrode. Such type of electrodes include a metal rod dipped in a solution of its own ions. Some examples are **Zn/Zn^{2+}, Ag/Ag^+, Cu/Cu^{2+}** etc.

2. Amalgam Electrodes. These are similar to metal-metal ion electrodes, but here the metal is used in the form of its amalgam with Hg. Amalgam is formed to modify the activity of the metal. **$(Zn\text{-}Hg)/Zn^{2+}$** is one of the common examples of this type.

3. Gas Electrodes. Such electrodes involve inert metals such as platinum dipped in a solution containing ions of the gaseous element. The arrangement is made in such a manner that gas and its ions are brought in contact at the surface of the inert metal. Some examples along with their electrode reactions are as follows:

Hydrogen electrode: **Pt, H_2(g)/H^+(aq)**

$H_2 \longrightarrow 2H^+ + 2e^-$ or $2H^+ + 2e^- \longrightarrow H_2$

Chlorine electrode : **Pt, Cl_2(g)/Cl^-(aq)**

$2Cl^- \longrightarrow Cl_2 + 2e^-$ or $Cl_2 + 2e^- \longrightarrow 2Cl^-$

Bromine electrode : **Pt, Br_2(g)/Br^-(aq)**

$2Br^- \longrightarrow 2Br_2 + 2e^-$ or $Br_2 + 2e^- \longrightarrow 2Br^-$

33.3 ELECTRODE POTENTIAL (E)

DEVELOPMENT OF ELECTRODE POTENTIAL

When a strip of metal (M) is brought in contact with a solution containing its own ions (M^{n+}), then either of the following three possible processes can take place:

(*i*) The metal ion M^{n+} may collide with the metallic strip and bounce back without any change.

(*ii*) The metal ion M^{n+} may collide with the strip, gain *n* electrons and get converted into a metal atom, *i.e.*, the ion is *reduced.*

$$M^{n+} + ne^- \longrightarrow M$$

(*iii*) The metal atom on the strip may lose *n* electrons and enter the solution as M^{n+} ion, *i.e.*, metal is *oxidised.*

$$M \longrightarrow M^{n+} + ne^-$$

The above changes have been shown in Fig. 33.2.

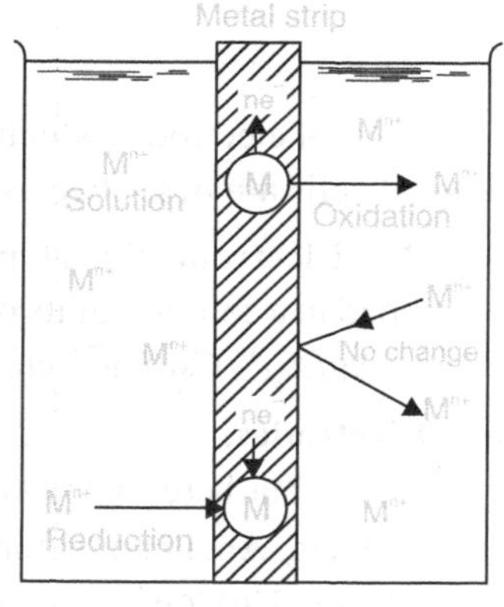

Fig. 33.2. Electrode equilibrium.

Now, if the metal has a relatively *high tendency to get oxidised*, its atoms would start losing electrons, change into positive ions and pass into the solution. The electrons lost accumulate in the metal strip and causes it to develop a *negative charge*. The negative charge developed on the strip does not allow metal atoms to continue losing electrons but it would reattract the metal ions from the solution in an attempt to neutralize its charge. Ultimately, a state of equilibrium will be established between the metal and its ions at the interface.

$$\underset{\text{(Metal strip)}}{M} \rightleftharpoons \underset{\text{(Solution)}}{M^{n+}} + \underset{\text{(In metal strip)}}{ne^-}$$

Similarly, if the metal ions have a relatively *greater tendency to get reduced,* they will accept electrons at the strip from the metal atoms and consequently, a *net positive* charge is developed on the metal strip. Ultimately, a similar equilibrium is established between the metal ions and the metal atoms at the interface.

$$\underset{\text{(Solution)}}{M^{n+}} + \underset{\text{(From metal strip)}}{ne^-} \rightleftharpoons \underset{\text{(Metal)}}{M}$$

The development of positive and negative charges on the metal strip has been shown in Fig. 33.3 and Fig. 33.4 respectively.

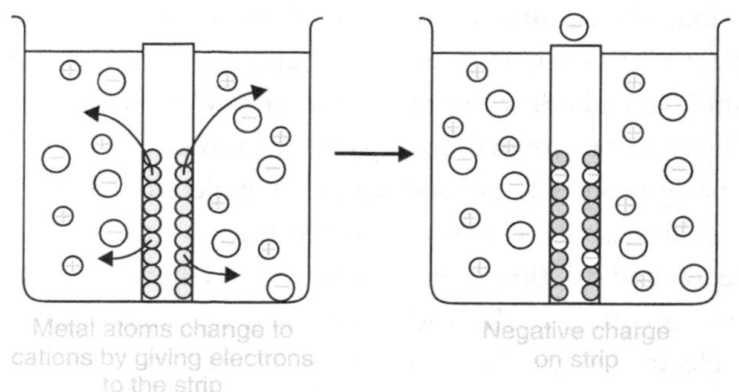

Fig. 33.3. Development of –ve charge on metal strip.

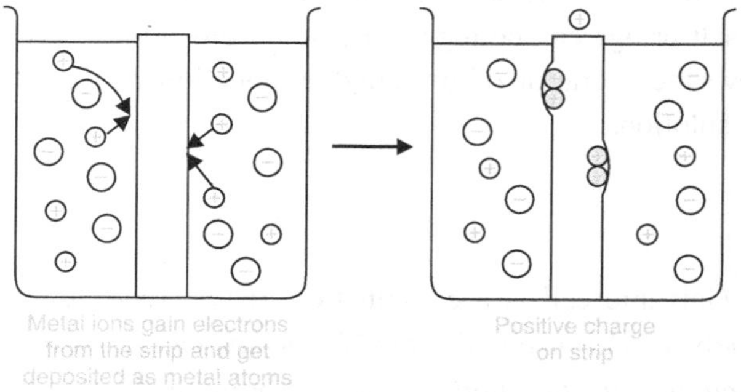

Fig. 33.4. Development of +ve charge on the metal strip.

Different metals have different tendencies to lose/gain electrons, and may develop different magnitude of negative or positive charge, which, unfortunately cannot be measured by any direct means. However, the separation of charges at equilibrium in either case, results in an electrical potential difference between the metal and the solution of its ions. This is known as **electrode potential.**

The exact potential difference at equilibrium depends on the **nature of metal, its ions, the concentration of ions** and the **temperature.**

According to the present IUPAC conventions, the half reactions are always written as **reduction half reactions** and their potentials are represented by reduction potentials. It may be noted that:

(*i*) Reduction potential (*tendency to gain electrons*) and oxidation potential (*tendency to lose electrons*) of an electrode are **numerically equal** but have **opposite signs.**

(*ii*) Reduction potential increases with an increase in the concentration of ions and decreases with a decrease in the concentration of the ions in solution.

STANDARD ELECTRODE POTENTIAL ($E^\ominus$)

The reduction potential of an electrode when the concentration of the ions in solution is *1 mol L^{-1}* and the temperature is 298 K is called **standard reduction potential ($E^\ominus_{red}$)** or simply **standard electrode potential ($E^\ominus$).**

In case of gas electrodes, the standard conditions chosen are 1 bar pressure and 298 K along with 1 M concentration of ions in solution.

The absolute value of $E^\ominus$ cannot be determined because once equilibrium is reached between the electrode and the solution in a half cell, no further displacement of charges can occur unless and until it is connected to another half cell with a different electrode potential. This difficulty is overcome by finding the electrode potentials of various electrodes relative to some reference electrode whose electrode potential is arbitrarily fixed. The common reference electrode used for this purpose is the *standard hydrogen electrode* (**SHE**) whose electrode potential is arbitrarily taken to be *zero.*

33.4 CELL POTENTIAL OR EMF OF THE CELL

The electrochemical cell consists of two half cells. The electrodes in these half cells have different electrode potentials. When the circuit is completed the loss of electrons occurs at the electrode having a lower reduction potential whereas the gain of electrons occurs at the electrode with higher reduction potential. The difference in the electrode potentials of the two electrodes of the cell is termed as **electromotive force** (abbreviated as **EMF**) or **cell voltage (E_{cell}).** Mathematically, it can be expressed as

$$\mathbf{EMF} = E_{red}\ (Cathode) - E_{red}\ (Anode) \text{ or simply as}$$

$$E_{cell} = E_{cathode} - E_{anode}$$

In the representation of a cell, the cathode is written on right hand side and the anode on left hand side. The EMF of a cell is also sometimes written as:

$$\mathbf{EMF} = E_{Right} - E_{Left} = E_R - E_L$$

EMF of a cell may be defined as the *potential difference between the two terminals of the cell when no current is drawn from it.* It is measured with the help of a *potentiometer* or a *vacuum tube voltmeter.*

The EMF of the cell depends on *nature of the reactants, concentration of the solutions in the two half cells,* and the *temperature.* The EMF of a cell under standard state conditions is called standard EMF and can be calculated from the standard electrode potentials of the two half cells.

$$E^\ominus_{cell} = E^\ominus_{cathode} - E^\ominus_{anode}$$

33.5 STANDARD HYDROGEN ELECTRODE (SHE)

Standard hydrogen electrode (Fig. 33.5) consists of a platinum wire sealed into a glass tube and carrying a platinum foil at one end. The platinum foil is coated with finely divided platinum. The electrode is placed in a beaker containing an aqueous solution of some acid having **one molar concentration of H^+ ions.** Hydrogen gas at 1 bar or 100 kPa pressure is continuously bubbled through the solution at a temperature of 298 K. The oxidation or reduction in the SHE takes place at the platinum foil. Hence, it can act as **anode** as well as **cathode** and may be represented as:

$$\mathbf{Pt,\ \tfrac{1}{2}H_2\ (1\ bar)/H^+(1\ M)}$$

or $$\mathbf{H^+(1\ M)/\tfrac{1}{2}H_2\ (1\ bar),\ Pt}\ \text{respectively.}$$

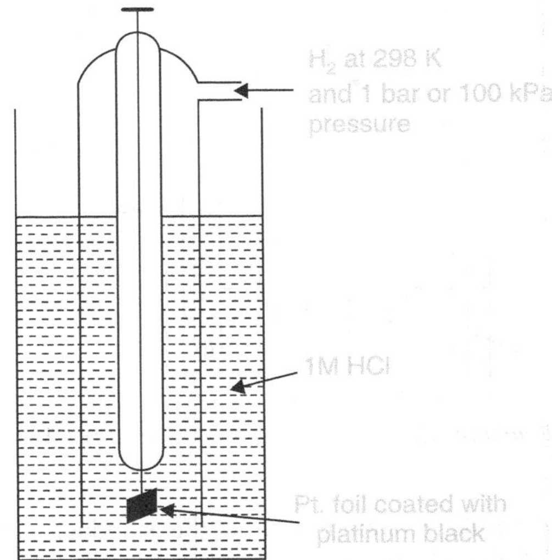

Fig. 33.5. Standard hydrogen electrode (SHE).

If the SHE acts as **anode** *then oxidation will take place at it as*

$$H_2(g) \rightleftharpoons 2H^+(aq) + 2e^-$$

If the SHE acts as **cathode** *then reduction will take place at it as*

$$2H^+(aq) + 2e^- \rightleftharpoons H_2(g)$$

33.6 MEASUREMENT OF STANDARD ELECTRODE POTENTIALS

In order to determine the standard electrode potential of an electrode, the electrode is connected to a standard hydrogen electrode (SHE) to constitute a cell. *If the electrode forms the negative terminal of the cell it is allotted a negative value of electrode potential and if it forms the positive terminal of the cell, it is allotted a positive value of electrode potential.* The potential difference between the electrodes is determined with the help of a *potentiometer or a vacuum tube voltmeter.* At the same time, the direction of flow of conventional current in the external circuit is also monitored with the help of a galvanometer. This helps us to know the positive and negative terminals of the cell since conventional current in an external circuit flows from the +ve terminal to the –ve terminal. The direction of the flow of electrons is opposite to that of conventional current. In an external circuit it occurs from *anode* to *cathode* and allows us to mark anode and cathode electrodes of the cell.

$$E^{\ominus}_{cell} = E^{\ominus}_{cathode} - E^{\ominus}_{anode}$$

Knowing the $E^{\ominus}_{cell}$ and electrode potential of one of the electrodes (SHE), that of the other can be calculated.

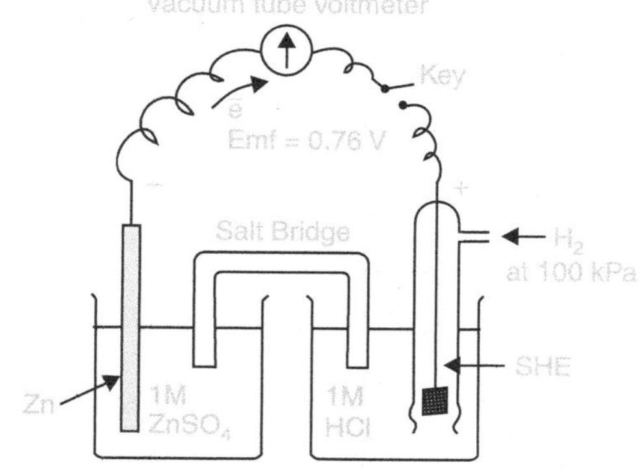

Cell Reaction: $Zn(s) + 2H^+(aq) \longrightarrow Zn^{2+}(aq) + H_2(g)$

Fig. 33.6. Determining $E^{\ominus}_{Zn^{2+}/Zn}$

For example, in order to find the standard electrode potential of zinc electrode, zinc electrode containing a 1 M concentration of Zn^{2+} ions is connected to a SHE as shown in Fig. 33.6. The voltmeter reading shows the potential difference ($E^{\ominus}_{cell}$) of 0.76 volts. The electron flow is found to be from the zinc electrode to the SHE. From this it follows that Zn acts as the anode while the SHE acts as the cathode. The net cell reaction is

$$Zn(s) + 2H^+(aq) \longrightarrow Zn^{2+}(aq) + H_2(g)$$

Now, $E^{\ominus}_{cell} = E^{\ominus}_{cathode} - E^{\ominus}_{anode}$

$$= E^{\ominus}_{H^+/\frac{1}{2}H_2} - E^{\ominus}_{Zn^{2+}/Zn},$$

$$0.76 = 0 - E^{\ominus}_{Zn^{2+}/Zn}$$

$$E^{\ominus}_{Zn^{2+}/Zn} = \mathbf{-0.76\ volts.}$$

Similarly, when standard copper electrode is coupled with a **SHE** the voltmeter reading shows a potential difference ($E^{\ominus}_{cell}$) of 0.34 volts. The electron flow is found to occur from the SHE towards the copper electrode as shown in Fig. 33.7. Therefore, the SHE in this cell acts as *anode* and the copper electrode acts as cathode. The cell reaction is

$$Cu^{2+}(aq) + H_2(g) \longrightarrow Cu(s) + 2H^+(aq)$$

$$E^{\ominus}_{cell} = E^{\ominus}_{cathode} - E^{\ominus}_{anode}$$

$$= E^{\ominus}_{Cu^{2+}/Cu} - E^{\ominus}_{H^+/\frac{1}{2}H_2}$$

$$0.34 = E^{\ominus}_{Cu^{2+}/Cu} - 0$$

$$E^{\ominus}_{Cu^{2+}/Cu} = \mathbf{0.34\ volts.}$$

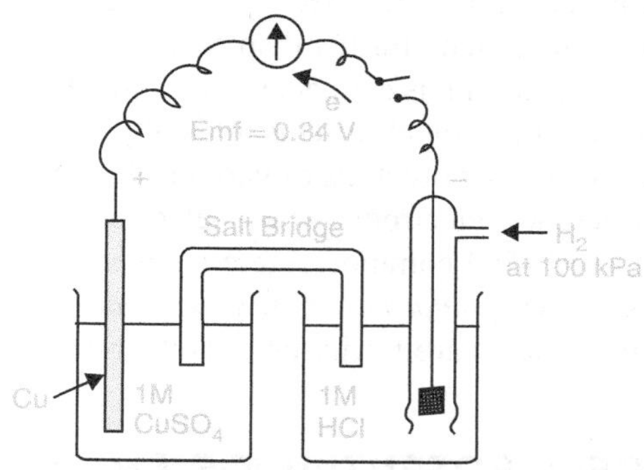

Cell Reaction: $H_2(g) + Cu^{2+}(aq) \longrightarrow Cu(s) + 2H^+(aq)$

Fig. 33.7. Determining $E^{\ominus}_{Cu^{2+}/Cu}$

33.7 ELECTROCHEMICAL SERIES

We have seen that different *metal/metal ion* combinations have different values of standard electrode potentials. The various elements can be arranged in order of increasing or decreasing values of their standard reduction potentials. *The arrangement of various elements in the order of increasing values of standard reduction potentials* is called **electrochemical series.** The electrochemical series, also called activity series consisting of some electrodes along with their respective reduction reactions has been given in Table 33.1.

Table 33.1. Standard Electrode Potentials

Elements	*Electrode Reaction*	$E^{\ominus}_{red}$ *(volts)*
	Oxidised Form + ne⁻ ⟶ Reduced Form	
Li	$Li^+(aq) + e^- \longrightarrow Li(s)$	– 3.05
K	$K^+(aq) + e^- \longrightarrow K(s)$	– 2.93
Ba	$Ba^{2+}(aq) + 2e^- \longrightarrow Ba(s)$	– 2.90
Ca	$Ca^{2+}(aq) + 2e^- \longrightarrow Ca(s)$	– 2.87
Na	$Na^+(aq) + e^- \longrightarrow Na(s)$	– 2.71
Mg	$Mg^{2+}(aq) + 2e^- \longrightarrow Mg(s)$	– 2.37
Al	$Al^{3+}(aq) + 3e^- \longrightarrow Al(s)$	– 1.66
Zn	$Zn^{2+}(aq) + 2e^- \longrightarrow Zn(s)$	– 0.76
Cr	$Cr^{3+}(aq) + 3e^- \longrightarrow Cr(s)$	– 0.74
Fe	$Fe^{2+}(aq) + 2e^- \longrightarrow Fe(s)$	– 0.44
	$H_2O(l) + e^- \longrightarrow \frac{1}{2}H_2(g) + OH^-(aq)$	– 0.41
Cd	$Cd^{2+}(aq) + 2e^- \longrightarrow Cd(s)$	– 0.40
Pb	$PbSO_4(s) + 2e^- \longrightarrow Pb(s) + SO_4^{2-}(aq)$	– 0.31
Co	$Co^{2+}(aq) + 2e^- \longrightarrow Co(s)$	– 0.28
Ni	$Ni^{2+}(aq) + 2e^- \longrightarrow Ni(s)$	– 0.25
Sn	$Sn^{2+}(aq) + 2e^- \longrightarrow Sn(s)$	– 0.14
Pb	$Pb^{2+}(aq) + 2e^- \longrightarrow Pb(s)$	– 0.13
H_2	**$2H^+ + 2e^- \longrightarrow H_2(g)$ (standard electrode)**	**0.00**
Cu	$Cu^{2+}(aq) + 2e^- \longrightarrow Cu(s)$	+ 0.34
I_2	$I_2(s) + 2e^- \longrightarrow 2I^-(aq)$	+ 0.54
Fe	$Fe^{3+}(aq) + e^- \longrightarrow Fe^{2+}(aq)$	+ 0.77
Hg	$Hg_2^{2+}(aq) + 2e^- \longrightarrow 2Hg(l)$	+ 0.79
Ag	$Ag^+(aq) + e^- \longrightarrow Ag(s)$	+ 0.80
Hg	$Hg^{2+}(aq) + 2e^- \longrightarrow Hg(l)$	+ 0.85
N_2	$NO_3^- + 4H^+ + 3e^- \longrightarrow NO(g) + 2H_2O$	+ 0.97
Br_2	$Br_2(aq) + 2e^- \longrightarrow 2Br^-(aq)$	+ 1.08
O_2	$O_2(g) + 2H_3O^+(aq) + 2e^- \longrightarrow 3H_2O$	+ 1.23
Cr	$Cr_2O_7^{2-} + 14H^+ + e^- \longrightarrow 2Cr^{3+} + 7H_2O$	+ 1.33
Cl_2	$Cl_2(g) + 2e^- \longrightarrow 2Cl^-(aq)$	+ 1.36
Au	$Au^{3+}(aq) + 3e^- \longrightarrow Au(s)$	+ 1.42
Mn	$MnO_4^-(aq) + 8H_3O^+(aq) + 5e^- \longrightarrow Mn^{2+}(aq) + 12H_2O(l)$	+ 1.51
F_2	$F_2(g) + 2e^- \longrightarrow 2F^-(aq)$	+ 2.87

Increase (upward): (*a*) Tendency for oxidation to occur (*b*) Power as reducing agent

Increase (downward): (*a*) Tendency for reduction to occur (*b*) Power as oxidising agent

APPLICATIONS OF THE ELECTROCHEMICAL SERIES

Some of the important applications of the electrochemical series have been discussed as follows:

1. Calculation of the standard EMF of a cell. The electrochemical series gives the standard reduction potentials of various electrodes. The electrode with higher reduction potential is taken as the cathode and the other as the anode. From this EMF° of the cell is calculated as:

$$EMF^\circ = E^\circ_{cathode} - E^\circ_{anode}$$

However, if the reaction taking place in the cell is also to be determined, the following steps are followed:

(*a*) Write reduction equations for both electrodes separately along with their standard reduction potentials.

(*b*) Balance each reaction with respect to the number of atoms of each kind and the electrical charges.

(*c*) Multiply the reactions by a suitable number so that the number of electrons involved in both half reactions are equal.

Caution

While multiplying the half cell reactions in calculations do not multiply the E° values with any such integer because it simply reflects the tendency of reduction.

(*d*) Subtract the equation for the reaction having the lower standard reduction potential from the other reaction having the higher standard reduction potential. The difference gives the standard EMF of the cell.

The electrode having higher standard reduction potential acts as the cathode and the other electrode acts as the anode. Now let us solve some numerical problems to understand the applications of this method in calculating standard EMF.

SOLVED EXAMPLES

Example 33.1. *Calculate the standard EMF of the given cell, also write the overall cell reaction.*

$$Zn \mid Zn^{2+}\ (1\ M) \parallel Ag^{+}\ (1\ M) \mid Ag$$

Standard electrode potential for Zn^{+2}/Zn is – 0.76 volts and for Ag^{+}/Ag is 0.80 volts.

Solution. It is clear from the cell, that silver electrode acts as a cathode and zinc electrode acts as an anode:

(*a*) $Ag^{+} + e^{-} \longrightarrow Ag \quad E^\circ = 0.80$ volts

$Zn^{+2} + 2e^{-} \longrightarrow Zn \quad E^\circ = -0.76$ volts

Multiply the first equation by 2 so as to equalise the number of electrons

(*b*) $2Ag^{+} + 2e^{-} \longrightarrow 2Ag \quad E^\circ = 0.08$ volts ...(*i*)

$Zn^{2+} + 2e^{-} \longrightarrow Zn \quad E^\circ = -0.76$ volts ...(*ii*)

(*c*) $EMF = E^\circ_{cathode} - E^\circ_{anode}$

$= 0.80 - (-0.76) = \mathbf{1.56\ volts.}$

The overall cell reaction can be obtained by subtracting equation (*ii*) from equation (*i*). *i.e.*,

$$2Ag^{+}(aq) + Zn(s) \longrightarrow 2Ag(s) + Zn^{+2}(aq).$$

Example 33.2. *Calculate the standard EMF of the cell in which the following reaction takes place.*

$$Zn(s) + Ni^{+2}(aq) \longrightarrow Zn^{+2}(aq) + Ni(s)$$

Given that $E^\circ_{Zn^{2+}/Zn} = -0.76$ *volts;*

$E^\circ_{Zi^{+2}/Ni} = -0.25$ *volts.*

Solution. The above reaction can be split into two half reactions as given below:

$Zn(s) \longrightarrow Zn^{+2}(aq) + 2e^{-}$ ... Oxidation

$Ni^{+2}(aq) + 2e^{-} \longrightarrow Ni(s)$ ... Reduction

At the zinc electrode oxidation takes place so it acts as the anode and at the nickel electrode reduction takes place so it acts as the cathode.

$$EMF^\circ = E^\circ_{cathode} - E^\circ_{anode}$$

$$= E^\circ_{Ni^{+2}/Ni} - E^\circ_{Zn^{2+}/Zn}$$

$$= -0.25 - (-0.76) = \mathbf{0.51\ volts.}$$

2. Comparison of the reactivities of metals. The metals which occupy higher position in the electrochemical series have lower reduction potentials. This indicates that ions of such metals will not be reduced to metals easily. On the contrary, such metals would be easily oxidized to their ions by losing electrons. Therefore, from the position of the metal in the electrochemical series, it is possible to predict its relative reactivity. The metal having a smaller reduction potential can displace metals having larger reduction potentials from the solutions of their salts. In other words, *the metal occupying higher position in the series can displace the metals lying below it from solutions of their salts.*

Thus, we may conclude that *the metals occupying higher positions in the electrochemical series are more reactive in displacing the other metals from the solutions of their salts.* For example, zinc lies above copper in the series and hence, it can displace copper from copper sulphate solution. Copper cannot displace zinc from zinc sulphate solution because it lies below zinc in the series and hence, it is less reactive.

3. Predicting the feasibility of a redox reaction. With the help of electromotive series we can predict whether a given redox reaction will take place or not. From a given equation, the substances undergoing oxidation and reduction are identified. The substance undergoing oxidation will act as an anode and the substance undergoing reduction will act as a cathode. The EMF of this hypothetical cell is calculated as:

$$EMF^\circ = E^\circ_{cathode} - E^\circ_{anode}$$

If the calculated EMF° is positive the given redox reaction will take place. If it is negative the given redox reaction will not take place.

SOLVED EXAMPLE

Example 33.3. *Predict whether the following redox reaction is feasible or not under standard conditions:*

$$Sn^{+2}(aq) + Cu(s) \longrightarrow Sn(s) + Cu^{+2}(aq)$$

Given that $E^\circ_{Sn^{+2}/Sn} = -0.136$ *volts and*

$$E^\circ_{Cu^{+2}/Cu} = 0.34 \text{ volts.}$$

Solution. The above reaction can be split into two half reactions as given below:

$$Cu(s) \longrightarrow Cu^{+2}(aq) + 2e^- \quad \text{... Oxidation}$$

$$Sn^{+2}(aq) + 2e^- \longrightarrow Sn(s) \quad \text{... Reduction}$$

At the copper electrode oxidation takes place so it acts as the anode and at the tin electrode reduction takes place so it acts as the cathode.

$$EMF = E^\circ_{cathode} - E^\circ_{anode}$$

$$= E^\circ_{Sn^{+2}/Sn} - E^\circ_{Cu^{2+}/Cu}$$

$$= -0.136 - 0.34 = -\mathbf{0.476\ volts.}$$

Since the value of the calculated EMF° is negative, the given reaction is not feasible.

4. To predict the reaction of a metal with dilute acids to liberate hydrogen gas. Some metals like Fe, Zn react with dil. acids like HCl, H_2SO_4 to liberate H_2 gas while some metals like Cu, Ag do not liberate H_2 gas with dil. HCl and dil. H_2SO_4. A prediction of the ability of a given metal to produce H_2 gas by its reaction with dilute acids can be made from a knowledge of the electromotive series. A chemical reaction between a metal and an acid can be represented in general as

$$\underset{\text{Metal}}{\mathbf{M}} + \underset{\text{From acid}}{n\mathbf{H}^+} \longrightarrow \mathbf{M}^{n+} + n/2\mathbf{H}_2$$

For the above reaction to occur, the E°_{red} of metal (M^{n+}/M) must be lower than that of hydrogen. Hence, it can be concluded that *metals which lie above hydrogen in the electromotive series can reduce H^+ ions to hydrogen and hence, liberate hydrogen gas on reaction with dil acids.* In other words, metals having negative reduction potentials can displace hydrogen from acids. For example, zinc ($E^\circ_{Zn^{+2}/Zn} = -0.76$ volt) lies above hydrogen in the series and can displace hydrogen from dilute acids, whereas copper ($E^\circ_{Cu^{+2}/Cu} = +0.34$ volts) lies below hydrogen in the series and cannot displace hydrogen from acids.

5. Relative Oxidising and Reducing Powers of Various Substances. A substance with a higher value of standard reduction potential have a greater tendency to undergo reduction. For example, F_2 has the highest reduction potential which means it is most easily reduced to F^- ions. In other words, F_2 is the best oxidising agent. Li^+ ion, on the other hand, has the lowest reduction potential. Hence Li^+ is the weakest reducing agent or conversely Li metal is the best reducing agent. Thus, it can be concluded that *substances with* ***higher reduction potentials*** *are* ***strong oxidising agents*** *while substances with* ***lower reduction potentials*** *are* ***strong reducing agents***.

ACTIVITY 33.1	**To set up a simple electro-chemical cell and measure its standard EMF.**

Requirements

Two beakers, U Tube, Copper and zinc rods, 1M $CuSO_4$ solution and 1M $ZnSO_4$ solution, KCl solution, Cotton plugs, Connecting wires, Digital voltmeter, Galvanometer.

Procedure

1. Set up the apparatus as shown in Fig. 33.8 using a digital voltmeter.
2. Complete the circuit and record the emf.
3. Now break the circuit and replace the voltmeter with a galvanometer. Again complete the circuit to note the direction of flow of conventional current. This will provide a clue about the +ve and –ve terminals. (*current flows from the +ve to the –ve terminal*).
4. Remove the salt bridge and observe if there is any flow of current.

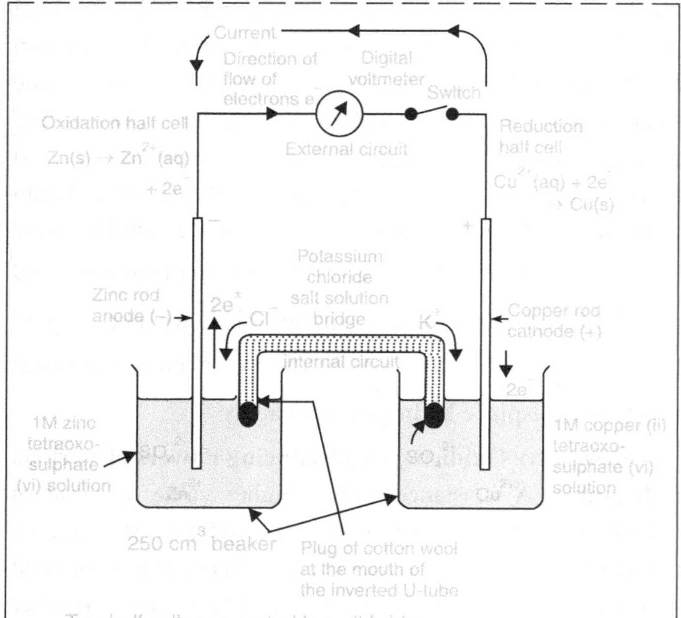

Fig. 33.8. A simple electrochemical cell.

5. Repeat the experiment by using the following electrochemical couples:

   (*a*) Zn/Zn^{2+}(1 M) and Ag/Ag^{+}(1 M)

   (*b*) Cu/Cu^{2+}(1 M) and Ag/Ag^{+}(1 M)

   Record EMF in each case.

   Now, (*i*) Draw the cell diagram for each pair of electrochemical couple.

   (*ii*) Record their standard EMF. Compare with theoretical values.

   (*iii*) Represent the cell in each case and write their net cell reaction.

33.8 APPLICATIONS OF ELECTRO-CHEMICAL CELLS

The importance of electrochemical cells or galvanic cells lies in their ability to provide us with a portable source of electrical energy. We have already studied that indirect redox reaction is, primarily, the basis of all the electrochemical cells. Quite often, we use the term **battery** to represent the arrangement of two or more galvanic cells connected in series.

However, in practice the redox reaction used should give the arrangement which fulfils the following requirements:

- It should be *light* and *compact*;
- Its voltage should not vary appreciably during its use;
- It should provide power for a longer period; and
- It should be rechargeable.

The galvanic cells can be broadly classified into two categories, namely; *primary cells* and *secondary cells.*

PRIMARY CELLS

This type of cells become dead over a period of time and the chemical reaction stops. They cannot be recharged or used again. Some common examples are *dry cell, mercury cell,* etc.

(*a*) **Dry Cell.** It is a compact form of ***Leclanche cell*** known after its inventor, a French chemist, G. Leclanche. In this cell, the *anode* consists of a *zinc container* while the *cathode* is a *graphite rod* surrounded by powdered MnO_2 and carbon. The space between the electrodes is filled with the paste of NH_4Cl and $ZnCl_2$. The arrangement is shown in Fig. 33.9.

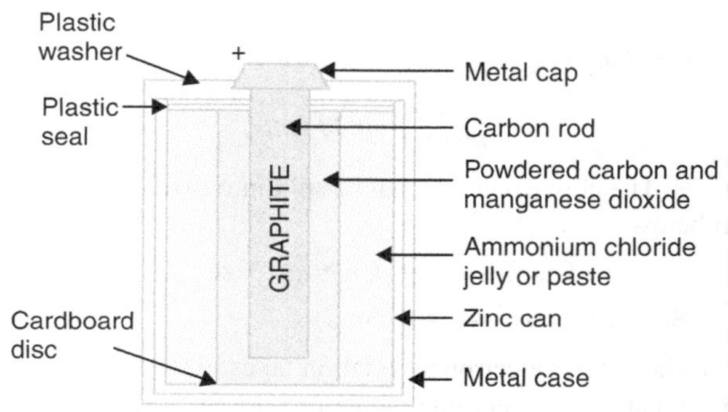

Fig. 33.9. A dry cell.

The reactions taking place at the electrodes are given in their simplified form as follows:

Cathode:

$$MnO_2 + NH_4^+ + e^- \longrightarrow MnO(OH) + NH_3$$

(*Oxidation state of Mn changes from + 4 to + 3*)

Anode:

$$Zn \longrightarrow Zn^{2+} + 2e^-$$

The zinc ions (Zn^{2+}) produced combine with ammonia liberated in the cathodic reaction to form diammine zinc (II) cation.

$$Zn^{2+} + 2NH_3 \longrightarrow [Zn(NH_3)_2]^{2+}$$

Dry cells do not have long life as NH_4Cl which is acidic, corrodes the zinc container even if the cell is not in use. The cell potential of dry cells lies in the range *1.25 V* to *1.5 V*.

(*b*) **Mercury Cell.** It is a miniature cell which is used these days to supply energy in watches, video cameras, hearing aids and other compact devices. In mercury cell the *anode* is a *zinc-mercury amalgam* and the *cathode* is a paste of *mercury(II) oxide* and *carbon.* Electrolyte is a moist paste of KOH-ZnO. The arrangement in its simple form is shown in Fig. 33.10.

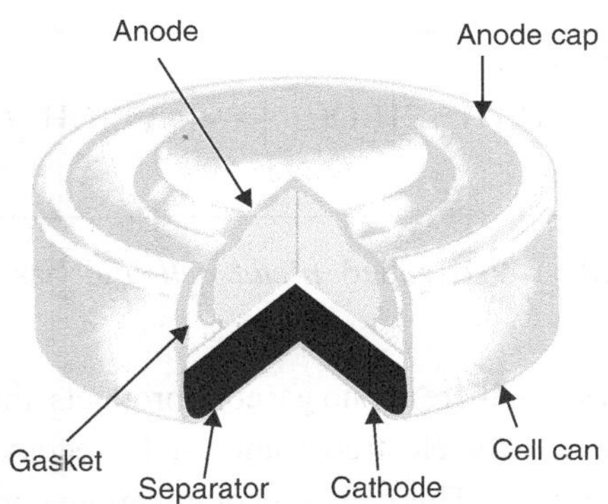

Fig. 33.10. Commonly used mercury cell. The reducing agent is zinc and the oxidising agent is mercury (II) oxide.

The operating voltage for mercury cell is ≃ 1.35 V and the cell reactions are as follows:

Anode:

$$\underset{\text{(Amalgam)}}{Zn(Hg)} + 2OH^- \longrightarrow ZnO(s) + H_2O(l) + 2e^-$$

Cathode:

$$HgO(s) + H_2O(l) + 2e^- \longrightarrow Hg(l) + 2OH^-$$

Net reaction:

$$\underset{\text{(Amalgam)}}{Zn(Hg)} + HgO(s) \longrightarrow ZnO(s) + Hg(l)$$

Such a cell has constant potential throughout its life.

SECONDARY CELLS

This type of cells can be recharged by passing *direct current* through them and can be used again and again. Some examples are lead-storage battery, nickel-cadmium storage cell, etc. Let us study the workings of a *lead storage cell.*

(*a*) **Lead-storage Battery.** It is the most frequently used battery in automobiles. It consists of six voltaic cells connected in series. In each cell the *anode* is made of *spongy lead* and the *cathode* is a grid of lead *packed with lead dioxide* (PbO_2). The electrolyte is an aqueous solution of H_2SO_4 which is 38% by mass. The reactions taking place in this type of cell can be represented as:

Anode:

$$Pb + SO_4^{2-} \longrightarrow PbSO_4 + 2e^-; \quad E^{\ominus}_{oxi} = 0.36 \text{ V}$$

Cathode:

$$PbO_2 + SO_4^{2-} + 4H^+ + 2e^- \longrightarrow PbSO_4 + 2H_2O; \quad E^{\ominus}_{red} = 1.69 \text{ V}$$

Net reaction:

$$Pb + PbO_2 + 4H^+ + 2SO_4^{2-} \longrightarrow 2PbSO_4 + 2H_2O; \quad E^{\ominus}_{cell} = 2.05 \text{ V}$$

During the working of the cell, the concentration of H_2SO_4 decreases as sulphate ions are consumed to form $PbSO_4$. The $PbSO_4$ precipitates and partially gets coated on both electrodes while the water formed dilutes the sulphuric acid. With the decrease in the concentration of H_2SO_4 the density of the solution also decreases. The condition of the battery can be easily checked by measuring the density of the solution.

To enhance the output of the cell, the anode and cathode plates are arranged in alternating manner and they are separated by sheets of insulating material. The anode and cathode plates are separately connected to each other so as to increase the electrode area in contact with the electrolyte solution. This increases the current delivering capacity of the cell. The *groups of electrodes* which constitutes *one cell* are shown in Fig. 33.11. The cells are further connected in series so as to increase the voltage of the battery. In a 6 volts battery there are 3 cells and in 12 volts battery there are 6 cells.

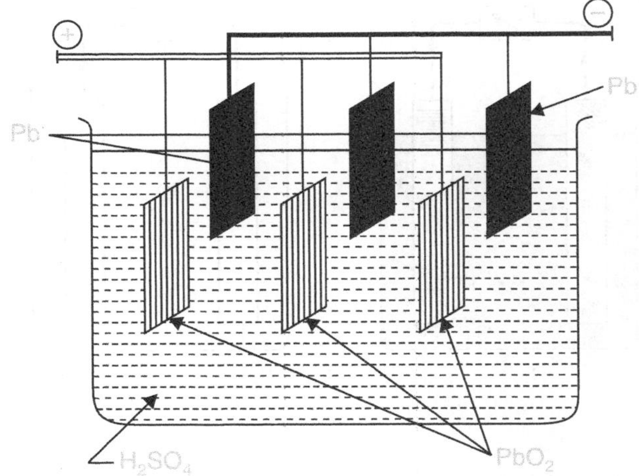

Fig. 33.11. Lead storage cell.

Recharging the battery. The battery can be recharged by connecting it to an external source of *direct current* with voltage greater than 12 V. It forces the electrons to flow in opposite direction resulting in the deposition of Pb on the anode and PbO_2 on the cathode.

During the *recharging* operation, the cell behaves as an *electrolytic cell*. The recharging reactions are:

Cathode (–ve):

$$PbSO_4(s) + 2e^- \longrightarrow Pb(s) + SO_4^{2-}(aq)$$

Anode (+ve):

$$PbSO_4(s) + 2H_2O(l) \longrightarrow PbO_2(s) + SO_4^{2-}(aq) + 4H^+(aq) + 2e^-$$

Net reaction:

$$2PbSO_4(s) + 2H_2O(l) \longrightarrow Pb(s) + PbO_2(s) + 4H^+(aq) + 2SO_4^{2-}(aq)$$

Such an operation becomes possible because the $PbSO_4$ formed during the discharge operation is solid and sticks to the electrodes. Therefore, it is in a position to lose or gain electrons during electrolysis.

The discharging and recharging of lead storage battery has been shown in Figs. 33.12 (*a*) and 33.12 (*b*) respectively.

In an automobile, the battery is discharged when the engine is started. While running, the engine powers the alternator which produces electrical energy sufficient enough to recharge the battery. Thus, the battery is constantly recharged as long as the automobile is being driven.

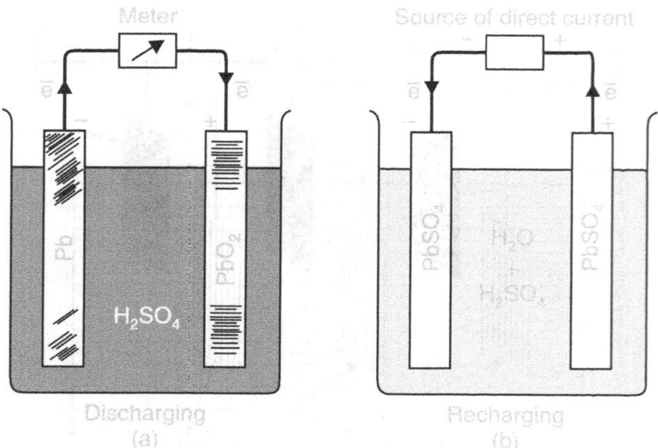

Fig. 33.12. Discharging and recharging of lead storage cell.

(*b*) **Nickel-cadmium Storage Cell.** It is another rechargeable cell. It consists of a *cadmium anode* and the cathode is made of a metal grid containing *nickel (IV) oxide.* These are immersed in KOH solution. The reactions occurring are:

Anode:

$$Cd(s) + 2OH^-(aq) \longrightarrow Cd(OH)_2(s) + 2e^-$$

Cathode:

$$NiO_2(s) + 2H_2O(l) + 2e^- \longrightarrow Ni(OH)_2(s) + 2OH^-(aq)$$

Net reaction:

$$Cd(s) + NiO_2(s) + 2H_2O(l) \longrightarrow Cd(OH)_2(s) + Ni(OH)_2(s)$$

The cell is also called ***nicad cell*** and has a *voltage* of ≈ **1.4 V.**

As is evident, there are no gaseous products, the products formed adhere to the electrodes and can be reconverted by a *recharging process*. This cell is becoming more popular these days and finds use in electronic watches and calculators.

33.9 FUEL CELLS

In recent years, scientists have designed cells which convert **chemical energy of a fuel** directly into **electrical energy.** Such cells are called **fuel cells.** These are the voltaic cells in which, the fuels such as H_2, CO, CH_4, C_3H_8, etc., are used to generate electrical energy without the intervention of thermal devices like boilers, turbines, etc.

The conventional method of conversion of chemical energy of a fuel into electrical energy involves combustion of a fuel to liberate heat. The heat energy produced is used to generate steam for spinning the turbines which are coupled to electrical generators. This process is approximately 40% efficient.

Fuel cells are designed in such a way that the materials to be oxidised and reduced at the electrodes are stored outside the cell and are constantly supplied to the electrodes. In fact, a fuel cell is a *flow battery* that continues to operate as long as the reactants from outside are fed into it. One of the most successful fuel cells uses the reaction of *hydrogen* and *oxygen* to form water and is known as H_2—O_2 fuel cell. The H_2—O_2 fuel cell is also called **Bacon cell** after the name of its inventor and it was used to power the *Apollo Space Missions*. The water vapours produced during the reaction were condensed and added to drinking water supply for the astronauts. The experimental arrangement is shown in Fig. 33.13.

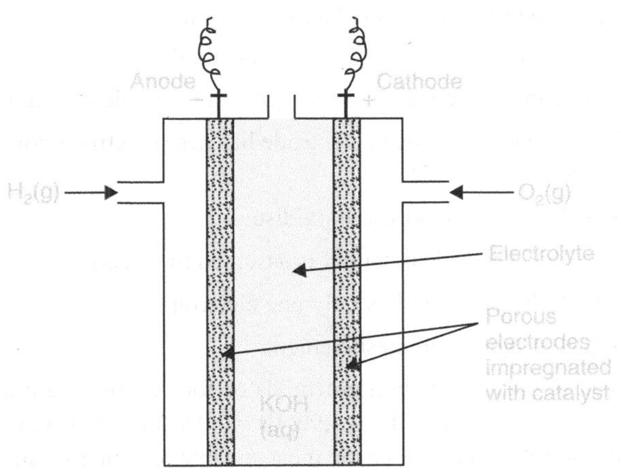

Fig. 33.13. A single bacon cell.

The cell consists of porous carbon electrodes which are impregnated with catalyst (Pt, Ag or CoO). Hydrogen and oxygen are bubbled through the electrodes into the electrolyte which is an aqueous solution of NaOH or KOH.

The electrode reactions are:

Anode:

$$[H_2(g) + 2OH^-(aq) \longrightarrow 2H_2O(l) + 2e^-] \times 2$$

Cathode:

$$O_2(g) + 2H_2O(l) + 4e^- \longrightarrow 4OH^-(aq)$$

Net reaction:

$$2H_2(g) + O_2(g) \longrightarrow 2H_2O(l)$$

The cell runs continuously as long as the gases hydrogen and oxygen are supplied at the temperature 525 K and 50 atm. pressure.

ADVANTAGES OF FUEL CELLS

Some prominent advantages of fuel cells are described as follows:

(*i*) ***Pollution Free.*** There is no harmful or objectionable product formed in fuel cells. Hence, they do not cause pollution problems.

(*ii*) ***High Efficiency.*** The efficiency of fuel cells is approximately 70–75%, which is much higher than the conventional cells.

THERMODYNAMIC EFFICIENCY

Thermodynamic efficiency of fuel cells is the ratio of the electrical free energy to the ΔH of the reaction. For example, for H_2—O_2 fuel cell

$$\eta = \frac{\Delta G}{\Delta H} = \frac{-nFE}{\Delta H} = \frac{-229 \text{ kJ}}{-242 \text{ kJ}}$$

$$= 0.095 \text{ or } 95\%$$

This theoretical thermodynamic efficiency, however, cannot be achieved because of the over potentials at the electrodes.

(*iii*) ***Continuous Source of Energy.*** Unlike conventional batteries, energy can be obtained from the fuel cell continuously so long as the supply of fuel is maintained.

The H_2/O_2 cell has been used for generating electrical power in the Apollo space programmes.

SUMMARY

- **Electrochemistry.** It is the branch of chemistry which deals with the process of conversion of chemical energy into electrical energy and vice versa.
- **Electrochemical Cell.** A device in which a redox reaction is carried out in such a way that its decrease in free energy during the process appears as electrical energy. In the cell the **anode** is the –ve terminal where oxidation occurs and the **cathode** is the +ve terminal where reduction occurs. The electron flow occurs from the anode to the cathode.
- **Electrode Potential.** The electrical potential difference between a metal and a solution of its own ions at equilibrium and at a given temperature.
- **Standard Electrode Potential (E°).** The electrode potential expressed under standard state conditions. The E° values are generally $E°_{red}$ values measured with reference to a standard hydrogen electrode whose E° is arbitrarily taken to be zero.
- **EMF of Cell or Cell Voltage ($E°_{cell}$).** Standard cell voltage is the difference in the standard electrode potentials of two terminals of a cell

$$E°_{cell} = E°_{cathode} - E°_{anode}$$

- **Electromotive Series.** The arrangement of various elements in order of increasing standard electrode potentials.

EVALUATE YOURSELF

I. Objective Type Questions

Select the most appropriate choice from the options given as (a), (b), (c) and (d) after each question:

1. Which instrument is used to measure standard electrode potential?
   (*a*) Electrometer (*b*) Voltmeter
   (*c*) Pyrometer (*d*) Galvanometer.
2. Standard electrode potentials of Ag^+/Ag is + 0.80 V and that of Cu^{2+}/Cu is + 0.34 V. These electrodes are connected through a salt bridge. The standard emf of the cell is
   (*a*) 0.46 V (*b*) – 0.46 V
   (*c*) 0.34 V (*d*) 1.14 V.
3. Aluminium displaces hydrogen from dilute HCl whereas silver does not. The EMF of the cell prepared by combining Al/Al^{3+} and Ag/Ag^+ is 2.46 V. The reduction potential of silver electrode is + 0.80 V. The reduction potential of aluminium electrode is
   (*a*) + 1.66 V (*b*) – 3.26 V
   (*c*) 3.26 V (*d*) – 1.66 V.
4. Consider the following E° values
   $$E^\circ_{Fe^{3+}/Fe^{2+}} = 0.77\text{ V}, \ E^\circ_{Sn^{2+}/Sn} = -0.14\text{ V}$$
   Under standard conditions, the standard potential for the reaction, $Sn(s) + 2Fe^{3+}(aq) \longrightarrow 2Fe^{2+}(aq) + Sn^{2+}(aq)$ is
   (*a*) 0.63 V (*b*) 1.40 V
   (*c*) 0.91 V (*d*) 1.68 V.
5. Standard electrode potential of three metals A, B and C are + 0.5 V, – 3.0 V and – 1.2 V respectively. The reducing power of these metals are:
   (*a*) B > C > A (*b*) A > B > C
   (*c*) C > B > A (*d*) A > C > B.
6. Which of the following is actually possible if the oxidation potentials of Zn = + 0.76 V and Ag = – 0.80 V?
   (*a*) $Zn + 2Ag^+ \longrightarrow Zn^{2+} + 2Ag$
   (*b*) $Zn^{2+} + 2Ag \longrightarrow Zn + 2Ag^+$
   (*c*) $2Zn + 2Ag \longrightarrow Zn^{2+} + 2Ag^+$
   (*d*) $Zn^{2+} + Ag^+ \longrightarrow Zn + Ag$.
7. Which of the following statements is true for the electrochemical Daniel cell?
   (*a*) Electrons flow from copper electrode to zinc electrode
   (*b*) Current flows from zinc electrode to copper electrode
   (*c*) Cations move towards copper electrode
   (*d*) Cations move towards zinc electrode.
8. The –ve terminal of an electrochemical cell is
   (*a*) anode (*b*) cathode
   (*c*) neither anode nor cathode (*d*) can be anode or cathode.
9. The standard hydrogen electrode has zero electrode potential because
   (*a*) Hydrogen is easiest to oxidise
   (*b*) This electrode potential is assumed to be zero
   (*c*) Hydrogen atom has only one electron
   (*d*) Hydrogen is lightest element.
10. The standard reduction potentials of four elements are given below: I = – 3.04 V, II = – 1.90, III = 0.0 V, IV = 1.90 V. Which of the following will be the most suitable reducing agent?
    (*a*) I (*b*) II
    (*c*) III (*d*) IV.

II. Fill in the Blanks

11. Complete the following sentences by supplying appropriate words:
    (*i*) The value E°_{OXI} of the standard hydrogen electrodes is ......
    (*ii*) Standard cell voltage should be ...... for the cell to work
    (*iii*) In H_2–O_2 fuel cell the electrolyte used is ......
    (*iv*) In dry cell the ...... rod acts as cathode
    (*v*) In mercury cell the electrolyte used is ......
    (*vi*) When lead storage battery is recharged it acts as ......
    (*vii*) *nicad* cell has a cell voltage equal to ......
    (*viii*) In H_2–O_2 fuel cell the cathode reaction is ......

III. Discussion Questions

12. What is meant by the term electrochemical cell? Give its mechanism by considering Zn—$CuSO_4$ reaction.
13. What is a salt bridge? What is its function? What alternative arrangement do you suggest in place of a salt bridge?
14. What is meant by the terms electrode potential and standard electrode potential?
15. Draw a well labelled diagram for a SHE. Give its one important use.
16. Draw a diagram for the galvanic cell which would have an overall chemical reaction as
    $$Zn + 2Ag^+ \longrightarrow Zn^{2+} + 2Ag$$
    Based on the above reaction, answer the following:
    (*i*) Write the reactions occurring at each electrode.
    (*ii*) In which directions do the electrons flow in the external circuit?
    (*iii*) Name the salt to be taken in the salt bridge.

(*iv*) How is the electrical neutrality maintained in the solutions of the two half cells?

(*v*) Label the anode and cathode.

Use data from Table 33.1 to find the standard EMF of the cell.

17. The net cell reactions of some electrochemical cells are given as follows. Give cell notations for all.

(*i*) $2Fe_{(aq)}^{3+} + 2Cl^-_{(aq)} \longrightarrow 2Fe^{2+}_{(aq)} + Cl_2(g)$

(*ii*) $Cd(s) + I_2 \longrightarrow Cd^{2+} + 2I^- (aq)$

(*iii*) $2Cr(s) + 3Cu^{2+} (aq) \longrightarrow 3Cu(s) + 2Cr^{3+} (aq)$

18. Using the electromotive series select a reducing agent that can convert

(*a*) Fe^{2+} to Fe (*b*) Ag^+ to Ag

(*c*) Al^{3+} to Al.

19. Can we store the solution of aqueous $CuSO_4$ in an iron vessel? Justify your answer.

20. Copper does not react with dilute HCl whereas zinc immediately produces H_2 gas.

21. If we have 1 M aqueous solution of $Pb(NO_3)_2$, which of the following vessels can be used to store it:

(*a*) Iron vessel (*b*) Zinc vessel

(*c*) Copper vessel.

22. Using the data of standard electrode potentials from Table 33.1, determine whether each of the following redox reactions will proceed spontaneously as written or not.

(*i*) $2Ag^+ + Fe \longrightarrow Fe^{2+} + 2Ag$

(*ii*) $2Cu + 2H^+ \longrightarrow H_2 + Cu^{2+}$.

23. (*i*) Explain what is meant by the cell representation $Cu(s) \mid Cu^{2+} (aq)\ (1.0M) \parallel Ag^+ (aq)\ (1.0M) \mid Ag(s)$.

(*ii*) Sketch and label such a cell to show its essential features.

(*iii*) Find out the $E°_{cell}$ using data from Table 33.1.

Answers

I. Objective Type Questions

1. (*b*)	**2.** (*a*)	**3.** (*c*)	**4.** (*c*)
5. (*a*)	**6.** (*a*)	**7.** (*c*)	**8.** (*a*)
9. (*b*)	**10.** (*a*).		

II. Fill in the Blanks

11. (*i*) Zero (*ii*) > 0
(*iii*) KOH (*iv*) graphite
(*v*) KOH–ZnO paste (*vi*) Electrolytic cell.
(*vii*) 1.4 V
(*viii*) $O_2 (g) + H_2O (l) + 4e^- \longrightarrow 4\overline{O}H_{(aq)}$.

SECTION-5

UNIT-34

ELECTROLYTIC CELLS

Specific Objectives

After studying this unit, the student should be able to:

- explain the operation of electrolytic cells
- illustrate the electrolysis of brine experimentally
- distinguish between electrolytic and electrochemical cells
- describe some uses of electrolysis in everyday life
- demonstrate an experiment to determine the quantity of metal deposited on an electrode
- state and explain Faraday's Laws of Electrolysis.

34.1 INTRODUCTION

In Unit 33, we have studied the use of redox reactions in electrochemical cells to produce electrical energy. Let us see the effect of opposing potential on the reversible electrochemical cell. To understand this, consider a Zn-$CuSO_4$ cell; having standard emf = 1.1 V. The cell is reversible with respect to its net cell reaction,

$$Zn(s) + Cu^{2+}(aq) \rightleftharpoons Zn^{2+}(aq) + Cu(s).$$

Let us now consider an arrangement in which an opposing potential is applied to Zn-$CuSO_4$ cell by coupling it to another cell (E_{ext}) through rheostat and galvanometer as shown in Fig. 34.1(*a*).

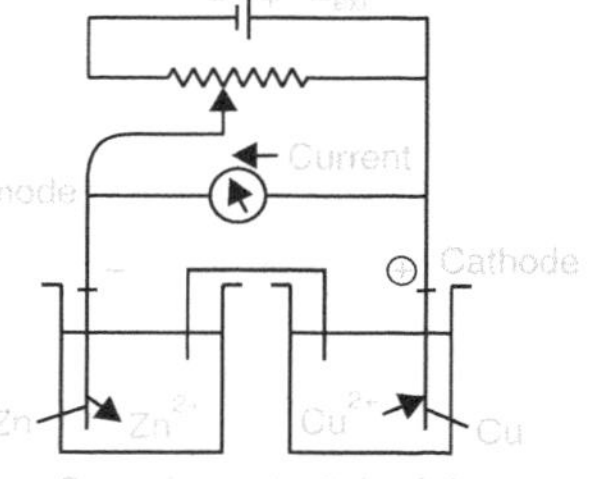

Opposing potential < 1.1
(i) Zn dissolves at anode and Cu deposits at cathode
(ii) Electrons flow from Zn to Cu and current flows from Cu to Zn.
Fig. 34.1(*a*)

In the beginning, the external opposing potential is less than 1.1 V. In this case the normal reaction occurs in the Zn-$CuSO_4$ cell and net current flows from the copper electrode to the zinc electrode. The opposing potential is now increased slowly and gradually. It is observed that the normal reaction continues to occur till the opposing potential reaches a value of 1.1 V [Fig. 34.1(*b*)]. At this stage the cell reaction stops and no current flows. Further increase in the opposing potential (beyond 1.1 V) again causes the chemical reaction to occur. But now the reaction occurs in the reverse direction forcing Zn^{2+} ion to get reduced and Cu atoms to get oxidised Fig. 34.1 (*c*).

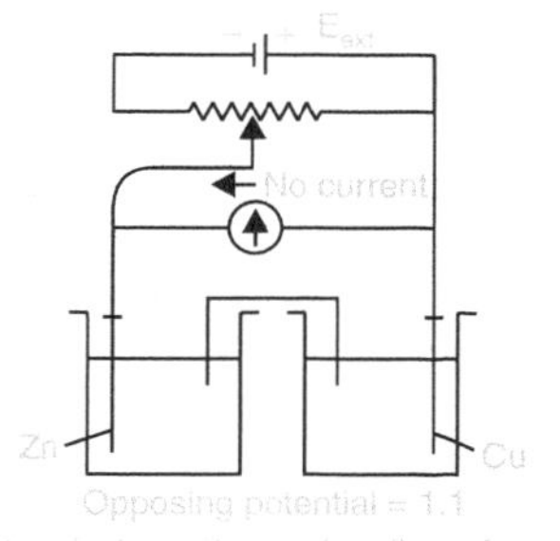

Opposing potential = 1.1
No chemical reaction and no flow of current
Fig. 34.1(*b*)

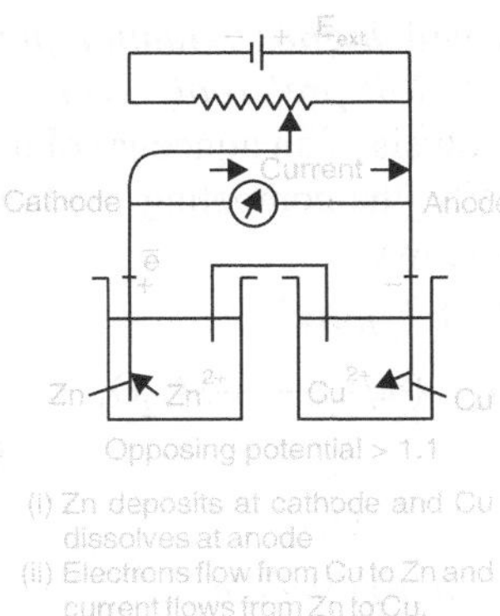

Fig. 34.1(c)

$$[Cu\,(s) + Zn^{2+}\,(aq) \longrightarrow Cu^{2+}\,(aq) + Zn\,(s)]$$

The flow of current in this case occurs from the Zn-electrode to the Cu-electrode. This process is non-spontaneous and involves the use of electrical energy to carry out a redox reaction. In this situation, the Zn-$CuSO_4$ cell starts acting as an ***electrolytic cell***. This phenomenon is also called *recharging* of the electrochemical cell. Let us now study different aspects of electrolysis and electrolytic cells.

34.2 ELECTROLYTIC CELLS AND ELECTROLYSIS

So far we have been discussing the process of conversion of chemical energy into electrical energy by means of electrochemical cell. In the reverse process, the passage of electricity through the electrolytes in their molten or dissolved state can cause chemical changes under suitable conditions. For example, the passage of electricity through acidified water results in the formation of hydrogen and oxygen gases. *The process of chemical decomposition of the electrolyte by the passage of electricity through its molten or dissolved state is* called **electrolysis.**

ELECTROLYTIC CELL

The device in which the process of electrolysis is carried out is called an electrolytic cell. It consists of:

(*i*) *Electrolytic tank*, which is made of some nonconducting materials like glass, wood or bakelite.

(*ii*) *Electrolyte* in its dissolved state or molten state.

(*iii*) *Source of electricity*; an electrochemical cell or battery.

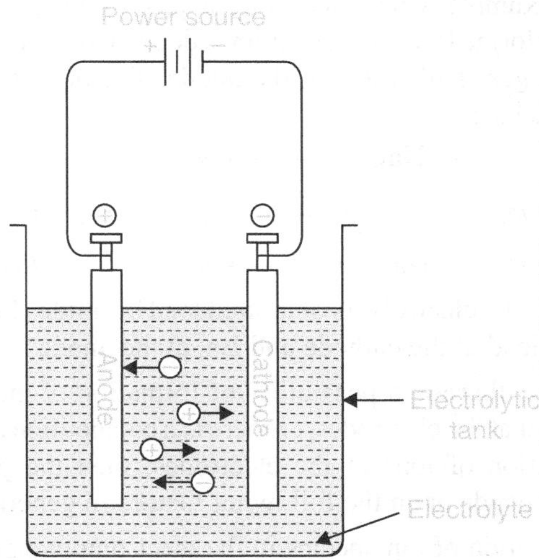

Fig. 34.2. Electrolytic cell.

(*iv*) Two *metallic rods*, suspended in the electrolyte and connected to the battery through conducting wires. These rods are called *electrodes*. The electrode connected to the negative terminal of the battery is called *cathode* while the other one which is connected to the positive terminal is called *anode*. The apparatus used as an electrolytic cell has been shown in Fig. 34.2.

34.3 MECHANISM AND CRITERIA OF PRODUCT FORMATION IN ELECTROLYTIC CELLS

The process of electrolysis can be explained on the basis of the *theory of ionisation. When an electrolyte is dissolved in water, it splits up into charged particles called ions*. The positively charged ions are called **cations** while the negatively charged ions are called **anions**. The ions are free to move about in aqueous solution. When electric current is passed through the solution, the ions respond to the applied potential difference and their movement is directed towards the oppositely charged electrodes. The *cations* move towards the *negatively charged electrode* while the *anions* move towards the *positively charged electrode*. The formation of products at the respective electrodes is due to **oxidation** (*loss of electrons*) at the **anode** and **reduction** (*gain of electrons*) at the **cathode.**

For example, when electricity is passed through molten sodium chloride [NaCl(*l*)], sodium is deposited at the cathode while Cl_2 gas is liberated at the anode. The process can be represented as:

$$NaCl(l) \longrightarrow Na^+ + Cl^-$$

Anode: $Cl^- \longrightarrow \frac{1}{2}Cl_2 + e^-$ *(Oxidation)*

Cathode: $Na^+ + e^- \longrightarrow Na$ *(Reduction)*

Similarly, electrolysis of molten lead bromide [$PbBr_2(l)$] produces lead at the cathode and Br_2 at the anode.

In case there is a possibility of formation of more than one product at the electrodes, or there is a competition between the liberation of ions at the electrodes, then the products formed depends upon the following factors in general:

- Position of ion/species in the electromotive series
- Concentration of ion/species
- Nature of electrodes.

One of the dominant factors which control the criterion of product formation at the electrodes is the values of electrode potentials of the species. Let us understand it as follows:

(*a*) **At the Cathode.** *Cathode involves a reduction process at its surface*. Therefore, for the different competing reduction processes, the one with *higher reduction potential will preferably take place*. For example, during the electrolysis of aqueous sodium chloride there is a possibility of the following reactions at the cathode:

Reduction of Na^+ ions

$$Na^+(aq) + e^- \longrightarrow Na(s); \quad E^\ominus_{red} = -2.71\ V$$

Or

Reduction of H_2O molecules

$$H_2O(l) + 2e^- \longrightarrow \tfrac{1}{2}H_2(g) + OH^-(aq); \quad E^\ominus_{red} = -0.41\ V$$

The reduction of water will preferably take place at the cathode because $E^\ominus_{red}$ of water is higher. Hence, the product of electrolysis of aqueous NaCl at the cathode will be H_2 gas instead of Na(*s*).

Similarly, during the electrolysis of an aqueous solution of copper(II)tetraoxosulphate(VI) reduction of Cu^{2+} ions will take place at the cathode in preference to the reduction of H_2O molecules because $E^\ominus_{Cu^{2+}/Cu}$ is greater than $E^\ominus_{H_2O/H_2}$.

$$Cu^{2+}(aq) + 2e^- \longrightarrow Cu(s); \quad E^\ominus = 0.34\ V$$

$$H_2O(l) + 2e^- \longrightarrow \tfrac{1}{2}H_2(g) + \bar{O}H(aq); \quad E^\ominus = -0.41\ V$$

(*b*) **At the Anode.** *Anode involves oxidation process at its surface*. Therefore, for the different competing oxidation processes, the one with ***higher oxidation potential*** (or *lower reduction potential) will preferably occur*. For example, if we carry out electrolysis of an aqueous solution of copper(II) tetraoxosulphate(VI), the competing oxidation processes at the anode are as follows:

Oxidation of SO_4^{2-} ions

$$2SO_4^{2-}(aq) \longrightarrow S_2O_8^{2-} + 2e^-; \quad E^\ominus_{oxi} = -2.01\ V$$

Or

Oxidation of water molecules

$$H_2O(l) \longrightarrow \tfrac{1}{2}O_2(g) + 2H^+(aq) + 2e^-; \quad E^\ominus_{oxi} = -1.23\ V$$

As oxidation potential of water is higher, the product formed at the anode will be O_2 gas instead of $S_2O_8^{2-}$ ion.

Similarly, if the electrolysis of copper sulphate solution is carried out using *copper electrodes*, then the process occurring at the anode will be the *oxidation of copper atoms to copper* ions instead of the oxidation of water because the oxidation potential of Cu is higher.

$$Cu(s) \longrightarrow Cu^{2+}(aq) + 2e^-; \quad E^\ominus_{oxi} = -0.34\ V$$

$$2H_2O(l) \longrightarrow \tfrac{1}{2}O_2(g) + 2H^+(aq) + 2e^-; \quad E^\ominus_{oxi} = -1.23\ V$$

Thus, in such a case the copper from the anode will go on dissolving into solution as Cu^{2+} ions while the Cu^{2+} ions from solution will go on depositing at the cathode as copper atoms.

The above discussion leads us to a general conclusion that for different competing reactions at the electrodes:

- **Cathodic reaction will be the one with higher $E^\ominus_{red}$ value**
- **Anodic reaction will be the one with higher $E^\ominus_{oxi}$ value**

It may be noted that in some cases the **unexpected results** are obtained due to *overvoltage**. For example, let us compare the oxidation potentials of Cl^- ion and water

---

*The transference of electrons from water is a *kinetically slow process* as it requires relatively large activation energy. The slowness of the electrode reaction creates some electrical resistance at the electrode surface. Hence, extra potential or voltage is required to provide activation energy and to overcome resistance in the case of oxidation of water molecules. **This extra voltage required for the oxidation of water is called overvoltage.** Due to overvoltage, the oxidation of Cl^- ions occurs in preference to H_2O.

$$H_2O(l) \longrightarrow \frac{1}{2} O_2(g) + 2H^+(aq) + 2e^-; \quad E^\ominus_{oxi} = -1.23\ V$$

$$Cl^-(aq) \longrightarrow \frac{1}{2} Cl_2(g) + e^-; \quad E^\ominus_{oxi} = -1.36\ V$$

Although the oxidation potential of H_2O is more than that of Cl^- ions, yet during the electrolysis of a *concentrated solution of sodium chloride*, the chloride ions oxidise in preference to H_2O molecules at the anode giving Cl_2 gas as the product.

EXAMPLES OF ELECTROLYSIS

In the light of the above discussion let us examine some examples of electrolytic reactions. Before we take up the actual examples, it is quite important to know about **inert electrodes** and **active electrodes**.

Inert electrodes are those electrodes which do not gain or lose electrons during the process. Two common examples are carbon (graphite) electrode and platinum electrode.

Active electrodes are those electrodes which themselves participate in the gain or loss of electrons.

1. Electrolysis of molten lead (II) bromide

Electrodes: Inert electrodes

Electrolyte: Molten lead (II) bromide

Molten lead (II) bromide ionises as

$$PbBr_2(l) \rightleftharpoons Pb^{2+} + Br^-$$

Reaction at anode (oxidation of Br^-)

$$2Br^- \longrightarrow Br_2 + 2e^-$$

Reaction at cathode (reduction of Pb^{2+})

$$Pb^{2+} + 2e \longrightarrow Pb$$

2. Electrolysis of a concentrated solution of sodium chloride (Brine solution)

Electrodes: Inert electrodes

Electrolyte: Sodium chloride dissolved in water. NaCl ionises in **aqueous** solution as

$$NaCl\ (l) \longrightarrow Na^+ + Cl^-$$

Cathode reaction: Reduction of H_2O occurs in preference to Na^+ ions as Na^+ is much higher on the electromotive series

$$2H_2O\ (l) + 2e^- \longrightarrow H_2(g) + 2\bar{O}H\ (aq)$$

Anode reaction: Oxidation of Cl^- ions occur in preference to H_2O because of a higher concentration of Cl^- ions and overvoltage

$$2Cl^-(aq) \longrightarrow Cl_2\ (g) + 2e^-$$

Thus, Cl_2 gas is produced at the anode and H_2 gas is liberated at the cathode. The pH of a solution increases due to an increase in the concentration of $\bar{O}H$ ions in solution.

3. Electrolysis of a very dilute solution of sodium chloride

Electrode: Inert electrodes.

Electrolyte: Very dilute NaCl solution.

NaCl ionises as

$$NaCl\ (s) \xrightarrow{\text{Water}} Na^+\ (aq) + Cl^-\ (aq)$$

Cathode reaction: Reduction of water occurs in preference to Na^+ ions

$$4H_2O\ (l) + 4e^- \longrightarrow 2H_2\ (g) + 4\ \bar{O}H\ (aq)$$

Anode reaction: Oxidation of $H_2O\ (l)$ occurs due to the relatively higher concentration of H_2O molecules as compared to Cl^- ions.

$$2H_2O \longrightarrow O_2 + 4H^+ + 4e^-$$

Here, H_2 is liberated at the cathode whereas O_2 is liberated at the anode. The pH of the solution does not change due to equal consumption of H^+ and $\bar{O}H$ ions. However, the concentration of NaCl gradually increases due to the decomposition of water.

4. Electrolysis of an aqueous solution of copper(II) tetraoxosulphate(VI) solution: using platinum electrodes

Electrodes: Inert electrodes (Pt)

Electrolyte: $CuSO_4$ dissolved in water

$CuSO_4$ ionises in aqueous solution as

$$CuSO_4\ (s) \xrightarrow{\text{water}} Cu^{2+}\ (aq) + SO_4^{2-}\ (aq)$$

Cathode reaction: Cu^{2+} undergoes reduction in preference to H_2O because H_2O lies above Cu^{2+} in the electrochemical series.

$$Cu^{2+}\ (aq) + 2e^- \longrightarrow Cu\ (s)$$

Anode reaction: H_2O gets oxidised in preference to SO_4^{2-} ions

$$2H_2O\ (l) \longrightarrow 4H^+ + O_2 + 4e^-$$

Thus, copper is deposited at the cathode and O_2 is liberated at the anode. The pH of the solution decreases gradually due to an increase in the concentration of H^+ ions. Then blue colour of the solution gradually fades due to the decrease in the concentration of Cu^{2+} ions.

5. Electrolysis of an aqueous solution of copper(II) tetraoxosulphate(VI) solution using copper electrodes

Electrodes: Active electrodes.

Electrolyte: $CuSO_4$ dissolved in water.

$CuSO_4$ ionises in aqueous solution as:

$$CuSO_4\ (s) \xrightarrow{\text{water}} Cu^{2+}\ (aq) + SO_4^{2-}\ (aq)$$

Cathode reaction: Cu^{2+} ions undergo reduction in preference H_2O molecules

$$Cu^{2+}(aq) + 2e^- \longrightarrow Cu(s)$$

Anode reaction: Cu atoms from the copper electrode get oxidised in preference to H_2O because Cu lies below H_2O in the activity series.

$$Cu(s) \longrightarrow Cu^{2+}(aq) + 2e^-$$

Thus, copper dissolves from anode and get deposited at the cathode. The mass of the cathode increases and that of the anode decreases. The blue colour of the solution does not fade because the concentration of Cu^{2+} ions in solution remains unaltered. The pH of the solution also does not change as the electrolysis proceeds.

34.4 COMPARISON BETWEEN ELECTROCHEMICAL CELL AND ELECTROLYTIC CELL

We have learnt that there are two types of cells namely: *electrochemical cells* and *electrolytic cells*. *Electrochemical cell is a device which converts chemical energy into electrical energy*. On the other hand, *electrolytic cell is a device which converts electrical energy into chemical energy*. The two cells also differ significantly with respect to the charges on the electrodes. For example, in electrochemical cell the anode is negative whereas in electrolytic cell, the anode is positive. Similarly, the cathode is positive in electrochemical cell whereas it is negative in the electrolytic cell. The main points of difference have been summed up as follows in Table 34.1.

Table 34.1. Differences between Galvanic Cell and Electrolytic Cell

Galvanic Cell	*Electrolytic Cell*
1. In galvanic cell, electrical energy is produced.	1. In electrolytic cell, electrical energy is consumed.
2. In galvanic cell, reaction taking place is spontaneous.	2. In electrolytic cell, reaction taking place is non-spontaneous.
3. The two half cells are set up in different containers and are connected through a salt bridge or porous partition.	3. Both electrodes are placed in the solution or molten electrolyte in the same container.
4. In galvanic cell, the anode is negative and the cathode is positive.	4. In electrolytic cell, the anode is positive and the cathode is negative.
5. The electrons move from the anode to cathode in external circuit.	5. The electrons are supplied by the external source. They enter through the cathode and come out the through the anode.

34.5 APPLICATIONS OF ELECTROLYSIS

Some of the important industrial applications of electrolysis are as follows.

ELECTROMETALLURGY

Many of the highly reactive metals such as the metals of group 1 or group 2 are extracted from their ores by electrolysis of their molten ores. This is because the chemical reduction of their compounds is either chemically not viable or highly expensive. Metals like sodium and magnesium are manufactured by the electrolysis of their molten chlorides. Pure aluminium is obtained at *VALCO* from a solution of its oxide in molten cryolite. Details of this extraction and that of some other metals is discussed in later units.

ELECTROREFINING OF METALS

Refining of the crude form of many metals such as copper, silver, aluminium, etc., is carried out by the process of electrolysis. In this process, the block of crude metal is made the anode while a thin sheet of pure metal is made the cathode and the electrolyte is an aqueous solution of some salt of the metal. As the electrolysis proceeds, metal from the anode dissolves and gets deposited in pure form at the cathode. The impurities settle down just below the anode in the form of **anode sludge** or **anode mud**.

MANUFACTURE OF CHEMICAL SUBSTANCES

Electrolysis is used in the manufacture of some important substances such as hydrogen, chlorine, sodium hydroxide, sodium oxochlorate (I) and oxygen.

34.6 ELECTROPLATING

It is another important industrial application of electrolysis. Electroplating is the art of coating a cheaper metal like iron with a layer of an expensive metal like gold, silver, etc. The purpose of electroplating is

- protection of cheaper metal like iron from corrosion
- beautification of articles like, earrings, bangles, rings, parts of wrist watches, etc.
- repair of the broken parts of delicate machinery where welding is not possible.

The **principle** of electroplating is similar to that of purification of metals by electrolysis. The article to be plated is thoroughly cleaned with aq H_2SO_4 and washed with distilled water. It is then made the cathode of the electrolytic cell. The anode is the pure sheet of metal to be coated or plated. The electrolyte is a solution of a salt of the metal to be plated. During electrolysis, the metal to be electroplated is transferred from the anode to the cathode. Experimental technique is similar to that discussed in Activity 34.1 D.

ACTIVITY 34.1 **To study the electrolysis of following compounds**

Requirements

(*a*) Acidified water or very dilute tetraoxosulphate(VI) acid.

(*b*) Dilute NaCl solution.

(*c*) Concentrated NaCl solution.

(*d*) Copper(II) tetraoxosulphate(VI) using copper electrodes.

A. Electrolysis of Acidified Water

1. Set up Hoffmann's voltameter as shown in Fig. 34.3 using either a D.C source of power or a battery. Fill the voltameter with water acidified with tetraoxosulphate(VI) acid by opening the taps at both ends. Close the taps after filling.

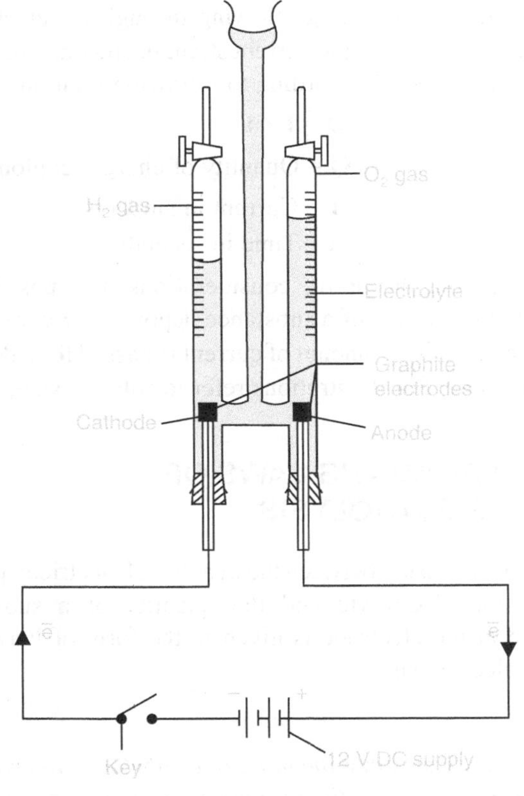

Fig. 34.3 Hoffmann's voltameter

2. Complete the circuit and let the current flow for five minutes.
3. Note the volumes of the gases at the electrodes.
4. Collect the gases in small glass tubes by holding them directly above the taps and closing them with your finger.
5. Test the cathodic gas by a burning splinter and anodic gas by a glowing splinter.

Record your observations and try to answer the following:

(*i*) Name the gases formed at the anode and cathode.

(*ii*) Write equations to represent the anodic and cathodic reaction.

(*iii*) What is the relationship between the volumes of the two gases? Justify.

(*iv*) What is the net reaction equation?

B. Electrolysis of Dilute NaCl Solution

1. Empty the voltameter, wash it thoroughly, rinse with distilled water and fill with very dilute aqueous sodium chloride solution. Complete the circuit and leave the current flowing for about 10 minutes.
2. Collect the gases, after noting their volumes.
3. After enough collection and noting the final volume of each gas, test the gas at the anode with a glowing splinter and that at the cathode with a burning splinter as in experiment *A*.

Record your observations and answer the following:

(*i*) Which ions will be present at each electrode and what are their sources?

Which species will be discharged at the anode? and at the cathode?

(*ii*) Give an explanation in each case.

(*iii*) Note the relative volume of gas in each arm as the electrolysis proceeds. What is the relationship?

(*iv*) Write an equation to represent the anodic reaction and the cathodic reaction.

Combine these equations with a net reaction equation.

(*v*) What will be the likely observation if the final solution is tested with moist litmus paper?

(*vi*) Give an explanation for the prediction, and test to verify it.

C. Electrolysis of Concentrated NaCl Solution

Repeat the experiment by using concentrated NaCl solution. Record your observations and answer the same questions (*i*) to (*vi*)

Note: *In this case you will test the gas at the anode with litmus paper and gas at the cathode with glowing splinter.*

D. Electrolysis of Copper(II)tetraoxosulphate(VI) Solution using Copper Electrodes

1. Set up the apparatus as shown in Fig. 34.4.

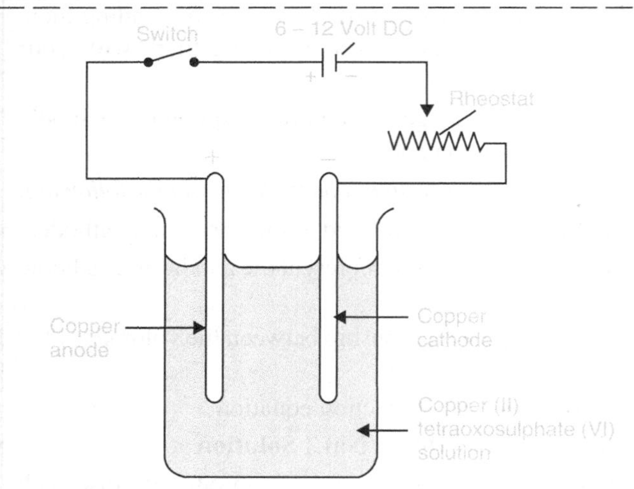

Fig. 34.4. Electrolysis of aqueous copper(II) tetraoxosulphate(VI) with copper electrodes

2. Note the mass of properly cleaned copper plates to be used as electrodes.
3. Close the switch and allow the current to flow for about 15 minutes.
4. Take out the electrodes, dry them and weigh them again.

Record the change in the mass of electrodes after the experiment. Try to provide answers to the following:

(*i*) Which electrode loses and which one gains weight? Why?

(*ii*) Write an equation to represent (*a*) the cathodic reaction (*b*) the anodic reaction.

(*iii*) With the help of a diagram, show how you would obtain a large piece of pure copper from a large piece of impure copper and a thin sheet of pure copper.

34.7 QUANTITATIVE ASPECT OF ELECTROLYSIS

We have already studied that during electrolysis the anodic reaction involve oxidation and the cathodic reaction involves reduction. For example, the cathodic reaction in the electrolysis of copper(II) tetraoxosulphate(VI) is

$$Cu^{2+} + 2e^- \longrightarrow Cu$$

This means that for the production of one mole of copper, 2 moles of electrons are needed. If the number of electrons available to copper ions are less than 2 moles, the amount of copper deposited at the cathode will also be less than 2 moles. Thus, the amount of product formed at the electrode depends on the number of electrons, which in turn depends on the quantity of electricity flowing through the electrolytic cell.

Now charge on one electron has been found to be 1.602×10^{-19} coulombs.

Charge on one mole of electrons

$$= 1.602 \times 10^{-19} \times 6.02 \times 10^{23} = 96488 \text{ C mol}^{-1}$$

This magnitude of charge (= 96488 C mol^{-1}) is called **Faraday constant** and is denoted by **F.** It is in honour of Michael Faraday who discovered the laws of electrolysis.

$$F = 96488 \text{ C mol}^{-1} \text{ or } \simeq 96500 \text{ C mol}^{-1}.$$

Charge on **n moles** of electrons (**Q**) is given by

$$\mathbf{Q = nF}$$

Thus, for **n** moles of electrons, the charge equal to **nF**. *i.e.*, $n \times 96500$ C of electricity has to be passed through the electrolyte. For example, during the electrolysis of aluminium chloride, 3 moles of electrons are required to producing 1 mole of aluminium.

$$Al^{3+} + 3e^- \longrightarrow Al$$

Therefore, charge equal to 3 F or 3×96500 C has to be passed through the electrolyte.

The quantity of charge flowing through the electrolyte can be calculated from the current strength and time for which the current is passed according to following relation.

$$\mathbf{Q = I \times t}$$

where, **Q** = Quantity of charge (coulombs)
I = Current in amperes
t = Time in seconds.

Thus, from the above considerations it is possible to calculate the amount of a substance deposited or evolved (if gas) when a definite amount of current is passed for a definite period of time. For illustrations refer to solved examples.

34.8 FARADAY'S LAWS OF ELECTROLYSIS

The relationship between the quantity of electricity passed through the electrolyte and the quantity of a substance liberated at the electrode is given in the form of Faraday's laws of electrolysis.

FARADAY'S FIRST LAW OF ELECTROLYSIS

This law states that *the mass of a substance liberated at the electrode is directly proportional to the quantity of electricity passed through the electrolyte.*

$m \propto Q$ Here, Q = quantity of electricity
$\propto I \times t$ I = current in amperes
$= Z \times I \times t$ t = time in seconds

Z = constant of proportionality called *electrochemical equivalent* (ECE)

If $I = 1$ ampere and $t = 1$ second, then $m = Z$

Thus, **electrochemical equivalent** *of a substance is the amount of substance liberated at the electrode when current of one ampere is passed through the electrolyte for one second.*

FARADAY'S SECOND LAW OF ELECTROLYSIS

This law states that the amounts of different substances liberated by the same quantity of electricity passing through their electrolytic solutions are directly proportional to their chemical equivalent masses (chemical equivalent mass of a metal can be obtained by dividing its atomic mass with the number of electrons required to reduce its cation). The law can also be stated as follows; *when same quantity of electricity is passed through different electrolytes connected in series then the masses of the substances liberated at the electrodes are in the ratio of their chemical equivalent masses* (atomic mass ÷ Number of electrons required to form the product) *or the ratio of their electrochemical equivalents.*

For example, if two electrolytic cells A (containing $AgNO_3$ solution) and B (containing $CuSO_4$ solution) are connected in series and the same quantity of electricity is passed through the cells, then the ratio of the mass of copper deposited at the cathode in electrolytic cell B (x g) to that of silver deposited in cell A (y g) is equal to the ratio of their chemical equivalent masses.

$$\frac{\textbf{Mass of Cu(x)}}{\textbf{Mass of Ag(y)}}$$

$$= \frac{\textbf{Chemical Equivalent Mass of Cu}}{\textbf{Chemical Equivalent Mass of Ag}} = \frac{Z_{Cu}}{Z_{Ag}}$$

Now, each copper (Cu^{2+}) ion requires 2 electrons to form Cu and each Ag^+ needs 1 electron to form Ag. Thus, the chemical equivalent mass of Cu is 63.5/2 and that of Ag is 108/1.

Thus, the ratio $\frac{x}{y} = \frac{63.5}{2 \times 108}$.

SOLVED EXAMPLES

Example 34.1. *The electrolysis of molten magnesium chloride is carried out with a current of 8.00×10^4 A. What mass of Mg and chlorine are produced in exactly 1 hour?*

Solution. The electrode reactions are

$$\underset{2\text{ mol}}{Mg^{2+}} + 2e^- \longrightarrow \underset{\substack{1\text{ mol}\\ \equiv 24\text{ g}}}{Mg}; \qquad 2\bar{Cl} \longrightarrow \underset{\substack{1\text{ mol}\\ \equiv 71\text{ g}}}{Cl_2} + \underset{2\text{ mol}}{2e^-}$$

Quantity of electricity passed = $I \times t$

$= 8.00 \times 10^4 \times 60 \times 60$ C

$= 2.88 \times 10^8$ C

Now, 2×96500 C produce Mg = 24 g

2.88×10^8 C produce $\quad Mg = \frac{24 \times 2.88 \times 10^8}{2 \times 96500}$

$= 3.58 \times 10^4$ g or **35.8 kg**

2×96500 C produce $\quad Cl_2 = 71$ g

2.88×10^8 C produce $\quad Cl_2 = \frac{71 \times 2.88 \times 10^8}{2 \times 96500}$

$= 1.059 \times 10^5$ g or **105.9 kg**

Example 34.2. *How long should a current of 1.5 ampere be passed through an electrolyte in order to deposit 1.0 g of Al when the electrode reaction is*

$$Al^{+3} + 3e^- \longrightarrow Al \text{ (At. mass of Al = 27)}$$

Solution. **Step 1.** *Calculation of number of coulomb required.*

From the electrode reaction, it is clear that

For depositing one mole of Al atoms the electrons required

= 3 mol

Charge on 3 mol of electrons = 3×96500 C

Now 1 mol of Al = 27 g

∴ Deposition of 27 g of Al requires charge

$= 3 \times 96500$ C

Deposition of 1.0 g of Al requires charge

$$= \frac{3 \times 96500}{27} \times 1.0 = \mathbf{10722.2\ C.} \qquad ...(i)$$

Step 2. *Calculation of time*

Current strength = 1.5 amp.

Let the time required be t sec

∴ No. of coulombs $= 1.5 \times t$ C ...(*ii*)

Equate (*i*) and (*ii*), $1.5\,t = 10722.2$

$$t = \frac{10722.2}{1.5} = 7148 \text{ sec} = \frac{7148}{60 \times 60} \text{ hrs}$$

= **1.98 hrs.**

Example 34.3. *Calculate the mass of hydrogen evolved by passing a current of 0.5 ampere for 40 minutes through acidified water.*

Solution. Current strength = 0.5 amp.

Time = 40 minutes = $40 \times 60 = 2400$ sec.

Total quantity of electricity passed $= 0.5 \times 2400$

$= 1200$ coulombs

Now the electrode reaction is $H^+ + e^- \longrightarrow \frac{1}{2} H_2$

Thus, formation of 1 mol (2.0 g) of hydrogen requires electrons = 2 mol.

$\therefore$ Formation of 2 g of H_2 requires charge $= 2 \times 96500$ C.

Now 2×96500 coulombs of electricity produces hydrogen $= 2.0$ g

1200 coulombs of electricity will produce hydrogen

$$= \frac{2 \times 1200}{2 \times 96500} = \mathbf{0.0124\ g.}$$

Example 34.4. *The electrolysis of a metal salt solution was carried out by passing a current of 4 amp for 45 minutes. It resulted in the deposition of 2.977 g of a metal. If atomic mass of the metal is 106.4 g mol^{-1}, calculate the charge on the metal cation.*

Solution. Let the charge on the metal ion be $n+$.

$\therefore$ The reduction half reaction is

$$\underset{(1\text{ mol})}{M^{n+}} + \underset{(n\text{ mol})}{ne^-} \longrightarrow \underset{(106.4\text{ g})}{M}$$

Quantity of electricity required for depositing 106.4 g of metal $= n \times 96500$ C

Quantity of electricity required for depositing 2.977 g of metal

$$= \frac{n \times 96500 \times 2.977}{106.4} = n \times 2700 \text{ C}$$

Quantity of electricity actually passed

$= 4 \times 45 \times 60$ C $= 10800$ C

Now, $10800 = n \times 2700$

$$\therefore \quad n = \frac{10800}{2700} = 4$$

Hence, charge on metal ion = **+ 4.**

SUMMARY

- **Electrolysis.** A process in which electrical energy is used to carry out a redox chemical reaction.
- **Electrolytic Cell.** A device which converts electrical energy into chemical energy.
- **Electrode Reactions.** Liberation of ionic species at the cathode involves ions with higher E°_{red} value whereas liberation of species at the anode involves ions with lower E°_{red} or higher E°_{oxi} value.
- **Faraday Constant.** (F) is equal to charge on 1 mol of electrons. Its numerical value is 96500 C mol^{-1}.

EVALUATE YOURSELF

1. Objective Type Questions

Select the most appropriate choice from the options given as (a), (b), (c) and (d) after each question:

1. The number of moles of electrons required to liberate one mole of an ion at an electrode is equal to
   (*a*) three times the charge on the ion
   (*b*) twice the charge on the ion
   (*c*) the charge on the ion
   (*d*) the number of electrons indicated in the electrode reaction.
2. The mass of a substance liberated during electrolysis is
   (*a*) directly proportional to the quantity of electricity flowing through it
   (*b*) inversely proportional to the quantity of electricity flowing through it
   (*c*) equal to the square of the quantity of electricity
   (*d*) equal to the reciprocal of the quantity of electricity.
3. The substances that would be formed when concentrated aqueous copper(II) chloride is electrolysed using inert electrodes are
   (*a*) chlorine and hydrogen (*b*) oxygen and copper
   (*c*) chlorine and copper (*d*) chlorine and ammonia.
4. In the electrolytic purification of crude copper which of the following are true?
   I. blocks of impure copper metal are used as the anode
   II. thin sheets of pure copper metal are used as the cathode
   III. impurities settle as anode mud
   IV. thin sheets of impure copper metal are used as the cathode.
   (*a*) I, II, III and IV (*b*) I, II and III
   (*c*) I and II (*d*) II and III.
5. Which of these occurs in the electrolysis of molten lead (II) bromide?
   I. electrons are lost at the cathode
   II. bromine is liberated at the anode
   III. lead is liberated at the cathode
   IV. electrons are gained at the anode.
   (*a*) II and III (*b*) I and IV
   (*c*) II and III (*d*) II and IV.
6. Which of the following occur during electrolysis?
   I. cations are discharged at the cathode
   II. cations are discharged at the anode
   III. electron gain occurs at the cathode
   IV. electron loss occurs at the anode.
   (*a*) I, II and III (*b*) I, III and IV
   (*c*) II, III and IV (*d*) I, II, III and IV.

7. Passage of electricity through dilute H_2SO_4 for 1 hr liberates a total of 224 cm^3 of H_2 at S.T.P. The strength of current would be
(*a*) 0.27 A (*b*) 0.54 A
(*c*) 0.042 A (*d*) 0.027 A.

8. The quantity of electricity required to deposit 1 mol of aluminium from a solution of $AlCl_3$ will be
(*a*) 0.33 F (*b*) 1 F
(*c*) 3 F (*d*) 1.33 F.

9. Two platinum electrodes are immersed in a $CuSO_4$ solution and a current is passed. Slowly the colour of the $CuSO_4$ disappears leaving behind a colourless solution. The colourless solution is
(*a*) water
(*b*) platinum sulphate solution
(*c*) copper hydroxide
(*d*) sulphuric acid solution.

10. The net charge required for the reduction of 1 mol of MnO_4^- to MnO_2 is
(*a*) 1 F (*b*) 3 F
(*c*) 6 F (*d*) 5 F.

II. Fill in the Blanks

11. Complete the following sentences by supplying appropriate words:
(*i*) During electrolysis of $CuSO_4$ solution using platinum electrodes the gas liberated at the anode is ....
(*ii*) In electrolytic cell the .... energy is converted into .... energy.
(*iii*) The electrolytic reaction is .... in nature.
(*iv*) When electrolysis of $CuSO_4$ solution is carried out by using copper electrodes, the pH of the solution ....
(*v*) Electrolysis is a process of .... of electrolyte by passage of electricity.
(*vi*) In the electrolysis of molten magnesium chloride, Cl_2 gas is produced at ....
(*vii*) In electroplating of a spoon by silver, the spoon is made .... electrode.

III. Discussion Questions

12. What is electrolysis? Describe the criteria for the formation of products at the respective electrodes.

13. Explain the following:
(*i*) Concentrated aqueous HCl can be used to clear oxidised copper (copper containing a coating of copper(II) oxide) without attacking the copper metal.
(*ii*) Sodium metal cannot be produced by the electrolysis of an aqueous solution of sodium chloride.

14. During electrolysis of a concentrated aqueous solution of NaCl the oxidation of Cl^- ion occurs at the anode in preference to oxidation of H_2O. Explain why?

15. Using $E^\ominus$ values explain the formation of products in electrolysis of:
(*i*) Aqueous $CuSO_4$ solution using platinum electrodes
(*ii*) Aqueous $CuSO_4$ solution using copper electrodes.

16. Predict the products of electrolysis of each of the following:
(*i*) An aqueous solution of $AgNO_3$ with silver electrodes.
(*ii*) An aqueous solution of $AgNO_3$ with inert electrodes.

17. What are the products of electrolysis of
(*i*) a dilute aqueous solution of tetraoxosulphate (VI) acid?
(*ii*) an aqueous acidified solution of copper (II) chloride?

18. A current of 4 A was passed for 1 hour through an acidified water voltameter and a copper voltameter, with copper anode and platinum cathode.
Calculate
(*i*) the volume of hydrogen at 27°C and 100 kPa that will be liberated in the water voltameter
(*ii*) the mass of copper that will be deposited on the cathode in the copper voltameter.
(F = 96500 C mol^{-1}, R = 8.31 Jmol^{-1} K^{-1},
Atomic mass = 63.50)

19. A steady current of 0.4 A was passed for 40 minutes through voltameters A and B, connected in series.
Voltameter A had silver electrodes dipped in aqueous silver trioxonitrate(V) solution. Voltameter B had a divalent metal M rod dipped in an aqueous solution of one of its salts.
At the end of the experiment the cathodes in A and B had increased in mass by 1.08 g and 0.16 g respectively.
[F = 96500 C mol^{-1}; Atomic mass (M) = 48]
(*a*) Calculate the:
(*i*) charge passed;
(*ii*) number of mol of electrons passed;
(*iii*) relative atomic mass of silver, assuming the silver ion to be Ag^+;
(*iv*) amount of metal M formed.
(*b*) Write the formula for the ion formed by M.
(*c*) (*i*) Write a half-equation for the reaction at each electrode.
(*ii*) Indicate the type of reaction in each case.

20. What is electroplating? What is its principle? Illustrate it with the help of an example.

Answers

I. Objective Type Questions

1. (*d*) **2.** (*a*) **3.** (*c*) **4.** (*b*)
5. (*c*) **6.** (*b*) **7.** (*b*) **8.** (*c*)
9. (*d*) **10.** (*d*).

II. Fill in the Blanks

11. (*i*) O_2 (*ii*) Electrical, chemical
(*iii*) Non-spontaneous (*iv*) does not change
(*v*) decomposition (*vi*) anode
(*vii*) Cathode.

III. Discussion Questions

18. H_2 = 1.87 dm^3; Cu = 4.74 g
19. (*a*) (*i*) 960 C, (*ii*) 9.95 × 10^{-3} mol, (*iii*) 108, (*iv*) 0.249; (*b*) M^{2+}.

SECTION-5

UNIT-35

CORROSION OF METALS

Specific Objectives

After studying this unit, the student should be able to:

- explain the concept of corrosion of metals
- state and describe methods of preventing corrosion.

35.1 CORROSION OF METALS

Metals are frequently used in almost all walks of our present day life. Bridges, railways, buildings, vehicles, industries, household articles, all involve the use of metals. But we commonly observe that certain metals (except those which are least reactive like Au, Pt, Pd etc.) are slowly eaten up on long exposure to the atmosphere. For example:

- silver gets tarnished,
- copper develops green coating on its surface,
- iron rusts, and
- aluminium, zinc, lead loses its lustre.

In fact, such metals react with the gases or moisture present in the environment to form undesirable compounds. This process in general is referred to as *corrosion of metals*. **Corrosion** may, thus, be defined as *the process of slow conversion of metals into their undesirable compounds (usually oxides) by reaction with moisture and other gases present in the atmosphere.*

CORROSION, AN OXIDATION-REDUCTION PROCESS

Corrosion of metals is basically an oxidation-reduction process. Most metals undergo slow oxidation with atmospheric oxygen and moisture to form an oxide layer on their surface. In certain metals like iron this oxide layer does not stick to the surface but gradually falls off exposing fresh metal surface to further corrosion. In this way, the metal is slowly eaten away weakening the structure in which it has been used.

Certain metals like aluminium, copper, chromium, zinc and lead react with atmospheric gases and moisture to form hydroxide, oxide or trioxocarbonate(IV) layers which generally stick to their surface and their surface get tarnished. This layer, however, becomes a sort of protective layer which prevent further corrosion. But at higher temperatures this layer can also crack exposing fresh surface to further corrosion.

Iron is generally used in buildings and other huge structures and rusting of iron (corrosion) causes damage to these structures. Sometimes this damage is so severe that it leads to the collapse of the structure. In this unit, we shall study the mechanism of corrosion of iron and the method used for its prevention.

FACTORS WHICH AFFECT CORROSION

The factors which affect the rate of corrosion are:

(*i*) **Reactivity of the metal.** The more active metals are more prone to corrosion.

(*ii*) **Presence of impurities.** Presence of impurities helps in setting up a corrosion cell and makes the corrosion to occur rapidly. For example, pure iron does not rust.

(*iii*) **Air and moisture.** Air and moisture are quite helpful in corrosion. The presence of gases like CO_2 and SO_2 in air makes it still rapid. For example, no rusting is caused if iron is kept in vacuum.

(*iv*) **Strains in metal.** Strains in metal also help in corrosion. For example, in iron articles, rusting is more pronounced on the areas having bends, dents, scratches, nicks and cuts.

(*v*) **Presence of electrolytes.** The presence of electrolytes also makes the corrosion process faster. For example, iron rusts more rapidly in saline water in comparison to pure water.

Let us understand the **mechanism of corrosion** by studying the most familiar example, the **rusting of iron.** Chemically rust is hydrated iron (III) oxide, $Fe_2O_3 . xH_2O$. It is generally caused by the moisture, CO_2 and O_2 in air. Rust is a non-sticking brown-coloured material which can be easily removed by scratching. There are a number of theories about the mechanism of rusting. The most widely accepted theory is *electrochemical theory* which will be discussed here.

ELECTROCHEMICAL THEORY OF RUSTING

According to this theory, the *impure iron* surface behaves like a small electrochemical cell in the presence of water containing dissolved oxygen or carbon dioxide. Such a cell is also called *corrosion cell* or *corrosion couple*. In these miniature corrosion cells, pure iron acts as an anode and the impure surfaces act as a cathode. Moisture having dissolved oxygen or carbon dioxide in it constitutes the electrolytic solution.

At anode, the oxidation of Fe atoms takes place. Thus, Fe atoms pass into solution as Fe^{2+} ions leaving behind electrons in the metal which are pushed into the cathodic area.

$$Fe \longrightarrow Fe^{+2} + 2e^-; \quad E^\ominus_{oxi} = -0.44\text{ V} \quad ...(i)$$

At cathode, the electrons are picked up by the H^+ ions which are produced either from H_2O or from H_2CO_3 (formed by dissolution of CO_2 in moisture).

$$CO_2 + H_2O \longrightarrow H^+ + HCO_3^-$$

$$2H^+(aq) + 2e^- \longrightarrow 2H$$

The H atoms thus formed reduce the dissolved oxygen as

$$2H + \tfrac{1}{2}\, O_2 \longrightarrow H_2O$$

The net reduction process at the *cathodic area* is

$$2H^+ + \tfrac{1}{2}\, O_2 + 2e^- \longrightarrow H_2O;\ E^\ominus_{red} = 1.23\text{ V} \ ...(ii)$$

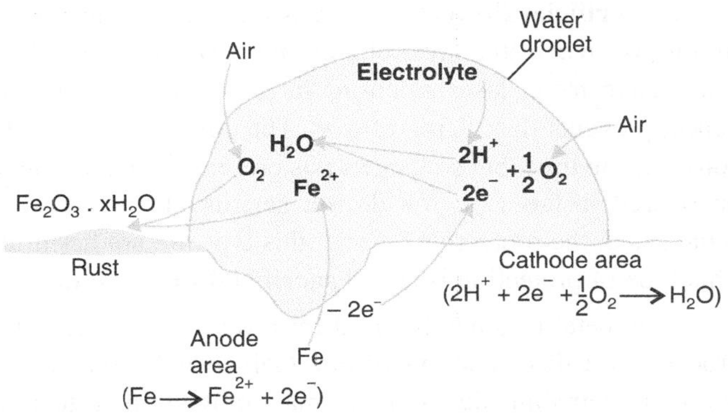

Fig. 35.1. Rusting of iron.

The net reaction of the corrosion cell can be obtained by adding equations (*i*) and (*ii*).

$$Fe + 2H^+ + \tfrac{1}{2}\, O_2 \longrightarrow Fe^{2+} + H_2O;\quad E^\ominus_{cell} = 1.67\text{ V}$$

The *iron (II)* ions so formed moves through the water and comes to the surface of the iron object where these are further oxidised to *iron (III)* by atmospheric oxygen to form rust which is hydrated iron (III) oxide.

$$2Fe^{2+} + \tfrac{1}{2}\, O_2 + 2H_2O \longrightarrow Fe_2O_3 + 4H^+$$

$$Fe_2O_3 + xH_2O \longrightarrow \underset{\text{(Rust)}}{Fe_2O_3.xH_2O}$$

35.2 PREVENTION OF CORROSION

There are several methods for protecting metals from corrosion (iron from rusting). Some of these methods are discussed as follows:

1. Barrier Protection. In this method, a barrier in the form of a film is introduced between iron and atmospheric oxygen and moisture. Barrier protection can be achieved by any of the following methods:

(*i*) *by painting the surface.*

(*ii*) *by coating the surface with a thin film of oil or grease.*

(*iii*) *by electroplating iron with some non-corrosive metal such as nickel, chromium, copper, etc.*

(*iv*) *Lamination with plastics*

All these methods do not involve any oxidation or reduction. Hence, these are called **non-redox methods.** However, in this type of protection, if scratches or cracks appear in the protective layer, then the surface of the iron may get exposed. Moisture and oxygen may then come into contact with iron in this region to trigger rusting. This rusting extends beneath the protective layer and eventually peels off the protective layer.

2. Sacrificial Protection. In this method, the surface of iron is covered with a layer of a *more active metal* like zinc. This active metal loses electrons in preference to iron and hence, prevents the rusting of iron. The covering metal gets consumed in time, but so long as it is present, even the nearly uncovered surfaces of iron do not get rusted. In this way, scratches in the protective layer of this type are not harmful. This type of protection is called **sacrificial protection.**

Zinc metal is generally used for protecting iron and the process is called **galvanization.** Galvanized iron sheets maintain their shine due to the formation of a thin protective layer of basic zinc carbonate, $ZnCO_3.Zn(OH)_2$, due to the reaction between zinc, oxygen, CO_2 and moisture in air.

Zinc, magnesium and aluminium powders dissolved in paints can also be applied as protective layers. The well known aluminium paint contains aluminium powder suspended in varnish.

3. Electrical Protection. This is also a case of sacrificial protection. In this method, the exposed surface of iron is protected by connecting it to some more active metal such as magnesium. The other metals which can be used for this purpose are aluminium, zinc, etc. The more active metal acts as an anode and loses electrons in preference to iron. The iron surface, acts as a cathode. This method is therefore, referred to as **cathodic protection** of iron. Underground water pipes or tanks made of iron are protected by connecting them to a more easily oxidisable metal as shown in Fig. 35.2.

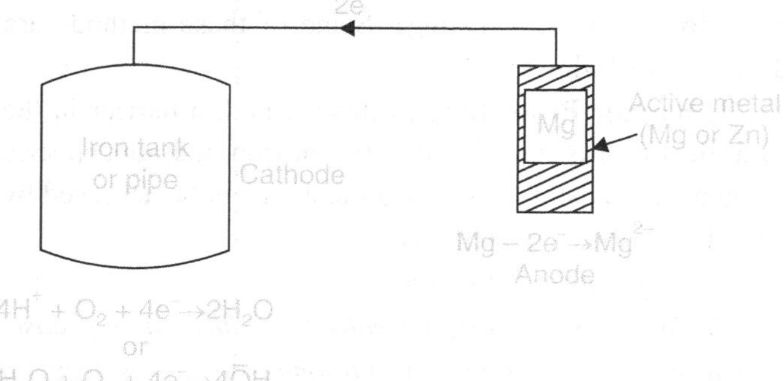

Fig. 35.2. Protection of underground pipe or tank.

4. Use of Anti-rust Solutions. The alkaline solutions of some phosphate and chromate salts act as anti-rust solutions. For example, when iron articles are dipped into a boiling and strongly alkaline solution of sodium phosphate, a protective insoluble film of iron phosphate is formed on them. This film protects the articles from rusting. The alkaline nature of such solutions decreases the availability of H^+ ions which facilitate the oxidation of iron to Fe^{2+}.

Why steel is preferred to iron in many instances?

Steel is an alloy of carbon and other metals like Cr, Ni, Mn, etc. Pure iron is soft whereas steel is much harder and has a high tensile strength. Moreover, iron is more prone to rusting but the presence of other non-corrosive metals in steel makes it quite resistant to corrosion/rusting. These qualities make steel more useful as a structural material as compared to iron.

ACTIVITY 35.1 — **To identify corroded metallic objects in your vicinity and estimate the cost of Repair/ Prevention**

Part I

1. Make a list of objects in your community that have corroded.
   - Try to know the metallic composition of each object.
   - Name the probable factors from the environment which might have caused the corrosion.
2. Investigate
   (*i*) any accident that the corrosion could have caused;
   (*ii*) the cost of repair of the object or prevention (or both) of the corrosion.
   Record your findings as follows.

Corroded object	Metallic composition	Probable cause of corrosion	Cost of repair, prevention or damage

3. Discuss your findings with your friends and teacher including possible accidents.
4. Discuss what types of maintenance work would be needed on (*i*) steel bridges, (*ii*) steel reinforced well tops, (*iii*) vehicle parts, (*iv*) sewing machines and (*v*) steel tools.

Part II

1. Visit an automobile welding or fitter's shop in your neighbourhood. Investigate areas on the vehicles where rust can be easily identified and/or concealed.

2. Find out the repair works being done by the welder and the cost of such repairs.
3. Record your findings using the following table:

Type of vehicle	Areas with rust	Repairs carried out	Cost

4. Write a short report on your findings and comments and read it to your class.

SUMMARY

- **Corrosion** of metal is the slow conversion of metals into their undesirable compounds by the action of gases in the atmosphere and moisture. Corrosion is basically a redox process. Corrosion causes severe damage to the structures made of metals. Corrosion in iron can be prevented by methods like barrier protection, sacrificial protection, electrical protection and by use of antirust solutions.

EVALUATE YOURSELF

I. Objective Type Questions

Select the most appropriate choice from the options given as (a), (b), (c) and (d) after each question:

1. The formula of rust is
   (*a*) FeO (*b*) Fe_3O_4
   (*c*) $Fe_2O_3 . xH_2O$ (*d*) $FeCO_3$.
2. Chromium metal kept at high temperature in air would
   (*a*) corrode quickly (*b*) corrode slowly
   (*c*) not corrode (*d*) form a thin protective layer.
3. Which of the following metals will not undergo corrosion?
   (*a*) Lead (*b*) Iron
   (*c*) Gold (*d*) Silver.
4. Corrosion of iron can be prevented by
   I. Painting the surface of article with an aqueous solution of brine.
   II. Electroplating the metal with some non-corroding metal.
   III. Covering the metal surface with more electronegative metal.
   (*a*) I, III (*b*) I
   (*c*) II, III (*d*) I, II.
5. Which statement is correct about corrosion?
   I. More active metals undergo corrosion more rapidly.
   II. Pure iron will develop rusting more quickly.
   III. Impure iron will rust more rapidly.
   IV. Surface of silver gets dull more quickly in the atmosphere of hydrogen sulphide gas.
   (*a*) I, III, IV (*b*) I, II
   (*c*) III (*d*) III, IV.

II. Fill in the Blanks

6. Complete the following sentences by supplying appropriate words:
   (*i*) Corrosion of metal is ...... process.
   (*ii*) Rusting of iron occurs ...... in saline water.
   (*iii*) Galvanised sheets have coating of ......
   (*iv*) In electrical protection of iron, the iron surface acts as ....
   (*v*) Painting the iron surface to protect it from rust is called ....

III. Discussion Questions

7. What is corrosion? Name some metals which do not undergo corrosion.
8. Describe the electrochemical mechanism for rusting of iron.
9. What are the factors which increase the rate of corrosion?
10. Describe means to prevent iron from rusting.
11. Iron does not rust when a coating of zinc over its surface cracks in galvanised iron pipes but it rusts rapidly when coating of tin done on its surface cracks. Explain.

Answers

I. Objective Type Questions

1. (*c*) **2.** (*d*) **3.** (*c*)
4. (*d*) **5.** (*a*).

II. Fill in the Blanks

6. (*i*) redox (*ii*) more rapidly
(*iii*) zinc (*iv*) cathode
(*v*) barrier protection.

SHS 3

SECTION 6

CHEMISTRY OF CARBON COMPOUNDS

Learning Objectives

After studying this section, the student should be able to:

1. show general awareness of source and preparation, structure and naming, physical properties and chemical reactions of carbon compounds;
2. apply chemical principles learnt to explain observed properties;
3. demonstrate knowledge of the characteristic tests to detect functional groups.

In this section, we shall study

Unit 36: Bonding in Carbon
Unit 37: Classification of Organic Compounds
Unit 38: Identification of Elements in Organic Compounds
Unit 39: Separation and Purification of Organic Compounds
Unit 40: Alkanes
Unit 41: Alkenes
Unit 42: Alkynes
Unit 43: Benzene
Unit 44: Alkanols
Unit 45: Carbonyl Compounds
Unit 46: Alkanoic Acid
Unit 47: Alkanoic Acid Derivatives: Amides
Unit 48: Alkanoic Acid Derivatives: Alkyl Alkanoate (Esters)

SECTION-6

UNIT-36

BONDING IN CARBON

Specific Objective

After studying this unit, the student should be able to:

- describe the electron structure of carbon.

36.1 ELECTRON STRUCTURE OF CARBON

Carbon is an element which can be found in the second period of the periodic table. Its atomic number is 6 and mass number is 12. It is represented thus $^{12}_{6}C$.

The ground state electronic configuration of carbon is:

$$_{6}\mathbf{C}\ \textbf{(ground state): } 1s^2\ 2s^2\ 2p_x^{\ 1}\ 2p_y^{\ 1}\ 2p_z^{\ 0}.$$

The carbon atom has 4 electrons in its valence shell and, therefore, it can attain a noble gas configuration either by losing or gaining or sharing 4 electrons. But the loss or gain of 4 electrons by the carbon atom to form highly charged C^{4+} or C^{4-} ions would require a very large amount of energy which is not ordinarily available during a chemical reaction. Therefore, carbon is unable to form ionic bonds and would participate only in the formation of covalent bonds.

The tendency of carbon atom to form covalent bonds only is also justified on the basis of its electronegativity value, which is 2.5. Since the carbon atom lies in group I of the periodic table, its electronegativity is neither very low nor very high. The electronegativities of other elements such as H (2.1), O (3.5), N (3.0), Cl (3.0) and S (2.5) which are generally linked with carbon in organic compounds are not very much different from that of carbon. Because of this small electronegativity difference between carbon and the other elements bonded to it, the bonds formed are predominantly covalent.

The electronic configuration of carbon suggests that it should be **bivalent** *i.e.*, should show a valency of two because of the presence of two half-filled $2p$ orbitals (*i.e.*, $2p_x$ and $2p_y$ orbitals) in its valence shell. All the known compounds of carbon confirm that carbon is tetracovalent. In order to account for tetravalency, it is believed that during the process of bond formation, which is an energy releasing process, the two electrons in the $2s$ orbital becomes unpaired because of the promotion of one electron into an empty $2p_z$ orbital. This is called the **excited state**. The electronic configuration of carbon in the excited state is:

$_6C$ (excited state) ; $1s^2$; $2s^1$. $2px^1$. $2py^1$. $2pz^1$

Carbon atom in the excited state has four half-filled orbitals ($2s^1\ 2p_x^1\ 2p_y^1 2p_z^1$). Since all these orbitals would be available for the formation of covalent bonds with other atoms, carbon would exhibit a covalency of four.

LEWIS SYMBOLS

The electrons involved in the process of combination are outer shell electrons, also called valence electrons. G.N. Lewis introduced simple notations to represent the valence electrons in an atom. These notations are called **Lewis symbols** or **electron dot symbols.** According to these notations,

(*i*) *the symbol of the element represents the nucleus as well as the electrons in the inner shells.*

(*ii*) *the electrons in the outer shell are represented by the dots surrounding the symbol.*

For example, the Lewis symbol of carbon is written as:

·Ċ·
 ·

The electrons in the valence shell are represented by dots (•) or by cross (X). The Lewis symbols for the other elements of 2nd period are:

Li̇ Ḃe ·Ḃ· ·N̈· ·Ö· :F̈· :N̈e:

Significance of the Lewis Symbols

The number of dots in the Lewis symbol represents the number of valence electrons. The common valencies of the elements can be deduced from the valence electrons. The common valency of the element is either *equal to the number of dots or 8 minus the number of dots.* For example, the common valencies of Li, Be, B and C are 1, 2, 3 and 4 respectively while those of N, O, F and Ne are 8 minus number of dots, *i.e.*, 3, 2, 1 and 0 respectively.

Lewis Representation of Simple Molecules and Polyatomic Ions

The Lewis dot structures provide a picture of bonding in simple molecules and polyatomic ions in terms of shared electron pairs. The following basic steps are generally used for writing Lewis dot structures.

(*i*) *Add the valence electrons of the combining atoms. This gives the total number of electrons for writing the structure.*

(*ii*) *For* ***polyatomic anions,*** *add one electron to the total number for each unit negative charge*. For example, in CO_3^{2-} ion, the total number of electrons are calculated as

Total electrons = ($4e^-$ from carbon)
+ 3 × ($6e^-$ from each O atom)
+ 2 ($1e^-$ for each negative charge)
= 4 + 3 × 6 + 2 × 1 = 24 electrons.

(*iii*) *For* ***polyatomic cations,*** *subtract one electron for each unit positive charge*.

(*iv*) *Write the skeleton structure by placing the least electronegative atom in the centre and more electronegative atoms on the terminal positions.* While doing this, some intelligent guess work is also required.

(*v*) *Now distribute the electrons properly as shared pairs in proportion to the total bonds.*

(*vi*) *After accounting for the shared pair of electrons for single bonds, the remaining electrons are utilised either for multiple bonding or for indicating the lone pairs.*

Illustrations *Let us write the Lewis dot structure for CO and HCN molecules*

1. Carbon monoxide (CO) molecule

Step I. Total number of valence electrons = ($4e^-$ from C) + ($6e^-$ from O) = $10e^-$

Step II. The skeletal structure is **C O**

Step III. Draw a single bond between the carbon and the oxygen atom (one shared pair) and complete the octet on the O atom. The remaining two electrons are the lone pair on the C atom.

:C : Ö: or :C—Ö:

Step IV. This does not complete the octet on the carbon atom. Hence, we have to resort to multiple bonding between carbon and oxygen to satisfy the octet.

The Lewis formula for carbon monoxide is therefore

:C ⋮⋮ O: or :C ≡ O:

2. Hydrogen cyanide (HCN) molecule

Step I. Total number of valence electrons
= ($1e^-$ from H) + ($4e^-$ from C) + ($5e^-$ from N)
= $10e^-$

Step II. The skeletal structure is **H C N**

Step III. Put one shared pair of electrons between H and C and one shared pair of electrons between C and N. The remaining electrons are lone pairs on C and N atoms.

H : C̈ : N̈:

Step IV. Since the octet of C and N are not complete, multiple bonding is required between C and N atoms. To complete their octets triple bonding is required between C and N atoms. Thus, the structure is

H : C ⋮⋮ N : or H—C ≡ N:

Let us now, write the Lewis structures of some more **molecules** and **polyatomic ions.**

(*i*) NF_3 :F̤̈:N̈:F̤̈: over :F̤: or :F̤̈—N̈—F̤̈: with :F̤̈: below N

Nitrogen trifluoride

(*ii*) CO_3^{2-} :Ö above; :Ö̤:C:Ö̤: or :Ö̤—C—Ö̤: with :Ö double bonded above C

Carbonate ion

(*iii*) SO_4^{2-} :Ö: above; :Ö̤··S··Ö̤: ; :Ö: below or :Ö̤—S—Ö̤: with :Ö: above and below S

Sulphate ion

(*iv*) H_2O_2 H :Ö̤:Ö̤:H or H—Ö̤—Ö̤—H

Hydrogen peroxide

(*v*) NO_2^- :O̤::N̈:Ö̤:⁻ or :O̤ = N̈—Ö̤:⁻

Nitrite ion

(*vi*) ClO_4^- :Ö: above; :Ö̤:Cl:Ö̤: ; :Ö: below or [:Ö̤—Cl—Ö̤: with :Ö: above and below Cl]

Perchlorate ion

(*vii*) H_2S H:S̤̈:H or H—S̤̈—H

Hydrogen sulphide

(*viii*) $SOCl_2$:C̈l: above; :C̤̈l:S::Ö: or :C̤̈l—S=Ö: with :C̈l: above S

Thionyl chloride

(*ix*) CS_2 :S̈::C::S̈: or :S̈=C=S̈:

Carbon disulphide

(*x*) BF_4^- [:F̈: above; :F̤̈:B:F̤̈: ; :F̤: below]⁻ or [F—B—F with F above and below B]⁻

Tetrafluoroborate ion

36.2 TYPES OF OVERLAPPING AND NATURE OF COVALENT BONDS

The formation of a covalent bond involves the overlapping of half-filled atomic orbitals. Covalent bonds can be classified into two depending upon the type of overlap. These are:

(*a*) *Sigma covalent bond,*

(*b*) *Pi covalent bond.*

(*a*) **Sigma (σ) bond.** *This type of covalent bond is formed by the* **axial overlap** *of half-filled atomic orbitals*. The atomic orbitals overlap along the inter-nuclear axis and involve *end to end* or *head on overlap.* The electron cloud formed as a result of an axial overlap is cylindrically symmetrical about its inter-nuclear axis. The electrons constituting sigma bond are called **sigma electrons.** There can be three types of axial overlap among *s* and *p*-orbitals as discussed below:

(*i*) **s-s overlap.** It involves mutual overlap of half-filled *s*-orbitals of the atoms approaching to form a bond. The bond formed is called **s-s σ bond.**

(*ii*) **s-p overlap.** It involves mutual overlap of a half-filled *s*-orbital of one atom with the half-filled *p*-orbital of another. The bond so formed is called **s-p σ bond.**

(*iii*) **p-p overlap.** It involves mutual overlap of half-filled *p*-orbitals of the two atoms. The bond so formed is called **p-p σ bond.**

The *s-s, s-p* and *p-p* overlaps have been shown diagrammatically in Fig. 36.1.

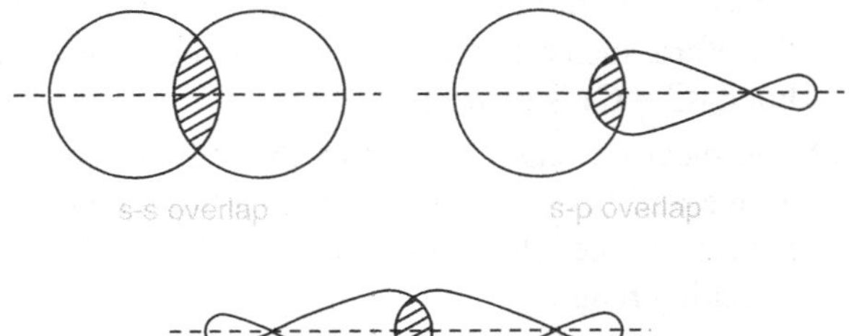

s-s overlap s-p overlap

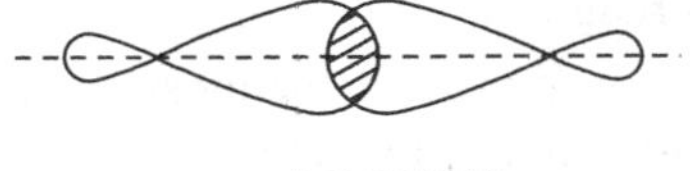

p-p overlap

Fig. 36.1. *s-s, s-p* and *p-p* overlap of atomic orbitals.

Strength of the three types of sigma bonds. The strength of the three types of sigma bonds varies as follows:

p-p > **p-s** > **s-s**

It is because of the fact that *p*-orbitals allow overlap to a greater extent as compared to *s*-orbitals. This means that *p-s* overlap is enhanced better than *s-s* overlap.

(*b*) ***Pi* (π) Bond.** *This type of covalent bond is formed by the* **lateral** or **sideways overlap** *of atomic orbitals.* The orbital overlap takes place in such a way that their axes are parallel to each other but perpendicular to the internuclear axis. The *pi* bond consists of two charge clouds above and below the plane of the atoms involved in the bond formation. The electrons involved in the π-bond formation are called **π-electrons.**

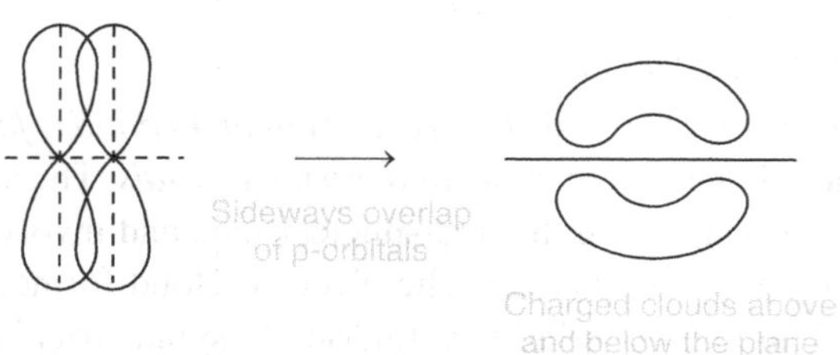

Fig. 36.2. Sideways overlap of *p*-orbitals.

SOME CHARACTERISTIC FEATURES OF π-BONDS

A *pi*(π) bond is constituted by sideways overlap of orbital perpendicular to the internuclear axis. Some characteristic features of π bonds are:

(*i*) All the atoms directly attached to the carbon atoms in a double bond lie in the same plane. For example, in $CH_2 = CH_2$ all the six atoms (2 carbon atoms and 4 hydrogen atoms) lie in the same plane.

(*ii*) Only the unhybridised *p*-orbitals perpendicular to the plane of the molecule form *pi* bonds.

(*iii*) Rotation of one CH_2 fragment with respect to the other interferes with maximum overlap of *p*-orbitals and, therefore, such rotation about a carbon-carbon double bond (C = C) is restricted.

(*iv*) The electron charge cloud of the π-bond is placed above and below the plane of bonding atoms. This results in the electrons being easily available to attacking reagents.

In general, π-bonds provide the most reactive centres in molecules containing multiple bonds.

It may be noted that:

(*i*) ***Sigma bond is stronger than pi bond.*** It is because of the fact that overlapping of atomic orbitals can take place to a greater extent during the formation of a sigma bond whereas overlapping of orbitals occurs to a smaller extent during the formation of a pi bond.

(*ii*) ***Pi bond between the two atoms is formed only in addition to a sigma bond.*** It is because of the fact that the atoms constituting a single bond prefer to form a strong sigma bond rather than a weak pi bond. Thus, pi bond is always present in molecules having multiple bonds, *i.e.*, double or triple bond. In other words, *a single bond cannot be a pi bond.*

(*iii*) The shape of a molecule is controlled by the sigma framework (orientations of sigma bonds) around the central atom. Pi bonds are superimposed on sigma bonds hence they simply modify the dimensions of the molecule.

36.3 HYBRID ORBITALS IN CARBON

The excited state electronic configuration of carbon is given as:

$${}_6C : 1s^2\ 2s^1\ 2p_x^1\ 2p_y^1 2p_z^1 \qquad ...(In\ excited\ state)$$

From this configuration there are four half-filled orbitals available for bond formation resulting in the formation of four covalent bonds.

It should be noted, however, that the four half-filled valence orbitals are not equivalent. There are three *p*-orbitals and one *s*-orbital. Hence, the four bonds expected would not be equivalent. The bond arising from the *s*-orbital would be different from the three bonds arising from the *p*-orbital. Actually carbon forms many compounds such as **carbon tetrachloride** in which all the four bonds are equivalent. To account for the equivalence of the four bonds, it is assumed that the four available orbitals of carbon, the 2*s* and three 2*p* orbitals are *mixed* or *hybridised* in a manner as to result in four equivalent orbitals.

The phenomenon of mixing of orbitals of the same atom having slightly different energies to form new orbitals which have identical shapes and equivalent energies is known as ***hybridisation.***

Carbon atom can undergo three types of hybridisation in its compounds.

sp^3 HYBRIDISATION (TETRAHEDRAL HYBRIDISATION)

In this type of hybridisation, four orbitals (one 2*s* and three 2*p*) of the excited carbon atom hybridise to form four orbitals of equivalent energy and same shape. Each orbital is *called sp^3-hybrid orbital.* The four hybrid orbitals are directed towards the corners of a regular tetrahedron. The hybridisation is called **tetrahedral hybridisation** and the angle between the hybridised orbitals is 109° 28′. (see Fig. 36.3)

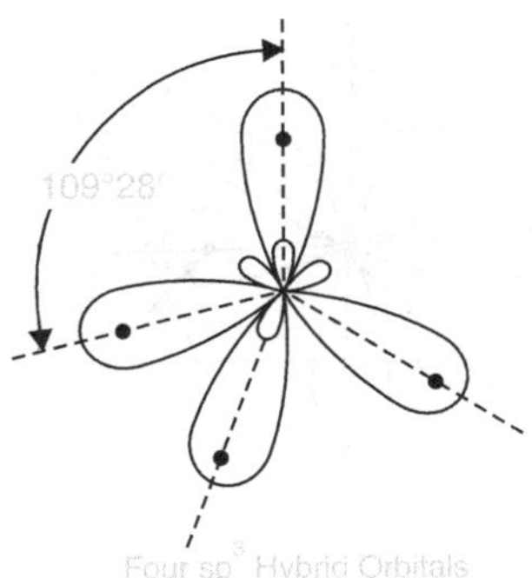

Fig. 36.3. sp^3 Hybrid orbitals.

Structure of Methane (CH_4)

Carbon in methane is sp^3 hybridised. Here, one orbital of 2*s*-sub-shell and three orbitals of 2*p*-sub-shell of excited carbon atom undergo hybridisation to form four sp^3 hybrid orbitals. The process which involves promotion of a 2*s*-electron followed by hybridisation is shown in Fig. 36.4.

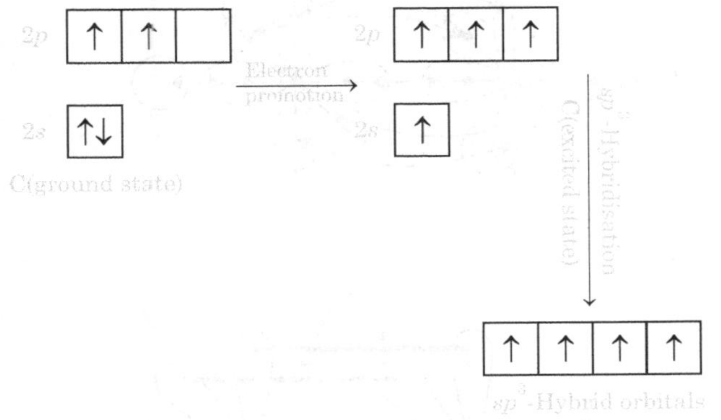

Fig. 36.4. sp^3 hybridisation of carbon.

The carbon atom lies at the centre of the tetrahedron while the four hydrogens occupy the four corners or vertices of the tetrahedron as shown in Fig. 36.5.

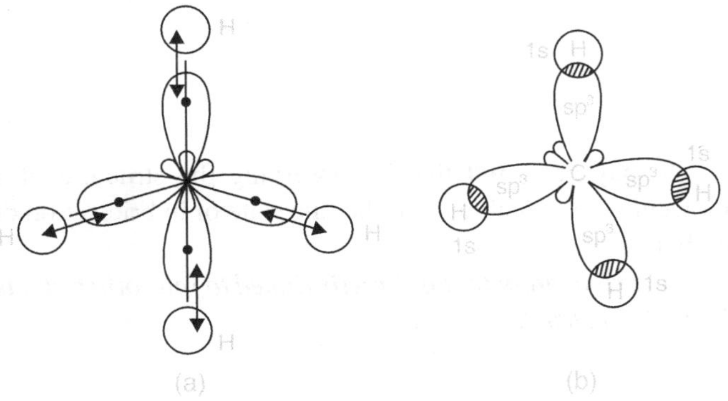

Fig. 36.5. Structure of methane molecule.

Structure of Ethane (C_2H_6)

In ethane molecule, both carbon atoms are sp^3 hybridised. The geometry around both carbon atoms is tetrahedral as shown in Fig. 36.6. C—C bond length is 1.54 Å and C—H bond length is 1.09 Å.

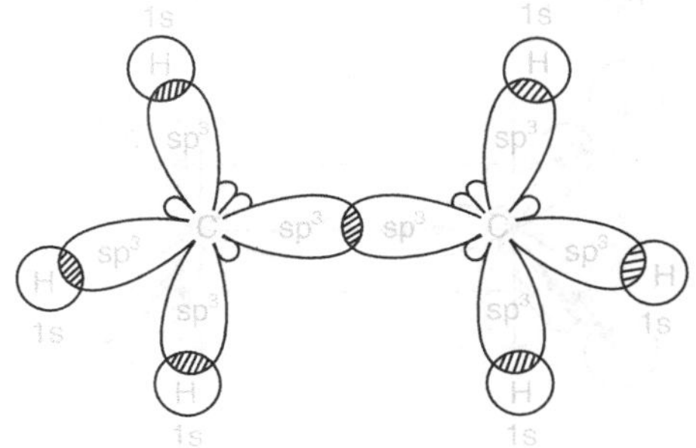

Fig. 36.6. Structure of ethane molecule.

sp^2 HYBRIDISATION (TRIGONAL HYBRIDISATION)

In this type of hybridisation, the 2*s* orbital and two out of the three 2*p* orbitals of excited carbon atom hybridise to form three sp^2 hybrid orbitals of equivalent energy and identical shape. Each orbital is called **sp^2-hybrid orbital.**

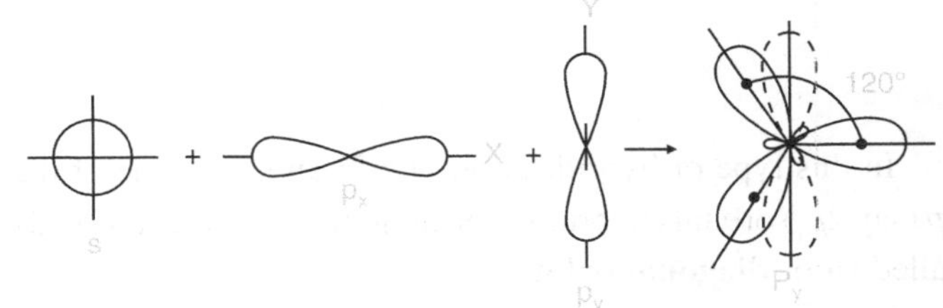

Fig. 36.7. Orientation of sp^2 hybrid orbitals.

The three sp^2 orbitals lie in one plane making an angle of 120° with each other. An sp^2 hybridised carbon is called a **trigonal** carbon atom and the hybridisation is known as **trigonal hybridisation.**

Unhybridised $2p_z$ orbital is oriented in a plane at right angle to the plane of three hybridised orbitals. (Fig. 36.8)

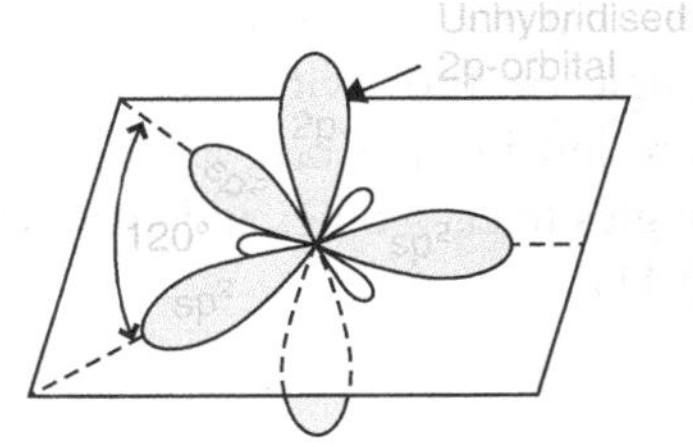

Fig. 36.8. sp^2-Hybrid orbitals in carbon atom.

Structure of Ethene ($CH_2 = CH_2$)

Both the carbon atoms in ethene are sp^2 hybridised. *All the atoms in ethene lie in the same plane.* The unhybridised *p*-orbital on each carbon is perpendicular to the plane of these atoms, and overlap sideways to form a π-bond between the two carbon atoms as shown in Fig. 36.9.

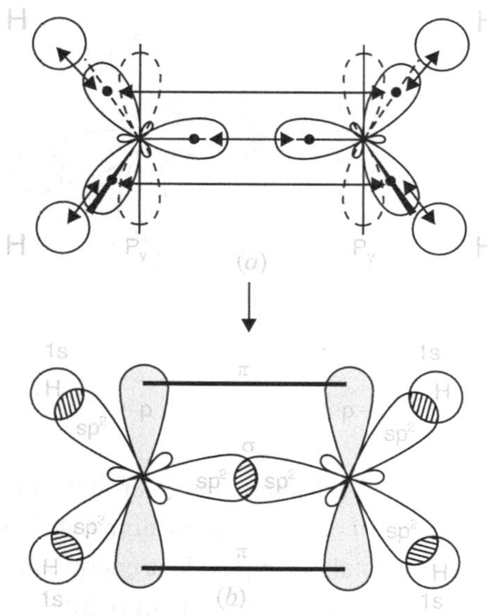

Fig. 36.9. The σ and π-bond in ethene.

sp-HYBRIDISATION OR DIAGONAL HYBRIDISATION (LINEAR HYBRIDISATION)

In this type of hybridisation, the 2*s* and only one of the 2*p* (say $2p_x$) orbitals hybridise forming two equivalent orbitals called ***sp*** or **diagonal orbitals.**

The remaining two 2*p* orbitals are left in their original state.

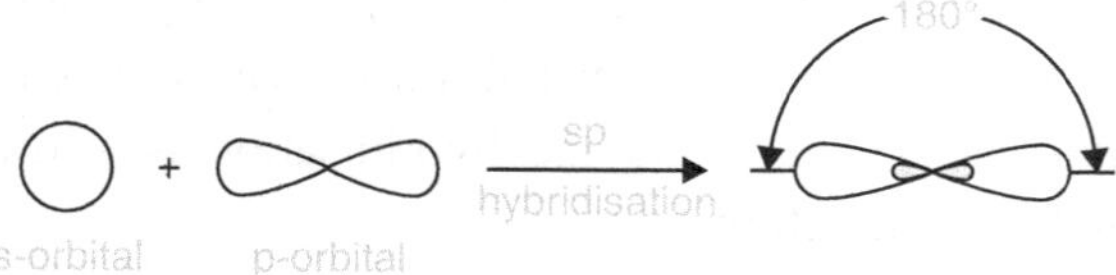

Fig. 36.10

The two *sp*-hybrid orbitals (each of which consists of two lobes, one big and one small) lie along a straight line and thus make an angle of 180° with each other. The two 2*p* orbitals which are left in their original state lie in different planes at right angles to each other and also to the hybridised orbitals (Fig. 36.11).

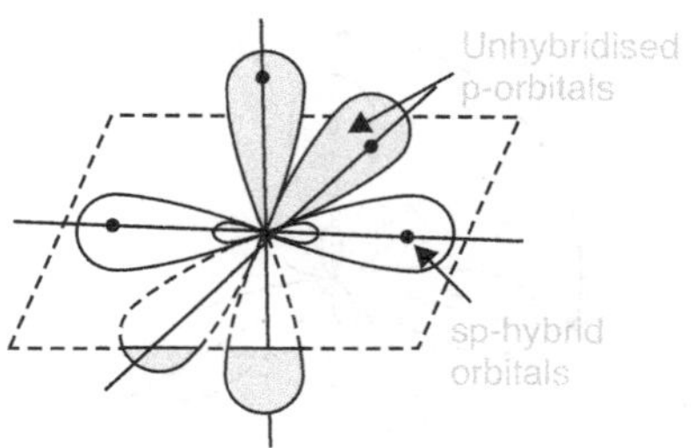

Fig. 36.11. *sp*-Hybrid orbitals in carbon atom.

Structure of Ethyne ($HC \equiv CH$)

Both the carbon atoms in ethyne are *sp*-hybridised.

The overlapping of orbitals has been shown in Fig. 36.12. The π cloud further merge to form a single cylindrical electron cloud along the internuclear axis (Fig. 36.12).

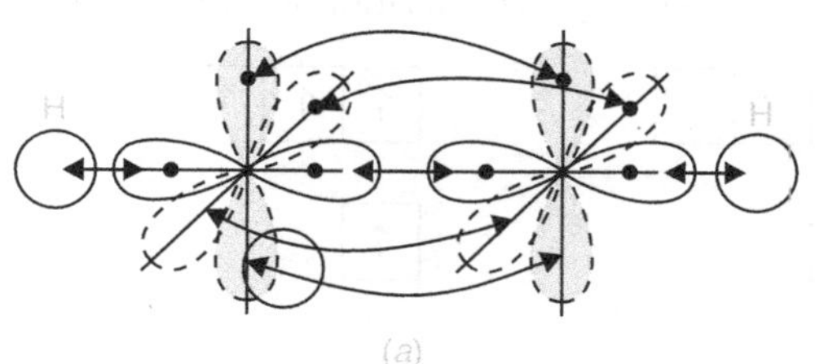

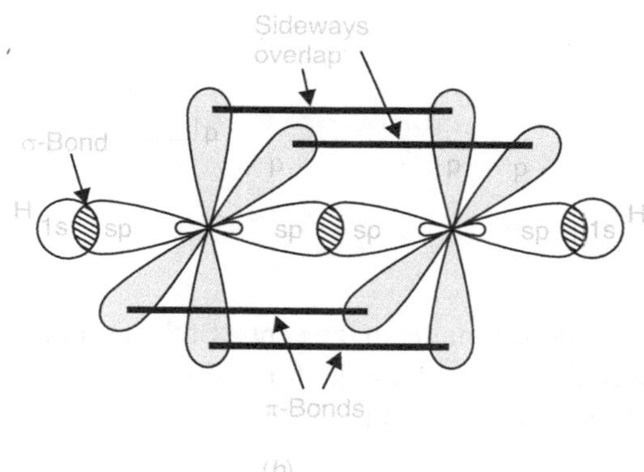

Fig. 36.12. The σ and π-bonds in ethyne. Cylindrical π-electron cloud in acetylene.

It may be **noted** that by observing the structure of a molecule it is possible to predict the state of hybridisation of carbon.

A carbon atom that is directly linked to four other atoms is sp^3-hybridised.

A carbon atom that is directly linked to three other atoms is sp^2-hybridised.

A carbon atom is sp-hybridised if it is directly linked to two other atoms.

Add To Your Knowledge

Effect of Hybridisation on Bonding Parameters

It may be mentioned here that state of hybridisation of carbon in organic molecules influences the bond length and bond strength. A hybrid orbital having greater *s*-character is closer to the nucleus and forms shorter and stronger bonds. Thus, *sp-hybrid orbitals have maximum s-character (among sp, sp^2 and sp^3 hybrid orbitals) and forms strongest bonds with shortest bond length.* Bond enthalpy and bond length of bonds formed by sp^2-hybrid orbitals lie between that of *sp* and sp^3-hybrid orbitals.

Decreasing bond dissociation energy →

Bond Energy (in kJ mol^{-1})	sp-s;	sp^2-s;	sp^3-s;	sp-sp;
	507	443	435	433

Bond Energy (in kJ mol^{-1})	sp^2-sp^2 ;	sp^3-sp^3
	383	347

Increasing bond length →

Bond length (in pm)	sp-sp;	sp^2-sp^2;	sp^3-sp^3
	138	148	154

The hybrid state of carbon atom also affects the electronegativity of the atom. Greater the *s*-character of the hybrid orbitals, greater is the electronegativity of the atom. Thus, *sp-hybridised carbon is more electronegative than sp^2 or sp^3-hybridised carbon atoms.*

SOLVED EXAMPLES

Example 36.1. *Give the hybridisation state of each carbon in the following molecules:*

(*i*) CH_3CHO (*ii*) $CH_2 = C = CH_2$

(*iii*) C_6H_6 (*iv*) $CH_2 = CH—CN$

(*v*) $(CH_3)_2CO$.

Solution. (*i*) $\overset{2}{C}H_3—\overset{1}{C}(H) = O$

C-1 is sp^2 hybridised

C-2 is sp^3 hybridised

(*ii*) $\overset{1}{C}H_2 = \overset{2}{C} = \overset{3}{C}H_2$

C-1 —sp^2

C-2 —sp

C-3 —sp^2

(*iii*) All carbons in benzene are sp^2 hybridised.

(*iv*) $\overset{3}{C}H_2 = \overset{2}{C}H—\overset{1}{C} \equiv N$

C-1 —sp

C-2 —sp^2

C-3 —sp^2

(*v*) $\overset{1}{C}H_3—\overset{2}{C}(=O)—\overset{3}{C}H_3$

C-1 —sp^3

C-2 —sp^2

C-3 —sp^3

Example 36.2. *Indicate the state of hybridisation of each carbon atom in the following molecule:*

$\overset{7}{C}H_3—\overset{6}{C}H(\text{cyclohexenyl})—\overset{5}{C}H_2—\overset{4}{C}(=O)—\overset{3}{C}H_2—\overset{2}{C} \equiv \overset{1}{C}H$

(ring carbons 1′ to 6′, with a double bond between C-2′ and C-3′; C-1′ attached to C-6)

Solution.

C-1	C-2	C-3	C-4	C-5	C-6	C-7
sp	*sp*	sp^3	sp^2	sp^3	sp^3	sp^3
C-1′	C-2′	C-3′	C-4′	C-5′	C-6′	
sp^3	sp^2	sp^2	sp^3	sp^3	sp^3	

Example 36.3. *What is the shape of the following molecule?*

(*a*) $CH_2 = O$ (*b*) CH_3NO_2 (*c*) $HC \equiv N$

Explain on the basis of hybridisation.

Solution.

(*a*) In $CH_2 = O$ *i.e.*, $H—C(=O)—H$, the carbon atom is sp^2 hybridised and hence it has a trigonal planar structure.

(*b*) In CH_3NO_2 the carbon atom is sp^3 hybridised and hence the molecule has a tetrahedral shape.

(*c*) In hydrocyanic acid the carbon is *sp* hybridised therefore the molecule is linear.

SUMMARY

- **Lewis Symbols.** It represents the valence electrons in an atom.
- **Sigma Covalent Bond.** It is formed by the axial overlapping or head on overlapping of half filled atomic orbitals.
- **Pi covalent bond** is formed by the lateral or sideways overlapping of atomic orbitals.
- Hybridisation is the phenomenon of mixing of orbitals of the same atom having slightly different energies to form new orbitals which have identical shapes and equivalent energies.
- Carbon undergoes three types of hybridisation *i.e.*, *sp*, sp^2 and sp^3

EVALUATE YOURSELF

I. Objective Type Questions

Select the most appropriate choice from the options given as (a), (b), (c) and (d) after each question:

1. The number of lone pairs of electrons in CS_2 molecule is/are:
   (*a*) one (*b*) two
   (*c*) three (*d*) four.
2. The correct order of decreasing strength of orbital overlap is
   (*a*) $s-s = s-p = p-p$ (*b*) $s-s > s-p > p-p$
   (*c*) $p-p > s-p > s > s$ (*d*) $s-p > p-p > s-s$.
3. Carbon atom in ethene molecule is
   (*a*) *sp* hybridised (*b*) sp^2 hybridised
   (*c*) sp^3 hybridised (*d*) *spd* hybridised.
4. Carbon atoms labelled as 1, 2, 3 in the structure below are hybridised respectively as $CH_3—CH={}^3CH—{}^2C{\equiv}{}^1CH$
   (*a*) *sp*, *sp* and sp^2 (*b*) *sp*, sp^2, sp^3
   (*c*) *sp*, sp^2 and sp^2 (*d*) sp^2, sp^2, sp^3.

II. Fill in the Blanks

5. Complete the following sentences by supplying appropriate words:
   (*i*) Carbon atom can show ...... hybridisation in ethyne.
   (*ii*) When ethyne changes to ethene hybridisation of carbon changes from ...... to ......
   (*iii*) ...... bond is stronger than ...... bond.
   (*iv*) Lewis dot structure of CO_2 will have ...... lone pairs of electrons.

III. Discussion Questions

6. Explain tetra covalency of carbon.
7. Explain briefly the sp^2 hybridisation.
8. Draw the structure of ethyne and explain it on the basis of hybridisation.
9. Differentiate between σ and π bonds.
10. What type of overlaps result in the formation of σ and π bonds?

Answers

I. Objective Type Questions

1. (*d*) **2.** (*c*) **3.** (*b*) **4.** (*a*).

II. Fill in the Blanks

5. (*i*) *sp* (*ii*) *sp*, sp^2 (*iii*) σ, π (*iv*) 4.

SECTION-6

UNIT-37

CLASSIFICATION OF ORGANIC COMPOUNDS

Specific Objective

After studying this unit, the student should be able to:

- classify organic compounds.

37.1 INTRODUCTION

The main attention of ancient chemists and alchemists was focused on the compounds obtained from *minerals i.e.*, non-living sources. Such compounds were known as **inorganic compounds.** In the eighteenth century, some progress was made in the investigation of substances such as sugar, gelatin, fats, oils etc., which were obtained from living organisms like plants and animals. Lavoisier was the first to establish that the compounds derived from animal and vegetable sources have almost similar composition of elements. Nearly all of them were composed of the elements carbon, hydrogen and oxygen. In view of the similarity of composition, *the compounds obtained from living organisms i.e., animals and plants,* were collectively called **organic compounds.** The study of organic compounds began to develop as a separate branch of chemistry known as **organic chemistry.**

In the early days, it was considered impossible to prepare organic compounds in the laboratory. Berzelius (1815) proposed that these compounds could only be produced by some mysterious force existing in living organisms. The laboratory synthesis of these compounds was not possible because of the absence of such a mysterious force. This simple generalisation was called **vital force theory.**

The vital force theory held its ground till 1828, when *Friedrich Wohler,* a German chemist succeeded in preparing an organic compound, *Urea* (NH_2CONH_2) from ammonium cyanate. In fact, he was trying to prepare ammonium cyanate by heating potassium cyanate and ammonium sulphate:

$$(NH_4)_2SO_4 + 2KCNO \longrightarrow K_2SO_4 + \underset{\text{Amm. Cyanate}}{2NH_4CNO} \xrightarrow{\text{Heat}} \underset{\text{Urea}}{2H_2N—CO—NH_2}$$

This accidental preparation of urea clearly indicated that *no mysterious force* was involved during the production of organic compounds. It also gave new direction to chemists who started working on the preparation of other organic compounds. Later on, *Kolbe* (1845) synthesised acetic acid and *Hennel* synthesised ethyl alcohol. Synthesis of these compounds gave a death blow to the vital force theory. At present, about 95 per cent of known organic compounds have been synthesised in the laboratory.

DEFINITION OF ORGANIC CHEMISTRY

Investigations about the structures of organic compounds gave way to a more precise definition of **organic chemistry.**

It is the *study of hydrocarbons (compounds of carbon and hydrogen elements) and other compounds regarded as derivatives of hydrocarbons.*

It may be noted that some organic compounds may also contain elements like nitrogen, oxygen, sulphur, phosphorus, halogen, etc., besides carbon and hydrogen.

SOURCES OF ORGANIC COMPOUNDS

Organic compounds are very abundantly distributed in nature and are also prepared by synthetic methods. The important sources of these compounds along with some products are given in the flow sheet below:

SOURCES

Plants	**Animals**	**Petroleum**	**Coal tar**
Sugar, starch, cellulose, dye, drugs, etc.	Proteins, fat, urea, enzyme, etc.	Petrol, kerosene oil, vaseline, paraffin wax, etc.	Benzene, toluene, xylene, phenol, perfumes, etc.

37.2 GENERAL CLASSIFICATION OF ORGANIC COMPOUNDS

Organic compounds are broadly grouped under open chains, carbocyclic rings and heterocyclic rings. The flow sheet of general classification is given below for ready reference.

ORGANIC COMPOUNDS
Hydrocarbons and their derivatives

Open Chain or Aliphatic Compounds
Compounds containing straight or branched chains of carbon atom skeletons with appropriate number of H atoms and functional groups. Some **examples** are: CH_4, C_2H_4, C_2H_2, C_2H_6

Cyclic Compounds
Compounds containing closed ring skeletons

Homocyclic or Carbocyclic Compounds
Ring skeleton of such compounds contains only carbon atoms

Heterocyclic Compounds
The ring skeletons of these compounds contain one or more atoms other than carbons

Alicyclic Compounds
These compounds show resemblance in properties with aliphatic compounds. Some **examples** are:

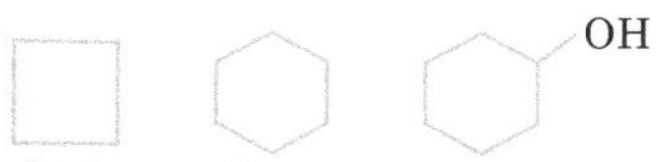

Cyclo butane, Cyclo hexane, Cyclo hexanol (OH)

Aromatic Compounds
Aromatic hydrocarbons and their derivatives They contain $(4n + 2)\pi$-electrons delocalised over their flat structures. Common **examples** are:

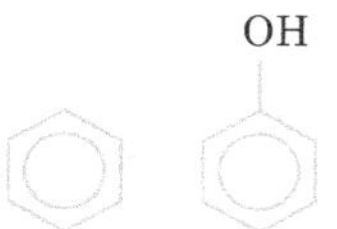

Benzene, Phenol (OH) — Carbocyclic aromatic

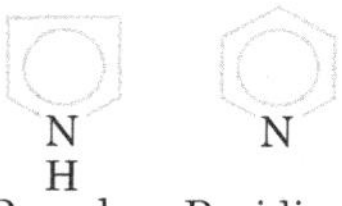

Pyrrole (N, H), Pyridine (N) — Heterocyclic aromatic

Alicyclic Type
Heterocyclic compounds have ring like structure with a heteroatom as a part of the ring. **Examples** are:

Tetra hydro furan (THF) (O), Piperidine (N)

SECTION-6

UNIT-38

IDENTIFICATION OF ELEMENTS IN ORGANIC COMPOUNDS

38.1 DETERMINATION OF COMPONENTS OF A GIVEN ORGANIC COMPOUND

Specific Objective

After studying this unit, the student should be able to:

- determine the components of a given organic compound.

The first step in the investigation of an organic compound after it has been obtained in a pure state is to carry out its qualitative analysis, *i.e.*, to detect the various elements present in it. The elements which commonly occur in organic compounds are carbon (present in all compounds), hydrogen (present in most of the compounds), oxygen, nitrogen, halogens, sulphur and in a few cases phosphorus and metals.

DETECTION OF CARBON AND HYDROGEN

Carbon must be present in a compound under examination if it is definitely known to be an organic compound (Fig. 38.1). Similarly, hydrogen too, is generally present. Both these elements can be tested together. A small quantity of the dry and powdered substance is mixed with freshly ignited cupric oxide. The mixture is heated in a well-dried hard glass test-tube provided with a delivery tube having a bulb at one end. The other end of the tube is dipped into lime water in another test tube. Carbon present in the compound is oxidised to carbon dioxide which turns lime water milky. The hydrogen present in the compound is oxidised to water which condenses on the cooler parts of the test-tube or in the bulb of the delivery tube. The drops turn white anhydrous copper sulphate placed in the bulb into blue.

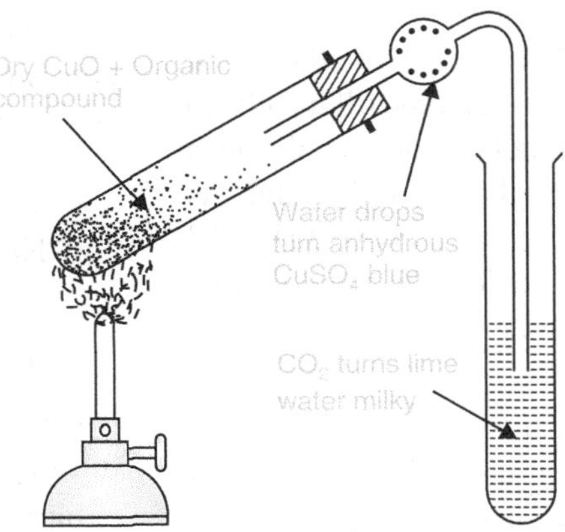

Fig. 38.1. Detection of carbon and hydrogen.

$$\underset{\text{From organic compound}}{C} + 2CuO \longrightarrow CO_2 \uparrow + 2Cu$$

$$CO_2 + \underset{\text{Lime water}}{Ca(OH)_2} \longrightarrow \underset{\text{Milky}}{CaCO_3 \downarrow} + H_2O$$

$$\underset{\text{From organic compound}}{2H} + CuO \longrightarrow H_2O + Cu$$

$$\underset{\text{White}}{CuSO_4} + 5H_2O \longrightarrow \underset{\text{Blue}}{CuSO_4 . 5H_2O}$$

If the compound to be analysed is a volatile liquid or a gas, then the vapour of the liquid or the gas itself is passed through heated copper oxide and the liberated gases are tested for carbon dioxide and water vapours.

DETECTION OF NITROGEN

Nitrogen in an organic compound is detected by the following tests:

(*i*) Sodalime Test

A pinch of an organic substance is heated strongly with *sodalime* (NaOH + CaO) in a test tube. The evolution of ammonia gives the indication of nitrogen.

$$\underset{\text{Acetamide}}{CH_3CONH_2} + [NaOH + CaO] \longrightarrow CH_3COONa + NH_3 \uparrow$$

Limitation. A large number of organic compounds such a *nitro and diazo* compounds do not liberate ammonia under these conditions.

(*ii*) Lassaigne's Test

This test is a confirmatory test for nitrogen in organic compounds. This test involves two steps:

(*a*) ***Preparation of Lassaigne's extract.*** A small piece of a sodium metal is heated gently in a fusion tube till it melts to a *shining globule*. At this stage, a small amount of substance is added and the tube is heated strongly. The red hot tube is plunged into distilled water contained in a china dish. The contents are boiled for some time, cooled and then filtered. The filterate is known as *sodium extract or Lassaigne's extract*.

(*b*) ***Test for nitrogen.*** The Lassaigne's extract is usually alkaline. If not, it may be made alkaline by the addition of a few drops of a dilute solution of sodium hydroxide. To a part of sodium extract a small amount of a freshly prepared ferrous sulphate solution is added and the contents are warmed. A few drops of ferric chloride solution are then added to the contents and the resulting solution is acidified with dilute hydrochloric acid. The appearance of a *prussian blue* colouration confirms the presence of nitrogen in the organic compound. The chemical reactions involved are as follows:

$$Na + \underset{\text{From organic compound}}{C + N} \xrightarrow{\text{Fuse}} NaCN$$

$$FeSO_4 + 2NaCN \longrightarrow Fe(CN)_2 + Na_2SO_4$$

$$Fe(CN)_2 + 4NaCN \longrightarrow \underset{\text{Sodium ferrocyanide}}{Na_4[Fe(CN)_6]}$$

$$3Na_4[Fe(CN)_6] + 4FeCl_3 \longrightarrow \underset{\substack{\text{Ferric ferrocyanide} \\ \text{(Blue colour)}}}{Fe_4[Fe(CN)_6]_3} + 12NaCl$$

If the organic compound contains both nitrogen and sulphur, sodium sulphocyanide may be formed during fusion. This reacts with ferric ions to form ferric sulphocyanide which gives a blood red colouration.

$$Na + \underset{\text{From organic compound}}{C + N + S} \longrightarrow \underset{\text{Sod. sulphocyanide}}{NaCNS}$$

$$3NaCNS + Fe^{3+} \longrightarrow \underset{\substack{\text{Ferric sulphocyanide} \\ \text{(Blood Red)}}}{Fe(CNS)_3} + 3Na^+$$

DETECTION OF SULPHUR

The presence of sulphur in the organic compound can be detected by the following tests:

(*i*) Oxidation Test

The organic compound is fused with a mixture of potassium nitrate and sodium carbonate. Sulphur, if present, gets oxidised to sodium sulphate.

$$3KNO_3 \longrightarrow 3KNO_2 + 3[O]$$

$$Na_2CO_3 + S + 3[O] \longrightarrow \underset{\text{Sodium sulphate}}{Na_2SO_4} + CO_2$$

The fused mass is extracted with water, the contents are boiled and filtered. The filtrate contains sodium sulphate. The filtrate is acidified with dilute hydrochloric acid and a solution of barium chloride is added. Formation of a white precipitate indicates the presence of *sulphur*.

$$BaCl_2(aq) + Na_2SO_4(aq) \longrightarrow \underset{\text{White ppt.}}{BaSO_4 \downarrow} + 2NaCl(aq)$$

(ii) Lassaigne's Test

Lassaigne's extract is prepared as described in the case of nitrogen. The extract contains sodium sulphide formed by the reaction between sulphur (present in the compound) and sodium.

$$2Na + S \longrightarrow Na_2S$$

The Lassaigne's extract is divided into two parts and the following tests are performed:

(*a*) ***Lead acetate test.*** One part of the extract is acidified with acetic acid and then lead acetate solution is added. Formation of a black precipitate confirms the presence of sulphur in the organic compound.

$$Na_2S + Pb(CH_3COO)_2 \longrightarrow \underset{\text{Black ppt.}}{PbS \downarrow} + 2CH_3COONa$$

(*b*) ***Sodium nitroprusside test.*** A few drops of sodium nitroprusside solution are added to another part of the Lassaigne's extract. The appearance of purple colouration confirms the presence of sulphur:

$$\underset{\text{Sodium extract}}{Na_2S} + \underset{\text{Sodium nitroprusside}}{Na_2[Fe(CN)_5NO]} \longrightarrow \underset{\text{Purple colouration}}{Na_4[Fe(CN)_5\ NO\ .\ S]}$$

DETECTION OF HALOGENS

The presence of halogens in an organic compound is detected by the following tests:

(i) Beilstein's Test

A copper wire flattened at one end is heated in the oxidising flame of a Bunsen burner. The heating is continued till it does not impart blue colour to the flame. The hot end of the copper wire is now touched with the organic substance and once again kept in the flame. The reappearance of green or blue colour indicates the presence of halogens in the organic compound.

Limitations. (*a*) Substances such as urea, thiourea, etc., do not contain halogens, but give this test.

(*b*) It does not tell as to which halogen is present in the organic compound.

(ii) Lassaigne's Test

A small amount of the compound is fused with metallic sodium as described earlier. The halogen present in the compound combines with sodium to form sodium halide which passes into the sodium extract.

$$\underset{\text{From compound}}{X} + Na \longrightarrow \underset{\text{Sodium halide}}{NaX}$$

(where X may be Cl, Br or I)

A part of the Lassaigne's extract is boiled with dilute nitric acid to expel all gases. Silver nitrate solution is added to the resulting solution.

If any of the halogen is present, a precipitate is formed.

$$NaX + AgNO_3 \longrightarrow \underset{\text{ppt.}}{AgX} + NaNO_3$$

(*a*) **A white precipitate soluble** in ammonium hydroxide solution indicates the presence of *chlorine* in the organic compound.

$$NaCl + AgNO_3 \longrightarrow \underset{\text{White ppt.}}{AgCl} + NaNO_3$$

(*b*) **A dull yellow precipitate** partially soluble in ammonium hydroxide solution indicates the presence of *bromine* in the organic compound.

$$NaBr + AgNO_3 \longrightarrow \underset{\text{Dull yellow ppt.}}{AgBr} + NaNO_3$$

(*c*) **A bright yellow precipitate,** completely insoluble in ammonium hydroxide solution indicates the presence of *iodine* in the organic compound.

$$NaI + AgNO_3 \longrightarrow \underset{\text{Bright yellow ppt.}}{AgI} + NaNO_3$$

Function of nitric acid. If nitrogen and sulphur are present along with the halogens in the organic compound, the Lassaigne's extract will contain sodium sulphide (Na_2S) and sodium cyanide (NaCN) along with sodium halide. Nitric acid decomposes sodium cyanide and sodium sulphide which otherwise form precipitate with silver nitrate and thus interfere with the test.

$$NaCN + HNO_3 \longrightarrow NaNO_3 + HCN \uparrow$$

$$Na_2S + 2HNO_3 \longrightarrow 2NaNO_3 + H_2S \uparrow$$

(iii) Special Test for Bromine and Iodine

Acidify the Lassaigne's extract with nitric acid and add a few drops of freshly prepared chlorine water. Free Br_2 and I_2 will be evolved. Shake this solution with carbon disulphide or carbon tetrachloride. The liberated halogens (Br_2 or I_2) will dissolve in it to give orange or brown colour for bromine and violet colour for iodine.

$$2NaBr + Cl_2 \longrightarrow 2NaCl + \underset{\text{(Brown colour in } CCl_4)}{Br_2}$$

$$2NaI + Cl_2 \longrightarrow 2NaCl + \underset{\text{(Violet colour in } CCl_4)}{I_2}$$

DETECTION OF PHOSPHORUS

Few organic compounds contain phosphorus. Its presence is detected by fusing the compound with an oxidizing mixture of sodium carbonate and potassium nitrate or with sodium peroxide alone when an alkali phosphate is formed. The fused mass is extracted with water and filtered. The filtrate containing sodium phosphate is heated with conc. HNO_3 and an excess of ammonium molybdate solution is added. A yellow ppt. or colouration is formed if phosphorus is present. The yellow precipitate is ammonium phosphomolybdate, $(NH_4)_3[PMo_{12}O_{40}]$ or $(NH_4)_3PO_4.12\ MoO_3$.

DETERMINATION OF EMPIRICAL AND MOLECULAR FORMULA ON THE BASIS OF GIVEN DATA

The empirical formula of a compound is the simplest formula which expresses the simple whole number ratio of the atoms of constituent elements present in the compound. For example, CH_2O is the empirical formula of acetic acid, it expresses that the simplest whole number ratio between carbon, hydrogen and oxygen atoms present in one molecule of acetic acid is 1 : 2 : 1. The empirical formula does not tell us the exact number of atoms of different elements present in one molecule of a substance. However, the exact number of atoms of different elements present in one molecule is always a simple multiple of the ratio of atoms in the empirical formula.

The formula which gives the actual number of atoms of various elements present in a molecule of a compound is termed the molecular formula.

Molecular formula of acetic acid

$= (CH_3COOH)$ or $(CH_2O) \times 2$ [here $n = 2$]

The value of 'n' can be obtained by dividing the molecular mass by empirical formula mass.

$$n = \frac{\text{Molecular mass}}{\text{Empirical formula mass}}$$

The **percentage composition** of the compound can be calculated from its empirical formula. For example,

Empirical formula of acetic acid $= CH_2O$

Empirical formula mass $= (12 + 2 + 16)$
$= 30$

$$\text{Percentage of C} = \frac{12}{30} \times 100 = 40.00$$

$$\text{Percentage of H} = \frac{2}{30} \times 100 = 6.66$$

$$\text{Percentage of O} = \frac{16}{30} \times 100 = 53.34$$

Thus, the simplest formula of a substance capable of expressing its percentage composition can be called its empirical formula.

Calculation of empirical formula. Empirical formula is calculated from the percentage composition. The steps involved in the calculation are as follows:

Step 1. The percentage of each element is divided by its atomic mass. This gives the relative number of different atoms present in the molecule.

Step 2. The relative numbers of different atoms obtained in Step 1 are divided by the lowest one amongst them as to get simple ratio of atoms present in the molecule.

Step 3. The values obtained in Step 2 may or may not be whole numbers. In case one or more values are fractional, these are multiplied by a suitable integer to get simplest ratio in whole numbers. Minor fractions are neglected.

Step 4. The symbols of each element present are written side by side in a line with the number of atoms as determined in Step 2 or Step 3 as subscripts to the lower corner of each. This gives the empirical or simplest formula.

Knowing the empirical formula, the molecular formula can be ascertained if the molecular mass of the substance is known. It may be the same as the empirical formula of the substance or an exact multiple of it.

Molecular formula $= n \times$ (Empirical formula)

The value of 'n' can be determined if molecular mass of the substance is known.

$$n = \frac{\text{Molecular mass}}{\text{Empirical formula mass}}$$

SOLVED EXAMPLES

Example 38.1. *An organic compound contains 40% C, 6.66% H and oxygen. Its vapour density is 30. Calculate its empirical and molecular formulas.*

Solution. Calculation of empirical formula:

Element	*Percentage*	*At. mass*	*Relative number of atoms*	*Simplest ratio*
Carbon	40.0	12	$\frac{40}{12} = 3.33$	$\frac{3.33}{3.33} = 1$
Hydrogen	6.66	1	$\frac{6.66}{1} = 6.66$	$\frac{6.66}{3.33} = 2$
Oxygen	53.34 (by difference)	16	$\frac{53.34}{16} = 3.33$	$\frac{3.33}{3.33} = 1$

Empirical formula $= CH_2O$

Empirical formula mass $= 12 + 2 + 16 = 30$

$\therefore$ Molecular mass $= 2 \times V.D = 2 \times 30 = 60$

Hence, $n = \frac{\text{Mol. mass}}{\text{Emp. mass}} = \frac{60}{30} = 2$

Molecular formula $= 2 \times$ Empirical formula $= 2 \times (CH_2O) = C_2H_4O_2$.

EVALUATE YOURSELF

I. Objective Type Questions

Select the most appropriate choice from the options given as (a), (b), (c) and (d) after each question:

1. In Lassaigne's test for detection of nitrogen, the blue colour is due to the compound
(*a*) $Na_4Fe(CN)_6$ (*b*) NaCN
(*c*) $Na_3[Fe(CN)_6]_4$ (*d*) $Fe_4[Fe(CN)_6]_3$.

2. The formula of compound which gives violet colour in Lassaigne's test for sulphur with sodium nitroprusside is
(*a*) $Na_4[Fe(CN)_5NOS]$ (*b*) $Na_2[Fe(CN)_5NOS]$
(*c*) $Na_2[Fe(CN)_5S]$ (*d*) $Na_4[Fe(CN)_6S]$.

3. In the Duma's method, the nitrogen present in organic compound gets converted to
(*a*) sodium cyanide (*b*) gaseous ammonia
(*c*) dinitrogen gas (*d*) ammonium sulphate.

4. The Lassaigne's extract is boiled with dil. HNO_3 before testing for halogens because
(*a*) Silver halides are soluble in HNO_3
(*b*) Na_2S and NaCN are decomposed by HNO_3
(*c*) Ag_2S is soluble in HNO_3
(*d*) AgCN is soluble in HNO_3.

5. The compound that does not give a blue colour in Lassaigne's test is
(*a*) aniline (*b*) glycine
(*c*) hydrazine (*d*) urea.

6. 59 g of an amide obtained from a carboxylic acid, RCOOH, liberated 17 g of ammonia upon heating with alkali. The acid is
(*a*) formic acid (*b*) acetic acid
(*c*) propionic acid (*d*) benzoic acid.

7. The Lassaigne's extract is boiled with dilute HNO_3 before testing for halogens because
(*a*) silver halide are soluble in HNO_3
(*b*) Na_2S and NaCN if present are decomposed by HNO_3
(*c*) Ag_2S is soluble in HNO_3
(*d*) AgCN is soluble in HNO_3.

8. Beilstein test for organic compounds is used to detect
(*a*) nitrogen (*b*) halogen
(*c*) sulphur (*d*) carbon.

9. 0.1914 g of an organic acid is dissolved in about 20 mL of water. 25 mL of 0.12 N NaOH is required for the complete neutralization of the acid solution. The equivalent mass of the acid is
(*a*) 65 (*b*) 64
(*c*) 63.80 (*d*) 62.50.

10. In quantitative estimation of hydrogen 0.9 g of an organic compound yielded 0.81 g of water. The percentage of hydrogen in the compound is
(*a*) 10 (*b*) 12
(*c*) 14 (*d*) 16.

11. The percentage of sulphur in organic compound whose 0.32 g produces 0.233 g of $BaSO_4$ is (At. mass of Ba = 137; S = 32)
(*a*) 1.0 (*b*) 10.0
(*c*) 23.5 (*d*) 32.4.

12. In the Lassaigne's test for N, the blue colour is due to the formation of
(*a*) $K_4[Fe(CN)_6]$ (*b*) sodium cyanide
(*c*) sodium ferrocyanide (*d*) ferri-ferrocyanide.

13. In Lassaigne's test, a blue colour is obtained when the organic compound contains nitrogen. Then blue colour is due to
(*a*) $K_4[Fe(CN)_6]$ (*b*) $Fe_4[Fe(CN)_6]_3$
(*c*) $Na_3[Fe(CN)_6]$ (*d*) $Cu_2[Fe(CN)_6]$.

14. The compound formed in the positive test for nitrogen with the Lassaigne solution of an organic compound is
(*a*) $Fe_4[Fe(CN)_6]_3$ (*b*) $Na_3[Fe(CN_6)]$
(*c*) $Fe(CN)_3$ (*d*) $Na_4[Fe(CN)_5COS]$.

II. Fill in the Blanks

15. Complete the following sentences by supplying appropriate words:
(*i*) The blue colour obtained during Lassaigne's test for nitrogen is due to ....
(*ii*) An organic compound containing nitrogen as well as sulphur gives ...... colouration when $FeCl_3$ solution is added to acidified Lassaigne's extract.

(*iii*) Duma's method is employed for the quantitative estimation of ....

(*iv*) Silver salt method is applied to find molecular masses of ....

(*v*) Molecular mass of an acid = Equivalent mass × ....

(*vi*) Benzene has same empirical formula as ....

(*vii*) Empirical formula of acetic acid is ....

III. Discussion Questions

16. Name any compound which does not contain halogen but responds to Beilstein test.
17. What is the function of fusing the organic compound with sodium metal?
18. A student was given the compound $H_2N—C_6H_4—SO_3H$ for elemental analysis. While performing Lassaigne's test for N, what colour will he get and why?
19. If silver nitrate solution is added to chlorobenzene will there be formation of white ppt.?
20. What is the colour of the solution, if Lassaigne's extract of the compound containing sulphur is mixed with sodium nitroprusside?
21. Do you expect sodium cyanide to show positive Lassaigne's test for N?
22. Why diazonium salts do not show Lassaigne's test?
23. How would you detect the presence of carbon and hydrogen in an organic compound?
24. Describe and explain Lassaigne's test for the detection of nitrogen, sulphur and halogens in an organic compound.
25. How will you proceed to detect the presence of halogens in an organic compound by Lassaigne's test?
26. (*i*) What is the relationship between empirical and molecular formula?

    (*ii*) How will you proceed to determine the molecular mass of a base by volumetric method?
27. Describe in brief the procedure for the estimation of phosphorus in organic compounds.
28. Why is an organic compound fused with sodium metal for testing nitrogen, sulphur and halogens?

Answers

I. Objective Type Questions

1. (*d*)	**2.** (*a*)	**3.** (*c*)
4. (*b*)	**5.** (*c*)	**6.** (*b*)
7. (*b*)	**8.** (*b*)	**9.** (*c*)
10. (*a*)	**11.** (*b*)	**12.** (*d*)
13. (*a*)	**14.** (*a*).	

II. Fill in the Blanks

15. (*i*) farri ferrocyanide (*ii*) blood red
(*iii*) nitrogen (*iv*) organic acid
(*v*) basicity (*vi*) ethyne
(*vii*) CH_2O.

III. Discussion Questions

16. *Urea; thiourea, etc.*

17. Sodium reacts with some of the elements present in organic compound and form corresponding water soluble salts.

18. *Blood red.*

19. *No, because the bond C—Cl is non ionisable.*

20. *Purple.*

21. *Yes. In fact during fusion of organic compound with sodium NaCN is formed if N is present.*

22. *Diazonium salts are unstable and lose nitrogen as N_2 gas on heating. Hence during fusion, no NaCN is formed in Lassaigne's extract due to the loss of nitrogen.*

SECTION-6

UNIT-39

SEPARATION AND PURIFICATION OF ORGANIC COMPOUNDS

Specific Objective

After studying this unit, the student should be able to:

- describe the separation and purification of an organic compound from a mixture of compounds.

39.1 SEPARATION AND PURIFICATION OF ORGANIC COMPOUNDS

Organic compounds when isolated from natural sources or prepared by organic reactions are seldom pure; they are usually contaminated with small amounts of other compounds which are produced along with the desired product. Before carrying out the qualitative and quantitative analysis of the organic compounds, in order to characterise them, it is very important of purify them. The various steps involved in the characterisation of an organic compound are:

(*a*) *Purification*

(*b*) *Qualitative analysis*

(*c*) *Quantitative analysis*

(*d*) *Determination of molecular mass*

(*e*) *Calculation of empirical and molecular formula*

(*f*) *Elucidation of the structure by various methods including chemical methods.*

In this unit, we shall consider some of the methods commonly employed for the purification, qualitative analysis and quantitative analysis of an organic compound.

PURIFICATION

The common techniques used for the purification of a particular compound are based on its nature and also on the nature of the impurities present in it. The various methods used for this purpose are:

1. *Crystallisation* 2. *Sublimation*

3. *Differential extraction* 4. *Distillation*
5. *Fractional distillation*
6. *Distillation under reduced pressure*
7. *Steam distillation* 8. *Chromatography.*

1. CRYSTALLISATION

Crystallisation is a process of solidification of a pure substance from its dissolved state. This is the most common way of purifying organic solids. This method is based upon differences in their solubility in a given solvent or a mixture of solvents. Since most organic compounds are not very soluble in water, other solvents such as alcohol, acetone, chloroform, ether, etc., are commonly used for this purpose. The process involves the steps given below:

(*i*) Selection of the Solvent

A small quantity of the substance is heated with 2–3 mL of the solvent under examination. When partial or complete dissolution has taken place, the solution is cooled. If a considerable proportion of the dissolved substance separates out in the form of crystals, the solvent is considered suitable. *Thus, a suitable solvent is one in which:*

(*a*) *The substance dissolves on heating readily and crystallises out on cooling.*

(*b*) *The solvent must not react chemically with the substance.*

(*c*) *It should not dissolve impurities or the impurities should be more soluble than the substance to be purified so that during crystallisation, the impurities are left in the mother liquor.*

Some examples are: *ethanol* can be used for purification of sugar containing impurities of common salt. Common salt does not dissolve in ethanol.

Similarly, a *benzoic acid* containing impurities of naphthalene can be purified by using *hot water* which dissolves only benzoic acid but not naphthalene.

(*ii*) Preparing the Solution

The impure substance is powdered and heated with the solvent in a flask. The amount of solvent should be just sufficient to dissolve the substance on heating. In the case of solvents with low boiling points, the flask is fitted with a water condenser or an air condenser to avoid the loss of solvent on heating (Fig. 39.1).

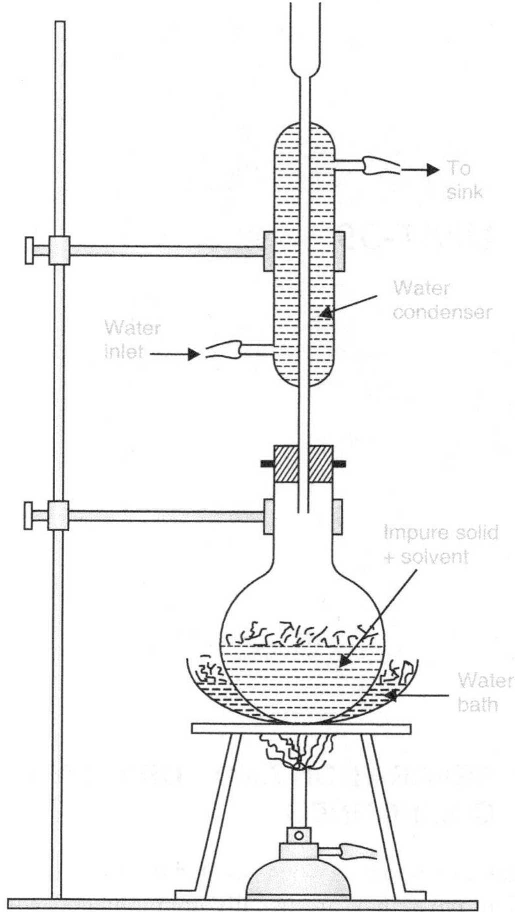

Fig. 39.1. Preparing the solution.

(*iii*) Filtration

The hot saturated solution prepared above is then filtered preferably through a fluted filter paper placed in a glass funnel (Fig. 39.2). The use of fluted filter paper makes the filtration rapid. If the organic substance crystallises during filtration, the filtration cannot be done with ordinary filter paper as the crystals will be formed on the filter paper itself and proper

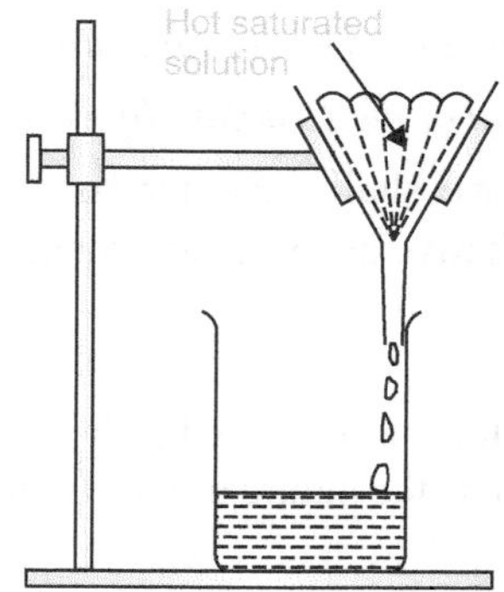

Fig. 39.2. Filtration through fluted filter paper.

filtration will not be possible. In such cases, the filtration is done with the help of *hot water funnel* (Fig. 39.3). The hot water that is circulated in the outer funnel keeps the solution hot and the crystals cannot be formed on the funnel itself.

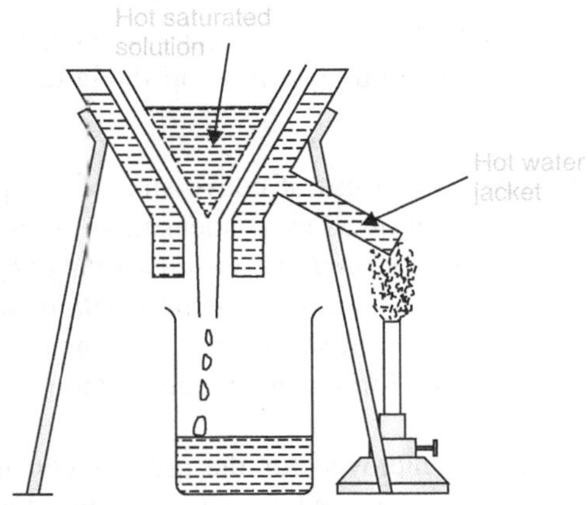

Fig. 39.3. A hot water funnel.

(*iv*) Crystallising

The hot filtrate is allowed to cool slowly in a beaker or a crystallising dish. In this way, large and beautiful crystals are obtained. Scratching the sides of the vessel often facilitates crystallisation.

(*v*) Separation and Drying of Crystals

The crystals are separated from the mother liquor by filtration with the help on a Buchner funnel and a suction pump (Fig. 39.4). The crystals obtained on the filter paper are washed two or three times with small quantities of the pure solvent. They are dried by pressing between folds of filter paper and are then left in a steam oven or an air oven for some time, if the substance is stable and high melting. The crystals are finally dried over sulphuric acid or calcium chloride in a vacuum desiccator.

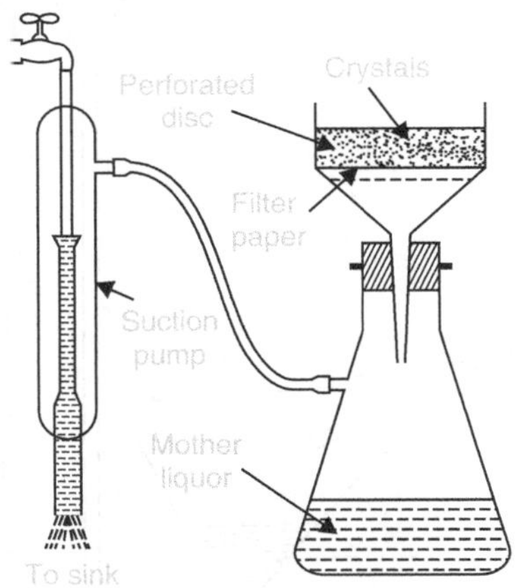

Fig. 39.4. Filtration with Buchner funnel and a suction pump.

Sometimes the crystals obtained are slightly coloured because of the presence of traces of impurities. In such cases, they are redissolved in a small amount of the solvent and a little of animal charcoal is added. The suspension thus obtained is boiled, filtered and recrystallised as described above.

FRACTIONAL CRYSTALLISATION

It is a process of separation of different components of a mixture by repeated crystallisation. For example, if the impure substance is a mixture of two compounds which do not have large difference in their solubilities, the process of fractional crystallisation is used for their separation. A saturated solution of the mixture is prepared and allowed to cool as usual. On cooling, the less soluble substance crystallises out first. The mother liquor is now very rich in the more soluble substance which separates out in the crystalline form on further concentration and cooling of the mother liquor. The less soluble substance separated in this way is usually contaminated with a little of the more soluble substance and *vice versa.* A complete separation of the substances is achieved by repeating the above process several times.

2. SUBLIMATION

Sublimation is a process of conversion of a solid into gas on heating without passing through the intermediate liquid and vice versa.

$$\text{Solid} \underset{\text{Cool}}{\overset{\text{Heat}}{\rightleftharpoons}} \text{Gas}$$

This process is used for the separation of volatile solids, which sublime on heating from the non-volatile solids. The impure substance is heated in a dish covered with a perforated filter paper over which an inverted funnel is placed (Fig. 39.5). The stem of the funnel is plugged with a little cotton. Vapours of the solid, which sublime, pass through the holes in the filter paper and condense on the cooler walls of the funnel. The non-volatile impurities are left behind in the dish.

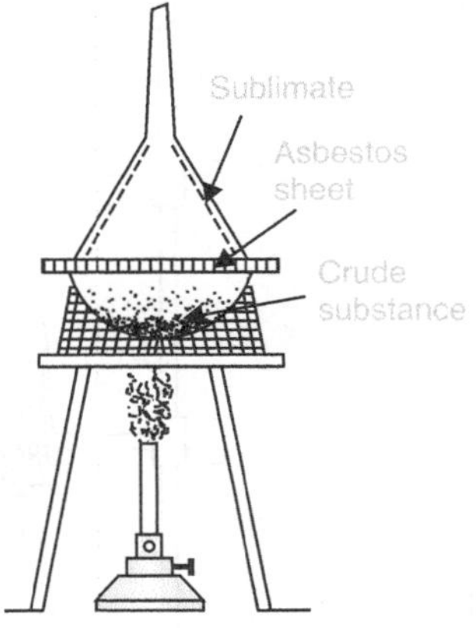

Fig. 39.5. Sublimation.

Iodine, camphor, naphthalene, benzoic acid, etc., are purified by this method.

3. DIFFERENTIAL EXTRACTION

Organic compounds, whether solids or liquids, can be recovered from aqueous solutions by shaking the solution in a separatory funnel with a suitable organic solvent which is immiscible with water but in which the organic compound is highly soluble. This process is known as **extraction** or **solvent extraction.** Some commonly employed solvents are, ether, benzene and chloroform. The process is carried out as follows:

The aqueous solution is mixed with a small quantity of the organic solvent in a separatory funnel (Fig. 39.6). The funnel is stoppered and its contents shaken for some time for the organic compound to move into the organic solvent. The mixture is then allowed to settle into two separate layers. The lower aqueous layer is run out by opening the tap and the solvent layer is collected separately. The whole process may be repeated to remove the organic compound completely from the aqueous solution. The compound is finally recovered from the organic solvent by distilling off the latter.

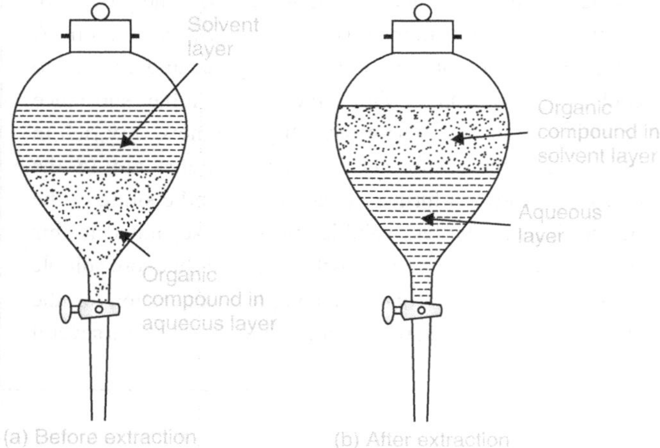

Fig. 39.6. Differential extraction

4. DISTILLATION

Distillation is a process of conversion of a liquid into vapours by heating followed by condensation of vapours so produced by cooling. *The method is used for the purification of liquids which boil without decomposition and are associated with non-volatile impurities.* The impure liquid is boiled and the vapours formed are condensed to get the pure liquid.

The apparatus employed for the process consists of a distillation flask having a side tube connected to a condenser. An air condenser which is a long glass tube is used for liquids with boiling points above 380 K and a water condenser in the case of other liquids. The neck of the flask is closed with a cork through which passes a thermometer and the bulb of the thermometer is kept just below the opening of the side tube. A receiver is attached to the lower end of the condenser (Fig. 39.7). The substance is placed in the distillation flask and one or two pieces of unglazed porcelain or glass bends are added to prevent bumping of the liquid during distillation. The flask is heated on a water bath or sand bath in the case of

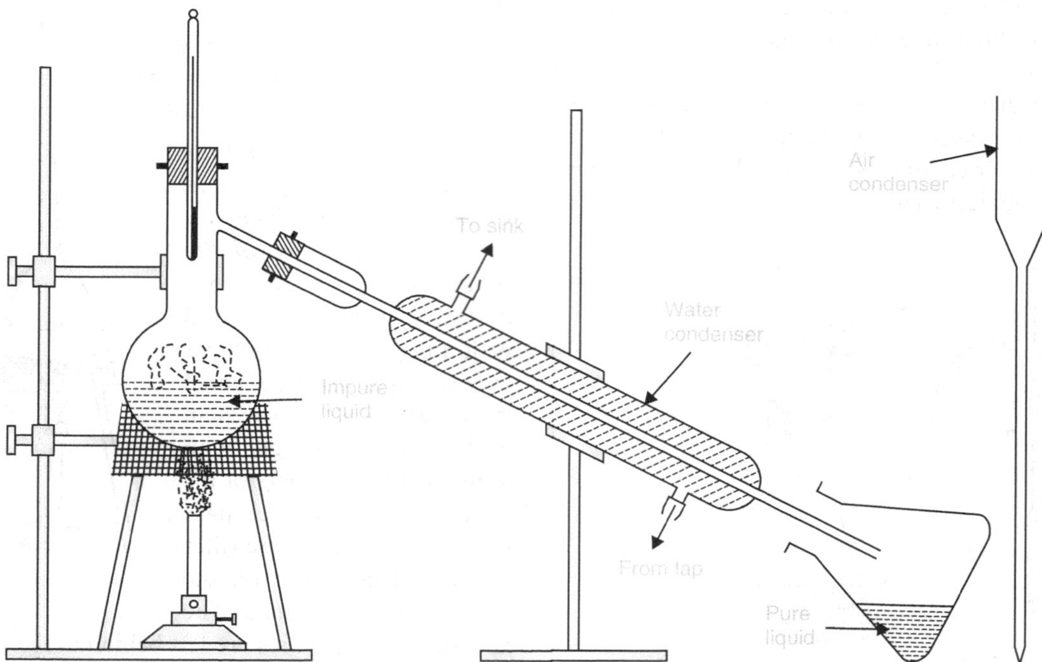

Fig. 39.7. Simple distillation

highly volatile and inflammable liquids, while it is heated directly with the flame when distilling high boiling liquids. The liquid changes into vapours which pass through the condenser and are condensed back into the liquid form. The liquid distillate is collected in the receiver while the impurities are left behind in the distillation flask.

5. FRACTIONAL DISTILLATION

This process is used to separate a *mixture of two or more miscible liquids having different boiling points.* It can be carried in two ways.

(*a*) By using an apparatus of simple distillation

When the two liquids have their boiling points wide apart, say about 40°C or more, the mixture may be separated by using simple distillation apparatus. The more volatile liquid (low b.p.) distils over first and is collected in a receiver. When the temperature begins to rise again, the receiver is disconnected. A new receiver is attached when the temperature becomes constant again. At this stage, the less volatile liquid distils over and is collected. The two fractions thus obtained are redistilled separately, a number of times, to ensure complete purity of the liquids.

(*b*) By using a fractionating column

When the boiling points of the two liquids are quite close (difference less than 30°C) the separation is effected by fitting the distillation flask with a *fractionating column* (Fig. 39.8). A **fractionating column** *is a long tube provided with obstructions to the passage of the vapours upwards and that to liquid downwards.* Various types of fractionating columns commonly used are shown in Fig. 39.9.

The fractionating column is fitted onto the neck of a round-bottomed flask containing the mixture of liquids (say A with b.p. 350 K and B with b.p. 360 K). The column itself is fitted with a thermometer and is attached to a condenser. On heating the flask the vapours obtained consist of more of A and less of B. As the vapours rise up the fractionating column, they condense partially and the condensed liquid flows down. Naturally, the vapours of B (having lower boiling point) condense more readily than the vapours of A (having lower boiling point). As a result, the vapours moving up get richer in A. The condensed liquid flowing down meets the ascending stream of vapours and in the process takes away more of B. This process is repeated throughout the length of

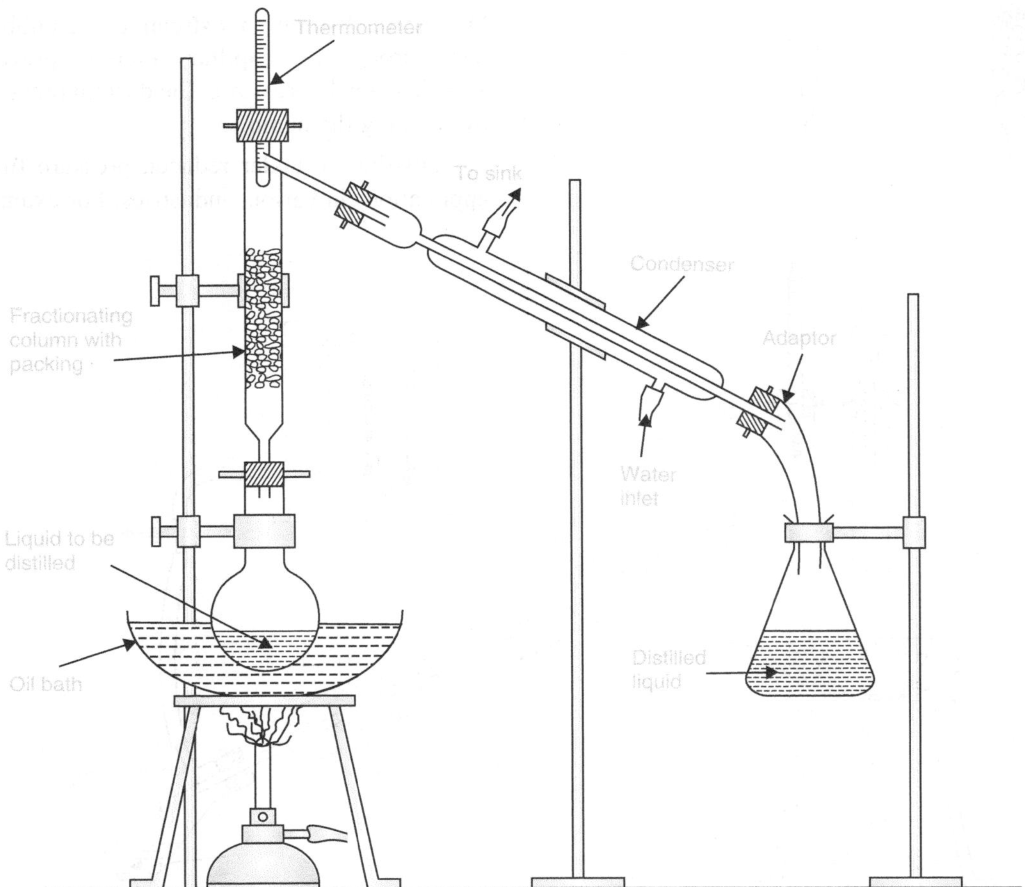

Fig. 39.8. Fractional distillation.

the fractionating column. Consequently, the vapours which escape from the top of the column into the condenser consist mostly of A. Thus, the distillate received is almost pure A whereas the liquid left behind in the flask is very rich in B. This process may be repeated to achieve the complete separation of liquids.

Fractional distillation has found remarkable application in modern industry especially in the distillation of petroleum, coaltar and crude alcohol.

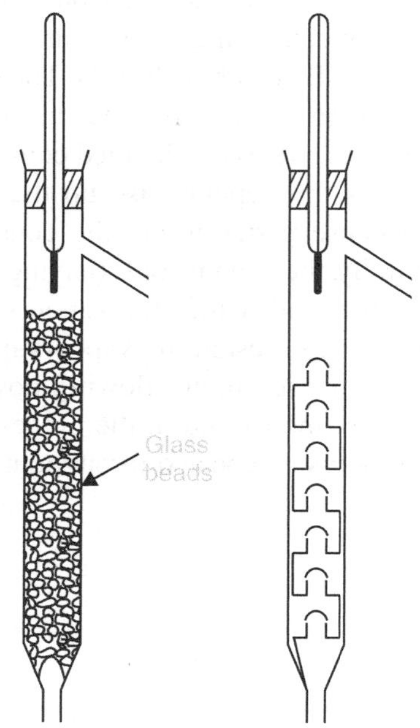

Fig. 39.9. Various types of fractionating columns.

6. DISTILLATION UNDER REDUCED PRESSURE

Principle. A liquid boils when its vapour pressure becomes equal to the applied pressure or external pressure. If applied pressure is decreased, the same liquid will now boil at a lower temperature. Thus, *this process is employed if the liquid has a tendency to decompose near its boiling point.* Under reduced pressure, the liquid will boil at a low temperature and the temperature of decomposition will not be reached. Glycerol, for example, boils with decomposition at 563 K but if the pressure it reduced to 12 mm Hg, it boils at 453 K without decomposition.

The apparatus used for distillation under reduced pressure is shown in Fig. 39.10. The liquid to be distilled is taken in a two necked flask called "*Claisen's flask*". It is fitted with a long capillary tube in the main neck and a thermometer in the side neck. The side tube of the second neck is connected to a condenser carrying a receiver at the other end. The receiver is further connected to a vacuum pump and a manometer which indicates the pressure. The whole apparatus is tested to be air-tight.

The flask is usually heated on an oil bath or sand bath. During the distillation a stream of air bubbles is allowed to pass through the capillary tube to prevent bumping by ensuring steady pressure. The desired pressure is maintained by working the pump.

Distillation under reduced pressure finds a number of applications in various industries. For example, *in the sugar*

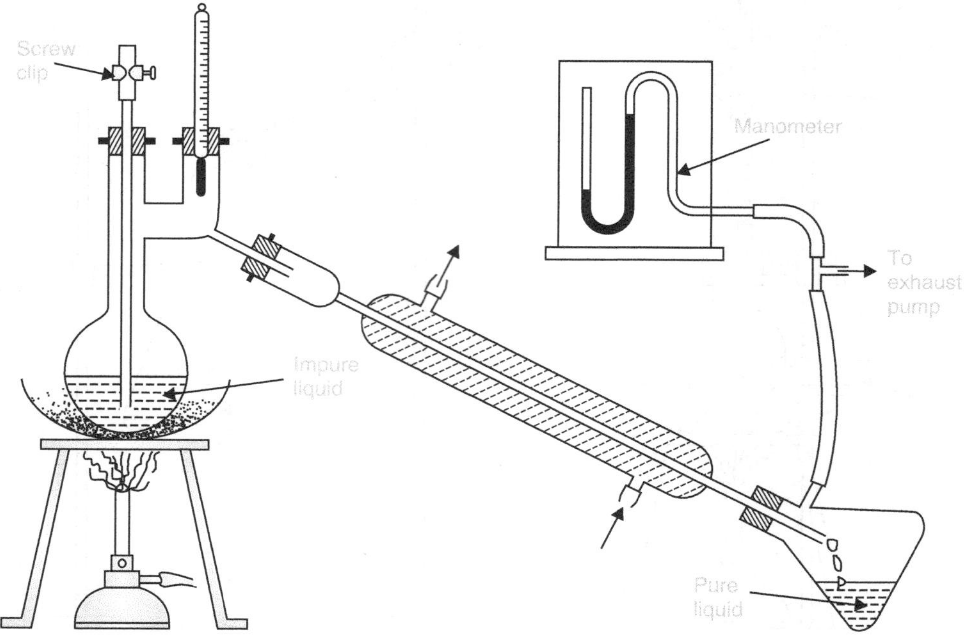

Fig. 39.10. Distillation under reduced pressure.

industry, the concentration of cane sugar juice and in the *soap industry*, the separation of glycerol from the spent-lye are carried out by this technique.

7. STEAM DISTILLATION

The process of steam distillation is employed in the purification of a substance from non-volatile impurities provided the substance **itself is volatile in steam** and **insoluble in water.** This method is based on the facts that:

(*i*) *A liquid boils at a temperature when its vapour pressure becomes equal to the atmospheric pressure.*

(*ii*) *The vapour pressure of a mixture of two immiscible liquids is equal to the sum of the vapour pressures of the individual liquids.*

In the actual process, steam is continuously passed through the impure organic liquid. Steam heats the liquid but itself gets practically condensed to water. After some time, the mixture of the liquid and water begins to boil because the vapour pressure of the mixture becomes equal to the atmospheric pressure. Obviously, this happens at a temperature which is lower than the boiling point of the substance or that of water. For instance, a mixture of aniline (b.p. 453 K, with decomposition), and water (b.p. 373 K), under normal atmospheric pressure, boils at 371 K. At this temperature the vapour pressure of water is 717 mm Hg and that of aniline is 43 mm Hg and, therefore, the total vapour pressure is equal to 760 mm Hg. Thus in steam distillation the liquid gets distilled at a temperature lower than its boiling point and any chances of decomposition is avoided.

The proportion of water and liquid in the mixture that distils over is given by the relation

$$\frac{w_1}{w_2} = \frac{p_1 \times 18}{p_2 \times M}$$

where w_1 and w_2 stand for the masses of water and the organic liquid that distils over; p_1 and p_2 represent the vapour pressures of water and the liquid at the distillation temperature and M is molecular mass of the liquid (molecular mass of water being 18).

The set up for a steam distillation is shown in Fig. 39.11. The impure liquid (or the solid, along with some water) is taken in a round-bottomed flask which is kept in a slightly slanting position. It is attached to a steam generator and steam is passed into the flask. The flask may be gently heated on a sand bath to prevent undue condensation of steam in it. The distillate consisting of water and the organic substance is collected in the receiver. The pure substance is separated from water with the help of a separatory funnel or by extraction with a suitable solvent.

8. CHROMATOGRAPHY

It is the most useful and modern technique of separation and purification of organic compounds. The method was first developed by ***Tswett*** in 1903 for the separation of coloured substances into individual components. Since then the method has undergone tremendous modifications. Nowadays, various types of chromatographic methods are in use to separate any mixture into its constituents irrespective of whether it is coloured or colourless.

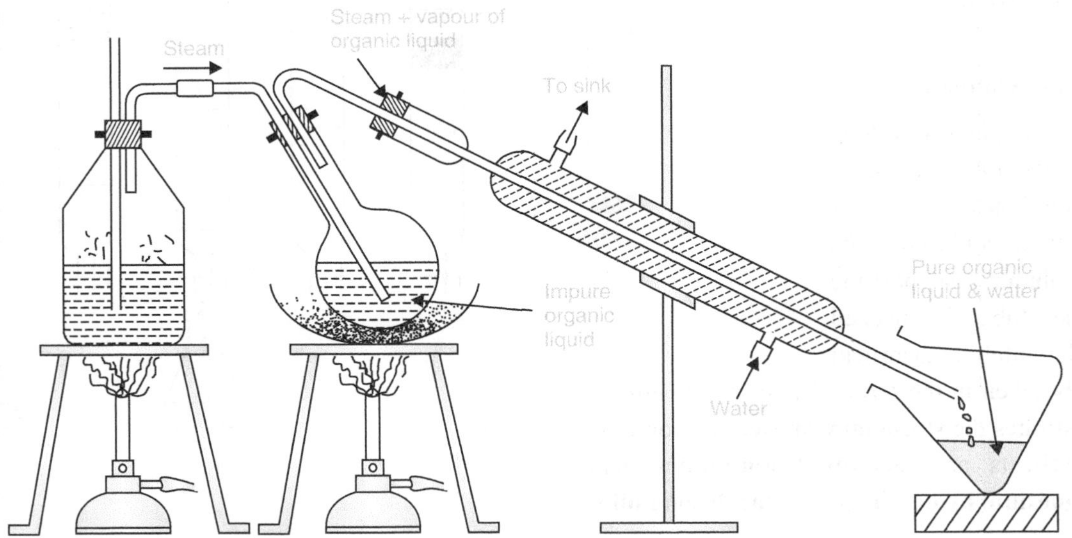

Fig. 39.11. Steam distillation.

Chromatography in general, may be described as *the technique of separating the constituents of a mixture by the differential movement of individual components through a stationary phase under the influence of a mobile phase.*

Different types of chromatographic techniques are employed depending upon the nature of the stationary and mobile phases.

Two of the important categories of chromatography are:

(A) *Adsorption chromatography*

(B) *Partition chromatography*

A. ADSORPTION CHROMATOGRAPHY

This category of chromatography is based on the principle of differential adsorption of various components of a mixture on a suitable adsorbent. Some components are more strongly absorbed than others. When the mobile phase is allowed to move over the stationary phase (adsorbent), the components of the mixture move by varying distances over the stationary phase. Two main types of chromatographic techniques based on the principle of differential adsorption are:

- *Column chromatography*
- *Thin layer chromatography.*

Column Chromatography

This is the simplest chromatographic method. In this technique, the mixture of substances or the substance to be purified is passed through the column of a suitable adsorbent. The constituents get adsorbed to different extents. In general, the following steps are involved (Fig. 39.12).

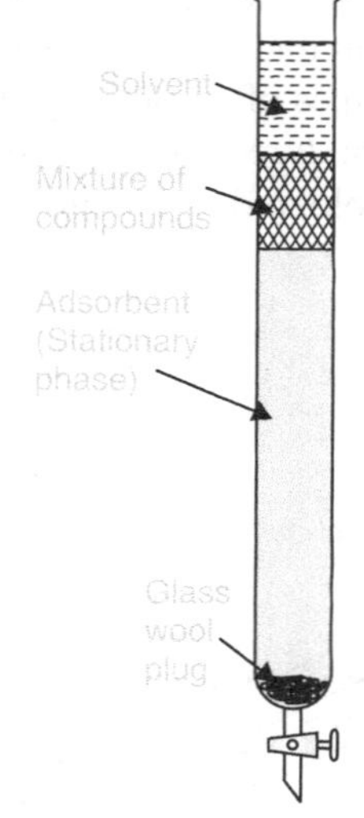

Fig. 39.12. Column chromatography.

(*i*) Preparation of Column

A column is prepared in a long burette like glass tube having a stopcock near the bottom. A plug of cotton or glass wool is placed at the bottom of the column to support the adsorbent. The tube is packed uniformly with suitable adsorbent which is usually taken in the *form of a slurry in petroleum ether.* This constitutes the **stationary phase.** The commonly employed adsorbents are: **activated aluminium oxide** (alumina), **magnesium oxide, silica gel, starch** and **Fuller's earth.** Nearly one-fourth of the tube is left empty. A loose plug of cotton or glass wool is then placed at the top of the adsorbent column.

(*ii*) Adsorption Process

The substance to be purified is added (Liquids are added as liquids but solids are added in the form of a *solution in some suitable solvent*) at the top of the column and allowed to pass slowly through it. As it passes through the column, the different components of the mixture get adsorbed. Different constituents of the mixture get adsorbed at different extents and forms bands at different parts of the column.

(*iii*) Elution

It is a process of extraction of the adsorbed components from the adsorbent with the help of suitable solvent called **eluent.** Eluent is added as soon as the last portion of the mixture to be separated enters the column. This acts as a **moving phase.** The eluents usually employed are *petroleum ether, carbon tetrachloride, benzene, alcohol,* etc., and their selection depends upon the relative solubilities of the components of the mixture in them. More than one eluent may be used in certain cases. The eluents dissolve out the different components selectively. The weakly adsorbed component is eluted more rapidly than a strongly adsorbed component. Such a progressive separation of a mixture is shown in Fig. 39.13. Different components of the mixture are collected in the form of different fractions in separate conical flasks. The eluent from each fraction is then distilled off to get the various components in pure form.

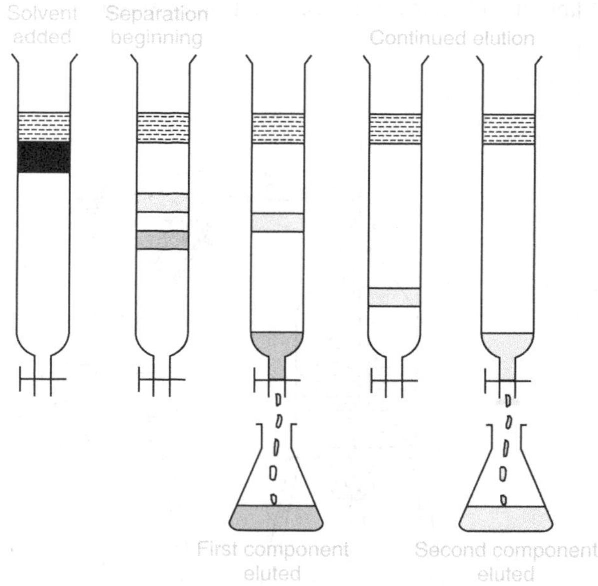

Fig. 39.13. Process of chromatography.

This technique is now widely used in research laboratories for the purification of different substances and for the separation of mixtures. For example, a mixture of an alcohol and a liquid hydrocarbon may be separated by using alumina as adsorbent and petroleum ether as eluent.

Thin Layer Chromatography (TLC)

Thin layer chromatography is another type of adsorption chromatography, which involves separation of the substances of a mixture over a thin layer of an adsorbent. The *various steps* involved are:

(*i*) *Preparation of TLC plate.* Two pencil lines are drawn across the width of the plate about 1 cm each from the top and bottom end. The lower pencil mark is called the *starting line* and the upper line is called the *finish line* or *solvent line.* A thin layer (about 0.2 mm thick) of an adsorbent (silica gel or alumina) is spread over a glass plate of suitable size. The plate is known as *thin layer chromatography plate* (TLC plate).

(*ii*) *Separation process.* The solution of the mixture to be separated is applied as a small spot about 2 cm from one end of the TLC plate. The glass plate is then placed in a closed jar containing the solvent (Fig. 39.14). As the solvent in the jar moves up, the components of the mixture move up along the plate to different distances depending on their degree of adsorption and separation takes place. The relative adsorption of each component of the mixture is expressed in terms of its retention factor *i.e.*, $\mathbf{R_f}$ **value** (Fig. 39.15).

$$R_f = \frac{\textit{Distance moved by the substance from base line}}{\textit{Distance moved by the solvent from base line}}$$

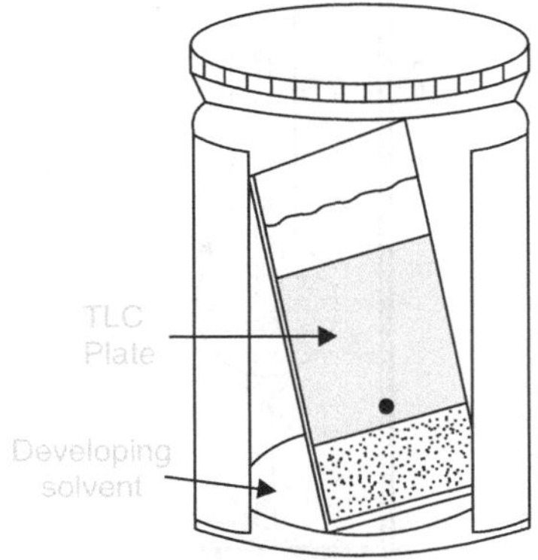

Fig. 39.14. Development chamber with TLC plate undergoing development.

For example, R_f *for component* $A = \frac{x_A}{y}$; R_f *for compound* $B = \frac{x_B}{y}$

Fig. 39.15. Thin layer chromatogram.

The spots of coloured compounds are visible on TLC plate due to their original colour. The spots of colourless compounds which are invisible to the eye can be detected by any of the following techniques.

(*a*) ***UV light.*** This method is used for certain organic compounds which produce fluorescence in *uv* light. The spots of such substance can be detected by placing it under *uv* lamp.

(*b*) ***Use of iodine vapours.*** In this method, the developed TLC plate is placed in a covered jar containing a few crystals of iodine. Spots of the compounds which adsorb iodine will become brown.

(*c*) ***Chemical methods.*** This method involves spraying of suitable chemical reagents on the TLC plate.

For example, amino acids may be detected by spraying the plate with ninhydrin solution. Similarly, aldehyde/ketones can be detected by spraying the plate with 2, 4-dinitrophenyl-hydrazine.

B. PARTITION CHROMATOGRAPHY

Partition chromatography is *liquid-liquid chromatography* unlike TLC and column chromatography which represents *solid-liquid* chromatography. In partition chromatography, the *stationary phase* as well as *mobile phase* are both ***liquids.***

Principle. Partition chromatography is based on *continuous differential partitioning (distribution) of components of a mixture between stationary and mobile phases*. One of the common example of partition chromatography is **paper chromatography.** In paper chromatography, a special quality paper known as chromatography *paper* is used. Paper is mainly cellulose, but cellulose does not constitute stationary phase. The ***stationary phase*** in paper chromatography is water, trapped or chemically bound to cellulose. The ***mobile phase*** is liquid mixture of two or more substances with water as one of the components.

Process. A suitable strip of chromatographic paper is taken and starting line is drawn across the width of the paper about 1 or 2 cm from the bottom. A *spot* of the mixture of components to be separated is applied on the starting line with the help of a fine capillary or a syringe. The spotted paper is then suspended in a suitable solvent (or a mixture of solvents) (Fig. 39.16). This solvent acts as the mobile phase. The solvent rises up the paper by capillary action and flows over the spot.

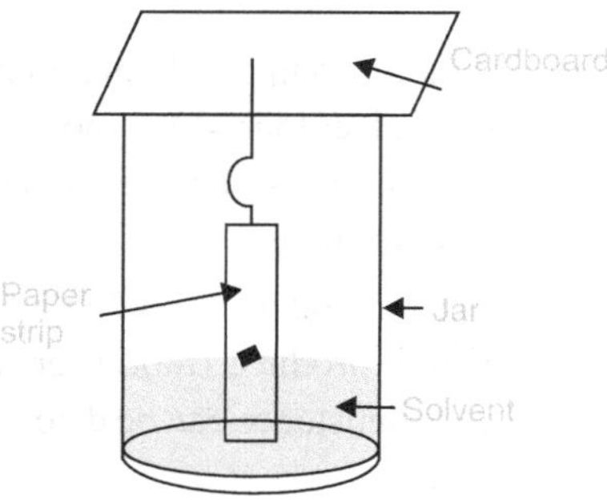

Fig. 39.16. Paper chromatography

The different component of the mixture travel through different distances depending upon their solubility in (or partioning between) the stationary and mobile phases. When the solvent reaches the top end the paper is taken out and allowed to dry. The paper strip so developed is known as a chromatogram. The spots of the separated coloured compounds are visible at different heights from the position of initial spot on the **chromatogram.** The spots of the separated colourless compounds may be observed either under ultraviolet light or by the use of an appropriate spray reagent as discussed under thin layer chromatography.

CRITERIA FOR PURITY OF ORGANIC COMPOUNDS

After purification, the test of purity of an organic compound is the next important step. A pure organic compound possesses characteristic physical properties such as refractive index, specific gravity, boiling point, crystalline structure and melting point. If a purified sample shows the same properties that the pure substance is known to possess, it may be considered as pure and no further purification is required. However, for practical purposes in the laboratory only melting point (in the case of solids) and boiling point (for liquids) are used as criteria for purity. *A pure organic solid has a definite and sharp (sudden, rapid and complete) melting point, while an impure substance has a lower and indefinite melting point.* The melting point of a solid may be defined as the temperature at which the solid and liquid states of the compound are in equilibrium with each other at an external pressure of 1 atmosphere.

Determination of melting point of a solid: A capillary tube, 5 to 6 cm long and 1 mm in diameter, is closed at one end and the dry powdered organic solid is inserted in it. The tube is tapped to accumulate the crystals at the bottom of the capillary tube. Filling is continued until the length of the packed solid is 3 to 4 mm.

The apparatus used for the determination of melting point is shown in Figs. 39.17(*a*), (*b*) and (*c*). The bulb of the thermometer is dipped in sulphuric acid bath and taken out.

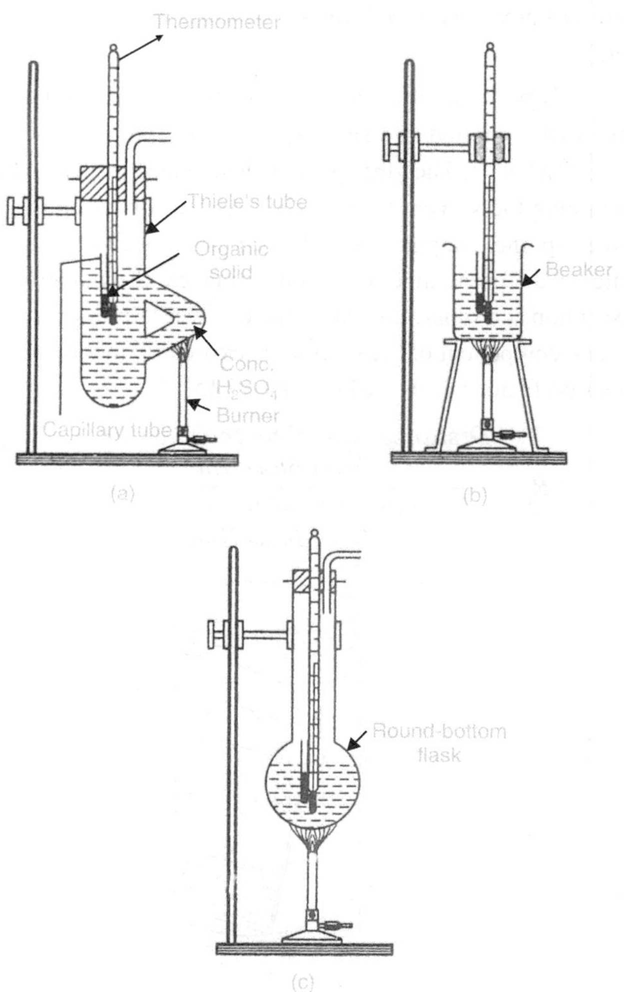

Fig. 39.17. Determination of melting point

The capillary tube is then placed along with the thermometer in such a manner so that the sealed end of the capillary is near the bulb. The capillary tube sticks to the thermometer due to surface tension and viscosity of the acid. The thermometer is now lowered in the bath and held in a position so as to keep half of the capillary tube out the bath. The bath is slowly heated to keep the temperature uniform throughout the bath. When the solid in the capillary tube just shows signs of melting, the heating is stopped. The temperature at which the substance just melts and becomes transparent is recorded. This is the melting point of the organic solid.

Thiele's tube or round bottom flask (50 mL) or a boiling tube can be used for the determination of melting point of an organic solid.

If the determined melting point is the same as that of the pure compound, the sample in question is also said to be a pure substance.

Mixed melting point: The melting point of two thoroughly mixed substances is called mixed melting point. This can also be used for ascertaining the purity of a compound.

The substance, whose purity is to be tested, is mixed with a pure sample of the same compound. The melting point of the mixture is determined. If the melting point of the mixture is sharp and comes out to be the same as that of pure compound, it means that the compound under test is pure. On the other hand, if the melting point of the mixture is less than the melting point of the pure compound, the compound in question is not pure.

Boiling point: The boiling point of a liquid is defined as the temperature at which the vapour pressure of the liquid is equal to the external pressure. Boiling points are also normally quoted at standard atmospheric pressure. A pure organic liquid boils at a fixed temperature which is characteristic of that substance.

If enough liquid is available, it can be distilled in a distillation apparatus and the constant temperature recorded by the thermometer is the boiling point (Fig. 39.18).

Boiling point is not very reliable as a test for purity as compared to the melting point for solids. There are many liquids which are miscible with other liquids and mixtures have fixed boiling points (azeotrope). Thus, other physical properties are being used these days for deciding the purity of liquids.

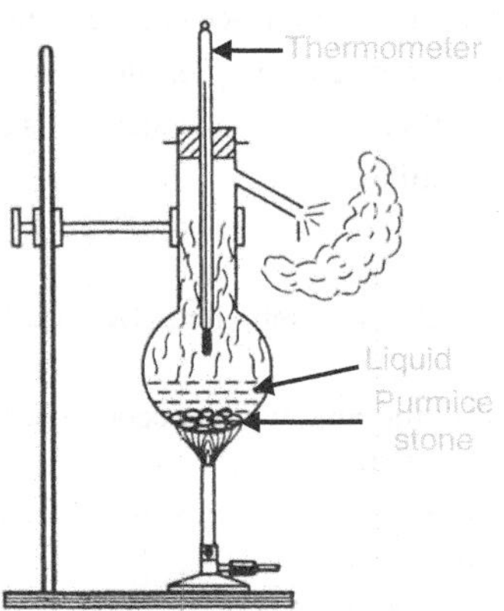

Fig. 39.18 Determination of boiling point

When the amount of the liquid is small, Siwoloff's method (capillary tube method) is used.

EVALUATE YOURSELF

I. Objective Type Questions

Select the most appropriate choice from the options given as (a), (b), (c) and (d) after each question:

1. The property which serves as a criterion of purity of an organic compound is
   (*a*) solubility in water (*b*) melting point
   (*c*) density (*d*) crystalline nature
2. Recrystallisation is used to purify
   (*a*) gases (*b*) liquids
   (*c*) solids (*d*) solids and liquid mixtures
3. In steam distillation of toluene, the pressure of toluene in vapour is
   (*a*) equal to pressure of barometer
   (*b*) less than pressure of barometer
   (*c*) equal to vapour pressure of toluene in simple distillation
   (*d*) more than the vapour pressure of toluene in simple distillation.
4. Fractional distillation is used to separate substances of different
   (*a*) melting points (*b*) boiling points
   (*c*) molar masses (*d*) melting and boiling points.
5. Chromatography is based on the principle of
   (*a*) solubility (*b*) differential extraction
   (*c*) preferential adsorption (*d*) absorption
6. Benzoic acid containing impurities of naphthalene can be purified by using
   (*a*) ethanol (*b*) hot water
   (*c*) air condenser (*d*) litmus paper.
7. A mixture of iodine can be separated from table salt by the method of
   (*a*) sublimation (*b*) fractional distillation
   (*c*) fractional crystallisation (*d*) steam distillation.

8. The concentration of cane juice can be done by
(*a*) distillation under reduced pressure
(*b*) the process of osmosis (*c*) steam distillation
(*d*) fractional distillation

II. Fill in the Blanks

9. Complete the following sentences by supplying appropriate words:
(*i*) An organic compound which decomposes below its boiling point can be purified by ....
(*ii*) Liquids whose boiling points are close to each other are separated by ....
(*iii*) Camphor can be purified by ....
(*iv*) An organic compound containing non-volatile impurities can be purified by steam distillation if it is .... and .... in water.
(*v*) A commonly used adsorbent in column chromatography is ....
(*vi*) In a chromatographic column, the weakly adsorbed component will elute .... than strongly adsorbed component.
(*vii*) The purity of the solid organic compound is determined in the laboratory by its ....
(*viii*) The purity of the liquid organic compound is determined in the laboratory by its ....
(*ix*) Impure organic compound's melting point is ....
(*x*) The organic compound insoluble in water having lower boiling point than water is purified by .... method.
(*xi*) In column chromatography, the process of isolating the separated zones by washing is known as ....
(*xii*) A mixture of common salt and naphthalene can be separated by ....
(*xiii*) Glycerol decomposes at its boiling point. It can be purified by ....
(*xiv*) Acetone (b.pt. 56°C) and methyl alcohol (b. p.t. 65.5°C) can be separated by ....
(*xv*) Raw juice is concentrated in factories by ....
(*xvi*) Two substances having different solubility in a particular solvent, can be separated by ....
(*xvii*) The boiling point of a liquid .... when pressure above the liquid is reduced.

III. Discussion Questions

10. Describe the method to separate two substances which have different solubilities in a solvent X.
11. What is the effect of pressure on the b.p. of a liquid? How is this fact utilized in the purification of compounds?
12. How will you separate two miscible liquids which differ in their b.p. by 10°? Give details of the method.
13. Under what conditions is a compound purified by steam distillation? Explain.
14. When do we use fractional crystallisation? Describe the method briefly.
15. Which technique can be used for the purification of iodine that contain traces of NaCl?
16. A liquid has a contamination of non-volatile impurities. Suggest method for its purification.
17. A substance has boiling point 350 K, but it starts decomposing near this temperature. Which type of distillation process is suitable for its purification?
18. What is the basic principle of chromatography? Explain briefly.
19. A liquid (10 mL) has three components A, B and C. Which technique is most suitable to separate A, B, C from such a small amount of mixture?
20. When is steam distillation most suitable for purification?
21. Suggest a method to separate a mixture of
(*i*) Camphor and NaCl (*ii*) Camphor and benzoic acid.
22. How can the mixture of kerosene oil and water be separated?
23. Discuss a brief chromatographic method for the purification of compounds.
24. How will you purify benzoic acid containing sodium chloride as impurity?
25. What is fractional distillation?
26. What is the principle of column chromatography?
27. Explain the principle of paper chromatography.
28. What is the difference between distillation, distillation under reduced pressure and steam distillation?

Answers

I. Objective Type Questions

1. (*b*) **2.** (*c*) **3.** (*b*)
4. (*b*) **5.** (*c*) **6.** (*b*)
7. (*a*) **8.** (*a*).

II. Fill in the Blanks

9. (*i*) distillation under reduced pressure
(*ii*) fractional distillation
(*iii*) sublimation (*iv*) volatile, insoluble
(*v*) alumina (*vi*) first
(*vii*) melting point (*viii*) boiling point
(*ix*) higher (*x*) steam distillation
(*xi*) elution (*xii*) sublimation
(*xiii*) steam distillation (*xiv*) fractional distillations
(*xv*) distillation under reduced pressure
(*xvi*) differential crystallisation
(*xvii*) lowers.

III. Discussion Questions

15. Sublimation. **16.** Distillation.
17. Distillation under reduced pressure.
19. Chromatography.
21.(*i*) Sublimation. Camphor sublimes leaving behind NaCl.
(*ii*) Treat the mixture with $NaHCO_3$. Sodium benzoate will form which is soluble. The camphor will be left undissolved.
22. Using separatory funnel.

SECTION-6

UNIT-40

ALKANES

Specific Objectives

After studying this unit, the student should be able to:

- describe the sources and characteristics of alkanes
- describe the nomenclature and isomerism of alkanes
- describe the preparation of and chemical reactions of alkanes
- identify the uses of alkanes and their contribution to the greenhouse effect.

40.1 INTRODUCTION

Alkanes are saturated hydrocarbons. Because of their low reactivity, they are also called ***paraffins*** (a Latin, word meaning *little affinity* or *reactivity*). The open chain (aliphatic) alkanes have the general formula C_nH_{2n+2} while cycloalkanes have the general formula C_nH_{2n}. All the carbon atoms in alkanes are sp^3-hybridized. Alkanes have tetrahedral structure around each carbon.

40.2 SOURCES OF ALKANES

Major sources of alkanes are:

1. **Petroleum.** Petroleum is a complex mixture of alkanes and other hydrocarbons. It usually contains alkanes having up to forty carbon atoms in their molecule. The natural wax found near petroleum wells is a mixture of higher solid hydrocarbons.
2. **Natural gas.** Methane is the major constituent of natural gas and occurs along with petroleum in the earth's sedimentary traps.
3. **Destructive distillation of wood and coal.** Methane is produced as one of the by-products in the manufacture of wood charcoal by destructive distillation of wood and coke by destructive distillation of coal.
4. **Anaerobic decomposition of organic matter** in the presence of bacteria liberates methane. For example, bio-gas produced by anaerobic decomposition of animal dung, excreta and plant wastes is rich in methane gas. Methane is produced in marshy lands by the bacterial decomposition of plant and animal matter under water and is observed bubbling through the surface. Since methane gas is produced in marshy lands, it is also known as **marsh gas.**

Decomposed wastes from rice fields and termites are other sources of methane.

5. Methane gas is also present in **coal mines** where it is known as **damp fire**. It is formed as a by-product during the formation of coal.

HOMOLOGOUS SERIES

Within a particular family, the compounds are further grouped in a number of series on the basis of structure. These series are known as homologous series.

Homologous series may be defined as a series of similarly constituted compounds in which the members possess similar chemical characteristics and the two consecutive members differ in their molecular formulae by —CH_2.

Characteristics of Homologous Series

The common characteristics of homologous series are as follows:

1. All the members of a series can be represented by the same general formula. For example, the general formula of alkanes is C_nH_{2n+2}.
2. Any two consecutive members differ in their formula by a common difference of —CH_2. Any two consecutive members of a homologous series differ in molecular mass by 14.
3. Different members in a series have a common functional group. For example, the members of the alkanol family have —OH group as their functional group.
4. The members in any particular family have almost identical chemical properties. Their physical properties such as melting point, boiling point, density, solubility, etc., show a regular gradation with the increase in the molecular mass.
5. The members of a particular series can be prepared almost by the identical methods known as *the general methods of preparation*.

The different members of a series are known as **homologues.** A few members of the alkane family are given below:

General Formula: C_nC_{2n+2}

Value of n	*Molecular Formula*	*Structural Formula*	*Name*
1.	CH_4	H | H—C—H | H	Methane
2.	C_2H_6	H H | | H—C—C—H | | H H	Ethane
3.	C_3H_8	H H H | | | H—C—C—C—H | | | H H H	Propane
4.	C_4H_{10}	H H H H | | | | H—C—C—C—C—H | | | | H H H H	Butane

40.3 NOMENCLATURE AND ISOMERISM

The first three members of this family have same common as well as IUPAC names. The higher members can exist as a number of isomers. The various isomers of butane and pentane are distinguished by the use of prefixes *n*- (for normal), *iso*- and *neo*-. Prefix *n*- is used for straight chain alkanes, *iso*- is used for the molecules containing $(CH_3)_2$CH— group and *neo*- is used for molecules containing $(CH_3)_3C$— group at one end of the chain. Some examples are given below:

Each of the first three members of alkanes; methane, ethane and propane has only one structural formula as shown below:

```
   H          H  H           H  H  H
   |          |  |           |  |  |
H—C—H     H—C—C—H       H—C—C—C—H
   |          |  |           |  |  |
   H          H  H           H  H  H
Methane      Ethane          Propane
```

Thus, *no chain isomerism is possible up to propane.*

The fourth member, that is, C_4H_{10}, can have **two** structural formulae as shown below:

```
  H  H  H  H            H     H     H
  |  |  |  |            |     |     |
H—C—C—C—C—H          H—C─────C─────C—H
  |  |  |  |            |     |     |
  H  H  H  H            H     |     H
                            H—C—H
                              |
                              H
Butane (n-Butane)       2-Methylpropane
(b.p. 273 K)            (iso-Butane)
                        (b.p. 261 K)
```

Replacement of hydrogen from first carbon atom of propane by —CH_3 gives *n*-butane while replacement of hydrogen from the second carbon atom by —CH_3 gives *iso*-butane.

Thus, *butane (C_4H_{10}) has two chain isomers n-butane and iso-butane.*

Similarly, the next hydrocarbon, C_5H_{12}, has three different structural formulae which have been represented below along with their names.

```
   H  H  H  H  H               H  H  H  H
   |  |  |  |  |               |  |  |  |
H— C— C— C— C— C —H         H— C— C— C— C —H
   |  |  |  |  |               |  |  |  |
   H  H  H  H  H               H  H  |  H
                                      |
                                   H— C —H
                                      |
                                      H
```

Pentane
(*n*-Pentane)
(b.p. 309 K)

2-Methylbutane
(*iso*-Pentane)
(b.p. 301 K)

```
          H
          |
       H— C —H
   H      |     H
   |      |     |
H— C ——— C ——— C —H
   |      |     |
   H      |     H
       H— C —H
          |
          H
```

2, 2-Dimethylpropane
(*neo*-Pentane)
(b.p. 282.5 K)

Thus, *pentane (C_5H_{12}) has three chain isomers; n-pentane, iso-pentane and neo-pentane.* These chain isomers differ in their physical properties such as boiling point.

From the above discussion, it follows that the number of chain isomers increases with an increase in the number of carbon atoms. For example, C_6H_{14} has **five,** C_7H_{16} has **nine,** C_8H_{18} has **eighteen** and $C_{10}H_{22}$ has 75 chain isomers. An alkane with sixty carbon atoms would have millions of chain isomers.

Some alkanes and their common as well as IUPAC names are given in the following table:

Molecular Formula	*Common Name*	*IUPAC Name*
CH_4	Methane	Methane
$CH_3—CH_3$	Ethane	Ethane
$CH_3—CH_2—CH_3$	Propane	Propane
$CH_3—CH_2—CH_2—CH_3$	*n*-Butane	Butane
CH_3 \ CH—CH_3 CH_3 /	*iso*-Butane	2-Methylpropane
$CH_3CH_2CH_2CH_2CH_3$	*n*-Pentane	Pentane
CH_3 \ $CHCH_2CH_3$ CH_3 /	*iso*-Pentane	2-Methylbutane
CH_3 | CH_3—C—CH_3 | CH_3	*neo*-Pentane	2, 2-Dimethylpropane

40.4 STRUCTURE OF ALKANES

The molecules of alkanes contain only single covalent bonds between carbon-carbon or carbon-hydrogen atoms. The formation of only single bonds requires that each carbon atom in an alkane be sp^3-hybridised. These hybrid orbitals of each carbon atom are directed towards the corners of a regular tetrahedron, the carbon atom being at the centre of the tetrahedron.

For example, let us study the **structure of methane** molecule.

STRUCTURE OF METHANE

The carbon atom in methane molecule is sp^3 hybridised. The four sp^3 hybrid orbitals of carbon overlap with 1*s* orbitals of four hydrogen atoms forming four C—H sigma bonds. Due to sp^3 hybridisation methane is a tetrahedral molecule. In methane molecule carbon is at the centre of the tetrahedron, whereas hydrogen atoms are at the corners. H—C—H bond angles are 109.5° and C—H bond length is 109 pm (Fig. 40.1).

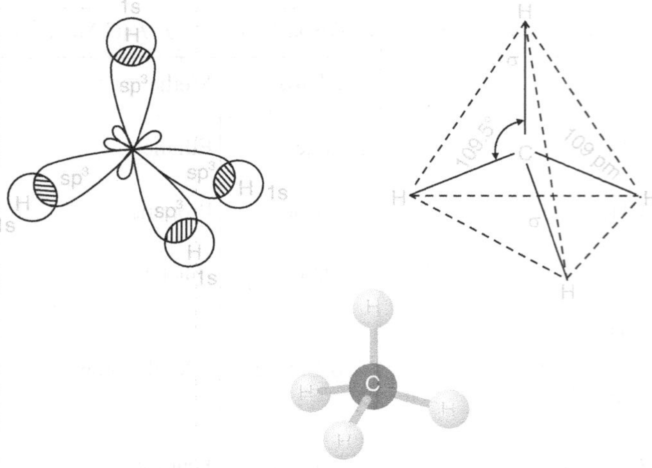

Fig. 40.1. Orbital picture of methane.

The carbon atoms in the alkane molecule with more than two carbon atoms do not lie along a straight line but form a zig-zag pattern. This is due to the sp^3-hybrid state of the carbon atoms. Due to the sp^3-hybrid state of carbon atoms the C—C—C bond angle in higher alkanes is expected to be around 109°. It is clear from the model of propane shown in Fig. 40.2.

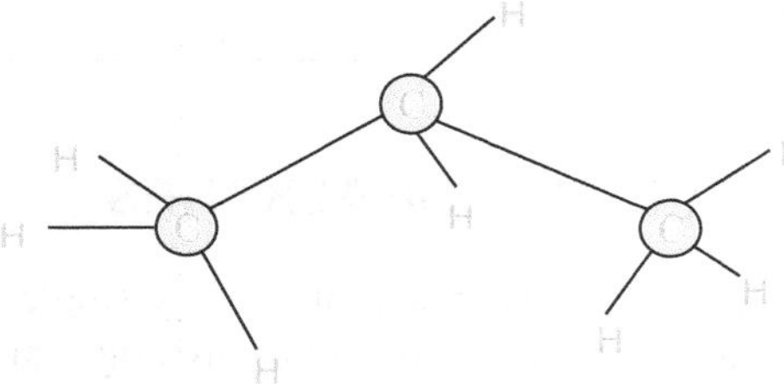

Fig. 40.2. Molecular model of propane.

All the carbon atoms in cycloalkanes are also sp^3-hybridised because each carbon atom is directly attached to four other atoms. Thus, alkanes have tetrahedral structure around each carbon. The average C—C and C—H bond distances are 154 pm and 110 pm respectively.

Structure of Some Cycloalkanes

Cyclobutane	Cyclopentane	Cyclohexane
C_4H_8	C_5H_{10}	C_6H_{12}

CLASSIFICATION OF CARBON ATOMS

The carbon atom in a molecule may be classified as primary (1°), secondary (2°), tertiary (3°) or quaternary (4°) depending upon whether it is directly linked to one, two, three or four carbon atoms respectively. Terminal carbon atoms are always primary. Primary, secondary, tertiary or quaternary carbon atoms in a molecule are denoted by the letters, p, s, t and q respectively. This is illustrated in the following formula:

$$\underset{p}{CH_3}-\underset{\underset{\underset{p}{CH_3}}{|}}{\overset{t}{CH}}-\underset{\underset{\underset{p}{CH_3}}{|}}{\overset{\overset{\overset{p}{CH_3}}{|}}{\underset{q}{C}}}-\overset{s}{CH_2}-\underset{\underset{\underset{s\quad p}{CH_2-CH_3}}{|}}{\overset{t}{CH}}-\overset{p}{CH_3}$$

There are six primary carbons, two secondary carbons, two tertiary carbons and one quaternary carbon atom in the above formula. It may be noted that hydrogen atoms in a molecule may also be termed as primary, secondary or tertiary hydrogens depending upon whether they are linked to primary, secondary, or tertiary carbon atom respectively.

SOLVED EXAMPLES

Example 40.1. *Write structures for each of the following compounds:*

(*i*) *3-Ethyl-3-methylpentane*

(*ii*) *3-Ethyl-4, 4-dimethylheptane.*

Solution. (*i*) $H_3C-CH_2-\underset{\underset{CH_2CH_3}{|}}{\overset{\overset{CH_3}{|}}{C}}-CH_2-CH_3$

(*ii*) $H_3C-CH_2-\underset{\underset{CH_2CH_3}{|}}{CH}-\underset{\underset{CH_3}{|}}{\overset{\overset{CH_3}{|}}{C}}-CH_2-CH_2-CH_3.$

Example 40.2. *Write IUPAC names of the following compounds:*

(*i*) $(C_2H_5)_4\,C$ (*ii*) $(CH_3)_3\,C\,CH_2\,CH(CH_3)_2$

(*iii*) $CH_3\,CH\,(C_2H_5)_2$.

Solution. (*i*) $\overset{1}{CH_3}\overset{2}{CH_2}-\underset{\underset{CH_2CH_3}{|}}{\overset{\overset{CH_2CH_3}{|}}{\overset{3}{C}}}-\overset{4}{CH_2}\overset{5}{CH_3}$

3, 3-Diethylpentane

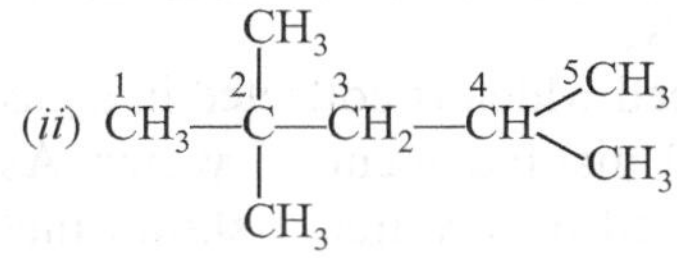

(*ii*) $\overset{1}{CH_3}-\overset{2}{C}(CH_3)_2-\overset{3}{CH_2}-\overset{4}{CH}<\overset{5}{CH_3}, CH_3$

2, 2, 4-Trimethylpentane

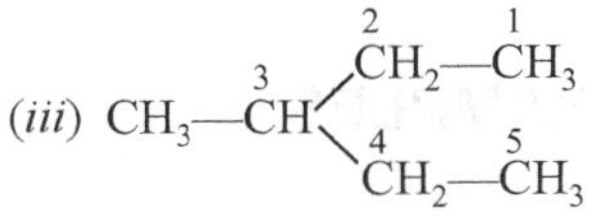

(*iii*) $CH_3-\overset{3}{CH}<(\overset{2}{CH_2}-\overset{1}{CH_3})(\overset{4}{CH_2}-\overset{5}{CH_3})$

3-Methylpentane

Example 40.3. *Write structures of different chain isomers of alkanes corresponding to the molecular formula C_6H_{14}. Also write their IUPAC names.*

Solution.

(*i*) $CH_3-CH_2-CH_2-CH_2-CH_2-CH_3$

n-Hexane

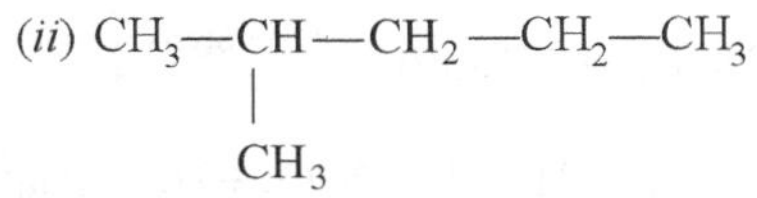

(*ii*) $CH_3-CH(CH_3)-CH_2-CH_2-CH_3$

2-Methylpentane

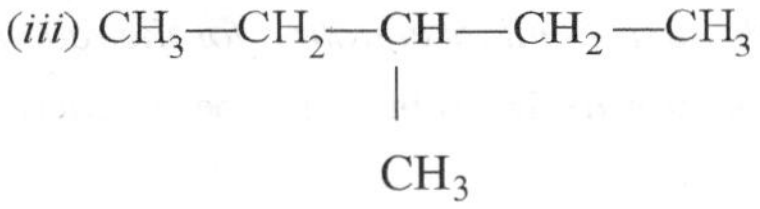

(*iii*) $CH_3-CH_2-CH(CH_3)-CH_2-CH_3$

3-Methylpentane

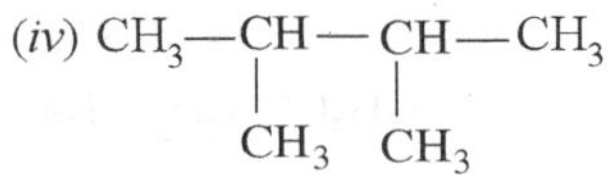

(*iv*) $CH_3-CH(CH_3)-CH(CH_3)-CH_3$

2, 3-Dimethylbutane

(*v*) $CH_3-C(CH_3)_2-CH_2-CH_3$

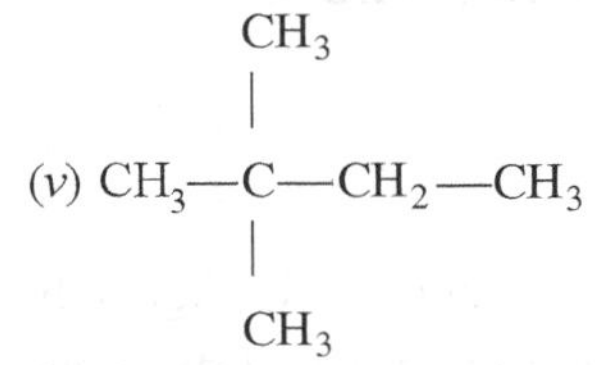

2, 2-Dimethylbutane.

40.5 PREPARATION OF ALKANES

Some of the general methods of preparation of alkanes are:

FROM ALKYL HALIDES (By Wurtz Reaction)

Alkyl halides are halogen derivatives of alkanes with general formula R—X. They can be used to prepare alkanes via the **Wurtz reaction:**

It involves the chemical reaction between alkyl halides (usually *bromides* and *iodides*) and metallic sodium in the presence of dry ether. The product is the symmetrical alkane containing twice the number of carbon atoms present in the alkyl halide.

$$2\,R-\mathbf{X} + 2Na \xrightarrow{\text{Dry ether}} \mathbf{R-R} + 2NaX$$

For example,

$$\underset{\text{Methyl iodide}}{2CH_3-I} + 2Na \xrightarrow{\text{Dry ether}} \underset{\text{Ethane}}{CH_3-CH_3} + 2NaI$$

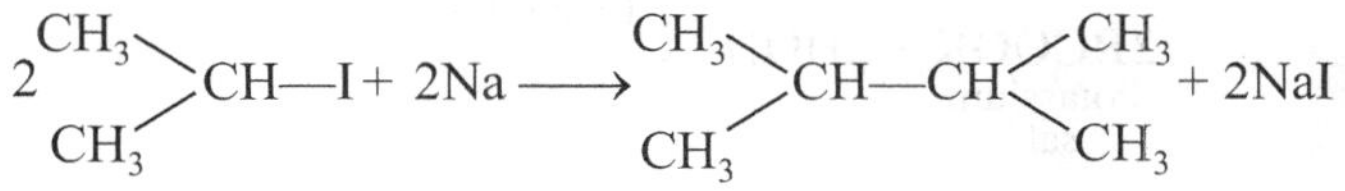

$$2\,(CH_3)_2CH-I + 2Na \longrightarrow (CH_3)_2CH-CH(CH_3)_2 + 2NaI$$

Limitations of Wurtz Reaction

- This method is not suitable to prepare alkanes containing odd number of carbon atoms.
- **Methane** *cannot be prepared by this method.*

FROM UNSATURATED HYDROCARBONS

Unsaturated hydrocarbons can be converted into alkanes by catalytic **hydrogenation.** The catalyst used may be nickel, platinum or palladium. These metals absorb dihydrogen gas on their surface and weaken the hydrogen-hydrogen bond. Platinum and palladium catalyse the reaction at room temperature while nickel requires higher temperature and pressure. The reaction is called **Sebatier and Senderen's reaction.**

$$\underset{\text{Ethene}}{CH_2=CH_2} + H_2 \xrightarrow[523-573\text{ K}]{\text{Ni}} \underset{\text{Ethane}}{CH_3-CH_3}$$

$$\underset{\text{Ethyne}}{CH\equiv CH} + 2H_2 \xrightarrow[523-573\text{ K}]{\text{Ni}} \underset{\text{Ethane}}{CH_3-CH_3}$$

FROM CARBOXYLIC ACIDS

(*a*) **Soda-lime Decarboxylation.** It involves the treatment of sodium salt of carboxylic acids with soda lime (NaOH and CaO in the ratio 3 : 1).

$$\underset{\text{Sodium salt of carboxylic acid}}{RCO\overset{\ominus}{O}\overset{\oplus}{Na}} + NaOH \xrightarrow[\text{Heat}]{\text{CaO}} \underset{\text{Alkane}}{\mathbf{RH}} + Na_2CO_3$$

$$\underset{\text{Sodium ethanoate}}{\mathbf{CH_3}CO\overset{\ominus}{O}\overset{\oplus}{Na}} + NaOH \xrightarrow[\text{Heat}]{\text{CaO}} \underset{\text{Methane}}{\mathbf{CH_4}} + Na_2CO_3$$

$$\underset{\text{Sodium propanoate}}{\mathbf{C_2H_5}CO\overset{\ominus}{O}\overset{\oplus}{Na}} + NaOH \xrightarrow[\text{Heat}]{\text{CaO}} \underset{\text{Ethane}}{\mathbf{C_2H_6}} + Na_2CO_3$$

The alkane formed has one carbon less than the sodium salt of the carboxylic acid.

In this reaction CaO does not participate in the reaction but helps in fusion of the reaction mixture.

(*b*) **Electrolytic Decarboxylation.** This reaction is also called **Kolbe's reaction.** In this reaction an aqueous solution of sodium or potassium salt of a carboxylic acid is electrolysed. Alkane is formed at the anode.

In general, the electrolytic decarboxylation can be represented as

$$\underset{\text{Potassium salt}}{2RCOOK} + 2HOH \xrightarrow{\text{Electrolysis}} \underset{\text{Alkane}}{R-R} + 2CO_2 + H_2 + 2KOH$$

This method is very useful for preparing ethane and higher members. *Methane cannot be prepared by this method* because the alkane is formed by the combination of two alkyl free radicals, therefore, the product should contain at least two carbon atoms.

ACTIVITY 40.1 Laboratory Preparation of Methane

Methane can be prepared in laboratory by heating a mixture of anhydrous sodium ethanoate and soda-lime. Soda-lime is a mixture of sodium hydroxide (NaOH) and calcium oxide (CaO) in the ratio 3 : 1.

$$\underset{\text{Sodium ethanoate}}{CH_3COONa} + \underset{\text{Sodium hydroxide}}{NaOH} \xrightarrow{CaO} \underset{\text{Methane}}{CH_4} + \underset{\text{Sodium carbonate}}{Na_2CO_3}$$

- Take a mixture of anhydrous sodium ethanoate and soda-lime in the ratio of 1 : 2 in a hard glass test-tube fitted with a delivery tube as shown in Fig. 40.3.

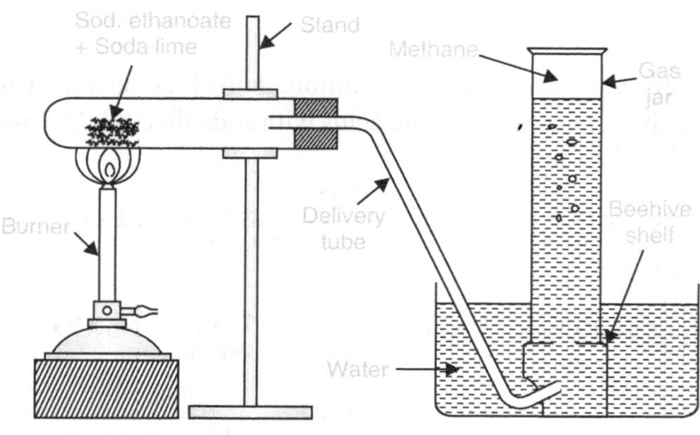

Fig. 40.3. Laboratory preparation of methane

- Heat the mixture slowly.

Methane gas is produced which is collected in a gas jar by the downward displacement of water. As methane gas is collected over water it shows that *methane is insoluble in water.*

SOLVED EXAMPLES

Example 40.4. *How would you obtain butane by (i) Wurtz reaction? (ii) Kolbe's electrolysis?*

Solution. (*i*) Ethyl bromide on reaction with sodium in the presence of dry ether yields butane.

$$2C_2H_5Br + 2Na \xrightarrow{\text{dry ether}} \underset{\text{Butane}}{CH_3CH_2CH_2CH_3} + 2NaBr$$

(*ii*) An aqueous concentrated solution of sodium propanoate on electrolysis yields butane at the anode.

$$2CH_3CH_2COONa(aq) + 2H_2O(l) \xrightarrow{\text{Electrolysis}} CH_3CH_2CH_2CH_3 + 2CO_2 + H_2 + 2NaOH$$

Example 40.5. *Sodium salt of which acid will be needed for the preparation of propane ? Write a chemical equation for the reaction.*

Solution. Sodium salt of **butanoic acid** will be needed for preparation of propane.

$$\underset{\text{Sodium butanoate}}{CH_3CH_2CH_2COO^-Na^+} + NaOH \xrightarrow{CaO} CH_3CH_2CH_3 + Na_2CO_3$$

40.6 PHYSICAL PROPERTIES OF ALKANES

1. PHYSICAL STATE

Alkanes contain only C—C and C—H sigma bonds which are almost non-polar due to very small difference of electronegativity between carbon and hydrogen atoms. Therefore, alkanes possess weak van der Waals' forces as the intermolecular forces. Due to the weak interparticle forces, the first four members (C_1 to C_4) are gases; the next thirteen members, (C_5 to C_{17}) are liquids while the higher members are waxy solids.

2. BOILING POINTS

Alkanes generally have low boiling points. *The boiling points of n-alkanes increase regularly with an increase in the number of carbon atoms.* This can be explained on the basis of interparticle forces. With the increase in the number of carbon atoms, the molecular mass and also the molecular size increases. This results in the increase in the magnitude of van der Waals' forces (London dispersion forces). Consequently, boiling point also increases. It is evident that there is an increase of 20–30 degrees for each carbon atom added to the chain. However, the first few members are exceptions where the increase in boiling point is relatively more.

Among the **isomeric alkanes,** *the branched chain isomers have relatively low boiling points as compared to their corresponding straight chain isomers.* The greater the branching of the chain, the lower the boiling point. This is due to the fact that branching of the chain makes the molecule more compact and brings it closer to a sphere. This decreases the surface area and hence the magnitude of interparticle London dispersion forces which leads to a decrease in boiling point. The boiling points of isomeric pentanes are given below:

Table 40.1. Physical Properties of the first 20 *n*-Alkanes

Alkane	*Number of carbon atoms (n)*	*Formula/ Physical State (at 20°C)*	*Boiling Point (K)*	*Melting Point (K)*	*Density ($g\ cm^{-3}$) at 20°C*
Methane	1	CH_4 (*g*)	109	90	0.55
Ethane	2	C_2H_6 (*g*)	184	90	0.51
Propane	3	C_3H_8 (*g*)	231	84	0.50
Butane	4	C_4H_{10} (*g*)	273	135	0.58
Pentane	5	C_5H_{12} (*l*)	309	143	0.63
Hexane	6	C_6H_{14} (*l*)	342	178	0.66
Heptane	7	C_7H_{16} (*l*)	371	182	0.68
Octane	8	C_8H_{18} (*l*)	399	216	0.70
Nonane	9	C_9H_{20} (*l*)	424	222	0.72
Decane	10	$C_{10}H_{22}$ (*l*)	447	243	0.73
Undecane	11	$C_{11}H_{24}$ (*l*)	469	247	0.74
Dodecane	12	$C_{12}H_{26}$ (*l*)	489	263	0.75
Tridecane	13	$C_{13}H_{28}$ (*l*)	508	268	0.76
Tetradecane	14	$C_{14}H_{30}$ (*l*)	527	279	0.76
Pentadecane	15	$C_{15}H_{32}$ (*l*)	544	283	0.77
Hexadecane	16	$C_{16}H_{34}$ (*l*)	560	291	0.77
Heptadecane	17	$C_{17}H_{36}$ (*l*)	576	296	0.76
Octadecane	18	$C_{18}H_{38}$ (*s*)	590	301	0.76
Nonadecane	19	$C_{19}H_{40}$ (*s*)	606	305	0.78
Eicosane	20	$C_{20}H_{42}$ (*s*)	616	310	0.79

$CH_3CH_2CH_2CH_2CH_3$
n-Pentane
(*no branching*)

$CH_3CHCH_2CH_3$ with CH_3 branch
iso-Pentane
(*one branch*)

Boiling points: 309 K 301 K

CH_3—C(CH_3)(CH_3)—CH_3
neo-Pentane
(*two branches*)

Boiling point: 282.5 K

3. VOLATILITY

Due to their low boiling points, lower alkanes are highly volatile. Volatility refers to the ability of a liquid to change into vapour. Among alkanes volatility decreases with increase in chain length. Among isomeric alkanes volatility increases with branching.

4. MELTING POINTS

The melting points of alkanes do not show a very smooth gradation with the increase in molecular mass. For example, let us look at the melting points of propane to *n*-octane.

Hydro-carbon	C_3H_8	C_4H_{10}	C_5H_{12}	C_6H_{14}	C_7H_{16}	C_8H_{18}
Melting Point (K)	84	135	143	178	182	216

It is evident that the increase in melting point is relatively more in moving from an alkane having an odd number of carbon atoms to the higher alkane while it is relatively less in moving from an alkane with an even number of carbon atoms to the higher alkane. This is because the intermolecular forces in a crystal depend not only upon the size of the molecule but also upon how well the molecules fit into the crystal lattice. The molecule which fit into the crystal lattice more easily due to their symmetrical structures usually have high lattice energy and therefore, high melting points.

For example, let us compare the melting points of isomeric pentanes.

$CH_3CH_2CH_2CH_2CH_3$
n-Pentane

$CH_3CHCH_2CH_3$ with CH_3 branch
iso-Pentane

Melting points: 143 K 113 K

$$CH_3-\underset{CH_3}{\overset{CH_3}{\underset{|}{\overset{|}{C}}}}-CH_3$$

neo-Pentane

Melting point: 256.4 K

Among these isomers *neo*-pentane has the highest melting point because its molecules are almost spherical and hence pack more closely.

5. SOLUBILITY

In keeping with the popular rule "*like dissolves like*" hydrocarbons are insoluble in polar solvent like water because they are predominantly non-polar in nature. However, they are soluble in non-polar solvents like ether, carbon tetrachloride, etc.

The liquid hydrocarbons themselves are good solvents for other non-polar organic molecules.

6. DENSITY

The densities of alkanes increase with increasing molecular masses but become constant at about 0.8 g cm^{-3}. This means that all alkanes are lighter than water.

ACTIVITY 40.2	**Drawing graphs of some physical properties of *n*-alkanes**

Using data from Table 40.1 draw graphs of

(*i*) boiling points against *n*

(*ii*) melting points against *n*

(*iii*) densities against *n*

Take *n*, the number of carbon atoms in an alkane on the horizontal axis.

SOLVED EXAMPLE

Example 40.6. *Arrange 2, 2-dimethylbutane, 3-methylpentane and n-hexane in the ascending order of their boiling point. Explain the basis for your order.*

Solution.

2, 2-Dimethylbutane < 3-methylpentane < *n*-hexane.

All the three are isomers having molecular formula C_6H_{14}. *n*-Hexane has the longest chain and therefore has the highest boiling point. 2, 2-Dimethylbutane is the most spherical and has the smallest surface area. In this case the van der Waals' forces will be weakest and so it has the lowest boiling point. 3-Methylpentane will have boiling point in between these two extremes.

40.7 SOME IMPORTANT TERMS

NUCLEOPHILES AND ELECTROPHILES

Organic reactions proceed by the attack of certain active species on the reacting molecule (generally called *substrate*). These active species are called attacking species. They may be classified into nucleophiles and electrophiles.

1. Nucleophiles

These are neutral or negatively charged species which are capable of donating a pair of electrons to some other molecules. These species act as **Lewis bases** and attack electron deficient centres (electrophilic centres) of organic molecules. Some examples are:

Charged nucleophiles:

$:\bar{C}l$, $:\bar{O}R$, $:\bar{C}N$, $:\bar{N}O_2$, $RCO\bar{O}:$, $:\bar{S}H$, $:\bar{O}H$, $:\bar{C}\equiv CR$, $:\bar{R}$

Neutral nucleophiles:

$H_2\ddot{O}:$, $R\underset{..}{\ddot{O}}H$, $\ddot{N}H_3$, $R\underset{..}{\ddot{O}}R$, $R\ddot{N}H_2$

2. Electrophiles

These are neutral or positively charged species capable of accepting a pair of electrons from another molecule. These species act as **Lewis acids** and attack the electron rich centres (nucleophilic centres) of an organic molecule. Some examples are:

Charged electrophiles:

$\overset{+}{H}$, $\overset{+}{Cl}$, $\overset{+}{Br}$, $Ar-\overset{+}{N}\equiv N$, $\overset{+}{N}O_2$, $\overset{+}{N}O$, $\overset{+}{R}$

Neutral electrophiles: BF_3, $AlCl_3$, SO_3.

FREE RADICALS

A free radical may be defined as *an atom or group of atoms which contains an unpaired electron.* Free radicals are formed by homolytic cleavage of a covalent bond. For example, in the presence of U.V. light, chlorine gives chlorine free radicals.

$$Cl:Cl \xrightarrow{U.V.} 2Cl^{\cdot}$$

If the unpaired electron is on carbon atom then they are called alkyl free radicals. Alkyl free radicals are classified as primary (1°), secondary (2°) or tertiary (3°) depending on the carbon atom carrying the unpaired electron.

$$CH_3—\dot{C}H_2 \qquad CH_3—\overset{H}{\overset{|}{\dot{C}}}—CH_3 \qquad CH_3—\overset{CH_3}{\overset{|}{\dot{C}}}—CH_3$$

Ethyl free radical (1°) — *iso*-Propyl free radical (2°) — *tert*-Butyl free radical (3°)

The order of stability of free radicals is 3° > 2° > 1°.

CLEAVAGE OR FISSION OF COVALENT BONDS

Organic reactions involve the breaking of bonds in the reacting molecules and formation of new bonds to give products. The breaking of a covalent bond can take place in two ways depending upon the *nature of the bond, nature of attacking reagent* and the *conditions of the reaction.*

1. Homolytic Cleavage or Homolytic Fission

In this type of cleavage each fragment formed as a result of the cleavage gets one electron from the shared pair of electrons.

This type of cleavage results in the formation of species with odd or unpaired electrons. Such species are called free radicals. Some examples of homolytic cleavage are given below:

$$Cl\cdot \vdots \cdot Cl \xrightarrow[\text{or } 525–725\ K]{h\nu} \dot{C}l + \dot{C}l$$

$$H—\overset{H}{\overset{|}{\underset{H}{\underset{|}{C}}}}\cdot \vdots \cdot H + \dot{C}l \longrightarrow H—\overset{H}{\overset{|}{\underset{H}{\underset{|}{C}}}}\cdot + H:Cl$$

Homolytic cleavage is favoured by conditions such as *non-polar nature of bond, high temperature or presence of high energy radiations.*

2. Heterolytic Cleavage or Heterolytic Fission

In this type of cleavage both the shared electrons of the covalent bond are taken away by one of the fragments (the other fragmentes does not get any of the shared electrons).

For example, $A \vdots :B \longrightarrow A^+ + :B^-$

This type of cleavage results in the formation of charged species. Heterolytic cleavage is favoured by conditions such as the *polar nature of a bond* and the *presence of polar solvent.*

40.8 CHEMICAL REACTIONS OF ALKANES

Alkanes are saturated hydrocarbons. They contain only strong C—C and C—H sigma bonds. Hence, alkanes are quite less reactive and are known as **paraffins** (*parum*—little, *affinis*—reactivity or affinity). Alkanes are generally inert towards acids, bases, oxidizing and reducing agents. Alkanes being saturated compounds can undergo only substitution reactions. Thus, **the typical reactions of alkanes are substitution reactions.**

Important reactions of alkanes are as follows:

HALOGENATION

Halogenation involves the replacement of one or more H atoms of alkane by a corresponding number of halogen atoms.

Chlorination can be carried out by treatment of alkane with chlorine, either in the presence of diffused sunlight or ultraviolet light or at high temperature (573–773 K). For example, when methane is chlorinated, we get a mixture of four different substitution products.

$$CH_4 + Cl_2 \xrightarrow[\text{or Heat}]{\text{Diffused sunlight}} \underset{\text{Chloromethane}}{CH_3Cl} + HCl$$

$$CH_3Cl + Cl_2 \longrightarrow \underset{\text{Dichloromethane}}{CH_2Cl_2} + HCl$$

$$CH_2Cl_2 + Cl_2 \longrightarrow \underset{\text{Trichloromethane}}{CHCl_3} + HCl$$

$$CHCl_3 + Cl_2 \longrightarrow \underset{\text{Tetrachloromethane}}{CCl_4} + HCl$$

Bromination, though similar to chlorination, occurs less rapidly.

Iodination is extremely slow and is reversible in nature

Fluorination of alkanes takes place with almost explosive violence to produce fluorinated compounds. It also involves rupture of some C—C bonds in the case of higher alkanes. The reaction can be made less violent by dilution of fluorine with nitrogen.

MECHANISM OF HALOGENATION OF ALKANES

Halogenation of alkanes is a chain reaction that takes place through **free-radicals.** Let us understand it by taking the example of chlorination of methane to give methyl chloride.

$$CH_4 + Cl_2 \xrightarrow{h\nu} CH_3Cl + HCl.$$

The substitution occurs through the following steps:

I. Chain initiation step. It involves the formation of attacking free radicals. Cl_2 molecules get excited by absorbing photons of light and undergo homolytic cleavage to produce chlorine free radicals.

$$Cl \cdot\cdot Cl \xrightarrow[\text{Homolysis}]{h\nu} \dot{Cl} + \dot{Cl}$$

II. Chain propagation step. Chlorine free radicals attack the methane molecules and abstract a hydrogen atom to form methyl free radicals.

$$(a)\ H{-}\underset{H}{\overset{H}{C}}{-}H + \dot{Cl} \longrightarrow H{-}\underset{H}{\overset{H}{C^{\bullet}}} + HCl$$

Methyl free radical in turn reacts with another molecule of chlorine to form CH_3Cl and chlorine free radical.

$$(b)\ H{-}\underset{H}{\overset{H}{C^{\bullet}}} + Cl{-}Cl \longrightarrow H{-}\underset{H}{\overset{H}{C}}{-}Cl + \dot{Cl}$$

The steps (*a*) and (*b*) are repeated again and again and are called **chain propagation steps**. In these steps, consumption of one free radical results in generation of another free radical and hence a chain reaction starts.

III. Chain termination step. This involves the combination of two free radicals

$$\dot{Cl} + \dot{Cl} \longrightarrow Cl_2$$

$$\dot{C}H_3 + \dot{Cl} \longrightarrow CH_3Cl$$

$$\dot{C}H_3 + \dot{C}H_3 \longrightarrow \underset{\text{Ethane}}{H_3C{-}CH_3}$$

SULPHONATION

Alkanes containing six or more carbon atoms on heating with fuming tetraoxosulphate (VI) acid for several hours yield the corresponding sulphonic acids.

$$\underset{\text{Hexane}}{C_6H_{14}} + H_2SO_4 \xrightarrow{SO_3} \underset{\text{Hexane sulphonic acid}}{C_6H_{13}SO_3H} + H_2O.$$

OXIDATION

Alkanes on burning in air or oxygen get completely oxidised to carbon (IV) oxide and water with the evolution of a large amount of heat.

$$CH_4 + 2O_2 \longrightarrow CO_2 + 2H_2O + \text{Heat}$$

$$C_4H_{10} + \frac{13}{2} O_2 \longrightarrow 4CO_2 + 5H_2O + \text{Heat}$$

Due to the evolution of a large amount of heat during combustion, alkanes are used as fuels.

The general chemical equation for combustion of alkanes may be written as:

$$C_nH_{2n+2} + \left(\frac{3n+1}{2}\right)O_2 \longrightarrow n\,CO_2 + (n+1)\,H_2O$$

In the presence of insufficient supply of oxygen, alkanes on combustion produce carbon (II) oxide, which is highly poisonous. Carbon (II) oxide combines with haemoglobin in the blood to form **carboxy haemoglobin** and makes it unfit for carrying oxygen. As a result, the person may die due to suffocation.

Do you know?

Carbon black is formed during incomplete combustion of alkanes with insufficient amount of air or oxygen. It is used in the manufacture of ink, printer ink, black pigments and as filters.

$$CH_4(g) + O_2(g) \xrightarrow[\text{Combustion}]{\text{Incomplete}} C(s) + 2H_2O(l).$$

ISOMERISATION

In this process, *n*-alkanes on heating with anhydrous aluminium chloride and hydrogen chloride get converted into branched chain alkanes.

$$\underset{n\text{-Hexane}}{CH_3(CH_2)_4CH_3} \xrightarrow[\Delta]{AlCl_3/HCl}$$

$$\underset{\text{2-Methylpentane}}{CH_3\overset{CH_3}{\overset{|}{C}}HCH_2CH_2CH_3} + \underset{\text{3-Methylpentane}}{CH_3CH_2\overset{CH_3}{\overset{|}{C}}HCH_2CH_3}$$

CRACKING OR PYROLYSIS

It is a process of decomposing higher hydrocarbons into lower hydrocarbons with low boiling points by strong heating. The process involves the cleavage of carbon-carbon and carbon-hydrogen bonds and results in the formation of lower hydrocarbons and is believed to be a free radical reaction.

The types of hydrocarbons formed depend upon the conditions employed for cracking. For example,

$$\underset{\text{Decane}}{CH_3(CH_2)_8CH_3} \xrightarrow{770\text{ K}} \underset{\text{Octane}}{C_8H_{18}} + \underset{\text{Ethylene}}{CH_2 = CH_2}$$

Oil gas or petrol gas is prepared by pyrolysis of kerosene oil or petrol. Similarly, gasoline can be prepared by pyrolysis of kerosene oil or fuel oil. For example, dodecane, a constituent of kerosene oil, on heating to 973 K in the presence of platinum, palladium or nickel, gives a mixture of heptane and pentene.

$$C_{12}H_{26} \xrightarrow{973\text{ K, Pd/Pt/Ni}} C_7H_{16} + C_5H_{10} + \text{other products}$$

40.9 USES OF ALKANES

Important uses of alkanes are as follows:

1. Petroleum and natural gas are largely mixtures of different alkanes. On refining, they give liquefied petroleum gas (LPG), gasoline, kerosene, diesel, furnace oil and wax which are used as **fuels.**
2. Some higher alkanes are used as **lubricating oils** and as **vaseline**.
3. Alkanes are used as starting materials for the **preparation of** many other useful **organic compounds.** For example, methane on chlorination gives chloromethane, dichloromethane, trichloromethane (chloroform) and tetrachloromethane.
4. Carbon formed during decomposition of methane is in the form of finely divided particles and is known as **carbon black.** Carbon black is used for making printer's ink and paints. It is also used in the rubber industry.
5. Alkanes are also used as **sources of hydrogen.**

GREENHOUSE EFFECT

Sun emits light of different wavelengths consisting of ultra-violet, visible and infra-red radiations. Out of these, the harmful ultra-violet (UV) radiations are largely absorbed by the ozone layer. The visible and infra-red radiations freely pass through the atmosphere and reach the surface of the earth. The infra-red rays, which come from the sun, are of short wavelength and pass through the atmosphere easily. The sunlight heats up the surface of the earth. The heated surface emits (or reflects) infra-red radiations of longer wavelength. These infra-red radiations (heat waves) are trapped by CO_2, water vapours and ozone in the atmosphere. Although these constituents are less than 1% of the earth's atmosphere, they absorb about 90% of the heat radiated from the earth's surface, causing the atmosphere to heat up. The heated atmosphere in turn warms the earth's surface to a higher temperature than it would have in the absence of an atmosphere. **The heating of earth due to trapped radiation is called greenhouse effect.**

40.10 CONTRIBUTION OF ALKANES TO THE GREENHOUSE EFFECT

Alkanes make a significant contribution to the greenhouse effect as discussed under:

1. Combustion of alkanes release large amount of carbon(IV) oxide into the atmosphere which causes greenhouse effect.
2. Anaerobic decomposition of plants and animal remains liberates methane into the atmosphere which is also responsible for greenhouse effect.
3. Alkanes are converted to chlorofluorocarbons (CFCs) which are used as propellants in aerosols and as refrigerant gases. CFCs are very efficient greenhouse gases. CFCs contribute about 15–20% to the greenhouse effect. Greenhouse warming effect of one molecule of a typical CFC molecule is about 1000 times that of a carbon(IV) oxide molecule.

SUMMARY

- Alkanes are saturated hydrocarbons. Their general formula is C_nH_{2n+2}.
- Members of a homologous series differ in molecular formula by —CH_2 and molecular mass by 14.
- Methane, CH_4 is the simplest alkane. It is also known as *damp-fire* and *marsh gas*.
- Alkanes with four or more carbon atoms exhibit chain isomerism.
- Petroleum, coal and natural gas are the major sources of alkanes.
- Unbranched alkanes are sometimes called normal alkanes.

- All the carbon atoms in alkanes are sp^3 hybridised.
- Methane molecule is tetrahedral in shape.
- Wurtz Reaction

$$2\,R—X + 2Na \xrightarrow{\text{Dry ether}} R—R + 2NaX$$

- Addition of hydrogen is known as **hydrogenation.**
- Alkanes have low melting and boiling points due to weak intermolecular forces.
- **Nucleophiles.** These are neutral or negatively charged species which are capable of donating a pair of electrons to some other molecule.
- **Electrophiles.** These are neutral or positively charged species which can accept a pair of electrons from some other species.
- **Free Radical** may be defined as an atom or group of atoms which contains an odd (or unpaired) electron. These are formed by homolytic cleavage of bond.
- Major use of alkanes is as fuel.
- Alkanes and the other compounds which are derived from them make a significant contribution towards the greenhouse effect. Some of these are CH_4, CO_2 and CFC's.

EVALUATE YOURSELF

I. Objective Type Questions

Select the most appropriate choice from the options given as (a), (b), (c) and (d) after each question:

1. The number of chain isomers possible for C_5H_{12} is

(*a*) 5 (*b*) 4

(*c*) 3 (*d*) 2.

2. The simplest alkane that can exhibit chain isomerism has the molecular formula

(*a*) C_3H_8 (*b*) C_2H_6

(*c*) C_5H_{12} (*d*) C_4H_{10}.

3. Which of the following alkanes exists in liquid state at room temperature?

(*a*) C_2H_6 (*b*) C_5H_{12}

(*c*) C_3H_8 (*d*) CH_4.

4. The C—C—C bond angle in propane is expected to be about

(*a*) 120° (*b*) 180°

(*c*) 109° (*d*) 90°.

5. The number of monochlorination products possible for *neo*-pentane is

(*a*) One (*b*) Two

(*c*) Three (*d*) Five.

6. Which of the following species is a nucleophile?

(*a*) BF_3 (*b*) H_3O^+

(*c*) H_2O (*d*) $AlCl_3$.

II. Fill in the Blanks

7. Complete the following sentences by supplying appropriate words:

(*i*) The simplest alkane is ....

(*ii*) The general formula of alkanes is ....

(*iii*) The number of structural isomers possible for C_4H_{10} is ....

(*iv*) Straight chain alkanes are known as ....

(*v*) The physical state of the alkane C_7H_{16} would be ...... at room temperature.

(*vi*) Chlorination of methane follows ...... mechanism.

(*vii*) Methane molecule has ...... geometry.

(*viii*) The IUPAC name of *iso*-pentane is ....

(*ix*) When sodium propanoate is heated with soda-lime, the gas produced is ....

(*x*) The full form of CFC is ....

III. Discussion Questions

8. What are alkanes? Why are they called paraffins?

9. Draw the structure of 2, 2, 4-trimethylhexane. What is the number of primary carbon atoms in it?

10. What is the general formula of cycloalkanes? Draw the structures of cyclobutane and cyclopentane.

11. An alkane has molecular mass 72. Draw the structures and give the IUPAC names of all the structural isomers of this alkane. Which of these has

(*i*) the highest boiling point?

(*ii*) the highest melting point?

(*iii*) the maximum volatility?

12. Give IUPAC names of the following compounds:

(*i*) $CH_3—CH_2—\underset{\underset{CH_2—CH_3}{|}}{CH}—CH\begin{matrix}CH_3\\CH_3\end{matrix}$

(*ii*) $CH_3—CH_2—\underset{\underset{C_2H_5}{|}}{CH}—CH_2—\underset{\underset{CH_3}{|}}{CH}—CH_2—CH_3$

(*iii*) Methylcyclopentane (cyclopentane ring bearing CH_3)

13. Explain and give two examples of each of the following:
(*i*) Free radicals (*ii*) Nucleophiles
(*iii*) Electrophiles.

14. Give chemical equations and state the products you would obtain when:
(*i*) Ethylbromide is treated with sodium metal in the presence of dry ether.
(*ii*) Ethane is burnt in the presence of sufficient oxygen.
(*iii*) Sodium ethanoate is heated with soda-lime.

15. List the important sources of alkanes.

16. Describe the mechanism of chlorination of methane.

17. Describe the laboratory preparation of methane.

18. Discuss the uses of alkanes. What is the contribution of alkanes to the greenhouse effect?

19. Explain why:
(*i*) alkanes with three or more than three carbon atoms contain zig-zag chains of carbon atoms.
(*ii*) characteristic reactions of alkanes are substitution reactions.

Answers

I. Objective Type Questions

1. (*c*) **2.** (*d*) **3.** (*b*)
4. (*c*) **5.** (*a*) **6.** (*c*).

II. Fill in the Blanks

7. (*i*) methane (*ii*) C_nH_{2n+2}
(*iii*) 2 (*iv*) normal alkanes
(*v*) liquid (*vi*) free radical chain
(*vii*) tetrahedral (*viii*) 2-methylbutane
(*ix*) ethane (*x*) chlorofluorocarbon.

SECTION-6

UNIT-41

ALKENES

Specific Objectives

After studying this unit, the student should be able to:

- describe the sources and characteristics of alkenes
- describe the nomenclature and isomerism of alkenes
- describe the preparation and chemical reactions of alkenes
- outline the uses of alkenes

41.1 INTRODUCTION

Alkenes are unsaturated hydrocarbons. Their molecules have two hydrogen atoms less than the corresponding members of alkanes. Thus, their general formula is C_nH_{2n}. The functional group in alkenes is carbon-carbon double bond (C = C). Alkenes are also known as ***olefins*** (derived from Greek word *olefiant* means *oil forming*). It is because of the fact that lower members of this series form oily products when treated with chlorine.

41.2 STRUCTURE OF ALKENES

The characteristic feature of alkenes is the presence of a carbon-carbon double bond.

In order to understand the structure of alkenes let us study the **structure of ethene** $CH_2 = CH_2$, the simplest member of this family. Each of the two carbon atoms of ethene is sp^2 hybridised and forms three sigma bonds, two with two hydrogens and the third with one another. They are all in the same plane. The *pi* bond between the two carbon atoms is perpendicular to this plane, as shown in Fig. 41.1, and is present both above and below the plane.

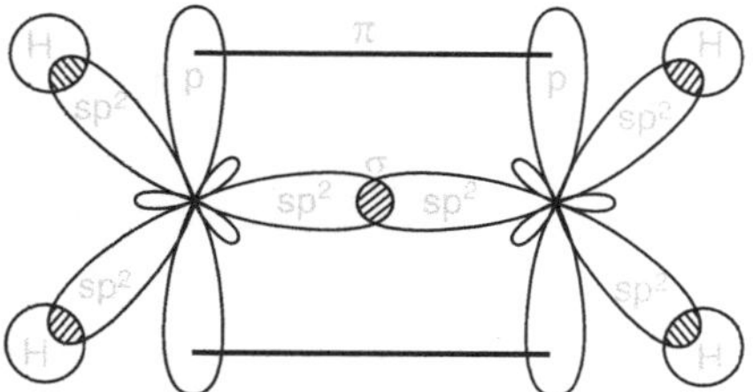

H 108 pm H
C=C
134 pm
H H

Fig. 41.1. Structure of Ethene.

C—C bond length in ethene is 134 pm which is much shorter than the C—C bond length in ethane (154 pm). C—H bond length in ethene is 108 pm.

Thus a carbon-carbon double bond comprises of a sigma bond and a *pi* bond. A *pi* bond is not as strong as a sigma bond. For example, the bond energies of a carbon-carbon single bond and a carbon-carbon double bond are respectively 347 and 605 kJ/mole. However, the *pi* bond is sufficiently strong to prevent rotation of the two carbon atoms with respect to one another. This restricted rotation about *pi* bonds gives rise to geometric isomerism in alkenes.

41.3 NOMENCLATURE AND ISOMERISM

In the **IUPAC system** the name of an alkene is derived by replacing the suffix ***-ane*** of the corresponding alkane by ***-ene.***

In **common system,** the name of an alkene is derived by replacing the suffix ***-ane*** of the corresponding alkane by ***-ylene.***

Molecular Formula	*Common Name*	*IUPAC Name*
$CH_2 = CH_2$	Ethylene	Ethene
$CH_3—CH = CH_2$	Propylene	Propene
$CH_3—CH_2—CH = CH_2$	α-Butylene	But-1-ene
$CH_3—CH = CH—CH_3$	β-Butylene	But-2-ene
$(CH_3)_2C = CH_2$	*iso*-Butylene	2-Methylpropene

ISOMERISM IN ALKENES

The simplest acyclic alkene which can exhibit isomerism is butene, C_4H_8. The possible isomers of C_4H_8 are:

$CH_2 = CHCH_2CH_3$ (But-1-ene) $\qquad$ $CH_3CH = CHCH_3$ (But-2-ene)

$CH_3C(CH_3) = CH_2$

2-Methylpropene

The three isomers shown above have different arrangement of atoms and hence are structural isomers. But-1-ene and but-2-ene are **position isomers** whereas But-1-ene and 2-methylpropene are **chain isomers.** However, experimentally it has been found that there are four isomers, having different physical properties. Careful study of the above isomers reveals that for but-2-ene, two isomers having different orientation of groups about the carbon-carbon double bond are possible as shown in Fig. 41.2.

These two isomers are stereoisomers.

H(H_3C)C = C(H)(CH_3) — *cis*-But-2-ene I

H(H_3C)C = C(CH_3)(H) — *trans*-But-2-ene II

Fig. 41.2. Geometric isomers of but-2-ene.

The stereoisomerism exhibited by alkenes due to the difference in the spatial arrangement of groups about the double bonded carbon atoms is called ***geometrical isomerism.***

Structures I and II are not interconvertible because rotation about carbon-carbon double bond is hindered. Therefore, they represent two different isomers of but-2-ene. Such isomers are called **geometric isomers**. The geometric isomers which have similar groups on same side are called *cis* (structure I) whereas the geometric isomers which have similar groups on opposite sides are called *trans* (structure II).

Conditions for Geometric Isomerism in Open Chain Compounds

1. *There should be a double bond in the molecule.*

2. *The two atoms or groups attached to each doubly bonded carbon atom should be different.* Thus, alkenes of the type *ab*C = C*ab* or *ab*C = C*cd* exhibit geometric or *cis-trans* isomerism. *If one of the two doubly bonded carbon atoms carries two identical groups then the molecule does not exhibit geometric isomerism.* Thus, compounds of the type *aa*C = C*ab* or *aa*C = C*bb* do not show geometric isomerism.

$(H)_2C = C(CH_3)(H)$ — Propene

$(CH_3)_2C = C(CH_3)(H)$ — 2-Methylbut-2-ene

No geometric isomerism

Geometric isomers differ in their physical properties such as melting point, boiling point, dipole moment, solubility, etc. Their chemical properties are similar but not identical.

It may be noted that:

- *cis*-isomer is more polar than the *trans*-isomer because in it the individual bond dipoles do not cancel each other. For example, dipole moment of *cis*-but-2-ene is 0.33 D whereas that of *trans*-but-2-ene is almost zero.

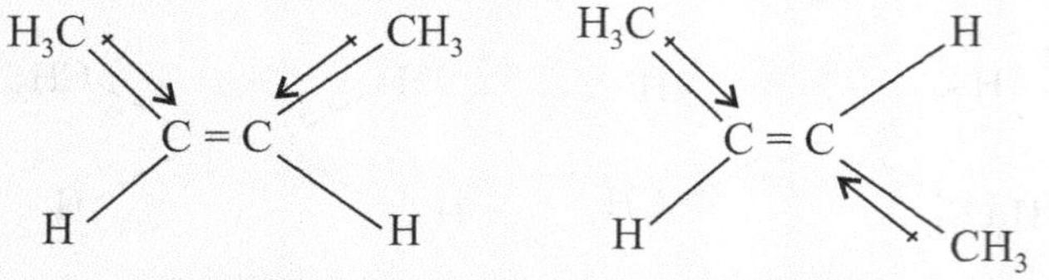

H_3C(H)C = C(H)CH_3 — *cis*-But-2-ene (μ = 0.33D)

H_3C(H)C = C(H)CH_3 (trans) — *trans*-But-2-ene (μ = 0)

- *cis*-isomer has greater boiling point than the *trans*-isomer due to greater polarity and consequent stronger interparticle forces. The boiling point of *trans*-but-2-ene is 274 K while that of *cis*-but-2-ene is 277 K.
- *trans*-isomer has higher melting point than the *cis*-isomer because its molecules are more closely packed in the solid state due to symmetry.

For example, melting point of maleic acid (*cis*-isomer) is 303 K while that of fumaric acid (*trans*-isomer) is 560 K.

H(COOH)C=C(H)(COOH) — Maleic acid, m. pt. 303 K

H(COOH)C=C(HOOC)(H) — Fumaric acid, m. pt. 560 K

- *trans*-isomer is more stable than the *cis*-isomer.

SOLVED EXAMPLES

Example 41.1. *Which of the following compounds can show geometric isomerism?*

(i) $(CH_3)_2C = C(CH_3)_2$

(ii) $CH_3CH = CHCH_2CH_3$

(iii) $CH_3CH_2CH_2CH = CH_2$

(iv) C_6H_5—$CH = CHCH_3$

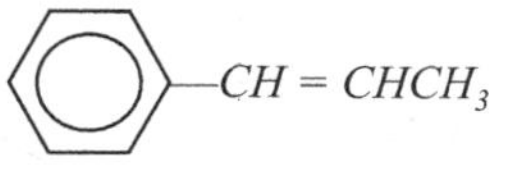

(v) C_6H_5—$CH = C(C_2H_5)_2$

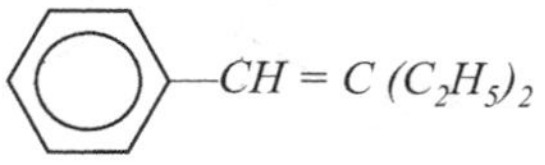

(vi) $CH_3C(Br) = C(Cl)CH_2CH_3$.

Draw the cis and trans-isomers of the compounds which can exhibit geometric isomerism and give their IUPAC names.

Solution. The compounds (*ii*), (*iv*) and (*vi*) can exhibit geometric isomerism because in these molecules each double bonded carbon has two different groups attached to it.

(*ii*) CH_3(H)C = C(CH_2CH_3)(H) — *cis*-Pent-2-ene; CH_3(H)C = C(H)(CH_2CH_3) — *trans*-Pent-2-ene

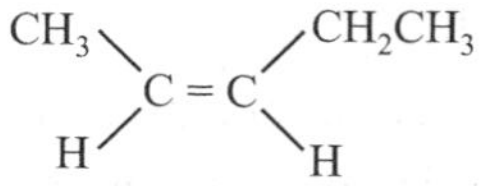

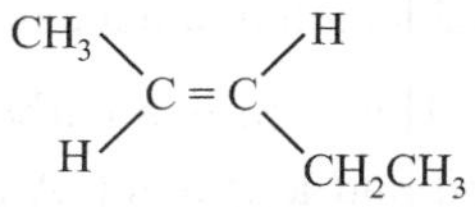

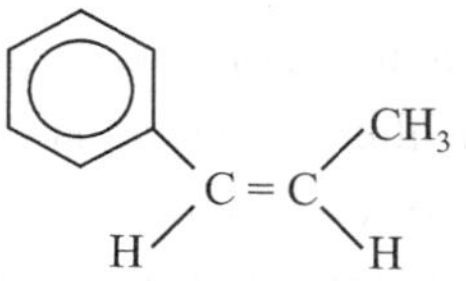

(*iv*) C_6H_5(H)C = C(CH_3)(H) — *cis*-1-Phenylprop-1-ene; C_6H_5(H)C = C(H)(CH_3) — *trans*-1-Phenylprop-1-ene

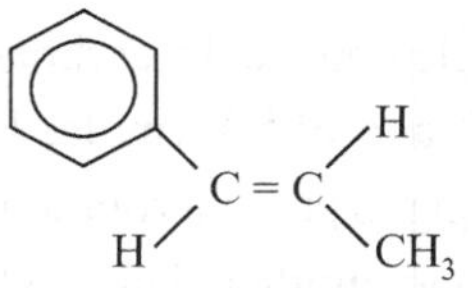

(*vi*) CH_3(Br)C = C(CH_2CH_3)(Cl) — *cis*-2-Bromo-3-chloropent-2-ene; CH_3(Br)C = C(Cl)(CH_2CH_3) — *trans*-2-Bromo-3-chloropent-2-ene

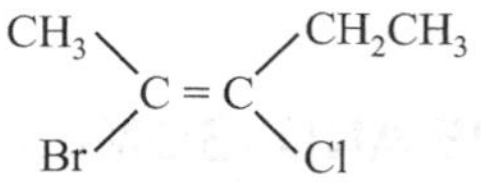

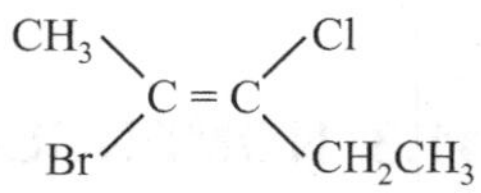

Example 41.2. *Write structures and IUPAC names of different structural isomers of alkenes corresponding to C_5H_{10}. Which of these can exhibit geometric isomerism?*

Solution.

(*a*) $CH_2 = CH—CH_2—CH_2—CH_3$
Pent-1-ene

(*b*) $CH_3—CH=CH—CH_2—CH_3$
Pent-2-ene

(*c*) $CH_3—C(CH_3) = CH—CH_3$
2-Methylbut-2-ene

(*d*) $CH_3—CH(CH_3)—CH = CH_2$
3-Methylbut-1-ene

(*e*) $CH_2 = C(CH_3)—CH_2—CH_3$
2-Methylbut-1-ene

The compound (*b*), pent-2-ene can exhibit geometric isomerism because it has two different atoms or groups attached to both doubly bonded carbon atoms.

Example 41.3. *Out of cis and trans 2, 3-dichlorobut-2-ene:*

(*i*) *Which is more polar?*

(*ii*) *Which has a higher boiling point?*

(*iii*) *Which has a higher melting point?*

Solution. (*i*) Out of *cis* and *trans* 2, 3-dichlorobut-2-ene, *cis*-isomer is more polar because in *trans*-isomer dipole moment of one C—Cl bond cancels the dipole moment of the other C—Cl bond whereas in the *cis*-isomer they do not cancel each other.

(*ii*) The boiling point of the *cis*-isomer is higher than that of *trans*-isomer because the interparticle forces in the *cis*-isomer are stronger due to greater polarity.

(*iii*) The melting point of the *trans*-isomer is higher than that of the *cis*-isomer because molecules of the *trans*-isomer due to their symmetry pack well in the solid state and hence come quite close to each other. So forces of attraction between them are quite strong.

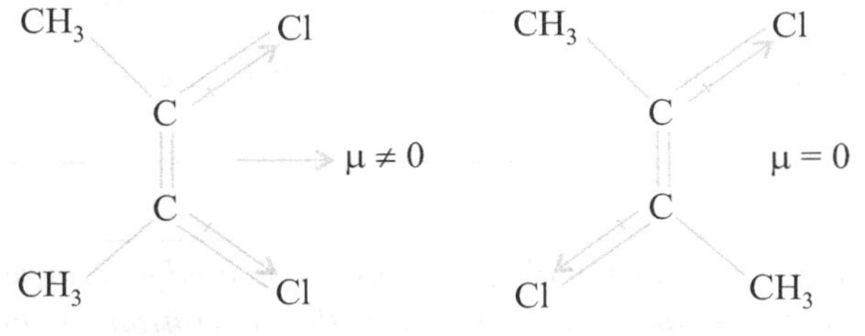

cis-2, 3-dichlorobut-2-ene   *trans*-2, 3-dichlorobut-2-ene

41.4 SOURCES OF ALKENES

INDUSTRIAL PREPARATION OF ALKENES

1. From Alkanes

Alkenes can be obtained **from alkanes by the process of cracking**. Alkanes for this purpose are obtained from petroleum or from natural gas. Cracking of ethane from petroleum gas yields ethene.

$$CH_3{-}CH_3 \xrightarrow[1\ atm]{1000\ K} CH_2{=}CH_2 + H_2$$

Higher alkanes may also be subjected to cracking to obtain ethene. Here propene is obtained as one of the by-products.

$$\underset{\text{Hexane}}{CH_3(CH_2)_4CH_3} \xrightarrow{1000\ K} \underset{40\%}{CH_2{=}CH_2} + \underset{20\%}{CH_3CH{=}CH_2} + \underset{15\%}{CH_4}$$

+ Other products

2. From Natural Gas

Alkenes can also be obtained by **cracking of natural gas**. This method is applied if the natural gas contains appreciable amounts of ethane, propane and butane.

$$CH_3{-}CH_3 \longrightarrow CH_2 = CH_2 + H_2$$

$$CH_3{-}CH_2{-}CH_3 \longrightarrow CH_2 = CH_2 + CH_4$$

$$CH_3{-}CH_2{-}CH_3 \longrightarrow CH_3{-}CH = CH_2 + H_2$$

$$CH_3{-}CH_2{-}CH_2{-}CH_3 \longrightarrow CH_3{-}CH = CH_2 + CH_4$$

Latex (Natural rubber) is a polymer of 2-methyl-1, 3-butadiene (*isoprene*). Thus, latex is an alkene polymer.

ACTIVITY 41.1 **Laboratory preparation of ethene by cracking of kerosene or wax**

- Take a spoonful of sand in a boiling tube and pour about 3 cm^3 of kerosene or melted wax into it.
- Set up the apparatus as shown in Fig. 41.3.

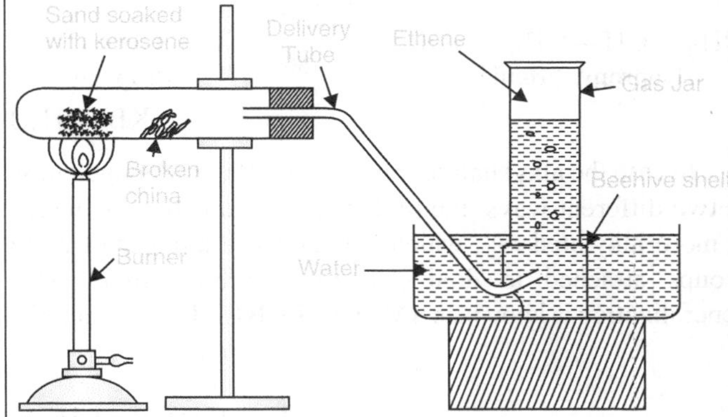

Fig. 41.3. Cracking of kerosene.

- Introduce few pieces of porcelain or broken china in the tube as shown in Fig. 41.3.
- Heat the porcelain pieces till they become red hot. Then heat the sand containing wax or kerosene. The hydrocarbons on heating change into vapours.
- When the vapours pass over heated porcelain, cracking takes place. To keep the porcelain hot, sand and porcelain are heated alternately.
- Collect the gas, produced as a result of cracking, in a gas jar by downward displacement of water. It mainly contains ethene.

PREPARATION OF ALKENES

1. From Alkyl Halides

Alkenes can be prepared from alkyl halides by treatment with an alcoholic solution of caustic potash (KOH) at about 353–363 K. This reaction is known as ***dehydrohalogenation of alkyl halides.***

$$\underset{\text{Ethyl bromide}}{H{-}\overset{\overset{H}{|\beta}}{\underset{\underset{H}{|}}{C}}{-}\overset{\overset{H}{|\alpha}}{\underset{\underset{Br}{|}}{C}}{-}H} + KOH\ (alcoholic)$$

$$\longrightarrow H_2C = CH_2 + KBr + H_2O$$

Similarly, treatment of 1-bromopropane or 2-bromopropane with *ethanolic solution of potassium hydroxide* produces propene.

$$\underset{n\text{-Propyl bromide}}{CH_3CH_2CH_2Br} + KOH \xrightarrow[353-363\ K]{C_2H_5OH} \underset{\text{Propene}}{CH_3CH=CH_2} + KBr + H_2O$$

$$\underset{\text{Isopropyl bromide}}{CH_3-\underset{Br}{CH}-CH_3} + KOH \xrightarrow[353-363\ K]{C_2H_5OH} \underset{\text{Propene}}{CH_3CH=CH_2} + KBr + H_2O$$

In case the alkyl halide can undergo dehydrohalogenation in two different ways, the preferred alkene is the one which is more alkylated *i.e.*, which carries more number of alkyl groups attached to double bonded carbon atoms. This generalization is known as **SAYTZEFF RULE.** For example,

$$\underset{\text{2-Bromo-3-methylbutane}}{CH_3-\overset{CH_3}{CH}-\underset{Br}{CH}-CH_3} \xrightarrow{\text{alc. KOH}} \underset{\substack{(I)\\ (Major\ product)}}{CH_3-\overset{CH_3}{C}=CH-CH_3} + \underset{\substack{(II)\\ (Minor\ product)}}{CH_3-\overset{CH_3}{CH}-CH=CH_2}$$

The alkene (I) is trialkylated whereas alkene (II) is monoalkylated. Here, alkene (I) is the major product.

2. From Alcohols

Alkenes can be prepared **by dehydration** (removal of water molecule) of alcohols. The two common methods for carrying out dehydration are to heat the alcohol with either ***alumina*** or a *mineral acid* such as ***tetraoxophosphate(V) acid (H_3PO_4)*** or ***conc. tetraoxosulphate(VI) acid (H_2SO_4).***

In dehydration reaction OH group is lost from the α carbon while H atom is lost from the β carbon atom creating a double bond between the α and β carbon atoms. Some examples are:

$$\underset{\text{Ethanol}}{H-\overset{H}{\underset{H}{C}}-\overset{H}{\underset{OH}{C}}-H} \xrightarrow[443\ K]{\text{Conc. } H_2SO_4} \underset{\text{Ethene}}{H_2C=CH_2} + \mathbf{H_2O}$$

$$\underset{\text{Propan-2-ol}}{CH_3-\underset{OH}{CH}-CH_3} \xrightarrow[443\ K]{\text{Conc. } H_2SO_4} \underset{\text{Propene}}{CH_3-CH=CH_2} + H_2O$$

$$\underset{\text{Propan-1-ol}}{CH_3CH_2CH_2OH} \xrightarrow[443\ K]{\text{Conc. } H_2SO_4} \underset{\text{Propene}}{CH_3-CH=CH_2} + H_2O$$

$$\underset{\text{Cyclohexanol}}{C_6H_{11}OH} \xrightarrow[373-383\ K]{H_3PO_4} \underset{\text{Cyclohexene}}{C_6H_{10}} + H_2O$$

MECHANISM OF DEHYDRATION

The dehydration of alcohols does not proceed in a single step but takes place through the formation of an intermediate carbocation. The various steps involved are:

$$H_2SO_4 \longrightarrow H^+ + HSO_4^-$$

(*i*) *Protonation of alcohol*

$$H-\overset{H}{\underset{H}{C}}-\overset{H}{\underset{H}{C}}-\ddot{O}H + H^+ \rightleftharpoons \underset{\text{Protonated alcohol}}{H-\overset{H}{\underset{H}{C}}-\overset{H}{\underset{H}{C}}-\overset{+}{\ddot{O}}H_2}$$

(*ii*) *Loss of water molecule*

$$H-\overset{H}{\underset{H}{C}}-\overset{H}{\underset{H}{C}}-\overset{+}{\ddot{O}}H_2 \rightleftharpoons \underset{\text{Carbocation}}{H-\overset{H}{\underset{H}{C}}-\overset{H}{\underset{H}{C^+}}} + H_2\ddot{O}$$

(*iii*) *Loss of H^+ ion by carbocation*

$$H-\overset{H}{\underset{H}{C}}-\overset{H}{\underset{H}{C^+}} \underset{}{\overset{H_2SO_4^-}{\rightleftharpoons}} \underset{\text{Ethene}}{H_2C=CH_2} + H_2SO_4$$

41.5 PHYSICAL PROPERTIES OF ALKENES

1. Physical State. The first three members of alkenes are gases, the next fourteen are liquids while the higher members are waxy solids. Ethene is a colourless gas with faint sweet smell. Higher alkenes are colourless and odourless.

2. Melting and Boiling Points. The melting and boiling points of alkenes are also quite low and increase regularly with increase in molecular mass. For every $—CH_2$ added, the boiling point increases by 20–30 K. Comparison of melting and boiling points of geometric isomers indicates that the *cis*-isomer has a higher boiling point than the corresponding

Table 41.1. Physical Properties of Some Alkenes

Name	*Structure*	*Carbons*	*Boiling Point (K)*	*Density (g cm^{-3})*
Ethene	$CH_2{=}CH_2$	2	169	
Propene	$CH_3CH{=}CH_2$	3	226	0.52
2-Methylpropene	$(CH_3)_2C{=}CH_2$	4	267	0.59
But-1-ene	$CH_3CH_2CH{=}CH_2$	4	268	0.59
trans-but-2-ene	$H_3C(H)C{=}C(H)CH_3$ (trans)	4	274	0.60
cis-but-2-ene	$H_3C(H)C{=}C(H)CH_3$ (cis)	4	277	0.62
3-Methyl-but-1-ene	$(CH_3)_2CH{-}CH{=}CH_2$	5	298	0.65
Pent-1-ene	$CH_3CH_2CH_2{-}CH{=}CH_2$	5	303	0.64
trans-Pent-2-ene	$H_3C(H)C{=}C(H)CH_2CH_3$ (trans)	5	309	0.65
cis-Pent-2-ene	$H_3C(H)C{=}C(H)CH_2CH_3$ (cis)	5	310	0.66

trans-isomer. On the other hand the melting point of the *trans*-isomer is generally higher.

For example, melting and boiling points of *cis* and *trans* isomers of but-2-ene are given below:

	cis-But-2-ene	*trans*-But-2-ene
	$H_3C(H)C{=}C(H)CH_3$ (cis)	$H_3C(H)C{=}C(H)CH_3$ (trans)
Melting point:	134 K	167 K
Boiling point:	277 K	274 K

3. Solubility. Like alkanes, alkenes are also insoluble in water. They are soluble in non-polar solvents such as benzene, hexane, CCl_4, etc.

4. Density. Densities of alkenes show regular increase with increase in molecular mass. However, the maximum density of alkenes is about 0.8.Thus, ***all alkenes are lighter than water.***

41.6 CHEMICAL PROPERTIES OF ALKENES

The functional group in alkenes is the carbon-carbon double bond. A carbon-carbon double bond consists of a strong sigma bond and a relatively weaker π bond. The typical reactions of alkenes are the reactions in which the π bond breaks and two strong sigma bonds are formed in its place.

$$>C{=}C< + X{-}Y \longrightarrow -\underset{X}{\overset{|}{\underset{|}{C}}}-\underset{Y}{\overset{|}{\underset{|}{C}}}-$$

A reaction in which two molecules combine to yield a single product is called an ***addition reaction.*** Most of the addition reactions in alkenes are initiated by electrophilic species because the loosely held π electrons in alkenes serve as source of electrons for the attacking species. Thus, ***the characteristic reactions of alkenes are electrophilic addition reactions.***

1. Addition of Hydrogen

Alkenes react readily with hydrogen in the presence of *finely-divided nickel, platinum* or *palladium* as catalyst. The process is known as **catalytic hydrogenation.** The ultimate products of this reaction are ***alkanes.***

$$\underset{\text{Ethene}}{CH_2 = CH_2} + H_2 \xrightarrow{\text{Catalyst}} \underset{\text{Ethane}}{CH_3—CH_3}$$

$$\underset{\text{Propene}}{CH_3—CH{=}CH_2} + H_2 \longrightarrow \underset{\text{Propane}}{CH_3—CH_2—CH_3}$$

Hydrogenation is used in the manufacture of **margarine** from *vegetable oils.*

Do You Know?

Regular consumption of large quantities of saturated fats can be dangerous to your health as it may cause cardiovascular diseases.

2. Addition of Halogens

Halogens, especially chlorine and bromine add onto *alkenes* readily to form dihalogen derivatives. The reaction is carried out in inert solvents like CCl_4. For example, ethene reacts with bromine dissolved in carbon tetrachloride to give 1, 2-dibromoethane.

$$\underset{\text{Ethene}}{H_2C{=}CH_2} + Br_2 \xrightarrow{CCl_4} \underset{\text{1, 2-Dibromoethane}}{H—CHBr—CHBr—H}$$

$$\underset{\text{Propene}}{CH_3CH{=}CH_2} + Br_2 \longrightarrow \underset{\text{1, 2-Dibromopropane}}{CH_3—CHBr—CHBr—H}$$

During this reaction the reddish brown colour of bromine is decolorized.

MECHANISM OF ADDITION OF BROMINE TO ALKENES

Addition of bromine to alkenes takes place through the following two steps:

Step I: The *pi* electrons of alkene attack the bromine molecule expelling bromide ion. A **bromonium ion** is formed.

$$H_2C{=}CH_2 + :\ddot{Br}—\ddot{Br}: \longrightarrow \underset{\text{Bromonium ion}}{H—C(H)—C(H)—H \text{ (bridged by } :Br:^{+})} + :\ddot{Br}:^{-}$$

Step II: Bromonium ion is then attacked by a bromide ion from the back side which results in opening of the bromonium ion and formation of the final product.

$$\text{Bromonium ion} + :\ddot{Br}:^{-} \longrightarrow \underset{\text{1, 2-Dibromoethane}}{H—CHBr—CHBr—H}$$

ADDITION OF UNSYMMETRICAL REAGENTS

Most of the addition reactions of alkenes are those with *unsymmetrical reagents* like HCl, HBr and H_2O, etc. When such reagents react with alkenes, one part of the molecule attaches itself to one carbon atom whereas the other part attaches itself to the second carbon atom of the double bond.

Some examples are as follows:

1. Addition of Halogen Acids

Alkenes react with halogen acids to form alkyl halides.

$$\underset{\text{Ethene}}{H_2C{=}CH_2} + HCl \longrightarrow \underset{\text{Chloroethane}}{H—CH_2—CHCl—H}$$

In case the *alkene* and the *reagent* are both unsymmetrical then two products are theoretically possible. For example, in the reaction between propene and HBr (in the dark and in the absence of peroxides) there can be two addition products. Experiments reveal that 2-bromopropane is the major product while 1-bromopropane is the minor product.

$$\underset{\text{Propene}}{CH_3CH = CH_2} + HBr \xrightarrow[\text{Absence of peroxides}]{\text{Dark}}$$

$$\underset{\text{2-Bromopropane (90\%)}}{CH_3—CH(Br)—CH_3} + \underset{\text{1-Bromopropane (10\%)}}{CH_3CH_2CH_2Br}$$

Such addition reactions are governed by the empirical rule which was given by the Russian Chemist *Markownikov* (1869) after studying a series of similar reactions. This rule is known as the **Markownikov's rule.** The rule states that:

During the addition across an unsymmetrical double bond, the negative part of the adding molecule attaches itself to the carbon atom carrying less number of hydrogen atoms.

$$H_3C—CH=CH_2 + \overset{\delta^+}{H}—\overset{\delta^-}{X} \longrightarrow H_3C—\underset{}{\overset{X}{\overset{|}{C}H}}—CH_3$$

Unsymmetrical double bond | Adding molecule | Addition product

MECHANISM OF ADDITION OF HALOGEN ACIDS TO ALKENES

The electrophilic addition of hydrogen halides to alkenes follows the ionic mechanism as given below:

Step I.

$$—C=C— + \overset{\delta^+}{H}—\overset{\delta^-}{X} \longrightarrow —\overset{+}{C}—\underset{H}{C}— + X^-$$

Step II.

$$—\overset{+}{C}—\underset{H}{C}— + X^- \longrightarrow —\underset{X}{C}—\underset{H}{C}—$$

Thus, addition of halogen acids to alkenes takes place through a carbocation intermediate.

We can explain the Markownikov's rule in terms of stability of the intermediate carbocation. Let us consider the addition of halogen acids to propene. In the step 1 there are two possibilities of formation of a carbocation.

$$CH_3—CH=CH_2 + H—X \xrightarrow{\text{Path I}} CH_3—\overset{+}{C}H—CH_3\ (2°) \xrightarrow{X^-} CH_3—\underset{X}{CH}—CH_3 \text{ (Major product)}$$

$$CH_3—CH=CH_2 + H—X \xrightarrow{\text{Path II}} CH_3—CH_2—\overset{+}{C}H_2\ (1°) \xrightarrow{X^-} CH_3CH_2CH_2X \text{ (Minor product)}$$

Path I involves the addition of the proton to C-1 and results in the formation of a 2° carbocation while path II involves the addition of the proton to C-2 and results in the formation of a 1° carbocation. Since a 2° carbocation is more stable it is formed at a faster rate. Hence, the reaction takes place mainly through path I resulting in the formation of same product as predicted by Markownikov's rule. Thus, in terms of stability of the intermediate, Markownikov's rule may be stated as:

Electrophilic addition to alkenes takes place via formation of a more stable intermediate carbocation.

It may be noted that an alkene which reacts via a more stable intermediate carbocation would be more reactive. Thus, *iso*-butylene would be more reactive than propene because the former reacts via a 3° carbocation as intermediate as compared to a 2° carbocation in the case of propene.

Decreasing **order of reactivity of alkenes** towards electrophilic addition is

$$R_2C=CR_2 > R_2C=CHR > R_2C=CH_2 > RCH=CHR > RCH=CH_2 > CH_2=CH_2$$

2. Reaction with Bromine Water (Addition of HOBr)

Alkenes react with bromine water to form bromoalkanols. The reaction involves the overall addition of HOBr.

$$\underset{\text{Ethene}}{H_2C=CH_2} + HOBr \xrightarrow{Br_2 + H_2O} \underset{\text{2-Bromoethanol}}{H—\underset{HO}{\overset{H}{C}}—\underset{Br}{\overset{H}{C}}—H}$$

$$H_3C(H)C{=}CH_2 + HOBr \longrightarrow H_3C{-}CH(OH){-}CH_2Br$$

Propene 1-Bromopropan-2-ol

During this reaction the reddish brown colour of bromine water is decolorized.

3. Addition of Tetraoxosulphate(VI) Acid (H_2SO_4)

Alkenes react with cold, conc. tetraoxosulphate(VI) acid to form alkyl hydrogen sulphates. The addition of tetraoxosulphate(VI) acid takes place in accordance with Markownikov's rule.

$$>C{=}C< + H_2SO_4 \longrightarrow -\overset{|}{\underset{H}{C}}-\overset{|}{\underset{OSO_3H}{C}}-$$

Alkene Alkyl hydrogen sulphate

$$CH_3{-}CH = CH_2 + H_2SO_4 \longrightarrow CH_3{-}\underset{OSO_3H}{CH}{-}CH_3$$

This reaction may be used for removal of alkenes from gaseous mixtures.

Alkyl hydrogensulphates on hydrolysis with water or dilute H_2SO_4 produce alcohols.

$$-\overset{|}{\underset{H}{C}}-\overset{|}{\underset{OSO_3H}{C}}- + H_2O \xrightarrow{\text{Warm}} -\overset{|}{\underset{H}{C}}-\overset{|}{\underset{OH}{C}}- + H_2SO_4$$

4. Addition of Water (H_2O)

Water adds to **alkenes** in the presence of mineral acids. In this case also the addition is governed by the Markownikov's rule. For example, propene, on reacting with water, gives propan-2-ol as the major product.

$$H_2C{=}CH_2 + H_2O \xrightarrow{H_2SO_4} H{-}\overset{H}{\underset{H}{C}}{-}\overset{H}{\underset{OH}{C}}{-}H$$

Ethene Ethanol

$$CH_3CH{=}CH_2 + H_2O \xrightarrow{H_2SO_4} CH_3\underset{OH}{CH}CH_3$$

Propene Propan-2-ol (Major product)

The reaction of alkenes with water to give an alcohol is known as **hydration** and is an industrially important method for the manufacture of alcohols.

MECHANISM OF HYDRATION OF ALKENES

Hydration of alkenes takes place through carbocations as intermediates.

$$H_2C{=}CH_2 \underset{}{\overset{H^+}{\rightleftharpoons}} H{-}\overset{H}{\underset{+}{C}}{-}\overset{H}{\underset{H}{C}}{-}H \quad \text{(Carbocation)}$$

$$\Big\Downarrow \text{Addition of } H_2O$$

$$H{-}\overset{H}{\underset{OH}{C}}{-}\overset{H}{\underset{H}{C}}{-}H \overset{\text{Loss of } H^+}{\rightleftharpoons} H{-}\overset{H}{\underset{\overset{+}{OH_2}}{C}}{-}\overset{H}{\underset{H}{C}}{-}H$$

SOME OTHER REACTIONS OF ALKENES

1. Hydroxylation

Alkenes react with cold dilute **neutral** or **alkaline** $KMnO_4$ solution to give *vicinal* diols (or glycols).

$$2KMnO_4 + H_2O \longrightarrow 2KOH + 2MnO_2 + 3[O]$$

$$H_2C{=}CH_2 + H_2O + [O] \longrightarrow H{-}\overset{H}{\underset{OH}{C}}{-}\overset{H}{\underset{OH}{C}}{-}H$$

Ethane-1, 2-diol (Ethylene glycol)

During the reaction the purple colour of potassium tetraoxomanganate(VII) solution disappears and a brownish precipitate of manganese(IV) oxide is formed.

2. Combustion

Alkenes, like other hydrocarbons, burn in the presence of oxygen or air to give carbon(IV) oxide and water. A large amount of heat is produced during this reaction.

$$2C_nH_{2n} + 3nO_2 \longrightarrow 2nCO_2 + 2nH_2O + \text{Heat}$$

$$C_2H_4 + 3O_2 \longrightarrow 2CO_2 + 2H_2O + \text{Heat}$$

In limited supply of oxygen, alkenes on combustion give carbon(II) oxide and carbon in addition to carbon(IV) oxide and water vapour.

3. Polymerization

The process of combination of a large number of molecules of same or different substances under suitable conditions to form a giant molecule with very high molecular mass is called **polymerization.** The product obtained by virtue of a polymerization reaction is called a **polymer.**

When ethylene is heated with traces of oxygen under pressure, it polymerizes to *polyethylene* or *polythene*

$$\underset{\text{Ethylene}}{n\,CH_2{=}CH_2} \xrightarrow[\text{High pressure, } 0.01\%\ O_2]{\text{High temp.}} \underset{\text{Polyethylene}}{-(-CH_2-CH_2-)_n-}$$

Similarly, propylene on polymerization gives polypropylene.

$$\underset{\text{Propene}}{n\,CH_2{=}\underset{\displaystyle CH_3}{\underset{|}{CH}}} \longrightarrow \underset{\text{Polypropylene}}{-(-CH_2-\underset{\displaystyle CH_3}{\underset{|}{CH}}-)_n}$$

TESTS FOR ALKENES

1. Bromine Water Test

When ethene is passed through bromine water, it discharges the reddish brown colour of bromine.

2. Potassium Permanganate Test (Baeyer's Test)

When ethene is passed through alkaline (or acidified) potassium permanganate ($KMnO_4$) solution, the **pink** colour of the solution gets discharged. This test is known as the Baeyer's test.

It may be noted that:

—*The above two tests are given by all the alkenes. Alkenes can be distinguished from alkanes by these tests.*

—*All alkynes also give these tests.*

Therefore, these two tests cannot be used to distinguish between ethene and ethyne.

SOLVED EXAMPLE

Example 41.4. *Explain why ethene decolourises bromine water whereas ethane does not.*

Solution. Ethene being an unsaturated hydrocarbon reacts with bromine to form the addition product which is colourless. On the other hand, ethane being a saturated hydrocarbon does not react with bromine.

$$H-\overset{\displaystyle H}{\overset{|}{C}}{=}\overset{\displaystyle H}{\overset{|}{C}}-H + Br_2 \longrightarrow H-\underset{\displaystyle Br}{\underset{|}{\overset{\displaystyle H}{\overset{|}{C}}}}-\underset{\displaystyle Br}{\underset{|}{\overset{\displaystyle H}{\overset{|}{C}}}}-H$$

41.7 USES OF ALKENES

1. Ethene is used in the manufacture of **polyethylene** or **polythene**—a plastic material. Similarly, propene is used in the manufacture of **polypropene**. These polymers are used in making plastic bags, pipes, electrical insulation, etc.
2. Ethene is used in the **preparation of solvents** like ethylene glycol, dioxane, etc. Ethylene glycol is also used as an **antifreeze in automobile radiators**.
3. Ethene is a plant hormone which controls **growth, seed germination** and **fruit development.** Therefore, ethene is used for **artificial ripening** of fruits, **flower maturation**, etc.
4. Alkenes are used as a starting material for the manufacture of many compounds such as alkyl halides, ethylene oxide, **ethanol** and **other alkanols**, etc.
5. 1, 3-Butadiene is used in the manufacture of **synthetic rubber**.

SUMMARY

- **Alkenes** are unsaturated hydrocarbons. Their general formula is C_nH_{2n}. The functional group in alkenes is a carbon-carbon double bond.
- **The simplest alkene that exhibits isomerism** has molecular formula C_4H_8.
- Alkenes exhibit chain, position and geometric isomerism.
- Rotation about C—C double bond is hindered due to the presence of a *pi* bond.
- Alkenes can be prepared on a large scale by the cracking of alkanes.
- The hybrid state of a double bonded carbon atom in an alkene is sp^2.
- C—C bond length in ethene is 134 pm.
- C—C double bond is stronger than C—C single bond.
- The reactivity of alkenes is due to the presence of weak *pi* bond in alkenes.

EVALUATE YOURSELF

I. Objective Type Questions

Select the most appropriate choice from the options given as (a), (b), (c) and (d) after each question:

1. Which of the following represents an unsaturated open chain compound?
   (*a*) C_4H_{10} (*b*) C_5H_{12}
   (*c*) C_6H_{12} (*d*) C_3H_8.
2. The hybrid state of carbon atoms in ethene is
   (*a*) sp^3, sp^3 (*b*) sp^2, sp^2
   (*c*) sp^3, sp^2 (*d*) sp, sp.
3. The numbers of sigma and *pi* bonds in propene are respectively
   (*a*) 8 and 1 (*b*) 8 and 2
   (*c*) 6 and 2 (*d*) 6 and 1.
4. How many isomeric alkenes (including geometric isomers) are possible for the molecular formula C_4H_8?
   (*a*) 3 (*b*) 4
   (*c*) 2 (*d*) 5.
5. Which of the following alkenes can exhibit geometric isomerism?
   (*a*) $(CH_3)_2C = C(CH_3)_2$ (*b*) $CH_3CH = CHCH_3$
   (*c*) $(CH_3)_2C = CH_2$ (*d*) $CH_3CH = CH_2$.
6. Which of the following statements is ***not*** correct about geometric isomers?
   (*a*) *trans*-isomer is more stable than *cis*-isomer
   (*b*) *trans*-isomer has higher melting point
   (*c*) *trans*-isomer is more polar
   (*d*) *cis*-isomer has higher boiling point.
7. The IUPAC name for the compound $CH_3CH_2CH = C(CH_3)_2$ is
   (*a*) 1, 1-dimethylbut-1-ene (*b*) 3-hexene
   (*c*) 2-methylpent-1-ene (*d*) 2-methylpent-2-ene.
8. Which of the following will decolorize a solution of bromine in CCl_4?
   (*a*) CH_4 (*b*) C_2H_6
   (*c*) C_2H_4 (*d*) $CH_3CH_2CH_3$.

II. Fill in the Blanks

9. Complete the following sentences by supplying the appropriate words:
   (*i*) The general formula of alkenes is ....
   (*ii*) Propene on reaction with HBr yields ...... as the major product.
   (*iii*) Hydration of ethene yields ....
   (*iv*) Structure of 3, 3-dimethylpent-1-ene is ....
   (*v*) 2-Bromobutane on reaction with an alcoholic solution of KOH yields ...... as the major product.

III. Discussion Questions

10. Which aliphatic hydrocarbon series is represented by the formula C_nH_{2n}? Write structural formulae for all the members of this series with four carbon atoms.
11. What is geometric isomerism? What are the conditions for geometric isomerism in open chain compounds?
12. Write the IUPAC names of the following compounds:
    (*i*) $CH_3CH{=}C(C_2H_5)_2$
    (*ii*) C_6H_5—$CH_2CH_2CH{=}CH_2$ (benzene ring)
    (*iii*) $CH_2{=}C(CH_3)—CH{=}CH_2$.
13. How does propene react with
    (*i*) HBr
    (*ii*) Water in the presence of dil H_2SO_4. Give chemical equations for the reactions.
14. Write the structural formula for 2-methylbut-2-ene. What is the major product obtained when HBr adds to this compound? Give the IUPAC name of the product.
15. Name the products A to F in the following reactions:
    (*i*) $CH_3—CH(Br)—CH_3 \xrightarrow[\text{Heat}]{\text{alc. KOH}} A \xrightarrow{H_2O/H^+} B$
    (*ii*) $CH_3—CH_2—CH_2Br \xrightarrow[\text{Heat}]{\text{alc. KOH}} C \xrightarrow{\text{HBr}} D$
    (*iii*) $CH_3CH_2OH \xrightarrow[\text{Heat}]{\text{Conc. } H_2SO_4} E \xrightarrow{Br_2/H_2O} F$.
16. List some important uses of alkenes. How would you distinguish between ethene and ethane?
17. Explain why:
    (*i*) Ethene is more reactive than ethane.
    (*ii*) Ethene undergoes electrophilic addition reactions.
    (*iii*) *cis*-but-2-ene has higher boiling point than *trans*-but-2-ene.

18. With the help of a balanced chemical equation in each case, show how ethene may be converted into:
(*i*) ethane-1, 2-diol
(*ii*) 2-bromoethanol
(*iii*) 1, 2-dibromoethane.

19. Discuss the mechanism of acid catalysed hydration of alkenes. How is this reaction used in the manufacture of ethanol?

Answers

I. Objective Type Questions

1. (*c*)	**2.** (*b*)	**3.** (*a*)	**4.** (*b*)
5. (*b*)	**6.** (*c*)	**7.** (*d*)	**8.** (*c*).

II. Fill in the Blanks

9. (*i*) C_nH_{2n}
(*ii*) 2-bromopropane
(*iii*) ethanol
(*iv*) $CH_3CH_2—C(CH_3)_2—CH=CH_2$
(*v*) but-2-ene.

SECTION-6

UNIT-42

ALKYNES

Specific Objectives

After studying this unit, the student should be able to:

- describe the sources and characteristic properties of alkynes
- describe the nomenclature and isomerism in alkynes
- describe the preparation and chemical reactions of alkynes
- outline the uses of alkynes.

42.1 INTRODUCTION

Alkynes are aliphatic unsaturated hydrocarbons. They are characterised by the presence of a *triple bond* ($C \equiv C$) between the carbon atoms in their molecules. Each member of the alkyne series has four hydrogen atoms less than the corresponding alkane. Thus, the general formula of alkynes is C_nH_{2n-2}. The first member of the series is acetylene and the family as a whole is also known as ***acetylenes.***

42.2 NOMENCLATURE AND ISOMERISM

The **IUPAC name** of an alkyne is derived from the name of the corresponding alkane by replacing suffix-*ane* by *yne*.

In **common nomenclature,** the name of the simplest alkyne is acetylene. The names of the higher alkynes are given as substituted acetylenes.

Some members of **homologous series of alkynes** are given below:

Molecular Formula	*Common Name*	*IUPAC Name*
$HC \equiv CH$	Acetylene	Ethyne
$CH_3—C \equiv CH$	Methylacetylene	Propyne
$CH_3CH_2—C \equiv CH$	Ethylacetylene	But-1-yne
$CH_3CH_2CH_2—C \equiv CH$	*n*-Propylacetylene	Pent-1-yne

Ethyne and propyne have only one structural formula and hence they do not exhibit isomerism. Butyne can have two structures as shown below:

Compounds having molecular formula C_nH_{2n} are alkenes. Thus, C_4H_8 and C_5H_{10} are alkenes. Compounds having molecular formula C_nH_{2n-2} are alkynes.

Thus, C_7H_{12} and C_2H_2 are alkynes.

$CH_3—CH_2—C \equiv CH$ (But-1-yne) $\quad$ $CH_3—C \equiv C—CH_3$ (But-2-yne)

But-1-yne and but-2-yne are **position isomers** as they differ in the position of the triple bond. The higher alkynes can exhibit **chain isomerism** in addition to position isomerism. For example, for pentyne, C_5H_8 the following isomeric alkynes are possible.

$CH_3—CH_2—CH_2—C \equiv CH$
Pent-1-yne
I

$CH_3—CH_2—C\equiv C—CH_3$
Pent-2-yne
II

$CH_3—CH(CH_3)—C \equiv CH$
3-Methylbut-1-yne
III

The compounds I and II are position isomers.

The compounds I and III or II and III are chain isomers.

SOLVED EXAMPLES

Example 42.1. *Write structures and IUPAC names of different isomers corresponding to the 5th member of the alkyne series.*

Solution. 5th member of the alkyne series has the molecular formula C_6H_{10}. The possible isomers are:

(*i*) $HC \equiv C—CH_2—CH_2—CH_2—CH_3$
Hex-1-yne

(*ii*) $CH_3—C \equiv C—CH_2—CH_2—CH_3$
Hex-2-yne

(*iii*) $CH_3—CH_2—C \equiv C—CH_2—CH_3$
Hex-3-yne

(*iv*) $HC \equiv C—CH(CH_3)—CH_2—CH_3$
3-Methylpent-1-yne

(*v*) $HC \equiv C—CH_2—CH(CH_3)—CH_3$
4-Methylpent-1-yne

(*vi*) $CH_3—C \equiv C—CH(CH_3)—CH_3$
4-Methylpent-2-yne

(*vii*) $HC \equiv C—C(CH_3)_2—CH_3$
3,3-Dimethylbut-1-yne

Example 42.2. *Classify the following compounds as alkanes, alkenes and alkynes:*

C_4H_8, C_5H_{12}, C_7H_{12}, C_5H_{10}, C_3H_8, C_2H_2.

Solution. Compounds having molecular formula C_nH_{2n+2} are alkanes. Thus, C_5H_{12} and C_3H_8 are alkanes.

Compounds having molecular formula C_nH_{2n} are alkenes. Thus, C_4H_8 and C_5H_{10} are alkenes.

Compounds having molecular formula C_nH_{2n-2} are alkynes. Thus, C_7H_{12} and C_2H_2 are alkynes.

42.3 STRUCTURE OF TRIPLE BOND

The simplest alkyne is ethyne. Both the carbon atoms in ethyne are *sp*-hybridised. The *sp*-hybrid orbital of one carbon atom overlaps axially with a similar orbital of the other carbon atom to form a C—C sigma bond. The remaining *sp*-hybrid orbital of each carbon atom overlaps axially with 1*s*-orbital of each of the two hydrogen atoms to form two C—H sigma bonds. H—C—C bond angle is 180°. Each carbon atom also has two unhybridised *p*-orbitals which are perpendicular to each other and also to the inter-nuclear axis. These *p*-orbitals undergo sideways overlap to form two *pi*-bonds (Fig. 42.1).

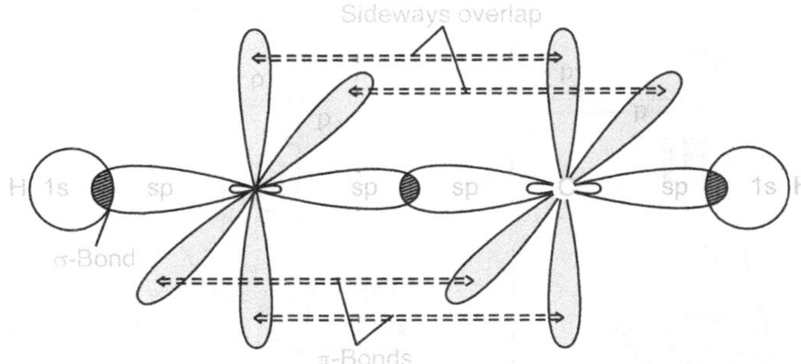

Fig. 42.1. Orbital diagram of ethyne.

Thus, a carbon-carbon triple bond consists of one sigma bond and two *pi*-bonds. The strength of C ≡ C bond is 823 kJ mol^{-1}. C ≡ C bond is stronger than C = C bond (605 kJ mol^{-1}) and C—C bond (347 kJ mol^{-1}). The bond length of a C ≡ C bond is 120 pm. The C ≡ C bond is shorter than the C = C bond (134 pm) which is also shorter than the C—C bond (154 pm).

Because of *sp*-hybridisation of the two carbon atoms, **ethyne molecule is linear**.

42.4 PREPARATION OF ALKYNES

Some of the important methods of preparation of alkynes are:

1. By the Action of Calcium Carbide and Water

Ethyne is prepared **on a large scale** by the reaction of water with calcium carbide.

$$\underset{\text{Calcium carbide}}{CaC_2} + 2H_2O \longrightarrow Ca(OH)_2 + \underset{\text{Ethyne}}{C_2H_2}$$

Calcium carbide is prepared by heating quicklime with coke.

$$\underset{\text{Quicklime}}{CaO} + \underset{\text{Coke}}{3C} \xrightarrow{2300\text{ K}} \underset{\text{Calcium carbide}}{CaC_2} + CO$$

Ethyne can also be prepared on **an Industrial scale** by pyrolysis of methane at a very high temperature of about 1800 K.

$$\underset{\text{Methane}}{2CH_4} \xrightarrow{\text{Heat}} \underset{\text{Ethyne}}{C_2H_2} + 3H_2$$

ACTIVITY 42.1	**Laboratory Preparation of Ethyne (Acetylene)**

Ethyne is prepared in the laboratory by the action of water on calcium carbide.

$$\underset{\text{Calcium carbide}}{CaC_2} + 2H_2O \longrightarrow Ca(OH)_2 + \underset{\text{Ethyne}}{C_2H_2}$$

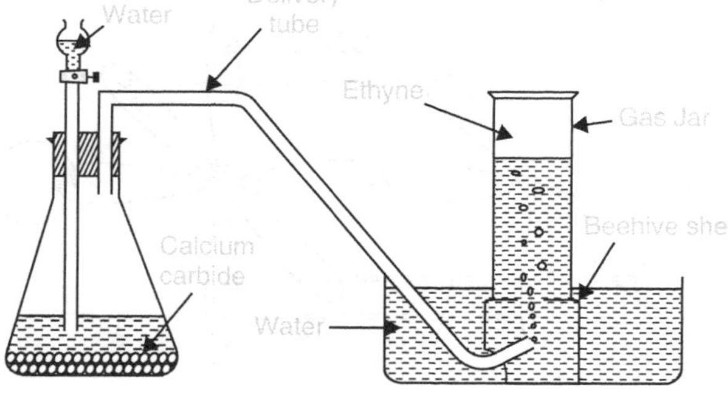

Fig. 42.2. Preparation of acetylene.

- Take small lumps of calcium carbide in a conical flask fitted with a dropping funnel and a side tube as shown in Fig. 42.2.
- Add water to calcium carbide, acetylene gas is produced.
- Collect the gas by downward displacement of water.

2. Dehydrohalogenation of Dihalides

Alkynes are prepared by dehydrohalogenation of *vicinal* dihalides, usually bromides, using an alcoholic solution of potassium hydroxide.

$$\underset{\text{1, 2-Dibromoethane}}{BrCH_2—CH_2Br} + 2KOH\,(alc) \xrightarrow{\text{Heat}} \underset{\text{Ethyne}}{HC{\equiv}CH} + 2KBr + 2H_2O$$

3. Formation of Higher Alkynes from Acetylene

Higher alkynes may be prepared by the action of alkyl halides on sodium ethynide. Sodium ethynide can be obtained from acetylene by the action of sodium or sodamide.

$$HC \equiv CH + \underset{\text{Sodamide}}{NaNH_2} \xrightarrow{\text{Diethyl ether}} \underset{\text{Sodium ethynide}}{HC \equiv \overset{-}{C}\overset{+}{Na}} + NH_3$$

$$2HC \equiv CH + 2Na \longrightarrow 2HC \equiv \overset{-}{C}Na^+ + H_2$$

Ethynides react with alkyl halides to form higher alkynes

$$HC \equiv \overset{-}{C}\overset{+}{Na} + \underset{\text{Ethyl bromide}}{C_2H_5Br} \longrightarrow \underset{\text{But-1-yne}}{HC \equiv CC_2H_5} + NaBr$$

Use of excess of $NaNH_2$ in the first step gives disodium ethynide.

$$HC \equiv CH + \underset{\text{Excess}}{2NaNH_2} \longrightarrow \underset{\text{Disodium ethynide}}{NaC \equiv CNa} + NH_3$$

$$NaC \equiv CNa + 2CH_3Br \longrightarrow \underset{\text{But-2-yne}}{CH_3C \equiv CCH_3} + 2NaBr$$

42.5 PHYSICAL PROPERTIES OF ALKYNES

1. Physical State

The first three alkynes are gases while those containing five to thirteen carbon atoms are liquids and higher alkynes are solids.

2. Melting and Boiling Points

The melting and boiling points of alkynes are quite low and increase regularly with increase in molecular mass. *Alkynes are less volatile than alkanes and alkenes.*

3. Solubility

Alkynes are insoluble in water but are soluble in organic solvents such as benzene, hexane, ether, carbon tetrachloride, etc.

4. Density

All alkynes are lighter than water. Their densities increase regularly with increase in molecular mass.

The physical properties of some alkynes are listed in Table 42.1.

Table 42.1. Physical Properties of Some Alkynes

Alkyne	*Number of Carbon atoms (n)*	*Molecular Formula*	*b.p. (K)*	*Density (g cm^{-3})*
Ethyne	2	$H—C≡C—H$	189	0.62
Propyne	3	$CH_3—C≡C—H$	250	0.67
But-1-yne	4	$CH_3—CH_2—C≡CH$	281	0.67
But-2-yne	4	$CH_3—C≡C—CH_3$	300	0.69
Pent-1-yne	5	$CH_3CH_2CH_2C≡CH$	313	0.70
Pent-2-yne	5	$CH_3CH_2C≡C—CH_3$	328	0.71
3-Methylbut-1-yne	5	$(CH_3)_2CH—C≡CH$	301	0.67
Hex-1-yne	6	$CH_3CH_2CH_2CH_2—C≡CH$	344	0.72
Hex-2-yne	6	$CH_3CH_2CH_2—C≡C—CH_3$	357	0.73
Hex-3-yne	6	$CH_3CH_2—C≡C—CH_2CH_3$	355	0.73
3, 3-Dimethylbut-1-yne	6	$(CH_3)_3C—C≡CH$	311	0.67

ACTIVITY 42.2 Plotting a graph of boiling points of Alkynes

Using data from Table 42.1 plot a graph of boiling points of *normal* 1-alkynes against the number of carbon atoms *n*. Take *n* along the horizontal axis.

42.6 CHEMICAL REACTIONS OF ALKYNES

Alkynes are unsaturated hydrocarbons. Like alkenes characteristic reactions of alkynes are electrophilic addition reactions. However, alkynes are less reactive than alkenes towards electrophilic addition reactions.

1. Addition of Hydrogen

Alkynes react readily with hydrogen in the presence of *finely-divided nickel, platinum* or *palladium* as catalyst. The process is known as **catalytic hydrogenation.** The ultimate product of this reaction is an ***alkane.***

$$\underset{\text{Ethyne}}{HC \equiv CH} + H_2 \xrightarrow{\text{Catalyst}} \underset{\text{Ethene}}{CH_2 = CH_2}$$

$$\text{Ethene} + H_2 \xrightarrow{\text{Catalyst}} \underset{\text{Ethane}}{CH_3—CH_3}$$

The hydrogenation of **alkynes** can be restricted to *alkenes* by employing suitable reaction conditions. For example, use of *Lindlar's catalyst* (Pd supported over $CaCO_3$ which is partially poisoned by lead acetate) can restrict the hydrogenation to the alkene stage.

$$\underset{\text{Propyne}}{CH_3C \equiv CH} + H_2 \xrightarrow[(CH_3COO)_2Pb]{Pd/CaCO_3} \underset{\text{Propene}}{CH_3CH = CH_2}$$

Reduction of alkynes to the alkene can also be carried out with sodium or lithium in liquid ammonia.

$$\underset{\text{Propyne}}{CH_3—C \equiv CH} \xrightarrow[+2H]{Na,\ NH_3(liq.)} \underset{\text{Propene}}{CH_3—CH = CH_2}$$

In case of non-terminal alkynes, the reduction to alkene can result in a *cis* or *trans* alkene. It is found that the reduction using Lindlar's catalyst results in a *cis*-alkene and the reduction using Na/liq. NH_3 leads to the formation of a *trans*-alkene.

$$\underset{\text{2-Butyne}}{CH_3—C \equiv C—CH_3} \xrightarrow[+H_2]{\text{Lindlar's Catalyst}} \underset{cis\text{-But-2-ene}}{(H)(H_3C)C = C(H)(CH_3)}$$

$$\underset{\text{2-Butyne}}{CH_3—C \equiv C—CH_3} \xrightarrow[+2H]{Na/liq.\ NH_3} \underset{trans\text{-But-2-ene}}{(CH_3)(H)C = C(H)(CH_3)}$$

2. Addition of Halogens

Halogens, especially chlorine and bromine add onto *alkynes* readily to produce a tetra-halogen derivatives. The reaction is carried out in inert solvent like carbon tetrachloride.

Due to the presence of two π-bonds, each molecule of the alkyne can react with two molecules of the halogen. For example, the addition of bromine to ethyne can be controlled to give 1, 2-dibromoethene or 1, 1, 2, 2-tetrabromoethane.

$$\underset{\text{Ethyne}}{H—C \equiv C—H} + Br_2 \xrightarrow{CCl_4} \underset{\text{1, 2-Dibromoethene}}{H—C(Br)=C(Br)—H}$$

$$\xrightarrow{+Br_2} \underset{\text{1, 1, 2, 2-Tetrabromoethane}}{H—CBr_2—CBr_2—H}$$

During the addition of bromine to alkynes, the reddish brown colour of bromine disappears.

3. Addition of Halogen Acids

In an alkynes two molecules of a halogen acid add across the triple bond to form a dihalogen derivative as product.

$$\underset{\text{Ethyne}}{H—C \equiv C—H} + HCl \longrightarrow \underset{\text{Chloroethene (Vinyl chloride)}}{H—C(H)=C(Cl)—H}$$

$$\xrightarrow{+HCl} \underset{\text{1, 1-Dichloroethane}}{H—CH_2—CCl_2—H}$$

4. Addition of Water (Hydration)

Alkynes undergo hydration in the presence of 60% H_2SO_4 and mercury(II) tetraoxosulphate (VI) as catalyst at about 330 K. The products are carbonyl compounds. In case of ethyne, ethanal is produced. The initially formed vinyl alcohol is unstable and rearranges to the more stable ethanal.

$$\underset{\text{Ethyne}}{H—C \equiv C—H} + H_2O \xrightarrow[H_2SO_4,\ 330\ K]{HgSO_4}$$

$$\left[\underset{\text{Vinyl alcohol (unstable)}}{H_2C=C(H)OH}\right] \xrightleftharpoons{\text{Rearrangement}} \underset{\text{Ethanal}}{H—CH_2—C(=O)—H}$$

Similarly, propyne reacts with water under similar conditions to give propanone.

$$\underset{\text{Propyne}}{CH_3C \equiv CH} + H_2O \xrightarrow[330\ K]{Hg^{2+}/H^+} \left[CH_3C(OH)=CH_2\right]$$

$$\xrightleftharpoons{\text{Rearrangement}} \underset{\text{Propanone}}{CH_3—C(=O)—CH_3}$$

5. Combustion

Alkynes burn with a luminous and sooty flame when heated in the presence of air to form carbon dioxide and water. The reaction is highly exothermic.

$$2C_2H_2 + 5O_2 \longrightarrow 4CO_2 + 2H_2O + \text{Heat}$$

The heat evolved during the reaction is employed as oxyacetylene flame for welding purposes.

Ethyne burns with a sooty flame because it has a very high percentage of carbon. Oxygen in the air is unable to oxidise all the carbon of ethyne to carbon dioxide. Due to unburnt carbon particles, the flame becomes sooty.

6. Oxidation with Alkaline $KMnO_4$ Solution

Alkynes are oxidised with alkaline $KMnO_4$ to a variety of products depending upon the structure. For example, ethyne is oxidized to oxalic acid.

$$\underset{\text{Ethyne}}{H—C \equiv C—H} + 4[O] \xrightarrow{\text{alk. } KMnO_4} \underset{\text{Oxalic acid}}{COOH—COOH}$$

During the reaction, the pink colour of $KMnO_4$ gets discharged. Hence, this reaction can be employed for detection of unsaturation in a molecule.

7. Polymerisation in Alkynes

When acetylene is passed through a red hot tube of iron or quartz it trimerises to form benzene

$$\underset{\text{Ethyne}}{3H—C \equiv C—H} \xrightarrow[873\ K]{\text{Red hot Fe tube}} \underset{\text{Benzene}}{C_6H_6}$$

This is the best route for entering from aliphatic to aromatic compounds.

This is an example of **cyclic trimerisation.**

Four molecules of acetylene polymerise under high pressure in the presence of nickel cyanide.

HC≡CH, CH≡CH, CH≡CH, HC≡CH (Acetylene) $\xrightarrow[\text{Pressure}]{Ni(CN)_2}$ Cyclooctatetraene

It is an example of **cyclic tetramerisation.**

Linear polymerisation of ethyne, under suitable conditions, yields polyethyne

$$2n\ HC \equiv CH \longrightarrow \text{—(—CH = CH—CH = CH—)}_n\text{—}$$
Polyethyne

Polyethyne is a high molecular mass polyene. Under special conditions this polymer conducts electricity. It can be used for making electrodes in batteries. Polyethyne electrodes are lighter and cheaper than metal electrodes.

In the presence of $CuCl/NH_4Cl$, *acetylene* gives vinyl acetylene and finally divinyl acetylene.

$$2H\text{—}C \equiv C\text{—}H \xrightarrow{CuCl,\ NH_4Cl} \underset{\text{But-1-en-3-yne}}{H\text{—}C \equiv C\text{—}CH = CH_2}$$

$$\xrightarrow[CH \equiv CH]{CuCl,\ NH_4Cl} \underset{\text{Divinyl acetylene}}{H_2C{=}CH\text{—}C{\equiv}C\text{—}CH{=}CH_2}$$

But-1-en-3-yne reacts with one mole of HCl to form **chloroprene** which is the starting material for the manufacture of **neoprene rubber**.

$$H\text{—}C{\equiv}C\text{—}CH{=}CH_2 + HCl \longrightarrow H_2C = \underset{\underset{Cl}{|}}{C}\text{—}CH = CH_2$$
Chloroprene
(2-chloro-1,3-butadiene)

8. Acidic Character of Alkynes

Terminal alkynes (1-alkynes) behave as very weak acids. They react with strong bases **like $NaNH_2$** to form acetylides. The acetylenic hydrogen atom is removed as a proton by the strong base leading to the formation of a stable acetylide.

$$H\text{—}C{\equiv}C\text{—}H + NaNH_2 \longrightarrow \underset{\text{(Sodium acetylide)}}{H\text{—}C \equiv \overset{-\ +}{CNa}} + NH_3$$

1-Alkynes also react with sodium metal to produce hydrogen gas.

$$\underset{\text{Ethyne}}{HC \equiv CH} + Na \longrightarrow \underset{\text{Sodium acetylide}}{HC \equiv \overset{-\ +}{CNa}} + \frac{1}{2} H_2$$

Acetylides react with alkyl halides to produce acetylene homologues.

$$HC{\equiv}\overset{-\ +}{CNa} + CH_3Br \longrightarrow \underset{\text{Propyne}}{HC{\equiv}C\text{—}CH_3}$$

$$HC{\equiv}\overset{-\ +}{CNa} + C_2H_5I \longrightarrow \underset{\text{But-1-yne}}{HC{\equiv}C\text{—}C_2H_5} + NaI$$

TESTS FOR ETHYNE

1. Bromine Water Test

When ethyne is passed through bromine water, the reddish brown colour of bromine water gets discharged. This test is also given by other unsaturated compounds such as ethene and other alkenes and alkynes.

2. $KMnO_4$ Test (Baeyer's Test)

When ethyne is passed through an alkaline (or acidified) solution of $KMnO_4$, the pink colour of $KMnO_4$ gets discharged. This reaction is also given by other unsaturated hydrocarbons such as ethene and other alkenes.

The above two reactions cannot be used to distinguish between an alkene and an alkyne.

3. Silver Nitrate Test

When ethyne is passed through an ammoniacal solution of silver nitrate it gives a **white precipitate.**

$$H\text{—}C{\equiv}C\text{—}H + 2[Ag(NH_3)_2]NO_3 \longrightarrow \underset{\text{White ppt.}}{Ag\text{—}C{\equiv}C\text{—}Ag} + 2NH_4NO_3 + 2NH_3$$

This reaction is not given by ethene and hence can be used to distinguish between ethene and ethyne.

4. Cuprous Chloride Test

When ethyne is passed through an ammoniacal solution of cuprous chloride it gives a **reddish brown precipitate.**

$$H\text{—}C{\equiv}C\text{—}H + 2[Cu(NH_3)_2]Cl \longrightarrow \underset{\text{Reddish brown ppt.}}{Cu\text{—}C{\equiv}C\text{—}Cu} + 2NH_4Cl + 2NH_3$$

This reaction can also be used to distinguish between ethene and ethyne. Ethene does not give reddish brown precipitate with ammoniacal cuprous chloride solution.

SOLVED EXAMPLES

Example 42.3. *How would you distinguish between (i) 1-Butyne and 2-Butyne (ii) Propene and Propyne?*

Solution. (*i*) Upon treatment with amm. solution of $AgNO_3$, 1-butyne would give a white ppt. whereas 2-butyne would not react.

(*ii*) Upon treatment with amm. solution of $AgNO_3$, propyne would give a white ppt. whereas propene would not react.

Example 42.4. *Suggest a method to separate a mixture of ethane, ethene and ethyne.*

Solution. The flow sheet for the separation is as follows:

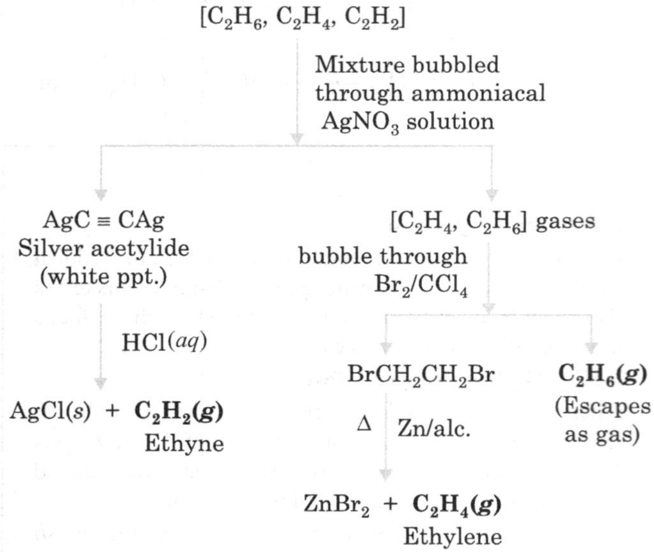

42.7 USES OF ALKYNES

Alkynes are used in the preparation of many other compounds. The **uses of ethyne**, the first member of alkynes are listed below:

1. Ethyne is used as the basic material for making many other organic compounds such as ethanoic acid, ethanal, acrylic acid, etc.
2. Ethyne is used for preparing starting material for many useful polymers. For example, vinyl chloride is the starting material for **PVC** whereas chloroprene is the starting material for **synthetic rubber neoprene**.
3. For illumination as in hawker's lamps.
4. For producing **oxy-acetylene flame** used in cutting and welding of iron.
5. For the **artificial ripening of fruits**.
6. Many organic **solvents such as trichloroethane** are prepared starting from ethyne.

SUMMARY

- Alkynes are the unsaturated hydrocarbons having the general formula C_nH_{2n-2}. They contain a carbon-carbon triple bond (—C≡C—) as the functional group.
- The simplest alkyne is ethyne (or acetylene).
- The simplest alkyne that can exhibit isomerism contains four carbon atoms.
- The carbon atoms in ethyne are *sp*-hybridized.
- Calcium carbide on reaction with water gives ethyne
$$CaC_2 + 2H_2O \longrightarrow Ca(OH)_2 + C_2H_2$$
- The first three alkynes exist in the gaseous state at room temperature.
- The characteristic reactions of alkynes are electrophilic addition reactions.
- Alkynes on complete hydrogenation yield alkanes. In this reaction 2 moles of hydrogen are consumed for each mole of alkyne.
- Lindlar's catalyst is palladium supported over $CaCO_3$, and partially poisoned by lead acetate.
- Ethyne decolorizes the colour of bromine water.
- Alkynes burn with a luminous, sooty flame due to the high percentage of carbon in them.
- Terminal alkynes or 1-alkynes exhibit some acidic character.

EVALUATE YOURSELF

I. Objective Type Questions

Select the most appropriate choice from the options given as (a), (b), (c) and (d) after each question:

1. The simplest alkyne which can exhibit chain isomerism has the molecular formula

   (*a*) C_4H_6 (*b*) C_5H_8
   (*c*) C_3H_4 (*d*) C_4H_8.

2. The IUPAC name of the compound, $(CH_3)_3CCH_2C{\equiv}CH$ is

   (*a*) 4, 4, 4-trimethylbut-1-yne
   (*b*) 2, 2-dimethylpent-4-yne
   (*c*) 4, 4-dimethylpent-1-yne
   (*d*) 1, 1, 1-trimethylbut-1-yne.

3. Which of the following represents an alkyne?

   (*a*) C_6H_{12} (*b*) C_6H_{14}
   (*c*) C_6H_6 (*d*) C_6H_{10}.

4. Which of the following compounds would give a white precipitate on reaction with ammoniacal silver nitrate solution?

   (*a*) $CH_3—C{\equiv}C—CH_3$
   (*b*) $CH_3—CH{\equiv}CH—CH_3$
   (*c*) $CH_3—CH_2—C{\equiv}CH$
   (*d*) $CH_3—CH_2—CH{=}CH_2$

5. Which of the following is the most acidic compound?
(*a*) $CH_3—CH_3$ (*b*) $HC≡CH$
(*c*) $CH_2=CH_2$ (*d*) $CH_3CH=CH_2$

II. Fill in the Blanks

6. Complete the following sentences by supplying appropriate words:
(*i*) In propyne molecule the total number of sigma bonds is ......
(*ii*) Calcium carbide is prepared by heating coke with ......
(*iii*) The physical state of propyne at room temperature is ......
(*iv*) One mole of ethyne on complete hydrogenation absorbs .... mole(*s*) of hydrogen.
(*v*) But-2-yne on reaction with sodium metal in the presence of liquid ammonia yields ....
(*vi*) Ethyne on hydration in the presence of dil. H_2SO_4 and mercuric ions yields ....

III. Discussion Questions

7. What are alkynes? Give the IUPAC and common name of the simplest alkyne.
8. Why does ethyne burn with a sooty flame?
9. Write structures and IUPAC names for all isomeric alkynes having molecular formula C_6H_{10}.
10. Discuss the structure of ethyne.
11. How would you obtain the following compounds from ethyne?
(*i*) Propyne
(*ii*) 1,1-dichloroethane
(*iii*) Chloroprene
12. How would you distinguish between ethane and ethyne? Give two tests.
13. Give one chemical test to distinguish between ethene and ethyne.
14. With the help of a labelled diagram, explain how ethyne is prepared in the laboratory. Give a chemical equation for the reaction that takes place.
15. Write a chemical equation for the combustion of ethyne.
16. Which of the following are unsaturated hydrocarbons?
C_4H_{10}, C_6H_{12}, C_5H_8, C_3H_6, C_2H_6.
17. Why does ethyne decolourise bromine water whereas ethane does not do so?
18. Complete the following equation:
$$CaC_2 + H_2O \longrightarrow$$
19. What happens when ethyne is passed through a red hot iron tube?
20. How is ethyne converted into divinyl acetylene?
21. Describe the following:
(*i*) Hydrogenation of but-2-yne to *cis*-but-2-ene.
(*ii*) Hydration of ethyne in the presence of dil. H_2SO_4 and mercuric ions.
22. Complete the following equations:
(*i*) $H—C \equiv C—H + 2HCl \longrightarrow$
(*ii*) $H—C \equiv C—H + 2Br_2 \longrightarrow$
(*iii*) $H—C \equiv C—H + 2H_2 \xrightarrow[200°C]{Ni}$
23. How is acetylene prepared from methane? Give four uses of ethyne.
24. Give four physical characteristics of ethyne.
25. You are given three jars containing ethane, ethene and ethyne. How would you identify which gas jar contains which gas?

Answers

I. Objective Type Questions

1. (*b*) **2.** (*c*) **3.** (*d*)
4. (*c*) **5.** (*b*).

II. Fill in the Blanks

6. (*i*) 8 (*ii*) quick-lime
(*iii*) gas (*iv*) 2
(*v*) *trans*-but-2-ene (*vi*) ethanal.

SECTION-6

UNIT-43

BENZENE

Specific Objectives

After studying this unit, the student should be able to:

- describe the structure and stability of benzene
- describe the reactions of benzene
- explain the differences between the reactivity of benzene and alkene towards certain reagents.

43.1 INTRODUCTION

Aromatic hydrocarbons or arenes are the compounds of carbon and hydrogen which contain at least one benzene type ring (hexagonal ring of carbons) in their molecules. We have hydrocarbons of the benzene series (containing one benzene ring), hydrocarbons of naphthalene series (two fused benzene rings) and so on. Names and structures of some common arenes are given below:

Benzene

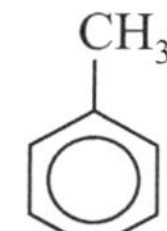

Methylbenzene (Toluene)

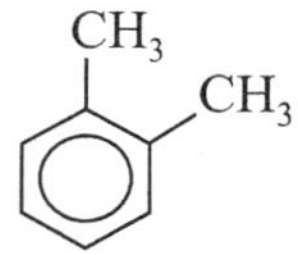

1, 2-Dimethylbenzene

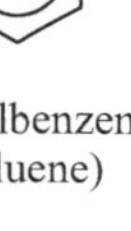

1, 3-Dimethylbenzene

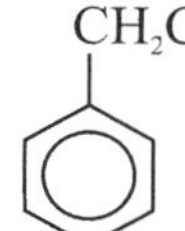

Ethylbenzene

Iso-propylbenzene (Cumene)

Naphthalene

1-Methylnaphthalene

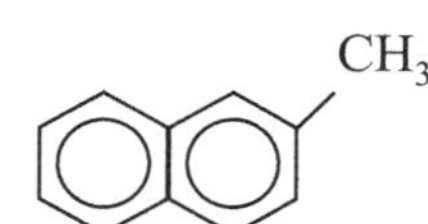

2-Methylnaphthalene

In this unit, you will study about the structure and properties of the simplest arene and benzene.

43.2 STRUCTURE OF BENZENE

The molecular formula of benzene is C_6H_6. It contains eight hydrogen atoms less than the corresponding parent hydrocarbon, *i.e.*, hexane (C_6H_{14}). It took several years to assign a structural formula to benzene because of its unusual stability and peculiar properties.

KEKULE STRUCTURE OF BENZENE

In 1865, Kekule proposed the first acceptable ring structure for benzene. In this structure there is a hexagonal ring of carbon atoms distributed in a symmetrical manner, with each carbon atom carrying one hydrogen atom. The fourth valence of carbon atoms is fulfilled by the presence of an alternate system of single and double bonds as shown:

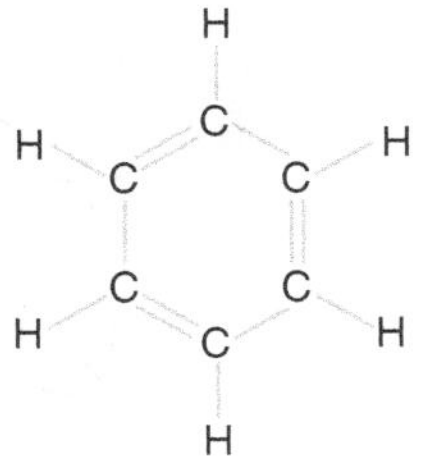

The above formula had many drawbacks as described below:

1. The presence of three double bonds should make the molecule highly reactive towards addition reactions. But contrary to this, benzene behaves like saturated hydrocarbons.
2. There should be two isomers of a 1, 2 disubstituted benzene as shown in Fig. 43.1. In one of the isomers, the bond between the substituted carbon atoms is a single bond while in the other it is a double bond. Actually only one 1, 2-disubstituted (or ortho) isomer is known.

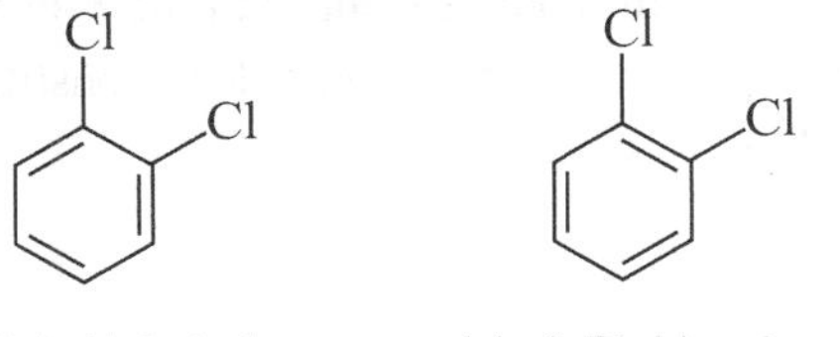

Fig. 43.1. Kekule isomers of 1, 2-Dichlorobenzene.

While the Kekule formula could not explain the difference in properties between benzene and alkenes based on his structure, he explained the lack of isomers as in Fig. 43.1 by postulating a rapid interchange in the position of the double bonds as follows:

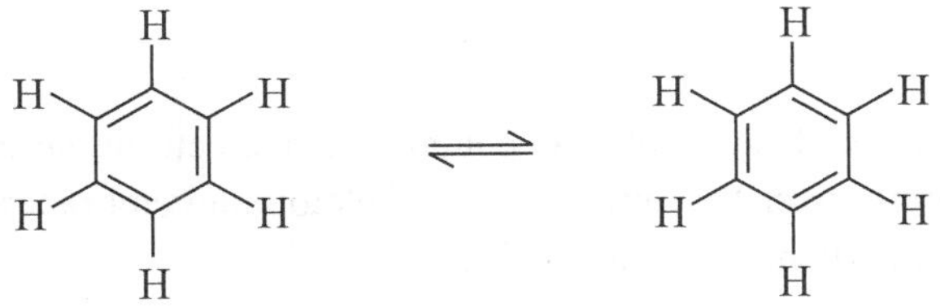

This structure came to be known as the Kekule's dynamic formula.

RESONANCE STRUCTURE OF BENZENE

X-ray studies indicate that all the carbon-carbon bonds in benzene are equivalent and have bond length 140 pm which is intermediate between C—C single bond (154 pm) and C=C bond (134 pm). This shows that double bonds in benzene differ from those of alkenes.

Structure of benzene can be explained on the basis of resonance. Benzene (C_6H_6) may be assigned following two structures A and B. Structures A and B have same arrangement of atoms and differ only in electronic arrangement. Any of these structures alone cannot explain all the properties of benzene.

Structures A and B are known as *resonating or canonical* structures of benzene. The actual structure of benzene is different from both A and B, and cannot be represented by conventional formulae. The actual structure of benzene lies somewhere in between A and B and may be represented as C, referred to as a **resonance hybrid.**

To indicate two structures which are resonance forms of the same compound, a double headed arrow is used as shown in Fig. 43.2.

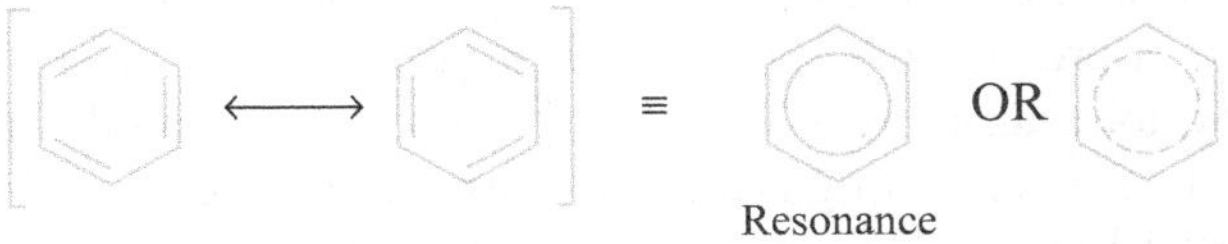

Resonance hybrid

(A) (B) (C)

Fig. 43.2. Resonance structure of benzene.

The resonance hybrid is more stable than any of the contributing (or canonical) structures. *The difference between the energy of the most stable contributing structure and the energy of the resonance hybrid* is known as **resonance energy.**

Since the contributing structures (A) and (B) are of exactly the same energy they make equal contribution to the resonance hybrid.

The resonance energy of benzene has been found to be 152 kJ/mole. The value of the resonance energy has been determined by studying the enthalpy of hydrogenation and enthalpy of combustion of benzene.

THE UNUSUAL STABILITY OF BENZENE

Reluctance of benzene to undergo alkene type reactions indicates that it must be unusually stable. The evidence for stability of benzene is obtained by comparing experimental and calculated values of enthalpies of hydrogenation of benzene. Benzene, cyclohexadiene and cyclohexene yield cyclohexane on hydrogenation.

Enthalpy of hydrogenation of cyclohexene is -120 kJ mol^{-1}

Enthalpy of hydrogenation of 1,4-cyclohexadiene is -240 kJ mol^{-1}

Thus, the calculated or expected value of enthalpy of hydrogenation of 1, 3, 5-cyclohexatriene is -360 kJ mol^{-1}

Cyclohexene $+ H_2 \longrightarrow$ Cyclohexane $\quad \Delta H = -120$ kJ mol^{-1} (Experimental)

1, 4-Cyclohexadiene $+ 2H_2 \longrightarrow$ Cyclohexane $\quad \Delta H = -240$ kJ mol^{-1} (Experimental)

1, 3, 5-Cyclohexatriene $+ 3H_2 \longrightarrow$ Cyclohexane $\quad \Delta H = -360$ kJ mol^{-1} (Expected)

Thus, the expected enthalpy of hydrogenation for benzene if it were to be represented hypothetically as 1, 3, 5- cyclohexatriene is -360 kJ mol^{-1}. The experimental value of enthalpy of hydrogenation of benzene has been found be -208 kJ mol^{-1}. Thus, 152 kJ mol^{-1} less energy is produced during hydrogenation of benzene than the expected for hypothetical 1, 3, 5-cyclohexatriene. In other words benzene molecule is more stable by 152 kJ mol^{-1} than 1, 3, 5-cyclohexatriene (Kekule benzene). This is the resonance energy of benzene. **The unusual stability of benzene makes it resistant to the usual addition reactions of alkenes.**

It is this stabilisation due to resonance which is responsible for the **aromatic character of benzene.**

ORBITAL STRUCTURE OF BENZENE

According to orbital structure, each carbon atom in benzene assumes sp^2-hybrid state. Each carbon has three sp^2-hybrid orbitals lying in one plane and oriented at an angle of 120°. There is one unhybridised *p*-orbital having two lobes lying perpendicular to the plane of the hybrid orbitals. Each carbon atom uses two hybrid orbitals for axial overlap with similar orbitals of two adjacent carbon atoms on either side to form C—C sigma bonds. The remaining one sp^2-hybrid orbital on each carbon atom overlaps axially with 1*s* orbital of hydrogen atom to form C—H sigma bond. The axial overlapping of hybrid orbitals to form C—C and C—H bonds has been shown in Fig. 43.3. As is clear, the framework of carbon and hydrogen atoms is *coplanar* with H—C—C or C—C—C bond angle as 120°.

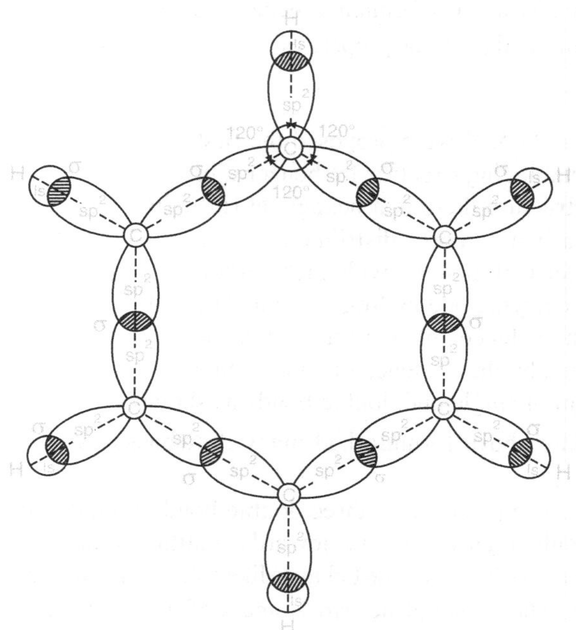

Fig. 43.3. Sigma bond formation in benzene.

The unhybridised *p*-orbital on each carbon atom can overlap to a small but equal extent with the *p*-orbitals of the two adjacent carbon atoms on either side to constitute π bonds as shown in Fig. 43.4.

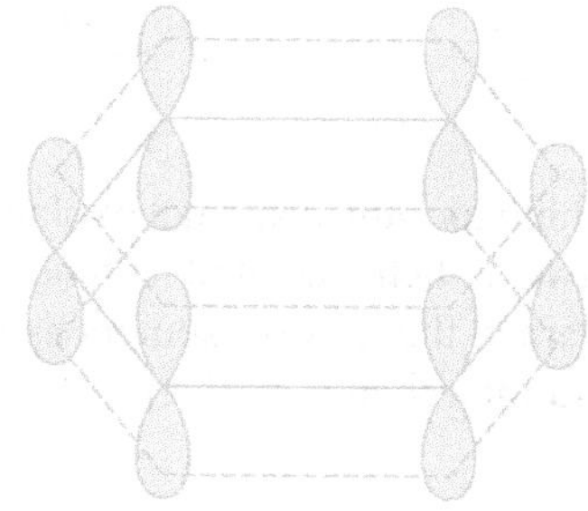

Fig. 43.4. Sidewise overlapping of *p*-orbitals.

The molecular orbital containing π electrons spreads uniformly over the entire carbon skeleton and embraces all the six carbons as shown in Fig. 43.5.

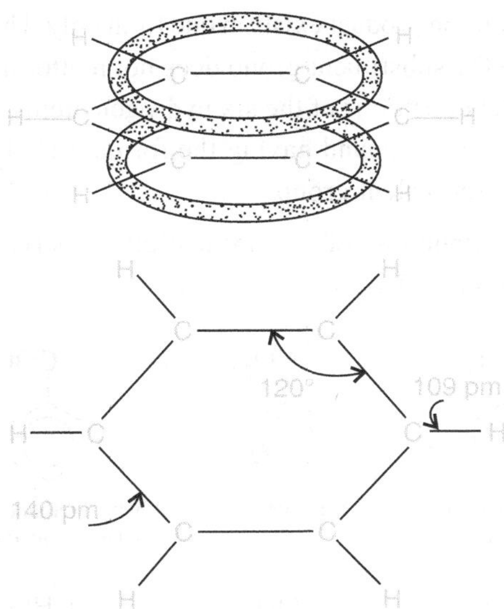

Fig. 43.5. Orbital picture of benzene.

This spreading of π electrons in the form of a ring of π-electrons above and below the plane of carbon atoms is called delocalisation of π-electrons. This delocalisation of π-electrons results in the decrease in energy, and hence, accounts for the stability of the benzene molecule.

C—C bond length in benzene is 140 pm and C—H bond length is 109 pm.

The delocalized structure of benzene also accounts for the X-ray data (all C—C bond lengths equal) and the absence of the type of isomerism shown in Fig. 43.1. Furthermore, molecular orbital theory predicts that those cyclic molecules which have alternate single and double bonds with $4n + 2$ ($n = 0, 1, 2, 3$ etc.) electrons in the delocalized π-cloud are particularly stable and have chemical properties different from other unsaturated hydrocarbons.

AROMATICITY OR AROMATIC CHARACTER

Benzene undergoes substitution reactions in spite of the high degree of unsaturation. This behaviour of benzene is referred to as aromaticity or aromatic character. Aromatic character of benzene can be explained on the basis of the resonance structure of benzene or on the basis of the orbital structure of benzene. In terms of resonance structure, benzene prefers to undergo substitution reactions because during addition reactions the resonance stabilised benzene ring would be destroyed. On the other hand, during substitution the ring structure remains intact.

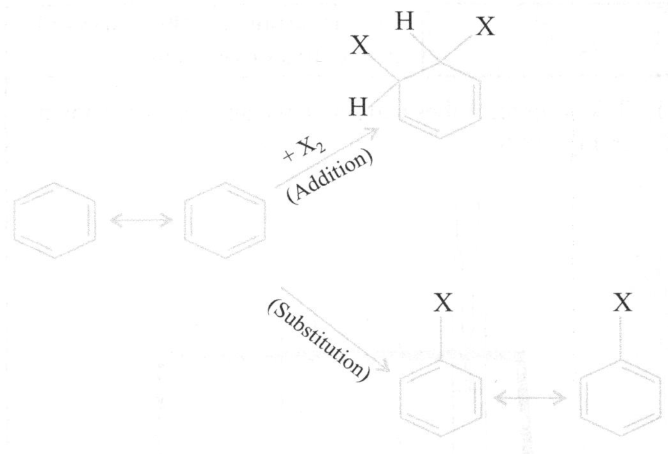

Add To Your Knowledge

HUCKEL RULE OR (4n + 2) RULE

According to this rule, a system possessing the following characteristics would behave as an aromatic compound.

- It should be planar.
- It should have complete delocalisation of π-electrons in the ring.
- It should have $(4n + 2)\pi$ electrons in the ring.

Thus, benzene molecule having six π-electrons (*aromatic sextet*) behaves as an aromatic compound.

43.3 PHYSICAL PROPERTIES OF BENZENE

The physical properties of benzene are as follows:

1. Benzene is a colourless liquid with characteristic odour.
2. It is highly toxic. It is easily absorbed through the skin and the inhalation of vapours of benzene may cause dizziness and nausea.
3. Its boiling point is 353.1 K and melting point is 278.3 K.
4. It is insoluble in water but is soluble in organic solvents.
5. It is highly inflammable and burns with a sooty flame.
6. It is lighter than water. Its density is 0.884 g/mL.

ACTIVITY 43.1	**Investigation of the Physical Properties of Benzene**

1. Take four test tubes and set up the apparatus as shown in Fig. 43.6.

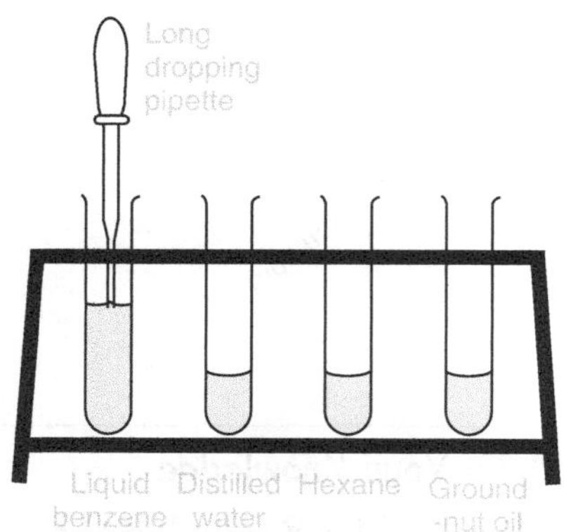

Fig. 43.6. Investigating physical properties of benzene.

- What is the colour of benzene?
- What is the physical state of benzene?

2. Add about 2 cm^3 of benzene in each of the other three tubes containing water, hexane and ground-nut oil and answer the following:
   - Is benzene miscible with water?
   - Is benzene miscible with hexane?
   - Does benzene dissolve ground-nut oil?
   - Is benzene a polar or a non-polar solvent?
   - Is benzene heavier than water?

43.4 CHEMICAL PROPERTIES OF BENZENE

As already mentioned benzene prefers to undergo substitution reactions in spite of the high degree of unsaturation. This is due to the high stabilisation of the benzene ring by resonance (or by delocalization of π electrons). The substitution reactions in benzene are initiated by electrophilies. *Thus, characteristic reactions of benzene are electrophilic substitution reactions.*

The molecule of benzene is symmetrical and the six carbon atoms as well as the hydrogen atoms occupy similar positions in the molecule. If one atom of hydrogen is substituted by a monovalent group or a radical (say methyl group), the resulting

monosubstitution product exists in one form only. The position assigned to the substituent group does not matter because of the equivalent positions of the six hydrogen atoms. Thus, we have only one compound having the formula C_6H_5X where X is some monovalent group.

Some examples of monosubstituted derivatives of benzene are given below:

CH_3	NO_2	COOH
Methyl benzene (Toluene)	Nitrobenzene	Phenylmethanoic acid (Benzoic acid)
NH_2	OH	CHO
Amino benzene (Aniline)	Phenol	Phenylmethanal (Benzaldehyde)

In addition to substitution reactions, benzene also undergoes addition reactions, but under more drastic conditions. Benzene and its homologues also undergo oxidation reactions and side chain substitution reactions.

I. ELECTROPHILIC SUBSTITUTION REACTIONS

1. Halogenation

Chlorine or bromine react with benzene in the presence of *Lewis acids* like ferric or aluminium salts of the corresponding halogen, which act as catalysts.

$$\text{Benzene} + Cl_2 \xrightarrow[300\text{–}320\ K]{\text{Anhy. } AlCl_3} \text{Chlorobenzene (Cl)} + HCl$$

$$\text{Benzene} + Br_2 \xrightarrow[300\text{–}320\ K]{\text{Anhy. } FeBr_3} \text{Bromobenzene (Br)} + HBr$$

> Chlorobenzene or bromobenzene can be converted into phenol on reaction with conc. NaOH solution at 625 K and 300 atm pressure.

Cl → OH: (i) NaOH(aq), 300 atm, 625 K; (ii) H^+

Phenol

2. Sulphonation

The replacement of hydrogen atom of benzene by a sulphonic acid group (—SO_3H) is known as **sulphonation.** The reaction is carried out by treating benzene with concentrated tetraoxosulphate (VI) acid containing dissolved sulphur(VI) oxide or with chlorosulphonic acid.

$$C_6H_6 + \text{Conc. } H_2SO_4 \xrightarrow[SO_3]{\approx 330\text{ K}} C_6H_5SO_3H + H_2O$$

Benzene → Benzene-sulphonic acid

$$C_6H_6 + ClSO_3H \longrightarrow C_6H_5SO_3H + HCl$$

Chlorosulphonic acid → Benzene-sulphonic acid

3. Nitration

A nitro group (—NO_2) can be introduced into a benzene ring using a *nitrating mixture* (a mixture of concentrated tetraoxosulphate(VI) acid and concentrated trioxonitrate(V) acid).

$$C_6H_6 + HNO_3 \xrightarrow[\approx 330\text{ K}]{H_2SO_4} C_6H_5NO_2 + H_2O$$

Benzene → Nitrobenzene

Nitrobenzene can be reduced with tin and conc. HCl to aminobenzene (aniline)

$$C_6H_5NO_2 + 6[H] \xrightarrow{Sn/HCl} C_6H_5NH_2 + 2H_2O$$

Aminobenzene (Aniline)

4. Friedel-Crafts Alkylation

On treatment with an alkyl halide in the presence of Lewis acids, such as *anhydrous aluminium chloride* as catalyst, benzene forms an alkylbenzene.

$$C_6H_6 + CH_3Cl \xrightarrow{\text{Anhyd. } AlCl_3} C_6H_5CH_3 + HCl$$

Benzene → Methyl-benzene (Toluene)

This reaction is called ***Friedel-Craft's alkylation.***

Directive Influence in Arenes

In benzene all the six hydrogens are equivalent. Therefore, any of the six positions can be occupied during the formation of monosubstituted products. However, when a monosubstituted product is to be converted into a disubstituted product, the substituent already present in the ring directs the incoming group to a particular position. This is referred to as *directive influence of the group.*

1. *The substituents or groups which direct the incoming group to ortho and para positions* are called ***ortho* and *para* directing groups.** Some examples are:

   —CH_3, —Cl, —OH, —Br, —NH_2, etc.

For example, if we carry out nitration of toluene, a mixture of 2 and 4 nitrotoluene is formed

$$\text{Toluene} \xrightarrow[H_2SO_4]{HNO_3} \text{2-Nitrotoluene} + \text{4-Nitrotoluene}$$

2. *The substituents or groups which direct the incoming group to meta position* are called **meta directing groups.** Some examples are:

—NO_2, —CN, —CHO, —COOH, —SO_3H, etc.

For example, nitration of nitrobenzene produces 1, 3-dinitrobenzene.

$$\text{Nitrobenzene} \xrightarrow[\text{Mixture}]{\text{Nitrating}} \text{1, 3-Dinitrobenzene}$$

II. ADDITION REACTIONS OF ARENES

Benzene and its homologues undergo some addition reactions characteristic of alkenes and alkynes, but under more drastic conditions (like higher temperature and pressure). This is due to the fact that these compounds are relatively more stable and behave like saturated hydrocarbons.

1. Addition of Hydrogen

Benzene reacts with hydrogen in the presence of a catalyst, *nickel* or platinum at 475–500 K to produce cyclohexane.

$$\text{Benzene} + 3H_2 \xrightarrow[475\text{–}500\ K]{Ni} \text{Cyclohexane}$$

2. Addition of Halogens

Benzene reacts with chlorine in sunlight and in the absence of substances like $AlCl_3$, $FeCl_3$ etc., to form benzene hexachloride (B.H.C.). Benzene hexachloride is an important pesticide.

$$\text{Benzene} + 3Cl_2 \xrightarrow{h\nu} \text{BHC}$$

BHC
Benzene hexachloride

Do You Know?

- *B.H.C.* is used as an insecticide.
- *B.H.C.* is also known as *Lindane*, *Gammexane* or *666*. Its IUPAC name is 1, 2, 3, 4, 5, 6-hexachloro-cyclohexane.

III. OXIDATION REACTIONS OF ARENES

1. Combustion

Arenes on combustion give carbon (IV) oxide and water, and large amount of heat is produced.

$$2C_6H_6 + 15O_2 \longrightarrow 12CO_2 + 6H_2O + \text{Heat}$$

Aromatic hydrocarbons burn with ***sooty flame*** due to the high carbon content of these compounds.

2. Controlled Oxidation

Among **arenes,** *benzene* is too stable to be oxidised even with hot $KMnO_4$ solution. The higher homologues, however, can be oxidised. During this reaction, the ring remains intact but the ***side alkyl chain*** is oxidised to a ***—COOH*** group irrespective of the length of the side chain. Some examples are:

$$\underset{\text{Toluene}}{C_6H_5CH_3} + 3[O] \xrightarrow[375\text{–}385\ K]{KMnO_4/H_2O} \underset{\text{Benzoic acid}}{C_6H_5COOH} + H_2O$$

Benzene, hexane and cyclohexane are inflammable and toxic.

ACTIVITY 43.2 Investigation of Some Chemical Properties of Benzene

1. Take liquid benzene, hexane and cyclohexane in three separate watch glasses.
2. With the help of a nickel spatula carry out combustion test with the three liquids and observe the appearance of flame in each case.

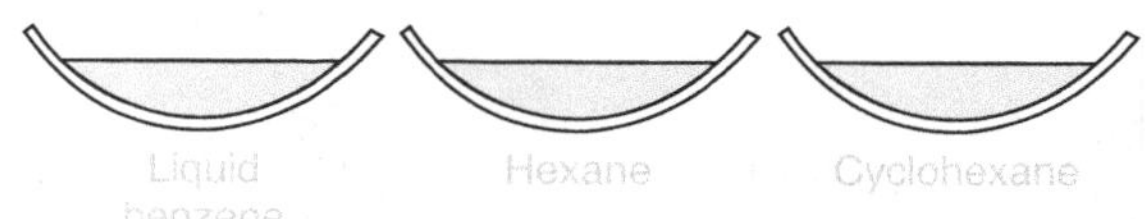

Fig. 43.7. Some liquids for study of combustibility.

3. Take four test tubes and set up the apparatus as shown in Fig. 43.8.

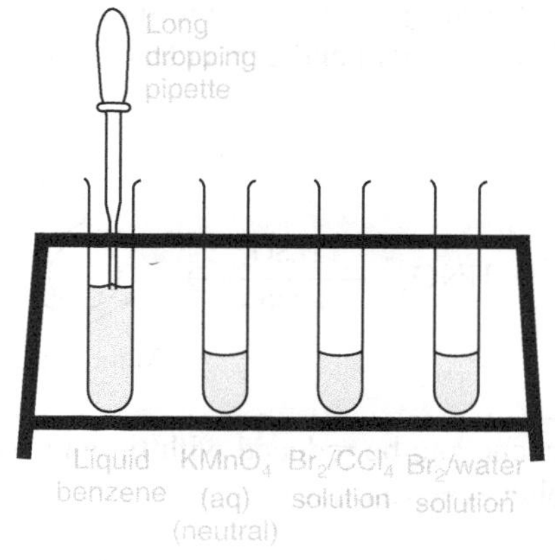

Fig. 43.8. Testing for chemical properties of benzene.

4. Add a few drops of benzene to the contents of each test tube. Warm the reaction mixture if no reaction occurs in the cold. Record your observations.

43.5 USES OF BENZENE

Important uses of benzene are as follows:

1. Benzene is used as solvent for extraction of fats and oils.

2. It is also used as motor fuel along with petrol.

3. Benzene is the starting material for the synthesis of various aromatic compounds which are used in the preparation of dyes, drugs, plastics, perfumes, etc.

Table 43.1. Comparison of Some Chemical Properties of Benzene with Alkanes, Alkenes and Alkynes

Chemical Reaction	*Benzene*	*Alkanes*	*Alkenes*	*Alkynes*
1. Combustion in air	Burns with a very smoky flame	Burn with luminous, non-smoky flame.	Burn with a smoky flame.	Burn with a very smoky flame.
2. Oxidation with acidified $KMnO_4$ solution	No reaction	No reaction	Purple colour of $KMnO_4$ solution is decolorized	Purple colour of $KMnO_4$ solution is decolorized
3. Reaction with bromine water	No reaction	No reaction	Solution of bromine is decolorized	Solution of bromine is decolorized
4. Catalytic addition of hydrogen	Yes	No	Yes	Yes

SUMMARY

- Benzene is an aromatic hydrocarbon. Its molecular formula is C_6H_6.
- Benzene is liquid at room temperature and has a characteristic odour.
- It is highly inflammable and toxic.
- C—C bond length in benzene is 140 pm which is intermediate between that of a single bond and a double bond.
- The cyclic structure of benzene was first proposed by Kekule in 1865.
- Benzene molecule possesses unusual stability and is resistant to addition reactions.
- Structure of benzene can be explained on the basis of resonance. The resonance energy of benzene is very high and is equal to 152 kJ mol^{-1}.
- All the carbon atoms in benzene are sp^2 hybridized.
- Benzene is immiscible with water but is soluble in organic solvents.
- The characteristic reactions of benzene are electrophilic substitution reactions.
- Benzene burns with a smoky flame due to its high carbon content.
- Benzene on catalytic hydrogenation absorbs three moles of hydrogen for each mole of benzene and yields cyclohexane as the product.
- Benzene does not decolorize $KMnO_4$ solution or bromine water or bromine in CCl_4.

EVALUATE YOURSELF

I. Objective Type Questions

Select the most appropriate choice from the options given as (a), (b), (c) and (d) after each question:

1. C—C bond length in benzene is about
   (*a*) 154 pm (*b*) 140 pm
   (*c*) 120 pm (*d*) 134 pm.
2. The structure of phenylmethanoic acid is

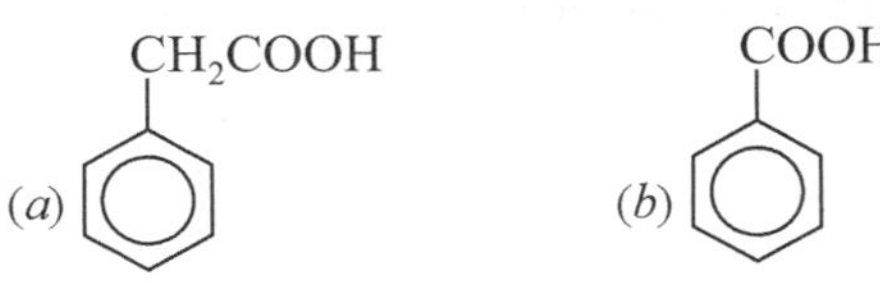

   (*c*) HCOOH (*d*) CH_3COOH.
3. How many monochloro derivatives are possible for benzene
   (*a*) 1 (*b*) 2
   (*c*) 4 (*d*) 6.
4. The hybrid state of carbon atoms in benzene is
   (*a*) sp^3 (*b*) sp
   (*c*) sp^2 (*d*) sp^3d.
5. In Friedel-Crafts alkylation of benzene, the catalyst used is
   (*a*) Nickel (*b*) Palladium
   (*c*) V_2O_5 (*d*) Anhydrous $AlCl_3$.

II. Fill in the Blanks

6. Complete the following sentences by supplying appropriate words:
   (*i*) Benzene burns with ...... flame.
   (*ii*) The physical state of benzene is ...... at room temperature.

(*iii*) Benzene reacts with chlorine in the presence of ...... to form benzene hexachloride.

(*iv*) Benzene on catalytic hydrogenation gives ....

(*v*) When benzene and water are taken in a test tube ...... forms the lower layer.

III. Discussion Questions

7. Explain why?

   (*i*) Benzene burns with a sooty flame

   (*ii*) Benzene undergoes substitution reactions more readily than addition reactions.

8. How does the orbital structure of benzene account for the equivalence of all C—C bonds in benzene?
9. Why is benzene extraordinarily stable though it contains three double bonds?
10. What do you understand by the aromatic character of benzene?
11. Discuss Kekule's structure of benzene. What is the evidence for and against the Kekule's structure?
12. Discuss the structure of benzene with particular reference to the orbital concept.
13. Discuss the resonance structure of benzene.
14. How does the resonance structure of benzene account for the equal C—C bond lengths in benzene?
15. What happens when:

    (*i*) Benzene is treated with a mixture of concentrated tetraoxosulphate(VI) acid and trioxonitrate(V) acid?

    (*ii*) Benzene is treated with an alkaline potassium permanganate solution (cold)?

    (*iii*) Benzene is treated with bromine in the presence of aluminium bromide as catalyst?

16. Draw the structure of

    (*i*) Nitrobenzene (*ii*) Phenylmethanal

    (*iii*) Methylbenzene (*iv*) Aminobenzene.

17. The calculated value of the enthalpy of hydrogenation of benzene, assuming its structure to be 1, 3, 5-cyclohexatriene, is about – 360 kJ mol^{-1} whereas the experimentally observed value for benzene has been found to be – 208 kJ mol^{-1}. What is the significance of the difference between the two values?
18. Name the reagent (*s*) and reaction condition(*s*) for the conversion of benzene to

    (*i*) Chlorobenzene (*ii*) Methylbenzene

    (*iii*) Nitrobenzene (*iv*) Benzenesulphonic acid

    (*v*) 1,3-Dinitrobenzene (*vi*) Cyclohexane

    (*vii*) Benzene hexachloride

    Write chemical equations in each case.

19. Suggest three different chemical tests to distinguish between benzene and cyclohexene.

Answers

I. Objective Type Questions

1. (*b*) **2.** (*b*) **3.** (*a*)

4. (*c*) **5.** (*d*).

II. Fill in the Blanks

6. (*i*) sooty (*ii*) liquid

(*iii*) ultraviolet light (*iv*) cyclohexane

(*v*) water.

SECTION-6

UNIT-44

ALKANOLS

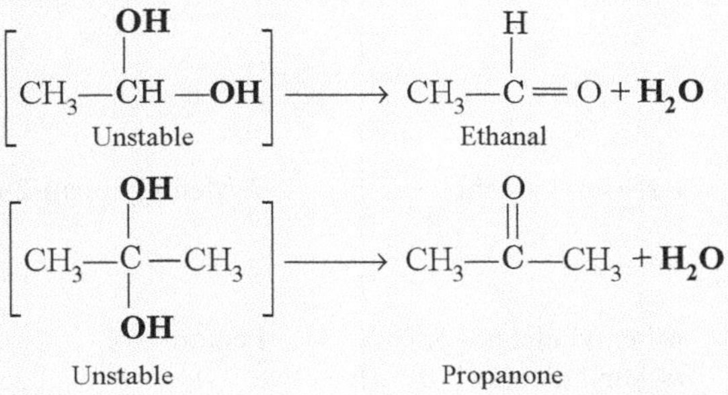

Specific Objectives

After studying this unit, the student should be able to:

- describe the preparation and properties of alkanols
- write the names and structures of given alkanols
- describe the chemical reactions of alkanols
- state some uses of alkanols.

Alcohols or alkanols are the compounds containing one or more hydroxyl groups (—OH). The alcohols contain the —OH group attached to an alkyl group.

44.1 CLASSIFICATION OF ALKANOLS

Alkanols are classified as *mono-*, *di-* and *trihydric* alcohols according to the number of —OH groups contained in their molecules. Some examples of mono-, di- and trihydric alcohols are as follows:

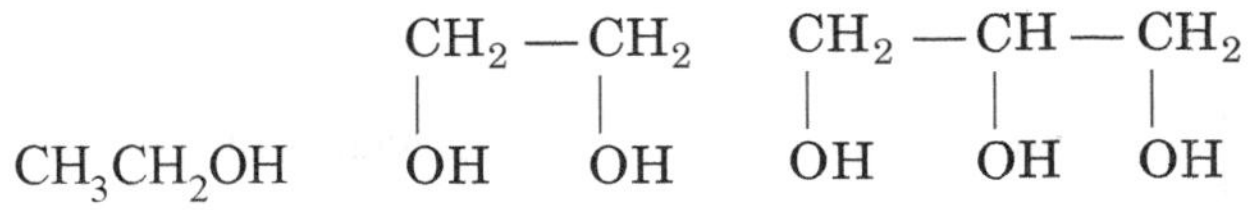

$$CH_3CH_2OH \qquad \underset{OH\quad\;\; OH}{CH_2 - CH_2} \qquad \underset{OH\quad\;\; OH\quad\;\; OH}{CH_2 - CH - CH_2}$$

Ethanol (Ethyl alcohol) (*Monohydric*) — Ethane-1, 2-diol (Ethylene glycol) (*Dihydric*) — Propane-1, 2, 3-triol (Glycerol) (*Trihydric*)

Alcohols containing more than one —OH group linked to the same carbon atom are highly unstable and lose water to form more stable aldehydes or ketones.

$$\left[CH_3—\overset{\mathbf{OH}}{\overset{|}{C}}H—\mathbf{OH}\right]_{\text{Unstable}} \longrightarrow \underset{\text{Ethanal}}{CH_3—\overset{H}{\overset{|}{C}}=O} + \mathbf{H_2O}$$

$$\left[CH_3—\underset{\mathbf{OH}}{\underset{|}{\overset{\mathbf{OH}}{\overset{|}{C}}}}—CH_3\right]_{\text{Unstable}} \longrightarrow \underset{\text{Propanone}}{CH_3—\overset{O}{\overset{\|}{C}}—CH_3} + \mathbf{H_2O}$$

Primary (1°), Secondary (2°) and Tertiary (3°) Alkanols

Monohydric alcohols are classified as primary, secondary or tertiary alcohols depending upon whether the hydroxyl group is attached to a primary, secondary or tertiary carbon atom. For example,

$$H_3C-\overset{H}{\underset{H}{\overset{|\,1^\circ}{\underset{|}{C}}}}-OH$$

Ethanol
(Primary)
or
(1° alcohol)

$$H_3C-\overset{CH_3}{\underset{H}{\overset{|\,2^\circ}{\underset{|}{C}}}}-OH$$

Propan-2-ol
(Secondary)
or
(2° alcohol)

$$H_3C-\overset{CH_3}{\underset{CH_3}{\overset{|\,3^\circ}{\underset{|}{C}}}}-OH$$

2-Methylpropan-2-ol
(Tertiary)
or
(3° alcohol)

Phenols. The compounds in which —OH group is directly attached to a benzene ring are known as phenols

OH
Phenol

OH, CH_3
2-Methylphenol

44.2 NOMENCLATURE

In the common system, alkanols are named as *alkyl alcohols*. In the IUPAC system, the names of saturated alcohols are derived from corresponding alkanes by replacing '*e*' of alkanes by '*ol*'. The general IUPAC name of alcohols is **alkanol.** The IUPAC names of a few members of this class are given in Table 44.1.

Table 44.1. IUPAC Names of Some Alkyl Alcohols

Formula	*Common Name*	*IUPAC Name*
CH_3OH	Methyl alcohol	Methanol
$CH_3—CH_2—OH$	Ethyl alcohol	Ethanol
$\overset{3}{C}H_3—\overset{2}{C}H_2—\overset{1}{C}H_2—OH$	*n*-Propyl alcohol	Propan-1-ol
$\overset{3}{C}H_3—\underset{OH}{\underset{\mid}{\overset{2}{C}H}}—\overset{1}{C}H_3$	*iso*-Propyl alcohol	Propan-2-ol
$\overset{4}{C}H_3—\overset{3}{C}H_2—\overset{2}{C}H_2—\overset{1}{C}H_2—OH$	*n*-Butyl alcohol	Butan-1-ol
$\overset{3}{C}H_3—\underset{CH_3}{\underset{\mid}{\overset{2}{C}H}}—\overset{1}{C}H_2—OH$	*iso*-Butyl alcohol	2-Methylpropan-1-ol
$\overset{1}{C}H_3—\underset{OH}{\underset{\mid}{\overset{2}{C}H}}—\overset{3}{C}H_2—\overset{4}{C}H_3$	*sec*-Butyl alcohol	Butan-2-ol
$H_3C—\overset{^1CH_3}{\overset{\mid}{\underset{^3CH_3}{\underset{\mid}{^2C}}}}—OH$	*tert*-Butyl alcohol	2-Methylpropan-2-ol
$CH_3CH_2CH_2CH_2CH_2OH$	*n*-Pentyl alcohol or *n*-Amyl alcohol	Pentan-1-ol

Some more examples of alcohols are given below:

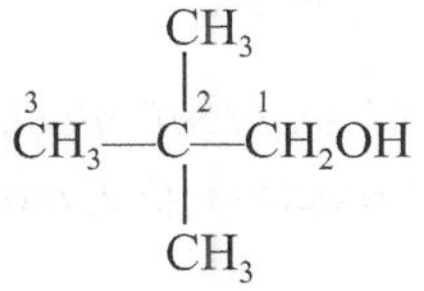

$$\begin{array}{ccc} & CH_3 & \\ & | & \\ \overset{3}{CH_3}- & \overset{2}{C} & -\overset{1}{CH_2OH} \\ & | & \\ & CH_3 & \end{array}$$

2, 2-Dimethylpropan-1-ol

$$\begin{array}{cccc} & \overset{4,5}{C_2H_5} & & \\ & | & & \\ CH_3- & \overset{3}{C} & -\overset{2}{CH} & -\overset{1}{CH_3} \\ & | & | & \\ & CH_3 & OH & \end{array}$$

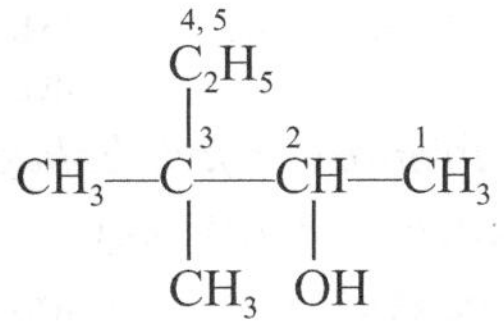

3, 3-Dimethylpentan-2-ol

$$H_2C = CH-CH_2OH$$

Prop-2-en-1-ol
(Allyl alcohol)

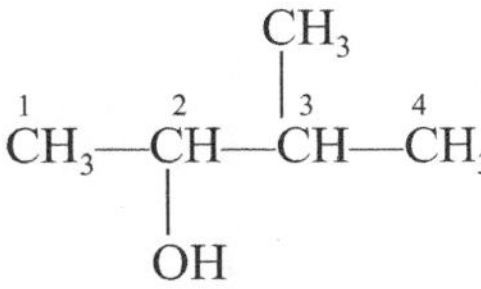

$$\begin{array}{cccc} & & CH_3 & \\ & & | & \\ \overset{1}{CH_3}- & \overset{2}{CH} & -\overset{3}{CH} & -\overset{4}{CH_3} \\ & | & & \\ & OH & & \end{array}$$

3-Methylbutan-2-ol

$$\begin{array}{ccccc} & & & CH_3 & \\ & & & | & \\ \overset{5}{CH_3}- & \overset{4}{CH_2}- & \overset{3}{CH} & -\overset{2}{C} & -\overset{1}{CH_2OH} \\ & & | & | & \\ & & CH_3 & CH_3 & \end{array}$$

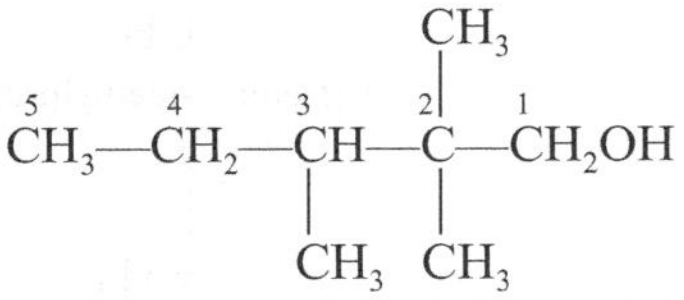

2, 2, 3-Trimethylpentan-1-ol

In the IUPAC system, dihydric alcohols are named as ***alkanediols*** and trihydric alcohols are named as ***alkanetriols.***

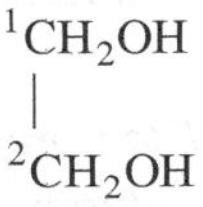

$$\begin{array}{c} {}^{1}CH_2OH \\ | \\ {}^{2}CH_2OH \end{array}$$

Ethane-1, 2-diol
(Ethylene glycol)

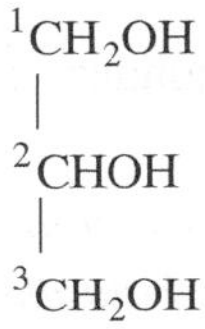

$$\begin{array}{c} {}^{1}CH_2OH \\ | \\ {}^{2}CHOH \\ | \\ {}^{3}CH_2OH \end{array}$$

Propane-1, 2, 3-triol
(Glycerol)

44.3 ISOMERISM IN ALKANOLS

Alkanols exhibit the following types of isomerism:

1. Chain Isomerism

Alkanols with four or more carbon atoms can exhibit this type of isomerism. For example, for C_4H_9OH the following chain isomers may be written.

$$CH_3CH_2CH_2CH_2OH$$

Butan-1-ol

$$\begin{array}{ccc} & CH_3 & \\ & | & \\ CH_3- & CH & -CH_2OH \end{array}$$

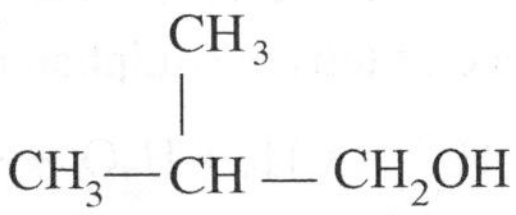

2-Methylpropan-1-ol

2. Position Isomerism

Alkanols with three or more carbon atoms can exhibit position isomerism. For example,

$$CH_3CH_2CH_2OH$$

Propan-1-ol

$$\begin{array}{ccc} CH_3- & CH & -CH_3 \\ & | & \\ & OH & \end{array}$$

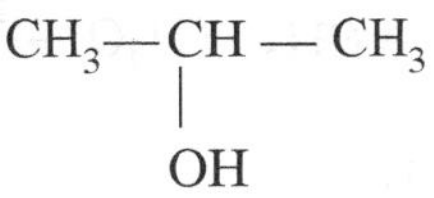

Propan-2-ol

3. Functional Isomerism

Alkanols with two or more carbon atoms can exhibit functional isomerism with ethers. Thus, ethers are the functional isomers of alkanols. For example,

$$CH_3CH_2CH_2CH_2OH$$

Butan-1-ol

$$CH_3CH_2-O-CH_2CH_3$$

Ethoxyethane

SOLVED EXAMPLES

Example 44.1. *Write IUPAC names for the following compounds:*

(*i*) $$\begin{array}{ccccc} & & & CH_3 & \\ & & & | & \\ CH_3- & CH- & CH- & C- & CH_3 \\ & | & | & | & \\ & CH_3 & OH & CH_3 & \end{array}$$

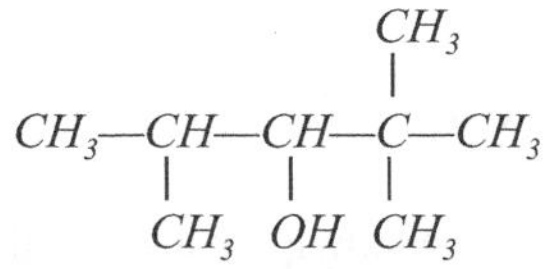

(*ii*) $$\begin{array}{cccccc} CH_3- & CH- & CH_2- & CH- & CH- & CH_2CH_3 \\ & | & & | & | & \\ & OH & & OH & C_2H_5 & \end{array}$$

(*iii*) $$\begin{array}{cccc} CH_3- & CH- & CH- & CH_3 \\ & | & | & \\ & OH & OH & \end{array}$$

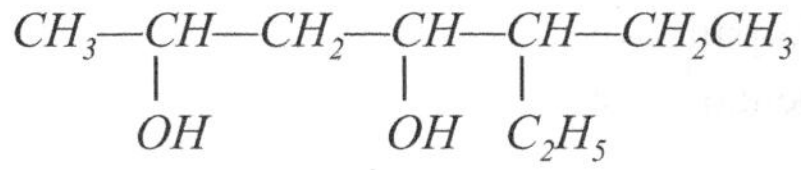

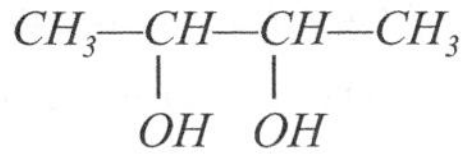

(*iv*) $HOCH_2-CHOH-CH_2OH$

(*v*) $$\begin{array}{ccc} & CH_3 & \\ & | & \\ C_6H_5- & C- & CH_2-CH_3 \\ & | & \\ & OH & \end{array}$$

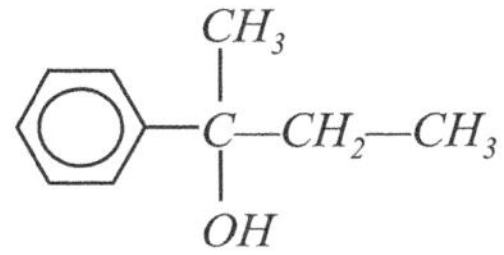

(*vi*) $C_6H_5-CH_2OH.$

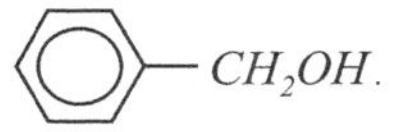

Solution. (*i*) 2, 2, 4-Trimethylpentan-3-ol

(*ii*) 5-Ethylheptane-2, 4-diol

(*iii*) Butane-2, 3-diol

(*iv*) Propane-1, 2, 3-triol

(*v*) 2-Phenylbutan-2-ol

(*vi*) Phenylmethanol.

Example 44.2. *Give structures and IUPAC names of all the isomeric alkanols having molecular formula* $C_4H_{10}O$.

Solution.

1. $CH_3CH_2CH_2CH_2OH$

Butan-1-ol

2. 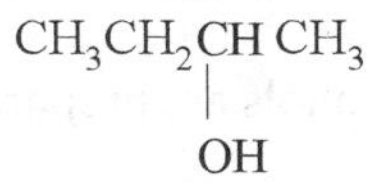 $$\begin{array}{cc} CH_3CH_2 & CH\ CH_3 \\ & | \\ & OH \end{array}$$

Butan-2-ol

3. 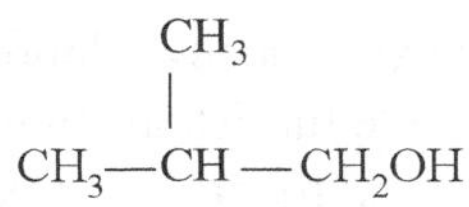 $$\begin{array}{ccc} & CH_3 & \\ & | & \\ CH_3- & CH & -CH_2OH \end{array}$$

2-Methylpropan-1-ol

4. 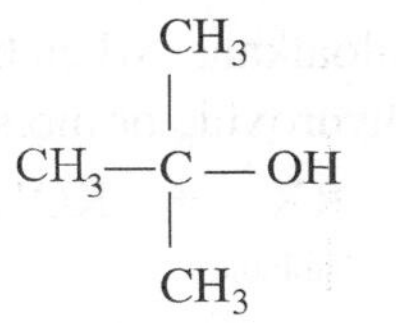 $$\begin{array}{ccc} & CH_3 & \\ & | & \\ CH_3- & C & -OH \\ & | & \\ & CH_3 & \end{array}$$

2-Methylpropan-2-ol

Example 44.3. *Give IUPAC names of the following compounds:*

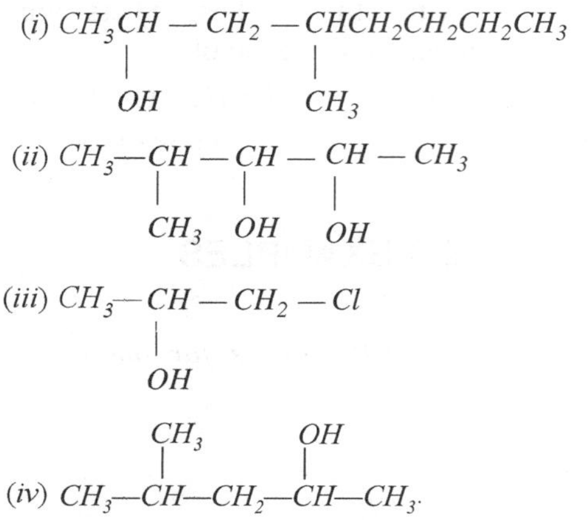

(*i*) $CH_3CH(OH)—CH_2—CH(CH_3)CH_2CH_2CH_2CH_3$

(*ii*) $CH_3—CH(CH_3)—CH(OH)—CH(OH)—CH_3$

(*iii*) $CH_3—CH(OH)—CH_2—Cl$

(*iv*) $CH_3—CH(CH_3)—CH_2—CH(OH)—CH_3$.

Solution.

(*i*) 4-Methyloctan-2-ol (*ii*) 4-Methylpentan-2, 3-diol
(*iii*) 1-Chloropropan-2-ol
(*iv*) 4-Methylpentan-2-ol.

Example 44.4. *Give structures of the following alcohols:*

(*i*) *2-Methylbutan-2-ol* (*ii*) *3-Methylpentan-2-ol*
(*iii*) *2, 5-Dimethylhexane-1, 3-diol.*

Solution.

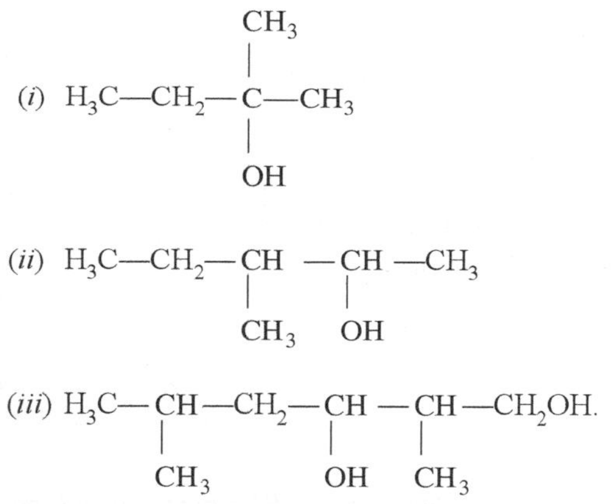

(*i*) $H_3C—CH_2—C(CH_3)(OH)—CH_3$

(*ii*) $H_3C—CH_2—CH(CH_3)—CH(OH)—CH_3$

(*iii*) $H_3C—CH(CH_3)—CH_2—CH(OH)—CH(CH_3)—CH_2OH$.

44.4 PREPARATION OF ALKANOLS

Alkanols are prepared by the following general methods:

1. BY THE HYDROLYSIS OF HALOALKANES

Haloalkanes when boiled with an aqueous solution of an alkali hydroxide or moist silver oxide furnish alkanols.

$$\underset{\text{Haloalkane}}{RX} + KOH(aq) \longrightarrow \underset{\text{Alkanol}}{ROH} + KX$$

$$C_2H_5Br + KOH(aq) \longrightarrow C_2H_5OH + KBr$$

$$\underset{\text{Bromoethane}}{C_2H_5Br} + \underset{\substack{\text{Moist}\\\text{silver oxide}}}{AgOH} \longrightarrow \underset{\text{Ethanol}}{C_2H_5OH} + AgBr$$

Primary alkyl halides give good yield of alcohols. However, tertiary alkyl halides, in this reaction give, mainly alkene due to dehydrohalogenation.

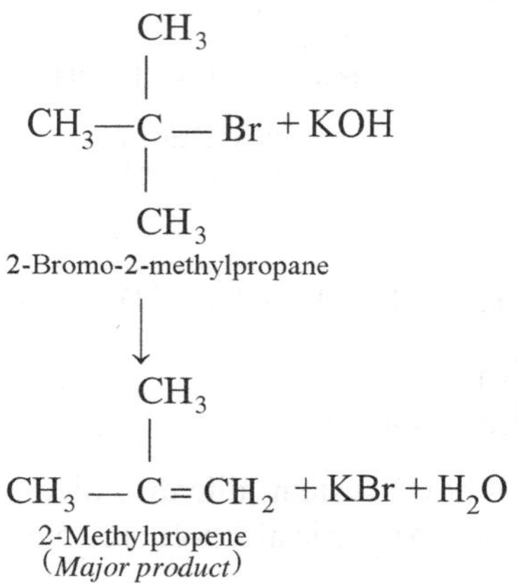

$$\underset{\text{2-Bromo-2-methylpropane}}{CH_3—C(CH_3)_2—Br} + KOH \longrightarrow \underset{\substack{\text{2-Methylpropene}\\(\text{Major product})}}{CH_3—C(CH_3)=CH_2} + KBr + H_2O$$

Secondary alkyl halides give a mixture of alkanol and alkene.

This method is not very useful for preparing alcohols because haloalkanes are themselves obtained from alcohols. However, the preparative utility of this method lies in the preparation of aromatic alcohols. For example,

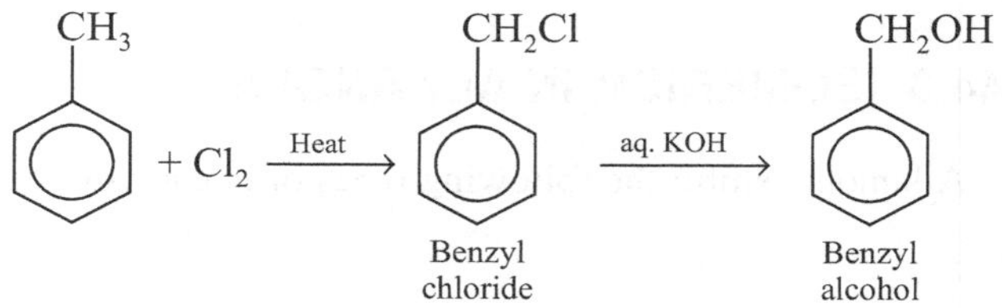

$$C_6H_5CH_3 + Cl_2 \xrightarrow{\text{Heat}} \underset{\text{Benzyl chloride}}{C_6H_5CH_2Cl} \xrightarrow{\text{aq. KOH}} \underset{\text{Benzyl alcohol}}{C_6H_5CH_2OH}$$

2. BY HYDRATION OF ALKENES

Alkenes are obtained by cracking of petroleum. They are easily converted to alkanols by the addition of water in presence of tetraoxosulphate(VI) acid.

$$\underset{\text{Ethene}}{H_2C=CH_2} + H_2O \xrightarrow{H_2SO_4} \underset{\text{Ethanol}}{CH_3—CH_2—OH}$$

In case of unsymmetrical alkenes, the addition takes place according to the Markownikov's rule.

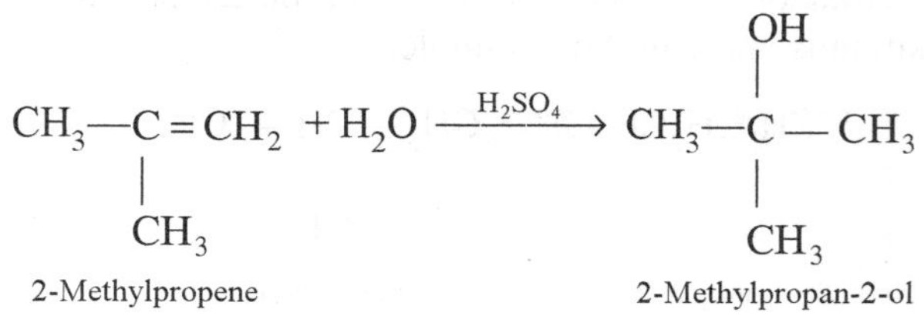

$$\underset{\text{2-Methylpropene}}{CH_3—C(CH_3)=CH_2} + H_2O \xrightarrow{H_2SO_4} \underset{\text{2-Methylpropan-2-ol}}{CH_3—C(OH)(CH_3)—CH_3}$$

3. BY OXO PROCESS

Alkenes react with carbon monoxide and hydrogen in the presence of octacarbonyl dicobalt $[Co(CO)_4]_2$ as catalyst and at high temperature and pressure to yield aldehydes. Catalytic hydrogenation of aldehydes gives primary alkanols.

$$\underset{\text{Ethene}}{CH_2 = CH_2} + CO + H_2 \xrightarrow[\text{High temp. and pressure}]{[Co(CO)_4]_2} CH_3—CH_2—CHO \xrightarrow{H_2/Ni} \underset{\text{Propan-1-ol}}{CH_3CH_2CH_2OH}$$

4. BY FERMENTATION OF CARBOHYDRATES

Formation of ethanol by the **fermentation of sugar** (obtained from mollases, grapes or beet) is one of the oldest methods. Sucrose is first converted to glucose and fructose with an enzyme *invertase*.

$$\underset{\text{Sucrose}}{C_{12}H_{22}O_{11}} + H_2O \xrightarrow{\text{Invertase}} \underset{\text{Glucose}}{C_6H_{12}O_6} + \underset{\text{Fructose}}{C_6H_{12}O_6}$$

Enzyme *zymase* then converts glucose and fructose to ethanol.

The enzyme zymase is present in yeast.

$$\underset{\text{Glucose or fructose}}{C_6H_{12}O_6} \xrightarrow{\text{Zymase}} 2C_2H_5OH + 2CO_2$$

The fermentation process is carried out under **anaerobic conditions** *i.e.,* in the absence of air. Carbon(IV) oxide released during fermentation keeps the fermentation mixture out of contact with air. If the fermentation mixture gets exposed to air, the oxygen of air oxidizes ethanol to ethanoic acid which makes the mixture sour.

In Africa, as in many other parts of the world ethanol is obtained from **starchy materials** such as *barley*, *millet*, *rice*, *maize*, *cassava*, *unripe banana* and *potatoes* with enzymes *diastase* and *maltase*.

$$\underset{\text{Starch}}{2(C_6H_{10}O_5)_n} + nH_2O \xrightarrow{\text{Diastase}} \underset{\text{Maltose}}{nC_{12}H_{22}O_{11}}$$

$$\underset{\text{Maltose}}{C_{12}H_{22}O_{11}} + H_2O \xrightarrow{\text{Maltase}} \underset{\text{Glucose}}{2C_6H_{12}O_6}$$

$$\underset{\text{Glucose}}{C_6H_{12}O_6} \xrightarrow{\text{Zymase}} \underset{\text{Ethyl alcohol}}{2C_2H_5OH} + 2CO_2$$

Do you know?

Enzyme diastase is obtained from germinated barley while enzyme maltase and zymase are obtained from yeast.

In the manufacture of ethanol by **fermentation of cellulose**, the cellulosic material is first hydrolysed to glucose by digesting it with dilute tetraoxosulphate(VI) acid and steam at a pressure of 6–7 atmospheres.

$$(C_6H_{10}O_5)_n + nH_2O \xrightarrow[\text{6–7 atm, Heat}]{\text{dil. } H_2SO_4} \underset{\text{Glucose}}{nC_6H_{12}O_6}$$

The excess acid is then neutralized with lime and the resulting solution is fermented with yeast. The enzyme *zymase* converts glucose to ethanol

$$C_6H_{12}O_6 \xrightarrow{\text{Zymase}} \underset{\text{Ethanol}}{2C_2H_5OH} + 2CO_2$$

Fermentation of carbohydrates yields ethanol containing water. This solution is concentrated by fractional distillation. By simple fractional distillation we can get, at the most, a solution containing 95 per cent ethanol. This 95 per cent ethanol is known as **rectified spirit.**

Ethanol in the form of alcoholic beverages is used for drinking. The alcohol used for drinking is heavily taxed. However, alcohol used for industrial purposes is not taxed. To prevent the misuse, the industrial alcohol is made unfit for drinking by the addition of 5–10 per cent methanol. The ethanol denaturated by methanol (and a little of pyridine and acetone), is called **methylated spirit.**

Ethyl alcohol may also be mixed with petrol. A mixture of 20% ethanol and 80% gasoline is used in internal combustion engines. Since alcohol does not mix with petrol, a third solvent benzene or ether or tetralin (tetra hydronaphthalene) is used as a solvent. Ethyl alcohol used for the purpose of generation of power in internal combustion engine is called **power alcohol.**

Add To Your Knowledge

100% Ethanol is known as **absolute alcohol.** It is prepared by azeotropic distillation of rectified alcohol by adding benzene.

SOLVED EXAMPLE

Example 44.5. *How will you prepare*

(*i*) *pentan-1-ol from 1-bromopentane and*

(*ii*) *pentan-2-ol from pent-1-ene?*

Solution. (*i*) By treating 1-bromopentane with an aqueous solution of potassium hydroxide.

$$\underset{\text{1-Bromopentane}}{CH_3CH_2CH_2CH_2CH_2Br} + KOH(aq) \longrightarrow \underset{\text{Pentan-1-ol}}{CH_3CH_2CH_2CH_2CH_2OH} + KBr$$

(*ii*) Pentan-2-ol can be prepared from pent-1-ene by hydration reaction.

$$\underset{\text{Pent-1-ene}}{CH_3CH_2CH_2CH=CH_2} + H_2O \xrightarrow{\text{dil. } H_2SO_4} \underset{\text{Pentan-2-ol}}{CH_3CH_2CH_2CH(OH)CH_3}$$

44.5 STRUCTURE OF ALKANOLS

The general formula of alkanols is R—OH, where R is some alkyl or substituted alkyl group. Since the oxygen atom of the hydroxyl group has two bond pairs and two lone pairs of electrons, therefore, the C—O—H bonds in alkanols is not linear. The C—O bond in alkanols is formed by overlap of sp^3-hybrid orbital of carbon with sp^3-hybrid orbital of oxygen. The C—O—H bond angle in alkanols is slightly less than the regular tetrahedral angle (109.5°). It is due to the repulsion of the lone pairs of oxygen on the bond pairs. The C—O bond length in alkanols is 142 pm.

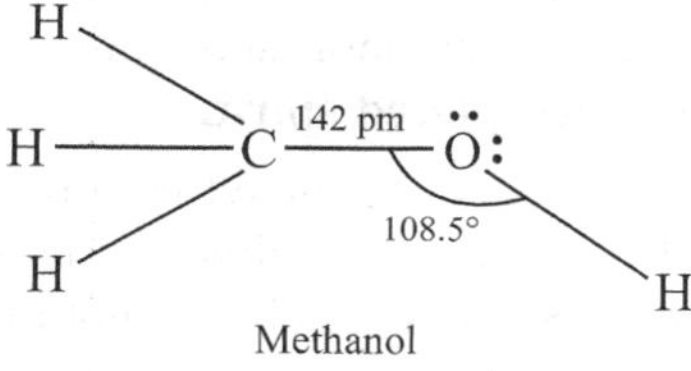

Methanol

As oxygen is more electronegative than carbon and hydrogen, therefore, C—O and O—H bonds in alkanols are polar bonds and hence alkanols possess a net dipole moment. Methanol has a dipole moment of 1.71 D.

44.6 PHYSICAL PROPERTIES OF ALKANOLS

1. The lower members of alkanols are colourless, volatile liquids with a characteristic alcoholic smell and burning taste whereas higher alkanols are odourless and tasteless.

   Higher alkanols having 12 or more carbon atoms are colourless waxy solids.

2. **Solubility.** The first three members are completely miscible with water. The solubility rapidly decreases with increase in molecular mass. The higher members are almost insoluble in water but are soluble in organic solvents like benzene, ether, etc.

   The solubility of lower alkanols is due to the existence of hydrogen bonds between water and the polar O—H group of alkanol molecules.

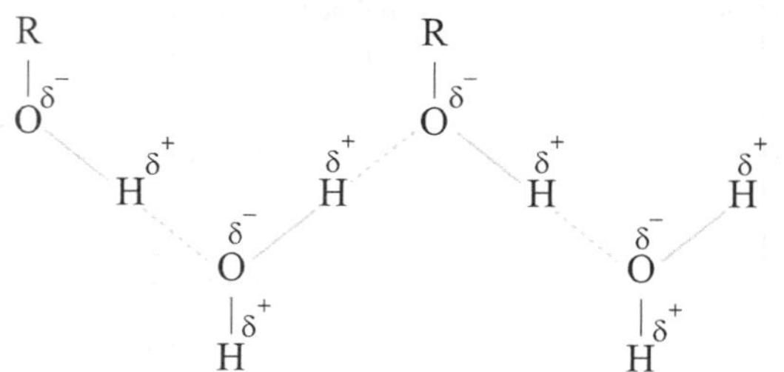

   ***The solubility of alkanols in water decreases with increase in molecular mass** because with increase in molecular mass, the non-polar alkyl group becomes predominant and masks the effect of the polar —OH group.*

   In addition, ***among the isomeric alkanols the solubility increases with branching of chain.***

3. **Boiling Points.** Boiling points of alcohols are much higher than those of alkanes, haloalkanes or ethers of comparable molecular masses. *This is because in alkanols the existing intermolecular hydrogen bonds require a large amount of energy to break.*

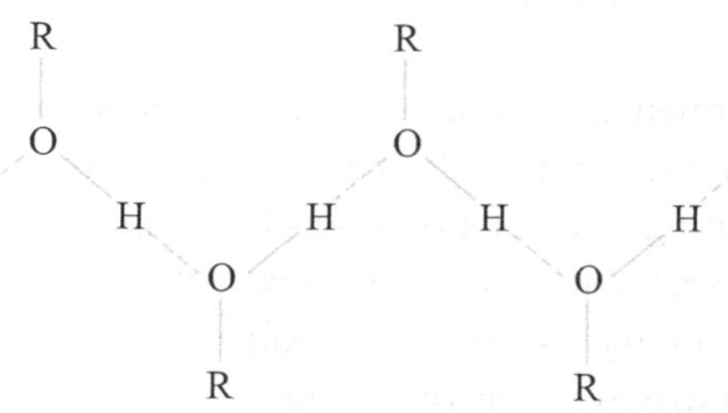

Compound	*Molecular Mass*	*Boiling Point*
CH_3CH_2OH (Ethanol)	46	351 K
CH_3OCH_3 (Methoxymethane)	46	297 K
CH_3Cl (Chloromethane)	50.5	249 K
$CH_3CH_2CH_3$ (Propane)	44	231 K

Among isomeric alkanols, the boiling point decreases with increase in branching in the alkyl group.

For isomeric alkanols the boiling points generally follow the order: 1° > 2° > 3°

4. Intoxicating Effect. Alcohols are known to have intoxicating effect. Methanol is poisonous and is not good for drinking purposes. It causes blindness. Ethanol on the other hand is used for drinking purposes. Sometimes it is rendered undrinkable by adding methanol.

SOLVED EXAMPLES

Example 44.6. *Explain why ethanol boils at higher temperature than the corresponding alkane.*

Solution. Ethanol (C_2H_5OH) contains O—H bond which gives rise to intermolecular H-bonding.

$$\cdots\underset{\underset{C_2H_5}{|}}{O}-H\cdots\cdots\underset{\underset{C_2H_5}{|}}{O}-H\cdots\cdots\underset{\underset{C_2H_5}{|}}{O}-H\cdots\cdots$$

As a result, ethanol has higher boiling point than ethane which has weak van der Waal's forces as intermolecular forces.

Example 44.7. *Explain why?*

(*i*) *tert-Butyl alcohol is much more soluble in water than n-butyl alcohol.*

(*ii*) *n-Hexanol is not soluble in water.*

Solution. (*i*) In *tert*-butyl alcohol, alkyl group is highly branched and approaches a sphere. As a result the non-polar part of the molecule becomes less prominent and hence it can have stronger intermolecular H-bonds with water molecules through its O—H group. On the otherhand, *n*-butyl alcohol has predominantly non-polar character.

(*ii*) In *n*-hexanol, alkyl group is quite large. As a result it resembles the hydrocarbons more closely and is not soluble in water. The extent of H-bonding between *n*-hexanol and water is very less.

Example 44.8. *Arrange the following sets of compounds in order of their increasing boiling points,*

(*i*) *Pentan-1-ol, butan-1-ol, butan-2-ol, ethanol, propan-1-ol, methanol*

(*ii*) *Pentan-1-ol, n-butane, pentanal, ethoxyethane.*

Solution.

(*i*) Methanol < ethanol < propan-1-ol < butan-2-ol < butan-1-ol < pentan-1-ol

(*ii*) *n*-Butane < ethoxyethane < pentanal < pentan -1-ol.

Example 44.9. *Arrange the following compounds in increasing order of their boiling point:*

(*i*) CH_3CH_2OH (*ii*) $HOCH_2CH_2OH$

(*iii*) CH_3CH_2Cl.

Solution. Alcohols have higher boiling points than alkyl halides due to *intermolecular hydrogen bonding*. Moreover dihydric alcohols undergo a greater extent of hydrogen bonding than monohydric alcohols due to the presence of two hydroxyl groups. Therefore, the boiling point of $HOCH_2\ CH_2OH$ is higher than that of CH_3CH_2OH while that of CH_3CH_2Cl is lowest due to the absence of H-bonding. Hence, the boiling points of the above given compounds increase in the order:

$$CH_3CH_2Cl < CH_3CH_2OH < HOCH_2CH_2OH.$$

44.7 CHEMICAL REACTIONS OF ALKANOLS

In alcohols, the —OH group is the functional group. Therefore, the chemical properties of alcohols generally involve the reactions of the —OH group. These can undergo **substitution** as well as **elimination** reactions. Alcohols can react both as nucleophiles and electrophiles. The bond between O–H is broken when they react as nucleophiles. The bond between C–O is broken when they react as electrophiles. Protonated alcohols react in this manner.

$$R—CH_2—OH + \overset{+}{H} \longrightarrow R—CH_2—\overset{+}{O}H_2$$

$$Br^- + \underset{\underset{R}{|}}{CH_2}—\overset{+}{O}H_2 \longrightarrow R—CH_2—Br + H_2O$$

(protonated alcohols as electrophiles)

Some common reactions of alcohols are as follows:

ACIDIC CHARACTER

Alcohols are very weak acids (Ka = $10^{-16} – 10^{-18}$), even feeble than water (Ka = 10^{-14}). They do not turn blue litmus red but when treated with active metals like sodium, potassium and aluminium they liberate hydrogen along with the formation of alkoxides.

$$2R—O—H + 2Na \longrightarrow \underset{\text{Sodium alkoxide}}{2R—ONa} + H_2$$

$$\underset{\text{Ethanol}}{2C_2H_5OH} + 2Na \longrightarrow \underset{\text{Sodium ethoxide}}{2C_2H_5—ONa} + H_2$$

Alcohols behave as acids because of the presence of the polar O—H group. Due to the greater electronegativity of the oxygen atom the shared pair between O and H is drawn towards the O.

The order of ***acidic strength*** among various types of alkanols is

primary > secondary > tertiary

The order can also be explained in terms of the electron releasing inductive effect of the alkyl groups. The greater the number of alkyl groups attached to the carbon carrying the —OH group, the lesser the polarity of the OH group and consequently the acidic character. In tertiary alcohols, the carbon atom carrying the —OH group is attached to three alkyl groups which confers a higher electron density on the oxygen atom. This means that the loss of H^+ is most difficult. Thus, tertiary alcohols are least acidic, whereas primary alcohols (with only one alkyl group) are most acidic. The secondary alcohols have intermediate acidic character.

$$R\rightarrow C(H)(H)\rightarrow OH > R\rightarrow C(R\downarrow)(H)\rightarrow OH > R\rightarrow C(R\downarrow)(R\uparrow)\rightarrow OH$$

Primary alcohol — Secondary alcohol — Tertiary alcohol

ACIDIC CHARACTER DECREASES →

Alkanols are weaker acids than water. This can be illustrated by the reaction of water with an alkoxide.

$$\underset{\text{Base}}{R{-}\ddot{\underset{..}{O}}{:}^-} + \underset{\text{Acid}}{H{-}\ddot{\underset{..}{O}}{-}H} \longrightarrow \underset{\text{Conjugate acid}}{R{-}O{-}H} + \underset{\text{Conjugate base}}{:\ddot{\underset{..}{O}}H^-}$$

This reaction shows that water is a better proton donor and hence a stronger acid than alkanols. Also, in the above reaction, we note that an alkoxide ion is a better proton acceptor than a hydroxide ion, which shows that alkoxides are stronger bases. Thus, sodium ethoxide is a stronger base than sodium hydroxide.

Due to the presence of lone pairs of electrons on oxygen, alkanols can also act as proton acceptors and hence behave as **Bronsted bases**.

$$\underset{\text{Alkanol}}{R{-}\ddot{\underset{..}{O}}{-}H} + H_2SO_4 \longrightarrow \underset{\text{Protonated alkanol}}{R{-}\overset{H}{\overset{|}{\underset{..}{O^+}}}{-}H} + HSO_4^-$$

Thus, **alkanols exhibit amphoteric character.**

REACTION WITH MONOCARBOXYLIC ACIDS (Esterification)

Alkanols react with monocarboxylic acids in the presence of conc. tetraoxosulphate(VI) acid to form esters. The function of conc. H_2SO_4 is to act as a protonating agent as well as a dehydrating agent. The reaction is called **esterification.**

$$\underset{\text{Acid}}{R{-}\overset{O}{\overset{\|}{C}}{-}\mathbf{OH}} + \underset{\text{Alkanol}}{\mathbf{H}{-}OR'} \overset{H_2SO_4}{\rightleftharpoons} \underset{\text{Ester}}{R{-}\overset{O}{\overset{\|}{C}}{-}OR'} + \mathbf{H_2O}$$

$$\underset{\text{Ethanoic acid}}{CH_3{-}\overset{O}{\overset{\|}{C}}{-}\mathbf{OH}} + \underset{\text{Ethanol}}{\mathbf{H}{-}OC_2H_5} \overset{H_2SO_4}{\rightleftharpoons} \underset{\text{Ethyl ethanoate}}{CH_3{-}\overset{O}{\overset{\|}{C}}{-}OC_2H_5} + \mathbf{H_2O}$$

The reaction is reversible and the equilibrium can be shifted to the right by removing water as soon as it is formed. During esterification water molecule is formed by combination of the —OH of the carboxylic acid and the H from the alkanol molecule.

REACTION WITH HYDROGEN HALIDES

Hydrogen halides react with alcohols to form alkyl halides. The alkanol may be primary, secondary or tertiary and the halogen acid may be HCl, HBr or HI.

$$ROH + HX \longrightarrow R{-}X + H_2O$$

$$CH_3CH_2OH + HCl \xrightarrow{ZnCl_2} C_2H_5Cl + H_2O$$

$$(CH_3)_2CH{-}OH + HBr \longrightarrow (CH_3)_2CH{-}Br + H_2O$$

$$(CH_3)_3C{-}OH + HI \longrightarrow (CH_3)_3C{-}I + H_2O$$

The order of reactivity of various alcohols towards this reaction is

tertiary > secondary > primary

The order of reactivity of various hydrogen halides is in the order

HI > HBr > HCl

ACIDIC DEHYDRATION (*Formation of Alkenes*)

When heated with conc. H_2SO_4 or H_3PO_4 alkanols undergo dehydration to form alkenes. The reaction with conc. H_2SO_4 is carried out at 443 K, whereas H_3PO_4 reacts at a higher temperature.

$$\underset{\text{Ethanol}}{H{-}\overset{\boxed{H}}{\underset{H}{C}}{-}\overset{\boxed{OH}}{\underset{H}{C}}{-}H} \xrightarrow[443\text{ K}]{\text{Conc. }H_2SO_4} \underset{\text{Ethene}}{H{-}\underset{H}{C}{=}\underset{H}{C}{-}H} + H_2O$$

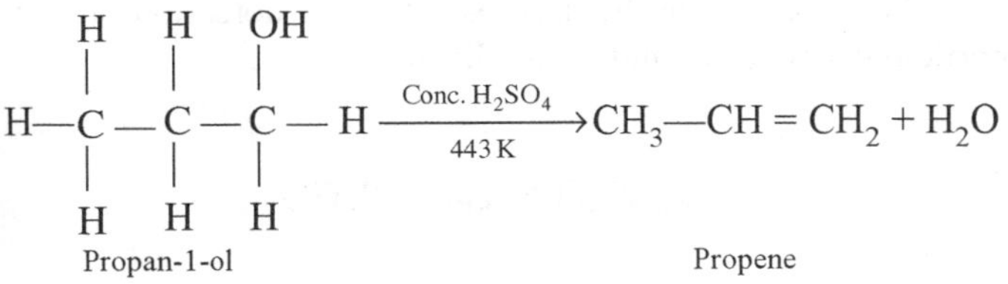

$$\underset{\text{Propan-1-ol}}{\mathrm{H{-}\overset{\overset{\displaystyle H}{|}}{\underset{\underset{\displaystyle H}{|}}{C}}{-}\overset{\overset{\displaystyle H}{|}}{\underset{\underset{\displaystyle H}{|}}{C}}{-}\overset{\overset{\displaystyle OH}{|}}{\underset{\underset{\displaystyle H}{|}}{C}}{-}H}} \xrightarrow[443\,K]{\text{Conc. } H_2SO_4} \underset{\text{Propene}}{CH_3{-}CH=CH_2} + H_2O$$

Secondary and tertiary alkanols are dehydrated under milder conditions. For example,

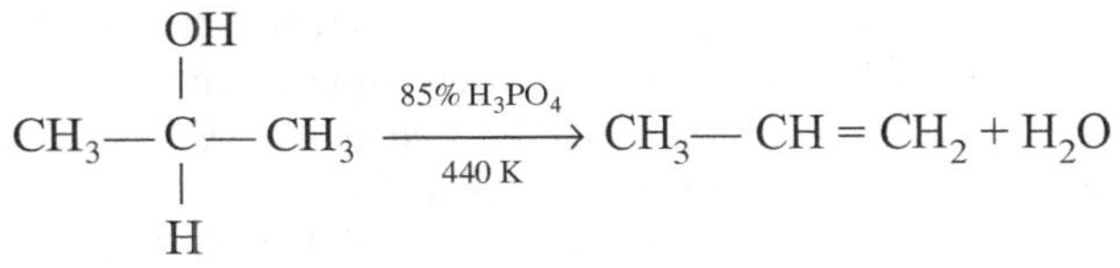

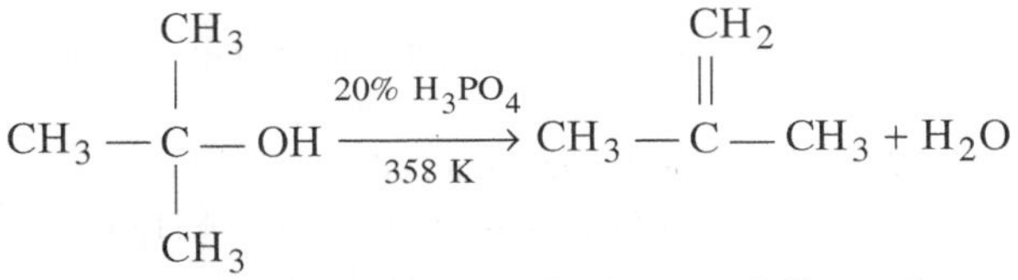

$$\mathrm{CH_3{-}\overset{\overset{\displaystyle OH}{|}}{\underset{\underset{\displaystyle H}{|}}{C}}{-}CH_3} \xrightarrow[440\,K]{85\%\ H_3PO_4} CH_3{-}CH=CH_2 + H_2O$$

$$\mathrm{CH_3{-}\overset{\overset{\displaystyle CH_3}{|}}{\underset{\underset{\displaystyle CH_3}{|}}{C}}{-}OH} \xrightarrow[358\,K]{20\%\ H_3PO_4} \mathrm{CH_3{-}\overset{\overset{\displaystyle CH_2}{||}}{C}{-}CH_3} + H_2O$$

The **ease of dehydration of alkanols** follows the order

$$3° > 2° > 1°$$

OXIDATION

Alcohols undergo oxidation with oxidising agents (in neutral or acidic or alkaline medium) such as **chromic anhydride or chromium(VI) oxide** (CrO_3), $K_2Cr_2O_7$, $KMnO_4$.

(*i*) ***A primary alkanol*** on oxidation with acidified $K_2Cr_2O_7$ first gives an aldehyde which on further oxidation gives an acid with the *same number of carbon atoms* as the original alcohol.

$$\underset{\text{Primary alkanol}}{\mathrm{R{-}\overset{\overset{\displaystyle H}{|}}{\underset{\underset{\displaystyle H}{|}}{C}}{-}OH}} + [O] \xrightarrow[-H_2O]{K_2Cr_2O_7/H^+} \underset{\text{Aldehyde}}{\mathrm{R{-}\overset{\overset{\displaystyle H}{|}}{C}=O}} \xrightarrow{+[O]} \underset{\text{Acid}}{\mathrm{R{-}\overset{\overset{\displaystyle OH}{|}}{C}=O}}$$

$$\underset{\text{Ethanol}}{\mathrm{CH_3{-}\overset{\overset{\displaystyle H}{|}}{\underset{\underset{\displaystyle H}{|}}{C}}{-}OH}} + [O] \xrightarrow[-H_2O]{K_2Cr_2O_7/H^+} \underset{\text{Ethanal}}{\mathrm{CH_3{-}\overset{\overset{\displaystyle H}{|}}{C}=O}} \xrightarrow{+[O]} \underset{\text{Ethanoic acid}}{\mathrm{CH_3{-}\overset{\overset{\displaystyle OH}{|}}{C}=O}}$$

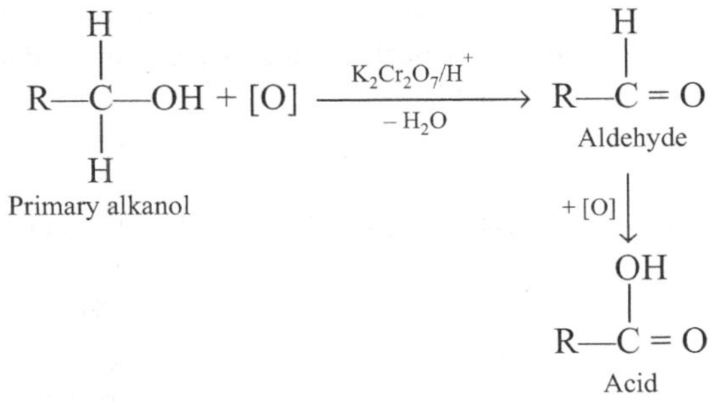

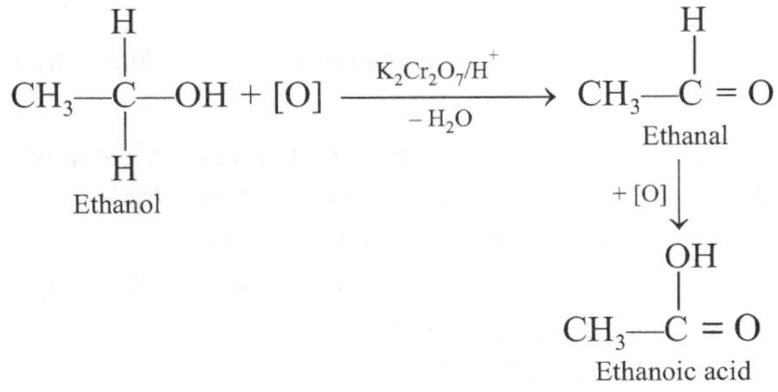

The orange colour of the reagent changes into green due to the formation of Cr^{3+} ions during the reaction.

(*ii*) ***A secondary alkanol*** on oxidation gives a ketone with the same number of carbon atoms as the original alcohol.

$$\underset{\text{Secondary alkanol}}{\mathrm{R{-}\overset{\overset{\displaystyle R}{|}}{\underset{\underset{\displaystyle H}{|}}{C}}{-}OH}} + [O] \xrightarrow[\text{or } CrO_3]{K_2Cr_2O_7/H^+} \underset{\text{Ketone}}{\mathrm{R{-}\overset{\overset{\displaystyle R}{|}}{C}=O}} + H_2O$$

$$\underset{\text{Propan-2-ol}}{\mathrm{CH_3{-}\overset{\overset{\displaystyle CH_3}{|}}{\underset{\underset{\displaystyle H}{|}}{C}}{-}OH}} + [O] \xrightarrow[\text{or } CrO_3]{K_2Cr_2O_7/H^+} \underset{\text{Propanone}}{\mathrm{CH_3{-}\overset{\overset{\displaystyle CH_3}{|}}{C}=O}} + H_2O$$

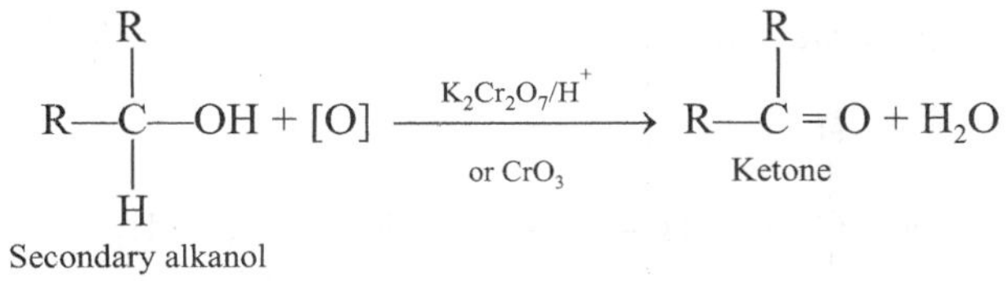

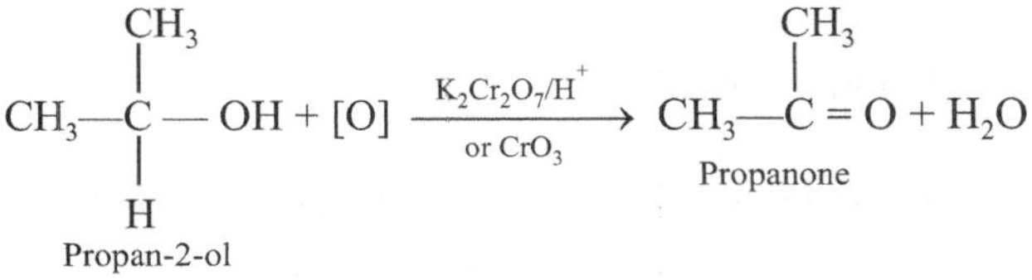

Ketones are resistant to further oxidation

(*iii*) ***Tertiary alkanols*** are not oxidised with mild oxidising agents in aqueous, alkaline or under neutral conditions. However, when oxidation is carried out under acidic conditions, tertiary alkanols undergo dehydration to form alkenes.

ACTIVITY 44.1

- Take about 2 ml of ethanol in a test tube
- Add a few drops of acidified (with dil. H_2SO_4) $K_2Cr_2O_7$ solution and warm.

What do you observe?

Primary alkanols can be oxidized to alkanals with **chromic anhydride (CrO_3)** in an anhydrous medium.

$$\underset{\text{Ethanol}}{CH_3CH_2OH} + [O] \xrightarrow{CrO_3 \text{ in } CH_3COOH} \underset{\text{Ethanal}}{CH_3CHO} + H_2O$$

ACTIVITY 44.2

- Take about 2 ml of ethanol in a test tube and add to it about 5% solution of chromium trioxide (CrO_3) in glacial acetic acid dropwise.

  During the reaction the orange colour of chromic anhydride disappears.
- Keep on adding the solution until the colour of chromic anhydride, no longer disappears.
- Observe the change in the smell of the reaction mixture before and after the reactions.

What do you observe?

DEHYDROGENATION (*Reaction with Hot Reduced Copper*)

Primary, secondary and tertiary alkanols give different products when their vapours are passed through a tube packed with reduced copper at 575 K.

(*i*) *A primary alkanol is dehydrogenated to aldehyde.*

$$\underset{\substack{\text{Primary alkanol}\\(1^\circ)}}{R\text{—}CH_2\text{—}O\text{—}H} \xrightarrow{Cu/575\,K} \underset{\text{Aldehyde}}{R\text{—}CH=O} + H_2$$

(*ii*) *A secondary alkanol is dehydrogenated to ketone.*

$$\underset{\substack{\text{Secondary alkanol}\\(2^\circ)}}{R\text{—}CH_2\text{—}OH} \xrightarrow{Cu/575\,K} \underset{\text{Ketone}}{R\text{—}CH=O} + H_2$$

(*iii*) *A tertiary alkanols* does not dehydrogenate due to absence of α-hydrogen. However, it *gets dehydrated to form an alkene.*

$$\underset{\text{2, 2-Dimethylpropan-2-ol}}{CH_3\text{—}C(CH_3)_2\text{—}OH} \xrightarrow{Cu/575\,K} \underset{\text{2-Methylpropene}}{CH_3\text{—}C(CH_3)=CH_2} + H_2O$$

IODOFORM REACTION

Alkanols containing a $CH_3\text{—}\underset{OH}{\underset{|}{CH}}\text{—}$structural unit on warming with sodium hydroxide solution and iodine yield a **pale yellow precipitate** of triiodomethane (iodoform).

$$CH_3\text{—}\underset{OH}{\underset{|}{\overset{H}{\overset{|}{C}}}}\text{—}R + 4I_2 + 6NaOH \xrightarrow{\text{Warm}} \underset{\text{Yellow ppt.}}{CHI_3}\downarrow + RCOONa + 5NaI + 5H_2O$$

This reaction can be used as a test for detection of this particular structural unit in an alkanol.

SOLVED EXAMPLES

Example 44.10. *Can we use sodium metal to remove traces of water from (i) benzene (ii) ethanol?*

Solution. (*i*) Sodium metal can be used to remove traces of water from benzene because it does not react with benzene.

(*ii*) Sodium metal cannot be used to remove traces of water from ethanol because it reacts with ethanol.

$$2C_2H_5OH + 2Na \longrightarrow 2C_2H_5ONa + H_2.$$

Example 44.11. *Which is a stronger acid ethanol or water and why?*

Solution. Water is a stronger acid than ethanol

(*i*) $C_2H_5OH + H_2O \rightleftharpoons \underset{\text{Ethoxide ion}}{C_2H_5\text{—}O^-} + H_3O^+$

(*ii*) $H_2O + H_2O \rightleftharpoons \underset{\text{Hydroxide ion}}{H\text{—}O^-} + H_3O^+$

Ethoxide ion is less stable than hydroxide ion due to the concentration of negative charge on the oxygen atom through electron releasing inductive effect of the ethyl group. Hence, the equilibrium in equation (*ii*) is more in the forward direction producing more H_3O^+ ions.

Example 44.12. *Among the isomeric alcohols having molecular formula C_4H_9OH, which is the strongest acid and which is the weakest acid?*

Solution. Among alcohols the order of acidic strength is $1° > 2° > 3°$.

∴ Among the isomeric alcohols having the molecular formula C_4H_9OH *tert*-butyl alcohol is the weakest acid and *n*-butyl alcohol is the strongest acid.

$$\underset{\substack{\text{Butan-1-ol }(1^\circ)\\(n\text{-Butyl alcohol})\\\textit{Strongest acid}}}{CH_3\text{—}CH_2\text{—}CH_2\text{—}CH_2\text{—}OH} \qquad \underset{\substack{\text{2-Methylpropan-2-ol }(3^\circ)\\(tert\text{-Butyl alcohol})\\\textit{Weakest acid}}}{CH_3\text{—}C(CH_3)_2\text{—}OH}$$

Example 44.13. *Name the reagents used in the following reactions:*

(*i*) *Oxidation of a primary alkanol to a carboxylic acid.*
(*ii*) *Oxidation of a primary alkanol to an aldehyde.*
(*iii*) *Dehydration of propan-2-ol to propene.*

Solution. (*i*) Acidified solution of $KMnO_4$ (or acidified $K_2Cr_2O_7$).
(*ii*) Hot reduced copper
(*iii*) Hot conc. H_2SO_4.

Tests for Alkanols

Alkanols can be identified by one of the following tests:

1. **Sodium Metal Test.** Alkanols can be distinguished from other organic compounds which do not contain acidic group by this test. A very small piece of sodium metal is added to the compound. **If hydrogen gas is produced with effervescence, it indicates the presence of alkanol.** It may be noted that alkanols cannot be distinguished from carboxylic acids by this test.
2. **Ester Test.** In this test, the compound to be tested is warmed with ethanoic acid and a few drops of conc. H_2SO_4. **Formation of sweet smelling vapours of some ester indicates the presence of alkanol.**
3. **Oxidation Test.** In this test, the compound to be tested is warmed with acidified $K_2Cr_2O_7$ solution. **If the colour changes from orange to green, this indicates the presence of some alkanol.**

   Similarly, the compound may be warmed with acidified potassium permanganate. **The change of colour from pink to colourless indicates the presence of some alkanol.**

44.8 USES OF ALKANOLS

Methanol and ethanol are the two most important alkanols. While methanol is largely manufactured from petroleum feed stock, ethanol is obtained from ethene or by fermentation of sugar and starch. Both have many uses in industry. **Methanol** is used as a solvent. It is also used in the preparation of ethanoic acid.

$$CH_3OH + CO \longrightarrow CH_3COOH$$

Ethan-1,2-diol is used in the manufacture of polymers. It is also used as an antifreeze.

Propane-1,2,3-triol is used in confectionery and cosmetics.

Uses of ethanol are given below in detail:

USES OF ETHANOL

1. Ethanol is a ***constituent of alcoholic beverages*** such as beer, wine, whisky and other liquors.
2. A mixture of ethanol and water has much lower freezing point than that of water. This mixture is known ***as antifreeze*** and is used in radiators of vehicles in cold countries.
3. It is also used ***as solvent*** for resins, fats, oils, fatty acids and hydrocarbons.
4. Ethanol is used to ***sterilise wounds and syringes.***
5. ***Ethanol is used in the manufacture of dyes, drugs and detergents.***
6. ***As a petrol substitute***. Ethanol up to 25% can be mixed with petrol and used as fuel in internal combustion engines.
7. Ethanol is used in the preparation of a large number organic compounds such as chloroform, iodoform, ethanoic acid, ethanal, etc.

SUMMARY

- General formula for monohydric alkanols is $C_nH_{2n+1}OH$
- Alkanols can exhibit chain, position and functional isomerism.
- Ethanol is prepared on a large scale by the hydration of ethene and fermentation of carbohydrates.
- C—O—H bond angle in alcohols is slightly less than the tetrahedral angle.
- Lower alcohols are soluble in water due to their ability to form hydrogen bonds with water molecules.
- Alkanols have relatively higher boiling points due to stronger intermolecular forces.
- Alkanols can behave as acids as well as bases.
- Alkanols are weaker acids than water.
- Alkanols on dehydration in the presence of conc. H_2SO_4, yield alkenes. The ease of dehydration of alkanols is $3° > 2° > 1°$.
- 1° and 2° alkanols can be oxidized with acidified $K_2Cr_2O_7$ solution or acidified $KMnO_4$ solution. 1° Alkanols yield carboxylic acids whereas 2° alkanols yield ketones.

EVALUATE YOURSELF

I. Objective Type Questions

Select the most appropriate choice from the options given as (a), (b), (c) and (d) after each question:

1. How many isomeric primary alkanols are possible for the molecular formula $C_4H_{10}O$?

   (*a*) 2 (*b*) 3 (*c*) 4 (*d*) 5.

2. The IUPAC name of the alkanol $(CH_3)_3CCH_2OH$ is

   (*a*) 2, 2-dimethylpentan-1-ol
   (*b*) 1, 1, 1-trimethylethanol
   (*c*) 2, 2-dimethylpropan-1-ol
   (*d*) 2, 2-dimethylpropan-2-ol.

3. Which of the following compounds is expected to be most viscous?

   (*a*) CH_3OH (*b*) CH_3CH_2OH

   (*c*) 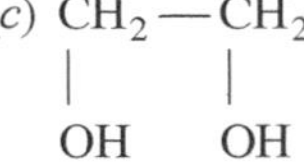

   (*d*) 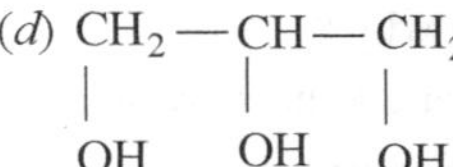

4. Which of the following alkanols is most acidic?

(*a*) CH_3CH_2OH (*b*) $CH_3CH(OH)CH_3$

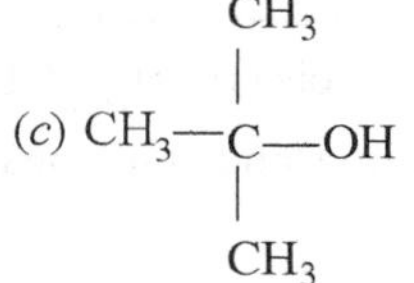

(*d*) $CH_3—CH(OH)—CH_2CH_3$

5. Which of the following compounds, on warming with sodium hydroxide solution and iodine, would yield a yellow precipitate?

(*a*) CH_3OH (*b*) $CH_3CH_2CH_2OH$

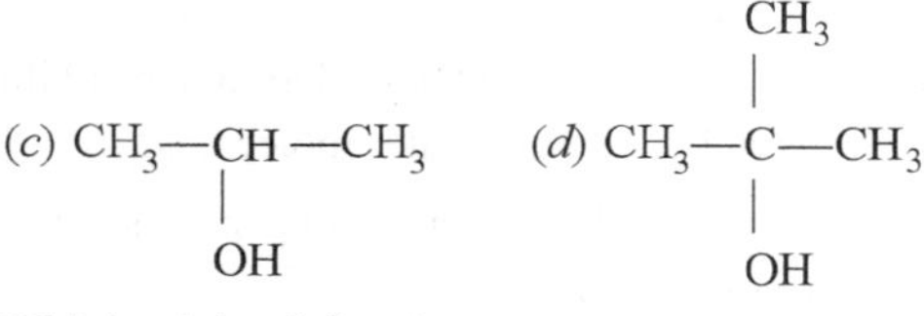

6. Which of the following compounds has the highest boiling point?

(*a*) CH_3CH_2OH (*b*) CH_3OCH_3

(*c*) CH_3Cl (*d*) $CH_3CH_2CH_3$

II. Fill in the Blanks

7. Complete the following sentences by supplying appropriate words:

(*i*) ...... are the functional isomers of alkanols.

(*ii*) Acid catalysed hydration of 2-methylpropene yields ......

(*iii*) Enzyme *diastase* is obtained from ......

(*iv*) 100% ethanol is known as ......

(*v*) When a small piece of sodium metal is added to a test tube containing ethanol, the gas produced is ......

(*vi*) Next homologue of CH_3OH is ......

(*vii*) IUPAC name of an alkanol having 2 carbon atoms is ....

III. Discussion Questions

8. Give one example of each of the following with its IUPAC name:

(*i*) A secondary monohydric alcohol

(*ii*) A dihydric alcohol

9. Give structures of all isomeric alkanols with formula $C_4H_{10}O$. Classify them as a primary, secondary or tertiary alkanol.

10. Give structures of the following alkanols:

(*i*) 2-methylpentan-2-ol

(*ii*) 2, 2-dimethylbutan-1-ol

(*iii*) 1, 2, 3-propanetriol.

11. Write all the possible isomeric alkanols having molecular formula C_3H_7OH and give their IUPAC names.

12. Explain why?

(*i*) lower alkanols are soluble in water

(*ii*) boiling point of butan-1-ol is higher than that of butane.

13. Suggest a chemical test to distinguish between the following pairs of compounds:

(*i*) ethanol and benzene

(*ii*) propan-1-ol and propan-2-ol

(*iii*) 2-methylpropan-2-ol and butan-1-ol.

14. Explain, how fermentation of starch yields ethyl alcohol?

15. How would you test for the presence of alcoholic group in an organic compound?

16. How does an alcohol react with sodium metal?

17. Give structural formula of *iso*-propyl alcohol. What is its IUPAC name?

18. What is the molecular formula of an alkanol containing five carbon atoms?

19. What is fermentation? How is ethanol prepared by fermentation of molasses?

20. What is denatured spirit? Name a poisonous alcohol which is added to ethyl alcohol to prepare denatured spirit.

21. Write the molecular formulae of the fourth and the fifth member of the homologous series of carbon compounds represented by the general formula $C_nH_{2n+1}OH$.

22. An organic compound 'A' is a constituent of wine and beer. This compound, on heating with potassium dichromate or oxygen forms another organic compound 'B'. Identify the compound 'A'. Write the chemical equation of the reaction that takes placed to form the compound 'B'. Name the compound 'B'.

23. How is ethanol prepared from ethene?

24. The sugar-cane juice mixed with yeast is kept in an air-tight pot. After a few days, this juice started to give a strong smell. Name and explain the process involved. Name the method used for separating the main product from this mixture.

[**Hint.** *The process involved is fermentation and the method used for separating the main product (ethanol) is fractional distillation.*]

25. Complete the following equations:

(*a*) $CH_3CH_2OH \xrightarrow{K_2Cr_2O_7/H^+}$

(*b*) $CH_3COOH + C_2H_5OH \xrightarrow{H^+}$

(*c*) $CH_2{=}CH_2 + H_2O \xrightarrow{H_3PO_4}$

Answers

I. Objective Type Questions

1. (*a*) **2.** (*c*) **3.** (*d*)

4. (*a*) **5.** (*c*) **6.** (*a*).

II. Fill in the Blanks

7. (*i*) Ethers (*ii*) 2-methylpropan-2-ol

(*iii*) germinated barley (*iv*) absolute alcohol

(*v*) hydrogen (*vi*) C_2H_5OH

(*vii*) ethanol.

SECTION-6

UNIT-45

CARBONYL COMPOUNDS

Specific Objective

After studying this unit, the student should be able to:

- identify the functional groups and names of carbonyl compounds.

45.1 FUNCTIONAL GROUPS OF CARBONYL COMPOUNDS

Aldehydes and ketones contain the most important functional group in organic chemistry—the carbonyl group, $>C=O$. Hence, they are known as carbonyl compounds. The carbonyl functional group is present in many important biological molecules like testosterone, progesterone, insect pheromones etc. It also occurs in the molecules in our eyes which are responsible for vision. It is also involved in manufacture of many important industrial chemicals, from plastics to solvents. The functional groups along with general formulae is given in Table 45.1.

Table 45.1. Names, Functional Groups and General Formulae of Alkanals and Alkanones

Name of the Family	*General Structural Formula*	*Functional Group*
1. Aldehydes or alkanals	$\begin{matrix} R \\ H \end{matrix}\!\!>C=O$	$H-\!\!\!>C=O$
2. Ketones or alkanones	$\begin{matrix} R \\ R \end{matrix}\!\!>C=O$	$>C=O$

Aldehydes and ketones are the *first oxidation* products of primary and secondary alcohols respectively. Aldehydes are obtained by replacing the two terminal hydrogen atoms of alkanes by the oxygen atom while ketones are obtained by replacing the two non-terminal hydrogen atoms by oxygen. The common feature of both aldehydes and ketones is the carbonyl group (>C = O) and, therefore, they are called **carbonyl compounds.**

45.2 STRUCTURE OF THE CARBONYL GROUP

Both aldehydes and ketones have a carbonyl group as the functional group. The carbonyl carbon is sp^2 hybridized and, thus, it has three sp^2 hybrid orbitals and one unhybridized *p*-orbital. It uses sp^2 hybrid orbitals to form three sigma bonds, one with oxygen atom and remaining two with two other atoms or groups (**R** or **H**). All these three sigma bonds lie in the same plane at the angle of 120°.

The unhybridized *p*-orbital of the carbonyl carbon forms a π bond with the oxygen atom via sideways overlap with half the filled *p*-orbital of oxygen atom as shown in Fig. 45.1.

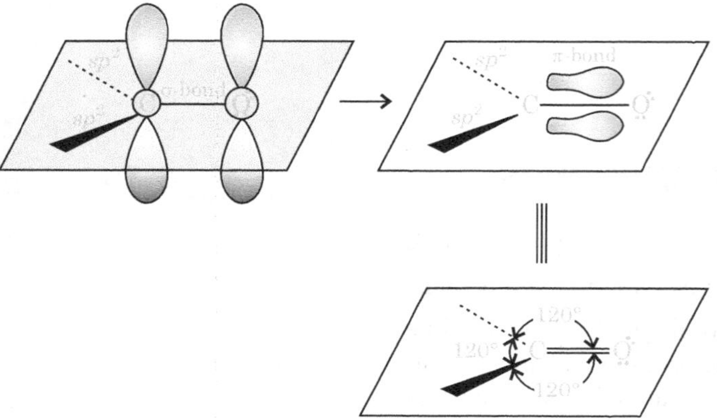

Fig. 45.1. Orbital diagram for the formation of a carbonyl group.

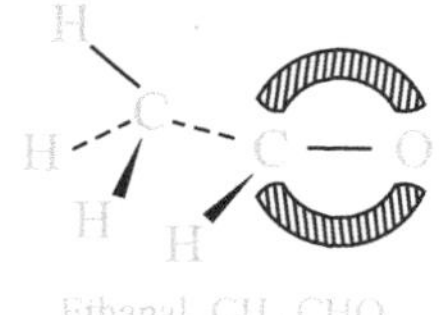

Since carbon and oxygen have different values of electronegativity, the bond between carbon and oxygen is polar. In fact, the electron density around the oxygen atom is increased which causes the development of a partial positive charge (δ^+) on carbon and a partial negative charge (δ^-) on the oxygen atom as shown in Fig. 45.2. Thus, the carbonyl carbon is an electrophillic species and the carbonyl oxygen a nucleophillic species.

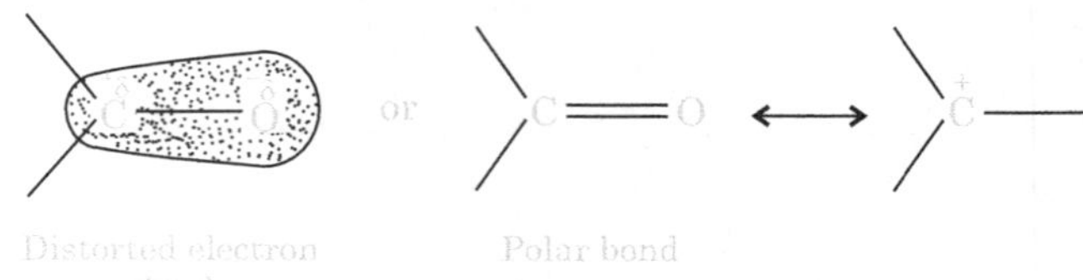

Fig. 45.2. Polarised electron cloud in carbonyl group.

The polar nature of the carbonyl group is confirmed by the fact that dipole moments of aldehydes and ketones (2.3 to 3.02 D) are quite high. The dipole moment of ethanal, propanone and diethyl ether are 2.72, 2.88 and 1.18 D respectively.

45.3 NAMES OF CARBONYL COMPOUNDS

ALDEHYDES OR ALKANALS

The trivial names of aldehydes are derived from the names of the corresponding carboxylic acids (oxidation products of aldehydes) by replacing the suffix *ic acid* with *aldehyde*. Position of substituents in aldehyde chain, if any, is indicated by Greek letters α, β, γ, δ etc. The α-carbon is the one which is directly attached to the *CHO* group.

$$\overset{\delta}{C}-\overset{\gamma}{C}-\overset{\beta}{C}-\overset{\alpha}{C}-CHO$$

In the **IUPAC system,** aldehydes are named by replacing '*e*' from the corresponding name of the alkane by the suffix '*al*'. Therefore, aldehydes are called ***alkanals.***

In case of higher aldehydes, the longest carbon chain is chosen and is numbered in such a way that the aldehyde group is given number one. The names of a few aldehydes are given in Table 45.2.

Table 45.2. IUPAC and Trivial Names of Some Alkanals

Bond Line Formulae, Formula	*Common Name*	*IUPAC Name*
HCHO	Formaldehyde	Methanal
CH_3CHO	Acetaldehyde	Ethanal

CH_3CH_2CHO	Propionaldehyde	Propanal
$CH_3CH_2CH_2CHO$	Butyraldehyde	Butanal
	Valeraldehyde	Pentanal
CH_3CHCHO with CH_3 on C-2	Isobutyraldehyde	2-Methyl-propanal
$CH_3\overset{\beta}{C}H-\overset{\alpha}{C}HCHO$ with Br on the β-carbon and CH_3 on the α-carbon	β-Bromo-α-methyl butyraldehyde	3-Bromo-2-methylbutanal
$CH_3\overset{\gamma}{C}H_2\overset{\beta}{C}H-\overset{\alpha}{C}HCHO$ with Cl on the β-carbon and Et on the α-carbon	β-Chloro-α-ethyl-valeraldehyde	3-Chloro-2-ethylpentanal
$CH_3CH_2\overset{\beta}{C}H\overset{\alpha}{C}H_2CHO$ with CH_3 on the β-carbon	β-Methylvaler-aldehyde	3-Methyl-pentanal

Alicyclic Aldehydes

In these compounds the —CHO group is attached to the ring and therefore the ring name is followed by the word *carbaldehyde*. For example:

Cyclopropane carbaldehyde; Cyclopentane carbaldehyde; Cyclohexane carbaldehyde

Aromatic aldehydes are of two types:

(*i*) Aldehydes in which the aldehydic group (—CHO) is directly attached to a benzene ring. The simplest member of the family is called *benzaldehyde*. Now, in case the substituents are attached in *ortho, meta* or *para* positions then such aldehydes are generally named as derivatives of benzaldehyde. Some examples are:

Vanillin (CHO, OCH_3, OH); *o*-Hydroxybenzaldehyde or Salicylaldehyde (CHO, OH); Cinnamaldehyde ($CH=CHCHO$)

(*ii*) In the second type of compounds, the aldehyde group is present in the side chain. For example,

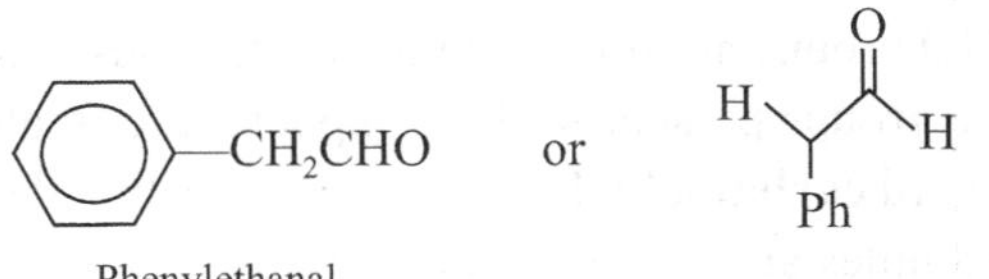

Phenylethanal

When a formyl group (—CHO) is linked to a ring, the ring name is followed by *carbaldehyde.* For example, benzaldehyde is also called,

Benzene carbaldehyde

However, the name benzaldehyde is accepted by the IUPAC.

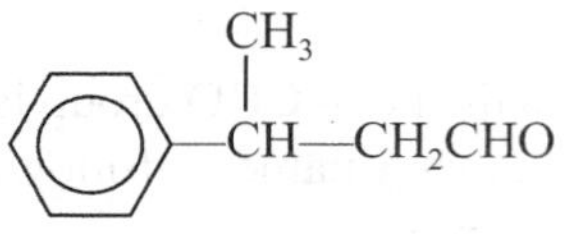

C_6H_5–CH(CH_3)–CH_2CHO

3-Phenylbutanal

or [bond line structure] or [bond line structure with Ph]

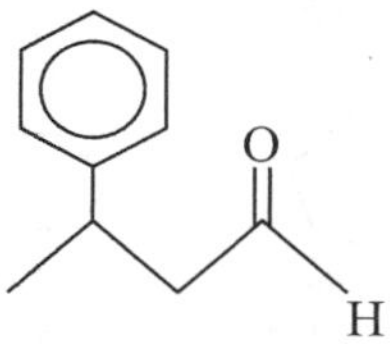

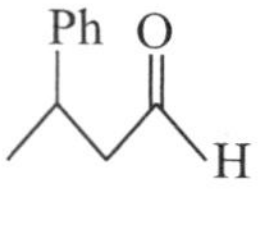

(Here Ph = phenyl group, C_6H_5)

The IUPAC names of some more aldehydes are discussed below in the form of the bond line formulae:

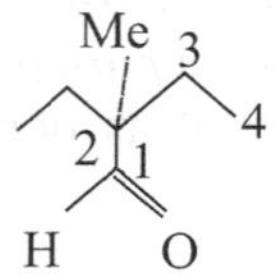

2-Ethyl-2-methylbutanal 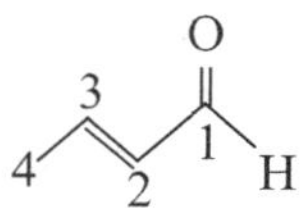 ; But-2-enal

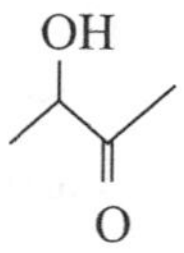

3-Hydroxy butan-2-one 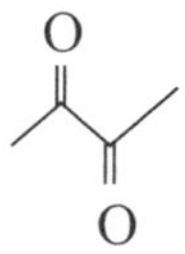 ; Butane-2, 3-dione

According to the 1993 recommendations, when three or more CHO groups are present in the compounds then all the aldehydic groups present in the compound are treated alike and the word **carbaldehyde** is used while writing the names. Some examples are discussed as follows:

(*i*) CH_2CHO–CH(CHO)–CH_2CHO

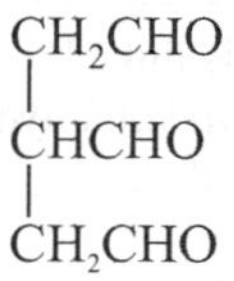

Propane-1, 2 3-tricarbaldehyde (Formerly 3-Aldopentane-1, 5-dial)

(*ii*) $OHCCH_2CH(CH_3)CH_2CHO$ (numbered 1 2 3 4 5)

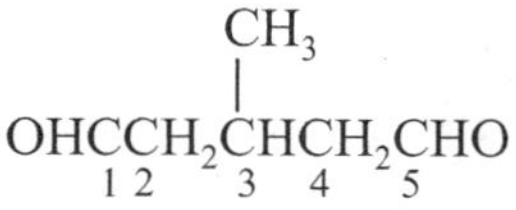

3-Methyl pentane-1, 5-dial

(*iii*) C_6H_{11}–CHO

Cyclohexane carbaldehyde

(*iv*) $OHC—^1CH(CHO)\overset{2}{C}H(CHO)—\overset{3}{C}H_2CHO$

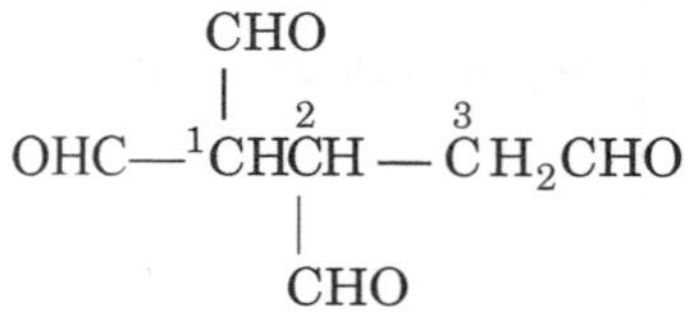

Propane-1, 1, 2, 3-tetra carbaldehyde

KETONES OR ALKANONES

Ketones are represented by the general formula RCOR′. They are termed as *simple ketones* if both R and R′ are the same alkyl groups. On the other hand, they are termed as *mixed ketones* if R and R′ are different. In the **common system, ketones** are named by using the names of alkyl groups attached to the carbonyl group followed by the word '*ketone*'.

For example, $CH_3COC_2H_5$ or [bond line structure]

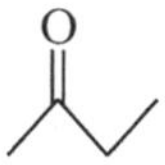

is *ethyl methyl ketone*.

In the case of substituted ketones the positions of substituents are indicated by the Greek letters, α, β, etc. (Trivial system) or numbers (IUPAC names).

According to the **IUPAC system** of naming compounds:

(*i*) *The longest carbon chain carrying the carbonyl group is considered as the parent chain.*

(*ii*) *The chain is numbered from one end such that the carbonyl carbon is given the lowest possible number.*

(*iii*) *The name of the ketone is obtained by replacing 'e' of the parent alkane by 'one'.*

(*iv*) *The positions of the carbonyl group and substituents are indicated by numbers. The names of some ketones are given in Table 45.3.*

Table 45.3. Common and IUPAC Names of Some Ketones

Bond Line Formulae, Formula	*Trivial Name*	*IUPAC Name*
 CH_3COCH_3	Acetone	Propanone
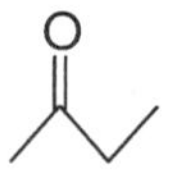 [bond line structure]	Ethyl methyl ketone	Butanone
—	—	—

Structure	Common name	IUPAC name
(skeletal structure)	Methyl propyl ketone	Pentan-2-one
(skeletal structure)	Diethyl ketone	Pentan-3-one
$CH_3CH(CH_3)COCH_3$	Methyl isopropyl ketone	3-Methyl-butan-2-one
$CH_3CH(CH_3)COCH(CH_3)CH_3$	Di-isopropyl ketone	2, 4-Dimethy-pentan-3-one
β α α′ β′ $CH_3CH(Cl)COCH_2CH_2Cl$	α, β′-Dichloro diethyl ketone	1,4-Dichloro-pentan-3-one
$(CH_3)_2CHCOCH(CH_3)_2$	Di-isopropyl ketone	2, 4-Dimethyl-pentan-3-one
(cyclohexanone with O and CH_3)	α-Methylcyclo-hexanone	2-Methyl-cyclohexanone
$(CH_3)_2C{=}CHCOCH_3$	Mesityl oxide	4-Methylpent-3-en-2-one

Examples of some aromatic ketones are given below:

$C_6H_5-CO-CH_3$ or Ph–CO–CH_3

Methyl phenyl ketone or Acetophenone

$C_6H_5-CH_2-CO-C_2H_5$ or Ph–CH_2–CO–CH_2CH_3

Ethyl benzyl ketone or 1-Phenylbutan-2-one

$C_6H_5-CO-C_6H_5$ or Ph–CO–Ph

Diphenyl ketone or Benzophenone

$C_6H_5-CH(CH_3)-CO-CH_2-C_6H_5$ or Ph–CH(CH_3)–CO–CH_2–Ph

1, 3-Diphenylbutan-2-one

SOLVED EXAMPLES

Example 45.1. *Name the following compounds according to the IUPAC system of nomenclature:*

(a) $CH_3CH(CH_3)CH_2CH_2CHO$

(b) $CH_3CH = CHCHO$

(c) $CH_3CH(CH_3)CH_2C(CH_3)_2COCH_3$

(d) $CH_3CH_2COCH(C_2H_5)CH_2CH_2Cl$

(e) $CH_3COCH_2COCH_3$.

Solution.

(*a*) 4-methyl pentanal

(*b*) But-2-enal

(*c*) 3, 3, 5-trimethyl hexan-2-one

(*d*) 6-chloro-4-ethyl hexan-3-one

(*e*) Pentane-2, 4-dione

Example 45.2. *Draw the structures of the following compounds:*

(a) 3-methylbutanal

(b) p-methylbenzaldehyde

(c) 4-chloropentan-2-one

(d) p, p-dihydroxybenzophenone

(e) p-nitropropiophenone

(f) 4-methylpent-3-en-2-one.

Solution.

(*a*) $CH_3-CH(CH_3)CH_2CHO$

(*b*) $H_3C-C_6H_4-CHO$

(*c*) $CH_3COCH_2CH(Cl)CH_3$

(*d*) $HO-C_6H_4-CO-C_6H_4-OH$

(e) 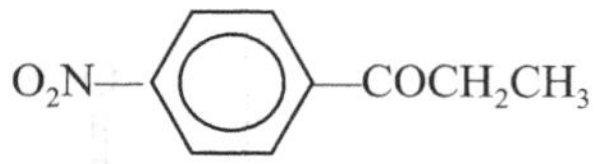 $O_2N-C_6H_4-COCH_2CH_3$

(f) $CH_3COCH=C(CH_3)-CH_3$

Example 45.3. *Write the structures of the following compounds:*

(a) α-Methoxypropionaldehyde

(b) 3-Hydroxybutanal

(c) 2-Hydroxycyclopentanecarbaldehyde

(d) 4-Oxopentanal

(e) Di-sec-butyl ketone

(f) 4-Fluoroacetophenone.

Solution. The structures are:

(a) $H_3C-CH(OCH_3)-C(=O)-H$

(b) $H_3C-CH(OH)-CH_2-C(=O)-H$

(c) 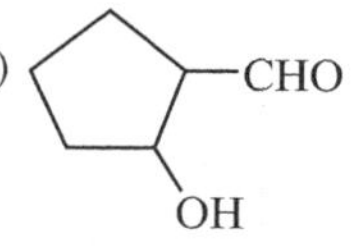 cyclopentane ring with CHO and adjacent OH

(d) $CH_3-C(=O)-CH_2-CH_2-CHO$

(e) $CH_3CH_2CH(CH_3)-C(=O)-CH(CH_3)-CH_2CH_3$

(f) 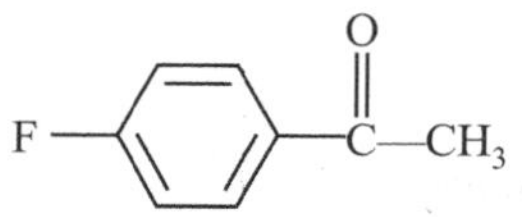 $F-C_6H_4-C(=O)-CH_3$

45.4 ISOMERISM IN ALDEHYDES AND KETONES

Aldehydes and Ketones exhibit the following type of isomerism:

(a) **Chain (nuclear) Isomerism.** Aldehydes with 4 or more carbon atoms and ketones with five or more carbon atoms show chain isomerism. For example:

C_4H_8O gives following isomers

$CH_3CH_2CH_2CHO$ and $CH_3-CH(CH_3)CHO$

Butanal 2-Methyl propanal

(b) **Position Isomerism.** Aromatic aldehydes and higher ketones give position isomers. For example:

(i) $CH_3CH_2CH_2C(=O)-CH_3$

Pentan-2-one

and $CH_3CH_2-C(=O)-CH_2CH_3$

Pentan-3-one

(ii) 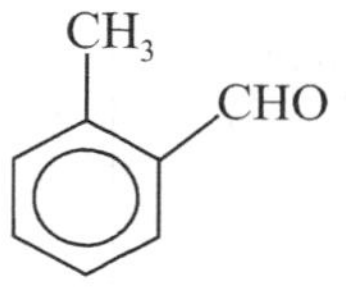 o-Tolualdehyde, 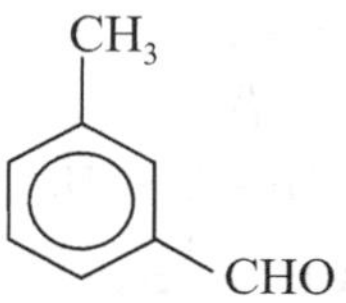 m-Tolualdehyde and p-Tolualdehyde ($CH_3-C_6H_4-CHO$)

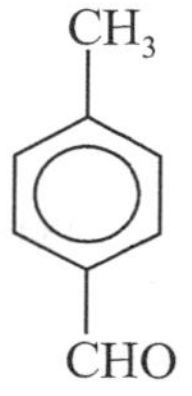

(c) **Functional Isomerism.** The general formula of aldehydes, ketones, unsaturated alcohols, oxiranes and oxolanes is $C_nH_{2n}O$

C_3H_6O has isomers as:

CH_3CH_2CHO (Propanal), $CH_3-C(=O)-CH_3$ (Propanone), $CH_2=CH-CH_2OH$ (Prop-2-en-1-ol),

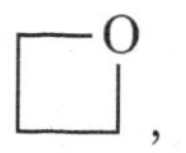

Oxobutane, 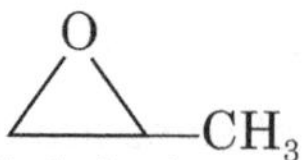 Methyl oxirane (CH_3)

(d) **Tautomerism.**

$$CH_3-C(=O)-H \rightleftharpoons CH_2=CH-OH$$

Keto Enol

SIMILARITIES AND DIFFERENCES BETWEEN C=C AND C=O BONDS

The important similarities and difference between the *carbonyl double bond (C = O) and the ethylenic double bond (C = C)* have been discussed as follows:

Similarities. C = O and C = C are similar in following aspects:

(*i*) Both consist of a σ and a π bond.

(*ii*) In both, carbon atom assumes sp^2 hybrid state.

(*iii*) Both have planar structures.

(*iv*) Both undergo addition reactions.

Differences. C = O and C = C bonds are different in following aspects:

(*i*) C = O bond is polar while C = C bond is non-polar.

(*ii*) Their bond lengths and bond strengths are different.

(*iii*) C = O bond gives nucleophilic additions while C = C bond gives electrophilic additions generally.

(*iv*) Geometric isomerism is possible in C = C bond but no such isomerism is possible in C = O bond.

SUMMARY

- **Alkanals** are known as aldehyde and have —CHO functional group.
- **Alkanones** are ketones with —C(|)= O functional group.
- **Isomers** are compound with same formula but different are physical or chemical properties.
- **Carbonyl group** >C=O group is known as the carbonyl group.

EVALUATE YOURSELF

I. Objective Type Questions

Select the most appropriate choice from the options given as (a), (b), (c) and (d) after each question:

1. The general formula of aldehydes is

(*a*) $C_nH_{2n}CHO$ (*b*) C_nH_{2n-2} CHO

(*c*) $C_nH_{2n+1}CHO$ (*d*) C_nH_nCHO.

2. Which of the following is an alicyclic aldehyde?

(*a*) 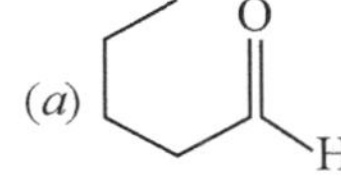

(*b*) cyclohexenyl—CHO

(*c*) C_6H_5—CHO

(*d*) C_6H_5—CH_2CHO

3. The common name of the compound, $(CH_3)_2CH$—CHO is

(*a*) isopropylaldehyde

(*b*) methylacetaldehyde

(*c*) isobutyraldehyde

(*d*) valeraldehyde.

4. Acrolein is

(*a*) an unsaturated ketone

(*b*) a ketoaldehyde

(*c*) prop-2-enal

(*d*) vinyl acetaldehyde.

5. Aldehydes and ketones are

(*a*) geometrical isomers

(*b*) optical isomers

(*c*) position isomers

(*d*) functional isomers.

6. The carbon atom of the carbonyl group in aldehydes and ketones is

(*a*) *sp* hybridised (*b*) sp^2 hybridised

(*c*) sp^3 hybridised (*d*) dsp^2 hybridised.

II. Fill in the Blanks

7. Complete the following sentences by supplying the appropriate words:

(*i*) CH_2=CH—OH is the enolic form of .....

(*ii*) $(CH_3)_2C=O$ and ..... are functional isomers.

(*iii*) The carbon of the carbonyl group is ..... hybridised.

(*iv*) In the IUPAC system, the suffix for ketones is ..... .

(*v*) H(C=O)-isopropyl and H, OH-substituted ... are .....

(*vi*) 2-pentanone and 3-pentanone are ..... isomers.

(*vii*) The C=O bond is ..... polar than the C=C bond.

III. Discussion Questions

8. Give the structures and IUPAC names of the following:

(*i*) Dibenzyl ketone (*ii*) Ethyl *neo*-pentyl ketone

(*iii*) Crotonaldehyde (*iv*) Phenyl acetaldehyde.

9. How many isomeric ketones are possible for 2, 2-Dimethyl propanal?

10. Vanillin is 4-hydroxy-3-methoxy *benzenecarbaldehyde*. Give its structure.

11. Write IUPAC names for the following:

(i) O, H (ii) O, Cl

(iii) CHO, CH_2Br (iv) Ph O, H

(v) O

12. Write IUPAC names and common names for the following:

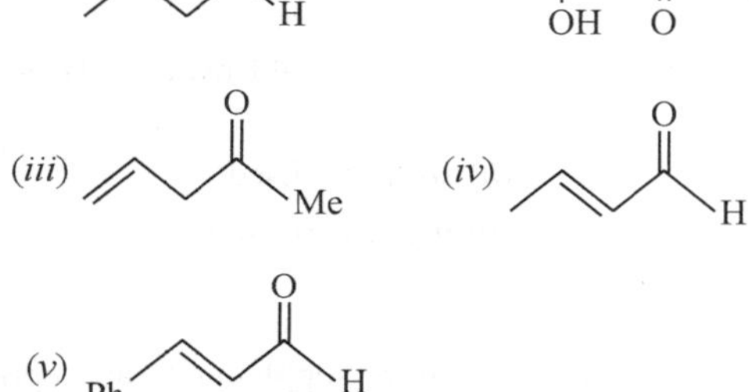

(v) Ph, O, H

13. Give structures and IUPAC names of all the carbonyl compounds having the molecular formula C_4H_8O.

14. Name the different kind of isomerism that is shown by the compound with formula C_4H_8O.

15. Write the correct IUPAC names for the following:

(i) Salicylaldehyde (ii) Acetylacetone
(iii) Vinyl acetone (iv) Benzyl phenyl ketone
(v) Cinnamaldehyde.

Answers

I. Objective Types Questions

1. (c) 2. (b) 3. (c) 4. (c)
5. (d) 6. (b).

II. Fill in the Blanks

7. (i) CH_3CHO (ii) O (iii) sp^2
(iv) one
(v) different compounds (vi) position (vii) more.

III. Discussion Questions

8. (i) $C_6H_5CH_2COCH_2C_6H_5$; *1, 3-Diphenyl-2-propanone*
(ii) $CH_3CH_2COCH_2C(CH_3)_3$; 5, 5-Dimethyl-2-hexanone
(iii) $CH_3CH = CH—CHO$; 2-Butenal
(iv) $C_6H_5CH_2CHO$; Phenyl ethanal.

9. *The molecular formula of 2, 2-Dimethyl propanal is $C_5H_{10}O$. The ketones possible are:*
(i) Pentan-2-one, (ii) Pentan-3-one,
(iii) 3-Methylbutan-2-one.

10. HO— —C(=O)—H, H_3CO

11. *(i) 3-Methylhexanal*
(ii) 4-Chloro-3-ethylpentan-2-one
(iii) 3-Bromomethyl benzaldehyde
(iv) 3-Phenylbutanal
(v) 2, 4-Dimethylpentan-3-one.

12. *(i) 3-Methyl butanal; isobutyraldehyde*
(ii) 3-Hydroxy butanal; β-Hydroxy butyraldehyde
(iii) Pent-4-en-2-one; Allyl methyl ketone
(iv) But-2-enal ; Crotonaldehyde
(v) 3-Phenylprop-2-enal; Cinnamaldehyde.

13. *It gives isomers, Butyraldehyde, iso-Butyraldehyde,*

Butan-2-one, 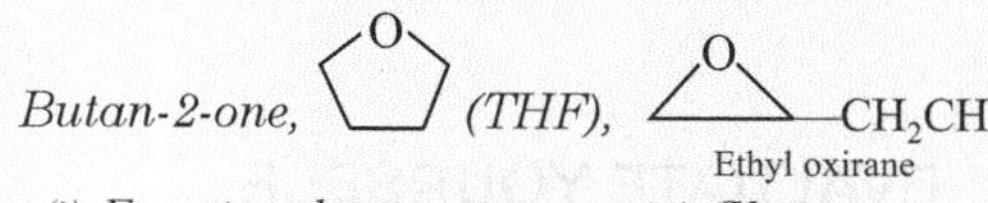

14. *(i) Functional (ii) Chain*
(iii) Tautomerism.

15. *(i) 2-Hydroxy benzaldehyde*
(ii) Pentane-2, 4-dione
(iii) Pent-4-en-2-one
(iv) Phenyl acetophenone
(v) 3-Phenylprop-2-enal.

SECTION-6

UNIT-46

ALKANOIC ACID

Specific Objectives

After studying this unit, the student should be able to:

- describe the sources, preparation and properties of alkanoic acids
- write the systematic names and structures of given alkanoic acids
- describe the uses of alkanoic acids.

Carboxylic acids are organic compounds containing the carboxyl group $-\overset{O}{\overset{\|}{C}}-OH$. These compounds possess sufficient acidic character and are called **carboxylic acids.** Many common chemicals such as citric acid (lemon juice), ethanoic acid (vinegar) are carboxylic acids. The carboxyl group is made up of a *carbonyl*, $>C=O$ and a *hydroxyl*, —OH group, hence, the name *carboxyl group* (*carb* from the carbonyl and *oxyl* from the hydroxyl). Carboxylic acids may be *aliphatic* (R—COOH) or *aromatic* (Ar—COOH) depending upon whether the —COOH group is attached to an aliphatic alkyl chain or aryl groups respectively.

Aliphatic monocarboxylic acids (containing one carboxyl group) are known as **fatty acids** because some of their higher members (C_{12}—C_{18}) *like palmitic acid* ($C_{15}H_{31}COOH$) and *stearic acid* ($C_{17}H_{35}COOH$) exist in natural fats as esters and are obtained via hydrolysis.

46.1 SOURCES, PREPARATION AND PROPERTIES OF ALKANOIC ACIDS

Many naturally occurring organic compounds are carboxylic acids. Some common examples are given in Table 46.1.

Table 46.1. Sources of Alkanoic Acids

Name	*Molecular structure*	*Major source*
methanoic acid (Latin word *formicus*) (formic acid)	HCOOH	ants, bees and nettles
ethanoic acid (Latin word acetum) (acetic acid)	CH_3COOH	vinegar
butanoic acid (Latin word butyrum) (butyric acid)	$CH_3CH_2CH_2COOH$	butter
hexanoic acid (caproic acid)	$CH_3(CH_2)_4COOH$	coconut oil, goat's milk
hexadecanoic acid (palmitic acid)	$CH_3(CH_2)_{14}COOH$	palm oil, olive oil, beef fat, cow's milk, lard
octadecanoic acid (stearic acid)	$CH_3(CH_2)_{16}COOH$	beef fat, olive oil, lard
dodecanoic acid (*vii*) valeric acid is found in plant valerian	$CH_3(CH_2)_{10}COOH$	coconut oil, palm kernel oil
cis-9-octadecanoic acid (oleic acid)	$CH_3(CH_2)_7\overset{H}{C}{=}\underset{9}{\overset{H}{C}}-(CH_2)_7COOH$	olive oil, plam oil, ground nut oil, lard, corn oil
cis, *cis*-9-12-octadecadienoic acid (linoleic acid)	$CH_3(CH_2)_4\underset{H}{C}{=}\underset{H}{C}-CH_2-\underset{H}{C}{=}\underset{H}{C}-(CH_2)_7COOH$	peanut oil, soyabean oil, corn oil, corn seed oil
2-hydroxypropanoic acid (lactic acid)	$CH_3-CH(OH)COOH$	breast milk, muscle, cow's milk
2-hydroxypropan-1,2,3 tricarboxylic acid (citric acid)	$HO-\overset{CH_2COOH}{\underset{CH_2COOH}{C}}-COOH$	citrus fruits, *e.g.*, lime, lemon, orange
2,3-dihydroxybutanedioic acid (tartaric acid)	$HOCH_2COOH$ | $HOCH_2COOH$	grapes, residue in akpeteshie distillation
phenylmethanoic acid (benzoic acid)	C_6H_5COOH	gums and balsams rasp berries food additives/preservatives
1,2-ethanedioic acid (oxalic acid)	HOOC—COOH	sweet potatoes, tomatoes, rhubarb
3-pyridinecarboxylic acid (nicotinic acid or niacin)	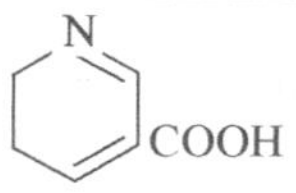 (pyridine ring with N) COOH	a vitamin B found in yeast, liver and green leaves, *e.g.*, kontomire

GENERAL METHODS OF PREPARATION

1. By Oxidation Reactions

(*i*) **By Oxidation of Primary Alcohols.** Carboxylic acids can easily be prepared by oxidation of primary alcohols with *acidified potassium permanganate* or *acidified potassium dichromate*.

$$R{-}CH_2OH + 2[O] \xrightarrow{KMnO_4/H^+} R{-}COOH + H_2O$$

$$\underset{\text{Ethyl alcohol}}{CH_3{-}CH_2OH} + 2[O] \xrightarrow{KMnO_4/H^+} \underset{\text{Acetic acid}}{CH_3COOH} + H_2O$$

(*ii*) **By Oxidation of Aldehydes.** Aldehydes on oxidation with usual oxidizing agents give carboxylic acids with same number of carbon atoms as in the aldehyde.

$$R{-}CHO + [O] \longrightarrow R{-}COOH$$

$$\underset{\text{Acetaldehyde}}{CH_3CHO} + [O] \longrightarrow \underset{\text{Acetic acid}}{CH_3COOH}$$

(*iii*) **By Oxidation of Methyl Ketones.** Methyl ketones can also be easily oxidized to carboxylate ion by hypohalite solutions. The salts of carboxylic acid so formed, can be hydrolysed to the corresponding carboxylic acid. *In this method, the acid formed has one carbon atom less than the parent ketone.*

$$\underset{\text{Methyl } n\text{-propyl ketone}}{CH_3CH_2CH_2{-}\overset{\overset{O}{\|}}{C}{-}CH_3} \xrightarrow{NaOI} \underset{\text{Sodium butyrate}}{CH_3CH_2CH_2CO\overset{-}{O}\overset{+}{N}a} + CHI_3$$

$$CH_3CH_2CH_2CO\overset{-}{O}\overset{+}{N}a + HCl \longrightarrow CH_3CH_2CH_2COOH + NaCl$$

2. From Alkyl Cyanides

Alkyl cyanides on hydrolysis with dilute acids or alkalies give carboxylic acids.

$$R{-}CN + 2H_2O \xrightarrow{H^+} R{-}COOH + \overset{+}{N}H_4$$

$$\underset{\text{Methyl cyanide}}{CH_3{-}CN} + 2H_2O \xrightarrow{\text{dil. HCl}} \underset{\text{Acetic acid}}{CH_3COOH} + NH_4Cl$$

This method serves as a very good synthetic method for the preparation of carboxylic acids.

3. By Carbonation of Grignard Reagents

Carboxylic acids can be obtained by carbonation of Grignard reagents. The reaction is carried out by bubbling CO_2 (dry ice is employed as a source of carbon dioxide) through ethereal solution of a Grignard reagent followed by subsequent hydrolysis with dil. acids.

$$RMgX + CO_2 \longrightarrow R{-}\overset{\overset{O}{\|}}{C}{-}OMgX \xrightarrow{H_2O/H^+} R{-}\overset{\overset{O}{\|}}{C}{-}OH + Mg(OH)X$$

$$\underset{\text{Methyl mag. bromide}}{CH_3MgBr} + CO_2 \longrightarrow \left[CH_3{-}\overset{\overset{O}{\|}}{C}{-}OMgBr\right] \xrightarrow{H_2O/H^+} \underset{\text{Acetic acid}}{CH_3{-}\overset{\overset{O}{\|}}{C}{-}OH} + Mg(OH)Br$$

This method gives carboxylic acid with one carbon more than the starting compound.

4. From Sodium Alkoxide and Carbon Monoxide

Heating of sodium alkoxide with CO under pressure yields sodium salt of fatty acids which on subsequent acidification gives acid.

$$\underset{\text{Sodium alkoxide}}{RONa} + CO \xrightarrow[\text{pressure}]{\text{high}} \underset{\text{Sodium salt of acid}}{RCOONa} \xrightarrow{HCl} \underset{\text{Fatty acid}}{RCOOH} + NaCl$$

5. From Olefins

Higher fatty acids can be obtained on large scale by heating an olefin with CO and steam under pressure at 570 K–675 K in the presence of phosphoric acid as catalyst.

$$\underset{\text{Ethylene}}{CH_2 = CH_2} + CO + H_2O \xrightarrow[H_3PO_4]{570-675\text{ K}} \underset{\text{Propanoic acid}}{CH_3CH_2COOH}$$

$$\underset{\text{Propene}}{CH_3CH = CH_2} + CO + H_2O \xrightarrow[H_3PO_4]{570-675\text{ K}} \underset{iso\text{-Butyric acid}}{(CH_3)_2CHCOOH}$$

Benzoic acid can be prepared by oxidation of alkyl benzenes with alkaline potassium permanganate, chromic anhydride or conc. nitric acid. The alkyl side chain gets oxidised to a —COOH group irrespective of the size of the chain. Many aromatic acids are obtained industrially by this method.

$$\text{Toluene } (C_6H_5CH_3) \xrightarrow[3[O]]{KMnO_4/OH^-} \text{Benzoic acid } (C_6H_5COOH) + H_2O$$

$$\text{Ethyl benzene } (C_6H_5CH_2CH_3) \xrightarrow[6[O]]{KMnO_4/OH^-} \text{Benzoic acid } (C_6H_5COOH) + CO_2 + 2H_2O$$

SOLVED EXAMPLES

Example 46.1. *How would you carry out the following transformations?*

(a) $CH_3CH_2CH_2Br$ to $CH_3CH_2CH_2COOH$

(b) $(CH_3)_2CHCl$ to $(CH_3)_2CHCOOH$.

Solution.

(a) $CH_3CH_2CH_2Br \xrightarrow[\text{ether}]{Mg} CH_3CH_2CH_2MgBr \xrightarrow{(i)\ CO_2\ (ii)\ H_3O^+} CH_3CH_2CH_2COOH$

(b) $(CH_3)_2CHCl \xrightarrow[\text{ether}]{Mg} (CH_3)_2CHMgCl \xrightarrow[(ii)\ H_3O^+]{(i)\ CO_2} (CH_3)_2CHCOOH$

Example 46.2. *Complete the following equations:*

(a) $(CH_3)_2CHCH_2Br \xrightarrow{KCN} (A) \xrightarrow{H_3O^+} (B)$

(b) $CH_3CH{=}CH_2 \xrightarrow{HCl} (A) \xrightarrow[(ii)\ H_3O^+]{(i)\ KCN} (B)$

Solution.

(a) (A) = $(CH_3)_2CHCH_2CN$; (B) = $(CH_3)_2CHCH_2COOH$

(b) (A) = $(CH_3)_2CHCl$; (B) = $(CH_3)_2CHCOOH$

PHYSICAL PROPERTIES OF ALKANOIC ACIDS

1. Physical State. The first three aliphatic acids are colourless liquids with pungent smell. The next six are oily liquids with an odour of rancid butter while the higher members are colourless, odourless waxy solids. Benzoic acid is a crystalline solid.

2. Solubility. The first four aliphatic members are soluble in water due to intermolecular hydrogen bonding with water molecules.

With increasing size of the alkyl group, the non-polar part of the molecule predominates thereby reducing the solubility in water. The higher members are practically insoluble in water.

3. Boiling Points. Carboxylic acids have quite high boiling points due to the presence of intermolecular hydrogen bonding which results in the formation of dimeric structures.

$$R{-}C({=}O\cdots H{-}O)({-}O{-}H\cdots O{=})C{-}R$$

Due to the dimeric structure, the effective molecular mass of the acid becomes double the actual mass. Hence, carboxylic acids have higher boiling points than alcohols and alkanes of comparable molecular masses. Moreover, the O—H bond in carboxylic acids is more polar than the O—H bond in alcohols. This is due to the electron withdrawing effect of the carbonyl group on the O—H. Hence, H-bonds in carboxylic acids are relatively stronger than those in alcohols.

4. Melting Points. In the first ten members of the homologous series, the **alternation effect** is observed. The alternation effect implies that the melting point of an acid with even number of carbon atoms is higher than the acid with odd number of carbon atoms above and below it. However, no such effect is observed in homologues with more than ten carbons. The alternation effect can be **explained** on the basis of the fact that in the carboxylic acids with even number of carbon atoms, the terminal methyl group and carboxylic group are on the opposite sides of a zig-zag carbon chain. Hence, they fit better in the crystal lattice and it results in stronger intermolecular forces. On the other hand, acids with odd number of C atoms have carboxyl and terminal methyl groups on the same side of a zig-zag carbon chain. Therefore, such molecules being relatively unsymmetrical, fit poorly in the crystal lattice. This causes weaker intermolecular forces and accounts for the relatively lower melting points.

CH_3—CH_2—CH_2—COOH (zig-zag)

Even number of C-atoms, fit better in crystal lattice, have higher m. pts (Terminal groups are on opposite side)

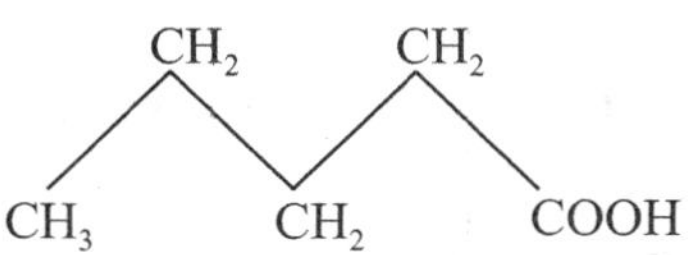

Odd number of C-atoms, fit poorly in crystal lattice, have lower m. pts. (Terminal groups are on same side)

The melting and boiling points of aromatic acids are usually higher than those of aliphatic acids of comparable molecular masses. This is presumably due to the fact that a planar benzene ring in these acids can pack closely in a crystal lattice than in zig-zag aliphatic acids.

The physical properties of some homologues of alkanoic acids are given in Table 46.2:

Table 46.2. Physical Properties of Some Homologues of Alkanoic Acids

Name	*Formula*	*Melting point/K (°C)*	*Boiling point/K(°C)*	*Solubility/g dm$^{-3}H_2O$*
methanoic acid	HCOOH	281(8)	374(101)	totally miscible (∞)
ethanoic acid	CH_3COOH	290(17)	391(118)	∞
propanoic acid	CH_3CH_2COOH	251(–22)	414(141)	∞
butanoic acid	$CH_3(CH_2)_2COOH$	267(–6)	437(164)	∞
pentanoic acid	$CH_3(CH_2)_3COOH$	239(–34)	460(187)	37
hexanoic acid	$CH_3(CH_2)_4COOH$	270(–3)	478(205)	10
octanoic acid	$CH_3(CH_2)_6COOH$	289(16)	512(239)	7
decanoic acid	$CH_3(CH_2)_8COOH$	304(31)	542(269)	2
dodecanoic acid	$CH_3(CH_2)_{10}COOH$	317(44)	498(225)	insoluble
tetradecanoic acid	$CH_3(CH_2)_{12}COOH$	327(54)	524(251)	insoluble
hexadecanoic acid	$CH_3(CH_2)_{14}COOH$	336(63)	542(269)	insoluble
octadecanoic acid	$CH_3(CH_2)_{16}COOH$	343(70)	560(287)	insoluble

CHEMICAL CHARACTERISTICS OF ALKANOIC ACIDS

In carboxylic acids, the functional group is the carboxyl group ($-\underset{\underset{O}{\|}}{C}-OH$). Carboxylic acids are a resonance hybrid of the following structures:

$$R-C(=\ddot{O}:)-\ddot{O}-H \;\;(I) \longleftrightarrow R-C(-O^{\ominus})=O^{\oplus}-H \;\;(II)$$

From these structures, it is clear that the carbonyl part of the carboxyl group does not have a double bond character but a reduced double bond character. Thus, it does not give the reactions of the carbonyl group. Also it is evident that the two contributing structures of the carboxylic acid are not equivalent, therefore, resonance stabilization in them is not much. Moreover, structure II involves charge separation, so this structure has higher energy and hence makes less contribution. It may be noted here that the oxygen atom of the —OH group has a positive charge in structure II, indicating its electron deficient nature. Hence, the shared pair of electrons of the O—H bond will be strongly be pulled towards oxygen and this makes the O—H bond quite polar. Thus, the reactions of carboxylic acids are characteristic of the carboxyl group and the alkyl group.

Now, let us study their chemical characteristics.

1. ACIDIC NATURE

Carboxylic acids are quite strong acids because of the presence of a polar O—H group. They ionize to give hydrogen ions and hence, behave as acids.

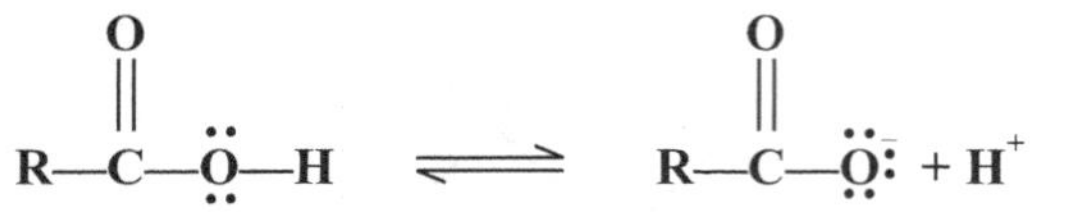

$$\underset{\text{Carboxylic acid}}{R—\overset{\overset{O}{\|}}{C}—\ddot{\underset{\cdot\cdot}{O}}—H} \rightleftharpoons \underset{\text{Carboxylate ion}}{R—\overset{\overset{O}{\|}}{C}—\ddot{\underset{\cdot\cdot}{O}}:^{-}} + H^{+} \quad ...(46.1)$$

Carboxylic acids behave as fairly strong acids. This can be explained as follows:

Carboxylic acids as well as the carboxylate ion are both stabilized by resonance. However, the carboxylate ion is better stabilized by resonance because its contributing structures are exactly identical. On the other hand, the contributing structures of a carboxylic acid involve charge separation. Since the carboxylate ion is better stabilized by resonance than a carboxylic acid, the equilibrium in Eqn. (46.1) would shift in the forward direction, *i.e.*, in favour of the ionized form. Hence, carboxylic acids behave as fairly strong acids. *They turn blue litmus red.* Some chemical reactions showing the acidic nature of carboxylic acids are:

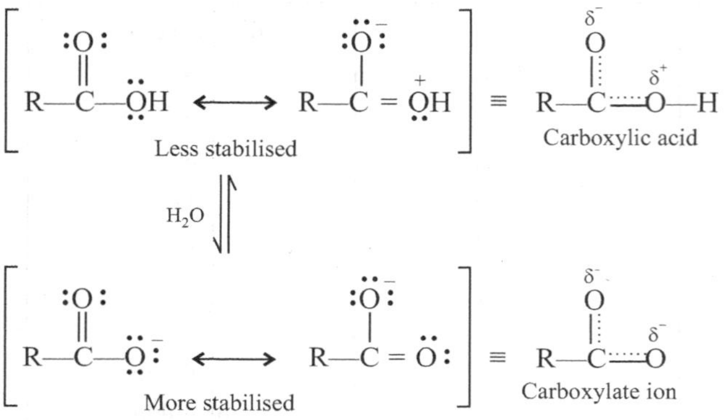

$$\left[R—\overset{\overset{:O:}{\|}}{C}—\ddot{O}H \longleftrightarrow R—\overset{\overset{:\ddot{O}:^{-}}{|}}{C}=\overset{+}{\ddot{O}}H\right] \equiv R—\overset{\overset{O^{\delta-}}{\vdots}}{C}\cdots\overset{\delta+}{O}—H$$

Less stabilised — Carboxylic acid

$\Big\Updownarrow H_2O$

$$\left[R—\overset{\overset{:O:}{\|}}{C}—\ddot{O}:^{-} \longleftrightarrow R—\overset{\overset{:\ddot{O}:^{-}}{|}}{C}=\ddot{O}:\right] \equiv R—\overset{\overset{O^{\delta-}}{\vdots}}{C}\cdots O^{\delta-}$$

More stabilised — Carboxylate ion

(*a*) **Reaction with Metals.** Carboxylic acids react with metals such as Na, K, Zn, etc., and liberate hydrogen gas.

$$2R—COOH + Zn \longrightarrow (RCOO)_2Zn + H_2$$

$$\underset{\text{Acetic acid}}{2CH_3COOH} + Zn \longrightarrow \underset{\text{Zinc acetate}}{(CH_3COO)_2Zn} + H_2$$

(*b*) **Reaction with Alkalies.** Carboxylic acids react with alkalies (NaOH, KOH) to form salt and water.

$$RCOOH + NaOH \longrightarrow RCOONa + H_2O$$

$$\underset{\text{Acetic acid}}{CH_3COOH} + NaOH \longrightarrow \underset{\text{Sodium acetate}}{CH_3COONa} + H_2O$$

(*c*) **Reaction with Bicarbonates and Carbonates.** Carboxylic acids react with bicarbonates and carbonates and produce brisk effervescence due to the liberation of CO_2.

$$RCOOH + NaHCO_3 \longrightarrow RCOONa + CO_2 + H_2O$$

$$\underset{\text{Acetic acid}}{CH_3COOH} + \underset{\text{Sodium bicarbonate}}{NaHCO_3} \longrightarrow \underset{\text{Sodium acetate}}{CH_3COONa} + CO_2 + H_2O$$

$$2CH_3COOH + Na_2CO_3 \longrightarrow 2CH_3COONa + CO_2 + H_2O$$

(*d*) **Reaction with Ammonia.** Carboxylic acid react with ammonia to form ammonium salts. Ammonium salts on heating give amides.

$$RCOOH + NH_3 \longrightarrow RCO\overset{-}{O}\overset{+}{N}H_4 \xrightarrow{\Delta} RCONH_2 + H_2O$$

$$\underset{\text{Acetic acid}}{CH_3COOH} + NH_3 \longrightarrow \underset{\text{Ammonium acetate}}{CH_3COONH_4} \xrightarrow{\Delta} \underset{\text{Acetamide}}{CH_3CONH_2} + H_2O$$

Comparison of acid strength of carboxylic acids and alcohols.

In order to understand the relative acidic strengths of carboxylic acids and alcohols, let us consider their dissociation to give H^+ ions.

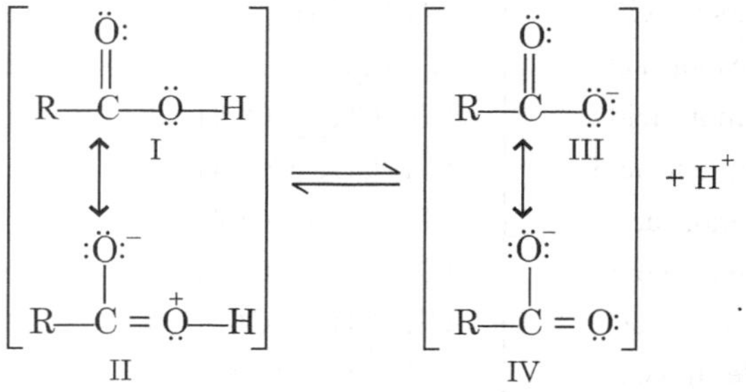

$$\left[\underset{\text{I}}{R—\overset{\overset{\ddot{O}:}{\|}}{C}—\ddot{O}—H} \updownarrow \underset{\text{II}}{R—\overset{\overset{:\ddot{O}:^{-}}{|}}{C}=\overset{+}{\ddot{O}}—H}\right] \rightleftharpoons \left[\underset{\text{III}}{R—\overset{\overset{\ddot{O}:}{\|}}{C}—\ddot{O}:^{-}} \updownarrow \underset{\text{IV}}{R—\overset{\overset{:\ddot{O}:^{-}}{|}}{C}=\ddot{O}:}\right] + H^{+} \quad ...(46.2)$$

$$R—\ddot{O}—H \rightleftharpoons R—\ddot{O}:^{-} + H^{+} \quad ...(46.3)$$

In Eqn. (46.2) though both the carboxylic acid and the carboxylate anion are resonance stabilised, the resonance stabilisation is larger in the carboxylate anion than in the carboxylic acid. The contributing structures of the carboxylate anion are identical but in the carboxylic acid, the contributing structures are non-equivalent. Therefore, the equilibrium lies more towards the right. On the other hand, in case of alcohols (Eqn. 46.3) neither alcohol nor alkoxide ion are stabilised by resonance and hence alcohols are weakly dissociated and behave as weaker acids than carboxylic acids.

Effect of substituents on the acid strength of carboxylic acids. *The factors which increase the stability of the carboxylate ion more than the carboxylic acid increase the acid strength of an acid while the factors that decrease the stability of a carboxylate ion decreases the acid strength.*

Electron withdrawing groups stabilize the carboxylate anion by dispersal of the negative charge and **increase** the strength of the acid. On the other hand, **electron releasing groups** increases the concentration of the negative charge, destabilize the carboxylate anion and decrease the strength of the acid.

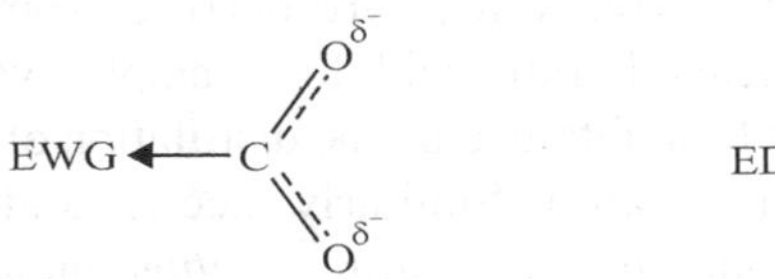

Electron withdrawing group (EWG withdraws electrons, stabilizes carboxylate anion and increases acidity.

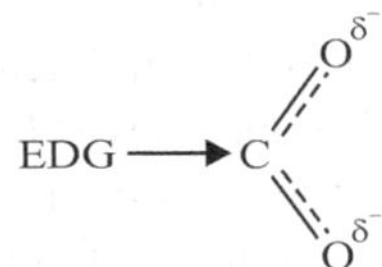

Electron donating group (EDG) gives electrons, destabilizes carboxylate anion and decreases acidity

Electron withdrawing groups such as the *halo group*, — NO_2, —*CN*, etc., increase the acidity of carboxylic acids whereas electron donating groups like *alkyl groups* decrease the acidity of carboxylic acids.

The effect of various substituents on the strength of acids has been further illustrated with the help of following examples.

1. The effect of **number of the substituents** is illustrated by the chloro substituted acetic acids. The acid strength increases from chloroacetic acid to trichloroacetic acid.

$$ClCH_2COOH < Cl_2CHCOOH < Cl_3CCOOH$$

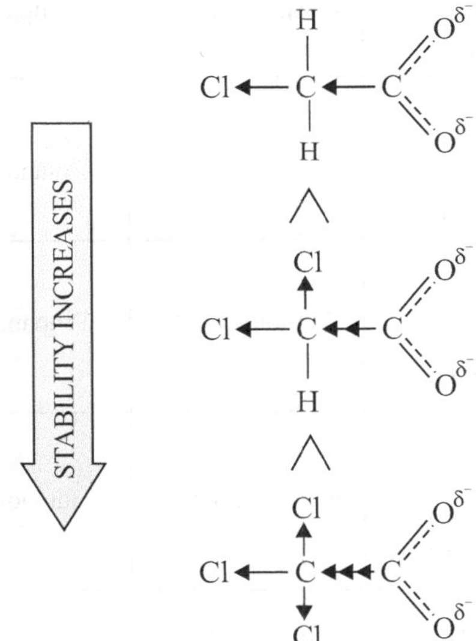

It is due to the fact that the increase in the number of chloro substituents on the α-carbon atom of acetic acid makes the electron withdrawing effect more pronounced and increases the stability of the corresponding conjugate base, *i.e.*, the carboxylate ion.

This causes an increase in the strength of their corresponding acids.

2. The effect of the **nature of the substituent** is illustrated by the various halo acetic acids. Their strength increases as shown below.

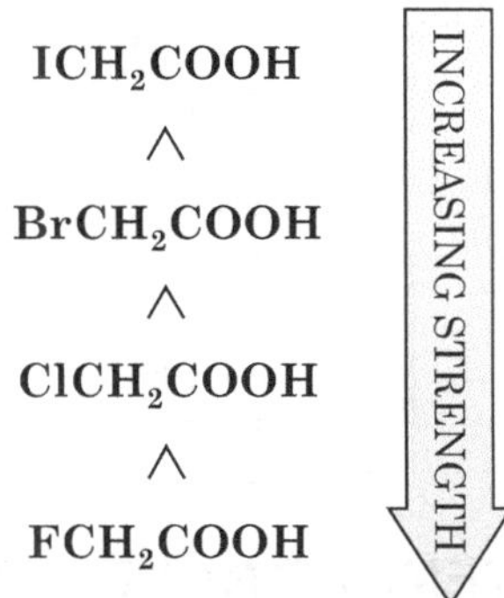

The above order is due to the fact that the electron withdrawing effect of halo-groups increase from

I < Br < Cl < F.

Consequently, the strength of the acids also increases.

3. The effect of the **position of the substituent** is illustrated by the acidity of α-*chloro* and β-*chloro* propionic acids.

$$\underset{\text{Cl}}{CH_3\overset{\alpha}{C}HCOOH} \quad > \quad \underset{\text{Cl}}{\overset{\beta}{C}H_2CH_2COOH}$$

The effect of the substituent decreases as its distance from the —COOH group increases. Thus, the electron withdrawing effect in β-chloropropionic acid is less pronounced because the —Cl group is relatively further away from the —COOH group. Thus, α-chloropropionic acid is a stronger acid than β-chloropropionic acid.

Measurement of the acidity of carboxylic acids. The acid strength of carboxylic acids is measured from their dissociation constant. For example,

$$RCOOH + H_2O \rightleftharpoons RCOO^- + H_3O^+$$

$$K_a = \frac{[RCOO^-][H_3O^+]}{[RCOOH]}$$

where K_a is the dissociation constant of the acid. The higher the value of the K_a the stronger the acid. As the numerical values of the K_a vary by large magnitudes, it is more convenient to express the acidic strength in terms of its pK_a value (= $-\log K_a$). Now, the *smaller the value of the* pK_a, *the stronger the acid and vice versa.* The K_a and pK_a values of some acids are given in Table 46.3.

Table 46.3. K_a and pK_a Values of Some Acids

Acid	K_a	pK_a
HCOOH	1.77×10^{-4}	3.75
CH_3COOH	1.8×10^{-5}	4.74
FCH_2COOH	2.6×10^{-3}	2.59
$ClCH_2COOH$	1.4×10^{-3}	2.86

$BrCH_2COOH$	1.3×10^{-3}	2.90
ICH_2COOH	6.7×10^{-4}	3.18
$Cl_2CHCOOH$	5.5×10^{-2}	1.26
Cl_3CCOOH	2.3×10^{-1}	0.64
CH_3CH_2COOH	1.3×10^{-5}	4.87
C_6H_5COOH	6.3×10^{-5}	4.20
$C_6H_5CH_2COOH$	4.9×10^{-5}	4.31
$HOCH_2COOH$	1.5×10^{-4}	3.83

46.2 SYSTEMATIC NAMES AND STRUCTURE OF ALKANOIC ACIDS

STRUCTURE OF THE CARBOXYL GROUP

In carboxylic acids, the carbon of the —COOH group is sp^2 hybridised. The carbon along with the two oxygen atoms lie in one plane and are separated by about 120°. The orbital structure of the —COOH group is shown below:

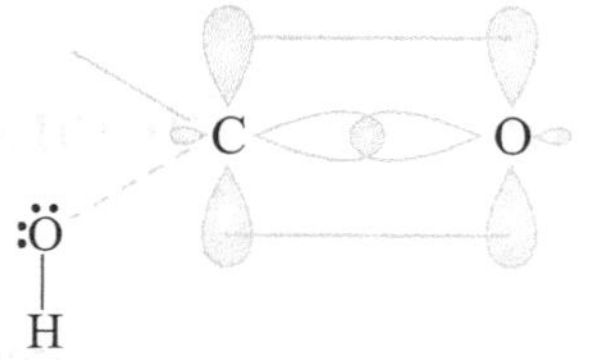

C O
:Ö
H

The carboxylic acid group, —C(=O)—ÖH, contains a —C(=O)— group and an —ÖH group. The π electrons of the —C(=O)— group are in conjugation with the lone pair of electrons on the —OH group. As a result of this, the —COOH group can be represented by different electronic structures called *canonical forms* or *mesomers*.

This phenomenon in which a compound/group can be written in two or more than two structures is known as resonance or mesomerism.

Resonance which results from the electron density around the oxygen of the O—H group decreases the displacement of electrons towards —C(=O)— group. The displacement of electrons due to resonance is known as mesomeric effect or resonance effect. The phenomenon of *resonance* or *mesomeric* effects in carboxylic acids helps to explain the acidic character of these compounds (see acidic nature).

—C(=Ö:)—Ö—H ⟷ —C(—Ö:⁻)=Ö⁺—H

(canonical structures)

NOMENCLATURE

The **common** or **trivial names** are derived from the source of individual acids. Formic acid, for example, was so named because it was first obtained by the distillation of ants (Latin: *fomica* meaning *ants*). Similarly, acetic acid has derived its name from vinegar (Latin: *acetum* meaning *vinegar*). Butyric acid is present in butter fat (Latin: *butyrum* meaning *butter*). The positions of substituents present is specified by Greek letters α, β, γ, etc. The carbon atom next to the carboxyl carbon is considered as α-carbon.

In the **IUPAC system,** the monocarboxylic acids are named as *alkanoic acids*. The name of the acid is derived by replacing the terminal '*e*-' of the corresponding alkane with **'-oic acid'.** Carboxyl carbon is always labelled number one in the chain of carbon atoms.

The trivial and IUPAC names of some common alkanoic acids are given in Table 46.4.

Table 46.4. The Trivial and IUPAC Names of Some Acids

Bond Line Formulae	*Common Name*	*IUPAC Name*
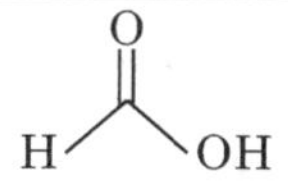 $HCOOH$	Formic acid	Methanoic acid
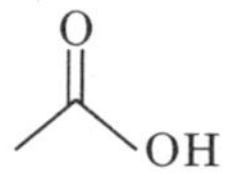 CH_3COOH	Acetic acid	Ethanoic acid
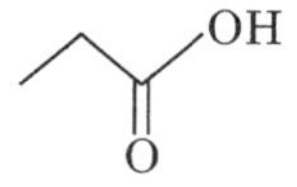 CH_3CH_2COOH	Propionic acid	Propanoic acid
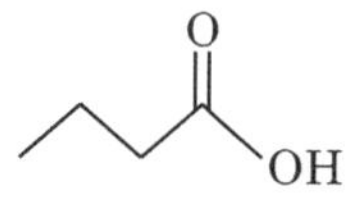 $CH_3CH_2CH_2COOH$	*n*-Butyric acid	Butanoic acid
$CH_3CH(CH_3)COOH$	Isobutyric acid	2-Methylpropanoic acid
$CH_3CH_2CH(Br)COOH$	α-Bromobutyric acid	2-Bromobutanoic acid

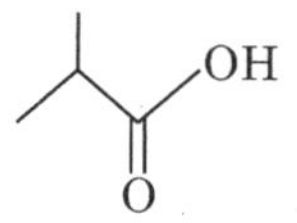

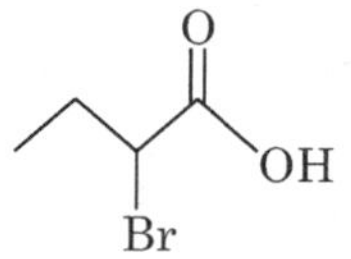

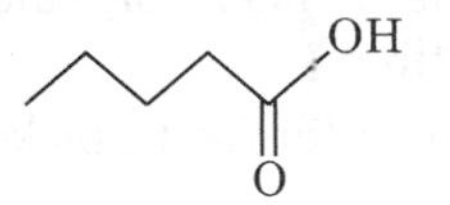 $CH_3(CH_2)_3COOH$	*n*-Valeric acid	Pentanoic acid
HOOC—COOH	Oxalic acid	Ethanedioic acid
$CH_2(COOH)_2$	Malonic acid	Propanedioic acid
CH_2COOH | CH_2COOH	Succinic acid	Butanedioic acid
COOH | $(CH_2)_3$ | COOH	Glutaric acid	Pentanedioic acid
COOH | $(CH_2)_4$ | COOH	Adipic acid	Hexanedioic acid
CH_2COOH | CH(COOH) | CH_2COOH	—	Propane-1, 2, 3-tricarboxylic acid
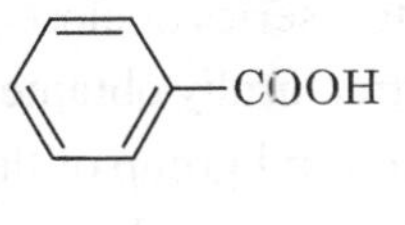	Benzoic acid	Benzene-carboxylic acid (phenyl methanoic acid)
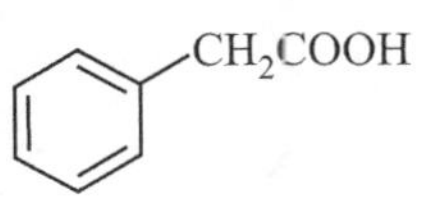	Phenylacetic acid	2-Phenylethanoic acid
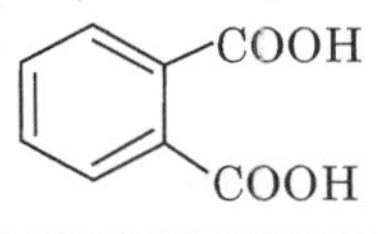	Phthalic acid	Benzene-1, 2-dicarboxylic acid

NOMENCLATURE OF POLYCARBOXYLIC ACIDS

In order to write the IUPAC names of polycarboxylic acids *i.e.*, when two or more carboxylic acid groups are present, the following rules are followed:

(*a*) **When Two COOH Groups are Present.** In such compounds the carbon atoms of both the carboxylic acid groups are to be counted in the chain. For example,

(*i*) $HOOC—CH_2—COOH$

Propanedioic acid

(*ii*) $CH_3\,\underset{2}{CH}(\overset{1}{COOH})\,\underset{3}{CH_2}\,\underset{4}{COOH}$

2-Methyl butanedioic acid

(*b*) **When Three or More COOH Groups are Present.** According to the 1993 recommendations when three or more COOH groups are present, the carbon atoms of the carboxylic acid group are not counted in the chain. Some examples are discussed as follow:

(*i*) $HOOC—\underset{1}{CH_2}—\underset{2}{CH}(COOH)—\underset{3}{CH_2}COOH$

Propane-1, 2, 3-tricarboxylic acid

(formerly 3-carboxy pentane dioic acid)

(*ii*) $HOOC\underset{4}{CH_2}\underset{3}{CH_2}\underset{2}{CH}(COOH)\underset{1}{CH_2}COOH$

Butane-1, 2, 4-tricarboxylic acid

(formerly 3-carboxy hexanedioic acid)

(*c*) **Some Typical Examples**

$\overset{2}{CH_2}\overset{1}{COOH}$
|
$^{3}CH—\overset{1}{CO}\overset{2}{CH_3}$
|
$_{4}CH_2\underset{5}{COOH}$

3-(1-oxoethyl) pentane-1, 5-dioic acid

CH_2COOH
|
$\underset{1}{HOOC}\underset{2}{CH_2}\underset{3}{CH}\underset{4}{CH_2}\underset{5}{COOH}$

3-(carboxy methyl) pentane-1, 5-dioic acid

ISOMERISM IN ALKANOIC ACIDS

(*i*) **Chain (nuclear isomerism):**

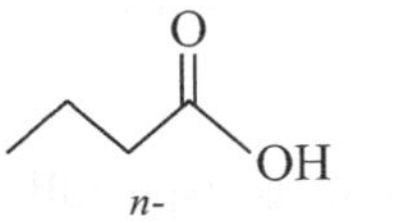
n-
and
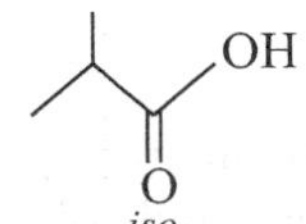
iso-

(*ii*) **Functional isomerism.** $C_nH_{2n}O_2$ represents saturated monocarboxylic acids, esters, hydroxy carbonyl compounds and hydroxy oxiranes. For example,

$C_2H_4O_2$ has isomers:

CH_3COOH, $HCOOCH_3$, $\underset{|}{CHO}$ CH_2OH , 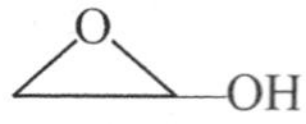

(*iii*) **Optical isomerism.** Carboxylic acids with one chiral atom can show optical isomerism.

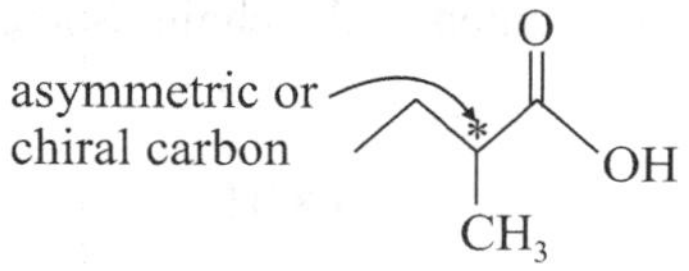
asymmetric or
chiral carbon
O
OH
CH3

46.3 SOME COMMERCIALLY IMPORTANT ALKANOIC ACIDS AND THEIR USES

Some commercially important carboxylic acids are *ethanoic acid, phenyl methanoic acid* and *benzoic acid.* Let us study their industrial preparation and uses.

1. ETHANOIC ACID (ACETIC ACID)

From ethyl alcohol. In this method, ethyl alcohol is first dehydrogenated with hot reduced copper to acetaldehyde. The acetaldehyde obtained is oxidized with air in the presence of manganous acetate as catalyst to acetic acid.

$$\underset{\text{Ethyl alcohol}}{C_2H_5OH} \xrightarrow[-H_2]{Cu/575\,K} \underset{\text{Acetaldehyde}}{CH_3CHO} \xrightarrow[\text{Manganous acetate}]{+[O]} \underset{\text{Acetic acid}}{CH_3COOH}$$

It can also be produced from carboxylation of methanol in the presence of *rhodium* catalyst.

$$CH_3OH + CO \xrightarrow{Rh} CH_3COOH$$

Acetic acid is also obtained in a dilute form (called *vinegar*) by the fermentation of ethyl alcohol with bacteria, *acetobacter* in the presence of air. This is a very old method for obtaining vinegar.

$$\underset{\text{Ethyl alcohol}}{CH_3CH_2OH + O_2} \xrightarrow{\text{Acetobacter}} \underset{\text{Acetic acid}}{CH_3COOH} + H_2O$$

Uses of Ethanoic Acid

1. *Solvent* for many organic chemicals especially in the textile, paint and colour making industry.
2. *Starting material* for the synthesis of many organic compounds, manufacture of rayon fibre.
3. *Preservative,* flavouring and tenderizer for foods, pickled foods.
4. Used in cooking (vinegar).

Uses of Phenylmethanoic Acid

1. *Disinfectant*
2. *Preservative* for foods, such as canned fruits and fruit juices; pharmaceuticals and cosmetics.
3. *Treatment of ringworm* as an ointment (Whitfield). In all the above uses the main work the acid does is to prevent and reduce growth of microbes or germs.
4. *Primary standard* in titrimetry to standardize alkalis.

Distinction between Alcohols and Carboxylic Acids

Property	*Alcohols*	*Carboxylic Acids*
1. Action on litmus	1. No action	1. Turn blue litmus red
2. Reaction with $NaHCO_3$	2. No reaction	2. Brisk effervescence, evolution of CO_2
3. Reaction with NaOH	3. No reaction	3. Form salts
4. Reaction with $FeCl_3$	4. No reaction	4. No reaction

DERIVATIVES OF CARBOXYLIC ACID

Functional derivatives of acids refers to a series of closely related organic compounds which can be structurally obtained by the replacement of the —*OH* of the carboxyl group of the acids by —*X*, —NH_2, —*OR′* or —*OCOR* respectively.

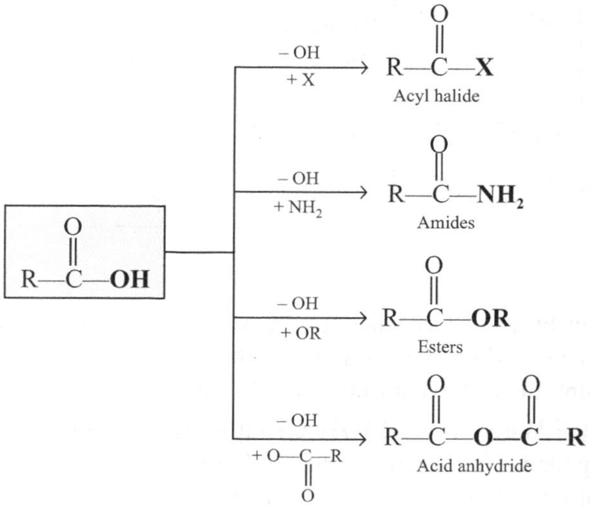
R—C—OH
– OH
+ X
R—C—X
Acyl halide
– OH
+ NH2
R—C—NH2
Amides
– OH
+ OR
R—C—OR
Esters
– OH
+ O—C—R
R—C—O—C—R
Acid anhydride

SUMMARY

- Monocarboxylic acids are called fatty acids.
- Boiling point of carboxylic acids is greater than corresponding alcohols.
- Carboxylic acid are liquids/solids at room temperature. Formic acid exists as a dimer in the vapour state and as a polymer in the liquid and solid state.
- Even membered carboxylic acids have markedly higher melting points than odd members preceeding or following it.
- Fatty acids are weaker than inorganic acids due to resonance stabilization in fatty acids.
- The presence of electron-withdrawing substituents, $—NO_2$, $—C_6H_5$, $—C \equiv N$, etc. enhance the acidic character while electron releasing groups such as Me, Et, etc. decrease the acidic strength.
- In esterification of acids, the rate of esterification follows the order
  (*i*) $CH_3OH > CH_3CH_2OH > (CH_3)_2CHOH > (CH_3)_3COH$
  (*ii*) $HCOOH > CH_3COOH > (CH_3)_2CHCOOH > (CH_3)_3CCOOH$.
- —COOH group is deactivating and *m*-directing. Deactivation by this group is so strong that aromatic acids do not undergo the Friedal-craft reaction.

EVALUATE YOURSELF

I. Objective Type Questions

Select the most appropriate choice from the options given as (a), (b), (c) and (d) after each questions:

1. Caproic acid is
   (*a*) $CH_3CH_2CH_2COOH$ (*b*) $CH_3(CH_2)_4COOH$
   (*c*) $HOOC\ CH_2COOH$ (*d*) $CH_3(CH_2)_3COOH$.
2. The general formula of carboxylic acids is
   (*a*) $C_nH_{2n+1}O_2$ (*b*) $C_nH_{2n-1}O_2$
   (*c*) $C_nH_{2n}O_2$ (*d*) C_nH_nCOOH.
3. The correct IUPAC name of the compound (HO–C(=O)–C(=O)–OH) is
   (*a*) dimethanoic acid (*b*) ethanedioic acid
   (*c*) oxalic acid (*d*) carboxymethanoic acid.
4. The melting and boiling point of carboxylic acids is greater than corresponding alcohols because
   (*a*) Carboxylic acids are more volatile than alcohols
   (*b*) Carboxylic acids are more soluble in water than alcohols
   (*c*) Carboxylic acids have stronger H bonds than alcohols
   (*d*) Carboxylic acids are oxidation products of alcohols.
5. All alkanoic acids contain the following group:
   (*a*) CHO (*b*) CH_3^- and $—\overset{O}{\overset{\|}{C}}—$
   (*c*) $—CONH_2$ (*d*) —COOH.
6. Carboxylic acids are acidic in character because
   (*a*) these dissolve in water
   (*b*) these produce H_3O^+ ions in aqueous solution
   (*c*) —C—OH bond is polar in character
   (*d*) —COOH group is a terminal group.
7. Butyric acid is correctly represented as

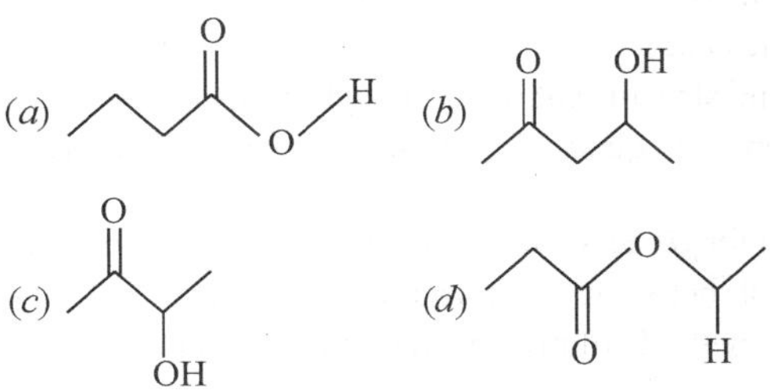

8. Which of the following is found in cow's milk?
   (*a*) linoelic acid (*b*) oleic acid
   (*c*) caproic acid (*d*) lactic acid.
9. (O, O, H) and (O, O, H) are
   (*a*) functional isomers (*b*) position isomers
   (*c*) chain isomers (*d*) tautomers.
10. The missing component is

+ (O, H) $\xrightarrow[\Delta]{H^+}$ (O, O) $+H_2O$

   (*a*) (O, H, OH) (*b*) (O, OH)
   (*c*) (OH, O) (*d*) (O, OH)

II. Fill in the Blanks

11. Complete the following sentences by supplying appropriate words:
   (*i*) Acetic acid reacts with ammonia to form ....
   (*ii*) Vinegar has an empirical formula ....
   (*iii*) Coconut oil contains ....

(*iv*) The IUPAC name of isobutyric acid is ....
(*v*) Heating ...... with alkaline $KMnO_4$ produces acetic acid.
(*vi*) The melting point of carboxylic acids is ...... than alcohols of corresponding molecular masses.
(*vii*) Carboxylic acids are ...... acidic than alcohols and turn ..... litmus ....

III. Discussion Questions

12. Give IUPAC names for the following compounds
(*a*) acetic acid (*b*) caproic acid
(*c*) valeric acid.
13. Draw structures of
(*a*) lactic acid (*b*) oxalic acid
(*c*) butyric acid.
14. Explain the structure of the carboxyl group.
15. Write chemical equations to show carboxylic acids are acidic in nature.
16. Give uses for (*i*) acetic acid and (*ii*) phenylmethanoic acid.
17. Distinguish between alcohols and carboxylic acids.
18. Give two methods of preparation of acetic acid.
19. Compare melting and boiling point of carboxylic acids with alcohols and alkanes of corresponding molecular masses.
20. Draw all the isomers of the compound with molecular formula $C_2H_4O_2$ and give their IUPAC names.

Answers

I. Objective Type Questions

1. (*b*) **2.** (*c*) **3.** (*b*)
4. (*c*) **5.** (*d*) **6.** (*b*)
7. (*a*) **8.** (*d*) **9.** (*c*)
10. (*c*).

II. Fill in the Blanks

11. (*i*) acetamide. (*ii*) CH_2O (CH_3COOH)
(*iii*) hexanoic acid (*iv*) 2-methyl propanoic acid
(*v*) acetaldehyde (*vi*) more
(*vii*) more, red, blue.

SECTION-6

UNIT-47

ALKANOIC ACID DERIVATIVES: AMIDES

Specific Objectives

After studying this unit, the student should be able to:

- describe the preparation and properties of amide
- describe the structure and names of amides.

AMIDES ($R—\overset{O}{\overset{\|}{C}}—NH_2$)

Acid amides are the derivatives of carboxylic acids in which the **—OH** part of the carboxylic acid group is replaced by the **—NH_2** group. These amides are known as primary amides because they have only one carbon atom attached to the nitrogen atom of the amide.

47.1 SOURCES, PREPARATION AND PROPERTIES OF AMIDES

SOURCES OF AMIDES

Amides occur naturally in plants and animals. For example, urea is found in the urine of mammals. A normal adult produces about 30 g of urea per day. Nicotinamide is found in many plants and animals as a source of vitamin B. Yeast, liver extract, kontomire and milk are important sources.

$H_2N—\overset{O}{\overset{\|}{C}}—NH_2$ (urea) $\qquad$ $C_5H_4N—\overset{O}{\overset{\|}{C}}—NH_2$ (nicotinamide)

Urea and nicotinamide are two naturally occurring amides

PREPARATION OF AMIDES

1. From Acid Chlorides or Acid Anhydrides. Amides are generally prepared by reaction of *ammonia* or *amines* with *acid chlorides* or *acid anhydrides*.

$$RCOCl + NH_3 \longrightarrow \underset{\text{Amide}}{RCONH_2} + HCl$$

$$\underset{\text{Acid anhydride}}{(RCO)_2O} + 2NH_3 \longrightarrow RCOONH_4 + \underset{\text{Amide}}{RCONH_2}$$

$$\underset{\text{Acetic anhydride}}{(CH_3CO)_2O} + 2NH_3 \longrightarrow \underset{\text{Amm. acetate}}{CH_3COONH_4} + \underset{\text{Acetamide}}{CH_3CONH_2}$$

$$\underset{\text{Acetyl chloride}}{CH_3COCl} + \underset{\text{Methyl amine}}{CH_3NH_2} \longrightarrow \underset{\text{N-Methyl acetamide}}{CH_3CONHCH_3} + HCl$$

2. From Ammonium Salts of Carboxylic Acids. Amides may also be prepared by the decomposition of ammonium salts of carboxylic acids.

$$RCO\overset{\ominus\oplus}{O}NH_4 \xrightarrow{\text{Heat}} RCONH_2 + H_2O$$

$$\underset{\text{Ammonium acetate}}{CH_3COONH_4} \xrightarrow{\text{Heat}} \underset{\text{Acetamide}}{CH_3CONH_2} + H_2O$$

3. From Nitriles. Alkanenitriles on partial hydrolysis produce amides.

$$\underset{\text{Alkanenitrile}}{R{-}C\equiv N} \xrightarrow[H_2O]{\text{Conc. HCl}} \underset{\text{Amide}}{R\overset{O}{\overset{\|}{C}}NH_2}$$

PHYSICAL PROPERTIES OF AMIDES

All amides except formamide are crystalline solids at room temperature. They have relatively high melting and boiling points. It is due to association of amide molecules by intermolecular hydrogen bonds.

Amides are associated by strong hydrogen bonds at both the polar oxygen of the C = O bond and the nitrogen of the —NH_2 group. The extent of hydrogen bonding appears to be greater in amides than in alkanoic acids.

Amides, especially the lower members, are very soluble in water due to hydrogen bond association. They are also soluble in common organic solvents.

(*a*) Hydrogen bonding in amides gives them relatively high melting and boiling points

(*b*) Amides dissolve in water by hydrogen bond association.

Hydrogen bonding in amides and with water change colour

Comparison of physical properties of amides and alkanoic acids

Name	*Structure*	*Relative molecular mass*	*Melting point/ K(°C)*	*Boiling point/ K(°C)*
methan-amide	$HCONH_2$	45	275 (2)	470 (197)
methanoic acid	$HCOOH$	46	282 (9)	374 (101)
ethanamide	CH_3CONH_2	59	355 (82)	494 (221)
ethanoic acid	CH_3COOH	60	300 (27)	391 (118)
propanamide	$CH_3CH_2CONH_2$	73	352 (79)	486 (213)
propanoic acid	CH_3CH_2COOH	74	251 (–22)	390 (117)

CHEMICAL PROPERTIES OF AMIDES

1. Amphoteric Character. Amides are very *feeble bases*. This is due to the fact that the lone pair of electrons on the nitrogen atom is involved in resonance with the carbonyl group as shown in the resonance structure II.

$$\underset{\text{I}}{R{-}\overset{\ddot{O}:}{\overset{\|}{C}}{-}\ddot{N}H_2} \longleftrightarrow \underset{\text{II}}{R{-}\overset{:\ddot{O}:^-}{\overset{|}{C}}{=}\overset{+}{N}H_2}$$

Thus, the electron pair on nitrogen is not easily available for protonation. Consequently, the basic character is considerably decreased. Amides can also exhibit a feeble acidic character. The amphoteric character of amides is illustrated by the following reactions:

(*i*) *Basic character*. In accordance with the resonance structure I already shown, it is evident that the nitrogen atom of the amide molecule has a lone pair of electrons. Therefore, it can act as a base. For example, acetamide (*a base*) reacts with hydrochloric acid (*an acid*) to form a salt.

$$\underset{\substack{\text{Acetamide}\\(Base)}}{CH_3CONH_2} + \underset{(Acid)}{HCl} \longrightarrow \underset{\substack{\text{Acetamide hydrochloride}\\(Salt)}}{CH_3CONH_2HCl}$$

(*ii*) *Acidic character*. In accordance with resonance structure II shown earlier, it is clear that the development of a positive character on the nitrogen atom facilitates the release of a proton. Thus, an amide can act as an acid. For example, acetamide (*an acid*) reacts with mercuric oxide (*a base*) to form mercury salt and water.

$$\underset{\substack{\text{Acetamide}\\(Acid)}}{2CH_3CONH_2} + \underset{\substack{\text{Mercuric oxide}\\(Base)}}{HgO} \longrightarrow \underset{(Salt)}{(CH_3CONH)_2Hg} + \underset{(Water)}{H_2O}$$

Similarly, acetamide reacts with metallic sodium to form sodium salt and hydrogen.

$$\underset{\text{Acetamide}}{CH_3CONH_2} + Na \longrightarrow \underset{\text{Sodium acetamide}}{CH_3CO\overset{-}{N}H\overset{+}{Na}} + \tfrac{1}{2}H_2$$

2. Reduction. Amides on *reduction* with *sodium in ethyl alcohol* or with *lithium aluminium hydride*, form primary amines. For example:

$$\underset{\text{Amide}}{RCONH_2} + 4[H] \xrightarrow{Na/C_2H_5OH} \underset{\text{Primary amine}}{RCH_2NH_2} + H_2O$$

$$\underset{\text{Benzamide}}{C_6H_5CONH_2} + 4[H] \xrightarrow{Na/C_2H_5OH} \underset{\text{Benzylamine}}{C_6H_5CH_2NH_2} + H_2O$$

3. Dehydration. Amides on heating with phosphorus pentoxide get dehydrated to form *nitriles*. For example:

$$\underset{\text{Amide}}{R\overset{\overset{O}{\|}}{C}NH_2} \xrightarrow[\text{Heat}]{P_2O_5} \underset{\substack{\text{Alkane}\\\text{nitrile}}}{RCN} + H_2O$$

$$\underset{\text{Acetamide}}{CH_3CONH_2} \xrightarrow[\text{Heat}]{P_2O_5} \underset{\text{Ethanenitrile}}{CH_3CN} + H_2O$$

4. Hydrolysis. On boiling with dilute acid or alkali, amides rapidly undergo hydrolysis. For example:

$$\underset{\text{Amide}}{RCONH_2} + H_2O \xrightarrow{\text{Dil. HCl}} \underset{\text{Carboxylic acid}}{RCOOH} + NH_4Cl$$

$$\underset{\text{Amide}}{RCONH_2} + H_2O \xrightarrow{\text{Dil. NaOH}} \underset{\text{Sodium salt}}{RCOONa} + NH_3$$

$$\underset{\substack{\text{Benzamide or}\\\text{phenylmethamide}}}{C_6H_5CONH_2} + H_2O \xrightarrow{\text{Dil. NaOH}} \underset{\text{Sodium benzoate}}{C_6H_5COONa} + NH_3$$

5. Reaction with Nitrous Acid. Amides react with nitrous acid to give carboxylic acids and nitrogen gas. Nitrous acid required is prepared *in situ* by reaction of $NaNO_2$ and HCl.

$$\underset{\text{Amide}}{R{-}\overset{\overset{O}{\|}}{C}{-}NH_2} + \underset{\text{Nitrous acid}}{HONO} \xrightarrow{(NaNO_2/HCl)} \underset{\text{Carboxylic acid}}{R{-}\overset{\overset{O}{\|}}{C}{-}OH} + N_2 + H_2O$$

6. Hoffmann's Bromamide Reaction. Amides (carrying primary nitrogen atom) react with bromine in the presence of alkali to form a *primary amine* carrying one carbon atom less than the parent amide.

$$\underset{\text{Amide}}{R{-}\overset{\overset{O}{\|}}{C}{-}NH_2} + 4NaOH + Br_2 \longrightarrow \underset{p\text{-Amine}}{R{-}NH_2} + 2NaBr + Na_2CO_3 + 2H_2O$$

$$\underset{\text{Acetamide}}{CH_3{-}\overset{\overset{O}{\|}}{C}{-}NH_2} + 4NaOH + Br_2 \longrightarrow \underset{\text{Methylamine}}{CH_3NH_2} + 2NaBr + Na_2CO_3 + 2H_2O$$

This reaction is useful in the preparation of a lower homologue from the higher one (descending of series).

EXPERIMENT TO TEST THE PRESENCE OF AMIDES

1. Take a pinch of ethanamide in a dry test tube.
2. Add 5–10 ml of 40% caustic soda (NaOH) solution to it.
3. To the mouths of test tube, attach the gas detector containing Nesseler's reagent (Fig. 47.1).
4. Warm the test tube on bunsen flame.

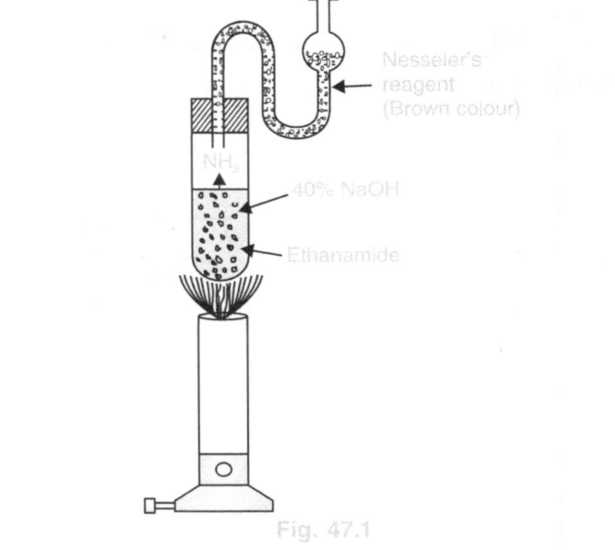

Fig. 47.1

Observations and Inference

A colourless gas with ammoniacal smell is evolved. On passing the gas through Nesseler's reagent, it turns it brown.

47.2 NOMENCLATURE AND STRUCTURE OF AMIDES

NOMENCLATURE

Acid amides are named by replacing the suffix *ic acid of the common name* or *oic acid of the IUPAC name* of the corresponding acid with the *suffix amide*. Some examples are given below:

Formula	*IUPAC Name*	*Common Name*
$HCONH_2$	Methanamide	Formamide
CH_3CONH_2	Ethanamide	Acetamide
CH_3CONH (N–CH_3)	N-methylethan-amide	N-methyl acetamide
$CH_3CH_2CONH_2$	Propanamide	Propionamide

STRUCTURE OF AMIDES

The general formula of amides is $R-\overset{\overset{\displaystyle O}{\|}}{C}-NH_2$.

The carbon of the amide group is sp^2 hybridised and hence the $-\overset{\overset{\displaystyle O}{\|}}{C}-NH_2$ group has a flat structure. The structure of ethanamide is given in Fig. 47.2.

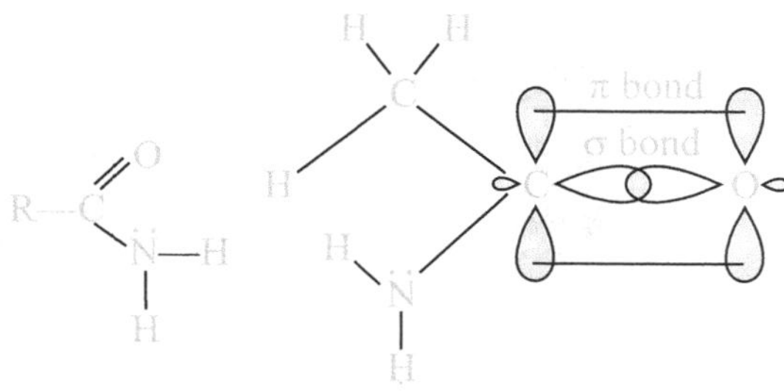

Fig. 47.2

SUMMARY

Important terms, facts and formulae

- Amides are derivatives of carboxylic acid.
- Formamide is a liquid at room temperature.
- Lower members are soluble in water due to H bonding.
- These have relatively high m.p and b.p due to H bonding.
- Amides have an amphoteric character.
- Amides evolve NH_3 gas on heating with 40% NaOH.
- Hoffman's Bromamide reaction is used to decrease the carbon chain from an amide to an amine.

EVALUATE YOURSELF

I. Objective Type Questions

Select the most appropriate choice from the options given as (a), (b), (c) and (d) after each question:

1. The correct IUPAC name of the compound $H_2N-C(=O)H$ is

   (*a*) Amino methanal (*b*) Amino acetaldehyde
   (*c*) Methanamide (*d*) Aldo amine.

2. $CH_3-C\equiv N \xrightarrow[H_2O]{\text{conc. HCl}}$ X. The compound X is

   (*a*) $CH_3CH_2NH_2$ (*b*) CH_3COOH
   (*c*) $CH_3-\overset{\overset{\displaystyle O}{\|}}{C}-CH_3$ (*d*) $CH_3-\overset{\overset{\displaystyle O}{\|}}{C}-NH_2$.

3. Which of the following amide is not a solid at room temperature?

   (*a*) CH_3CONH_2 (*b*) $CH_3CH_2CONH_2$
   (*c*) $HCONH_2$ (*d*) $CH_3CH_2CH_2CONH_2$

4. Amides are

   (*a*) Only acidic in character
   (*b*) Only Basic in character
   (*c*) Neutral in character
   (*d*) Amphoteric in character.

5. Acetamide produce 1° amine with

   (*a*) NaOH (*b*) HCl
   (*c*) $NaOH/Br_2$ (*d*) H_2O.

6. A colourless gas evolved by the reaction of propanamide with $NaNO_2/HCl$ is

   (*a*) ammonia (*b*) dinitrogen
   (*c*) dioxygen (*d*) propane.

7. On heating acetamide with phosphorus pentaoxide the reaction produces
(*a*) an amine (*b*) an acid
(*c*) an alcohol (*d*) a nitrile.

8. $CH_3C{\equiv}N \xrightarrow{\text{conc. HCl}} CH_3CONH_2 \xrightarrow[Br_2]{NaOH} P$. The compound P is
(*a*) CH_3COOH (*b*) $CH_3CH_2NH_2$
(*c*) CH_3NH_2 (*d*) NH_3.

9. Reduction of an amide, $RCONH_2$ with Na/C_2H_5OH produces
(*a*) an alcohol (*b*) an amine
(*c*) an aldehyde (*d*) an acid.

10. $A \xrightarrow{NH_3} CH_3CONH_2 \xrightarrow[HCl]{NaNO_2} B$

The compound A and B in the reaction are respectively
(*a*) CH_3COOH, CH_3NH_2 (*b*) CH_3COCl, CH_3OH
(*c*) CH_3COOCH_3, CH_3NH_2 (*d*) CH_3COCl, CH_3COOH

II. Fill in the Blanks

11. Complete the following sentences by supplying the appropriate words:
(*i*) Amides are feeble ....
(*ii*) Formamide is a .... at room temperature.
(*iii*) The reaction of acetamide with nitrous acid is accompanied with the evolution of ....
(*iv*) The reaction of an amide with $NaOH/Br_2$ to produces an amine with one carbon .... than the parent compound is known as ....
(*v*) Acetamide is .... in nature.

III. Discussion Questions

12. Explain why amides are amphoteric in character.
13. How are amides obtained from acid chlorides and acid anhydrides?
14. Name the amides up to three carbon atoms in a chain and discuss their physical state giving reasons.
15. How will you convert acetamide to
(*i*) acetic acid (*ii*) methylamine?
16. Write chemical equations to show that amides are both acidic as well as basic in character.

Answers

I. Objective Type Questions

1. (*c*)	**2.** (*d*)	**3.** (*c*)	**4.** (*d*)
5. (*c*)	**6.** (*d*)	**7.** (*d*)	**8.** (*c*)
9. (*b*)	**10.** (*d*).		

II. Fill in the Blanks

11. (*i*) basic and feebly acidic
(*ii*) liquid
(*iii*) dinitrogen gas
(*iv*) less, Hoffman's Bromamide reaction
(*v*) amphoteric.

SECTION-6

UNIT-48

ALKANOIC ACID DERIVATIVES: ALKYL ALKANOATE (ESTERS)

Specific Objectives

After studying this unit, the student should be able to:

- describe the sources, preparation and properties of alkanoates
- describe the nomenclature and structure of alkyl alkanoates
- describe the uses of alkanoates.

ESTERS ($R\overset{O}{\overset{\|}{C}}OR'$)

Esters are the derivatives of the carboxylic acids in which the —OH part of the carboxylic acid group has been replaced by the —OR group where, R may be an *alkyl* or an *aryl* group.

48.1 SOURCES, PREPARATION AND PROPERTIES OF ALKANOATES

Alkyl alkanoates are found widely in nature. Short carbon chain simple alkyl alkanoates exist as liquids and have a characteristic pleasant odour. They occur in essential oils, many fruits and flowers and are sometimes called **fruit essences** because of their pleasant odours.

More complex alkyl alkanoates are found in fats and oils and some waxes.

Fats and oils contain one, two or three alkanoate groups which generally do not have pleasant odours. It is worth noting that there is a mixture of different alkanoate derivatives of propane-1, 2, 3-triol and long-chain fatty acids in fats and oils. Table 48.1 tabulates some common and direct sources of some alkyl alkanoates which give pleasant or fruity smells. It includes some waxes which do not have a fruity smell and are formed from long carbon chain fatty acids. Esters and their mixtures are prepared and used as artificial flavours in many foods, drinks and perfumes.

Fig. 48.1. Synthetic essence

Fig. 48.2. Honeycombs—a source of alkyl alkanoates

Table 48.1. Examples of Sources and Uses of Alkyl Alkanoates

Name	*Formula*	*Source and/or uses*
ethyl ethanoate (ethyl acetate)	$CH_3COOC_2H_5$	wines, brandy, pineapple, synthetic flavours
pentyl ethanoate (amyl acetate)	$CH_3COOC_5H_{11}$	occurs naturally in apples and bananas and can be synthesized
ethyl butanoate (ethyl butyrate)	$C_3H_7COOC_2H_5$	component of synthetic pineapple, bananas, strawberry and raspberry essences
octyl butanoate	$C_3H_7COOC_8H_{17}$	occurs naturally in the seeds of parsnip
nonyl octanoate (*n*-nonylcoprylate)	$CH_3(CH_2)_6COOC_9H_{19}$	present in orange oil
ethyl methanoate (ethyl formate)	$HCOOC_2H_5$	raspberry essence
3-methylbutyl ethanoate	$CH_3COOCH_2CH(CH_3)_2$	pear essence
menthyl ethanoate (menthyl acetate)	CH_3 $CH_3—CH$ H H $CH_3COO—$ H H H H H H CH_3	found in peppermint oil and tablets (has an alicyclic ring)
Waxes beeswax	$C_{15}H_{31}COOC_{31}H_{63}$	found in honeycombs of bees—used in polishes
hexadecyl heptadecanoate	$C_{16}H_{33}COOC_{16}H_{33}$	head of whale—used in cosmetic ointments and creams

PREPARATION OF ALKANOATES

1. Acylation. Esters are prepared by the *acylation of alcohols* or *phenols*. The acylating agents can be any of the following:

Carboxylic acid/H_2SO_4 or *Acyl chloride* or *Acid anhydride*

(*i*) **Condensation of alcohols with carboxylic acid**

This reaction involves **esterification** of alkanols by alkanoic acid

$$RCOOH + R'OH \underset{\text{heat}}{\overset{\text{conc. } H_2SO_4}{\rightleftharpoons}} RCOOR' + H_2O$$

$$e.g., \underset{\text{ethanoic acid}}{CH_3COOH(l)} + \underset{\text{propan-1-ol}}{CH_3CH_2CH_2OH(l)} \underset{\text{heat}}{\overset{\text{conc. } H_2SO_4}{\rightleftharpoons}} \underset{\text{propyl ethanoate}}{CH_3COOCH_2CH_2CH_3(l)} + H_2O(l)$$

LABORATORY METHODS OF PREPARATION OF ALKYL ALKANOATES

Alkyl alkanoates can be prepared in the laboratory from alkanoic acids. The reaction is called **esterification**. In a one-step esterification reaction with alkanols, acid catalysts are used. Yields using this method are not as good as those obtained using alkanoyl chlorides and alkanoic acid anhydrides.

A pleasant fruity smell accompanies the formation of the simple alkyl alkanoates in all the preparations. *The formation of fruity odour in esterification reaction is a test for ester or alkanoate formation.* Phosphoric acid, H_3PO_4, can be used instead of tetraoxosulphate (VI) acid, H_2SO_4. It produces less decomposition at high temperatures.

(*ii*) **Esterification through acid derivatives**

$$\underset{\text{dry}}{RCOOH} \xrightarrow[\text{or } PCl_5]{SOCl_2} \underset{\text{alkanoyl chloride}}{RCOCl} + HCl \xrightarrow{R'OH} RCOOR' + HCl$$

$$e.g.\ \underset{\text{ethanoic acid}}{\underset{\text{dry}}{CH_3COOH(l)}} \xrightarrow{SOCl_2} CH_3COCl(l) + HCl(g) \xrightarrow{C_2H_5OH} \underset{\text{ethyl ethanoate}}{CH_3COOC_2H_5(l)}$$

$$RCOOH \xrightarrow{NaOH(aq)} \underset{\text{isolate solid}}{RCOONa} \xrightarrow{RCOCl} \underset{\text{alkanoic anhydride}}{RCOOCOR} \xrightarrow[\text{heat}]{R'OH} RCOOR' + RCOOH$$

$$e.g.\ \underset{\text{ethanoic acid}}{CH_3COOH(l)} \xrightarrow{NaOH(aq)} CH_3COONa(s) \xrightarrow{CH_3COCl(l)} \underset{\text{ethanoic acid anhydride}}{(CH_3CO)_2O}$$

$$(CH_3CO)_2O \xrightarrow[\text{heat}]{CH_3CH_2OH} \underset{\text{ethyl ethanoate}}{CH_3COOCH_2CH_3} + CH_3COOH$$

2. By Reaction of Acids with Diazomethane. Acids on being treated with an ethereal solution of diazomethane yield methyl esters.

$$\underset{\text{Acid}}{RCOOH} + \underset{\text{Diazo methane}}{CH_2N_2} \longrightarrow \underset{\text{Methyl ester}}{RCOOCH_3} + N_2$$

3. By Tischenko Reaction. When aldehydes containing α-hydrogen atoms are treated with aluminium ethoxide. They undergo condensation to produce esters.

$$\underset{\text{Acetaldehyde}}{CH_3-\overset{O}{\overset{\|}{C}}-H} + O=\overset{H}{\overset{|}{C}}-CH_3 \xrightarrow{Al(OC_2H_5)_3} \underset{\text{Ethyl acetate}}{CH_3-\overset{O}{\overset{\|}{C}}-O-\underset{H}{\underset{|}{\overset{H}{\overset{|}{C}}}}-CH_3}$$

PHYSICAL PROPERTIES OF ALKANOATES

1. Physical State. Esters are colourless, oily liquids with a characteristic fruity odour. The odour of most flowers and fruits is due to the esters present in them. The characteristic tastes and odours of different esters find applications in the manufacture of artificial flavouring and perfuming agents. The flavours of some of the esters are given below:

Ester	*Flavour*	*Ester*	*Flavour*
n-Pentyl ethanoate	Banana	Amyl butyrate	Apricot
Octyl ethanoate	Orange	Isobutyl methanoate	Raspberry
Ethyl butanoate	Pineapple	Benzyl ethanoate	Jasmine

2. Solubility. Esters are sparingly soluble in water but are quite miscible in organic solvents like alcohols and ethers. In fact, most of the esters are themselves very good solvents for plastics and nitrocellulose.

3. Boiling Points. The boiling points of esters are always less than the corresponding carboxylic acids because esters do not form hydrogen bonds.

CHEMICAL PROPERTIES

1. Hydrolysis. Esters are hydrolysed slowly by water at boiling temperature. The reaction is catalysed by small amount of *acid* or *base*. The basic hydrolysis is also known as **saponification.** It is because of the fact that the esters with high molecular mass acids (C_{12}—C_{17}) give soap on hydrolysis with a base. **Soaps** are sodium or potassium salts of carboxylic acids with high molecular mass (C_{12}—C_{17}).

$$\underset{\text{Ethyl ethanoate}}{CH_3COOC_2H_5} \xrightarrow[\text{Acidic hydrolysis}]{H_2O/HCl} \underset{\text{Acetic acid}}{CH_3COOH} + \underset{\text{Ethanol}}{C_2H_5OH}$$

$$\underset{\text{Ethyl ethanoate}}{CH_3COOC_2H_5} \xrightarrow[\text{Basic hydrolysis}]{NaOH} \underset{\text{Sod. acetate}}{CH_3CO\overset{-}{O}\overset{+}{Na}} + \underset{\text{Ethanol}}{C_2H_5OH}$$

2. Reduction. Esters are reduced to alcohols by the reducing agents like (*sodium/ethanol*) or (*lithium aluminium hydride*).

$$R-\overset{O}{\overset{\|}{C}}-OR' \xrightarrow[\text{or } LiAlH_4]{Na/C_2H_5OH} RCH_2OH + R'OH$$

$$\underset{\text{Methyl ethanoate}}{CH_3COOCH_3} \xrightarrow[4[H]]{Na/C_2H_5OH} \underset{\text{Ethanol}}{CH_3CH_2OH} + \underset{\text{Methanol}}{CH_3OH}$$

3. Reaction with Ammonia. Esters on treatment with alcoholic ammonia yield *acid amides*. This reaction is known as *ammonolysis of esters.*

$$\underset{\text{Ester}}{RCOOR'} + NH_3 \longrightarrow \underset{\text{Acid amide}}{RCONH_2} + \underset{\text{Alcohol}}{R'OH}$$

$$\underset{\text{Ethyl ethanoate}}{CH_3COOC_2H_5} + NH_3 \longrightarrow \underset{\text{Ethanamide}}{CH_3CONH_2} + \underset{\text{Ethanol}}{C_2H_2OH}$$

4. Reaction with Phosphorus Pentachloride. Esters are converted into acid chlorides and alkyl halides by heating with phosphorus pentachloride.

$$RCOOR' + PCl_5 \longrightarrow RCOCl + R'Cl + POCl_3$$

$$\underset{\text{Ethyl benzoate}}{C_6H_5COOC_2H_5} + PCl_5 \longrightarrow \underset{\text{Benzoyl chloride}}{C_6H_5COCl} + C_2H_5Cl + POCl_3$$

5. Alcoholysis. An ester on refluxing with a large excess of an alcohol in the presence of a little acid or alkali, undergoes exchange of alcohol residues, *i.e., alkoxy parts* as shown below:

$$\underset{\text{Ester}}{RCOOR'} + \underset{\text{Alcohol}}{R''OH} \overset{H^+}{\rightleftharpoons} RCOOR'' + R'OH$$

$$\underset{\text{Ethyl ethanoate}}{CH_3COOC_2H_5} + \underset{\text{Methanol}}{CH_3OH} \overset{H^+}{\rightleftharpoons} \underset{\text{Methyl ethanoate}}{CH_3COOCH_3} + \underset{\text{Ethanol}}{C_2H_5OH}$$

This reaction is known as *alcoholysis* or *trans-esterification*.

6. Reaction with Grignard's Reagent. All esters except the esters of formic acid react with Grignard's reagent to give tertiary alcohols. However, the esters of formic acid, *i.e., alkyl formates* form secondary alcohols.

$$\underset{\text{Ethyl acetate}}{CH_3-\overset{\overset{O}{\|}}{C}-OC_2H_5} \xrightarrow{CH_3MgI} \left[CH_3-\overset{\overset{\ominus\ \oplus}{OMgI}}{\underset{CH_3}{C}}-OC_2H_5\right] \xrightarrow{-C_2H_5OMgI} \underset{\text{Acetone}}{CH_3-\underset{CH_3}{C}=O}$$

$$\xrightarrow{CH_3MgI} \left[CH_3-\overset{\overset{\ominus\ \oplus}{OMgI}}{\underset{CH_3}{C}}-CH_3\right] \xrightarrow{HOH} \underset{\textit{tert}\text{-Butyl alcohol}}{CH_3-\overset{OH}{\underset{CH_3}{C}}-CH_3}$$

Similarly, if we start with ethyl formate ($HCOOC_2H_5$) and methyl magnesium iodide, the product will be *isopropyl alcohol.*

$$H-\overset{\overset{O}{\|}}{C}-O-C_2H_5 + CH_3MgI \longrightarrow H-\overset{OMgI}{\underset{CH_3}{C}}-OC_2H_5 \xrightarrow{-C_2H_5OMgI} H-\overset{\overset{O}{\|}}{C}-CH_3 +$$

$$\xrightarrow{CH_3MgI} H-\overset{OMgI}{\underset{CH_3}{C}}-CH_3 \xrightarrow{HOH} H-\overset{OH}{\underset{CH_3}{C}}-CH_3$$

48.2 NOMENCLATURE AND STRUCTURE OF ALKYL ALKANOATES

The esters are named by writing the name of the *alkyl* or *aryl* group of the —OR′ part of the molecule before the name of the parent acid and replacing the suffix *ic acid* of the (*IUPAC* or *common*) name of the acid by *ate.* some examples are:

Formula	*IUPAC Name*	*Common Name*
$HCOOC_2H_5$	Ethyl methanoate	Ethyl formate
$CH_3COOC_2H_5$	Ethyl ethanoate	Ethyl acetate
$CH_3COOC_6H_5$	Phenyl ethanoate	Phenyl acetate
$\overset{\gamma}{\underset{4}{C}}H_3\overset{\beta}{\underset{3}{C}}H_2—\underset{\underset{Br}{\vert 2}}{\overset{\overset{H}{\vert}}{C}}—\underset{1}{C}OOCH_3$	Methyl-2-bromo-butanoate	Methyl, α-bromobutyrate

The traditional name ester has its origin in the shortened old German word 'essigother' which means 'acetic ether' or pleasant smell.

The names of some aromatic esters are given below:

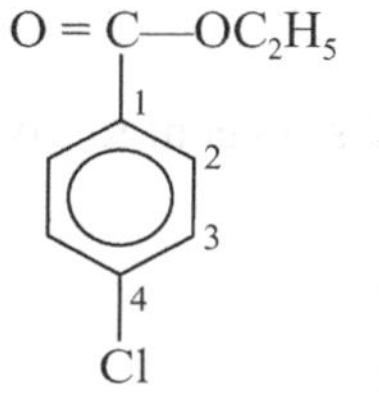

$O=C—OC_2H_5$ attached at position 1 of a benzene ring, Cl at position 4 — Ethyl 4-chloro benzoate

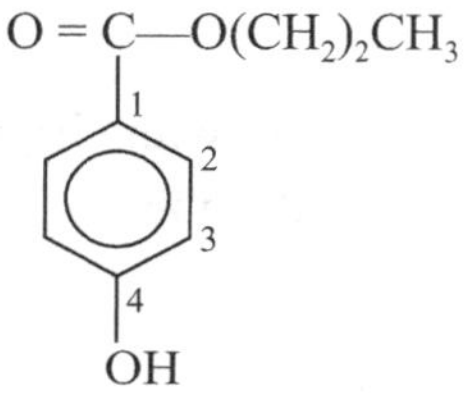

$O=C—O(CH_2)_2CH_3$ attached at position 1 of a benzene ring, OH at position 4 — *n*-Propyl 4-hydroxy benzoate

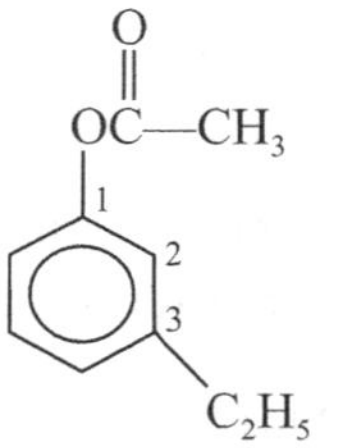

$\overset{\overset{O}{\Vert}}{O}C—CH_3$ attached at position 1 of a benzene ring, C_2H_5 at position 3 — (3-Ethyl phenyl) ethanoate

STRUCTURE OF ALKYL ALKANOATES

The alkyl alkanoates, RCOOR′, contain the carboxyl $—C\langle^{O}_{O}$ (C=O and C–O) functional group. The carbon of — C=O group is sp^2 hybridised.

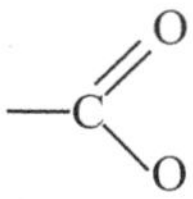

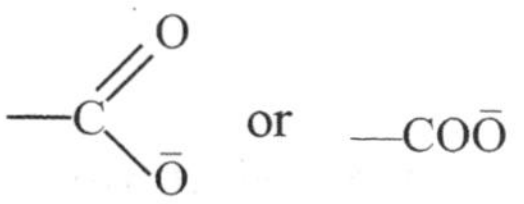

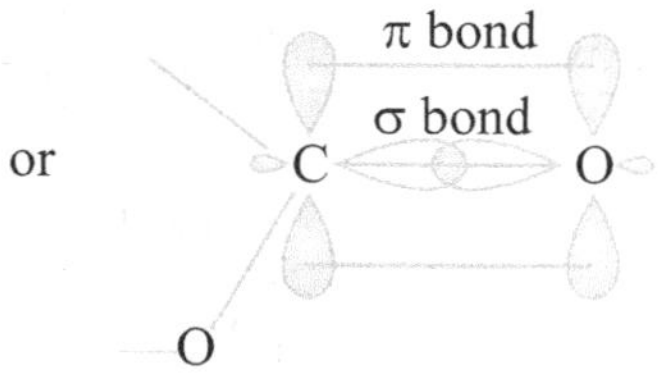

$—C\langle^{O}_{\bar{O}}$ or $—CO\bar{O}$

or [orbital diagram: C—O with π bond and σ bond]

π bond

σ bond

48.3 USES OF ALKANOATES

1. Soap making. The alkyl alkanoates of higher fatty acids like palmitic, stearic or oleic acids are used in the process of soap making. The reaction is known as **saponification.**

The saponification process can be represented generally for fats and oils which contain one or more ester groups.

$$\begin{array}{l} CH_3(CH_2)_{16}\,COOCH_2 \\ \qquad\qquad\qquad\quad\vert \\ CH_3(CH_2)_{16}\,COOCH + 3NaOH(aq) \xrightarrow{heat} \\ \qquad\qquad\qquad\quad\vert \\ CH_3(CH_2)_{16}\,COOCH_2(l) \end{array}$$

tristearin (oil in palm oil)

$$3CH_3(CH_2)_{16}\,COO^-Na^+(aq) + \begin{array}{l} CH_2OH \\ \vert \\ CHOH \\ \vert \\ CH_2OH(l) \end{array}$$

soap (sodium stearate or sodium octadecanoate) (glycerol – propane-1, 2, 3-triol)

2. Flavouring and perfuming agents. Esters or alkanoates have characteristic odours and are therefore used in the manufacture of artificial flavouring and perfuming agents (see Table 48.1 for details).

3. As plasticizers. Many esters are used as plasticizers *i.e.*, substances which are added to thermoplastics to soften them so as to be reused. Some common examples are, *di-n-*butyl pthalate, triceryl phosphate, etc.

4. As solvents. Esters are used as solvents due to their ability to dissolve various greases.

- Methyl acetate is used as solvents for many oils and resins.
- Ethyl acetate is used as extraction solvent in the food processing and pharmaceutical industry.
- Butyl acetate is used as solvent for inks and as a cleaning liquid for surfaces.

SUMMARY

Important terms, facts and formulae

- Esters have a pleasant smell
- Esters are derivatives of alkanoic acids
- In lab alkanoates are prepared by the reaction of ethyl alcohol with acetic acid in the presence of H_2SO_4. The reaction is known as esterification.
- Esters are sparingly soluble in water
- The b.p. of esters is lower than their corresponding acids
- The reaction of animal fats or vegetable oil with NaOH is known as saponification.

EVALUATE YOURSELF

I. Objective Type Questions

Select the most appropriate choice from the options given as (a), (b), (c) and (d) after each question:

1. Alkaline hydrolysis of an ester is called
   (*a*) Neutralisation (*b*) Esterification
   (*c*) Polymerisation (*d*) Saponification.
2. The reverse of esterification is called
   (*a*) alcoholysis (*b*) hydrolysis
   (*c*) acidolysis (*d*) neutralisation.
3. The correct IUPAC name of $CH_3—O—\overset{\overset{O}{||}}{C}—H$ is
   (*a*) Methoxy aldehyde (*b*) Methoxymethanol
   (*c*) Methyl methanoate (*d*) Methyl formate
4. Pineapple flavour is due to which of the following esters
   (*a*) $CH_3COOC_2H_5$ (*b*) CH_3COOCH_3
   (*c*) $CH_3CH_2COOC_2H_5$ (*d*) $CH_3CH_2CH_2COOC_2H_5$.
5. The boiling point of $CH_3CH_2COOCH_3$ is ...... than $CH_3CH_2CH_2COOH$
   (*a*) greater (*b*) less
   (*c*) same (*d*) unpredictable.
6. During esterification of carboxylic acid with alcohol which bond of the carboxylic acid undergoes clevage?
   (*a*) C—C (*b*) C = O
   (*c*) C—O (*d*) O—H.
7. Heating sodium formate with soda lime (NaOH/CaO) produces
   (*a*) Ethane (*b*) dihydrogen
   (*c*) Formic acid (*d*) Calcium formate.
8. When acetaldehyde is heated with aluminium ethoxide it produces
   (*a*) acetic acid (*b*) methyl formate
   (*c*) ethyl acetate (*d*) ethanol.
9. The correct IUPAC name of

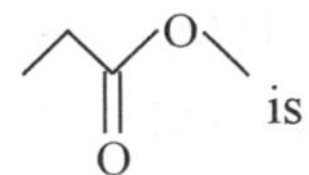

   is
   (*a*) ethyl methanoate (*b*) methyl ethanoate
   (*c*) methyl propanoate (*d*) methoxy propanone.
10. When an ester is refluxed with a large excess of an alcohol in the presence of little and or alkali, it is known as
   (*a*) hydrolysis (*b*) acidolysis
   (*c*) trans esterification (*d*) saponification.

II. Fill in the Blanks

11. Complete the following sentences by supplying appropriate words:
   (*i*) The functional group of esters is ....
   (*ii*) The reaction of acetic acid with ethyl alcohol in the presence of conc. H_2SO_4 is known as ....
   (*iii*) Reaction of .... with diazomethane produces methyl-acetate.
   (*iv*) Benzyl ethanoate has flavour of ....
   (*v*) The boiling point of esters is always .... than corresponding carboxylic acid.
   (*vi*) The basic hydrolysis of esters is known as ....
   (*vii*) Heating ethyl acetate with Grignard's reagent will give ....

III. Discussion Questions

12. Briefly explain the process of soap making.
13. Discuss the physical properties of alkyl alkanoates.
14. Discuss the acidic and basic hydrolysis of esters and also write chemical equations for them.
15. Discuss the physical state of esters and give example of two esters having characteristic flavours.
16. What is esterification? Discuss with chemical equations.
17. Write short notes on
   (*i*) Tischenko reaction (*ii*) Alcoholysis.

18. Give the IUPAC names of the following:

(*i*) 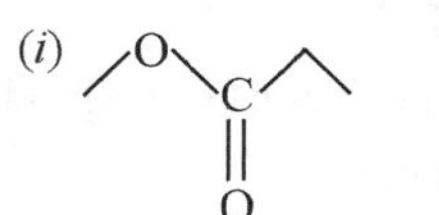

(*ii*) Phenyl acetate

(*iii*) $CH_3CH_2COOC_2H_5$

(*iv*) $CH_3CH_2-C(=O)-O-CH_2CH_3$

19. Complete the following reactions:

(*i*) $CH_3COOC_2H_5 + NH_3 \longrightarrow$

(*ii*) $CH_3COOC_2H_5 + CH_3MgI \longrightarrow$

(*iii*) $CH_3COOC_2H_5 + CH_3OH \xrightarrow{H^+}$

(*iv*) $CH_3COOC_2H_5 \xrightarrow{Na/C_2H_5OH}$

20. Briefly explain the process of soap making. Write a chemical equation for the process.

Answers

I. Objective Type Questions

1. (*d*) 2. (*b*) 3. (*c*) 4. (*d*) 5. (*b*) 6. (*c*) 7. (*b*) 8. (*c*) 9. (*c*) 10. (*c*).

II. Fill in the Blanks

11. (*i*) $-\overset{\overset{O}{\|}}{C}-OR$ (*ii*) esterification (*iii*) CH_3COOH (*iv*) Jasmine (*v*) less (*vi*) saponification (*vii*) tert-butyl alcohol.

SHS 4

SECTION 1

CHEMISTRY INDUSTRY AND ENVIRONMENT

General Objectives

After studying this section, the student should be able to:

1. recognize the important role chemical industries play in the lives of people.
2. be aware of the chemical industries in Ghana and understand what they do.
3. be aware of the mineral resources in Ghana and understand how they are exploited and used.

In this section, we shall study

SECTION-1

UNIT-49

CHEMICAL INDUSTRY

Specific Objectives

After studying this unit, the student should be able to:

- explain the terms 'industry' and 'chemical industry'
- explain what a chemical plant is.

49.1 CHEMICAL INDUSTRY

Chemical Industry refers to that which is involved in the business of making different products from raw materials through chemical reactions. Generally, the raw materials used in the chemical industry are oil, natural gas, metals, minerals, coal, salt and last but not the least air and water.

Chemical Industries are very important for the economy of any country. This is because these chemical industries supply the farmers *pesticides* and *fertilizers* which are essential for crop growing. In this way chemical industries contribute to agriculture and food self sufficiency in every country.

Apart from the direct contribution to agriculture, the chemical industry contributes indirectly to almost every sector of every economy. In our everyday life, we use a lot of industrial products and most of them are related to the chemical industry in some way or the other.

- Chemical industry produces the fibers and dyes which are used in the textile industries.
- It supplies the synthetic sweeteners and synthetic flavours which are used by food manufacturing companies.
- Chemical industry indirectly helps the pharmaceutical industry and health care industry by providing the essential chemical components.
- The plastic requirement of the packaging industry and the artificial rubber requirement of the automobile industry are also met by the very same chemical industries.

One can look at the chemical industry from many points of view.

(*i*) Some people think of it in terms of huge factories or oil refining plants giving out smoke and polluting the environment.

(*ii*) Others see it in terms of facts and figures, profits and losses.

(*iii*) The people who work in the industry see it as the source of their livelihoods.

Inspite of the different opinions, a chemical industry is of great importance to us. In our daily lives we make use of chemicals. They are in the food we eat, or rather we should say that chemicals *are* the food we eat. They are the clothes we wear, the medicines we take and the immense variety of articles that we use. In simple words, knowledge of their behaviour has been put to good use, sometimes to bad; and often the effects of their use have been entirely unexpected. Hence it is impossible to avoid the influence of the chemical industry on individuals, on the environment and on the economy of a country. One of the signs of the economic development of a country is the state of its chemical industry. One reason for this is that the industry can take essentially simple, and often cheap raw materials and turn them into much more valuable items. For example, the Haber process employed for the preparation of ammonia. This exploits the free supply of nitrogen in the atmosphere and converts it into ammonia, and then into fertilizers, upon which the efficient supply of food has come to rely.

CASE STUDY: GHANA

Distinction Between Chemical Industry and Chemical Plant

A chemical industry is often comprised of units or machinery where the chemical reactions are carried out. It is known as a chemical plant. The details of some chemical plants in Ghana is given in section 49.2. The chemical plants together constitutes the chemical industry of Ghana.

Add To Your Knowledge

It may not be out of place to mention here that, very often it is the poorer nations that are the sources of the original chemicals, *i.e.*, the *feedstock*. Traditionally these nations export the chemicals for processing in the sophisticated chemical plants of developed countries. The developed nations make the most money out of this arrangement because the final product can be sold at a much greater profit than the feedstock. It is for this reason that underdeveloped countries have tried to develop their own chemical industries.

World Chemical Industries are growing presently at a considerable rate after a slow down in the year 2000 and 2001. If we refer back to the origin of the World Chemical Industries we will be astonished by the fact that use of chemicals and chemical process started in 7000 BC. In those early days of civilization, some artisans from the middle east refined alkali and limestone for producing glass. After them people from china invented black powder which was nothing but a chemical explosive. In 1635, different chemicals were produced for tanning in Massachusetts.

But, the large scale World Chemical Industries with proper form and infrastructure came into existence in the 19th century. The last half of the 19th century experienced tremendous advancement in the field of Organic Chemistry which gave the World Chemical Industries a massive boost. The World Chemical Industries started their journey of development in real terms.

In 1850, World Chemical Industries managed to produce synthetic dyes which were being used in the textile industries. In 1890, production of sulphuric acid, caustic soda and chlorine started at a mass level by the World Chemical Industries. Then came two revolutionary Chemical Products. The first one was Rayon which was made from wood fiber and which changed the total scenario of the textile industry. The second one's impact was even larger. It was Synthetic Fertilizers which led to the Green Revolution in agriculture resulting in drastic improvement in agricultural crop yield.

In this way World Chemical Industries developed over the years and continued to contribute immensely to mankind.

FUTURE OF THE CHEMICAL INDUSTRY IN GHANA

The future of the chemical industry in Ghana is bright. A number of factors point to this.

- It is likely that people in Ghana will be able to turn more of major raw materials into finished goods for export. At present, Ghana, like many other developing countries, is not deriving maximum benefit from major raw materials since the country is unable to turn them into secondary or finished goods with added value.
- Certain species of timber are now exported as sawn timber only. It is no longer exported in the log form. Some of the sawn timber is also exported as self-assembly furniture, but with mounting deforestation problems the forest must be managed correctly with a regrowth programme for hard woods.
- The chocolate industry turns cocoa into chocolate, cocoa butter and other products.
- Aluminium and its products are in demand on the world market. There is a high probability that this industry will be linked to the development of other industries. Some of the industries that can be developed are the chemical, petroleum, iron and steel industries. The mining of bauxite will also be expanded.

- Ghana possesses very large renewable solar salt production potential along its 500 km coastline. Effective exploitation would enable Ghana supply the entire subregion with salt. Unfortunately, dispute over right of exploitation of the Ada-Songhor salt concession has prevented production to begin as at April, 2005.
- An integrated aluminium industry leads to a demand for caustic soda to refine bauxite into alumina. Caustic soda in turn is derived from salt. Thus salt extraction has a great potential for the future. Apart from being a source of raw material, it has a potential as a high foreign exchange earner. The industry can achieve this aim by exporting to West African countries. Apart from Burkina Faso which imports from Ghana, the others depend on salt imports from Europe and elsewhere.
- Other developments include further improvement of mining methods so that more efficient mining can take place. The use of microbes has already been mentioned.
- Recycling methods, particularly of scrap metals and paper, will have an increasing importance. The following are some of the advantages of recycling:
  (*i*) It keeps down the cost of production of some goods since it is cheaper to use a waste product than to use fresh raw material.
  (*ii*) It ensures that some natural resources (*e.g.*, trees for manufacturing paper) are used more economically.
  (*iii*) It helps in checking solid waste pollution.

The future of the industry is not without problems. One of the very important problems is the shortage of imported inputs in all the industrial sectors that rely wholly or partially on imported materials. For example, it is not safe to forecast an increase in mineral production on the basis of the five year period production figures. Only manganese ore witnessed a continuous rise in production during the period under study. Note the phenomenal decrease in bauxite production in 2003 (*i.e.*, from 653,634 mt in 2002 to 492,544 mt in 2003) and the unstable world unit price for each mineral. Low prices and low production seriously affects national income from mineral exports.

RAW MATERIALS OF INDUSTRY

- The main raw materials of the industry are fossil fuels (coal, petroleum and natural gas), air, water, salt, limestone, sulphur, chlorine and some special raw materials such as phosphates.
- Raw materials are converted into products that nearly always need further processing before consumers can use them, *e.g.*, petroleum and its products, such as petrol and plastics.
- The less readily consumable products are called **primary products**, the readily consumable products are called **secondary products** while the more readily consumable products are called **tertiary products**, *e.g.*, from petrol to ethene to plastics.
- An average chemical product has to pass through many stages before reaching the market or consumer.
- Industries compete with each other because the same product can often be made in more than one way.
- Many chemicals compete for the same use. Thus there is always an on-going research to find the best method for manufacturing a particular chemical and how best it can be used to improve the quality of products. This approach is known as research and development (R and D). This is why it is said that '*change is the rule in the chemical industry*'.

49.2 CHEMICAL PLANTS

The design of a chemical plant is sometimes as much an art as a science. The **overall aim** is to increase production while minimising costs, and at the same time keeping a good margin for safety. It is the chemical engineer who has the task of balancing a number of criteria to ensure the most efficient running. For example, chemicals tend to react faster at high temperatures. However, if the reaction is exothermic and involves an equilibrium, a high temperature will hinder the formation of products. Often a compromise must be reached. Even if a high temperature is best, for chemical reasons and for reasons of economy it may be best to work at a lower temperature. It may cost more to provide the energy to increase the temperature than is returned by the profit on the sale of the extra chemical produced.

Similarly, gaseous reactions may be most effective at high pressures, but it is far more costly to build a plant to withstand high pressures. Indeed, a high pressure plant is also more expensive to run.

Energy costs are one of the most important variables in the design and running of a plant. Large chemical sites are often run as *integrated* concerns. This means that, for example, the heat from an exothermic reaction in one part of the plant may be used to produce steam that drives turbines, compressors or pumps used in another part of the plant. Likewise an endothermic reaction may be used to cool fluid, which in turn cools gases from an exothermic reaction.

Running a chemical plant

There are many different types of chemical plant, but they usually use one of the following two methods of production.

(*i*) *Making chemical continuously*

(*ii*) *Making a chemical in batches.*

Continuous reactors are the best when there is large and dependable demand for a product. For example, oil refineries work on a continuous basis. Crude oil is fed into the distillation towers, and the various products are directed into the catalytic crackers, polymerization reactors and so on. On the other hand, if the requirement of the chemical is less, then a speciality chemicals firm makes small amounts of chemicals for research establishments in industry or universities. Thus, a chemical would be made in small batches.

The economics and physical scale of batch and continuous methods will be different, but they have many features in common. In particular, safety is of absolute importance, as is quality control.

With the advances in automation it is possible to run a huge chemical plant with no more than a dozen workers. Temperature and pressure changes can be measured by instruments directly connected to computers. The computers can be programmed to control pumps, heaters and the like so that the plant is kept running under the best conditions.

The way the reactors are controlled has a direct bearing on safety. Dangerous situations can be caused if the sensors respond too slowly, or not at all. Especially, exothermic reactions that are not cooled properly can increase their rate of reaction extremely quickly. If gases are produced, a sudden rise in pressure may fracture pipes and cause fires or explosions. Similarly, if the plant is not regularly inspected for metal fatigue, leaks can give rise to a poisonous cloud of gas escaping into the atmosphere, or to an explosion taking place.

Add To Your Knowledge

Two accidents, one in India at Bhopal in 1984, the other in England at Flixborough in 1974, illustrate the damage that can be caused if something goes wrong. At Bhopal a cloud of poisonous gas released from a chemical works proved fatal to a large number of people in the surrounding area, and caused severe harm to many more. The cause of the accident has not been finally decided. There is evidence that the company was to blame for lack of care in monitoring the running of the plant, but the company claims that the accident was the result of sabotage. At Flixborough, caprolactam was made for use in Nylon manufacture. A leak in a pipe allowed vapour to escape into the atmosphere. The loss of vapour was detected late and the entire plant was destroyed in a huge explosion.

SOME CHEMICAL PLANTS IN GHANA

Crude Oil Refinery

Crude oil, like many other natural resources, is not found in all parts of the world. For example, at present, Nigeria and, to a lesser extent, the Cote d'Ivoire are the only countries in West Africa that produce oil. Offshore oil deposits have been found in Ghana, but oil wells have yet to be sunk to obtain the oil in commercial quantities.

Ghana imports both refined and unrefined crude oil, much of it from Nigeria.

Composition of crude oil

Petroleum (or crude oil) comes from Latin words meaning 'oil from rock'. It has already been stated that crude oil is a complex mixture. On heating to 500 K (~230°C), crude oil can yield at least 500 different compounds.

Crude oil refining means separating crude oil into usable and useful fractions.

Initial separation is based on the different boiling points (and volatilities) of the compounds present in the crude oil mixture.

Fractional distillation or **fractionation** is used to separate the petroleum into fractions. One fraction differs from another in the temperature range over which its components boil. The boiling ranges and composition of the various fractions are shown in Fig. 49.1.

Industrial refining of crude oil

Solids such as sand are separated from the impure crude oil by filtration. The filtrate of crude oil and water is allowed to stand in a reservoir for some time. The oil separates as the upper layer and is then pumped through pipes to a refinery such as the one at Tema in Ghana (Fig. 49.1).

Fig. 49.1. Part of Tema oil refinery (TOR).

At the refinery, the oil is vaporized in a furnace at 570 K (~300°C). The hot vapour is pumped into a **fractionating tower** such as the one shown in Fig. 49.2.

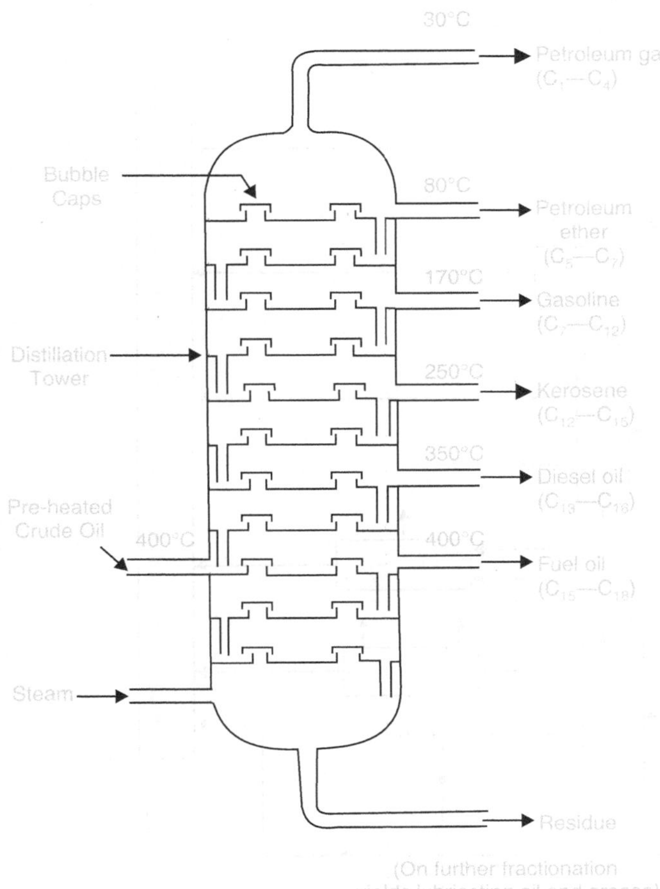

Fig. 49.2. Refining of petroleum.

The fractionating tower is about 45 m high; it contains 30 to 40 perforated horizontal trays, regularly spaced throughout its height. The trays provide a large surface area on which evaporation and condensation can take place. Each tray contains a number of **bubble caps**. The bubble caps force the vapour through the liquid which has condensed on the tray. This results in continuous condensation and re-evaporation.

The vaporization of the crude oil allows the more volatile fractions to rise up the tower. Since the temperature of the vapour falls as it goes up the tower (*i.e.*, there is a temperature gradient), the less volatile fractions condense first, near the base. The more volatile fractions, which contain molecules of low molecular mass and are gases at room temperature, are collected from the top of the tower. Fractions with middle range boiling temperatures and molecular mass condense in the middle of the tower. The process goes on continuously, vapour being ejected from the furnace and the various fractions being removed for storage or further processing.

All the fractions are redistilled before they become usable and useful products. The composition of the fractions varies from refinery to refinery, since the crude oil varies in its composition and the methods of distillation differ in their detail.

Flow diagram of a typical refinery

The image given on the next page is a schematic *flow diagram* of a typical oil refinery that depicts the various *unit* processes and the flow of intermediate product streams that occurs between the inlet crude oil feedstock and the final end products. The diagram depicts only one of the literally hundreds of different oil refinery configurations. The diagram also does not include any of the usual refinery facilities providing utilities such as steam, cooling water, and electric power as well as storage tanks for crude oil feedstock and for intermediate products and end products.

There are many process configurations other than that depicted above. For example, the *vacuum distillation* unit may also produce fractions that can be refined into endproducts such as: spindle oil used in the textile industry, light machinery oil, motor oil, and steam cylinder oil. As another example, the vacuum residue may be processed in a *coker unit* to produce petroleum coke.

Refineries also produce pure chemicals, called feedstocks, from crude oil. These feedstocks are sold to the petrochemical industries, where a great variety of products are made.

Feedstock	*Products*	
Methane (Natural Gas)	Fertilizers	Polyethylene
Ethylene	Plastics	Explosives
Propylene	Rayon	Synthetic Rubber
Butylene	Vinyl	Polystyrene
Naphthene	Nylon	Dyes
	Dacron	Drugs
	Teflon	Antifreeze

GOLD

Mining of gold in Ghana began long before the arrival of the Europeans. The metal was extracted from alluvial materials deposited on the banks and bottoms of rivers. Ghana's pre-independence name of 'Gold Coast' shows the importance of the metal in the economy and history of the country.

Gold is an inert metal and, it is always found uncombined and in association with copper and silver. The extraction of the metal is done from the concentrated ore.

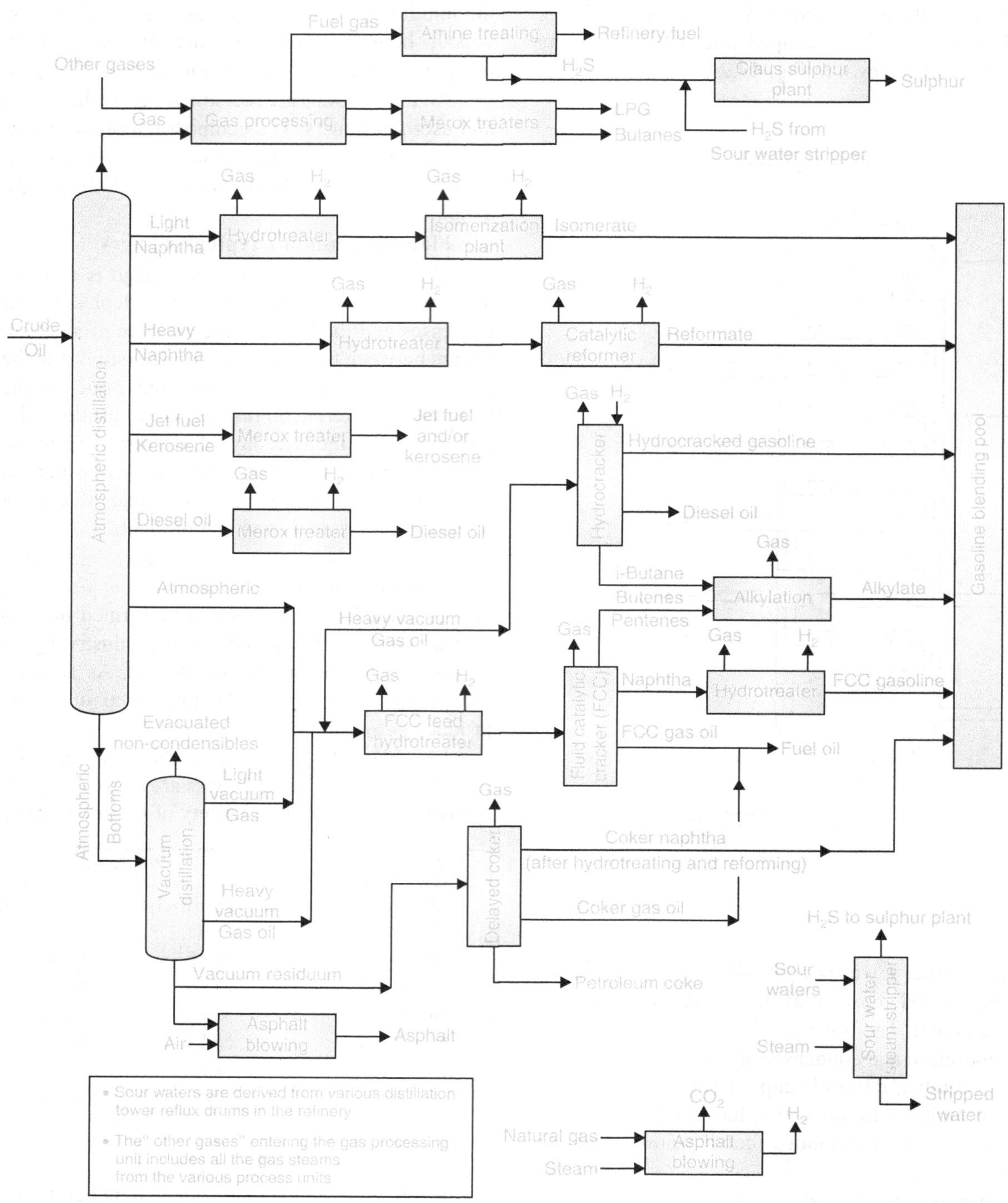

To concentrate the ore, it is ground up into a very fine powder. It is then dissolved by leaching in sodium cyanide solution in the presence of oxygen and water. A soluble gold complex is formed.

$$4Au(s) + 8NaCN(aq) + O_2(g) + 2H_2O(l) \longrightarrow 4NaAu(CN)_2(aq) + 4NaOH(aq)$$

The gold reacts with the cyanide solution, leaving the impurities behind. Products are a gold-rich slurry. The slurry is passed through a series of thickeners in a filtration process. Addition of metallic zinc dust to the concentrated filtrate precipitates the metal:

$$2NaAu(CN)_2(aq) + Zn(s) \longrightarrow 2Au(s) + Na_2Zn(CN)_4(aq)$$

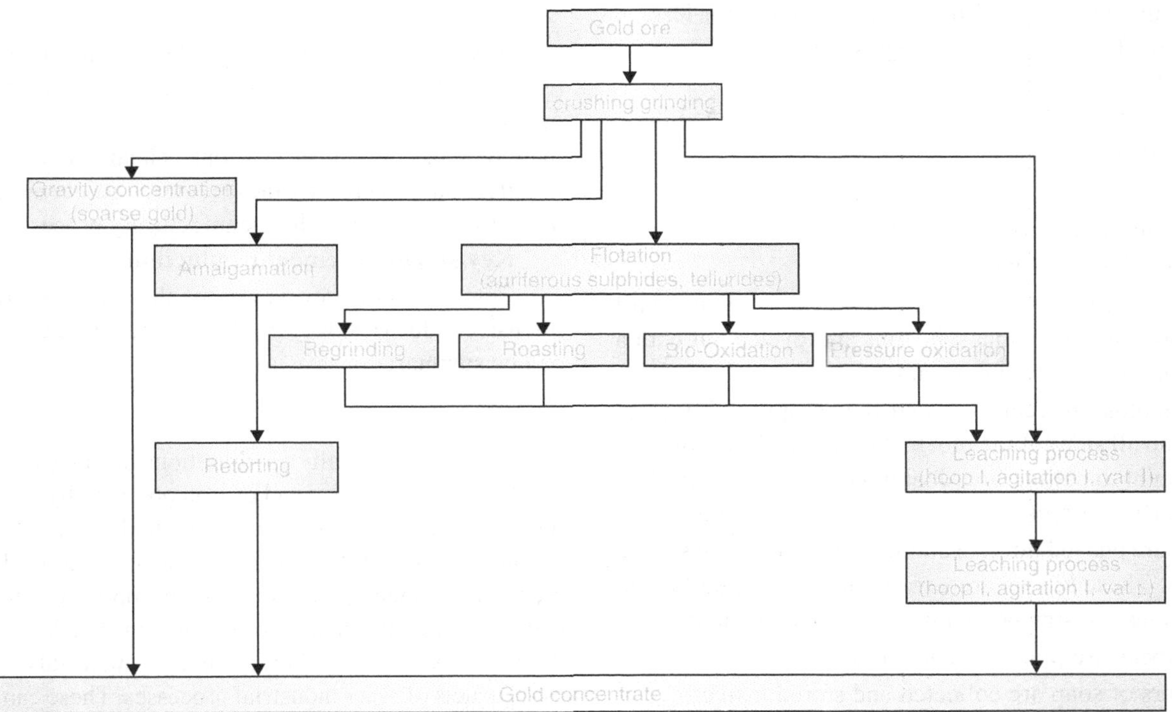

Refining of Gold

The metal is refined by electrolysis. The anode is impure gold. The cathode is a strip of pure gold coated with a thin layer of graphite. The electrolyte is an aqueous solution of gold trioxonitrate(V) and trioxonitrate(V) acid. The gold ions are reduced at the cathode on electrolysis:

$$Au^+ + e^- \longrightarrow Au$$

Crude mining and extraction of gold and diamond, is carried out on a small scale in Ghana. It is popularly called '**Galamsey**'. The ores or tailings, *i.e.*, waste from previous ore treatment, is first concentrated by washing several times with water in a pan or other suitable container. This helps to remove slime (mud and other filth).

The product is then poured on a jute bag or any wide mesh cloth laid on an inclined table and repeatedly washed with water to remove the lighter gangue. The heavier gold material, trapped in the sack or cloth, is washed off into a pan with water and then decanted. The solid gold product is rubbed well with mercury to form an amalgam leaving the remaining gangue.

In the final stage, the gold/mercury amalgam is put in a white handkerchief and squeezed hard to let out as much mercury as possible. The gold product looks white due to mercury contamination. This is washed with water in a pan. The heavier mercury falls to the bottom and the gold remains on top. The gold is harvested (collected) ready to be sold. The gold may also be recovered by distillation.

The gold produced by both **Ashanti Goldfields Company Limited** and **'Galamsey' operators** is not totally pure. Purer gold is obtained by refining their products outside Ghana.

Soap Making

Soap making is an activity that the Rural Enterprises Project (REP) supports through running training sessions on how it's made and business training for startup businesses run out of the Business Advisory Centre (BAC).

Process of Making Bar Soap in Ghana:

1. The materials used are (coconut oil, buckets, scoops, caustic soda, water, soda ash, perfume, mixing spoons, hydrometer, rubber gloves, moulds, plastic bags, tables, and cutting equipment.
2. The moulds are set on the table and lined with plastic.
3. A solution is made by mixing caustic soda (NaOH) and water. Measure out the volume and ensure correct level of acidity (pH) with the hydrometer.

The solution of caustic soda and water is corrosive! Gloves should be used when handling this liquid

4. Mixing 10 L of coconut oil, 1 scoop of perfume, 1 scoop of soda ash, and the measured volume of caustic soda and water into a large mixing bowl.

5. The contents are mixed for about 10 minutes or less, in the same direction at an average speed.

Caution

Don't mess around there, or you will spoil the batch!

6. The contents are poured into the moulds (caution, heavy lifting!)
7. These are left to dry for approximately 2.5–3 hours until they are white in colour. The tops are leveled out with a scrapper as the cooling occurs.
8. The moulds are removed, and the scrapper is used to clean up all sides of the block of soap. The block of soap is brought to cutting board, and pushed through to slice it into 10 long rectangles (caution, very difficult!)
9. These are pushed through another cutter to create square blocks. These blocks are carried back to the table, and all 6 sides are stamped with the logo (time intensive).
10. The blocks are placed outside to dry in the sun.
11. The bars of soap are collected and stored inside.
12. Bars of soap are distributed to various sellers to sell at the market and in villages.

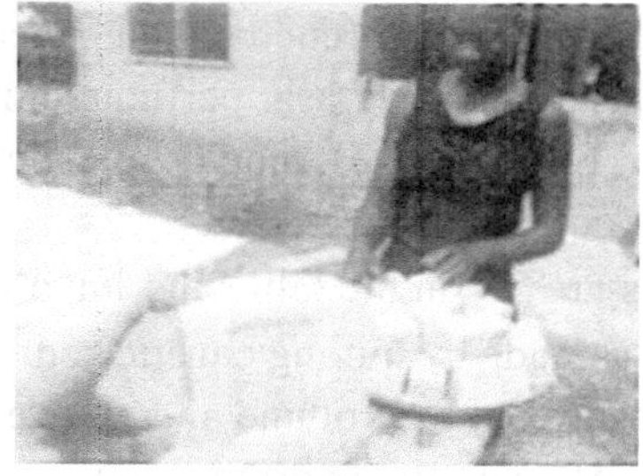

CEMENT INDUSTRY

Portland cement was first produced and patented in 1824 by a British stonemason. Today around 1700 million tonnes of cement are used every year, with different types manufactured to meet various chemical and physical requirements. To produce these requires a clear understanding and careful control of the manufacturing processes.

Key stages in cement production

Cement production is one of the world's most energy intensive industries. Key production stages as used in Ghana can be summarized as:

1. Raw Materials

These are generally combinations of limestone, shells or chalk, and shale, clay, sand or iron ore, usually mined from a quarry close to the plant where they undergo reduction using primary and secondary crushers. When the reduced materials reach the cement plant they are proportioned to create a cement of specific chemical composition. Much work is being done on the use of alternative raw materials—often the by-products of other industrial processes. These can minimize the effects of quarrying, reduce the impact of the cement plant on the local environment and enable the cement industry to become a major player in materials recycling.

There are two basic methods used in Portland cement production—**wet** and **dry**.

In the dry process dry materials are proportioned, ground to a powder, blended and fed into the kiln dry.

The wet process involves adding water to the proportioned raw materials and completing the grinding and blending operations in slurry form. Increasingly alternative fuels are being used. These include materials such as scrap tyres, processed sewage sludge and packaging waste.

2. Cooling/Finish Grinding

Clinker is discharged from the lower end of the kiln and transferred to various types of coolers. Cooled clinker is combined with gypsum and ground to a fine powder in a ballmill to produce the final grade cement.

EXTRACTIVE METALLURGY OF ALUMINIUM

Aluminium is the most abundant metal on Earth.

It occurs in nature in form of aluminium oxide and other combined forms. The ore containing aluminium compound, which is commercially used in extractive metallurgy is called *bauxite*. ***Bauxite*** is a hydrated aluminium oxide. The extraction of aluminium is expensive in terms of energy consumption owing to the large amount of electrical energy required. Between 13000 and 17000 kWh are needed to produce 1 tonne of aluminium. For example, the smelter at Tema, Ghana, owned by the Volta Aluminium Company, uses 2,760,000 MWh to produce nearly 200,000 tonnes of aluminium annually. The company's power lines are connected to the Akosombo power station of the Volta River hydroelectric dam and the site of the company allows it to obtain relatively cheap and plentiful electrical energy otherwise the electrolysis operation would be too expensive.

Aluminium production makes a major contribution to the economies of some countries, particularly the members of the International Bauxite Association (IBA).

Extraction of aluminium from bauxite is carried out in three stages:

- **Ore dressing:** cleaning ore by means of separation of the metal containing mineral from the waste (gangue).
- **Chemical treatment of bauxite** for converting the hydrated aluminium oxide to pure aluminium oxide.

At this stage crushed and ground bauxite is mixed with hot sodium hydroxide solution, which dissolves the aluminium hydroxide, forming a solution of sodium aluminate.

The residual impurities (oxides of silicon, iron and titanium), which is called "red mud", are separated from the sodium aluminate solution.

The solution is then treated in precipitator tanks, where aluminium trihydrate precipitates from the solution.

The aluminium trihydrate after separation from the sodium hydroxide is converted into pure aluminium oxide by heating to 1800F (1000°C).

- **Reduction of aluminium from aluminium oxide by the electrolytic process.**

As aluminium oxide is a very poor conductor of electricity its electrolysis is carried out in a bath of molten cryolite (mineral, containing sodium aluminium fluoride—Na_3AlF_6).

The technology is called **Hall-Heroult process.**

Schematically the process is presented in the picture:

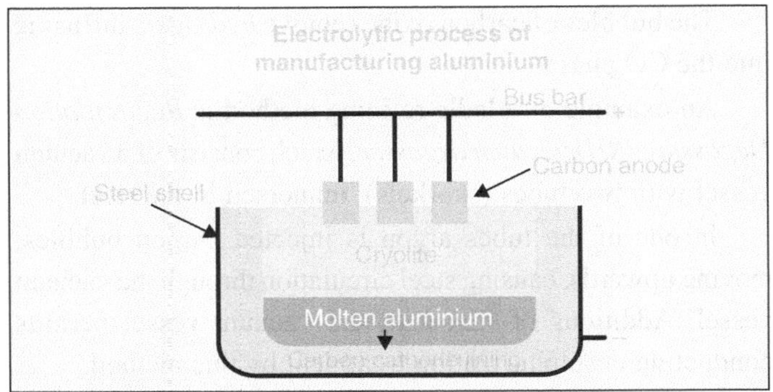

The electrolytic cell for aluminium production consists of a pot with *carbon* lining, serving as negative electrode (cathode) and positive electrodes (anodes), connected to the current conductor (bus bar). The anodes are immersed into the bath of molten cryolite.

The aluminium oxide is added to the cryolite and dissolved in it. When electric current passes between the anode and the cathode through the cryolite, aluminium oxide decomposes to metallic aluminium which is deposited at the cathode and oxygen liberated at the anode.

The molten aluminium is periodically tapped from the furnace into a crucible and cast into ingots.

STEEL MAKING

Steel is an alloy of iron, containing up to 2% of *carbon* (usually up to 1%).

Steel contains lower (compared to pig iron) quantities of impurities like phosphorus, sulphur and silicon.

Steel is produced from pig iron by processes involving reducing the amounts of carbon, silicon and phosphorus.

The main steel making methods are *basic oxygen process* (basic oxygen furnace, basic oxygen converter) and *electric-arc furnace.*

1. Basic Oxygen Process (BOP)

The Basic Oxygen Process is the most powerful and effective method of steel manufacturing.

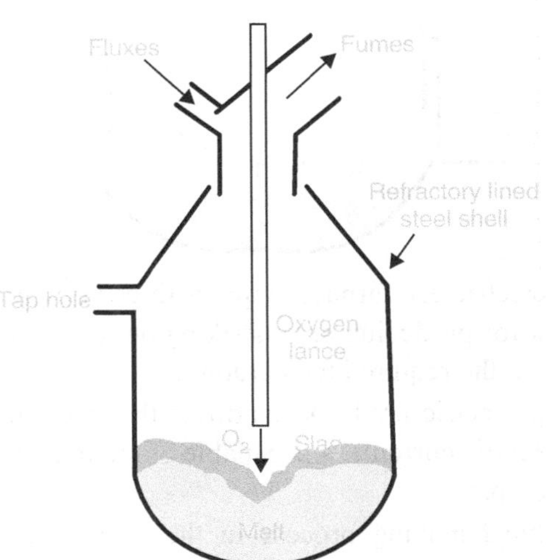

The scheme of the *Basic Oxygen Furnace (BOF)* is presented in the picture above.

Typical basic oxygen converter has a vertical steel shell lined with refractory lining.

The furnance can to rotate about its horizontal axis on trunnions. This rotation is necessary for charging raw materials and fluxes, sampling the melt and pouring the steel and the slag out of the furnace.

The Basic Oxygen is equipped with a water cooled oxygen lance for blowing oxygen into the melt.

The basic oxygen converter uses no additional fuel. The pig iron impurities (carbon, silicon, manganese and phosphorus) serve as fuel.

The steel making process in the oxygen converter consists of:

- *Charging steel scrap*
- *Pouring liquid pig iron into the furnace*
- *Charging fluxes*
- *Oxygen blowing*
- *Sampling and temperature measurement*
- *Tapping the steel to a ladle*
- *De-slogging.*

The iron impurities oxidize evolving heat necessary for the process.

The forming oxides and sulphur are absorbed by the slag.

The oxygen converter has a capacity up to 400 t and production cycle of about 40 min.

2. Electric-arc Furnace

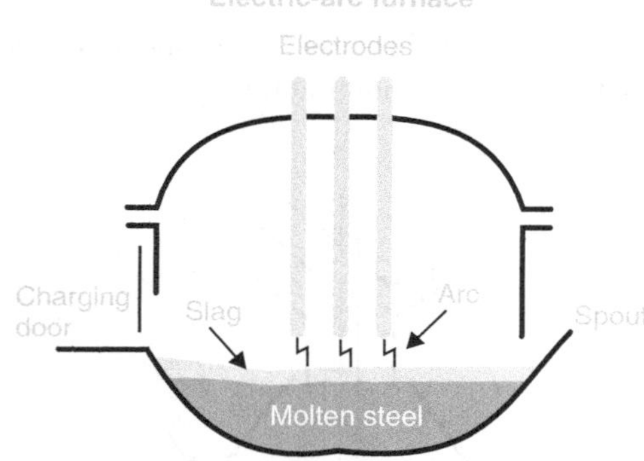

The electric-arc furnace employs three vertical *graphite electrodes* for producing arcs, striking on to the charge and heating it to the required temperature.

As the electric-arc furnace utilizes the external origin of energy (electric current), it is capable of melting up to 100% of steel scrap.

The steel making process in the electric-arc furnace consists of:

- *Charging scrap metal, pig iron, limestone*
- *Lowering the electrodes and starting the power (melting)*
- *Oxidizing stage.*

At this stage the heat, produced by the arcs, causes oxidation of phosphorus, silicon and manganese. The oxides are absorbed into the slag. By the end of the stage the slag is removed.

- *De-slagging*
- *Reducing stage.*

New fluxes (lime and anthracite) are added at this stage for formation of *basic reducing slag.*

The function of this slag is refining of the steel from sulphur and absorption of oxides, formed as a result of *deoxidation.*

- *Tapping*
- *Lining maintenance.*

The *advantages* of the electric-arc furnance are as follows:

- Unlimited scrap quantity may be melt;
- Easy temperature control;
- Deep *desulphurization;*
- Precise alloying.

Ladle refining (ladle metallurgy, secondary refining)

Ladle refining is post steel making technological operations performed in the ladle prior to casting with the purposes of *desulphurization*, degassing, temperature and chemical homogenization, *deoxidation* and others. Ladle refining may be carried out at atmospheric pressure, in a vacuum and may involve heating, gas purging and stirring.

Sulphur refining (desulphurization) in the ladle metallurgy is performed by addition of fluxes (CaO, CaF_2 and others) into the ladle and stirring the steel together with the slag, absorbing sulphur.

In the production of high quality steel the operation of *vacuum treatment* in the ladle is widely used.

Vacuum causes the chemical reaction below within the molten steel:

$$[C] + [O] = CO$$

This reaction results in the reduction of the quantity of oxide inclusions.

The bubbles of carbon oxide remove *hydrogen*, diffusing into the CO phase.

An example of a ladle refining method is *Recirculation Degassing (RH) vacuum degasser*, which consists of a vacuum vessel with two tubes (snorkels), immersed in the steel.

In one of the tubes argon is injected. Argon bubbles, moving upwards, causing steel circulation through the vacuum vessel. Additions of fluxes in the vacuum vessel permits conducting desulphurization treatment by this method.

USE OF FOREIGN RAW MATERIALS

The government of a country may sometimes act as a factor in influencing the siting of an industry in spite of better economic pulls elsewhere. This is done with a view to giving employment to people of the area where the industry is located. It may also be located to win the people's political support or to check the movement of the area's labour force to the urban centres. Ghana's old Matchet Factory at Bibiani was sited for these reasons. Currently, it is the policy of governments in West Africa, for example, to promote the establishment of cottage industries through tax concessions. This is one of the attempts to check the concentration of industries and people in the urban areas.

Finally, another way of maximizing profit or producing low-cost but quality products is to avoid waste. Uses and hence markets must be found for the waste products—they should be turned into by-products.

Factors that influenced the location of some specific individual chemical industries will now be discussed.

Volta Aluminium Company (VALCO) Tema

The Volta aluminium company obtains electrical energy supply from the hydroelectric power at Akosombo. The energy supply is cheap and regular because Tema is only about 80 kilometres from Akosombo. VALCO at present imports its raw materials. Alumina is one of them. Alumina is extracted from bauxite. Although there are bauxite deposits in Ghana, VALCO does not use this raw material at present. This importation increases production cost and hence the aluminium product is expensive. On the other hand, the situation of the industry at Tema has the advantage that imported raw materials can easily be received and products exported. The industry aims at both internal and external markets. With an assured market, the aluminium sheets products need not be stored for long periods, hence much space is not needed by the industry.

AngloGold Ashanti Limited

There is no question about the marketability of processed gold and its importance in international trade. The industry is sited at Obuasi because this is one area in the country where the concentration of gold, more than 10 parts in a million, is considered adequate for commercial exploitation. It has one of the richest gold ore deposits in the world. Good security is needed to prevent stealing of the precious metal. The Akosombo hydroelectric plant supplies electrical power.

Fig. 49.3 (a). Volta aluminium company (VALCO), Tema

Fig. 49.3 (b). Kumasi jute factory

Panbros salt industry, Accra

This industry is situated at Weija, a suburb of Accra. The location of the industry is at an open space near the sea. The open space provides an area for the evaporation of the sea water in pans. The sea-water is an abundant raw material. Thus the industry does not face the problem of raw material transportation to the factory. The use of the salt for preservation and flavouring of food means it has a ready market. Transportation of the products to market areas is not a problem because of its location.

BASIC CHEMICAL REACTIONS TO TRANSFORM RAW MATERIALS TO PRODUCTS

Saponification

Saponification is the *hydrolysis of esters of higher fatty acids in alkaline medium*

$$\underset{\text{Ester}}{RCOOR} + NaOH(aq) \longrightarrow \underset{\text{Sodium alkanoate}}{RCOONa} + \underset{\text{Alcohol}}{R'OH}$$

The alkaline hydrolysis of oils/fats, which are triesters of glycerol and higher fatty acids produces glycerol and sodium salts of fatty acids which are called **soaps.** The reaction used in **soap making is:**

$$\begin{array}{l} CH_2OCOC_{17}H_{35} \\ | \\ CHOCOC_{17}H_{35} + 3NaOH(aq) \\ | \\ CH_2OCOC_{17}H_{35} \end{array}$$

Glyceryl tristearate

$$\longrightarrow \begin{array}{l} CH_2OH \\ | \\ CHOH + 3C_{17}H_{35}COONa \\ | \\ CH_2OH \end{array}$$

Glycerol   Sodium stearate (soap)

Electrolysis

It is a *process of decomposition of electrolyte in fused state or dissolved state by the passage of electricity.* This process is utilised in a large number of industrial preparations. For *example,* production of aluminium is carried out by the electrolysis of molten alumina (Al_2O_3) dissolved in cryolite (Na_3AlF_6).

$$Al_2O_3(l) \xrightarrow[(Na_3AlF_6)]{\text{Electrolysis/}} 2Al(l) + 3O_2(g)$$

Similarly, electrolysis of brine solution (NaCl) is used for the production of caustic soda (NaOH)

$$2NaCl(aq) + 2H_2O(l) \xrightarrow{\text{Electrolysis}} 2NaOH(aq) + H_2(g) + Cl_2(g)$$

Fermentation

Fermentation is a process of breaking down complex organic material into smaller fragments by the action of living organisms which act as catalysts. Some common *examples* are souring and curdling of milk, putrefaction of meat, conversion of fruit juices into wines, etc.

Alcohol is produced in the brewing industry by the action of the enzyme ***Zymase*** on glucose/fructose found in sweet fruits. Zymase is secreted by the microorganism called ***Yeast.***

$$\underset{\text{Glucose/Fructose}}{C_6H_{12}O_6} \xrightarrow{\text{Zymase}} \underset{\text{Alcohol}}{2C_2H_5OH} + 2CO_2$$

Solvent Extraction

Solvent extraction is a technique used to separate a component of a mixture by making use of its different solubilities in two immiscible solvents. Immiscible solvents cannot dissolve in each other, and so they form two layers when mixed. The desired component in a reaction mixture dissolves well in one solvent and is insoluble or only slightly soluble in the other at a particular temperature. When the reaction mixture is shaken well in the two immiscible solvents, the desired product is (dissolved) extracted into one solvent whilst the other components or impurities are (dissolved) extracted into the other solvent.

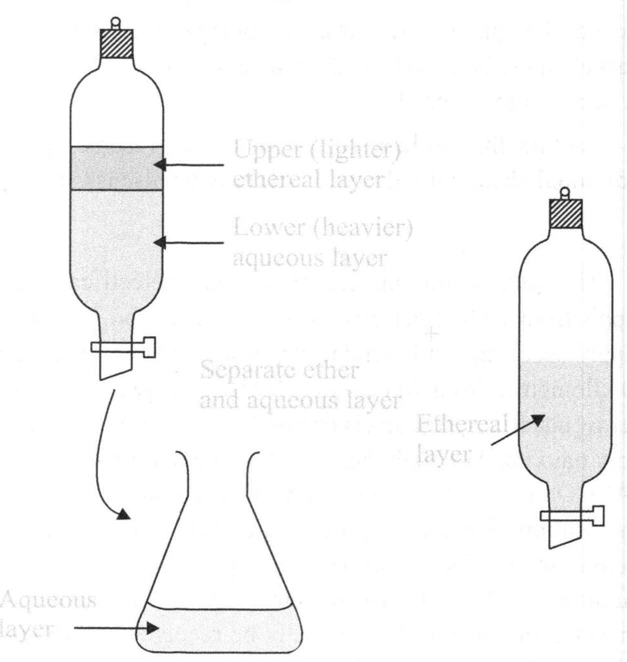

Solvent extraction is very useful in extracting natural organic products from their natural sources such as tree barks, leaves, roots, stems, and animal parts.

Caffeine is a heterocyclic base (alkaloid) found in cocoa, tea, coffee beans, cola nuts and drinks. It is an organic compound which is far more soluble in tetrachloromethane (CCl_4) than water. The solvent extraction technique can therefore be used to separate it from its aqueous solutions.

Precipitation

The process of precipitation is employed in the extraction of metals from their ores. For example, in the refining of gold, the ore is ground up into a very fine powder. It is then dissolved by leaching in sodium cyanide solution in the presence of oxygen and water. A soluble gold complex is formed.

$$4Au(s) + 8NaCN(aq) + O_2(g) + 2H_2O(l) \longrightarrow 4NaAu(CN)_2\,(aq) + 4NaOH(aq)$$

The gold reacts with the cyanide solution, leaving the impurities behind. Products are a gold-rich slurry. The slurry is passed through a series of thickeners in a filtration process. Addition of metallic zinc dust to the concentrated filtrate **precipitates** the metal:

$$2NaAu(CN)_2(aq) + Zn(s) \rightarrow 2Au(s) + Na_2Zn(CN)_4(aq)$$

Group Activity

Visit a chemical plant in Ghana, write a report on the kind and sources of raw materials; finished products, by-product and the basic chemical process/or processes used to convert raw materials into products, sources of energy to the chemical plants and benefits the industry gives to the community in which the plant is cited.

In groups present reports to the class for discussions.

EVALUATE YOURSELF

I. Objective Type Questions

1. Petroleum is found
   (*a*) in the atmosphere
   (*b*) in the arctic ocean
   (*c*) deep under the surface of the earth
   (*d*) on the surface of the earth
2. Petroleum is formed by chemical changes in
   (*a*) inorganic matter
   (*b*) animal matter only
   (*c*) vegetable matter only
   (*d*) both vegetable and animal matter
3. Petroleum is a mixture of
   (*a*) alkanes (*b*) cycloalkanes
   (*c*) aromatic hydrocarbons (*d*) all of the above
4. The fraction obtained between temperature 200–300°C during fractional distillation of crude petroleum is
   (*a*) paraffin wax (*b*) heavy oil
   (*c*) kerosene oil (*d*) naphtha
5. Kerosene oil is a mixture of
   (*a*) aliphatic hydrocarbons (*b*) aromatic hydrocarbons
   (*c*) saturated hydrocarbons (*d*) alicyclic hydrocarbons
6. Which of the following fractions of petroleum has the lowest boiling point?
   (*a*) gasoline (*b*) kerosene oil
   (*c*) diesel oil (*d*) heavy oil
7. Gold dissolves in sodium cyanide to form
   (*a*) NaAuCN (*b*) $Na_2Au(CN)_4$
   (*c*) $NaAu(CN)_2$ (*d*) $Au(CN)_3$
8. Saponification is
   (*a*) hydrolysis of esters of higher fatty acids in a medium with pH = 5.0
   (*b*) hydrolysis of esters of higher fatty acids in a neutral medium with pH =7.0
   (*c*) hydrolysis of esters of acetic acid with conc. NaOH
   (*d*) hydrolysis of esters of higher fatty acids in alkaline medium
9. The metal used for precipitation of gold from its leached solution is
   (*a*) Cu (*b*) Ag
   (*c*) Ba (*d*) Zn
10. Fermentation of Glucose with Zymase produces
   (*a*) fructose (*b*) ethyl alcohol
   (*c*) methyl alcohol (*d*) gluconic acid

II. Fill in the Blanks

11. Complete the following sentences by supplying appropriate words:
   (*i*) Each tray in the fractionating tower contains ....
   (*ii*) Petroleum coke is obtained by processing vacuum residue in ....
   (*iii*) Refineries produce pure chemical called ....
   (*iv*) The leaching reagent used in the extraction of gold is ....
   (*v*) The oil used in the manufacture of bar soap in Ghana is ....
   (*vi*) Aluminium is extracted from the ore ....
   (*vii*) The anode used in the electrolytic cell of Hall-Heroult process is ....
   (*viii*) Cryolite is chemically ....
   (*ix*) Steel is an alloy of iron containing carbon ....
   (*x*) Alcohol is produced from Fructose by the process of ....

III. Discussion Questions

12. What is saponification? Discuss with an example.
13. How is Aluminium extracted in Hall-Heroult's method?
14. Discuss the production of steel by BOP method.
15. What raw materials are used in the production of cement?
16. Explain the terms
   (*i*) Fermentation (*ii*) Leaching
   (*iii*) Refining.
17. Discuss briefly the process of making bar soap.
18. What is Ladle refining?
19. Explain the process of solvent extraction with an example.
20. What is clinker? How is cement obtained from clinker?

Answers

I. Objective Type Questions

1. (*c*) **2.** (*d*) **3.** (*d*)
4. (*c*) **5.** (*b*) **6.** (*a*)
7. (*c*) **8.** (*d*) **9.** (*d*)
10. (*b*).

II. Fill in the Blanks

11. (*i*) bubble caps (*ii*) coker unit
(*iii*) feed stocks (*iv*) sodium cyanide
(*v*) coconut oil (*vi*) bauxite
(*vii*) carbon (*viii*) Na_3AlF_6
(*ix*) 1% – 2% (*x*) fermentation.

SECTION-1

UNIT-50

EXTRACTION OF METALS

Specific Objectives

After studying this unit, the student should be able to:

- outline the properties and reactivity of metals
- explain the term mineral/ore
- identify the different types of mineral deposits in Ghana
- identify ores for gold, aluminium, iron and manganese
- outline the extraction of gold and aluminium from their ores
- describe the extraction of iron from its ore
- outline the economic importance of iron to the people of Ghana.

50.1 INTRODUCTION

It is a well-known fact that elements are the building blocks of various kinds of materials. Although there are only about 114 elements known, they constitute the entire matter on this earth. The elements are classified into three categories, namely: *metals, non-metals* and *metalloids.*

Metals are generally *lustrous, solids, malleable, ductile* and *good conductors of heat* and *electricity.* They constitute about two-third of the known elements. Some of the metals which find common use in our daily life are: iron, copper, aluminium, lead, gold, silver, mercury, etc.

50.2 PROPERTIES AND REACTIVITY OF METALS

PROPERTIES OF METALS

The characteristic properties of metals are summarized in Table 50.1.

Table 50.1. Properties of Metals

Physical Properties	
1. *State*	Metals are solids at ordinary temperature. *Exception. Mercury* is a liquid.
2. *Lustre*	They possess lustre or shine.
3. *Malleability and Ductility*	Metals are generally malleable and ductile.
4. *Hardness*	Metals are generally hard. *Alkali metals* are exceptions.

5. *Density*	They have high densities.
6. *Conductivity*	Metals are good conductors of heat and electricity.
7. *Melting and boiling points*	They usually have high melting and boiling points.
Chemical Properties	
1. *Action with mineral acids*	Metals generally react with dilute mineral acids to liberate H_2 gas.
2. *Nature of oxides*	They form basic oxides. For example, Na_2O, MgO, etc. These oxides are ionic in nature.
3. *Combination with hydrogen*	Metals generally do not combine with hydrogen. However, Li, Na, Ca, etc. form unstable hydrides. For example, LiH, NaH, CaH_2, etc. These hydrides are ionic in character.
4. *Combination with halogens*	They combine with halogens to form well defined and stable crystalline solids. For example, NaCl, KBr, etc.
5. *Electrochemical behaviour*	Metals are electropositive in character. They form cations in solution and are deposited on the cathode when electricity is passed through their solution.
6. *Oxidising or reducing behaviour*	Metals behave as reducing agents. This is *because* of their tendency to lose electrons. $Na \longrightarrow Na^+ + e^-$

REACTIVITY SERIES OR ACTIVITY SERIES

By observing the relative reactivities of different metals in their reactions with oxygen, dilute acids, water and chlorine, the metals have been arranged in order of their chemical reactivities as given in Table 50.2. Reactivities of certain metals are also compared by studying displacement reactions. A more reactive metal displaces a less reactive metal from the solution of its salt. Such a list is called the **activity series** or **reactivity series.** Thus, ***the activity series is an arrangement of metals in the order of decreasing reactivity.*** In the reactivity series, the most reactive metal is placed at the top whereas the least reactive metal is placed at the bottom. From this series it becomes possible to predict the **general** behaviour and relative chemical activity of the common metals.

Table 50.2. Reactivity Series of Metals

Elements	*Symbol*	
Potassium	K	**Most Reactive**
Sodium	Na	↓
Calcium	Ca	
Magnesium	Mg	
Aluminium	Al	
Zinc	Zn	
Iron	Fe	
Nickel	Ni	
Tin	Sn	
Lead	Pb	
Hydrogen	H	
Copper	Cu	
Mercury	Hg	
Silver	Ag	
Platinum	Pt	**Least Reactive**
Gold	Au	

Metals occupying higher position in the series have more tendency to lose electrons and are more reactive. The metals at the bottom of the series are least reactive.

50.3 MODES OF OCCURRENCE OF METALS

Elements occur in nature in two states: *native state* and *combined state*.

1. NATIVE STATE

The metals are said to be in a native state if they are found in their elementary form. Generally, less active metals are found in their native state. The common examples of metals which occur in their native state are gold, silver, copper, platinum, etc.

2. COMBINED STATE

The metals are said to occur in the combined state if they are found in nature in the form of their compounds. Generally, the reactive metals occur in the form of their compounds.

In the combined state the metals are found in the crust of the earth as *oxides, carbonates, sulphides, silicates, phosphates*, etc. *The naturally occurring chemical substances in the earth's crust which can be obtained by mining* are known as **minerals.** Minerals extracted from the crust of the earth are not pure but instead they are associated with a large number of earthly, rocky and silicious impurities. *The impurities associated with the minerals* are collectively known as **gangue** or **matrix.**

Every mineral of a metal cannot be used for its extraction. Economic factors or availability of the mineral may be a hinderance. *The mineral from which the metal can be economically and conveniently extracted* is called **ore.** For example, the earth's crust contains aluminium in the form of two well-known minerals, *bauxite* $(Al_2O_3.2H_2O)$ and *china clay* $(Al_2O_3.2SiO_2.2H_2O)$, but the extraction of aluminium is cheaper and easy from bauxite. Hence, the ore of aluminium is *bauxite.* Similarly, minerals of copper are *copper glance* (Cu_2S), *cuprite* (Cu_2O), *malachite* ($CuCO_3$. $Cu(OH)_2$), *copper pyrites* ($CuFeS_2$), etc., but the ore of copper is *copper pyrites.* Thus, it can be concluded that ***all the ores are minerals but all the minerals are not ores.***

Examples of some important ores are given in Table 50.3.

Table 50.3. Metals and Their Ores

Metal	*Ore*	*Composition*
Potassium	Sylvine	KCl
Sodium	Rock-salt	$NaCl$
Calcium	Limestone Gypsum	$CaCO_3$ $CaSO_4 . 2H_2O$
Magnesium	Magnesite Dolomite	$MgCO_3$ $MgCO_3 . CaCO_3$
Aluminium	Bauxite Diaspore Kaolinite (a form of clay)	$Al_2O_3 . 2H_2O$ $Al_2O_3 . H_2O$ $Al_2(OH)_4 . Si_2O_5$
Titanium	Rutile Ilmenite	TiO_2 $FeTiO_3$
Zinc	Zinc blende or Sphalerite Calamine Zincite	ZnS $ZnCO_3$ ZnO
Iron	Haematite Iron pyrite *(Fool's gold)* Magnetite	$Fe_2O_3 . xH_2O$ FeS_2 Fe_3O_4
Tin	Cassiterite or Tin stone	SnO_2
Chromium	Chromite	$FeCr_2O_4$
Manganese	Pyrolusite	MnO_2
Lead	Galena	PbS
Copper	Chalcopyrite or Copper pyrites Malachite Cuprite Copper glance	$CuFeS_2$ $CuCO_3 . Cu(OH)_2$ Cu_2O Cu_2S
Mercury	Cinnabar	HgS
Silver	Argentite	Ag_2S
Gold	Native metal	Au

It may be noted that the metals at the top tend to occur more commonly as chlorides, carbonates and sulphates. Metals in the middle commonly occur as oxides and those which are lower in the series occur as sulphides. The metals at the bottom such as silver, platinum and gold exist in their native state.

A few metals in decreasing order of their abundance in the earth's crust are listed as follows:

Aluminium, Iron, Calcium, Sodium, Potassium, Magnesium, Titanium

DECREASING ABUNDANCE IN EARTH'S CRUST

SOLVED EXAMPLES

Example 50.1. *What type of metals are likely to exist in a native state in nature? Give some examples of such metals.*

Solution. The less reactive metals are likely to exist in a native state. These metals lie below hydrogen in the activity series. Some examples are gold, silver, platinum, copper, etc.

Example 50.2. *Differentiate between "minerals" and "ores".*

Solution. Minerals are the naturally occurring chemical substances in the earth's crust which can be extracted by mining. The minerals from which a metal can be extracted economically and conveniently are called ores. For example, minerals of copper are copper glance (Cu_2S), cuprite (Cu_2O) and copper pyrites ($CuFeS_2$) but its ore is copper pyrites. All ores are minerals but all minerals are not ores.

Example 50.3. *Predict the modes of occurrence of the following three types of metals:*

(i) Highly reactive (such as sodium).

(ii) Moderately reactive (such as iron).

(iii) Noble metal (such as gold, platinum).

Solution. (*i*) Highly reactive metals occur in a combined state generally as their chlorides, sulphates or carbonates.

(*ii*) Moderately reactive metals generally occur as their oxides.

(*iii*) Noble metals exist in a free state or native state.

CASE STUDY: GHANA

50.4 MINERAL DEPOSITS IN GHANA

Utilization of Ghana's mineral resources is managed by the Mineral Commission of Ghana. Important mineral deposits in Ghana and their locations in Ghana are discussed as follows:

1. METALS

The major metallic minerals of Ghana are gold, bauxite, iron and manganese. At present, gold, bauxite and manganese are being mined in Ghana.

Gold: There is great potential of gold exploration in Ghana. It is the main focus of majority of foreign and local exploration companies in the country. The main gold deposits in Ghana are at Obuasi, Tarkwa, Prestea, Damang, Bogoso, Aboso, Bibiani, Kenyase and Ntronang-Abrem.

Bauxite: Ghana has large deposits of good quality bauxite in the Western, Eastern and Ashanti regions. The four major bauxite deposits in Ghana are ; Sefwi-Bekwai (Awaso), Kibi, Ejuanema and Aya-Nyinahin deposits.

Iron: There are three main iron ore deposits in Ghana. These are: Shieni, Oppong Mansi and Pudo.

Manganese: There are two types of manganese ores which occur in Ghana. These are oxides and carbonates. The major deposits are at Nsuta in the Western region.

Manganese occurs in seven of the ten regions of Ghana, in the Central, Eastern, Western, Ashanti, Northern, Upper-west and Upper-east.

2. PRECIOUS STONES

Diamonds: Ghana's diamond mining industry has produced industrial grade gems since the 1920's. These are mainly mined from the Birim and the Bonza diamond fields. Diamonds found in Ghana are found in alluvial gravel. More than 11 million carats of proven and probable reserves are located about seventy miles North-west of Accra. Currently one million carats of diamonds is produced in a year. Ghana is the 9th largest producer of diamonds in the world.

3. INDUSTRIAL MINERALS

Limestone ($CaCO_3$). There are four limestone deposits in Ghana. These are Buipe, Nauli Oterkpolu and Bongo-Da deposits.

Dolomite ($CaCO_3$ $MgCO_3$). The main dolomite deposit is located at the Buipe-Baka area in the Northern region.

Solar Salt. Ghana possesses one of the largest proven renewable solar salt production potential along the entire coastline stretching over a distance of over 500 km. Effective exploitation would enable the country to meet the needs of the entire sub-region.

At present Ghana's total annual production of salt is about 250,000 tonnes.

A massive development of the salt industry within the catchment area of the Songhor Lagoon, in the Ada Traditional Area, is to take off soon.

50.5 EXTRACTION OF METALS—METALLURGY

As already mentioned only a few most unreactive metals occur in nature in a free state. Most of the metals occur in nature in oxidized form as their compounds. Therefore, metals are generally extracted by subjecting their ores to reduction by chemical methods or by electrolytic methods.

The process of extracting pure metal from its ore is known as metallurgy.

Since the nature of the ore and also the properties of different metals are different, it is not possible to have a universal scheme which may be applicable to all the metals. However, some common steps involved in the metallurgical operations are:

I. *Crushing and grinding of the ore.*

II. *Concentration or benefication of the ore.*

III. *Preliminary treatment of the concentrated ore.*

IV. *Reduction.*

V. *Purification or refining of crude metal.*

I. CRUSHING AND GRINDING OF THE ORE

Most of the ores are obtained from the crust of the earth in the form of huge lumps. These lumps have to be converted into powdered form so that the chemical changes which have to take place at the later stages may become convenient. The huge lumps are broken into small pieces in *jaw crushers.* They are further *pulverised in stamp mill* or *ball mill.*

II. CONCENTRATION OR BENEFICATION OF THE ORE

The ore obtained from the earth's crust is associated with rocky and silicious impurities. It is quite essential to get rid of these impurities so that they may not cause any interference in the process of extraction. *The removal of unwanted materials such as sand, clays, etc., from the pulverised ore* is called **concentration, dressing** or **benefication of the ore.** The benefication of the ore is carried out by any of the following methods depending upon the nature of the ore and also the impurities present in the ore:

1. Hydraulic Washing, Levigation or Gravity Separation Method. This method is usually applicable to *oxide ores* in which the ore particles are heavier than the impurities. The powdered ore is washed with a running stream of water. The lighter impurities are washed away leaving behind the heavier ore particles.

Native metals such as gold can be separated from sand and gravel by shaking and washing of earth with water in a pan or sieve. The process is known as **panning**.

2. Froth Floatation Process. This process is generally used for the concentration of *sulphide ores*. The finely powdered ore and *water* are taken in a tank. Additional reagents such as *pine oils, fatty acids,* etc., are added to the mixture. These reagents increase the nonwettability of the mineral particles and are known as **collectors.** The contents are kept agitated by the blast of air. As a result of agitation the froth is produced.

The ore particles are preferentially wetted by the oil and are carried to the surface by the foam. The gangue material, which is preferentially wetted by water sinks to the bottom of the tank. The foam at the surface of the tank is transferred to the other tank where it is washed with water to recover the ore particles.

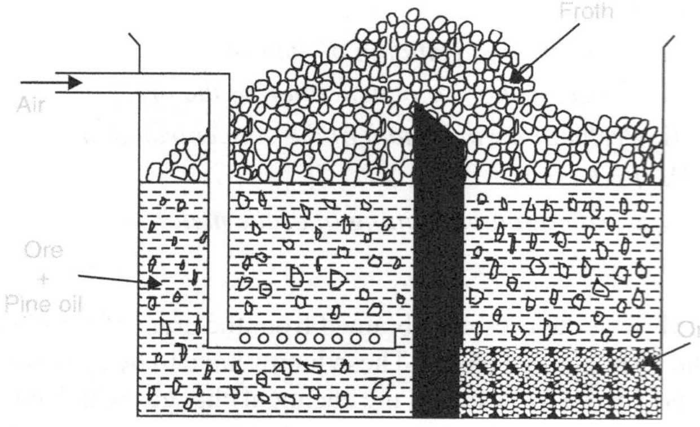

Fig. 50.1. Froth floatation process.

3. Magnetic Separation of Impurities: This method is usually employed when either the ore or the gangue is capable of being attracted by a magnetic field. For example, tungstates of iron and manganese from tin stone are separated by this method. The powdered ore is dropped over the belt revolving around the rollers, one of which is magnetic. The magnetic roller attracts the magnetic part of the ore and they are collected in the form of a heap near it. The non-magnetic part of the ore flies off and forms a heap away from the impurities.

4. Leaching: It is a chemical method for the concentration of the ore. In this process the powdered ore is dissolved selectively in acids, bases or other suitable reagents. The impurities remain undissolved as *sludge*. The solution of ore is filtered and the ore is recovered by precipitation or crystallisation.

III. PRELIMINARY TREATMENT OF THE CONCENTRATED ORE

The process of extraction of metal from the concentrated ore depends upon the nature of the ore as well as the nature of impurities present in the ore. Before the concentrated ore is subjected to final metallurgical operations in order to get the metal in the free state, **preliminary chemical treatment** may be necessary. The objective of this preliminary chemical treatment is:

(*a*) *to get rid of impurities which would cause difficulties in the later stages*; and

(*b*) *to convert the ore into oxide of the metal* because it is easier to reduce an oxide than the carbonate or sulphide.

The processes employed for preliminary treatment are *calcination* and *roasting*.

1. Calcination

It is a process of heating the ore in a limited supply of air below its melting point. The process involves:

— the removal of volatile impurities,

— the removal of moisture,

— the decomposition of any carbonate ore into oxide.

$$Fe_2O_3.xH_2O \longrightarrow Fe_2O_3 + xH_2O(g)$$
$$ZnCO_3 \longrightarrow ZnO + CO_2$$
$$CuCO_3 . Cu(OH)_2 \longrightarrow 2CuO + H_2O + CO_2$$

2. Roasting

It is the process of heating the ore in excess supply of air below its melting point. This process is employed when oxidation of the ore is required. As a result of roasting,

— moisture is driven away,

— volatile impurities are removed,

— the impurities like sulphur, phosphorus, arsenic are removed as their oxides,

— the ore undergoes oxidation to form metal oxide or sulphate.

$$2PbS + 3O_2 \longrightarrow 2PbO + 2SO_2$$
$$2ZnS + 3O_2 \longrightarrow 2ZnO + SO_2$$

It is advantageous to roast a sulphide ore to the oxide before reduction because metal oxides can be reduced to metal by carbon and hydrogen much more easily than sulphides.

IV. REDUCTION

After the preliminary treatment, the ore may be subjected to a reduction process by one of the following methods depending upon its nature:

1. Smelting or Reduction with Carbon. In this process, the roasted or calcined ore is mixed with suitable quantity of coke or charcoal (which act as a *reducing agent*) and is heated to a high temperature above its melting point. During reduction, an additional reagent is also added to the ore to remove the impurities still present in the ore. This additional reagent is called **flux.** *Flux combines with the impurities to form a fusible product called* **slag.**

$$\textit{Flux} + \textit{Impurities} \longrightarrow \textit{Slag}$$

The selection of flux depends upon the nature of impurities. If impurities are *acidic* in nature, the *flux is basic, such as, lime (CaO)*. On the other hand, for *basic impurities*, an *acidic flux such as silica* (SiO_2) is used.

$$\underset{\text{(Basic impurity)}}{CaO} + \underset{\text{(Acidic flux)}}{SiO_2} \longrightarrow \underset{\text{(Slag)}}{CaSiO_3}$$

2. Reduction with Aluminium. Certain metal oxides such as Cr_2O_3 and Mn_3O_4 are not easily reduced with carbon. In such cases aluminium is used as a reducing agent because it is more electropositive than chromium or manganese. *The process of reduction of oxides with aluminium* is called **aluminothermy.** Some examples are:

$$Cr_2O_3 + 2Al \longrightarrow Al_2O_3 + 2Cr$$
$$3Mn_3O_4 + 8Al \longrightarrow 4Al_2O_3 + 9Mn$$

The chemical methods are suitable for reduction of compounds of metals which are in the middle of the activity series.

3. Auto-reduction. Certain metals are obtained from their ores by simple roasting without using any reducing agent. For example, mercury is directly obtained by roasting its ore cinnabar (HgS) in air.

$$HgS(s) + O_2(g) \longrightarrow Hg(l) + SO_2(g)$$

or

$$2HgS(s) + 3O_2(g) \longrightarrow 2HgO(s) + 2SO_2(g)$$
$$2HgO(s) + HgS(s) \longrightarrow 3Hg(l) + SO_2(g)$$

4. Electrolytic Reduction. The highly electropositive elements such as *alkali metals, alkaline earth metals* and aluminium cannot be extracted by carbon reduction methods. They are extracted by the electrolysis of their fused salts. *The process of extraction of metals by the use of electrolysis* is called **electrometallurgy.**

For example, sodium metal is extracted by electrolysis of molten sodium chloride containing other salts as impurities.

$$NaCl(l) \longrightarrow Na^+ + Cl^-$$

At cathode: $Na^+ + e^- \longrightarrow Na$

At anode: $2Cl^- \longrightarrow Cl_2 + 2e^-$

5. Displacement Method (Hydrometallurgy). Some metals like gold and silver are extracted from their concentrated ores by *leaching*. They are dissolved in suitable reagents like acids or bases leaving behind insoluble impurities. The metal is recovered from the solution by displacement with some more electropositive metal such as zinc. For example, silver ore is leached with dilute solution of sodium cyanide. Silver dissolves forming a complex, sodium dicyanoargentate (I). The solution is further treated with scrap zinc which displaces silver from complex.

$$Ag_2S + 4NaCN \rightleftharpoons \underset{\text{Sod. dicyanoargentate (I)}}{2\,Na[Ag(CN)_2]} + Na_2S$$

$$2Na[Ag(CN)_2] + Zn \longrightarrow \underset{\text{Sod. tetracyano-zincate (II)}}{Na_2\,[Zn(CN)_4]} + 2Ag\downarrow$$

On the basis of reactivity we can group the metals into the following three categories:

(*i*) Metals of low reactivity.

(*ii*) Metals of medium reactivity.

(*iii*) Metals of high reactivity.

Different reduction processes are used for obtaining the metals falling in each category.

The relation between the reduction process employed and the position of the metal in the activity series is depicted in Table 50.4.

Table 50.4. Position of the Metal in the Activity Series and the Related Reduction Process

Position in the Activity Series	*Element*	*Reduction Process*
Metals found near the top of the activity series	**K, Na, Ca, Mg, Al**	**Electrolysis**
Metals found in the middle of the activity series	**Zn, Fe, Pb**	**Reduction using carbon or some other reducing agent**
Metals found in the lower regions of the activity series	**Cu, Hg, Ag**	**Reduction by heat alone**
Metals found at the bottom of the activity series	**Pt, Au**	**Found in native state**

V. REFINING OF METALS

The process of purification of impure metals is known as refining.

Depending upon the nature of the metal and the nature of the impurities present different methods are applied for the refining of metals. Some of the commonly used methods are discussed below:

1. Distillation. Volatile metals like zinc and mercury are purified by this method. The pure metal distils over and is condensed in a receiver. The non-volatile impurities are left behind in the retort.

2. Liquation. The method is used for easily fusible metals like bismuth, tin and lead. The crude metal is placed on the sloping hearth of a furnace and heated gently until the metal melts and flows down, leaving behind the infusible impurities which remain stuck to the floor of the hearth.

3. Electrolytic Refining or Electro-refining. This method is based on the phenomenon of electrolysis. Many metals like ***copper, silver, tin, gold, zinc, nickel and chromium*** are purified by electro-refining.

The impure metal is made the anode whereas a thin sheet of pure metal is made the cathode. Electrolyte is a solution of some salt of the metal. On passing electricity, the metal from the anode goes into solution as ions due to oxidation while the pure metal gets deposited at the cathode due to the reduction of metal ions. The insoluble impurities settle down below the anode as **anode mud** whereas the soluble impurities go into solution. The reactions taking place at the two electrodes may be represented as:

At cathode: $M^{n+} + ne^- \longrightarrow M$

At anode: $M \longrightarrow M^{n+} + ne^-$

The various steps involved in the extraction of pure metals from their respective ores are summarized in Fig. 50.2.

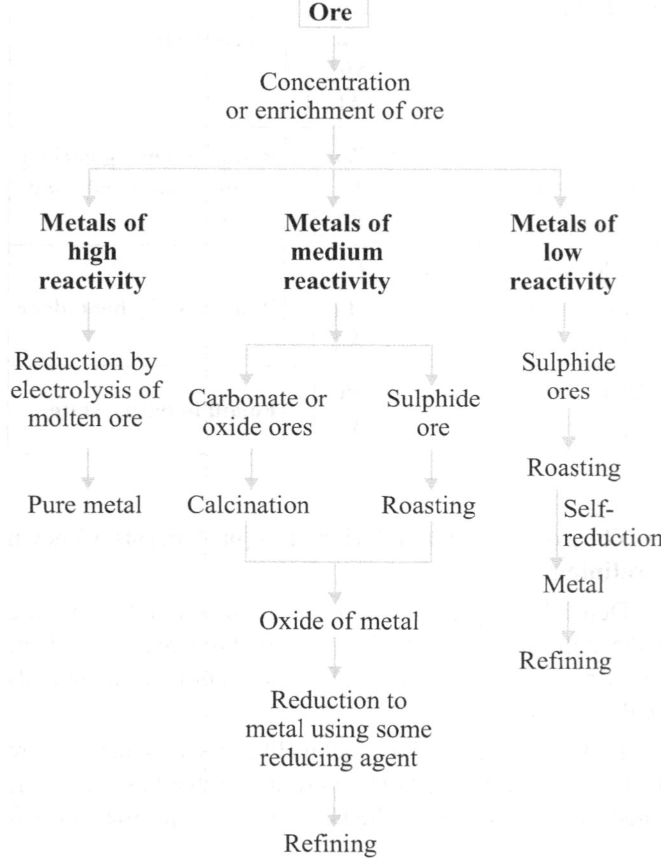

Fig. 50.2. Steps involved in the extraction of metals from ores.

50.6 EXTRACTION OF GOLD

Mining of gold in Ghana began long before the arrival of the Europeans, mainly by panning by hand. The metal was extracted from alluvial sand and gravel deposited on the banks and beds of rivers.

The use of machines in the mining of the metal started in about 1880 in the Tarkwa area and extended to Obuasi in 1898. Ghana's pre-independence name of 'Gold Coast' shows the importance of the metal in the economy and history of the country.

OCCURRENCE

Since gold is very unreactive it occurs in a free state in nature. The primary sources of gold are:

(*i*) *Alluvial gravel.* It is found in the beds of certain rivers. It is referred to as *alluvial gold.*

(*ii*) *Auriferous quartz.* Gold also occurs in the veins of quartz. This is called *relf gold.*

EXTRACTION

The various steps involved in the extraction of gold are:

1. The concentrated ore is *roasted* to remove all the oxidisable impurities.

2. Treatment with KCN

The roasted ore is then treated with a solution of sodium cyanide or potassium cyanide for some days. In the presence of atmospheric oxygen, gold dissolves in the form of a complex.

$$4Au + 8NaCN + 2H_2O + O_2 \longrightarrow \underset{\text{Sodium dicyanoaurate (I)}}{4Na[Au(CN)_2]} + 4NaOH$$

3. Precipitation of Gold. From the solution the metal is precipitated by adding zinc shavings.

$$2Na[Au(CN)_2] + Zn \longrightarrow \underset{\text{Sodium tetracyanozincate (II)}}{Na_2[Zn(CN)_4]} + 2Au \downarrow$$

The process is called ***Mac Arthur Forest Cyanide Process.***

4. Purification. Gold obtained in the above process is impure and contains metals such as silver and copper as impurities. ***Removal of silver and copper impurities from impure gold is known as parting.*** It is achieved by boiling impure gold with conc. H_2SO_4. Silver and copper dissolve in hot conc. H_2SO_4 while the gold remains as such. Parting can also be done with conc. HNO_3. Purification of gold can also be done by electrolysis.

The anode is impure gold. The cathode is a strip of pure gold coated with a thin layer of graphite. The electrolyte is an aqueous solution of gold (III) chloride or gold (III) trioxonitrate(V) and trioxonitrate(V) acid. The gold ions are reduced at the cathode on electrolysis:

$$Au^{+3} + 3e^- \longrightarrow Au$$

Crude mining and extraction of gold and diamond is carried out on a small scale in Ghana. It is popularly called '**Galamsey**'. Gold bearing ores or tailings, *i.e.*, waste from previous ore treatment, is first concentrated by washing several times with water in a pan or other suitable container to remove mud and other fifth.

The gold-bearing product is then poured on a jute bag or any wide mesh cloth laid on an inclined table. It is then repeatedly washed with water to remove the lighter gangue. The heavier gold material, trapped in the sack or cloth, is washed off into a pan with water and then decanted. The solid gold product is rubbed well with mercury to form an amalgam leaving the remaining gangue.

In the final stage, the gold amalgam is put in a white cloth and squeezed hard to let out as much mercury as possible. The gold product thus obtained is white due to the mercury contamination. This is washed with water in a pan. The heavier mercury falls to the bottom and the gold remains on top. The gold is harvested (collected) ready to be sold. The gold may also be recovered by distillation.

The gold produced by both Ashanti Goldfields Company Limited and 'Galamsey' operators is not very pure. Pure gold is obtained by refining outside Ghana.

USES OF GOLD

1. For preparing ornaments.
2. For electrogilding (*i.e.*, gold plating by electrolysis).
3. For filling teeth, and in switch contacts.
4. For making compounds of gold.
5. For preparing alloys.

Do You Know?

Pure gold is soft and is generally hardened by adding silver or copper to it for the purpose of making ornaments. The weight of gold in gold ornaments is expressed in terms of '**carats**'. Pure gold is taken as 24 carats. 18 carat gold means that it contains 18 parts by weight of gold in 24 parts by weight of the given alloy. The percentage of gold in this sample may be calculated as:

Percentage of gold in 18 carat gold

$$= \frac{18}{24} \times 100 = 75\%$$

Most jewellery is made out of 22 carat gold.

50.7 EXTRACTION OF ALUMINIUM

Aluminium is one of the elements of Group 13 of the periodic table. Although aluminium is a reactive metal according to the electrochemical series, it is rendered unreactive due to the formation of a film of oxide on its surface.

OCCURRENCE

Aluminium is quite a reactive element and hence does not occur in nature in the native form. Aluminium is the *most abundant metal* and the *third most abundant element* in the earth's crust. The important ores of aluminium are:

(*i*) *Bauxite*, $Al_2O_3.2H_2O$
(*ii*) *Cryolite*, $Na_3\,AlF_6$
(*iii*) *Feldspar*, $KAlSi_3O_8$
(*iv*) *Mica*, $KAlSiO_{10}\,(OH)_2$.

EXTRACTION (From bauxite)

Aluminium metal is extracted from bauxite in a two stage process.

Stage 1. Involves the extraction of alumina (Al_2O_3) from bauxite.

Stage 2. Involves extraction of pure aluminium from Al_2O_3 by its electrolysis in molten cryolite [Na_3AlF_6].

1. Purification of Bauxite. Bauxite contains SiO_2, iron oxides and titanium (*IV*) oxide as impurities.

Bauxite is digested with a hot concentrated (45%) solution of sodium hydroxide at about 473–523 K and 35–36 bar pressure.

Alumina dissolves to form sodium tetrahydroxoalu-minate (III), $Na[Al(OH)_4]$ leaving behind iron oxide and TiO_2.

$$Al_2O_3 + 2NaOH + 3H_2O \longrightarrow 2Na[Al(OH)_4]$$

Silica (SiO_2) also dissolves in sodium hydroxide to form soluble sodium trioxosilicate (IV), Na_2SiO_3.

$$SiO_2 + 2NaOH \longrightarrow 2Na_2SiO_3 + H_2O$$

The impurities are filtered out and CO_2 is bubbled through the filtrate containing sodium tetrahydroxo-aluminate (III) and sodium trioxosilicate (IV). At this stage the solution may be seeded with freshly precipitated aluminium hydroxide. Aluminium hydroxide precipitates leaving behind sodium trioxosilicate (IV) in solution. This is filtered and the precipitate of $Al(OH)_3$ is heated at 1473 K to obtain pure alumina.

$$Na[Al(OH)_4] + CO_2 \longrightarrow NaHCO_3 + Al(OH)_3 \downarrow$$

$$2Al(OH)_3 \xrightarrow{1473\,K} Al_2O_3 + 3H_2O$$

2. Electrolysis of Pure Alumina (Hall-Heroult Process). Aluminium is obtained from alumina by the process of electrolysis. This method is known as **Hall-Heroult process.** Purified alumina is dissolved in molten cryolite (Na_3AlF_6) and is electrolysed in an iron tank lined with carbon (Fig. 50.3). *Carbon lining serves as cathode* while a number of *carbon rods* dipping in the fused electrolyte serve as *anode.* Cryolite improves the electrical conductivity of the cell as alumina is a poor conductor. Moreover, cryolite lowers the melting point of the mixture to about 1250 K. Other impurities such as CaF_2, NaF and AlF_3 may also be added. The temperature of the electrolyte is maintained between 1200–1300 K.

On passing electric current, aluminium is liberated at the cathode and gets collected at the bottom of the tank from where it is removed. Oxygen liberated at the anode combines with the carbon of the anode to produce carbon monoxide which either burns or escapes out. The reactions taking place during electrolysis are:

At cathode: $Al^{3+}\text{ (melt)} + 3e^- \longrightarrow Al\ (l)$

At anode: $C(s) + O^{2-}\text{ (melt)} \longrightarrow CO(g) + 2e^-$

$C(s) + 2O^{2-}\text{ (melt)} \longrightarrow CO_2(g) + 4e^-$

Since during electrolysis, the carbon electrodes get consumed, they have to be replaced periodically. For each kg of aluminium about 0.5 kg carbon is burnt away.

The extraction of aluminium requires large amount of electrical energy. About 15,000 kWh are needed to produce 1 tonne of aluminium. For example, the smelter at Tema, Ghana, owned by the Volta Aluminium Company, uses about 2,760,000 MWh electricity to produce nearly 200,000 tonnes of aluminium annually. The company receives power from the Akosombo power station of the Volta River hydroelectric dam. Due to its proximity to the power station, the company receives sufficient electrical energy at a relatively cheap rate.

USES OF ALUMINIUM

Aluminium is a soft metal with a density of 2.7 g cm^{-3}. Aluminium can be easily extruded through dies to form various shapes.

1. It is used for making angles used in windows.
2. It is a good conductor of electricity. Since it is not as good a conductor as copper, thicker cables of aluminium are used for transmission of electricity.
3. Aluminium forms many useful alloys *e.g., magnalium* (Al and Mg), *duralumin* (Al, Cu, Mg and Mn). Aluminium alloys are used in aircraft and other transportation vehicles.
4. In the form of finely-divided powder, aluminium is used in antirust paints.
5. Aluminium foil is used for wrapping cigarettes, confectionery, etc.
6. Aluminium is used to produce metals such as chromium and manganese from their ores (*aluminothermic process*)

$$Cr_2O_3 + 2Al \longrightarrow Al_2O_3 + 2Cr$$

$$3MnO_2 + 4Al \longrightarrow 2Al_2O_3 + 3Mn$$

7. Aluminium utensils are extensively used for household purposes.

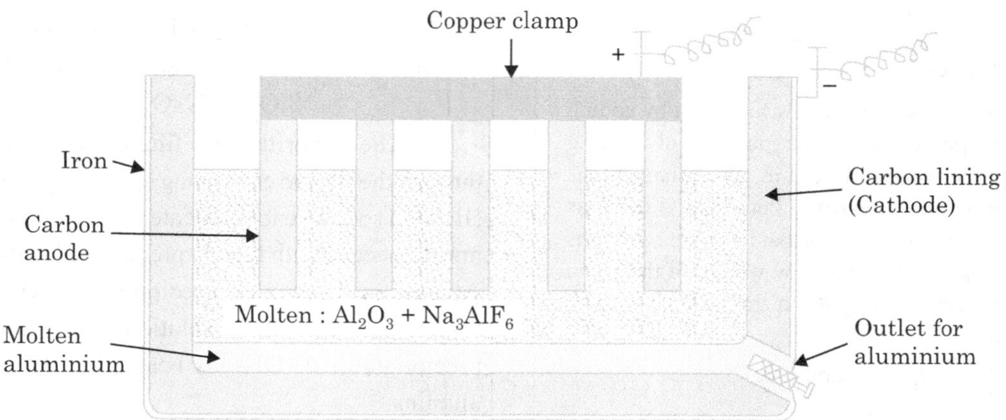

Fig. 50.3. Electrolytic cell for production of aluminium.

50.8 EXTRACTION OF IRON

Iron is the second most abundant metal occurring in the earth's crust. It is an element of the first transition series.

OCCURRENCE

The important ores of iron are:

1. *Haematite*, Fe_2O_3 (red oxide of iron)
2. *Magnetite*, Fe_3O_4 (magnetic oxide of iron)
3. *Limonite*, $Fe_2O_3.3H_2O$ (hydrated oxide of iron)
4. *Iron pyrites*, FeS_2
5. *Siderite or Spathic ore*, $FeCO_3$.

The cast iron is generally extracted from haematite (Fe_2O_3). The various steps involved in the process are as follows:

1. Concentration. The ore is crushed with the help of jaw crushers into small pieces of about 5 cm size. The crushed ore is washed with a stream of water whereby lighter sand particles are washed away and the heavier ore particles settle down.

2. Calcination. The concentrated ore is heated strongly in the presence of air. This is called calcination. During calcination the following changes take place:

- Moisture is driven out.
- Sulphur, arsenic and phosphorus impurities are expelled as their volatile oxides.
- Carbonate ore changes into oxide ore.

$$FeCO_3 \longrightarrow FeO + CO_2$$

- Ferrous oxide changes into ferric oxide.

$$4FeO + O_2 \longrightarrow 2Fe_2O_3$$

3. Smelting. After calcination the ore is subjected to reduction with carbon in a blast furnace (Fig. 50.4). The process is called **smelting.** Blast furnace is made up of steel, lined on the inside with fire resistant bricks. It has a cup and cone arrangement for the introduction of charge at the top, and at the base it has:

(*i*) a tapping hole for removing molten iron from time to time.

(*ii*) an arrangement for the introduction of hot air.

(*iii*) an outlet for slag.

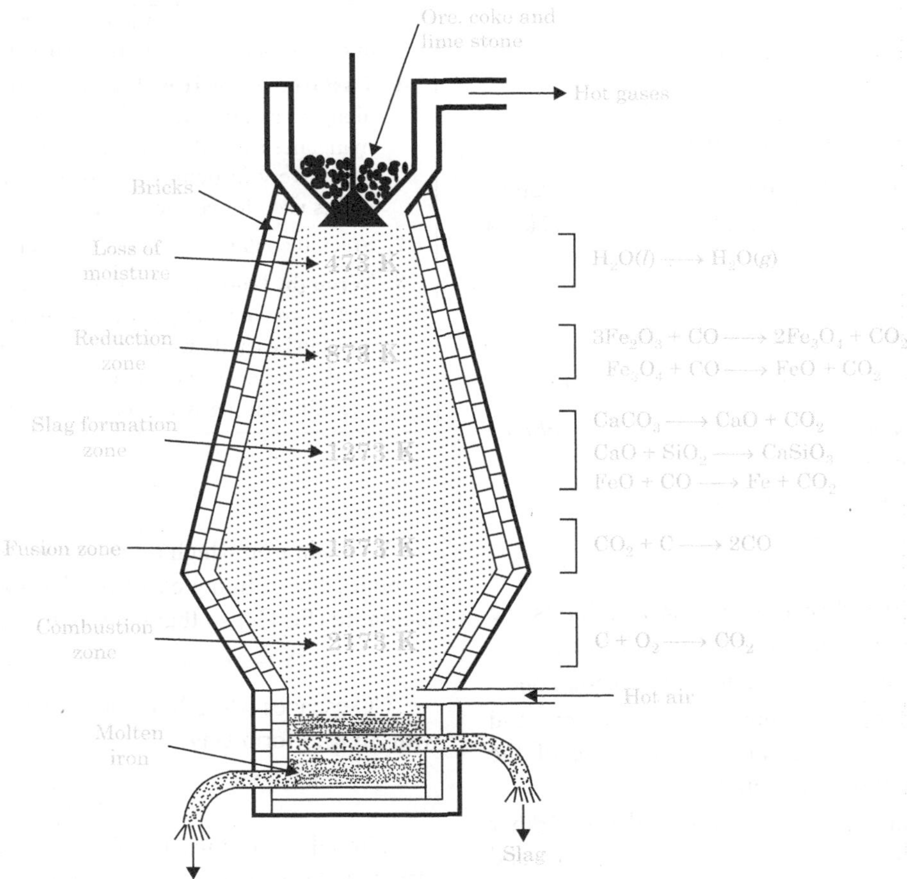

Fig. 50.4. Blast furnace for the manufacture of iron.

The calcined ore, coke and lime stone are mixed in the ratio 8 : 4 : 1 and are fed into the furnace through a cup and cone arrangement. Blast of hot air at about 1000 K is blown in through narrow pipes (*tuyeres*) at the base of the furnace. Coke here serves as a fuel as well as a reducing agent while lime is a flux. Reactions which take place in the furnace are as follows:

(*i*) Coke burns at the base to produce CO_2 which rises up.

$$C + O_2 \longrightarrow CO_2, \Delta H = -406 \text{ kJ mol}^{-1}$$

The reaction is exothermic and temperature here is about 2173 K. This region is called the **combustion zone.**

(*ii*) As CO_2 rises up it is reduced to CO with the coke.

$$CO_2 + C \longrightarrow 2CO, \Delta H = +173 \text{ kJmol}^{-1}$$

Due to the endothermic nature of this reaction, the temperature in this region falls to 1475–1575 K. In this region *Fe_2O_3, if present, gets reduced to iron by hot coke and the spongy iron produced in the upper region gets melted.* This is known as the **fusion zone.**

(*iii*) Near the top of the furnace, where temperature is about 873 K, *the oxides of iron* (Fe_2O_3 and Fe_3O_4) *are reduced to iron carbon (II) oxide.*

$$Fe_2O_3 + CO \longrightarrow 2FeO + CO_2$$

$$3Fe_2O_3 + CO \longrightarrow 2Fe_3O_4 + CO_2$$

$$Fe_3O_4 + 4CO \longrightarrow 3Fe + 4CO_2$$

(*iv*) In the middle portion of the furnace where temperature is about 1073 to 1273 K, FeO is reduced to iron while coke is oxidized to CO.

$$CO_2 + C \longrightarrow 2CO$$

$$FeO + CO \longrightarrow Fe + CO_2$$

$$FeO + C \longrightarrow Fe + CO$$

Lime stone decomposes to produce CaO which combines with silica (impurity) to form slag.

$$CaCO_3 \longrightarrow CaO + CO_2$$

$$\underset{\text{Quick lime}}{CaO} + SiO_2 \longrightarrow \underset{\text{Slag}}{CaSiO_3}$$

Molten slag is lighter than molten iron and forms the upper layer. This region is called **slag formation zone.**

Iron formed moves down and melts in the fusion zone.

Molten iron dissolves in it some carbon, silicon, and phosphorus and forms the lower layer at the base of the furnace. Iron thus formed is called **pig iron.**

Pig iron contains about 4% of carbon and many other impurities such as phosphorus, silicon, sulphur and manganese.

Iron works are located close to coal fields so that abundant coal is accesible at a cheap rate. Coal is used as a reducing agent and as fuel.

Pig iron is converted into cast iron by heating molten pig iron with scrap iron and coke in specially designed furnaces. A blast of hot air is blown through the mixture. Cast iron contains about 3% carbon. It slightly expands on solidifying and hence reproduces the shape of the mould. Cast iron is very hard and brittle. Its melting point is about 1473 K.

CAST IRON FROM IRON PYRITES

Iron pyrites (FeS_2) is not directly used as an ore for the extraction of iron because it contains quite high percentages of sulphur. Due to the high content of sulphur it is primarily used for the manufacture of H_2SO_4. For this purpose it is burnt in air to get SO_2:

$$4FeS_2 + 11O_2 \longrightarrow 2Fe_2O_3 + 8SO_2 \uparrow$$

SO_2 is converted into SO_3 which is absorbed in water to get H_2SO_4. The residual Fe_2O_3 formed during burning of FeS_2 is reduced to iron in a blast furnace as already discussed.

COMMERCIAL FORMS OF IRON

There are three commercial forms of iron which differ from each other mainly in their carbon content.

1. **Cast Iron or Pig Iron.** It contains about 2 to 5% of carbon, along with impurities such as sulphur, silicon, phosphorus, manganese, etc. It is the least pure form of iron. It is brittle and is very difficult to weld. It is resistant to corrosion and is used for sewage pipes.
2. **Steel.** It contains 0.1 to 1.5% carbon and other impurities.
3. **Wrought Iron.** It is the purest form of iron and contains carbon and other impurities less than 0.2%. It is malleable and can be easily welded. This type of iron is structurally weak and hence cannot be magnetised permanently. It is used for making wires, chains, electromagnets, etc.

CONVERSION OF CAST IRON INTO WROUGHT IRON

Wrought iron which is comparatively a pure form of iron is obtained by refining cast iron. The cast iron is heated with haematite (Fe_2O_3) so that the impurities are oxidized.

$$Fe_2O_3 + 3C \longrightarrow 2Fe + 3CO$$

CO escapes while MnO present as an impurity combines with SiO_2 to form slag.

$$MnO + SiO_2 \longrightarrow \underset{\text{(slag)}}{MnSiO_3}$$

Wrought iron thus formed contains about 0.2 – 0.5% carbon and traces of P and Si in the form of slag.

Wrought iron is malleable and ductile. Its melting point is about 1400°C.

STEEL

The amount of carbon present in steel is nearly intermediate between that contained in cast and wrought iron. The various steels contain carbon from 0.1 to 1.5%. The increase in carbon content in the steel decreases its ductility and increases its tensile strength. The following methods are commonly used for the production of steel:

1. *Bessemer Process*
2. *The Basic Oxygen Process*
3. *Open Hearth Process*
4. *The Electric Arc Process*
5. *The High Frequency Induction Process.*

All these methods are based on the oxidation of impurities in the pig iron.

Two of these methods are discussed below:

1. Bessemer Process. The *Bessemer process* involves blowing a strong blast of air through the molten pig iron.

The Bessemer process is accomplished in large pear-shaped iron vessels called *converters* which are lined on the inside with silica bricks and can hold upto 50 tons of pig iron at a time. The converter is mounted on horizontal pivots around which it can be tilted. The converter is filled with molten pig iron at a temperature of about 1473 K and air is blown into the chamber. Oxygen reacts with the impurities and raises the temperature to 2173 K. The impurities are oxidised by air.

$$Si + O_2 \longrightarrow SiO_2$$
$$2Mn + O_2 \longrightarrow 2MnO$$
$$C + O_2 \longrightarrow CO_2$$

SiO_2 is slagged off with lime and MnO with silica in the form of their respective silicates.

$$CaO + SiO_2 \longrightarrow CaSiO_3$$
$$SiO_2 + MnO \longrightarrow MnSiO_3$$

The slag floats on the surface whereas purified iron being denser than impure iron sinks to the bottom.

The Bessemer process produces iron containing less than 0.3 per cent carbon. If it is desired to obtain steel, the air blast is either shut off before all the carbon has burnt out, or a definite amount of pig iron rich in carbon is added to the iron produced in the converter. Air is again blown through it for a short period in order to mix the ingredients.

2. The Basic Oxygen Process. The Basic Oxygen process is a development of the Bessemer Process. In this process pure oxygen is used instead of air.

Initially, the furnace is charged with hot molten pig iron from the blast furnace lime and scrap steel. A jet of oxygen is blown onto the surface of the metal at a great speed through a water cooled pipe.

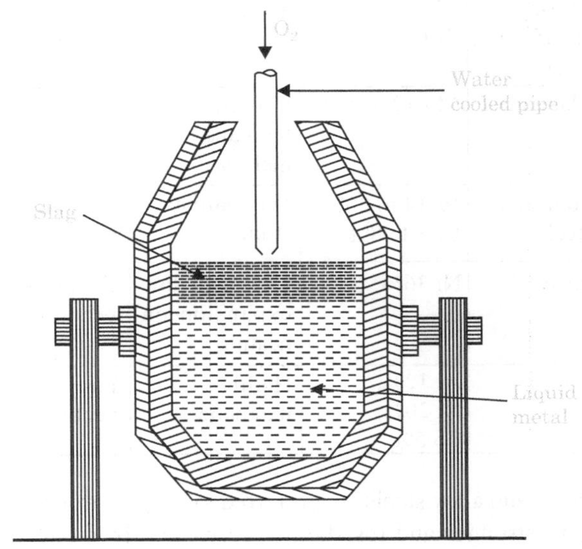

Fig. 50.5. The Basic Oxygen process.

The oxygen oxidizes the impurities rapidly. Since the reactions taking place in the furnace are highly exothermic, the heat evolved keeps the contents of the furnace in the molten state. When steel of desired composition is obtained, the oxygen is turned off and molten steel is poured by tilting the furnace.

ALLOY STEELS

The properties of steel can be modified by the addition of other metals. *When steel is alloyed with at least one metal like nickel, chromium, vanadium, manganese, cobalt, tungsten, etc., an alloy steel is obtained.* Some of these alloy steels are given in Table 50.5.

Table 50.5. Some Alloy Steels

Alloy Steel	*Special Components*	*Chief Properties*	*Uses*
1. Nickel steel	Ni 3.5%	Hard, flexible, rust resistant	For cables armour plates
2. Stainless steel	Cr 18%, Ni 8%	Does not rust or corrode	For household utensils, shaving blades, watch cases
3. Chrome-vanadium steel	Cr 1% V 0.15%	Tenacious and load bearing	For axles, springs and cog-wheels
4. Manganese steel	Mn 12 to 15%	Extremely hard and high melting	For rock crusher, burglar proof safes
5. Tungsten steel	W 14 to 20% Cr 3 to 8%	Very hard and strong	For cutting tools, springs
6. Invar	Ni 36%	Extremely low expansion on heating	For clock pendulums
7. Alnico	Al 12%, Ni 20%, Co 5%	Highly magnetic	For permanent magnets

Steel and alloy steels are preferred to the pure iron because they are stronger and resistant to corrosion. Iron undergoes corrosion (rusting).

SOLVED EXAMPLES

Example 50.4. *Giving examples, differentiate between 'roasting' and 'calcination'.*

Solution. Roasting is the process of heating the ore in excess supply of air below its melting point. This process is applied for sulphide ores. As a result sulphide ore changes into oxides. Calcination is the process of heating the ore in limited supply of air below its melting point. This process is applied to oxide and carbonate ores.

For example, *zinc blende* (sulphide ore of zinc) is roasted whereas *haematite* (oxide ore of iron) is subjected to calcination.

$$2ZnS + 3O_2 \xrightarrow{\text{Heat}} 2ZnO + SO_2 \quad \text{...Roasting}$$

$$Fe_2O_3.xH_2O \xrightarrow{\text{Heat}} Fe_2O_3 + xH_2O \quad \text{...Calcination}$$

Example 50.5. *How is 'cast iron' different from 'pig iron'?*

Solution. Cast iron contains about 3% carbon whereas pig iron contains about 4% carbon. Pig iron is harder than cast iron.

50.9 LOCAL METHOD OF EXTRACTION OF IRON

1. Arc Method. This is the modern method used in GHANA. This method involves smelting **scrap iron** at a high temperature. Scrap iron is any discarded iron material. The high temperature is provided by an electric arc produced when a pair of high voltage oppositely charged electrodes are brought near each other.

2. In a typical local production of iron, a mixture of iron ore, oyster shells and charcoal or palm kernel shells is loaded into a kiln or furnace. Fire is then set at the bottom of the furnace. Bellows are used to blow air continuously through the lower parts of the furnace.

The charcoal burns to produce CO_2.

$$C(s) + O_2(g) \longrightarrow CO_2(g)$$

As the CO_2 rises up the heap in the furnace, it reacts with more charcoal to produce carbon (II) oxide, CO.

$$CO_2(g) + C(s) \longrightarrow 2CO(g)$$

CO reduces iron ore, either FeO or Fe_2O_3, to iron metal.

$$FeO(s) + CO(g) \longrightarrow Fe(l) + CO_2(g)$$

$$Fe_2O_3(s) + 3CO(g) \longrightarrow 2Fe(l) + 3CO_2(g)$$

The oyster shells contain limestone which is a source of $CaCO_3$. When heated, $CaCO_3$ decomposes producing calcium oxide and CO_2.

$$CaCO_3(s) \longrightarrow CaO(s) + CO_2(g)$$

The calcium oxide combines with silica (SiO_2) removing this impurity in the form of slag.

$$CaO(s) + SiO_2(s) \longrightarrow \underset{\text{Calcium trioxosilicate(IV)}}{CaSiO_3(s)}$$

The molten iron flows to the bottom of the furnace where it is collected.

50.10 CORROSION OF METALS

The process of slow conversion of metals into their undesirable compounds (usually oxides) by reaction with moisture and other gases present in the atmosphere is called ***corrosion of metals.***

For example,

(*i*) Iron when exposed to moist air, gets corroded and a layer of reddish brown flaky substance, known as rust, is formed on its surface. Rust is chemically hydrated iron (III) oxide, $Fe_2O_3.xH_2O$.

(*ii*) Copper metal on exposure to moist air gets coated with greenish white powdery substance which is basic copper carbonate, $CuCO_3.Cu(OH)_2$.

(*iii*) Silver on exposure to air, loses it shine and its surface becomes black due to formation of silver sulphide, Ag_2S.

The metals occupying higher position in the activity series undergo corrosion rapidly. Only a few metals such as gold, platinum, palladium, etc., which are at the bottom of the activity series, are resistant to corrosion.

Sometimes corrosion of metals is an advantage because it prevents the metal underneath from further corrosion. For example, aluminium when exposed to air becomes coated with a layer of aluminium oxide which protects the metal underneath from further corrosion.

Now let us study corrosion of iron (rusting) in some detail:

RUSTING OF IRON

The corrosion of iron with moist air is known as ***rusting***. When an iron object is exposed to air in the presence of moisture, it gets corroded and a reddish brown flaky substance, which is known as rust, is formed. Rust is hydrated iron (III) oxide ($Fe_2O_3 . xH_2O$). Unlike the oxide layers of other metals such as aluminium and chromium, rust does not stick to the surface and does not protect the metal from further corrosion. Rust once formed, causes more and more rusting until the whole of the metal is eaten up. Thus, rusting weakens structures that are made up of iron.

$$4Fe + 3O_2 \xrightarrow{\text{Moisture}} \underset{\substack{\text{Hydrated iron (III) oxide} \\ \text{(Rust)}}}{2Fe_2O_3 . xH_2O}$$

The two conditions necessary for the rusting of iron are:

1. Presence of moisture, and
2. Presence of oxygen.

Conditions for rusting can be investigated by the following simple experiment.

ACTIVITY 50.1	**Investigation of Conditions for Rusting**

Take three test tubes and label them as A, B and C. Add a few clean iron nails to each of them. To the tube A add tap water. To the tube B add anhydrous calcium chloride and cork it. To the tube C add boiled distilled water and then add about 1 cm^3 oil and cork it. Leave the test tubes as such for a few days.

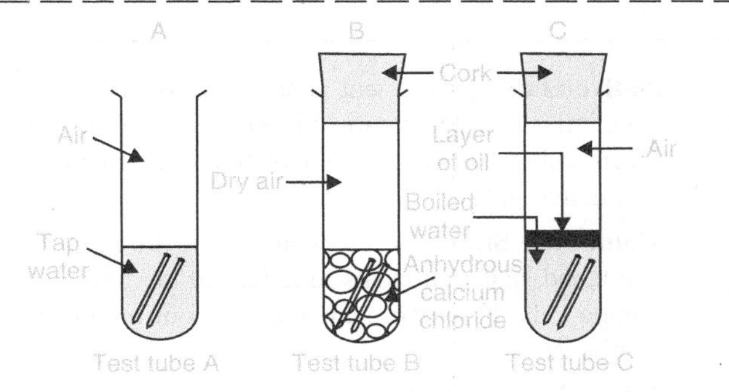

Fig. 50.6. Investigation of conditions for rusting.

What do you abserve?

After a few days it is observed that the nails in the tube A rust but the nails in the tubes B and C do not rust. In tube A both air and water are present. In tube B nails are exposed only to air while in the tube C nails are exposed to water but the layer of oil prevents the air from dissolving in water. Thus, presence of water and air both is essential for rusting.

The following factors further catalyse the process of rusting.

1. *Presence of carbon dioxide,*
2. *Presence of acids,*
3. *Presence of impurities in the iron.*

Due to rusting, iron object loses its strength.

Rusting can be prevented by alloying iron with other metals and non-metals. For example, stainless steel is an alloy of iron which contains about 0.05% carbon, 8% nickel and 18% chromium. It is quite hard and is resistant to corrosion.

Add To Your Knowledge

Anodising is a process of forming a thick layer of aluminium oxide on the surface of aluminium so as to protect it from corrosion.

Aluminium which has been exposed to the atmosphere is covered with a thin layer of aluminium oxide, Al_2O_3 which protects the metal from further corrosion. In order to protect the aluminium even more, it is possible to increase the thickness of the oxide layer to about 10^{-5} m by anodising. The aluminium is anodised by making it the anode during electrolysis of dilute sulphuric acid. Oxygen, released at the anode combines with the aluminium and thickens the oxide layer.

The oxide layer at this stage readily absorbs dyes. By carrying out the electrolytic anodising process in the presence of dyes, the anodised material can be coloured attractively.

SUMMARY

- **Native State.** An element is said to exist in a native state if it is found in nature in its elementary form. Less reactive elements such as gold, platinum, etc., are found in a native state.
- **Combined State.** An element is said to exist in a combined state if it exists in nature in the form of its compounds. Reactive elements occur in nature in a combined state.
- **Minerals.** A mineral is a naturally occurring material obtained by mining from the earth's crust that contains metal in its native state or combined state.
- **Ore.** A mineral from which metal can be extracted conveniently and economically is called an ore.
- **Gangue or Matrix.** The earthly and silicious impurities associated with the ore are known as gangue or matrix.
- **Metallurgy.** The process of extraction of pure metal from one of its ores is known as metallurgy.
- **Concentration.** *The process of removal of earthly impurities from the pulverized ore is called **concentration** or **benefication** of ore.* This is also sometimes called **dressing** of ore.
- **Calcination.** *It is the process of heating the ore in a limited supply of air below its melting point.* This process is employed for *oxide ores* and *carbonate ores.*
- **Roasting.** *It is the process of heating the ore in excess supply of air below its melting point.* This process is employed for *sulphide ores.*
- **Smelting.** In this process, the roasted or calcined ore is mixed with a suitable quantity of coke or charcoal (which act as a *reducing agent*) and is heated to a high temperature above its melting point. During reduction, an additional reagent is also added to the ore to remove the impurities still present in the ore. This additional reagent is called **flux.** *Flux combines with the impurities to form a fusible product called* ***slag.***
- **Reduction with Aluminium.** *The process of reduction of oxides with aluminium is called* **aluminothermy.**
- **Electrolytic Reduction.** The highly electropositive metals such as alkali metals, alkaline earth metals and aluminium are extracted by electrolysis of their fused salts.
- **Liquation.** This process is applied for purification of metals having low melting points, such as tin and lead.
- **Distillation.** Metals having low boiling points, such as mercury and zinc, can be purified by distillation.
- **Electro-refining.** Metals such as copper, silver, aluminium and gold are refined by process of electrolysis. In this method impure metal is made anode, while a thin strip of pure metal is made cathode and solution of some salt of the metal is used as electrolyte.
- Gold occurs in native form in alluvial sand or gravel.
- Gold is extracted by hydrometallurgy.
- **Aluminium** is the most abundant metal in the earth's crust.

  Its most important ore is **bauxite** ($Al_2O_3 . 2H_2O$)
- Bauxite is concentrated by a leaching process. Aluminium is obtained by electrolysis of pure alumina dissolved in molten cryolite. The process is known as **Hall-Heroult process.**
- ***Iron is the second most abundant metal in the earth's crust***
- The important ores of iron are:

  (*i*) **Haematite, Fe_2O_3** (red oxide of iron)

  (*ii*) Magnetite, Fe_3O_4 (Magnetic oxide of iron).

EVALUATE YOURSELF

I. Objective Type Questions

Select the most appropriate choice from the options given as (a), (b), (c) and (d) after each question:

1. The impurities associated with the ore after mining are collectively called

   (*a*) flux (*b*) slag
   (*c*) minerals (*d*) gangue.

2. An ore after levigation is found to have acidic impurities. Which of the following can be used as flux during the smelting operation?

   (*a*) H_2SO_4 (*b*) $CaCO_3$
   (*c*) SiO_2 (*d*) Both $CaCO_3$ and SiO_2.

3. In which of the following minerals is aluminium not present?

   (*a*) Fluorspar (*b*) Cryolite
   (*c*) Feldspar (*d*) Mica.

4. The process in which metal oxide is reduced to metal by aluminium is called

   (*a*) smelting (*b*) aluminothermy
   (*c*) hydrothermy (*d*) no specific name.

5. Which of the following metals can be extracted by smelting?

   (*a*) Aluminium (*b*) Magnesium
   (*c*) Iron (*d*) None of these.

6. In the froth-floatation process for benefication of the ores, the ore particles float because

   (*a*) they are light
   (*b*) their surface is not easily wetted by water
   (*c*) they bear electrostatic charge
   (*d*) they are insoluble.

7. Which of the following benefication processes is used for the mineral bauxite?
   (*a*) Froth floatation (*b*) Leaching
   (*c*) Liquation (*d*) Magnetic separation.
8. Heating an ore in the absence of air below its melting point is called
   (*a*) leaching (*b*) roasting
   (*c*) smelting (*d*) calcination.

II. Fill in the Blanks

9. Complete the following sentences by supplying appropriate words:
   (*i*) Froth floatation process is generally employed for ...... ores.
   (*ii*) The most abundant metal in the earth's crust is ......
   (*iii*) The purification of crude metal is referred to as ....
   (*iv*) The process of reduction of oxides by aluminium is known as ....
   (*v*) The process of removal of gangue from ore is known as ....
   (*vi*) Aluminium is obtained from Al_2O_3 by ...... reduction.
   (*vii*) In the metallurgical process for electrorefining of the metal, the anode is made of ...... metal.
   (*viii*) Slag is a product formed during smelting by combination of ...... and impurities.

III. Discussion Questions

10. (*a*) Name the three most abundant metals in the decreasing order of abundance.
    (*b*) Name any two metals which exist in a native state.
    (*c*) Define the following:
    (*i*) mineral (*ii*) ore
    (*iii*) gangue (*iv*) metallurgy.
11. What is the significance of leaching in the extraction of aluminium?
12. What type of ores are concentrated by hydraulic washing?
13. Which process is generally used for the benefication of sulphide ores?
14. An ore has impurities which are attracted by a magnet. Suggest a process for its benefication.
15. What is the difference between calcination and roasting?
16. What name is given to a carbon reduction process for extracting a metal?
17. When is electrolytic reduction applied for getting metal?
18. What is flux and in which operation is it used?
19. Explain the terms: *mineral* and *ore*. Suggest the formula of the following minerals:
    (*i*) Bauxite (*ii*) Haematite
    (*iii*) Magnetite.
20. Describe the froth floatation process.
21. Describe the magnetic separation method for the concentration of ores.
22. What are ores? How do they differ from minerals? Justify the statement that 'all the ores are minerals but all the minerals are not ores'.
23. What is meant by leaching? Give reactions involved during the leaching of alumina from bauxite ore.
24. Describe the process of extraction of gold. What are the important uses of gold?
25. Describe how pure aluminium oxide is obtained from its ore.
26. Explain why aluminium, though an electropositive metal, finds extensive use as a structural material.
27. Give important uses of aluminium.
28. What is magnalium?
29. Give the name and formula of the flux used during the extraction of iron.
30. What is difference between pig iron and cast iron? How is pig iron converted into cast iron?
31. Name the most pure form of iron. What is the percentage of carbon in it?
32. Give equations for the reactions taking place in different regions of the blast furnace during extraction of iron from haematite.
33. Name any two ores of iron.
34. What are the components of stainless steel?
35. What is rust? List two ways by which rusting of iron can be prevented.
36. Explain why the surface of some metals acquires a dull appearance when exposed to air for a long time.
37. What are the factors which catalyse the process of rusting? Explain, why rusting is considered harmful?
38. Which metals do not corrode easily?
39. Define corrosion. Name three metals which when exposed to air gets corroded. Identify the substance formed after corrosion of these metals.

Answers

I. Objective Type Questions

1. (*d*)	**2.** (*b*)	**3.** (*a*)
4. (*b*)	**5.** (*c*)	**6.** (*b*)
7. (*b*)	**8.** (*d*).	

II. Fill in the Blanks

9. (*i*) sulphide (*ii*) aluminium
(*iii*) refining (*iv*) aluminothermy
(*v*) concentration (*vi*) electrolytic
(*vii*) impure (*viii*) flux.

SECTION-1

UNIT-51

EXTRACTION OF CRUDE OIL AND PETROLEUM PROCESSING

Specific Objectives

After studying this unit, the student should be able to:

- identify the chemical elements and compounds found in crude oil
- outline the formation of crude oil from biological sources
- demonstrate understanding of the classification of crude oil
- describe how crude oil is extracted from an oil well
- describe the fractional distillation of crude oil
- explain cracking and reforming of organic compounds
- outline the uses of the fractions obtained from crude oil distillation
- outline the sources and uses of petrochemicals
- explain octane number and its importance to the petroleum industry.

Petroleum is also called *rock oil* (*petra* = rock, *oleum* = oil). It is a dark coloured oily liquid which occurs at various depths below the surface of the earth. The oil derived from oil wells is generally called **mineral oil** or **crude oil**.

51.1 COMPOSITION OF PETROLEUM

Petroleum is essentially a mixture of hydrocarbons particularly ***alkanes***, ***cycloalkanes*** and ***aromatic hydrocarbons***. The tarry residue which is left when the volatile components have been boiled off, consists of ***resins*** and ***asphaltenes***. This non-volatile component contains very large organic molecules. Apart from hydrocarbons it may also contain compounds containing ***oxygen***, ***nitrogen*** and ***sulphur***. The exact composition of crude oil varies from one place to another and also from one oil field to another.

51.2 FORMATION OF CRUDE OIL

According to a modern view, crude oil has been produced as a result of bacterial decomposition of the remains of animals and plants (*zooplankton, algae,* etc.) under high pressure which got buried under the sea, millions of years ago. The hydrocarbons formed by the decay of microorganisms rise through porous rocks until they are trapped by impervious rocks. If the region is uplifted during the passage of time, the oil trap will be found to lie under land. Oil in these deposits generally floats over salt water. The oil in the petroleum field is generally covered with a gaseous mixture known as **natural gas.** Natural gas mainly contains methane, ethane, propane and butane.

51.3 CLASSIFICATION OF CRUDE OIL

I. SWEET AND SOUR CRUDE OIL

On the basis of sulphur content crude oil is classified as sweet crude oil or sour crude oil.

1. **Sweet Crude Oil.** Petroleum is considered sweet if it contains less than 0.5% sulphur. Sweet crude oil contains small amounts of hydrogen sulphide and carbon dioxide. The term sweet originates because the low level of sulphur provides the oil with a mildly sweet taste and pleasant smell.
2. **Sour Crude Oil.** Petroleum is considered sour if it contains more than 0.5% sulphur.

II. LIGHT AND HEAVY CRUDE OIL

On the basis of wax content and viscosity, crude oil is classified as light or heavy crude oil.

1. **Light Crude Oil.** It is the crude oil with low wax content. It is less viscous.
2. **Heavy Crude Oil.** It is the crude oil with higher wax content. It is more viscous and therefore is more difficult to transport and pump.

Do You Know?

- Sweet crude oil commands a higher price than sour oil because it has fewer environmental problems and requires less refining.
- Light crude oil is more desirable than heavy crude oil because it produces higher yields of gasoline, kerosene and high quality diesel.

III. CLASSIFICATION ON THE BASIS OF API GRAVITY

The **American Petroleum Institute gravity or API gravity**, is a measure of how heavy or light a petroleum liquid is compared to water. If API gravity of petroleum is greater than 10, it is lighter and floats on water, if less than 10, it is heavier and sinks. API gravity is used to compare the relative densities of petroleum liquids.

For example, if one petroleum liquid floats on another and is therefore less dense, it has a greater API gravity.

Crude oil is classified as light, medium or heavy according to its measured API gravity.

1. **Light Oil** has an API gravity higher than 31.1° API.
2. **Medium Oil** has an API gravity between 22.3° API and 31.1° API.
3. **Heavy Oil** has an API gravity below 22.3° API.

Do You Know?

- Crude oil with an API gravity between 40° and 45° commands the highest price.
- Crude oil with API gravity less than 10° API is referred to as extra heavy oil or bitumen.

IV. CLASSIFICATION OF CRUDE OIL BY GEOGRAPHIC LOCATION

The oil industry also classifies crude oil by the geographic location of its origin. Some classes of crude on this basis are given below:

1. **West Texas Intermediate (WTI)** is the reference used for North American Crude oil. It is very high quality sweet light oil.
2. **Brent Crude or Brent Blend** is the benchmark used for crude oil from the North Sea. Brent Blend is a light crude oil though not as light as WTI.
3. **Dubai-Oman** is the benchmark used for the crude oil from Middle East
4. **OPEC Basket** is the reference used for crude oils from various OPEC (The Organisation of the Petroleum Exporting Countries) countries.

Thus, crude oils from different sources have different qualities and hence command different prices.

Now, can you answer *why there are differences in prices, for the same quantity of crude oil from different sources?*

51.4 MINING OF PETROLEUM

Petroleum is obtained by drilling holes in the earth's crust and sinking pipes into it as shown in Fig. 51.1.

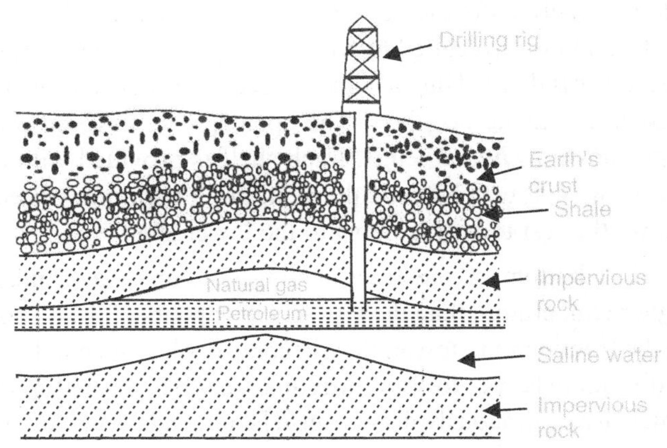

Fig. 51.1. Mining of petroleum.

When the pipe reaches the oil deposit, natural gas comes out with a great pressure and for a time the crude oil comes out by itself due to the pressure of the gas over oil. This is the primary recovery process. Gradually the pressure of the gas decreases. At this stage, the crude oil is extracted with the help of lift pumps. The is the secondary recovery process.

In West Africa, Nigeria and, to a lesser extent Cote d' Ivoire are the only countries that produce oil.

Ghana meets its oil requirements by importing refined as well as crude oil. Much of the oil is imported from Nigeria.

51.5 DISTILLATION OF CRUDE OIL

The crude petroleum, obtained by mining, is a dark-coloured viscous liquid having an unpleasant smell due to the presence of sulphur containing compounds. It also contains impurities of sand and brine. Since crude oil is a mixture of hydrocarbons, it cannot be used as such as a fuel or as a base compound for producing other useful substances. The crude oil is subjected to refining by continuous fractional distillation. *This continuous fractional distillation of crude oil into different useful fractions and the removal of undesirable impurities is called **refining** or **processing of petroleum**.* The different fractions, thus, obtained contain hydrocarbons of different chain lengths.

Solids such as sand are separated from the impure crude oil by filtration. The filtrate is allowed to stand in a reservoir for sometime. The oil separates as the upper layer and is then pumped through pipes to a refinery (such as the Tema Oil Refinery in Ghana).

The crude oil is heated to 400°C in coiled pipes in a gas heated furnace and the vapours, thus, obtained are introduced into the fractionating tower. The fractionating tower is made of steel and is also called ***bubble tower*** (shown in Fig. 51.2). The tower is divided into a number of compartments by means of shelves having openings. The openings are covered with caps known as ***bubble caps***. Each shelf is provided with an overflow pipe which keeps the liquid to a certain level and allows the rest to trickle down to the lower shelf.

The bottom of the fractionating column is at a much higher temperature than the top. As the vapours of the oil rise up the fractionating tower, they become cooler and condense in different shelves to form different fractions of the liquids which are continuously drawn out. The uncondensed vapours pass out of the tower at the top. The fractions, thus, obtained consist of a mixture of hydrocarbons having close ranges of boiling point. The actual number of fractions and their boiling point ranges depend upon their commercial uses and the source of petroleum.

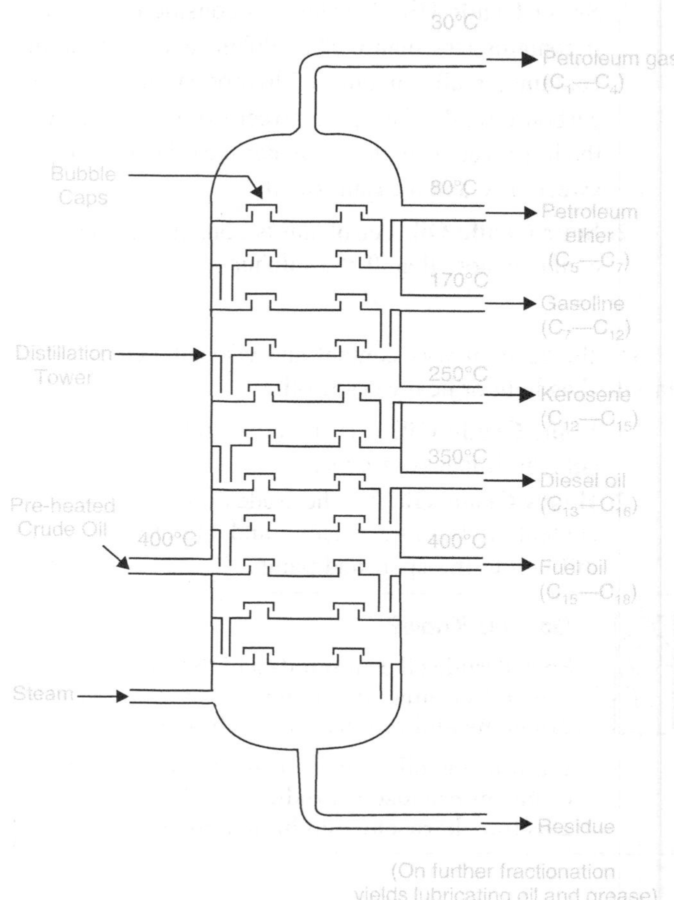

Fig. 51.2. Refining of petroleum.

The various components obtained from the fractional distillation of petroleum, from the base upwards are: *asphalt, lubricating oil, paraffin wax, fuel oil, diesel, kerosene, gasoline (petrol)* and *petroleum gas*. Except for asphalt, lubricating oil and paraffin wax, all other fractions readily burn and are used as fuels. Asphalt, lubricating oil and paraffin wax are obtained by further fractionation of residual oil which collects at the bottom of the fractionating tower. *The major fractions obtained during fractional distillation along with their uses and composition are given in Table 51.1.*

Table 51.1. Typical Fractionation of Petroleum and Uses of Various Fractions

Fraction	*Boiling Range (°C)*	*Uses*
1. Petroleum gas (C_1—C_4)	Below 30°	*Gaseous fuel, LPG, production of carbon black, hydrogen and carbon, monoxide (used to make) ammonia and methanol)*
2. Petroleum ether (C_5—C_7)	30—80°	*As solvent in dry cleaning.*
3. Gasoline (C_7—C_{12})	80—170°	*Motor fuel.*
4. Kerosene (C_{12}—C_{15})	170—250°	*Household fuel, illuminant fuel, jet engine fuel.*
5. Diesel oil (C_{13}—C_{16})	250—350°	*Fuel for diesel engines, in cracking.*
6. Fuel oil (C_{15}—C_{18})	350—400°	*Furnace fuel in industries*
7. Lubricating oils (C_{16} onward)	above 400°	*Lubrication of machinery.*
8. Paraffin wax (C_{20} onward)	non-volatile solids	*Candles, water-proofing, vaseline, fabrics.*
9. Petroleum coke or asphalt		*Artificial asphalt, fuel, electrodes.*

PETROLEUM GAS

Petroleum gas is a mixture of butane, propane and ethane. ***The main constituent of petroleum gas is butane.*** Petroleum gas burns readily in the presence of air with a blue flame, producing a large amount of heat. Butane, propane and ethane are gases under ordinary pressure. However, butane can be easily liquefied under pressure. *The petroleum gas which has been liquefied under pressure is called Liquefied Petroleum Gas (LPG).* Petroleum gas is supplied in liquid form so that a cylinder of even small volume may contain an appreciable amount of the gas. A strong smelling substance ***(Ethyl mercaptan,*** C_2H_5SH***)*** is added to LPG gas cylinders to help in the detection of gas leakage. The gas used for domestic cooking is called Liquefied Petroleum Gas (LPG) because it is present in liquid form in the cylinders.

Do You Know?

Asphaltenes

The non-volatile tarry residue left after distillation of residue from the fractionating tower contains compounds with large molecules. These are *resins* and *asphaltenes*.

Asphaltenes consist primarily of carbon, hydrogen, nitrogen, oxygen and sulphur. Asphaltenes are defined operationally as *n*-heptane insoluble and toluene soluble component of a carbonaceous material such as crude oil. They constitute the non-volatile, high molecular mass fraction of petroleum. Moreover, since they are insoluble in heptane, they remain in solid form even in the crude. Asphaltenes are present in dispersed form in the crude oil. They are stabilised by resins which keep them dispersed in oil.

Asphaltenes create many production problems because of their depositional effect in the production equipment. Asphaltenes in the form of distillation products from refineries are used as '*tar-mats*' on roads.

SOLVED EXAMPLES

Example 51.1. *The boiling points of three components A, B and C of petroleum are 120°C, 70°C and 250°C respectively. If the mixture of three is fractionally distilled, which of the three will be obtained at the bottom of the distillation column?*

Solution. During fractional distillation of a mixture, the component having the highest boiling point is obtained at the bottom of the column. In the given mixture, the component C has the highest boiling point and hence will be obtained at the bottom of the distillation column.

Example 51.2 *A mixture contains three components P, Q and R with boiling points 150°C, 250°C and 350°C respectively. If this mixture is subjected to fractional distillation, in which order would you get these substances?*

Solution. During fractional distillation of a mixture, the component having the highest boiling point is obtained at the bottom of the column whereas the component having the least boiling point is obtained at the top of the column. In the given mixture the component R, having the highest boiling point will be obtained at the bottom of the column. The component Q having intermediate boiling point will be obtained in the middle of the column while the component P, having the lowest boiling point, will be obtained at the top of the column. Thus, these substances will be obtained (from bottom to top) in the order R, Q and P.

ACTIVITY 51.1 **Petroleum fractions are highly inflammable**

Collect samples of petroleum ether, gasoline, kerosene oil and diesel oil.

- Study the ignition behaviour of these fractions by burning a little of each fraction.
- Observe the ease with which each fraction ignites and observe the flame.
- Observe the viscous nature of each fraction.

Why are the first fractions such as petroleum ether and gasoline easiest to ignite?

Why do the higher fractions such as kerosene and diesel burn with the smokiest flames?

Suggest a reason why the fractions become increasingly viscous.

51.6 NATURAL GAS

Natural gas is a naturally occurring mixture of simple aliphatic hydrocarbons containing 1–4 carbon atoms and a small amount of hydrogen. Among aliphatic hydrocarbons, it mainly contains methane CH_4 (up to 97%) with a little ethane (C_2H_6) propane (C_3H_8) and butane (C_4H_{10}). Natural gas occurs deep under the earth either alone or along with petroleum deposits. Thus, during petroleum mining natural gas is always obtained as a byproduct, but there are some wells that produce only natural gas. After extraction from wells, the gas is purified to remove impurities such as water and sulphur and then sent through pipelines for use.

USE OF NATURAL GAS

1. Natural gas is a near ideal ***fuel*** and is used as the fuel for domestic and industrial purposes.
2. Natural gas is used as fuel for automobiles in the form of ***compressed natural gas (CNG).***
3. Natural gas is a rich ***source of hydrogen*** gas which is needed in the manufacture of fertilizers.

   For obtaining hydrogen, natural gas is reacted with superheated steam in the presence of nickel as catalyst

$$CH_4 + H_2O \xrightarrow[\text{Catalyst}]{1100\text{ K}} CO + 3H_2$$

51.7 CRACKING AND REFORMING OF ORGANIC COMPOUNDS

We know that the requirement of petrol or gasoline is increasing day by day and at the same time the availability of petroleum is falling short of our requirement. Therefore, in order to meet the ever-increasing demand of petrol and to overcome the crisis due to less availability of petroleum, scientists have been successful in developing some processes to convert other fractions of petroleum into gasoline. Let us familiarise ourselves with some of these processes.

1. CRACKING

It is a process of decomposing higher hydrocarbons of the petroleum fraction into lower hydrocarbons with low boiling points by strong heating.

The process involves the cleavage of carbon-carbon and carbon-hydrogen bonds and results in the formation of molecules of lower hydrocarbons. By this process, the higher boiling fractions of petroleum, like fuel oil, are converted into gasoline. Thus, cracking can be employed to improve the yield of gasoline. The types of hydrocarbons formed depend upon the conditions employed for cracking. In **thermal cracking** the higher hydrocarbons are heated to a high temperature, whereby lower hydrocarbons are formed. For example,

$$\underset{\text{Decane}}{CH_3(CH_2)_8CH_3} \xrightarrow{775\text{ K}} \underset{\text{Octane}}{C_8H_{18}} + \underset{\text{Ethylene}}{CH_2 = CH_2}$$

Similarly, octane can produce other hydrocarbons by cleavage of C—C bonds as shown below:

$$\underset{n\text{-Octane}}{C_8H_{18}} \xrightarrow{\text{cracking}} \underset{\text{Hexane}}{C_6H_{14}} + \underset{\text{Ethene}}{C_2H_4}$$

$$\underset{n\text{-Octane}}{C_8H_{18}} \xrightarrow{\text{cracking}} \underset{\text{Butane}}{C_4H_{10}} + \underset{\text{Ethene}}{C_2H_4} + \underset{\text{Methane}}{CH_4} + C$$

Cracking can also be carried out at lower temperatures by the use of suitable catalysts like silica and alumina. This type of cracking is referred to as **catalytic cracking.**

Cracking is generally employed for the production of good quality gasoline, from heavier fractions of petroleum.

2. REFORMING

Reforming is a process of conversion of straight-chain alkanes into ring molecules—cycloalkanes and arenes. The reaction is carried out in the presence of suitable catalyst like, ***palladium, platinum or nickel.*** The reaction involves *dehydrogenation*, *cyclisation* and even *isomerisation* in some cases. The reaction is generally carried out at high temperatures. For example, the conversion of *n*-hexane to benzene as shown below:

$$\text{Hexane} \xrightarrow[(-H_2)]{670\text{ K, Pt}} \text{Cyclohexane} \xrightarrow[(-3H_2)]{670\text{ K, Pt}} \text{Benzene}$$

Hexane Cyclohexane Benzene

Reforming converts low-grade gasoline to high-grade gasoline.

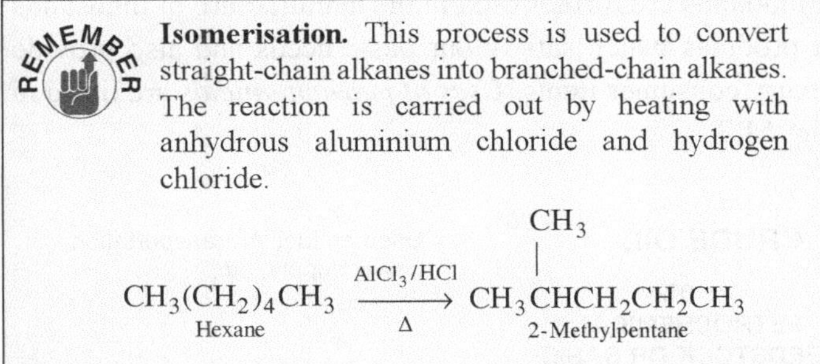

Isomerisation. This process is used to convert straight-chain alkanes into branched-chain alkanes. The reaction is carried out by heating with anhydrous aluminium chloride and hydrogen chloride.

$$\underset{\text{Hexane}}{CH_3(CH_2)_4CH_3} \xrightarrow[\Delta]{AlCl_3/HCl} \underset{\text{2-Methylpentane}}{CH_3CH(CH_3)CH_2CH_2CH_3}$$

51.8 QUALITY OF GASOLINE—OCTANE NUMBER

Gasoline is used as a fuel in the internal combustion engines of motor cars and aeroplanes. In such engines, a mixture of gasoline vapours and air is compressed by a piston within the cylinder of the engine and the compressed mixture is ignited by a spark from the *spark plug*. As the compressed fuel-air mixture burns, considerable expansion is caused. This results in the movement of piston and the transmission of the force to the wheel of the motor car or the propeller of the aeroplane. The efficiency of the engine depends upon the extent to which the fuel-air mixture is compressed at the time of ignition. The greater the compression, the greater is the efficiency. But the increase in compression beyond a certain point results in the sudden and irregular burning of fuel mixture which causes jerk against the piston and gives rise to violent metallic sound. *This rattling metallic noise produced due to irregular burning of fuel mixture in the cylinder* is known as **knocking.** Knocking reduces the power of the engine and also causes damage to the piston and the cylinder.

The maximum compression that can be attained in a given engine without any knocking taking place depends upon the composition of the gasoline used as fuel. Therefore, it is a matter of importance to know the relative *"anti-knock quality"* of the fuel. For this purpose, a scale involving the term **octane number** was introduced by *Edger* in 1929.

By carrying out tests on individual hydrocarbons, it has been found that **n-heptane** knocks maximum when used as a fuel. Thus, due to its very poor anti-knock quality, it has been assigned *octane number,* **zero.** On the other hand, **iso-octane** (2, 2, 4-trimethylpentane) was found to have the best anti-knock property. Therefore, it has been assigned *octane number* **100.**

$CH_3CH_2CH_2CH_2CH_2CH_2CH_3$; (*octane number* = 0)

n-Heptane

$CH_3-C(CH_3)_2-CH_2-CH(CH_3)-CH_3$; (*octane number* = 100)

iso-Octane

All other fuels are graded between these two extreme values (0–100) by comparing with suitable mixture of above mentioned compounds taken as standard. For example, if a fuel has octane number 85, it implies that it has same anti-knock quality as the mixture of 85 parts *iso*-octane and 15 parts *n*-heptane. Thus, **octane rating** of a fuel may be defined as *the percentage by volume of iso-octane in the mixture of iso-octane and n-heptane which has the same anti-knock qualities as the fuel under test.*

It may be noted that for aeroplanes the fuel used should have octane number 100, while for motor vehicles the fuel with octane number of 80 is rated as good fuel. The '*premium*' grade petrol used in Ghana has octane number of 91 or above.

Influence of Structure of Hydrocarbons on the Octane Number

A study of the octane numbers of different hydrocarbons has revealed that the value of the octane number depends upon the structure of the hydrocarbon.

- *Straight chain hydrocarbons have very low value of octane number. The value decreases further with increase in the length of the chain.*
- *Branching of the chain increase the value of the octane number.*
- *Cycloalkanes have higher value of octane number than the corresponding straight chain alkanes.*
- *Aromatic hydrocarbons have generally very high value octane numbers.*

Gasoline Additives or Anti-knocking Agents

In internal combustion engines, the problem of knocking can be checked either by *using a fuel of higher octane number* or by *adding certain compounds to gasoline which decrease knocking*. The compounds which are added to gasoline to improve its octane number are called **anti-knocking agents.**

Tetraethyl lead (TEL), $\mathbf{(C_2H_5)_4Pb}$ is added to gasoline (~0.01%) to reduce knocking. Gasoline containing tetraethyl lead is called **ethyl gasoline** or **leaded gasoline.**

Use of lead anti-knock agents in the vehicles causes lead pollution of the environment due to release of volatile lead compounds. Lead compounds, especially organic lead compounds, are highly toxic. They affect the central nervous system of human beings and also retard the mental development of their children. Many countries including Ghana have stopped using lead anti-knock agents because of lead pollution. The storage and use of leaded petrol has been banned in Ghana since January 2004. TOR started producing unleaded petrol in October 2003. To minimize knocking unleaded gasolines contain methyl *tert*-butylether (MTBE). It minimizes knocking and increases the oxygen content of gasoline making the fuel burn cleaner.

$$H_5C_2—\underset{\underset{C_2H_5}{|}}{\overset{\overset{C_2H_5}{|}}{Pb}}—C_2H_5 \qquad\qquad H_3C—O—\underset{\underset{CH_3}{|}}{\overset{\overset{CH_3}{|}}{C}}—CH_3$$

Tetraethyl lead (TEL) Methyl *tert*-butylether (MTBE)

Pollution problem. A gasoline-engine discharges various substances into the atmosphere. These are *carbon(IV) oxide, water, carbon(II) oxide, nitrogen oxides, unburnt hydrocarbons*, etc. Some of these substances are highly poisonous. Their emission into the atmosphere, poses a serious pollution problem to mankind. The leaded gasoline also produces some poisonous compounds of lead along with other above mentioned gases. The pollution problem is getting aggravated with increasing use of automobiles.

SOLVED EXAMPLE

Example 51.3. *A sample of gasoline has the same knocking characteristics as a mixture of iso-octane and n-heptane in the ratio 3 : 1 by volume. What is the octane number of the gasoline sample?*

Solution. Percentage of *iso*-octane (by volume) in the given sample of gasoline $= \frac{3}{4} \times 100 = 75\%$

Hence, the octane number of the given sample of gasoline is 75.

51.9 SOURCES AND USES OF PETROCHEMICALS

Petroleum serves as a source of a large number of chemicals. **These chemicals obtained from petroleum are known as petrochemicals.** Thousands of organic compounds are made from the refined products of petroleum. These compounds are further used in the manufacture of thousands of products which satisfy our basic needs and also provide luxury consumer items. ***Uses of petrochemicals*** are listed in Fig. 51.3.

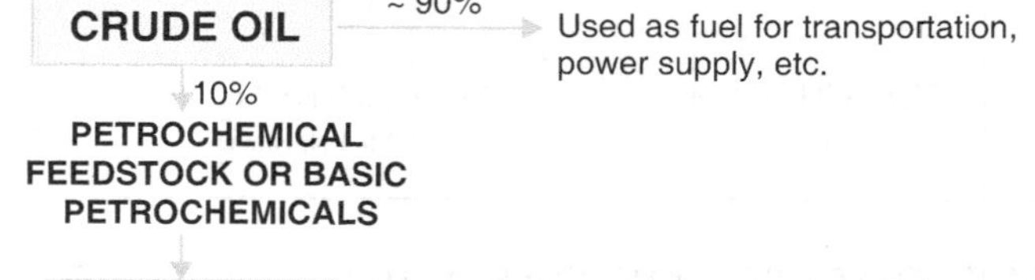

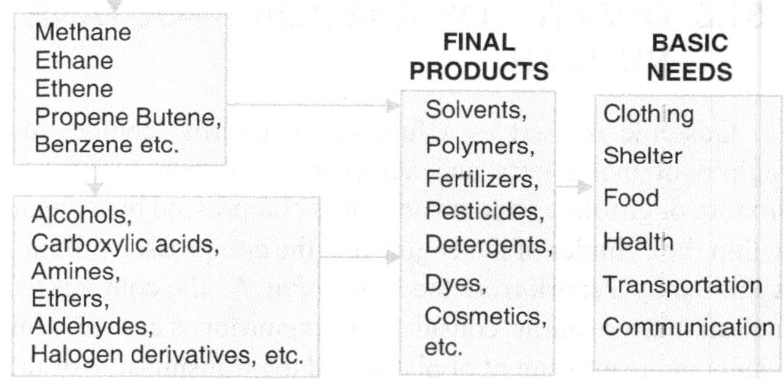

Fig. 51.3. Petrochemicals and their uses

Thus, the source of ethanol, ethane-1, 2-diol and ethanoic acid is ethene which in turn is obtained from petroleum.

Similarly, methanol, freon-12, chloroform and CCl_4 are obtained from methane

Group Discussion

Name three petrochemicals and state their sources and one use of each of them.

Some petrochemicals derived from methane and ethene are given below in the form of a flow sheet:

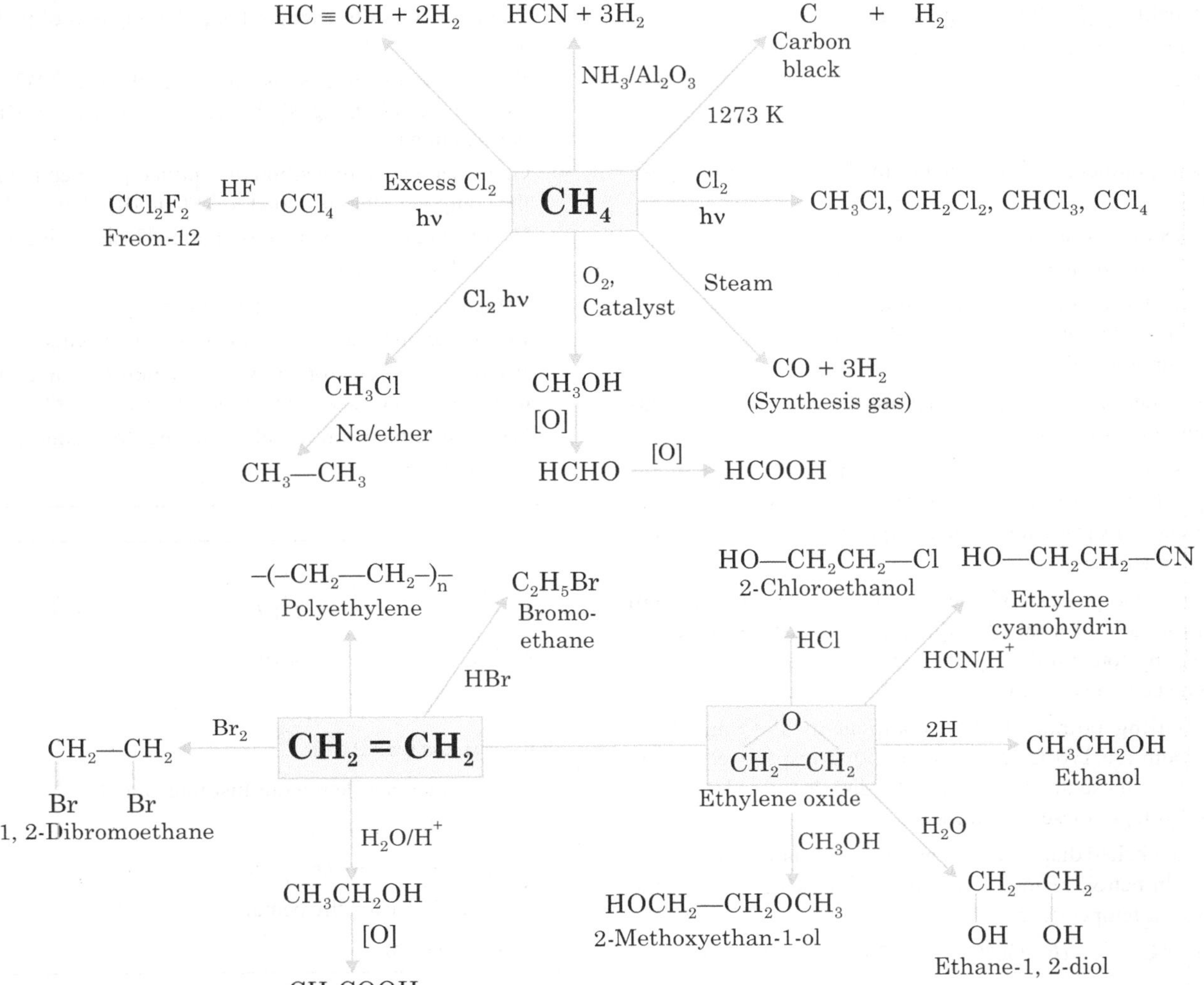

EVALUATE YOURSELF

I. Objective Type Questions

Select the most appropriate choice from the options given as (a), (b), (c) and (d) after each question:

1. Which of the following commands the highest price in the oil industry?
   (*a*) Light sweet crude oil
   (*b*) Heavy sweet crude oil
   (*c*) Light sour crude oil
   (*d*) Heavy sweet crude oil.
2. Light oil has API gravity
   (*a*) less than 31.1° (*b*) between 22.3 and 31.1°
   (*c*) less than 22.3° (*d*) higher than 31.1°.
3. Separation of crude oil into various fractions is possible because the fractions have different
   (*a*) melting points (*b*) densities
   (*c*) boiling points (*d*) viscosity.
4. Which one of the following is not a petrochemical?
   (*a*) methanol (*b*) ethanol
   (*c*) benzene (*d*) glass.
5. Which of the following fractions of petroleum has the highest boiling range?
   (*a*) Kerosene (*b*) Fuel oil
   (*c*) Gasoline (*d*) Petroleum ether.

II. Fill in the Blanks

6. Complete the following sentences by supplying the appropriate words:
   (*i*) The percentage of sulphur in sweet crude oil is ....
   (*ii*) The full form of API gravity is ....

(*iii*) API gravity of heavy oil is ....
(*iv*) Quality of gasoline is expressed in terms of ....
(*v*) IUPAC name of *iso*-octane is ....
(*vi*) Octane number of heptane is ....

III. Discussion Questions

7. What is petroleum? How was it formed? What is its composition?
8. Discuss the classification of crude oil by
   (*i*) sulphur content (*ii*) API gravity.
9. Explain the classification of crude oil as light and heavy. Explain why light and sweet crude oils are preferred to heavy and sour crude oils.
10. How is petroleum mined? What are the two main elements present in petroleum?
11. Discuss the classification of crude oil on the basis of the geographic location of its source. Explain why there are differences in price for the same quantity of crude oil from different sources.
12. Describe the process of fractional distillation of petroleum with the help of a labelled diagram of a distillation tower. Name any four products obtained. State the temperature at which each one of them is obtained.
13. The boiling points of three components P, Q and R of petroleum are 170°C, 250°C and 350°C respectively. If the mixture of these three is fractionally distilled, in which order would you get these substances?
14. Draw a labelled diagram of a fractional distillation tower used to obtain petrol. Name the products obtained at each of the following temperatures:
    (*a*) 170°C (*b*) 250°C (*c*) 360°C (*d*) 450°C
15. Out of gasoline and kerosene oil which is more volatile?
16. Point out one difference for the following pair of fuels: petrol and kerosene oil.
17. What is petroleum gas and how is it obtained? What are the uses of petroleum gas? What is the main constituent of petroleum gas?
18. Give chemical composition of liquified petroleum gas. Name the compound that is added to LPG to detect its leakage.
19. How is domestic cooking gas obtained? Why is it called LPG?
20. Write short notes on:
    (*i*) cracking (*ii*) reforming
    (*iii*) octane number (*iv*) anti-knock agents.
21. What is the relation between structures of constituent hydrocarbons of gasoline on its octane number?
22. What are petrochemicals? Name three petrochemicals and state their sources.

Answers

I. Objective Type Questions

1. (*a*) **2.** (*d*) **3.** (*c*)
4. (*d*) **5.** (*b*).

II. Fill in the Blanks

6. (*i*) less than 0.5%
(*ii*) American Petroleum Institute Gravity
(*iii*) below 22.3°
(*iv*) octane number
(*v*) 2, 2, 4-trimethylpentane
(*vi*) 0 (zero).

SECTION-1

UNIT-52

ENVIRONMENTAL POLLUTION

Specific Objectives

After studying this unit, the student should be able to:

- explain pollution
- describe natural air pollution
- describe human activities that cause pollution
- describe atmospheric events such as acid-rain, greenhouse effect and ozone depletion
- describe the effects of air pollution
- describe the sources of water pollution
- describe the sources of land pollution.

52.1 INTRODUCTION

The study of various phenomena related to the *environment* has gained prime importance in last two decades. *Environmental pollution* has become one of the most important issues of global concern. In the recent years, the terms *ozone depletion, greenhouse effect, acid rains,* etc., have become hot topics of discussion among socially and educationally advanced people. Whenever we talk about these hot problems, we tend to curse industrialisation and related developments in transportation, refrigeration, air conditioning and the manufacture of all kinds of chemicals. Chemistry plays a significant role in our environment, which literally means surroundings. Large number of chemical phenomena are associated with the species present in the environment. *The branch of chemistry which deals with the study of various chemical phenomena occurring in the environment is called* **environmental chemistry.**

52.2 ENVIRONMENTAL POLLUTION

Environmental pollution, in general, may be described as the contamination of the environment with harmful wastes arising mainly from human activities. These activities release certain materials which produce pollution. The various systems of the environment that can be polluted are as follows:

(*i*) **Air.** *Air comprises a blanket of gaseous layer* that extends to a height of about 1600 km above the surface of the earth. It consists of gases like N_2, O_2, $H_2O(g)$ as *major components,* Ar, CO_2 as *minor components* and nobel gases, oxides of nitrogen (NO_x), oxides of sulphur (SO_x), and H_2, O_3 as trace components. This part of environment is called the **atmosphere** and it protects life on earth from harmful radiations coming from outer space and the sun.

(*ii*) **Water.** This comprises about 75% of the earth's surface and includes all the sources of water like oceans, rivers, lakes, glaciers, ground water, etc. This part is commonly called **hydrosphere.**

(*iii*) **Land.** It refers to the earth's solid crust containing the outer mineral cover like soil, rocks, mountains, etc. Soil is the upper most part of the earth's crust which contains weather rocks and organic matter. This part is commonly called **lithosphere.**

Environmental pollution may be defined as *direct or indirect changes as a result of human activities, which adversely influence the biological and non-biological equilibrium of the environment.* The environmental pollution may be caused by:

Chemical agents (gases, pesticides, particulates),

Physical agents (heat, noise, radiation) or

Biological agents (micro-organisms).

Pollution produces adverse effects on living as well as non-living materials. For example, it can:

- cause illness or even death in case of humans and animals.
- retard the growth of plants and may cause them severe injury.
- result in corrosion of metals, marble and decolourise paints.

Do You Know?

Some Episodes of Environmental Pollution

(*i*) About 4000 people died in England in 1952 due to respiratory diseases caused by the London smog.

(*ii*) In Japan, thousands of people died during the 1953–1960 period by the consumption of fish from the polluted Minamta Bay.

(*iii*) In 1984 about 3500 people died in Bhopal (India) due to leakage of methyl isocyanate vapours. Very large number of people are still suffering from respiratory, lung, eye and throat diseases.

(*iv*) Radioactive emission from the use of the nuclear weapons in Hiroshima and Nagasaki during the second world war resulted in heavy loss of human lives.

(*v*) Consumption of polluted wheat in Iraq resulted in the loss of about 500 human lives in 1972.

(*vi*) Acid rains in Greece and Italy caused damage to historical marble statues. Sulphur dioxide emissions are effecting the beauty of Tajmahal in Agra (India).

(*vii*) Thousands of people lost their lives in the Ukrain area of the Soviet Union due to world's worst nuclear disaster.

1. POLLUTANT

Any substance or species produced either by a natural source or by human activity, which produces adverse effect on the environment is called a **pollutant**. Some common examples of pollutants are carbon monoxide, oxides of sulphur (SO_2, SO_3), oxides of nitrogen, pollens, dust particles, etc.

It may be noted that the name pollutant is given to a substance which is already present in nature but human activities has increased its concentration beyond the desirable limit so that it becomes harmful to the natural environment. Quite often pollutants are thought to be highly toxic substances only, but even the substances which are normally considered harmless can cause pollution if they are present in undesirable concentrations or in the wrong places at the wrong time. For example, nitrates are added to the soil to increase plant growth but excessive concentration of nitrates in drinking water can be toxic.

2. CONTAMINANT

A substance which does not occur in nature but is introduced by human activity into the atmosphere is called **contaminant.** It may or may not be harmful to the living organism or the non-living component.

3. SOURCE

The site from which the pollutants or contaminants originate is called **source.** The source is particularly important because its knowledge helps us to develop the methods to eliminate pollutants.

4. SINK

The material or medium which consumes or interacts with a long lived pollutant is called **sink.** For example, a marble wall acts as a sink for atmospheric sulphuric acid because of the reaction:

$$CaCO_3 + H_2SO_4 \longrightarrow CaSO_4 + H_2O + CO_2$$

Similarly, the oceans act as sink for atmospheric carbon dioxide and other water soluble gases. Ground water and subsoil water also act as sinks for pesticides used in agriculture.

5. RECEPTOR

Anything that is affected by the pollutants is called **receptor.** For example, human beings are the receptor of photochemical smog which cause irritation in the eyes and also breathing problems.

6. THRESHOLD LIMIT VALUE (TLV)

This indicates the permissible limit of a pollutant in the atmosphere to which a healthy worker is exposed during 8 hours a day or 40 hours a week for a life time without any adverse effects. TLVs are determined by experimentation on animals, by use of medical knowledge, epidemiological surveys and environmental studies.

For example, TLV of CO is 40 ppm and that of CO_2 is 5000 ppm. But TLV for the poisonous gas phosgene is only 0.1 ppm.

The environmental pollution process may be summarized as follows. A pollutant originates from source and gets transported by air or water or it is dumped by humans on land. Some of the pollutants may be assimilated (absorbed) or chemically changed by the environment while the rest build up their concentrations slowly and gradually and become harmful to the environment. This has been shown in the flow sheet in Fig. 52.1.

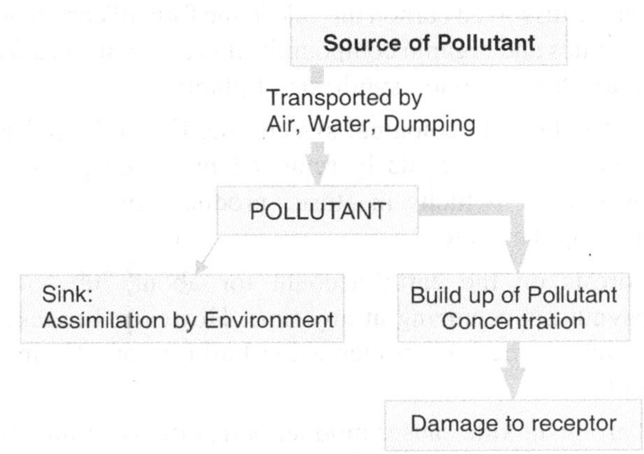

Fig. 52.1. Flow sheet of environmental pollution process

52.3 AIR POLLUTION

Air in the atmosphere is never pure and clean. It contains many chemical substances which arise because of *natural* or *human activity*. These substances include:

- *gases* like CO, CO_2, SO_2, NO_2, H_2S, NO, hydrocarbons (C_xH_y) along with many other volatile organic compounds and
- *suspended particulate* matter such as dust, smoke, fumes, and also some *radioactive material.*

The presence of these substances in low concentration does not cause any major problem but when their concentration becomes high they produce harmful effects and become *pollutants*. For example, SO_2 is a pollutant when its concentration in the atmosphere exceeds the normal limit of 2×10^{-4} ppm. Similarly, CO becomes a pollutant when its concentration exceeds the normal limit of 1×10^{-4} ppm.

Air pollution may thus, be referred to as *any condition in which the chemical substance already present in air or those which are added or formed in air, gradually build up their concentration in the atmosphere to such an extent that it becomes undesirable and produce harmful effects on man and his environment.*

Air pollution can be caused both by natural as well as human activities.

NATURAL ACTIVITIES

Three main **natural activities** responsible for air pollution are:

- ***Wild fires.*** These release oxides of carbon CO, CO_2 into the air through combustion.
- ***Volcanic Eruptions.*** The volcanic eruptions release CO_2, SO_2, SO_3, HCl, H_2S, water vapours, particulate matter, etc., into the air.
- ***Duststorms.*** Windblown dust or windstorms also cause the release of particulate matter in the air.

HUMAN ACTIVITIES

The **human activities** which cause air pollution are:

- ***Burning of octane, fossil fuels like*** coal, oil in limited supply of oxygen to release CO and $H_2O(g)$.
- ***Vehicular exhausts*** release CO, oxides of nitrogen.
- ***Combustion of sulphur rich ores*** which release oxides of sulphur in the air.

Let us discuss major air pollutants, their source, their harmful effects and the control.

52.4 MAJOR AIR POLLUTANTS

Major air pollutants can be divided into two categories, namely: *gaseous pollutants* and *particulate pollutants.*

A. GASEOUS POLLUTANTS

1. Carbon Monoxide (CO)

Carbon monoxide is a colourless, poisonous, lethal gas, which is one of the most serious air pollutants.

Sources:

(*a*) ***Natural activities.*** Major amount of CO in the atmosphere comes from natural sources as given below:

- From forest fires and volcanic activities.
- From combustion of methane (which is produced by the bacterial decay of organic matter) into CO.
- From synthesis and decay of chlorophyll (a green pigment of plants).

(*b*) ***Human activities.***

- *Incomplete combustion of carbonaceous matter in an automobile engine and also in defective furnaces. This produces major chunks of CO in the air.* For example, complete combustion of octane (C_8H_{18}) requires 12.5 mole of oxygen per mole of octane

$$C_8H_{18} + 12.5\ O_2 \longrightarrow 8CO_2 + 9H_2O$$

(*Complete combustion*)

However, if we have only 10 mol of O_2 then we shall get 3 mole of carbon dioxide and 5 mol of CO.

$$\underset{\text{Octane}}{C_8H_{18}} + 10O_2 \longrightarrow \underset{(\textit{Incomplete combustion})}{3CO_2 + 5CO + 9H_2O}$$

- *Incomplete combustion of fossil fuels, agricultural and slash matter.*
- *Reaction of carbon dioxide and carbon at high temperatures in industrial processes*

$$CO_2 + C \longrightarrow CO$$

- *Dissociation of carbon dioxide at high temperatures*

$$2CO_2 \longrightarrow 2CO + O_2$$

Toxic effects of CO

Toxic effect of CO arises from its capacity to bind itself to the haemoglobin present in red blood corpuscles (RBC) of the blood more strongly (about 200 times) than oxygen. As a result of this, the amount of haemoglobin available in the blood for the transport of oxygen to the body cells decreases. Thus CO cause obstruction in the normal functioning of the blood which further produces disorders in normal metabolism due to less O_2 level. The presence of CO in blood can cause mental retardation, respiratory problems, muscular weakness, dizziness and even death depending on its concentration in the blood.

The **maximum permissible** concentration of CO in the atmosphere is **40 ppm** for an exposure of **6–8 hours.** Higher concentrations of CO or longer exposure may start causing adverse effects. At concentrations greater than 100 ppm difficulties in breathing starts which produces headache and/or dizziness. Concentrations greater than 750 ppm causes acute oxygen-starvation which is referred to an ***asphixiation*** or ***anorexia*** which leads to coma or even death.

Sinks of CO

Human activities release CO into the atmosphere and its concentration is estimated to double every 5 years. However, the actual increase in CO in the atmosphere is less because of some natural sinks.

(*i*) Soil is the major sink for CO. Some microorganisms in the soil remove CO from the atmosphere.

(*ii*) Some of the species in the atmosphere such as hydroxyl $(HO\cdot)$ and perhydroxyl $(HO\dot{O})$ radicals, atomic oxygen and ozone help in the oxidation of CO to CO_2 and consequently, remove it from the atmosphere.

2. Carbon Dioxide (CO_2)

Carbon dioxide (CO_2) is a natural constituent of the atmosphere and is vital to all forms of plant life. Normally, it forms about 0.03% by volume of the atmosphere.

Sources:

It is released mainly into the atmosphere by the combustion of fossil fuels (coal, oil etc.) in factories and also at homes. Carbon dioxide is also produced by the biological decay of plants. It is also given out by plants and animals in the process of respiration.

Sink

The ocean is the main sink for carbon dioxide. It contains most of the dissolved carbon dioxide in the form of carbonates, bicarbonates and organic compounds. It is also removed from the atmosphere by photosynthesis of plants.

In a balanced ecological system, the CO_2 released into the atmosphere is regularly removed by green plants for photosynthesis which in turn produce oxygen, thus maintaining the cycle.

Forests on the earth account for about 70% of all photosynthesis occurring at any time. However, because of deforestation, there is an increasing build up of CO_2 in the atmosphere.

Carbon dioxide causes mild narcotic effects, stimulation of respiratory centres and leads to asphixiation. The increasing concentration of carbon dioxide also changes climatic conditions especially by raising the global temperature. The phenomenon is known as the **greenhouse effect.**

3. Oxides of Sulphur (SO_x)

The two common oxides of sulphur SO_2 and SO_3 are represented by the formula SO_x. These are probably the most harmful of the gaseous pollutants.

Sources:

(*a*) ***Natural activities.*** SO_2, the common oxide of sulphur is released into the atmosphere through volcanic eruptions. It accounts for about 67% of the total SO_2 present in the atmosphere. The remaining 33% comes from human activity.

(*b*) ***Human activities.*** The human activities which release SO_2 into the atmosphere are as follows:

- Combustion of sulphur-bearing fuels such as coal and oil in thermal plants.
- Roasting and smelting of sulphide ores such as $CuFeS_2$, PbS, Cu_2S, ZnS, etc.
- A part of SO_2 undergoes photolytic and catalytic oxidation in the atmosphere to form SO_3 ($SO_2 + O_2 \longrightarrow SO_3$). Thus, SO_2 present in the atmosphere is generally contaminated with SO_3. The SO_3 so formed gets converted into H_2SO_4 in the

presence of moisture. This acid comes down from the atmosphere in the form of ***sulphuric acid rain.***

Toxic effects of SO_x

- Both SO_2 and SO_3 are strongly irritating to the respiratory tract of humans and animals. It has been reported that even at lower concentrations ($\approx$ 5 ppm), SO_2 causes cough, shortness of breath and spasm of the larynx (voice box). It also causes acute irritation to the membrane of the eyes resulting in tears and redness. SO_3 is more harmful even at a concentration of 1 ppm.
- SO_2 also produces *leaf injuries* (called *necrotic blotching*) to broad leaved plants and grasses.
- SO_2 also produces corrosive effect on building materials such as lime stone, marble, roof slate etc., and metals such as iron, steel and aluminium. It also causes deterioration of fabrics (cotton, rayon), paper and leather.

Sinks of SO_x

The oxides of sulphur are converted into H_2SO_4 which combine with lime stone minerals. In this way SO_x are considerably removed from the atmosphere. Thus, marble structures, lime stone and geological sources of ammonia act as **sink** for atmospheric SO_x.

4. Oxides of Nitrogen (NO_x)

Nitrogen and oxygen forms five oxides. However, *three* of these five oxides, namely ; *nitrogen dioxide* (NO_2), *nitric oxide* (NO) and *nitrous oxide* (N_2O) are present in the atmosphere in significant amounts. Of these three, only two oxides (NO and NO_2) are atmospheric pollutants.

Sources:

(*a*) ***Natural activities.*** Significant amounts of NO_x, mainly NO is released into the atmosphere by natural bacterial action.

- Small amounts of NO_2 is produced in the soil by microbiological processes.
- Lightening discharge also results in the formation of NO and NO_2 from N_2 and O_2 present in the atmosphere.

(*b*) ***Human activities.*** It involves the production of oxides of nitrogen from N_2 and O_2 present in the atmosphere during the combustion of carbonaceous matter such as coal, oil and natural gas at high temperature (exceeding 1200°C) in industries.

$$N_2 + O_2 \rightleftharpoons 2NO \; ; \quad 2NO + O_2 \rightleftharpoons 2NO_2$$

- These oxides are also produced during combustion of gasoline or diesel in automobiles.
- Chemical industries involved in the manufacture of H_2SO_4 and HNO_3 produce oxides of nitrogen as by-products which are released into atmosphere.

It may be noted that natural activity releases about 5×10^3 tonnes of NO_x into the air mainly as NO. On the other hand, human activity release about 5×10^7 tonnes of NO_x every year all over the world. The concentration of NO_x is about 10–100 times greater in urban areas (because of industrialisation and automobile pollution) as compared to rural areas.

Toxic effects of oxides of nitrogen

The main harmful effects of oxides of nitrogen are:

- NO is biologically less active and less toxic than NO_2. Like CO, it binds to haemoglobin and decreases the oxygen transport efficiency of blood.
- Inhalation of the nitrogen oxides by humans results in *pulmonary odema* and haemorrhage.
- The oxides of nitrogen cause damage to plants. Exposure of plants to NO_x causes leaf spotting and breakdown of plant tissues. Excessive concentration (10 ppm) of NO causes decrease in the rate of photosynthesis.
- Sunlight reacts with nitrogen dioxide to produce highly active oxygen atoms.

$$NO_2 \xrightarrow{\text{Sunlight}} NO + O$$

The active oxygen attacks traces of hydrocarbons in the air and produces irritants which constitute *photochemical smog*, which is a big health hazard.

- Acute exposure to NO_2 can also cause damage to human health. The smaller concentration of NO_2 leads to the inflammation of lung tissues. NO_2 has irritating effects on mucous membrane and higher doses of NO_2 may cause bronchitis and respiratory problems. Excessive exposure can even cause death.
- The nitrogen oxides also cause fading of dyes used in textiles.

Sinks of NO_x

Many natural processes acts as sink for oxides of nitrogen. These oxides are inherently unstable and decompose to N_2 and O_2 after some time. Therefore, the concentration of nitrogen oxides in the atmosphere tends to remain low. Some natural processes and photochemical reactions also take care of NO_x in the atmosphere and serve as sinks. The end product of these reactions is HNO_3. The probable mechanism for the change of NO_x to nitric acid are:

(*i*) $$2NO + O_2 \longrightarrow 2NO_2$$
$$3NO_2 + H_2O \longrightarrow 2HNO_3 + NO$$
$$4NO_2 + 2H_2O + O_2 \longrightarrow 4HNO_3$$

(*ii*) $$NO_2 + O_3 \longrightarrow NO_3 + O_2$$
$$NO_3 + NO_2 \longrightarrow N_2O_5$$
$$N_2O_5 + H_2O \longrightarrow 2HNO_3$$

5. Hydrogen Sulphide

The main **natural sources** of H_2S are volcanic activity and biological decay of protein matter in stagnant water and swamps. A large number of industries such as petroleum refining, manufacture of paper, manufacture of sulphur dyes, tanning, etc., also produce considerable amounts of H_2S.

Toxic effect of H_2S

H_2S gas is extremely toxic to humans. In higher atmospheric concentrations, it can be fatal. The reaction of H_2S with essential proteins is primarily responsible for its toxic effects.

6. Hydrocarbons

These are the chemical compounds of carbon and hydrogen only. The hydrocarbons involved in air pollution are those which are volatile or gases under ordinary conditions. The hydrocarbons in air by themselves alone cause no harmful effects. However, they undergo chemical reactions in the presence of sunlight and nitrogen oxides forming photochemical oxidants.

The natural processes such as bacterial decomposition of organic matter, forest fires and vegetation are important sources of hydrocarbons in the environment. Gasoline fuelled vehicles and industrial emissions account for 1/6th of all hydrocarbons in the atmosphere.

B. PARTICULATE POLLUTANTS

In general, the term particulate refers to **finely divided solid particles or liquid droplets suspended in air.** The size of particulates varies from about 0.0002 μ to about 500 μ (1 μ = 10^{-4} cm). In polluted air, the number of particulate particles is as high as 10^{11} per m^3. In urban areas the mass level of particulates varies from 60 μ g to 2 × 10^3 μ g per cubic metre.

Some Common Particulates and their Sources

1. Soot. The most common particulate injected into the atmosphere through human activity is **soot** which is produced by incomplete combustion of carbonaceous fossil fuels such as coal, fuel oil, natural gas, wood, etc., in insufficient supply of oxygen.

2. Metal Particles. These are released by various metal-finishing operations. When SO_2 gas present in a polluted atmosphere gets absorbed on to the microparticles of toxic metals, it renders the particles highly toxic.

3. Metal Oxides. Metal oxides are generated by combustion of fuels containing metallic compounds. For example, the particulate Fe_3O_4 is produced by the combustion of coal containing iron pyrites (FeS_2).

$$3FeS_2 + 8O_2 \longrightarrow Fe_3O_4 + 6SO_2$$

4. Lead Salts. Lead used to be a major air pollutant emitted by automobile exhaust. Earlier lead tetraethyl [$Pb(C_2H_5)_4$] was added to gasoline to improve its antiknock property. Suitable amounts of $C_2H_4Cl_2$ and $C_2H_4Br_2$ are added to gasoline along with $Pb(C_2H_5)_4$. The volatile lead salts get eliminated through the exhaust. The complete reaction may be represented as follows:

$$2Pb(C_2H_5)_4 + 32O_2 + C_2H_4Br_2 + C_2H_4Cl_2 \longrightarrow 20CO_2 + 24H_2O + PbCl_2 + PbBr_2$$

This problem has now been overcome by using unleaded gasoline in vehicles.

5. Fly Ash. It originates from the combustion of high-ash fossil fuels. It contains partially burnt particles of the fuels. This pollutant is released into the atmosphere from thermal power plants, smelters and mining operations.

6. Asbestos Dust. It originates from industrial units manufacturing asbestos sheets, gaskets, ropes, etc. Asbestos flooring and asbestos insulations also contribute towards asbestos dust in the atmosphere. The pollutant is also released into the atmosphere during mining and processing operations.

7. Solid Hydrocarbons. These are emitted from petroleum refineries and comprise of *paraffins, olefins* and *aromatics*. The polynuclear aromatic hydrocarbons such as benzo (α-) pyrene, chrysene and dibenzanthracenes are carcinogens. These compounds are readily adsorbed on the surface of soot particles present in the atmosphere and thus become serious health hazards.

8. Dust Particulates. Dust particulates originate from natural, domestic, industrial or agricultural sources. These are thrown into the atmosphere by volcanic eruptions, blowing of dust by wind, mining operations, etc. Depending upon the sources, they may contain a number of harmful pollutants such as Pb, Cd, Hg and hydrocarbons. These particles do not remain dispersed in air but gradually settle down on the surface because of gravitational forces.

9. Acid Mist. *Sulphuric acid mist* is produced when SO_3 present in the atmosphere comes into contact with moisture. Similarly, *nitric acid mist* is produced when oxides of nitrogen, *viz.*, NO and NO_2, undergo a series of reactions in the atmosphere. For example,

$$2NO + O_2 \longrightarrow 2NO_2$$

$$2NO_2 + H_2O + \tfrac{1}{2}O_2 \longrightarrow 2HNO_3$$

or $$2NO_2 + O_3 + H_2O \longrightarrow 2HNO_3 + O_2$$

Harmful Effects of Particulates

1. Effect on Human Beings. The particulate pollutants seriously affect the human respiratory system and cause several respiratory diseases. The particles with small size are more harmful in this context. The larger particles are easily checked by the hairlike cilia in the nose while the smaller particles reach the lungs easily and get accumulated there. Exposure to particulates for long periods of time causes *scarring* or *fibrosis* of the lung lining. This type of disease is well known in industrial settings and is termed as *pneumoconiosis.*

The particulates thus become the carriers of these toxic substances from the atmosphere to the humans and cause big health hazards. They cause *bronchitis, bronchial asthma, emphysema* and *lung cancer* and various *allergic reactions.*

2. Effect on Visibility. Particulates in the atmosphere cause scattering and absorption of sunlight and reduce the visibility. They also produce a blurring effect on vision. The reduced visibility is dangerous particularly for aircraft and motor vehicle drivers.

3. Effects on Materials. The adverse effects of particulates on materials include *corrosion of metals* (when the atmosphere is humid), *erosion* and *soiling of buildings, sculptures* and *painted surfaces* and *soiling of clothes and draperies.* Particulate matter also damages electronic equipment through chemical or mechanical action on electrical contact.

4. Effects on Plants. Particulates get deposited on the leaves of the plants thereby blocking the stomata and retarding the transpiration of minerals from the soil. They also hinder the intake of CO_2 from the air thereby causing obstruction in photosynthesis.

The various air pollutants and the sources from which they originate, are given in Table 52.1.

52.5 CONTROL OF AIR POLLUTION

Air pollution can be controlled either at *source* or *at the dispersal stage* of various pollutants. Although, the problem of controlling air pollution at source is not easy, it can be reduced to some extent by taking steps such as:

- *use of appropriate raw materials,*
- *selection of proper sites for industrial units,* and
- *use of modern techniques for efficient reduction of emissions and effluents.*

Different methods for the control of various air pollutants are described as follows:

1. Control of CO. Some of the measures which can be adopted for control of CO pollution are:

(*a*) *Use of anti-pollution exhaust catalyst in automobiles.*

(*b*) *Modification and improvements in internal combustion engine.*

(*c*) *Use of other fuels* such as *CNG* (Compressed natural gas), *LPG* (Liquefied petroleum gas) and *LHG* (Liquefied hydrogen gas) which are almost pollution free.

(*d*) *Alternative power sources such* as *solar energy, fuel cells*, etc.

2. Control of CO_2. CO_2 levels can be maintained by reducing its release from industrial units, controlling vehicular exhaust, etc. *Hydrogen is being looked into as a possible pollutionless fuel for the future.*

3. Control of NO_x Pollution. (*i*) Power plants and industrial units emit large amounts of NO_x. Such emissions can be reduced by almost 90% with the help of a two step combustion process as given below:

- In the first step, the fuel is burnt at high temperature in a limited supply of air. Under these conditions only a limited quantity of NO is formed.
- In the second step, the unburnt fuel is burnt at low temperature in excess air. Since the formation of NO requires higher temperature, only a small quantity of NO is formed at the lower temperature in the second step.

(*ii*) *Acidic scrubbing of flue gases.* The large amounts of nitrogen oxides emitted from power plants and industrial units can also be removed by scrubbing the flue gases with sulphuric acid.

(*iii*) NO_x in vehicular exhaust is converted to N_2 and trace amounts of NH_3 with the help of an anti-pollution exhaust catalyst (finely divided platinum).

Table 52.1. Air Pollutants and their Sources

Pollutants	*Major Sources*	*Maximum permissible exposure in ppm during an average period of 8 hours per day*
CO	Incomplete combustion of carbonaceous matter in automobile engines and defective furnaces, incomplete combustion of agricultural and slash matter, volcanic eruptions, forest fires.	50
SO_2	Combustion of sulphur-bearing fuels such as coal and oil, volcanic eruptions.	5
NO, NO_2	Combustion of fuel, interaction of N_2 and O_2 of the atmosphere at high temperatures.	5
H_2S	Biological decay of organic matter, oil refineries.	20
Cl_2	Industrial units producing paper, plastics, chlorinated hydrocarbons, dyes, chlorochemicals, etc.	1
Hydrocarbons	Combustion of fuel in automobiles, refineries, anaerobic bacterial decomposition of organic matter, natural gas.	1000
Chlorofluoro-carbons	Industrial units manufacturing refrigerants, fire fighting agents, their use as solvents.	1030
Pesticides	Application of pesticides and their volatilization from soil, water, treated surfaces and industrial units producing pesticides.	—
Particulates	Volcanic eruptions, fly ash, smelting and mining operations, smoke from incomplete combustion, dust from crushers and grinders.	—
$PbCl_2$, $PbBr_2$	Addition of $(C_2H_5)_4Pb + C_2H_4Cl_2 + C_2H_4Br_2$ to gasoline to improve its antiknock property.	—
CO_2	Combustion of fossil fuels, bacterial decay of plants/animals.	5000

4. Control of SO_2 Pollution. The process of removal of SO_2 is known as flue gas desulphurisation (FGD). Sulphur dioxide from flue gases is commonly removed by using chemical scrubbers consisting of a slurry of $CaCO_3$ (lime stone) when SO_2 is absorbed forming $CaSO_3$.

$$CaCO_3 + SO_2 \longrightarrow CaSO_3 + CO_2$$

If the flue gases happen to contain SO_3, $CaSO_4$ is formed:

$$CaCO_3 + SO_3 \longrightarrow CaSO_4 + CO_2$$

5. Control of Particulates. The most effective and efficient method of control of particulates is based on the use of **electrostatic precipitator.** It is based on the principle that particles of all sizes *acquire electrical charge when exposed to a high potential electric field.* The *flue gases* are passed through a tall chamber fitted with two electrodes. A high voltage potential difference is applied through the electrodes. The particulates settle down and get collected in a reservoir. In this way about 99% of particulate matter gets removed from flue gases.

52.6 ACID RAIN

Rain occurs when water vapours condense in clouds and fall to the earth. As it begins to fall, it is neutral (pH = 7). While it travels through the air, it dissolves floating chemicals and washes down particles that are suspended in air. In clean air, rain picks up materials that occurs naturally such as dust, pollen, some CO_2 and other chemicals produced by lightening or volcanic activities. These substances make the rain slightly acidic (pH ≈ 6). This level of acidity is not dangerous.

However, when the rain falls through *polluted air*, it comes across chemicals such as gaseous oxides of sulphur (SO_x), oxides of nitrogen (NO_x), mists of hydrochloric acid and phosphoric acid, etc. These substances dissolve in falling rain making it more acidic than normal with pH ranging between (5.6 – 3.5). In some cases, the pH lowers to the extent of 2. This leads to *acid rain*. Similarly, acid fog is formed when chemical pollutants dissolve in moist air. Now-a-days, the term **acid rain** is used to describe all types of *precipitation, namely; rain, snow, fog and dew which are more acidic than normal.*

CHEMISTRY OF ACID RAIN

Natural processes such as *volcanic eruptions, forest fires* and *bacterial decomposition of organic matter* produce oxides of sulphur and nitrogen. Man-made sources such as *power plants, smelting plants, industrial plants, burning of coal* and *gasoline* also release sulphur dioxide, nitrogen oxides and acidic soots. Sulphur dioxide and nitrogen dioxide interact with water vapours in the presence of sunlight to form sulphuric acid and nitric acid mist.

$$SO_2 + H_2O \longrightarrow \underset{\text{Trioxosulphate (IV) acid}}{H_2 + SO_3}$$

$$SO_2 + \underset{\text{Ozone}}{O_3} \longrightarrow SO_3 + O_2;$$

$$SO_3 + H_2O \longrightarrow \underset{\text{Tetraoxosulphate (VI) acid}}{H_2SO_4}$$

$$2NO + O_2 \longrightarrow 2NO_2;$$

$$2NO_2 + H_2O \longrightarrow \underset{\text{Trioxonitrate (V) acid}}{HNO_3} + \underset{\text{Dioxonitrite (III) acid}}{HNO_2}$$

The H_2SO_4 and HNO_3, thus, formed remain as vapour at high temperatures and begin to condense slowly as the temperature falls. These acids mix with rain or snow, on its way down to the earth and make it sufficiently acidic.

HARMFUL EFFECTS OF ACID RAIN

Some of the significant ill effects of acid rain are as follows:

1. Damage to Animals. Most of the aquatic animals cannot survive when the pH is less than 4. Some species of fish, such as salmon, die even when the pH is less than 5.5. Certain species of algae and zooplankton are eliminated at pH less than 6. A reduction in the zooplankton and bottom fauna ultimately affects the food availability for fish population.

2. Damage to Plants. Acidic water is dangerous for plants. Leaf pigments are decolourised. Acid also affects the green pigment (chlorophyll) in plants. Agricultural productivity is also decreased. Several non-woody plants, such as barley, cotton and fruit trees like apple, pear, etc., are severely affected by acid rain.

3. Material Damage. Metallic surfaces exposed to acid rain are readily corroded. *Textile fabrics, paper* and *leather products* lose their material strength or disintegrate on exposure to acid rain.

Building materials such as *lime stone, marble, dolomite, mortar* and *slate* are weakened on reaction with acid rains because of the formation of soluble compounds. Thus, acid rain is dangerous for historical monuments.

$$CaCO_3 + H_2SO_4 \longrightarrow CaSO_4 + H_2O + CO_2.$$

Effect of acid rain on lakes with lime stone deposits

Although acid rain adversely affects buildings made of lime stone, marble, etc., it does not affect the lakes with lime stone deposits severely. It is because of the fact that the acid rain gets diluted largely by the lake water and is also neutralised by the lime stone present in the lake. Hence, its effect on aquatic life is almost nullified.

52.7 GREENHOUSE EFFECT AND GLOBAL WARMING

The term *Greenhouse effect* has its origin from the practice of encasing vegetation in glass chambers to protect them from frost particularly in cold countries. It was observed that there was a continued rise in temperature in such chambers even when the temperature outside remained low. This enabled the warming up of vegetation inside the chamber, resulting in good plant growth.

WHAT HAPPENS IN A TYPICAL GREENHOUSE?

The transparent roof and wall of the glass chamber allows sun rays to pass through and heat up the soil and plants in the chamber. The warm soil and plants emit infra-red radiation. Since glass is opaque to infra-red radiations. It partly absorbs and partly reflect these radiation back to the soil and plants. In this way the energy of the sun remains trapped in the greenhouse. This causes a rise in temperature inside the chamber.

GREENHOUSE EFFECT AROUND THE EARTH

An effect similar to the one in a glass chamber is responsible for keeping the earth's surface warmer than it would otherwise be. The atmosphere around the earth acts like the glass of a greenhouse chamber just as the glass in a greenhouse holds the sun's warmth inside. The atmosphere around the earth traps the sun's heat near the earth's surface and keeps it warm. This is called the ***natural greenhouse effect*** and it is responsible for maintaining the temperature and making the earth a perfect habitat for life. The gases present in the atmosphere which cause greenhouse effect are referred to as greenhouse gases. The various greenhouse gases are: *Carbon dioxide, Water vapours, Chlorofluorocarbons and Oxides of nitrogen.* Out of these, the most significant role is played by CO_2.

The greenhouse gases in the atmosphere form a thick cover around the earth. The earth receives a large amount of energy from the sun. The sun emits *u.v. radiations, visible light* and *infrared radiations*. Of these, the harmful u.v. radiations are absorbed by the ozone layer in the stratosphere. The visible and I.R. radiations pass through the atmosphere and reach the earth's surface. About 75% of solar energy which reaches the earth's surface is absorbed by it causing an increase in its temperature. The additional 25% of energy is reflected back into the atmosphere. The earth also radiates heat from its heated surface in the form of I.R. radiations. Carbon dioxide and water molecules present in the air are transparent to u.v. radiations but opaque to I.R. radiation. Thus, they do not allow I.R. radiations emitted by earth to escape.

The heat thus absorbed by the cover of carbon dioxide and water vapour around the earth does not go to the upper atmosphere and is partly radiated back to the surface of the earth making it warmer. In this way, the additional heat is kept within the lower atmosphere causing the warming of the earth.

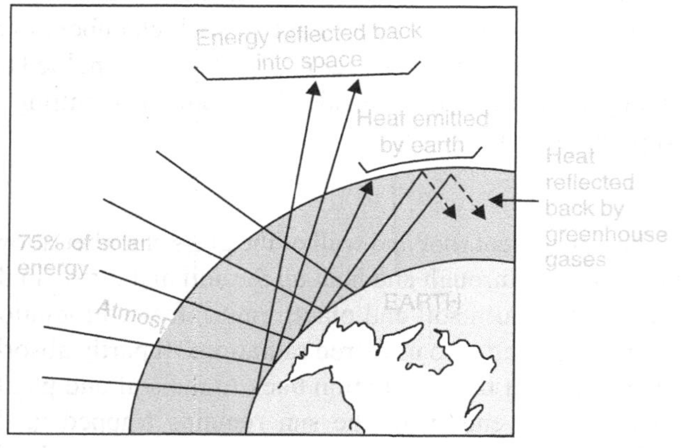

Fig. 52.2. Greenhouse-like effect on the earth.

The heating of the atmosphere due to absorption of infra-red radiations by carbon dioxide and other gases is called **greenhouse effect.** In fact, the atmosphere, and the natural greenhouse effect has kept the temperature of the earth constant for centuries.

HARMFUL EFFECTS OF GREENHOUSE EFFECT

If the amount of CO_2 in the atmosphere becomes more than 0.03%, the natural greenhouse balance will get disturbed. In this situation more and more heat will be trapped by the CO_2 cover and will be subsequently be radiated back to earth causing excessive heating of the earth's atmosphere. In recent years, quantities of CO_2 and other greenhouse gases in the atmosphere have increased significantly because of the use of chemical fertilizers, deforestation and excessive burning of fossil fuels. This excessive amount of CO_2 cannot be removed from the atmosphere by plants and ocean waters. The gradual increase in the concentration of CO_2 and other greenhouse gases has in turn increased the greenhouse effect. This poses a serious threat to mankind as it leads to a slow and gradual increase in the temperature of the earth's environment, a phenomenon referred to as **global warming.**

In the last 50 years the average temperature of the earth has increased by 1°C. If the concentration of carbon dioxide in the atmosphere continues rising, it is estimated that the average temperature of atmosphere may rise by 3–6°C. This in turn will have following disastrous consequences:

1. The high temperature of the atmosphere may melt polar ice caps releasing enormous amounts of water. The level of the sea is likely to rise by about 100 m sinking most coastal areas and causing large scale destruction.
2. The high temperature may reduce crop production, thereby causing famines.
3. The high temperature will reduce the work efficiency of human beings.
4. Tropical rains and hurricans will become more frequent and also more stronger causing more devastation.
5. The change in ocean temperature will adversely affect aquatic life.
6. Increasing concentration of greenhouse gases are expected to cause cooling of the stratosphere. This may happen because most of the *thermal IR radiation* will be absorbed at lower altitute and little of them will be left to warm the stratosphere. Cooling of the stratosphere would mean an enhanced greenhouse effect *i.e.*, greater global warming.

52.8 OZONE LAYER AND ITS DEPLETION

FORMATION AND BREAKDOWN OF OZONE

The upper stratosphere which is about 50 km above the earth's surface consists of a considerable amount of ozone (O_3) which protects us from the harmful ultraviolet (u.v) radiation ($\lambda \approx 255$ nm) coming from the sun. The ozone in the upper stratosphere is a product of u.v. radiation acting on dioxygen (O_2) molecules according to the following reactions.

$$O_2(g) + \text{u.v.} \longrightarrow O(g) + O(g)$$

$$O(g) + O_2(g) \xrightleftharpoons{\text{u.v.}} O_3(g)$$

Ozone is energetically unstable. It absorbs u.v. radiation to give dioxygen along with the liberation of heat which raises the temperature of the stratosphere.

$$O_3(g) \xrightleftharpoons{\text{u.v.}} O_2(g) + O$$

$$O + O_3(g) \longrightarrow O_2(g) + \text{Heat}$$

In this way, the dynamic equilibrium existing between the production and decomposition of ozone molecules causes the ozone layer to absorb u.v. radiation and prevent most of them from reaching the earth's surface.

DEPLETION OF OZONE LAYER

The equilibrium between formation and decomposition of ozone has been upset by the influx of several substances into the atmosphere which react with ozone to destroy it. The rate at which ozone is being destroyed is much faster than the rate at which it is being formed. The factors which accelerate the ozone depletion are as follows.

1. Effect of Chlorofluorocarbons (Freons)

A very serious threat to the existence of ozone in the stratosphere comes from the use of chlorofluorocarbons (CFC) such as *Freon*-1 ($CFCl_3$), *Freon*-12 (CCl_2F_2), *Freon*-22 ($CHCl_2F$), *Freon*-113 (CCl_2FCClF_2) and *Freon*-114 ($CClF_2CClF_2$), etc. CFC's find a wide and varied applications due to their properties like *non-corrosiveness, non-flammability, low toxicity* and *high chemical stability*. They are used as refrigerants, propellants in aerosol sprays, foaming agents in plastic manufacturing, fire extinguishing agents and solvents for cleaning electronic and metallic components, etc.

Once these CFCs enter the atmosphere by gradual evaporation from their source, they do not react with any substance due to their chemical inertness and thus cannot be eliminated from the atmosphere. They upfloat the atmosphere unchanged and eventually go to the stratosphere.

In the stratosphere, they absorb the u.v. radiation and break up liberating chlorine free radicals. The chlorine free radicals react with O_3 to form chlorine monoxide radicals ($Cl\dot{O}$) which further combines with atomic oxygen to form O_2.

$$CF_2Cl_2(g) \xrightarrow[\text{Break up}]{\text{u.v.}} \dot{C}F_2Cl + \underset{\text{Chlorine radicals}}{\dot{C}l}$$

$$\dot{C}l + O_3 \longrightarrow Cl\dot{O} + O_2$$

$$Cl\dot{O} + O \longrightarrow O_2 + \dot{Cl}$$

$$\text{Net process} \quad O_3 + O \longrightarrow O_2 + O_2$$

Thus, we find that chlorine radicals which converts O_3 to O_2 is regenerated at the end. A single chlorine radical can, thus, destroy millions of O_3 molecules which eventually causes depletion of ozone. This depletion of the ozone layer over the south pole is referred to as the ozone hole. It was first reported in 1980 by scientists working in Antarctica.

2. Effect of Depletion of Ozone Layer

Depletion of the ozone layer poses a severe threat to mankind. As a result of the decreased concentration of ozone in the stratosphere, the influx of u.v. radiation reaching the surface of the earth would increase. Medical science have pointed out that every 1% decrease in ozone concentration would lead to a 2% increase in risk to skin cancer (*melanoma*) in humans due to exposure to u.v. radiation.

- u.v. radiation leads to ageing of the skin, cataract, sunburns and also tend to damage the immune system which may lead to increased viral infections.
- Aquatic animals and aquatic plants also get damaged by their excessive exposure to u.v. radiation. U.V. radiation kill many micro phytoplanktons and damage the fish productivity.
- It has also been reported that plant proteins get easily affected by u.v. radiations which leads to the harmful mutation of cells. It also increases evaporation of surface water through the stomata of the leaves and decreases the moisture content of the soil. Increased u.v. radiations damages paints and fibres, causing them to fade faster.

52.9 SMOGS

The word *smog* is obtained from the combination of the words *smoke* and *fog*. It is one of the best known example of **air pollution** that occurs in many cities throughout the world. The word *smog* was initially used to describe a smoke-fog condition. However, nowadays, the name depends upon the composition or the mode of its formation. There are two types of smogs as described below:

CLASSICAL SMOG

It is also known as ***sulphurous smog*** or ***London smog***. This type of smog was first noticed in **London** in 1952 and it resulted in the loss of many human lives. It occurs in cool and humid conditions and is generally formed in the early morning hours of winter months. Its formation is primarily initiated by a mixture of *particulates, gaseous oxides of* *sulphur and moisture present in the atmosphere*. The **source** of classical smog is the combustion of industrial and household fuels (coal and petroleum). Because of the presence of SO_2 and carbon (soot) particles, classical smog has a **reducing character.** Classical smog produces more dangerous effects and causes severe lung and throat irritation.

PHOTOCHEMICAL SMOG

This type of smog is distinctly different from the **classical smog.** It is also called ***Los Angeles smog*** as it was first noticed in Los Angeles in 1950. It is formed in warm, dry and sunny climate. It results by the photochemical reactions between oxides of nitrogen and hydrocarbons produced by automobiles and factories. Hence it is known as **photochemical smog.** It is formed around mid day of summer months when the intensity of solar radiation is very high. The principal constituents of photochemical smog are *ozone* and *oxides of nitrogen*. This type of smog is **oxidising in character** because of the presence of O_3, NO_2 and some other photochemical oxidants.

HARMFUL EFFECTS OF PHOTOCHEMICAL SMOG

Ozone, nitrogen oxide and organic compounds (PAN, $CH_2{=}O$, $CH_2{=}CHCHO$) are the main constituents of photochemical smog each of which produces hazardous effects.

- It causes irritation to the eyes and affect the respiratory tract of human beings.
- Ozone causes cracks in rubber materials and is also harmful to fabric, crops and ornamental plants.
- The toxic nature of photochemical smog cause coughing, wheezing, bronchial constrictions and irritation to the respiratory mucous system.
- It attacks newly grown leaves and causes bronzing and glazing of their surfaces.
- The brownish colour of photochemical smog due to NO_2 reduces visibility. NO_2 also produces throat and eye irritation and leads to several chronic diseases of the throat, eyes, lungs and heart.

52.10 WATER POLLUTION

Water constitutes about 80% of the earth's surface. It comprises all kinds of sources of water such as oceans, rivers, lakes, glaciers, polar ice caps and ground water. This part of the earth which has water is called **hydrosphere.** Most of the earth's water (about 97%) is contained in oceans which is unfit for human consumption due to the high salt content. World total supply of fresh water that also includes glaciers and polar ice caps is about 2.5% only.

On the average, man consumes about 2 L of water daily. About 70% of the human body is water. Survival of plants is not possible without water. Water in abundance is required for *domestic use, agricultural activities* and *industries*. Human activities have polluted much of this limited quantity of water. Rain water on its way down to earth also brings a number of air pollutants which mix with water on the ground and pollute it. **Water pollution** refers to *the presence of any foreign substance (organic, inorganic, radioactive or biological) in water which produce harmful effect and decrease the usefulness of water*.

MAJOR WATER POLLUTANTS AND THEIR SOURCES

The various water pollutants are as follows:

1. Sewage and other oxygen-demanding wastes.
2. Infectious or disease causing agents.
3. Plant nutrients.
4. Synthetic organic chemicals.
5. Inorganic minerals and chemical compounds.
6. Suspended solids or sediments.
7. Radioactive substances.
8. Thermal discharges.
9. Oil.
10. Industrial wastes.

Let us now discuss these pollutants in brief.

1. Sewage and other Oxygen-demanding Wastes

Sewage from municipalities and other oxygen-demanding waste from industry and agriculture comprise of *organic matter* which undergoes oxidation in rivers under the action of micro-organisms. If the amount of sewage and other waste is relatively small, the dissolved oxygen (DO) is sufficient to decompose it and the products formed, CO_2, SO_4^{2-}, NH_4^+ or NO_3^- ions do not cause much of a pollution. If, however, the amount of sewage and other wastes is heavy, then the dissolved oxygen (DO) becomes insufficient and oxidation of organic matter is incomplete. The incomplete oxidation results in the formation of amines, methane, H_2S, etc., which produces a wormy smell. Thus, *sewage and other oxygen-demanding wastes* become water **pollutants** on two accounts:

(*i*) *They give rise to compounds which have an extremely offensive odour*.

(*ii*) *They deplete the dissolved oxygen (DO) from water which is rendered harmful because water devoid of dissolved oxygen cannot sustain aquatic life*. It affects or even kills fish and other aquatic living beings.

2. Infectious or Disease Causing Agents

These are the various ***pathogenic microorganisms*** which may enter water along with sewage or other wastes. These microbes (mainly bacteria and viruses) can cause various diseases such as cholera, typhoid, dysentery, gastroenteritis, polio, hepatitis, etc.

3. Plant Nutrients

The presence of plant nutrients in lakes and slow-moving waters supports a high population of aquatic plants which on decay produce disagreeable odour. Besides this, these plants also deplete dissolve oxygen (DO) making the survival of aquatic life problematic.

The enrichment of water by nutrients is known as ***eutrophication.*** Lakes and slow-moving waters age through eutrophication and over periods of several millennia, they get converted into swamps and marshes.

4. Synthetic Organic Chemicals

These include pesticides, detergents and other industrial chemicals. These chemicals when present in water can act as poisons for plants, animals and humans. These chemicals enter the hydrosphere either by losses during their transport and usage or by accidental or intentional disposal of wastes from manufacturing units.

5. Inorganic Minerals and Chemical Compounds

These include various metals and metallic compounds released from human activities or from natural minerals. These pollutants enter the water bodies from municipal and industrial waste waters and *mine runoff.* The waste which consists of sulphate and nitrate ions make water acidic. Most of these inorganic compounds particularly, those of heavy metals such as cadmium, mercury, chromium, lead and silver, etc., are toxic. These substances get attached to the tissues of aquatic organisms, produce physiological poisoning and are, therefore capable of killing the living organisms in water bodies. Alkalies discharged by industries such as textiles, tanneries, paper, etc., can also destroy aquatic life. The main constituents of **mine drainage** are sulphuric acid and iron compounds. These cause corrosion of metals and concrete and are fatal to fish.

In the recent years the levels of fluorides in drinking water has considerably increased in different parts of India, causing **fluorosis** (an incurable bone disease). The toxic elements which are commonly present in municipal and industrial waste along with their adverse effect are given in Table 52.2.

Table 52.2. Toxic Elements Commonly Present in Municipal and Industrial Waste Waters

Element	*Sources*	*Adverse Effects*
Arsenic	Pesticides, chemical wastes, mining by-product.	Enzyme-inhibitor, carcinogenic
Beryllium	Nuclear power and space industries, coal.	Toxic, carcinogenic.
Boron	Industrial wastes, detergent formulations, coal.	Toxic to some plants.
Cadmium	Industrial discharge, metal plating, Ni-Cd batteries, mining waste.	Causes high blood pressure, kidney malfunctioning, anaemia, disorder of the bone marrow.
Chromium	Metal plating industries.	Cr(VI) is carcinogenic
Zinc	Metal plating industries, industrial waste.	Toxic to plants.
Copper	Metal plating industries, mining.	Not very toxic to animals, toxic to plants and algae.
Fluorine (F^- ion)	Natural geological sources, industrial waste.	Causes bone damage, mottled teeth.
Lead	Plumbing, mining, coal, gasoline.	Causes anaemia, kidney malfunctioning, nervous disorder.
Mercury	Pesticides, coal, industrial waste, mining.	Highly toxic.

6. Suspended Solids or Sediments

Suspended solids in water are mainly *sand, silt* and *minerals eroded* from land. Solid particles settle in reservoirs and dams and thus reduce their water storage capacity. The suspended particles in water bodies also block the sunlight required by bottom vegetation for photosynthesis and thus reduce the availability of food to fish.

7. Radioactive Substances

Radioactive substances can be carried into water from *nuclear power plants, wastes of uranium and thorium during their mining and refining processes* and also from *medical and scientific institutions* which utilise radioactive materials. These substances may cause radioactivity in living organisms and produce harmful effects.

8. Thermal Discharges

Large amounts of water are used for cooling purposes in thermal and nuclear power plants. Water is also used as a coolant in many industries. Cooling water is thus discharged at a raised temperature. The increased temperatures have several adverse effects on water. The rates at which chemical reactions occur increase considerably. This results in faster assimilation of water wastes and hence faster depletion of dissolved oxygen which affects aquatic life. The density and viscosity of water decrease with increase in temperature. This results in faster settling of suspended solids. The rate of evaporation also increases appreciably with an increase in temperature. This results in waste of water in the form of its vapour and so on.

9. Oil

Oil and oil wastes enter rivers and other water bodies from different sources such as *oil refineries, storage tanks, automobile waste oil, petrochemical plants* and *industrial effluents.* Normal tanker operations and spillage from oil tankar accidents cause marine pollution and shore contamination.

Since oil is insoluble in water, it floats and spreads rapidly into a thin layer. The oil layer on the surface of water reduces the DO levels in water as oxygen transfer from atmosphere is prevented. At sea, oil layer is responsible for the death of birds. The oil penetrates the bird feathers thereby affecting their insulation and buoyancy. The birds experience difficulty in floating and flying. Oil may be driven to shores through wind and tides where it is accumulated and pose aesthetic problems.

10. Industrial Wastes

Industrial wastes and effluents contaminate water in the following respects:

(*i*) **Heavy Metals.** Metals such as Cd, Pb and Hg may be present in industrial or mining waste. These metals can prove poisonous to humans. Cadmium and mercury can cause kidney damage, and lead poisoning can cause damage to the kidneys, liver, brain and the central nervous system. All of these metals are cumulative poisons because the body does not excrete them and their concentration builds up.

(*ii*) **Detergents and Fertilisers.** These may contain phosphates as additives. The addition of phosphorus to water, in the form of the phosphate anion PO_4^{3-}, encourages the formation of algae, which reduces the dissolved oxygen concentration of water. This process, known as **eutrophication,** impedes the development of higher life forms, such as fish.

(*iii*) **Acid-polluted Water (pH < 3).** This is deadly to most forms of aquatic life. Water downstream from a mine may be contaminated by acid mine drainage, the result of microbial oxidation of discarded waste material at the mine site. Acid mine water principally contains sulphuric acid produced by the oxidation of iron pyrites (FeS_2). Industrial wastes and acid rain may also contribute to the acidity of natural waters.

(*iv*) **Polychlorinated Biphenyls (PCBs).** These chemicals are recent additions to the list of contaminants of water. PCBs are generally used as fluids in transformers and capacitors because of their high stabilities. PCBs are resistant to oxidation and their release into the environment causes skin disorders in humans. They are reported to be carcinogenic.

Add To Your Knowledge

International Standards for Drinking Water

The quality of water is of vital concern for mankind since it is directly linked with human welfare. There are some international standards for drinking water, which must always be obeyed if water is to be used for drinking purposes. These are:

Fluoride. Soluble fluoride is often added to drinking water to bring it up to a concentration of 1 ppm or 1 mg dm^{-3}. This concentration is within agreed safety limits because it protects teeth against decay. However, concentrations of fluoride ions above 10 ppm (mg dm^{-3}) are poisonous and are harmful to bones and teeth.

Lead. The limit for the concentration of lead ions in drinking water is 50 ppb (μg dm^{-3}). If water is relatively acidic and lead pipes are used for water transport, then the water is likely to get contaminated with lead.

pH. The pH of drinking water should be between 5.5 and 9.5. A decrease in the pH of the water increases the solubility of metal ions.

Other metals. The maximum recommended levels of common metals in drinking water, set by the European Environment Commission (EEC), are as follows.

Metal	Zn	Fe	Mn	Cu	Cd	Al
Max Conc. (ppm)	5	0.2	0.05	3	0.005	0.2

Sulphate. Sulphate is harmless at moderate levels, but excessive sulphate (>500 ppm) is thought to have a laxative effect.

Nitrate. Excess nitrate in drinking water can lead to *methemoglobinemia* ('blue-baby' syndrome). It also may be linked to stomach cancer, although this link has not been proved. The EEC have set a maximum limit of 50 ppm for the nitrate ion in drinking water.

CONTROL OF WATER POLLUTION

Some of the steps which are helpful for the control of water pollution are described as follows:

Some examples are:

1. Waste water treatment techniques can be applied before the polluted water enters a river, lake or pool. Available waste water treatment processes can be physical, chemical or biological. **Physical processes** comprise *screening, sedimentation, floatation* and *filtration*. Commonly used **chemical processes** are *precipitation, coagulation* and *disinfection*. The **biological processes** are *biological filtration* and the *activated sludge process*. In particular cases, processes such as *carbon adsorption, oxidation* and *reduction, ion-exchange, reverse osmosis, electrodialysis*, etc., are also used.

2. Septic tanks should be used for each house. This will reduce the flow of municipal sewage and human excreta into a river, lake or pool.

3. Rivers, lakes, etc., should not be used for bathing and washing purposes. In this way, water sources do not get polluted with detergents and germs.

4. Too much use of pesticides which are not biodegradable should be avoided. These are highly toxic substances.

5. Efforts should be made to increase the use of *low grade* or *polluted water*. Some examples are: *Treatment of domestic sewage for industrial cooling; reuse of water in mining and similar industries where water availability is less*. These efforts will save fresh water from pollution.

52.11 LAND POLLUTION

Any factor which deteriorates the quality, texture and mineral content of the soil or which disturbs the biological balance of the organisms in the soil is referred to as **Land or Soil pollutant.** Pollution in the soil has adverse effect on plant growth. Soil pollution is associated with

1. Indiscriminate use of fertilizers
2. Indiscriminate use of pesticides
3. Dumping of large quantities of waste materials
4. Deforestation.

COMMON SOIL POLLUTANTS

1. Indiscriminate Use of Fertilizers

Fertilizers contaminate the soil with impurities which come from the raw materials used for their manufacture. For instance, As, Pb and Cd present in traces in rock phosphate mineral get transferred to superphosphate fertilizers. Since the metals are not degradable, their accumulation in the soil (due to excessive use of phosphate fertilizers) above their toxic levels becomes an indestructible poison for crops. The quantity of vegetables and crops grown on soil which is constantly treated with fertilizers over the years, goes down. For example, the indiscriminate use of NPK fertilizers reduces the protein content of wheat, maize, grains, etc., grown on that soil. The carbohydrate quality of such crops also gets degraded. Excess potassium content in soil decreases vitamin C and carotene content in vegetables and fruits. The vegetables and fruits grown on over-fertilized soil are more prone to attacks by insects and diseases.

2. Indiscriminate Use of Pesticides

Pesticides are substances which are used to kill or block the reproduction processes of unwanted organisms. Most of the pesticides can be put into one of the following three categories:

(*a*) ***Insecticides.*** These are used to control insects and thus help to curb diseases and protect crops. The important insecticides are *chlorinated hydrocarbons, DDT, BHC, aldrin, malathion*, etc. They are stable in the environment and toxic to insects in small amounts, but much less so to humans.

(*b*) ***Herbicides.*** These are used to kill weeds. Some common weed killers which were used earlier were, $NaClO_3$, $NaAsO_3$, etc. However, because of their toxic effect on mammals their use have been restricted. Now-a-days, organic herbicides such as *triazines* are widely used as herbicides.

(*c*) ***Fungicides.*** These are used to check the growth of fungi. Fungi are the plants without chlorophyll and cannot use solar energy. They grow at the expense of living organisms or they live on decaying matter. Organic compounds of mercury are frequently used as fungicides.

The remnants of the pesticides used on pests may get adsorbed by the soil particles which may contaminate root crops grown in that soil. Through the consumption of these crops, the remnants may enter human biological systems affecting them adversely. Most of the pesticides are quite stable and their biodegradation may take weeks and even months. Pesticides not only bring toxic effect on humans and animals but also decrease the fertility of the soil.

3. Dumping of Large Quantities of Waste Materials

Solid waste include domestic refuse and discarded solid materials such as those from commercial, industrial and agricultural operations. They contain increasing amounts of paper, cardboards, plastics, glass, packing material and toxic or otherwise hazardous substances. A large number of heavy metals get deposited into the soils surrounding smelting industries. These effluents in the long run pollute the soil because the chemicals present in the waste are absorbed by the soil. This eventually alters the chemical and biological properties of the soil.

4. Deforestation

It results in floods and cause soil erosion. During the past few years quite a vast green land has been converted into deserts. In India, the Thar desert spread annually over 12000 hectares of productive land. Roots of grasses are an excellent binding material and keep the soil intact. Overgrazing, over-cropping and improper tilling accelerate soil erosion.

CONTROL OF SOIL POLLUTION

The following steps have been suggested to control soil pollution:

1. The use of chemical fertilizers can be reduced by applying bio-fertilizers and manures. Biological methods of pest control can also reduce the use of pesticides and thereby minimise soil pollution.

2. Recycling and recovery of materials appears to be a reasonable solution for reducing soil pollution. Materials such as paper, glass and some kinds of plastics can be recycled. This would decrease the volume of refuse and help in the conservation of natural resources. For example, the recovery of one tonne of paper can save 17 trees.

3. Control of land loss can be attempted through restoring forest and grass cover to check soil erosion and floods. Crop rotation or mixed cropping can improve the fertility of the land.

4. Proper methods should be adopted for the disposal of solid wastes. The simplest and most widely used technique of solid waste management is to bury the waste in locations situated away from residential areas. While adopting this practice, environmental and aesthetic considerations must be taken before selecting the dumping sites. Acidic and alkaline wastes should be first neutralised, the insoluble material if biodegradable should be allowed to degrade under controlled conditions before their disposal. *Incineration* of wastes is expensive, leaves a huge residue and adds to air pollution. *Pyrolysis*, a process of combustion in the absence of oxygen on the material burned under a controlled atmosphere of oxygen, is an alternative to incineration. The gas and liquid thus obtained can be used as fuels. Pyrolysis of carbonaceous wastes like firewood, coconut, palm waste, corn combs, cashew shell, rice husk, paddy straw and saw dust, yields charcoal along with products like tar, methyl alcohol, acetic acid, acetone and fuel gas. Dung should be used to produce "gobar gas" and good manure through gobar gas plants.

52.12 INDUSTRIAL WASTE

Industrial solid wastes can be placed into two categories as *biodegradable* and *non-degradable wastes.*

- *Biodegradable* wastes are generated by cotton mills, food processing units, paper mills, and textile factories, etc.
- *Non-biodegradable* wastes are generated by *thermal power plants* which produce fly ash, *Integrated iron and steel plants* which produce blast furnace slag and steel melting slag. *Metallurgical industries* such as aluminium, zinc and copper produce anode mud and tailings. *Fertilizer industries* produce gypsum. Hazardous waste such as inflammables, composite explosives or highly reactive substances are produced by *industries dealing in metals, chemicals, drugs, pharmaceuticals, dyes, pesticides, rubber goods, etc.*

  The disposal of non-biodegradable industrial solid waste, if not done by a suitable method, may cause serious threat to the environment. Recent developments and new innovations have led to different uses of waste material. Now-a-days, fly ash and slag from the steel industry is utilized by the cement industry. Large quantities of toxic wastes are usually destroyed by controlled incineration. Whereas small quantities are burnt along with factory garbage in open bins. Moreover, solid waste if not managed effectively, affects the components of the environment.

RECYCLING OF WASTE

- Fuel obtained from plastic waste has high octane rating. It contains no lead and is known as "green fuel".
- Due to recent developments made in the chemical and textile industries, clothes will be made from recycled plastic waste. These will be available soon in the global textile market.
- Our cities and towns face endless hours of power cuts. We can also see piles of rotting garbage here and there. There is a good news that we can get rid of both of these problems simultaneously. A multinational company has developed a technology where electricity is produced from garbage. The company has set up a pilot plant, where after removing ferrous metals, plastic, glass, paper, etc., from the garbage, it is mixed with water. It is then cultured with bacterial species to produce methane, commonly known as biogas. The remaining product is used as manure and the biogas is used to produce electricity.

SUMMARY

- **Pollutant.** Substance produced by a natural source or by human activity, which produces adverse effect on the environment.
- **Environmental Chemistry.** The branch of chemistry which deals with the study of the various chemical reactions occurring in the environment.
- **Air Pollution.** Any condition of the atmosphere in which the presence of certain substances beyond certain concentrations produce harmful effect on man and his environment.
- **Primary Pollutants.** Pollutants which are emitted directly from the sources.
- **Secondary Pollutants.** Pollutants which are formed in the atmosphere by chemical interaction among primary pollutants.
- **Major Air Pollutants.** Major air pollutants are: CO, SO_2, SO_3, NO_2, NO, CEC, H_2S, particulates like dust, smoke, lead salts, fly ash, asbestos dust, acid mist.
- **Ozone Layer.** Ozone layer is present in the stratosphere which protect the earth from harmful u.v. radiations of the sun. The *CEC and oxide of nitrogen* pose a danger to the ozone layer and becomes the cause for its depletion.
- **Greenhouse Effect.** Heating of the atmosphere due to the absorption of infrared radiation by carbon dioxide and other gases.
- **Smog.** Smog is a mixture of smoke and fog of particulates. Principal constituents of classical smog are SO_2 and carbon soot. It has a reducing nature.
- **Photochemical Smog.** It is formed by photochemical reactions involving solar radiation. The principal constituents are O_3, NO_2 and some photo-chemical oxidants.

EVALUATE YOURSELF

I. Objective Type Questions

Select the most appropriate choice from the options given as (a), (b), (c) and (d) after each question:

1. What is not correct about the greenhouse effect?
   (*a*) It results in global warming
   (*b*) Carbon dioxide is one of the main chemical species responsible for it
   (*c*) It results in lowering of the levels of the ocean over the years
   (*d*) CH_4, O_3, CFC also contribute to the greenhouse effect.
2. Ozone in the stratosphere is depleted by
   (*a*) CF_2Cl_2 (*b*) C_9F_{16}
   (*c*) $C_6H_6Cl_6$ (*d*) C_6F_6.
3. A smog is essentially caused by the presence of
   (*a*) O_2 and O_3 (*b*) O_2 and N_2
   (*c*) O_3 and N_2 (*d*) Oxides of S and N.
4. Which of the following is responsible for the depletion of the ozone layer in the upper strata of the atmosphere?
   (*a*) Polyhalogens (*b*) Ferrocene
   (*c*) Fullerencs (*d*) Freons.
5. Pick up the correct statement
   (*a*) CO which is a major pollutant resulting from combustion of fuels in automobiles plays a major role in photochemical smog.
   (*b*) Classical smog has an oxidising character while photochemical smog has a reducing character.
   (*c*) Photochemical smog occurs in day time whereas classical smog occurs in morning hours.
   (*d*) The level of ozone in the atmosphere goes down during the formation of smog.
6. High concentration of fluoride is poisonous and harmful to bones and teeth at levels over
   (*a*) 1 ppm (*b*) 3 ppm
   (*c*) 5 ppm (*d*) 10 ppm.
7. Which of the following is not a greenhouse gas?
   (*a*) CO_2 (*b*) CCl_2F_2
   (*c*) O_3 (*d*) N_2.
8. Which of the following does not cause water pollution?
   (*a*) Automobile exhaust (*b*) Plant nutrients
   (*c*) Oxygen demanding wastes
   (*d*) Disease causing agents.

9. Lung diseases are four times more in urban areas than rural areas. This is due to presence of which of the following in the atmosphere?
   (*a*) CO_2 (*b*) SO_2
   (*c*) N_2 (*d*) O_2.
10. In the stratosphere decomposition and formation of ozone is continuous. Which free radical retards the formation of O_3?
   (*a*) $\dot{C}H_3$ (*b*) Cl
   (*c*) $\dot{F}$ (*d*) Cl_2.

II. Fill in the Blanks

11. Complete the following sentences by supplying the appropriate words:
   (*i*) pH of acid rain lies below ....
   (*ii*) The part of the environment which supports life is called ....
   (*iii*) The part of the atmosphere where temperature increases with increasing altitude is/are ....
   (*iv*) The oxide of carbon which has no role to play in photo-chemical smog is ....
   (*v*) Lead pollution in the atmosphere is caused mainly by ....
   (*vi*) Main constituents of acid rain are ....
   (*vii*) The oxide of nitrogen that is not present in the atmosphere is ....
   (*viii*) Incomplete combustion of fuel, or carbon compounds produces ...... which is toxic to humans.

III. Discussion Questions

12. How does environmental pollution affect humans, plants and other materials?
13. What are the natural and human sources of CO? What is sink for CO?
14. What are particulate pollutants? What is their role in air pollution? What measures can be taken to control it?
15. Name major air pollutants.
16. Describe the harmful effects of particulates in air.
17. What is greenhouse effect? Name the important gases responsible for greenhouse effect.
18. (*i*) What is acid rain? What are its adverse effects?
   (*ii*) Why do lakes with lime stone deposits experience less adverse effect from acid rain?
19. What are smogs? Give harmful effects of smog.
20. What name is given to the region of the atmosphere where ozone is present? What is the function of the ozone layer?
21. What are the major causes of water pollution?
22. Describe the water pollution caused by
   (*i*) plant nutrients, (*ii*) oil,
   (*iii*) thermal discharges.
23. Name the common soil pollutants.
24. Do you observe any soil pollution in your neighbourhood? What efforts will you make to control the soil pollution?
25. Discuss various steps to control soil pollution.

Answers

I. Objective Type Questions

1. (*c*)	**2.** (*a*)	**3.** (*d*)
4. (*d*)	**5.** (*c*)	**6.** (*d*)
7. (*d*)	**8.** (*a*)	**9.** (*b*)
10. (*b*).		

II. Fill in the Blanks

11. (*i*) 5, 6 (*ii*) biosphere
(*iii*) Stratosphere/ Thermosphere
(*iv*) CO (*v*) $Pb(C_2H_5)_4$ or leaded gasoline
(*vi*) H_2SO_4, HNO_3, HNO_2, etc.
(*vii*) N_2O_3, N_2O_5 (*viii*) CO.

SECTION-1

UNIT-53

BIOTECHNOLOGY

Specific Objectives

After studying this unit, the student should be able to:

- describe the concept of biotechnology
- describe biotechnology processes that gives products for human use
- describe biotechnology processes that provide services to humans
- visit a traditional (indigenous) industrial facility
- describe an industrial visit.

53.1 BIOTECHNOLOGY

The term 'biotechnology' is derived from the fusion of biology and technology. It utilizes the biological agents along with its components for generating useful products for mankind. The area covered under this is very vast and the techniques involved are highly divergent.

Biotechnology consists of "the controlled use of biological agents such as micro-organisms or cellular components, for beneficial use."

—U.S. National Science Foundation

APPLICATIONS OF BIOTECHNOLOGY

(*a*) Food and Drink (*b*) Treatment of Waste
(*c*) Genetic Engineering (*d*) Medical Products/pharmaceuticals
(*e*) Mining (*f*) Fuel.

A. Food and Drink

Biotechnology has great applications in the development of different kinds of food and beverages. It helps in improving the quality of food, producing fermented food and in the preservation of food.

Food includes meat (from healthy animals), poultry, eggs, fruits and vegetables, fermented food that includes, yoghurt and buttermilk.

Drink includes milk, fruit juice, alcoholic beverages (wine, beer).

Fermented Foods

A variety of foods are produced by fermentation mainly in Asia and Africa. Natural growth of certain microbes in some cases on food improve their flavour, texture and nutritional value.

Advantages of Fermented Foods

1. Improved flavour.
2. Elimination of undesirable flavours.
3. Improvement in the texture of food.
4. Enhanced nutritional value.
5. Increased digestibility.

Cheese Production

It involves the three main steps (*i*) coagulum formation (*ii*) separation of curd from whey (*iii*) ripening of cheese.

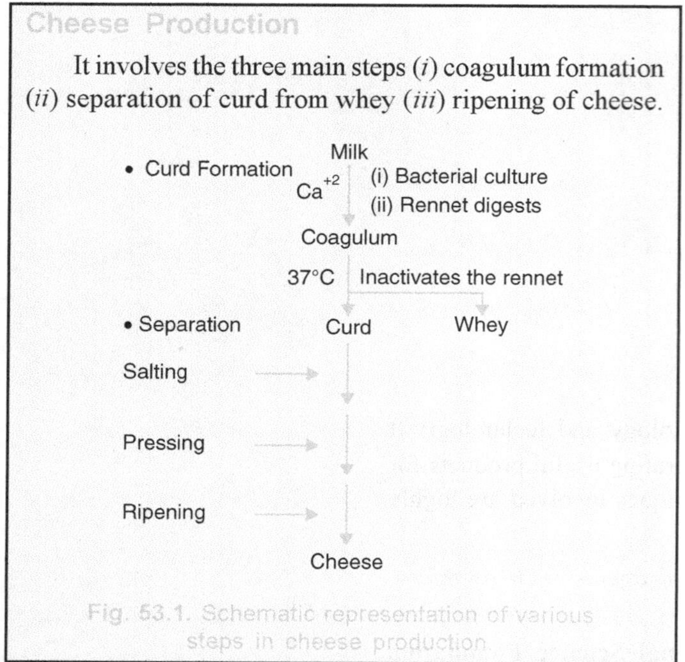

Fig. 53.1. Schematic representation of various steps in cheese production.

Use of Biotechnology in Dairy, Fruit Juice and Brewing Industry

Improved quality of milk is produced now-a-days. Different ways of preserving the milk are available like pasteurization of milk.

Enzymes are used in fruit juices to reduce the viscosity, cloudiness, increase yield and enhance the flavour. Similarly enzymes are utilized in preparing alcoholic beverages.

B. Treatment of Waste

Any product, by product or residue that cannot be used profitably is called **waste.** A waste product is regarded as a **pollutant** when it damages the environment.

Wastes may be (*i*) *biological* (*ii*) *chemical* or (*iii*) *physical* in nature.

Human activities have increased by many orders the amount of wastes and have also generated a very wide variety of novel wastes and pollutants.

The aim of biotechnology is to develop such processes and products, which minimize the damage to the environment. This is a major step in adding value to the quality of life.

The wastes may be **biodegradable**, **recalcitrant** or a mixture of both. Some recalcitrant components of wastes like heavy metals, some man-made organic compounds etc., may interfere with biodegradation processes by inhibiting the growth of micro-organisms.

The various approaches to waste treatment may be grouped into the following five categories:

(*i*) biofilters (gases)
(*ii*) landfills (solids)
(*iii*) burning or incineration (solids)
(*iv*) aerobic digestion (liquid)
(*v*) anaerobic digestion (liquid).

Contribution of Biotechnology to Waste Treatment and Environmental Management

Biotechnology has made several contributions to waste treatment and environmental management. The various categories are summarized below:

- Technologies for degradation and conversion of readily biodegradable wastes.
- Cleaner technologies of production which generate less number of pollutants and much less pollution.
- Safer products which are far less polluting.
- Sensitive and rapid detection techniques for a variety of pollutants.

C. Genetic Engineering

DNA or gene cloning is taken to include both the development of recombinant DNA as well as their cloning in a suitable host.

A clone consists of a sexual progeny of a single individual or cell while the process/technique of producing a clone is called **cloning**. As a result all the individuals of a clone have the same genetic constitution, which is also identical with that of the individual from which the clone was derived. It shows that, the genomes present in members of a single clone are also identical.

The gene or DNA cloning produces a large number of copies of the gene or DNA being cloned. Similarly, the term recombinant DNA technology is used as a synonym for DNA or gene cloning. A rather popular term for these activities is **Genetic Engineering.**

Steps in Gene Cloning

The entire procedure of gene cloning or recombinant DNA technology may be classified into the following **six steps:**

(*i*) *Production and isolation of DNA fragments to be cloned.*

(*ii*) *Insertion of isolated gene in a suitable vector to obtain recombinant DNA.*

(*iii*) *Introduction of recombinant DNA into a suitable organism or cell called* ***host*** *(transformation).*

(*iv*) *Selection of the transformed host cells and identification of the clone containing the desired gene.*

(*v*) *Expression of introduced gene in the host.*

(*vi*) *Where needed, transfer and expression of the gene into another organism.*

The recombinant DNA molecules produced have one of the following three objectives:

1. To obtain a large number of copies of specific DNA fragments.
2. To recover large quantities of the protein produced by the concerned gene.
3. To integrate the gene in question into the chromosome of a target organism where it expresses itself.

To achieve all this, the DNA segments are integrated into an autonomously replicating DNA molecule called a **vector**.

So all the steps involved in the transformation of a suitable host with recombinant DNA and cloning of the transformed cells is called **DNA cloning** or **gene cloning**.

D. Medical Products/Pharmaceuticals

Biotechnology contributes a lot in the healthcare and pharmaceuticals sector. These can be grouped as follows:

1. Disease prevention
2. Disease detection
3. Therapeutic agents
4. Correction of genetic disease
5. Fertility control
6. Forensic medicine.

Prevention of disease is most desirable, most convenient and a highly effective approach to health which can be achieved by vaccination or immunisation using biological products/preparations known as vaccines.

Disease Treatment

Treatment of disease utilizes a wide variety of preparations of both biological or non-biological origins.

Products from Non-recombinant Organisms

Therapeutic agent from non-recombinant organisms may originate from the following systems.

(*i*) Micro-organisms

(*ii*) Plant tissue cultures

(*iii*) Animal cell cultures.

Product from Recombinant Organisms

Genetic engineering has enabled the production of a large number of recombinant proteins. Genetic engineering or recombinant DNA technology is used in micro-organisms, cultured animal cells and more recently in plants.

Genetically engineered micro-organisms are used in the production of the following products.

(*i*) Human insulin (humulin).

(*ii*) Human growth hormone (somato statin).

(*iii*) Interferons

(*iv*) Growth factors

(*v*) Monoclonal antibodies

(*vi*) Artificial tissues/organs

(*vii*) Therapeutic oligonucleotides.

Disease Diagnosis

It includes:

1. *Microscopic examination.*
2. *Culture of the specimen.*
3. *Immunological assays for specific antigens present on the surface of pathogens.*
4. *Detection and measurement of pathogen specific antibodies produced by the patient in response to attack by pathogen.*

E. Mining

Biotechnology also contributes in the process of **mining.**

Microbes can be employed to recover valuable metals from low grade ores and also from dilute solutions for which the conventional metallurgical processes are uneconomical and are generally polluting.

In 1947, it was discovered that the bacterium **Thiobacillus ferrooxidans** induces leaching of copper sulphate from the ores. Other species of bacteria, for example, **Thiobacillus thiooxidans, Thiobacillus acidophilus** and **Thiobacillus arganoparus** were also discovered to have similar capabilities although they are of lesser importance.

These bacteria require extremely low pH for optimum growth and activity and derive their energy needs by oxidizing metal sulphides. Metals like copper, zinc and iron are recovered by this method.

Another bacterium **Leptospirillum ferrooxidans** oxidizes ferrous ions and is also involved in microbial ore leaching. **Microbial ore leaching** consists of the processes involving bacteria, mainly **T. ferrooxidans** that leads to leaching of metal sulphates from low grade ores containing insoluble metal sulphides.

F. Fuels

The use of biological agents (or their components) for the purpose of generating energy is known as fuel bio-technology. Some sources of energy *e.g.,* biomass and sunlight are convenient to be used as fuels. Some more *examples* are: methane, ethanol, butanol, biodiesel and hydrogen.

Petroleum oil or crude oil is pumped out of holes. The crude oil is transported in tankers or pipelines to a refinery where it is distilled to separate various types of molecules on the basis of their boiling points. This processing of oil yields various types of fuels for example gases (lowest boiling point), petrol, kerosene, etc.

Useful Features of Biofuels

The biofuels have a number of desirable features, which are briefly summarised below.

1. Most of the biofuels are derived from biomass, which is renewable, low cost and locally available entailing little or no commitment of foreign exchange.
2. In general, they lead to relatively low CO_2 emission than do fossil fuels.
3. They do not contribute to environmental pollution due to gases like SO_2.
4. The substrate is often a waste including municipal waste. Use of such materials for biofuel production not only generates a more valuable product from low cost substrate but also helps in cleaning up the environment.

53.2 PRODUCTS OBTAINED USING BIOTECHNOLOGY

Microbes have been employed for product generation, like wine, bread etc, for thousands of years. The use of microbes to obtain a product of economic value constitutes **industrial microbiology**. Any process mediated by or involving micro-organisms in which a product of economic value is obtained is called **fermentation**. Microbial biomass production has been developed into an industrial activity to obtain protein rich food.

A. FOOD AND DRINK

- **Single Cell Protein**

Biomass produced by unicellular and multicellular organisms like bacteria, yeast, filamentous fungi and algae is processed and used as human food or animal feed supplement.

This biomass is called Single Cell Protein (SCP) as it is rich in protein.

The micro-organisms used for SCP production must be

(*i*) non-pathogenic to plants, animals and man

(*ii*) of good nutritional value

(*iii*) easily and cheaply produced

(*iv*) toxin free.

A variety of substrates are used for SCP production. These range from inorganic carbon (CO_2), industrial effluents, confectionary effluents (whey etc.,), cellulosic wastes (like straw etc.) to high cost materials like starch hydrolysate.

Process of SCP Production from any Micro-organism or Substrate would have the Following Basic Steps

1. Provision of a carbon source.
2. Addition to the carbon source, a source of nitrogen, phosphorus and other nutrients needed to support growth.
3. Prevention of contamination.
4. Selected M.O. is inoculated in a pure state
5. SCP processes are highly aerobic so proper aerotion should be there.
6. The microbial biomass is recovered from the medium.
7. Processing of the biomass for enhancing its usefulness and/or stability.

B. DAIRY PRODUCTS

Milk drawn from a healthy animal contains several hundred to thousand micro-organisms. In the dairy industry various fermented milk products are produced by inoculating pasteurized milk with known cultures of micro-organisms referred to as **Starter Culture.**

Some of the important fermented milk products are butter milk, yoghurt, butter and various kinds of cheese.

A simple flowsheet for production of plain yoghurt is found below:

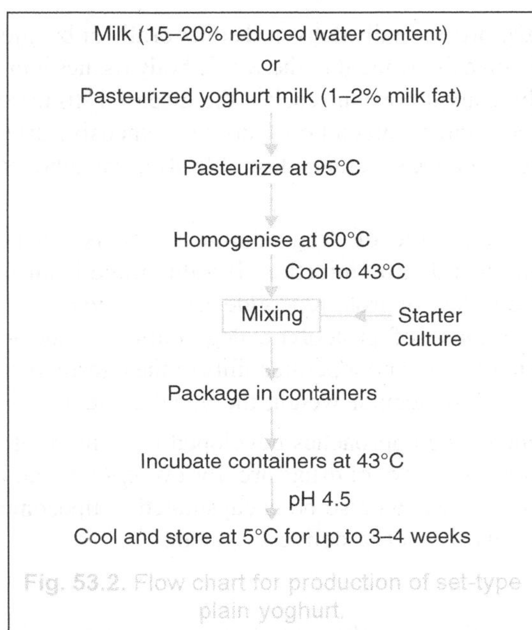

Fig. 53.2. Flow chart for production of set-type plain yoghurt.

Buttermilk is obtained from pasteurized skim milk cultured with lactic acid and aroma producing organisms. **Streptococcus lactis** may be used as starter culture.

The term buttermilk is also used for a phospholipid rich fluid fraction obtained as a by-product during the churning of cream in butter manufacture. However, cultured buttermilk is a viscous, cultured, fluid milk containing a characteristic pleasing aroma and flavour.

C. BAKING AND BREWING

Baking Process

Let us take the example of bread making to illustrate the baking process in food. Micro-organisms are useful in two main ways in bread making:

(*i*) *They may produce gas to leaven or raise the dough giving the bread or any food the desired loose and porous texture.*

(*ii*) *They may produce desirable flavouring substances.*

Dough is usually leavened by bread yeasts which ferment the sugars in the dough and produce mainly CO_2 and alcohol. This can also be done by the addition of ammonium bicarbonate. Alcohols, acids, esters and aldehydes are the products that may be added to give desirable flavours.

Although the interior of the loaf does not require to reach 100°C during baking, the heat serves to kill the yeasts, inactivate their enzymes, expands the gas present and set the structure of the loaf. Baking besides producing the appearance of the loaf also produces desirable flavours.

Brewing

Beer and alcohol are the principal malt beverages produced and consumed in our country. They are made of malt hops, yeasts, water and malt adjuncts.

Brewing of Beer

The manufacture of beer as an example of the brewing process. The different steps involved are:

(*a*) **Malting.** In the preparation of malt, barley grains are soaked or steeped at 10–15°C and allowed to germinate over 5–7 days.

(*b*) **Mashing.** The purpose of mashing is to make valuable portions of the malt and malt adjuncts as soluble as possible. The main mash is prepared by mixing the ground malt with water at 38–50°C. To this are added cooked starchy malt adjuncts in water, after cooking under steam pressure.

(*c*) **Fermentation.** A special beer yeast of the bottom type, a chain of saccharomyces is used for the inoculation of the cooled wort. During fermentation, the yeast converts the sugar in the wort mainly to alcohol and carbon dioxide.

(*d*) **Aging or Maturing.** The young or green beer is stored or lagered in vats at about 0°C for several weeks to several months, during which period precipitation of proteins, yeast and other undesirable substances takes place and the beer becomes clear and matured.

D. MEDICAL PRODUCTS

A large number of human genes have been encoded and hence pharmaceutically valuable proteins have been cloned and expressed in micro-organisms. Initially *E. coli* was used as the host for reasons of ease in cloning. But yeast is fast becoming the host of choice for the production of recombinant proteins.

Several of the recombinant proteins are used for the treatment of the *diabetes mellitus* (insulin protein), *dwarfism* [(protein-human growth hormone hGH)], etc. Many other useful recombinant proteins are in advanced stages of development.

1. Production of Human Insulin (Humulin)

Human insulin is a dimer comprising one chain of 21 amino acids (A chain) and the other of 30 amino acids (B chain). Both chains A and B are derived from a single polypeptide chain called **preproinsulin**, which has the following sequences:

(1) an N′ terminal leader.

(2) the B chain sequence.

(3) the C-chain that links the A and B chains.

(4) C-terminal A chain.

The preproinsulin folds spontaneously to yield proinsulin and chains A and B becomes linked by two disulphide bridges. This is followed by cleavage of the leader and the C-chain sequences leaving the mature insulin molecule. It is important that the insulin molecule is not glycosylated making its production in *E. coli* feasible.

Flow Sheet of Production of Mature Insulin

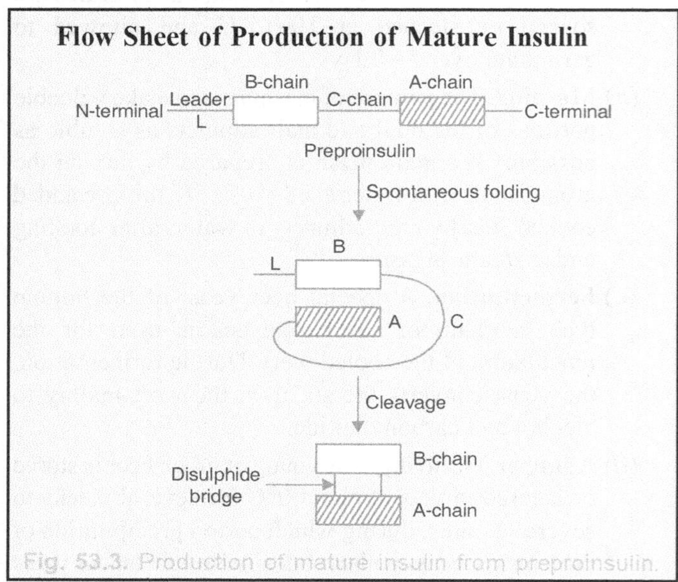

Fig. 53.3. Production of mature insulin from preproinsulin.

2. Drugs

Drug designing aims at designing drugs, which selectively and specifically fits into the critical sites of the target molecules, thereby inactivating the latter.

The target molecule may be

(*a*) a receptor (45% of cases)

(*b*) an enzyme (20% of cases)

(*c*) ion channel (5% of cases)

(*d*) DNA (2% of cases)

(*e*) nuclear receptor (2% of cases).

The aim of drug designing is to develop highly efficient drugs which have little or no side effects.

The three main steps in drug designing are:

(*i*) detailed knowledge (including the three-dimensional structure) of the critical sites of target molecules.

(*ii*) designing of drug molecules which will specifically fit into and bind to these critical sites and synthesis of such molecules.

(*iii*) evaluation of the interaction of the synthesized drugs with the target molecules, and further modifications in the former to make them 'safe' for medical use.

Drugs are normally delivered either orally or by injection. They become distributed in the whole body tissues and fluids and only a small portion reaches the diseased tissue/organ. This necessitates a much larger dose of expensive drugs and may often produce several undesirable effects in other organs/tissues.

The oral route of drug administration is much more desirable than that by injection. But this route is unsuitable for the new class of protein/peptide drugs due to poor uptake. This is because of proteolytic degradation in the gastro-intestinal tracts and poor permeability of the intestinal mucosa to these high molecular weight therapeutic agents.

Some of the approaches developed for a more efficient and targeted delivery of drugs are, for *example,* nasal, bucal, rectal etc. Drugs can also be encapsulated in liposomes and polymers used as drug delivery systems.

3. Antibiotics

Antibiotics are the metabolites having preferential antimicrobial activity. Therefore they are widely used for curing of human ailments caused by micro-organisms.

The antibiotic penicillin was discovered by Fleming in 1929 but its commercial production could commence only during the early 1940s. Antibiotic compounds are used either in their natural form or as semisynthetic derivatives; the latter are usually produced by isolating the antibiotic nucleus and subjecting it to chemical modification.

Steps in the Production of Antibiotics:

- **Inoculum.** A high yielding strain is a prerequisite for antibiotic production. Constant strain improvement is an integral part of production. The improved strains are kept in long term storage.

- **Fermenter.** Antibiotics are generally produced in stainless steel fermenters used in batch or fed-batch mode. Agitation is mostly done by impellers. Water cooling is done to maintain the temperature. Sterile air is supplied as per need.

In most cases, it is critical to prevent contamination. The final stage fermenter is preferably used for antibiotic production for the longest possible period.

- **Production Media.** Antibiotic production employs a variety of media, a different one for each stage of production. Seed stage and production media are used. A typical production media has about 10% (*w*/*v*) solids. Generally yields are much higher on complex media. In some cases a suitable precursor is provided.

For example for penicillin-G production, phenylacetic acid is used as a precursor.

In contrast the seed stage medium is devised for rapid growth and to prevent antibiotic production.

In production processes, the production fermenter is run in batch mode in which nutrient is added continuously throughout the fermentation to enhance the duration of antibiotic production. Antifoaming agents are added to prevent foaming.

- **Recovery.** At the end of the final production stage of incubation, the froth contains only a low concentration of the antibiotic.

The initial step in antibiotic recovery is separation of cells from the froth; this is usually achieved by filtration or centrifugation. But in some cases whole froth is used for extraction.

Antibiotic isolation and purification employs solvent extraction, ion exchange ultrafiltration etc.

4. Vaccines

Prevention of diseases is the most desirable, most convenient and highly effective approach to health. This is achieved by vaccination or immunization using biological preparations called **vaccines.**

Vaccines represent an invaluable contribution of biotechnology as they provide protection against such diseases for which effective cures are not yet available.

- **An Ideal Vaccine.** An ideal vaccine or vaccination protocol should have the following features:

(*i*) *It should not be toxic or pathogenic. i.e., it should be safe.*

(*ii*) *It should have very low levels of side effects.*

(*iii*) *It should not cause problems in individuals with an impaired immune system.*

(*iv*) *It should not contaminate the environment.*

(*v*) *The vaccine should be cheap so that it is generally affordable.*

Conventional vaccines consist of the whole pathogenic organisms, which may be either killed or live (attenuated).

- **Recombinant Vaccines.** It contains either a protein or a gene encoding a protein of a pathogen.

Various steps involved are:

1. Identify a protein that is both immunogenic and critical for a pathogen.
2. Gene encoding where the identified protein is isolated.
3. The gene is integrated into a suitable expression vector and introduced into a suitable host where the protein is produced in large quantities.
4. The protein is then isolated and purified from the host cells.
5. It is used for the preparation of vaccine.

The host organism used for expression of immunogenic proteins may be any one of the following:

1. **A microorganism:** The yeast for the expression of hepatitis B surface antigen (HBsAg) used as vaccine against hepatitis B virus.
2. **Cultured animal cells,** for example HBsAg expressed in CHO (Chinese hamster ovary).
3. **Transgenic plants.**

- **DNA Vaccines.** Now-a-days, vaccines based on DNA are being developed. These are regarded as the third revolution in vaccines. It involves gene encoding and the relevant immunogenic protein is isolated, cloned and then integrated into a suitable expression vector. This preparation is introduced into the individual to be immunised.

E. CHEMICALS

1. Ethanol

The use of ethanol as a fuel for internal combustion engines, either alone or in combination with other fuels has been given much attention mostly because of its possible environmental and long term economic advantages over fossil fuels.

Ethanol is mostly used to reduce emissions in gasoline. It is sometimes blended with gasoline to produce E 10 fuel. (10% ethanol and 90% gasoline), but it is also used in higher concentrations such as E 85 or E 95.

It is generally produced by fermentation from biomass and its products like sugarcane, sugarbeet molasses and usual grains.

Ethanol is produced by enzymatically converting the starch in grains to fermentable sugars followed by metabolism of the sugars to ethanol by distiller's yeast Saccharomyces cerevisial as shown below:

$$\underset{\text{Starch}}{(C_6H_{10}O_5)_n} + nH_2O \xrightarrow[\text{Enzymes}]{\text{Amylolytic}} \underset{\text{Glucose}}{nC_6H_{12}O_6}$$

$$\underset{\text{Glucose}}{C_6H_{12}O_6} \xrightarrow{\text{Saccharomyces cerevisial}} \underset{\text{Ethanol}}{2C_2H_5\text{—}OH} + 2CO_2$$

The application of yeast is due to their homofermentative nature, which yields mostly a single product ethanol in relatively high concentrations. Under anaerobic conditions, these yeasts normally produce small amounts of byproducts such as *glycerol, methanol* and *fusel* oil.

- Most regions of the world have traditionally produced alcohol from locally available substrates.
- The annual world production of ethanol is over 37 billions litres.
- Ethyl alcohol has long been used as an automotive fuel in two ways:

First it replaces gasolines and secondly it is an effective "octane booster" when mixed with gasoline.

2. Polymer

A polymer is a compound of high molecular mass formed by the combination of large numbers of small molecules. The small molecules which constitute the *repeating units* in a polymer are called **monomer units.**

As polymers are *single, giant molecules, i.e.*, big size molecules, they are also called **macromolecules.**

Polymers are classified into two types:

Natural polymers and *Synthetic polymers.*

1. Natural Polymers. *Polymers found in nature, mostly from plants* and *animal sources*, are called **natural polymers.** A few examples are:

(*a*) *Polysaccharides*

(*b*) *Proteins*

(*c*) *Nucleic Acids*

(*d*) *Natural Rubber.*

2. Synthetic Polymers. The polymers which are prepared in the laboratory are referred to as **synthetic polymers** or man-made polymers. Some examples of synthetic polymers are *polyethylene, polystyrene, PTFE, synthetic rubber, nylon, PVC, bakelite, teflon, orlon,* etc.

F. AGRICULTURES AND HORTICULTURE

- **Nitrogen fixation**

A number of micro-organisms are able to use molecular nitrogen in the atmosphere as their source of nitrogen.

The conversion of molecular nitrogen into ammonia is known as **nitrogen fixation.**

Two types of micro-organisms are involved in this process:

1. **Non symbiotic** micro-organisms, those living freely and independently in the soil.
2. **Symbiotic** micro-organisms, those living in the roots of plants.

The essential reactants in the bacterial nitrogen fixation process are:

- The **nitrogenase enzyme complex**. It consists of two components and neither is active without the other. Component I is nitrogenase and II is nitrogenase reductase.

Component I is known as **MoFe protein** and II is a smaller molecule known as **Fe protein**. Both molecules contain sulphur.

- A strong reducing agent such as ferredoxin or flavodoxin.
- ATP
- A regulating system for NH_3 production and utilization.
- A system that protects the nitrogen fixing system from inhibition by molecular oxygen.

The overall biochemical reaction for nitrogen fixation can be expressed as:

$$N_2 + 6e^- + 12ATP + 12H_2O \xrightarrow{\text{Nitrogenase complex}} 4H^+ + 12Pi + 12ADP + 2NH_4^+$$

G. GENETIC ENGINEERING IN DOMESTIC ANIMALS AND CROPS

Now-a-days genetic engineering or modifying the genome constitution of animals and plants is being done to provide various uses for mankind.

Vector system is used to modify the gene of interest in the host organism.

Various transgenic animals are produced such as transgenic mice, sheep, goat, fish.

Transgenic plants like tomato, rice, sugarcane etc. are produced to increase yield.

Applications of Transgenic Plants

1. It provides tools in the study of plant molecular biology, regulation of gene action.
2. Specific genes have been transferred into plants to improve their agronomic and other features.
3. Genes for resistance to various biotic stresses have been engineered to generate transgenic plants resistant to insects.
4. Several gene transfers have been aimed at improving the product quality.

   For example, protein or lipid quality.
5. It aims at producing novel biochemicals like interferon, insulin etc.

A transgenic animal contains in its genome a gene or genes introduced by one or other technique of transfection (introduction of DNA segment into an animal cell).

Objective of Gene Transfer in Animals

1. Majority of gene transfers aim at studies on promoter function, reporter gene expression, regulation of gene expression, functions of transferred genes/DNA sequences etc.
2. Genes have been transferred and expressed into cultured cell lines to obtain the proteins encoded by them.
3. Genetic modification of animals may be aimed at improving their milk, meat, wool etc. production.
4. A special case of gene transfer aims at eliminating the symptoms and miseries of genetic diseases.

   In this approach, normal and functional copies of the defective gene are introduced into the patient (called gene therapy).
5. Specific transgenic animal strains or lines are created to fulfil specialized experimental and/or biomedical needs. Example is "**knock out**" mice strain in which specific genes have been replaced or knocked out by their disrupted counterparts through a process of homologous recombination.

H. FUEL

1. Biogas (Methane)

- It is a mixture of gases. (about 60% methane) produced by anaerobic bacterial digestion of organic matter.
- It consists mainly of methane (CH_4 60%) and CO_2 (40%) plus traces of hydrogen (H_2) and several other gases.
- It is produced by anaerobic degradation of a variety of organic materials, ranging from simple sugars to polymers like cellulose and nucleic acids by a community of organisms.
- Anaerobic digestion occurs in nature in the sediments of lakes and ponds, and in the rumen of cattle.
- The **substrate** usually employed for the biogas generation is a waste product of industrial, agricultural, domestic and municipal origin.
- A **digester** is needed to carry out all the processes. Its design mainly depends on the type of waste to be handled and the level of operation *i.e.*, small rural or large industrial operation.
- Several hundred species of miro-organisms are involved in the anaerobic digestion and biogas production. It includes:

(*i*) Hydrolytic and fermentative bacteria.

(*ii*) Methanogenic bacteria.

$$4H_2 + CO_2 \longrightarrow CH_4 + 2H_2O;$$
$$\Delta G = -139 \text{ kg/mole.}$$

(*iii*) Acetogenic bacteria.

Process of Biogas Production

- This plant is based on the excreta of cattle and other farm animals, which contains about 20% of inorganic particles.
- The level of dust particles is reduced.
- The feeding rate is 3,500 kg dung/day. Calcium ammonium nitrate is added.
- In addition to cowdung, human excreta and kitchen waste can also be used.
- Optimal temperature for biogas production is between 35–38°C.
- pH is around 7. Under favourable conditions, the biogas yield may be up to 60 P/kg of dung.

Advantages

1. The technology is cheaper.
2. Recovery of product is spontaneous.
3. Dilute waste materials can be used as substrate.
4. Organic pollutants are removed from the environment.
5. Asceptic conditions are not needed for operation.

2. Gasohol (Ethanol-gasoline)

- Ethanol produced by micro-organisms, for example saccharomyces **cerevisial** from biomass is called **bioethanol.**
- Bioethanol is the most widely used biofuel for transport purposes.

Production of Bioethanol

At present there are three important routes for producing bioethanol.

(*i*) from starch or sugar crops.

(*ii*) from cellulose following enzymatic hydrolysis.

(*iii*) from cellulose following chemical hydrolysis.

The process for producing ethanol using cellulose is done by micro-organisms ***Trichoderma reesei*** and ***S. cerevisiae.*** The fermentation of hexoses produce alcohol.

Thus, the conversion of cellulose to ethanol involves the following four steps:

(*i*) production of cellulose enzyme
(*ii*) hydrolysis of cellulose
(*iii*) hexose fermentation
(*iv*) pentose fermentation.

The cellulosic biomass must be suitably pretreated. For example, a brief exposure to high temperature in combination with acid or ammonia is done for this purpose.

- Recovery of ethanol from the fermentation froth is by distillation which exploits the difference in boiling points of ethanol (87°C) and water (100°C). As a result, when a water ethanol mixture having < 95% ethanol is heated, the vapour has a greater concentration of ethanol than the liquid phase.

Thus, a dilute ethanol water mixture can be repeatedly distilled to obtain more concentrated (up to 95%) ethanol solution.

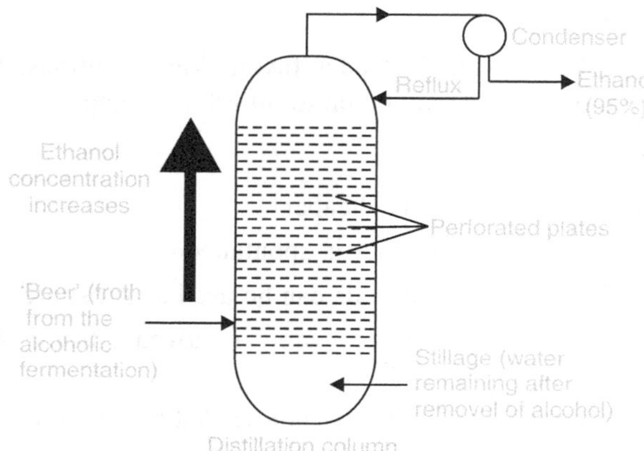

Fig. 53.4

53.3 BIOTECHNOLOGY SERVICES

A. MINING

Extraction of metals by bioleaching

Microbial ore leaching consists of the processes involving bacteria, mainly *Thiobacillus ferrooxidans,* that leads to leaching of metal sulphates from low grade ores containing insoluble metal sulphides.

The metal solubilizing action of bacteria is both direct as well as indirect. **In direct conversion,** the metal sulphide is itself converted to metal sulphate as follows:

$$CuS + 2O_2 \xrightarrow[\text{T. ferrooxidans}]{} CuSO_4.$$

In contrast to this, indirect conversion is based on the reduction of ferric sulphate to ferrous sulphate.

$$CuS + Fe_2(SO_4)_3 \xrightarrow{\text{T. ferrooxidans}} CuSO_4 + FeSO_4 + S\downarrow$$

T. ferrooxidans also oxidizes sulphur to H_2SO_4 and converts $FeSO_4$ to $Fe_2(SO_4)_3$.

$$S_8 + 12O_2 + 8H_2O \xrightarrow{\text{T. ferrooxidans}} 8H_2SO_4.$$

$$2FeSO_4 + H_2SO_4 \xrightarrow{\text{T. ferrooxidans}} Fe_2(SO_4)_3 + H_2.$$

Limitations of Microbial Ore Leaching

1. The desired metal is recovered as a dilute solution of its salts and not as elemental metal. This makes a recovery process from the solution essential.
2. The micro-organisms must be kept viable by providing appropriate conditions.
3. It is a relatively slow process.

Merits of Using Microbes for Ore Leaching

1. It does not require high energy inputs.
2. It can be applied both on small and large scales.
3. The process can be used to extract a variety of metals.
4. The process can be applied **in situ.** *i.e.*, in the mine itself.
5. It does not produce toxic or noxious gases as by-products.

Applications of Microbial Ore Leaching

1. Extraction of metals from ores. For example, copper as their sulphates.
2. Partial leaching of mineral concentrates to remove impurities.
3. Conversion of minerals into a form suitable for processing by other means.

B. TREATMENT OF WASTE

(*i*) Oil spills (*ii*) Paper waste
(*iii*) Domestic waste (*iv*) Toxic waste
(*v*) Agricultural waste (*vi*) Sewage treatment.

The environment consists of the sum of all the factors outside an organism. It consists of both biotic as well as abiotic factors. Any unwanted material present in the environment is called **waste.** Waste may be biological, chemical or physical in nature.

Sources of Wastes and Pollutants

(*i*) **Manufacturing.** It includes chemical industry, oil refinery/spill, fertilizer industry wastes. These activities generate a wide variety of wastes depending on the nature of raw materials and products, as well as the design and the operation of the processes involved.

(*ii*) **Agriculture and Dairy.** These activities produce crop residues, which are biodegradable. Most of the pesticides used are nonbiodegradable. In addition, plastics, copper present in feed additives and wastes arising from use of fossil fuels are recalcitrant to biodegradation.

(*iii*) **House Building and Domestic Activities.** It generates biodegradable and non-biodegradable wastes. It include stone, asbestos, fly ash, sewage, garbage, plastic, paper, tins and bottles etc.

- Treatment of Liquid Wastes

Liquid wastes are treated by either aerobic or anaerobic digestion in large centralized, water reclamation or sewage treatment units.

- Treatment of Solid Wastes

Solid wastes can be sorted out into biodegradable and non-biodegradable components; the latter may be recycled while the former may be (*i*) either incinerated or (*ii*) used for anaerobic digestion.

The entire solid waste is usually dumped into pits, called **landfill.**

1. Aerobic Reactors or Digesters

Biological treatment of settled sewage may be based on fixed film digester or dispersed growth digester under aerobic conditions.

These are basically of two types:

(*i*) Trickling filter digesters

(*ii*) Rotating biological contractors

(*i*) Trickling Filter Digester

These reactors are either rectangular or circular and consist of 3 m deep filters made up of 25–100 mm diameter chunks of crushed rock or ceramic bricks. The micro-organisms are present as a layer called biofilm on the surface of the filter particles. With time, as the biofilm grows thick and peels off the filter particles, a new biofilm forms due to the microbial growth.

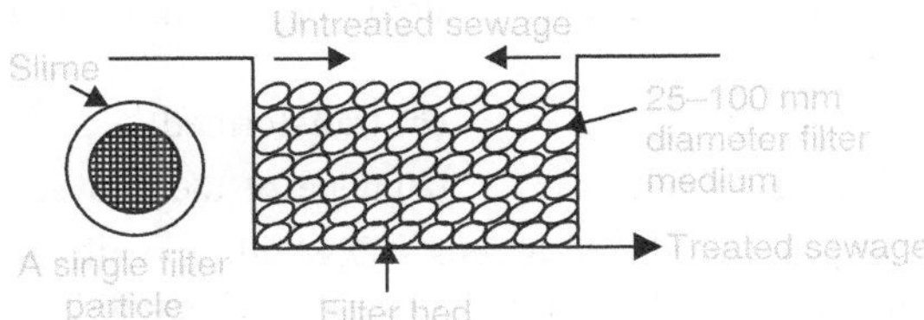

Fig. 53.5. The filter medium is crushed rock or 'blast furnace slag' and each filter particle is covered with a slime composed of the micro-organisms forming a biofilm.

Arrow indicates the direction of waste water flow.

(*ii*) Rotating Biological Contractors

It consists of set of discs about 2–3 m in diameter oriented vertically to the direction of waste water flow. The biofilm is adhered to the disc surface, which keeps on rotating at a speed of 1–2 rpm.

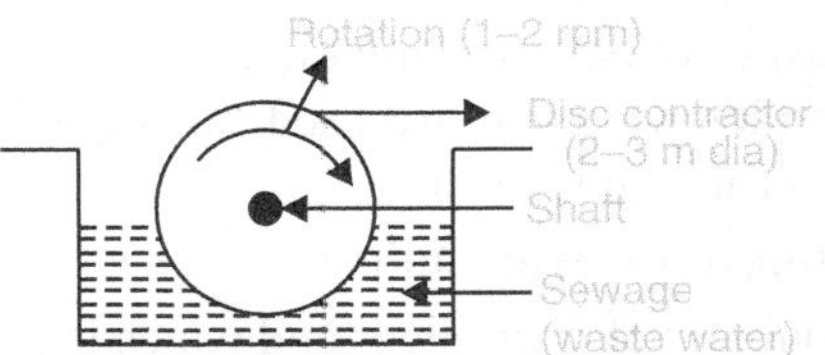

Fig. 53.6. Representation of rotating biological contractor: the biofilm is present on the surface of the disc. The waste water flows at right angles to the disc. A series of discs is used in the contractor.

2. Dispersed Growth Digesters

In such processes, the microbial population is dispersed throughout the sewage being treated. The **activated sludge process** is a typical example; it uses a large aerated vessel for a large scale oxidation of liquid wastes. The micro-organisms are distributed as flocs, throughout the waste water being treated.

3. Enzyme Technology

(*i*) Biological washing powders or Detergents

(*ii*) Pectinase in fruit juices.

(*i*) Detergents

Detergents represent the largest industrial application of enzymes amounting to 25–30% of the total sales of enzymes. The enzymes used in detergents must be cost-effective, safe to use and be able to perform the task in the presence of ionic and non-ionic detergents.

Enzymes constitute only 0.4–0.8% crude enzyme by weight. The main enzymes used are **proteases**, **α-amylases** and sometimes **cellulases**.

- **Proteases** are used to digest proteins present in blood stains, milk, grass etc. and also in association with dirt; therefore they help in the removal of dirt as well. Only serine proteases are suitable for use in detergents. These enzymes are produced by ***Bacillus licheniformis*** and ***Bacillus amyloliquefaciens***.

  Proteases are packed inside dust-free granules coated with wax materials made from paraffin oil.

- **α-Amylase.** It is used to digest the starch present in association with dirt and stains. They are produced by ***B. licheniformis.***

- **Cellulases.** It is produced by fungi and are used for washing cotton fabrics. The enzyme digests the small fibres raised from the fabric without damaging the major fibres of the fabric. This restores the fabric to "a new" condition and also removes soil particles by digesting the associated cellulase.
- **Lipases.** Which are suitable for detergent use have been identified and are used for digestion of lipids present in stains or dirt.

(*ii*) Pectinase in Fruit Juices

The cloudiness of fruit juices and wines is primarily due to pectins. It exhibits various degrees of methyl esterification and are usually associated with other plant polymers and even cell debris.

The pectins are digested by pecteolytic enzymes prepared from **Aspergillus niger.**

These enzyme preparations are mixtures of the following enzymes and activities:

(*i*) **polygalacturonase** (random digestion of pectin)

(*ii*) **pectin esterase** (removal of methyl esters and release of methanol)

(*iii*) **pectin lyase** (cleaves pectins into oligosaccharides even without pectin esterase action).

(*iv*) **hemicellulose** (a mixture of enzymes degrading hemicellulose).

These enzymes work best in an optimum pH between 4–5 and optimum temperature below 50°C.

EVALUATE YOURSELF

I. Objective Type Questions

Select the most appropriate choice from the options given as (a), (b), (c) and (d) after each question:

1. All the information needed to make an entire organism is in a molecule called
   (*a*) DNA (*b*) a protein
   (*c*) a cell (*d*) a gene.
2. What controls inherited traits in an organism such as eye, hair, or skin colour?
   (*a*) meiosis (*b*) a bacterium
   (*c*) a cell (*d*) a gene.
3. Examples of biotechnology in animals include
   (*a*) animal cloning (*b*) cancer research
   (*c*) human tissue replacement
   (*d*) all of the above.
4. A transgenic organism is one that
   (*a*) has genes inserted into it from another organism
   (*b*) has genes created through mitosis
   (*c*) normally occurs in nature
   (*d*) none of the above.
5. Clones are
   (*a*) organisms from which genes are removed
   (*b*) genes that cause the production of twins
   (*c*) genetically identical organisms
   (*d*) organisms which produce a clone protein.
6. Bioreactors are
   (*a*) pieces of machinery that extract genes
   (*b*) organisms engineered to make specific compounds
   (*c*) nuclear power plants that use biotechnology
   (*d*) none of the above.
7. Goals of biotechnology in plants include making plants
   (*a*) stronger (*b*) more productive
   (*c*) more nourishing (*d*) all of the above.
8. Biotechnology may help some people in developing countries who need
   (*a*) vaccinations and disease prevention
   (*b*) better nutrition
   (*c*) more food
   (*d*) all of the above.
9. In humans, genes control all of the following except
   (*a*) nutrition (*b*) eye colour
   (*c*) skin colour (*d*) inherited diseases.
10. Biotechnology can be used in all of the following except
    (*a*) plants (*b*) animals
    (*c*) human medicine (*d*) replication of cells.
11. The most commonly-used bacterium for gene transfer in plants is
    (*a*) Bacillus thuringiensis
    (*b*) Escherichia coli
    (*c*) Agrobacterium
    (*d*) none of the above.
12. When a gene is transferred into an organism, the organism is called
    (*a*) transgenic (*b*) transformed
    (*c*) all of the above (*d*) none of the above.
13. What are inserted into a cells's DNA along with genes of interest to visually identify transgenic character?
    (*a*) marker genes (*b*) cultures
    (*c*) clones (*d*) bioreactors.

II. Fill in the Blanks

14. Complete the following sentences by supplying appropriate words:
    (*i*) The production of cheese involves separation of ...... from whey.
    (*ii*) A clone consists of ...... of a single individual or cell.
    (*iii*) A/An ...... is used to introduce recombinant DNA into cells.
    (*iv*) Genetic engineering has enabled the production of large numbers of ...... proteins.
    (*v*) Leaching of copper sulphate from the ores is possible with ...... ferroxidans bacteria.
    (*vi*) Biomass produced by unicellular organisms is called ......
    (*vii*) The pasteurized milk with known cultures of micro-organisms is referred to as ......
    (*viii*) Dough is ...... by bread yeast which ...... the sugars in dough to produce ...... and ......
    (*ix*) Human insulin is a dimer comprising of two chains A and B which are derived from a single polypeptide chain called ......
    (*x*) The micro-organism used for SCP must be of ....
    (*xi*) Recombinant vaccines contains ...... encoding a protein of a ....
    (*xii*) The process of producing ethanol using cellulose is done by the micro-organism ....

III. Discussion Questions

15. Explain briefly the application of biotechnology in the industries of
    (*i*) Mining (*ii*) Medical products.
16. Outline the various steps involved in cheese production.
17. What is Genetic Engineering? Explain briefly.
18. What are Biofuels? Give useful features of biofuels.
19. Explain the following terms
    (*i*) SCP (*ii*) Starter culture
    (*iii*) Brewing.
20. How is Humulin produced? Explain.
21. Give the main steps involved in the process of drug designing.
22. What is an ideal vaccine and a recombinant vaccine? Clearly differentiate between the two.
23. What is the objective of gene transfer in animals?
24. Explain briefly the various types of fuels used.
25. Discuss two methods for treatment of solid wastes.

Answers

I. Objective Type Questions

1. (*a*)	**2.** (*d*)	**3.** (*d*)
4. (*a*)	**5.** (*c*)	**6.** (*b*)
7. (*d*)	**8.** (*d*)	**9.** (*a*)
10. (*d*)	**11.** (*c*)	**12.** (*c*)
13. (*a*).		

II. Fill in the Blanks

(*i*) curd (*ii*) sexual progeny
(*iii*) vector (*iv*) recombinant
(*v*) thiobacillus (*vi*) single cell protein
(*vii*) starter culture
(*viii*) leavened, ferments, CO_2, alcohol
(*ix*) preproinsulin (*x*) good nutritional valve
(*xi*) gene, pathogen origen
(*xii*) Trichoderma reesei.

SECTION-1

UNIT-54

GLASS AND ITS USES

Specific Objective

After studying this unit, the student should be able to:

- describe raw materials used in making glass.

54.1 GLASS

Glass is a *transparent or translucent amorphous supercooled solid solution of silicates and borates.* The most common silicates present are those of potassium, calcium and lead. It is not a true compound because its composition is not fixed. An approximate formula for ordinary glass may be given as:

$$R_2O.MO.6SiO_2$$

where R = Na or K and M = Ca, Ba, Zn or Pb

SiO_2 component in the above general formula can be replaced by Al_2O_3, B_2O_3 or P_2O_5.

Some of the common **physical characteristics** of different types of glass are as follows:

- Glass is generally a transparent or sometimes a translucent non-crystalline solid.
- Glass does not have a definite melting point. On heating, it gradually softens and can be moulded into any desired shape. It is this property which makes glass a useful material for making articles of different shapes and sizes.
- It is poor conductor of heat and electricity.
- It has sufficiently high viscosity (>1013 poise).
- Glass can incorporate colouring material preserving transparency.
- Glass melts into a liquid at very high temperature.

54.2 RAW MATERIALS FOR MAKING GLASS

Glass is completely a vitrified product which has no definite composition. The major ingredients which form over 90% of all types of glasses are *Lime* ($CaCO_3$), *Sand* (SiO_2) and *soda ash* (Na_2CO_3). The raw materials used in the manufacture of glass can be broadly divided into the following categories:

1. Acidic oxides
2. Basic oxides
3. Cullets and
4. Colouring matter.

1. ACIDIC OXIDES

A large variety of acidic oxides can be used. The choice depends upon the quality of glass to be made. The various acidic oxides that can be used are as follows:

- **Sand (SiO_2).** It forms silicates on fusion with the other ingredients. The sand used in glass manufacture should be pure and free from iron as it imparts colour to the glass. The sand is washed before it is used in order to make it free from impurities. The sand used should not be too fine as it makes the reaction violent. It should not be too coarse either because it makes the reaction slow.
- **Borax ($Na_2B_4O_7.10H_2O$).** It is a minor ingredient which supplies B_2O_3. Besides its high fluxing power, borax lowers the coefficient of expansion of glass and adds to its chemical durability.
- **Phosphorus(V) Oxide, P_2O_5.** It is used as Na_2HPO_4 or $Ca_3(PO_4)_2$ and is added in tableware glass for imparting bright appearance.
- **Arsenious oxide, As_2O_3.** It is generally used to remove air bubbles from the glass.

2. BASIC OXIDES

Some of the basic oxides used are as follows:

- **Limestone ($CaCO_3$).** It provides CaO. Sometimes burnt dolomite containing CaO and MgO is employed as a substitute of lime.
- **Feldspar.** It is a naturally occurring mineral with a formula $Na_2O.\ Al_2O_3.\ 6SiO_2$ (soda feldspar) or $K_2O.\ Al_2O_3.\ 6SiO_2$ (Potash feldspar). It is cheap, pure and easily fusible. Feldspar is used as a flux, *i.e.*, to lower the melting point, and to retard denitrification of glass.
- **Lead oxide.** Litharge (PbO) or red lead (Pb_3O_4) is used for the production of heavy and high quality flint glass, which is used in optical instruments, tableware and decorative articles.

3. CULLET

It is the crushed glass from imperfect or defective articles or their trimmings. It represents the utilisation of waste and it can easily be melted.

4. COLOURING MATTER

Coloured glasses are obtained by adding certain metallic oxides or salts in the fused mass. Various colouring agents and the corresponding shades are grouped as follows:

Red : CdO (1.4%), Se (1%)
Ruby red : (*i*) Colloidal gold or ruby gold, (*ii*) Cu_2O
Light yellow : Cerium oxide (2%), Titanium oxide (TiO_2)
Deep blue : Cobalt oxide (CoO) (0.1%)
Greenish blue : Copper (II) oxide, CuO
Emerald green : Cr_2O_3 (0.15%), CoO (0.001%)
Amber : FeS (0.3%)
Black : $MnO_2 + Fe_2O_3$
Yellow : CdS

54.3 MANUFACTURE OF GLASS

The manufacturing process of glass consists of four major operations: (1) *Melting*, (2) *Shaping*, (3) *Annealing*, (4) *Finishing*. Each operation is discussed briefly as follows:

1. **Melting.** The ingredients called *batch materials* are mixed in the appropriate proportion and heated to fusion in a furnace. Many designs of glass furnace are in use. The two most commonly used furnaces are: (*i*) Pot furnace and (*ii*) Tank furnace.

(*i*) **Pot Furnace.** In this furnace, the charge is fused in fire clay pots. The pots may be opened or closed. The closed pots are used when the glass is to be protected from the products of combustion.

The batch materials are put in the pots. They are placed in a circle inside a furnace and heated by burning producer gas around them (Fig. 54.1). When the fusion is complete the pots are removed from the furnace and the fused plastic mass is taken out for shaping. *Pot furnace is employed for the production of high quality glass, since the charge remains protected from the products of combustion.*

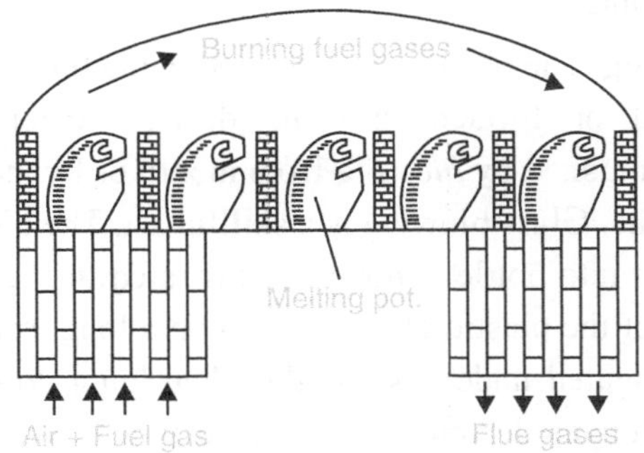

Fig. 54.1. Pot furnace

(*ii*) **Tank Furnace.** It consists of a large rectangular tank built of fire clay blocks. The batch materials are fed into the tank and producer gas is used as a fuel in the furnace. (Fig. 54.2).

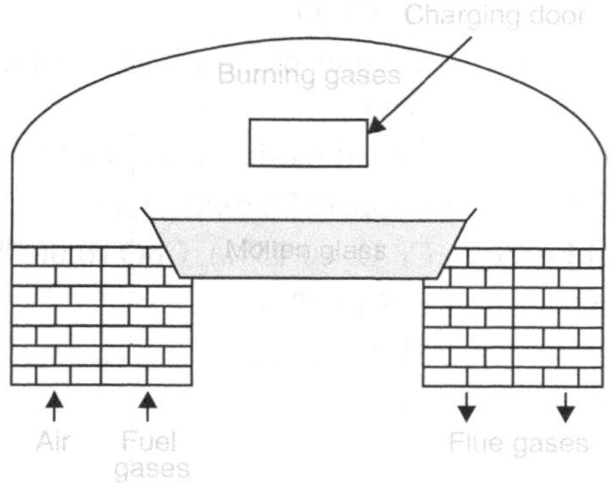

Fig. 54.2. Tank furnace

The charge is heated at 1400°–1500°C for 10–12 hours. The chemical reactions involved in both furnaces are:

$$Na_2CO_3 + SiO_2 \longrightarrow Na_2SiO_3 + CO_2$$

$$2Na_2SO_4 + 2SiO_2 \longrightarrow 2Na_2SiO_3 + O_2 + 2SO_2$$

$$CaCO_3 + SiO_2 \longrightarrow CaSiO_3 + CO_2$$

At 1400°C silica also in silicates of calcium and sodium fuses thus $Na_2SiO_3 + CaSiO_3 + 4SiO_2 \longrightarrow$

$$\underbrace{Na_2SiO_3 \, . \, CaSiO_3 \, . \, 4SiO_2}_{\text{Glass}}$$

During the melting, a lot of frothing is caused owing to the evolution of gases like CO_2, SO_2, O_2, etc. When the frothing subsides, the temperature is raised and the molten glass is allowed to stand for some time. This is called refining, and its objective is to form a homogeneous mass free from gas bubbles and bits of undissolved material or batch stones. Tank furnace is a *continuous process* and usually employed for the production of large quantities of only one variety of glass at a time.

2. **Shaping.** The plastic glass formed in the furnace is next shaped or formed into the desired articles. It is accomplished by *blowing* from mouth or by means of a machine. Glass blowing is a skillful art. The blowing of a glass into bottle is done as illustrated in Fig. 54.3. A lump of the plastic glass is taken on a long iron pipe. It is elongated under its weight when hung downwards. The elongated lump is introduced into a mould and is inflated by blowing air into it from the mouth. On cooling, the bottle is taken out by removing the two-halves of the mould.

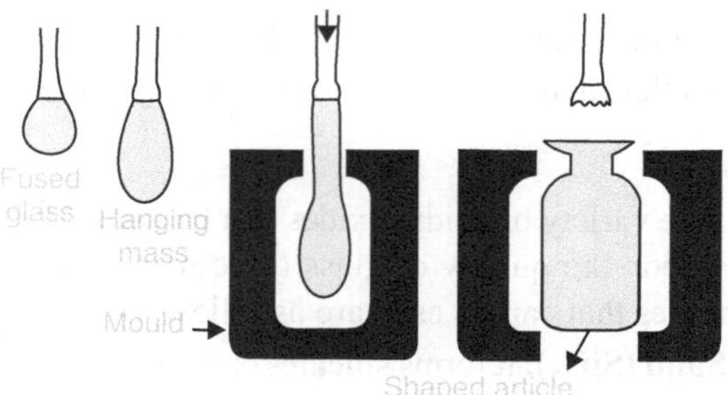

Fig. 54.3. Shaping of glass lump

3. **Annealing.** It is a process of cooling slowly the newly shaped articles. If they are cooled quickly they become brittle on account of the high internal strain. Annealing allows the molecules to arrange themselves in such a way that there is no internal strain when the mass is cooled.

Annealing is done in a tunnel like oven called *lehr* which is 50 to 60 feet long. At one end the temperature is a little below the softening point of glass, *i.e.*, 500–600°C and it gradually falls along the length of the oven. At the other end the temperature is almost the same as the room temperature. Immediately after shaping, the articles are introduced into the *lehr* at the hotter end and travel towards the cooler end by means of a moving belt. It takes a few hours for the articles to move along through the tunnel. Some high quality glasses require long annealing.

4. **Finishing.** The articles obtained from the lehr are subjected to a number of operations such as cleaning, polishing, grinding, rounding edges, etc., for bringing them to a useable form.

54.4 DIFFERENT VARIETIES OF GLASS

Some important types of glasses along with their uses are described briefly as follows:

1. **Lime Glass.** This glass involves the use of lime or limestone along with the alkali carbonates (Na_2CO_3 and CO_3). This is further classified into:

(*i*) Soft Glass and (*ii*) Hard Glass.

(*i*) *Soft Glass.* It is made by fusing soda ash, sand and limestone together. It is believed to have an approximate composition $Na_2O.CaO.5SiO_2$ (or $Na_2O.\ CaO.6SiO_2$). It fuses at a low temperature and cannot stand mechanical injuries. It is used for making bottles, tumblers, window panes, cheap glassware, etc.

(*ii*) *Hard Glass.* It is obtained by fusing potassium carbonate and limestone together. It is understood to have a formula $K_2O.CaO.5SiO_2$ (or $K_2O.CaO.6SiO_2$). It is harder and stronger than the soda glass. It is used for making apparatus capable of standing high temperature and pressure.

2. **Flint or Lead Glass.** This is made from potassium carbonate, lead oxide or red lead and sand. This is heavy, shining and extraordinarily durable. This is used for making optical instruments like lenses, prisms, etc. and cut glass articles, like objects of decoration, imitation diamonds or jewels.
3. **Borosilicate Glass.** In this glass a part of the silica is replaced by boric oxide (B_2O_3). This is a super cooled solution of various borates and silicates and hence its name borosilicate glass. This is made by fusing a mixture of sand, lime, borax and alkali carbonates. Sometimes a little alumina (Al_2O_3) and zinc oxide are also added. The former makes the glass more durable and the latter adds to its thermal and chemical resistance. **Pyrex glass** and **Jena glass** are two well-known trade marks of borosilicate glasses, which are widely used for making laboratory apparatus, thermometers, kitchenware, tableware, insulators, etc.
4. **Crooks' glass.** It is a special type of glass containing cerium oxide. It does not allow the passage of ultraviolet rays. It is used for making lenses.

ETCHING OF GLASS

Glass is attacked by hydrofluoric acid. HF and glass react to form a soluble compound hydrofluoric silicic acid (H_2SiF_6). This property is used in the etching of glass.

The glass to be etched is coated with a thin layer of wax and the design to be produced is scratched with a needle or blade. An aqueous solution of hydrofluoric acid is applied to the exposed part. After some time it is placed in water and the wax is removed from the surface. The marks made by blade get engraved on the exposed parts.

54.5 USES OF GLASS

We have learned that different types of glasses can be prepared by adjusting the chemical composition of the batch material. Different types of glasses are prepared for specific uses. Some of the common uses of glass are as follow:

1. Glass in Construction

- Soda glass is used in making window panes, mirrors and internal glazed partitions.
- Safety type glass, tempered glass and laminated glass are used in buildings.
- Glass-fibre insulations are common in roofs and walls.

2. Glass in Decoration and Fabric

Coloured glasses and painted glasses are used for manufacturing fancy articles, artificial gems, and decorating fabric.

3. Glassware

Household articles are generally made from soft glass. Optical instruments, cut glass objects are made from flint glass or lead crystal glass. Laboratory ware are made of borosilicate glasses.

4. Mirrors and Lenses

Mirrors are generally made from soft glass. Lenses for spectacles, microscopes, telescopes, cameras, etc., are made from crooks' glass. Sunglasses, however, are made from photo-chromatic glasses.

SUMMARY

- **Glass.** Glass is a transparent supercooled solid solution of silicates and borates. It is a pseudo solid and non-crystalline in nature. The *raw material* used for the production of glass consists of acidic oxides, basic oxides colouring matter and cullet. It is manufactured by heating the ***batch material*** in a furnace. The molten glass so obtained is blown into desired article/product and is then subjected to annealing followed by finishing.

EVALUATE YOURSELF

I. Objective Type Questions

Select the most appropriate choice from the options given as (a), (b), (c) and (d) after each question:

1. The approximate formula of ordinary glass is $R_2O.MO.6SiO_2$. In this R can be

(*a*) Al (*b*) Ba

(*c*) Ra (*d*) Na.

2. Which property is not associated with glass?

(*a*) crystalline nature (*b*) amorphous nature

(*c*) pseudo solid (*d*) super cooled liquid.

3. Which of the following is not a raw material for ordinary soda glass?

(*a*) $CaCO_3$ (*b*) SiO_2

(*c*) Na_2CO_3 (*d*) $PbCO_3$.

4. The formula of Borax is

(*a*) $NaBO_3$ (*b*) Na_3BO_3

(*c*) $Na_2B_4O_7.10H_2O$ (*d*) $Na_4B_2O_7.10H_2O$.

5. Addition of CdS to the glass melts imparts ...... colour to the glass
(*a*) Blue (*b*) Red
(*c*) Brown (*d*) Yellow.
6. Which oxide is present in flint glass?
(*a*) PbO (*b*) NO_2
(*c*) RbO (*d*) As_2O_3.

II. Fill in the Blanks

7. Complete the following sentences by supplying appropriate words:
(*i*) The halogen acid used for etching of glass is ..... .
(*ii*) Variety of glass which does not allow *u.v* rays is ..... .
(*iii*) Soft glass is generally made by fusing ..... .
(*iv*) Flint glass is generally made of ..... .
(*v*) Pyrex glass is a kind of ..... .

III. Discussion Questions

8. What is glass? List some of its common physical characteristics.
9. Name the raw materials for the preparation of glass.
10. Name the basic steps used for the manufacture of glass.
11. Describe the tank furnace for the manufacture of glass. Give various chemical reactions taking place.
12. Explain the following terms.
Hard glass, Soft glass, Pyrex glass, Flint glass
13. Discuss the etching of glass.

Answers

I. Objective Type Questions

1. (*b*)	2. (*a*)	3. (*d*)
4. (*c*)	5. (*d*)	6. (*a*).

II. Fill in the Blanks

7. (*i*) HF (*ii*) Crooks' glass
(*iii*) soda ash, sand and lime stone
(*iv*) K_2CO_3, PbO, PbO_2, SiO_2
(*v*) borosilicate glass.

SECTION-1

UNIT-55

CEMENT AND ITS USES

Specific Objectives

After studying this unit, the student should be able to:

- describe raw materials used in clinker and cement production
- describe process of cement production
- describe the environmental impact of cement production
- describe the uses of cement.

55.1 CEMENT

Cement is a general name given to a powdered material which initially have plastic flow when mixed with water, but sets into a hard solid structure in several hours with varying degree of strength and bonding properties. It was discovered in 1824 by an English Mason Joseph Aspdine.

Chemically, cement is a *fine ground mixture of calcium aluminates and silicates of varying composition which hardens into a rigid solid mass after treatment with water*. It is called **portland cement** because of the resemblance of the hardened mass to ***portland rock***, a famous building stone in England.

Composition of Portland Cement

Portland Cement is a dirty greyish heavy powder containing calcium aluminates and silicates. The silicates and aluminates which form more than 90% of cement are:

- Tricalcium silicate ($3CaO.SiO_2$)
- Dicalcium silicate ($2CaO.SiO_2$)
- Tricalcium aluminate ($3CaO.Al_2O_3$)
- Tetra calcium alumino ferrite ($4CaO.Al_2O_3.Fe_2O_3$)

Of these, tricalcium silicate is the most important.

55.2 RAW MATERIAL FOR CLINKER PRODUCTION

The essential *raw materials* for the production of cement are **limestone** and **clay** which supply all the *four principle ingredients*; CaO, Al_2O_3, SiO_2 and Fe_2O_3. Calcium oxide and iron oxide are obtained from ***calcareous materials*** like limestone, chalk, calcium carbonate sluge, etc. while silica and alumina are obtained from ***siliceous or argillaceous materials*** like clay, blast furnace slag, slate, etc. The composition of normal portland cement is as follows:

Constituents	*Per cent*
Lime (CaO)	62
Silica (SiO_2)	22.5
Alumina (Al_2O_3)	7.5
Magnesia (MgO)	2.0
Iron Oxide (Fe_2O_3)	2.5
Sulphur trioxide (SO_3)	1.25
Alkalis (Na_2O, K_2O) from clay	0.5

55.3 PRODUCTION OF CEMENT

FORMATION OF CEMENT CLINKERS

The raw materials, **lime** and **clay** are mixed and subjected to the following processes.

1. Physical Treatment of Raw Materials

It involves mixing and making the raw materials ready for chemical treatment. Two common processes employed in the physical treatment are **dry process** and **wet process**.

(*a*) Dry Process: In this process, the calcareous and argillaceous materials are ***mixed*** in the ratio 3 : 1, ***crushed, dried, sieved*** and then ***ground to a fine powder***. The mixture is now ready for the chemical treatment in the kiln.

(*b*) Wet Process: In this process, the rocky matter is ***crushed*** and water is sprinkled over it. The resulting materials is ground in the wet condition to form a ***slurry***. The slurry, which contains 35–40% of water, is stored in large tanks equipped with mechanical stirrers. It is then fed into the kiln.

2. Chemical Treatment

It involves ***calcination*** of the mixture of **limestone** and **clay**. *Dry mixture* from the dry process or the *slurry* from the wet process is fed into the rotary kiln (Fig. 55.1) for calcination.

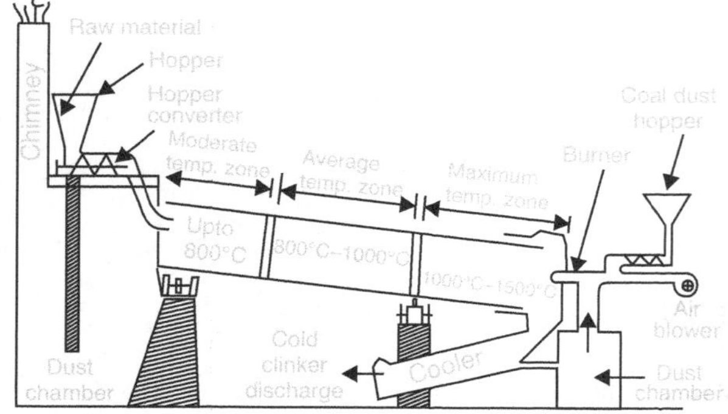

Fig. 55.1. Rotary kiln for manufacture of portland cement.

Rotary kiln consists of a steel cylinder lined with fire bricks. It is 45–120 *m* long and has about 3 *m* internal diameter. The kiln is inclined at an angle of about 15 degrees to the horizontal and it rotates on its axis at the rate of 1/2 to 1 revolution per minute. The charge is introduced at the upper end and it travels down as the kiln rotates.

The charge is heated by burning ***fuels*** like ***pulverized coal*** or, ***fossil fuel***, or ***oil*** or ***natural gas***. The hot fuel gases are made to enter at the lower end of the kiln with the help of a blower. As the charge moves forward, it meets higher temperatures till it reaches the lower end where the temperature is highest (around 1400°C). The charge takes 2 to 3 hours to cover the journey in the kiln. Reactions taking place in the rotary kiln in three major zones are as follows:

(*a*) *Maximum Temperature Zone (1000°C–1500°C)*: The main reactions between lime, alumina and silica takes place near the lower end of the kiln resulting in the formation of calcium silicates and aluminates as shown below.

$$2CaO + SiO_2 \longrightarrow 2CaO.SiO_2$$

Dicalcium silicate

$$3CaO + Al_2O_3 \longrightarrow 3CaO.Al_2O_3$$

Tricalcium aluminate

$$3CaO + SiO_2 \longrightarrow 3CaO.SiO_2$$

Tricalcium silicate

$$4CaO + Al_2O_3 + Fe_2O_3 \longrightarrow 4CaO.Al_2O_3.FeO_3$$

Tetracalcium alumino ferrite

(*b*) *Moderate Temperature Zone (up to 800°C)*: At 100°C, the removal of free moisture takes place. At 500°C, kaolin (Al_2O_3,$2SiO_2$.$2H_2O$) present in clay breaks up into amorphous silica and alumina.

(*c*) *Average Temperature Zone (up to 800°C–1500°C)*: Limestone decomposes to form lime and carbon dioxide.

$$CaCO_3 \longrightarrow CaO + CO_2$$

Moreover, due to the high temperature in this zone, about 20–30% of mass melts and combines with solids to form pebbles (which vary in size) and are called **cement clinkers.**

The burnt gases (containing a good amount of dust) are made to pass through a dust chamber, where most of the dust is retained and gases are allowed to escape through a chimney.

The hot clinkers from the rotary kiln go on to coolers which consist of a series of tubes parallel to the kiln in which air is forced through. As a result of this, the clinkers cool down while the air gets heated. The hot air is then used for the combustion of the fuel as an economy measure.

FORMATION AND PACKING OF CEMENT

The cooled **clinkers** are mixed with 3-5% of **gypsum $CaSO_4.2H_2O$**, along with certain *dispersing* and *water proofing agents* and then ground to a fine powder. Since pure finely powdered, clinkers set very rapidly, therefore, gypsum has to be added to retard the setting of cement. The powdered material is passed through fine sieves and finally it is sent to the automatic packing machines where it is packed in jute or polythene bags.

Group Discussion

To explore the composition of raw materials used in making cement

1. Collect the samples of raw materials used for the production of cement, limestone, clay and coal.
2. Explore the chemical composition of each of them using literature.
3. Name the principal ingredients of cement.
   - Name the ingredients supplied by limestone
   - Name the ingredients supplied by clay.
   - What is the function of coal in cement manufacture?

SETTING OF CEMENT

When cement is mixed with water and left for some time, it becomes a hard mass. This is known as **setting of cement**. The setting of cement is a complicated process and the exact mechanism of this process is not known. It is believed that various aluminates and silicates present in the cement form hydrates with water which separates in the form of gels.

$$3CaO.Al_2O_3 + 6H_2O \xrightarrow{\text{Hydration}} \underset{\text{Colloidal gel}}{3CaO.Al_2O_3.6H_2O}$$

$$2CaO.SiO_2 + xH_2O \xrightarrow{\text{Hydration}} \underset{\text{Colloidal gel}}{2CaO.SiO_2.xH_2O}$$

$$4CaO.Al_2O_3.Fe_2O_3 + 6H_2O \xrightarrow{\text{Hydration}} \underset{\text{Colloidal gel}}{3CaO.Al_2O_3.6H_2O + Fe_2O_3.CaO}$$

At the same time some $Ca(OH)_2$ and $Al(OH)_3$ are formed as precipitates due to hydrolysis.

$$3CaO.SiO_2 + H_2O \xrightarrow{\text{Hydrolysis}} Ca(OH)_2 + 2CaO.SiO_2$$

$$3CaO.Al_2O_3 + 6H_2O \xrightarrow{\text{Hydrolysis}} 3Ca(OH)_2 + 2Al(OH)_3$$

Gypsum combines with tricalcium aluminate to form calcium sulpho-aluminate.

$$3CaO.Al_2O_3 + 3CaSO_4 + 2H_2O \longrightarrow 3CaO.Al_2O_3.3CaSO_4.2H_2O$$

This reaction slows down the setting.

The gels formed start losing water partly by evaporation and partly by forming hydrates with unhydrated constituents. This results in the formation of a hard mass. $Ca(OH)_2$ binds the particles of calcium silicate together, while $Al(OH)_3$ fills the interstices resulting in hardening of the mass.

Setting of cement is an exothermic process. Hence, cement structures have to be cooled during setting by sprinkling water.

55.4 ENVIRONMENTAL IMPACT OF CEMENT PRODUCTION

Although cement is very useful in buildings and construction work, its cement production has adverse effect on the environment in the following way. One of the major pollutants released from cement plants is ***cement dust***. Its harmful effects are:

- It increases the concentration of particulate matter in the air.
- The heavy metals present in cement dust cause disturbances in various metabolic activities which further retards the process of growth.
- It also causes breathing problems and lungs infections.
- Direct contact of cement dust with skin become the cause of many allergies and skin diseases.
- Cement kiln-dust gets accumulated on the soil and researchers have reported that it changes the soil alkalinity which is harmful for crops.

USES OF CEMENT

1. Cement on a very large scale is used in building construction. For this purpose it is generally used in the form of a ***thick paste*** of cement, sand and water in appropriate proportions.
2. In the form of **concrete**, it used for the construction of floors, and roads. **Concrete** is a mixture of *cement, sand, gravel or small pieces of stone* along with *water*. It sets into an exceedingly hard structure.
3. In the form of **reinforced concrete** it is used in building roofs and bridges, etc. Reinforced concrete is the concrete which is made more stronger by

embeding into it steel frames or wire netting. The structures made of reinforced concrete are extremely strong because in these structures the compression strength of cement is coupled with the tensile strength of steel.

4. It is also used in making *asbestos sheets*.

SUMMARY

- **Cement** is a fine mixture of calcium aluminates and silicates of varying composition which sets into a rigid solid mass after treatment with water. The raw materials required for production of cement are **lime** and **clay.** For production of cement, the raw materials are crushed and their mixture is subjected to calcination at 1400–1600°C to get *cement clinkers*. The clinkers are mixed with gypsum and other additives and ground to get cement.

EVALUATE YOURSELF

I. Objective Type Questions

Select the most appropriate choice from the options given as (a), (b), (c) and (d) after each question:

1. Cement is
   (*a*) Mixture of various elements
   (*b*) Single compound of definite composition
   (*c*) Mixture of aluminates and silicates of alkali metals
   (*d*) Mixture of calcium aluminates and silicates.
2. The credit of discovery of portland cement goes to
   (*a*) Rutherford (*b*) Joseph Aspdin
   (*c*) Joseph Pristlay (*d*) Linus Pauling.
3. Which of the following is not a constituent of Portland cement?
   (*a*) CaO (*b*) Fe_2O_3
   (*c*) Al_2O_3 (*d*) B_2O_3.
4. The colour of portland cement is
   (*a*) Greyish white (*b*) Greyish black
   (*c*) Bluish white (*d*) White.
5. In the rotary kiln the product that we get is
   (*a*) Clinkers (*b*) Concrete
   (*c*) Gypsum (*d*) Powdered cement.
6. The formula of gypsum is
   (*a*) $CaSO_4.\frac{1}{2}H_2O$ (*b*) $CaSO_4$
   (*c*) $CaSO_4.2H_2O$ (*d*) $CaSiO_3$.

II. Fill in the Blanks

7. Complete the following sentences by supplying appropriate words:
   (*i*) The most important silicate component of cement is ......
   (*ii*) The charge introduced in the lime kiln consists ......
   (*iii*) The temperature in the rotary kiln ...... as we move down the kiln.
   (*iv*) The formation of silicates/aluminates in the kiln occurs at the temperature range of ......
   (*v*) Concrete is a name given to ......

III. Discussion Questions

8. What is cement? Name its important raw materials.
9. Describe briefly the production of clinkers from the raw materials.
10. How is cement produced from clinkers? What is the function of adding gypsum to it?
11. Describe the chemical reactions involved in a rotary kiln.
12. What do you understand by the setting of cement?
13. Give important uses of cement.
14. Explain the terms concrete and reinforced concrete.

Answers

I. Objective Type Questions

1. (*d*) **2.** (*b*) **3.** (*b*)
4. (*b*) **5.** (*a*) **6.** (*c*).

II. Fill in the Blanks

7. (*i*) $3CaO.SiO_2$ (*ii*) finely ground lime and clay
(*iii*) increases (*iv*) 1000–1500°C
(*v*) mixture of cement, sand and gravels.

SHS 4

SECTION

2

BASIC BIOCHEMISTRY

General Objectives

After studying this section, the student should be able to:

1. recognize important applications of functional group chemistry in the manufacture and use of organic compounds.
2. show awareness of the general impact of organic products on our lives and the environment e.g., medicines, clothing, shelter, plastics and detergents.

In this section, we shall study

Unit 56: Proteins

Unit 57: Carbohydrates

Unit 58: Fats and Oils

Unit 59: Synthetic Polymers

SECTION-2

UNIT-56

PROTEINS

Specific Objectives

After studying this unit, the student should be able to:

- describe the sources, preparation and properties of proteins
- describe the nomenclature of amino acids
- describe proteins as a natural polymer
- describe the uses of proteins

56.1 INTRODUCTION

Proteins are the complex organic substances which are the basis of protoplasm and are found in all living organisms. The name protein (Greek, *proteios means first*) was introduced by Mudlar (1839) because of the prime importance of such substances to animal life. In human beings, proteins constitute about 18% by mass of the body.

In general, proteins are polymers of α-amino acids. The amino acid units in proteins are held by peptide (**—CONH—**) linkages. Polymeric products of α-amino acids with molecular mass up to 10,000 are called *polypeptides* while those having molecular mass higher than 10,000 are considered as *proteins*. However, there is no sharp demarcation between polypeptides and proteins.

SOURCES OF PROTEINS

Proteins are found in animal as well as plant foods. Animal foods such as ***meat, poultry, fish, eggs, milk and cheese*** contain proteins with many essential amino acids. Plant products such as ***beans, groundnuts, cashew nut, cereals (maize and wheat) and pulses*** are also good protein food, though they contain fewer essential amino acids.

56.2 AMINO ACIDS

These are amino substituted carboxylic acids. Of the various amino acids, the α-amino acids or 2-amino acids are quite important because they are the building blocks of peptides and proteins. In α-amino acids, the amino group is present on the α-carbon atom (*i.e.*, C atom next to COOH group). Thus, they can be represented by the general formula,

$$R{-}CH(NH_2){-}COOH$$

Group R—is different for different α-amino acids.

About 20 of the α-amino acids have been identified as the constituents of most of animal and plant proteins. The formulae of such amino acids along with their common and abbreviated names have been given in Table 56.1.

CLASSIFICATION OF AMINO ACIDS

The amino acids containing equal number of amino and carboxyl groups are known as **neutral amino acids.** The amino acids which contain more of carboxyl groups than amino groups are known as **acidic amino acids** while those containing more of amino groups are called **basic amino acids.** For example, *glutamic acid* and *aspartic acid* are acidic amino acids while *lysine* is basic amino acid. *Alanine* is a neutral amino acid.

Add To Your Knowledge

ESSENTIAL AND NON-ESSENTIAL AMINO ACIDS

Out of the 20 amino acids which are required for protein synthesis, the human body can synthesize 10. The amino acids which the body can synthesize are called **non-essential amino acids.** The remaining 10 amino acids which the body is not able to synthesize are called **essential amino acids.** The essential amino acids must be supplied to our bodies through our diet because they are required for proper health and growth. The deficiency of essential amino acids may cause diseases like *Kwashiorkor* in which the water balance of the body is disturbed.

STRUCTURE OF AMINO ACIDS

It has been found by spectroscopic means that the —COOH and —NH_2 groups of amino acids do not exist as such but they react with each other to form an *internal salt* which is also called a **zwitter ion.** In the formation of a zwitter ion, a proton from the —COOH part of the molecule is released and attaches itself to the —NH_2 part to form a dipolar ion as shown below:

$$H_2\ddot{N}{-}CH(R){-}C(=O){-}\ddot{O}{:}\,\text{(H)} \longrightarrow H_3\overset{\oplus}{N}{-}CH(R){-}C(=O){-}\ddot{O}{:}^{\ominus}$$

Zwitter ion

Zwitter ion is a neutral species but carries both positive and negative charges.

PROPERTIES OF AMINO ACIDS

Amino acids are usually colourless, crystalline and high melting solids. They are moderately soluble in water. They are insoluble in organic solvents such as ether, benzene and petroleum ether. In *acidic solution* amino acids exist as *cations* and migrate towards the cathode in an electric field whereas in basic solutions they exist as *anions* and migrate towards the anode. At an intermediate pH, however, they exist as a **zwitter ion** (a dipolar ion) and do not migrate towards either electrode. This pH is known as the **isoelectric** point of the α-amino acid.

$$\underset{\text{Cation}}{H_3\overset{+}{N}{-}CH(R){-}COOH} \underset{H_3O^+}{\overset{OH^-}{\rightleftharpoons}} \underset{\text{Zwitter ion}}{H_3\overset{+}{N}{-}CH(R){-}CO\overset{-}{O}}$$

$$\underset{\text{Zwitter ion}}{H_3\overset{+}{N}{-}CH(R){-}CO\overset{-}{O}} \underset{H_3O^+}{\overset{OH^-}{\rightleftharpoons}} \underset{\text{Anion}}{H_2N{-}CH(R){-}CO\overset{-}{O}}$$

Different amino acids have different isoelectric points. The isoelectric point of an amino acid depends upon the functional groups present in the amino acid. Neutral amino acids have isoelectric point in the pH range of 5.6–6.3.

At isoelectric point the amino acids have least solubility in water. This property of amino acids is exploited in the separation of different α-amino acids obtained by hydrolysis of proteins.

α-Amino acids exhibit chemical reactions characteristic of a primary —NH_2 group and a —COOH group.

NAMING THE AMINO ACIDS

Amino acids are known by their common names. However, they can also be named by following the IUPAC rules. In protein or peptides, the amino acid units are represented by a three letter symbol or a one letter symbol. For example, the abbreviation for glycine and alanine are *gly, ala* respectively. The names, structures, and abbreviations of the various α-amino acids are given in Table 56.1.

Table 56.1. Structure and Names of Some α-Amino Acid

General Structure

$$H_2N—\underset{\underset{R}{|}}{CH}—COOH$$

	Common Name	Three Letter Symbol	One Letter Symbol	– R
	Neutral amino acids			
1.	Glycine	*Gly*	G	—H
2.	Alanine	*Ala*	A	$—CH_3$
3.	***Valine**	*Val*	V	$—CH(CH_3)_2$
4.	***Leucine**	*Leu*	L	$—CH_2—CH(CH_3)_2$
5.	***Phenylalanine**	*Phe*	F	$—CH_2—C_6H_5$ (benzene ring)
6.	Serine	*Ser*	S	$—CH_2—OH$
7.	Tyrosine	*Tyr*	Y	$—CH_2—C_6H_4—OH$ (benzene ring)
	Acidic amino acids			
8.	Aspartic acid	*Asp*	D	$—CH_2—COOH$
9.	Glutamic acid	*Glu*	E	$—CH_2—CH_2—COOH$
	Basic amino acids			
10.	***Lysine**	*Lys*	K	$—CH_2—(CH_2)_3—NH_2$

ACTIVITY 56.1 **Study of α-Amino Acids**

Draw structures of the following α-amino acids

(*i*) Alanine (*ii*) Phenylalanine

(*iii*) Tyrosine (*iv*) Aspartic acid

(*v*) Lysine (*vi*) Glutamic acid

Give the IUPAC names of these α-amino acids. Use the information from Table 56.1.

SOLVED EXAMPLE

Example 56.1. *Give reasons for the following:*

On electrolysis in acidic solution amino acids migrate towards the cathode while in alkaline solution they migrate towards the anode.

Solution. In acidic medium, amino acids exist as cations $(R—\underset{\underset{^{+}NH_3}{|}}{CH}—COOH)$ and in basic medium they exist as anions $(R—\underset{\underset{NH_2}{|}}{CH}—COO^-)$. Therefore, on electrolysis in an acidic medium they migrate towards the cathode while in an alkaline medium they migrate towards the anode.

56.3 FORMATION OF PEPTIDES AND PROTEINS

Peptides are the products formed by the condensation of two or more amino acids through their amino and carboxylic acid groups involving the elimination of water molecules. They may be classified as ***dipeptides, tripeptides, tetrapeptides,*** etc., depending upon whether the number of amino acid molecules taking part in condensation is *two, three* or *four* respectively. When the number of such amino acids is more than ten, the product is called a **polypeptide.** A polypeptide having molecular mass more than 10000 *u* is called a **protein.** Thus, **polypeptides and proteins are condensation polymers of α-amino acids. The linkage (—CO—NH—) which unites various amino acid units in a peptide molecule is** called **peptide linkage** or **peptide bond.**

$$H_2NCH(R)-C(=O)-\boxed{OH + H}-N(H)-CH(R)COOH$$

Amino acids

$$\xrightarrow[-H_2O]{\text{Condense}}$$

$$H_2NCH(R)-\underbrace{C(=O)-N(H)}_{\text{Peptide Bond}}-CH(R)-COOH$$

A Dipeptide

A simple convention is used to write the structure and name of a peptide. The amino acid unit having the free —NH_2 group is called **N-terminal end** whereas the amino acid unit with the free —COOH group is called **C-terminal end.** The structure is written with the *N-terminal end to the left* and *C-terminal end to the right.* The base name of the peptide is taken from the C-terminal amino acid unit. Other amino acid units are taken to be substituents of this acid and the suffix *ine* of their name is replaced with *yl.* In case of polypeptides and proteins, the abbreviated names of amino acid units are used. Let us write the structure of a tripeptide formed from *glycine, alanine* and *serine.*

$$H_2N-CH_2-C(=O)-N(H)-CH(CH_3)-C(=O)-NH-CH(CH_2OH)-C(=O)-OH$$

Glycylalanylserine (*Name*) Gly.Ala.Ser or G–A–S (*Abbreviated name*)

A section of a polypeptide with different amino acid units is given below:

$$\sim HN-CH(R)-\mathbf{CONH}-CH(R')-\mathbf{CONH}-CH(R'')-CO\sim$$

Polypeptide

HYDROLYSIS OF PROTEINS

Polypeptides and proteins can be hydrolysed by dilute acids or enzymes. The ultimate product of hydrolysis of proteins is a mixture of α-amino acids.

When food is digested, the proteins present in it are broken into constituent amino acid molecules. During digestion the peptide linkage that joins the amino acids in proteins gets hydrolysed. **Hydrolysis of proteins takes place in the stomach and small intestine** and the amino acids produced in the process are absorbed from the intestine by the blood. These amino acids are then regrouped to form specific proteins in the cells of our body.

Hydrolysis of tripeptide glycylalanylglycine (gly.ala.gly) is shown below:

$$H_2N-CH_2-C(=O)-NH-CH(CH_3)-C(=O)-NH-CH_2-C(=O)-OH$$

(gly. ala. gly)

$$\downarrow \text{dil HCl}$$

$$\underset{\text{glycine}}{H_2N-CH_2-COOH} + \underset{\text{alanine}}{H_2N-CH(CH_3)-COOH} + \underset{\text{glycine}}{H_2N-CH_2-COOH}$$

56.4 CLASSIFICATION OF PROTEINS

CLASSIFICATION BASED UPON MOLECULAR STRUCTURE

Proteins can be classified into two broad classes on the basis of their molecular structure:

Fibrous proteins and *Globular proteins.*

1. Fibrous Proteins

Such proteins have thread like molecules which tend to lie side by side to form fibres. The molecules are held together at many points by hydrogen and disulphide bonds. They are generally insoluble in water.

2. Globular Proteins

Such proteins have molecules which are folded into compact units that often approach spheroidal shapes. The areas of contact between the molecules are small. Therefore, the intermolecular forces here are comparatively weak. These proteins are soluble in water or aqueous solutions of acids, bases or salts.

CLASSIFICATION BASED UPON THE CORRESPONDING HYDROLYSIS PRODUCTS

On the basis of chemical composition or the nature of products of hydrolysis, proteins are classified into three categories.

1. Simple Proteins

These proteins on hydrolysis give a mixture of α-amino acids only. Some examples are *albumins* in the white section of an egg, *oxyzenin* in rice, *keratin* in hair, nails, horns and *glutenin* in wheat.

2. Conjugated Proteins

These proteins on hydrolysis give a non-protein part in addition to the α-amino acids. This non-protein portion is *called the* ***prosthetic group.*** The main function of the prosthetic group is to control the biological functions of the protein. Some examples of conjugate proteins along with their corresponding prosthetic groups are given below in tabular form:

Name of the Protein	*Prosthetic Group*
Nucleoproteins	Nucleic acids
Glycoproteins	Sugars
Lipoproteins	Lipids such as lecithin
Phosphoproteins	Phosphoric acid residues
Chromoproteins	Some colouring matter
Metalloproteins.	Some metal.

3. Derived Proteins

These are the degradation products obtained by partial hydrolysis of simple or conjugated proteins with acids, alkalies or enzymes. Some examples are, proteoses, peptones and polypeptides.

Proteins ⟶ Proteoses ⟶ Peptones ⟶ Polypeptides

56.5 STRUCTURE OF PROTEINS

The structure of proteins is generally studied at four different levels, *i.e., primary, secondary, tertiary* and *quaternary* structures as discussed below:

1. PRIMARY STRUCTURE OF PROTEINS

It refers to the sequence in which amino acids are arranged in proteins. Each protein has a specific sequence of amino acid units which is critical to its biological activity. The change of just one amino acid in the sequence can alter the biological activity. For example, *haemoglobin* a protein responsible for carrying oxygen in the blood consists of 574 amino acid units. The change of just one amino acid in the sequence produces ineffective haemoglobin (which is found in patients suffering from *sickle cell anaemia*).

—Val—His—Leu—Thr—Pro—**Glu**—Glu—Lys—

(*Normal haemoglobin*)

—Val—His—Leu—Thr—Pro—**Val**—Glu—Lys

(*Sickle cell haemoglobin*)

The procedure for the determination of the primary structure of a protein is quite complex. Such procedure was first developed by *Frederic Sanger* in 1953, who determined the amino acid sequence in *insulin*.

2. SECONDARY STRUCTURE OF PROTEINS

It refers to the arrangement of polypeptide chains giving rise to a particular shape, which arises as a result of hydrogen bonding.

The two common secondary structures are α-**helix** and β-**pleated sheet**.

3. TERTIARY STRUCTURE OF PROTEINS

It refers to the definite geometric pattern in which the entire protein molecule folds up to produce a specific shape. It arises due to folding and superposition of various secondary structural elements. The two important tertiary structures of proteins are ***fibrous structure*** and ***globular structure.*** At normal pH and temperature, each protein will take a shape that is energetically most stable. This shape is called **native state of the protein.**

4. QUARTERNARY STRUCTURE OF PROTEINS

Many proteins exist as aggregates of two or more polypeptide chains. ***The overall structure of the protein which arises due to specific spatial arrangement of multiple sub-units is called quarternary structure.*** For example, the quarternary structure of haemoglobin involves four polypeptide chains.

SOLVED EXAMPLE

Example 56.2. *What type of linkages are responsible for the formation of,*

(*i*) *Primary structure of proteins?*

(*ii*) *α-Helix formation?*

Solution.

(*i*) Primary structure of proteins arises due to ***peptide bonds*** (—CO—NH— bonds) between various constituent of amino acids.

(*ii*) ***Hydrogen bonds*** between —C(=O)— and —N(H)— groups of the same polypeptide chain.

56.6 DENATURATION OF PROTEINS

All proteins in a living system possess definite configuration and biological activity. *A protein found in a living system with definite configuration and biological activity is called a* ***native protein.*** When a protein in its native form is subjected to physical change such as change in temperature, addition of solvents miscible with water or chemical change such as addition of acids, alkalies or salts, the hydrogen bonds are disturbed. As a result, globules unfold and helix get uncoiled and protein loses its biological activity and gets coagulated. The proteins in this state are said to be **denatured.**

The process that brings about changes in physical as well as biological properties of the proteins is called ***denaturation of proteins****.*

Some important features of denaturation are

(*i*) *Chemically, denaturation does not change the primary structure but changes the secondary and tertiary structures of proteins.*

(*ii*) *Denaturation of proteins can be caused by changes in pH, temperature or by adding other denaturating agents.*

(*iii*) *Denatured protein loses its biological activity.*

(*iv*) *During denaturation, the protein molecule uncoils from an ordered and specific conformation into a more random conformation leading to precipitation. Thus, denaturation leads to increase in entropy.*

(*v*) *Denaturation may be reversible or irreversible.*

The most common example of irreversible denaturation of proteins is coagulation of the albumin present in the white part of an egg. When an egg is boiled, the soluble globular proteins present in it denature resulting in the formation of insoluble fibrous proteins.

Another example of irreversible denaturation is **curdling of milk** which is caused due to the formation of lactic acid by bacteria present in milk.

In the case of reversible denaturation, the protein recovers its original properties and biological activity when the disruptive agent is removed. The reverse of denaturation is called **renaturation.**

56.7 ENZYMES

Enzymes *are the important group of globular proteins which act as biological catalysts in living systems.* They may be simple or conjugated proteins. Some enzymes can be non-proteins also. Enzymes increase the rate of biochemical reactions by providing an alternative path of lower activation energy. About 3000 different kinds of enzymes have been identified so far. Some common enzymes along with their functions have been given in Table 56.2.

Table 56.2. Some Common Enzymes

Enzymes	*Reaction Catalysed*
Amylase	Starch $\longrightarrow$ Glucose
Maltase	Maltose $\longrightarrow$ Glucose
Lactase	Lactose $\longrightarrow$ Glucose + Galactose
Invertase	Sucrose $\longrightarrow$ Glucose + Fructose
Pepsin	Proteins $\longrightarrow$ Amino acids
Trypsin	Proteins $\longrightarrow$ Amino acids

56.8 USES OF PROTEINS

USES OF PROTEINS FOR FOOD

Proteins are an important component of our food. Proteins build up, maintain and replace the tissues in our body. Our muscles, our organs, and our immune systems are made up mostly of protein. Our body uses the protein we eat to make large numbers of specialized protein molecules that have specific functions.

The best sources of proteins are poultry, fish, eggs, nuts, dairy products (such as milk, cheese, curd), seeds and legumes.

Proteins, in our food, supply the essential amino acids which our body is unable to synthesize. When we eat food that contains proteins, the digestive juices in the stomach and intestines break down the proteins present in food into amino acids. The amino acids can then be used to give the energy that our body needs to perform various functions.

FUNCTIONS OF PROTEINS

Some of the different forms of proteins and their important functions in the human body are given below:

1. Proteins are needed for **growth** and **repair** of the body. Growing children need proteins in large quantities for making new tissues. Proteins such as muscle, skin, hair and other tissues ***constitute the bulk of the body's non-skeletal structure.***
2. Some proteins ***as hormones*** regulate many body functions. For example, the hormone **insulin** is a protein. It regulates sugar level in the blood.
3. Some proteins ***as enzymes*** catalyse or help in biochemical reactions. For example, **pepsin** and **trypsin.**
4. Some proteins act as ***antibodies,*** and protect the body from the effect of invading species or substances.

5. ***Transport proteins*** carry different substances in the blood to different tissues. For example, **haemoglobin** is a transport protein. It carries oxygen from the lungs to the tissues and carbon dioxide from the tissues to the lungs.
6. ***Contractile proteins*** help in contraction of muscle and other cells of our body. **Myosin** is an example of a contractile protein.
7. Proteins also serve as a **source of energy** when the carbohydrates are not available in the food, especially during long fasting or starvation.
8. Proteins act as buffers and maintain appropriate pH required for many biochemical reactions.

56.9 TEST FOR PROTEINS

1. BIURET TEST

To the dispersion of the substance to be tested (say 5% solution of egg albumin) add about 2 ml of NaOH solution. Now add 4–5 drops of 1% $CuSO_4$ solution.

Bluish violet colouration indicates the presence of protein.

2. MILLON'S TEST

This test is given by proteins containing phenolic amino acids. *Gelatin does not give this test.*

To 1–2 ml of egg albumin dispersion add 2 drops of Millon's reagent.

White ppt. which changes to brick red on boiling, confirms the presence of proteins.

TOXIC

Millon's Reagent is prepared by dissolving 5 g each of $HgNO_3$ and $Hg(NO_3)_2$ in 100 ml of dil. HNO_3.

Caution: Millon's reagent is poisonous. Avoid contact with skin and do not inhale fumes.

3. NINHYDRIN TEST

Take about 2 ml of egg albumin dispersion in a test-tube and add 1–2 ml of Ninhydrin solution. Boil the contents.

Intense blue or purple colouration confirms the presence of proteins.

Ninhydrin Solution is prepared by dissolving 0.1 g of ninhydrin in about 100 ml of distilled water. This solution is unstable and can be kept only for two days.

SUMMARY

- **α-Amino Acids.** The organic compounds containing a –COOH group and an amino group at the α-carbon atom. These are the building blocks of proteins and peptides.
- **Zwitter ion** or **dipolar ion** structure of α-amino acids

$$H_3\overset{+}{N}-\underset{\substack{|\\R}}{CH}-COO^-$$

- **Proteins.** The complex nitrogenous organic molecules which are essential for the growth and maintenance of life.
- Proteins are **condensation polymers** of α-amino acids. In proteins various amino acids are joined by **peptide linkage.**
- Hydrolysis product of peptides and proteins is α-amino acids
- **Denaturation.** It is the process that brings about changes in the physical as well as biological properties of the protein.
- **Enzymes.** The biological catalysts produced by living cells which catalyse the biochemical reactions in living organisms. Most enzymes are globular proteins.
- Presence of proteins in a food item can be detected by
  — Biuret test
  — Millon's test
  — Ninhydrin test.

EVALUATE YOURSELF

1. Objective Type Questions

Select the most appropriate choice from the options given as (a), (b), (c) and (d) after each question:

1. Which of the following is not a source of proteins?
   (*a*) fish (*b*) potato
   (*c*) egg (*d*) milk.
2. The number of naturally occurring α-amino acids is about
   (*a*) 10 (*b*) 50
   (*c*) 40 (*d*) 20.
3. The structure of phenylalanine is

$$C_6H_5-CH_2-\underset{\substack{|\\NH_2}}{CH}-COOH$$

Its IUPAC name is
(*a*) 2–amino–1–phenylpropanoic acid
(*b*) 1–phenyl–2–aminopropanoic acid

(*c*) 3–phenyl–2–aminopropanoic acid

(*d*) 2–amino–3–phenylpropanoic acid.

4. The correct statement in respect of the protein haemoglobin is that it

(*a*) maintains blood sugar level

(*b*) acts as an oxygen carrier in the blood

(*c*) forms antibodies and offers resistance to diseases

(*d*) functions as a catalyst for biological reactions.

5. Which one of the following structures represents a peptide chain?

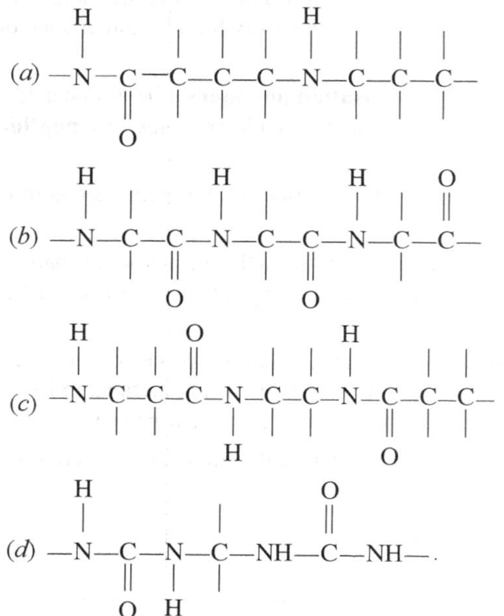

6. The helical structure of protein is stabilized by

(*a*) hydrogen bonds (*b*) ether bonds

(*c*) peptide bonds (*d*) dipeptide bonds.

7. The presence of an amino acid or protein will change the colour of ninhydrin solution from red to

(*a*) green (*b*) yellow

(*c*) blue (*d*) orange.

II. Discussion Questions

8. Give examples of two natural polymers.

9. What is the final product obtained by hydrolysis of proteins?

10. Draw structures of glycine and alanine, and give their IUPAC names.

11. List any five sources of proteins.

12. Discuss the following:

(*i*) zwitter ion structure of amino acids

(*ii*) peptide linkage

(*iii*) isoelectric point.

13. List four important physical properties of α-amino acids.

14. Explain briefly:

(*i*) primary structure of proteins

(*ii*) denaturation of proteins

(*iii*) enzymes

15. List any four functions of proteins in the human body.

16. Describe two tests to detect the presence of proteins in a food item.

Answers

I. Objective Type Questions

1. (*b*)	**2.** (*d*)	**3.** (*a*)
4. (*b*)	**5.** (*b*)	**6.** (*a*)
7. (*c*).		

SECTION-2

UNIT-57

CARBOHYDRATES

Specific Objectives

After studying this unit, the student should be able to:

- identify the sources and properties of carbohydrates
- describe the nomenclature and structure of carbohydrates
- describe carbohydrate as a polymer
- describe the uses of carbohydrates.

57.1 INTRODUCTION

Carbohydrates constitute an important class of compounds like glucose, fructose, sucrose, starch, cellulose, etc., which play a vital role in our everyday life. They are the ultimate source of most of our food. We clothe ourselves with cellulose in the form of cotton, rayon and linen. We build furniture and houses from cellulose in the form of wood. Thus, carbohydrates provide us with the basic necessities of life, food, clothing and shelter.

Carbohydrates are also known as **saccharides.** Originally, the name carbohydrate was given to the compounds conforming to the general formula $C_x(H_2O)_y$, and they were considered to be hydrates of carbon. However, this definition could not hold ground for long because many compounds like *formaldehyde* (CH_2O), *acetic acid* ($C_2H_4O_2$), etc., conform to the formula $C_x(H_2O)_y$ but they do not exhibit the characteristic properties of carbohydrates.

According to the modern definition **carbohydrates** are defined as *polyhydroxy aldehydes or polyhydroxy ketones or the compounds that yield such compounds on hydrolysis.*

57.2 CLASSIFICATION OF CARBOHYDRATES

Carbohydrates are classified into three major categories depending upon their behaviour towards hydrolysis:

1. MONOSACCHARIDES

These are carbohydrates which cannot be hydrolysed to simpler carbohydrates. About 20 monosaccharides are known to occur in nature. **Glucose** and **fructose** are common examples.

2. OLIGOSACCHARIDES

These are the carbohydrates which on hydrolysis give two to ten units of monosaccharides. Accordingly, they may be further divided into *di, tri* or *tetrasaccharides* depending upon the actual number of monosaccharide units formed by the hydrolysis of a particular oligosaccharide.

Disaccharides give two units of monosaccharides on hydrolysis. The two monosaccharide units obtained on hydrolysis of a disaccharide may be the same or different. Common examples are **sucrose** and **maltose**. Both have molecular formula $C_{12}H_{22}O_{11}$.

Sucrose on hydrolysis gives one molecule of glucose and one molecule of fructose whereas maltose on hydrolysis gives two molecules of glucose only.

Trisaccharides give three units of monosaccharides on hydrolysis. Raffinose, $C_{18}H_{32}O_{16}$ is a common example.

Tetrasaccharides give four units of monosaccharides on hydrolysis. Stachyose, $C_{24}H_{42}O_{21}$ is a common example.

3. POLYSACCHARIDES

These are the carbohydrates which are polymeric molecules and can be hydrolysed to give large number of monosaccharide units. The commonly occurring polysaccharides have the general formula $(C_6H_{10}O_5)_n$. The common examples are **starch**, **glycogen** and **cellulose**.

It may be noted that the carbohydrates which are sweet in taste are collectively called **sugars** while those which are not sweet are called **non-sugars.** *Monosaccharides and disaccharides are* **sugars** *but polysaccharides are* **non-sugars.**

The relative degree of sweetness of various sugars is given below in tabular form:

Sugar	*Lactose*	*Maltose*	*Galactose*	*Glucose*	*Sucrose*	*Fructose*
Relative Sweetness	16	32	32	74	**100**	173

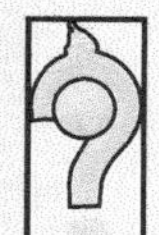

Do you know?

Besides carbohydrates, some other compounds are far more sweet. For example,

- **Saccharin** is about 500 times as sweet as sucrose.
- **Aspartame,** a peptide, is about 160 times sweeter than sucrose.

REDUCING AND NON-REDUCING CARBOHYDRATES

The carbohydrates may also be classified as **reducing** and **non-reducing** sugars. The carbohydrates which can reduce Tollen's reagent or Fehling's or Benedict's solution are classified as reducing sugars, while those which do not reduce these reagents are called non-reducing sugars.

Reducing sugars contain free aldehyde or ketonic group. All monosaccharides are reducing sugars. Disaccharides may be reducing or non-reducing. If the carbonyl groups of both the monosaccharides are involved in linkage, the disaccharide is non-reducing. On the other hand, if one of the carbonyl groups is free, the disaccharide is reducing. **Sucrose is a non-reducing sugar while maltose is a reducing sugar.**

SOURCES OF CARBOHYDRATES

Glucose occurs in *sweet fruits* such as grapes, mangoes, oranges, etc. *Honey* is also rich in glucose. In a combined state it is present in maltose, sucrose, starch, cellulose, etc.

Fructose is found in *ripe fruits* and *honey*.

Sucrose. Major sources of sucrose are *sugarcane* and *sugarbeet*

Lactose is present in *milk*.

Starch. Major sources of starch are *wheat, rice, cassava, root tubers* such as potatoes, *legumes* and *vegetables*.

Cellulose. Cellulose is present in *cotton*, *wood* and *jute*.

57.3 MONOSACCHARIDES

As pointed out earlier, these are the simplest carbohydrates and cannot be hydrolysed to give still simpler carbohydrates. Monosaccharides containing an *aldehyde group* are called **aldoses** while those containing *ketonic group* are called **ketoses.** They can be further classified into different categories depending upon the number of carbon atoms. For example,

Monosaccharides with three carbon atoms are called trioses.

Monosaccharides with four carbon atoms are called tetroses.

Monosaccharides with five, six and seven carbon atoms are called pentoses, hexoses and heptoses respectively.

Table 57.1. Different Types of Monosaccharides

Carbon atoms	*General term*	*Aldehyde*	*Ketone*
3	Triose	Aldotriose	Ketotriose
4	Tetrose	Aldotetrose	Ketotetrose
5	Pentose	Aldopentose	Ketopentose
6	Hexose	Aldohexose	Ketohexose

Some examples of various types of monosaccharides are given below:

$$\begin{array}{c} CHO \\ | \\ H-C-OH \\ | \\ H-C-OH \\ | \\ H-C-OH \\ | \\ CH_2OH \end{array} \qquad \begin{array}{c} CHO \\ | \\ H-C-OH \\ | \\ HO-C-H \\ | \\ H-C-OH \\ | \\ H-C-OH \\ | \\ CH_2OH \end{array} \qquad \begin{array}{c} CH_2OH \\ | \\ C=O \\ | \\ HO-C-H \\ | \\ H-C-OH \\ | \\ H-C-OH \\ | \\ CH_2OH \end{array}$$

Ribose **(Aldopentose)** — Glucose **(Aldohexose)** — Fructose **(Ketohexose)**

GLUCOSE (DEXTROSE OR GRAPE SUGAR), $C_6H_{12}O_6$

Glucose occurs in nature in a free state as well as in combined state. In a free state, **it occurs in sweet fruits (such as grapes, mangoes, oranges) and honey.** Ripe grapes contain about 20% glucose and hence the name grape sugar. In a combined state, glucose is present in di- and polysaccharides. Maltose and starch on hydrolysis yield only glucose.

Preparation of Glucose

1. ***By Hydrolysis of Cane-sugar.*** In the laboratory, glucose can be prepared by hydrolysis of cane-sugar in the presence of alcohol using dilute hydrochloric acid. Glucose and fructose are formed in equal amounts. Glucose, being less soluble in ethyl alcohol than fructose, crystallizes out.

$$C_{12}H_{22}O_{11} + H_2O \xrightarrow{H^+} \underset{\text{Glucose}}{C_6H_{12}O_6} + \underset{\text{Fructose}}{C_6H_{12}O_6}$$

2. ***By Hydrolysis of Starch.*** Glucose is obtained, **on commercial scale,** by hydrolysis of starch by boiling it with dilute H_2SO_4 acid at 393 K under a pressure of 2–3 atm.

$$\underset{\text{Starch}}{(C_6H_{10}O_5)_n} + nH_2O \xrightarrow[393\text{ K, 2–3 atm.}]{H^+} \underset{\text{Glucose}}{nC_6H_{12}O_6}$$

Physical Properties of Glucose

1. Glucose is a white crystalline solid. Its **melting point is 419 K.**
2. It is sweet in taste.
3. It is very soluble in water due to its ability to form hydrogen bonds with water. Glucose is insoluble in organic solvents such as ether.

Structure of Glucose

Glucose is an aldohexose. It is monomer of many of the larger carbohydrates such as starch and cellulose. The reactions of glucose indicate that its molecule contains one primary (—CH_2OH) and four secondary (—CHOH) hydroxyl groups. Glucose is assigned the following structure:

$$\begin{array}{c} H\diagdown\;\;\diagup\!\!\!\!=O \\ C \\ H-\!\!\!-\!\!\!-OH \\ HO-\!\!\!-\!\!\!-H \\ H-\!\!\!-\!\!\!-OH \\ H-\!\!\!-\!\!\!-OH \\ CH_2OH \end{array}$$

Glucose also exists in two cyclic forms, α-D-glucose and β-D-glucose. The cyclic structures of glucose are shown below:

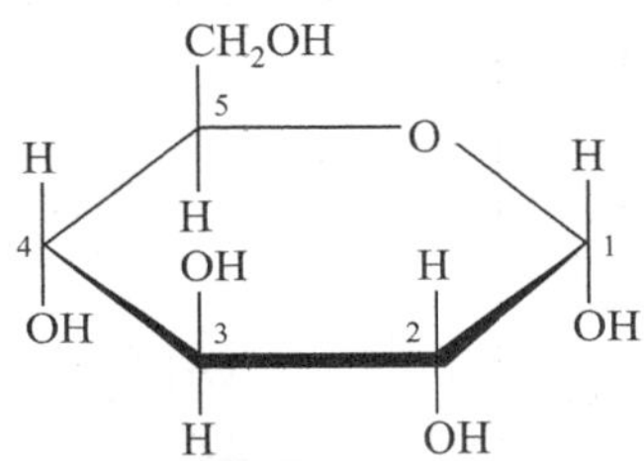

α-D-Glucose

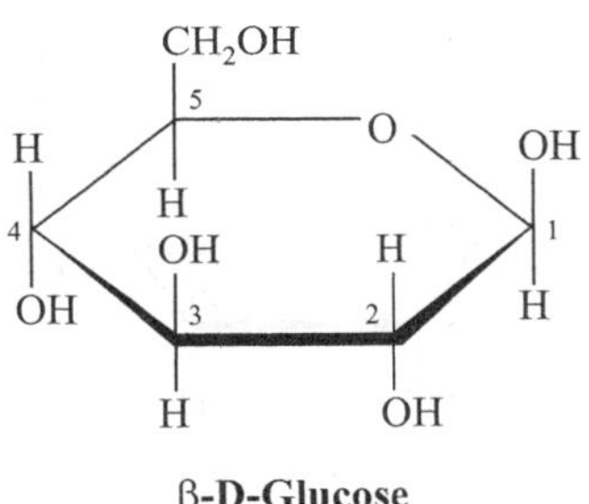

β-D-Glucose

FRUCTOSE (LAEVULOSE), $C_6H_{12}O_6$

Fructose is a ketohexose. It is obtained along with glucose by the hydrolysis of sucrose. It is found in nature in **ripe fruits** and **honey.**

Physical Properties of Fructose

1. Fructose is a white crystalline solid. Its melting point is 377 K.
2. It is highly soluble in water. It is insoluble in organic solvents such as ether.
3. It is very sweet in taste and is used in making sweets.

Structure of Fructose

Fructose has the molecular formula $C_6H_{12}O_6$ and on the basis of its reactions it was found to contain a ketonic functional group at C—2 and six carbon atoms in a straight chain as in the case of glucose. Its open chain structure may be written as:

Fructose

Fructose also exists in two cyclic forms, α-D-fructose and β-D-fructose.

The two cyclic structures of fructose are represented by Haworth structures as given:

α-D-Fructose

β-D-Fructose

57.4 DISACCHARIDES

These are the carbohydrates which give two units of monosaccharides on hydrolysis with dilute acids or enzymes. Some examples are: Sucrose, maltose and lactose.

$$\underset{\text{Sucrose}}{C_{12}H_{22}O_{11}} \xrightarrow{H_2O/H^+} \underset{\text{Glucose}}{C_6H_{12}O_6} + \underset{\text{Fructose}}{C_6H_{12}O_6}$$

$$\underset{\text{Lactose}}{C_{12}H_{22}O_{11}} \xrightarrow{H_2O/H^+} \underset{\text{Galactose}}{C_6H_{12}O_6} + \underset{\text{Glucose}}{C_6H_{12}O_6}$$

$$\underset{\text{Maltose}}{C_{12}H_{22}O_{11}} \xrightarrow{H_2O/H^+} \underset{\text{Glucose}}{2C_6H_{12}O_6}$$

This implies that a disaccharide is formed by the condensation of two monosaccharide units. The two monosaccharide units in a disaccharide are joined together by an oxide (or ether) linkage formed by the loss of a water molecule. Such a linkage between two monosaccharide units through oxygen atom is called **glycosidic linkage.**

Now, let us study some disaccharides in some details.

SUCROSE (CANE-SUGAR), $C_{12}H_{22}O_{11}$

Sucrose is the most widely occurring disaccharide. It is found in all photosynthetic plants. **It is obtained commercially from sugar cane or sugarbeets.** On hydrolysis with dilute acids or the enzyme *invertase*, 1 mole of sucrose gives 1 mole of glucose and 1 mole of fructose.

$$\underset{\text{Sucrose}}{C_{12}H_{22}O_{11}} + H_2O \xrightarrow{H^+} \underset{\text{Glucose}}{C_6H_{12}O_6} + \underset{\text{Fructose}}{C_6H_{12}O_6}$$

Sucrose is a ***non-reducing sugar.*** Haworth (1927) suggested the following structure for sucrose.

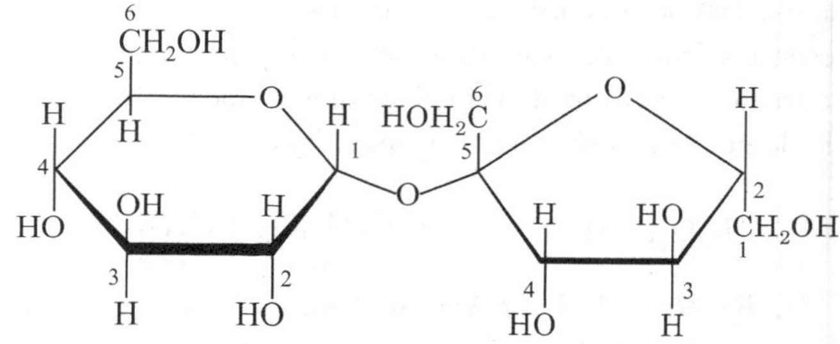

Haworth's Representation of Sucrose

MALTOSE, $C_{12}H_{22}O_{11}$

When starch is hydrolysed by the enzyme *diastase*, maltose is formed as one of the products.

$$\underset{\text{Starch}}{2(C_6H_{10}O_5)_n} + nH_2O \xrightarrow{\text{Diastase}} \underset{\text{Maltose}}{nC_{12}H_{22}O_{11}}$$

On hydrolysis with dilute acids, one mole of maltose yields 2 moles of glucose.

$$C_{12}H_{22}O_{11} + H_2O \xrightarrow{H^+} 2C_6H_{12}O_6$$

Maltose Glucose

Maltose is a ***reducing sugar.*** In maltose the two glucose units are linked through ***α-glycosidic linkage*** between C—1 of one glucose unit and C—4 of the other. The free aldehyde group can be produced at C—1 of the second glucose in solution. Hence, maltose shows reducing properties.

The structure of maltose is given below:

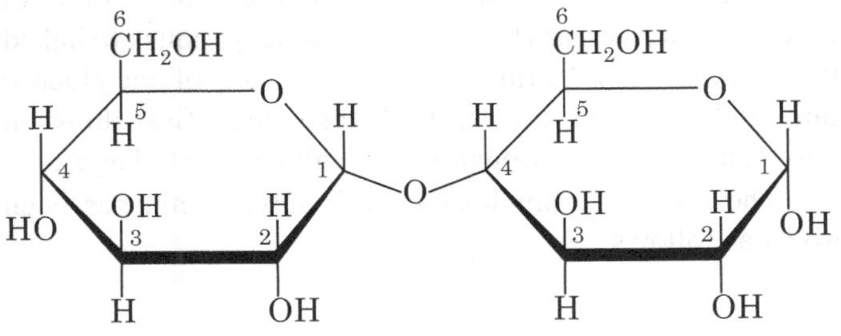

α-D-Glucose unit α-D-Glucose unit

Maltose

LACTOSE, (MILK SUGAR), $C_{12}H_{22}O_{11}$

Lactose is present in **milk** and is also known as milk sugar. On hydrolysis with dilute acids one mole of lactose yields 1 mole of glucose and one mole of galactose

$$\underset{\text{Lactose}}{C_{12}H_{22}O_{11}} \xrightarrow[+H_2O]{H^+} \underset{\text{Glucose}}{C_6H_{12}O_6} + \underset{\text{Galactose}}{C_6H_{12}O_6}$$

Lactose is a ***reducing sugar.*** In lactose, glucose and galactose units are linked through ***β-glycosidic linkage*** between C—1 of galactose and C—4 of glucose unit. The structure of lactose is given below:

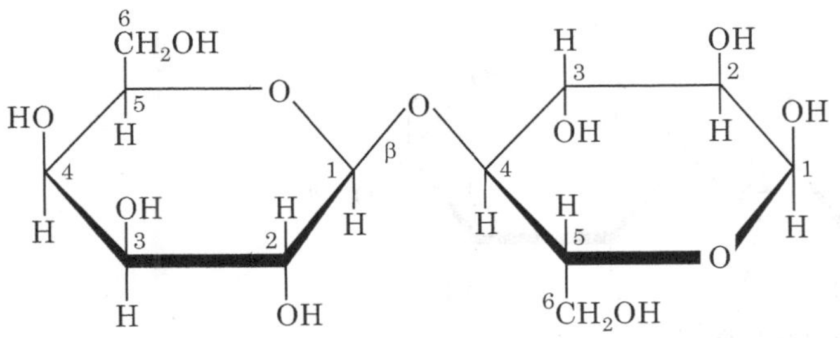

β-D-Galactose unit β-D-Glucose unit

Lactose

ACTIVITY 57.1 Tests for Reducing Sugars

1. FEHLING'S TEST

Take 2 cm^3 of an aqueous solution of a carbohydrate (nearly 5%) and add 1–2 cm^3 each of Fehling's solution A and Fehling's solution B. Keep the test tube in a boiling water bath.

Reddish ppt. indicates the presence of a reducing sugar.

Preparation of Fehling's Solution

Fehling's Solution A. Dissolve 17.5 g of $CuSO_4$ in 250 cm^3 of distilled water containing few drops of H_2SO_4.

Fehling's Solution B. Dissolve 8.65 g of sodium potassium tartarate and 35 g NaOH in 150 cm^3 of distilled water.

2. BENEDICT'S TEST

To 1–2 cm^3 of an aqueous solution of a carbohydrate in a test tube add 1–2 cm^3 of Benedict's reagent. Keep the test tube in a boiling water bath.

Reddish ppt. indicates the presence of a reducing sugar.

Preparation of Benedict's Reagent

Dissolve 17.3 g of an sodium citrate and 9 g of anhydrous Na_2CO_3 in about 80 cm^3 of distilled water. Heat if necessary. Dissolve 1.73 g of copper sulphate in 100 cm^3 of water. Mix the two and make the volume 250 cm^3 by adding water.

3. TOLLEN'S TEST

Take 2–3 cm^3 of an aqueous solution of carbohydrate in a test tube. Add to it 2–3 cm^3 of Tollen's reagent. Keep the test tube in a boiling water bath for 10 minutes.

A shining silver mirror indicates the presence of a reducing carbohydrate.

Preparation of Tollen's Reagent

Add NaOH solution to $AgNO_3$ solution. Then add NH_4OH solution dropwise till the ppt. just dissolved. The clear solution obtained is Tollen's reagent.

4. IODINE TEST (FOR STARCH ONLY)

To the aqueous suspension of the sample, add 1–2 drops of iodine solution.

Appearance of a blue colouration indicates the presence of starch.

57.5 POLYMERS OF CARBOHYDRATES—POLYSACCHARIDES

These are the carbohydrates which are made of large numbers of monosaccharide units. They are naturally occurring condensation **polymers** in which large numbers of monosaccharide units are joined together through glycosidic linkages. They are *colourless, tasteless amorphous powders*. They play vital roles in plant and animal life.

Let us study the structure of three important polysaccharides **starch, cellulose** and **glycogen.**

1. STARCH (AMYLUM), $(C_6H_{10}O_5)_n$

It is the most abundant source of carbohydrate in human diet. It is a major food reserve material of plants and occurs mainly in seeds, fruits, tubers and roots. The important **sources of starch** are *wheat, rice, cassava, cocoyam maize, potatoes, legumes* and *other vegetables*. Prepared foods such as *fufu, akple, omotuo, tapioca* and *kokonte* have a high starch content.

Starch is a polymer of D-glucose. It consists of two components:

(*a*) ***Amylose,*** a water soluble fraction, and a linear polymer of D-glucose units. It gives blue colour with iodine.

(*b*) ***Amylopectin,*** a water insoluble fraction, and consists of linear as well as branched chain polymers of D-glucose.

Natural starch contains approximately 15–20% of amylose and 80–85% of amylopectin. Amylopectin does not give blue colour with iodine. A molecule of amylose may contain 200–1000 glucose units whereas a molecule of amylopectin may contain 2000–3000 glucose units. In amylose as well as amylopectin the D-glucose units are linked through α-glycosidic linkages between C—1 of one glucose unit and C—4 of the next glucose unit. Branching in amylopectin occurs through $C_1 - C_6$ glycosidic linkage.

The section of amylose as well amylopectin has been given as follows:

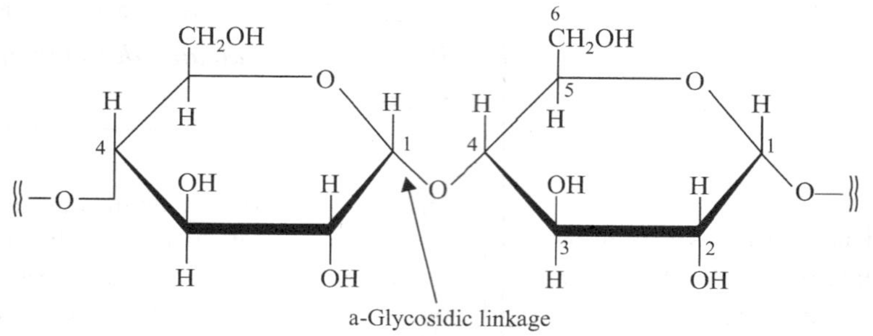

a-Glycosidic linkage

Amylose (Linear Polymer)

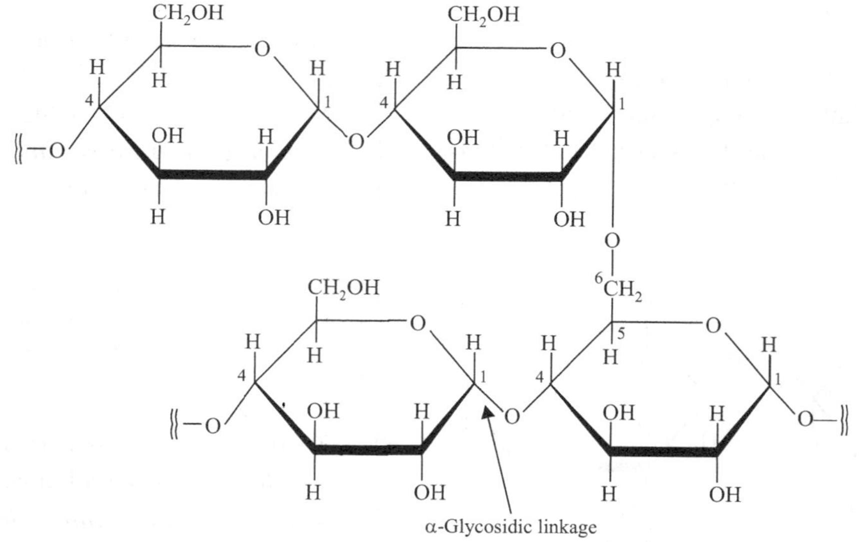

α-Glycosidic linkage

Amylopectin

On hydrolysis with dilute acids or enzymes, starch breaks down into maltose and finally glucose.

$$\underset{\text{Starch}}{(C_6H_{10}O_5)_n} \xrightarrow[+H_2O]{\text{Diastase}} \underset{\text{Maltose}}{C_{12}H_{22}O_{11}} \longrightarrow \underset{\text{Glucose}}{C_6H_{12}O_6}$$

Starch is a *white amorphous powder. It is tasteless* and is *insoluble in cold water.*

2. CELLULOSE, $(C_6H_{10}O_5)_n$

Cellulose occurs exclusively in plants. It is the most abundant organic substance in the plant kingdom. It is a chief structural material of cell walls of all plants. It is also the chief component of **cotton, wood** and **jute**. Wood contains 45–50% while cotton contains 90–95% cellulose. It may be noted that over 50% of the total organic matter of the living world is cellulose.

Structurally, *cellulose is a linear polymer of glucose* units joined by β-glycosidic linkage *between C—1 of one glucose unit and C— 4 of the next glucose unit* as shown below:

Cellulose

The chains are held together by hydrogen bonds between glucose units of adjacent strands. The chains are so arranged as to constitute bundles. This lends rigidity to its structure.

Cellulose does not reduce Tollen's reagent or Fehling's solution. It is not fermented by yeast. Cellulose cannot be hydrolysed easily. However, when it is heated with dilute sulphuric acid, under pressure, it undergoes hydrolysis and yields glucose.

Add To Your Knowledge

- Cellulose forms many useful products when treated with suitable chemical reagents. Some of the important products are:
- ***Celluloid,*** which is used in films.
- ***Gun cotton,*** which is an explosive.
- ***Cellulose acetate*** is used in plastics, for wrapping films and nail polishes.
- ***Methyl cellulose*** is used in cosmetics, pastes and for fabric sizing.
- ***Ethyl cellulose*** is used in plastic coats and films.

On treatment with concentrated sodium hydroxide, cellulose forms a transluscent mass which imparts a silky lustre to cotton. The process is called *mercerisation* and cotton so produced is called mercerised cotton.

3. GLYCOGEN

Glycogen is the principal reserve of carbohydrates in animals. The molecular structure of glycogen is similar to amylopectin. That is why it is also known as animal starch. However, one main difference between glycogen and amylopectin is that the average chain length in glycogen is 10–14 while that in amylopectin is 25–30 glucose units.

Glycogen is present in liver, muscles and brain. When the body needs glucose, enzymes break glycogen into glucose.

In addition to starch and cellulose, a number of other polysaccharides are used as food components. These are the **gums** and **pectins. Gums** are polysaccharides made up of more than one type of monosaccharide. They are used for thickening and improvement of texture in the food industry. **Pectins** are found in fruit skins and are extracted by boiling. Jelly preparations contain pectin dissolved in a fruit juice. The pectin causes jelly to set into a semi-solid.

57.6 USES OF CARBOHYDRATES

Some important uses of carbohydrates are discussed below:

1. **Glucose** is widely used in the food industry in the production of fruit drinks and sweets.
2. **Glucose** is soluble in water and is absorbed rapidly into the blood stream. Therefore, it is useful for sick people and sportsmen who need instant energy.
3. **Fructose** is the sweetest sugar. It is used as a sweetening agent.
4. **Sucrose** is commonly used as table sugar. It is also used as a sweetening agent.
5. **Sucrose** is used to prepare glucose and fructose by hydrolysis. Sucrose is also used in the manufacture of ethanol by fermentation.
6. **Starch** can be used for the production of ethanol by hydrolysis and fermentation. It is extensively used as food.
7. **Starch** is used for stiffening cloth after laundering. Its suspension in water is used as an adhesive.
8. **Cellulose** is used for the production of paper.

57.7 FUNCTIONS OF CARBOHYDRATES

The carbohydrates perform many important functions in living bodies.

1. ***They act as biofuels to provide energy for functioning of living organisms.***

In the human system, all the carbohydrates except cellulose can serve as a source of energy. Starch and various sugars which are taken as food are first hydrolysed to glucose by the enzymes present in the digestive system.

$$\underset{\text{Starch}}{2(C_6H_{10}O_5)_n} + nH_2O \xrightarrow{\text{Amylase}} \underset{\text{Maltose}}{nC_{12}H_{22}O_{11}}$$

$$\underset{\text{Maltose}}{C_{12}H_{22}O_{11}} + H_2O \xrightarrow{\text{Maltase}} \underset{\text{Glucose}}{2C_6H_{12}O_6}$$

Glucose on slow oxidation to carbon dioxide and water in the presence of enzymes liberates large amounts of energy which is used by the body for carrying out various functions.

$$C_6H_{12}O_6 + 6O_2 \longrightarrow 6CO_2 + 6H_2O + 2832 \text{ kJ}$$

In order to fulfil emergency requirements, our body also stores some of the carbohydrates as glycogen in the liver. Glycogen on hydrolysis gives glucose.

It may be noted that cellulose cannot be hydrolysed in our body because enzymes required for its hydrolysis are not present in our body. However, grazing animals are capable of hydrolysing cellulose to glucose. In these animals cellulolytic bacteria present in the rumen, break down cellulose with the help of the enzyme ***cellulase*** and is subsequently digested and converted into glucose.

2. ***They act as constituents of cell membrane of plants and bacteria.***
3. ***D-ribose and 2-deoxy-D-ribose are present in nucleic acids.***

SOLVED EXAMPLES

Example 57.1. *What are reducing and non-reducing sugars?*

Solution. Carbohydrates which reduce Tollen's reagent and Fehling solution are called reducing sugars, while those which do not reduce these are called non-reducing sugars. For example, glucose, fructose, maltose and lactose are reducing sugars while sucrose is a non-reducing sugar.

Example 57.2. *Draw open chain structures of an aldopentose and an aldohexose.*

Solution.

Ribose (Aldopentose):

```
      CHO
       |
  H—C—OH
       |
  H—C—OH
       |
  H—C—OH
       |
     CH2OH
```

Ribose
(Aldopentose)

```
       CHO
        |
   H—C—OH
        |
  HO—C—H
        |
   H—C—OH
        |
   H—C—OH
        |
      CH2OH
```

Glucose
(Aldohexose)

Example 57.3. *Amylose and cellulose are both straight chain polysaccharides containing only D-glucose units. What is the structural difference between the two?*

Solution. In amylose, D-glucose units are joined together by *α-glycosidic linkages* involving C—1 of one glucose molecule and C—4 of the next glucose molecule. In cellulose, D-glucose units are joined together by *β-glycosidic linkages* between C—1 of one molecule and C—4 of the next glucose molecule.

SUMMARY

- **Carbohydrates** are the polyhydroxy aldehydes or polyhydroxy ketones or the compounds which can be hydrolysed to such compounds.
- Carbohydrates are classified as **monosaccharides**, **oligosaccharides** or **polysaccharides.**
- The carbohydrates which can reduce Tollen's reagent, Fehling's solution or Bededict's solution are known as **reducing carbohydrates.**
  Glucose, fructose and lactose are reducing sugars. Sucrose is a non reducing sugar.
- Carbohydrates which are sweet in taste are known as **sugars.** Fructose is the sweetest sugar.
- Sucrose on hydrolysis gives glucose and fructose.
- Lactose on hydrolysis gives glucose and galactose.
- Maltose, starch and cellulose on hydrolysis give only glucose.

- All monosaccharides and disaccharides are soluble in water.
- Starch, cellulose and glycogen are examples of **polysaccharides**.
- Carbohydrates are the major source of energy for functioning of living beings. They are the constituents of the cell membrane of plants.

EVALUATE YOURSELF

I. Objective Type Questions

Select the most appropriate choice from the options given as (a), (b), (c) and (d) after each question:

1. Which of the following is an example of an aldohexose?
(*a*) Ribose (*b*) Fructose
(*c*) Sucrose (*d*) Glucose.
2. Maltose on hydrolysis gives
(*a*) α-D-glucose (*b*) α and β-D-glucose
(*c*) glucose and fructose (*d*) fructose only.
3. Fructose has/is
(*a*) bitter taste (*b*) sweet taste
(*c*) tasteless (*d*) nothing certain.
4. The constituent units of sucrose are
(*a*) lactose and glucose (*b*) glucose and fructose
(*c*) galactose and glucose (*d*) glucose and maltose.
5. Which disaccharide is present in milk?
(*a*) Maltose (*b*) Galactose
(*c*) Sucrose (*d*) Lactose.
6. Which of the following is a ketohexose?
(*a*) Fructose (*b*) Maltose
(*c*) Glucose (*d*) Ribose.
7. The linkage that holds monosaccharide units together in a polysaccharide is called
(*a*) peptide linkage (*b*) glycoside linkage
(*c*) ester linkage (*d*) ionic linkage.
8. Lactose on hydrolysis yields
(*a*) glucose (*b*) fructose
(*c*) glucose and fructose (*d*) glucose and galactose.
9. Which of the following carbohydrates is not a reducing sugar?
(*a*) Sucrose (*b*) Lactose
(*c*) Glucose (*d*) Fructose.

II. Fill in the Blanks

10. Complete the following sentences by supplying appropriate words:
(*i*) Sucrose on hydrolysis yields.............
(*ii*) .... is the sweetest sugar.
(*iii*) .... (carbohydrate) is not digested in our body.
(*iv*) Two monosaccharide units in a disaccharide are held by .... linkage
(*v*) Molecular formula of fructose is ....

III. Discussion Questions

11. What are carbohydrates? How are they classified? Give examples. What are the important sources of glucose and fructose?
12. What are reducing and non-reducing sugars?
13. Give an example of an aldohexose and an aldotriose.
14. What is the name given to the linkage which holds together two monosaccharides units in a polysaccharide?
15. Give the structure of:
(*i*) sucrose (*ii*) α-D-fructose
(*iii*) maltose.
16. Out of amylose and amylopectin which form of starch is soluble in water?
17. Glucose or sucrose are soluble in water but cyclohexane or benzene (simple six membered ring compounds) are insoluble in water. Explain.
18. What products are expected when lactose is hydrolysed?
19. Draw structures for α-D-fructose and β-D-glucose.
20. What are the hydrolysis products of?
(*i*) sucrose (*ii*) lactose
21. What is glycogen?
22. Describe the functions of carbohydrates in living organisms.
23. List some important uses of carbohydrates.
24. Draw the structure of cellulose. What is the product of hydrolysis of cellulose?

Answers

I. Objective Type Questions

1. (*d*) **2.** (*a*) **3.** (*b*)
4. (*b*) **5.** (*d*) **6.** (*a*)
7. (*b*) **8.** (*d*) **9.** (*a*).

II. Fill in the Blanks

10. (*i*) glucose and fructose (*ii*) Fructose
(*iii*) cellulose (*iv*) glycoside
(*v*) $C_6H_{12}O_6$.

SECTION-2

UNIT-58

FATS AND OILS

Specific Objectives

After studying this unit, the student should be able to:

- identify sources and properties of fats and oils
- describe fats and oils as alkylalkanoates
- describe the extraction of fats and oils
- describe the preparation of soap from fats and oils
- compare soapy and soapless detergents
- outline some uses of fats and oils.

58.1 INTRODUCTION

Fats and oils are important constituents of a balanced diet for human beings. Fats are solids while oils are liquids at room temperature. A substance which behaves as an oil in a tropical country such as Ghana may behave as a fat in a country with colder climate. However, fats and oils have the same basic structure. Fats and oils are generally insoluble in water but are soluble in organic solvents.

58.2 GENERAL STRUCTURE OF FATS AND OILS

Fats and oils are triesters of glycerol with long chain fatty acids. Each molecule of fat is composed of one molecule of glycerol and three molecules of fatty acids. A molecule of fat may be represented by the general formula,

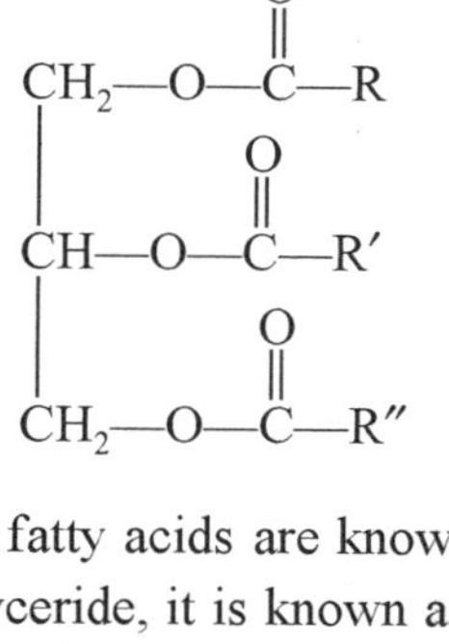

$$\begin{array}{l} CH_2-O-\overset{\overset{\large O}{\|}}{C}-R \\ | \\ CH-O-\overset{\overset{\large O}{\|}}{C}-R' \\ | \\ CH_2-O-\overset{\overset{\large O}{\|}}{C}-R'' \end{array}$$

These triesters of glycerol and fatty acids are known as **triglycerides**. If all the three acid groups are the same in a triglyceride, it is known as **simple glyceride** and if the acid groups are different, it is called a **mixed glyceride**. Glycerides consisting predominantly of saturated fatty acids are solid at room temperature while those with a high proportion of unsaturated acids are usually liquid at room temperature and are called **oils.**

$$\begin{array}{l} CH_2-O-\overset{O}{\overset{\|}{C}}-C_{15}H_{31} \\ | \\ CH-O-\overset{O}{\overset{\|}{C}}-C_{15}H_{31} \\ | \\ CH_2-O-\overset{O}{\overset{\|}{C}}-C_{17}H_{33} \end{array} \qquad \begin{array}{l} CH_2-O-\overset{O}{\overset{\|}{C}}-C_{17}H_{33} \\ | \\ CH-O-\overset{O}{\overset{\|}{C}}-C_{17}H_{33} \\ | \\ CH_2-O-\overset{O}{\overset{\|}{C}}-C_{17}H_{33} \end{array}$$

Tripalmitin
(A saturated fat)

Triolein
(A unsaturated fat or oil)

The fats containing saturated fatty acids are also called saturated fats and the fats containing unsaturated fatty acids are called unsaturated fats. Saturated fats have higher melting point than unsaturated fats. *Saturated fatty acids occur in animal fats, margarine and butter. Unsaturated fatty acids occur in vegetable oils like ground-nut oil, palm kernel oil, olive oil, corn oil, cotton seed oil, mustard oil, etc.* Because of the presence of double bonds in them, unsaturated fats are more reactive than saturated fats. *Saturated fats being more stable tend to accumulate and get stored under the skin.* Therefore, persons consuming more amounts of saturated fat gain weight and become obese. Excessive use of saturated fats leads to many diseases such as hypertension.

Vegetable and marine oils can be hardened and turned into solid fats by hydrogenation in the presence of some appropriate catalyst such as nickel. This process converts most of the unsaturated fatty acids into saturated fatty acids.

Among saturated fatty acids palmitic acid ($C_{15}H_{31}COOH$) is widely present in fats. **Oleic acid ($C_{17}H_{33}COOH$), an unsaturated fatty acid, is the most widely distributed in nature.** In most fats from animal as well as plant origin it forms 30 per cent or more of the total fatty acids.

ESSENTIAL FATTY ACIDS

Cells, in our body, can synthesize most of the fatty acids that it needs, from the carbohydrates. However, a few polyunsaturated fatty acids cannot be synthesized in the body. *The fatty acids which cannot be synthesized in the body are known as essential fatty acids (EFA). Linoleic acid* and *linolenic acid* are examples of essential fatty acids.

$$CH_3(CH_2)_4-CH{=}CH-CH_2-CH{=}CH-(CH_2)_7COOH$$

Linoleic Acid

$$CH_3CH_2CH{=}CHCH_2CH{=}CHCH_2-CH{=}CH_3CH_2COOH$$

Linolenic Acid

Essential fatty acids are present in large amounts in many vegetable oils. A diet rich in essential fatty acids (linoleic acid and linolenic acid) can be obtained by eating plenty of vegetable seed oils. *Essential fatty acids are drastically reduced during hydrogenation of oils.* For example, hydrogenation of groundnut oil reduces essential fatty acids from 28 to 2 per cent.

Some common food items which are rich in fats and the type of fatty acids they contain are given in Table 58.1.

58.3 SOURCES OF OILS AND FATS

Oils and fats are widely distributed in both plants and animals.

In plants oils and fats are stored in their seeds, roots and fruits. Ground nuts, cotton seeds, palm kernels, castor beans, olives are rich in fats and oils. These are rich in unsaturated fats.

Table 58.1. Major Fatty Acids in some Fats

Types of Fat	*Major Fatty Acid*	*Formula of the Fatty Acid*
1. Butter, Cream or milk	Butyric acid	$CH_3-CH_2-CH_2-COOH$
2. Coconut oil	Octanoic acid	$CH_3-(CH_2)_6-COOH$
3. Animal fat	Oleic acid and	$CH_3-(CH_2)_7-CH{=}CH-(CH_2)_7-COOH$
	Palmitic acid	$CH_3-(CH_2)_{14}-COOH$
4. Plant fats (Seed Oils)	Oleic acid,	$CH_3-(CH_2)_7-CH{=}CH-(CH_2)_7-COOH$
	Palmitic acid and Essential fatty acids	$CH_3-(CH_2)_{14}-COOH$

Animals are also an important source of fats. Milk and milk products such as butter, cream, cheese are rich in fats. Meat and eggs also contain fat. Creamish white fat found in pig meat (pork) is called ***lard. Tallow***, the fat from sheep and cattle is used largely for making soap.

58.4 EXTRACTION OF FATS AND OILS

Fats and oils are extracted from natural sources by the following processes:

1. **Rendering.** The animal tissues containing the fat are heated dry or with water until the fat melts and can be removed.
2. **Pressing.** Oils are obtained from seeds by crushing between steel rollers and then pressing in a *hydraulic press*.
3. **Solvent Extraction.** It is often applied to the residue after pressing or rendering for complete removal of oil or fat. The solvents used include petroleum ether and benzene.

58.5 PROPERTIES OF OILS AND FATS

PHYSICAL PROPERTIES

1. Oils and fats are liquids or solids having a greasy feel. When pure, they are colourless, odourless and tasteless.
2. They are insoluble in water but soluble in organic solvents such as ether, chloroform and benzene.
3. They have a lower density than water and consequently float on the surface when mixed with water.

CHEMICAL PROPERTIES

Fats and oils are triesters of glycerol with saturated and unsaturated fatty acids. Their reactions are those of ester groups in triplicate and carbon-carbon double bonds.

1. **Hydrolysis.** They are readily hydrolysed by heating with acids or alkalies or superheated steam. When boiled with sodium or potassium hydroxide solution, the hydrolysis products are glycerol and sodium or potassium salts of long-chain fatty acids. The latter are called **soaps** and alkaline hydrolysis is known as **saponification.**

$$\begin{array}{l} CH_2\text{—O—}\overset{O}{\overset{\|}{C}}\text{—R} \\ | \\ CH\text{—O—}\overset{O}{\overset{\|}{C}}\text{—R}' \\ | \\ CH_2\text{—O—}\overset{O}{\overset{\|}{C}}\text{—R}'' \end{array} + 3NaOH \longrightarrow \begin{array}{l} CH_2OH \\ | \\ CHOH \\ | \\ CH_2OH \end{array} + \begin{array}{l} R\text{—}\overset{O}{\overset{\|}{C}}\text{—}\overset{-}{O}\overset{+}{Na} \\ R'\text{—}\overset{O}{\overset{\|}{C}}\text{—}\overset{-}{O}\overset{+}{Na} \\ R''\text{—}\overset{O}{\overset{\|}{C}}\text{—}\overset{-}{O}\overset{+}{Na} \end{array}$$

Oil or Fat (triglyceride) — Glycerol — Soap

2. **Hydrogenation.** On catalytic hydrogenation, hydrogen adds across the carbon-carbon double bonds of the acid components of the triglycerides. This results in the formation of saturated glycerides which are solid fats at room temperature. This hydrogenation process is called **hardening of oils.** For example,

$$\begin{array}{l} CH_2O\overset{O}{\overset{\|}{C}}(CH_2)_7CH{=}CH(CH_2)_7CH_3 \\ | \\ CHO\overset{O}{\overset{\|}{C}}(CH_2)_7CH{=}CH(CH_2)_7CH_3 \\ | \\ CH_2O\overset{O}{\overset{\|}{C}}(CH_2)_7CH{=}CH(CH_2)_7CH_3 \end{array} + 3H_2 \xrightarrow[\text{Ni, }\Delta]{\text{Pressure}}$$

Triolein (An oil from palm oil)

$$\begin{array}{l} CH_2O\overset{O}{\overset{\|}{C}}(CH_2)_{16}CH_3 \\ | \\ CHO\overset{O}{\overset{\|}{C}}(CH_2)_{16}CH_3 \\ | \\ CH_2O\overset{O}{\overset{\|}{C}}(CH_2)_{16}CH_3 \end{array}$$

Tristearin (A solid fat)

Partial hydrogenation of vegetable oils is used for the manufacture of **margarine** and other hardened oils.

Add To Your Knowledge

For the **manufacture of margarine**, oils such as groundnut oil and palm oil are treated to remove undesirable material, bleached and heated. Hydrogen gas at a pressure of about 4 atmospheres is passed through the hot oil in the presence of nickel as catalyst. By controlling the extent of hydrogenation the oil is hardened to the desired level. The hardened oil is deodorized and then blended with milk. The product is then mixed with salt, vitamins, flavouring agents and preservatives to obtain margarine. Margarine is used as a substitute for butter.

58.6 TESTS FOR OILS AND FATS

SOLUBILITY TEST

This test is based on the fact that oils and fats are soluble in organic solvents such as chloroform, ethanol, etc., but are insoluble in water. Shake a small amount of the given sample with 5 cm^3 each of water, ethanol and chloroform in three test tubes. Observe the solubility and draw inferences.

Test tube	*Solvent*	*Observation*	*Inference*
1.	Water	(*i*) Sample is immiscible	Oil or fat present.
		(*ii*) Sample is miscible	Oil or fat absent.
2.	Alcohol	(*i*) Sample forms lower layer, which *dissolves on heating.*	Oil or fat present.
		(*ii*) Sample *does not dissolve* even on heating.	Oil or fat absent.
3.	Chloroform	(*i*) Sample is *miscible.*	Oil or fat present.
		(*ii*) Sample is *immiscible.*	Oil or fat absent.

TRANSLUSCENT SPOT TEST

Press a little of the substance in the folds of a filter paper. On unfolding the filter paper, the appearance of a **transluscent or a greasy spot** on the filter paper indicates the presence of fat or oil. The spot grows larger on heating and drying the filter paper.

ACROLEIN TEST

Heat a little of the sample with some crystals of $KHSO_4$ in a test-tube. **A pungent irritating odour of acrolein** confirms the presence of fat or oil.

$$\text{Oil or fat} \xrightarrow{\Delta} \text{Glycerol} + \text{Fatty acid}$$

$$\underset{\text{Glycerol}}{\begin{array}{c} CH_2OH \\ | \\ CHOH \\ | \\ CH_2OH \end{array}} \xrightarrow[\Delta]{KHSO_4} \underset{\text{Acrolein}}{\begin{array}{c} CH_2 \\ \| \\ CH \\ | \\ CHO \end{array}} + 2H_2O$$

58.7 SOAPS AND SYNTHETIC DETERGENTS

The word 'detergent' means '*cleansing agent*' and so *the detergents are substances which remove dirt and have cleansing action in water.* According to this definition of detergents, soap is also a detergent and has been used for more than two thousand years. There are two types of detergents:

1. *Soapy detergents or soaps*
2. *Non-soapy detergents or soapless soaps.*

Soap

A soap is a sodium or potassium salt of some long chain carboxylic acids (fatty acid).

Sodium salts of fatty acids are known as **hard soaps** and potassium salts of fatty acids are known as **soft soaps.** Hard soaps are prepared from cheap oils and fats and sodium hydroxide. They contain free alkali and are used for washing purposes. Soft soaps are prepared from good oils and potassium hydroxide. They do not contain free alkali, produce more lather and are used as toilet soaps, shaving creams and shampoos.

A soap has a large non-ionic hydrocarbon group and an ionic COO^-Na^+ group. So for simplicity the structure of soap can be represented as

COO^-Na^+

Soap Molecule

Some examples of soaps are: *sodium stearate,* $C_{17}H_{35}COO^-Na^+$, *sodium palmitate,* $C_{15}H_{31}COO^-Na^+$ *and sodium oleate,* $C_{17}H_{33}COO^-Na^+$.

Soaps of metals other than sodium and potassium are usually water-insoluble and do not find application as a cleansing agent. Therefore, hard water, which contains salts of magnesium and calcium, reacts with soap to form magnesium salt of fatty acid and calcium salt of fatty acid.

$$\underset{\text{Soap}}{2C_{17}H_{35}COONa} + \underset{\text{(Hardness of water)}}{MgCl_2} \longrightarrow \underset{\text{White ppt.}}{(C_{17}H_{35}COO)_2Mg \downarrow} + 2NaCl$$

$$2C_{17}H_{35}COONa + \underset{\text{(Hardness of water)}}{CaCl_2} \longrightarrow \underset{\text{White ppt.}}{(C_{17}H_{35}COO)_2Ca \downarrow} + 2NaCl$$

These calcium and magnesium salts of fatty acids are insoluble in water and separate as curdy white precipitate. Thus, a lot of soap is wasted if water is hard.

ACTIVITY 58.1	**Study of Action of Soap in Hard Water**

- Take two test tubes and label them as A and B.
- In test tube A, put 10 cm^3 of distilled water (or rain water) and in test tube B put 10 cm^3 of hard water.
- Add 5 drops of soap solution to both tubes. Shake the test tubes vigorously for an equal period of time.

What do you observe?

It is observed that more foam is formed in the test tube containing distilled water. In the test tube containing hard water less foam is formed and at the same time a ***curdy white precipitate*** is formed.

Note. If hard water is not available, it can be prepared by adding small amount of calcium chloride or magnesium sulphate to ordinary water.

PREPARATION OF SOAP

Soap is prepared by heating oil or fat of vegetable or animal origin with a calculated quantity of concentrated sodium hydroxide solution (caustic soda solution). Hydrolysis of fat takes place and a mixture of sodium salts of fatty acids and glycerol is formed.

$$\begin{array}{l} CH_2O.COH_{15}H_{31} \\ | \\ CHO.COC_{15}H_{31} + 3NaOH \\ | \\ CH_2O.COC_{15}H_{31} \end{array} \longrightarrow \begin{array}{l} CH_2OH \\ | \\ CHOH + 3C_{15}H_{31}COONa \\ | \\ CH_2OH \end{array}$$

Fat or oil (Tripalmitin) — Glycerol — Soap (Sodium palmitate) or Sodium hexadecanoate

Soap which is formed as a result of alkaline hydrolysis of oil or fat is separated from the solution by the addition of sodium chloride (common salt). Salt decreases the solubility of soap which is released from the solution. The process is known as **salting out.** The crude soap that separates out is called **grained soap.** Now, soap being lighter than water floats on its surface from where it is removed. The lower aqueous layer is called **lye.**

After the removal of soap the solution which is left behind contains glycerol and sodium chloride. Glycerol is then recovered from this mixture as it is an important chemical which finds use in drugs, paints, cosmetics and explosives.

TRADITIONAL SOAP MAKING

In Ghana and other parts of Western Africa, soap was prepared by traditional methods long before the modern methods were available. In the traditional methods, ashes from the burning of plantain peels, cocoa husk and wood are used in place of alkali. The ashes provide potassium hydroxide. Oils or fats are obtained from vegetable sources such as palm, coconut, cocoa, etc.

The oil is heated with dry ash or its aqueous solution, with constant stirring. In the traditional soap, glycerol is not recovered. The product thus obtained is usually coloured due to impurities and is not as good as the soap manufactured by modern methods.

ACTIVITY 58.2	**Laboratory Preparation of Soap**

- Take about 25 cm^3 of castor oil or vegetable oil in a beaker. To this add about 50 cm^3 of 20 per cent sodium hydroxide solution slowly with constant shaking.
- Heat the mixture slowly to boil and let it continue boiling for about ten minutes.
- Remove the beaker from the burner and add about 5 g of common salt to the beaker.
- Allow the mixture to cool.

What do you observe?

After sometime a solid crust of soap will be seen floating in the beaker.

DETERGENTS

These are also called **synthetic detergents** or **syndets** or **soapless soaps** or just **detergents**. So in our routine language when we say detergent, it means the synthetic detergent. These synthetic detergents are designed in such a way that while using them problems do not arise with the hardness of water just like they arise with the use of soap. They do not form insoluble calcium and magnesium salts with hard water and can be used for washing even with hard water.

A ***synthetic detergent*** *is the sodium salt of a long chain benzene sulphonic acid or the sodium salt of a long chain alkyl hydrogen sulphate.*

Like soaps they contain an ionic group such as a sulphonate group, $S\overset{-}{O}_3\overset{+}{Na}$ or a sulphate group, $OS\overset{-}{O}_3\overset{+}{Na}$ and a long chain hydrocarbon which is a non-ionic group.

$SO_3^-Na^+$

A Detergent Molecule

PREPARATION OF SYNTHETIC DETERGENTS

Synthetic detergents are prepared by reacting hydrocarbons from petroleum with conc. tetraoxosulphate (VI) acid and converting the product into its sodium salt. Examples of synthetic detergents are sodium *p*-dodecyl benzenesulphonate and sodium lauryl sulphate.

$C_{12}H_{25}-C_6H_4-\overset{-}{SO_3}\overset{+}{Na}$ $\quad$ $C_{12}H_{25}-O-SO_2-\overset{-}{O}\overset{+}{Na}$

Sodium *p*-dodecyl benzenesulphonate $\quad$ Sodium lauryl sulphate

Add To Your Knowledge

Washing powders available in the market contain about 15 to 30 per cent detergents by weight and the remaining part is other chemicals which are added to it. These are:

- Tetraoxophosphate(V) compounds which prevent formation of insoluble compounds with Ca^{2+} ions and fatty acids
- Bleaching agents to remove stains and to produce whiteness in clothes.
- Perfumes
- Fluorescents
- Additives to prevent corrosion of metals, and to maintain the powdery texture of the detergent.

ADVANTAGES OF SYNTHETIC DETERGENTS OVER SOAPS

Synthetic detergents are widely used these days as cleansing agents. Synthetic detergents have the following advantages over soaps:

1. Synthetic detergents can be used even in hard water whereas some of the soap gets wasted if water is hard.
2. Synthetic detergents can be used even in acidic medium as they are the salts of strong acids and are not decomposed in acidic medium.
3. Synthetic detergents have a stronger cleansing action than soaps.
4. Synthetic detergents are more soluble in water than soaps.
5. Synthetic detergents are prepared from the hydrocarbons obtained from petroleum. This saves vegetable oils which are otherwise used in preparation of soap.

ACTIVITY 58.3 Comparison of Action of Soap and Detergent in Hard Water

- Take two test tubes and label them as **A** and **B.** Add 10 cm^3 of hard water in each tube.
- Now add 5 drops of soap solution to test tube **A** and 5 drops of detergent solution to test tube **B**.
- Shake both test tubes vigorously for the same period.

What do you observe?

It is observed that more foam is formed in test tube B (the test tube containing detergent solution).

In the test tube containing soap, less foam is formed and at the same time some curdy precipitate is formed.

DISADVANTAGES OF SYNTHETIC DETERGENTS

Some of the early synthetic detergents were non-biodegradable, that is they were not broken down by bacteria. This problem has now been overcome. *It has been found that if straight chain hydrocarbon is used in the detergent instead of a branched chain hydrocarbon then the detergent is biodegradable. All the soaps, on the other hand, are biodegradable.*

Another problem is with the use of certain tetraoxophosphate(V) additives which are added to the detergent. These additives act as nutrients for algae which grows rapidly and consumes most of the dissolved oxygen in water. This rapid algal growth is referred to as **eutrophication.** Due to lack of oxygen the fish and the other aquatic animals die. But action has now been taken to resolve this problem also.

A comparison between properties of soaps and detergents is given in Table 58.2.

Table 58.2. Differences between Soaps and Detergents

Soaps	*Synthetic Detergents*
1. Soaps are sodium salts of higher fatty acids.	1. Synthetic detergents are sodium alkyl sulphates or sodium alkyl benzene sulphonates with alkyl group having more than ten carbon atoms.
2. Soaps are prepared from natural oils and fats.	2. Synthetic detergents are prepared from the hydrocarbons of petroleum.
3. Soaps form insoluble salts (curdy white ppt.) with calcium and magnesium ions which are present in hard water and hence, cannot be used in hard water.	3. Calcium and magnesium salts of detergents are soluble in water and, therefore, no curdy white precipitates are obtained in hard water and hence, they can be used even in hard water.
4. Soaps cannot be used in acidic medium as they are decomposed into carboxylic acids in acidic medium.	4. They can be used in acidic medium as they are the salts of strong acids and are not decomposed in acidic medium.
5. Soaps are biodegradable.	5. Some of the synthetic detergents are not biodegradable.

CLEANSING ACTION OF SOAPS AND DETERGENTS

Now, let us see how soaps and detergents help in cleansing. A molecule of soap is made up of two parts—a non-polar part consisting of a long chain of 12–18 carbon atoms and a polar part, —COO^-Na^+. The polar end is water soluble and is thus **hydrophilic** whereas the hydrocarbon part is insoluble in water and is thus **hydrophobic.** The hydrocarbon part is, however, soluble in non-polar solvents and is sometimes called **lipophilic**. In a soap solution, the hydrocarbon portions of several soap molecules huddle together whereas the polar ends project outward into the polar solvent, water. **These spherical clusters of hundreds of molecules (or ions) are called micelles.** The soap micelles are negatively charged due to the presence of carboxylate ions at the surface. Repulsion between similarly charged micelles keeps them dispersed in solution.

Mostly the dirt is held to any surface such as cloth by the oil or grease which is present there. Now since the oil and grease are not soluble in water, the dirt particles cannot be removed by simple washing of the cloth with water. However, when soap is applied, the non-polar hydrocarbon part of the soap molecules dissolves in oil droplets while the polar —COO^-Na^+ groups remain attached to the water molecules. In this way, each oil droplet gets surrounded by negative charge (Fig. 58.1).

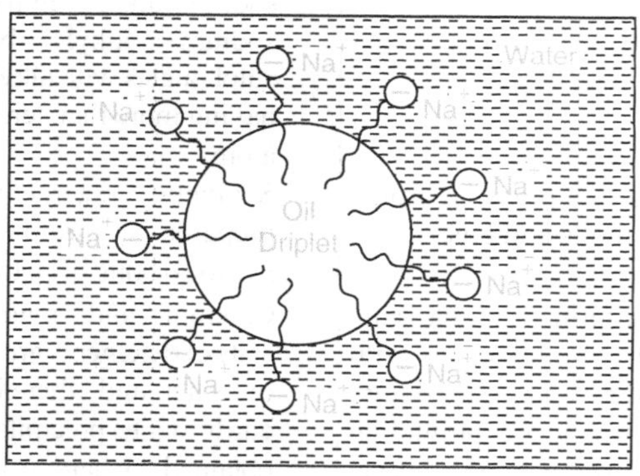

Fig. 58.1. Cleansing action of soap.

These negatively charged oil droplets cannot coalesce due to repulsion between similar charges and a stable emulsion is formed. These oil droplets (containing dirt particles) can be washed away with water along with dirt particles. **So the action of soap or detergent is to emulsify oil or grease.** This loosens the solid particles of dirt and they are easily rinsed away.

ACTIVITY 58.4	**Study of Action of Soaps and Detergents in Emulsification**

- Take three test tubes and add 10 cm^3 of water to each of them. Label them as **A**, **B** and **C**.
- Now add 2 drops of mustard oil to each of the three test tubes.
- Add a few drops of soap solution to test tube B and a few drops of detergent solution to test tube C.
- Shake vigorously all the three test tubes, for the same period of time and then keep them undisturbed for some time.

What do you observe?

It is observed that oil layer separates out quickly in the tube A whereas in tubes B and C stable emulsion is formed.

Soap or detergent helps in cleansing in another way. Not only does it emulsify oil or grease but ***it also lowers the surface tension of water.*** As a result of this water wets things more effectively. Now, let us see how this happens. When water is added on to the surface of the cloth the water molecules tend to stay as close to each other as possible because of the strong forces of attraction (hydrogen bonding) for each other and do not wet the cloth properly. If some soap solution is added to this water, the polar end of soap dissolves in water and the non-polar hydrocarbon end remains away from the water. Thus, soap molecules arrange themselves between the water molecules on the surface of water and decrease the forces of attraction between the water molecules. This reduces the surface tension of water. Water molecules can now spread on the surface of cloth and can make it wet effectively.

SOLVED EXAMPLE

Example 58.1. *Give an example of a soap and an example of a synthetic detergent. Which of them can be used in acidic medium and why?*

Solution. An example of a soap is sodium stearate, $C_{17}H_{35}COONa$. An example of a synthetic detergent is sodium *p*-dodecyl benzene sulphonate.

$$C_{12}H_{25}-C_6H_4-\overset{-}{S}O_3\overset{+}{N}a$$

Synthetic detergents can be used in acidic medium as they are salts of strong acids and hence are not decomposed in acidic medium.

58.8 USES OF FATS AND OILS

The uses of fats and oils are as follows:

1. Oils and fats are important ***constituent of food***. They are used as food mainly to provide energy.
2. Oils are used for the ***preparation of margarine*** and ***soaps***.
3. Some oils such as cod liver oil find ***medicinal*** application.
4. Fats and oils are also used in the manufacture of ***cosmetics*** such as cold creams, lip-sticks, lotions, etc.
5. Certain drying oils such as linseed oil are used in oil-based ***paints***.
6. Some oils are used for ***lubrication of machine parts***.

58.9 FUNCTIONS OF FATS IN THE BODY

1. *The main function of fats in the body is to provide energy.* By supplying energy, fats save proteins from being used for energy and allow them to perform their more important role of building and repairing tissues. Fats on oxidation provide almost twice as much energy as that given by carbohydrates. The fats provide on oxidation about 37 kJ of energy per gram as compared to about 17 kJ of energy per gram of carbohydrates. ***Fats*** yield more energy than carbohydrates because fats contain a less percentage of oxygen and a higher percentage of carbon and hydrogen as compared with carbohydrates. Fats can also be stored in the body for subsequent use. When we consume food which has more energy than is required by the body for performing various functions, the excess food is deposited under our skin in the form of subcutaneous fat.
2. In addition to supplying energy, *fats also help in forming the structural material of cells and tissues such as the cell membrane.*
3. *Fats also carry the fat soluble vitamins A, D, E and K into the body and help in the absorption of these vitamins in the intestines.*
4. Fat stored under skin protects animals from cold because it is a poor conductor of heat.

SUMMARY

- **Fats and oils** are triesters of glycerol with long chain fatty acids.
- Animal fats are rich in saturated fats whereas vegetable oils are rich in unsaturated fats.
- Fats and oils are insoluble in water but are soluble in organic solvents such as benzene, hexane, ether, petroleum ether, etc.
- **Saponification.** The alkaline hydrolysis of fats and oils is called saponification. The products of saponification are glycerol and sodium or potassium salts of long chain fatty acids (soaps).
- **Hydrogenation of oils** transforms them into fats.
- Oil can be extracted from plant seeds by pressing and solvent extraction.
- ***Soaps.*** These are sodium or potassium salts of long chain fatty acids.
- ***Synthetic Detergents.*** These are sodium salts of long chain benzene sulphonic acids or long chain alkyl hydrogen sulphates.
- Synthetic detergents can be used in acidic medium or in hard water.
- The cleansing action of soaps and detergents is based on the presence of both hydrophobic and hydrophilic groups in their molecules which helps to emulsify the oily dirt which can then be rinsed out.

EVALUATE YOURSELF

I. Objective Type Questions

Select the most appropriate choice from the options given as (a), (b), (c) and (d) after each question:

1. Fatty acids are
   (*a*) Unsaturated dicarboxylic acids
   (*b*) Long-chain alkanoic acids
   (*c*) Aromatic carboxylic acids
   (*d*) Aromatic dicarboxylic acids.
2. Fats and oils are
   (*a*) monoesters of glycerol (*b*) diesters of glycerol
   (*c*) triesters of glycerol (*d*) diesters of glycol.
3. Liquid oils can be converted to solid fats by
   (*a*) Hydrogenation (*b*) Saponification
   (*c*) Hydrolysis (*d*) Oxidation of double bonds.
4. Alkaline hydrolysis of oils and fats is known as
   (*a*) Saponification (*b*) Rancidification
   (*c*) Diazotization (*d*) Hydrogenation.

5. Sodium or potassium salts of fatty acids are known as
   (*a*) Carbohydrates (*b*) Soaps
   (*c*) Non-soapy detergents (*d*) Proteins.

II. Fill in the Blanks

6. Complete the following sentences by supplying appropriate words:
   (*i*) Triglycerides containing unsaturated fatty acids are ...... at room temperature and are called ....
   (*ii*) The products of saponification of fats are soaps and ....
   (*iii*) Hard soaps are ...... salts of long chain fatty acids.

III. Discussion Questions

7. What are fats and oils? How do they differ from each other? Is it possible to convert oil into fat?
8. Give the structure of a saturated fat and an unsaturated fat.
9. Write down the structure of
   (*i*) propane-1, 2, 3-triol (*ii*) hexadecanoic acid
   (*iii*) sodium octadecanoate.
10. What are the various steps involved in the extraction of oil from groundnuts?
11. What are the uses of fats and oils?
12. What are the major functions of fats and oils in the human body?
13. What is a detergent? Name two types of detergents.
14. What are soaps? Why does it not produce lather easily in hard water?
15. What are hard and soft soaps?
16. What is the basic structure of a soap molecule?
17. Why is sodium chloride added during the manufacture of soap from oils?
18. Why is soap not suitable for washing clothes when water is hard?
19. How do synthetic detergent differ from soaps? What is the basic structure of a synthetic detergent?
20. Why are synthetic detergents called soapless soaps?
21. What are synthetic detergents? Give one example of a synthetic detergent.
22. What are the advantages of synthetic detergents over soaps?
23. How is soap prepared in the laboratory?
24. Name the raw materials required for the manufacture of soap.
25. What are the advantages of synthetic detergents?
26. Explain the cleansing action of soap.
27. Explain why detergents are preferred over soaps?
28. Explain the water pollution by the use of synthetic detergents.
29. Name the two ways by which soaps or detergents help in cleansing.
30. Would you be able to check if water is hard by using a detergent?

Answers

I. Objective Type Questions

1. (*b*) **2.** (*c*) **3.** (*a*) **4.** (*a*)
5. (*b*).

II. Fill in the Blanks

6. (*i*) liquid, oils
(*ii*) glycerol
(*iii*) sodium.

SECTION-2

UNIT-59

SYNTHETIC POLYMERS

Specific Objectives

After studying this unit, the student should be able to:

- describe synthetic polymers
- describe addition and condensation polymerization
- describe how the properties of polymers depend on their structural features
- describe ways of modifying the properties of polymers
- describe plastics and resins
- explain the chemical tests for plastics
- state the uses of polymers.

59.1 INTRODUCTION

Polymers are the chief products of the modern chemical industry and a backbone of our present society. Daily life without the discovery and varied applications of polymers would not have been easier and colourful. The materials made of polymers find multiple uses and applications in all walks of our life. They have influenced our day-to-day life to such an extent that it is impossible to get through the day without using a material based on polymers. Common examples of these include.

Plastic dishes, cups, non-stick, pans; automobile tyres and seat covers, plastic bags, rain coats, plastic pipes and fitting; radio, TV and computer cabinets; wide range of synthetic fibres for clothing, synthetic glues, flooring materials and materials for biomedical and surgical operations.

The word **polymers** means "many parts" (Greek: *poly* means *many* and *merors* means *parts*). ***A polymer is a compound of high molecular mass formed by the combination of large number of small molecules.*** The small molecules which constitute the *repeating units* in a polymer are called **monomer units.** The process by which the monomers are transformed into polymer is called **polymerisation.** For example, polyethylene is a polymer which is obtained by the polymerisation of ethylene. The ethylene molecules are referred to as monomer units.

$$\underset{\text{Ethylene}}{n\,CH_2=CH_2} \xrightarrow{\text{Polymerisation}} \underset{\text{Repeating unit}}{n \left(CH_2-CH_2 \right)}$$

$$\longrightarrow \underset{\text{Polythene polymer}}{\left(CH_2-CH_2 \right)_n}$$

$$\underset{\text{Hexa methylene diamine}}{nNH_2(CH_2)_6NH_2} + \underset{\text{Adipic acid}}{nCOOH(CH_2)_4\ COOH} \xrightarrow{\text{Polymerisation}}$$

$$\left(NH(CH_2)_6-NH-\overset{\overset{O}{\|}}{C}-(CH_2)_4-\overset{\overset{O}{\|}}{C}\right)_n$$

Nylon 6, 6

As polymers are *single, giant molecules, i.e.*, big size molecules, they are also called **macromolecules.**

HOMOPOLYMERS AND COPOLYMERS

Polymers are divided into two broad categories depending upon the nature of the repeating structural units. These are *homopolymers* and *co-polymers*.

The polymer formed from one kind of monomers is called **homopolymer** while the polymer formed from more than one kind of monomer unit is called **copolymer** or **mixed polymer.** For **example,** *polyethylene* is an example of a homopolymer whereas *Buna-S* rubber which is formed from 1,3-butadiene ($CH_2 = CH—CH = CH_2$) and styrene ($C_6H_5CH = CH_2$) is an example of a copolymer.

59.2 CLASSIFICATION OF POLYMERS

Polymers are classified in a number of ways depending upon one criterion or the other as described below.

On the basis of source or origin, the polymers are classified into two types:

Natural polymers and *Synthetic polymers.*

1. Natural Polymers. *Polymers found in nature, mostly in plants* and *animals* are called **natural polymers.** A few examples are:

(*a*) ***Polysaccharides.*** Starch and cellulose are very common examples of polysaccharides. They are the polymers of glucose. Starch is the main food reserve of plants while cellulose is the main structural material of plants.

(*b*) ***Proteins.*** These are the polymers of α-amino acids. They are the building blocks of animal cells. They constitute an indispensable part of our food. Wool, natural silk, leather, etc., are proteins.

(*c*) ***Nucleic Acids.*** These are the polymers of various *nucleotides.* RNA and DNA are common examples.

(*d*) ***Natural Rubber.*** Substance obtained from latex is a polymer of 2-methyl buta-1, 3-diene (*isoprene*).

It may be noted that polymers like polysaccharides, nucleic acids, proteins, etc., which control various life processes in plants and animals are also called **biopolymers.**

2. Semi Synthetic Polymers. These are mostly derived from naturally occurring polymers by carrying out chemical modifications. For example,

$$\text{Cellulose} + \underset{\text{Acetic anhydride}}{(CH_3CO)_2O} \xrightarrow{H_2SO_4} \text{Cellulose diacetate}$$

Cellulose diacetate is used in making *threads, films, glasses,* etc.

3. Synthetic Polymers. The polymers which are prepared in the laboratory are referred to as **synthetic polymers** or man-made polymers. Some examples of synthetic polymers are *polyethylene, polystyrene, PTFE, synthetic rubber, nylon, PVC, bakelite, teflon, orlon,* etc.

The synthetic polymers can be further classified as those made up of monomers and comonomer units.

The polymers are also classified based on:

(*i*) structure (*ii*) molecular forces

(*iii*) modes of polymerisation.

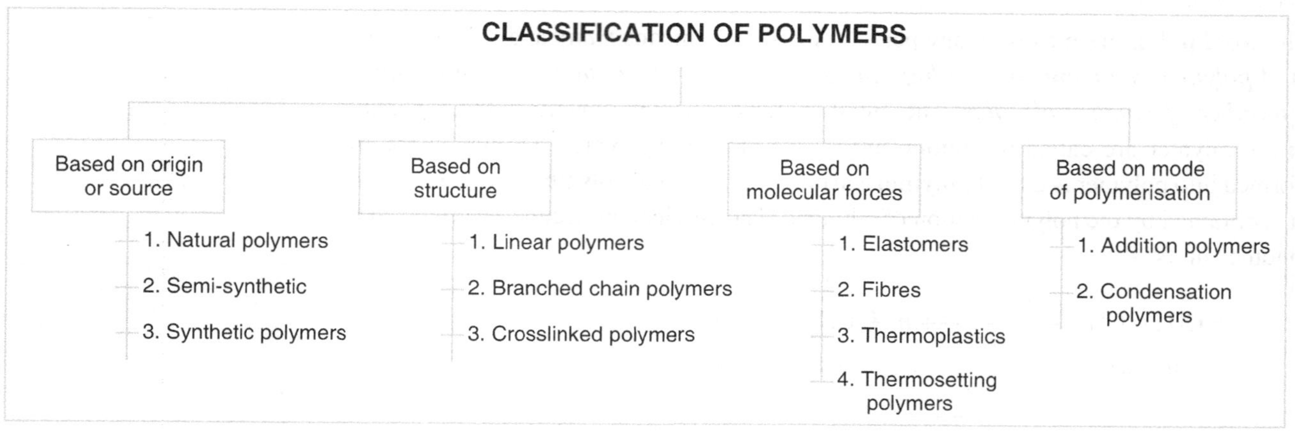

59.3 ADDITION AND CONDENSATION POLYMERISATION

1. Addition Polymers. When the monomer units are repeatedly added to form long chains without the elimination of any by-product molecules, the product formed is called *addition polymer* and the process involved is called *addition polymerisation*. The monomer units are unsaturated compounds and are usually the derivatives of alkenes. The molecular formula and hence the molecular mass of the addition polymer is an *integral multiple* of that of the monomer units. Some **examples** of addition polymerisation are:

(*i*) $nCH_2=CH_2 \longrightarrow (CH_2—CH_2)_n$

Ethylene → Polyethylene or polyethene

(*ii*) $nCH_2=CH(CH_3) \longrightarrow (CH_2—CH(CH_3))_n$

Propylene → Polypropylene

(*iii*) $nCH_2=CH(CN) \longrightarrow (CH_2—CH(CN))_n$

Acrylonitrile → Orlon

(*iv*) $nCH_2=CH(Cl) \longrightarrow (CH_2—CH(Cl))_n$

Vinyl chloride → Poly vinyl chloride (PVC)

2. Condensation Polymers. In this type of polymers, the monomers react together with the elimination of a simple molecule like H_2O, ROH or NH_3, etc. The reaction is called condensation and the product formed is called **condensation polymer.**

As the process involves the elimination of by-product molecules, the molecular mass of the polymer is not the integral multiple of the monomer units. For examples, *nylon-6,6* is a condensation polymer of hexamethylene diamine and adipic acid.

$$n\,H_2N—(CH_2)_6—NH_2 + n\,HOOC—(CH_2)_4—COOH$$

Hexamethylene diamine + Adipic acid

$$\downarrow$$

$$(—NH—(CH_2)_6—NH—CO—(CH_2)_4—CO—)_n + 2nH_2O$$

Nylon-6, 6

Some other **examples** of condensation polymers are:

Dacron (polyester): A polymer of *ethylene glycol* and *terepthalic acid.*

Bakelite: A polymer of *phenol* and *formaldehyde.*

Polyurethanes. These are the condensation polymers of *toluene-2, 6-diisocyanate* and *ethylene glycol.* During polymerisation a low boiling liquid such as Freon-11 are added to the reaction mixture. The heat liberated during polymerisation causes vaporisation of a low boiling liquid producing bubbles which converts viscose polymer to form *polyurethane foams.*

$$R—N=C=O + ROH \rightarrow RNH—C(=O)—OR$$

alkyl isocyanate + alcohol → a urethane

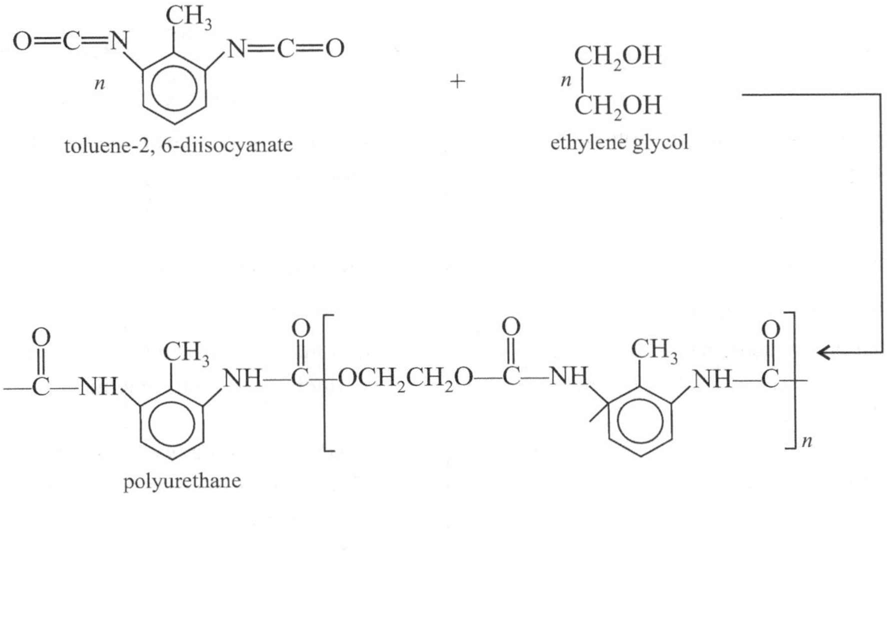

$$n\ O=C=N—C_6H_3(CH_3)—N=C=O + n\ CH_2OH—CH_2OH$$

toluene-2, 6-diisocyanate + ethylene glycol

$$\downarrow$$

$$—C(=O)—NH—C_6H_3(CH_3)—NH—[C(=O)—OCH_2CH_2O—C(=O)—NH—C_6H_3(CH_3)—NH—C(=O)]_n—$$

polyurethane

59.4 PROPERTIES OF POLYMERS ON THE BASIS OF THEIR STRUCTURAL FEATURES

There are several factors which determine the physical as well as chemical properties of a polymer, for example, nature of chain packing, type of molecules forces etc. These factors influence, the crystallinity of polymers, which in turn, along with other factors influence properties such as melting point, tensile strength, viscosity, toughness, etc. **Based on their structure,** *i.e.*, how the monomers are linked to each other we have the following types of polymers:

1. *Linear polymers,*
2. *Branched chain polymers,* and
3. *Cross-linked polymers.*

1. Linear Polymers. These are the polymers in which monomeric units are linked together to form long straight chains. The polymeric chains are stacked over one another to give a well packed structure. As a result of close packing, such polymers have *high densities, high tensile strength* and *high melting points.* Common **examples** of such type of polymers are *polyethylene, nylons* and *polyesters.*

Fig. 59.1 Linear polymer.

2. Branched Chain Polymers. In this type of polymers, the monomeric units are linked to constitute long chains (called the *main-chain*). There are side chains of different lengths which constitute branches. Branched chain polymers are irregularly packed and thus have *low density, low tensile strength* and *low melting point* as compared to linear polymers. *Amylopectin* and *glycogen* are common examples of this type.

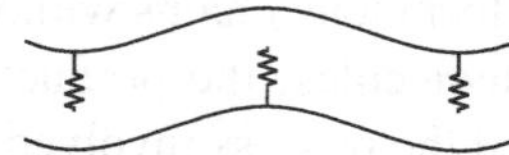

Fig. 59.2 Branched chain polymer.

3. Cross-linked or Network Polymers. In this type of polymers, the monomeric units are linked together to constitute a three-dimensional network. The links involved are called cross-links. Cross-linked polymers are *hard, rigid* and *brittle* because of their network structure. Common examples of this type of polymers are *bakelite, melamine formaldehyde resin,* etc.

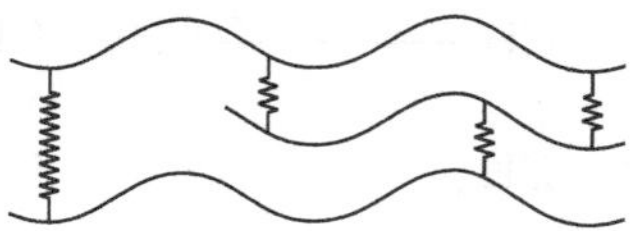

Fig. 59.3 Cross-linked polymer.

Low Density (LDPE) and High density (HDPE) polyethylene

Polyethylene is a branched chain polymer. This means that the polymer chains are not be in parallel but instead form a tangled mass (Fig. 59.4). Now since the chains cannot be close together, the density of polyethylene is low. It is therefore called low-density

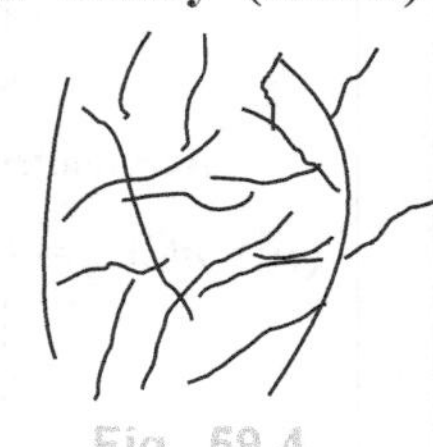

Fig. 59.4

High Density Polythene (HDPE)	*Low Density Polythene (LDPE)*
$nCH_2=CH_2 \xrightarrow[\text{6–7 atm, } TiCl_4 + Al(C_2H_5)_3]{\text{333–343 K}} \text{+}CH_2\text{—}CH_2\text{+}_n$ (Polythene)	$nCH_2=CH_2 \xrightarrow[\text{1000–1500 atm, Traces of } O_2]{\text{350–570 K}} \text{+}CH_2\text{—}CH_2\text{+}_n$ (Polythene)
• It consists of linear chains polymers molecules	• It consist of branched chain structure of polymer molecules
• Density ≈ 0.97 g cm^{-3}	• Density ≈ 0.92 g cm^{-3}
• Melting point 403 K	• Melting point 383 K
• It is transluscent polymer.	• It is transparent polymer
• It is chemically inert, having relatively greater toughness and high tensile strength than LDPE	• It is also chemically inert with moderate tensile strength and high toughness
• Used in the manufacture of *containers, pipes, bottles, toys, bags,* etc.	• Used as *packing material* in the form of thin film or sheet, also as *insulation for electrical wires and cables.*

polyethylene (or LDPE). LDPE has a lower tensile strength, low melting point ($\simeq$ 130ºC)

Linear chains of polyethylene *i.e.,* chains with no branching can be close together and hence result in increase in density. Such a polymer is called high density polyethylene (HDPE). It was first produced in 1953 in Germany by Karl Ziegler.

$$nC_2H_4 \xrightarrow[50°C,\ 1.5\ atom]{(C_2H_5)_3\ Al/TiCla} \text{+}CH_2\text{—}CH_2\text{+}_n$$

HDPE

The HDPE chains form ordered structures, and therefore, it has higher density than the LDPE. Its melting point and tensile strength is also higher than LDPE.

The molecular forces like van der Waals' forces and hydrogen bonding existing in the monomeric units affect many mechanical properties of polymers such as tensile strength, toughness, elasticity, etc. Although these intermolecular forces are present in simple molecules also, their effect is less significant in them as compared to that in macromolecules. It is because of the fact that in polymers there is a combined effect of these forces all along the length of the chains. Obviously, the longer the chain, the more intense is the effect of intermolecular forces.

On the basis of the magnitude of **inter-molecular forces** present in polymers, we have,

Elastomers, Fibers, Thermoplastics, and *Thermosetting polymers.*

1. Elastomers. These are *the polymers in which the polymer chains are held up by the weakest attractive forces.* They are amorphous polymers having high degree of elasticity. The weak forces permit the polymer to be stretched out about 10 times their normal length but they return to their original position when the stretching force is withdrawn. In fact, these polymers consist of randomly coiled molecular chains having few cross-links. When the stress is applied, these randomly coiled chains straighten out and the polymer gets stretched. As soon as the stretching force is released, the polymer regains its original shape because the weak forces do not allow the polymer to remain in the stretched form.

2. Fibers. *These are the polymers which have quite strong interparticle forces such as H-bonds.* They have high tensile strength and high modulus. They are thread-like polymers and can be woven into fabrics. Nylon, dacron, silk are some common examples of this type of polymers. The H-bonds in nylon-66 have been shown in Fig. 59.5.

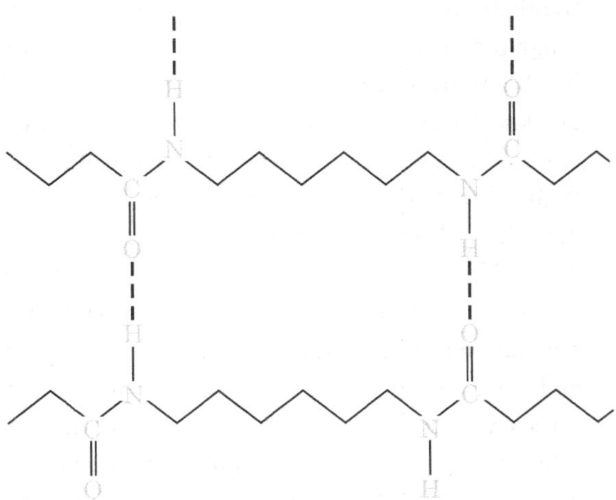

Fig. 59.5. H-Bonding in nylon-66.

3. Thermoplastics. *There are the polymers in which the interparticle forces of attraction are in between those of elastomers and fibers.* The polymers can be easily moulded into desired shapes by heating and subsequent cooling to room-temperature. There is no cross-linking between the polymer chains. In fact, thermoplastic polymers soften on heating and become fluids but on cooling they become hard. They are capable of undergoing such reversible changes on heating and cooling repeatedly. Common examples of thermoplastics are *polyethene, polystyrene, polyvinyl chloride,* etc.

4. Thermosetting Polymers. These are the *polymers which become hard and infusible on heating.* They are normally made from *semi-fluid substances* with low molecular masses, by heating in a mould. Heating results in *excessive cross-linking between the chains* forming a three dimensional network of bonds as a consequence of which a *non-fusible* and *insoluble* hard material is produced. Bakelite is a common example of a thermosetting polymer.

In short, a **thermoplastic material** *can be remelted time and again without change, while a* **thermosetting material** *undergoes a permanent change upon melting and thereafter sets to a solid which cannot be remelted.*

59.5 MODIFICATION OF PROPERTIES OF POLYMERS

Plasticizers, fillers and dyes are often added in various quantities to modify the physical properties of plastics. For example, it is a common feature among many plastics that they do not soften to a workable extent on heating.

(*a*) Synthetic resins and cellulose derivatives are horny tough materials. Such plastics can be easily softened by the addition of some organic compounds which are known as **plasticizers.** For example, polyvinyl chloride (PVC) is extremely hard and brittle. It was found that it often decomposed before it could be moulded. However, it was discovered in 1926 that the addition of certain substances like, di-*n*-butylphthalate, makes it soft and workable. The plasticizing effect is due to the solubilization action and an accompanying reduction in intermolecular forces which permits free movement of molecules relative to each other. In simple words plasticizers are molecules which get in between the chains, allowing them to slide over one another more easily.

Some **important** plasticizers are:

(*i*) *Triceryl phosphate* (*ii*) *Dimethyl pthalate*

(*iii*) *Triphenyl phosphate* (*iv*) *Camphor.*

(*b*) The moulding of a polymer can also be achieved by using the thermoplastic or thermosetting polymers. The method of moulding is to use an expansion method. In this method either a chemical reaction is done or a gas is bubbled into the polymer. For example,

(*i*) **Polyurethane foam** is made by mixing isocyanates and diols. The isocyanate reacts to give carbon dioxide gas, which produces foaming. The polyurethane foams are used as filling for furniture.

(*ii*) Similarly **expanded polystyrene** is made by bubbling gas into the polymer. Pentane has been used for this purpose. However, CFCs (Chlorofluoro carbons) have been increasingly used for making polymer foams.

(*c*) **Vulcanisation of rubber.** The natural as well as synthetic rubber are very soft and sticky. These are hardened by a process known as **vulcanisation.** Charles Goodyear found that when rubber is heated with sulphur, it becomes hard and more flexible. He called the heating of rubber with sulphur as *vulcanisation*, (The name is coined after *vulcan, the Roman god of fire.*)

Heating rubber with sulphur causes cross-linking of polymer chains through disulphide bonds. Thus, the individual chains which were entangled together in the rubber now get together into a giant molecule through vulcanisation. As the polymer has double bonds, the chains have bends and kinks that prevent them from forming a tightly packed crystalline polymer. When rubber is stretched, the chains straighten out along the direction of the pull. Cross-linking prevents the polymer from being torn when it is stretched. When the stress is removed it springs back to its original shape and size. This is shown in Fig. 59.6.

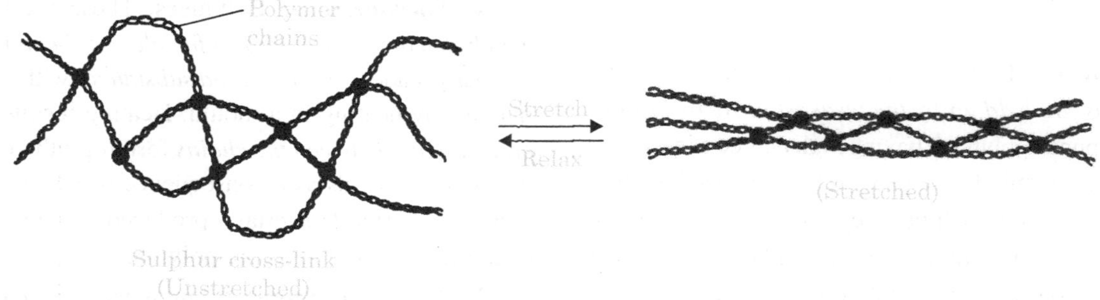

Fig. 59.6. Vulcanised rubber.

59.6 RESINS AND PLASTICS

(*i*) **Resins.** Resins are amorphous organic solids or sticky organic liquids or semi-solids which usually have a typical metallic lustre and are often transparent or translucent. Natural resins are secreted or flow out of plants (and some insects) when they are cut or wounded. They can harden when exposed to the air to form brittle, non-crystalline solids. Resins are insoluble in water but soluble in ether, ethanol and other organic solvents. Examples of natural resins are amber (fossil resin) and shellac.

(*ii*) **Plastics.** A plastic is a substance which is capable of being moulded whereas a resin lacks this property. Typical plastics on heating become soft but do not give mobile melts. In contrast, resins on heating give mobile melts. Plastics usually have much higher molecular mass than resins. Nevertheless, this differentiation is not very clear and the two terms are often used interchangeably.

59.7 CHEMICAL TESTS ON PLASTICS

Identification of an unknown sample of a plastic, rubber or a fiber may be required for a variety of reasons. Many of these polymers carry their trade name only and their detailed structure is not disclosed. In this age of competition, a manufacturer may like to develop a product similar to that of another producer. In such cases, it may be necessary to analyse the polymer sample and identify it.

Some common methods are discussed as follows:

1. Visual Examination. A visual examination of the sample may reveal useful information. For example, the presence of a hard, inflexible flash line would indicate a thermoset moulded material. Similarly, the presence of a gate scar would indicate an injection moulded material. The physical form of the sample, *i.e.*, whether granules, film, sheet or fiber and its flexibility or rigidity would give some indication of its identity. The colour of the product can also be used for identification of a thermoset material. For example, a pastel shade would rule out a phenolic and probably indicate that it is a urea or melamine-formaldehyde plastics material.

2. Cutting. The cutting of a plastics material with a pen knife can also provide some information. For example, one can differentiate between crystal clear cellulose acetate and poly styrene. Thermoset and thermoplastic material can also be differentiated in this way. The ivory like cut of casein and cast phenolics is noteworthy.

3. Heating tests. A small amount of the material to be tested is taken in a spoon type of spatula and heated on a small bunsen flame. The ease of burning, whether the burning continues after removal from the flame, colour of flame and so on, all give an indication of the possible identity of the material. If the material explodes or burns away rapidly, it is possibly of a cellulose nitrate composition.

A second heating test can be conducted with the help of a **clean copper wire.** The wire is first heated in a clear bunsen flame and then touched with a small quantity of the material. It is heated again and the colour of the flame is noted. Blue and green colours indicate the presence of halogens in the composition, that is, chlorine, fluorine and rarely, bromine. Presence of poly vinyl chloride or its co-polymers, poly vinylidine chloride, Polytetrafluoroethylene, chlorinated rubber, rubber hydrochloride or cellulose acetate containing a plasticizer like tricresyl phosphate is indicated by this test.

A third heating test is carried out by heating a small sample in a hard glass tube. The gas evolved is condensed in another tube and very carefully smelled. This can then be compared with the gas evolved from a known polymer.

4. Fusion Test. The metallic sodium fusion test, can show the presence or absence of nitrogen and halogens. Similarly, the potassium nitrate/potassium carbonate fusion test will indicate the presence or absence of phosphorus in the material under examination. Thus, one can easily distinguish between Nylon and Terylene.

Identification of Polymers

From the preliminary tests, a great deal of information can be gathered regarding the possible identity of the unknown polymer. If the polymer does not show rubber like elasticity, an elastomer is ruled out. Further, if on heating it does not melt or flow, a thermosetting polymer is indicated. If it does melt, a thermoplastic polymer is indicated.

59.8 PREPARATION AND USES OF POLYMERS

1. Polythene or Polyethylene. This is an addition polymer of ethene. Two types of polythenes namely; *high density polythene* and *low density polythene*, are being produced these days using different conditions for polymerisation.

$$\underset{\text{ethene}}{nCH_2 = CH_2} \longrightarrow \underset{\text{polythene}}{\left[CH_2—CH_2 \right]_n}$$

Characteristics and uses. It is used in film wraps and bags for packaging, water pipes, coating of telephones, electric wires and cables.

2. Polypriopylene. The monomer units are **propylene molecules.** It is generally manufactured by passing propylene through *n-hexane* (inert solvent) containing Ziegler-Natta catalyst (a mixture of triethyl aluminium and titanium chloride).

$$\underset{\text{Propylene}}{nCH(CH_3)=CH_2} \xrightarrow[TiCl_3]{Al(C_2H_5)_3} \underset{\text{Polypropylene}}{\left(CH(CH_3)—CH_2 \right)_n}$$

Characteristics and uses. It is harder, stronger and lighter than polyethene. It is used in:

packing of textile material and food, liners of bags, gramophone records, ropes, carpet fibres, etc.

3. Polystyrene or Styron. The monomer units are **styrene molecules.** It is prepared by free radical polymerisation of styrene in the presence of benzoyl peroxide.

$$\underset{\text{Styrene}}{nCH(C_6H_5) = CH_2} \xrightarrow{(C_6H_5CO)_2O} \underset{\text{Polystyrene}}{\left(CH(C_6H_5)—CH_2 \right)_n}$$

Characteristics and uses. It is a white thermoplastic material which is transparent and floats on water. It is used for making *toys, combs, model construction kits, ceiling tiles, packing for delicate articles* and *lining material for refrigerators* and *TV cabinets.*

Polystyrene is sold under the name *styrofoam* or *styron.*

4. Teflon or Poly Tetrafluoro Ethylene (PTFE). The monomer unit is **Tetrafluoro-ethylene.** It is prepared by heating tetrafluoroethylene under pressure in the presence of ammonium peroxo-sulphate $[(NH_4)_2S_2O_8]$.

$$\underset{\text{Tetrafluoroethylene}}{nCF_2 = CF_2} \xrightarrow[\text{Heat, pressure}]{(NH_4)_2S_2O_8} \underset{\text{Teflon}}{\left(CF_2 - CF_2 \right)_n}$$

Characteristics and uses. It is a very tough material and is resistant towards heat, action of acids or bases. It is a bad conductor of electricity. It is used in:

coating utensils to make them non-sticking, making seals and gaskets which can withstand high pressures, insulations for high frequency electrical installations.

5. Poly Monochloro Trifluoro Ethylene (PCTFE). The monomer unit are **chloro-trifluoro ethylene** molecules. It is prepared by the polymerisation of monochlorotrifluoro ethylene

$$\underset{\text{Chlorotrifluoro ethylene}}{nClFC = CF_2} \longrightarrow \underset{\text{PCTFE}}{\left(ClFC - CF_2 \right)_n}$$

The properties of PCTFE are similar to teflon. However, this is relatively less resistance to heat and chemicals.

6. Poly Vinyl Chloride (PVC). The monomer units are **vinyl chloride** molecules. It is prepared by heating vinyl chloride in an inert solvent in the presence of dibenzoyl peroxide.

$$\underset{\text{Vinyl chloride}}{nCH_2 = \underset{\underset{Cl}{|}}{CH}} \xrightarrow{\text{Dibenzoyl peroxide}} \underset{\text{PVC}}{\left(CH_2 - \underset{\underset{Cl}{|}}{CH} \right)_n}$$

Characteristics and uses. PVC is a hard horny material. However, it can be made to acquire any degree of pliability by the addition of a *plasticizer.* It is resistant to chemicals as well as heat. It is used for making:

rain coats, hand bags, toys, hosepipes, gramophone records, electrical insulations and *floor covering.*

7. Neoprene. This is synthetic rubber which resembles natural rubber in its properties. It is obtained by polymerisation of **chloroprene** (2-chloro-1, 3-butadiene) in the presence of potassium persulphate.

$$\underset{\text{Chloroprene}}{nCH_2 = \underset{\underset{Cl}{|}}{C} - CH = CH_2} \xrightarrow{K_2S_2O_8} \underset{\text{Neoprene}}{\left(CH_2 - \underset{\underset{Cl}{|}}{C} = CH - CH_2 \right)_n}$$

Characteristics and uses. Neoprene is superior to natural rubber in its stability to aerial oxidation and also in its resistance to oils and other solvents. It is generally used for making:

hoses, shoeheels, stoppers and *belts.*

8. Buna-S. It is a copolymer of *1,3-butadiene* and *styrene.* It is obtained by the polymerisation of butadiene and styrene in the ratio of 3 : 1 in the presence of sodium. It is also known as *styrene butadiene rubber* **(SBR).**

$$\underset{\text{Butadiene}}{nCH_2 = CH - CH = CH_2} + \underset{\text{Styrene}}{nC_6H_5CH_5 = CH_2} \xrightarrow{\text{Na, Heat}} \underset{\text{Buna-S}}{\left(CH_2 - CH = CH - CH_2 - \underset{\underset{C_6H_5}{|}}{CH} - CH_2 \right)_n}$$

In Buna-S, *Bu* stands for butadiene; *na* for sodium which is the polymerising agent and S stands for *styrene.*

Characteristics and uses. SBR has slightly less tensile strength than natural rubber. It is used in the manufacture of:

automobile tyres, rubber soles, belts, hoses, etc.

9. Terelene. It is a polymer obtained by the condensation reaction between *ethylene glycol* and *terephthalic acid*

$$\underset{\text{Ethylene glycol}}{nHO - CH_2 - CH_2 - OH} + \underset{\text{Terephthalic acid}}{nHO - \overset{\overset{O}{||}}{C} - C_6H_4 - \overset{\overset{O}{||}}{C} - OH} \xrightarrow[-2nH_2O]{425\text{–}475\ K} \left(O - CH_2 - CH_2 - O - \overset{\overset{O}{||}}{C} - C_6H_4 - \overset{\overset{O}{||}}{C} \right)_n$$

Poly (ethylene glycol terephthalic) or *Terelene*

Characteristics and uses. Terelene is resistant to the action of chemical and biological substances and also to abrasion. It has a low moisture absorbing power. As such it is widely used in making wash and wear fabrics. The polyester textile fibers made from the polymer are marketed under the trade name *terelene* or *dacron*.

It is used as a blend with cotton and wool in clothing. It is also used in *seat belts* and *sails*. The polymer is also used as **mylar** in the preparation of *films, magnetic recording tapes* and for *packing frozen food*. Dacron (and teflon) tubes are *good substitutes for human blood vessels in heart by-pass operations*.

10. Glyptal or Alkyd Resin. Glyptal is a general name of all polymers obtained by condensation of *di-basic acids*, and *polyhydric alcohols*. The simplest glyptal is *poly (ethylene glycol phthalate)* which is obtained by the condensation of *ethylene glycol* and *phthalic acid*.

$$\underset{\text{Ethylene glycol}}{n\text{HO—CH}_2\text{—CH}_2\text{—OH}} + \underset{\text{Phthalic acid}}{n\text{HO—}\overset{\text{O}}{\overset{\|}{\text{C}}}\text{—C}_6\text{H}_4\text{—}\overset{\text{O}}{\overset{\|}{\text{C}}}\text{—OH}} \xrightarrow{-2n\text{H}_2\text{O}} \underset{\text{Poly (ethylene glycol phthalate)}}{\left[\text{—O—H}_2\text{C—CH}_2\text{—O—}\overset{\text{O}}{\overset{\|}{\text{C}}}\text{—C}_6\text{H}_4\text{—}\overset{\text{O}}{\overset{\|}{\text{C}}}\text{—}\right]_n}$$

Characteristics and uses. These are three dimensional cross-linked polymers. Poly (ethylene glycol phthalate) dissolves in suitable solvents and the solution on evaporation leaves a tough and non-flexible film. Thus, it is used in:

adherant paints, lacquers and *building materials* like *asbestos and cement*.

11. Nylon-6,6. It is a polymer of *adipic acid* (1, 6-hexanedioic acid) and *hexamethylene diamine* (1, 6-diaminohexane)

$$\underset{\text{Hexamethylene diamine}}{n\,\text{H}_2\text{N—(CH}_2)_6\text{—NH}_2} + \underset{\text{Adipic acid}}{n\text{HO}\overset{\text{O}}{\overset{\|}{\text{C}}}\text{—(CH}_2)_4\text{—}\overset{\text{O}}{\overset{\|}{\text{C}}}\text{OH}} \xrightarrow[\text{Heat}]{\approx 525\text{ K}} \underset{\text{(Nylon-66)}}{\left(\overset{\text{H}}{\overset{|}{\text{N}}}\text{—(CH}_2)_6\text{—}\overset{\text{H}}{\overset{|}{\text{N}}}\text{—}\underset{\text{O}}{\underset{\|}{\text{C}}}\text{—(CH}_2)_4\text{—}\overset{\text{O}}{\overset{\|}{\text{C}}}\right)_n} + 2n\text{H}_2\text{O}$$

Characteristics and uses. Nylon-66 (read as *nylon-six-six*) can be cast into a sheet or fibers by spinning devices. Nylon fibers have high tensile strength. They are tough and resistant to abrasion. They are also somewhat elastic in nature. Nylon finds uses in:

making bristels and brushes, carpets and *fabrics* in the *textile industry, elastic hosiery* in the form of *crinkled nylon*.

12. Nylon 6,10. It is a polymer of hexamethylene diamine (six carbon atoms) and sebacoyl chloride (ten carbon atoms)

$$n\text{H}_2\text{N(CH}_2)_6\text{—NH}_2 + n\text{Cl—}\overset{\text{O}}{\overset{\|}{\text{C}}}\text{(CH}_2)_8\text{—}\overset{\text{O}}{\overset{\|}{\text{C}}}\text{—Cl} \xrightarrow{\text{heat}} \underset{\text{Nylon 610}}{\left(\text{HN—(CH}_2)_6\text{—}\overset{\text{H}}{\overset{|}{\text{N}}}\text{—}\overset{\text{O}}{\overset{\|}{\text{C}}}\text{—(CH}_2)_8\text{—}\overset{\text{O}}{\overset{\|}{\text{C}}}\right)_n} + 2n\text{HCl}$$

13. Nylon-6 (or Perlon). It is obtained from the monomer *caprolactum*. Caprolactam is obtained from *cyclohexane* according to the reaction sequence given below:

$$\underset{\text{Cyclohexane}}{\text{C}_6\text{H}_{12}} \xrightarrow{\text{Oxidation}} \underset{\text{Cyclohexanone}}{\text{C}_6\text{H}_{10}\text{=O}} \xrightarrow[-\text{H}_2\text{O}]{\text{NH}_2\text{OH}} \underset{\text{Oxime}}{\text{C}_6\text{H}_{10}\text{=NOH}} \xrightarrow[\text{Beckmann rearran-gement}]{\text{H}_2\text{SO}_4} \underset{\text{Caprolactam}}{\text{(CH}_2)_5\text{C(=O)NH}}$$

Caprolactam on heating with traces of water hydrolyses to *6-amino caproic acid* which on continued heating undergoes self-condensation and polymerises to give *nylon-6.*

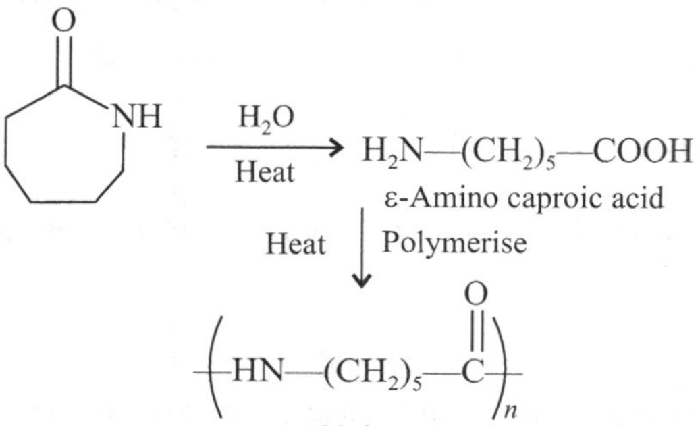

$$\text{Caprolactam} \xrightarrow[\text{Heat}]{H_2O} H_2N\text{—}(CH_2)_5\text{—}COOH$$

ε-Amino caproic acid

Heat ↓ Polymerise

$$\left(HN\text{—}(CH_2)_5\text{—}\overset{O}{\overset{\|}{C}} \right)_n$$

Nylon-6

Nylon-6 is **used** in the manufacture of *tyre cords, fabrics* and *ropes*.

14. Phenol-Formaldehyde Resin (Bakelite). These are made by the reaction of *phenol* and *formaldehyde* in basic medium. The reaction involves the formation of methylene bridges in the *ortho, para* or the *ortho* as well as the *para* position as shown in the following reactions.

Characteristics and uses. Bakelite is a cross-linked thermosetting polymer. Soft bakelites with low degree of polymerisation are used as *bonding glue for laminated wooden planks, in the preparation of varnishes and lacquers.*

High degree of polymerisation leads to the formation of hard bakelite which is used for making combs, fountain pen barrels, gramophone records, electrical goods, formica table tops and many other products. Sulphonated bakelites are used as ion-exchange resins for softening of hard water.

The reaction starts with the initial formation of *o*-and / or *p*-hydroxymethylphenol derivatives, which further react with phenol to form compounds having rings joined to each other through – CH_2 groups. The initial product could be a linear product - ***Novolac*** used in paints.

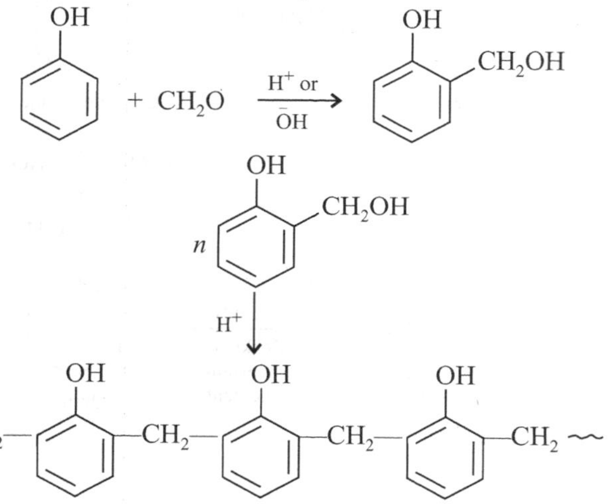

$$C_6H_5OH + CH_2O \xrightarrow[\bar{O}H]{H^+ \text{ or}} o\text{-}HOC_6H_4CH_2OH$$

$$n \; o\text{-}HOC_6H_4CH_2OH \xrightarrow{H^+} \sim CH_2\text{—}C_6H_3(OH)\text{—}CH_2\text{—}C_6H_3(OH)\text{—}CH_2\text{—}C_6H_3(OH)\text{—}CH_2 \sim$$

Novolac

Novolac on heating with formaldehyde undergoes cross linking to form an infusible solid mass called bakelite.

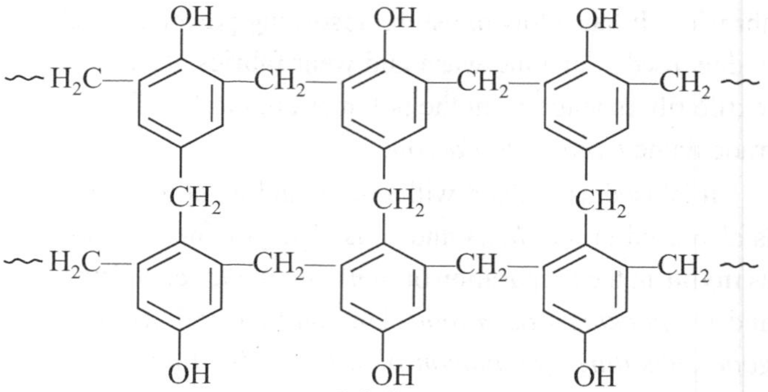

Bakelite

15. Melamine Formaldehyde Resin. *Melamine* and *formaldehyde* copolymerise to give another polymer.

Characteristics and uses. Melamine polymer is quite hard and is **used** in making plastic crockery under the trade name *melmac*. Cups, plates and other articles made from melamine polymer do not break on being dropped.

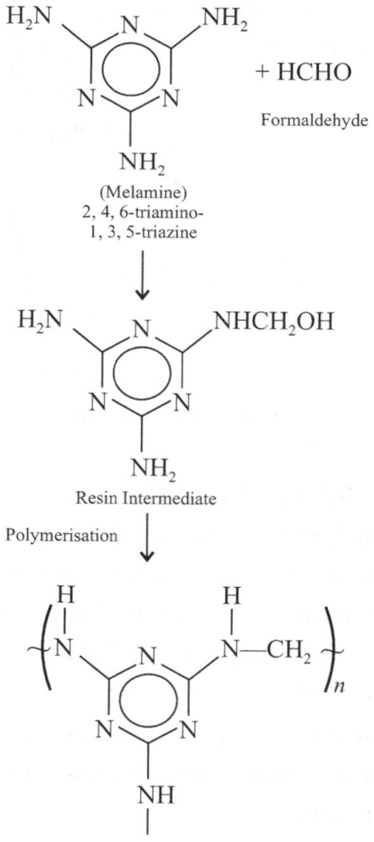

$$\text{Melamine} + HCHO \text{ (Formaldehyde)}$$

(Melamine) 2, 4, 6-triamino-1, 3, 5-triazine

↓

$$H_2N\text{—}C_3N_3(NH_2)\text{—}NHCH_2OH$$

Resin Intermediate

Polymerisation ↓

$$\left(\overset{H}{N}\text{—}C_3N_3(NH\text{—})\text{—}\overset{H}{N}\text{—}CH_2 \right)_n$$

Melamine Polymer

16. Natural Rubber. Rubber is a natural polymer and possesses elastic properties. It is obtained from the rubber tree. When the bark of the tree is cut, a sticky white liquid, *latex*, oozes out. It is a suspension of rubber particles in water.

Natural rubber is a linear polymer of 2-methyl-1, 3-butadiene (isoprene). It is also called *cis*-1,4-poly isoprere. On an average it contains 5000 isoprene units. All the double bonds in natural rubber are *Cis*. Rubber is a water proof material.

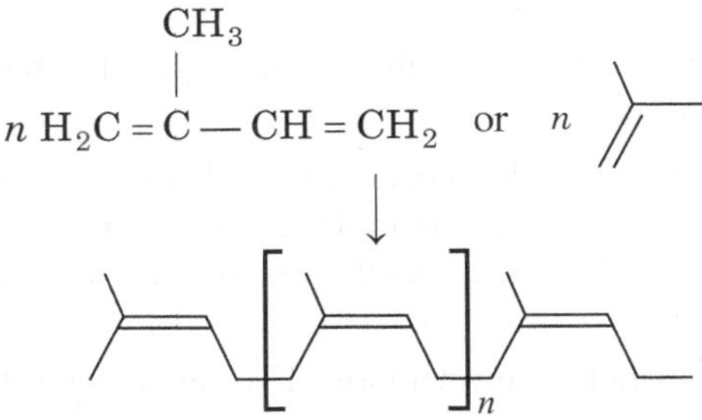

$$n\ H_2C=\underset{|}{\overset{CH_3}{C}}-CH=CH_2 \quad \text{or} \quad n$$

Cis-poly (2-methyl-1, 3-butadiene)
natural rubber

The *cis*-poly isoprene molecule consists of various chains held by weak van der Waals interactions and has a coiled structure. Thus it can be stretched like a spring and exhibits elastic properties. Gutta-percha (*getah* means gum and *percha* means tree) is a naturally occurring isomer of rubber in which all the double bonds are trans. Like rubber, gutta percha is exuded by certain trees. It is harder and more brittle than rubber.

SYNTHETIC RUBBER

To meet human needs, scientists have started preparing synthetic rubbers. Besides having similar properties as natural rubbers they are tougher, more flexible and more durable than natural rubber. They are capable of getting stretched to twice its length. However, it returns to its original shape and size as soon as the external stretching force is released. Synthetic rubbers have been made by the polymerisation of dienes other than isoprene. The polymerisation is carried out in the presence of a Zeigler-Natta catalyst. For example, Polymerisation of 1, 3-butadiene

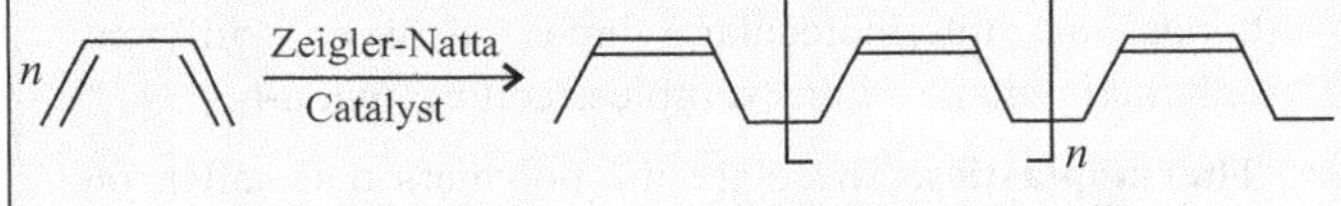

n Zeigler-Natta Catalyst →

Cis-poly (1, 3-butadiene)
(synthetic rubber)

PREPARATION OF SYNTHETIC RUBBERS

Neoprene or polychloroprene is formed by the free radical polymerisation of chloroprene.

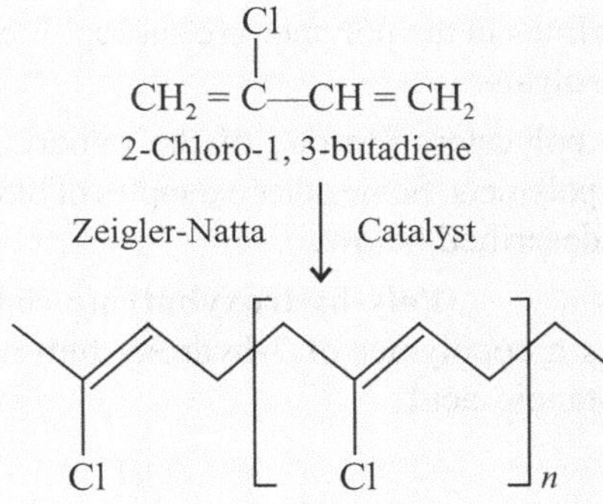

$$CH_2=\overset{Cl}{\overset{|}{C}}-CH=CH_2$$

2-Chloro-1, 3-butadiene

Zeigler-Natta ↓ Catalyst

Cl Cl n

neoprene

It has superior resistance to vegetable and mineral oils. It is used for manufacturing of conveyer belts, gaskets and hoses.

BIODEGRADABLE POLYMERS

These are polymers that can be broken into small segments by enzyme-catalysed reactions. The required enzymes are produced by microorganisms. It is a known fact that the carbon-carbon bonds of chain growth polymers are inert to enzyme-catalysed reactions, and hence they are non-biodegradable. To make such polymers biodegradable we have to insert certain bonds in the chains so that these can be easily broken by enzymes. Now when such polymers are buried as waste, microorganisms present in the ground can degrade the polymer.

One of the best method of making a polymer biodegradable is by inserting a hydrolysable ester group into the polymer.

For **example** if acetal is added to an alkene undergoing radical polymerisation, the ester group will be inserted into the polymer.

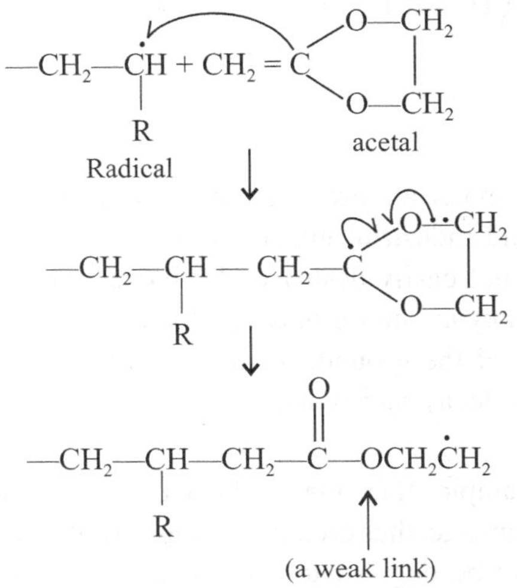

$$-CH_2-\underset{\underset{R}{|}}{\dot{C}H}+CH_2=C\begin{matrix}O-CH_2\\ \quad|\\ O-CH_2\end{matrix}$$

Radical   acetal

↓

$$-CH_2-\underset{\underset{R}{|}}{CH}-CH_2-C\begin{matrix}O-\dot{C}H_2\\ \quad|\\ O-CH_2\end{matrix}$$

↓

$$-CH_2-\underset{\underset{R}{|}}{CH}-CH_2-\overset{\overset{O}{\|}}{C}-OCH_2\dot{C}H_2$$

↑
(a weak link)

The weak links in the polymer are susceptible to enzyme catalysed hydrolysis.

Aliphatic polyesters are one of the important class of biodegradable polymers. Some other examples of biodegradable polymers are described below:

(*i*) **PHBV (Poly-hydroxybutrate-co-β-hydroxy valerate).** It is a copolymer of 3-hydroxy butyric acid and 3-hydroxypentanoic acid.

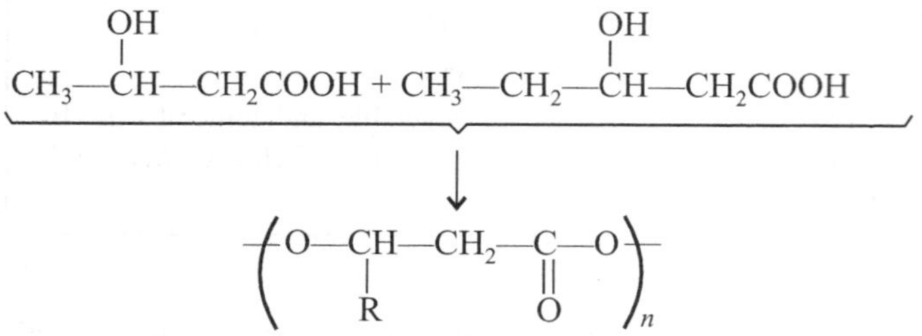

$$CH_3-\underset{\displaystyle OH}{CH}-CH_2COOH + CH_3-CH_2-\underset{\displaystyle OH}{CH}-CH_2COOH$$

$$\downarrow$$

$$\left(O-\underset{\displaystyle R}{CH}-CH_2-\underset{\displaystyle O}{\overset{\|}{C}}-O\right)_n$$

PHBV
($R = CH_3, C_2H_5$)

PHBV is used in orthopaedic devices and controlled drug release. The drug put in PHBV capsule is released after this polymer is degraded by enzymatic action. It can also be degraded by bacterial action.

(*ii*) **Poly glycolic acid and poly lactic acid.** These are also biodegradable polymers and are used for post operative stiches. These are bioabsorbable structures.

(*iii*) **Nylon-2-Nylon-6.** It is an alternating polyamide copolymer of glycine ($H_2N—CH_2—COOH$) and amino caproic acid ($H_2N—(CH_2)_5COOH$) and is biodegradable.

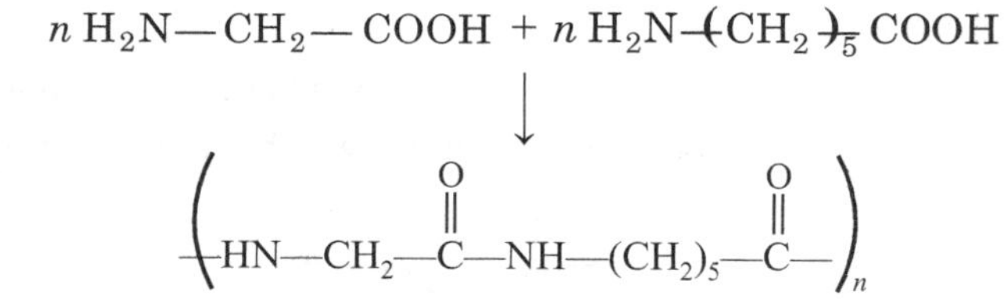

$$n\,H_2N-CH_2-COOH + n\,H_2N\!\left(CH_2\right)_5COOH$$

$$\downarrow$$

$$\left(HN-CH_2-\overset{\displaystyle O}{\overset{\|}{C}}-NH-(CH_2)_5-\overset{\displaystyle O}{\overset{\|}{C}}-\right)_n$$

Nylon-2,6

DISADVANTAGES OF POLYMERS

Environmental Pollution by Plastics

Plastic materials are used to make a large number of domestic and industrial materials. Most of the plastics are strong and not easily destroyed by bacteria when disposed after use. They are non-biodegradable. When disposed of on the surface of the ground, inside the ground or elsewhere, they do not decay and remain there for about one hundred years.

For example, they may collect together in drains and block the drainage, thus creating sewage problems. The drains may have to be dug out to remove the non-biodegradable plastic materials such as polythene.

Plastics can be burnt at refuse dumps, but the products of burning can be hazardous. Burning of PVC plastics, which contain chlorine, produces pungent smelling hydrogen chloride gas and chlorine gas which are poisonous to humans and plants.

Pollution by plastics may be reduced by reduction in plastic use, more use of biodegradable plastics, incineration and recycling.

SUMMARY

- **Polymer.** A substance with giant molecules having high molecular mass.
- **Polymerisation.** The process in which large number of small molecules combine to form a giant molecule or a macromolecule. The small molecules are called *monomers*.
- **Homopolymer.** A polymer formed by combination of only one type of monomers.
- **Copolymer.** A polymer formed by combination of two or more different types of monomers.
- **Addition Polymers.** The polymers formed by combination of large number of monomers without the elimination of any smaller molecules.
- **Cross-linked Polymers.** Monomers cross-link to form a network. These are hard and rigid.
- **Condensation Polymers.** During formation of such polymers, elimination of smaller molecules such as H_2O, NH_3, etc., takes place.
- **Polythene.** It is a polymer of ethene.
- **Poly vinylchloride.** It is formed by the polymerisation of vinylchloride ($CH_2 = CH—Cl$).
- **Phenol-methanal plastics.** It is commercially known as Bakelite.
- **Elastomers.** In this type of polymers, the polymeric chains are held by weak intermolecular forces. For example, rubber.
- **Fibres.** These are thread like polymers having hydrogen bonds as intermolecular forces between different polymeric chains. For examples, cotton, nylon-66.
- **Thermoplastics.** These are the polymers that soften on heating and can be moulded into any shape. Intermolecular forces in them are intermediate between fibres and elastomers.
- **Thermosetting Polymers.** These polymers are very hard, infusible and have extensive cross links. For example, bakelite.

- **Plastic.** A polymer which is readily deformable and can be moulded into any shape.
- **Vulcanisation.** Heating of natural rubber with sulphur to make it strong.
- **Resiliency Property.** The property of returning to the original shape after distortion within elastic limit.
- **Isoprene.** 2-Methyl-1, 3-butadiene. Monomer of natural rubber.
- **Plasticizer.** A substance that when added to a thermoplastic improves its workability.
- **Caprolactam.** Starting material for Nylon-6, a synthetic polyamide fibre.
- **Wool.** A natural polyamide fibre.
- **Poly Urethane.** It is formed by the reaction of di-isocyanate with a polyester having hydroxy groups on ends. These are used as leather substitute.

EVALUATE YOURSELF

I. Objective Type Questions

Select the most appropriate choice from the options given as (a), (b), (c) and (d) after each question:

1. Polymers are
(*a*) micromolecules (*b*) macromolecules
(*c*) sub-micromolecules (*d*) none of these.

2. Bakelite is
(*a*) addition polymer (*b*) elastomer
(*c*) thermoplastic (*d*) thermosetting.

3. The S in Buna-S refers to
(*a*) sodium (*b*) sulphur
(*c*) styrene (*d*) just a trade name.

4. The repeating units of PTFE is
(*a*) $Cl_2CH—CH_3$ (*b*) $F_2C=CF_2$
(*c*) $F_3C—CF_3$ (*d*) $FClC=CF_2$.

5. The inter-particle forces between linear chains in Nylon-66 are
(*a*) H-bonds (*b*) covalent bonds
(*c*) ionic bonds (*d*) unpredictable.

6. Nylon-66 is a polyamide of
(*a*) vinylchloride and formaldehyde
(*b*) adipic acid and methyl amine
(*c*) adipic acid and hexamethylene diamine
(*d*) formaldehyde and malamine.

7. Which of the following is not a condensation polymer?
(*a*) Glyptal (*b*) Nylon-66
(*c*) Dacron (*d*) PTFE.

8. Which of the following is a condensation polymer?
(*a*) Polystyrene (*b*) Neoprene
(*c*) PAN
(*d*) Poly (ethylene glycol phthalate).

9. The monomer of PVC is
(*a*) ethylene (*b*) tetrafluoroethylene
(*c*) chloroethene (*d*) none of the above.

10. Which of the following polymers is a copolymer?
(*a*) Polypropylene (*b*) Nylon-66
(*c*) PVC (*d*) Teflon.

11. Which of the following polymers is a homopolymer?
(*a*) Bakelite (*b*) Nylon-66
(*c*) Terylene (*d*) Neoprene.

12. Which of the following types of polymers has the strongest interparticle forces?
(*a*) Elastomers (*b*) Thermoplastics
(*c*) Fibers (*d*) Thermosetting polymers.

13. Bakelite is obtained from phenol by reacting with
(*a*) ethanal (*b*) methanal
(*c*) vinyl chloride (*d*) ethylene glycol.

14. Polymer used in bullet proof glass is
(*a*) PMMA (*b*) lexan
(*c*) nomex (*d*) kevlar.

15. Nylon-6 is made from
(*a*) 1, 3-Butadiene (*b*) chloroprene
(*c*) adipic acid (*d*) caprolactam.

16. $F_2C=CF_2$ is a monomer of
(*a*) teflon (*b*) glyptal
(*c*) nylon-6 (*d*) buna-5.

17. Soft drinks and baby feeding bottles are generally made up of
(*a*) polyester (*b*) polyurethane
(*c*) polyurea (*d*) polyamide
(*e*) polystyrene.

18. Nylon threads are made of
(*a*) polyvinyl polymer (*b*) polyester polymer
(*c*) polyamide polymer (*d*) polyethylene polymer.

19. $\left(NH(CH_2)_6—NHCO(CH_2)_4CO \right)_n$ is a
(*a*) homopolymer (*b*) Copolymer
(*c*) addition polymer (*d*) thermosetting polymer.

20. Which is not a polymer?
(*a*) Sucrose (*b*) enzyme
(*c*) Starch (*d*) teflon.

II. Fill in the Blanks

21. Complete the following sentences by supplying the appropriate words:
(*i*) On the basis of their origin polymers are classified as ...... and ...... polymers.
(*ii*) The three natural fibres are ......
(*iii*) Natural rubber is a polymer of ......
(*iv*) Phenol formaldehyde resin is commonly called ....
(*v*) Nylon-6 is a polymer of ....
(*vi*) The monomer units of PAN is ......
(*vii*) The monomer units of PMMA is ....
(*viii*) In Buna-S, S stands for ......
(*ix*) Lucite is a polymer of ......
(*x*) The starting material of PCTFE is ......

III. Discussion Questions

22. Explain the following terms with examples:
(*i*) Polyester (*ii*) Polyamide
(*iii*) Natural polymers (*iv*) Synthetic polymers.
23. Give the classification of polymers:
(*i*) On the basis of their mode of synthesis; and
(*ii*) On the basis of the nature of forces between the macromolecules.
Give suitable examples in each case.
24. What is a polyamide? How is Nylon-6 synthesised?
25. How does vulcanised rubber differ from natural rubber?
26. Define the terms:
Thermoplastic polymers, Thermosetting polymers, Fibers, and Elastomers.
27. Give the starting material of each of the following:
(*i*) PMMA (*ii*) PCTFE
(*iii*) PAN (*iv*) PVC.
28. What is the difference between nylon-6 and nylon-66?
29. Give the structures of monomers of bakelite. How is it formed? To which class, *thermosetting* or *thermoplastic* does it belong? Give reasons.
30. Write the names and structures of the monomer of each of the following. Give one use of each of the polymers.
(*i*) Terelene (*ii*) Neoprene
(*iii*) Natural rubber (*iv*) Lucite.
31. What do you understand by linear, branched chain and cross-linked polymers?
32. Given examples of polymers belonging to the following categories. Give their structures:
(*i*) Polyamide (*ii*) Polyhaloolefins
(*iii*) Polyolefins (*iv*) Polyesters
(*v*) Polyacrylates.
33. Arrange the following polymers in increasing order of their intermolecular forces. Also classify them as addition and condensation polymers. Nylon-6, Neoprene, PVC.
34. Mention which of the following are addition polymers?
(*i*) Terylene (*ii*) Nylon-66
(*iii*) Neoprene (*iv*) Teflon.
35. Write the information required for the following polymers:
(*i*) Bakelite: materials required for preparation
(*ii*) PVC: monomer unit
(*iii*) Synthetic rubber: monomer unit.
36. Give one method of synthesis of each of the following:
(*i*) Buna-S (*ii*) Teflon
(*iii*) Nylon-66.
37. What is the difference between homopolymer and co-polymer? Give one example.
38. Define any two plasticizers.
39. How are polymers classified into different categories on the basis of intermolecular forces? Give one example of a polymer of each of these categories.
40. How do double bonds in rubber molecule influence their structure and reactivity?
41. Differentiate between addition and condensation polymers based on the mode of polymerisation. Give one example of each type.

Answers

I. Objective Type Questions

1. (*b*)	**2.** (*d*)	**3.** (*c*)	**4.** (*b*)
5. (*a*)	**6.** (*c*)	**7.** (*d*)	**8.** (*d*)
9. (*c*)	**10.** (*b*)	**11.** (*d*)	**12.** (*d*)
13. (*b*)	**14.** (*b*)	**15.** (*d*)	**16.** (*a*)
17. (*e*)	**18.** (*c*)	**19.** (*b*)	**20.** (*a*).

II. Fill in the Blanks

21. (*i*) natural, synthetic (*ii*) wool, cotton and silk
(*iii*) isoprene (*iv*) bakelite
(*v*) caprolactam (*vi*) acrylonitrile
(*vii*) methylmethacrylate (*viii*) sulphur
(*ix*) methylmethacrylate (*x*) chlorotrifluoroethylene.

APPENDIX 1

IUPAC NOMENCLATURE FOR SOME INORGANIC SUBSTANCES AS ADOPTED BY WAEC

EXAMPLES OF RECOMMENDED NAMES

BINARY COMPOUNDS

Common name	Formula	WAEC adopted IUPAC name
nitrous oxide	N_2O	nitrogen(I) oxide or dinitrogen monoxide
nitric oxide	NO	nitrogen(II) oxide or nitrogen monoxide
nitrogen dioxide	NO_2	nitrogen(IV) oxide or nitrogen dioxide
nitrogen tetraoxide	N_2O_4	dinitrogen tetraoxide*
nitrogen pentoxide	N_2O_5	nitrogen(V) oxide
phosphorus decaoxide	P_4O_{10}	tetraphosphorus decaoxide**
sodium hydride	NaH	sodium hydride
triplumbic tetraoxide	Pb_3O_4	trilead tetraoxide***
lead monoxide	PbO	lead(II) oxide
lead dioxide	PbO_2	lead(IV) oxide
hydrogen fluoride	HF	hydrogen fluoride
water	H_2O	water

* dinitrogen tetraoxide is so called to distinguish it from nitrogen(IV) oxide of which it is the dimer.

** Where P_2O_5 is mentioned it should be referred to as phosphorus(V) oxide.

*** This compound behaves as though the formula is $2PbO.PbO_2$ and so it is sometimes called dilead(II) lead(IV) oxide.

Common name	Formula	WAEC adopted IUPAC name
hydrogen sulphide	H_2S	hydrogen sulphide
ammonia	NH_3	ammonia
calcium acetylide/carbide	CaC_2	calcium dicarbide
potassium oxide	K_2O	potassium oxide
potassium peroxide	K_2O_2	potassium peroxide
potassium superoxide	KO_2	potassium superoxide
phosphorus trichloride	PCl_3	phosphorus(III) chloride
phosphorus pentachloride	PCl_5	phosphorus(V) chloride
carbon dioxide	CO_2	carbon(IV) oxide or carbon dioxide
carbon monoxide	CO	carbon(II) oxide or carbon monoxide
sulphur dioxide	SO_2	sulphur (IV) oxide or sulphur dioxide
sulphur trioxide	SO_3	sulphur (VI) oxide or sulphur trioxide
manganous oxide	MnO	manganese(II) oxide
manganic oxide	Mn_2O_3	manganese(III) oxide
manganese dioxide	MnO_2	manganese(IV) oxide or manganese dioxide
trimanganic tetraoxide	Mn_3O_4	trimanganese tetraoxide
ferrous oxide	FeO	iron(II) oxide
ferric oxide	Fe_2O_3	iron(III) oxide
aluminium carbide	Al_4C_3	aluminium carbide
hydrogen chloride	HCl	hydrogen chloride
hydrated hydrogen ion	H_3O^+	oxonium ion or hydronium ion or hydroxonium ion
borohydride ion	BH_4^-	tetrahydridoborate(III) ion
ammoniacal cuprous ion	$Cu(NH_3)_2^+$	diamminecopper(I) ion
zinc ammonium ion	$Zn(NH_3)_4^{2+}$	tetraamminezinc(II) ion
silver ammonium ion	$Ag(NH_3)_2^+$	diamminesilver(I) ion
aluminium hydride ion	AlH_4^-	tetrahydridoaluminate(III) ion
ammonium ion	NH_4^+	ammonium ion
ferrocyanide	$Fe(CN)_6^{4-}$	hexacyanoferrate(II) ion
ferricyanide	$Fe(CN)_6^{3-}$	hexacyanoferrate(III) ion
cuprammonium ion	$Cu(NH_3)_4^{2+}$	tetraaminecopper(II) ion
aluminate ion	*$Al(OH)_4^-$ $Al(H_2O)_2(OH)_4^-$	tetrahydroxoaluminate(III) ion tetrahydroxodiaquaaluminate(III)

* Where AlO_2^- is met, it should be called dioxoaluminate(III) ion.

OXOANIONS

Common name	Formula	WAEC adopted IUPAC name
hydrogen carbonate ion	HCO_3^{2-}	hydrogentrioxocarbonate(IV) ion
hypochlorite ion	ClO^-	oxochlorate(I) ion
perchlorate ion	ClO_4^-	tetraoxochlorate(VII) ion

Common name	Formula	WAEC adopted IUPAC name
permanganate ion	MnO_4^-	tetraoxomanganate(VII) ion
manganate ion	MnO_4^{2-}	tetraoxomanganate(VI) ion
dichromate ion	$Cr_2O_7^{2-}$	hexaoxo-μ-oxo-dichromate(VI) ion
chromate ion	CrO_4^{2-}	tetraoxochromate(VI) ion
thiosulphate ion	$S_2O_3^{2-}$	thiosulphate ion or trioxosulphur-sulphate(VI) ion or trioxothio-sulphate(VI) ion
tetrathionate ion	$S_4O_6^{2-}$	tetrathionate or tetraoxodithioperoxo disulphate(VI) ion or tetraoxodisulphur-μ-peroxodisulphate(VI) ion
persulphate ion	$S_2O_8^{2-}$	persulphate or hexaoxo-μ-peroxodisulphate(VI) ion
sulphate ion	SO_4^{2-}	tetraoxosulphate(VI) ion
sulphite ion	SO_3^{2-}	trioxosulphate(IV) ion
nitrate ion	NO_3^-	trioxonitrate(V) ion
nitrite ion	NO_2^-	dioxonitrate(III) ion
carbonate ion	CO_3^{2-}	trioxocarbonate(IV) ion

ACIDS

Common name	Formula	WAEC adopted IUPAC name
nitric acid	HNO_3	trioxonitrate(V) acid
nitrous acid	HNO_2	dioxonitrate (III) acid
hydrochloric acid	$HCL(aq)$	hydrochloric acid
sulphuric acid	H_2SO_4	tetraoxosulphate(VI) acid
sulphurous acid	H_2SO_3	trioxosulphate(IV) acid
carbonic acid	H_2CO_3	trioxocarbonate(IV) acid

BASES

Common name	Formula	WAEC adopted IUPAC name
ferric hydroxide	$Fe(OH)_3$	iron(III) hydroxide
ammonium hydroxide	$NH_3(aq)$	aqueous ammonia/ammonia solution
sodium hydroxide	$NaOH$	sodium hydroxide
ferrous hydroxide	$Fe(OH)_2$	iron(II) hydroxide

SALTS

Common name	Formula	WAEC adopted IUPAC name
sodium sulphate	Na_2SO_4	sodium tetraoxosulphate(VI)
sodium carbonate	Na_2CO_3	sodium trioxocarbonate(IV)
sodium nitrate	$NaNO_3$	sodium trioxonitrate(V)

Common name	Formula	WAEC adopted IUPAC name
cupric sulphate	$CuSO_4$	copper(II) tetraoxosulphate(VI)
sodium chloride	$NaCl$	sodium chloride
	Hydrated salts	
sodium carbonate decahydrate	$Na_2CO_3.10H_2O$	sodium trioxocarbonate(IV) decahydrate
ferrous sulphate heptahydrate	$FeSO_4.7H_2O$	iron(II) tetraoxosulphate(VI) heptahydrate
	Double salts	
potash alum	$KAl(SO_4)_2.12H_2O$	aluminium potassium bisulphate (tetraoxosulphate (VI)) dodecahydrate
ferrous ammonium sulphate	$Fe(NH_4)_2(SO_4).6H_2O$	diammonium iron(II) bisulphate (tetraoxosulphate(VI)) hexahydrate
	Basic salts	
bismuth oxychloride	$BiOCl$	bismuth(III) chloride oxide
basic lead carbonate	$Pb_2CO_3(OH)_2$	dilead(II) trioxocarbonate(IV) dihydroxide

APPENDIX 2

NOMENCLATURE OF ORGANIC COMPOUND

The term nomenclature means the system of naming of organic compounds. In the early stages of development of organic chemistry, the organic compounds were named after the source from which they were prepared. For example, the name *acetic acid* was derived from *acetum* (Latin: *acetum* means vinegar). Similarly, the name *formic acid* was derived from *formicus* (means *red ants*) because the compound was obtained by the distillation of red ants. Methyl alcohol was named *wood-spirit* as it was obtained from the destructive distillation of wood. These names of organic compounds are called **common names** or **trivial names.** The common names were assigned to the organic compounds at the whim and fancy of the discoverer and there was no systematic basis. In many cases, even the same compound was given more than one name. For example, methane was named as *marsh gas* as well as *damp fire.*

Towards the end of nineteenth century the number of organic compounds known was so large that it became difficult to remember their individual names. In order to rationalise the system of nomenclature, a comprehensive system of nomenclature known as the **Geneva system** was developed. There was further improvement in the Geneva system by the *International Union of Chemists* **(IUC System).**

The IUC system has been further revised by the *International Union of Pure and Applied Chemistry* in the light of practical difficulties. This revised system of nomenclature has been accepted all over the world and is abbreviated as **IUPAC system** of nomenclature. The rules discussed here are based on the recommendations made by the IUPAC in 1979 and 1993.

NOMENCLATURE OF ALIPHATIC HYDROCARBONS

A. Straight Chain Hydrocarbons

The name of a straight chain aliphatic hydrocarbon, in general, may be divided into two parts; ***word root*** and ***suffix.*** Word root designates the number of carbon atoms in the chain. For chains containing one to four carbon atoms, the special word roots are used and for chains of five or more carbon atoms Greek number roots are used. IUPAC word roots for a few carbon chains are given in Table 1. The general word root for any carbon chain is **alk-.**

Table 1. Word Roots for Carbon Chain Length

Chain Length	*Word Root*	*Chain Length*	*Word Root*
C_1	Meth-	C_6	Hex-
C_2	Eth-	C_7	Hept-
C_3	Prop-	C_8	Oct-
C_4	But-	C_9	Non-
C_5	Pent-	C_{10}	Dec-

In order to derive the IUPAC name, *a suffix is added to the word root to indicate saturation or unsaturation in the molecule.* The suffixes, thus used are called **primary suffixes.** The primary suffixes are given in Table 2.

Table 2. Primary Suffixes

Class of Compound	*Primary Suffix*	*General Name*
Saturated	*—ane*	Alkane
Unsaturated ($>C=C<$)	*—ene*	Alkene
Unsaturated ($—C\equiv C—$)	*—yne*	Alkyne

IUPAC names of alkanes containing carbons 1–10 are given in Table 3.

Table 3. Names of Alkanes

Molecular Formula	*IUPAC Name*
1. CH_4	**Methane**
2. CH_3CH_3	**Ethane**
3. $CH_3CH_2CH_3$	**Propane**
4. $CH_3CH_2CH_2CH_3$	**Butane**
5. $CH_3CH_2CH_2CH_2CH_3$	**Pentane**
6. $CH_3CH_2CH_2CH_2CH_2CH_3$	**Hexane**
7. $CH_3CH_2CH_2CH_2CH_2CH_2CH_3$	**Heptane**
8. $CH_3CH_2CH_2CH_2CH_2CH_2CH_2CH_3$	**Octane**
9. $CH_3CH_2CH_2CH_2CH_2CH_2CH_2CH_2CH_3$	**Nonane**
10. $CH_3CH_2CH_2CH_2CH_2CH_2CH_2CH_2CH_2CH_3$	**Decane**

NAMES OF SOME HIGHER MEMBERS OF ALKANE SERIES			
$C_{11}H_{24}$	Undecane	$C_{16}H_{34}$	Hexadecane
$C_{12}H_{26}$	Dodecane	$C_{17}H_{36}$	Heptadecane
$C_{13}H_{28}$	Tridecane	$C_{18}H_{38}$	Octadecane
$C_{14}H_{30}$	Tetradecane	$C_{19}H_{40}$	Nonadecane
$C_{15}H_{32}$	Pentadecane	$C_{20}H_{42}$	Icosane

B. Branched Chain Hydrocarbons

In branched chain compounds all the carbon atoms are not present in a straight chain. Some of the carbon atoms in these compounds are present as side chains. These carbon atoms in side chains constitute **alkyl groups** or **alkyl radicals** and are expressed as **prefixes** in the IUPAC name. An alkyl group is obtained from an alkane by removal of a hydrogen. Since the general formula of alkanes is C_nH_{2n+2}, the general formula for an alkyl group is C_nH_{2n+1}. An alkyl group is generally represented by R—. The alkyl group is named by replacing suffix *ane* from the name of the corresponding alkane by *yl*. The position of the hydrogen atom removed is indicated by giving it the lowest possible number from the chain end. The names and formulae of some common alkyl groups are given in Table 4.

A branched chain hydrocarbons is named using the following general rules:

Rule 1. Longest Chain Rule

Select the longest continuous chain of carbon atoms as the parent chain. If some carbon-carbon multiple bond is present, the parent chain must contain the carbon atoms involved in it. The number of carbon atoms in the parent chain determines the *word root.* The carbon atoms which are not included in the parent chain are considered as alkyl substituents and determine the *prefixes.*

```
CH3—CH2—CH2—CH2—CH—CH3
                |
                CH2—CH3
```

Prefix: *methyl*
Word root: *hept*
p. suffix: *ane*

```
        CH3
        |
CH3—CH—C—CH2—CH3
        ||
        CH2
```

Prefixes: *ethyl; methyl*
Word root: *but*
p. suffix: *ene*

If two equally long chains are possible, the chain with the maximum number of side chains is selected as the parent chain. For example,

```
                  CH3
                  |
CH3—CH2—CH—CH—CH3
            |
            CH—CH3
            |
            CH3
```

Correct

Prefixes: *ethyl; dimethyl*
word root: *pent*
p. suffix: *ane*

```
                  CH3
                  |
CH3—CH2—CH—CH—CH3
            |
            CH—CH3
            |
            CH3
```

Wrong

Rule 2. Lowest Number or Lowest Sum Rule

The selected parent chain is numbered using Arabic numerals and the positions of the alkyl groups are indicated by the number of the carbon atom to which the alkyl group is attached. ***The numbering is done in such a way that the substituted carbon atoms have the lowest possible numbers.***

Table 4. Some Common Alkyl Groups

Parent Alkane	*Formula (R—H)*	*Alkyl Group R—*	*Name*
Methane	CH_4	CH_3—	Methyl
Ethane	CH_3CH_3	CH_3CH_2—	Ethyl
Propane	$CH_3CH_2CH_3$	$CH_3CH_2CH_2$—	1-Propyl (*n*-Propyl)
		CH_3CHCH_3	2 - Propyl (*iso*-Propyl)

When series of locants containing the same number of terms are compared term by term, that series is lowest which contains the lowest number on the occasion of first difference. Some examples are given to illustrate the rule.

$$^{1}CH_3{-}^{2}\underset{\substack{|\\ CH_3}}{CH}{-}^{3}CH_2{-}^{4}CH_3$$

2-Methylbutane
(Correct)

$$^{4}CH_3{-}^{3}\underset{\substack{|\\ CH_3}}{CH}{-}^{2}CH_2{-}^{1}CH_3$$

3-Methylbutane
(Wrong)

$$^{1}CH_3{-}^{2}\overset{\substack{CH_3\\ |}}{\underset{\substack{|\\ CH_3}}{C}}{-}^{3}CH_2{-}^{4}\overset{\substack{CH_3\\ |}}{CH}{-}^{5}CH_3$$

2, **2**, 4-Trimethylpentane
(Correct)

$$^{5}CH_3{-}^{4}\overset{\substack{CH_3\\ |}}{\underset{\substack{|\\ CH_3}}{C}}{-}^{3}CH_2{-}^{2}\overset{\substack{CH_3\\ |}}{CH}{-}^{1}CH_3$$

2, **4**, 4-Trimethylpentane
(Wrong)

$$\overset{1}{CH_3}{-}\underset{\substack{|\\ CH_3}}{\overset{2}{CH}}{-}\underset{\substack{|\\ CH_3}}{\overset{3}{CH}}{-}\overset{4}{CH_2}{-}\underset{\substack{|\\ CH_3}}{\overset{5}{CH}}{-}\overset{6}{CH_3}$$

2, **3**, 5-Trimethylhexane
(Correct)

$$\overset{6}{CH_3}{-}\underset{\substack{|\\ CH_3}}{\overset{5}{CH}}{-}\underset{\substack{|\\ CH_3}}{\overset{4}{CH}}{-}\overset{3}{CH_2}{-}\underset{\substack{|\\ CH_3}}{\overset{2}{CH}}{-}\overset{1}{CH_3}$$

2, **4**, 5-Trimethylhexane
(Wrong)

$$\overset{\substack{CH_3\\ |}}{\overset{9}{CH_2}}{-}\underset{\substack{|\\ CH_3}}{\overset{8}{CH}}{-}\overset{\substack{CH_3\\ |}}{\overset{7}{CH}}{-}\overset{6}{CH_2}{-}\overset{5}{CH_2}{-}\overset{4}{CH_2}{-}\overset{3}{CH_2}{-}\underset{\substack{|\\ CH_3}}{\overset{2}{CH}}{-}\overset{1}{CH_3}$$

2, 7, 8-Trimethyldecane

2, 7, 8-Trimethyldecane is the correct name of the above compound. **3**, 4, 9-Trimethyldecane is a wrong name although it has the lowest sum of numbers. The first name is correct because 2 is lower than 3 when we compare the locants in the two names, term by term.

In case of unsaturated hydrocarbons, the carbon atoms involved in the multiple bond should get the lowest possible number. For example,

$$^{4}CH_3{-}^{3}\underset{\substack{|\\ CH_3}}{CH}{-}^{2}CH={}^{1}CH_2$$

3-Methylbut-1-ene
(Correct)

$$^{1}CH_3{-}^{2}\underset{\substack{|\\ CH_3}}{CH}{-}^{3}CH={}^{4}CH_2$$

2-Methylbut-3-ene
(Wrong)

The name of the compound, in general, is written in the following sequence:

(Position of substituents)—(Prefixes) (Word root) (p. suffix)

The above sequence is illustrated in the following examples

$$^{1}CH_3{-}^{2}CH_2{-}^{3}\overset{\substack{CH_3\\ |}}{CH}{-}^{4}CH_2{-}^{5}CH_3$$

Here, prefix is *methyl*; the methyl group is at carbon number 3 in the parent chain. Word root is *pent* and primary suffix is *ane*. Hence, the name is: ***3-Methylpentane***

$$^{4}CH_3{-}^{3}\overset{\substack{CH_3\\ |}}{CH}{-}^{2}C\equiv{}^{1}CH$$

Here, prefix is *methyl*; word root is *but*; primary suffix is *yne*. Methyl group is at the carbon number 3 and the triple bond is carbon number 1. Thus, the name is:

3-Methylbut-1-yne.

Rule 3. Use of Prefixes Di, Tri, etc.

If the compound contains more than one similar alkyl group, their positions are indicated separately. The name of the substituent is not repeated but an appropriate numerical prefix, di, tri, etc., is attached to the name of the substituent. The positions of the substituents are separated by commas.

For example,

$$\overset{\displaystyle CH_3}{} \quad \overset{\displaystyle CH_3}{}$$
$$^{1}CH_3—^{2}CH—^{3}CH—^{4}CH_2—^{5}CH_3$$

2,3-Dimethylpentane

$$^{5}CH_3—^{4}CH_2—^{3}\overset{CH_3}{\underset{CH_3}{C}}—^{2}\underset{CH_3}{CH}—^{1}CH_3$$

2, 3, 3-Trimethylpentane

Rule 4. Alphabetical Arrangement of Prefixes

If there are different alkyl substituents present in the compound, their names are written in alphabetical order. However, the numerical prefixes such as di, tri, etc., are not considered for the alphabetical order. For example,

$$^{1}CH_3—^{2}\underset{CH_3}{CH}—^{3}\overset{C_2H_5}{\underset{CH_3}{C}}—^{4}CH_2—^{5}CH_3$$

3-Ethyl-2,3-dimethylpentane

Naming different alkyl substituents at the equivalent positions. If two different alkyl groups are located at the equivalent positions, then numbering of the chain is done in such a way that the alkyl group which comes first in alphabetical order gets the lower position. For example, if ethyl and methyl groups are present at equivalent positions, then the carbon bearing the ethyl group should get the lower number as illustrated in the following example:

$$^{1}CH_3—^{2}CH_2—^{3}\overset{C_2H_5}{CH}—^{4}\overset{CH_3}{CH}—^{5}CH_2—^{6}CH_3$$

3-Ethyl-4-methylhexane

In the light of the above rules, let us write the IUPAC names of some aliphatic hydrocarbons.

$$^{6}CH_3—^{5}CH_2—^{4}\overset{C_2H_5}{\underset{CH_3}{C}}—^{3}CH_2—^{2}\underset{CH_3}{CH}—^{1}CH_3$$

4-ethyl-2,4-dimethexane

$$^{1}CH_3—^{2}\overset{CH_3}{\underset{CH_3}{C}}—^{3}CH_3$$

2,2-Dimethylpropane

$$CH_3—^{3}CH—^{4}CH_3$$
$$CH_3—\underset{2}{C}=\underset{1}{C}H_2$$

2,3-Dimethylbut-1-ene

NOMENCLATURE OF COMPOUNDS CONTAINING FUNCTIONAL GROUPS

In case some functional group (other than C = C and C ≡ C) is present in molecule, it is indicated by adding a ***secondary suffix*** after the primary suffix. The terminal **'e'** of the primary suffix is generally removed before adding the secondary suffix. *The terminal* **'e'** *of the primary suffix is removed if it is followed by a suffix beginning with* **'a', 'i', 'o', 'u' or 'y'.**

Some functional groups (such as —OR) are indicated by the prefixes. The various classes of organic compounds, their functional groups, the secondary suffixes (or prefixes) used to indicate them along with their general IUPAC names are given in Table 5. It may be mentioned here that the groups $—NO_2$, —OR, —F, —Cl, —Br and —I are considered as substituents and are indicated by the ***prefixes*** whereas the groups —CHO, —CO—, —COOH, —COCl, $—CONH_2$, —COOR, $—NH_2$, —CN and —OH, are considered as functional groups and are indicated by ***suffixes.***

Name of an organic compound containing functional group is derived through the following steps:

Step 1. *Select the longest continuous chain of the carbon atoms as* **parent chain.** *The selected chain must include the carbon atoms involved in the functional groups like —COOH, —CHO, —CN etc., or those which carry the functional groups like —OH,* $—NH_2$, *—Cl,* $—NO_2$, *etc. The number of carbon atoms in a parent chain decides the* **word root.**

Step 2. *In order to decide the* **primary suffix,** *the presence of carbon-carbon multiple bond in chain is observed.*

Step 3. *The functional group present is identified. This enables the selection of an appropriate* **secondary suffix** *or* **prefix.**

Table 5. Different Functional Groups and Corresponding Suffixes and Prefixes

Class of Compounds	*Functional Group Structure*	*General Formula*	*IUPAC Group Prefix (p) or IUPAC Secondary Suffix (s)*	*General IUPAC Name*
(1)	*(2)*	*(3)*	*(4)*	*(5)*
1. Alcohols	—OH	R—OH	—ol (s)	Alkanol
2. Ethers	—O—R	R—O—R	Alkoxy (p)	Alkoxyalkane
3. Ketones	—C(=O)—	R—C(=O)—R	—one (s)	Alkanone
4. Primary amines	$—NH_2$	$R—NH_2$	Amino-(p)	Aminoalkane
5. Halides	—X X = F, Cl Br, 1	R—1	Halo-(p)	Haloalkane
6. Nitro compounds	—N(=O)(→O) $(—NO_2)$	$R—NO_2$	Nitro-(p)	Nitroalkane
7. Aldehydes	—C(=O)—H(—CHO)	R—CHO	—al (s)	Alkanal
8. Carboxylic acids	—C(=O)—OH	R—COOH	—oic acid (s)	Alkanoic acid
9. Amides	$—C(=O)—NH_2$	$R—C(=O)—NH_2$	—amide (s)	Alkanamide
10. Acyl halides	—C(=O)—X X Cl, Br	R—C(=O)—X	—oyl halide	Alkanoyl halide
11. Esters	—C(=O)—OR′	R—C(=O)—O—R′	Alkyl (p) oate (s)	Alkyl alkanoate
12. Acid anhydride	—C(=O)—O—C(=O)—	R—C(=O)—O—C(=O)—R	—oic anhydride (s)	Alkanoic anhydride
13. Cyanides	—CN	R—CN	—nitrile (s)	Alkanenitrile

Step 4. *The carbon atoms of the parent chain are numbered in such a way that the carbon atom of the functional group gets the lowest possible number. In case the functional group does not have the carbon atom, then the carbon atom of the parent chain attached to the functional group should get the lowest possible number.*

Step 5. *The name of the compound is then arrived at by arranging prefixes and suffixes along with their positions as follows:*

Prefixes—word root—primary suffix —secondary suffix

Let us now apply various rules to write the names of some organic compounds.

SOLVED EXAMPLES

Example 1. *Give the IUPAC name of the following compound:*

$$CH_3{-}CH_2{-}\underset{\displaystyle CH_2OH}{\underset{|}{CH}}{-}CH_2{-}\underset{\displaystyle CH_2{-}CH_3}{\underset{|}{CH}}{-}CH_3$$

Solution. 1. The longest chain containing functional group is of six carbon atoms. Therefore, the *word root* is **hex.**

$$CH_3{-}CH_2{-}\underset{\displaystyle {}_1CH_2OH}{\underset{|}{{}^2CH}}{-}{}^3CH_2{-}\underset{\displaystyle {}_5CH_2{-}{}_6CH_3}{\underset{|}{{}^4CH}}{-}CH_3$$

2. There is no multiple bond in it. Hence, the *primary suffix* is **ane.**

3. The functional group is —OH. Hence, *secondary suffix* is **—ol.**

4. The chain is numbered as shown. The functional group —OH is on carbon 1. Moreover, there is a methyl group on carbon 4 and ethyl group on carbon 2.

5. The IUPAC name is, therefore, **2-Ethyl-4-methylhexan-1-ol.**

Example 2. *Give the IUPAC name of the following compound:*

$$CH_3{-}CH_2{-}\underset{\displaystyle CH_3}{\underset{|}{CH}}{-}CH=CH{-}CH_2{-}CHO$$

Solution.

$$\overset{7}{C}H_3{-}\overset{6}{C}H_2{-}\underset{\displaystyle CH_3}{\underset{|}{\overset{5}{C}H}}{-}\overset{4}{C}H=\overset{3}{C}H{-}\overset{2}{C}H_2{-}\overset{1}{C}HO$$

1. The longest chain containing functional group is of seven carbon atoms. Therefore, *word root* is **hept.**

2. A C = C double bond is present in the molecule. Thus, the *primary suffix* is **ene.**

3. The *secondary suffix* is **al** because of the presence of —CHO group.

4. The chain is numbered as shown so that carbon atom of the —CHO group gets number 1. The methyl group is present on carbon 5 while the position of double bond is 3. Thus, IUPAC name is **5-Methylhept-3-en-1-al.**

MORE EXAMPLES OF COMPOUNDS CONTAINING ONE FUNCTIONAL GROUP

$${}^4CH_3{-}{}^3CH={}^2CH{-}\overset{\displaystyle O}{\overset{\|}{{}^1C}}{-}OH$$

But-2-en-1-oic acid

$${}^4CH_3{-}{}^3CH={}^2\overset{\displaystyle OCH_3}{\overset{|}{C}}{-}{}^1CH_3$$

2-Methoxybut-2-ene

$${}^3CH_2={}^2\overset{\displaystyle CH_3}{\overset{|}{C}}{-}\overset{\displaystyle O}{\overset{\|}{{}^1C}}{-}OC_2H_5$$

Ethyl 2-methylprop-2-enoate

$${}^4CH_3{-}{}^3\overset{\displaystyle CH_3}{\overset{|}{CH}}{-}{}^2\overset{\displaystyle CH_3}{\overset{|}{CH}}{-}{}^1CH_2NH_2$$

2, 3-Dimethylbutan-1-amine

$${}^1CH_3{-}\overset{\displaystyle O}{\overset{\|}{{}^2C}}{-}\underset{\displaystyle C_2H_5}{\underset{|}{{}^3CH}}{-}\underset{\displaystyle CH_3}{\underset{|}{{}^4CH}}{-}{}^5CH_3$$

3-Ethyl-4-methylpentan-2-one

$${}^4CH_3{-}\underset{\displaystyle CH_3}{\underset{|}{{}^3CH}}{-}{}^2CH_2{-}\overset{\displaystyle O}{\overset{\|}{{}^1C}}{-}Br$$

3-Methylbutanoyl bromide

$^{3}CH_3—^{2}CH(C_6H_5)—^{1}CH_2—OH$

2-Phenylpropan-1-ol

$C_6H_5—^{2}CH(^{3}CH_3)—^{1}CHO$

2-Phenylpropanal

$^{1}CH_3—^{2}C(CH_3)_2—^{3}CH(CH_3)—^{4}CH(I)—^{5}CH_2—^{6}CH_3$

4-Iodo-2, 2, 3-trimethylhexane

$^{1}CH_3—^{2}C(=O)—^{3}CH(^{4}CH_3)CH_3$

3-Methylbutan-2-one

$^{6}CH_3—^{5}CH_2—^{4}CH(Cl)—^{3}CH(CH_3)—^{2}C(CH_3)(OH)—^{1}CH_3$

4-Chloro-2, 3-dimethylhexan-2-ol

APPENDIX 3

SOME USEFUL CONVERSION FACTORS

Common Units of Mass and Weight

1 pound = 453.59 grams

1 pound = 453.59 grams = 0.45359 kilogram,

1 kilogram = 1000 grams = 2.205 pounds

1 gram = 10 decigrams = 100 centigrams

= 1000 milligrams

1 gram = 6.022×10^{23} atomic mass units or u,

1 atomic mass unit = 1.6606×10^{-24} gram

1 metric tonne = 1000 kilograms = 2205 pounds

Common Units of Volume

1 quart = 0.9463 litre, 1 litre = 1.056 quarts

1 litre = 1 cubic decimetre = 1000 cubic centimetres

= 0.001 cubic metre

1 millilitre = 1 cubic centimetre = 0.001 litre

= 1.056×10^{-3} quart

1 cubic foot = 28.316 litres = 29.902 quarts

= 7.475 gallons

Common Units of Energy

1 joule = 1×10^7 ergs

1 thermo chemical calorie* = 4.184 joules

= 4.184×10^7 ergs = 4.129×10^{-2} litre-atmospheres

= 2.612×10^{19} electron volts

1 ergs = 1×10^{-7} joule = 2.3901×10^{-8} calorie

1 electron volt = 1.6022×10^{-19} joule

= 1.6022×10^{-12} erg = 96.487 kJ/mol

1 litre-atmosphere = 24.217 calories

= 101.32 joules = 1.0132×10^9 ergs

1 British thermal unit = 1055.06 joules

= 1.05506×10^{10} ergs = 252.2 calories

Common Units of Length

1 inch = 2.54 centimetres (exactly)

1 mile = 5280 feet = 1.609 kilometres

1 yard = 36 inches = 0.9144 metre

1 metre = 100 centimetres = 39.37 inches

= 3.281 feet = 1.094 yards

1 kilometre = 1000 metres = 1094 yards = 0.6215 mile

---

* The amount of heat required to raise the temperature of one gram of water from 14.5°C to 15.5°C

1 Angstrom = 1.0×10^{-8} centimetres = 0.10 nanometer
= 1.0×10^{-10} metre = 3.937×10^{-9} inch

Common Units of Force* and Pressure

1 atmosphere = 760 millimetres of mercury
= 1.013×10^5 Pascals
= 14.70 pounds per square inch

1 bar = 10^5 Pascals

1 torr = 1 millimetre of mercury

1 Pascal = 1 kg/m.s^2 = 1 N/m^2

Temperature

SI Base Unit: Kelvin (K)

0 K = – 273.15°C

? K = °C + 273.15°

?°F = 1.8(°C) + 32°

$$? \text{°C} = \frac{F - 32°}{1.8}$$

Note: All other units are per particle and must be multiplied by 6.022×10^{23} to be strictly comparable.

* Force: 1 newton (N) = 1 kg.m/s^2, *i.e.,* the force that, when applied for 1 second, gives a 1-kilogram mass a velocity of 1 metre per second.

The Modern Periodic Table of Elements

PERIODS GROUPS

This numbering system is used by the international Union of Pure and Applied Chemistry (IUPAC)

Key: 12 Mg Magnesium 24.305 (2, 8, 2) — Atomic number; Electrons in each energy level; Element symbol; Average atomic mass

TRANSITION METALS (groups 3–12)

PERIODS	1	2	3	4	5	6	7	8	9	10	11	12	13	14	15	16	17	18
1	1 **H** Hydrogen 1.0079 (1)																	2 **He** Helium 4.0026 (2)
2	3 **Li** Lithium 6.941 (2, 1)	4 **Be** Beryllium 9.0122 (2, 2)											5 **B** Boron 10.81 (2, 3)	6 **C** Carbon 12.011 (2, 4)	7 **N** Nitrogen 14.007 (2, 5)	8 **O** Oxygen 15.999 (2, 6)	9 **F** Fluorine 18.998 (2, 7)	10 **Ne** Neon 20.179 (2, 8)
3	11 **Na** Sodium 22.990 (2, 8, 1)	12 **Mg** Magnesium 24.305 (2, 8, 2)											13 **Al** Aluminium 26.982 (2, 8, 3)	14 **Si** Silicon 28.086 (2, 8, 4)	15 **P** Phosphorus 30.974 (2, 8, 5)	16 **S** Sulphur 32.06 (2, 8, 6)	17 **Cl** Chlorine 35.453 (2, 8, 7)	18 **Ar** Argon 39.948 (2, 8, 8)
4	19 **K** Potassium 39.098 (2, 8, 8, 1)	20 **Ca** Calcium 40.08 (2, 8, 8, 2)	21 **Sc** Scandium 44.958 (2, 8, 9, 2)	22 **Ti** Titanium 47.90 (2, 8, 10, 2)	23 **V** Vanadium 50.941 (2, 8, 11, 2)	24 **Cr** Chromium 51.996 (2, 8, 13, 1)	25 **Mn** Manganese 54.938 (2, 8, 13, 2)	26 **Fe** Iron 55.847 (2, 8, 14, 2)	27 **Co** Cobalt 58.933 (2, 8, 15, 2)	28 **Ni** Nickel 58.71 (2, 8, 16, 2)	29 **Cu** Copper 63.546 (2, 8, 18, 1)	30 **Zn** Zinc 65.38 (2, 8, 18, 2)	31 **Ga** Gallium 69.72 (2, 8, 18, 3)	32 **Ge** Germanium 72.59 (2, 8, 18, 4)	33 **As** Arsenic 74.922 (2, 8, 18, 5)	34 **Se** Selenium 78.96 (2, 8, 18, 6)	35 **Br** Bromine 79.904 (2, 8, 18, 7)	36 **Kr** Krypton 83.80 (2, 8, 18, 8)
5	37 **Rb** Rubidium 85.468 (2, 8, 18, 8, 1)	38 **Sr** Strontium 87.62 (2, 8, 18, 8, 2)	39 **Y** Yttrium 88.906 (2, 8, 18, 9, 2)	40 **Zr** Zirconium 91.22 (2, 8, 18, 10, 2)	41 **Nb** Niobium 92.906 (2, 8, 18, 12, 1)	42 **Mo** Molybdenum 95.94 (2, 8, 18, 13, 1)	43 **Tc** Technetium (97) (2, 8, 18, 14, 1)	44 **Ru** Ruthenium 101.07 (2, 8, 18, 15, 1)	45 **Rh** Rhodium 102.91 (2, 8, 18, 16, 1)	46 **Pd** Palladium 106.4 (2, 8, 18, 18)	47 **Ag** Silver 107.87 (2, 8, 18, 18, 1)	48 **Cd** Cadmium 112.41 (2, 8, 18, 18, 2)	49 **In** Indium 114.82 (2, 8, 18, 18, 3)	50 **Sn** Tin 118.69 (2, 8, 18, 18, 4)	51 **Sb** Antimony 121.75 (2, 8, 18, 18, 5)	52 **Te** Tellurium 127.60 (2, 8, 18, 18, 6)	53 **I** Iodine 126.90 (2, 8, 18, 18, 7)	54 **Xe** Xenon 131.30 (2, 8, 18, 18, 8)
6	55 **Cs** Cesium 132.91 (2, 8, 18, 18, 8, 1)	56 **Ba** Barium 137.33 (2, 8, 18, 18, 8, 2)	71 **Lu** Lutetium 174.97 (2, 8, 18, 32, 9, 2)	72 **Hf** Hafnium 178.49 (2, 8, 18, 32, 10, 2)	73 **Ta** Tantalum 180.95 (2, 8, 18, 32, 11, 2)	74 **W** Tungsten 183.85 (2, 8, 18, 32, 12, 2)	75 **Re** Rhenium 186.21 (2, 8, 18, 32, 13, 2)	76 **Os** Osmium 190.2 (2, 8, 18, 32, 14, 2)	77 **Ir** Iridium 192.22 (2, 8, 18, 32, 15, 2)	78 **Pt** Platinum 196.09 (2, 8, 18, 32, 17, 1)	79 **Au** Gold 196.97 (2, 8, 18, 32, 18, 1)	80 **Hg** Mercury 200.59 (2, 8, 18, 32, 18, 2)	81 **Tl** Thallium 204.37 (2, 8, 18, 32, 18, 3)	82 **Pb** Lead 207.2 (2, 8, 18, 32, 18, 4)	83 **Bi** Bismuth 208.98 (2, 8, 18, 32, 18, 5)	84 **Po** Polonium (209) (2, 8, 18, 32, 18, 6)	85 **At** Astatine (210) (2, 8, 18, 32, 18, 7)	86 **Rn** Radon (222) (2, 8, 18, 32, 18, 8)
7	87 **Fr** Francium (223) (2, 8, 18, 32, 18, 8, 1)	88 **Ra** Radium 226.03 (2, 8, 18, 32, 18, 8, 2)	103 **Lr** Lawrencium (260) (2, 8, 18, 32, 32, 9, 2)	104 **Rf** Rutherfordium (261) (2, 8, 18, 32, 32, 10, 2)	105 **Db** Dubnium (262) (2, 8, 18, 32, 32, 11, 2)	106 **Sg** Seaborgium (263) (2, 8, 18, 32, 32, 12, 2)	107 **Bh** Bohrium (264) (2, 8, 18, 32, 32, 13, 2)	108 **Hs** Hassium (265) (2, 8, 18, 32, 32, 14, 2)	109 **Mt** Meitnerium (266) (2, 8, 18, 32, 32, 15, 2)	110 **Ds** Darmstadtium (269) (2, 8, 18, 32, 32, 16, 2)	111 **Rg** Roentgenium (272)	112 **Uub** Ununbium (277)	—	114 **Uuq** Ununquadium (289)	—	116 **Unh** Unhexium (292)		

Series														
Lanthanide Series	57 **La** Lanthanum 138.91 (2, 8, 18, 18, 9, 2)	58 **Ce** Cerium 140.12 (2, 8, 18, 20, 8, 2)	59 **Pr** Praseodymium 140.91 (2, 8, 18, 21, 8, 2)	60 **Nd** Neodymium 144.24 (2, 8, 18, 22, 8, 2)	61 **Pm** Promethium (145) (2, 8, 18, 23, 8, 2)	62 **Sm** Samarium 150.4 (2, 8, 18, 24, 8, 2)	63 **Eu** Europium 151.96 (2, 8, 18, 25, 8, 2)	64 **Gd** Gadolinium 157.25 (2, 8, 18, 25, 9, 2)	65 **Tb** Terbium 158.93 (2, 8, 18, 27, 8, 2)	66 **Dy** Dysprosium 162.50 (2, 8, 18, 28, 8, 2)	67 **Ho** Holmium 164.93 (2, 8, 18, 29, 8, 2)	68 **Er** Erbium 167.26 (2, 8, 18, 30, 8, 2)	69 **Tm** Thulium 168.93 (2, 8, 18, 31, 8, 2)	70 **Yb** Ytterbium 173.04 (2, 8, 18, 32, 8, 2)
Actinide Series	89 **Ac** Actinium (227) (2, 8, 18, 32, 18, 9, 2)	90 **Th** Thorium 232.04 (2, 8, 18, 32, 18, 10, 2)	91 **Pa** Protactinium 231.04 (2, 8, 18, 32, 20, 9, 2)	92 **U** Uranium 238.03 (2, 8, 18, 32, 21, 9, 2)	93 **Np** Neptunium 237.06 (2, 8, 18, 32, 22, 9, 2)	94 **Pu** Plutonium (244) (2, 8, 18, 32, 24, 8, 2)	95 **Am** Americium (245) (2, 8, 18, 32, 25, 8, 2)	96 **Cm** Curium (247) (2, 8, 18, 32, 25, 9, 2)	97 **Bk** Berkelium (247) (2, 8, 18, 32, 27, 8, 2)	98 **Cf** Californium (251) (2, 8, 18, 32, 28, 8, 2)	99 **Es** Einsteinium (254) (2, 8, 18, 32, 29, 8, 2)	100 **Fm** Fermium (257) (2, 8, 18, 32, 30, 8, 2)	101 **Md** Mendelevium (258) (2, 8, 18, 32, 31, 8, 2)	102 **No** Nobelium (259) (2, 8, 18, 32, 32, 8, 2)

INDEX

A

B

C

D

E

F

G

H

I

K

L

M

N

W

Z

LOGARITHMS

	0	1	2	3	4	5	6	7	8	9	1	2	3	4	5	6	7	8	9
10	0000	0043	0086	0128	0170						5	9	13	17	21	26	30	34	38
						0212	0253	0294	0334	0374	4	8	12	16	20	24	28	32	36
11	0414	0453	0492	0531	0569						4	8	12	16	20	23	27	31	35
						0607	0645	0682	0719	0755	4	7	11	15	18	22	26	29	33
12	0792	0828	0864	0899	0934						3	7	11	14	18	21	25	28	32
						0969	1004	1038	1072	1106	3	7	10	14	17	20	24	27	31
13	1139	1173	1206	1239	1271						3	6	10	13	16	19	23	26	29
						1303	1335	1367	1399	1430	3	7	10	13	16	19	22	25	29
14	1461	1492	1523	1553	1584						3	6	9	12	15	19	22	25	28
						1614	1644	1673	1703	1732	3	6	9	12	14	17	20	23	26
15	1761	1790	1818	1847	1875						3	6	9	11	14	17	20	23	26
						1903	1931	1959	1987	2014	3	6	8	11	14	17	19	22	25
16	2041	2068	2095	2122	2148						3	6	8	11	14	16	19	22	24
						2175	2201	2227	2253	2279	3	5	8	10	13	16	18	21	23
17	2304	2330	2355	2380	2405						3	5	8	10	13	15	18	20	23
						2430	2455	2480	2504	2529	3	5	8	10	12	15	17	20	22
18	2553	2577	2601	2625	2648						2	5	7	9	12	14	17	19	21
						2672	2695	2718	2742	2765	2	4	7	9	11	14	16	18	21
19	2788	2810	2833	2856	2878						2	4	7	9	11	13	16	18	20
						2900	2923	2945	2967	2989	2	4	6	8	11	13	15	17	19
20	3010	3032	3054	3075	3096	3118	3139	3160	3181	3201	2	4	6	8	11	13	15	17	19
21	3222	3243	3263	3284	3304	3324	3345	3365	3385	3404	2	4	6	8	10	12	14	16	18
22	3424	3444	3464	3483	3502	3522	3541	3560	3579	3598	2	4	6	8	10	12	14	15	17
23	3617	3636	3655	3674	3692	3711	3729	3747	3766	3784	2	4	6	7	9	11	13	15	17
24	3802	3820	3838	3856	3874	3892	3909	3927	3945	3962	2	4	5	7	9	11	12	14	16
25	3979	3997	4014	4031	4048	4065	4082	4099	4116	4133	2	3	5	7	9	10	12	14	15
26	4150	4166	4183	4200	4216	4232	4249	4265	4281	4298	2	3	5	7	8	10	11	13	15
27	4314	4330	4346	4362	4378	4393	4409	4425	4440	4456	2	3	5	6	8	9	11	13	14
28	4472	4487	4502	4518	4533	4548	4564	4579	4594	4609	2	3	5	6	8	9	11	12	14
29	4624	4639	4654	4669	4683	4698	4713	4728	4742	4757	1	3	4	6	7	9	10	12	13
30	4771	4786	4800	4814	4829	4843	4857	4871	4886	4900	1	3	4	6	7	9	10	11	13
31	4914	4928	4942	4955	4969	4983	4997	5011	5024	5038	1	3	4	6	7	8	10	11	12
32	5052	5065	5079	5092	5105	5119	5132	5145	5159	5172	1	3	4	5	7	8	9	11	12
33	5185	5198	5211	5224	5237	5250	5263	5276	5289	5302	1	3	4	5	6	8	9	10	12
34	5315	5328	5340	5353	5366	5378	5391	5403	5416	5428	1	3	4	5	6	8	9	10	11
35	5441	5453	5465	5478	5490	5502	5515	5527	5539	5551	1	2	4	5	6	7	9	10	11
36	5563	5575	5587	5599	5611	5623	5635	5647	5658	5670	1	2	4	5	6	7	8	10	11
37	5682	5694	5705	5717	5729	5740	5752	5763	5775	5786	1	2	3	5	6	7	8	9	10
38	5798	5809	5821	5832	5843	5855	5866	5877	5888	5899	1	2	3	5	6	7	8	9	10
39	5911	5922	5933	5944	5955	5966	5977	5988	5999	6010	1	2	3	4	5	7	8	9	10
40	6021	6031	6042	6053	6064	6075	6085	6096	6107	6117	1	2	3	4	5	6	8	9	10
41	6128	6138	6149	6160	6170	6180	6191	6201	6212	6222	1	2	3	4	5	6	7	8	9
42	6232	6243	6253	6263	6274	6284	6294	6304	6314	6325	1	2	3	4	5	6	7	8	9
43	6335	6345	6355	6365	6375	6385	6395	6405	6415	6425	1	2	3	4	5	6	7	8	9
44	6435	6444	6454	6464	6474	6484	6493	6503	6513	6522	1	2	3	4	5	6	7	8	9
45	6532	6542	6551	6561	6571	6580	6590	6599	6609	6618	1	2	3	4	5	6	7	8	9
46	6628	6637	6646	6656	6665	6675	6684	6693	6702	6712	1	2	3	4	5	6	7	7	8
47	6721	6730	6739	6749	6758	6767	6776	6785	6794	6803	1	2	3	4	5	5	6	7	8
48	6812	6821	6830	6839	6848	6857	6866	6875	6884	6893	1	2	3	4	4	5	6	7	8
49	6902	6911	6920	6928	6937	6946	6955	6964	6972	6981	1	2	3	4	4	5	6	7	8

	0	1	2	3	4	5	6	7	8	9	1	2	3	4	5	6	7	8	9
50	6990	6998	7007	7016	7024	7033	7042	7050	7059	7067	1	2	3	3	4	5	6	7	8
51	7076	7084	7093	7101	7110	7118	7126	7135	7143	7152	1	2	3	3	4	5	6	7	8
52	7160	7168	7177	7185	7193	7202	7210	7218	7226	7235	1	2	2	3	4	5	6	7	7
53	7243	7251	7259	7267	7275	7284	7292	7300	7308	7316	1	2	2	3	4	5	6	6	7
54	7324	7332	7340	7348	7356	7364	7372	7380	7388	7396	1	2	2	3	4	5	6	6	7
55	7404	7412	7419	7427	7435	7443	7451	7459	7466	7474	1	2	2	3	4	5	5	6	7
56	7482	7490	7497	7505	7513	7520	7528	7536	7543	7551	1	2	2	3	4	5	5	6	7
57	7559	7566	7574	7582	7589	7597	7604	7612	7619	7627	1	2	2	3	4	5	5	6	7
58	7634	7642	7649	7657	7664	7672	7679	7686	7694	7701	1	1	2	3	4	4	5	6	7
59	7709	7716	7723	7731	7738	7745	7752	7760	7767	7774	1	1	2	3	4	4	5	6	7
60	7782	7789	7796	7803	7810	7818	7825	7832	7839	7846	1	1	2	3	4	4	5	6	6
61	7853	7860	7868	7875	7882	7889	7896	7903	7910	7917	1	1	2	3	4	4	5	6	6
62	7924	7931	7938	7945	7952	7959	7966	7973	7980	7987	1	1	2	3	3	4	5	6	6
63	7993	8000	8007	8014	8021	8028	8035	8041	8048	8055	1	1	2	3	3	4	5	5	6
64	8062	8069	8075	8082	8089	8096	8102	8109	9116	8122	1	1	2	3	3	4	5	5	6
65	8129	8136	8142	8149	8156	8162	8169	8176	8182	8189	1	1	2	3	3	4	5	5	6
66	8195	8202	8209	8215	8222	8228	8235	8241	8248	8254	1	1	2	3	3	4	5	5	6
67	8261	8267	8274	8280	8287	8293	8299	8306	8312	8319	1	1	2	3	3	4	5	5	6
68	8325	8331	8338	8344	8351	8357	8363	8370	8376	8382	1	1	2	3	3	4	4	5	6
69	8388	8395	8401	8407	8414	8420	8426	8432	8439	8445	1	1	2	2	3	4	4	5	6
70	8451	8457	8463	8470	8476	8482	8488	8494	8500	8506	1	1	2	2	3	4	4	5	6
71	8513	8519	8525	8531	8537	8543	8549	8555	8561	8567	1	1	2	2	3	4	4	5	5
72	8573	8579	8585	8591	8597	8603	8609	8615	8621	8627	1	1	2	2	3	4	4	5	5
73	8633	8639	8645	8651	8657	8663	8669	8675	8681	8686	1	1	2	2	3	4	4	5	5
74	8692	8698	8704	8710	8716	8722	8727	8733	8739	8745	1	1	2	2	3	4	4	5	5
75	8751	8756	8762	8768	8774	8779	8785	8791	8797	8802	1	1	2	2	3	3	4	5	5
76	8808	8814	8820	8825	8831	8837	8842	8848	8854	8859	1	1	2	2	3	3	4	5	5
77	8865	8871	8876	8882	8887	8893	8899	8904	8910	8915	1	1	2	2	3	3	4	4	5
78	8921	8927	8932	8938	8943	8949	8954	8960	8965	8971	1	1	2	2	3	3	4	4	5
79	8976	8982	8987	8993	8998	9004	9009	9015	9020	9025	1	1	2	2	3	3	4	4	5
80	9031	9036	9042	9047	9053	9058	9063	9069	9074	9079	1	1	2	2	3	3	4	4	5
81	9085	9090	9096	9101	9106	9112	9117	9122	9128	9133	1	1	2	2	3	3	4	4	5
82	9138	9143	9149	9154	9159	9165	9170	9175	9180	9186	1	1	2	2	3	3	4	4	5
83	9191	9196	9201	9206	9212	9217	9222	9227	9232	9238	1	1	2	2	3	3	4	4	5
84	9243	9248	9253	9258	9263	9269	9274	9279	9284	9289	1	1	2	2	3	3	4	4	5
85	9294	9299	9304	9309	9315	9320	9325	9330	9335	9340	1	1	2	2	3	3	4	4	5
86	9345	9350	9355	9360	9365	9370	9375	9380	9385	9390	1	1	2	2	3	3	4	4	5
87	9395	9400	9405	9410	9415	9420	9425	9430	9435	9440	0	1	1	2	2	3	3	4	4
88	9445	9450	9455	9460	9465	9469	9474	9479	9484	9489	0	1	1	2	2	3	3	4	4
89	9494	9499	9504	9509	9513	9518	9523	9528	9533	9538	0	1	1	2	2	3	3	4	4
90	9542	9547	9552	9557	9562	9566	9571	9576	9581	9586	0	1	1	2	2	3	3	4	4
91	9590	9595	9600	9605	9609	9614	9619	9624	9628	9633	0	1	1	2	2	3	3	4	4
92	9638	9643	9647	9652	9657	9661	9666	9671	9675	9680	0	1	1	2	2	3	3	4	4
93	9685	9689	9694	9699	9703	9708	9713	9717	9722	9727	0	1	1	2	2	3	3	4	4
94	9731	9736	9741	9745	9750	9754	9759	9763	9768	9773	0	1	1	2	2	3	3	4	4
95	9777	9782	9786	9791	9795	9800	9805	9809	9814	9818	0	1	1	2	2	3	3	4	4
96	9823	9827	9832	9836	9841	9845	9850	9854	9859	9863	0	1	1	2	2	3	3	4	4
97	9868	9872	9877	9881	9886	9890	9894	9899	9903	9908	0	1	1	2	2	3	3	4	4
98	9912	9917	9921	9926	9930	9934	9939	9943	9948	9952	0	1	1	2	2	3	3	4	4
99	9956	9961	9965	9969	9974	9978	9983	9987	9991	9996	0	1	1	2	2	3	3	3	4

ANTILOGARITHMS

	0	1	2	3	4	5	6	7	8	9	1	2	3	4	5	6	7	8	9
.00	1000	1002	1005	1007	1009	1012	1014	1016	1019	1021	0	0	1	1	1	1	2	2	2
.01	1023	1026	1028	1030	1033	1035	1038	1040	1042	1045	0	0	1	1	1	1	2	2	2
.02	1047	1050	1052	1054	1057	1059	1062	1064	1067	1069	0	0	1	1	1	1	2	2	2
.03	1072	1074	1076	1079	1081	1084	1086	1089	1091	1094	0	0	1	1	1	1	2	2	2
.04	1096	1099	1102	1104	1107	1109	1112	1114	1117	1119	0	1	1	1	1	2	2	2	2
.05	1122	1125	1127	1130	1132	1135	1138	1140	1143	1146	0	1	1	1	1	2	2	2	2
.06	1148	1151	1153	1156	1159	1161	1164	1167	1169	1172	0	1	1	1	1	2	2	2	2
.07	1175	1178	1180	1183	1186	1189	1191	1194	1197	1199	0	1	1	1	1	2	2	2	2
.08	1202	1205	1208	1211	1213	1216	1219	1222	1225	1227	0	1	1	1	1	2	2	2	3
.09	1230	1233	1236	1239	1242	1245	1247	1250	1253	1256	0	1	1	1	1	2	2	2	3
.10	1259	1262	1265	1268	1271	1274	1276	1279	1282	1285	0	1	1	1	1	2	2	2	3
.11	1288	1291	1294	1297	1300	1303	1306	1309	1312	1315	0	1	1	1	2	2	2	2	3
.12	1318	1321	1324	1327	1330	1334	1337	1340	1343	1346	0	1	1	1	2	2	2	2	3
.13	1349	1352	1355	1358	1361	1365	1368	1371	1374	1377	0	1	1	1	2	2	2	3	3
.14	1380	1384	1387	1390	1393	1396	1400	1403	1406	1409	0	1	1	1	2	2	2	3	3
.15	1413	1416	1419	1422	1426	1429	1432	1435	1439	1442	0	1	1	1	2	2	2	3	3
.16	1445	1449	1452	1455	1459	1462	1466	1469	1472	1476	0	1	1	1	2	2	2	3	3
.17	1479	1483	1486	1489	1493	1496	1500	1503	1507	1510	0	1	1	1	2	2	2	3	3
.18	1514	1517	1521	1524	1528	1531	1535	1538	1542	1545	0	1	1	1	2	2	2	3	3
.19	1549	1552	1556	1560	1563	1567	1570	1574	1578	1581	0	1	1	1	2	2	3	3	3
.20	1585	1589	1592	1596	1600	1603	1607	1611	1614	1618	0	1	1	1	2	2	3	3	3
.21	1622	1626	1629	1633	1637	1641	1644	1648	1652	1656	0	1	1	2	2	2	3	3	3
.22	1660	1663	1667	1671	1675	1679	1683	1687	1690	1694	0	1	1	2	2	2	3	3	3
.23	1698	1702	1706	1710	1714	1718	1722	1726	1730	1734	0	1	1	2	2	2	3	3	4
.24	1738	1742	1746	1750	1754	1758	1762	1766	1770	1774	0	1	1	2	2	2	3	3	4
.25	1778	1782	1786	1791	1795	1799	1803	1807	1811	1816	0	1	1	2	2	2	3	3	4
.26	1820	1824	1828	1832	1837	1841	1845	1849	1854	1858	0	1	1	2	2	3	3	3	4
.27	1862	1866	1871	1875	1879	1884	1888	1892	1897	1901	0	1	1	2	2	3	3	3	4
.28	1905	1910	1914	1919	1923	1928	1932	1936	1941	1945	0	1	1	2	2	3	3	4	4
.29	1950	1954	1959	1963	1968	1972	1977	1982	1986	1991	0	1	1	2	2	3	3	4	4
.30	1995	2000	2004	2009	2014	2018	2023	2028	2032	2037	0	1	1	2	2	3	3	4	4
.31	2042	2046	2051	2056	2061	2065	2070	2075	2080	2084	0	1	1	2	2	3	3	4	4
.32	2089	2094	2099	2104	2109	2113	2118	2123	2128	2133	0	1	1	2	2	3	3	4	4
.33	2138	2143	2148	2153	2158	2163	2168	2173	2178	2183	0	1	1	2	2	3	3	4	4
.34	2188	2193	2198	2203	2208	2213	2218	2223	2228	2234	1	1	2	2	3	3	4	4	5
.35	2239	2244	2249	2254	2259	2265	2270	2275	2280	2286	1	1	2	2	3	3	4	4	5
.36	2291	2296	2301	2307	2312	2317	2323	2328	2333	2339	1	1	2	2	3	3	4	4	5
.37	2344	2350	2355	2360	2366	2371	2377	2382	2388	2393	1	1	2	2	3	3	4	4	5
.38	2399	2404	2410	2415	2421	2427	2432	2438	2443	2449	1	1	2	2	3	3	4	4	5
.39	2455	2460	2466	2472	2477	2483	2489	2495	2500	2506	1	1	2	2	3	3	4	5	5
.40	2512	2518	2523	2529	2535	2541	2547	2553	2559	2564	1	1	2	2	3	4	4	5	5
.41	2570	2576	2582	2588	2594	2600	2606	2612	2618	2624	1	1	2	2	3	4	4	5	5
.42	2630	2636	2642	2649	2655	2661	2667	2673	2679	2685	1	1	2	2	3	4	4	5	6
.43	2692	2698	2704	2710	2716	2723	2729	2735	2742	2748	1	1	2	3	3	4	4	5	6
.44	2754	2761	2767	2773	2780	2786	2793	2799	2805	2812	1	1	2	3	3	4	4	5	6
.45	2818	2825	2831	2838	2844	2851	2858	2864	2871	2877	1	1	2	3	3	4	5	5	6
.46	2884	2891	2897	2904	2911	2917	2924	2931	2938	2944	1	1	2	3	3	4	5	5	6
.47	2951	2958	2965	2972	2979	2985	2992	2999	3006	3013	1	1	2	3	3	4	5	5	6
.48	3020	3027	3034	3041	3048	3055	3062	3069	3076	3083	1	1	2	3	4	4	5	6	6
.49	3090	3097	3105	3112	3119	3126	3133	3141	3148	3155	1	1	2	3	4	4	5	6	6

	0	1	2	3	4	5	6	7	8	9	1	2	3	4	5	6	7	8	9
.50	3162	3170	3177	3184	3192	3199	3206	3214	3221	3228	1	1	2	3	4	4	5	6	7
.51	3236	3243	3251	3258	3266	3273	3281	3289	3296	3304	1	2	2	3	4	5	5	6	7
.52	3311	3319	3327	3334	3342	3350	3357	3365	3373	3381	1	2	2	3	4	5	5	6	7
.53	3388	3396	3404	3412	3420	3428	3436	3443	3451	3459	1	2	2	3	4	5	6	6	7
.54	3467	3475	3483	3491	3499	3508	3516	3524	3532	3540	1	2	2	3	4	5	6	6	7
.55	3548	3556	3565	3573	3581	3589	3597	3606	3614	3622	1	2	2	3	4	5	6	7	7
.56	3631	3639	3648	3656	3664	3673	3681	3690	3698	3707	1	2	3	3	4	5	6	7	8
.57	3715	3724	3733	3741	3750	3758	3767	3776	3784	3793	1	2	3	3	4	5	6	7	8
.58	3802	3811	3819	3828	3837	3846	3855	3864	3873	3882	1	2	3	4	4	5	6	7	8
.59	3890	3899	3908	3917	3926	3936	3945	3954	3963	3972	1	2	3	4	5	5	6	7	8
.60	3981	3990	3999	4009	4018	4027	4036	4046	4055	4064	1	2	3	4	5	6	6	7	8
.61	4074	4083	4093	4102	4111	4121	4130	4140	4150	4159	1	2	3	4	5	6	7	8	9
.62	4169	4178	4188	4198	4207	4217	4227	4236	4246	4256	1	2	3	4	5	6	7	8	9
.63	4266	4276	4285	4295	4305	4315	4325	4335	4345	4355	1	2	3	4	5	6	7	8	9
.64	4365	4375	4385	4395	4406	4416	4426	4436	4446	4457	1	2	3	4	5	6	7	8	9
.65	4467	4477	4487	4498	4508	4519	4529	4539	4550	4560	1	2	3	4	5	6	7	8	9
.66	4571	4581	4592	4603	4613	4624	4634	4645	4656	4667	1	2	3	4	5	6	7	9	10
.67	4677	4688	4699	4710	4721	4732	4742	4753	4764	4775	1	2	3	4	5	7	8	9	10
.68	4786	4797	4808	4819	4831	4842	4853	4864	4875	4887	1	2	3	4	6	7	8	9	10
.69	4898	4909	4920	4932	4943	4955	4966	4977	4989	5000	1	2	3	5	6	7	8	9	10
.70	5012	5023	5035	5047	5058	5070	5082	5093	5105	5117	1	2	4	5	6	7	8	9	11
.71	5129	5140	5152	5164	5176	5188	5200	5212	5224	5236	1	2	4	5	6	7	8	10	11
.72	5248	5260	5272	5284	5297	5309	5321	5333	5346	5358	1	2	4	5	6	7	9	10	11
.73	5370	5383	5395	5408	5420	5433	5445	5458	5470	5483	1	3	4	5	6	8	9	10	11
.74	5495	5508	5521	5534	5546	5559	5572	5585	5598	5610	1	3	4	5	6	8	9	10	12
.75	5623	5636	5649	5662	5675	5689	5702	5715	5728	5741	1	3	4	5	7	8	9	10	12
.76	5754	5768	5781	5794	5808	5821	5834	5848	5861	5875	1	3	4	5	7	8	9	11	12
.77	5888	5902	5916	5929	5943	5957	5970	5984	5998	6012	1	3	4	5	7	8	10	11	12
.78	6026	6039	6053	6067	6081	6095	6109	6124	6138	6152	1	3	4	6	7	8	10	11	13
.79	6166	6180	6194	6209	6223	6237	6252	6266	6281	6295	1	3	4	6	7	9	10	11	13
.80	6310	6324	6339	6353	6368	6383	6397	6412	6427	6442	1	3	4	6	7	9	10	12	13
.81	6457	6471	6486	6501	6516	6531	6546	6561	6577	6592	2	3	5	6	8	9	11	12	14
.82	6607	6622	6637	6653	6668	6683	6699	6714	6730	6745	2	3	5	6	8	9	11	12	14
.83	6761	6776	6792	6808	6823	6839	6855	6871	6887	6902	2	3	5	6	8	9	11	13	14
.84	6918	6934	6950	6966	6982	6998	7015	7031	7047	7063	2	3	5	6	8	10	11	13	15
.85	7079	7096	7112	7129	7145	7161	7178	7194	7211	7228	2	3	5	7	8	10	12	13	15
.86	7244	7261	7278	7295	7311	7328	7345	7362	7379	7396	2	3	5	7	8	10	12	13	15
.87	7413	7430	7447	7464	7482	7499	7516	7534	7551	7568	2	3	5	7	9	10	12	14	16
.88	7586	7603	7621	7638	7656	7674	7691	7709	7727	7745	2	4	5	7	9	11	12	14	16
.89	7762	7780	7798	7816	7834	7852	7870	7889	7907	7925	2	4	5	7	9	11	13	14	16
.90	7943	7962	7980	7998	8017	8035	8054	8072	8091	8110	2	4	6	7	9	11	13	15	17
.91	8128	8147	8166	8185	8204	8222	8241	8260	8279	8299	2	4	6	8	9	11	13	15	17
.92	8318	8337	8356	8375	8395	8414	8433	8453	8472	8492	2	4	6	8	10	12	14	15	17
.93	8511	8531	8551	8570	8590	8610	8630	8650	8670	8690	2	4	6	8	10	12	14	16	18
.94	8710	8730	8750	8770	8790	8810	8831	8851	8872	8892	2	4	6	8	10	12	14	16	18
.95	8913	8933	8954	8974	8995	9016	9036	9057	9078	9099	2	4	6	8	10	12	15	17	19
.96	9120	9141	9162	9183	9204	9226	9247	9268	9290	9311	2	4	6	8	11	13	15	17	20
.97	9333	9354	9376	9397	9419	9441	9462	9484	9506	9528	2	4	7	9	11	13	15	17	20
.98	9550	9572	9594	9616	9638	9661	9683	9705	9727	9750	2	4	7	9	11	13	16	18	20
.99	9772	9795	9817	9840	9863	9886	9908	9931	9954	9977	2	5	7	9	11	14	16	18	20